Image Processing and Analysis

Image Processing and Analysis

Stan Birchfield
Clemson University

CENGAGE
Learning™

Australia • Brazil • Mexico • Singapore • United Kingdom • United States

CENGAGE Learning®

Image Processing and Analysis,
First Edition

Stan Birchfield

Product Director, Global Engineering:
  Timothy L. Anderson

Senior Content Developer: Mona Zeftel

Associate Media Content Developer:
  Ashley Kaupert

Product Assistant: Teresa Versaggi

Marketing Manager: Kristin Stine

Director, Higher Education Production:
  Sharon L. Smith

Content Project Manager: D. Jean Buttrom

Production Service: RPK Editorial Services,
  Inc.

Copyeditor: Shelly Gerger-Knechtl

Proofreader: Lori Martinsek

Indexer: Shelly Gerger-Knechtl

Compositor: SPi Global

Senior Art Director: Michelle Kunkler

Cover and Internal Designer: Ramsdell
  Design, LLC

Cover Image: Jessica Birchfield

Chapter Opener Images: Jessica Birchfield
  and Stan Birchfield

Intellectual Property
  Analyst: Christine Myaskovsky
  Project Manager: Sarah Shainwald

Text and Image Permissions Researcher:
  Kristiina Paul

Manufacturing Planner: Doug Wilke

For product information and technology assistance, contact us at
Cengage Learning Customer & Sales Support, 1-800-354-9706.
For permission to use material from this text or product,
submit all requests online at **www.cengage.com/permissions**.
Further permissions questions can be emailed to
permissionrequest@cengage.com.

Library of Congress Control Number: 2016952392
ISBN: 978-1-285-17952-0

Cengage Learning
20 Channel Center Street
Boston, MA 02210
USA

Cengage Learning is a leading provider of customized learning solutions with employees residing in nearly 40 different countries and sales in more than 125 countries around the world. Find your local representative at **www.cengage.com**.

Cengage Learning products are represented in Canada by

Nelson Education Ltd.

To learn more about Cengage Learning Solutions, visit **www.cengage.com/engineering**.

Purchase any of our products at your local college store or at our preferred online store **www.cengagebrain.com**.

Unless otherwise noted, all items © Cengage Learning.

Printed in Canada
Print Number: 01   Print Year: 2016

BRIEF CONTENTS

CONTENTS

PREFACE

The seeds of this book were sown two decades ago when, as a graduate student, I took a course in *digital image processing* and a separate course in *computer vision*. The former course was taught in the electrical engineering department, whereas the latter was taught in the computer science department. The former followed the traditional approach of beginning with 1D signal processing, then moving to 2D image processing, Fourier transforms, filtering, and compression. The latter, on the other hand, began with image formation and edge detection, then covered segmentation, classification, stereo, and motion. Not only did the two courses from different departments cover distinct topics, but they also relied upon different underlying mathematical foundations, and they seemed to have non-overlapping goals: one course was more concerned with manipulating images as they existed, whereas the other course focused more on how the images were formed and how they related to the world. Overall, the experience left me with the distinct impression that the two fields have little in common.

Nothing could be further from the truth. Despite the fact that the fields of digital image processing and computer vision have traditionally been taught as separate courses—sometimes in separate departments—with little attention paid to their relationship, the two are in fact inseparable. Just as electricity and magnetism are taught together, or statics and dynamics, or algorithms and data structures, so too should image processing and computer vision. Over the past decade or so, it has become increasingly apparent that the overlap between these two fields can no longer be ignored, regardless of their distinct histories.

The title of this book was deliberately chosen to emphasize the seamless overlap between the two fields. Instead of *Image Processing and Computer Vision,* which could lead to the false impression that the two fields have little in common, the present title suggests that the two topics are intertwined and interrelated—two sides of the same coin. The term *Image Processing* is self-explanatory and carries the well-understood meaning, whereas the term *Image Analysis* is used to encompass all of computer vision while relaxing the often implicit restriction upon input modality (images taken by an optical camera). Together they form the dual field of *Image Processing and Analysis.*

Purpose

This book offers a comprehensive introduction to both of these exciting fields, in a format that is as accessible as possible. The text is designed for use in a senior-level undergraduate or first-year graduate course in computer science, electrical engineering, or related field of study. It should also serve as a useful reference for researchers and practitioners due to its emphasis upon real-world problems, practical algorithms, and implementation issues. The book covers hundreds of algorithms and techniques that are used every day in research and industry. It presents both the underlying mathematical concepts and principles behind these techniques, as well as detailed descriptions of the actual steps involved (in the form of pseudocode) in implementing the most commonly used algorithms. Throughout, an attempt has been made to keep the presentation accessible to all levels of readers by keeping the explanations as simple as possible and focusing on the core concepts. The book assumes some knowledge of probability, linear algebra, signal processing, programming, and algorithms and data structures. However, even readers deficient in these areas should be able to digest the essentials of the material without too much additional effort.

Selecting material for a book of this scope has been no easy task. In a rapidly-changing field such as computer vision, it is possible for a book to be obsolete even before it is published. Therefore, to help to maximize the relevance of the book, topics were selected according to the following criteria. First, any topic described in a research paper receiving at least a thousand citations was considered important enough to be included. Secondly, any algorithm or method that is widely used in industry, regardless of publication status, was deemed worthy of inclusion due to its practical relevance. Finally, foundational material was selected when it seemed necessary (or at least helpful) to understand other concepts. No doubt these principles were not applied perfectly: space limitations did not permit all topics to receive the attention they deserve, and some topics or papers may have been inadvertently overlooked. Nevertheless, it is hoped that this principled approach has resulted in a text that remains relevant for years to come (or at least until the ink dries on the page—or whatever is the digital equivalent).

The twin goals of comprehensiveness and accessibility are in tension with one another. Comprehensiveness involves both breadth of topics and depth of coverage. Some readers will no doubt find fault with my attempt to cover as many topics as I could with as much detail as I could. I confess to being guilty as charged. Indeed, I have intentionally painted both with a broad brush that covers tremendous ground for such a short book, as well as with a fine brush that insists upon mathematical and intellectual rigor wherever possible. To make the fire hose drinkable, however, I have tried my best to provide gentle introductions, to introduce topics in a graduated manner from simple to complex, to motivate the work with real-world examples, and to frequently bring the reader back to the "big picture" so as not to get lost in all the details. Nevertheless, my working assumption has been that, when in doubt, more information is better than less information, since the reader can always skip over material but cannot easily insert new material; I hope readers will agree with this philosophy.

One unique feature of the text is its approach to mathematical derivations and proofs. A common practice is to follow a long derivation with the result. By the time the result is reached, however, the reader has often become hopelessly lost in the details so as to forget the importance of the problem being addressed, and oftentimes readers are not even interested in the cumbersome process required to obtain the result, desiring only the result itself. To address this problem, I have, for the most part, followed the approach of presenting the result first, followed by the derivation later. This inversion of order serves both types of readers: anyone not interested in the derivation can simply skip it, whereas anyone curious about the details has access to them either at the time of reading or later for reference.

Organization

Two approaches for teaching image processing and computer vision are common: one begins with convolution and filtering, whereas the other starts with image formation, and in particular projective geometry. Neither of these approaches, however, is easiest for students: convolution and filtering are not the first operations that one considers when using an image editing program, and projective geometry involves abstract mathematical concepts that are intimidating for first timers. Moreover, these topics do not provide an underlying foundation for later topics, thus sometimes leaving students disappointed when they realize that their effort to master the math early in the course does not pay off when they encounter later topics that do not leverage the same mathematical concepts.

In this book a different approach is used. After the first two introductory chapters, students are presented immediately with extremely simple algorithms that allow them to appreciate the process of manipulating 2D image data. In other words, students are not asked to wade through complicated mathematics before experiencing the joy and wonder of seeing the result of image transformations, and no pretense is made that a single underlying mathematical theory will guide them through the rest of the book. Rather, the math is woven through the chapters as necessary.

For example, converting an RGB image to grayscale is something that everyone should learn in the first week, and simple approximate algorithms are sufficient for nearly all practical applications. In contrast, the correct formula for conversion requires a great deal of math and several advanced concepts in order to properly describe the process. A judgment call must be made therefore, whether to burden the reader up front with all the details, or to delay introducing such an important topic until halfway through the book. In this book, the dilemma is resolved by presenting the simple algorithm up front, then delaying the more advanced algorithm until the proper prerequisite material has been covered. While the resulting fragmentation is admittedly suboptimal, it far surpasses the alternatives. To help minimize the impact of such fragmentation, footnotes are liberally sprinkled throughout the text to point out connections between topics as an aid to the reader.

Another reason it is so difficult to organize this material is that these fields are nonlinear webs of knowledge rather than a linear sequence of topics building on one another. Neither image processing nor computer vision easily lends itself to a linear progression, and neither field requires a single underlying mathematical foundation. A technique like graph cuts, for example, can be used to solve a variety of problems, so a judgment call must be made as to which problem with which to associate it, or whether to assign it a separate section. Mean-shift filtering is closely related to bilateral filtering, but mean-shift segmentation belongs with other segmentation algorithms that have no relationship to bilateral filtering. Grayscale morphology is closely related to binary morphology, but there are many other algorithms for binary images that do not have analogs with grayscale imagery. The approach taken here is to make the math subservient to the problems being solved, with chapters and sections organized (with few exceptions) by problems to be solved rather than by the tools used to solve them. Again, footnotes help to connect the material in different sections.

Roughly speaking, the book begins with image processing and ends with computer vision, but I have deliberately tried to avoid introducing any artificial barriers between sections due simply to differences in their respective histories or communities. The book can be divided into three major areas:

- **Basic Concepts (Chapters 1-2).** An overview of the field, including motivating applications, along with some basic concepts in storing and accessing image data (Chapter 1). Natural vision systems, followed by image formation and acquisition, and a fairly detailed look at imaging modalities and electromagnetic radiation (Chapter 2).
- **Image Processing (Chapters 3-9).** A variety of practical, easy-to-understand algorithms requiring little to no mathematical background for transforming images (Chapter 3) and processing binary images (Chapter 4). Spatial- and frequency-domain filtering (Chapters 5 and 6), along with approaches for detecting edges and features (Chapter 7). Finally, compression (Chapter 8) and color representations (Chapter 9).
- **Image Analysis (Chapter 10-13).** The three core problems of computer vision / image analysis. First, techniques for segmenting dense pixels and fitting models to sparse data (Chapters 10 and 11). Then, methods for classifying pixels and images (Chapter 12).

Finally, problems involving multiple images, such as stereopsis, optical flow, camera calibration, and 3D reconstruction, along with the mathematics of projective geometry (Chapter 13).

There is enough material in the book for a two-course sequence on image processing and computer vision. The chapters follow a logical progression from simpler to more advanced topics, and there is inevitably some dependence between them. Nevertheless, the book has been designed to support a variety of different course types and academic schedules. Each chapter is relatively self-contained, so that it should be easy to select the chapters of interest without worrying whether important prerequisite material has been skipped. Within each chapter, the simple concepts are presented first, followed by more advanced ones, thus providing flexibility in picking and choosing which topics to cover, depending on the goals of the course. In fact, in many cases the chapters can even be covered out of order, as necessitated by the interests of the instructor or the needs of the practitioner.

Instructor Resources

A variety of resources are available to instructors via Cengage Learning's secure, password-protected Instructor Resource Center. These resources include the **Instructor's Solution Manual**, providing complete solutions to all problems from the text, as well as **Lecture Note PowerPoint slides, algorithmic pseudocode** and **processed images** in PowerPoint slides. To access these resources, please visit https://login.cengage.com.

I am deeply indebted to Carlo Tomasi, my Ph.D. advisor at Stanford, who introduced me to computer vision as only he could and at a time when the field was transitioning from the old world of running handcrafted algorithms on a handful of images to the new world of running machine-learned algorithms on millions of images. He taught me the importance of paying attention to detail and of always striving for excellence. Without his wise counsel and guidance, it is doubtful that I would ever have gotten involved in this field in the first place. Thanks are also in order to Robert Gray, who introduced me to the world of image processing at Stanford.

Many individuals read early versions of the manuscript and provided extremely valuable feedback. In particular, I wish to thank the following reviewers: Zekeriya Aliyazicioglu (California Polytechnic State University at Pomona), Saeid Belkasim (Georgia State University), Eliza Y. Du (Indiana University-Purdue University Indianapolis), Roger S. Gaborski (Rochester Institute of Technology), Arthur A. Goshtasby (Wright State University), Artyom Grigoryan (University of Texas at San Antonio), K. R. Rao (University of Texas at Arlington), Michael C. Roggemann (Michigan Technological University), Ezzatollah Salari (University of Toledo), Min C. Shin (University of North Carolina at Charlotte), and Jane Zhang (California Polytechnic State University at San Luis Obispo), as well as Serge Belongie (Cornell Tech and Cornell University), Raffay Hamid, Ashley Feniello, Greg Shirakyan, and several anonymous reviewers.

Some details of the algorithms were influenced by discussions with Yujie Dong, Satyajeet Bhide, and Michael Gillam. Zhengyou Zhang provided helpful discussions regarding calibration, and long-ago discussions with Chris Bregler shaped my presentation of Lucas-Kanade. Thanks to Ross Girshick for contributing one of the figures. Some of the problems at the end of the chapters were writted by Xueting Yu and Edwin Weill, whose assistance is greatly appreciated. Many thanks to all my former students, especially Neeraj Kanhere, Guang Zeng, Shrinivas Pundlik, Zhichao Chen, Vidya Murali, Bryan Willimon, Xiaoxia Huang, Ninad Pradhan, Brian Peasley, J. P. Kwon, Douglas Dawson, and Kalaivani Sundararajan, from whom I learned a great deal. Without the encouragement of Joshua Tarbutton to continue and complete this work when I was ready to give up, this book would likely not have happened. Thanks also to Olaf Hall-Holt, who has always been ready to provide encouragement when needed.

It has been a joy to work with the folks at Cengage Learning. I wish to express my gratitude for all the hard work of the Global Engineering team at Cengage Learning for their dedication to this new book: Timothy Anderson, Product Director; Mona Zeftel, Senior Content Developer; D. Jean Buttrom, Content Project Manager; Kristin Stine, Marketing Manager; Elizabeth Brown and Brittany Burden, Learning Solutions Specialists; Ashley Kaupert, Associate Media Content Developer; Teresa Versaggi and Alexander Sham, Product Assistants; Kristiina Paul, Text and Image Permissions Researcher, and Rose Kernan of RPK Editorial Services, Inc. They have skillfully guided every aspect of this text's development and production to successful completion. In addition, both Clemson and Microsoft provided extremely supportive environments.

Finally, this book would not have been possible without the loving support of my patient wife, MeMe, who, along with our children, sacrificed a great deal to allow me the time to work on this project. I hope I can make it up to them one day.

Stan Birchfield
Redmond, Washington

CHAPTER 1
Introduction

Vision is, without a doubt, our most dominant sense. With our eyes, we are able to navigate through complicated environments, detect and recognize the faces of our friends, and identify items to purchase on a shelf at a store. We see an object and reach for it, without even appreciating the immense complexity of the sensing task we have just performed to determine not only *what* object we are looking at but also *where* it is located. Indeed, it is nothing short of a miracle that we are able to process the signals resulting from visual stimuli on our retinas in order to make sense of the world around us. This pervasive reliance on vision has formed metaphors that permeate our daily vocabulary, such as:

"*Seeing* is believing." "A *picture* is worth a thousand words." "Our company needs a *vision* statement." "Don't you *see* what this means?" "They are like the *blind* leading the blind."

Inspired by the success of natural vision systems such as our own, it has long been the goal of scientists and engineers to harness the power of imagery to accomplish otherwise impossible tasks. Achieving this goal requires programming a computer to extract meaningful information from images, or, more generally, to use a computer to manipulate images in order to make the data that they contain more useful. The aim of this book is to introduce the basic concepts and algorithms necessary to prepare you to understand and use the algorithms for accomplishing this ambitious goal.

This is truly an exciting time to be studying this field. Not that long ago, such manipulation was restricted to researchers and specialists in the field, but nowadays any of us can acquire and manipulate digital imagery, given the ease with which we can snap a digital photograph or scan a document, and given the increasing levels of computational performance available. The fact that you are reading this indicates that you probably have at least some desire to understand in a deeper way the underlying principles and techniques for taking advantage of this newfound opportunity, so welcome to the fascinating world of images!

1.1 Image Processing and Analysis

This book covers both digital image processing and digital image analysis, where the adjective "digital" can safely be omitted these days since essentially all images are now available in digital form. In this chapter we will discover what exactly is meant by these two terms, how they relate to one another, and how their principles and algorithms are used every day in real-world systems. We will also cover some basic concepts regarding images, such as the representations used and the different types of images.

Since the fields of image processing and image analysis overlap significantly in their concepts, methods, and aims, it is difficult to know exactly where to draw the line between them. Nevertheless, a view that many have found helpful is to distinguish algorithms based on the type of output they produce. According to this view, **image processing** is the field of study in which algorithms operate on input images to produce output images, whereas **image analysis** is the field of study in which algorithms operate on images to extract higher-level information. In other words, an image processing algorithm outputs another image, whereas an image analysis algorithm outputs a nonimage type of data structure. Another way to think about the division is to consider image processing algorithms to be low-level in nature, whereas image analysis algorithms are more high-level, although not all algorithms are easily classified in this manner.

Three primary problems of image processing are shown in Figure 1.1. The first, known as **enhancement**, involves transforming an input image into another image so as to improve its visual appearance. An example of enhancement is to brighten an originally dark image, or to increase the contrast of an image to make the details more visible. Another example is to detect the intensity edges of an image in order to highlight the boundaries of objects, or to colorize a grayscale image (usually with false colors, known as *pseudocolors*) to make the different data values more distinguishable to a human observer. **Restoration**, the second problem, has as its purpose to restore an image that has been corrupted by some type of noise. The corruption may have been caused by noise introduced by the sensor, noise added during the transmission of the signal, or noise introduced by some external process. The third problem, **compression**, involves storing an image with fewer bits than are required by the original signal, while affecting viewing quality of the decompressed image as little

Figure 1.1: Three example problems of image processing. Top: A dark image, an image corrupted by noise, and a clean image. Bottom: the results of contrast enhancement, image restoration, and compression. The latter shows intentionally poor quality to better illustrate the effects of the operation.

Enhancement   Restoration   Compression

as possible. Compression algorithms can be applied either to a still image or to a video sequence. To solve these three types of problems, image processing utilizes concepts such as image transformations, linear and nonlinear filtering, and frequency-domain processing.

Three primary problems of image analysis are shown in Figure 1.2. **Segmentation** is the process of determining which pixels in an image belong together, that is, which pixels are projections of the same object in the scene. Segmentation can be viewed as a *bottom-up process* in which pixels are grouped together based upon low-level, local properties of the pixels and their neighbors, without any model of the particular object in the scene that produced the group of pixels. In contrast, the problem of **classification** involves determining which pixels in an image belong to a model that has been created beforehand. Classification is a *top-down process*, relying upon a human trainer or some other system to facilitate the creation of the model to which the pixels will be compared. If you have ever seen the display on a digital camera outlining the faces of all the people, that is the result of classification. The third problem, **shape from X**, aims to recover the three-dimensional (3D) structure of the scene using any of a variety of techniques (hence the "X"), such as stereo, video, shading, or texture. To solve these three types of problems, image analysis utilizes concepts such as linear algebra, statistical analysis, projective geometry, and function optimization.

The goal of image analysis is for the computer to be able "to see," because algorithms analyze images in order to extract useful information about the world. In this sense, image analysis is nearly synonymous with both machine vision and computer vision. **Machine vision** typically refers to systems in an industrial setting in which the placement of the camera and lighting conditions can be controlled, and the scene being viewed by the camera is, for the most part, two-dimensional (2D), such as parts on a conveyor belt. **Computer vision**, on the other hand, refers to systems operating on images taken in unstructured settings, such as those taken by ordinary people in everyday life using their personal digital cameras, or by a mobile robot navigating through unknown territory. We will often use these three terms interchangeably, since the distinction between them is too subtle to be important in most contexts. Nevertheless, as summarized in Table 1.1, it is proposed here to use the term image analysis to encompass techniques applicable to images from any type of sensor, optical or otherwise, whereas machine and computer vision refer to techniques that are applied to images obtained by a traditional camera capturing visible light.

Although the set of six core problems above is not necessarily exhaustive, it is truly remarkable how many problems that arise in practice are instances of one of them.

Figure 1.2: Three example problems of image analysis. Top: input images. Bottom: From left to right, the results of color-based segmentation, human face detection (a type of classification), and 3D reconstruction.

 Segmentation   Classification   Shape from X

TABLE 1.1: Comparison between image processing and analysis, machine and computer vision.

	environment	sensor	algorithm	output
image processing	any	any	low-level (2D)	another image
image analysis	any	any	low- to high-level	nonimage
machine vision	industrial	camera	low-level (2D)	nonimage
computer vision	everyday	camera	mid- to high-level	nonimage

For example, thresholding an image is really a segmentation problem, because the goal is to determine regions in the image whose pixels belong together. Edge detection is a type of classification problem, because the goal is to determine whether each pixel is an edge pixel or not. Tracking involves matching a query set of pixels in the current image to a model of the target captured in previous images, which is an inherent classification problem.[†] Image inpainting is a type of image restoration because it aims to reconstruct missing data in the image, and computed tomography is a type of shape from X because its goal is to determine the 3D structure of the object being viewed. And so on.

1.2 History and Related Fields

Some perspective can be gained by looking at a brief history. Image processing was born in the mid-1960s due to the convergence of two phenomena: First, the space program began to transmit priceless images of the moon back to earth, which happened to be distorted; and secondly, digital computers were becoming powerful enough to perform useful tasks such as removing that distortion. Before the decade was over, a wave of other applications of image processing began to assert themselves, such as medical imaging, remote sensing, and document image analysis. In the 1970s commercially viable machine vision systems were introduced to inspect manufactured parts for defects, a thriving industry that continues today. The 1980s saw the expansion of machine vision systems into the transportation industry, among other areas, as images were processed automatically to detect vehicles on the highway in a variety of weather and lighting conditions. Meanwhile, researchers in computer vision were laying the foundation of solutions to many problems in the field in the 1970s and 1980s. Although fundamental breakthroughs were achieved throughout the 1990s, it was not until the mid-2000s that computer vision began to impact commercial products, thanks to the convergence of faster processing, inexpensive sensors, and the availability of large amounts of training data. Today the computer vision market is booming, with application areas multiplying faster than developers are able to tackle them.

Interwoven throughout this history is a rich interplay between image processing, image analysis, and several closely related fields, such as those illustrated in Figure 1.3:

Vision science. Scientists in **psychophysics** study the relationship between physical stimuli and the resulting perceptual sensations that they cause. Since the mid-19th century, such scientists have spent considerable effort to understand how the human visual system operates by studying its reaction to different types of scenes and environmental conditions. This has given rise to heated debates between schools of thought like structuralism, gestalt psychology, ecological optics, and constructivism. Some of the more well-known figures are Hermann von Helmholtz (1821–1894), who conducted some of the first psychophysical experiments, Max Wertheimer (1880–1943), one of the main proponents of *gestalt psychology* which emphasized grouping as the key to visual perception, and J. J. Gibson (1904–1979), one of the most influential researchers in visual perception of

---

[†] This connection is particularly evident in the recent interest in "tracking by detection" approaches. (Note that detection is a type of classification.)

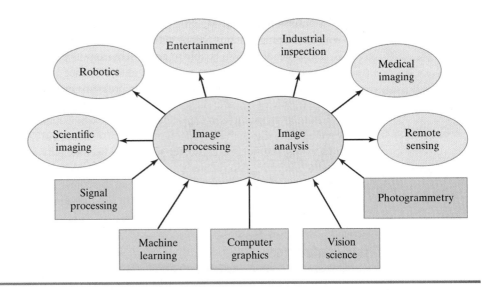

Figure 1.3: Image processing and analysis, along with related fields (bottom rectangles) and sample applications (top ovals).

the 20th century who contributed to the training of airplane pilots, in which they learned to orient themselves via visual cues on the ground. More recently, there has been a flurry of activity in cross-disciplinary work, in which psychophysical researchers apply computational models to more precisely characterize the operations of the human visual system, while computer scientists use techniques inspired by psychophysical models to propose new computational algorithms. The field at the intersection between these two approaches, known as *vision science*, has been important in establishing principles regarding perceptual quality for applications such as image compression. Nevertheless, while much progress has been made, the actual workings of the human vision system remain largely a mystery.

Photogrammetry. As its name suggests, photogrammetry involves making metric measurements from photographs. Starting in the mid-19th century, sophisticated techniques were developed to facilitate the creation of accurate, detailed 3D terrain maps using images captured by cameras mounted on kites, balloons, and aircraft. Before the advent of digital computers, such calculations were carried out meticulously by hand by carefully measuring the image coordinates of points on high-resolution photographs, then using the machinery of projective geometry to infer 3D coordinates. Many of these techniques, such as triangulation and bundle adjustment, are still widely used today in automated 3D reconstruction systems.

Signal processing. With the advent of electronic forms of communication near the turn of the 20th century, such as radio, telephone, radar, and television, the need to process these one-dimensional (1D) electronic signals became important. The field of signal processing, and later digital signal processing, is concerned with filtering signals in order to reduce the effects of noise, enhance the information that is present, or make better use of the available bandwidth. The origins of image processing lie in the extension of one-dimensional digital signal processing techniques to two-dimensional images.

Computer graphics. While the goal of computer vision is to infer a model of the world from sensor data, the goal of computer graphics is the exact opposite: to create an image from a model of the world. As such, the two fields overlap in their shared use of the mathematics of geometric optics, in particular projective geometry. In recent years, there has been a surge of interest in applications that intersect both fields, such as augmented reality, urban and archaeological site modeling, medical visulization, facial animation, teleimmersion, and telecollaboration. Motion capture of actors, as well as the automatic computation of optical flow, is used to produce a variety of special effects for movies, such as retiming, artificial motion blur, image-based animation, and non-photorealistic rendering.

Machine learning. Machine learning is a branch of artificial intelligence concerned with developing systems whose output improves as more empirical data are provided; that is, a learning algorithm is able to generalize from its experience. A tight connection exists between machine learning and image analysis since two of the main areas of machine learning map directly into two of the main problems in image analysis. Image segmentation is an example of *unsupervised learning*, which aims to find clusters in data, whereas classification is an example of *supervised learning*, which makes decisions based upon labeled training data.[†] Since the introduction of the first successful face detection algorithms in the mid-1990s, the field of computer vision has been heavily influenced by the paradigm of machine learning. Many long-standing elusive problems are now beginning to be tractable by providing large amounts of training data to sophisticated machine learning algorithms that extract the desired underlying properties of the signals.

1.3 Sample Applications

Due to the explosion in the use of image processing and analysis over the past several decades, it is not difficult to find a myriad of real-world applications in which these technologies are used every day. Chances are pretty good, for example, that you or someone you know snapped a picture or video with your smartphone recently, which was subsequently compressed using techniques from image processing. In the same manner, machine vision is a thriving, mature, and growing multibillion dollar industry that is used to improve the quality of manufactured products. Although computer vision has only recently begun to find profitable niches, these application areas will inevitably multiply over the coming decades as the technology matures to handle the difficult issues that arise in unstructured settings. Some of the more important application areas of these fields are highlighted in Figure 1.4.

Industrial inspection. Machine vision systems are commonly used to inspect manufactured parts for defects, particularly in the semiconductor industry where the sensed semiconductor wafer is compared with a model template to detect defects. Similar systems are also used to identify missing components or broken traces on printed circuit boards, missing pills in pharmaceutical packaging, defects in fiber bundles, errors in packaging labels, or missing tamper bands on consumer products. Other systems inspect and measure machined parts such as automotive engines to ensure alignment and tolerance specifications are met, while yet others inspect food to identify foreign objects accidentally dropped in bread loaves, diseased corn, or blemishes in fried potatoes.

Document image analysis. Another mature application area is the automated analysis of documents. The postal service routinely uses optical character recognition (OCR) technology to automatically read the characters and numerals printed on envelopes to sort the mail. Similar methods have been used to build reading machines for the blind and automated license plate recognition (ALPR) systems to read the license plate numbers of vehicles from high-resolution cameras controlled by an external trigger. Comparing the captured image with a template is also one way that vending machines are able to verify that a dollar bill inserted in the machine is genuine. Using similar techniques, the now ubiquitous QR (quick response) codes are two-dimensional bar codes that are capable of being quickly read by smartphones to reveal product data or other information.

Transportation. Cameras mounted on poles on the side of the road are used to automatically determine the volume of traffic and occupancy of the roadway by measuring pixel

---

[†] The third area of machine learning, *reinforcement learning*, is not as obviously related, although it has been used in tracking and interactive systems.

Figure 1.4: Sample applications. From left to right, top to bottom: industrial inspection, optical character recognition, tracking vehicles on a highway, detecting a drowning person at the bottom of a pool, photgrammetry, detecting tree roots in an underground image, medical imaging, robotic assembly, moviemaking.

changes in the video relative to a background image. Cameras installed at intersections are used to determine the presence of vehicles in individual lanes in order to control the traffic signal or to automatically take a snapshot of any driver who illegally runs a red light. Other systems count the number of vehicles passing through an intersection, inspect railroad tracks for fatigue or corrosion, or detect stray, fallen parts on an airport runway for safety purposes. Thermal infrared cameras are increasingly being used for their insensitivity to shadows, rain, glare, or other environmental conditions. Cameras are also being deployed on vehicles themselves, with integrated computer vision algorithms automatically detecting the headlights of oncoming vehicles, pedestrians in front of the vehicle, and inadvertent lane departure. They are also used to automate parallel parking.

Security and surveillance. Biometric devices are used to read fingerprints, recognize irises, and identify faces for contact-free access control. X-ray security scanners at airports are able to detect banned objects in luggage, while full body scanners utilize the backscatter X-ray or millimeter waves. Security cameras installed around the perimeter of high-security areas, within public areas, or around places of business are primarily used for manual viewing either during an incident or afterward, although efforts have been made to automate the detection of intruders. Such cameras are also used to track people through shopping areas to determine purchasing habits and product interest. Underwater video cameras continuously watch for motionless people at the bottom swimming pools to alert lifeguards to save them from drowning.

Remote sensing. Information regarding the earth is collected by acquiring and processing data from multiple spectral bands obtained by sensors on aircraft flying over specific locations or satellites orbiting the planet. Some of the goals of remote sensing are to identify land features, measure the amount of vegetation, locate ore deposits, measure the temperature of land and water, and estimate changes in sea level. The large number of images collected continuously from orbiting satellites over the past several decades provides a long-term record of changes to the natural landscape due to either natural or human causes.

Scientific imaging. Scientists in a variety of fields use imaging to study and measure phenomena of interest. Biologists track live cells in time-lapse microscopy images, bio-image informaticists analyze cells using light or electron microscopy, and horticultural-ists estimate tree health by measuring the growth of roots using cameras in underground minirhizotron tubes. Space scientists use triangulation from cameras that are miles apart to estimate wind direction, while astronomers use speckle imaging techniques to increase the resolution of ground-based telescopes for viewing faint stars and other distant phenomena. Chemists use atomic force microscopes (AFMs) to view extremely fine details such as the chemical bonds linking atoms in a molecule. For studying the flow of liquids or gases, par-ticle imaging velocimetry (PIV) provides scientists with the instantaneous velocity profile of the flow field. Most scientific imaging is still done manually, with basic low-level image processing and analysis routines aiding the human viewer in conducting measurements.

Medical imaging. One of the largest areas of active research is medical imaging, in which images of the human body are captured using a variety of imaging modalities to detect tumors, diagnose diseases, verify whether a bone has been broken, and view neural activity in the brain to identify the region that is responsible for a certain type of processing. These different types of images are registered so that they may be overlaid on one another to create 3D models to aid visualization. Medical imaging is also used to guide surgery, and images captured from tiny images on the end of a catheter allow physicians to see the block-age of arteries and other phenomena that would be difficult to sense otherwise. Although much effort has been spent automating medical image analysis, the images are primarily interpreted by a trained professional, as in scientific imaging.

Robotics. Commercial industrial robotic systems use machine vision for quality inspec-tion during operations such as parts feeding, manufacturing assembly, arc welding, and automatic wire bonding. Computer vision also plays an important role in mobile robotics systems that navigate an environment, follow a certain person, or build a map of a building. Computer vision systems are just now beginning to reach levels of robustness that allow them to be deployed in real unmanned systems operating in unstructured environments, whether in smaller robotic systems or larger autonomous vehicles.

Entertainment. With the proliferation of cameras on smart phones, an emerging area is that of *computational photography*, in which specialized optics and image processing can be used to produce high dynamic range images, all-focus images, or high-resolution mosaics and panoramas. Another application area is human-computer interaction, in which natural user interfaces allow a user to control a computer by sending the user's gestures from a camera. With the increasing availability of inexpensive depth sensors, such applications are becoming mainstream. In the sports world, cameras are used to display the location of first down markers in American football, the pitch speed in baseball, the angle of the shot in basketball, the location of the puck in hockey, the identities of boats in sailing races, and other metadata to enhance the visual experience for viewers. In the moviemaking industry, computer vision techniques are now the standard way of combining computer generated imagery (CGI) with live action footage by tracking features in the video to determine the camera motion.

1.4 Image Basics

In preparation for the material in the rest of the book, we now consider some of the basic concepts in storing and representing images, as well as some of the conventions that we will use. When a camera (or alternative imaging device like those we will see in the next chapter) forms an **image** of the scene, it captures in some way a likeness of the scene. In fact, the word *image* comes from the Latin word (*imago*) meaning "likeness," which is why when we look in the mirror we say that we see an image of ourselves, because the person inside the mirror looks just like us. Because the image captured by a camera is usually a digitized version of some two-dimensional sensory input, it is appropriately called a **digital image**.

1.4.1 Accessing Image Data

At its most basic level, then, a digital image is simply a discrete two-dimensional array of values, much like a matrix. We use *width* to refer to the number of columns in the image, and *height* to refer to the number of rows, so that the dimensions of the image are *width* by *height*, represented as *width* × *height*, and the **aspect ratio** is width divided by height, or *width* / *height*. Each element of the array is known as a **pixel**, which is short for "picture element." Pixel values are accessed by a pair of coordinates (x, y), where x and y are nonnegative integers. For a grayscale image I, the value v of the pixel at coordinates (x, y) is given by

$$v = I(x, y) \tag{1.1}$$

Sometimes we will find it more convenient to represent pixel coordinates using a vector. According to the standard convention, each vector is vertically oriented, while its transpose is horizontally oriented:

$$\mathbf{x} = \begin{bmatrix} x \\ y \end{bmatrix} = [x \ y]^\mathsf{T} = (x, y) \tag{1.2}$$

where the boldface indicates a vector, and the superscript $^\mathsf{T}$ indicates the transpose operator. We will use the vector and coordinate notation interchangeably, so that $I(x, y) = I(\mathbf{x})$. In the case of a color image, each pixel contains multiple values, which we represent as another vector, $\mathbf{v} = I(\mathbf{x})$, that contains the values of the different color channels, e.g., $\mathbf{v} = (v_{red}, v_{green}, v_{blue})$.

For accessing the pixels we adopt the convention that the positive x axis points to the right and the positive y axis points down, so that x specifies the column and y specifies the row, as depicted in Figure 1.5. We also assume zero-based indexing, so that the top-left pixel is at $(0, 0)$. Other conventions are possible, but this coordinate system has the advantage that it is closely tied to the way images are typically stored in memory, and, although in 2D this is a left-handed coordinate system, in 3D the right-hand rule causes the z axis to point toward the scene along the camera's optical axis, which is convenient when performing 3D reconstruction.

Despite the fact that an image is actually a 2D array, it is stored in memory as a 1D array. Sometimes images are stored in **column major order**, that is, the first column is stored, then the second column, then the third column, and so on until the last column. More commonly, however, they are stored in **row major order**, also known as raster scan order. Hearkening back to the days when images were displayed on a cathode ray tube (CRT) by an electron gun scanning the tube one row at a time, a **scanline** is one row of an image; **raster scan order** therefore refers to storing the first row, then the second row, then the third row, and so on until the final row. Since a CRT display always begins its scan at the top-left corner of the image and proceeds downward, this is the historical basis for setting the origin at the top-left pixel.

The elements of this 1D array have indices $0, 1, 2, \ldots, n - 1$, where $n = width \cdot height$ is the number of pixels in the image, and the dot $(\cdot)$ indicates ordinary multiplication. If we let i refer to the index of this 1D array, then the first pixel at $(x, y) = (0, 0)$ has the 1D index $i = 0$. Assuming row major order, the second pixel at $(1, 0)$ has the index $i = 1$, the third pixel at $(2, 0)$ has the index $i = 2$, and so on. If the pixels are stored contiguously, then the last pixel of the first row at $(width - 1, 0)$ has the index $i = width - 1$, while the first

Figure 1.5: Top: Image as a 2D array, showing the 1D index of each pixel. Bottom: Internal representation of image as a 1D array using row major order.

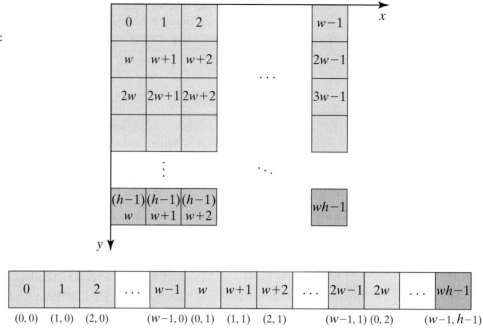

pixel of the second row at $(0, 1)$ has the index $i = width$. From this, it is easy to see that the 1D index can be obtained from the 2D coordinates as follows:

$$i = y \cdot width + x \tag{1.3}$$

and the inverse relationship is given by

$$x = \mathrm{mod}(i, width) = i - y \cdot width \tag{1.4}$$

$$y = \lfloor i \,/\, width \rfloor \tag{1.5}$$

where mod (a, b) is the modulo operator that returns the remainder of a divided by b, and the floor operator $\lfloor c \rfloor$ returns the largest integer that is less than or equal to c.

1.4.2 Image Types

Several types of images exist. In a **grayscale image**, the value of each pixel is a scalar indicating the amount of light captured. These values are quantized into a finite number of discrete levels called **gray levels**. If b is the number of bits used to store each pixel value (called the **bit depth**), then 2^b is the number of gray levels, which we shall refer to as *ngray*. Usually there are eight bits (one byte) per pixel, so that $ngray = 2^b = 2^8 = 256$. Therefore, in an 8-bit grayscale image, a pixel whose value is 0 represents black, whereas a pixel whose value is 255 represents white. All the bits of a black pixel are 0, whereas all the bits of a white pixel are 1, so using hexadecimal notation these values are 00 and FF, respectively. Some specialized applications such as medical imaging require more quantization levels (e.g., 12 or 16 bits per pixel) to increase the dynamic range that can be captured, but we will generally assume 8 bits per pixel to simplify the presentation; the extension to larger bit depths is straightforward.

In an **RGB color image**, the pixel values are triples containing the amount of light captured in the three color channels: red, green, and blue. Color images, therefore, usually require 24 bits per pixel, or one byte for each of the three color channels. For an RGB color image, a black pixel has hexadecimal value 000000, while a white pixel has value FFFFFF. Although the bytes could be stored in the order of red-green-blue (*RGB*), with blue as the lowest-order byte, most frame buffers and frame grabbers adopt the reverse convention in which the order is blue-green-red (*BGR*), so that red is stored as 0000FF. The values for the different color channels are usually stored in an **interleaved** manner, that is, all three values for one pixel are stored before the three values of the next pixel, as in $B_0 G_0 R_0 B_1 G_1 R_1 B_2 G_2 R_2 \cdots B_{n-1} G_{n-1} R_{n-1}$, where the subscript is the pixel index. An alternate approach is to store the color channels in a **planar** manner, so that the red, green, and blue channels are stored as separate one-byte-per-pixel images, as in $B_0 B_1 B_2 \cdots B_{n-1} G_0 G_1 G_2 \cdots G_{n-1} R_0 R_1 R_2 \cdots R_{n-1}$. Either way, sometimes a fourth value is associated with each pixel, called the **alpha value** or the **opacity**, which is used for blending multiple images, as in $B_0 G_0 R_0 A_0 B_1 G_1 R_1 A_1 B_2 G_2 R_2 A_2 \cdots B_{n-1} G_{n-1} R_{n-1} A_{n-1}$, in which case 32 bits are associated with each pixel; an alpha value of 00 indicates complete transparency, whereas an alpha value of FF indicates that the color is fully opaque.

Although grayscale and RGB color images are used for capture and display, the processing of images leads to several additional types. First, there is the **binary image**, which arises from applying a propositional test to each pixel. The most common test is that of thresholding, in which case each pixel in the output image receives the logical value ON or OFF (or equivalently TRUE or FALSE, respectively) depending upon whether the value of the input pixel is above or below a given threshold. These logical values can be stored using one bit per pixel, (0 for OFF or 1 for ON), or they can be stored using one byte per pixel, where their values are usually 0 (hexadecimal 00) or 255 (hexadecimal FF). Although this latter practice is somewhat wasteful, it is often more convenient for both display and processing. We adopt the convention that OFF is displayed as black, whereas ON is displayed as white, when the binary image is displayed as an image; we reverse this convention when graphically depicting algorithms, where OFF is displayed as white, and a color such as blue or orange is used for ON. This minor inconsistency arises naturally from the fact that, although black is the color of a blank screen, white is the color of a blank piece of paper.

Another type of image is the **real-valued image**, or **floating-point image**, in which each pixel contains a real number, at least conceptually. In practice, the number is stored in the computer as an IEEE single- or double-precision floating point number, in which case the number requires 32 or 64 bits, respectively, to be stored. A single-precision number can represent any integer in the range $\left[-2^{24}, 2^{24}\right]$ exactly, and it can represent any real number in the approximate range $\left[-10^{38}, 10^{38}\right]$ with an accuracy of about 10^{-7}. A double-precision number can represent any integer in the range $\left[-2^{53}, 2^{53}\right]$ exactly, and it can represent any real number in the approximate range $\left[-10^{308}, 10^{308}\right]$ with an accuracy of about 10^{-16}. Unlike *signal processing*, which often involves numerically delicate operations that require double-precision, for *image processing* it is difficult to find situations for which single-precision is not sufficient. In fact, an increasingly common format stores images using half-precision, which requires just 16 bits per pixel. A half-precision number can represent any integer in the range $\left[-2^{11}, 2^{11}\right]$ exactly, and it can represent any real number in the range $\left[-65535, 65535\right]$ with an accuracy of about 0.001. These numbers, which are summarized in Table 1.2, arise from the general rule that if e and s are the number of exponent and significand bits, respectively, then the range of exact integers is $\left[-2^{s+1}, 2^{s+1}\right]$, the entire range is $\left[-2^{2^{e-1}}, 2^{2^{e-1}}\right]$, and the accuracy is $s \log_{10} 2$. Floating-point images are useful not only to store the results of arithmetic operations, but also for high dynamic range images and radiance maps.

Some image processing algorithms output an **integer-valued image** in which the value of each pixel is an integer. Integer-valued images arise whenever it is necessary to store

precision	number of bits				range		accuracy
	sign	exponent	significand	total	integers	reals	
half	1	5	10	16	$[-2^{11}, 2^{11}]$	$[-10^4, 10^4]$	10^{-3}
single	1	8	23	32	$[-2^{24}, 2^{24}]$	$[-10^{38}, 10^{38}]$	10^{-7}
double	1	11	52	64	$[-2^{53}, 2^{53}]$	$[-10^{308}, 10^{308}]$	10^{-16}

TABLE 1.2: Half-, single-, and double-precision floating point representations.

negative numbers or somewhat arbitrarily large numbers. For example, to label each pixel with the region to which it belongs, we obviously cannot store the result in a grayscale image if there are more than 256 regions. Similarly, the subtraction of two images, which will in general contain negative numbers, cannot be stored in a grayscale image. Although in practice an integer-valued image uses a finite number of bits per pixel (usually 32 or 64), these values are large enough that the chance of overflowing the buffer is usually not a practical concern. A 32-bit integer, for example, can represent all integers between approximately -10^9 and 10^9, and a 64-bit integer can represent the integers from approximately -10^{19} to 10^{19}, both of which are extremely large ranges.

Finally, images can have multiple **channels**. We have already seen, for example, that an RGB color image is an 8-bit image with three channels. Similarly, after transforming from RGB color space to another color space, the result can be stored as a multichannel image, either real-valued or 8-bit. Another common multichannel image type is a **complex-valued image**, which arises from computing the Fourier transform of an image. A complex-valued image contains two floating-point values for each pixel, one for the real component and one for the imaginary component. Similarly, a multichannel integer-valued image might store the (x, y) coordinates of another pixel associated with each pixel, or a set of regions to which the pixel might belong.

These image types are summarized in Table 1.3. In this book we shall exercise care to maintain the distinction between the different types in order to support applications for which speed and memory considerations warrant this extra level of detail. Real-time applications tend to squeeze the result into as few bits as possible, so that grayscale and RGB color images are commonly used not only for capture and display, but also for holding results that may conceptually be considered integers or real values. The reason for this is that, although memory itself is cheap, processing time is greatly affected by the amount of memory used, due to the relatively high cost of cache misses and page swaps. Although the type of image should either be clear from the context or mentioned explicitly, when in doubt it will always be safe to assume (if computation is not an issue) the most general model, namely that of a multichannel floating-point image. Such a model is flexible enough to hold all of the image types mentioned (grayscale, RGB color, binary, integer, real, complex, and other color spaces), as well as any others that you will ever encounter.

	grayscale	RGB color	binary	integer-valued	real-valued	complex-valued
channels	1	3	1	1	1	2
bit depth	8	24	1	32/64	32/64	64/128
value range	$\{0, \dots, 255\}$	$\{0, \dots, 255\}^3$	$\{0, 1\}$	$\mathbb{Z}$	$\mathbb{R}$	$\mathbb{R}^2$

TABLE 1.3: Common image types, shown with the number of channels, the most commonly encountered bit depth (number of bits per pixel), and the set of possible values. In the final three columns this set is conceptual only, since the integers $\mathbb{Z}$ and real numbers $\mathbb{R}$ are infinite sets.

1.4.3 Conceptualizing Images

We normally think of an image as a picture. That is, if we display the image so that the brightness of each tiny region on the screen or page is proportional to the value of a pixel, then the representation is easily interpreted by viewing it. There are several other ways to conceptualize an image, however, as shown in Figure 1.6, each of which provides additional insight into the algorithmic processing of images.

At its most basic level a digital image is stored in the computer as a discrete array of values, which can be visualized either by considering the raw pixel values themselves arranged in a 2D lattice, or equivalently as a height map, or 3D surface plot, where the height of each point is the value of the pixel. Alternatively, an image can be considered as a function that returns the value given the coordinates of a pixel. In this case, $I(x, y)$ means to evaluate the function at the position (x, y). If x and y are restricted to nonnegative integers in the domain of the image, then the function is equivalent to accessing a 2D array. However, if we expand the domain of each axis of the function to the entire set of real numbers, then it allows us to capture the values of the image even when accessed out of bounds. For example, $I(-1, -1)$ makes no sense when I is viewed as a 2D array, because $(-1, -1)$ would cause a memory access violation; but when viewed as a function, $I(-1, -1)$ yields a value that is computed from the nearby pixels, e.g., the value of the nearest pixel. Similarly, the parameters to $I(2.5, 3.5)$ would have to be rounded if the image were accessed as an array, but as a function we can define an appropriate interpolation function to compute values between pixels.

Another way to conceptualize an image is as a set of pixels. In its most general form, this set contains triplets of values capturing both the coordinates and values of the pixels. For example, the grayscale image

$$I = \begin{bmatrix} 3 & 8 & 0 \\ 2 & 9 & 4 \end{bmatrix} \tag{1.6}$$

can be represented as $\{(0, 0, 3), (1, 0, 8), (2, 0, 0), (0, 1, 2), (1, 1, 9), (2, 1, 4)\}$. However, this representation is most commonly used for binary images, where the set is

Figure 1.6: Different ways to visualize an image: as a picture, as a height map, as an array of values, as a function, as a set, as a graph, and as a vector. The 5 × 4 array is a small portion of the image; the set contains the coordinates of all pixels in the array whose value is greater than 80; and the weights of the edges in the graph are the absolute differences between values in the array.

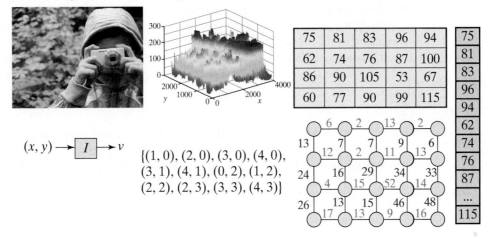

usually simplified to contain just the coordinates of those pixels whose value is on, that is, $\{(x, y) : I(x, y) = \text{ON}\}$. For example, the binary image

$$I = \begin{bmatrix} 1 & 0 & 1 \\ 1 & 1 & 1 \\ 1 & 0 & 1 \end{bmatrix} \tag{1.7}$$

can be represented as the set

$$\{(0, 0), (2, 0), (0, 1), (1, 1), (2, 1), (0, 2), (2, 2)\}. \tag{1.8}$$

An image can also be viewed as a graph, where each pixel of the image is a vertex in the graph, and each edge in the graph connects two pixels that are adjacent in the image. The weight associated with each edge is usually some measure of the similarity or dissimilarity in value between the two pixels. For example, the image in Equation (1.6) can be represented as a graph with 6 vertices and 7 edges, where the weights of the edges are given by the absolute difference between neighboring pixels:

Occasionally it is useful to view an image as a matrix. Since a matrix is a 2D array of values, this representation is easy to imagine. The only difficulty is that, for historical reasons, the conventions for matrices and images are different. Matrix entries are accessed using one-based indexing, so the top-left entry is at position (1, 1) rather than (0, 0). Also, matrices are indexed first by their row, then by their column, so the entry just to the right of the top-left entry is at position (1, 2), and an $m \times n$ matrix has m rows and n columns (as opposed to a $w \times h$ image, which has w columns and h rows). To avoid confusion, we will use boldface to indicate matrices, and we will access matrix entries using subscripts. Thus, if $\mathbf{A}$ is an $m \times n$ matrix whose $(i, j)^{\text{th}}$ entry is given by a_{ij}, we will write

$$\mathbf{A}_{\{m \times n\}} = \begin{bmatrix} a_{11} & a_{12} & \cdots & a_{1n} \\ a_{21} & a_{22} & \cdots & a_{2n} \\ \vdots & \vdots & \ddots & \vdots \\ a_{m1} & a_{m2} & \cdots & a_{mn} \end{bmatrix} \tag{1.9}$$

where the braces in the subscript of the matrix indicate its dimensions.

Finally, it is sometimes useful to view the image as a vector, which is obtained by either concatenating the columns of the image or by concatenating the rows and transposing the result. Adopting the latter approach, if we let $v_i = I(x, y)$ be the value of the pixel at (x, y) according to the 1D indexing of Equation (1.3), then the resulting vector is given by

$$\mathbf{v} = \begin{bmatrix} v_0 & v_1 & v_2 & \cdots & v_{n-1} \end{bmatrix}^{\mathsf{T}} \tag{1.10}$$

where n is the number of pixels in the image. This vector is a point in an n-dimensional space, so if we let each pixel take on a real value for simplicity, then $\mathbf{v} \in \mathbb{R}^n$. The vector notation allows us to imagine linear transformations of the image that involve multiplying the vector by an $m \times n$ matrix $\mathbf{T}$ on the left-hand side to produce a new vector $\mathbf{v}' = \mathbf{T}\mathbf{v}$:

$$\mathbf{v}'_{\{m \times 1\}} = \mathbf{T}_{\{m \times n\}} \mathbf{v}_{\{n \times 1\}} \tag{1.11}$$

If $\mathbf{T}$ is the $n \times n$ identity matrix $\mathbf{I}_{\{n \times n\}}$, then the input is unchanged: $\mathbf{v}' = \mathbf{v}$. More interestingly, $\mathbf{T}$ may be defined appropriately to translate or rotate the image, perform bilinear interpolation, downsample, upsample, crop, or extend past the borders as needed. Other linear operations, such as convolution and the Fourier transform, can also be represented in this way, as we shall see later in the book.

1.4.4 Mathematical Prerequisites and Notation

To successfully master the material in this book, it is necessary to be at least somewhat familiar with three areas of mathematical study. First, it is important to be comfortable with the basic concepts of linear algebra, such as matrices, vectors, matrix multiplication, and solving linear systems. Secondly, it is helpful to have some familiarity with probability and statistics, so that you know what is meant by joint probability, conditional probability, or a probability distribution function (PDF). Finally, the work will be easier if you already have been exposed to signal processing, so that discrete signals, convolution, and the Fourier transform are not entirely new concepts. Having said that, this book aims to ease the transition as much as possible by explaining concepts at an elementary level, so that having some deficiencies in these areas should not prevent anyone from progressing through the material and digesting most of it.

Because image processing and analysis are at the intersection of a number of different mathematical traditions, developing a clear and consistent notation is a challenge. The goal of this book has been to strike a balance between using notation that is internally consistent on the one hand, while at the same time maintaining consistency with existing conventions whenever possible. The result is the following set of notational conventions, which are used throughout the book almost everywhere. This list may not be interesting upon first reading, but you may find it helpful to refer to it from time to time as you progress through the book. On a few occasions these conventions are violated in order to adhere to existing widely established conventions, but the context should make the meaning clear wherever this occurs.

g, ψ	Lowercase Latin or Greek characters indicate scalars
g, ψ	Lowercase Latin or Greek characters also indicate functions of one variable
G, Ψ	Uppercase Latin or Greek characters indicate functions of more than one variable
$\mathcal{A}$	Uppercase calligraphic Latin characters indicate sets
$g(x)$	Either the 1D function g evaluated at x, or the function g itself
$G(x, y)$	Either the 2D function G evaluated at (x, y) or the function G itself
$g(\cdot)$	The function evaluated at some value, where the variable name is unimportant or obvious
$G(\cdot, \cdot)$	The function evaluated at some pair of values, where the variable names are unimportant or obvious
$g[x]$	Brackets indicate a discrete array indexed by nonnegative integers
$\dot{g}, \ddot{g}$	First and second derivatives of function
$a = b$	The variable a is equal to b
$a \equiv b$	The variable a is defined to be equal to b
$\mathbf{g}, \boldsymbol{\psi}$	Boldface lowercase Latin or Greek characters indicate vectors
$\mathbf{G}, \boldsymbol{\Psi}$	Boldface uppercase Latin or Greek characters indicate matrices
$\mathbf{G} = [g_{ij}]$	The ij^{th} element of matrix $\mathbf{G}$ is given by g_{ij}
$\mathbf{g}^{\mathsf{T}}$	Transpose of vector $\mathbf{g}$
$\mathbf{FG}$	Matrix multiplication
$\cdot$	Central dot indicates ordinary multiplication (also used for divergence)

:	Colon means either a range, as in $1{:}10$, or "such that," as in $\{x{:}x < 0\}$
$\circledast$	Asterisk with a circle indicates convolution
$\mathbb{R}, \mathbb{R}^n, \mathbb{R}^{m \times n}$	Set of real numbers, set of vectors of n real numbers, set of $m \times n$ real matrices
$\mathbb{Z}, \mathbb{Z}_{a:b}$	Set of integers, set of integers from a to b, inclusive
$O(\cdot)$	Big O notation for asymptotic running time of algorithms
$*$	Asterisk indicates ordinary multiplication (only used in pseudocode)
$==$	Long equal sign indicates test for equality (pseudocode)
$\leftarrow$	Assignment (pseudocode)
$\leftarrow_+$	Assignment with addition; same as $+=$ in C/C++/Java (pseudocode)
$\leftarrow_-$	Assignment with subtraction; same as $-=$ in C/C++/Java (pseudocode)

1.4.5 Programming

It has been said that a person does not really know anything until he or she is able to write it down. In a similar way, a person does not really understand an algorithm until he or she is able to implement it. Therefore, the best way to learn image processing and analysis is by programming real algorithms on real images. To aid the reader in this endeavor, this book provides detailed pseudocode for many of the algorithms presented. Although the pseudocode may not be very interesting upon first reading, you will likely find it indispensable when you desire to acquire a deeper understanding of any given technique by implementing it yourself. The pseudocode has been written to balance between precision on the one hand and readability on the other. If you are proficient at a programming language, it should not be difficult to translate the pseudocode into actual working code.

By far the most common language used in learning image processing and analysis is MATLAB, or its open-source alternative, Octave. MATLAB has a clean syntax, is very easy to use, is interpreted rather than compiled, and comes with built-in visualization capabilities, an editor, and a debugger. In industry, however, the need for efficient computation requires the use of a lower-level language like C or C++, for which the most widely used library is OpenCV. OpenCV has extensive capabilities for loading and displaying images, connecting to cameras, and performing basic operations, as well as advanced algorithms like face detection and camera calibration. OpenCV also has bindings to other languages such as Python and Java for more rapid prototyping. Other libraries include CImg, vxl, ImageJ, and dozens of others. More information about these tools and libraries can easily be found by searching online.

1.5 Looking Forward

With these basics under our belt, we are now ready to begin tackling the topics of image processing and analysis. As we do so, one word of caution is in order. In other fields of study, we are accustomed to dealing with convergent problems. A **convergent problem** is one in which there is a single unique solution, and the more one studies the problem the more one learns about it. In contrast, as pointed out by a well-known economist [Schumacher, 1973], a **divergent problem** has no correct solution, and the more it is studied the more the answers seem to contradict one another. Image analysis, and to a lesser extent image processing, are full of divergent problems for which there is not a single unique solution but rather a variety of different solutions, each with its own merits and shortcomings. Therefore, do not be surprised if, when faced with a particular problem, you try the leading algorithms, only to discover that they fail miserably and that a completely different (and oftentimes far simpler) approach outperforms them all in the particular context in which you are working.

Image analysis is a young field, and the solutions are elusive. While progress will undoubtedly continue over the coming decades to produce practical systems that process imagery to provide useful information, this will happen by continually questioning existing techniques and exploring new ones. Therefore whether you are a student, researcher, or practitioner, put on your creativity cap and be ready to think outside the box and try new approaches. After all, image analysis is for the most part a bag of tricks, so feel free to select whatever tricks you find in the bag, as well as any new tricks you develop on your own, in order to solve the problems that you encounter.

1.6 Further Reading

This chapter has presented an overview of image processing and analysis, along with their relationship to machine and computer vision. A variety of alternative overviews of one or more of these fields can be found in various textbooks. Burger and Burge [2008] provide an easy-to-read introduction to the field of image processing, while Gonzalez and Woods [2008] present a more detailed treatment of the subject. For computer vision, Shapiro and Stockman [2001] provide an introduction, whereas Forsyth and Ponce [2012] cover the subject at an advanced level, and Szeliski [2010] provides a readable treatment with a helpful summary of the latest research. Machine vision is covered thoroughly by Davies [2005]. A combined treatment of the fields can be found in the introductory text of Umbaugh [2010] or the more comprehensive book of Sonka et al. [2008]. For more historical texts, the classic books of Rosenfeld and Kak [1982], Jain [1989], Pratt [1991], Jain et al. [1995], or Castleman [1995] on image processing; or the classic works of Marr [1982], Ballard and Brown [1982], Horn [1986], or Nalwa [1993] on computer vision can be consulted. For learning about 3D computer vision, Trucco and Verri [1998] provide an easy-to-read treatment, while Hartley and Zisserman [2003] is the definitive resource. A myriad of monographs or edited works on more specialized topics can also be found but are too numerous to list here.

The latest research can be found in a variety of conferences and journals. The leading conferences in image processing are International Conference on Image Processing (ICIP) and International Conference on Pattern Recognition (ICPR), while the leading journal is *IEEE Transactions on Image Processing*. The leading conferences in computer vision are Computer Vision and Pattern Recognition (CVPR), International Conference on Computer Vision (ICCV), and European Conference on Computer Vision (ECCV), while the leading journals are *IEEE Transactions on Pattern Analysis and Machine Intelligence* (PAMI) and *International Journal of Computer Vision* (IJCV). The leading venues for medical imaging research are *IEEE Transactions on Medical Imaging* and *Medical Image Analysis*.

PROBLEMS

1-1 Define image processing and image analysis.

1-2 Even though machine vision and computer vision are nearly synonymous, there are some subtle distinctions between them. List at least two of these differences.

1-3 Image analysis, as defined in this book, is very closely related to computer vision. What is the key difference?

1-4 Image processing, as defined in this book, produces an output image from an input image. What are the two primary purposes for such output images?

1-5 Another way to categorize the information in this book would be in terms of low-, mid-, and high-level vision. Explain how you would map image processing, image analysis, machine vision, and computer vision into these alternative categories.

1-6 List three basic image processing problems and three basic problems in image analysis.

1-7 Skim the table of contents to identify at least one topic for each of the six basic problems mentioned in the previous question (list the chapter and/or section number for each, along with the title). Can you identify a topic that overlaps more than one basic problem? Can you identify a topic that does not fit into any of the basic categories?

1-8 Explain the statement, "Computer vision is the inverse of computer graphics."

1-9 The three main problems in machine learning are unsupervised learning, supervised learning, and reinforcement learning. Relate any two of these to the main problems in image analysis.

1-10 Provide the make and model of an automobile that processes images from one or more cameras permanently mounted on the vehicle, and explain the purpose of the processing. Search the Web if needed.

1-11 Give an example of an application that you have used personally in the past month that involves image processing and/or analysis.

1-12 List three psychologists whose work has been influential in understanding the human visual system.

1-13 Give a real-world example of technology using each of the following fields: (a) photogrammetry, (b) signal processing, (c) computer graphics, and (d) machine learning.

1-14 Search the Web for job openings in computer vision. List three jobs that you found, along with the qualifications needed to apply.

1-15 Suppose we have the following image: $I = \begin{bmatrix} 4 & 5 & 2 \\ 1 & 3 & 8 \end{bmatrix}$. Using the conventions of this book, what are the values of $I(0, 1)$, $I(1, 1)$, and $I(2, 1)$?

1-16 Suppose an image has 640 columns and 480 rows and is stored in row-major order. Convert the coordinates $(x, y) = (38, 52)$, $(592, 241)$, and $(33, 0)$ to 1D indices. Conversely, convert the following 1D indices to (x, y) coordinates: $i = 8092$, 24061, and 38190.

1-17 Equations $(1.3) - (1.5)$ apply to an image stored in row-major order. Write the equivalent expressions to convert between 2D coordinates and 1D indices for an image stored in column-major order.

1-18 Suppose the following 1D array of bytes in memory stores a 2×2 color image (in blue-green-red order): 52, 68, 31, 133, 192, 88, 255, 208, 32, 233, 161, 25.

a. Assuming that the image is stored in interleaved format, convert to planar format. What are the RGB values of the pixel at location $(1, 1)$?

b. Assuming that the image is stored in planar format, convert to interleaved format. What are the RGB values of the pixel at location $(0, 1)$?

1-19 Suppose the following 1D array of bytes in memory stores 8 consecutive pixels of a binary image: 0, 0, 0, 255, 255, 0, 255, 0. Show how to store these pixels in a single packed byte.

1-20 For increased fidelity, medical images are often stored using more than 8 bits. Suppose you needed to store a 12-bit-per-pixel grayscale image. Would you try to pack 3 pixels into 2 bytes to avoid wasted bits? Why or why not?

1-21 Convert the following grayscale image to set notation: $\begin{bmatrix} 112 & 195 & 48 \\ 97 & 203 & 125 \end{bmatrix}$.

1-22 Convert the following 3 × 2 binary image back to array notation: $\{(1, 0), (0, 1), (2, 1)\}$.

1-23 Consider the following 2D array, which has 3 columns and 2 rows: $\begin{bmatrix} 167 & 30 & 245 \\ 41 & 127 & 87 \end{bmatrix}$.

a. If the array is an image I, what is the value of $I(1, 1)$?

b. If it is a matrix $\mathbf{A}$, what is the value of a_{11}?

c. In which case would you write the dimensions as 2 × 3? As 3 × 2?

1-24 List three mathematical prerequisites for studying the material in this book.

1-25 Explain the difference between a convergent problem and a divergent problem.

1-26 Briefly explain why computer vision is so difficult.

1-27 Download a software library (e.g., OpenCV), and write a program to load an image from a file and display it in a window.

CHAPTER 2
Fundamentals of Imaging

Before delving into specific techniques for image processing and analysis, in this chapter we consider some of the fundamental concepts of imaging. Because most of the images you will encounter will already be digitized, it is not necessary to understand all the details presented in this chapter before proceeding through the rest of the book. Nevertheless, familiarizing yourself at least somewhat with the material of this chapter will better prepare you to appreciate the subtle distinctions that you will encounter, as well as to make it easy to refer back to this chapter later as needed. This chapter provides a quick tour of natural vision systems, with particular attention paid to the human visual system. Afterwards we proceed to the topics of image formation and acquisition, such as the pinhole camera model, lenses, sampling, and quantization, which are followed by a survey of alternative imaging modalities. Finally, a detailed look at the electromagnetic spectrum is presented.

2.1 Vision in Nature

We begin with a tour of natural vision systems, starting with a single photoreceptor, followed by the human visual system, and finally the visual systems of various animals. Since image processing is concerned with producing a new image with improved visual quality over the original, having some knowledge of human visual perception is necessary, and understanding how animals are able to achieve amazingly robust behavior with little computational power can yield inspiration for developing our own digital image analysis algorithms.

2.1.1 A Single Photoreceptor

The miracle of vision starts with the amazingly complex system of a single **photoreceptor**, which is the most fundamental component of any natural vision system. When a single photon of light hits a single photoreceptor, it sets off a wonderfully complex chain of events that leads to the surprising ability to see. In a nutshell, the events are as follows. The absorption of a photon causes a change in shape of a small organic molecule called *retinal*. This change in shape, in turn, causes the larger protein (called *rhodopsin*) holding the retinal molecule to change shape and bind itself to another protein (*transducin*), which causes another molecule (*GTP*) to bind to it. This newfound combination then goes around cutting any instances of another type of molecule (*cGMP*) that it encounters, thus reducing its concentration. This reduction in concentration causes an ion channel to close, which then reduces the flow of positively charged sodium ions into the cell. The resulting imbalance of charge causes another channel to close, which reduces the concentration of calcium ions. Since calcium is required by the neurotransmitter, this reduced level of calcium causes the neurotransmitter to slow down, which therefore indicates to the next cell that a photon has been absorbed.[†] A set of enzymes then goes to work to restore the rhodopsin to its original shape, to resynthesize *c*GMP, to restore the concentration of sodium ions, and so forth, so that the process can begin all over again. As you can see, the complexity involved in even this most basic step of vision, that of sensing a single photon, is quite impressive. In fact, some scientists call such a system "irreducibly complex," because if any one of the many components does not function properly, then the entire system fails, and the process of vision cannot even begin.

2.1.2 Human Visual System

Individual photoreceptors are more useful if they are, in turn, packaged into an even larger system. In this section we consider the human visual system.

Structure of the Eyes

As shown in Figure 2.1, the human eyeball is approximately spherical in shape, covered by a transparent layer (called the **cornea**) in front and an opaque layer (called the **sclera**, the white part of the eye) everywhere else. Light rays enter the eyeball through the cornea, where they are bent before they pass through the **aqueous humor**, where they are bent again. These rays then pass through the small aperture known as the **pupil**, whose size is controlled by muscles attached to the **iris**, the colored circular region surrounding the pupil whose circular boundary with the sclera is known as the **limbus**. From the pupil the rays are bent yet again by the **lens**, whose thickness is controlled by the **ciliary muscle** in a process known as **accommodation**. The lens provides only about a third of the refractive power of the eyeball, the rest being achieved by the cornea and aqueous humor. Nevertheless, the accommodation of the lens is needed to focus the light to form an image on the **retina** at the back of the eyeball. After absorbing a photon, the photoreceptors in the retina are nourished by the layer between the sclera and retina called the **choroid**.

The retina consists of two types of photoreceptors. The 100 million or so **rods** are sensitive to low levels of light and able to generate a detectable photocurrent from as little as a single photon. The 6 million or so **cones** respond to normal, everyday light levels at which the rods are saturated. Color vision is possible because of the three types of cones, namely L-, M-, and S-cones, which respond primarily to long-, middle-, and short-wavelength

---

[†] One of the ironies of vision (at least in vertebrates) is that in its resting state the neurotransmitter of a photoreceptor constantly emits a signal, so that it is actually the *lack* of a signal that indicates the absorption of a photon. That is, unlike most ordinary sensory receptors (including invertebrate photoreceptors), which become *depolarized* in response to a stimulus, vertebrate photoreceptors become *hyperpolarized*.

Figure 2.1 Cross section of
the human eyeball.

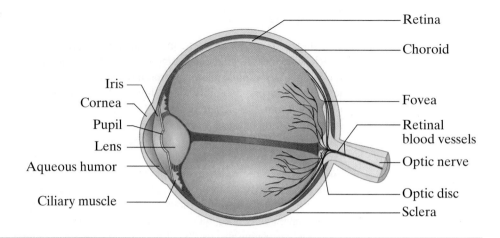

Figure 2.1 Cross section of
the human eyeball.

light, respectively. Colloquially these are known as red, green, and blue cones, although
in fact the peak sensitivity of the three types is closer to yellow, green, and violet, as
shown in Figure 2.2. Rods contain a protein called **rhodopsin**, while cones contain three
different proteins called **photopsins**. According to the property of **univariance**, when
a photopigment absorbs a photon of light, it generates the same response no matter the
wavelength of the photon. That is, although a photoreceptor is more sensitive to some
wavelengths than others and therefore is more likely to absorb some wavelengths than
others, once the photon is absorbed, all information about the wavelength is lost. It is for
this reason that different types of cones are needed for color vision.

The rods and cones are not distributed equally throughout the retina. As shown in
Figure 2.3, no photoreceptors are present in the **optic disc**, also known as the blind spot.
Otherwise, cones exist throughout the retina but are concentrated more heavily in the **fovea**,
the central pit responsible for the greatest visual acuity. The fovea is within the **macula
lutea**, the highly pigmented yellow spot near the center of the retina that absorbs harmful
ultraviolet light to protect the retina. In the fovea, the cones are tightly packed to form a
regular sampling array, with the centers of adjacent cells spaced approximately 2.5 μm,
which is about the same as the spacing between pixels on a typical camera sensor. The
fovea senses only about 5° of the visual field, which is slightly more than the width of both

Figure 2.2 Relative sensitivity of the S-, M-,
and L-cones of the human visual system to
different wavelengths. These functions are
also known as the cone fundamentals. Based
on data from http://www.cvrl.org.

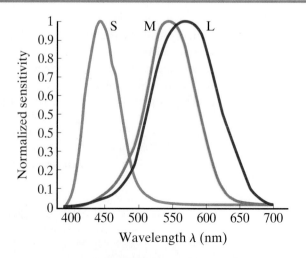

Figure 2.3 Distribution of cones and rods in the retina. Based on B. A. Wandell. *Foundations of Vision*. Sunderland, Mass., Sinauer Associates, Inc., 1995.

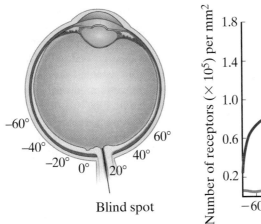

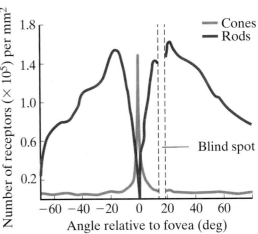

your thumb joints when held together at arm's length. The fovea is responsible for detailed pattern recognition, which can be demonstrated by attempting to read a book while fixating on two fingers covering the page; while you will have a general sense from your periphery that there are words on the page, it will be nearly impossible to recognize them. The central portion of the fovea, known as the **foveola**, is one-fifth the size of the fovea, or 1° of the visual field, covering approximately the width of the nail of your index finger at arm's length. No rods exist in the foveola, which explains why to view a faint star at night it is better to look slightly away from it, so that light from the star falls on the rods rather than on the cones. Except for the blind spot and foveola, rods are present throughout the retina, reaching their highest density about 20 degrees from the center.

In the fovea, S-cones account for approximately 6% of the cones, as shown in Figure 2.4, although like rods they are completely absent from the foveola. The increased spacing between S-cones (compared with M- and L- cones) matches the increased blurring of short wavelengths (compared with the blurring of longer wavelengths) due to chromatic aberration in the lens. The ratio of the number of M- to L-cones is highly variable among individuals, making it difficult to distinguish between the roles of these two types of cones. It is important to note that, even though the S-cones are less numerous in the retina, they are more sensitive to light than are L- and M-cones. As a result, it would be wrong to conclude that short wavelengths are less important, because in fact humans are able to distinguish between different shades of blue as well as they are between different shades of other colors.

Figure 2.4 An actual photoreceptor mosaic, 1.25° from the center, pseudocolored to show the different types of cones: L (red), M (green), S (blue). From Hofer et al. [2005].

Light does not land directly on the rods or cones but instead first passes through a layer of cells, the details of which are examined below. Such an arrangement is called an **inverted retina**. While at first glance this approach appears counterintuitive, there is in fact a good reason for this design. An inverted retina is needed so that the **retinal pigment epithelium (RPE)**, which is attached to the choroid, is able to replenish the damaged photoreceptors. Since the RPE is opaque, if the retina were not inverted then the RPE would have to be in front of the photoreceptors, which would block the light and make it impossible for us to see at all. Moreover, in the fovea the cones are elongated, so that light falls directly on the cones without passing through other cells in the central portion of the retina, and hence the inverted retina does not have much effect on the detailed, central vision anyway. And even in the periphery where the light passes through additional cells, these cells are for the most part transparent.

Each of the two eyeballs is approximately 24 mm in diameter, and the spacing between them (called the **interpupillary distance**, or IPD) is approximately 60 to 70 mm. When a person views a point in the scene, the eyes are said to be **fixated** on the point, and the horizontal angle between the axes of the eyes is known as the **vergence angle**. The point of fixation projects onto the retina at the same location (that is, directly in the center of

Figure 2.5 Retinal disparity is defined as the distance between corresponding points on the two retinas, after the retinas have been overlaid on top of one another and rotated so that their optical axes are coincident. Based on B. A. Wandell. *Foundations of Vision*. Sunderland, Mass., Sinauer Associates, Inc., 1995.

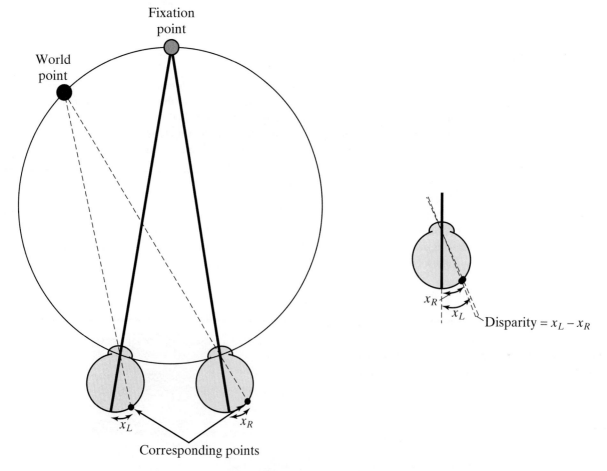

the fovea) in both eyes. For all other points in the scene, the light rays project onto different locations on each retina depending on their location in the scene. The lateral shift between these locations is known as the **disparity** of the point. The locus of points in the scene that yield zero disparity is known as the **horopter**, and the theoretical horopter is the **Vieth-Müller circle**, which passes through the two lens centers and the fixation point, as shown in Figure 2.5. Because the disparity is related to the distance (or depth) to the point, computing the disparity is a key step toward determining the distance to a point, and hence to 3D perception. Stereo vision involves establishing correspondence between grayscale patterns in the two images to determine the depth to each point in the scene. The resulting fused image is known as a **Cyclopean image** (after the famous one-eyed mythical Greek monster), because the fused image almost seems to result from an additional sensor in the center of the head. Only points within a small area, known as **Panum's fusional area**,[†] around the fixation point are fused into the Cyclopean image.[‡] Outside this area the brain retains both images, a situation known as **diplopia** ("double vision"). To experience these phenomena, hold your finger in front of your face while focusing on the finger, and you will see one 3D finger that is fused from both images (the Cyclopean image); then without moving your finger focus on the scene in the distance, and you will see two fingers (diplopia).

Because of the fovea, your eyes do not look at a scene by staring at one spot for a long period of time. Instead, they jump erratically from one spot to another to allow high-resolution imaging of various parts of the scene in order to build a full 3D mental picture. For example, as you read this page, your eyes are not fixating on a single point, nor are they moving with a continuous motion. Rather they jump from one word to another, giving your brain enough time to read the word (and surrounding text) and move on. These rapid movements are called **saccades**,[§] and they are the fastest movements made by the human body. When a person looks at a photograph, for example, these saccades cause the person's eye to jump from one location to another; the path of these movements is known as a **scan path**, an example of which is shown in Figure 2.6.

The Visual Pathway

Once light has been captured by the photoreceptors (rods and cones), the information is processed and transmitted via **neurons**, or nerve cells. A neuron contains many **dendrites** (inputs) and one **axon** (output). The outputs from many neurons are tied to the inputs of other neurons via connections known as **synapses** to form a **neural network**. As shown in

Figure 2.6 A scan path records the path traversed by a person when viewing a photograph, shown is a scan path of a photograph, showing that the viewer focused primarily upon the facial features and objects of interest, in order to build a complete mental model.

Courtesy of J. Kevin O'Regan

[†] Section 13.2.1 (p. 625).

[‡] The Cyclopean image, along with Cyclopean coordinates, is covered in Chapter 13.

[§] Pronounced *seh-KAHD*.

Figure 2.7, the photoreceptors connect directly to the inputs of either rod or cone **bipolar cells**, whose outputs then connect to the inputs of the **ganglion cells**. All visual signals must pass through these cells, since the outputs of the ganglion cells actually form the **optic nerve**, which exits the retina at the optic disc. An indirect pathway is also present due to the **horizontal cells**, which are connected to the receptors, as well as **amacrine cells**, which are connected to the bipolar and ganglion cells — neither has an identifiable axon.

Human vision is **foveated**, with the rods and surrounding cones contributing to the **ambient** (peripheral) component of vision that senses motion and continuity, while the cones in the fovea contribute to the **focal** (foveal) component of vision used to recognize detailed patterns such as text or faces. The ratio of rods to ganglion cells is approximately 100:1 ("100 to 1"), meaning that the information from the rods is greatly compressed and aggregated. For this reason, rods do not provide good spatial acuity but instead trade high acuity for a high signal-to-noise ratio. On the other hand, the ratio of cones to ganglion cells, in the fovea at least, is 1:3, meaning that information from one cell maps to multiple ganglion cells. It is for this reason that foveal cones provide high spatial acuity.

The optic nerve (composed of the axons of the ganglion cells) causes the 5-degree-wide blind spot at the optic disc from 15 to 20 degrees from the center, on the side of the nose. At the **optic chiasm** the optic nerve splits: One half of the bundle goes to the left, while the

Figure 2.7 After a photoreceptor absorbs a photon, the information is passed through several layers of cells before exiting via the optic nerve. Surprisingly, the light passes through these same layers of cells before landing on the photoreceptors.

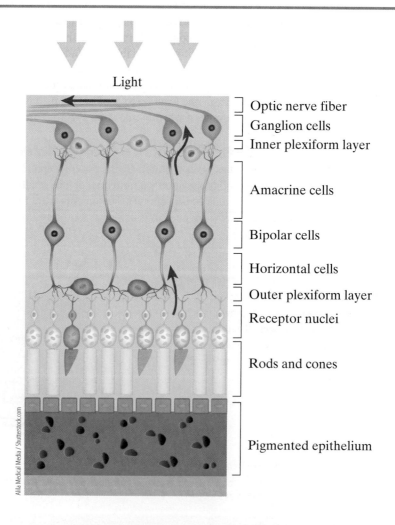

Light

Optic nerve fiber
Ganglion cells
Inner plexiform layer

Amacrine cells

Bipolar cells

Horizontal cells

Outer plexiform layer

Receptor nuclei

Rods and cones

Pigmented epithelium

Alila Medical Media / Shutterstock.com

other half goes right, both terminating in the **lateral geniculate nucleus (LGN)**. A common myth is that the left half of your brain processes data from your right eye, while the right half processes data from your left eye. In reality, the left half of your brain processes data from the right half of your visual field (containing information from the left half of both retinas), while the right half of your brain processes data from the left half of your visual field (containing information from the right half of both retinas). This is necessary for binocular stereo processing, so that the brain has access to information from both eyes in order to establish correspondence for disparity computation. Nevertheless, processing in the LGN is primarily monocular, with binocular processing reserved for the next step.

Information passes from the LGN to the **visual cortex**, which is at the rear of the brain. The visual cortex is composed of several different stages, called V1 through V5. Area V1 is also known as the **primary visual cortex**, or **striate cortex**, while V5 is known as MT (middle temporal). Early cells in V1 are essentially locally tuned Gabor filters[†] that extract spatiotemporal features such as spatial frequency, orientation, temporal frequency, and motion direction. Beyond this, the actual inner workings of the visual cortex are largely a mystery. The currently accepted theory is that there are two streams of processing, namely, the **dorsal stream**, whose purpose is to analyze motion, locations of objects, and tracking, and the **ventral stream**, which is responsible for object recognition and representation. These streams are known as the "where pathway" and the "what pathway," respectively.

Human Visual Perception

The human visual system can respond to levels of light ranging an astounding 14 orders of magnitude.[‡] The eye, however, cannot process this entire dynamic range simultaneously but instead adapts using the different types of photoreceptors and by adjusting the size of the pupil. As shown in Figure 2.8, at low light levels the rods dominate, and the resulting monochromatic vision in these conditions is known as **scotopic** vision, from the Greek word for "darkness." At normal to higher light levels, the rods become saturated so that their responses are not meaningful, and the cones take over. The resulting color vision is known as **photopic** vision, from the Greek word for "light." In between, there is a small range when both rods and cones respond, known as **mesopic** vision. It may be helpful to think of these three types of vision as being applicable, respectively, to starlight, sunlight, and moonlight, but keep in mind that photopic vision also includes most normal viewing conditions, such as indoor lighting.

Figure 2.8 Scotopic, mesopic, and photopic vision at different light levels. While the human visual system is capable of sensing light in approximately a range of 10^{14} overall (from 10^{-6} to 10^8 cd/m²), light can be sensed in a range of 10^3, at any particular state of adaptation.

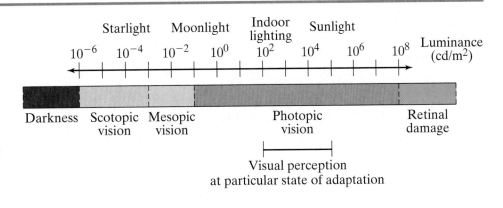

[†] Gabor filters are covered in Section 6.6.7 (p. 321).

[‡] To get a sense of the enormity of this range, imagine a device that could measure distance not only in kilometers but also in nanometers!

The **luminous efficiency function (LEF)** captures the relative sensitivity of the visual system to different wavelengths. As shown in Figure 2.9, the **photopic LEF** corresponds to normal light levels where the cones dominate due to the saturation of the rods, while the **scotopic LEF** corresponds to low light levels where the rods dominate due to the lack of sensitivity of the cones. The difference in peak wavelength is called the **Purkinje effect** and explains why objects appear to have a more bluish tint as the light dims. Not surprisingly, the scotopic LEF closely matches the rod spectral sensitivity function (SSF),[†] and the photopic LEF can be well approximated as a weighted combination of the cone fundamentals of Figure 2.2.

At a particular state of adaptation, human vision can discern luminances across a range of about 1000 to 1, depending on conditions. A good monitor can reproduce luminances in a range of about 500 to 1, while a range of only 100 to 1 is possible from the reflectance of paper. If two different shades of gray are placed adjacent to one another, a person can discern the difference between them if their luminances differ by approximately at least 1%, which is called the **just-noticeable difference (JND)**; otherwise, their brightness is perceived to be the same. While this number is a helpful rule of thumb, it is important to keep in mind that it is only a rough approximation to the actual behavior, which is quite complex depending upon spatial frequency, temporal frequency, and overall light intensity.

The visual **receptive field** of a neuron is the retinal area in which light influences the neuron's response. The receptive fields of neurons in the visual cortex are optimized for extracting efficient information using sparse coding constraints, where the learned receptive fields arise from exposure to natural images. These neurons, like Gabor filters, perform local spatial frequency analysis to form edge and line detectors that respond to luminance information in the proper orientation and polarity. Unlike photoreceptors, which respond to absolute levels of light intensity, these later neurons produce outputs that are independent of the overall level because they respond to contrast (or change in light), in a process known as **lateral inhibition**. Similarly, the retinal ganglion cells exhibit center-surround receptive fields, in the shape of the Laplacian of Gaussian.[‡]

2.1.3 Animal Vision

In addition to the human visual system, nature provides us with an astonishing array of diverse vision systems. Studying such systems provides us with a fresh dose of humility when we learn how effortless it is for a simple low-level animal with very little processing

Figure 2.9 Photopic and scotopic luminous efficiency functions (LEFs). Based on data from http://www.cvrl.org.

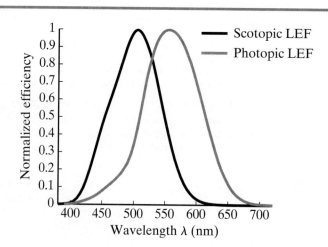

[†] SSFs are covered in Section 9.2.1 (p. 405).

[‡] The Laplacian of Gaussian is covered in detail in Section 5.4.1 (p. 242).

Figure 2.10 The common housefly has the fastest visual response of any animal, leading to extreme maneuverability in flight. Tiny flying robots (such as this one from Centeye) have been inspired to mimic the housefly's navigation ability based on optic flow.

power to robustly extract information from an image when we find it difficult to do the same even with powerful latest-generation computers. Studying such systems can also provide us with confidence that the particular problem that we are addressing can, in fact, be solved, if an existing animal demonstrates that particular capability. As a result, such systems can provide inspiration for designing artificial machines. This imitation of natural systems, known as **biomimicry** (or *biomimetics*), is an important approach to discovering novel solutions in both software and hardware.

One example is the common housefly, which—like other insects—has compound eyes, as shown in Figure 2.10. In a compound eye, the photoreceptors are arranged in small groups called *ommatidia*. Each ommatidium views the world from a different direction, yielding a mosaic of images providing a fairly low-resolution representation of the scene. Even so, the fly has the fastest visual response in the animal kingdom, which is achieved by the photoreceptors physically contracting a tiny amount in response to light. Such mechanical response, in contrast to the chemical response of our own visual system, is extremely fast, and it is one of the reasons that the fly has the most maneuverable flight system. Flies maneuver by detecting optic flow, which is the relative motion of the surrounding environment projected onto the eye, thus inspiring flying robotic systems that weigh just a few grams and can avoid obstacles using optic flow algorithms embedded on a tiny vision chip. Note that the ability of a fly to land effortlessly on a seemingly untextured surface proves that texture is always present in the world.

If the housefly wins the award for the fastest visual response, the hawk (and other raptors) wins for the highest visual acuity. A hawk, shown in Figure 2.11, can see a rabbit from a mile away, which is about 8 times better than human vision. With today's aerial imagery, satellite images, and megapixel video cameras, similarly impressive resolutions are possible. Tigers and other cats also have excellent eyesight. Like most predators, their eyes are in

Figure 2.11 Raptors, such as this hawk, have the highest visual acuity of any animal. Megapixel video cameras with similar ability are now commercially available.

Figure 2.12
Predators such as this tiger have two eyes facing forward, so that it can estimate the distance to its intended prey via stereo vision. Prey such as this rabbit have eyes on the sides of the head, providing a much wider field of view to detect danger.

front of their head, as shown in Figure 2.12; this overlap in the visual field of the two eyes enables the predator to estimate the distance to its target by means of stereo vision. The defenseless rabbit, on the other hand, has two eyes on the sides of its head, with very little overlap in their fields of view. While this makes the rabbit unable to perceive depth from stereo vision, it gives the creature a much wider field of view overall, thus enabling it to perceive when it is being threatened by a predator.

Sometimes the best way to extract useful information from light is to filter special wavelengths. Pit vipers, for example, have a heat-sensing pit organ between each eye and nostril to detect infrared light, as shown in Figure 2.13. Modern forward-looking infrared (FLIR) cameras also detect heat, making it much easier to find people or machinery in all weather conditions. At the other end of the spectrum, bees use ultraviolet filters to help them see their target when pollinating flowers, as shown in Figure 2.14. Since many flowers have low reflectance of ultraviolet light near the center, these filters simplify the detection of the flower center, providing a convenient natural landing pattern for the creature. Similarly, most birds can see four different color bands, similar to modern multispectral imaging equipment used for applications ranging from astronomy to medicine.

The mantis shrimp, shown in Figure 2.15, has arguably the most sophisticated eye in the animal kingdom. It can see 12 color dimensions (roughly half of the bands for visible light and half for ultraviolet). It has 4 filters to tune the pigments, it sees several planes of polarized light, and it can distinguish between left and right circularly polarized light. It

Figure 2.13 The loreal pit between the eye and nostril on a pit viper leads to an organ that detects heat via infrared light. Forward-looking infrared (FLIR) cameras detect warm bodies by examining the infrared portion of the spectrum, as seen in this thermal image.

Figure 2.14 Bees have ultraviolet filters enabling them to detect the flower center, which is helpful for pollinization. The middle image shows the flower (left) as it appears to a bee. Ultraviolet cameras are also used to detect heavenly bodies, such as the sun (right).

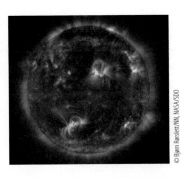

can even convert linearly polarized light into circularly polarized light. The most surprising aspect of the shrimp is that it lives in the ocean at depths of 40 meters, where light is only a dim, filtered blue. The purpose of this fancy vision system, then, is not to detect the surroundings so much as it is for communication: The shrimp communicate with each other by fluorescing their spots. Similarly, cephalopods such as squid, octopus, and cuttlefish have tiny hairlike membranes (called *microvilli*) in their photoreceptor cells that are oriented perpendicular to one another, enabling the animals to detect differences in polarized light, and they are able to produce polarized light patterns on their skin as a means of communication.

One of the most unique eyes is that of the lobster, shown in Figure 2.16. The eye of the lobster (as well as other long-bodied decapods) focuses not by refraction but by reflection. That is, instead of using a lens, light is focused using a honeycomb-like arrangement of mirror-lined tubes. These tiny facets are perfectly square and from a distance look like tiny graph paper. These square tubes are on a spherical surface, with flat shiny mirrors on the sides of the tubes. This precise geometrical arrangement allows the eye to focus parallel light rays from any direction. This principle is the inspiration behind a new generation of astronomical telescope that focuses X-rays using reflection, since no practical lens can focus such high-frequency waves.

The eyes of extinct trilobites have the amazing property that their calcite lenses are shaped almost exactly as needed to minimize lens aberration, as shown in Figure 2.17. In the 17th century, two different lens designs were developed by Descartes[†]

Figure 2.15 The mantis shrimp has arguably the most sophisticated eye of all, which can detect 11 different color bands and has sophisticated machinery to deal with polarized light.

[†] René Descartes (DAY-cart) (1596–1650) was not only an influential figure in mathematics but also the father of modern philosophy, credited with the well-known saying, "I think therefore I am." The Cartesian coordinate system is named after him.

Figure 2.16 The lobster eye focuses by reflection, not refraction, and is the inspiration for a new generation of telescope. Based on Denton, M.J., *Nature's Destiny: How the Laws of Biology Reveal Purpose in the Universe*, ch. 15, The Free Press, New York/London, 1998

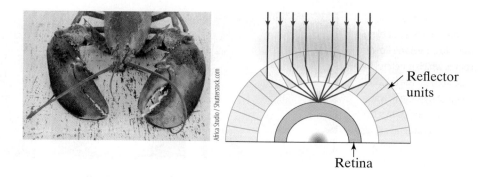

and Huygens[†] according to optical principles to minimize lens aberration. As it turns out, the lenses of trilobites form an internal doublet structure that follows one or the other of these designs, depending upon the type of trilobite.

Other creatures that have unusual eyes are shown in Figure 2.18. The brittle star secretes calcite crystals that form microlenses, so its whole body is composed of little eyes. Similarly, the scallop contains an array of eyes around its opening. The nautilus eye is unique in that it has no lens at all and therefore cannot focus light. The jumping spider has eight eyes in total: The two largest eyes in the front and center have four retinas stacked in layers, allowing the spider to judge distance by a technique called depth from defocus; the smaller eyes are called *ocelli* and feed into a distinct visual pathway. Other creatures, such as certain lizards and frogs, have an extra third eye called a *parietal eye* that contains a small lens and retina between their primary two eyes.

2.2 Image Formation

We now consider the process by which an image is formed on the surface of a sensor, focusing our attention primarily upon the case of standard optical cameras capturing visible light; alternative imaging systems are discussed in a later section.

Figure 2.17 The doublet structure of the trilobite lens follows the shape necessary to minimize lens aberration. Depending upon the type of trilobite, this shape is essentially identical to those deduced by Descartes and Huygens using the geometrical principles of optics. Based on E. N. K. Clarkson and R. Levi-Setti. Trilobite eyes and the optics of Descartes and Huygens. *Nature* 254: 663-667, 1975.

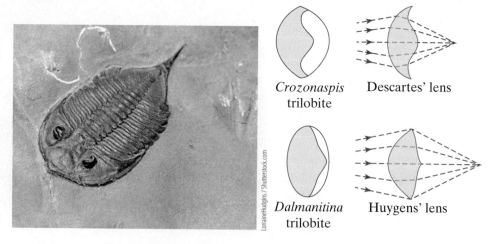

Crozonaspis trilobite   Descartes' lens

Dalmanitina trilobite   Huygens' lens

[†] Christiaan Huygens (HIGH-guns) (1629–1695) was an influential Dutch mathematician and scientist who discovered Saturn's rings as well as its first moon.

Figure 2.18 Top: The entire body of the brittle star (left) is covered with little eyes (middle). The scallop (right) has eyes all around its opening. Bottom: The nautilus eye (left) has no lens and therefore cannot produce a focused image. The jumping spider (middle) has extra little eyes called *ocelli*. Creatures such as this frog (right) contain a third light-sensitive spot (the tiny blue dot) called a *parietal eye* between the two main eyes.

Ethan Daniels / Shutterstock.com, Lucent Technologies'Bell Labs / Science Source, Stephen Frink / Photodisc / Getty Images, bluehand / Shutterstock.com, Tomatito / Shutterstock.com, Kenneth M. Highfill / Science Source

2.2.1 Light and the Electromagnetic Spectrum

From basic physics you may recall that light is an electromagnetic wave traveling through space. The **wavelength** λ (measured in meters) of such a wave is the distance between successive peaks in the sinusoid, while the **frequency** ν (measured in hertz) is inversely related to the wavelength. That is, λ times ν is the speed[†] of light in the medium. Visible light ranges in wavelength from about 380 nm to about 720 nm, or equivalently, 0.38 μm to 0.72 μm. A nanometer is one billionth of a meter, or $10^{-9} = .000000001$ meters, so the wavelength of visible light is about a hundred times smaller than the diameter of a human hair, as shown in Figure 2.19. These short wavelengths explain why vision systems are able to achieve such accurate measurements of the world.

Figure 2.19 The wavelength of visible light is about 1/100th the diameter of a human hair.

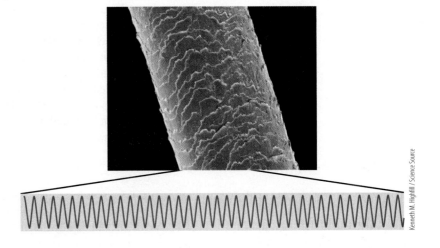

Kenneth M. Highfill / Science Source

[†] Technically the *phase velocity*.

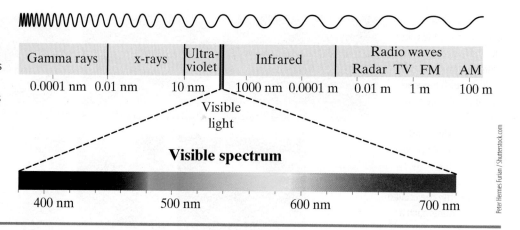

Figure 2.20 The electromagnetic spectrum consists of gamma rays and X-rays at one end, and radio waves and microwaves at the other end. The visible spectrum is between about 380 and 720 nm.

The visible spectrum is a tiny part of the entire **electromagnetic spectrum**, depicted in Figure 2.20. Since the color violet has the highest frequency (shortest wavelength) among visible light, waves with slightly higher frequency than that of violet are called "beyond violet," or **ultraviolet** (UV). Beyond ultraviolet light are X-rays and gamma rays. At moderate to high frequencies, an electromagnetic wave can alternatively be viewed as a stream of particles called **photons**, where a photon is a single quantum of light containing an amount of energy that is proportional to the frequency. This is known as the theory of **wave-particle duality**. The amount of energy in a photon is given by h times ν, where the proportionality factor $h \approx 6.626 \times 10^{-34}$ watts seconds squared $(\mathrm{W} \cdot \mathrm{s}^2)$ is Planck's constant.[†] The fact that energy increases with frequency means that a high-frequency photon contains much more energy than a visible light photon, which explains why X-rays and gamma rays are so dangerous.

On the other end of the visible spectrum, light whose frequency is slightly lower (wavelength slightly longer) than that of red is known as being "below red," or **infrared** (IR). Roughly speaking, three types of infrared light can be distinguished. **Near infrared (NIR)** light consists of wavelengths only slightly longer than that of visible light. Such light is prevalent in sunlight as well as indoor light sources, motivating camera manufacturers to insert filters on the inside of consumer cameras to reduce the influence of these wavelengths. At the same time, *night vision* is made possible by shining invisible near infrared light on the scene and removing the infrared filter from the camera to increase its sensitivity to this range of wavelengths, even when the scene appears to the unaided human eye to be dark. **Mid infrared (MIR)** light, also known as **thermal infrared**, has much longer wavelengths than visible light. Thermal infrared cameras do not require any artificial illumination but instead sense the electromagnetic radiation emitted by objects in the scene. Such cameras are typically expensive due to their need to cool the electronics to avoid confusing these thermal emissions with those of the device itself. **Far infrared (FIR)** light is used primarily in astronomical applications. Beyond infrared, the electromagnetic spectrum contains microwaves (used in radar), and radio waves.

2.2.2 Plenoptic Function

Whenever light is present in a scene, the light rays bounce around the environment in a complicated manner as they repeatedly reflect off the surfaces in the scene. These rays carry all the information necessary to form an image of the scene at any point in space. Imagine,

---

[†] Max Planck (1858–1947) revolutionized science with his proposal of quantum theory.

for example, standing in a well-lit room full of various objects and looking at an empty area of one of the walls. At each point on the wall, rays of light that have interacted with surface points throughout the scene impinge the wall, so that there is enough information in that bundle of rays to produce a sharp, well-focused image of the room. In fact, there is enough information to produce many sharp, well-focused images of the room, each from a different point of view. However, the reason you will not see an image on the wall is because what you are looking at is the sum of all those images. Or, stated another way, an image is displayed on the wall, but it is so extremely defocused that it is unrecognizable.

The many bundles of light rays in the room are modeled by the **plenoptic function**. The plenoptic function (the prefix *plen-* comes from the Greek word meaning "complete") specifies the radiance[†] along all light rays in a region of space, that is, along all light rays passing all locations (x, y, z) in all directions (θ, ϕ), where θ and ϕ are two angles that uniquely specify the direction of a ray in 3D space. In other words, the plenoptic function models all images of the scene that could be taken if a camera were placed at any possible viewing position and any viewing angle. The plenoptic function is typically considered to be five-dimensional (5D) but could be extended by including other parameters, such as time, wavelength, polarization, or instantaneous phase, if desired.

In an area of free space, the values in the 5D plenoptic function are not independent of one another. Radiance is a measure of the energy along a ray of light, and radiance is defined so that its value does not change along a ray traveling through free space. As a result, the plenoptic function is equal if evaluated at any two location-directions $(x_1, y_1, z_1, \theta, \phi)$ and $(x_2, y_2, z_2, \theta, \phi)$ such that the ray along the direction (θ, ϕ) passes through the two points (x_1, y_1, z_1) and (x_2, y_2, z_2) unimpeded. This observation motivates the definition of the **light field**, which is a 4D version of the plenoptic function in free space. The most common parameterization of the light field, called the *light slab representation*, uses two points (x_1, y_1) and (x_2, y_2), each on a different parallel plane, which can be thought of as the collection of perspective images of one plane from a point on the other plane, as shown in Figure 2.21.

2.2.3 Pinhole Camera

To form a recognizable image, the light rays must be constrained somehow. One way to do this is to construct an empty, opaque box that is so tight that no light can enter the box. Then, a small hole the size of a pin is pierced into one side of the box, which allows light to enter the box only through the hole. The hole is called a **pinhole**, and the camera is therefore called a **pinhole camera**. A pinhole camera will cause a faint image to be projected onto the inner wall of the box opposite the pinhole, as shown in Figure 2.22. The image, however, is trapped inside the box. To view the image, one must either replace the

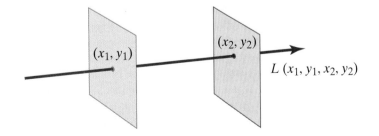

Figure 2.21 The light field is a 4-dimensional function of the radiance over position and direction. Shown is the light slab representation of the light field, in which each ray of light passes through two parallel planes.

(x_1, y_1)

(x_2, y_2)

$L\,(x_1, y_1, x_2, y_2)$

[†] The term *radiance* is precisely defined later in the chapter, but it basically refers to the amount of energy in a light ray.

Designua / Shutterstock.com, 19th Century Dictionary Illustration / Public Domain

Figure 2.22 In a pinhole camera, light rays pass through the tiny aperture and form an upside-down image on the opposite wall. A camera obscura was an early form of pinhole camera in which light rays pass through the small aperture, reflect off the mirror, and form an image on the top horizontal surface near the rear of the enclosed box.

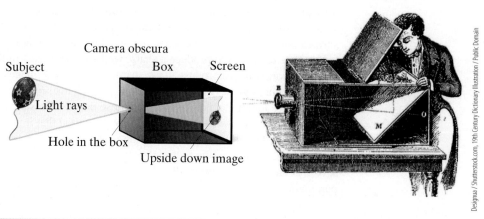

side of the box with a translucent material and drape a dark cloth over the viewer (an early primitive type of camera called the *camera obscura*, which literally means, "dark room"), or the inner wall must be lined with a photosensitive material in order to capture the image for viewing at a later time. Either way, the pinhole camera can be thought of as a way of sampling the plenoptic function at the 3D location of the pinhole, allowing the light rays at all angles θ and ϕ to uniquely determine the image formed, subject only to the field of view of the camera.

Geometrically, an ideal pinhole camera consists of a point and (typically) a plane. The point, known as the **focal point**, is the pinhole through which all rays of light pass. The plane, known as the **image plane**, is the sensor surface on which the image is formed. The line through the focal point perpendicular to the image plane is known as the **optical axis**, and the distance from the focal point to the image plane along this line is the **focal length**.

A pinhole camera forms images via **perspective projection**: Light rays from the source reflect off the surface of an object in the scene, travel through the focal point (also called the **center of projection**), then land on the image plane. Consider a right-hand coordinate system so that its origin is the focal point, the positive z axis points toward the world along the optical axis, and the y axis points vertically, parallel to the columns of the image, as shown in Figure 2.23. If we let (x_w, y_w, z_w) be the 3D coordinates of the world point, and (x, y) the 2D coordinates of its projection onto the image, it is easy to see from the figure that the two triangles in the y-z plane are *similar* because their angles are equivalent, leading us to conclude from what we know about similar triangles that the ratios of the lengths of their sides are equal: $y/f = y_w/z_w$, where f is the focal length and z_w is the depth of the point (i.e., distance to the point from the focal point). By symmetry, in the x-z plane we have $x/f = x_w/z_w$. Rearranging yields the coordinates of the point where the imaging ray lands on the image plane:

$$x = f\frac{x_w}{z_w} \tag{2.1}$$

$$y = f\frac{y_w}{z_w} \tag{2.2}$$

Although in reality the image is upside down, these equations exactly describe the right-side up image that would be created on a virtual image plane at a distance of f in front of the focal point, rather than at a distance of f behind it. This mathematical trick not only

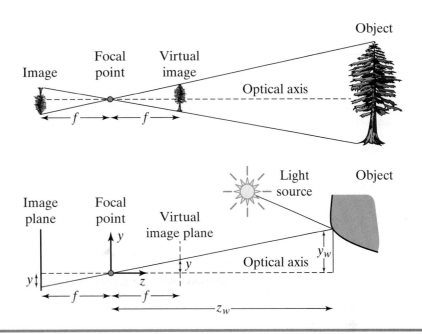

Figure 2.23 Perspective projection caused by a pinhole camera, showing the focal point (pinhole), image plane, focal length, and optical axis. The light rays emitted by the light source reflect off the surface in the world and pass through the aperture to form an upside-down image on the image plane. This is mathematically equivalent to producing a rightside-up image on the virtual image plane in front of the focal point.

simplifies the equations by obviating any need for a minus sign but also makes it easier to visualize the imaging process by removing any need to imagine the image upside down. Notice that perspective projection involves a loss of dimension in going from 3D to 2D: Since any point in space along the light ray will project to the same point on the image plane, it is impossible to recover the third dimension (distance from the camera) without additional information.

Orthographic projection occurs when all the light rays are parallel to the image plane. In that case the nonlinear equations above become linear: $x = x_w$, $y = y_w$. Orthographic projection is an approximation of perspective projection in the unlikely scenario that the scene being viewed is far from the camera, close to the optical axis, and no bigger than the camera's sensor in size. A more realistic approximation is **scaled orthographic projection**, which is orthographic projection with a single uniform scaling factor. It is easy to see that if the objects in the scene vary little in depth relative to their distance from the camera, then the distance z_w to all points in the scene can be approximated with a constant z_0, leading to

$$x = f \frac{x_w}{z_w} \approx \alpha x_w \tag{2.3}$$

$$y = f \frac{y_w}{z_w} \approx \alpha y_w, \tag{2.4}$$

where $\alpha \equiv f/z_0$ is the scaling factor. Like orthographic projection, scaled orthographic projection is linear. Scaled orthographic projection of a scene is mathematically equivalent to the orthographic projection of the scene onto a plane parallel to the image plane, followed by perspective projection of all the points in that plane, where this last step is simply a uniform scaling, as shown in Figure 2.24. Scaled orthographic projection, also known as **weak perspective projection**,[†] is a reasonable approximation when either the depth varies little over the scene or the scene lies close to the optical axis. These two sufficient

---

[†] Some authors distinguish *weak perspective* from *scaled orthographic* by allowing nonuniform scaling in the former.

Figure 2.24 Perspective, weak perspective (scaled orthographic), and paraperspective projection models. Based on V. S. Nalwa. *A Guided Tour of Computer Vision*. Reading, MA: Addison-Wesley, 1993; S. E. Palmer. *Vision Science: Photons to Phenomenology*. Cambridge, Mass.: The MIT Press, 1999.

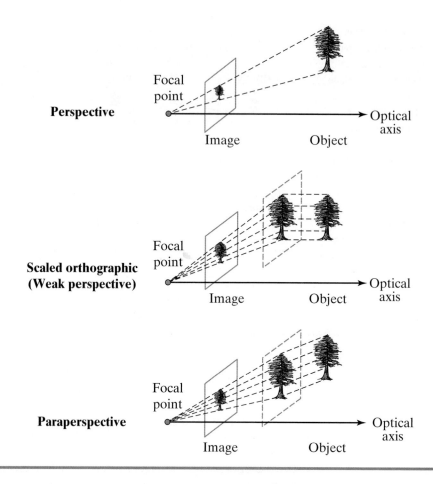

conditions are interrelated: An object near the optical axis may vary more in depth, while an object far from the optical axis must vary less in depth to achieve the same amount of error in approximation due to the simplified imaging model. Note that a long focal length restricts the field of view, causing the image to be formed by light rays that are nearly parallel to one another and hence ensuring that all visible objects are close to the optical axis. For this reason scaled orthographic projection is a good approximation when viewing distant objects with a zoom lens, as long as the camera remains at a roughly constant distance from the scene over time.

Affine projection is a generalization of scaled orthographic projection in which the light rays remain parallel to each other but are not required to be parallel to the optical axis. A special case of affine projection is **paraperspective projection**, in which light rays are projected in parallel along the direction from the focal point to the centroid of the object of interest onto a plane parallel to the image plane. The points on this plane then undergo perspective projection, which is mathematically equivalent to uniform scaling since the planes are parallel.

2.2.4 Camera with Lens

The ideal pinhole model remains important because even real cameras with lenses are well modeled mathematically as pinhole cameras, once the distortions due to the lens are accounted for. However, pinhole cameras themselves are not very practical because the tiny hole does not let in much light. Although it is possible to build a real pinhole camera that works, such a camera must be allowed to remain unmoved for several minutes in front of a perfectly static scene in order to gather enough photons for a quality image to be formed.

To build a more practical imaging device, the pinhole is replaced with a **lens**. A simplified lens consists of two spherical surfaces joined so that the centers of the spheres are collinear with the centers of the surfaces. Such a spherical lens has three basic parameters: the radii r_1 and r_2 of the two surfaces, and the refractive index n.[†] We make the common assumption known as **Gaussian optics** in which all light rays are **paraxial**, that is, they form small angles with respect to the optical axis. For such rays, assuming that the thickness of the lens is negligible, the relationship between the lens parameters and the focal length f is given by the **lens maker's formula**, also known as the **thin lens formula**:

$$\frac{1}{f} = (n - 1)\left(\frac{1}{r_1} - \frac{1}{r_2}\right) \tag{2.5}$$

which follows the *Cartesian sign convention* in which light travels from left to right, and the sign of r_1 or r_2 is positive if the surface makes the light rays more convergent, or negative if it makes them more divergent. A surface that bulges out is called *convex*, whereas a surface that curves inward is *concave*, so r_1 is positive if the left surface is convex, whereas r_2 is positive if the right surface is concave, as shown in Figure 2.25. The **Gaussian lens formula** specifies the distance from the lens to the image s_i for an object at a distance of s_o from the lens:

$$\frac{1}{f} = \frac{1}{s_o} + \frac{1}{s_i} \tag{2.6}$$

Figure 2.25 Thin lens, thick lens, and double Gauss lens.

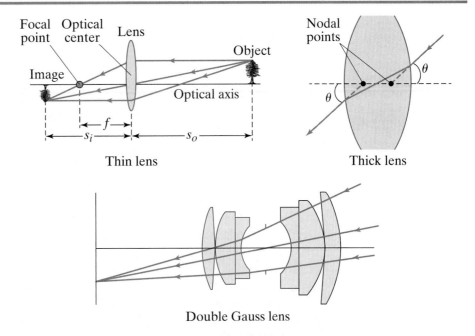

Thin lens

Thick lens

Double Gauss lens

[†] If the surrounding medium is not air (which has a refractive index very close to 1), then n is the ratio of the refractive index of the material and the refractive index of the surrounding medium.

If the thickness τ of the lens is not negligible, then we have a thick lens, in which the relationship between the lens parameters and the focal length are given by the **thick lens formula**:

$$\frac{1}{f} = (n-1)\left(\frac{1}{r_1} - \frac{1}{r_2} + \frac{n-1}{n}\cdot\frac{\tau}{r_1 r_2}\right) \tag{2.7}$$

The Gaussian (paraxial) optical behavior of a thick lens is specified by three pairs of **cardinal points** along the optical axis. Of these, the front and rear **focal points** are the same as the focal points on either side of a thin lens. The front and rear **nodal points** are such that a light ray aimed at one of them will emerge from the other one at the same angle as the incoming ray. Similarly, a ray of light crossing the front *principal plane* at a certain distance from the optical axis appears to emerge from the rear principal plane at the same distance; these principal planes are defined as passing through the **principal points** and perpendicular to the optical axis. If the surrounding medium on both sides is the same, then each nodal point coincides with the nearest principal point. If, in addition, the lens is thin, the front and rear nodal/principal points coincide at the center of the lens. The principal point of a thin lens is therefore the center of the lens. More commonly, the term **principal point** refers to the intersection of the line passing through the center of the lens and the image plane.

   Real lenses are neither thick nor thin but rather **compound**, in which several simple lens elements are combined to improve performance. One of the most successful and common compound lenses is the **double Gauss lens**, which consists of six simple lenses arranged in a nearly symmetric relationship. The **optical power** of a lens is the inverse of the focal length, measured in **diopters** (inverse meters). To a good approximation, the overall optical power of a compound lens system is simply the addition of the optical power of the individual lens elements.

   The **aperture** of a camera is the opening through which light rays enter the lens on their way to the sensor. The ratio, f/d, of the focal length f to the diameter d of the aperture is a dimensionless quantity called the **f-number**. Since the area of a circular aperture is proportional to the square of the diameter, if the diameter is decreased by $\sqrt{2}$, the amount of light is reduced by a factor of 2. The aperture setting is measured in **f-stops**, where a stop is a power of 2. That is, reducing the aperture by one stop means reducing the amount of light by a factor of 2, while increasing the aperture by one stop means increasing the amount of light by a factor of 2. The sensor size and focal length determine the angular **field of view (FOV)**, which is given by 2θ, where $\tan\theta = d/2f$. The aperture determines the **depth of field (DOF)**, which is the range of distances in the scene that form acceptably sharp images. That is, at distances other than the focused distance, a point will project onto the image not as a point but rather as a blur spot, with the amount of lens blur defined by

$$c = \frac{d|s_1 - s_2|}{s_2} \tag{2.8}$$

where s_1 is the distance to the plane that is in focus, s_2 is the distance to the point, and c is the diameter of the **circle of confusion**. While this formula makes the convenient assumption that a point is imaged as a circle, more generally the impulse response, or **point spread function (PSF)**, specifies the shape that a point will take on the image plane. The Fourier transform of the PSF is the **optical transfer function (OTF)**, whose magnitude is the **modulation transfer function (MTF)**. Photographers refer to the aesthetic quality of the blur as the **bokeh**.[†]

[†] Pronounced BOH-keh.

Figure 2.26 An undistorted image, barrel distortion, and pincushion distortion.

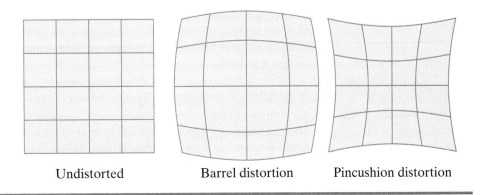

Undistorted   Barrel distortion   Pincushion distortion

Lenses exhibit various types of **aberrations**, which include any deviation of the performance of a lens from ideal. For example, **distortion** arises from the fact that the light rays do not necessarily follow straight lines when passing through the lens. The two most common types of distortion are barrel distortion and pincushion distortion, as shown in Figure 2.26. Distortion is more noticeable for lenses with small focal lengths (called **fisheye lenses**), which are designed to capture a wide field of view. Camera calibration usually includes nonlinear terms to account for such distortion. This bending of the light is often different for different wavelengths due to *material dispersion*, leading to **chromatic aberration**. Chromatic aberration is usually reduced by adjoining multiple lenses in an *achromatic doublet*, or by postprocessing the image inside the camera. When non-paraxial light rays enter the lens from the side, they reflect and scatter inside the lens, producing **lens flare**, which manifests itself as bright spots on the image due to a light source (such as the sun) that is outside the field of view.

Another aberration is known as **vignetting**[†] which is the darkening of an image away from the center. There are several types of vignetting, illustrated in Figure 2.27. *Optical vignetting* is caused by the fact that off-axis rays may not travel through all the lens elements in a complex lens. *Mechanical vignetting* is caused by obstruction of the light rays by external camera elements (such as the lens hood). *Pixel vignetting* is caused by the angular sensitivity of digital sensors. *Natural vignetting* is caused by the dependence of light intensity on the angle θ that the incoming ray makes with the optical axis. For any given pixel the irradiance[‡] E falling on the sensor after passing through a simple lens is proportional to the radiance L:

$$E = L \frac{\pi}{4} \left(\frac{d}{f} \right)^2 \cos^4 \theta \tag{2.9}$$

Figure 2.27 Optical vignetting.

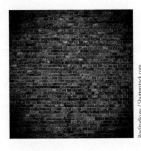

RoyStudio.eu / Shutterstock.com

[†] Pronounced vin-YET-ing.
[‡] Irradiance, the radiant power landing on a surface, is discussed in more detail later.

where the proportionality constant is related not only to the f-number f/d, but also to the angle θ. If we let $E_0 \equiv L\frac{\pi}{4}\left(\frac{d}{f}\right)^2$ be the on-axis $(\theta = 0)$ irradiance on the sensor for a given scene radiance L, and let E be the irradiance at a different point on the sensor, then their relationship is given by

$$\frac{E}{E_0} = \cos^4\theta \tag{2.10}$$

Since the light intensity decreases according to the cosine of the angle raised to the fourth power, this is known as the **cosine fourth law**. In practice this law is not important since most modern cameras are designed to compensate for this effect using, for example, a graduated neutral density filter that reduces the amount of light in the center of the lens to balance the effects of the law. The effects of most types of vignetting are negligible when using apertures smaller than an f-number of 8, represented as $f/8$.

2.2.5 A Simplified Imaging Model

Despite the intricate details of electromagnetic waves, radiometry, and lenses, a simple imaging model that provides a reasonable approximation for many tasks specifies the irradiance E on the image sensor as the product of a lighting function Λ and a reflectance function R:

$$E(x, y, \lambda) = \Lambda(x, y, \lambda)R(x, y, \lambda) \tag{2.11}$$

where $E(x, y, \lambda)$ is the irradiance at a point (x, y) on the sensor at wavelength λ. In this model, the lighting function Λ models the light source(s) and all interreflections and shadows, not according to their location in 3D space, but rather according to the light rays collected at each point on the sensor. Similarly, the reflectance function R models how much light at wavelength λ incident on the surface seen by the point (x, y) is reflected toward the sensor. Reflectance values vary from 0 (complete absorber) to 1 (perfect reflector). In other words, if every object in the scene were perfectly diffuse then $E(x, y, \lambda) = \Lambda(x, y, \lambda)$ for every point on the sensor and every wavelength. This imaging model is closely related to the notion of **intrinsic images**, in which multiple images of a static scene under different imaging conditions can be used to estimate the reflectance or other properties in the scene, as illustrated in Figure 2.28.

Figure 2.28 Intrinsic images are a mid-level description of scenes determined by decomposing an image into constituent components, such as an illumination image and a reflectance image. Based on Y. Weiss, "Deriving intrinsic images from image sequences," *Proceedings of the International Conference on Computer Vision*, pages 68-75, July 2001.

Input	=	Illumination	×	Reflectance

  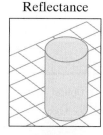

2.3 Image Acquisition

Once an image has formed on the surface of a sensor as an irradiance function, the information must be converted by the sensor into a digital image, which is then transmitted or stored. The steps involved in this process are the focus of this section.

2.3.1 Sampling and Quantization

Let $s(\lambda)$, where $0 \leq s(\lambda) \leq 1$, be the sensitivity of the sensor to a particular wavelength λ. Then the image pixel value $I(x, y)$ can be modeled as the integration of the irradiance function over the area of the pixel and over all wavelengths, after first multiplying by the sensitivity function:

$$I(x, y) = \varphi \left(\int \int \int E(x', y', \lambda')s(\lambda') \, dx' \, dy' \, d\lambda' \right) \qquad (2.12)$$

where the primes indicate dummy variables. The integrals over wavelength and sensor position, in addition to an integration over time (which is not shown), perform the work of **sampling** to convert the continuous irradiance function into a discrete function defined only over the rectangular lattice of integer (x, y) coordinates. **Quantization** then assigns a discrete gray level to every pixel in order to represent its value in digital form. However, to avoid an artifact known as *false contouring*,[†] it is important not only that there is a sufficient number of gray levels but also that they are meaningfully spaced. This is accomplished by applying a nonlinear mapping known as **gamma compression**, described in detail below, prior to quantization. The function φ includes both gamma compression and quantization, along with any sensor artifacts like blooming or noise.

2.3.2 Gamma Compression

The basic idea of gamma compression is shown in Figure 2.29. The raw measurement of light obtained by the sensor is transformed by a nonlinear mapping before transmission, storage, and/or manipulation. This step transforms the linear, physical light intensity into a perceptually uniform quantity, so that the pixel values in a digital image are not (in most cases) directly proportional to the amount of light collected by the sensor.

To understand gamma compression, we need to go back in time to consider an important fact of a now-obsolete technology. Cathode ray tubes (CRTs), which were the prevailing display technology for three-quarters of a century, have the curious property that the intensity of the light displayed on the screen is nonlinearly related to the applied voltage. More specifically, the transfer function of a CRT display follows a power law, in which the displayed intensity L (representing radiance or luminance) is proportional to the voltage V raised to some power:

$$L = cV^{\gamma} + b \qquad (2.13)$$

Figure 2.29 Linear light intensities are gamma compressed by the camera into perceptually uniform quantities, which are then gamma expanded by the display.

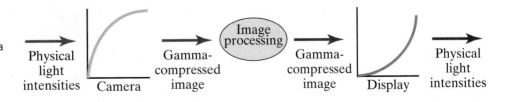

Physical light intensities → Camera → Gamma-compressed image → Image processing → Gamma-compressed image → Display → Physical light intensities

[†] Also known as *banding*, a form of *posterization*.

where γ is the exponent of the power function, and the constants b and c are the *blacklevel* and *contrast*, respectively, of the CRT display. If the monitor is adjusted properly so that its blacklevel is zero (i.e., the black pixels just barely emit light), then $b = 0$, leading to a simpler formulation:

$$L = cV^{\gamma} \tag{2.14}$$

Because of the widespread use of the Greek letter gamma (γ) for the exponent, this function is known as a **gamma function**. Figure 2.30 shows several plots of this function for different values of gamma, assuming $c = 1$ for simplicity. If $\gamma > 1$, the function is convex (curves upward) and is known as *gamma expansion*; if $\gamma < 1$, the function is concave (curves downward) and is known as *gamma compression*; if $\gamma = 1$, the function is linear.

CRT displays have a typical value of $\gamma_d \approx 2.2$, where the subscript indicates that this is the gamma of the display. To counter this effect and to simplify the electronics, video engineers decided many years ago that the voltage inside a display should be proportional not to the intensity of light being displayed but rather to the intensity raised to the power of γ_c, where $\gamma_c \approx 1/\gamma_d$. Cameras were therefore designed to encode the image according to $V = L_i^{\gamma_c}$, where L_i is the incoming light intensity, while displays produced light according to $L = V^{\gamma_d} = L_i^{\gamma_c \gamma_d}$. Images encoded in such a way are said to be *gamma compressed*, and if the gammas are inverses of each other $(\gamma_d = 1/\gamma_c)$ then they cancel each other $(L = L_i)$ so that the intensity displayed is the same as the intensity captured.

In practice, while the exponents used by the camera and display are nearly inverses of each other, they are not exactly so. In fact, when the image is expected to be viewed in lighting conditions different from those under which it was captured, the compression exponent γ_c is intentionally designed so that the product $\gamma_c \gamma_d$ is not 1. The reason for this choice is a perceptual phenomenon known as **simultaneous contrast** in which the human visual system's ability to discern contrast decreases in dark surroundings, as depicted in Figure 2.31. As a result, if a scene is captured in a bright outdoor setting but the resulting image (or movie) is viewed in a dim room or dark theater, it will lack contrast if displayed at the same intensity level. For this reason, various television and movie industry standards specify the *viewing gamma* (the product of the camera and display gammas) to be between 1.0 and 1.2, so that the viewing experience is subjectively correct even though it is not necessarily mathematically correct. Today, video cards typically also have a lookup table (LUT) that provides an additional adjustment, and the viewing gamma is defined to take this into account as well.

Figure 2.30 Gamma function with different values of γ.

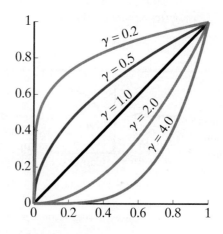

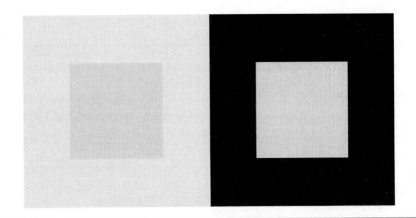

Figure 2.31 Simultaneous contrast. The pixels inside the middle squares have the same luminance, but the pixels on the right appear brighter due to its surroundings. Therefore, if an image is displayed at the correct luminance in a dimmer environment than the one in which it was captured, it will appear to be lacking in contrast.

In addition to this curious property of CRT displays, there is another more foundational reason that television engineers chose to introduce gamma compression many decades ago. Due to an amazing coincidence, the compression of light intensity according to $\gamma_c \approx 1/\gamma_d$ closely models the way in which the human visual system perceives light. In other words, although the doubling of the amount of light produces a *physical intensity* equal to twice the original, the *perceived intensity* is not increased linearly but rather nonlinearly according to a function quite similar to $L_i^{\gamma_c}$. Gamma compression therefore transforms a linear light intensity into a nonlinear quantity that is perceptually uniform. As a result, additive noise introduced in the transmission of a gamma-compressed analog video signal has minimal impact on visual perception, because a constant amount of additive noise affects the perceived intensity by an equal amount regardless of the overall signal value. Without gamma compression, however, additive noise in dark regions would produce a much more noticeable (and objectionable) effect on the viewing experience than noise added to bright regions. In the digital age this phenomenon still applies, particularly in the case of lossy compression which introduces additive noise to the image. JPEG and MPEG compression, for example, should always be performed on the nonlinear, perceptually uniform gamma-compressed signal rather than on the original linear signal, in order to minimize unacceptable artifacts. For the same reason, gamma compression prior to quantization results in a more effective use of the finite number of digital codes available. It is important therefore to view gamma compression not simply as an unfortunate relic necessary to maintain backward compatibility with now-obsolete CRT displays, but rather as an essential part of the image digitization and transmission process based upon timeless characteristics of human visual perception.

To see this connection between gamma compression and human visual perception, recall from our discussion on the human visual system that two luminances[†] can be discerned if their difference is at least 1% (approximately). That is, the **contrast threshold** of the human visual system is approximately $\frac{\Delta L}{L} = 1\% = 0.01$. Also recall that for image reproduction purposes the range of luminances is about 100 to 1. Now suppose we were to digitize these luminances in equally spaced intervals, in increments of 0.01, from 1 to 100, resulting in 9901 digital codes representing the luminance values of $1.00, 1.01, 1.02, \ldots, 99.98, 99.99, 100.00$. The drawback of this approach would be that, while the consecutive codes at the lower end of the scale, say 1 and 1.01, are discernible, consecutive codes at the higher end of the scale, say 99.99 and 100.00, are not discernible at all. The reason for this is that $(100.00 - 99.99)/100.00 = 0.01/100 = 0.0001 = 0.01\%$, which is much less than 1%, so that many codes at the higher end of the scale would be

---

[†] Luminance, which is radiance multiplied by the sensitivity of the sensor, is discussed in more detail later.

wasted. On the other hand, if we were to digitize luminances in equally spaced intervals in increments of 1.00, the digital codes of 1, 2, 3, ..., 98, 99, 100 would yield a barely discernible difference at the higher end but unacceptably large differences at the lower end, leading to objectionable false contouring. This is because $(2 - 1)/1 = 1.0 = 100\%$, which is much greater than 1%.

A more effective use of the digital codes occurs when consecutive codes correspond to relative luminance differences of approximately 1%. If we let v be the gray level and L be the luminance, then this is expressed as

$$0.01\Delta v = \frac{\Delta L}{L} \tag{2.15}$$

or

$$\frac{\Delta v}{\Delta L} = \frac{1}{0.01L} \tag{2.16}$$

where $\Delta v = 1$ is understood to be the difference between two consecutive gray levels, and ΔL is the difference between the corresponding luminances. If we let φ be the function that maps luminances to gray levels, i.e., $v = \varphi(L)$, the derivative of this function is $d\varphi/dL \approx \Delta v/\Delta L$. Therefore, the function f is the integral of the above expression, or

$$\varphi(L) = \int \frac{1}{0.01L} = 100 \log L \tag{2.17}$$

where log is the natural logarithm. This expression tells us that the desired nonlinear function that maps linear intensity to a perceptually uniform value is logarithmic. Table 2.1 compares this nonlinear coding with the two linear coding attempts just described. Linear coding requires approximately $100/0.01 = 10000$ gray levels, or 14 bits, to cover the 100:1 range with an increment of 0.01, while an increment of 1.0 requires $100/1 = 100$ gray levels, or 7 bits. In contrast, if the codes are spaced nonlinearly according to a ratio of 1.01, then only $(\log 100)/(\log 1.01) = 463$ gray levels, or 9 bits, are needed. The common 8-bit format, which owes its popularity to the widespread practice of grouping 8 bits into a byte in a digital computer, is sufficient for about a 50:1 ratio, roughly equivalent to traditional broadcast-quality television.

The assumption of a constant 1% threshold in Equation (2.15) is known as **Weber's law**. Although Weber's law is a good model of the transfer function of some cortical cells, it is not

linear 14-bit codes			linear 7-bit codes			logarithmic 9-bit codes		
gray level	L	$\Delta L/L$	gray level	L	$\Delta L/L$	gray level	L	$\Delta L/L$
00000000000000	1.00	1.00%	00000	1.00	100.00%	000000000	1.00	1.00%
00000000000001	1.01	1.00%	00001	2.00	50.00%	000000001	1.01	1.00%
00000000000010	1.02	1.00%	00010	3.00	33.33%	000000010	1.02	1.00%
⋮	⋮	⋮	⋮	⋮	⋮	⋮	⋮	⋮
10011010101010	99.98	0.00%	1100001	98.00	1.02%	111001100	98.01	1.00%
10011010101011	99.99	0.00%	1100010	99.00	1.01%	111001101	99.00	1.00%
10011010101100	100.00	0.00%	1100011	100.00	1.00%	111001110	100.00	1.00%

TABLE 2.1 Logarithmic coding is a more efficient use of the available bits than linear coding because it results in successive codes that differ by the contrast threshold of 1% across the entire range of luminances. In contrast, linear 14-bit coding waste bits in bright regions where successive gray levels look identical, and linear 7-bit coding produces objectionable artifacts in dark regions.

an accurate model of human visual perception over all luminances.[†] As it turns out, a more accurate mapping between physical light intensity and perceived light intensity is obtained with a power-law function, known as **Stevens' power law**:

$$\varphi(L) = cL^{\gamma_h} \tag{2.18}$$

where $\gamma_h \approx 0.5$. Here we see the amazing coincidence that this gamma of the human visual system γ_h is nearly the same as the inverse of the CRT display gamma, γ_d, because $1/2.2 \approx 0.45$. Therefore, the gamma compression of a camera produces nearly the same mapping as that of the human visual system, which justifies our saying that gamma-compressed signals are perceptually encoded. Note that the power-law function with $\gamma < 1$ performs a similar operation to that of a logarithm function, since they both have similar concave shapes. While the linear quantity L is referred to as the *luminance* as we saw earlier, the nonlinear quantity that captures human perception on a uniform scale is known as **lightness**.

One drawback of the gamma compression function $\varphi(L) = cL^{\gamma}$ is that its slope is infinite at $L = 0$, leading to high amplification of noise in dark regions of the image. To overcome this problem, it is common practice to modify the function by specifying a linear section for values below some threshold τ:

$$\varphi(L) = \begin{cases} mL & \text{if } L \leq \tau \\ (1 + \epsilon)L^{\gamma} - \epsilon & \text{otherwise} \end{cases} \tag{2.19}$$

where the slope m and offset ϵ are set to ensure that the value and first derivative of the two sections of the function match at the point $L = \tau$:

$$m = \frac{\gamma\tau^{\gamma-1}}{\tau^{\gamma}(\gamma - 1) + 1} \tag{2.20}$$

$$\epsilon = \frac{1}{\tau^{\gamma}(\gamma - 1) + 1} - 1 \tag{2.21}$$

The nonlinear transfer function obtained by modifying gamma compression is uniquely specified by the parameters γ and τ. There are two widely used standards that offer slightly different variations of gamma compression by choosing different values for these two parameters. **Rec. 709**,[‡] the standard for high-definition television (HDTV) that was first approved in 1990, uses $\gamma = 0.45 \approx 1/2.222$ and $\tau = 0.018$, leading to $m = 4.5$ and $\epsilon = 0.099$:

$$\varphi_{709}(L) \equiv \begin{cases} 4.5L & \text{if } 0 \leq L < 0.018 \\ 1.099L^{0.45} - 0.099 & \text{if } 0.018 \leq L < 1 \end{cases} \tag{2.22}$$

where the intensity L has been normalized to be in the range of 0 to 1.[§] Six years after the approval of Rec. 709, **sRGB** was developed to standardize the RGB color space used for still images for display on computer monitors and printers. sRGB uses $\gamma = 1/2.4 \approx 0.417$ and $\tau \approx 0.0031308$, leading to $m = 12.92$ and $\epsilon = 0.055$:

$$\varphi_{sRGB}(L) \equiv \begin{cases} 12.92L & \text{if } 0 \leq L \leq 0.0031308 \\ 1.055L^{(1/2.4)} - 0.055 & \text{if } 0.0031308 < L \leq 1 \end{cases} \tag{2.23}$$

[†] In fact, if b is the bit depth, it is easy to see from Table 2.1 that the ratio of the highest luminance to the lowest non-zero luminance is $(1.01)^n$, where $n = 2^{b-1}$. For $b = 9$ bits, this yields $(1.01)^{511} \approx 162$, which is a reasonable number. But for $b = 16$, the logarithmic model yields $(1.01)^{65535} \approx 10^{283}$, which is more than the number of atoms in the universe.

[‡] Formally known as ITU-R Recommendation BT.709.

[§] Rec. 2020, used for UHDTV, uses the same transfer function as Rec. 709. However, in 12-bit mode, the precision is increased to $\tau = 0.0181$ and $\epsilon = 0.0993$.

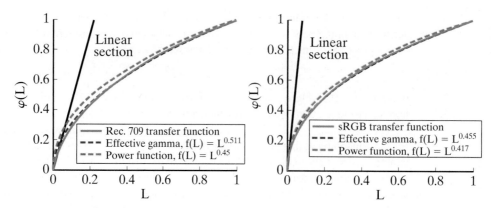

Figure 2.32 Rec. 709 gamma function (left), and sRGB gamma function (right). In both cases, the effective gamma function closely follows the modified gamma function, while the power function (using the same exponent as the modified gamma function but without the linear segment) is noticeably different. The dashed black line indicates the linear section, which is valid only for $L \leq \tau$.

as shown in Figure 2.32. Either way, the resulting nonlinear quantity is known as the **grayscale value** (or simply *gray level*), and since the nonlinear transfer function is designed to correspond closely to the power function of Steven's law, generally speaking *lightness* and *value* can be thought of as synonyms. Note that in the case of a color image, Equation (2.23) is applied to the color channels separately, with L replaced by R, G, and B after normalizing to the range of 0 to 1.

It is important to note that modifying the gamma function by inserting a linear section changes the **effective gamma** of the function. The effective gamma is defined as the exponent of the gamma function (*without* the linear section) that best fits the curve (*with* the linear section). As a result, although the exponent of the Rec. 709 transfer function is 0.45, its effective gamma is closer to 0.511. Similarly, although the exponent of sRGB is approximately 0.417, its effective gamma is closer to 0.455. In fact, both of these standards were designed by first specifying the desired effective gamma, then determining the exponent that best approximates that function. Because Rec. 709 is intended for viewing in dim environments, it was designed for a 1.125 viewing gamma, which is achieved using an effective camera gamma of $1/1.955555 \approx 0.511$, since $2.2/1.955555 = 1.125$, assuming a CRT gamma of 2.2. Empirically, this effective gamma of 0.511 is achieved pretty well using an exponent of $1/2.222222 = 0.45$. sRGB, on the other hand, was designed for a viewing gamma of 1.0, because it is intended for typical office environments. With a CRT gamma of 2.2, this yields $1/2.2 = 0.454545$ effective camera gamma, which empirically is achieved with an exponent of $1/2.4 = 0.416666$.

Having presented the concept of gamma compression in some detail, it is only appropriate to caution the reader that not all cameras perform gamma-compression. That is, some high-end cameras offer the possibility of storing the raw non-gamma compressed image, using a large number of bits per pixel in order to prevent false contouring. Raw images are useful for some computer graphics work, as well as for measuring the actual radiance of the scene. Nevertheless, unless you have good reason to believe otherwise, you should always assume that an image has been gamma compressed, especially with 8-bit-per-pixel-per-channel images.

2.3.3 CCD and CMOS Sensors

The light that falls onto the image plane is sampled by the sensor to produce values. These days, nearly all cameras are digital. The two most common digital sensors are CCD (charge-coupled device) or CMOS (complementary metal-oxide semiconductor). Each consists of a dense array of photodiodes (typically spaced 1 to 5 μm apart) that convert light photons to electrons. In a **CCD sensor**, these electrons are collected and stored in local potential wells during exposure time and then read out by transferring electrons down the line of potential wells until they reach the readout register known as the *horizontal shift register*. Electrons in the horizontal shift register are then transferred one at a time to an amplifier that converts the collected electrons into a voltage. In a **CMOS sensor**, transistors next to each photodiode convert the electrons to a voltage.

CCD sensors dominated the digital camera industry for two decades before CMOS sensors began to gain in popularity in the late 1990s. As a result, the CCD sensor is a more mature technology and produces superior image quality overall. In particular, in low light conditions CMOS sensors produce grainy images because the photons that land on the transistors next to the photodiodes are wasted. These transistors thus reduce the **fill factor** of CMOS sensors, which is the percentage of the pixel that collects light, as shown in Figure 2.33. However, CMOS sensors have now progressed to the point that the images captured by CMOS and CCD sensors in bright light settings are nearly indistinguishable in quality. The primary advantages of CMOS sensors is that they are less expensive to produce, consume less power, and are smaller and lighter. Moreover, CMOS sensors provide more flexibility: because CMOS pixel values can be read individually, a subset of the pixels called a region of interest (ROI) can be read from a CMOS sensor without reading the entire image, as opposed to the CCD sensor which requires an entire line to be read out. CMOS sensors can also achieve a wide dynamic range by resetting individual pixel wells when they near their capacity. Because the CMOS manufacturing process is identical to that used to produce processors and memory, CMOS sensors can include circuitry directly on the same chip to perform image processing operations such as stabilization or compression.

There are two dominant types of CCDs. A **full-frame CCD** has 100% fill factor because the entire sensor surface area collects light, but this type of CCD requires a mechanical shutter to prevent light from striking the photodiodes as the charge is read out. An **interline transfer CCD**, on the other hand, consists of masked columns between the photodiodes; when an exposure has ended, the charge is transferred from each photodiode to the adjacent masked column, which is then used to transfer the charge. Because an interline transfer CCD can read out the image even while light continues to strike the photodiodes, it has an *electronic shutter* which leads to much faster shutter speeds than are possible with a mechanical shutter.

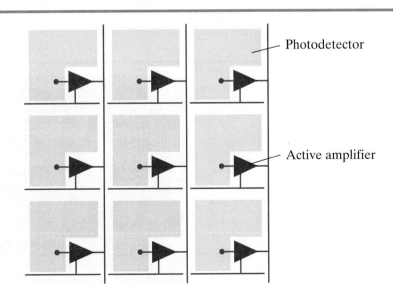

Figure 2.33 Fill factor is the percentage of the pixel on the physical sensor that captures light. A full frame CCD has 100% fill factor, whereas an interline transfer CCD has significantly less fill factor. Based on http://www.siliconimaging.com/cmosfundamentals.htm, image from photobit.

Photodetector

Active amplifier

Video cameras are either progressive scan or interlaced. An **interlaced camera** divides each image *frame* of a video sequence into two alternating *fields* — an odd field and an even field. The odd field consists of the odd rows of the image, while the even field consists of the even rows of the image. By displaying the two fields in succession, the effective frame rate of the display is doubled while taking advantage of the human visual system's tendency to blur information temporally. If the camera captures the two fields at different times, then a moving object will be shifted between the two fields, even within the same image frame. Interlaced cameras were the prevailing technology for analog video cameras, with 60 Hz (or 50 Hz, depending upon the country) power circuitry driving 60 Hz (or 50 Hz) field refresh rates, leading to effectively 30 Hz (or 25 Hz) frame rates for broadcast television. A **progressive scan camera**, on the other hand, captures the entire image frame simultaneously, without dividing it into fields. With the move toward digital formats and standards, progressive scan cameras are more popular, and interlaced cameras are all but obsolete.

To capture color, two approaches are common. Professional and other high-end cameras, where quality is more important than cost, use an optical device known as a *trichroic prism* to split the incoming light beam into the three separate beams of differing wavelengths, which are then sensed by three separate CCD sensors, one for each color channel. Together, the device is known as a **three-CCD (3CCD) camera**. A less expensive approach is to cover the sensor with a **color filter array (CFA)** (or *color filter mosaic*), which filters the wavelengths of the incoming light differently for the individual pixels. A common CFM is the **Bayer filter**, which blocks all but green light for alternating pixels throughout the sensor in a checkerboard pattern; of the remaining pixels, red and blue light filters are placed over alternating rows, as shown in Figure 2.34. With a Bayer filter, each pixel senses only one of the three colors. The remaining colors are estimated using a **demosaicking algorithm** that interpolates missing colors based on the colors sensed by neighboring pixels. Having no such filters, a monochrome camera is responsive to a wider range of frequencies. Figure 2.35 shows the spectral sensitivity functions (SSFs) for typical CCD color and monochrome video cameras.

Cameras contain a number of controls. The shutter speed, along with the aperture, controls the amount of light allowed to strike the sensor. **Automatic gain control (AGC)** causes the camera to automatically adjust its exposure time to ensure that the output level remains relatively constant, which can yield widely differing values, even for consecutive image frames, when light sources enter or exit the scene. **White balance** is the proper adjustment of the relative intensities of the primary colors needed to ensure that the colors are captured properly.

Figure 2.34 A Bayer color filter placed over an image sensor is an inexpensive way to sense color using a single sensor. Green light is sensed by half the pixels in the checkerboard pattern, with the remaining pixels sensing blue or red in alternating rows.

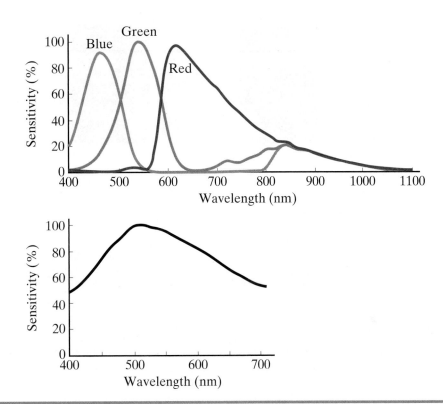

Figure 2.35 Spectral
sensitivities of a typical CCD
color (top) and monochrome
(bottom) video camera.
Note that, in contrast to the
human eye, CCD sensors are
sensitive to near infrared
light, up to approximately
1000 nm. Cameras typically
include an infrared filter
that cuts off frequencies
greater than about 700 nm.
Based on http://www.
theimagingsource.com/
downloads/fwcamspecwp.en.

There are three main sources of noise in an image sensor. **Shot noise** occurs at extremely low light levels due to the statistical nature of the discrete number of photons arriving in any given length of time. Another source is **sensor noise**, which includes the *fixed pattern noise* that arises due to differences in the individual pixel properties, as well as *transfer noise*, *quantization noise*, and the *dark current* that flows through a photosensitive device even when no photons are entering, leading to non-zero values regardless of the light level. The third source is **readout (or amplifier) noise**, which is added uniformly to the image by the amplifier used to convert electrons to voltage.

Other degradations in the image are possible. When the number of incoming photons exceeds the capacity of the photodiode to hold charge, the excess charge leaks out of the saturated photodiode into neighboring photodiodes, resulting in bright vertical streaks, an artifact called **blooming**. Although blooming has traditionally been a problem for CCD sensors, nearly all modern CCD sensors have anti-blooming protection to prevent the charges from overflowing, and CMOS sensors have never been affected. **Glare** is the presence of a bright light source that interferes with the ability to discern detail in the image. Another degradation is **motion blur**, which occurs when an object moves rapidly relative to the exposure time, causing streaking in the image in the direction of object motion, regardless of the sensor type. A related problem is the **rolling shutter effect**, which occurs in CMOS sensors that read out pixels sequentially rather than simultaneously, so that different pixels capture light entering the lens at different times. In the case of an analog video signal, **line jitter** refers to the random horizontal shift of rows of the image due to the inability of the phase-locked loop (PLL) circuitry to detect the start of the active line period perfectly. Similarly, if the signal is interlaced, then alternating lines of the image frame can show a moving object captured at different times, as shown in Figure 2.36.

Figure 2.36 Interlacing. A half-black, half-white piece of paper is translated. Left: no motion, edge is sharp. Middle: motion, edge is blurred. Right: zoomed in. The length of each horizontal bar is the distance traveled in 1/60 of a second.

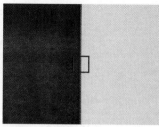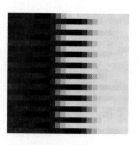

2.3.4 Transmission and Storage

The digital image captured by the sensor is transmitted to the computer and either processed live or stored for later processing. Traditional video cameras transmit the video signal using analog waveforms, the three standards being **NTSC** (used in North America and Japan), **SECAM** (used in France and the former Soviet Union), and **PAL** (used most everywhere else). Based on an AC voltage frequency of 60 Hz, NTSC transmits even and odd image fields at 60 Hz, leading to an effective frame rate of 30 frames per second (fps).[†] Specified originally in 1941 for black-and-white television and augmented in 1953 to include color, the NTSC format was discontinued as a broadcast signal in 2009 to make room for digital video transmission. Also interlaced, PAL transmits image fields at 50 Hz for an effective frame rate of 25 fps. NTSC, PAL, and SECAM use an aspect ratio of 4:3, resulting in an image resolution of approximately 640 × 480 and 768 × 576, respectively. The former is known as VGA resolution, while the latter is (after taking into account the non-square aspect ratio of older displays) 4CIF. All three analog standards are known as **composite video** because the luminance and chrominance information is combined into a single signal, whereas **component video** separates the individual color channels into individual signals, resulting in higher fidelity. Component video is typically used in production studios and other high-end applications.

The original digital replacement for analog video was Rec. 601,[‡] but this has been replaced by Rec. 709, which is the **high-definition television (HDTV)** standard mentioned earlier. The term HDTV encompass a variety of different resolutions, but they typically use an aspect ratio of 16:9. Common HDTV formats are 1080i or 1080p, containing image frames of size 1920 × 1080 either interlaced or progressive, respectively, and 720p, containing 1280 × 720 progressive scan images. More recently, **Ultra-high-definition television (UHDTV)**, defined by Rec. 2020, includes 3840 × 2160 and 7680 × 4320 video (known as 4K and 8K, respectively).

To take advantage of the human visual system's insensitivity to color changes, the color information in digital video is often downsampled. Because the nonlinear version of chrominance is known as **chroma** (just as the nonlinear version of luminance is **luma**), this is known as **chroma subsampling**. The nomenclature is rather nonintuitive, but as summarized in Table 2.2, $J{:}a{:}b$ means that chroma is downsampled by $h = \frac{J}{a}$ and $v = \frac{2a}{a+b}$ in the horizontal and vertical directions, respectively, so that the total amount of downsampling is $hv = \frac{2J}{a+b}$, where J is nearly always equal to 4.[§] For example, with 4:2:0 subsampling (the most common format), a 2 × 2 window of pixels contains 4 bytes of luma data (assuming

---

[†] Actually the frame rate of the standard was modified to $30/1.001 \approx 29.97$ fps when color was introduced to reduce interference between the chroma subcarrier and the accompanying audio signal.

[‡] Formerly known as CCIR 601.

[§] A special case in the notation occurs when $b = 1$ but $a \neq 1$ (the final 2 columns of the table), in which $v = 4$ and therefore $hv = 4h$; otherwise $v = 4$ would require a or b to be negative.

	4:4:4	4:2:2	4:2:0	4:1:1	4:1:0	4:4:0	4:4:1	4:2:1
horizontal chroma downsampling	1	2	**2**	4	4	1	1	2
vertical chroma downsampling	1	1	**2**	1	2	2	4	4
total chroma downsampling	1	2	**4**	4	8	2	4	8
number of luma bytes in 2 × 2 window	4	4	**4**	4	4	4	4	4
number of chroma bytes in 2 × 2 window	8	4	**2**	2	1	4	2	1

TABLE 2.2 The nomenclature for chroma subsampling is $J{:}a{:}b$, where J is nearly always equal to 4. From top to bottom, the rows are $h = 4/a$, $v = 2a/(a + b)$, $hv = 8/(a + b)$, 4, and $8/hv$; the final two columns show the special case when $v = 4$. Boldface is used to indicate that 4:2:0 is the most common case.

an 8-bit image) and 2 bytes of chroma data[†] since the chroma data is downsampled by 2 in both directions). To allow for filter overshoot and undershoot, video standards typically do not allow pixels to use all the available values but instead reserve a certain amount of **headroom** and **footroom**, so that black has a value of 16, and white has a value of 235. (While luma ranges from 16 to 235, chroma ranges from 16 to 240.)

A digital image is stored as an array of values in memory or as a sequence of bytes in a file. A large number of file formats exist. Some of the most common include **PNM**, a barebones format for uncompressed images used by researchers that includes **PGM** (for grayscale) and **PPM** (for color); **BMP**, a simple format widely used for its connection to the Windows operating system; **GIF**, an unusual format that supports multiple images for animation but only a limited color palette, making it suitable to simple shapes and logos; **PNG**, an open-source successor to GIF that supports lossless compression; **TIFF**, a flexible format with an extremely wide range of options, making it important for high-end manipulation of photographs but limiting its support in other applications such as Web browsers; **JPEG**, a widely-used format that makes use of lossy compression to reduce the file size; and **JPEG 2000**, a successor to JPEG that never gained widespread acceptance. The **EXIF** file format is increasingly being used, rather than the original **JFIF** format, to store JPEG files in order to allow metadata to be stored with the image, such as when and where the image was captured, the settings of the camera, the color space, and so forth. Another format is **OpenEXR**, which is used in the movie industry for high-dynamic range images using 16- or 32-bit floating point numbers. For video, one option is to store the video as a sequence of JPEG frames, known as **M-JPEG**. More common file formats include the historic **MPEG-1** format, or the more recent **MPEG-2**, **MPEG-4**, **AVI**, and **QuickTime** formats, which come with a dizzying array of choices for the **codec** (compressor-decompressor).[‡] The foundational video compression standard is **H.261**, which is used by MPEG-1 and forms the basis for all later standards. The more recent standards are **H.262** (used by MPEG-2 and DVD discs) and the ubiquitous **H.264** (used by MPEG-4, Blu-ray discs, streaming Internet video, and HDTV broadcasts).

2.4 Other Imaging Modalities

Now that we have spent considerable effort explaining the imaging process for a standard optical camera, in this section we consider several alternate imaging modalities to help appreciate the great diversity of techniques for gathering images, as well as the peculiar properties of each.

[†] That is, 1 byte for C_B and 1 byte for C_R, see Section 9.5.3 (p. 427).
[‡] Compression and decompression are discussed in more detail in Chapter 8.

2.4.1 Consumer Imaging: Catadioptric, RGBD, and Light-Field

We mentioned earlier that some animals, like lobsters, focus light not by using lenses, but rather by using mirrors. In a similar way, cameras can be made that either focus or bend light using mirrors. A standard optical imaging system using a lens is called **dioptric**, while a camera that uses mirrors is called **catoptric**. Putting these two together, a system that uses both lenses and mirrors is called **catadioptric**. One of the most widely used catadioptric imaging systems is the **omnidirectional** sensor, in which a camera points upward at a hyperbolic (or parabolic) mirror, which allows the camera to see 360 degrees around the scene, as shown in Figure 2.37. Such a camera system has an effective focal point at the focus of the hyperboloidal-shaped (or paraboloid-shaped) mirror, so that it is called a **central panoramic camera**. The resulting image is donut-shaped, with approximately half the pixels wasted.

Another useful sensor is the **RGBD camera** that captures not only the RGB values for each pixel but also the depth of the scene point from the camera plane. Such sensors operate by either time-of-flight, stereo processing after projecting an invisible, infrared texture onto the scene to simplify the correspondence problem, or shape from shading. Currently the most popular sensors are the Microsoft Kinect, Asus Xtion, and Intel RealSense, which are revolutionizing robotics and user interfaces due to their richer capturing modality. An entirely different approach is achieved by a **light-field camera** (also known as a plenoptic camera) that samples the light field using an array of microlenses; by tracing the rays of light using the appropriate computation, the image can be refocused after it has been captured, or it can be viewed as a 3D stereoscopic image whose appearance changes with the viewing angle.

2.4.2 Medical Imaging: CAT, PET, MRI, and Sonar

Medical applications use a variety of imaging technologies. One of the most well-known is **X-ray radiography**, which produces images by transmission rather than reflection. A generator emits X-rays toward an object of interest (such as part of a person's body), and a detector measures the photons that make it to the other side, rather than being absorbed by the object. The high amount of calcium in bones, for example, along with their high

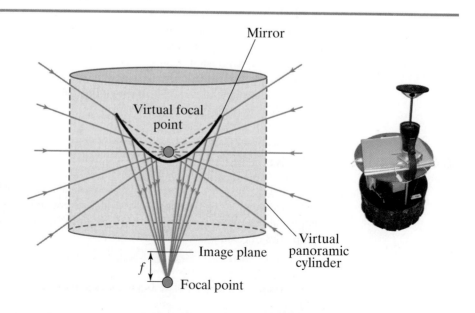

Figure 2.37 An omnidirectional camera can be achieved by attaching a hyperbolic or parabolic mirror to an upward-facing camera. Based on Valdir Grassi Junior and Jun Okamoto Junior, Development of an omnidirectional vision system, J. Braz. Soc. Mech. Sci. & Eng. vol. 28 no. 1 Rio de Janeiro Jan./ Mar. 2006, http://www.scielo.br/scielo.php?script=sci_arttext&pid=S1678-58782006000100007

density, causes them to absorb X-rays, which explains why bone structure is revealed so prominently in an X-ray. X-ray technology can also be used to capture 3D structure as an array of slices. The word *tomography* refers to imaging by slices, so this approach is known as **computed tomography (CT)**. The term *axial* refers to the horizontal plane through the human body when standing upright, and since the slices are parallel to this plane, the approach is also known as **computed axial tomography (CAT)**, so that CT scan and CAT scan are essentially synonymous. A patient is enclosed in a ring of scintillation detectors, and the X-ray emitting tube is rotated around the patient, collecting an image for each slice. Reconstruction of the patient's body is then obtained using algorithms such as filtered back projection, or iterative reconstruction.

Another common technique is **magnetic resonance imaging (MRI)**, which is safer than X-ray because it does not use ionizing radiation. Soft tissue in the human body contains water, and MRI uses powerful magnets to align the hydrogen nuclei (that is, protons) in these water molecules. A radio frequency (RF) signal pulse at the resonance frequency is emitted that systematically alters the alignment of the nuclei by flipping the spin of the protons. As the nuclei return to their original state, their motion generates an RF signal which is detected by receiver coils. MRI is widely used for medical diagnosis, and its extension called **functional MRI (fMRI)** uses MRI technology to detect change in magnetization between oxygen-rich and oxygen-poor blood to measure brain activity.

Positron emission tomography (PET) is another technique for nuclear imaging. A patient is injected with a radioactive isotope which, as it undergoes positron emission decay, emits a positron. The positron travels a short distance, decelerates, and then interacts with an electron. Both the electron and positron are annihilated, emitting a pair of gamma photons in opposite directions in the process, which are detected by a scintillator. Unlike CT or MRI, PET can detect details at the level of molecular biology.

Fluoroscopy is a way of obtaining real-time images of a patient using an X-ray image intensifier to convert the X-rays on the sensor to visible light for viewing by a radiologist. A popular fluoroscopy technique is **digital subtraction angiography**, in which a contrast medium has been injected into a structure; by subtracting the precontrast image, an enhanced image is obtained which enables a physician to more easily see the blood vessels for catheters and vascular imaging. Another technique is **fluorescence in situ hybridization (FISH)** which is used to detect DNA sequences on chromosomes using fluorescent probes that bind to certain parts of chromosomes.

Finally, **ultrasound** does not use electromagnetic radiation at all but rather sound waves, which are longitudinal and require a medium for transmission. These broadband sound waves are reflected by the tissue, allowing real-time imaging of moving structures with no ionizing radiation. Ultrasound imaging is widely used for observing babies in the womb, as well as elastography, which is measuring the elastic properties of soft tissue.

2.4.3 Remote Sensing: SAR and Multispectral

Cameras are often attached to aircraft or satellites for remote sensing of the earth for applications in meteorology, agriculture, surveillance, and geology. To enable detailed sensing of the terrain in all weather conditions, these cameras typically sense multiple frequencies simultaneously. A **multispectral** sensor senses a small number of frequencies, typically 5 to 7, while a **hyperspectral** sensor senses a much larger range of frequencies. Due to the larger number of frequencies, it is often not possible to build a 2D array that yields an image directly. Instead, either a **whiskbroom** sensor is used, in which a rotating mirror scans one pixel at a time, or a **pushbroom** sensor, which is a 1D linear array perpendicular to the direction of travel. Comparing the two alternatives, a pushbroom sensor is smaller, lighter, consumes less power, and has high reliability because it has no mechanical parts.

The **Landsat program** is the longest-running program to gather satellite imagery of the earth's surface, beginning in the 1970s and continuing to the present day. The latest version of Landsat uses a whiskbroom multispectral scanner with 8 spectral bands and an opto-mechanical sensor to collect information about earth from space, with calibration used to convert raw sensed values to absolute units of radiance. Another satellite imaging program is **SPOT (Système Pour l'Observation de la Terre)**, which uses a pushbroom camera consisting of a linear array of CCDs to collect 5 spectral bands. The SPOT sensor is able to collect more photons than Landsat, so it has a higher signal-to-noise ratio. The **AVIRIS (Airborne Visible InfraRed Imaging Spectrometer)** instrument, which uses a hyperspectral sensor capable of collecting radiance in 224 contiguous spectral bands from 400 to 2500 nm, is a more recent sensor mounted on aircraft for measuring the Earth's surface atmosphere. In remote sensing, it is common to call the raw digital values from an uncalibrated sensor **digital numbers (DNs)**, to distinguish them from physically meaningful quantities such as radiance or reflectance.

Synthetic aperture radar (SAR) illuminates the scene with radio waves whose wavelength ranges from one meter to millimeters. The received echo waveforms are detected and processed to form an image. SAR is usually mounted on a moving platform with a single beam-forming antenna attached to an aircraft or spacecraft. SAR is an advanced form of side-looking airborne radar (SLAR), which is essentially a virtual phased array. Related to SAR is **ultra-wideband radar**, whose signals are defined as having a bandwidth exceeding 500 MHz or 20% the center frequency of radiation and are sometimes used for through-the-wall imaging.

2.4.4 Scientific Imaging: Microscopy

A **micrograph** is an image obtained by connecting a camera to a microscope or similar device to obtain a magnified image. An optical microscope, also known as a **light microscope**, uses visible light and a system of lenses to focus the image. Some forms of light microscopy are **bright field microscopy**, in which the light shines below the sample, yielding a dark sample on a bright background; **phase contrast microscopy**, which exploits phase shifts that occur when light passes through media, thus avoiding the need to stain the specimen and allowing for *in vivo* imaging; and **fluorescent microscopy**, which illuminates the specimen with a nearly monochromatic light to excite fluorescent stains or proteins. Most fluorescent microscopes use **epifluorescence**, in which reflected light from the specimen combines with the emitted light, yielding a high signal-to-noise ratio. To reduce the out-of-focus light and improve the contrast, the recent approach of **light sheet microscopy** has been gaining in popularity. Another advanced approach is that of a **confocal microscope**, which uses point illumination and a beam splitter to allow 2D or 3D imaging of the object with increased contrast and resolution. Further improvements in resolution are achieved using electron microscopes such as a **scanning electron microscope (SEM)**, which scans the surface using beams of electrons, or a **scanning tunneling microscope (STM)**, which uses quantum tunneling.

2.5 A Detailed Look at Electromagnetic Radiation

You may know that there are three ways to transfer heat energy. If you pick up a pan from the stove, it will feel hot to the touch because of **conduction**. If you sit in front of a rotating fan, the fan will cool your skin due to the movement of the air, known as **convection**. Both of these methods require the source responsible for heat transfer to be nearby. In contrast, if you stand outside on a sunny day, you will feel warmth from the sun, even though the sun is millions of miles away. This form of energy transfer, known as **electromagnetic radiation**, has nothing to do with the surrounding air, and it explains why the front side of your body is heated while the back side is not when you stand in front of a campfire.

Even though it may seem that we have already treated the imaging process with a fair amount of detail, to understand light on an even deeper level we have to consider precisely what is meant by this third form of energy transfer. The energy in electromagnetic radiation is carried solely by **electromagnetic waves**, which are perfectly capable of traveling through a vacuum. In the following discussion we provide a more detailed description of these waves, the energy carried by them, and the ways in which they interact with the world around us.

2.5.1 Transverse Electromagnetic Waves

It is a fundamental principle of physics that a time-varying electric field causes, or *induces*, a time-varying magnetic field, and vice versa. Mathematically, this coupling is described in a set of equations known as **Maxwell's equations**.[†] Let E and B be the 3D vector electric and magnetic fields, respectively. In a vacuum containing no electric charge (imagine space through which the sun's rays travel), Maxwell's equations in differential form are

$$\nabla \cdot E = 0 \qquad \nabla \times E = -\frac{\partial B}{\partial t} \tag{2.24}$$

$$\nabla \cdot B = 0 \qquad \nabla \times B = \epsilon_0 \mu_0 \frac{\partial E}{\partial t} \tag{2.25}$$

where $\partial/\partial t$ is the partial derivative with respect to time, ∇ is the "dell" operator, and ϵ_0 and μ_0 are fundamental constants of nature, namely the **permittivity** and **permeability** of free space, respectively. The operator $\nabla\cdot$ ("dell dot") is the **divergence** of a vector field, while $\nabla \times$ ("dell cross") is the **curl**.

In their most general form, Maxwell's equations succinctly capture almost everything we know about electricity and magnetism, making them foundationally important for electric circuits, transmission lines, radio transmission, antenna design, communications, fiber optics, microwave ovens, waveguides, and sensing. While the details of these equations are beyond our scope of interest, what is important to note is that electric and magnetic fields are tightly coupled when they vary in time. This coupling, in fact, is the basis for the term **electromagnetism**. As E changes, the non-zero value of $\partial E/\partial t$ modifies the magnetic field; the changing value of B, in turn, modifies the electric field. From the first-order equations above, it is easy to derive second-order equations called the **homogeneous electromagnetic wave equations**:

$$\nabla^2 E = \epsilon \mu \frac{\partial^2 E}{\partial t^2} \tag{2.26}$$

$$\nabla^2 B = \epsilon \mu \frac{\partial^2 B}{\partial t^2} \tag{2.27}$$

where the **Laplacian** operator $\nabla^2 = \nabla\cdot\nabla$ is the divergence of the **gradient**.[‡] In these equations we have replaced ϵ_0 and μ_0 with the constants ϵ and μ appropriate for the medium,

---

[†] James Clerk Maxwell (1831–1879) is widely considered one of the greatest physicists of all time.
[‡] The gradient is covered in more detail in Section 5.3 (p. 234), while the Laplacian is discussed in Section 5.4.1 (p. 242).

to emphasize that the equations are applicable to any simple (linear, isotropic,[†] and homogeneous), non-conducting medium containing no electric charge, which includes all media through which light travels, such as empty space, air, water, glass, or a variety of plastics. For non-metallic materials, $\mu \approx \mu_0$ to an accuracy of one part in a billion, so we may safely assume $\mu = \mu_0$. We will discuss the effects of ϵ later in this section when we consider refraction.

Solutions to the homogeneous electromagnetic wave equations are electromagnetic waves. An electromagnetic wave consists of oscillating, coupled electric and magnetic fields propagating at a speed (*phase velocity*) of $\frac{1}{\sqrt{\epsilon\mu}}$. The electric and magnetic fields are in-phase sinusoids (meaning they reach their maxima together) along the direction of propagation, and they are perpendicular not only to each other but also to the direction of propagation, so that E, B, and the propagation direction form a right-handed orthogonal set, as shown in Figure 2.38. For this reason such a wave is called *transverse*, hence the name **transverse electromagnetic (TEM) wave**. The **polarization** of the wave is described by the orientation of the fields. If the fields retain their direction as the wave travels, then the wave is said to be *linearly polarized*. Alternatively, if their direction changes (rotates) as the wave travels, then the wave is *elliptically polarized* (with *circular polarization* as a special case). In the latter case the handedness of the wave is the direction of the change in orientation, namely, whether the orientation rotates clockwise or counterclockwise.

2.5.2 Radiometry and Photometry

To precisely describe the amount of energy transferred by electromagnetic radiation, several distinct quantities must be carefully defined. We begin by considering the simple example of a old-fashioned 60-watt incandescent light bulb radiating energy. The basic unit of energy is the joule (J), and the basic unit of power is the watt (W), which is defined as one joule per second. Therefore, this particular light bulb consumes 60 joules of energy per second, or equivalently, 60 watts of power. If all of this power were used for light (which it is not), then we would say that the **radiant flux** for the bulb is 60 watts. The integration of the radiant flux over a certain amount of time yields the total energy radiated by the light during that time, called **radiant energy** and measured in joules.

Oftentimes we are interested in the radiant flux in a particular direction. This quantity is known as **radiant intensity**, measured in watts per steradian $(\mathrm{W} \cdot \mathrm{sr}^{-1})$. Just as the radian is the unit of measure for a 2D angle, the *steradian* (sr) is the unit of measure for a 3D angle, also known as a *solid angle*.[‡] Whereas a half-circle spans π radians, and a complete circle

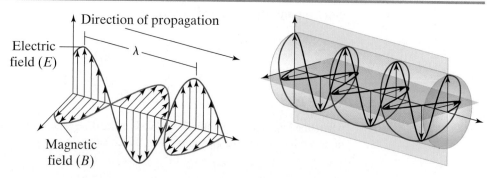

Figure 2.38 Left: A transverse electromagnetic (TEM) wave with linear polarization. Right: A TEM wave with circular polarization. Based on D. K. Cheng. *Field and Wave Electromagnetics*. Addison Wesley, second edition, 1989.

Direction of propagation

Electric field (E)

λ

Magnetic field (B)

[†] The term **isotropic** (from the Greek *isos*, equal, + *tropos*, way) means "the same in all directions." The opposite term is **anisotropic**.

[‡] The prefix *ster-* comes from the Greek word for *solid*, which is where we also get the word *stereo*.

spans 2π radians (the circumference of a unit circle), a hemisphere spans 2π steradians, and an entire sphere spans 4π steradians (the surface area of a unit sphere). Integrating the radiant intensity over all possible angles yields the radiant flux.

Now imagine a sphere centered at the light bulb. As the radius of the sphere increases, a point on the sphere gets farther from the light bulb and therefore receives less and less radiant intensity, because a constant solid angle yields a larger area on the sphere as the radius increases. This reduction motivates defining a new quantity, **radiance**, as the radiant intensity divided by the cross-section area, measured in watts per steradian per square meter $(W \cdot sr^{-1} \cdot m^{-2})$. Equivalently, radiance is the power per unit foreshortened area emitted into a unit solid angle. Along a ray of light emanating from the source, the radiance remains constant at all distances from the source. Therefore, radiance can be thought of as the amount of light along a ray traveling in any direction at any point in space.

Now suppose the light falls on a surface in the scene, such as the image sensor. The amount of radiant power that lands on a portion of the surface is called the **irradiance**, measured in watts per square meter $(W \cdot m^{-2})$. While radiance is a directional quantity, irradiance is not. Instead, it is the integration of the radiance of all incoming rays on an infinitesimal surface patch, after first considering the reduction in intensity due to **foreshortening**, which reduces the amount of incident light on the patch based on the angle between the surface normal and the incoming light ray. For example, if the light rays are parallel to the surface (that is, perpendicular to the surface normal), then the irradiance is zero because no light hits the surface.

These are the basic quantities of **radiometry**, which is the measurement of electromagnetic radiation, summarized in Table 2.3. Radiometry captures the rate at which light energy is emitted or absorbed when such power is sufficiently high that these quantities can be treated as continuous values, and when the light can be assumed to travel in straight lines according to geometrical optics. Radiometry is applicable not only to visible light but also to infrared, ultraviolet, and shorter wavelengths such as microwaves and radio waves. However, at high frequencies like X-rays and gamma rays, it is more appropriate to talk about individual photons, since the corpuscular nature of light makes the continuous quantities of radiometry less applicable.

Each quantity defined above also has a spectral version, namely **spectral radiant flux**, **spectral radiant intensity**, **spectral radiance**, **spectral irradiance**, and so forth. The spectral versions are normalized by wavelength, thus capturing the corresponding quantity per wavelength. If any spectral version is integrated over all wavelengths, it yields the non-spectral version. The per wavelength contribution to any radiometric quantity is known as the **spectral power distribution (SPD)**.

Related to radiometry is **photometry**, which is the measurement of electromagnetic radiation after weighting each wavelength by the sensitivity of the human eye to that wavelength. For example, a radio wave might have large radiometric values but zero photometric values, since the human eye is not sensitive to radio waves. The basic unit in

radiometry	photometry	meaning
radiant energy $(W \cdot s)$	luminous energy $(lm \cdot s)$	energy
radiant flux (W)	luminous flux (lm)	power
radiant intensity $(W \cdot sr^{-1})$	luminous intensity $(lm \cdot sr^{-1})$	power in a direction
radiance $(W \cdot sr^{-1} \cdot m^{-2})$	luminance $(lm \cdot sr^{-1} \cdot m^{-2})$	power along ray
irradiance $(W \cdot m^{-2})$	illuminance $(lm \cdot m^{-2})$	power incident on surface

TABLE 2.3 Quantities of radiometry and photometry. Note that $W \cdot s = J$, $lm \cdot sr^{-1} = cd$, $lm \cdot m^{-2} = lux$.

photometry is the *candela* (cd), which is roughly the power emitted by one candle in any particular direction. Other units include the *lumen* (lm), which is a candela times a steradian, and *lux*, which is a lumen per square meter. By considering the sensitivity of the human eye, each radiometric quantity has a corresponding photometric quantity, with *radiant* replaced by *luminous*, and *watt* replaced by *lumen*.[†] Thus in photometry we have **luminous energy** instead of radiant energy, **luminous flux** instead of radiant flux, **luminous intensity** instead of radiant intensity, **luminance** instead of radiance, **illuminance** instead of irradiance, and so forth. Be sure not to confuse any of the precisely defined radiometric or photometric terms described here with the subjective term **brightness**, which refers to the perceptual sensation of light.

2.5.3 Blackbody Radiators

Electromagnetic radiation is closely connected with temperature. In our everyday experience, light sources such as incandescent bulbs, fire, and the sun are usually hot. Such light sources can be closely approximated as idealized objects known as **blackbody radiators**.[‡] It may seem strange to use the term *blackbody* to refer to a brightly shining object, but the name stems from the fact that such an object absorbs all incident EM radiation, just as a completely black object absorbs all incident light. A blackbody radiator is in thermal equilibrium with its surroundings, so that it emits and absorbs the same amount of EM radiation at any given wavelength (otherwise it would increase or decrease in temperature). **Planck's law** expresses the spectral radiance of a blackbody radiator as a function of wavelength:

$$L_\lambda(T) = \frac{2hc^2}{\lambda^5 \left(e^{\frac{hc}{\lambda kT}} - 1 \right)}$$

where h is the same Planck's constant mentioned earlier,[§] $k \approx 1.38 \times 10^{-23}$ joules per kelvin $(J \cdot K^{-1})$ is Boltzmann's constant,[¶] and T is the absolute temperature. Blackbody radiance is plotted as a function of wavelength in Figure 2.39. The important aspect of this equation to note is that the amount of EM radiation emitted at any wavelength is based solely upon the constant temperature of the blackbody. This power per surface area is given by **Stefan's law** (derived from Planck's law) as σT^4 in watts per square meter $(W \cdot m^{-2})$, where $\sigma = 2\pi^5 k^4 / 15 c^2 h^3$ is Stefan's constant.

Also derived from Planck's law is **Wien's displacement law**,[*] which says that the peak wavelength at which the most energy is radiated is inversely proportional to the temperature: $\lambda_{peak} = b/T$, where $b \approx 2.9 \times 10^6$ nm · K is Wien's displacement constant. At room temperature $(27°C \approx 81°F)$, the peak wavelength is 9.7 μm, which is very much in the far infrared band (recall that the visible spectrum ends around 0.7 μm). Because the energy is highly concentrated around the peak (approximately 70% of the energy emitted is between one half and twice the peak wavelength), the amount of energy in the visible band is effectively zero. As the temperature of the blackbody increases, its intensity and frequency also increase, causing the peak wavelength to move from infrared to red to orange to yellow to white to blue to ultraviolet. Lower temperatures (red through yellow) are known as warm colors, while higher temperatures (bluish white) are known as cool colors. The difference

---

[†] For monochromatic light of 555 nm, 1 watt equals 683 lumens; at other wavelengths the conversion factor is multiplied by the photopic luminous efficiency function (LEF) described in Figure 2.9.

[‡] Ludwig Boltzmann (1844–1906) was an Austrian physicist.

[§] Section 2.2.1 (p. 33).

[¶] Not all light sources are well modeled as blackbody radiators; light-emitting diodes (LEDs), for example, do not waste as much energy on heat because they do not emit significant amounts of infrared light.

[*] Wilhelm Wien (1864–1928) was a German physicist.

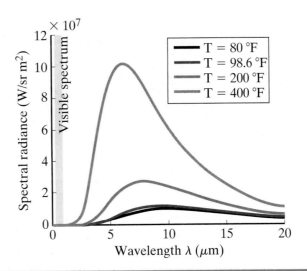

between two light sources can be measured by taking the difference between the micro reciprocal degree (**mired**), defined as $10^6/T$, for each.

2.5.4 Interaction with a Surface

When an electromagnetic wave impinges on a surface, it interacts with the surface in one of three ways: the energy is either absorbed, reflected, or transmitted, as illustrated in Figure 2.40. Absorption turns the electromagnetic energy into other forms of energy, such as heat. Reflection and transmission, which allow the light to continue its journey, are more relevant to our purposes.

Reflection

An opaque surface is one that only absorbs and reflects light, with no transmission. At the extremes, there are two kinds of opaque surfaces, as shown in Figure 2.41. A **specular** surface is one in which incoming light from one direction reflects in only one direction. A specular surface is a good model for a completely smooth plane boundary. This is why if you take a piece of reflective material (a metal like aluminum or silver) and smooth it, you will get

Figure 2.40 An EM wave interacts with a surface in one of three ways: absorption, transmission, or reflection.

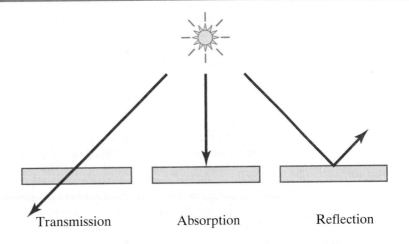

Figure 2.41 In specular reflection, incident light reflects in a single direction (as in a mirror). In diffuse reflection, incident light reflects in all directions equally.

Specular Reflection   Diffuse Reflection

a mirror. When you look at a mirror, the light you see depends very much on the direction from which you are viewing. The reflection of a mirror is easy to model because the reflected radiance is equal to the incoming radiance, and the angle of incidence equals the angle of reflection. Glossy surfaces, such as the body of a polished automobile, are not pure mirrors but nevertheless reflect the incoming light in a small set of directions around the specular angle.

At the other extreme is a **diffuse**, or matte, surface, which reflects light in many directions due to either a rough surface shape or internal scattering of the light by molecules below the surface. An ideal diffuse surface that reflects equal luminance in every direction is called **Lambertian**.[†] Although the intensity of the light reflected from such a surface is proportional to the cosine of the angle between the surface normal and the direction of the incoming light ray, known as **Lambert's cosine law**, this effect is canceled by the foreshortening of the apparent area of the viewer. As a result, a Lambertian surface appears equally bright when viewed from any direction.

For a diffuse surface, the **reflection coefficient** ρ is the ratio of the reflected radiation to the incident radiation. It is a unitless number that ranges from 0 to 1, with 0 meaning that none of the radiation is reflected, and 1 meaning that all of it is reflected. The average reflection coefficient is known as the **albedo** (from the Latin *albus*, meaning "white") of the surface, which also ranges from 0 to 1. A surface with albedo of 1 looks white, whereas albedo of 0 looks black. Albedo is often used to describe the surface of the earth in different terrains: dark soil, for example, has an albedo as low as 0.05, while fresh snow has an albedo as high as 0.95.

Most real-world surfaces are not perfectly specular or diffuse. The intensity of light seen from everyday surfaces can usually be modeled as a weighted combination of the two extreme phenomena. The **Phong reflection model** is a widely-used empirical model that combines specular and diffuse reflection terms, along with a term for ambient light, to model the energy due to **interreflections** of other surfaces in the scene. Interreflections occur when light bounces off one surface, then another, then another, and so on, which cause regions to be illuminated that otherwise would be in shadow. More generally, a surface can be modeled by the **bidirectional reflectance distribution function (BRDF)**, which is defined as the ratio of the radiance to the irradiance of a surface, measured in sr^{-1}:

$$BRDF(\omega_i, \omega_r) = \frac{L_r(\omega_r)}{E_i(\omega_i)} \tag{2.28}$$

where E_i is the incoming irradiance, L_r is the outgoing (reflected) radiance, and ω_i and ω_r are the incoming and outgoing directions, respectively.[‡] In the case of an ideal diffuse

---

[†] Johann Heinrich Lambert (1728–1777), was a versatile Swiss scientist and mathematician.

[‡] The directions are specified by two numbers indicating the angles in spherical coordinates, $\omega_i = (\theta_i, \phi_i)$ and $\omega_r = (\theta_r, \phi_r)$.

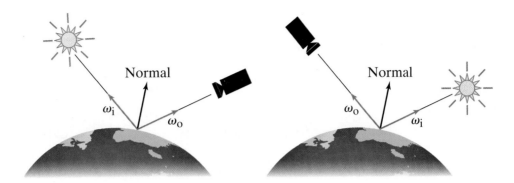

Figure 2.42 The bidirectional reflectance distribution function (BRDF) is a model of the reflectance of a surface as a function of the incoming and outgoing directions. According to the Helmholtz reciprocity principle, the BRDF outcome does not change if the direction of the light ray is reversed.

surface, the BRDF is a single number ρ_b, which is related to the diffuse reflectance by a factor of π: $\rho = \pi \rho_b$. The BRDF is symmetric in the incoming and outgoing directions, so that $BRDF(\omega_i, \omega_r) = BRDF(\omega_r, \omega_i)$, which is known as the **Helmholtz reciprocity principle**,[†] shown in Figure 2.42.

Transmission

Light that is neither absorbed nor reflected by the surface is transmitted through it. If light is transmitted through the material without being scattered, then the material is **transparent**. On the other hand, a **translucent** material also allows light to pass, but the internal structure of the material causes scattering of the light rays. Either way, when the light hits the boundary of the surface, its speed changes from $\frac{1}{\sqrt{\mu_0 \epsilon_1}}$ to $\frac{1}{\sqrt{\mu_0 \epsilon_2}}$, where ϵ_1 and ϵ_2 are the permittivities of the two media, assuming (as before) that $\mu_1 = \mu_2 = \mu_0$.[‡] As a result, the angle of the light with respect to the surface changes according to **Snell's law of refraction**, illustrated in Figure 2.43, which says that at the interface between two dielectric media the ratio of the sines of the angles is equal to the ratio of these speeds:

$$\frac{\sin \theta_t}{\sin \theta_i} = \sqrt{\frac{\epsilon_1}{\epsilon_2}} = \frac{n_1}{n_2} \tag{2.29}$$

where θ_i is the angle between the incident light ray and the surface normal, θ_t is the angle between the transmitted light ray and the surface normal, and $n_1 = \sqrt{\epsilon_1/\epsilon_0}$ and $n_2 = \sqrt{\epsilon_2/\epsilon_0}$ are the indices of refraction. This bending of the light according to Snell's law is the basic principle behind a lens.

When light passes from a higher permittivity to a lower one, that is, when $n_1 > n_2$, a surprising possibility arises. Rearranging the terms reveals

$$\sin \theta_t = \frac{n_1}{n_2} \sin \theta_i \tag{2.30}$$

[†] Hermann von Helmholtz (1821–1894) made important contributions to diverse areas of science, including visual perception, color science, electrodynamics, and thermodynamics.

[‡] Note that the speed of the light is not constant. Rather, it is the speed of light *in a vacuum* that is the well-known fundamental constant in nature: $c = \frac{1}{\sqrt{\epsilon_0 \mu_0}} = 299,792,458$ meters per second.

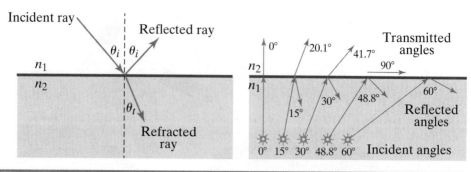

Figure 2.43 Snell's law of refraction (left) and total internal reflection (right). Note that 48.8 degrees is the angle of total internal reflection for water (assuming index of refraction = 1.33). Based on http://www.timbercon.com/Total-Internal-Reflection.html

so that the right hand side of the equation can be greater than 1, in which case θ_t does not exist at all! This is called **total internal reflection**, and it occurs when θ_i is greater than a certain *critical angle* that depends on the ratio n_1/n_2. Total internal reflection is the mechanism behind fiber optics, in which a transparent core is surrounded by a material with a lower index of refraction, allowing light to pass through the fiber by continuous reflections with almost no loss. A natural example of this phenomenon is the mineral *ulexite*, which contains fibrous compact veins that act as fiber optic cables, transmitting light from one surface to the other. When the rock is polished on both sides and placed on top of, say, a newspaper, the words seem to leap to the top of the stone, as shown in Figure 2.44 — hence the nickname "television stone."

Other Phenomena

Our brief tour in this section has only begun to explore the rich and complex capabilities of electromagnetic waves. For example, when the dimensions of objects are small compared with the wavelength of light, the wavelike properties of light become important, allowing light to bend around the edges of an object or through tiny slits, causing **diffraction** and **interference** between different waves. There are two types of mathematical models commonly used to model diffraction: *Fresnel diffraction* describes what happens when the light wave is near the object, while *Fraunhofer diffraction* applies to plane waves at a distance.[†] Another phenomenon is **iridescence**, which occurs when a surface appears to change color due to the viewing angle, as in soap bubbles, peacock feathers, or some butterfly wings. When the material is anisotropic so that a ray of light is split into two rays, we have **birefringence**, a phenomenon exploited by a variety of applications. *Circular*

Figure 2.44 Ulexite (television stone) is a naturally occurring rock with internal veins that act like fiber optic cables.

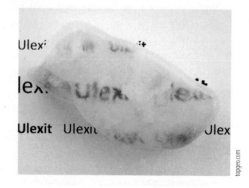

_____

[†] Pronounced fray-NELL and FROWN-hoof-uh.

birefringence is caused by applying a magnetic field that changes the relative speed between left and right circularly polarized waves, known as the *Faraday effect*. **Luminescence** is the emission of light by a method other than heat, such as by a light-emitting diode (LED), which relies on *electroluminescence*. Other forms of luminescence include **fluorescence**, in which the wavelength of light changes upon reflection, and **phosphorescence**, when there is a time delay between the absorption and emission. **Scintillation** is the twinkling or flickering of a light source, including the flash of light produced by an ionizing event, used in the sensing mechanism of positron emission tomography (PET). As you can see from these examples, electromagnetic waves exhibit a variety of phenomena, many of which are useful in practical applications, beyond the simplifying assumptions that we normally consider when studying light rays.

2.6 Further Reading

Excellent overviews of the human visual system can be found in the works of Wandell [1995] and Palmer [1999]. Another excellent reference on the inner workings of vision is that of Hubel [1988]. A recent survey on the human visual system is provided by Krüger et al. [2013]. Pioneering work in this field is too numerous to cite, but the research studies on receptive fields by Hubel and Wiesel [1962] and Olshausen and Field [1996] are particularly well known and relevant. The irreducible complexity of a photoreceptor is well argued by Behe [1996], while the explanation of the need for an inverted retina can be found in Gurney [1999]. The cone fundamentals in Figure 2.2 come from the data provided online by Stockman and colleagues.[†]

The plenoptic function is due to Adelson and Bergen [1991], while the light field was independently proposed by and Levoy and Hanrahan [1996] and Gortler et al. [1996] (where it was called the "lumigraph"); all of these can be traced to Gibson's ambient optic array [Gibson, 1966] and even earlier to the integral camera of Lippmann near the turn of the 20th century. And in fact the term "light field" itself was coined by Gershun [1939]. Further detail regarding the sampling of the light field can be found in Ng [2006].

Standard image formation and acquisition is treated in any computer vision book, such as Forsyth and Ponce [2012] or Szeliski [2010]. Gamma compression is described in detail by Poynton [2003, 1998], who also maintains an online Gamma FAQ.[‡] A complementary treatment of gamma compression can be found in the work of Stokes et al. [1996]. A description of how pixels are stored in a frame buffer, including resolution, color channels, and so forth, can be found in Glassner [1990]. Pawley [2006] provides a detailed overview of the imaging process in the context of microscopy. Vignetting is described by Goldman [2010], while intrinsic images are due to Barrow and Tenenbaum [1978] and investigated by Weiss [2001], Tappen et al. [2005], and Grosse et al. [2009]. For recent work on intrinsic images, see Imber et al. [2014]. Further information regarding CCD sensing and high dynamic range imaging can be found in Debevec and Malik [1997]. Another important consideration that we did not have space to consider involves atmospheric effects; see, for example, the dark channel prior of He et al. [2009].

Electromagnetic waves are the subject of any standard text on electromagnetism, such as Cheng [1989]. Light and the electromagnetic spectrum are covered in standard physics texts, such as Gettys et al. [1989]. Radiometry and photometry are difficult subjects to grasp, with subtle differences between the terms being extremely difficult to perceive for the uninitiated. For more information on the subjects, the reader may wish to consult the *Illumination Fundamentals* booklet from the Lighting Research Center.[§]

[†] http://www.cvrl.org.
[‡] http://www.poynton.com/PDFs/GammaFAQ.pdf.
[§] http://www.opticalres.com/lt/illuminationfund.pdf.

PROBLEMS

2.1 What is unusual about vertebrate photoreceptors, compared with most sensory receptors?

2.2 Explain the purpose of an inverted retina in the human eyeball. Give two additional reasons why the inverted retina does not cause significant distortion in the image.

2.3 List three parts of the human eyeball that refract light.

2.4 What are the actual names of the three types of cones, which are colloquially called red, green, and blue?

2.5 Define horopter.

2.6 Explained what is meant by foveated vision.

2.7 How do we know that the original signal captured by the rods is compressed by subsequent cells before leaving the eyeball?

2.8 True or false: Your right eye is mapped to the left half of your brain, while your left eye is mapped to the right half of your brain.

2.9 Match each term on the left with with the lighting condition on the right.

scotopic vision	sunlight
photopic vision	moonlight
mesopic vision	starlight

2.10 Cones do not work in the dark, because they are not sensitive enough. What about the converse: Do rods produce meaningful signals in everyday well-lit conditions? Why or why not?

2.11 Draw a labeled diagram of the human visual system, including at least ten parts indicated in bold in the text.

2.12 Why would it be tempting to conclude that short (blue) wavelengths are less important to the human visual system? Why is this conclusion false?

2.13 Suppose the following pairs of numbers indicate the luminances of the left and right halves of a piece of paper (ignore units): 100/101, 200/201, 300/301, 150/160, 250/260, 350/360. Which can be discerned?

2.14 What is a receptive field?

2.15 Which cells in the visual pathway transform the signal similar to the Laplacian of Gaussian (LoG)?

2.16 The axons of which cells comprise the optic nerve?

2.17 What is unique about the lobster eye?

2.18 True or false: The speed of light is constant no matter what medium it is passing through.

2.19 Which has a longer wavelength: red light or blue light?

2.20 Suppose a lens is made of a high-index plastic whose index of refraction is 1.74. If the speed of light is approximately $3 \cdot 10^8$ m/s in a vacuum, what is the speed of light (phase velocity) as it passes through the lens?

2.21 What is the plenoptic function?

2.22 Describe the essential elements of a pinhole camera.

2.23 What are the wavelengths of visible light?

2.24 Which has a longer wavelength: radio waves or X-rays? Which is more dangerous, and why?

2.25 You are sitting at a stoplight listening to 102.1 FM on your old-fashioned radio, getting a weak signal. You wish to roll the car to improve the signal. How far must you roll to move one wavelength? What is the ratio of this wavelength to that of green light?

2.26 Is scaled orthographic projection more appropriate for a zoom lens or a fisheye lens? Explain your answer.

2.27 Suppose we have a symmetric thin lens composed of two sections of a sphere glued together, where the radii of both sides are equal. Explain why $\left(\frac{1}{r_1} - \frac{1}{r_2}\right)$ in Equation (2.5) is not equal to zero.

2.28 Apply the nonlinear transfer function from both Rec. 709 and sRGB in Equations (2.22) and (2.23) to the values $L = 0.2, 0.4, 0.6$, and 0.8. Compute the difference for each value as a percentage of the answer for Rec. 709.

2.29 Will gamma compression become obsolete now that CRT displays are obsolete? Why or why not?

2.30 Specify the two sections of the modified gamma function with exponent $\gamma = 0.5$ and threshold $\tau = 0.1$.

2.31 Explain the idea of effective gamma.

2.32 What is the name of the most popular color filter array (CFA)?

2.33 Explain the difference between a field and a frame of video.

2.34 List some similarities and differences between CCD and CMOS sensors.

2.35 How much is a CMOS sensor affected by blooming?

2.36 Why does black have the value 16 and not 0 in a digital image?

2.37 Explain what is meant by a Lambertian surface. What is albedo? Which is more likely to be Lambertian: a piece of cloth or a shiny piece of metal?

2.38 Which radiometric quantity is appropriate for a ray of light? What is the corresponding photometric quantity?

2.39 Explain why a thermal infrared camera is able to measure the heat emanating from people and animals.

2.40 Suppose I am standing on the shore looking at a body of water. If the water has an index of refraction of 1.33, at what angle will I experience total internal reflection? How does the answer change if I am underwater looking up? In both cases, express the angle with respect to the vertical axis.

2.41 A light field, which is 4D, can be represented as a 2D array of tiny 2D images. Indeed, this is the representation used by a light field camera. Explain how a 2D array of microlenses placed in front of the image sensor might be able to accomplish this.

2.42 Mathematically show the two sufficient conditions for scaled orthographic projection to closely approximate perspective projection. (*Hint*: Show from Equations (2.1)–(2.2) that bounding the error $\left| f\frac{x_w}{z_w} - f\frac{x_w}{z_0} \right| < \epsilon$ for some nominal depth z_0 implies $\frac{x}{z_0}\frac{|\delta_z|}{z_0 + \delta_z} \leq \frac{\epsilon}{f}$, where $z = z_0 + \delta_z$, and ϵ is a constant. Then interpret the result in terms of the two sufficient conditions.)

2.43 Suppose a person is standing in front of a pinhole camera so that their face occupies a certain width in the image. If the person moves laterally so that the perpendicular distance to the camera is maintained, does the face width in the image change? Why or why not?

2.44 Consider two thin lenses, both symmetric and made of the same material. If one lens has twice the focal length of the other, what is the relationship between their radii?

2.45 In the case of a thin lens, what condition is necessary in order for a distinction between the focal points and nodal points to be important?

2.46 Suppose a camera has an f-number of 8. What is the aperture, expressed as a function of f? What is the aperture of a camera whose light gathering ability is twice as great, also expressed as a function of f?

2.47 List the four types of vignetting. Under what conditions are they important?

2.48 List the three ways of transferring energy. Of these, which one can travel through a vacuum?

2.49 What is the name of the set of equations that underlie all applications using electromagnetism?

2.50 Derive the homogeneous vector wave equations in Equations (2.26)–(2.27) from Maxwell's equations in Equations (2.24)–(2.25). (*Hint*: It is *not* necessary that you understand what the divergence and curl operators actually do. Simply take the curl of (2.24), and apply the fact that the curl operator is linear. You will need the vector identity $\nabla \times \nabla \times E = \nabla(\nabla \cdot E) - \nabla^2 E$, and similarly for B.)

CHAPTER 3
Point and Geometric Transformations

In this chapter we discuss some of the simplest ways to transform an image into another image. These transformations fall into one of two types. In a *point transformation*, a pixel's value is changed solely based upon its original value, without changing its location within the image. We consider several types of point transformations, such as graylevel transformations in which the pixel values are scalars, multispectral transformations that operate on images with multiple channels, and multi-image transformations that operate on more than one image. In a *geometric transformation*, the location of a pixel changes from the input image to the output image, but the value of the pixel does not change. Our discussion of geometric transformations includes both simple transformations that involve a one-to-one mapping from input to output pixels as well as more complex transformations that require interpolation.

3.1 Simple Geometric Transformations

We begin by considering simple geometric transformations in which the output pixel is dependent only upon a single input pixel. More general geometric transformations are considered later in the chapter.

3.1.1 Flipping and Flopping

Perhaps the simplest geometric transformation is to reflect the image about a horizontal or vertical axis passing through the center of the image, as shown in Figure 3.1. If the axis is horizontal, then the transformation **flips** the image upside down; whereas if it is vertical, then the transformation **flops** the image to produce a right-to-left mirror image. A flip followed by a flop (or equivalently, a flop followed by a flip, since order does not matter here) is referred to as a **flip-flop**.[†]

Consider, for example, the following 3×3 grayscale image and its reflections about the horizontal and vertical axes:

$$\begin{bmatrix} 128 & 78 & 174 \\ 181 & 48 & 77 \\ 109 & 49 & 138 \end{bmatrix} \xrightarrow{\text{FLOP}} \begin{bmatrix} 174 & 78 & 128 \\ 77 & 48 & 181 \\ 138 & 49 & 109 \end{bmatrix}$$

$$\downarrow \text{FLIP}$$

$$\begin{bmatrix} 109 & 49 & 138 \\ 181 & 48 & 77 \\ 128 & 78 & 174 \end{bmatrix}$$

In this example, flipping swaps the first and last rows, whereas flopping swaps the first and last columns. More generally, recall that for an arbitrarily sized *width* $\times$ *height* image, the columns are $x = 0, 1, \ldots, width - 1$, while the rows are $y = 0, 1, \ldots, height - 1$. Thus, all pixels in the first row have coordinates $(x, 0)$, while all pixels in the last row have coordinates $(x, height - 1)$; all pixels in the second row have coordinates $(x, 1)$, while all pixels in the penultimate (next-to-last) row have coordinates $(x, height - 2)$; and so on. Therefore, flipping involves swapping row y with row $height - 1 - y$, while flopping involves swapping column x with column $width - 1 - x$, as illustrated in Figure 3.2. If we let (x, y) be the coordinates of an input pixel, and (x', y') the coordinates of the corresponding output pixel, then the relationship between these coordinates can be expressed mathematically as follows:

$$x' = x \qquad\qquad y' = height - 1 - y \qquad (\text{flip}) \qquad (3.1)$$

$$x' = width - 1 - x \qquad y' = y \qquad\qquad (\text{flop}) \qquad (3.2)$$

Using these equations, the transformations are expressed as functions that define the mapping between each input pixel $I(x, y)$ and its corresponding output pixel $I'(x', y')$:

$$I'(x, height - 1 - y) = I(x, y) \qquad (\text{flip, forward mapping}) \qquad (3.3)$$

$$I'(width - 1 - x, y) = I(x, y) \qquad (\text{flop, forward mapping}) \qquad (3.4)$$

where in each case I is the input image and I' is the output image. These equations, known as **forward mappings**, instruct how to compute the destination coordinates from the source coordinates. Rearranging the equations yields **inverse mappings**, in which the source coordinates are computed from the destination coordinates:

Figure 3.1 An image, and the result of flipping, flopping, and flip-flopping.

Image

Flip

Flop

Flip-flop

[†] The terms *flip* and *flop* are widely used in the graphics community; the term *flip-flop* is introduced here as a natural consequence.

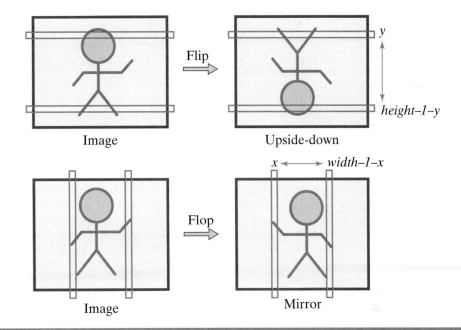

Figure 3.2 To flip an image (turn it upside-down), swap each row *y* with its corresponding row *height* − 1 − *y*. Similarly, to flop an image (produce a mirror version), swap each column *x* with its corresponding column *width* − 1 − *x*.

$$I'(x', y') = I(x', height - 1 - y') \quad \text{(flip, inverse mapping)} \quad (3.5)$$

$$I'(x', y') = I(width - 1 - x', y') \quad \text{(flop, inverse mapping)} \quad (3.6)$$

Although both the forward and inverse mappings are equivalent to each other in the simple case of flipping and flopping, the distinction between them is important with more complicated transformations, as we shall see later.

Throughout this book we will be presenting a variety of different algorithms to process images. To capture these algorithms precisely we present them using **pseudocode**. Pseudocode (which literally means "false code") is a compromise between the two extreme alternatives of explaining an algorithm in human language (which leads to ambiguity) and providing actual working code (in which uninteresting details obscure the important steps). Pseudocode allows us to precisely express the steps of an algorithm in a manner detailed enough to aid implementation, but independently of any particular programming language.

The pseudocode for the algorithms to flip and flop an image using the forward mapping are shown as the procedures FLIPIMAGE and FLOPIMAGE in Algorithms 3.1 and 3.2, respectively. Each of these procedures takes one input parameter, namely the image *I*, which can be of any type (grayscale, RGB, floating-point, or otherwise), and each procedure produces exactly one output, namely the upside-down or mirror-reversed image *I'*. The comments, which are set apart by a right-facing triangle (➤), help to explain what each line of the pseudocode is doing, although these particular procedures are so simple that there is really no need for comments at this point. Line 1 allocates memory to store the output image; we will often omit this step and assume that memory has already been allocated. Line 2 indicates a "for loop" over all the pixels in the image. In this line the image is treated as a set, so that $(x, y) \in I$ means a pixel in the image. Line 2 is therefore a compact way of saying, "for each pixel in image *I*," which in many programming languages would be expressed as two separate for loops, one over *x* and another over *y*:

> **for** $y \leftarrow 0$ **to** $height - 1$ **do**
>
> > **for** $x \leftarrow 0$ **to** $width - 1$ **do**

ALGORITHM 3.1 Flip an image by reflecting about a horizontal axis

FlipImage (I)

Input: image I of size $width \times height$
Output: upside-down image I'

1	$I' \leftarrow$ AllocateImage($width, height$)	➤ Allocate memory for output image.
2	**for** $(x, y) \in I$ **do**	➤ For each pixel in input image,
3	$\quad I'(x, height - 1 - y) \leftarrow I(x, y)$	set corresponding pixel in output image.
4	**return** I'	➤ Return output image.

ALGORITHM 3.2 Flop an image by reflecting about a vertical axis

FlopImage (I)

Input: image I of size $width \times height$
Output: mirror-reversed image I'

1	$I' \leftarrow$ AllocateImage($width, height$)	➤ Allocate memory for output image.
2	**for** $(x, y) \in I$ **do**	➤ For each pixel in input image,
3	$\quad I'(width - 1 - x, y) \leftarrow I(x, y)$	set corresponding pixel in output image.
4	**return** I'	➤ Return output image.

or equivalently using a single for loop over the 1D index using Equation (1.3):

$$\textbf{for } i \leftarrow 0 \textbf{ to } width \cdot height - 1 \textbf{ do}$$

However, we use the set notation $(x, y) \in I$ to simplify the presentation and to emphasize that the order in which the pixels are processed in this case does not matter. In Line 3 the image is treated as a 2D array, so that $I(x, y)$ yields the value of pixel (x, y). This value is then copied to a different location in the output image I'. The left arrow ($\leftarrow$) denotes the setting of a variable, performing the same role in our pseudocode as the equal sign ($=$) in most programming languages. This notation should help to avoid confusion, because incrementing a variable, for example, will be written in our pseudocode as $x \leftarrow x + 1$ (or $x \leftarrow_+ 1$) rather than $x = x + 1$, which is not a valid mathematical statement. Once all pixels have been transformed, the resulting image is returned in Line 4.

3.1.2 Rotating by a Multiple of 90 Degrees

Another important geometric transformation is to rotate the image. Later in the chapter we will consider arbitrary rotation angles, but for now let us limit ourselves to rotations that are multiples of 90 degrees about the center of the image, which simplifies the problem considerably by ensuring that the transformation is a one-to-one mapping from pixels in the input image to those of the output image. Figure 3.3 shows the result of rotating an image by multiples of 90 degrees. Note that rotating an image by 180 degrees is equivalent to a flip-flop.

Consider, for example, the clockwise 90-degree rotation of a $width \times height$ image, as illustrated in Figure 3.4, where each pixel (x, y) in the input image maps to the pixel (x', y') in the output image. From the figure it is not hard to see that the dimensions of the $new\text{-}width \times new\text{-}height$ output image are swapped with respect to those of the input image: $new\text{-}width = height$ and $new\text{-}height = width$; and that $I(x, y)$ maps to $I'(height - 1 - y, x)$, so that $x' = height - 1 - y$ and $y' = x$, or

Figure 3.3 An image rotated by 0, +90, −90, and 180 degrees.

0°

+90°

−90°

180°

$$I'(height - 1 - y, x) = I(x, y) \tag{3.7}$$

which is the forward mapping. Alternatively, we can rewrite the correspondence as $x = y'$ and $y = height - 1 - x'$, leading to the inverse mapping:

$$I'(x', y') = I(y', height - 1 - x') \tag{3.8}$$

As with flipping and flopping, these two approaches are equivalent since rotating about a multiple of 90 degrees is a one-to-one mapping.

EXAMPLE 3.1

Rotate the following 3×2 grayscale image clockwise by 90 degrees, using both forward and inverse mapping approaches:

$$I = \begin{bmatrix} 105 & 90 & 35 \\ 228 & 207 & 52 \end{bmatrix}$$

Solution

Considered as a forward mapping, Equation (3.7) indicates that the pixel at $I(0,0)$ maps to $I'(2 - 1 - 0, 0) = I'(1, 0)$; the pixel at $I(1,0)$ maps to $I'(2 - 1 - 0, 1) = (1, 1)$; the pixel at $I(2,1)$ maps to $I'(2 - 1 - 1, 2) = I'(0, 2)$; and so forth. Considered as an inverse mapping, Equation (3.8) indicates that the pixel at $I'(0, 0)$ is mapped from $I(0, 2 - 1 - 0) = I(0, 1)$; the pixel at $I'(1, 0)$ is mapped from $I(0, 2 - 1 - 1) = I(0, 0)$; the pixel at $I'(1, 2)$ is mapped from $I(2, 2 - 1 - 1) = I(2, 0)$; and so forth. Either way, the result is as follows:

$$\begin{bmatrix} 105 & 90 & 35 \\ 228 & 207 & 52 \end{bmatrix} \xrightarrow{\text{ROTATE CLOCKWISE}} \begin{bmatrix} 228 & 105 \\ 207 & 90 \\ 52 & 35 \end{bmatrix}$$

Figure 3.4 To rotate an image clockwise by 90 degrees, the pixel (x, y) in the input image is mapped to (x', y') in the output image. From the drawing, it is easy to see that $x' = new\text{-}width - 1 - y = height - 1 - y$, and $y' = x$.

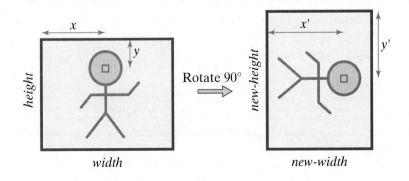

Similar reasoning can be applied to other multiples of 90 degrees. If we let $90m$ be the clockwise rotation angle, where m is an integer, then only four possible cases exist: A clockwise rotation of 90 degrees (the case we just considered in detail) occurs if the remainder of dividing m by 4 is 1, a counterclockwise rotation of 90 degrees occurs if the remainder is 3, a rotation of 180 degrees occurs if the remainder is 2, and no rotation occurs if the remainder is 0. Note that the image dimensions are swapped when the rotation is either clockwise 90° or counterclockwise 90°, whereas the image dimensions are unchanged when the rotation is either 0° or 180°. The pseudocode for these cases, using inverse mapping, is shown in Algorithm 3.3, where we have used the notation of modulo arithmetic to specify the remainder, that is, $\mathrm{mod}(m,4)$ is the remainder of dividing m by 4.

3.1.3 Cropping an Image

Figure 3.5 shows the result of **cropping** a smaller region out of a larger image. The region to be cropped is specified by a rectangle, where the coordinates of the top-left pixel in the rectangle are given by (*left*, *top*), and the coordinates of the bottom-right pixel just outside the rectangle are given by (*right*, *bottom*). In other words, we adopt the common convention that the rectangle specified by the four parameters *left*, *top*, *right*, and *bottom* includes the pixels in the sets $x \in \{left, \ldots, right-1\}$ and $y \in \{top, \ldots, bottom-1\}$, so that the pixel (*left*, *top*) is included but the pixel (*right*, *bottom*) is excluded. One advantage of this convention is that the number of pixels in the rectangle is given simply by $(right - left) \cdot (bottom - top)$. The pseudocode of this procedure, which unlike the previous transformations considered is not one-to-one, is shown in Algorithm 3.4.

3.1.4 Downsampling and Upsampling

Another common operation is to **downsample** an image to produce a smaller image than the original. In general it is advised to smooth the image before downsampling to avoid aliasing artifacts, as discussed later in the chapter. For now, however, let us simply discard a subset of the pixels. For example, to downsample by a factor of 2, every other column and row are discarded:

$$I'(x, y) = I(2x, 2y) \qquad (\text{downsample by two}) \qquad (3.9)$$

ALGORITHM 3.3 Rotate an image by a multiple of 90 degrees

RotateImageByMultipleOf90Degrees(I, m)

Input: image I of size *width* $\times$ *height*, signed integer m indicating the number of 90-degree turns
Output: image I' of size *new-width* $\times$ *new-height*, which is I rotated by $90m$ degrees clockwise

```
1   case mod(m,2) of
2        0: new-width ← width, new-height ← height          ➤ Image dimensions remain the same.
3        1: new-width ← height, new-height ← width          ➤ Image dimensions are swapped.
4   I' ← AllocateImage(new-width, new-height)
5   for (x', y') ∈ I' do
6        case mod(m,4) of
7             0:   I'(x', y') ← I(x', y')                    ➤ no rotation
8             1:   I'(x', y') ← I(y', height − 1 − x')       ➤ 90 degrees clockwise
9             2:   I'(x', y') ← I(width − 1 − x', height − 1 − y')  ➤ 180 degrees
10            3:   I'(x', y') ← I(width − 1 − y', x')        ➤ 90 degrees counterclockwise
11  return I'
```

Figure 3.5 An image and an automobile cropped out of the region of the image indicated by the red rectangle.

Image   Cropped region

Similarly, an image can be **upsampled** to produce a larger image than the original. For best results, interpolation should be performed between pixel values to avoid pixelization artifacts, as discussed later. For now, however, simply replicate each pixel a certain number of times. For example, to upsample by a factor of 2, each row and column is copied twice:

$$I'(x, y) = I\left(\left\lfloor \frac{x}{2} \right\rfloor, \left\lfloor \frac{y}{2} \right\rfloor\right) \qquad \text{(upsample by two)} \qquad (3.10)$$

where the floor operator ensures that the input image is accessed by integer coordinates. Results of repeatedly downsampling and upsampling by a factor of two in both directions are presented in Figure 3.6.

ALGORITHM 3.4 Crop an image

CROPIMAGE(*I*, *left*, *top*, *right*, *bottom*)

Input: image *I*, rectangle with corners (*left*, *top*) and (*right*−1, *bottom*−1)
Output: cropped image *I*' of size *new-width* × *new-height*

1   *new-width* ← *right* − *left*
2   *new-height* ← *bottom* − *top*
3   *I*' ← ALLOCATEIMAGE(*new-width*, *new-height*)
4   **for** $(x', y') \in I'$ **do**
5     $I'(x', y') \leftarrow I(x' + left, y' + top)$
6   **return** *I*'

Figure 3.6 LEFT: An image and the result of downsampling by a factor of 2 and 4, respectively, in each direction. RIGHT: A cropped region and the result of upsampling by a factor of 2 and 4, respectively, in each direction.

3.2 Graylevel Transformations

A geometric transformation, as we have just seen, changes a pixel's location without changing its value. A complementary idea is that of a **point transformation**, which changes a pixel's value without changing its location. To be a point transformation, it is required that the mapping be independent of the pixel's coordinates as well as of the coordinates and values of all other pixels. In this respect point transformations can be considered as a special case of spatial-domain filtering, which we consider in detail later.[†] Spatial-domain filtering removes the latter restriction, allowing the output values to be dependent upon the values of other pixels.

The simplest type of point transformation is a **graylevel transformation**, which transforms a grayscale input image into a grayscale output image:

$$I'(x, y) = f(I(x, y))$$ (3.11)

where f is a function that maps the gray level of a pixel in the input image I to the gray level of a pixel in the output image I'. Graylevel transformations are used for many purposes, such as contrast enhancement, nonlinearity correction, and binarization. If we let $\mathbb{Z}_{0:255} = \{0, 1, 2, \ldots, 255\}$ be the set of integers between 0 and 255, inclusive, then for an 8-bit grayscale image, f is a mapping from an element of the set $\mathbb{Z}_{0:255}$ to another element of the same set, represented mathematically as

$$f \colon \mathbb{Z}_{0:255} \to \mathbb{Z}_{0:255}$$ (3.12)

and depicted graphically in Figure 3.7. Note that f is not dependent upon the coordinates x or y themselves but only upon the value of the image pixel.

Pseudocode to perform a graylevel transformation is presented as the generic procedure TRANSFORMGRAYLEVELS in Algorithm 3.5. The procedure takes two input parameters: an image I and a function f[‡]. For each pixel in the image, the function f is called with the pixel's gray level, and the return value of the function is then stored at the same location in the output image. After all pixels have been transformed, the resulting image is returned. Since the pixels' locations do not change, and therefore $x' = x$ and $y' = y$, there is no need to distinguish between the forward and inverse mappings, as is done with geometric

Figure 3.7 A graylevel transformation maps input gray levels to output gray levels. Based on http://www.unit.eu/cours/ videocommunication/Point_Transformation_ histogram.pdf

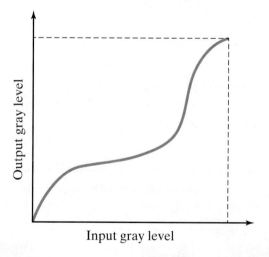

Output gray level

Input gray level

[†] Chapter 5 (p. 215).

[‡] In code f would be considered a *function object* or "functor."

ALGORITHM 3.5 Transform gray levels of an image

TRANSFORMGRAYLEVELS(I, f)

Input: grayscale image I, graylevel mapping f
Output: transformed image I'

1 **for** $(x, y) \in I$ **do**
2      $I'(x, y) \leftarrow f(I(x, y))$
3 **return** I'

transformations. Note that other variations of the same basic idea can be achieved by modifying the procedure to suit the particular needs at hand. For example, instead of returning the output image by value, as is shown here, it may be preferable to pass a previously allocated array into the procedure to store the output. Or, instead of using a separate output image to hold the result, an alternative would be to store the result *in place*, that is, to overwrite the pixel's value using $I(x, y) \leftarrow f(I(x, y))$. Such variations, which are common to all the pseudocode presented in this book, are left as exercises for the reader.

Graylevel transformations are also known as *intensity transformations*, but the former term has the slight advantage in emphasizing that the mapping is between discrete sets of values, as well as being more accurate when images are already gamma-compressed, in which case the values do not (strictly speaking) represent intensity anyway.[†] Nevertheless, the transformations presented in this section are easily extended to integer or floating-point images by simply removing the discretization, or to RGB images by applying the transformation to each color channel separately.

3.2.1 Arithmetic Operations

A useful class of graylevel transformations is the set of arithmetic operations, depicted in graphical form in Figure 3.8. The identity transformation maps each gray level to itself, thus rendering the image unchanged, whereas inversion reverses the gray levels to create a photographic negative. Addition and multiplication, as their names imply, simply add or multiply, respectively, a constant to each gray level, and the gain-bias transformation combines these two operations. Except for inversion, all of these transformations are **monotonically nondecreasing**, meaning that the ordering of gray levels does not change as a result of the mapping. That is, if $z_1 \geq z_2$, then $f(z_1) \geq f(z_2)$. When the transformation

Figure 3.8 Arithmetic graylevel transformations. From left to right: identity, inversion, addition (bias), multiplication (gain), and gain-bias transformation, where saturation arithmetic prevents the output from exceeding the valid range. Note that the slope remains 1 under addition, while the mapping passes through the origin under multiplication.

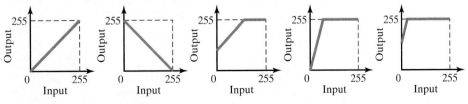

[†] Recall from Section 2.3.2 (p. 43) that gamma-corrected gray levels represent *lightness*, which is perceptually uniform; whereas *intensity* is proportional to the power in the electromagnetic wave, as explained in Section 2.5.2 (p. 58).

is monotonically non-decreasing, then the relative values of pixels remain the same, i.e., if one pixel is brighter than another in the input image, then it remains brighter in the output image.

Let us consider these operations in more detail. Apart from the identity function, the simplest arithmetic operation is to invert each pixel's value by subtracting it from the maximum gray level, which in the case of an 8-bit image leads to

$$I'(x, y) = 255 - I(x, y) \tag{3.13}$$

resulting in an image that looks like the photographic negative of the input image. An example of graylevel inversion is shown in Figure 3.9, where the input and output gray levels of the pixels in a 5×5 window are

$$\begin{bmatrix} 7 & 21 & 25 & 38 & 76 \\ 27 & 45 & 58 & 88 & 155 \\ 28 & 46 & 96 & 163 & 216 \\ 40 & 55 & 123 & 216 & 226 \\ 42 & 55 & 94 & 173 & 201 \end{bmatrix} \xrightarrow{\text{INVERTGRAYLEVELS}} \begin{bmatrix} 248 & 234 & 230 & 217 & 179 \\ 228 & 210 & 197 & 167 & 100 \\ 227 & 209 & 159 & 92 & 39 \\ 215 & 200 & 132 & 39 & 29 \\ 213 & 200 & 161 & 82 & 54 \end{bmatrix}$$

Note that for each pixel, the input and output values sum to 255.

Another arithmetic operation is to add a constant number, say b, to each pixel's value:

$$I'(x, y) = I(x, y) + b \tag{3.14}$$

which is one way to brighten a dark image. If I' were a floating-point or integer-valued image, we would have nothing more to say. But for grayscale images with a finite number of bits devoted to each pixel, we must concern ourselves with the possibility of **overflow**, which occurs when the result is too large to fit into those bits. For example, adding 75 to an 8-bit pixel whose value is 200 would result in the value 275, which would exceed the storage capacity of the pixel. The solution to this problem is to use **saturation arithmetic** in which results are **clamped** to the nearest valid value, leading to $200 + 75 = 255$, which is obviously only valid if the plus sign is interpreted as saturation addition with a valid range of 0 to 255. As a result, we add an extra test to ensure that no value greater than 255 will attempt to be stored:

$$I'(x, y) = \min(I(x, y) + b, 255) \tag{3.15}$$

Figure 3.9 An 8-bit grayscale image (left), and the inverted image obtained by subtracting each pixel from 255 (right).

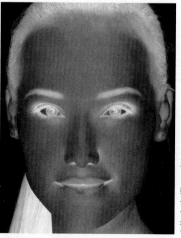

Moreover, if b is allowed to be negative (in which case the image will be darkened), an additional test is needed to ensure that negative results are clamped at zero. To handle the general case in which the sign of b is not known beforehand, clamping must occur at both ends:

$$I'(x, y) = \min(\max(I(x, y) + b, 0), 255) \tag{3.16}$$

Note that with saturation arithmetic, addition and subtraction are not necessarily inverses of each other.

| **EXAMPLE 3.2** | Compute the result of adding either 100 or -100 to the following 3×3 grayscale image, using saturation arithmetic: |

$$I = \begin{bmatrix} 216 & 171 & 174 \\ 134 & 214 & 97 \\ 52 & 5 & 212 \end{bmatrix}$$

Solution   For $b = 100$, add 100 to each pixel and clamp the result at 255. For $b = -100$, subtract 100 from each pixel and clamp the result at 0. These operations yield

$$I + 100 = \begin{bmatrix} 255 & 255 & 255 \\ 255 & 255 & 197 \\ 152 & 105 & 255 \end{bmatrix}$$

$$I - 100 = \begin{bmatrix} 116 & 71 & 74 \\ 34 & 114 & 0 \\ 0 & 0 & 112 \end{bmatrix}$$

An alternative approach to brightening or darkening an image is to multiply each pixel by a positive value:

$$I'(x, y) = cI(x, y) \tag{3.17}$$

where $c > 0$ is a constant. As before, we must apply a minimum, $I'(x, y) = \min(cI(x, y), 255)$, to prevent the result from exceeding the number of bits allowed for storage. If, in addition, c is a floating-point value and I' is a grayscale image, then we also must round the result to the nearest integer before storing.

Combining multiplication and addition yields the transformation

$$I'(x, y) = cI(x, y) + b \tag{3.18}$$

where the constant c is called the **gain** and b is called the **bias**. Recall from the previous chapter that a standard television or computer monitor has two controls called the **contrast** and **blacklevel** (or brightness).[†] Mathematically, the gain-bias transformation is identical to that of the contrast and blacklevel controls, as can be seen from comparing Equation (3.18) with Equation (2.13). Note that c plays the role of the contrast, while b governs the blacklevel. Together these two parameters are useful to increase both the overall gray levels as well as the contrast of a dark image. Figure 3.10 shows the effects of applying gain and bias to an image, using the pseudocode presented in Algorithm 3.6. Note that multiplication, which is implicit in the equation, is explicitly denoted using the asterisk (*) symbol in the

---

[†] Section 2.3.2 (p. 43).

Figure 3.10 Improving image quality by applying gain or bias to an image. From left to right: Original image, brightened image by adding a constant value ($b = 50$), higher contrast image by multiplying a constant value ($c = 2.5$). Source: Movie *Hoop Dreams*

ALGORITHM 3.6 Apply gain and bias to a grayscale image, using saturation arithmetic

APPLYGAINANDBIAS(I, b, c)

Input: grayscale image I, constants b, c
Output: grayscale image I' with increased brightness and contrast

1   **for** $(x, y) \in I$ **do**
2       $I'(x, y) \leftarrow \text{MIN}(\text{MAX}(\text{ROUND}(I(x, y)*c + b), 0), 255)$
3   **return** I'

pseudocode to better reflect the appearance of actual code. Also note that addition and multiplication are special cases of this procedure, when $c = 1$ and $b = 0$, respectively. Grayscale inversion is also a special case, when $b = 255$ and $c = -1$, but in that case the procedure can be simplified since rounding and saturation arithmetic are not needed.

EXAMPLE 3.3

Compute the output of Algorithm 3.6 on the following 3×3 grayscale image, with $b = 50$ and $c = 2$, using saturation arithmetic:

$$I = \begin{bmatrix} 216 & 171 & 174 \\ 134 & 214 & 97 \\ 52 & 5 & 212 \end{bmatrix}$$

Solution

For each pixel, we simply multiply the value by c and add b, then clamp the result:

$$\text{APPLYGAINANDBIAS}(I, 50, 2) = \begin{bmatrix} 255 & 255 & 255 \\ 255 & 255 & 244 \\ 154 & 60 & 255 \end{bmatrix}$$

3.2.2 Linear Contrast Stretching

A closely related transformation specifies a line segment that maps gray levels between $g_{\min}$ and $g_{\max}$ in the input image to the gray levels $g'_{\min}$ and $g'_{\max}$ in the output image according to a linear function. Called **linear contrast stretch**, this transformation is given by

$$I'(x, y) = \frac{g'_{\max} - g'_{\min}}{g_{\max} - g_{\min}}(I(x, y) - g_{\min}) + g'_{\min} \tag{3.19}$$

It is easy to verify that that the minimum input value maps to the minimum output value, that is, $I(x, y) = g_{min}$ maps to $I'(x, y) = g'_{min}$. Similarly, the maximum input value maps to the maximum output value, that is, $I(x, y) = g_{max}$ yields $I'(x, y) = g'_{max}$. As a consequence, if some values in the input image are less than g_{min} or greater than g_{max}, then it is necessary to clamp the output of Equation (3.19) using $\min(\max(\cdot, g'_{min}), g'_{max})$.

One widely used application of linear contrast stretching is to display an integer-valued or floating-point image. Most displays require pixel values to be in the range of 0 to 255 for all color channels. Therefore, to maximize the output contrast, we usually take advantage of the full output range by setting $g'_{min} = 0$ and $g'_{max} = 255$, in which case Equation (3.19) simplifies to

$$I'(x, y) = \text{Round}\left(255 \cdot \frac{I(x, y) - g_{min}}{g_{max} - g_{min}}\right) \tag{3.20}$$

where g_{min} and g_{max} are the minimum and maximum of the values in the image, and rounding has been included for clarity. (Oftentimes, floating-point values are between 0.0 and 1.0.) It is easy to see that this equation is identical to the gain-bias transform of Equation (3.18), with $c = 255/(g_{max} - g_{min})$ and $b = -255 \cdot g_{min}/(g_{max} - g_{min})$.

A **piecewise linear contrast stretch**, formed by combining several of these line segments, can model any graylevel transformation with arbitrary precision, given enough line segments. Figure 3.11 illustrates linear contrast stretching and piecewise linear contrast stretching.

EXAMPLE 3.4

Apply the piecewise linear contrast stretch shown in Figure 3.12 to the following 3×3 grayscale image:

$$\begin{bmatrix} 73 & 56 & 3 \\ 15 & 188 & 239 \\ 82 & 45 & 64 \end{bmatrix}$$

Solution

From the figure, we see that the mapping contains 3 linear segments. In the first segment, any value less than 50 is mapped to 100. In the second segment, values between 50 and 100 are linearly mapped to the range 100 to 200. In the third segment, values between 100 and 255 are linearly mapped to the range 200 to 255. Plugging these values into Equation (3.19), the mapping can be expressed as

$$I'(x, y) = \begin{cases} 100 & \text{if } I(x, y) \le 50 \\ 2(I(x, y) - 50) + 100 & \text{if } 50 < I(x, y) \le 100 \\ \frac{55}{155}(I(x, y) - 100) + 200 & \text{if } 100 < I(x, y) \end{cases} \tag{3.21}$$

The value 64, for example, falls into the second category and therefore is mapped to $2(64 - 50) + 100 = 128$. The value 188 falls into the third category and is mapped to $(55/155) \cdot 88 + 200 = 231$ (after rounding). Applying the same procedure to the other values yields the following output image:

$$\begin{bmatrix} 146 & 112 & 100 \\ 100 & 231 & 249 \\ 164 & 100 & 128 \end{bmatrix}$$

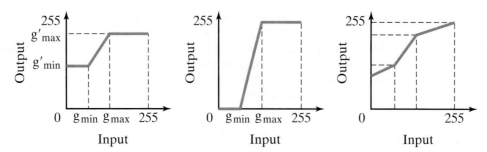

Figure 3.11 Left: Linear contrast stretching maps all the gray levels between g_{min} and g_{max} to the range g'_{min} to g'_{max}. Middle: If $g'_{min} = 0$ and $g'_{max} = 255$, then the full output range is used. Right: A piecewise linear contrast stretch can model any graylevel transformation with arbitrary precision.

3.2.3 Analytic Transformations

In addition to arithmetic operations, graylevel transformations can be specified using **analytic functions** such as the logarithm, exponential, or power functions, as shown in Figure 3.13:

$$I'(x, y) = \log(I(x, y)) \tag{3.22}$$

$$I'(x, y) = \exp(I(x, y)) \tag{3.23}$$

$$I'(x, y) = (I(x, y))^\gamma \tag{3.24}$$

where γ is a nonnegative real number and the necessary rounding and clamping are omitted for brevity. The logarithm is useful for squeezing images with a high dynamic range, such as the magnitude of the Fourier transform,[†] into a grayscale image with a small bit depth (typically 8 bits per pixel). The exponential is the inverse of the logarithm. The power function, as we saw in the previous chapter, is also known as the *gamma function* and is useful for gamma expansion ($\gamma > 1$) or gamma compression ($\gamma < 1$). Most grayscale images are already gamma-compressed, in which case gamma expansion can be applied to convert it back into a radiance map by reversing the effects of gamma compression. On the other hand, if the image is in raw format, then gamma compression can be used to compress the radiance map into a perceptually uniform grayscale image.[‡]

Figure 3.12 An example piecewise linear graylevel mapping.

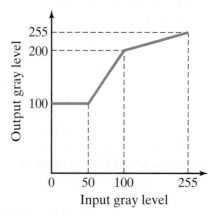

[†] See Section 6.3.3 (p. 293) for an example.
[‡] As explained in Section 2.3.2 (p. 43), $\gamma \approx 2$ for expansion, while $\gamma \approx 0.5$ for compression.

Figure 3.13 Analytic graylevel mapping. From left to right: logarithm, exponential, gamma expansion $(\gamma = 2)$, and gamma compression $(\gamma = 0.5)$. All transformations are monotonically nondecreasing.

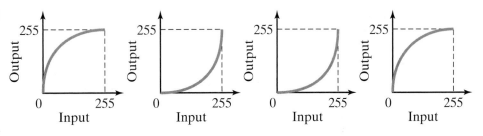

3.2.4 Thresholding

Another important graylevel transformation is **thresholding**, which takes a grayscale image and sets every output pixel to 1 if its input gray level is above a certain threshold, or to 0 otherwise:

$$I'(x, y) = \begin{cases} 1 & \text{if } I(x, y) > \tau \\ 0 & \text{otherwise} \end{cases} \tag{3.25}$$

where τ is the threshold. The result is a binary image that, for some images at least, separates the foreground object(s) from the background, as shown in Figure 3.14. By convention, foreground pixels are labeled ON, or 1, while background pixels are labeled OFF, or 0. Because thresholding produces a binary image as output, it is also known as **binarization**. Even though only one bit per pixel is needed to store a binary image, it is often more convenient to use one byte per pixel, setting the nonzero output values to 255 instead of 1, and storing the output as a grayscale image; this approach has the advantage that it allows the output to be displayed without scaling, and it also creates a bitwise mask in which all the bits in each pixel agree with each other, which can be used in masking.

3.2.5 Other Transformations

Several other transformations are worth mentioning. **Density slicing** (or *graylevel slicing* or *intensity slicing*), assigns all gray levels within a certain range to a certain value. For example, all gray levels between 128 and 164 might be mapped to 255. Figure 3.15 shows two versions of density slicing, one in which all gray levels outside this range are mapped to black (zero), and one in which all gray levels outside this range remain unchanged. Either way, density slicing is a useful way of highlighting some feature of interest that generates a relatively narrow range of values in the image. If multiple features generate multiple non-overlapping ranges, then multiple slices can be combined, so that all gray levels within one

Figure 3.14 An 8-bit grayscale image (left), and the binarized result obtained by thresholding with $\tau = 150$ (right).

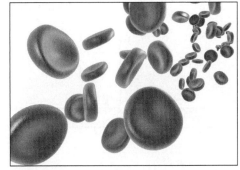

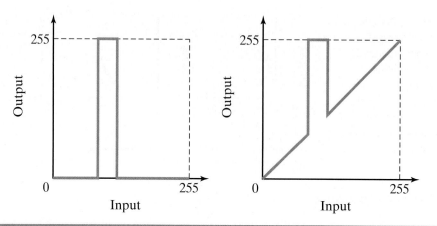

Figure 3.15 Density slicing maps a range of input gray levels to a specific output gray level. All other input gray levels are either mapped to zero (left) or remain unchanged (right).

range map to one value, while all gray levels within another range map to a different value. We will see an example of density slicing in the next section.

Quantization, which discards one or more of the lower-order bits, is a staircase function as shown in Figure 3.16, with the number of stairs equal to 2^b, where b is the number of bits per pixel retained. Since the higher-order bits typically contain more useful information than the lower-order bits, quantization is an easy way to reduce the storage requirements of an image without making it unrecognizable. For example, a pixel whose value is 163 (that is, 10100011 in binary) in an 8-bit-per-pixel image can be stored as 160 (that is, 10100000 in binary) in a 6-bit-per-pixel image. Although some information has been lost, 160 is similar enough to 163 that for some purposes recognizability will not be significantly affected. Too much quantization, however, can seriously degrade the quality of the image, and there are better ways to achieve high compression ratios anyway,[†] so quantization should be used with care. An example of an image with varying levels of quantization is shown in Figure 3.17.

Bit-plane slicing is another transformation that is sometimes mentioned in the context of graylevel transformations. A *bit plane* is a binary image whose value at each pixel is the same as the appropriate bit in the corresponding pixel of the original image. As a result, the number of bit planes is equal to the number of bits per pixel. By convention these bit planes are called 0 through $b - 1$, where b is the number of bits per pixel: Bit plane 0 is associated with the lowest-order bit, bit plane 1 with the second-lowest-order bit, and so on through bit plane $b - 1$ with the highest-order bit. For an 8-bit-per-pixel image, $b = 8$, so the highest-order bits of all the pixels are stored in bit plane 7. Figure 3.18 shows the transformations

Figure 3.16 Quantization discards the lower-order bits via a staircase function, with the number of stairs determined by the number of bits retained. From right to left: Only 1 bit is retained, so the gray levels in the dark half (less than 128) map to 0, while the gray levels in the bright half (above 127) map to 128 (binary: 10000000); 2 bits are retained, so gray levels are mapped to either 0, 64, 128, or 192; 3 bits are retained, so all gray levels are mapped to either 0, 32, 64, 96, 128, 160, 192, or 224; all 8 bits are retained (no quantization, the identity function).

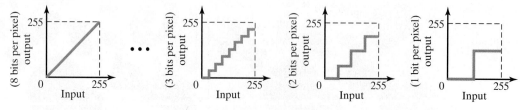

[†] Compression is covered in Chapter 8 (p. 355).

Figure 3.17 From left-to-right and top-to-bottom: An original 8-bit-per-pixel image of a fire hydrant and its quantized versions with 7, 6, 5, 4, 3, 2, and 1 bit per pixel. Note that the image is quite recognizable with as few as 3 bits per pixel.

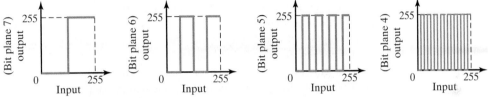

Figure 3.18 Bit-plane slicing transformations for the four highest-order bit planes. The transformation for bit plane 7 is identical, apart from scaling, to thresholding with a value of 127.

for bit planes 4 through 7 (that is, the four highest-order bits), and Figure 3.19 shows the bit plane slices of an image. Note that bit plane 7 is identical to the result of thresholding the image with $\tau = 127$ and multiplying by 255, and the lower-order bit planes resemble noise. Bit planes are not very useful on their own, but an image can often be represented with some fidelity by retaining the higher-order bit planes while discarding the lower-order bit planes, which is effectively quantization.

3.2.6 Lookup Tables

If the same transformation is to be performed many times, it is often more computationally efficient to compute the transformation beforehand and store it as a **lookup table (LUT)**. A lookup table is an array specifying the output gray level for any input gray

Figure 3.19 From left-to-right and top-to-bottom: Bit planes 7, 6, 5, 4, 3, 2, 1, and 0 of the hydrant image. Note that the higher-order bit planes bear some resemblance to the original image, while the lower-order bit planes appear as noise. Bit plane 7 is identical, apart from scaling, to the quantization with 1 bit per pixel. The image reconstructed from the highest four bit planes (that is, 4 through 7) is shown in the bottom-left of Figure 3.17.

level. To transform an image, one simply needs to look up the appropriate output value in the LUT:

$$I'(x, y) = lut[I(x, y)] \tag{3.26}$$

where *lut* is a one-dimensional array of 256 bytes (assuming an 8-bit image as input), and the bracket operator is used to select an element of the array. Any discrete graylevel transformation can be implemented using a LUT.

3.3 Graylevel Histograms

A **histogram** is a simple but powerful technique for capturing the statistics of any type of data. The space in which the data reside is divided into bins, and the histogram records the number of occurrences in each bin. Throughout this book we will encounter a variety of different types of histograms, but in this section we focus on the **graylevel histogram**, which is a histogram of image gray levels. The space in which the gray levels reside is the discrete set of values $\{0, 1, \ldots, 255\}$, and this space is divided into 256 bins, one for each gray level. The graylevel histogram is a one-dimensional array that stores for each gray level the number of pixels having that value. If n_ℓ is the number of pixels in the image with gray level ℓ, then the histogram h is an array with 256 values specified by

$$h[\ell] = n_\ell, \ \ell = 0, \ldots, 255$$

The graylevel histogram can be thought of as a summary of image data that captures only *which* gray levels occur, not *where* they occur. That is, all spatial information is discarded.

The computation of the graylevel histogram of an image is straightforward, as shown in Algorithm 3.7. After initializing each element of the array to zero, all the pixels in the image are visited, and each time a particular gray level is encountered, the appropriate element of the array is incremented. When the procedure has completed, each element of the array therefore stores the count of pixels for that particular gray level.

The **normalized histogram** $\bar{h}[\ell]$ is computed from the histogram by simply dividing each value by the total number of pixels in the image. That is, if we let $n \equiv width \cdot height$, then $\bar{h}[\ell] = h[\ell]/n$ for $\ell = 0, \ldots, 255$. The normalized histogram is the **probability density function (PDF)**[†] capturing the probability that any pixel drawn at random from the image has a particular gray level, so $\sum_{\ell=0}^{255} \bar{h}[\ell] = 1$. Note that while $h[\ell]$ is an integer, $\bar{h}[\ell]$ is a floating-point value. As shown in Algorithm 3.8, the computation of the normalized histogram requires just a single pass through the image.

ALGORITHM 3.7 Compute the graylevel histogram of an image

COMPUTEHISTOGRAM(I)

Input: grayscale image I
Output: graylevel histogram h

1   **for** $\ell \leftarrow 0$ **to** 255 **do**
2       $h[\ell] \leftarrow 0$
3   **for** $(x, y) \in I$ **do**
4       $h[I(x, y)] \leftarrow_+ 1$ ➤ $h[I(x,y)] \leftarrow h[I(x,y)] + 1$
5   **return** h

[†] Since the image is discrete, $\bar{h}$ is technically a *probability mass function (PMF)*, but the distinction is not important for our purposes.

ALGORITHM 3.8 Compute the normalized graylevel histogram of an image

COMPUTENORMALIZEDHISTOGRAM(I)

Input: grayscale image I
Output: normalized graylevel histogram $\bar{h}$

1   $h \leftarrow$ COMPUTEHISTOGRAM(I)
2   $n \leftarrow$ width $*$ height
3   **for** $\ell \leftarrow 0$ **to** 255 **do**
4        $\bar{h}[\ell] \leftarrow h[\ell]/n$
5   **return** $\bar{h}$

EXAMPLE 3.5

Compute the histogram and normalized histogram of the following 4×3 3-bit grayscale image:

$$I = \begin{bmatrix} 7 & 4 & 2 & 0 \\ 4 & 2 & 4 & 5 \\ 3 & 3 & 5 & 6 \end{bmatrix}$$

Solution

Because it is a 3-bit image, there are only $2^3 = 8$ possible gray levels. To compute the histogram, we simply count the number of times that each gray level appears in the image. The value 0 appears once, 1 appears not at all, 2 appears twice, 3 appears twice, 4 appears three times, and so forth. To compute the normalized histogram, each value in the histogram is divided by the total number of pixels in the image, which in this case is $4 \cdot 3 = 12$. The results are

$$\text{histogram: } h = \begin{bmatrix} 1 & 0 & 2 & 2 & 3 & 2 & 1 & 1 \end{bmatrix}$$

$$\text{normalized histogram: } \bar{h} = \begin{bmatrix} 0.083 & 0.00 & 0.167 & 0.167 & 0.250 & 0.167 & 0.083 & 0.083 \end{bmatrix}$$

3.3.1 Interpreting Histograms

Graylevel histograms provide an easy way of visualizing the statistical properties of an image. The histogram of an image that is too dark, for example, will have large values for the bins with small indices and small values for bins with large indices. On the other hand, the reverse will be true for the histogram of an image that is too bright. If the image is overexposed, then a large number of pixels will be saturated at 255, which will be visible as a large spike in the histogram at $h[255]$. If the image has low contrast, then all the values in the histogram will be concentrated in a relatively narrow range, whereas a high contrast image will have values spread throughout the entire range. Some examples of images and their histograms are shown in Figure 3.20.

3.3.2 Histogram Equalization

Increasing the contrast of a low-contrast image involves distributing the pixel values of an image more evenly across the range of allowable values. We saw two approaches to achieve this effect in previous sections, namely, gain-bias modification and linear contrast stretching. The drawback of these methods is that they require someone to manually specify the parameters of the transformation. In this section we consider an alternate approach known

Figure 3.20 Images and their histograms. From left to right: an image with high contrast and many dark or bright pixels, a dark image with low contrast, and another high contrast image with good exposure. Note the spikes at 255 in the first and last images, indicating pixel saturation.

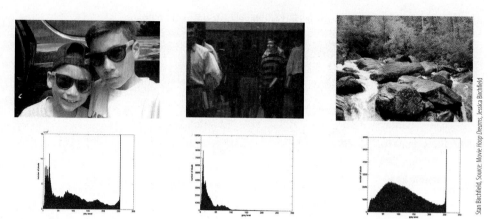

Stan Birchfield, Source: Movie Hoop Dreams, Jessica Birchfield

as **histogram equalization**, which is completely automatic, extremely simple to implement, and parameter-free.

Histogram equalization first converts the PDF (captured by the normalized histogram) to a **cumulative distribution function (CDF)** by computing the **running sum** of the histogram:

$$\bar{c}[\ell] = \sum_{k=0}^{\ell} \bar{h}[k], \quad \ell = 0, \ldots, 255 \tag{3.27}$$

The running sum can be computed efficiently by initializing the first element of the array according to $\bar{c}[0] = \bar{h}[0]$ and then updating $\bar{c}[\ell] = \bar{c}[\ell - 1] + \bar{h}[\ell]$ for each gray level ℓ. Once the CDF has been computed, a pixel with gray level ℓ is simply transformed to $\ell' = \text{Round}(255 \cdot \bar{c}[\ell])$. The algorithm is straightforward, as shown in Algorithm 3.9, and an example is shown in Figure 3.21. Note that since the integral of a PDF is always 1, the CDF always evaluates to 1 at the largest value, and thus $\bar{c}[255] = 1$. As a result, the output ℓ' is in the range from 0 to 255 as desired.

ALGORITHM 3.9 Perform histogram equalization on an image

HistogramEqualize(I)

Input: grayscale image I
Output: histogram-equalized grayscale image I' with increased contrast

1   $\bar{h} \leftarrow$ ComputeNormalizedHistogram(I)
2   $\bar{c} \leftarrow$ RunningSum$(\bar{h})$
3   **for** $(x, y) \in I$ **do**
4          $I'(x, y) \leftarrow$ Round$(255 * \bar{c}[I(x, y)])$
5   **return** I'

RunningSum(a)

Input: 1D array a of *length* values
Output: 1D running sum s of array

1   $s[0] \leftarrow a[0]$
2   **for** $k \leftarrow 1$ **to** $length - 1$ **do**
3          $s[k] \leftarrow s[k - 1] + a[k]$
4   **return** s

Figure 3.21 The result of histogram equalization applied to an image. The increase in contrast is noticeable. The normalized histogram of the result is much flatter than the original histogram, but it is not completely flat due to discretization effects. Source: Movie *Hoop Dreams*.

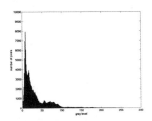
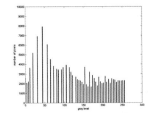

EXAMPLE 3.6	Perform histogram equalization on the image of Example 3.5.

Solution

The normalized histogram of the image was given by the previous example:

$$\bar{h} = \begin{bmatrix} 0.083 & 0.00 & 0.167 & 0.167 & 0.250 & 0.167 & 0.083 & 0.083 \end{bmatrix}$$

The cumulative histogram is obtained by computing the running sum of the normalized histogram:

$$\bar{c} = \begin{bmatrix} 0.083 & 0.083 & 0.250 & 0.417 & 0.667 & 0.834 & 0.917 & 1.00 \end{bmatrix}$$

Since it is a 3-bit image, the maximum gray level is $ngray = 2^3 - 1 = 7$, so the mapping from input to output gray levels is determined by $\text{ROUND}(7 \cdot \bar{c}[\ell])$. For example, gray level 2 maps to $\text{ROUND}(7 \cdot 0.250) = \text{ROUND}(1.75) = 2$, while gray level 4 maps to $\text{ROUND}(7 \cdot 0.667) = \text{ROUND}(4.67) = 5$. This results in the following mapping: $0 \to 1, 1 \to 1, 2 \to 2, 3 \to 3, 4 \to 5, 5 \to 6, 6 \to 6$, and $7 \to 7$, which yields the following result:

$$\begin{bmatrix} 7 & 4 & 2 \\ 4 & 2 & 4 \\ 3 & 3 & 5 \end{bmatrix} \xrightarrow{\text{HISTOGRAMEQUALIZE}} \begin{bmatrix} 7 & 5 & 2 \\ 5 & 2 & 5 \\ 3 & 3 & 6 \end{bmatrix}$$

Why does such a simple algorithm work? In other words, what is Line 4 (the heart of the algorithm) in HISTOGRAMEQUALIZE doing? To gain some intuition, consider the example shown in Figure 3.22, in which we assume that the gray levels are continuous for simplicity. To emphasize their continuous nature, we will use z instead of ℓ to designate a pixel value. The desired PDF $p'(z)$, which is the normalized histogram of the gray levels of the output image, should be flat. That is, given some constant δ, the value $\int_{a'}^{a'+\delta} p'(z)dz$ should be the same for any gray level a'. Since the algorithm uses the CDF $q(z)$ of the original histogram to transform gray levels, this transformation is visualized

Figure 3.22 Why histogram equalization works. In this example, the histogram of the original image is heavily weighted toward darker pixels. If we let the CDF be the mapping from the old gray level to the new one, the new PDF is flat and therefore weights all gray levels equally. This is because any interval of width δ in the new histogram captures the same number $(\delta/255)$ of pixels in the original image. In this example the area within each orange region is identical. Note that discretization effects have been ignored for this illustration.

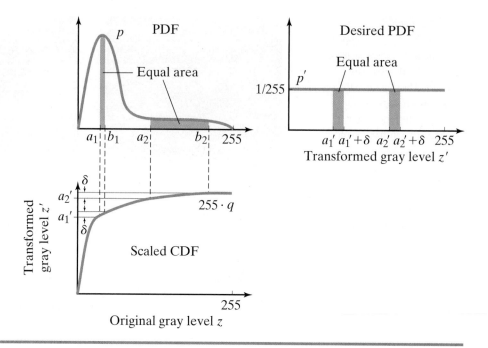

Figure 3.22 Why histogram equalization works. In this example, the histogram of the original image is heavily weighted toward darker pixels. If we let the CDF be the mapping from the old gray level to the new one, the new PDF is flat and therefore weights all gray levels equally. This is because any interval of width δ in the new histogram captures the same number $(\delta/255)$ of pixels in the original image. In this example the area within each orange region is identical. Note that discretization effects have been ignored for this illustration.

in the lower-left plot of the figure, with gray level z along the horizontal axis transforming to the new gray level $z' = 255 \cdot q(z)$ along the vertical axis. Since q is the integral of p, the area under the PDF for any interval corresponding to an output interval of δ is $\int_{q^{-1}(a')}^{q^{-1}(a'+\delta)} p(z)\, dz = q(q^{-1}(a'+\delta)) - q(q^{-1}(a')) = a' + \delta - a' = \delta$. In other words, equally spaced intervals of width δ along the axis of the new PDF capture equal numbers of pixels in the original PDF. The CDF thus provides a simple means of ensuring that an equal number of pixels contribute to an equally spaced interval in the output. Of course this analysis assumes that the gray levels are continuous—in practice the algorithm only produces an approximately flat output because of discretization effects.

An alternative, and slightly more mathematical, explanation is as follows. The goal is to apply a mapping $z' = f(z)$ to convert the image I, whose normalized histogram is $\bar{h}$, to the image I', whose normalized histogram is $\bar{h}'$. To maximize contrast, we want the normalized histogram to be flat, that is, $\bar{h}'(z) = 1$, over all possible gray levels z. From basic probability theory, we know that if $p(x)$ is a PDF over x and if $y = f(x)$ is a transformation of the random variable x, then the PDF $q(y)$ of the result is given by

$$q(y) = p(x) \cdot \left| \frac{dx}{dy} \right| \tag{3.28}$$

Let the transformation f be the CDF of the input histogram:

$$z' = f(z) = \bar{c}(z) = \int_0^z \bar{h}(\zeta)\, d\zeta \tag{3.29}$$

where we assume $0 \le z \le 1$ and $0 \le z' \le 1$ for simplicity. Using Leibniz's integral rule from calculus, the derivative is calculated as

$$\frac{dz'}{dz} = \frac{df(z)}{dz} = \frac{d}{dz}\left[\int_0^z \bar{h}(\zeta)\, d\zeta \right] = \bar{h}(z) \tag{3.30}$$

Substituting z for x, z' for y, $\bar{h}$ for p, and $\bar{h}'$ for q into Equation (3.28) leads to

$$\bar{h}'(z') = \bar{h}(z) \cdot \left|\frac{dz}{dz'}\right| = \bar{h}(z) \cdot \frac{1}{\bar{h}(z)} = 1 \tag{3.31}$$

which is indeed the desired uniform distribution.

3.3.3 Histogram Matching

Instead of attempting to flatten the histogram as much as possible, sometimes the goal is to transform the histogram in a specific way. In this more general case, we wish to modify the image so that its histogram closely matches that of a given reference histogram. For example, we may have two images taken under different lighting conditions, and we wish to modify them so that they appear as if they had been taken under similar conditions. This procedure, called **histogram matching** (also known as *histogram specification*), is based on the same principle as that of histogram equalization.

Suppose we are given an image I with normalized graylevel histogram $\bar{h}$, along with a reference normalized graylevel histogram $\bar{h}_{ref}$. Our goal is to apply a transformation to I to obtain an image I' whose normalized graylevel histogram is $\bar{h}_{ref}$. From Equation (3.29), we can apply the following transformations to flatten both $\bar{h}$ and $\bar{h}_{ref}$:

$$z_1' = \bar{c}(z) = \int_0^z \bar{h}(\zeta)d\zeta \tag{3.32}$$

$$z_2' = \bar{c}_{ref}(z') = \int_0^{z'} \bar{h}_{ref}(\zeta)d\zeta \tag{3.33}$$

The first equation transforms the gray level z in the input image to the gray level z_1' in a histogram-equalized image. The second equation transforms the gray level z' in the output image to the gray level z_2' in a histogram-equalized image. To make sure that z maps to z', we simply set $z_1' = z_2'$, leading to the desired transformation:

$$z' = f(z) = \bar{c}_{ref}^{-1}(\bar{c}(z)) \tag{3.34}$$

In other words, the inverse of the CDF of $\bar{h}_{ref}$ is applied to the CDF of $\bar{h}$ to yield the output gray level. The transformation is illustrated in Figure 3.23.

The pseudocode for histogram matching is presented in Algorithm 3.10. Although the idea is conceptually as simple as histogram equalization, the procedure is considerably

Figure 3.23 Histogram matching. Given the CDF $\bar{c}$ of the original image and the desired CDF $\bar{c}_{ref}$, histogram matching transforms an original gray level z to a new gray level z' by finding the value of z' such that $\bar{c}(z) = \bar{c}_{ref}(z')$. As before, discretization effects are ignored in this illustration.

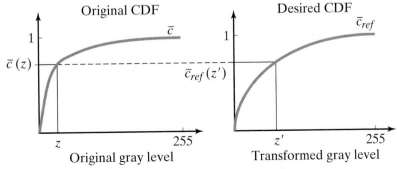

ALGORITHM 3.10 Perform histogram matching on an image

HistogramMatch(I, h_{ref})

Input: grayscale image I, reference normalized graylevel histogram $\bar{h}_{ref}$
Output: grayscale image I' whose normalized graylevel histogram closely matches $\bar{h}_{ref}$

1 $\bar{h} \leftarrow$ ComputeNormalizedHistogram(I) ▶ Compute the normalized histogram of the image,
2 $\bar{c} \leftarrow$ RunningSum$(\bar{h})$ then compute the CDF of the image,
3 $\bar{c}_{ref} \leftarrow$ RunningSum$(\bar{h}_{ref})$ as well as the desired CDF.
4 **for** $\ell \leftarrow 0$ **to** 255 **do**   ▶ For each possible gray level ℓ, set $f[\ell] \approx \bar{c}_{ref}^{-1}[\bar{c}(\ell)]$.
5        $\ell' \leftarrow 255$ This is done by finding ℓ' such that $\bar{c}_{ref}[\ell'] \approx \bar{c}[\ell]$,
6   **repeat**   and setting $f[\ell] = \ell'$. To handle discretization effects,
7              $f[\ell] \leftarrow \ell'$ ℓ' is set to the maximum possible gray level,
8              $\ell' \leftarrow\_ 1$ then repeatedly decremented
9   **while** $\ell' \geq 0$ AND $\bar{c}_{ref}[\ell'] > \bar{c}[\ell]$ until $\bar{c}_{ref}[\ell'] \leq \bar{c}[\ell]$.
10 **for** $(x, y) \in I$ **do**   ▶ Once the mapping f has been determined,
11       $I'(x, y) \leftarrow f[I(x, y)]$ it is applied to all pixels in the image.
12 **return** I'

more complicated due to the need to invert the reference CDF. During initialization in Lines 1–3, the two cumulative histograms, $\bar{c}$ and $\bar{c}_{ref}$, are computed. Then, for each gray level ℓ, we find in Lines 5–9 the index ℓ' such that $\bar{c}_{ref}[\ell'] \approx \bar{c}[\ell]$. Because these are discrete histograms, an exact equality is not possible. Instead, we set ℓ' to the maximum possible gray level, then decrement it until $\bar{c}_{ref}[\ell'] < \bar{c}[\ell]$. This yields the largest value i such that $\bar{c}_{ref}[\ell'] \leq \bar{c}[\ell]$. (Alternatively, we could set ℓ' to the smallest value such that $\bar{c}_{ref}[\ell'] \geq \bar{c}[\ell]$.) The result is stored in a 1D array f, which is the histogram matching function. After all possible gray levels have been processed, f is applied as a lookup table. Notice that in the pseudocode $\bar{c}$ and $\bar{c}_{ref}$ could be replaced by c and c_{ref}, since the normalizations cancel when the histograms are compared.

3.4 Multispectral Transformations

While grayscale images are useful for many applications, even more compelling are **multispectral images**, which store multiple values for each pixel by capturing the amount of light in different *bands* of the electromagnetic spectrum. These bands typically include portions of either the visible spectrum or the infrared or ultraviolet regions, but may include other regions as well. All the pixel values for a given band are known as a *channel*, so that the collection of channels makes up the image. The most common multispectral image is the RGB image containing 3 channels corresponding to the red, green, and blue regions of the spectrum; other sources of multispectral images are remote sensing devices on satellites or aircraft as we discussed in the previous chapter.

A mapping in which either the input or output (or both) are multispectral images is called a **multispectral transformation**. In its most general form, this type of transformation can be represented as a mapping from a vector of values to another vector of values. If we let $I_i(x, y)$ be the value of the ith band in image I, and let $I'_j(x, y)$ be the value of the jth band in image I', then the transformation of pixel (x,y) is

$$[I'_1(x, y) \quad \dots \quad I'_n(x, y)] = f([I_1(x, y) \quad \dots \quad I_m(x, y)]) \tag{3.35}$$

where the input image I has m bands, the output image I' has n bands, and f is a mapping from an m-element vector to an n-element vector.

3.4.1 RGB Transformations

One such multispectral transformation is to convert an RGB image to a grayscale image, which involves a mapping from a three-channel image $(m = 3)$ to a single-channel image $(n = 1)$. The simplest approach one might consider would be to average the three values:

$$I'(x, y) = \frac{1}{3}(I_R(x, y) + I_G(x, y) + I_B(x, y)) \tag{3.36}$$

where I_R, I_G, and I_B are the red, green, and blue channels of the input image. However, the resulting image will not look correct, because the human visual system is not equally sensitive to all frequencies. A significant improvement is obtained by increasing the weight of the green channel, since the human visual system is more sensitive to green than to the other bands:

$$I'(x, y) = \frac{1}{4}(I_R(x, y) + 2I_G(x, y) + I_B(x, y)) \tag{3.37}$$

which has the advantage that the division by 4 can be easily and efficiently implemented as a bitwise shift to the right by two. This simple equation, shown as pseudocode in Algorithm 3.11, is actually a decent approach to use in many practical applications, although we shall discuss a more accurate method later in the book that is preferred if computational time is not an issue.[†]

The reverse operation, namely to convert a one-byte-per-pixel grayscale image to a three-byte-per-pixel image with RGB color channels, is straightforward and involves simply replicating the values:

$$I'_R(x, y) = I(x, y) \quad I'_G(x, y) = I(x, y) \quad I'_B(x, y) = I(x, y) \tag{3.38}$$

as shown in Algorithm 3.12. Note that the resulting RGB image will still look like a grayscale image when displayed, because the color information has been lost. Nevertheless, converting to RGB is useful for overlaying results on the image (see Figure 1.2), as well as for pseudocoloring, described later.

At first glance it may not be obvious why the reverse transformation does not use the same coefficients as the forward transformation. That is, why is the reverse transformation not $I'_R \leftarrow I$, $I'_G \leftarrow 2I$, and $I'_B \leftarrow I$, or something similar? The following example illustrates why such an approach would be fundamentally wrong.

ALGORITHM 3.11 A simple, approximate RGB to grayscale conversion algorithm

$\textsc{RgbToGraySimple}(I_R, I_G, I_B)$

Input: RGB image with channels I_R, I_G, and I_B
Output: grayscale image I'

1   **for** $(x, y) \in I_R$ **do**
2       $I'(x, y) \leftarrow (I_R(x, y) + 2 * I_G(x, y) + I_B(x, y))/4$
3   **return** I'

[†] Section 9.5.6 (p. 428).

EXAMPLE 3.7

Suppose a pixel has RGB values $R = 126$, $G = 222$, $B = 94$. Convert the pixel to grayscale according to Equation (3.37), then convert back to RGB using both Equation (3.38) and the non-replicating transformation. Is this new RGB value consistent with the grayscale value in both cases?

Solution

Applying Equation (3.37) we have

$$\text{grayscale} = \frac{1}{4}(126 + 2(222) + 94) = 166$$

Now, according to Equation (3.38) this grayscale value converts back to RGB as follows:

$$R = 166 \quad G = 166 \quad B = 166$$

Note that this result is not the same as the original pixel's RGB values, because the color information has been lost. Nevertheless, the values are internally consistent, because if we apply Equation (3.37) again, we arrive at the same grayscale value:

$$\text{grayscale} = \frac{1}{4}(166 + 2(166) + 166) = 166$$

which is what we want.

However, if we apply the non-replicating transformation we get

$$R = 166 \quad G = 2(166) = 332 \quad B = 166$$

Here we immediately see a problem, because G does not even fit into 8 bits. Moreover, when we convert back to grayscale we get

$$\text{grayscale} = \frac{1}{4}(166 + 2(332) + 166) = 249$$

which is much brighter than the original pixel.

ALGORITHM 3.12 Grayscale to RGB conversion

GRAYTORGB(I)

Input: grayscale image I
Output: RGB image with channels I'_R, I'_G, and I'_B

1 **for** $(x, y) \in I$ **do**
2      $I'_R(x, y) \leftarrow I(x, y)$
3      $I'_G(x, y) \leftarrow I(x, y)$
4      $I'_B(x, y) \leftarrow I(x, y)$
5 **return** I'_R, I'_G, I'_B

Various typical point transformations of an RGB image are displayed in Figure 3.24. In the first row, the three separate color channels are shown by setting, in turn, the other color channels to zero. For example, the red channel is visualized by setting

$$I'_R(x, y) = I_R(x, y) \quad I'_G(x, y) = 0 \quad I'_B(x, y) = 0 \tag{3.39}$$

and similarly for green and blue. The second row shows a different way of visualizing these channels, where grayscale images were formed by simply copying the values from the appropriate color channel:

$$I'_R(x, y) = I_R(x, y) \quad I'_G(x, y) = I_R(x, y) \quad I'_B(x, y) = I_R(x, y) \tag{3.40}$$

Figure 3.24 TOP: Original RGB image and three separate color channels. MIDDLE: RGB image obtained by swapping the color channels, and same three color channels of original image shown as grayscale images. BOTTOM: RGB image obtained by swapping the color channels in the reverse order, and three different grayscale transformations of the original image. See text for details.

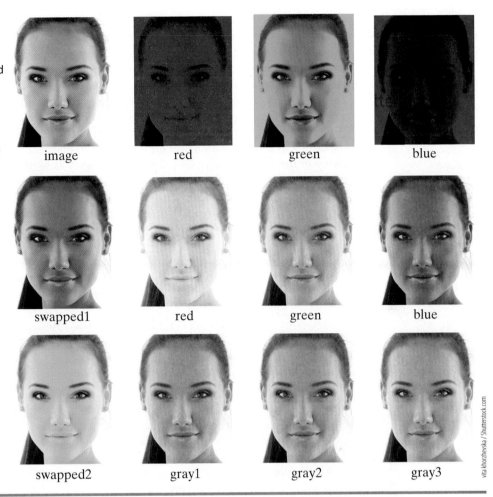

image   red   green   blue

swapped1   red   green   blue

swapped2   gray1   gray2   gray3

vita khorzhevska / Shutterstock.com

and similarly for green and blue. Notice that skin contains significantly more red than green, and significantly more green than blue, which explains why the red image is so much brighter than the other two. Just below the original image in the first column, the purple face (swapped1) was obtained by swapping the three color channels:

$$I'_R(x, y) = I_G(x, y) \quad I'_G(x, y) = I_B(x, y) \quad I'_B(x, y) = I_R(x, y) \tag{3.41}$$

Since red is the dominant color in the original image, blue is the dominant color in the result, leading to the purplish appearance. Similarly, swapping the color channels in the reverse order leads to a greenish appearance, shown as the light green face (swapped2). Finally, the bottom row displays three different versions of RGB to grayscale conversion of the original image. The average of the three channels, Equation (3.36), is too dark (gray1), but there is not much difference between the recommended simple conversion of Equation (3.37) (gray2) and the more correct version (gray3) described later.[†]

3.4.2 Pseudocolor

The transformation of Algorithm 3.12 is not the only way to produce an RGB image from a grayscale image. Instead of creating a colorless image, an alternate approach is to assign

---

[†] Section 9.5.6 (p. 428).

an RGB value to each gray level based upon some criterion other than simply replicating the gray level three times. Because this process assigns colors not according to their actual appearance in the real world, the resulting image is known as a *false color* image, or **pseudocolor** image. One of the most common ways of pseudocoloring is to assign each range of values to a different RGB color — another form of density slicing that we considered in the previous section. You have probably seen the result of density slicing on weather satellite display, such as Geostationary Operational Environmental Satellite (GOES) infrared images. A geostationary satellite is one whose rotation is the same as the earth's so that its position over the earth remains constant, enabling continuous monitoring of the same location. GOES satellite imagery is used to monitor severe weather conditions such as hurricanes and tornadoes, and to estimate rainfall for flash flood warnings. GOES data includes both infrared images and water vapor images. In GOES infrared images, the mapping from 8-bit gray levels to the temperature τ in degrees Celsius is[†]

$$\tau(x, y) = \begin{cases} 57 - \frac{1}{2}I(x, y) & \text{if } I(x, y) \le 176 \\ 145 - I(x, y) & \text{otherwise} \end{cases} \tag{3.42}$$

where a higher gray level means a lower temperature. This temperature mapping can be used to create a variety of different pseudocolorings, such as the following density slicing:

$$I'(x, y) = \begin{cases} I(x, y) & \text{if } I(x, y) \le 130 \\ \text{green} & \text{if } 130 < I(x, y) \le 170 \\ \text{yellow} & \text{if } 170 < I(x, y) \le 210 \\ \text{red} & \text{if } 210 < I(x, y) < 225 \end{cases} \tag{3.43}$$

so that red regions are the coldest, indicating clouds that are higher in elevation. Similarly, brighter values in the water vapor image indicate more moisture, so that yellow and red areas indicate rain. Figure 3.25 shows GOES infrared and water vapor images, along with the pseudocolor results using the same mapping.

Another popular remote sensing program is Landsat, which was mentioned in the previous chapter. Since the program's inception, the Landsat hardware has undergone several revisions. Landsat 5, first launched in 1984 and decommissioned in 2013, used a Thematic Mapper (TM) sensor to collect 7 spectral bands: 3 visible channels, 3 near and mid infrared channels, and 1 thermal channel. These 7 bands are shown in Figure 3.26.[‡] Since the first three bands correspond approximately to the blue, green, and red wavelengths, an RGB image of the area can be produced by combining these three images with appropriate weights (doubling the green and red). The first image of the last row shows the result of this combination.

Looking closely at the RGB image, you will notice that the peninsula in the northwest corner is sparsely populated, covered mostly with rolling hills of vegetation. This same area appears dark in the green and red images (bands 2 and 3) but bright in the near infrared images (bands 4 and 5). This is because live green plants absorb light in the longer wavelengths of the visible spectrum for photosynthesis, while they reflect near infrared light because their energy level per photon is too low to be useful for synthesizing organic molecules. As a result, a simple way to detect the presence of living, healthy, green vegetation is to compute the ratio of the red value to the near infrared value for each pixel, which is called the **Ratio Vegetation Index (RVI)**:

$$I'_{RVI}(x, y) = \frac{I_{IR}(x, y)}{I_{Red}(x, y)} \tag{3.44}$$

[†] From http://www.goes.noaa.gov/ECIR3.html

[‡] Notice that the thermal channel's spatial resolution is significantly less than that of the other channels.

Figure 3.25

Pseudocolor display using density slicing of GOES infrared (top) and water vapor (bottom) images.

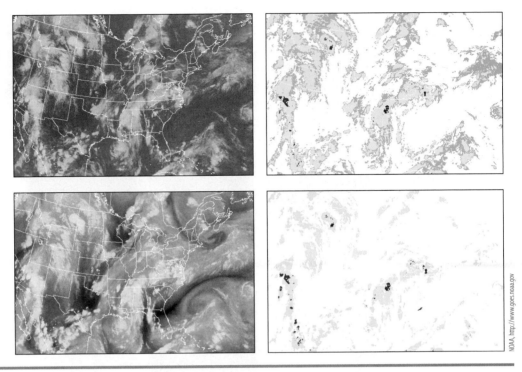

NOAA, http://www.goes.noaa.gov

where I_{Red} is the spectral reflectance (ratio of reflected to incoming radiation) in the visible red band (approximately 580 to 680 nm) and I_{IR} is the spectral reflectance in the near infrared band (approximately 720 to 1100 nm). One drawback of the RVI is that it ranges from 0 to infinity, which is inconvenient. A more common approach, therefore, is to compute the difference between the bands normalized by the sum of the bands. This **Normalized Difference Vegetation Index (NDVI)** is functionally equivalent to the RVI:

$$I'_{NDVI}(x, y) = \frac{I'_{RVI}(x, y) - 1}{I'_{RVI}(x, y) + 1} = \frac{I_{IR}(x, y) - I_{Red}(x, y)}{I_{IR}(x, y) + I_{Red}(x, y)} \tag{3.45}$$

Since both of the input values are nonnegative, the NDVI value is between -1 and $+1$. High values (0.3 to 1.0) indicate live, green vegetation. Soils are generally in the range of 0.1 to 0.2, water is near zero or slightly negative, and clouds and snow have values less than water. The bottom row of Figure 3.26 shows the NDVI computed using bands 3 and 4, along with a pseudocolor output showing the vegetation, soil, and water detected by this simple approach.

3.4.3 Chromakey

Another simple but popular multispectral technique is **chromakeying**, which is widely used in the movie and broadcasting industry to separate foreground from background for the purpose of blending multiple images. For example, a weather forecaster stands and points in front of a blue screen, and all the pixels that do not contain blue are placed on top of a map, making it look like the forecaster (who, of course, is not allowed to wear blue) is pointing at the map. This is the traditional way of broadcasting weather forecasts on television, and it is similar to the way live action is blended with animation in movies.

While in theory the color can be anything (purple or orange, for example), in practice one of the three color channels is usually chosen. Historically blue was used, because blue is the

Figure 3.26 7 bands of a Landsat image of the San Francisco Bay Area. The bottom row shows the RGB image obtained by combining bands 1, 2, and 3; the NDVI calculated using bands 3 and 4; and the pseudocolored image obtained by density slicing on the NDVI (blue indicates water, green indicates vegetation, and tan indicates soil). Notice the vegetation occurs outside the city itself in Marin County (upper peninsula in the image) and several parks (namely, the Presidio, Golden Gate Park, and San Bruno Mountain State Park on the lower peninsula). Source: http://glcf.umd.edu, http://glcf.umd.edu/data/landsat/

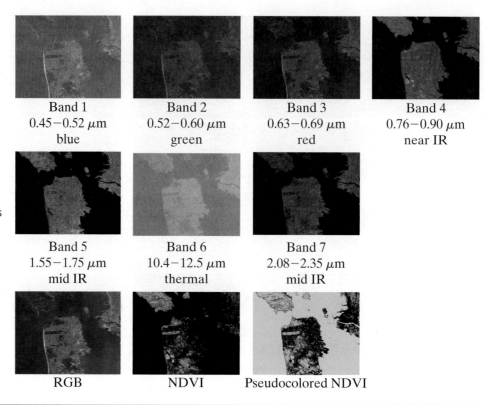

farthest color from human skin color and because high-contrast film that was sensitive to only blue was widely available. The result, therefore, was known as **bluescreening**. More recently, the widespread use of digital video cameras that are more sensitive to green than to red or blue, combined with the use of digital video formats that also emphasize green over blue, have led to the dominance of using a green backdrop, known as **greenscreening**.

In the simplest possible implementation, a binary mask is produced from an RGB image by thresholding the amount of the particular color (green, for example):

$$M(x, y) = \begin{cases} 1 & \text{if } I_G(x, y) > \tau \\ 0 & \text{otherwise} \end{cases} \tag{3.46}$$

where τ is a threshold. Better results are obtained if, instead of thresholding the color value, the value itself is retained and scaled between 0 and 1. Known as the *alpha value* (discussed in more detail later in this chapter), this approach leads to fewer artifacts around the foreground / background edges.

3.5 Multi-Image Transformations

A **multi-image transformation** involves two or more input images. The multispectral transformations that we have just seen are a special case in which the different images are the individual spectral bands. More generally, though, multi-image transformations include images taken of the same scene at different times, or of different scenes entirely. The two basic types of multi-image transformations are arithmetic and logical operations, which are covered in this section, after which we describe several applications of these basic ideas.

3.5.1 Arithmetic Operations

Earlier in the chapter we saw how to add a constant to an image, subtract a constant from an image (or subtract an image from a constant), and multiply an image by a constant. In such cases an output image was produced from a single input image and a parameter specifying the constant. In a similar manner, arithmetic operations can be used, without an additional parameter, to produce an output image from multiple input images. These images must have the same dimensions, because the arithmetic operator is applied to pixels at the same (x, y) coordinates in each image.

EXAMPLE 3.8

Compute the sum of the following 3×3 grayscale images:

$$I_1 = \begin{bmatrix} 216 & 171 & 174 \\ 134 & 214 & 97 \\ 52 & 5 & 212 \end{bmatrix} \quad I_2 = \begin{bmatrix} 72 & 134 & 106 \\ 68 & 89 & 23 \\ 189 & 91 & 212 \end{bmatrix}$$

Solution

The solution is obtained by straightforward addition without saturation, storing the result as an integer or floating-point value:

$$I_1 + I_2 = \begin{bmatrix} 288 & 305 & 280 \\ 202 & 303 & 120 \\ 241 & 96 & 424 \end{bmatrix}$$

Unlike the case of a single input image, multi-image transformations do not typically use saturation arithmetic but rather store the result as an integer-valued or floating-point image. However, when multiple operations are combined, the result is often stored directly into an 8-bit grayscale or 24-bit RGB output image without having to employ a floating-point image. For example, the **absolute difference** between two images computes for each pixel location the absolute value of the difference between the pixels:

$$I'(x, y) = |I_1(x, y) - I_2(x, y)| \tag{3.47}$$

Since the difference between two 8-bit pixels ranges from -255 to 255, the absolute difference ranges from 0 to 255, and therefore no loss of information is incurred when storing the result. Similarly, the weighted average of two images, where the weights sum to 1, results in pixel values in the same range as the original. This operation, known as **linear interpolation,**[†] is conveniently written using a single parameter:

$$I'(x, y) = \eta I_1(x, y) + (1 - \eta)I_2(x, y) \tag{3.48}$$

where $0 \le \eta \le 1$. Note that the operation produces a convex combination of the two inputs because the output is guaranteed to lie inclusively between them:

$$\min(I_1(x, y), I_2(x, y)) \le I'(x, y) \le \max(I_1(x, y), I_2(x, y))$$

for any pixel coordinates (x, y).

[†] Also known as first-order Lagrange interpolation.

EXAMPLE 3.9

Compute (a) the absolute difference between the two images in the previous example and (b) the weighted average, with weights 0.1 and 0.9.

Solution

If we apply rounding to the weighted average, both results fit into an 8-bit grayscale image:

(a)
$$|I_1 - I_2| = \begin{bmatrix} 144 & 37 & 68 \\ 66 & 125 & 74 \\ 137 & 86 & 0 \end{bmatrix} \tag{3.49}$$

(b)
$$\text{Round } 0.1I_1 + 0.9I_2) = \begin{bmatrix} 86 & 138 & 113 \\ 75 & 102 & 30 \\ 175 & 82 & 212 \end{bmatrix} \tag{3.50}$$

3.5.2 Logical Operations

The standard logical, or Boolean, operators AND, OR, XOR (exclusive or), and NOT (complement) produce a 0 or 1 as output, given 0s or 1s as input, as shown in Table 3.1. These operators apply naturally, therefore, to binary images. For example, given two binary images in which 1s indicate the objects of interest, the logical AND produces an image containing 1s where the objects intersect. Similarly, the logical OR produces an image containing 1s where either of the objects appears, and the logical NOT produces an image containing 1s where the objects do *not* appear.

EXAMPLE 3.10

Apply the logical operators to the following two binary images:

$$I_1 = \begin{bmatrix} 1 & 1 & 0 \\ 1 & 1 & 0 \\ 1 & 1 & 0 \end{bmatrix} \quad I_2 = \begin{bmatrix} 0 & 0 & 0 \\ 0 & 1 & 1 \\ 0 & 1 & 1 \end{bmatrix}$$

Solution

The results are straightforward:

$$\underbrace{\begin{bmatrix} 0 & 0 & 0 \\ 0 & 1 & 0 \\ 0 & 1 & 0 \end{bmatrix}}_{I_1 \text{ AND } I_2} \underbrace{\begin{bmatrix} 1 & 1 & 0 \\ 1 & 1 & 1 \\ 1 & 1 & 1 \end{bmatrix}}_{I_1 \text{ OR } I_2} \underbrace{\begin{bmatrix} 1 & 1 & 0 \\ 1 & 0 & 1 \\ 1 & 0 & 1 \end{bmatrix}}_{I_1 \text{ XOR } I_2} \underbrace{\begin{bmatrix} 0 & 0 & 1 \\ 0 & 0 & 1 \\ 0 & 0 & 1 \end{bmatrix}}_{\text{NOT } I_1} \underbrace{\begin{bmatrix} 1 & 1 & 1 \\ 1 & 0 & 0 \\ 1 & 0 & 0 \end{bmatrix}}_{\text{NOT } I_2}$$

It is also common to apply logical operators to a pair of images in which one is a binary image while the other is a regular grayscale or RGB image. In such cases the binary image is interpreted as a *mask*, with 1s indicating the pixels of interest, and 0s indicating the pixels not of interest. The most common operator is AND, which sets all the pixels in the

a	b	a AND b	a	b	a OR b	a	b	a XOR b	a	NOT a
0	0	0	0	0	0	0	0	0	0	1
0	1	0	0	1	1	0	1	1	1	0
1	0	0	1	0	1	1	0	1		
1	1	1	1	1	1	1	1	0		

TABLE 3.1 The truth tables of the standard logical operators.

image to 0 if the corresponding pixel in the binary image is 0 and leaves the remaining pixels intact. Equivalently, we can think of first extending each mask pixel by the number of bits per image pixel before applying the logical operator bitwise, as shown in the following example.

| EXAMPLE 3.11 | Apply the following binary mask M to the RGB image I below (of a gray square around a green dot), where pixel values are specified in hexadecimal notation: |

$$I = \begin{bmatrix} 888888 & 888888 & 888888 \\ 888888 & 00FF00 & 888888 \\ 888888 & 888888 & 888888 \end{bmatrix} \quad M = \begin{bmatrix} 0 & 0 & 0 \\ 0 & 1 & 1 \\ 0 & 1 & 1 \end{bmatrix}$$

Solution

First, extend the mask to have the same bit depth as the image, replicating the 1s or 0s as many times as needed:

$$M' = \begin{bmatrix} 000000 & 000000 & 000000 \\ 000000 & FFFFFF & FFFFFF \\ 000000 & FFFFFF & FFFFFF \end{bmatrix}$$

A bitwise AND between the image and extended mask retains the 4 pixels in the lower-right corner, setting the others to 0:

$$\text{MASKIMAGE}(I,M) = \begin{bmatrix} 000000 & 000000 & 000000 \\ 000000 & 00FF00 & 888888 \\ 000000 & 888888 & 888888 \end{bmatrix}$$

3.6 Change Detection

Suppose a stationary camera (mounted on a tripod, for example) observes a scene containing one or more moving objects. By comparing image frames in the video sequence, the moving objects (foreground) can be separated from the stationary objects (background). Although the problem of foreground / background segmentation is discussed more thoroughly in Chapter 10 , this approach of subtracting image frames is so simple and easy to implement that there is no need to delay introducing this powerful and widely used technique. There are two basic variations on the theme, depending upon whether a reference frame is available. Both variations are covered in this section.

3.6.1 Frame Differencing

One way to detect motion, known as **frame differencing**, is to compare successive image frames in the video sequence. The key insight is that, because the camera is stationary, the background pixel values should not change much, whereas the values of the pixels containing the moving foreground will change considerably. Because we are not interested in whether the pixel gets brighter or darker, but rather only in the amount of difference, it is sufficient to compute the absolute difference between image frames. The simplest approach is to use two successive frames to compute a **difference image**:

$$I'(x, y) = |I_t(x, y) - I_{t-1}(x, y)| > \tau \tag{3.51}$$

ALGORITHM 3.13 Compute the double difference between three consecutive image frames

$\textsc{FrameDifferenceDouble} \left(I_{t-1}, I_t, I_{t+1}, \tau\right)$

Input: successive images I_{t-1}, I_t, and I_{t+1}, and threshold τ
Output: binary image indicating the moving regions

1   **for** $(x, y) \in I_t$ **do**
2         $d_1 \leftarrow |I_{t-1}(x, y) - I_t(x, y)|$
3         $d_2 \leftarrow |I_{t+1}(x, y) - I_t(x, y)|$
4         $I'(x, y) \leftarrow 1$ **if** $d_1 > \tau$ AND $d_2 > \tau$ **else** 0
5   **return** I'

ALGORITHM 3.14 Compute the triple difference between three consecutive image frames

$\textsc{FrameDifferenceTriple} \left(I_{t-1}, I_t, I_{t+1}, \tau\right)$

Input: successive images I_{t-1}, I_t, and I_{t+1}, and threshold τ
Output: binary image indicating the moving regions
1   **for** $(x, y) \in I_t$ **do**
2         $d_1 \leftarrow |I_{t-1}(x, y) - I_t(x, y)|$
3         $d_2 \leftarrow |I_{t+1}(x, y) - I_t(x, y)|$
4         $d_3 \leftarrow |I_{t+1}(x, y) - I_{t-1}(x, y)|$
5         $I'(x, y) \leftarrow 1$ **if** $d_1 + d_2 - d_3 > \tau$ **else** 0
6   **return** I'

where I_t is the frame at time t, I_{t-1} is the previous frame, and τ is some threshold. Unfortunately, this two-frame approach suffers from the **double-image problem**, that is, the difference image will contain foreground pixels not only where the foreground object is located in the current frame but also where it was in the previous frame. This problem can be solved using a third frame by applying the logical AND to the two difference images computed using the two pairs of adjacent frames. The result, known as the **double-difference image** (or *three-frame difference*), is given by

$$I' = |I_t - I_{t-1}| > \tau \ \text{ AND } \ |I_{t+1} - I_t| > \tau \qquad (3.52)$$

where I_{t+1} is the next frame in the sequence, and the pixel coordinates have been omitted for brevity. An alternate approach is to combine the absolute differences from all three image pairs using addition and subtraction prior to thresholding:

$$I'(x, y) = \left(|I_{t-1} - I_t| + |I_{t+1} - I_t| - |I_{t-1} - I_{t+1}|\right) > \tau \qquad (3.53)$$

which we call the **triple-difference image**. These two procedures are shown as Algorithms 3.13 and 3.14, and the results can be seen in Figure 3.27, where the absolute differences from the individual color channels have been combined to improve results over grayscale processing.

3.6.2 Background Subtraction

A **background image** is a reference image that does not contain any foreground objects. If a background image is available, then the foreground can be separated from the background

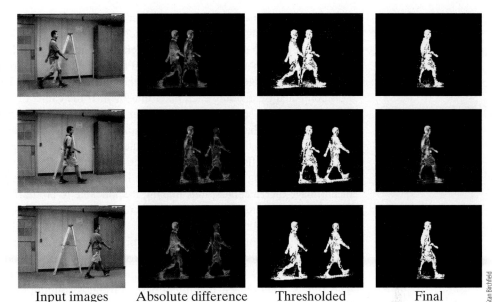

Figure 3.27 Detecting a moving object by frame differencing. LEFT COLUMN: Three image frames from a video sequence. SECOND COLUMN: The absolute difference between pairs of frames. THIRD COLUMN: Thresholded absolute difference. RIGHT COLUMN: Final result using double difference (top), triple difference (middle), and thresholded triple difference (bottom) methods.

Input images   Absolute difference   Thresholded   Final

with just two images rather than three. This technique, known as **background subtraction**, is straightforward:

$$I'(x, y) = |I(x, y) - B(x, y)| \tag{3.54}$$

where I is the image and B is the background image.

One way to obtain a background image is to remove all foreground objects from the scene before taking the picture, or similarly, to take the picture before changing the scene. This latter approach is used in **digital subtraction angiography (DSA)**, where a reference image is captured of a blood vessel before injecting it with dye to increase contrast. In many applications, however, it is not possible to exercise so much control over the environment. In such cases the best approach is to compute a *mean image*, or *average image* by adding successive images in a video sequence to each other and then dividing by the number of images:

$$I'(x, y) = \frac{1}{n} \sum_{i=1}^{n} I_i(x, y) \tag{3.55}$$

ALGORITHM 3.15 Compute the mean of a set of images

COMPUTEMEANIMAGE $(\mathcal{I})$

Input: set of n images $\mathcal{I} = \{I_i\}_{i=1}^{n}$
Output: mean image
1 **for** $(x, y) \in I_1$ **do**
2      $S(x, y) \leftarrow 0$
3   **for** $i \leftarrow 1$ **to** n **do**
4          $S(x, y) \leftarrow_+ I_i(x, y)$
5      $I'(x, y) \leftarrow S(x, y)/n$
6 **return** I'

where I_i is the ith image and n is the number of images. Since the order does not matter, we can assume the images are collected in a set $\mathcal{I} = \{I_i\}_{i=1}^n$. The procedure requires iterating through the images of this set and, for each pixel, adding the value of the pixel in the ith image to the sum obtained so far. To avoid the problem of overflow, an integer-valued (or floating-point) image S is needed to hold the sum for each pixel. The desired mean image I' is computed by dividing each pixel in S by n. If the camera is stationary and the foreground objects are small and moving often enough, then the mean image will contain the appearance of the background of the scene, without any foreground objects. Figure 3.28, for example, shows the mean image computed from successively larger sets of images from a traffic camera. After 100 frames of video (approximately 3 seconds), the vehicles have disappeared, leaving only the background. It is easy to modify the procedure to update the mean image incrementally, thus making it applicable to arbitrarily long video sequences without having to store all the images. The result of using this mean image as a background image is shown in Figure 3.29.

One advantage of background subtraction over frame differencing is that it separates the foreground objects even when they cease moving for a period of time. On the other hand, one drawback is that objects that remain stationary for a very long time prevent the detection of other objects that might pass in front. A solution to this problem is to adaptively update the background, so that stationary objects blend with the background over time and the background image adapts to changing lighting conditions. Another issue is the distraction caused by slightly moving background objects, such as trees waving in the wind, which can be handled by using more sophisticated probabilistic models of pixel colors.[†]

Figure 3.28 TOP: Five images from a video sequence. BOTTOM: Each column shows the mean image obtained using all the images up to and including the one above it. As time progresses the moving objects disappear, leaving only the background.

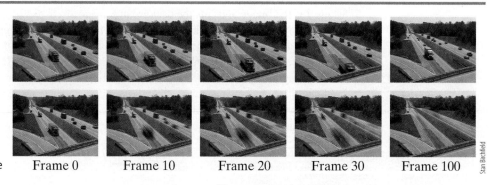

Frame 0   Frame 10   Frame 20   Frame 30   Frame 100

Stan Birchfield

Figure 3.29 Background subtraction. From left to right: the background image, the current image, the absolute difference between the image and the background, and the thresholded absolute difference.

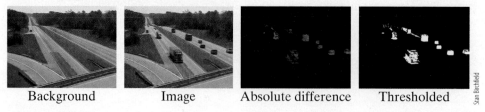

Background   Image   Absolute difference   Thresholded

Stan Birchfield

[†] See Problem 3.40.

3.7 Compositing

Another application of multi-image transformations is **digital compositing**, which is widely used in the movie industry to blend live action with computer graphics or to blend different areas of a computer graphic scene rendered by different pieces of software. If you have ever watched a movie with special effects, it is almost certain that you witnessed the results of digital compositing of 2D images. We begin with the simplest approach, namely dissolving, followed by compositing with binary masks, and then compositing with alpha values.

3.7.1 Dissolving

Suppose we have two input images I_A and I_B that are the same size as each other. The weighted combination between them is given by

$$I'(x, y) = w_A I_A(x, y) + w_B I_B(x, y) \tag{3.56}$$

where w_A and w_B are scalar weights. If we restrict $w_A + w_B = 1$, then the output is a convex combination of the inputs, as mentioned earlier: $I'(x, y) = \eta I_A(x, y) + (1 - \eta)I_B(x, y)$, where $\eta = w_A/(w_A + w_B)$. When $\eta = 0$ the output is identical to the second image, when $\eta = 1$ it is identical to the first image, and for other values the output is a blend of the two. By varying η from 1 to 0, the first image slowly **dissolves** into the second. This simple algorithm is often used to transition from one scene to another in movies, as shown in Figure 3.30.[†]

3.7.2 Compositing with Binary Masks

Now suppose that the two images are accompanied by binary masks M_A and M_B that define the support of the pixels. That is, $M_A(x, y) = 1$ wherever the value $I_A(x, y)$ is valid, and $M_A(x, y) = 0$ wherever the value $I_A(x, y)$ is invalid; and similarly for M_B and I_B. Invalid pixels can be ignored.

Since the masks are binary, exactly four cases exist for any given pixel. For each of these cases, one or more choices are available for the output pixel, as shown in Table 3.2. For example, if $M_A(x, y) = M_B(x, y) = 0$, then the output pixel mask $M'(x, y)$ must be 0, because both inputs are invalid. In that case, the value of the output RGB pixel $I'(x, y)$ is irrelevant, which is indicated by the dot ($\cdot$) in the table. If, on the other hand, $M_A(x, y) = 0$ but $M_B(x, y) = 1$, then the output pixel mask can be 0 (invalid), or it can be 1 (valid) with the output RGB pixel set to the only valid input RGB pixel, namely $I_B(x, y)$. If $M_A(x, y) = M_B(x, y) = 1$, then three choices exist, because the pixel value can be selected from I_A, from I_B, or from neither.

Multiplying the number of entries in the third column of the table, there are exactly $3 \cdot 2 \cdot 2 \cdot 1 = 12$ possible ways of combining two images with binary masks. These 12 compositing operations are called the **Porter-Duff operators** and are illustrated in Figure 3.31. Of these operations, three are trivial, namely COPY I_A, COPY I_B, and CLEAR; one is commutative, because I_A XOR $I_B = I_B$ XOR I_A; and four are noncommutative. The latter include I_A OVER I_B, which places the first image over the second; I_A IN I_B, which copies the first image only where it lies inside the second; I_A OUT I_B (short for I_A "held out by" I_B), which places the first image only where it lies outside the second; and I_A ATOP I_B, which combines the images within the second mask. The reverse versions are obtained by swapping the two operands. Table 3.3 shows the formulas for these compositing operations, which are obtained by inspection from the figure (using the simplest possible formulas for I' by ignoring invalid pixels outside the mask M').

[†] If the images are not already aligned, then the dissolve is accompanied by a warp to align the images, which is called morphing, described in Section 3.9.7 (p. 126). Screenshots from the 2011 movie "The Adventures of Tintin."

Figure 3.30 Two examples of dissolving one image into another. Source: Screenshots by WETA Digital Ltd. – © 2011 Paramount Pictures. 'The Adventures of Tintin'

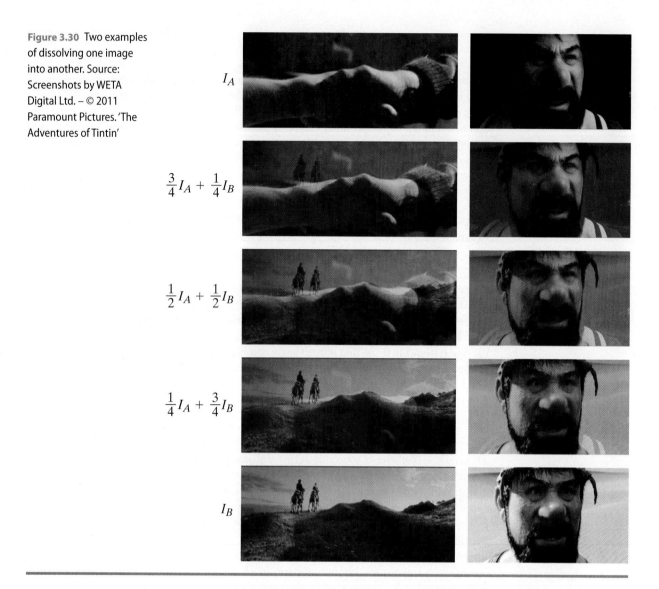

I_A

$\dfrac{3}{4}I_A + \dfrac{1}{4}I_B$

$\dfrac{1}{2}I_A + \dfrac{1}{2}I_B$

$\dfrac{1}{4}I_A + \dfrac{3}{4}I_B$

I_B

The result of applying these binary compositing operators to a pair of images is shown in Figure 3.32. We only show the forward operators, in which the tree is I_A and the house is I_B; the reverse operators are left as an exercise. Note that some operators copy invalid pixels (i.e., pixels outside the mask) to the output image, even though they are still considered invalid (because they remain outside the output mask).

$M_A(x, y)$	$M_B(x, y)$	$M'(x, y)$	$I'(x, y)$
0	0	0	$\cdot$
0	1	0,1	$\cdot, I_B(x, y)$
1	0	0,1	$\cdot, I_A(x, y)$
1	1	0,1,1	$\cdot, I_A(x, y), I_B(x, y)$

TABLE 3.2 The four cases for any given pixel in compositing images with binary masks. For each case, the choices available for the output pixel are given in the last two columns, with a dot ($\cdot$) meaning that the RGB value is irrelevant since the mask is zero. Each entry in the M' column is paired with an entry in the I' column.

Operation	I'	M'
CLEAR	0	0
COPY I_A	I_A	M_A
I_A OVER I_B	$I_A \wedge M_A + I_B \wedge M_B \wedge \neg M_A$	$M_A + M_B$
I_A IN I_B	I_A	$M_A \wedge M_B$
I_A OUT I_B	I_A	$M_A \wedge \neg M_B$
I_A ATOP I_B	$I_A \wedge M_A + I_B \wedge \neg M_A$	M_B
I_A XOR I_B	$I_A \wedge M_A \wedge \neg M_B + I_B \wedge \neg M_B \wedge \neg M_A$	$M_A \wedge \neg M_B + M_B \wedge \neg M_A$

TABLE 3.3 Formulas for compositing two images with binary masks. The formulas for the reverse versions of the non-commutative operations are easily obtained by swapping the operands. The caret $(\wedge)$ symbol refers to logical AND, the angle $(\neg)$ refers to logical NOT, and the plus $(+)$ symbol indicates logical OR (which is equivalent to addition in these formulas due to mutual exclusion between the terms).

3.7.3 Compositing with Alpha Channels

One of the problems with binary masks is that they produce harsh, unnatural edges around the boundaries of objects when compositing two images, as evident from the figure. A better way is to associate with each image an **alpha channel** (sometimes called an **opacity map**) instead of a binary mask. The alpha channel is the same size as the image, and each pixel in the alpha channel is (conceptually at least) a floating-point value, typically between 0 and 1, with 1 meaning that the associated RGB pixel is opaque and 0 meaning that it is transparent (or, equivalently, invisible or invalid). Pixels between 0 and 1 indicate varying degrees of opacity.

Let the alpha channels of the first input, second input, and output be given by α_A, α_B, and α', respectively, and let the RGB images be I_A, I_B, and I', as before. The general rule for compositing is to compute a convex combination of the RGB images and a weighted combination of the alpha channels:

$$I'(x, y) = \eta(x, y)I_A(x, y) + (1 - \eta(x, y))I_B(x, y) \tag{3.57}$$

$$\alpha'(x, y) = \phi_A(x, y)\alpha_A(x, y) + \phi_B(x, y)\alpha_B(x, y) \tag{3.58}$$

Figure 3.31 The twelve binary compositing operations. The first column shows the original two images. Columns 2 through 5 show the noncommutative operations, with the order of operands reversed in the two rows. The final column shows the CLEAR (top) and XOR (bottom) operations. In all cases the display shows the RGB image after applying the mask, i.e., I' AND M', with black pixels indicating a mask value of 0.

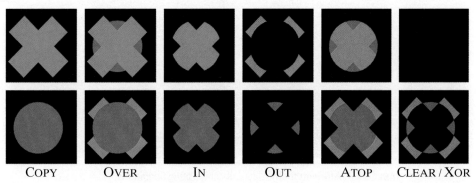

COPY   OVER   IN   OUT   ATOP   CLEAR / XOR

Figure 3.32 Common binary compositing operations applied to a pair of masked images. The top two rows show, from left to right: Original image I_A and mask M_A, original image I_B and mask M_B, and image I' and mask M' resulting from the four operations OVER, IN, OUT, and ATOP, respectively. The bottom row shows the result of ANDing each image with each mask.

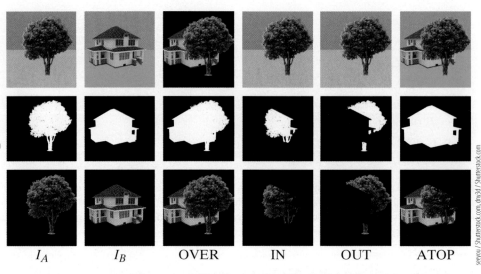

$$I_A \qquad I_B \qquad \text{OVER} \qquad \text{IN} \qquad \text{OUT} \qquad \text{ATOP}$$

seeyou / Shutterstock.com, dny3d / Shutterstock.com

where the fraction is given by

$$\eta(x, y) = \frac{\phi_A(x, y)\alpha_A(x, y)}{\phi_A(x, y)\alpha_A(x, y) + \phi_B(x, y)\alpha_B(x, y)} = \frac{\phi_A(x, y)\alpha_A(x, y)}{\alpha'(x, y)} \tag{3.59}$$

and the coefficients ϕ_A and ϕ_B are determined by the operation, as listed in Table 3.4.

By substituting the coefficients from Table 3.4 into Equations (3.57), (3.58), and (3.59), it is easy to derive the actual formulas for any of the 12 possible compositing operations. The formulas for the most common operators are given in Table 3.5, where the similarity with Table 3.3 should be evident by replacing M_A with α_A, the logical AND ($\wedge$) operator with multiplication, and the logical NOT ($\neg$) operator with "one minus". As a simple example, consider the computation I_A OVER I_B, where I_B is a background image with $\alpha_B(x, y) = 1$ everywhere. In that case, $\alpha'(x, y) = 1$ everywhere, so that the output image is a simple convex combination of the two images according to the alpha channel of I_A:

$$I'(x, y) = \alpha_A I_A + (1 - \alpha_A)I_B \tag{3.60}$$

as expected.

Operation	ϕ_A	ϕ_B
I_A OVER I_B	1	$1 - \alpha_A$
I_B OVER I_A	$1 - \alpha_B$	1
I_A IN I_B	α_B	0
I_B IN I_A	0	α_A
I_A OUT I_B	$1 - \alpha_B$	0
I_B OUT I_A	0	$1 - \alpha_A$
I_A ATOP I_B	α_B	$1 - \alpha_A$
I_B ATOP I_A	$1 - \alpha_B$	α_A
I_A XOR I_B	$1 - \alpha_B$	$1 - \alpha_A$

TABLE 3.4 Coefficients for the different compositing operations.

Operation	I'	α'
I_A OVER I_B	$\dfrac{1}{\alpha'}(\alpha_A I_A + (1 - \alpha_A)\alpha_B I_B)$	$\alpha_A + (1 - \alpha_A)\alpha_B$
I_A IN I_B	I_A	$\alpha_A \alpha_B$
I_A OUT I_B	I_A	$\alpha_A(1 - \alpha_B)$
I_A ATOP I_B	$\alpha_A I_A + (1 - \alpha_A)I_B$	α_B
I_A XOR I_B	$\dfrac{1}{\alpha'}(\alpha_A(1 - \alpha_B)I_A + (1 - \alpha_A)\alpha_B I_B)$	$\alpha_A(1 - \alpha_B) + \alpha_B(1 - \alpha_A)$

TABLE 3.5 Formulas for compositing two images with alpha channels for the most common operations. The formulas for the reverse versions are easily obtained by swapping the operands.

3.7.4 Using Premultiplied Alphas

As evident from Figure 3.32, the image must first be multiplied by the mask before displaying, to avoid the possibility of displaying invalid pixels. Therefore, since the image must be multiplied by the alpha channel anyway, to save computation it is often preferred to *premultiply* the image by the alpha channel before processing. In this approach, all compositing operations are performed on premultiplied images, so that when it is necessary to display the image, no further multiplication is necessary. Although at first glance the resulting savings may not seem all that significant, this approach is actually widely used in the entertainment industry — keep in mind that a single movie contains hundreds of thousands of image frames at very high resolution, so any possible reduction in computation cannot be dismissed. An additional advantage to using premultiplied images is that they are more easily interpretable visually because they do not show arbitrary colors for invalid pixels, which was also noted in the figure.

The premultiplied image values are defined as

$$\bar{I}_A(x, y) \equiv \alpha_A(x, y)I_A(x, y) \tag{3.61}$$

$$\bar{I}_B(x, y) \equiv \alpha_B(x, y)I_B(x, y) \tag{3.62}$$

$$\bar{I}'(x, y) \equiv \alpha'(x, y)I'(x, y) \tag{3.63}$$

By substituting Equations (3.57) and (3.58) into Equation (3.63), it is easy to show that

$$\bar{I}'(x, y) = \phi_A(x, y)\bar{I}_A(x, y) + \phi_B(x, y)\bar{I}_B(x, y) \tag{3.64}$$

$$\alpha'(x, y) = \phi_A(x, y)\alpha_A(x, y) + \phi_B(x, y)\alpha_B(x, y) \tag{3.65}$$

Note that the first equation is actually simpler than Equation (3.57), and the second equation is copied from Equation (3.58) for completeness. When compositing it is important to always keep track of whether the images have been premultiplied by the alpha channel to avoid introducing artifacts.

3.8 Interpolation

As a 2D array of values, a digital image can be thought of as a sampling of an underlying continuous, gamma-compressed intensity function. The problem of **interpolation** is to estimate this underlying continuous function by computing pixel values at any real-valued coordinate pair in the image plane. In order to be a true interpolation function, the estimated continuous function must coincide with the sampled data at the sample points, although it is sometimes desirable to relax this requirement, as we shall see. In this section we describe several methods of interpolation, beginning with the simplest.

3.8.1 Nearest Neighbor Interpolation

Let I be a grayscale image defined over discrete locations $x = 0, \ldots, width - 1$ and $y = 0, \ldots, height - 1$, and let $\hat{I}$ be the continuous underlying function defined for any real-valued coordinate location (x, y). For now, let us require $\hat{I}$ to be a true interpolation function, so $\hat{I}(x, y) = I(x, y)$ whenever x and y are integers that are within the image bounds. The simplest approach to interpolation, called **nearest neighbor** interpolation, returns the gray level of the pixel nearest the coordinates:

$$\hat{I}(x, y) \equiv I(\min(\max(\text{ROUND}(x), 0, width - 1)), \min(\max(\text{ROUND}(y), 0, height - 1))) \quad (3.66)$$

although the notation is simplified considerably if bounds checking can be assumed:

$$\hat{I}(x, y) \equiv I(\text{ROUND}(x), \text{ROUND}(y)) \quad (3.67)$$

Note that even though (x, y) are real numbers, with nearest-neighbor interpolation the result is still a discrete gray level, $\hat{I}(x, y) \in \mathbb{Z}_{0:255}$.

3.8.2 Bilinear Interpolation

A more accurate approach is **bilinear interpolation**. As the name implies, bilinear interpolation is a 2D extension of 1D linear interpolation. If f is a 1D function, then recall that linear interpolation at a point $f(x)$ computes a weighted average of the two nearest samples,[†] as illustrated in Figure 3.33:

$$\hat{f}(x) \equiv (1 - \alpha)f(x_0) + \alpha f(x_0 + 1) \quad (3.68)$$

where $x_0 \equiv \lfloor x \rfloor$ is the index of the nearest pixel to the left and $\alpha \equiv x - x_0, 0 \leq \alpha < 1$ is the fractional distance between the real value x and the integer x_0. To verify that this is a true interpolation, note that $\hat{f}(x) = I(x_0)$ when $\alpha = 0$, and $\hat{f}(x) = f(x_0 + 1)$ when $\alpha = 1$, so that the interpolation passes through the samples, as required.

Similarly, bilinear interpolation computes an appropriately weighted average of the four nearest pixels, i.e., the pixels in the surrounding 2×2 neighborhood, as shown in the figure. The nearest pixel up and to the left of the point is given by (x_0, y_0), where $x_0 \equiv \lfloor x \rfloor$ and $y_0 \equiv \lfloor y \rfloor$ are integers. Again, the fractional part is what remains after subtracting the coordinates of the upper-left pixel coordinates: $\alpha_x \equiv x - x_0, \alpha_y \equiv y - y_0, 0 \leq \alpha_x, \alpha_y < 1$. The interpolated value is then the weighted average of the four nearby pixels:

$$\hat{I}(x, y) = \overline{\alpha}_x \overline{\alpha}_y I_{00} + \alpha_x \overline{\alpha}_y I_{10} + \overline{\alpha}_x \alpha_y I_{01} + \alpha_x \alpha_y I_{11} \quad (3.69)$$

where $\overline{\alpha}_x \equiv 1 - \alpha_x$, and $\overline{\alpha}_y \equiv 1 - \alpha_y$, and we define $I_{00} \equiv I(x_0, y_0)$, $I_{10} \equiv I(x_0 + 1, y_0)$, $I_{01} \equiv I(x_0, y_0 + 1)$, $I_{11} \equiv I(x_0 + 1, y_0 + 1)$. To verify that this is a true interpolation, note that if x and y are integers, then $x_0 = x$, $y_0 = y$, and $\alpha_x = \alpha_y = 0$, so that $\hat{I}(x, y) = I(x, y)$; and so on. The pseudocode is provided in Algorithm 3.16.

3.8.3 Bicubic Interpolation

Now suppose we want to interpolate between two adjacent data points for which we know not only their values but also their derivatives:

$$f_0 \equiv f(x_0), \quad f_1 \equiv f(x_0 + 1), \quad \dot{f}_0 = \frac{df(x)}{dx}\bigg|_{x_0}, \quad \dot{f}_1 = \frac{df(x)}{dx}\bigg|_{x_0+1} \quad (3.70)$$

[†] Section 3.5.1 (p. 99).

Figure 3.33 TOP: Linear interpolation $\hat{f}(x)$ at an arbitrary point x of a discrete function f is computed as the weighted average of the two nearby sampled values, namely $f(x_0)$ and $f(x_0 + 1)$, where $x_0 = \lfloor x \rfloor$. BOTTOM: Bilinear interpolation $\hat{I}(x, y)$ at a point (x, y) of a discrete image I is computed as the weighted average of the four nearby gray levels, namely I_{00}, I_{10}, I_{01}, and I_{11}. The alternating white and shaded regions indicate the extent of the sampled pixels in the continuous domain.

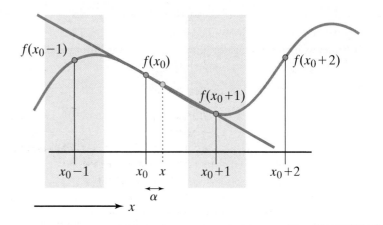

1D function

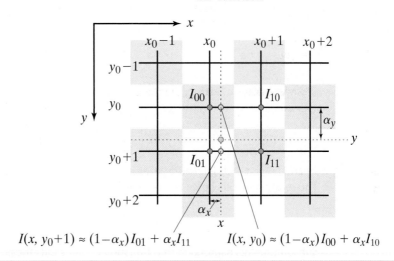

2D function

$$I(x, y_0+1) \approx (1-\alpha_x)I_{01} + \alpha_x I_{11} \qquad I(x, y_0) \approx (1-\alpha_x)I_{00} + \alpha_x I_{10}$$

That is, an estimate of the derivative $\dot{f}$ of the underlying continuous function is available at the sampled points x_0 and $x_0 + 1$. In this case, more accurate results (at greater computational expense) are achieved via **cubic interpolation**, which is the third-order function that passes through the sampled values and also satisfies the derivatives. It is not difficult to show[†] that cubic interpolation is given by

$$\hat{f}(x) \equiv \underbrace{(2\alpha^3 - 3\alpha^2 + 1)}_{h_0} f_0 + \underbrace{(\alpha^3 - 2\alpha^2 + \alpha)}_{h_0{'}} \dot{f}_0 + \underbrace{(-2\alpha^3 + 3\alpha^2)}_{h_1} f_1 + \underbrace{(\alpha^3 - \alpha^2)}_{h_1{'}} \dot{f}_1 \qquad (3.71)$$

where h_0, h_0', h_1, and h_1' are the **cubic Hermite splines** shown in Figure 3.34 alongside the coefficient functions used in linear interpolation, namely, $1 - \alpha$ and α. Note that $\hat{f}(x) = f_0$

[†] Problem 3.32

ALGORITHM 3.16 Perform bilinear interpolation on an image at a point

INTERPOLATEBILINEAR(I, x, y)

Input: image, floating-point coordinates (x, y)
Output: weighted average of gray levels of nearest four pixels

1    $x_0 \leftarrow$ FLOOR (x)
2    $y_0 \leftarrow$ FLOOR (y)
3    $\alpha_x \leftarrow x - x_0$
4    $\alpha_y \leftarrow y - y_0$
5    $\overline{\alpha}_x \leftarrow 1 - \alpha_x$
6    $\overline{\alpha}_y \leftarrow 1 - \alpha_y$
7   **return** $\overline{\alpha}_x * \overline{\alpha}_y * I(x_0, y_0) + \alpha_x * \overline{\alpha}_y * I(x_0+1, y_0) + \overline{\alpha}_x * \alpha_y * I(x_0, y_0+1) + \alpha_x * \alpha_y * I(x_0+1, y_0+1)$

and $\hat{f}(x) = \dot{f}_0$ when $\alpha = 0$, while $\hat{f}(x) = f_1$ and $\hat{\dot{f}}(x) = \dot{f}_1$ when $\alpha = 1$, as desired; where

$$\hat{\dot{f}}(x) \equiv \frac{d\hat{f}}{dx} = (6\alpha^2 - 6\alpha)f_0 + (3\alpha^2 - 4\alpha + 1)\dot{f}_0 + (-6\alpha^2 + 6\alpha)f_1 + (3\alpha^2 - 2\alpha)\dot{f}_1 \qquad (3.72)$$

Several ways of estimating the derivative are possible. The family of **cardinal splines** approximates the derivative as

$$\dot{f}_i \approx \frac{1 - \tau}{2}(f_{i+1} - f_{i-1}) \qquad (3.73)$$

where $0 \leq \tau \leq 1$ is a tension parameter controlling the extent of the derivative's influence. The most common choice is $\tau = 0$, leading to

$$\dot{f}_i \approx \frac{1}{2}(f_{i+1} - f_{i-1}) \qquad (3.74)$$

Substituting this expression into Equation (3.71) yields the popular **Catmull-Rom spline**:

$$\hat{f}(x) = \frac{1}{2}((-\alpha^3 + 2\alpha^2 - \alpha)f_b + (3\alpha^3 - 5\alpha^2 + 2)f_0$$
$$+ (-3\alpha^3 + 4\alpha^2 + \alpha)f_1 + (\alpha^3 - \alpha^2)f_2) \qquad (3.75)$$

Figure 3.34 Linear interpolation coefficient functions (left), and cubic Hermite splines (right). Note the similarity: h_0 and h_1 are smooth versions of the linear interpolation coefficient functions.

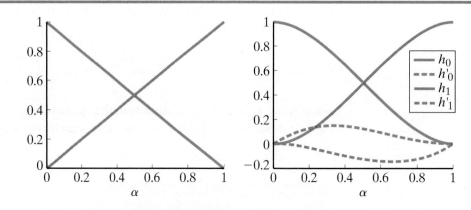

or in matrix form:

$$\hat{f}(x) = \frac{1}{2}\begin{bmatrix} \alpha^3 & \alpha^2 & \alpha & 1 \end{bmatrix} \begin{bmatrix} -1 & 3 & -3 & 1 \\ 2 & -5 & 4 & -1 \\ -1 & 0 & 1 & 0 \\ 0 & 2 & 0 & 0 \end{bmatrix} \begin{bmatrix} f_\flat \\ f_0 \\ f_1 \\ f_2 \end{bmatrix} \quad (\text{Catmull-Rom}) \quad (3.76)$$

where we define $\flat \equiv -1$ so that $f_\flat \equiv f(x_0 - 1)$, and $f_2 \equiv f(x_0 + 2)$. Note that, whereas linear interpolation requires two nearby sampled values, Catmull-Rom spline interpolation requires four. Catmull-Rom spline interpolation is the most common variety of cubic interpolation, and we shall use the terms interchangeably from now on.

An alternate way of visualizing linear and cubic interpolation is shown in Figure 3.35. An *interpolation kernel* $k(x)$ is defined over the continuous domain, and the interpolated value is the weighted sum of the neighboring samples, where the weights are chosen by centering the interpolation kernel over the desired position x, that is, $\hat{f}(x) = \sum_i k(i - \alpha)f_i$. In linear interpolation, for example, $k(x)$ is the triangle function:

$$k(x) = \begin{cases} 1 - |x| & \text{if } |x| \leq 1 \\ 0 & \text{otherwise} \end{cases} \quad (\text{linear}) \quad (3.77)$$

so that

$$\hat{f}(x) = \sum_{i=0}^{1} k(i - \alpha)f_i = k(-\alpha)f_0 + k(1 - \alpha)f_1 \quad (3.78)$$

$$= (1 - |-\alpha|)f_0 + (1 - |1 - \alpha|)f_1 = (1 - \alpha)f_0 + \alpha f_1 \quad (3.79)$$

Figure 3.35 Top: Linear (left) and cubic (right) 1D interpolation kernels. The dashed line indicates $k(x) = 0$ to emphasize that the cubic interpolation kernel contains negative values. Bottom: Interpolation involves shifting the kernel so that it is centered at the desired position x, then the neighboring samples are combined using the weights from the kernel.

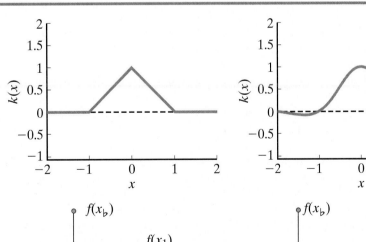

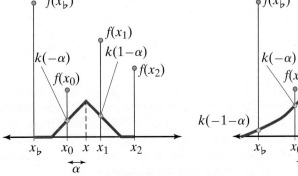

which is equivalent to Equation (3.68). For cubic interpolation, $k(x)$ is defined as

$$k(x) = \frac{1}{2} \cdot \begin{cases} 3|x|^3 - 5|x|^2 + 2 & \text{if } |x| \le 1 \\ -|x|^3 + 5|x|^2 - 8|x| + 4 & \text{if } 1 < |x| < 2 \quad \text{(cubic)} \\ 0 & \text{otherwise} \end{cases} \tag{3.80}$$

so that the cubic interpolation function evaluated at position x is computed as

$$\hat{f}(x) = \sum_{i=-1}^{2} k(i - \alpha)f_i = k(-1 - \alpha)f_{\flat} + k(-\alpha)f_0 + k(1 - \alpha)f_1 + k(2 - \alpha)f_2 \tag{3.81}$$

$$= \frac{1}{2}(((-1-\alpha)^3 + 5(-1-\alpha)^2 + 8(-1-\alpha) + 4)f_{\flat}$$

$$+ (-3(-\alpha)^3 - 5(-\alpha)^2 + 2)f_0$$

$$+ (3(1-\alpha)^3 - 5(1-\alpha)^2 + 2)f_1$$

$$+ (-(2-\alpha)^3 + 5(2-\alpha)^2 - 8(2-\alpha) + 4)f_2) \tag{3.82}$$

With a bit of algebraic simplification, this expression can be shown to be equivalent to Equation (3.75).

Bicubic interpolation is an extension of this basic idea to 2D, in which the weighted sum of the values of the 16 pixels in the surrounding 4×4 neighborhood are computed:

$$\hat{I}(x, y) = \sum_{i=-1}^{2} \sum_{j=-1}^{2} k_{ij} \alpha_x^{i+1} \alpha_y^{j+1} \tag{3.83}$$

where the weights k_{ij} are determined by a linear combination of the 16 pixel values:

$$\begin{bmatrix} k_{\flat\flat} \\ k_{0\flat} \\ k_{1\flat} \\ k_{2\flat} \\ k_{\flat0} \\ k_{00} \\ k_{10} \\ k_{20} \\ k_{\flat1} \\ k_{01} \\ k_{11} \\ k_{21} \\ k_{\flat2} \\ k_{02} \\ k_{12} \\ k_{22} \end{bmatrix} = \frac{1}{4} \begin{bmatrix} 0 & 0 & 0 & 0 & 0 & 4 & 0 & 0 & 0 & 0 & 0 & 0 & 0 & 0 & 0 & 0 \\ 0 & 0 & 0 & 0 & -2 & 0 & 2 & 0 & 0 & 0 & 0 & 0 & 0 & 0 & 0 & 0 \\ 0 & 0 & 0 & 0 & 4 & -10 & 8 & -2 & 0 & 0 & 0 & 0 & 0 & 0 & 0 & 0 \\ 0 & 0 & 0 & 0 & -2 & 6 & -6 & 2 & 0 & 0 & 0 & 0 & 0 & 0 & 0 & 0 \\ 0 & -2 & 0 & 0 & 0 & 0 & 0 & 0 & 0 & 2 & 0 & 0 & 0 & 0 & 0 & 0 \\ 1 & 0 & 0 & 0 & 0 & 0 & -1 & 0 & 0 & -1 & 1 & 0 & 0 & 0 & 0 & 0 \\ -2 & 5 & -6 & 0 & 0 & 0 & 2 & 1 & 0 & -4 & 5 & -1 & 0 & 0 & 0 & 0 \\ 1 & -3 & 4 & 0 & 0 & 0 & -1 & -1 & 0 & 3 & -4 & 1 & 0 & 0 & 0 & 0 \\ 0 & 4 & 0 & 0 & 0 & -10 & 0 & 0 & 0 & 8 & 0 & 0 & 0 & -2 & 0 & 0 \\ -2 & 0 & 0 & 0 & 5 & 0 & -4 & 0 & -6 & 2 & 5 & 0 & 0 & 1 & -1 & 0 \\ 4 & -10 & 12 & 0 & -10 & 25 & -22 & 4 & 12 & -22 & 12 & -5 & 0 & 4 & -5 & 1 \\ -2 & 6 & -8 & 0 & 5 & -15 & 16 & -4 & -6 & 12 & -9 & 5 & 0 & -3 & 4 & -1 \\ 0 & -2 & 0 & 0 & 0 & 6 & 0 & 0 & 0 & -6 & 0 & 0 & 0 & 2 & 0 & 0 \\ 1 & 0 & 0 & 0 & -3 & 0 & 3 & 0 & 4 & -1 & -4 & 0 & 0 & -1 & 1 & 0 \\ -2 & 5 & -6 & 0 & 6 & -15 & 12 & -3 & -8 & 16 & -9 & 4 & 0 & -4 & 5 & -1 \\ 1 & -3 & 4 & 0 & -3 & 9 & -9 & 3 & 4 & -9 & 7 & -4 & 0 & 3 & -4 & 1 \end{bmatrix} \begin{bmatrix} I_{\flat\flat} \\ I_{0\flat} \\ I_{1\flat} \\ I_{2\flat} \\ I_{\flat0} \\ I_{00} \\ I_{10} \\ I_{20} \\ I_{\flat1} \\ I_{01} \\ I_{11} \\ I_{21} \\ I_{\flat2} \\ I_{02} \\ I_{12} \\ I_{22} \end{bmatrix} \tag{3.84}$$

where $I_{\flat\flat} = I(x_0 - 1, y_0 - 1)$, and so forth. Deriving this 16×16 matrix is left as an exercise, but basically they are the coefficients necessary to ensure that the interpolated function maintains continuity in its values, first derivatives, and cross derivatives. Bicubic interpolation is illustrated in Figure 3.36, and the pseudocode is provided in Algorithm 3.17. Compared with bilinear interpolation, bicubic interpolation preserves finer detail but requires more computational expense.

Figure 3.36 Bicubic interpolation at a point (x, y) is a weighted average of the 16 nearby gray levels.

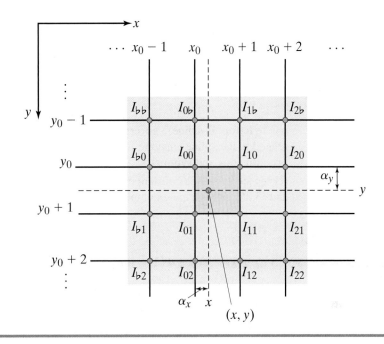

3.8.4 Keys Filters

Bicubic interpolation can be improved upon in two ways: first, by reducing its computational expense, and secondly, by relaxing the requirement that the weighting function $k(x)$ be a true interpolation. (Relaxing this requirement tends to produce more visually pleasing results in practice than true interpolation.) To implement both of these improvements, we turn to the **cubic convolution filter**:

$$\hat{f}(x) \equiv \sum_{i=-1}^{2} k(i - \alpha) f(x_0 + i) \tag{3.85}$$

where, as before, $x_0 \equiv \lfloor x \rfloor$ and $\alpha \equiv x - x_0$. This filter is applied to the image along the rows, then along the columns (or vice versa) in a separable manner, thus achieving an efficient

ALGORITHM 3.17 Perform bicubic interpolation (slow version)

INTERPOLATEBICUBICSLOW (I, x, y)

Input: image, floating-point coordinates (x, y)
Output: weighted average of graylevel values of nearest 16 pixels
1   $x_0 \leftarrow$ FLOOR (x)
2   $y_0 \leftarrow$ FLOOR (y)
3   $\alpha_x \leftarrow x - x_0$
4   $\alpha_y \leftarrow y - y_0$
5   Compute $k_{ij}, i = -1, \ldots, 2, j = -1, \ldots, 2$ using Equation (3.84)
6   **return** $\sum_{i=-1}^{2} \sum_{j=-1}^{2} k_{ij} \alpha_x^{i+1} \alpha_y^{j+1}$

approximation to 2D bicubic interpolation.[†] The kernel $k(x)$ is a piecewise cubic spline function specified by two parameters, b and c:

$$k(x) = \frac{1}{6} \begin{cases} (12 - 9b - 6c)|x|^3 + (-18 + 12b + 6c)|x|^2 + (6 - 2b) & \text{if } |x| < 1 \\ (-b - 6c)|x|^3 + (6b + 30c)|x|^2 + (-12b - 48c)|x| + (8b + 24c) & \text{if } 1 \le |x| < 2 \\ 0 & \text{otherwise} \end{cases} \quad (3.86)$$

where b (for "B-spline") governs the amount of smoothing, and c (for "cardinal spline") is related to the spline tension.[‡] It is easy to verify that, for any values of b and c, the kernel is continuous and symmetric about the origin, has a continuous first derivative, and preserves the value of a constant input signal, $\sum_{x=-1}^{2} k(x) = 1$. Combining Equations (3.85) and (3.86) yields a more complete expression:

$$\hat{f}(x) = \sum_{i=-1}^{2} k(i - \alpha)f_i$$

$$= k(-1 - \alpha)f_b + k(-\alpha)f_0 + k(1 - \alpha)f_1 + k(2 - \alpha)f_2$$

$$= \frac{1}{6}((-b - 6c)(1 + \alpha)^3 + (6b + 30c)(1 + \alpha)^2 + (-12b - 48c)(1 + \alpha) + (8b + 24c))f_b$$

$$+ \frac{1}{6}((12 - 9b - 6c)\alpha^3 + (-18 + 12b + 6c)\alpha^2 + (6 - 2b))f_0$$

$$+ \frac{1}{6}((12 - 9b - 6c)(1 - \alpha)^3 + (-18 + 12b + 6c)(1 - \alpha)^2 + (6 - 2b))f_1$$

$$+ \frac{1}{6}((-b - 6c)(2 - \alpha)^3 + (6b + 30c)(2 - \alpha)^2 + (-12b - 48c)(2 - \alpha) + (8b + 24c))f_2$$

$$= \frac{1}{6}\begin{bmatrix} \alpha^3 & \alpha^2 & \alpha & 1 \end{bmatrix} \begin{bmatrix} -b - 6c & 12 - 9b - 6c & -12 + 9b + 6c & b + 6c \\ 3b + 12c & -18 + 12b + 6c & 18 - 15b - 12c & -6c \\ -3b - 6c & 0 & 3b + 6c & 0 \\ b & 6 - 2b & b & 0 \end{bmatrix} \begin{bmatrix} f_b \\ f_0 \\ f_1 \\ f_2 \end{bmatrix} \quad (3.87)$$

where $f_i \equiv f(x_0 + i)$, as before.

The two parameters b and c govern the type of filter, allowing us to generate any smoothly fitting piecewise cubic filter. The resulting filter space is illustrated in Figure 3.37, with an overlay of the subjective quality as assessed by image processing experts on sample images. The vertical line along the left of the figure at $c = 0$ contains the well-known family of **B-splines**, of which the **uniform cubic B-spline** ($b = 1, c = 0$) is the most famous:

$$\hat{f}(x) = \frac{1}{6}\begin{bmatrix} \alpha^3 & \alpha^2 & \alpha & 1 \end{bmatrix} \begin{bmatrix} -1 & 3 & -3 & 1 \\ 3 & -6 & 3 & 0 \\ -3 & 0 & 3 & 0 \\ 1 & 4 & 1 & 0 \end{bmatrix} \begin{bmatrix} f_b \\ f_0 \\ f_1 \\ f_2 \end{bmatrix} \quad (3.88)$$

Similarly, the horizontal line along the bottom of the figure at $b = 0$ contains the family of **cardinal splines** that we considered in the previous section:

$$k(x) = \begin{cases} (2 - c)|x|^3 + (c - 3)|x|^2 + 1 & \text{if } |x| \le 1 \\ -c|x|^3 + 5c|x|^2 - 8c|x| + 4c & \text{if } 1 < |x| < 2 \\ 0 & \text{otherwise} \end{cases} \quad (3.89)$$

[†] Convolution and separability are covered in detail in Chapter 5.

[‡] If τ is the spline tension described in the previous section, then $c = \frac{1}{2}(1 - \tau)$.

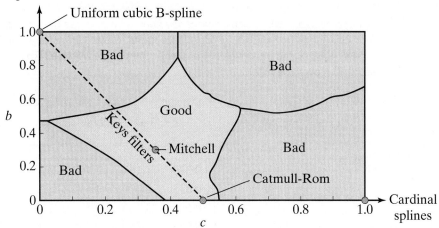

Figure 3.37 The space of smoothly fitting piecewise cubic filters, governed by the parameters b and c. The true interpolation filters are the cardinal splines, which satisfy $b = 0$, while the B-spline cubics satisfy $c = 0$. The Keys filters satisfy $b + 2c = 1$, with special cases being the Mitchell filter ($b = 1/3, c = 1/3$), the Catmull-Rom spline ($b = 0, c = 0.5$), and the standard uniform cubic B-spline ($b = 1$, $c = 0$). The color indicates subjective quality as assessed by image processing experts, with the "bad" regions exhibiting anisotropy, excessive ringing, blocking, aliasing, or smoothing artifacts. Based on D. P. Mitchell and A. N. Netravali. Reconstruction Filters in computer graphics. *Computer Graphics* (SIGGRAPH), 22(4):221-228, June 1988.

$$\hat{f}(x) = [\alpha^3 \quad \alpha^2 \quad \alpha \quad 1] \begin{bmatrix} -c & 2-c & c-2 & c \\ 2c & c-3 & 3-2c & -c \\ -c & 0 & c & 0 \\ 0 & 1 & 0 & 0 \end{bmatrix} \begin{bmatrix} f_b \\ f_0 \\ f_1 \\ f_2 \end{bmatrix} \tag{3.90}$$

These splines incorporate no smoothing and hence are true interpolation filters, that is, $\hat{f}(x) = f(x)$ whenever x is an integer, which is easily seen from Equation (3.90) because $\hat{f}(x) = f_0$ when $\alpha = 0$.

From the figure notice that the visually pleasing filters tend to lie on or near the line

$$b + 2c = 1. \tag{3.91}$$

Any filter satisfying this constraint is called a **Keys filter**. Substituting Equation (3.91) into Equations (3.86)–(3.87) reveals that Keys filters are parameterized as

$$k(x) = \frac{1}{6} \begin{cases} (3 + 12c)|x|^3 + (-6 - 18c)|x|^2 + (4 + 4c) & \text{if } |x| < 1 \\ (-1 - 4c)|x|^3 + (6 + 18c)|x|^2 + (-12 - 24c)|x| + (8 + 8c) & \text{if } 1 \le |x| < 2 \\ 0 & \text{otherwise} \end{cases} \tag{3.92}$$

$$\hat{f}(x) = \frac{1}{6}[\alpha^3 \quad \alpha^2 \quad \alpha \quad 1] \begin{bmatrix} -1 - 4c & 3 + 12c & -3 - 12c & 1 + 4c \\ 3 + 6c & -6 - 18c & 3 + 18c & -6c \\ -3 & 0 & 3 & 0 \\ 1 - 2c & 4 + 4c & 1 - 2c & 0 \end{bmatrix} \begin{bmatrix} f_b \\ f_0 \\ f_1 \\ f_2 \end{bmatrix} \tag{3.93}$$

where $0 \le c \le 0.5$.

The Catmull-Rom spline is not only a cardinal spline but also a Keys filter, occurring when $b = 0$ and $c = \frac{1}{2}$, which is seen by substituting $c = \frac{1}{2}$ into Equation (3.93) and comparing with Equation (3.76) . Although it is beyond our scope to prove this, Catmull-Rom achieves third-order convergence with respect to the Taylor series approximation of the original signal, which is the best of any known filter. Catmull-Rom is therefore a popular filter that in practice tends to produce acceptable results without noticeable artifacts.

Nevertheless, the most widely used Keys filter is the **Mitchell filter** (also known as the **Mitchell-Netravali filter**), defined as $b = 1/3, c = 1/3$:

$$k(x) \approx \begin{cases} 1.167|x|^3 - 2|x|^2 + 0.889 & \text{if } |x| < 1 \\ -0.389|x|^3 + 2|x|^2 - 3.333|x| + 1.778 & \text{if } 1 \le |x| < 2 \\ 0 & \text{otherwise} \end{cases} \tag{3.94}$$

$$\hat{f}(x) \approx \begin{bmatrix} \alpha^3 & \alpha^2 & \alpha & 1 \end{bmatrix} \begin{bmatrix} -0.389 & 1.167 & -1.167 & 0.389 \\ 0.833 & -2 & 1.5 & -0.333 \\ -0.5 & 0 & 0.5 & 0 \\ 0.055 & 0.890 & 0.055 & 0 \end{bmatrix} \begin{bmatrix} f_b \\ f_0 \\ f_1 \\ f_2 \end{bmatrix} \tag{3.95}$$

From the figure, it can be seen that the Mitchell filter lies even further from the objectionable regions than Catmull-Rom. In practice the Mitchell filter achieves a nice compromise between not enough smoothing and too much smoothing, making it a popular choice for image upsampling.

3.8.5 Lanczos Interpolation

Another important method is **Lanczos interpolation**,[†] whose interpolation kernel is the well-known **sinc** function multiplied by a truncated sinc function:[‡]

$$k(x) = \begin{cases} (\text{sinc } x) \cdot (\text{sinc } \frac{x}{a}) & \text{if } -a < x < a \\ 0 & \text{otherwise} \end{cases} \tag{3.96}$$

where

$$\text{sinc } x \equiv \frac{\sin \pi x}{\pi x} \tag{3.97}$$

and a is an integer specifying the number of positive zero crossings to include. The first factor has zeros at integer values of x, while the second factor is the bell-shaped window function (known as the *Lanczos window*) that reaches zero at $x = \pm a$.

To interpolate a 1D discrete signal f, the weighted sum of the $2a$ values of the signal is computed, using the values of the interpolation kernel as weights. Typically $a = 2$ or 3, where the kernel is known as Lanczos-2 or Lanczos-3, respectively. Figure 3.38 shows an example using the Lanczos-2 kernel, where $x = 9.2$, $x_0 \equiv \lfloor x \rfloor = 9$, and $\alpha = x - x_0 = 0.2$. The value $f(x_0)$ is multiplied by $k(-\alpha)$, the value $f(x_0 - 1)$ is multiplied by $k(-\alpha - 1)$, the value $f(x_0 + 1)$ is multiplied by $k(-\alpha + 1)$, and the value $f(x_0 + 2)$ is multiplied by $k(-\alpha + 2)$. More generally,

$$\hat{f}(x) = \sum_{i=-a+1}^{a} k(i - \alpha)f(x_0 + i) \tag{3.98}$$

[†] Cornelius Lanczos (1893–1974) made a number of important contributions to math and physics, including several numerical algorithms.

[‡] Pronounced "sink", the function in Equation (3.97) is the so-called *normalized* version. The unnormalized verison, $\frac{\sin x}{x}$, omits π.

Figure 3.38 Interpolation of a 1D signal. Here the signal shown by the vertical lollipops is evaluated at $x = 9.2$. The interpolation function is the smooth curve (Lanczos-2 in this case). The 4 green circles indicate the values of the interpolation function that are elementwise multiplied by the corresponding signal values, and then summed to yield the interpolated value.

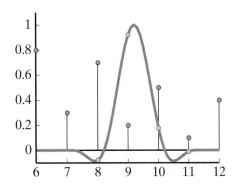

Like the cubic convolution filters, Lanczos interpolation is applied horizontally, then vertically (or vice versa) to obtain an approximation to 2D interpolation.

One must be careful to distinguish between Lanczos *interpolation*, which we consider here, and Lanczos *filtering*, which we consider later.[†] In the latter, the rationale for using the sinc function is that it is the ideal low-pass filter, which makes sense in the context of filtering an image by convolving with a large kernel that has been sampled at integer positions. In interpolation, however, the kernel has only 4 or 6 values (depending upon whether $a = 2$ or 3), and the kernel is sampled at non-integer positions.

Figure 3.39 shows the 1D interpolation kernels defining the different filters of this section, and Figure 3.40 compares the results of these filters on an example 1D signal. Note that the first six filters are true interpolations because they pass through all the sample points of the original signal, whereas the last two introduce some amount of smoothing. Although the negative lobes of Catmull-Rom, Mitchell, and Lanczos filters cause ringing in the signal, which is theoretically undesirable, this behavior actually improves visual appearance by

Figure 3.39 Various interpolation kernels, including those that introduce a small amount of smoothing. The "bad filter" is at $b = 0$, $c = 1$. Note that the last two kernels are not true interpolation functions, because they do not satisfy $k(\pm 1) = k(\pm 2) = 0$.

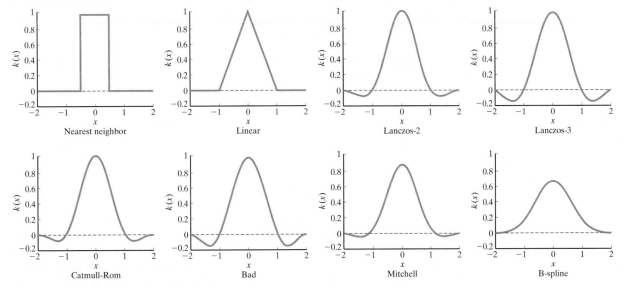

[†] Section 6.4.1 (p. 296).

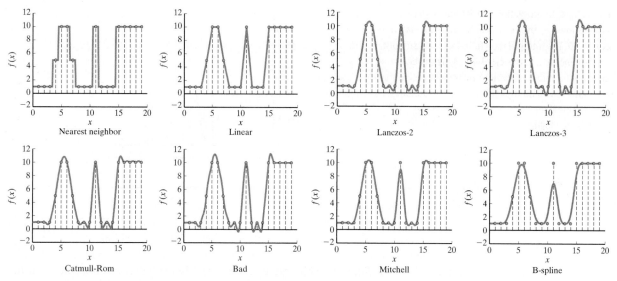

Figure 3.40 Comparison of 1D interpolation methods, some with smoothing, on an example signal. Overall the Catmull-Rom, Mitchell, and Lanczos-2 methods do the best job of providing a smooth fit to the signal without excessive overshoot or ringing.

increasing acutance (edge contrast) in the image in a manner similar to the Mach bands phenomenon.[‡] Note that nearest neighbor and linear methods produce sharp transitions, whereas the higher-order methods produce smoother results. The B-spline filter smooths too much, while the "bad filter" at $b = 0, c = 1$ causes excessive overshoot, or ringing. A good compromise is achieved by either the Mitchell or Lanczos-2 filters.

3.9 Warping

Armed with the ability to interpolate pixel values, in this section we consider arbitrary geometric transformations from real-valued coordinates (x, y) to real-valued coordinates (x', y'):

$$I'(x', y') = I(x, y) \tag{3.99}$$

where the **mapping function** $f : \mathbb{R}^2 \to \mathbb{R}^2$ specifies the transformation, or **warping**, from the input coordinates to the output coordinates:

$$(x', y') = f(x, y) \tag{3.100}$$

Since f is a function of two inputs that produces two outputs, another way to write the equation is to split it into two components:

$$x' = f_x(x, y) \tag{3.101}$$

$$y' = f_y(x, y) \tag{3.102}$$

where $f(x, y) = (f_x(x, y), f_y(x, y))$.

Earlier in the chapter we discussed the distinction between forward and inverse mapping of coordinates. Unlike the simple transformations considered there, transformations involving real-valued coordinates do not involve a one-to-one mapping between the pixels of the input and output images. As a result, some pixels in the output image may not be touched by the

---

[‡] Section 6.4.3 (p. 303).

transformation, because the pixels that would have mapped to them are, in fact, out of bounds in the input image. At the same time, multiple pixels in the input image sometimes map to the same pixel in the output image due to discretization effects (imagine viewing a slanted plane, as illustrated in Figure 3.41). To ensure that all pixels in the output image are set exactly once, it is important not to loop over the pixels in the input, performing the forward transformation, but rather to loop over the pixels in the output, performing the inverse transformation:

$$(x, y) = f^{-1}(x', y') \tag{3.103}$$

where interpolation is used to calculate $I(x, y)$. In this manner, the coordinates (x', y') in the output are guaranteed to be integers, even though the coordinates (x, y) in the input may be real-valued.

3.9.1 Downsampling and Upsampling Revisited

In Equations (3.9)–(3.10) we showed how to downsample or upsample an image by a factor of 2, without averaging the pixel values. In the more general case, let s_x and s_y be real-valued scaling factors in the horizontal and vertical directions, respectively:

$$I'(x', y') = I(s_x x', s_y y') \qquad (\text{downsample}) \tag{3.104}$$

$$I'(x', y') = I(x'/s_x, y'/s_y) \qquad (\text{upsample}) \tag{3.105}$$

where $s_x, s_y \geq 1$. If $s_x = s_y$, then the transformation preserves the aspect ratio of the original image.

Although downsampling can be achieved by simply discarding pixels that are not needed (nearest-neighbor interpolation), it is better to compute output pixel values as an average of the neighboring input pixels to avoid aliasing.[†] Any of the interpolation methods can be used, with best results obtained by introducing a small amount of smoothing (i.e., not using a true interpolation filter). For example, if $s_x = s_y = 2$, then the output pixel is computed as the weighted average of the pixel and its two immediate neighbors. This is achieved for two alternative techniques by setting $\alpha = 0$ in Equation (3.88) and Equation (3.95), respectively:

$$\hat{f}(x) = 0.167f(x-1) + 0.666f(x) + 0.167f(x+1) \qquad (\text{uniform cubic B-spline}) \tag{3.106}$$

$$\hat{f}(x) = 0.055f(x-1) + 0.889f(x) + 0.055f(x+1) \qquad (\text{Mitchell}) \tag{3.107}$$

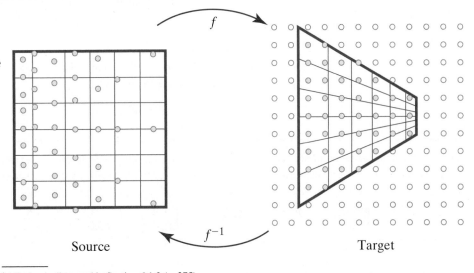

Figure 3.41
A frontoparallel plane in the input is warped to a slanted plane in the output. The inverse transformation guarantees that every pixel in the output receives a value, whereas the forward transformation leads to some pixels not receiving values while others receive multiple values. Based on Burger and Burge: W. Burger and M. J. Burge. *Digital Image Processing: An Algorithmic Introduction Using Java.* Springer, 2008.

Source

Target

f

f^{-1}

[†] Aliasing is discussed in Section 6.1.3 (p. 275).

When s_x, $s_y > 2$, additional neighboring pixels must be averaged, requiring techniques that will be explored in Chapter 5.[†]

Similarly, although upsampling can simply replicate pixels (nearest-neighbor interpolation), to avoid blocking artifacts it is best to use a higher-order form of interpolation, such as the Mitchell filter. For example, with $s_x = s_y = 2$ and substituting $\alpha = 0.5$ into Equation (3.95), the Mitchell filter applies

$$\hat{f}(x) = -0.035f(x-1.5) + 0.535f(x-0.5) + 0.535f(x+0.5) - 0.035f(x+1.5) \tag{3.108}$$

There is a fundamental limit to how much upsampling can be performed, since we cannot recreate detail that was not present in the original image, and thus upsampling necessarily involves hallucinating information. Nevertheless, for higher scaling factors, more sophisticated algorithms can be used to recover sharp edges and textures while suppressing artifacts; such algorithms are beyond our scope but are mentioned later as additional reading.

3.9.2 Euclidean Transformations

A number of different primitive geometric transformations are illustrated in Figure 3.42. *Translation*, for example, involves shifting the image by a certain amount t_x horizontally and t_y vertically:

$$x' = x + t_x \tag{3.109}$$
$$y' = y + t_y \tag{3.110}$$

whose inverse mapping is easy to obtain:

$$\begin{bmatrix} x \\ y \end{bmatrix} = f^{-1}(x', y') = \begin{bmatrix} x' - t_x \\ y' - t_y \end{bmatrix} \tag{3.111}$$

where interpolation is necessary to compute $I(x, y)$ if t_x and t_y are not both integers. Although out-of-bounds values are typically set to the nearest pixel, as mentioned above, another possibility is to wrap the pixels around the other opposite side of the image, which is sometimes used in the case of a landscape background.

Another transformation is to *rotate* the image by a clockwise angle θ:

$$\begin{bmatrix} x' \\ y' \end{bmatrix} = \underbrace{\begin{bmatrix} \cos\theta & -\sin\theta \\ \sin\theta & \cos\theta \end{bmatrix}}_{\mathbf{R}} \begin{bmatrix} x \\ y \end{bmatrix} \tag{3.112}$$

or $\mathbf{x}' = \mathbf{R}\mathbf{x}$, where $\mathbf{x}' = \begin{bmatrix} x' & y' \end{bmatrix}^\mathsf{T}$ and $\mathbf{x} = \begin{bmatrix} x & y \end{bmatrix}^\mathsf{T}$. This equation specifies rotation about the origin in the upper-left corner of the image. Usually, however, we want to rotate about some point $\mathbf{c} = \begin{bmatrix} c_x & c_y \end{bmatrix}^\mathsf{T}$ inside the image:

$$\mathbf{x}' = \mathbf{R}(\mathbf{x} - \mathbf{c}) + \mathbf{c} \tag{3.113}$$

Figure 3.42 Various geometric transformations applied to a square. From left to right: Identity, translation, rotation, uniform scaling, non-uniform scaling, shear, and projective transformation. Note that all transformations but the last one preserve parallel lines.

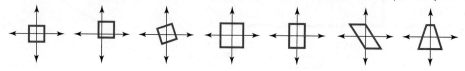

[†] Subsampling is also discussed in detail in Section 7.1.1 (p. 329).

Since a rotation matrix has the property that its inverse is its transpose, $\mathbf{R}^{-1} = \mathbf{R}^\mathsf{T},$ all we need to do to invert the equation is transpose the matrix, which means swapping the sign on the sine:

$$\begin{bmatrix} x \\ y \end{bmatrix} = \underbrace{\begin{bmatrix} \cos\theta & \sin\theta \\ -\sin\theta & \cos\theta \end{bmatrix}}_{\mathbf{R}^\mathsf{T}} \left(\begin{bmatrix} x' \\ y' \end{bmatrix} - \begin{bmatrix} c_x \\ c_y \end{bmatrix} \right) + \begin{bmatrix} c_x \\ c_y \end{bmatrix} \tag{3.114}$$

Unless θ happens to be a multiple of 90 degrees, interpolation will need to be performed, because x and y will not be integers for arbitrary values of θ.

Conveniently, translation and rotation can be combined into a single **Euclidean transformation**: $\mathbf{x}' = \mathbf{R}(\mathbf{x} - \mathbf{c}) + \mathbf{c} + \mathbf{t} = \mathbf{R}\mathbf{x} + \tilde{\mathbf{t}}$:

$$\begin{bmatrix} x' \\ y' \end{bmatrix} = \underbrace{\begin{bmatrix} \cos\theta & -\sin\theta \\ \sin\theta & \cos\theta \end{bmatrix}}_{\mathbf{R}} \begin{bmatrix} x \\ y \end{bmatrix} + \begin{bmatrix} \tilde{t}_x \\ \tilde{t}_y \end{bmatrix} \tag{3.115}$$

where $\tilde{\mathbf{t}} \equiv \begin{bmatrix} \tilde{t}_x & \tilde{t}_y \end{bmatrix}^\mathsf{T} = -\mathbf{R}\mathbf{c} + \mathbf{c} + \mathbf{t}$. This can be viewed as either a rotation about $\mathbf{c}$ followed by a translation by $\mathbf{t}$, or equivalently as a rotation about the origin followed by a translation by $\tilde{\mathbf{t}}$. Euclidean transformations preserve the shape and scale of an object.

By appending a 1 to each point, we obtain the 3×1 vectors $\begin{bmatrix} x & y & 1 \end{bmatrix}^\mathsf{T}$ and $\begin{bmatrix} x' & y' & 1 \end{bmatrix}^\mathsf{T},$ which are the **homogeneous coordinates** of the points. Using homogeneous coordinates, Equation (3.115) can be rewritten as a single matrix multiplication:

$$\begin{bmatrix} x' \\ y' \\ 1 \end{bmatrix} = \begin{bmatrix} \cos\theta & -\sin\theta & \tilde{t}_x \\ \sin\theta & \cos\theta & \tilde{t}_y \\ 0 & 0 & 1 \end{bmatrix} \begin{bmatrix} x \\ y \\ 1 \end{bmatrix} \tag{3.116}$$

which is convenient mathematically. Homogeneous coordinates are powerful representations that have many uses in computer vision and are covered in more detail later in the book.[†]

3.9.3 Similarity Transformations

Uniform scaling is achieved by multiplying all the coordinates by the same scalar k:

$$x' = kx \tag{3.117}$$
$$y' = ky \tag{3.118}$$

Similarity transformations are a superset of Euclidean transformations, because they include not only translations and rotations, but also uniform scaling:

$$\begin{bmatrix} x' \\ y' \\ 1 \end{bmatrix} = \begin{bmatrix} k\cos\theta & -k\sin\theta & k\tilde{t}_x \\ k\sin\theta & k\cos\theta & k\tilde{t}_y \\ 0 & 0 & 1 \end{bmatrix} \begin{bmatrix} x \\ y \\ 1 \end{bmatrix} \tag{3.119}$$

Similarity transformations also preserve shape, but not necessarily size.

[†] See Section 13.4.1 (p. 654).

3.9.4 Affine Transformations

A 2×2 matrix R is a rotation matrix if and only if its determinant is 1, i.e., $\det(\mathbf{R}) = 1$. Removing this constraint to allow for any arbitrary, invertible 2×2 matrix leads to an **affine transformation**:

$$\begin{bmatrix} x' \\ y' \end{bmatrix} = \begin{bmatrix} a_{11} & a_{12} \\ a_{21} & a_{22} \end{bmatrix} \begin{bmatrix} x \\ y \end{bmatrix} + \begin{bmatrix} a_{13} \\ a_{23} \end{bmatrix} \tag{3.120}$$

which can be rewritten as a single matrix multiplication using homogeneous coordinates:

$$\begin{bmatrix} x' \\ y' \\ 1 \end{bmatrix} = \begin{bmatrix} a_{11} & a_{12} & a_{13} \\ a_{21} & a_{22} & a_{23} \\ 0 & 0 & 1 \end{bmatrix} \begin{bmatrix} x \\ y \\ 1 \end{bmatrix} \tag{3.121}$$

To warp an image using an affine transformation, the matrix must be inverted:

$$\begin{bmatrix} x \\ y \\ 1 \end{bmatrix} = \frac{1}{a_{11}a_{22} - a_{12}a_{21}} \begin{bmatrix} a_{22} & -a_{12} & a_{23}a_{12} - a_{22}a_{13} \\ -a_{21} & a_{11} & -a_{23}a_{11} + a_{21}a_{13} \\ 0 & 0 & a_{11}a_{22} - a_{12}a_{21} \end{bmatrix} \begin{bmatrix} x' \\ y' \\ 1 \end{bmatrix} \tag{3.122}$$

Affine transformations include rotation, translation, and uniform scaling, since similarity transformations are a special case. In addition, affine transformations can change the scale nonuniformly:

$$x' = s_x x \tag{3.123}$$

$$y' = s_y y \tag{3.124}$$

as well as produce something called *shear*, in which one coordinate is shifted by an amount proportional to the other coordinate:

$$x' = x + ay \tag{3.125}$$

$$y' = y \tag{3.126}$$

where s_x, s_y, and a are scalars. Together, these possibilities mean that a square can warp into a rectangle or parallelogram. Note that lines that are parallel to each other in the original remain parallel in the output.

3.9.5 Projective Transformations

A 2D **projective transformation** relaxes the constraint that the bottom row of the matrix be $\begin{bmatrix} 0 & 0 & 1 \end{bmatrix}^T$, leading to an invertible 3×3 matrix $\mathbf{H}$ known as a **homography**:

$$\begin{bmatrix} x' \\ y' \\ 1 \end{bmatrix} \propto \underbrace{\begin{bmatrix} h_{11} & h_{12} & h_{13} \\ h_{21} & h_{22} & h_{23} \\ h_{31} & h_{32} & h_{33} \end{bmatrix}}_{\mathbf{H}} \begin{bmatrix} x \\ y \\ 1 \end{bmatrix} \tag{3.127}$$

With projective transformations, homogeneous coordinates become considerably more difficult to visualize and understand, because the left-hand side is not necessarily equal to, but rather is proportional to, the right-hand side. If we use λ to represent this proportionality constant, then we can rewrite the equation as

$$\lambda \begin{bmatrix} x' \\ y' \\ 1 \end{bmatrix} = \underbrace{\begin{bmatrix} h_{11} & h_{12} & h_{13} \\ h_{21} & h_{22} & h_{23} \\ h_{31} & h_{32} & h_{33} \end{bmatrix}}_{\mathbf{H}} \begin{bmatrix} x \\ y \\ 1 \end{bmatrix} \tag{3.128}$$

where $\lambda = h_{31}x + h_{32}y + h_{33}$. Note that if the bottom row of the matrix is $\begin{bmatrix} 0 & 0 & 1 \end{bmatrix}^{\mathsf{T}}$ as before, then $\lambda = 1$ for all possible values of x and y, and we are back to an affine transformation. On the other hand, if $\lambda \neq 1$, then the output vector has a value in the last entry that is not equal to 1. To compute the true (x', y'), then, we must divide by λ:

$$x' = \frac{\lambda x'}{\lambda} = \frac{h_{11}x + h_{12}y + h_{13}}{h_{31}x + h_{32}y + h_{33}} \tag{3.129}$$

$$y' = \frac{\lambda y'}{\lambda} = \frac{h_{21}x + h_{22}y + h_{23}}{h_{31}x + h_{32}y + h_{33}} \tag{3.130}$$

EXAMPLE 3.12

Apply the following projective transformation to the point $(1,2)$:

$$\mathbf{H} = \begin{bmatrix} 7 & 3 & 2 \\ 2 & 4 & 8 \\ 1 & 3 & 2 \end{bmatrix} \tag{3.131}$$

Solution

The homogeneous coordinates of the point are obtained by appending a 1 to $(1,2)$, leading to $\begin{bmatrix} 1 & 2 & 1 \end{bmatrix}^{\mathsf{T}}$. Multiplying by the matrix $\mathbf{H}$ yields:

$$\begin{bmatrix} \lambda x' \\ \lambda y' \\ \lambda \end{bmatrix} = \begin{bmatrix} 7 & 3 & 2 \\ 2 & 4 & 8 \\ 1 & 3 & 2 \end{bmatrix} \begin{bmatrix} 1 \\ 2 \\ 1 \end{bmatrix} = \begin{bmatrix} 15 \\ 18 \\ 9 \end{bmatrix} \tag{3.132}$$

so that $\lambda = 9$ and therefore $(x', y') = \left(\frac{15}{9}, \frac{18}{9}\right) = \left(\frac{15}{9}, 2\right)$.

If $\lambda = 0$ then it is not possible to divide by it. Instead of this being an error, however, the resulting point is known as a **point at infinity**, which is closely related to the concept of a **vanishing point**. Points at infinity, and projective transformations in general, are covered in more detail later in the book.[†] Projective transformations subsume affine transformations, so they are capable of performing translation, rotation, non-uniform scaling, and shear. Projective transformations also have the ability to transform parallel lines into intersecting lines, such as occurs in a picture of railroad tracks captured from a low vantage point. Examples of various geometric transformations applied to an image are shown in Figure 3.43.

3.9.6 Arbitrary Warps

Projective (and therefore the special cases of affine, similarity, and Euclidean) warps can be captured succinctly in 3×3 matrix transformations using homogeneous coordinates. Such an approach can be used, for example, to warp the image to provide a bird's-eye view given a perspective image of a planar ground (e.g., road or railroad tracks). Similarly, such a transformation can be used to intentionally distort the image so that when the image is projected onto a flat surface at an angle, it appears undistorted (a procedure known

---

[†] Section 13.4 (p. 654).

Figure 3.43 Various geometric warpings applied to an image.

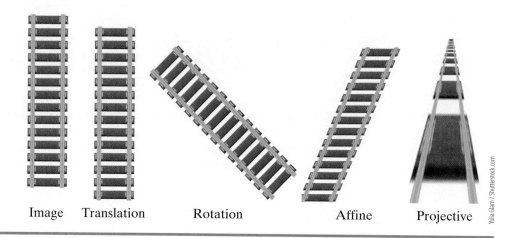

Image   Translation   Rotation   Affine   Projective

as **keystone correction**). Beyond projective warps, however, the mapping function f in Equation (3.103) can be any desired function, thus providing increased flexibility to allow an image that has been geometrically distorted in a nonlinear way (e.g., because of lens aberration or a catadioptric sensor) to be unwarped to provide a perspective view, or to define a spiral transform that swirls the image about its center to distort the image to create a visually pleasing effect, or to perform radial distortion to increase or decrease the number of pixels devoted to the center. The details of such warps are left as exercises to the reader.

3.9.7 Image Registration and Morphing

Finally, to close this chapter we mention that when two images are provided as input, it is sometimes desirable to align them so that they can be overlaid on top of one another. This alignment is known as **image registration**. Depending on the type of scene and the relationship between the images, the warp needed for image registration can vary from simple to complex. For example, if two images are taken of a planar scene from different locations, then they are related by a homography, which can be estimated by numerical techniques given correspondences between the two images. However, if an airplane flies over non-flat terrain to take pictures, the resulting images cannot be perfectly aligned using such a 3×3 matrix, because trees or buildings will cause 3D parallax effects, which in turn will cause misalignment between pixels on those surfaces when the images are registered based on the ground plane motion.

If two images are not only registered but also dissolved into each other, we say they are **morphed**. A common example is to morph one person's face into another by specifying a warping function that maps the corresponding features of one person's face (eyes, nose, and mouth corners) into the corresponding features of the other person's face, and to blend the pixel values over some number of image frames to provide the appearance of a continual blending from one image to the other when the sequence of frames is played back as a video. The warping of such nonlinear, complex surfaces as faces requires transformations that are local rather than global in nature.

3.10 Further Reading

Most of the material in this chapter, such as graylevel transformations, graylevel histograms, histogram equalization, and histogram matching, can be found in any image processing textbook, such as Burger and Burge [2008] or Gonzalez and Woods [2008]. Multispectral transformations such as NDVI are described in the remote sensing literature, such as Campbell and Wynne [2011] or Lillesand et al. 2007]. Another popular approach for dealing with multispectral data in remote sensing (but beyond our scope) is known as the Tasseled Cap Transformation, which is due to Kauth and Thomas [1976]. For a thought-provoking discussion regarding the use of pseudocoloring, consult the work of Borland and Taylor [2007].

Background subtraction can be traced to the early work of Jain and Nagel [1979], which also addressed the problem of updating the background image once objects begin moving. The double-difference image for frame differencing is originally due to Kameda and Minoh [1996]. A notable investigation of the topic of background subtraction was conducted by Toyama et al. [1999]. One of the most popular background subtraction algorithms is based on mixtures of Gaussians and is due to Stauffer and Grimson [2000], which was later updated by Zivkovic and van der Heijden [2006]. Additional approaches to background subtraction are that of Javed et al. [2002], which combines color and gradient information, and the texture-based approach of Heikkila and Pietikainen [2006].

Anyone interested in learning more about digital compositing is encouraged to pick up the delightful book by Brinkmann [2008], which explains techniques widely used in the entertainment industry, as well as the terms flipping and flopping as they are commonly used in the computer graphics community. The classic paper on compositing is that of Porter and Duff [1984], which remains a readable and self-contained introduction. Another short introduction to compositing is the paper by Thompson [1990].

Cubic convolution interpolation filters were developed by Keys [1981], who also showed that the Catmull-Rom spline (without mentioning it by name) minimizes the reconstruction error. This work was later extended by Mitchell and Netravali [1988], who demonstrate that a small amount of smoothing improves the subjective quality of the reconstructed image — the Mitchell filter is a result of this study. A comparison of various interpolation methods is found in Turkowski and Gabriel [1990]. Another excellent description of 1D and 2D geometric transformations, warps, and interpolation methods can be found in the Burger and Burge book [2008]. For more recent work on sophisticated upsampling algorithms, consult Shan et al. [2008], HaCohen et al. [2010], or Kopf et al. [2007].

PROBLEMS

3.1. Given the following 4×4 grayscale image, perform the following operations: flip, flop, flip-flop, invert, and rotate clockwise by 90 degrees.

$$\begin{bmatrix} 176 & 94 & 201 & 219 \\ 37 & 161 & 16 & 88 \\ 71 & 129 & 177 & 81 \\ 41 & 198 & 107 & 19 \end{bmatrix}$$

3.2. Based on Algorithms 3.1 and 3.2, write pseudocode to flip-flop an image.

3.3. Modify Algorithm 3.1 to perform the flipping in place, so that the output and input images occupy the same memory.

3.4. Modify the pseudocode of Algorithm 3.3 to rotate the image in place, using only enough additional storage to hold one pixel value at a time. (Only implement rotate clockwise by 90 degrees.)

3.5. Use 8-bit saturation arithmetic to compute the following: (a) $52 + 200$, (b) $86 + 199$, (c) $30 - 50$, and (d) $32 + 11$. Then repeat with 4-bit saturation arithmetic.

3.6. Using the 8-bit image in Problem 3.1, perform the following arithmetic operations, clamping where necessary: add 65, subtract 85, multiply by 2.

3.7. Compute the (a) sum, (b) difference, and (c) absolute difference of the following two 8-bit images, using saturation arithmetic to store the result in another 8-bit image:

$$I_1 = \begin{bmatrix} 19 & 171 & 91 & 68 \\ 123 & 99 & 74 & 195 \\ 85 & 71 & 208 & 18 \\ 241 & 212 & 189 & 68 \end{bmatrix} \qquad I_2 = \begin{bmatrix} 106 & 97 & 190 & 5 \\ 81 & 64 & 183 & 82 \\ 71 & 200 & 251 & 94 \\ 181 & 76 & 9 & 18 \end{bmatrix}$$

3.8. Suppose you want to display the following floating-point image. Perform a linear contrast stretch to convert the pixels to 8 bits, mapping the smallest value to 0 and the largest value to 255:

$$\begin{bmatrix} 0.327 & 0.945 & 0.559 & 0.381 \\ 0.181 & 0.252 & 0.080 & 0.950 \\ 0.240 & 0.399 & 0.737 & 0.148 \\ 0.986 & 0.170 & 0.246 & 0.447 \end{bmatrix}$$

3.9. Determine the lookup table (LUT) for the piecewise linear graylevel mapping given in Figure 3.12. (To avoid excessive writing, provide entries only for the gray levels divisible by 25.)

3.10. Generate the LUT for the operation $I'(x, y) = \eta \cdot (\frac{1}{\eta} I(x, y))^{2.0}$, assuming I and I' are 4-bit grayscale images. Set η to ensure that the maximum value of I maps to the maximum value of I'.

3.11. Threshold the image in Problem 3.1 with $\tau = 155$.

3.12. Compute bit plane 7 and bit plane 4 for the image in Problem 3.1.

3.13. Compute the histogram, normalized histogram, and cumulative normalized histogram for the following 4-bit image:

$$\begin{bmatrix} 5 & 8 & 3 & 7 \\ 1 & 3 & 3 & 9 \\ 6 & 8 & 2 & 7 \\ 4 & 1 & 0 & 9 \end{bmatrix}$$

3.14. Implement histogram equalization in your favorite programming language. Run your code on the image in Problem 3.13, as well as on a low-contrast image of your own.

3.15. Implement histogram matching, and run your code on the image in Problem 3.13, using the reference histogram $h_{ref} = \begin{bmatrix} 1 & 1 & 1 & 0 & 2 & 2 & 2 & 0 & 4 & 3 \end{bmatrix}$. Check by hand to verify that the output is correct.

3.16. If histogram equalization is applied twice to an image, that is, if it is applied to the result of histogram equalization, will the second application change the image or not? Explain why or why not.

3.17. Compute the grayscale equivalent for each of the following RGB triplets using both Equations (3.36) and (3.37), rounding to the nearest whole number: (a) (128,128,128), (b) (64,245,198), and (c) (255,253,128). Also, (d) for each color, compute the difference in grayscale between the two conversions.

3.18. Implement the RGB-to-grayscale conversion approaches of Equations (3.36) and (3.37), and run on a few sample images. Also run the RGB-to-grayscale conversion of some existing software application. Can you tell a difference in quality between the various outputs?

3.19. An alternative to the NDVI of Equation (3.45) is the Infrared Percentage Vegetation Index (IPVI), given by

$$I'_{IPVI}(x, y) = \frac{I_{IR}(x, y)}{I_{IR}(x, y) + I_{Red}(x, y)} \tag{3.133}$$

Show that IPVI is functionally equivalent to NDVI, in that it is simply a linear transformation. Compute the transformation, and state the range of IPVI. What are the thresholds for detecting live green vegetation in IPVI?

3.20. What temperature (in degrees Celsius) is represented by a pixel in a GOES infrared image with a value of 200?

3.21. Suppose you have an application for which you want to apply background subtraction. List some difficulties that might prevent the output from being perfect.

3.22. Compute the (a) double difference and (b) triple difference between the following 3 consecutive frames, using the threshold $\tau = 40$:

$$I_1 = \begin{bmatrix} 18 & 168 & 94 & 67 \\ 120 & 97 & 78 & 198 \\ 83 & 70 & 208 & 17 \\ 238 & 208 & 189 & 68 \end{bmatrix} \quad I_2 = \begin{bmatrix} 21 & 168 & 92 & 71 \\ 122 & 71 & 191 & 227 \\ 83 & 212 & 16 & 187 \\ 240 & 216 & 188 & 68 \end{bmatrix} \quad I_3 = \begin{bmatrix} 20 & 171 & 92 & 70 \\ 76 & 193 & 39 & 228 \\ 209 & 20 & 20 & 194 \\ 241 & 210 & 190 & 73 \end{bmatrix}$$

3.23. Explain the concept of dissolving and how it is used in the movie/film industry for digital compositing.

3.24. In addition to the 12 binary Porter-Duff operators (and their alpha-channel equivalents), other compositing operators are possible. One of these is SCREEN, defined as

$$I_A \text{ SCREEN } I_B = 1 - (1 - I_A)(1 - I_B)$$

where the pixel values are assumed to range between 0 and 1 to simplify the equation.

(a) What does the screen operator do? (*Hint*: Assume I_A and I_B are the foreground and background, respectively, and examine what happens when the images are at their minimal or maximal values.)

(b) Implement the screen operator and test it on two images, each with constant alpha channel of 0.5 for every pixel.

3.25. Another pair of compositing operators is dodging and burning. *Dodging* brightens certain pixels in an image, while *burning* darkens the pixels; in both cases the pixels are specified by a second binary image. One way to implement these operators is to add or subtract a constant, say 128, to every pixel in the image where the binary image is ON, using saturation arithmetic. Write the pseudocode for these two operators.

3.26. Suppose your computer has saturation arithmetic built-in, that is, $a + b = \min(a + b, 255)$ for byte operations, and so forth. (Such logic is common for CPUs with SIMD instructions such as MMX/SSE.) How would you modify your pseudocode in the previous problem?

3.27. The 12 Porter-Duff operators are as follows:

(a)	CLEAR	(e)	I_B OVER I_A	(i)	I_B OUT I_A
(b)	COPY I_A	(f)	I_A IN I_B	(j)	I_A ATOP I_B
(c)	COPY I_B	(g)	I_B IN I_A	(k)	I_B ATOP I_A
(d)	I_A OVER I_B	(h)	I_A OUT I_B	(l)	I_A XOR I_B

Apply these operators to the following two images with masks:

$$I_A = \begin{bmatrix} 132 & 231 & 227 \\ 237 & 105 & 238 \\ 193 & 59 & 128 \end{bmatrix} \quad M_A = \begin{bmatrix} 255 & 255 & 0 \\ 255 & 255 & 0 \\ 0 & 0 & 0 \end{bmatrix} \quad I_B = \begin{bmatrix} 43 & 79 & 116 \\ 56 & 246 & 184 \\ 36 & 119 & 162 \end{bmatrix} \quad M_B = \begin{bmatrix} 0 & 0 & 0 \\ 0 & 255 & 255 \\ 0 & 255 & 255 \end{bmatrix}$$

Show only the values of the valid pixels; indicate invalid pixels using an X.

3.28. One of the 12 Porter-Duff operators is equivalent to $(I_A$ IN $I_B)$ OVER I_B. Which one is it?

3.29. Using the same values for I_A and I_B as in Problem 3.27, but with the following opacity values, compute (a) I_A OVER I_B, (b) I_A IN I_B, and (c) I_A ATOP I_B.

$$\alpha_A = \begin{bmatrix} 1.0 & 0.5 & 0 \\ 0.5 & 0.4 & 0 \\ 0 & 0 & 0 \end{bmatrix} \quad \alpha_B = \begin{bmatrix} 0 & 0 & 0 \\ 0 & 0.2 & 0.6 \\ 0 & 0.6 & 1.0 \end{bmatrix}$$

3.30. The four operators in Figure 3.32 are not commutative. Briefly describe what the result of each operator would be if the order of the operands were reversed.

3.31. Given the following image, use bilinear interpolation to compute the value at (a) (0.1, 0.7), (b) (1.2, 0.5), (c) (1.3, 1.6), and (d) (2.8,1.7).

$$\begin{bmatrix} 232 & 177 & 82 & 7 \\ 241 & 18 & 152 & 140 \\ 156 & 221 & 67 & 3 \end{bmatrix}$$

3.32. Show that Equation (3.71) passes through the values f_0 and f_1 and maintains the derivatives $\dot{f}_0$ and $\dot{f}_1$.

3.33. Derive the 16×16 bicubic interpolation matrix in Equation (3.84), using the central difference operator for derivatives.

3.34. The general form of a 1D symmetric cubic filter is given by

$$k(x) = \begin{cases} a_3|x|^3 + a_2|x|^2 + a_1|x| + a_0 & \text{if } |x| < 1 \\ a_3'|x|^3 + a_2'|x|^2 + a_1'|x| + a_0' & \text{if } 1 \le |x| < 2 \\ 0 & \text{otherwise} \end{cases} \quad (3.171)$$

Show that if the value and derivatives are enforced to be continuous everywhere, and the coefficients sum to 1, then the 8 parameters are reduced to 2, leading to Equation (3.86).

3.35. Why is the Mitchell filter generally preferred over Catmull-Rom?

3.36. It can be shown that the Mitchell filter in Equation (3.95) is a linear combination of Catmull-Rom in Equation (3.76) and uniform cubic B-spline in Equation (3.88) Find the weighting coefficients.

3.37. List a popular filter for downsampling, and another for upsampling.

3.38. Find two images of a fairly static scene taken by a camera that panned and zoomed between the images. Manually click on two corresponding points on both images to compute the scale and rotation between them, then use one of these corresponding points to compute the translation. Construct a similarity transform, and apply the transform to one of the images to bring the two images into approximate alignment.

3.39. Set up a stationary camera and capture a short video sequence of some moving foreground objects while the background remains stationary. Compute the mean image of the sequence to generate a background image, then use background subtraction to segment the foreground objects. Display the foreground objects that result from thresholding the absolute difference between the current image(s) and the background image.

3.40. Implement the popular background subtraction approach that uses mixtures of Gaussians to model the background pixel colors, explained in Stauffer and Grimson [2000].

CHAPTER 4
Binary Image Processing

In the previous chapter we examined simple operations to transform one image into another. In this chapter we continue the investigation by focusing on the special case of binary images. Our goal in this chapter is to cover well-known algorithms that are widely used in practice, while at the same time laying a foundation for later chapters. Binary images provide a natural excuse to introduce mathematical morphology, which is an approach to image processing that views a binary image as a subset of the plane, as well as other mathematical concepts such as the eigendecomposition of a matrix, which is useful for determining the orientation of a set of points in the plane. Unlike the previous chapter, the operations in this chapter are not restricted to those in which the output pixel is determined by a single input pixel.

4.1 Morphological Operations

Recall that thresholding a grayscale image, or thresholding the result of an operation like background subtraction, produces a binary image in which foreground pixels have the value of 1, while background pixels have the value of 0. Logically, these values can be considered as ON or OFF, respectively. Figure 4.1 shows an example grayscale image of objects on a conveyor belt, along with a thresholded version. Such images are common in manufacturing environments, where machine vision techniques play an important role in ensuring that the parts being manufactured are without defects, are sorted into

Figure 4.1 Left: A grayscale image of several types of fruit on a dark conveyor belt. Right: A binary image resulting from thresholding. The white pixels are ON and indicate the foreground, while the black pixels are OFF and indicate the background.

Stan Birchfield

the proper bins, and so on. For such applications, it is often important to isolate the foreground objects from the background, as well as to compute properties of the objects for the purpose of classifying and manipulating them. Unfortunately, as seen from the figure, thresholding does not produce a perfect separation between the foreground and background. Although we will consider algorithms for automatically computing a threshold value later,[†] on a difficult image like this one, the result will be noisy no matter what threshold value is chosen.

Morphological operations provide a powerful way to clean up such noise. The word *morphology* means "the study of shape," and **mathematical morphology** is an entire branch of mathematics that has been developed to process images by considering the shape (or form) of the pixel regions. Mathematical morphology models binary images as sets by considering foreground (ON) pixels as subsets of the image plane. In the next chapter, we shall extend this concept to grayscale images, but for this chapter we focus on binary mathematical morphology.

4.1.1 Binary Image as a Set

A binary image is generally thought of as an array of values such that $I(x,y)$ returns 1 or 0 for each pixel location (x,y). Alternatively, as explained earlier,[‡] a binary image can be represented as the set of coordinates of all the foreground (1-valued, or ON) pixels. These two representations are equivalent, and we can freely convert between them as desired.

EXAMPLE 4.1

Write the set representation of the following binary image:

$$I = \begin{bmatrix} 1 & 0 & 1 \\ 0 & 1 & 1 \\ 0 & 0 & 0 \end{bmatrix}$$

Solution

The set representation contains the coordinates of all the pixels with a value of 1. Using our coordinate system convention in which the x axis points to the right, the y axis points down, and the origin is the top-left pixel, this yields $\{(0, 0), (1, 1), (2, 0), (2, 1)\}$.

[†] Chapter 10 (p. 444).
[‡] Section 1.4.3 (p. 13).

EXAMPLE 4.2

A 3×3 binary image is represented by $\{(0, 1), (0, 2), (1, 2), (2, 2), (2, 1)\}$. What is the array representation?

Solution

The array representation is obtained by placing a 1 at every location contained in the set, and 0 everywhere else:

$$I = \begin{bmatrix} 0 & 0 & 0 \\ 1 & 0 & 1 \\ 1 & 1 & 1 \end{bmatrix}$$

The two fundamental set operators are *union* and *intersection*. If A and B are sets of points in the plane, then the union of A and B is the set containing all points that are in either A or B (or both), while the intersection of A and B is the set containing all points that are in both A and B:

$$A \cup B \equiv \{\mathbf{z} : \mathbf{z} \in A \text{ or } \mathbf{z} \in B\} \qquad \text{(union)} \qquad (4.1)$$

$$A \cap B \equiv \{\mathbf{z} : \mathbf{z} \in A \text{ and } \mathbf{z} \in B\} \qquad \text{(intersection)} \qquad (4.2)$$

where $\mathbf{z} \in \mathbb{R}^2$ is a point in the plane. Additional operators are the *translation* of the set A by the vector $\mathbf{b}$, denoted $A_{\mathbf{b}}$; the *reflection* of the set B about its origin (the flip-flop operator we saw in the previous chapter), denoted $\check{B}$; the logical *complement* $\neg A$ containing all the points not in A, which is equivalent to replacing each 1 in the array with 0 and each 0 with 1; and the *difference* $A \setminus B$ between two sets containing all the points in A that are not in B. If we let $\mathbf{a} = (a_x, a_y) \in A$ be a point in the first set, and let $\mathbf{b} = (b_x, b_y) \in B$ be a point in the second set, then these operators are summarized as follows:

$$A_{\mathbf{b}} \equiv \{\mathbf{z} : \mathbf{z} = \mathbf{a} + \mathbf{b}, \mathbf{a} \in A\} \qquad \text{(translation)} \qquad (4.3)$$

$$\check{B} \equiv \{\mathbf{z} : \mathbf{z} = -\mathbf{b}, \mathbf{b} \in B\} \qquad \text{(reflection)} \qquad (4.4)$$

$$\neg A \equiv \{\mathbf{z} : \mathbf{z} \notin A\} \qquad \text{(complement)} \qquad (4.5)$$

$$A \setminus B \equiv \{\mathbf{z} : \mathbf{z} \in A, \mathbf{z} \notin B\} = \mathbf{A} \cap \neg \mathcal{B} \qquad \text{(difference)} \qquad (4.6)$$

where $-\mathbf{b} = (-b_x, -b_y)$ indicates that the sign of both the x and y coordinates have changed. The application of these operators is shown in Figure 4.2, from which it is easy to see that $\mathbf{z} \in A \cap B$ implies $\mathbf{z} \in A \cup B$, since any point in the intersection must also be in the union. As a result, the intersection is a subset of the union: $A \cap B \subseteq A \cup B$.

In addition, it is obvious that any point not in the union of A and B is also not in A and not in B; similarly, any point not in the intersection of A and B is either not in A or not in B. These are known as **De Morgan's laws**:

$$\neg(A \cup B) = \neg A \cap \neg B \qquad (4.7)$$

$$\neg(A \cap B) = \neg A \cup \neg B \qquad (4.8)$$

Figure 4.2 Set operators. The first two columns show two sets A and B in blue. Then, from left to right, shown are the union, intersection, shift of A by some amount $\mathbf{b}$ (not related to B), reflection about the origin (assumed to be at the center), complement, and set difference.

$$\mathcal{A} \qquad \mathcal{B} \qquad A \cup B \qquad A \cap B \qquad A_{\mathbf{b}} \qquad \check{B} \qquad \neg A \qquad A \setminus B$$

4.1.2 Minkowski Addition and Subtraction

The coordinates of two points can be added easily using vector addition, $\mathbf{a} + \mathbf{b} = (a_x + b_x, a_y + b_y)$, which leads naturally to a definition for the sum of two sets. The **Minkowski addition**[†] of two sets $\mathcal{A}$ and $\mathcal{B}$ is defined as the set of points resulting from all possible vector additions of elements of the two sets:

$$\mathcal{A} \oplus \mathcal{B} \equiv \{\mathbf{z} : \mathbf{z} = \mathbf{a} + \mathbf{b}, \mathbf{a} \in \mathcal{A}, \mathbf{b} \in \mathcal{B}\} \tag{4.9}$$

$$= \bigcup_{\mathbf{b} \in \mathcal{B}} \{\mathbf{a} + \mathbf{b} : \mathbf{a} \in \mathcal{A}\} = \bigcup_{\mathbf{b} \in \mathcal{B}} \mathcal{A}_{\mathbf{b}} \tag{4.10}$$

In other words, Equation (4.9) says that $\mathcal{A} \oplus \mathcal{B}$ is the set $\{\mathbf{z}\}$ such that for each point $\mathbf{z}$ in the set there is some $\mathbf{a}$ in $\mathcal{A}$ and some $\mathbf{b}$ in $\mathcal{B}$ whose sum is $\mathbf{z}$. Equivalently, Equation (4.10) says that $\mathcal{A} \oplus \mathcal{B}$ is the union of the sets resulting from translating $\mathcal{A}$ by each element of $\mathcal{B}$.

The **Minkowski subtraction** of two sets is defined in an analogous manner:

$$\mathcal{A} \ominus \mathcal{B} \equiv \{\mathbf{z} : \mathbf{z} - \mathbf{b} \in \mathcal{A}, \forall \mathbf{b} \in \mathcal{B}\} \tag{4.11}$$

$$= \bigcap_{\mathbf{b} \in \mathcal{B}} \{\mathbf{a} + \mathbf{b} : \mathbf{a} \in \mathcal{A}\} = \bigcap_{\mathbf{b} \in \mathcal{B}} \mathcal{A}_{\mathbf{b}} \tag{4.12}$$

where the symbol $\forall$ means "for all." That is, Equation (4.11) says that $\mathcal{A} \ominus \mathcal{B}$ is the set $\{\mathbf{z}\}$ such that for each point $\mathbf{z}$ in the set and for all $\mathbf{b}$ in $\mathcal{B}$, the point $\mathbf{z} - \mathbf{b}$ is in $\mathcal{A}$. Equivalently, Equation (4.12) says that it is the intersection of the sets resulting from translating $\mathcal{A}$ by each element of $\mathcal{B}$.

These two operations are best understood by example. Although Minkowski addition and subtraction are defined for arbitrary point sets in the plane (and can be extended beyond the plane, as we shall see in the next chapter), our goal is to apply these concepts to images, that is, to discrete point sets in a square lattice in which the point coordinates are integers. Our example, therefore, highlights this case.

EXAMPLE 4.3

Compute the Minkowski addition of the two discrete sets $\mathcal{A}_1$ and $\mathcal{B}$ shown in Figure 4.3, and the Minkowski subtraction of the two discrete sets $\mathcal{A}_2$ and $\mathcal{B}$ shown in the same figure.

Solution

The first set contains just two points: $\mathcal{A}_1 = \{(0, 0), (1, 0)\}$. The second set contains four points: $\mathcal{B} = \{(0, 0), (1, 0), (0, 1), (0, 2)\}$. Minkowski addition is computed by adding each element of $\mathcal{B}$ to each element of $\mathcal{A}_1$:

$$
\begin{array}{ccccc}
\overbrace{}^{\in \mathcal{A}_1} & & \overbrace{}^{\in \mathcal{B}} & & \overbrace{}^{\in \mathcal{A}_1 \oplus \mathcal{B}} \\
(0, 0) & + & (0, 0) & = & (0, 0) \\
(0, 0) & + & (1, 0) & = & (1, 0) \\
(0, 0) & + & (0, 1) & = & (0, 1) \\
(0, 0) & + & (0, 2) & = & (0, 2) \\
(1, 0) & + & (0, 0) & = & (1, 0) \\
(1, 0) & + & (1, 0) & = & (2, 0) \\
(1, 0) & + & (0, 1) & = & (1, 1) \\
(1, 0) & + & (0, 2) & = & (1, 2)
\end{array}
$$

[†] The mathematician Hermann Minkowski (1864–1909) was a teacher of Albert Einstein.

Since one point is repeated, the result is the set containing the 7 unique points: $\mathcal{A}_1 \oplus \mathcal{B} = \{(0,0), (1,0), (0,1), (0,2), (2,0), (1,1), (1,2)\}$. The figure shows that this set of points is the union of all the points in $\mathcal{A}_1$ when $\mathcal{A}_1$ is translated to every point in $\mathcal{B}$.

Minkowski subtraction is less intuitive. Nevertheless, it is easy to show that $(0,0)$ is in $\mathcal{A}_2 \ominus \mathcal{B}$ because, for every element of $\mathcal{B}$, their difference is in $\mathcal{A}_2$. On the other hand, the point $(1,0)$ is not in $\mathcal{A}_2 \ominus \mathcal{B}$, because for the point $\mathbf{b} = (0,2) \in \mathcal{B}$, the difference is not in $\mathcal{A}_2$:

$$
\begin{array}{ll}
\overbrace{}^{\in \mathcal{B}} & \overbrace{}^{\in \mathcal{B}} \\
(0,0) - (0,0) = (0,0) \in \mathcal{A}_2 & (1,0) - (0,0) = (1,0) \in \mathcal{A}_2 \\
(0,0) - (1,0) = (-1,0) \in \mathcal{A}_2 & (1,0) - (1,0) = (0,0) \in \mathcal{A}_2 \\
(0,0) - (0,1) = (0,-1) \in \mathcal{A}_2 & (1,0) - (0,1) = (1,-1) \in \mathcal{A}_2 \\
\underbrace{(0,0) - (0,2) = (0,-2) \in \mathcal{A}_2}_{\in \mathcal{A}_2 \ominus \mathcal{B}} & \underbrace{(1,0) - (0,2) = (1,-2) \notin \mathcal{A}_2}_{\notin \mathcal{A}_2 \ominus \mathcal{B}}
\end{array}
\tag{4.13}
$$

The figure shows that the set $\mathcal{A}_2 \ominus \mathcal{B}$ is the intersection of the sets of points resulting from translating $\mathcal{A}_2$ to every point in $\mathcal{B}$.

Properties

The Minkowski operators have a number of interesting properties. From the previous example, it is clear that Minkowski addition grows the foreground $\mathcal{A}_1$, whereas Minkowski subtraction shrinks the foreground $\mathcal{A}_2$. Formally, as long as the second set $\mathcal{B}$ contains the origin (which is nearly always the case in practice), then Minkowski addition is **extensive**, which means its output is a superset of the input, while Minkowski subtraction is **anti-extensive**, which means its output is a subset of the input. Similarly, if $\mathcal{A}_1$ is a subset of $\mathcal{A}_2$, then the result of either operator on $\mathcal{A}_1$ will be a subset of the result of the same operator on $\mathcal{A}_2$, a property known as **increasing**. Other properties are fairly easy to show, which are listed here for reference:

$$\mathcal{A} \oplus \mathcal{B} = \mathcal{B} \oplus \mathcal{A} \qquad \text{(commutativity)} \tag{4.14}$$

$$\mathcal{A} \ominus \mathcal{B} = \neg\mathcal{B} \ominus \neg\mathcal{A} \qquad \text{(non-commutativity)} \tag{4.15}$$

$$(\mathcal{A} \oplus \mathcal{B}) \oplus \mathcal{C} = \mathcal{A} \oplus (\mathcal{B} \oplus \mathcal{C}) \qquad \text{(associativity, separability)} \tag{4.16}$$

$$(\mathcal{A} \ominus \mathcal{B}) \ominus \mathcal{C} = \mathcal{A} \ominus (\mathcal{B} \oplus \mathcal{C}) \qquad \text{(separability)} \tag{4.17}$$

$$(\mathcal{A} \oplus \mathcal{B}) \oplus \mathcal{C} = (\mathcal{A} \oplus \mathcal{C}) \oplus \mathcal{B} \qquad \text{(order does not matter)} \tag{4.18}$$

$$(\mathcal{A} \ominus \mathcal{B}) \ominus \mathcal{C} = (\mathcal{A} \ominus \mathcal{C}) \ominus \mathcal{B} \qquad \text{(order does not matter)} \tag{4.19}$$

$$\mathcal{A} \oplus \mathcal{B} \subseteq \mathcal{A} \ \text{ if}(0,0) \in \mathcal{B} \qquad \text{(extensivity)} \tag{4.20}$$

$$\mathcal{A} \ominus \mathcal{B} \subseteq \mathcal{A} \ \text{ if}(0,0) \in \mathcal{B} \qquad \text{(anti-extensivity)} \tag{4.21}$$

$$\mathcal{A}_1 \oplus \mathcal{B} \subseteq \mathcal{A}_2 \oplus \mathcal{B} \ \text{ if } \mathcal{A}_1 \subseteq \mathcal{A}_2 \qquad \text{(increasing)} \tag{4.22}$$

$$\mathcal{A}_1 \ominus \mathcal{B} \subseteq \mathcal{A}_2 \ominus \mathcal{B} \ \text{ if } \mathcal{A}_1 \subseteq \mathcal{A}_2 \qquad \text{(increasing)} \tag{4.23}$$

$$\mathcal{A} \oplus (\mathcal{B} \cup \mathcal{C}) = (\mathcal{A} \oplus \mathcal{B}) \cup (\mathcal{A} \oplus \mathcal{C}) \qquad \text{(parallelism)} \tag{4.24}$$

$$\mathcal{A} \ominus (\mathcal{B} \cup \mathcal{C}) = (\mathcal{A} \ominus \mathcal{B}) \cap (\mathcal{A} \ominus \mathcal{C}) \qquad \text{(parallelism)} \tag{4.25}$$

$$(\mathcal{A} \cap \mathcal{B}) \ominus \mathcal{C} = (\mathcal{A} \ominus \mathcal{C}) \cap (\mathcal{B} \ominus \mathcal{C}) \tag{4.26}$$

$$(\mathcal{A} \cup \mathcal{B}) \oplus \mathcal{C} = (\mathcal{A} \oplus \mathcal{C}) \cup (\mathcal{B} \oplus \mathcal{C}) \tag{4.27}$$

$$\neg(\mathcal{A} \oplus \mathcal{B}) = \neg\mathcal{A} \ominus \mathcal{B} \qquad \text{(duality)} \tag{4.28}$$

$$\neg(\mathcal{A} \ominus \mathcal{B}) = \neg\mathcal{A} \oplus \mathcal{B} \qquad \text{(duality)} \tag{4.29}$$

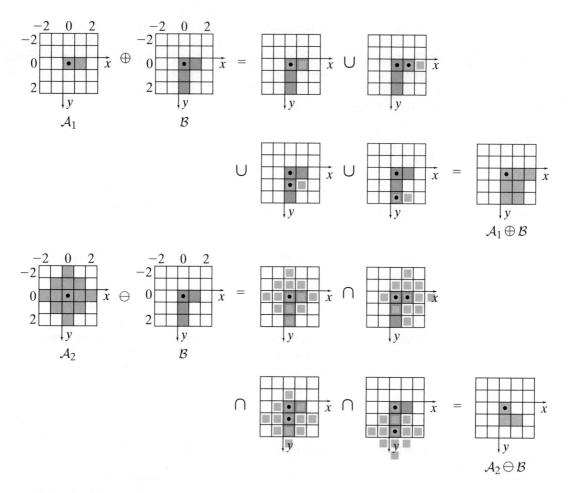

Figure 4.3 Minkowski addition and subtraction, from Equations (4.10) and (4.12) , respectively. For both operations, the first set ($\mathcal{A}_1$ or $\mathcal{A}_2$) is translated so that its origin is placed at each element of the second set ($\mathcal{B}$). The union of the blue cells then yields $\mathcal{A}_1 \oplus \mathcal{B}$, while their intersection yields $\mathcal{A}_2 \ominus \mathcal{B}$. In other words, each ON (blue) cell in $\mathcal{A}_1 \oplus \mathcal{B}$ is ON (blue) in *at least one* of the intermediate results, while each ON (blue) cell in $\mathcal{A}_2 \ominus \mathcal{B}$ is ON (blue) in *all* of the intermediate results. Colored cells are ON (value 1); white cells are OFF (value 0); and all pixels outside the 5 $\times$ 5 image are assumed to be OFF. The small black dots indicate the origins of the coordinate systems. (As explained later, this is the "center-in" approach.)

One particularly interesting property is that of **duality**. Two operators Ψ and Ψ' are said to be duals of each other with respect to complementation if $\neg\Psi(\mathcal{A}) = \Psi'(\neg\mathcal{A})$. It is not hard to see that Minkowski addition and subtraction are *duals* of each other with respect to complementation, because the former grows the foreground $\mathcal{A}$ (or, equivalently, shrinks the background $\neg\mathcal{A}$), while the latter shrinks the foreground $\mathcal{A}$ (or, equivalently, grows the background $\neg\mathcal{A}$). In other words, for a given $\mathcal{B}$ the same result is achieved if we complement the output or if we complement the input and change the operator. Duality is a powerful tool for proving properties.

Swapping the Order of the Operands

Due to the commutative property, it is not surprising that Minkowski addition can also be computed by leaving $\mathcal{A}$ stationary and instead shifting $\mathcal{B}$, whereas in the above example we left $\mathcal{B}$ stationary and shifted $\mathcal{A}$. Similarly, by combining the commutativity and duality properties, it is easy to show that Minkowski subtraction can be computed by leaving $\neg\mathcal{A}$ stationary and instead shifting $\neg\mathcal{B}$:

$$\mathcal{A} \oplus \mathcal{B} = \mathcal{B} \oplus \mathcal{A} \tag{4.30}$$

$$= \{\mathbf{z} : \mathbf{z} = \mathbf{a} + \mathbf{b}, \mathbf{a} \in \mathcal{A}, \mathbf{b} \in \mathcal{B}\} \tag{4.31}$$

$$= \bigcup_{\mathbf{a} \in \mathcal{A}} \{\mathbf{a} + \mathbf{b} : \mathbf{b} \in \mathcal{B}\} = \bigcup_{\mathbf{a} \in \mathcal{A}} \mathcal{B}_{\mathbf{a}} \tag{4.32}$$

$$\mathcal{A} \ominus \mathcal{B} = \neg\mathcal{B} \ominus \neg\mathcal{A} \tag{4.33}$$

$$= \{\mathbf{z} : \mathbf{z} - \mathbf{a} \notin \mathcal{B}, \forall \mathbf{a} \notin \mathcal{A}\} \tag{4.34}$$

$$= \bigcap_{\mathbf{a} \notin \mathcal{A}} \{\mathbf{a} + \mathbf{b} : \mathbf{b} \notin \mathcal{B}\} = \bigcap_{\mathbf{a} \notin \mathcal{A}} \neg\mathcal{B}_{\mathbf{a}} \tag{4.35}$$

Computations involving these swapped operands are illustrated in Figure 4.4.

Center-In Versus Center-Out

All the formulations of Minkowski addition and subtraction that we have seen so far are examples of what we call the "center-in" approach, in which the center of one set is placed only at certain locations, and the output is determined by examining the entire set (applying

Figure 4.4 Alternate view of Minkowski addition and subtraction from Equations (4.32) and (4.35), which are obtained by swapping the order of the operands using the properties $\mathcal{A}_1 \oplus \mathcal{B} = \mathcal{B} \oplus \mathcal{A}_1$ and $\mathcal{A}_2 \ominus \mathcal{B} = \neg\mathcal{B} \ominus \neg\mathcal{A}_2$, respectively. For addition, the second set ($\mathcal{B}$) is translated so that its origin remains within the first set ($\mathcal{A}_1$), and the result is the union of all the ON (orange) cells. For subtraction, the second set ($\mathcal{B}$) is translated so that its origin remains outside the first set ($\mathcal{A}_2$), and the result is the intersection of all the OFF (non-orange) cells. In other words, each ON (blue) cell in $\mathcal{A}_1 \oplus \mathcal{B}$ is ON (orange) in *at least one* of the intermediate results, while each ON (blue) cell in $\mathcal{A}_2 \ominus \mathcal{B}$ is OFF (non-orange) in *all* of the intermediate results. Equivalently, the result of subtraction is ON everywhere except where the pixel is ON (orange) in at least one of the intermediate results. (This is also the "center-in" approach.)

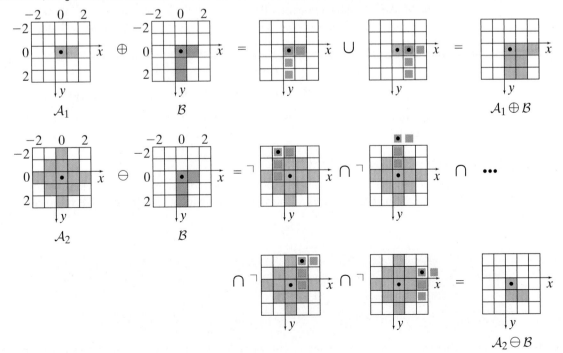

$\mathcal{A}_2 \ominus \mathcal{B}$

either union or intersection). In contrast, the "center-out" approach performs a test at each location on all the elements of the set, then updates the value of the center element. In other words, the former approach treats the center element (i.e., the element at the origin) as an input to decide whether to even look at the other elements, whereas the latter approach treats the center element as the output that aggregates information from the other elements. All formulations of Minkowski addition or subtraction can be classified as either center-in or center-out.

The center-out approach to Minkowski addition finds the set of points $\mathbf{z}$ such that *at least one* point in the reflected $\check{\mathcal{B}}$, when centered at $\mathbf{z}$, intersects $\mathcal{A}$. Similarly, Minkowski subtraction is the set of points $\mathbf{z}$ such that *all* the points in the reflected $\check{\mathcal{B}}$, when centered at $\mathbf{z}$, intersect $\mathcal{A}$:

$$A \oplus B = \{\mathbf{z} : \check{\mathcal{B}}_{\mathbf{z}} \cap \mathcal{A} \neq \emptyset\} \tag{4.36}$$

$$A \ominus B = \{\mathbf{z} : \check{\mathcal{B}}_{\mathbf{z}} \subseteq \mathcal{A}\} \tag{4.37}$$

The two approaches are summarized in Table 4.1. Despite the (perhaps surprising) need for a reflection, the center-out approach is more intuitive and easier to visualize than the center-in approach, because the set $\check{\mathcal{B}}$ is translated by every possible $\mathbf{z}$, and a test is performed on the entire set. This is illustrated in Figure 4.5, where the equivalence of the center-in and center-out approaches is obvious by comparing with Figures 4.3 and 4.4.

4.1.3 Dilation and Erosion

Minkowski addition and subtraction lead naturally to the two fundamental morphological operators called **dilation** and **erosion**. As shown in previous examples, Minkowski addition grows (or "dilates") the region by increasing its size, whereas Minkowski subtraction shrinks (or "erodes") the region by decreasing its size. Examining Figure 4.5, however, reveals a problem: Minkowski subtraction $\mathcal{A} \ominus \mathcal{B}$ yields the set of locations at which the *reflected* set $\check{\mathcal{B}}$ fits entirely within $\mathcal{A}$, whereas it is more natural to want the set of locations at which the *original* set $\mathcal{B}$ fits entirely within $\mathcal{A}$; as a result, the erosion operator $\check{\ominus}$ is defined to be Minkowski subtraction after first reflecting the second operand. Minkowski addition, on the other hand, exhibits no such reflection in its output, and therefore the dilation operator $\oplus$ is identical to Minkowski addition:

$$A \oplus B \equiv A \oplus B = \{\mathbf{z} : \mathbf{z} = \mathbf{a} + \mathbf{b}, \mathbf{a} \in \mathcal{A}, \mathbf{b} \in \mathcal{B}\} \quad \text{(dilation)} \tag{4.38}$$

$$A \check{\ominus} B \equiv A \ominus \check{\mathcal{B}} = \{\mathbf{z} : \mathbf{z} + \mathbf{b} \in \mathcal{A}, \forall \mathbf{b} \in \mathcal{B}\} \quad \text{(erosion)} \tag{4.39}$$

where the right-most equations are copied from Equations (4.9) and (4.11), changing the sign in the latter.

Structuring Elements

Up to now the two operands have been treated more or less equally. Typically, however, the first set, which we call I, is a binary image containing tens of thousands of pixels, whereas

	$\mathcal{A} \oplus \mathcal{B}$	$\mathcal{A} \ominus \mathcal{B}$
center-in	$\{\mathbf{z} : \mathbf{z} = \mathbf{a} + \mathbf{b}, \mathbf{a} \in \mathcal{A}, \mathbf{b} \in \mathcal{B}\}$ $= \bigcup_{\mathbf{b} \in \mathcal{B}} \mathcal{A}_{\mathbf{b}} = \bigcup_{\mathbf{a} \in \mathcal{A}} \mathcal{B}_{\mathbf{a}}$	$\{\mathbf{z} : \mathbf{z} - \mathbf{b} \in \mathcal{A}, \forall \mathbf{b} \in \mathcal{B}\}$ $= \bigcap_{\mathbf{b} \in \mathcal{B}} \mathcal{A}_{\mathbf{b}} = \bigcap_{\mathbf{a} \notin \mathcal{A}} \neg \mathcal{B}_{\mathbf{a}}$
center-out	$\{\mathbf{z} : \check{\mathcal{B}}_{\mathbf{z}} \cap \mathcal{A} \neq \emptyset\}$	$\{\mathbf{z} : \check{\mathcal{B}}_{\mathbf{z}} \subseteq \mathcal{A}\}$

TABLE 4.1 All formulations of Minkowski addition and subtraction can be classified into one of two categories, center-in or center-out.

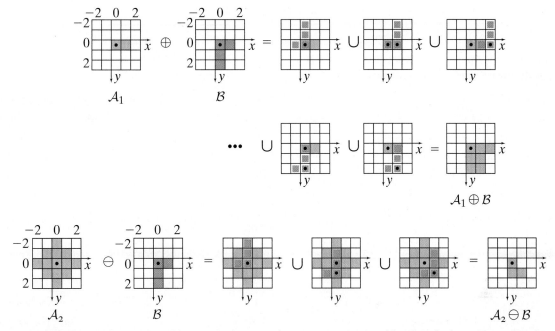

Figure 4.5 Yet another view of Minkowski addition and subtraction, from Equations (4.36) and (4.37). For both operations, the reflected second set $\check{B}$ is translated throughout the space. For addition, the output is the union of the locations of the center of $\check{B}$ when *at least one* element of $\check{B}$ overlaps $\mathcal{A}_1$. For subtraction, the output is the union of locations of the center of $\check{B}$ when *all* elements of $\check{B}$ overlap $\mathcal{A}_2$. (This is the "center-out" approach.)

the second set B is a much smaller binary mask called a **structuring element (SE)**.[†] From Equations (4.32) and (4.35), the center-in formulation of dilation and erosion involves translating B across the image, performing a test on the central pixel, and updating the output of all the pixels under B:

$$I \oplus B = \bigcup_{\mathbf{z} \in I} B_{\mathbf{z}} \qquad \text{(dilation, center-in)} \qquad (4.40)$$

$$I \stackrel{\vee}{\ominus} B = \bigcap_{\mathbf{z} \notin I} \neg \check{B}_{\mathbf{z}} \qquad \text{(erosion, center-in)} \qquad (4.41)$$

as shown in Algorithm 4.1. In the case of dilation, the output is initially set to ON everywhere. Then, for every location (x,y), the structuring element B is overlaid on I centered at (x,y), and the value $I(x,y)$ is tested. If $I(x,y)$ is ON, then the output I' is set to ON wherever B is ON. Ignoring border effects, I' will be the same size as I. Similarly, in the case of erosion, the output is initially set to ON everywhere. For every location (x,y), the structuring element B is overlaid on I centered at (x,y), and the value $I(x,y)$ is tested. If $I(x,y)$ is OFF, then the output I' is set to OFF wherever B is ON. In the code, note that $x' = 0, \ldots, w_B - 1$, and $y' = 0, \ldots, h_B - 1$, where w_B and h_B are the width and height of B, respectively. Typically the size of B is odd, in which case the floor of the half-width and half-height simplify to $\left\lfloor \frac{w_B}{2} \right\rfloor = \frac{w_B - 1}{2}$ and $\left\lfloor \frac{h_B}{2} \right\rfloor = \frac{h_B - 1}{2}$. The notation $v' \leftarrow_{\text{OR}} v$ in Line 5 is a shorthand

---

[†] Since SEs are usually stored as arrays rather than sets, from now on we use the non-calligraphic notation B rather than $\mathcal{B}$ for clarity; the two are mathematically equivalent.

ALGORITHM 4.1 Dilate or erode an image with a structuring element (center-in approach)

DILATE-CENTERIN (I, B)

Input: binary image I and structuring element B
Output: binary image I' from dilating I with B

1	$I' \leftarrow$ OFF	➤ Initialize the output $I'(x, y) \leftarrow$ OFF for all (x, y).
2	**for** $(x, y) \in I$ **do**	➤ For each ON-pixel in the input I,
3	**if** $I(x, y) ==$ ON **then**	overlay B at the pixel,
4	**for** $(x', y') \in B$ **do**	➤ and set all pixels overlapping an ON-pixel in B to ON.
5	$I'(x + x' - \lfloor \frac{w_B}{2} \rfloor, y + y' - \lfloor \frac{h_B}{2} \rfloor) \leftarrow_{\text{OR}} B(x', y')$	➤ (i.e., $I' \leftarrow I'$ OR B.)
6	**return** I'	

ERODE-CENTERIN (I, B)

Input: binary image I and structuring element B
Output: binary image I' from eroding I with B

1	$I' \leftarrow$ ON	➤ Initialize the output $I'(x, y) \leftarrow 1$ for all (x, y).
2	**for** $(x, y) \in I$ **do**	➤ For each OFF-pixel in the input I,
3	**if** $I(x, y) ==$ OFF **then**	overlay $\check{B}$ at the pixel,
4	**for** $(x', y') \in B$ **do**	and set all pixels overlapping an OFF-pixel in $\check{B}$ to OFF.
5	$I'(x - x' + \lfloor \frac{w_B}{2} \rfloor, y - y' + \lfloor \frac{h_B}{2} \rfloor) \leftarrow_{\text{AND}}$ NOT $B(x', y')$	➤ (i.e., $I' \leftarrow I'$ AND $\neg \check{B}$.)
6	**return** I'	

way of saying $v' \leftarrow v'$ OR v, and similarly for $v' \leftarrow_{\text{AND}} v$, much like the operators $|=$ and $\&=$ in the C programming language.

From Equations (4.36) and (4.37) , the center-out formulation of dilation and erosion involves translating B (or $\check{B}$) across the image, performing a test at each pixel, and outputting a value in the center:

$$I \oplus B = \{\mathbf{z} : \check{B}_\mathbf{z} \cap I \neq \emptyset\} \qquad \text{(dilation, center-out)} \qquad (4.42)$$

$$I \check{\ominus} B = \{\mathbf{z} : B_\mathbf{z} \subseteq I\} \qquad \text{(erosion, center-out)} \qquad (4.43)$$

as shown in Algorithm 4.2. In the case of dilation, for every location (x, y) the reflected structuring element $\check{B}$ is overlaid on I, and the corresponding output value $I'(x, y)$ is set to ON if there is an ON-pixel in I under *at least one* of the ON-pixels in $\check{B}$. The erosion $I \check{\ominus} B$ is computed similarly. For every location (x, y) the structuring element B is overlaid on I, and the corresponding output value $I'(x, y)$ is set to ON if there is an ON-pixel in I under *all* of the ON-pixels in $\check{B}$. In both cases, the reflection of B is accomplished by changing the signs in Line 4. It is no doubt ironic that, whereas the dilation (Minkowski addition) operator has avoided all mention of reflection until now, its center-out computation actually requires reflection, whereas the implementation of the erosion operator, which we might have expected to involve reflection, in fact requires none. For clarity, the situation is summarized in Table 4.2. Nevertheless, in case it is not already clear why dilation requires a reflection of the structuring element, an additional example is provided in Figure 4.6.

Using Dilation and Erosion for Noise Removal

The most common use of dilation and erosion is to remove noise. In this application, erosion and dilation are morphological filters that either remove anything smaller than the structuring

ALGORITHM 4.2 Dilate or erode an image with a structuring element (center-out approach)

DILATE-CENTEROUT (I, B)

Input: binary image I and structuring element B
Output: binary image I' from dilating I with B

1	**for** $(x, y) \in I$ **do**	▸ For each pixel in the input I,
2	$\quad any \leftarrow$ OFF	initialize the output value to OFF.
3	$\quad$ **for** $(x', y') \in B$ **do**	▸ Overlay $\breve{B}$ on I, and if any ON-pixel in $\breve{B}$
4	$\quad\quad$ **if** $B(x', y')$ AND $I(x - x' + \lfloor \frac{w_B}{2} \rfloor, y - y' + \lfloor \frac{h_B}{2} \rfloor)$ **then**	overlaps an ON-pixel in I,
5	$\quad\quad\quad any \leftarrow$ ON	then set the pixel in the output
6	$\quad I'(x, y) \leftarrow any$	to ON, else set it to OFF.
7	**return** I'	

ERODE-CENTEROUT (I, B)

Input: binary image I and structuring element B
Output: binary image I' from eroding I with B

1	**for** $(x, y) \in I$ **do**	▸ For each pixel in the input I,
2	$\quad all \leftarrow$ ON	initialize the output value to ON.
3	$\quad$ **for** $(x', y') \in B$ **do**	▸ Overlay B on I, and if any ON-pixel in B overlaps
4	$\quad\quad$ **if** $B(x', y')$ AND NOT $I(x + x' - \lfloor \frac{w_B}{2} \rfloor, y + y' - \lfloor \frac{h_B}{2} \rfloor)$ **then**	an OFF-pixel in I,
5	$\quad\quad\quad all \leftarrow$ OFF	then set the pixel in the output
6	$\quad I'(x, y) \leftarrow all$	to OFF, else set it to ON.
7	**return** I'	

element (erosion) or fill any gaps / holes smaller than the structuring element (dilation), and therefore the structuring element is usually a discrete approximation to a circular disk. The two most common structuring elements are the **elementary structuring elements**, which include a 3×3 cross of 1s (known as B_4) and a 3×3 array of all 1s (known as B_8):

$$B_4 = \begin{bmatrix} 0 & 1 & 0 \\ 1 & 1 & 1 \\ 0 & 1 & 0 \end{bmatrix} \qquad B_8 = \begin{bmatrix} 1 & 1 & 1 \\ 1 & 1 & 1 \\ 1 & 1 & 1 \end{bmatrix} \tag{4.44}$$

depicted graphically in Figure 4.7. The symmetry of such structuring elements means that there is no difference between Minkowski subtraction and erosion, because the structuring element is equal to its reflection. Also, the size and shape of the structuring element greatly simplify the implementation. In the case of B_4, for example, the code in the outer for loop

Operator	Center-in	Center-out
Minkowski addition	No	Yes
Minkowski subtraction	No	Yes
Dilation	No	Yes
Erosion	Yes	No

TABLE 4.2 Whether reflection of the second argument (structuring element) is needed for both center-in and center-out approaches for all operators. Note that Minkowski addition and dilation agree on both, since the operators are identical. Similarly, Minkowski subtraction and erosion disagree on both, since the definition of the latter already includes a reflection.

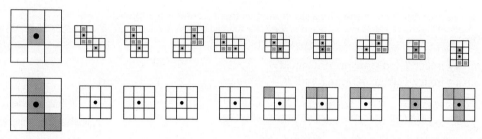

Figure 4.6 A simple example illustrating the need for reflecting the structuring element in "center-out" dilation. Left column: A 3 × 3 binary image with a single ON pixel in the center (top), and an L-shaped 3 × 3 structuring element (bottom). Remaining columns: As the structuring element slides past the image (top), dilation sets the pixel in the output (bottom) corresponding to the central pixel of the structuring element to ON if there is overlap between the image and the structuring element. In all cases, colored pixels are ON, whereas white pixels are OFF. As can be seen, if an asymmetric structuring element is not first reflected, then the resulting dilation will exhibit an undesirable reflection.

(Lines 2–6) can be replaced with a single line, leading to Algorithm 4.3. Similar code can be written using B_8. Note, therefore, that the most widely used cases of dilation and erosion are extremely easy to implement.

Properties

It should be obvious that dilation and erosion inherit the properties of the Minkowski operators. All the properties in Equations (4.14)–(4.29) are also true for dilation and erosion by simply replacing $\ominus$ with $\check{\ominus}$ everywhere. The only exception is duality, which requires an additional reflection:

$$\neg(A \oplus B) = \neg A \,\check{\ominus}\, \check{B} \qquad (\text{duality}) \tag{4.45}$$

$$\neg(A \,\check{\ominus}\, B) = \neg A \oplus \check{B} \qquad (\text{duality}) \tag{4.46}$$

That is, dilation and erosion are duals of each other with respect to complementation and reflection, which is illustrated in Figure 4.8.

One of the more interesting properties is separability:

$$A \oplus (B \oplus C) = (A \oplus B) \oplus C \qquad (\text{separability}) \tag{4.47}$$

$$A \,\check{\ominus}\, (B \oplus C) = (A \,\check{\ominus}\, B) \,\check{\ominus}\, C \qquad (\text{separability}) \tag{4.48}$$

In other words, if we dilate an image by a structuring element B, then dilate it again with a different structuring element C, the result is the same as if we had dilated the image by the structuring element resulting from dilating B with C. A similar argument holds for erosion. As an example, B_8 can be decomposed as follows:

$$B_8 = \begin{bmatrix} 1 & 1 & 1 \\ 1 & 1 & 1 \\ 1 & 1 & 1 \end{bmatrix} = \begin{bmatrix} 0 & 0 & 0 \\ 1 & 1 & 1 \\ 0 & 0 & 0 \end{bmatrix} \oplus \begin{bmatrix} 0 & 1 & 0 \\ 0 & 1 & 0 \\ 0 & 1 & 0 \end{bmatrix} = \begin{bmatrix} 0 & 1 & 0 \\ 0 & 1 & 0 \\ 0 & 1 & 0 \end{bmatrix} \oplus \begin{bmatrix} 0 & 0 & 0 \\ 1 & 1 & 1 \\ 0 & 0 & 0 \end{bmatrix} \tag{4.49}$$

since dilation is commutative. By dilating (or eroding) with the horizontal structuring element, then dilating (or eroding) with the vertical structuring element, significant computation

Figure 4.7 The two most common structuring elements.

$$B_4 \qquad\qquad B_8$$

ALGORITHM 4.3 Dilate or erode an image with a B_4 structuring element (center-out approach)

DILATE_B4(I)

Input: binary image I
Output: binary image I' from dilating I with the B_4 structuring element

1 **for** $(x, y) \in I$ **do**
2      $I'(x, y) \leftarrow I(x, y)$ OR $I(x - 1, y)$ OR $I(x + 1, y)$ OR $I(x, y - 1)$ OR $I(x, y + 1)$
3 **return** I'

ERODE_B4(I)

Input: binary image I
Output: binary image I' from eroding I with the B_4 structuring element

1 **for** $(x, y) \in I$ **do**
2      $I'(x, y) \leftarrow I(x, y)$ AND $I(x - 1, y)$ AND $I(x + 1, y)$ AND $I(x, y - 1)$ AND $I(x, y + 1)$
3 **return** I'

can be avoided. This approach, which can be applied to rectangular structuring elements of all 1s of any size, is particularly beneficial for large structuring elements.

Another interesting property is parallelism, which allows us to dilate or erode piece by piece, then combine the results by union or intersection:

$$A \oplus (B \cup C) = (A \oplus B) \cup (A \oplus C) \tag{4.50}$$

$$A \check{\ominus} (B \cup C) = (A \check{\ominus} B) \cap (A \check{\ominus} C) \tag{4.51}$$

For example, if we had three processors, we could decompose B_8 as follows:

$$B_8 = \begin{bmatrix} 1 & 1 & 1 \\ 1 & 1 & 1 \\ 1 & 1 & 1 \end{bmatrix} = \begin{bmatrix} 1 & 1 & 1 \\ 0 & 0 & 0 \\ 0 & 0 & 0 \end{bmatrix} \cup \begin{bmatrix} 0 & 0 & 0 \\ 1 & 1 & 1 \\ 0 & 0 & 0 \end{bmatrix} \cup \begin{bmatrix} 0 & 0 & 0 \\ 0 & 0 & 0 \\ 1 & 1 & 1 \end{bmatrix}$$

perform dilation (or erosion) using a different structuring element with each processor, then combine the results using union (or intersection).

Figure 4.8 The duality of dilation and erosion, namely, $\neg(\mathcal{A} \oplus \mathcal{B}) = \neg\mathcal{A} \check{\ominus} \check{\mathcal{B}}$. Note that out-of-bounds pixels receive the value of their closest neighbor.

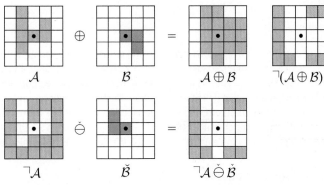

$\mathcal{A}$ $\mathcal{B}$ $\mathcal{A} \oplus \mathcal{B}$ $\neg(\mathcal{A} \oplus \mathcal{B})$

$\neg\mathcal{A}$ $\check{\mathcal{B}}$ $\neg\mathcal{A} \check{\ominus} \check{\mathcal{B}}$

Alternative Definitions and Notation

The reader is cautioned that the notation used for dilation and erosion varies among authors. The notation used in this book is summarized as follows:

$$\left(\text{Mink. addition/dilation}\right)\; A \oplus B = \bigcup_{\mathbf{b}\in B} A_{\mathbf{b}} \qquad A \check{\oplus} B = \bigcup_{\mathbf{b}\in B} A_{-\mathbf{b}} \;\left(\text{reflected dilation}\right) \qquad (4.52)$$

$$\left(\text{Mink. subtraction}\right)\; A \ominus B = \bigcap_{\mathbf{b}\in B} A_{\mathbf{b}} \qquad A \check{\ominus} B = \bigcap_{\mathbf{b}\in B} A_{-\mathbf{b}} \;\left(\text{erosion}\right) \qquad (4.53)$$

By retaining the standard notation for Minkowski addition and subtraction while introducing the notation $\check{\ominus}$ for erosion, the distinction between these concepts is clearly maintained. Many authors, such as Gonzalez and Woods [2008] and Davies [2005], reuse the Minkowski subtraction notation for erosion, leading to ambiguity in interpreting the symbol $\ominus$. Other authors, such as Serra [1982] and Soille [2003], not only use different notation but also define dilation to include a reflection ("reflected dilation"), in the same way that erosion includes a reflection. These differences are summarized in Table 4.3, where the top rows show how each author defines each term, using our notation to describe the definition; and the bottom rows show the notation used by these authors. When the structuring element is not symmetric, maintaining these distinctions is important.

4.1.4 Building Large Structuring Elements

We have just seen[†] that an SE is *separable* when it can be written as the dilation of two SEs. The dilation or erosion of an image with a separable SE is equivalent to two successive dilations or erosions, respectively, with the component SEs, as shown in Equations (4.46)–(4.47). This property can be used to greatly decrease the number of computations needed when an SE is separable. For example, a 3×3 SE of all 1s is the dilation of a vertical SE of all 1s with a horizontal SE of all 1s:

$$B_8 = \begin{bmatrix} 1 & 1 & 1 \\ 1 & 1 & 1 \\ 1 & 1 & 1 \end{bmatrix} = \underbrace{\begin{bmatrix} 1 \\ 1 \\ 1 \end{bmatrix}}_{L_{3,(0,1)}} \oplus \underbrace{\begin{bmatrix} 1 & 1 & 1 \end{bmatrix}}_{L_{3,(1,0)}} \qquad (4.54)$$

		This Book	**GW, Davies, SHB, SS**	**Serra**	**Soille**
definition	dilation	$A \oplus B$	$A \oplus B$	$A \check{\oplus} B$	$A \check{\oplus} B$
	erosion	$A \check{\ominus} B$	$A \check{\ominus} B$	$A \check{\ominus} B$	$A \check{\ominus} B$
notation	dilation	$A \oplus B$	$A \oplus B$	$A \oplus \check{B}$	$\delta_B(A)$
	erosion	$A \check{\ominus} B$	$A \ominus B$	$A \ominus \check{B}$	$\varepsilon_B(A)$

TABLE 4.3 Comparison of the definition and notation for dilation and erosion among various authors (from left-to-right: Gonzalez and Woods [2008], Davies [2005], Sonka et al. [2008], Shapiro and Stockman [2001], Serra [1982], and Soille [2003]). The top rows present the definitions of the operators by the various authors with respect to our notation. All authors agree on the definitions, except that the last two authors introduce a reflection in the definition of dilation. The bottom rows present the notation used by the various authors. Ignoring the last column, all authors agree on the notation for dilation, whereas our notation for erosion differs from others' by making the reflection explicit in the operator symbol. (Keep in mind, however, that in a "center-out" implementation, it is actually dilation, not erosion, that reflects the structuring element, even using the definitions in the middle two columns. See Table 4.2.)

[†] Section 4.1.2 (p. 134).

where $L_{r,(x,y)}$ indicates a straight line SE in the direction of (x,y) with length r. A variety of shapes of arbitrary size can be composed from this basic definition of a line, as shown in Figure 4.9:

$$\text{Square}_r = L_{r,(1,0)} \oplus L_{r,(0,1)} \tag{4.55}$$

$$\text{Rectangle}_{w,\ell} = L_{w,(1,0)} \oplus L_{\ell,(0,1)} \tag{4.56}$$

$$\text{Diamond}_r = L_{r-1,(1,1)} \oplus L_{r-1,(1,-1)} \oplus \text{Diamond}_2 \tag{4.57}$$

$$\text{Octagon}_r = L_{r-1,(1,1)} \oplus L_{r-1,(1,-1)} \oplus L_{r,(1,0)} \oplus L_{r,(0,1)} \tag{4.58}$$

$$\text{Octagon}'_r = L_{r,(1,1)} \oplus L_{r,(1,-1)} \oplus L_{r,(1,0)} \oplus L_{r,(0,1)} \tag{4.59}$$

where Diamond_2 is identical to B_4 in which the 1s are arranged in a plus sign. For the two types of octagon SEs, the following ordering relation can be shown to hold for all r: $\text{Octagon}_r \subset \text{Octagon}'_r \subset \text{Octagon}_{r+1} \subset \text{Octagon}'_{r+1}$.

4.1.5 Opening and Closing

One problem with dilation and erosion is that they inadvertently grow or shrink a binary object. For example, with a 3×3 structuring element, dilation extends the object by 1 pixel around the border of the object, whereas erosion removes (in addition to noise) a 1-pixel-thick border around the object. The solution to this problem lies in two additional morphological operations called **opening** (which is erosion followed by dilation) and **closing** (which is dilation followed by erosion):

$$A \circ B = (A \ominus B) \oplus B \qquad \text{(opening)} \tag{4.60}$$

$$A \bullet B = (A \oplus B) \ominus B \qquad \text{(closing)} \tag{4.61}$$

Figure 4.9 Composing large structuring elements by dilating smaller ones. From top to bottom: Square$_3$, Rectangle$_{5,3}$, Diamond$_3$, Octagon$_3$, and Octagon$'_3$.

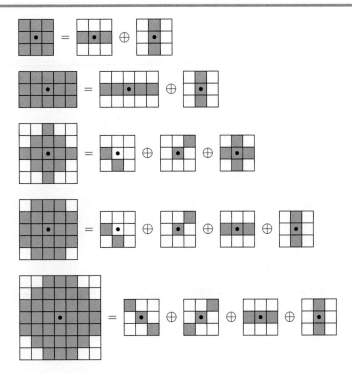

Equivalently, the opening of a set is the union of all the SEs that fit entirely within the set, whereas the closing of a set is the intersection of all translations of the complement of the SE such that it contains the set:

$$A \circ B = \bigcup_{\mathbf{z}} \{B_{\mathbf{z}} : B_{\mathbf{z}} \subseteq A\} \qquad \text{(opening)} \qquad (4.62)$$

$$A \bullet B = \bigcap_{\mathbf{z}} \{ \neg B_{\mathbf{z}} : A \subseteq \neg B_{\mathbf{z}} \} = \neg \left(\bigcup_{\mathbf{z}} \{B_{\mathbf{z}} : B_{\mathbf{z}} \subseteq \neg A\} \right) \quad \text{(closing)} \qquad (4.63)$$

It is easy to show that opening and closing are also duals of each other with respect to complementation and reflection:

$$\neg(A \bullet B) = (\neg A \circ \check{B}) \qquad \text{(duality)} \qquad (4.64)$$

$$\neg(A \circ B) = (\neg A \bullet \check{B}) \qquad \text{(duality)} \qquad (4.65)$$

The open and close operators are **idempotent**, meaning that repeated applications of opening or closing do nothing:

$$(A \circ B) \circ B = A \circ B \qquad \text{(idempotence)} \qquad (4.66)$$

$$(A \bullet B) \bullet B = A \bullet B \qquad \text{(idempotence)} \qquad (4.67)$$

It is easy to see that openings are anti-extensive, whereas closings are extensive.

If the foreground pixels are visualized as land and the background pixels as water, features can be defined, such as lakes, bays, channels, capes, isthmuses, and islands. Figure 4.10 illustrates some binary regions with these features, along with the results of dilation, erosion, closing, and opening. For simplicity we ignore discretization and imagine the structuring element in the shape of a circular ball (or disk). If the ball is rolled on the outer contour of the object, the result of dilation is the set of all points whose enclosing contour is defined by the path of the center of the ball. This operation fills the lakes, bays, and channels. Similarly, if the ball is rolled on the inner contour of the object, the result of erosion is the set of all points whose enclosing contour is defined by the path of the center of the ball. This operation removes capes, isthmuses, and islands. Note that dilation increases

Figure 4.10 Top: Dilation of a binary region by a circular structuring element can be visualized as rolling the disk along the outside of the region; the result is enclosed by the path of the center of the disk. The right column shows the result of closing (dilation followed by erosion), which fills lakes, bays, and channels. Bottom: Erosion can be visualized as rolling the disk along the inside of the region; the result is again enclosed by the path of the center of the disk. The right column shows the result of opening (erosion followed by dilation), which removes capes, isthmuses, and islands.

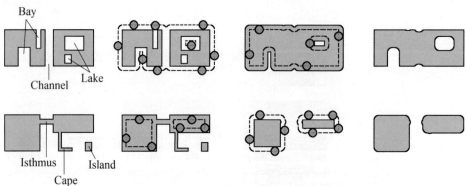

the size of the region, whereas erosion decreases the size of the region. By following one operation after the other, the undesired features can be filtered without changing the overall size of the region. An example of applying various morphological operations to a binary image (obtained by thresholding a background subtraction result) is shown in Figure 4.11.

In addition to **morphological openings**, which we have just seen, there are also **algebraic openings**, which cannot be written as a single erosion followed by a dilation. The **area opening**, for example, removes all regions whose area (in terms of the number of pixels) is smaller than a given threshold. The area opening is equivalent to the union of all morphological openings with connected SEs whose size (in terms of the number of pixels) is equal to the given threshold. Similarly, there are **algebraic closings**. The **area closing**, for example, is the dual of area opening, given by the intersection of all morphological closings with connected SEs whose size is equal to the threshold. Another algebraic opening is the **parametric opening**, which is the union of all morphological openings by all subsets of the SE whose size is equal to the given threshold. Such operators are mathematically interesting but not widely used.

4.1.6 Hit-Miss Operator

Erosion can be used to define a simple approach to shape detection in a binary image. Suppose we have a shape B_{ON} that we wish to detect in an image A. The erosion $A \hat{\ominus} B_{\text{ON}}$ will find all places in the image where B_{ON} fits entirely within the foreground of A (meaning that every ON pixel in B_{ON} lines up with an ON pixel in A), but it will lead to spurious detections as well. To avoid these spurious detections, use another structuring element B_{OFF} that contains the pixels that we do not want to be ON. These could be the inverse $\neg B_{\text{ON}}$ of the shape, the outer boundary $(B_{\text{ON}} \oplus B_4) - B_{\text{ON}}$ of the shape, or some other set of pixels. To detect the shape in the image, the **hit-miss operator**[†] uses erosion to find all the places in the image where B_{ON} matches the foreground and B_{OFF} matches the background:

$$A \circledast (B_{\text{ON}}, B_{\text{OFF}}) \equiv (A \hat{\ominus} B_{\text{ON}}) \cap (\neg A \check{\ominus} B_{\text{OFF}}) \qquad \text{(hit-miss operator)} \qquad (4.68)$$

The hit-miss operator provides a simple approach to object detection, for which more robust techniques will be considered in Chapter 12.

Notice that the hit-miss operator takes two arguments, the second of which is a pair of SEs. We can combine these two SEs into a single ternary SE that holds one of three values for each pixel:

$$B \equiv \begin{cases} \text{ON} & \text{if } B_{\text{ON}} = \text{ON} \\ \text{OFF} & \text{if } B_{\text{OFF}} = \text{ON} \\ \text{DONT-CARE} & \text{otherwise} \end{cases} \qquad (4.69)$$

Figure 4.11 A binary image and the result of morphological operations: Erode, dilate, open, and close. Erosion removes salt noise but shrinks the foreground. Dilate fills pepper noise but expands the foreground. Opening and closing removes the respective types of noise while retaining the overall size of the foreground.

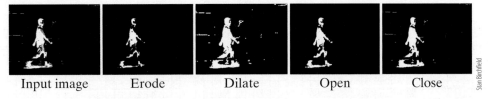

Input image   Erode   Dilate   Open   Close

[†] The hit-miss operator is also known as the *hit-and-miss transform* or *hit-or-miss transform*, but the latter is inaccurate since the parts are combined by conjunction rather than disjunction. Similarly, *operator* is more appropriate than *transform* because the coordinate frame does not change.

allowing us to write Equation (4.68) in a more compact form:

$$A \circledast B \qquad \text{(hit-miss operator with ternary argument)}$$

Figure 4.12 shows an example of the hit-miss operator applied to detect a simple smiley face pattern in an image. The foreground pattern contains the eyes and mouth, while the background pattern contains the negation of the foreground—except that the corner pixels are considered "don't care." The foreground pattern is found in the image at three places, in addition to several spurious readings along the bottom of the image. Notice that the detection in the upper-right is of a face for which a nose was added. The background pattern is also found in three places, including a detection of a face with a pixel missing from the mouth. When logically ANDed together, only the two unmodified faces are detected. Notice that the DONT-CARE pixels do not have any effect on any of the detections. Figure 4.13 shows the hit-miss operator applied to another image for the purpose of detecting a particular letter ("e") in an image of an English sentence.

4.1.7 Thinning

One application of the hit-miss operator is morphological **thinning**. Thinning a binary image involves removing pixels from the foreground (that is, setting pixels to OFF) while maintaining as much as possible the structure and connectivity of the foreground regions. At its core, thinning takes an image and a ternary SE and removes all the points detected by the SE:

$$I \boxminus B \equiv I - (I \circledast B) = I \cap \neg(I \circledast B) \tag{4.70}$$

where the subtraction operator performs set differencing by setting all pixels that are ON in $I \circledast B$ to OFF in the output. Typically we want to thin an image using several ternary SEs, organized as either an ordered or unordered set. An ordered set is known as a *sequence*, in which case thinning applies the elements of the sequence in order, one at a time, to the output of the previous element. If we let $\mathcal{B} = (B_1, B_2, \ldots, B_m)$ be a sequence of m ternary SEs, then the thinning operator is defined as

$$I \boxminus \mathcal{B} \equiv ((I \boxminus B_1) \boxminus B_2) \cdots \boxminus B_m \qquad (\text{sequential}) \tag{4.71}$$

Figure 4.12 Hit-miss operator. TOP: The foreground pattern B_{ON}, the background pattern B_{OFF}, and the ternary representation B with X indicating DONT-CARE. BOTTOM: From left-to-right: The image A within which to search for the pattern, the negation $\neg A$ of the image, the erosion of the image with the foreground pattern, the erosion of the negated image with the background pattern, and the final result, which shows two successful detections of the smiley face.

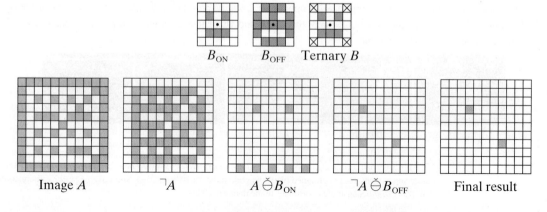

$$B_{\text{ON}} \qquad B_{\text{OFF}} \qquad \text{Ternary } B$$

Image A \qquad $\neg A$ \qquad $A \ominus B_{\text{ON}}$ \qquad $\neg A \ominus B_{\text{OFF}}$ \qquad Final result

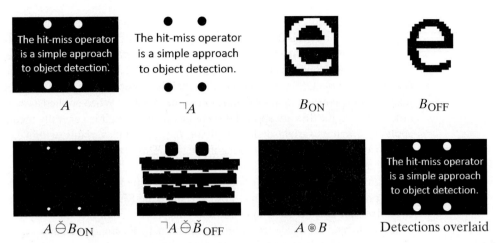

Figure 4.13 The hit-miss operator applied to a binary image, with white indicating ON and black indicating OFF. TOP: From left-to-right: The image, the inverted image, the foreground pattern, and the background pattern. BOTTOM: From left-to-right: The erosion of the image with the foreground pattern, the erosion of the inverted image with the background pattern, the result of hit-miss, and the outlines of the detections overlaid on the original image.

On the other hand, if we let $\mathcal{B} = \{B_1, B_2, \ldots, B_m\}$ be an unordered set of m ternary SEs, then the thinning operator first flags all pixels that are detected by the SEs, then removes those pixels:

$$I \boxminus \mathcal{B} \equiv I - ((I \circledast B_1) \cup (I \circledast B_2) \cup \cdots \cup (I \circledast B_m)) \quad \text{(nonsequential)} \quad \text{(4.72)}$$

Either way, the equations above constitute one iteration of thinning with the set/sequence. Applying multiple iterations is straightforward:

$$I \boxminus^n \mathcal{B} \equiv \underbrace{(((I \boxminus \mathcal{B}) \boxminus \mathcal{B})) \cdots \boxminus \mathcal{B}}_{n \text{ iterations}} \quad \text{(4.73)}$$

where n is the number of iterations. If the procedure is run until convergence, the computation is represented as $I \boxminus^\infty \mathcal{B}$, so that repeated applications have no effect: $(I \boxminus^\infty \mathcal{B}) \boxminus \mathcal{B} = I \boxminus^\infty \mathcal{B}$.

Figure 4.14 shows 8 ternary SEs that are commonly used for morphological thinning. The two basic SEs are

$$B_{\text{EDGE}} = \begin{bmatrix} 0 & 0 & 0 \\ X & 1 & X \\ 1 & 1 & 1 \end{bmatrix} \quad B_{\text{CORNER}} = \begin{bmatrix} X & 0 & 0 \\ 1 & 1 & 0 \\ X & 1 & X \end{bmatrix}$$

and the other 6 SEs are rotated versions of these. As a rough approximation, B_{EDGE} and its varieties detect points that are on an interior edge of a region, or that are protruding out of the region. Similarly, B_{CORNER} and its varieties detect points that are on a corner or on a thin diagonal line.

Figure 4.14 Structuring elements commonly used for morphological thinning. Colored pixels are ON, white pixels are OFF, and x indicates DONT-CARE. The top row shows the 4 edge SEs, while the bottom row shows the 4 corner SEs.

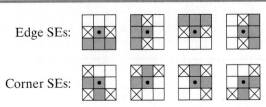

Figure 4.15 shows the process of thinning a binary image using these 8 SEs as a sequence. The first SE is applied to the image, matching 5 pixels along the top that are then removed. Afterward, the second SE is applied, and no pixels are matched. The third SE is then applied, and 8 pixels along the bottom are matched and removed. The process continues until convergence, at which point the result is an approximation of the skeleton of the image, a topic covered in more detail later in the chapter.[†] When applied as a sequence, the order in which the SEs are applied does in fact matter, and it is generally recommended to apply B_{EDGE} and its variants before applying B_{CORNER} and its variants, within each iteration, so as to avoid unnecessarily shrinking the region. This is the approach taken in the example shown in the figure.

Figure 4.16 shows the process of thinning the same binary image using the same SEs as the previous figure. This time, however, the edge SEs are applied as a set, so that all edge pixels are first detected and removed. This process continues until convergence. Once no more changes are made to the image, the corner SEs are applied as a sequence, beginning with the first one and continuing until convergence. This approach produces a thinner skeleton than the approach in the previous example which, as we shall see later, is an improvement because the goal is usually to obtain the thinnest skeleton possible. Note from the figure that the corner SEs cannot be applied as a set, because then two adjacent pixels would be detected and deleted (i.e., the pixel just below the hashed pixel in the next-to-last graphic), leading to a disconnected skeleton.

4.1.8 Thickening

Thickening is the opposite of thinning. Thickening a binary image involves adding pixels to the foreground (that is, setting pixels to ON) while maintaining as much as possible the

---

Figure 4.15 Morphological thinning of a binary image using the SEs of Figure 4.14 treated as a sequence. The first SE matches pixels (indicated by hashed squares) along the top of the region, which are then removed. The second SE matches no pixels, while the third SE matches pixels along the bottom, which are then removed. This process continues until convergence, yielding an approximate skeleton of the original image.

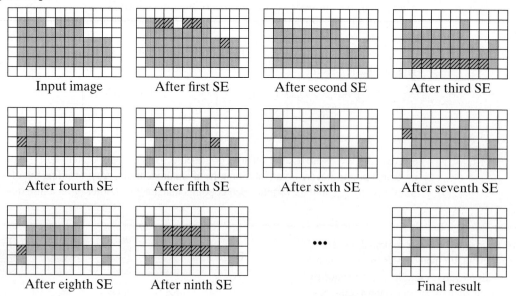

Input image	After first SE	After second SE	After third SE
After fourth SE	After fifth SE	After sixth SE	After seventh SE
After eighth SE	After ninth SE	•••	Final result

† Section 4.5.1 (p. 195).

Figure 4.16 Morphological thinning of the same binary image using the same SEs as the previous example. In this case, however, the edge SEs are applied repeatedly as a set until convergence, before applying the corner SEs repeatedly as a sequence. In the first iteration pixels along the top, right, bottom, and left of the region are removed by the edge SEs. In the second iteration, 9 additional pixels are removed by the edge SEs. In the final iteration a single pixel is removed by one of the corner SEs, thus producing a thinner skeleton than in the previous example.

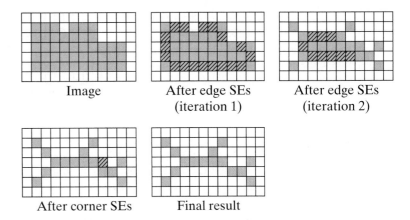

Image   After edge SEs (iteration 1)   After edge SEs (iteration 2)

After corner SEs   Final result

overall shape of the foreground regions. At its core, thickening takes an image and a ternary SE and adds all the points detected by the SE:

$$I \boxplus B \equiv I \bigcup (I \circledast B) \qquad (4.74)$$

where the union operator sets all pixels that are ON in $I \circledast B$ to ON in the output. Using X to represent DONT-CARE, commonly used SEs for thickening are

$$B_{\text{GROW-SE}} = \begin{bmatrix} 1 & 1 & X \\ 1 & 0 & X \\ 1 & X & 0 \end{bmatrix} \quad B_{\text{GROW-SW}} = \begin{bmatrix} X & 1 & 1 \\ X & 0 & 1 \\ 0 & X & 1 \end{bmatrix}$$

and the rotated versions of these, which are all shown in Figure 4.17. $B_{\text{GROW-SE}}$ grows the region in the southeast (bottom-right) direction, $B_{\text{GROW-SW}}$ grows in the southwest (bottom-left) direction, and so on. Note that the central pixel of thickening SEs are OFF, whereas the central pixel of thinning SEs are ON.

The process of morphological thickening of a binary image is shown in Figure 4.18. For brevity we simply show the input and output of the procedure, leaving the detailed steps as an exercise for the reader. Note that the thickening process fills the concavities of the region, resulting in an approximation to the convex hull, discussed later.[†] Thickening is not as sensitive as thinning to the order in which the SEs are applied, so the distinction between set and sequence is less important. Comparing this figure with Figure 4.16, notice that the thinning SEs are designed to retain 8-connectedness of the foreground, while the thickening SEs are designed to thin 4-connected foregrounds. For example, if the original region in Figure 4.18 were first thinned to an 8-connected region, thickening would have no effect because the SEs would not match anything. We will discuss 4- and 8-connectedness in more detail in the next section.

Figure 4.17 Structuring elements commonly used for morphological thickening.

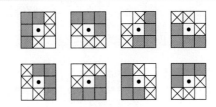

Figure 4.18 Morphological thickening of a binary image using the SEs in Figure 4.17. Shown are the original image (left) and the final result after convergence (right). The thickened result is an approximation to the convex hull.

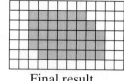

Input image   Final result

An alternate approach to thickening the foreground is to thin the background. That is, the complement of the image is taken, thinning is applied until convergence, and the result is then complemented again. This approach is illustrated in Figure 4.19, from which it is clear that this approach does not yield the convex hull—rather it retains the approximate shape of the original region. One drawback of this approach is the lack of meaningful convergence. As we saw in Figure 4.18, thickening converges to a solution in a finite number of iterations, without regard to the size of the background in which the foreground is embedded. In this alternate approach, however, if the foreground were embedded in a much larger background, then the background would be repeatedly thinned until it reached the borders of the image, at which point the background size would be greatly reduced, thus causing the foreground region to have enlarged well beyond its original size.

4.2 Labeling Regions

Once the noise in a binary image has been removed (or at least reduced) using morphological operations, the next step is often to find regions of pixels in the image. Each region is a set of connected pixels and is assigned a unique label to distinguish it from other regions. In this section we consider basic definitions, followed by several algorithms for labeling regions.

4.2.1 Neighbors and Connectivity

The morphological algorithms of the previous section are the first algorithms we have considered that use neighbors of pixels to compute a result. A pixel $\mathbf{q} = (q_x, q_y)$ is a **neighbor** of pixel $\mathbf{p} = (p_x, p_y)$ if $\mathbf{q}$ is in the **neighborhood** of $\mathbf{p}$, denoted $\mathbf{q} \in \mathcal{N}(\mathbf{p})$, where $\mathcal{N}$ is the neighborhood function. The most common neighborhoods used in image processing are shown in Figure 4.20:

Figure 4.19 Morphological thickening of the same binary image by thinning the background using the thinning SEs in Figure 4.14. Top: The background (obtained as the complement of the original image), and the output after each iteration of morphological thinning. Bottom: The original image and the thickened output after each iteration, obtained by complementing the image above it.

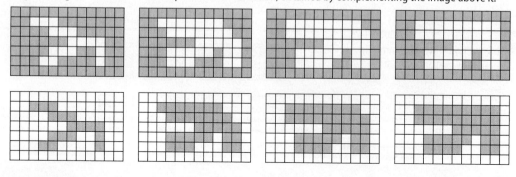

Figure 4.20 Commonly used neighborhoods. From left to right: $\mathcal{N}_4$, $\mathcal{N}_8$, and $\mathcal{N}_D$.

- The 4-neighborhood, denoted by $\mathcal{N}_4$, consists of the four pixels to the left, right, above, and below the pixel.[†]
- The D-neighborhood, denoted by $\mathcal{N}_D$, consists of the four pixels diagonal from the pixel.
- The 8-neighborhood, denoted by $\mathcal{N}_8$, consists of the eight closest pixels. That is, $\mathcal{N}_8 = \mathcal{N}_4 \cup \mathcal{N}_D$.

Note that the structuring element B_4 that we encountered earlier is a set consisting of the central pixel and its 4-neighbors, whereas B_8 consists of the central pixel and its 8-neighbors.

Any algorithm that uses neighbors must decide what to do when those neighbors do not exist. For example, in the case of dilation or erosion, when the structuring element is centered near the boundary of the image, some of its elements will extend past the image, placing them out of bounds. Although for simplicity the pseudocode presented in the previous section ignores this detail, in a real implementation a decision must be made how to handle these out-of-bounds pixels. While there is no agreed-upon solution for this problem, several common approaches are the following:

- *Do not process boundary pixels.* Keep the SE in bounds at all times and set the output pixels near the boundary to zero (or some other arbitrary value). This is the fastest and simplest solution and is acceptable if you do not care what happens near the border.
- *Resize the SE.* Near the border, shrink the SE so that it does not extend past the image border. For example, if we have a 3×3 SE of all ones, we could use a 2×3 SE of all ones near the left and right border, a 3×2 SE near the top and bottom borders, and 2×2 SEs (with the center placed appropriately) near the four corners.
- *Pad values outside the image.* The most common approaches to padding are to *replicate* (that is, out of bounds pixels are assigned the value of the nearest pixel in the image), *reflect* (image values are mirror-reflected about the image border), *wrap* (image values are extended in a periodic wrap, which is what the discrete Fourier transform does implicitly), and *set to constant*. Note that this last option was used in the previous section, where we assumed that out-of-bounds pixels were OFF.

The type of neighborhood determines the type of adjacency. Two pixels are said to be **adjacent** if they have the same value (i.e., either ON or OFF in the case of a binary image) and if they are neighbors of each other. Pixels are said to be **connected** (or contiguous) if there exists a path between them, where a **path** is defined as a sequence of pixels $\mathbf{p}_0, \mathbf{p}_1, \ldots, \mathbf{p}_{n-1}$ such that $\mathbf{p}_{i-1}$ and $\mathbf{p}_i$ are adjacent for all $i = 1, \ldots, n-1$. A **region** in an image is therefore a set of connected pixels. Not surprisingly, the two most common adjacencies are 4-adjacency and 8-adjacency, illustrated in Figure 4.21:

- Two pixels $\mathbf{p}$ and $\mathbf{q}$ in an image I are *4-adjacent* if $I(\mathbf{p}) = I(\mathbf{q})$ and $\mathbf{q} \in \mathcal{N}_4(\mathbf{p})$;
- Two pixels $\mathbf{p}$ and $\mathbf{q}$ in an image I are *8-adjacent* if $I(\mathbf{p}) = I(\mathbf{q})$ and $\mathbf{q} \in \mathcal{N}_8(\mathbf{p})$.

In addition, the notion of 4- and 8-neighbors can be combined to yield *m*-adjacency ("mixed adjacency"):

- Two pixels $\mathbf{p}$ and $\mathbf{q}$ in an image I are *m-adjacent* if $I(\mathbf{p}) = I(\mathbf{q})$ and ($\mathbf{q} \in \mathcal{N}_4(\mathbf{p})$ or ($\mathbf{q} \in \mathcal{N}_D(\mathbf{p})$ and $\mathcal{N}_4(\mathbf{p}) \cap \mathcal{N}_4(\mathbf{q}) = \emptyset$)).

[†] In cellular automata theory, $\mathcal{N}_4$ is known as the von Neumann neighborhood, while $\mathcal{N}_8$ is the Moore neighborhood. J. von Neumann (1903–1957) was a pioneer in many areas and is perhaps most famous for the von Neumann architecture used in almost all computers today. E. F. Moore (1925–2003), the inventor of the Moore finite state machine, was a pioneer of artificial life.

Figure 4.21 A binary region and the 4-, 8-, and *m*-adjacency of its pixels. Note that *m*-adjacency removes the loops that sometimes occur with 8-adjacency.

Region   4-adjacency   8-adjacency   *m*-adjacency

In other words, two pixels are *m*-adjacent if they are either 4-adjacent or 8-adjacent and there is not another pixel that is 4-adjacent to both of them. For all adjacencies, the neighborhood relations are symmetric, so that $\mathbf{q} \in \mathcal{N}(\mathbf{p})$ if and only if $\mathbf{p} \in \mathcal{N}(\mathbf{q})$ for any neighborhood $\mathcal{N}$.

Discretization introduces a subtle complication in dealing with adjacency. According to the **Jordan curve theorem**, any simple closed curve (called a *Jordan curve*) divides the plane into the interior region bounded by the curve and the exterior region which is the complement of the interior region unioned with the curve. Therefore, any continuous path from one region to the other intersects the curve. With discretization, however, it is easy to demonstrate scenarios in which a path from the background to the foreground does not intersect the discretized curve if the same adjacency is used for both foreground and background. The well-known solution to this problem is to use 4-adjacency for the foreground and 8-adjacency for the background, or vice versa.

4.2.2 Floodfill

Floodfill, also called **seed fill**, is the problem of coloring all the pixels that are connected to a *seed pixel* with some desired new color (where "color" refers to either an RGB triplet or gray level or binary value, depending on the image type). Several algorithms exist for performing floodfill, with the recursive version being particularly easy to explain. As shown in Algorithm 4.4, the algorithm takes the coordinates of a seed pixel $\mathbf{p} = (x, y)$, a new color, and an image, and it sets the colors of the pixels. (The pseudocode uses non-bold-face *p* (rather than $\mathbf{p}$) to represent the pixel to emphasize that it is just like any other variable.) The seed pixel is examined and set to the new color after first storing the original color. The neighbors (either 4- or 8-), in turn, are examined and set to the new color if they are equal to the original color under the seed pixel. The neighbors of these pixels are then examined and set in the same manner, with the process recursively repeating until no neighboring pixels share the original seed pixel color. No value is returned, since the pixels are modified in place. Although this algorithm is simple to understand, it is never used in practice because, not only is recursion computationally inefficient due to the overhead of making function calls, but, more importantly, recursive floodfill will cause the stack to be overrun, because it is not uncommon for floodfilled regions to contain tens of thousands of pixels. And of course a stack overrun will cause the program to crash.

A more computationally efficient approach overcomes these problems by using a dynamic array of pixels called the *frontier*. In the initialization, the original color of the seed pixel is grabbed, the seed pixel is colored with the new color, and the coordinates of the pixel are pushed onto the initially empty frontier. Then the algorithm repeatedly pops a pixel off the frontier and expands all the adjacent pixels (the neighbors that still have the original color), where expansion involves setting the pixel to the new color and pushing its coordinates onto the frontier. The algorithm terminates when the frontier is empty. In the pseudocode of Algorithm 4.5, the algorithm performs a depth-first search if the frontier is implemented as a stack, because PUSH and POP operate on the same end of the array. Alternatively, if the stack is replaced by a queue, then PUSH and POP operate on opposite ends of the array, and the FIFO (first-in-first-out) operations cause a breadth-first search

ALGORITHM 4.4 Perform floodfill on an image (stack-unfriendly version using recursion)

FLOODFILLRECURSIVE(I, p, *new-color*)

Input: image I, seed pixel p, and new color
Output: all pixels in I connected to p are colored *new-color*

1	**if** $I(p) \neq$ *new-color* **then**	If pixel is already new color, then terminate this branch in the recursion.
2	*orig-color* $\leftarrow$ $I(p)$	
3	$I(p) \leftarrow$ *new-color*	Otherwise, set pixel to new color, and recursively call the procedure on any neighbor having the same color.
4	**for** $q \in \mathcal{N}(p)$ **do**	
5	**if** $I(q) ==$ *orig-color* **then**	
6	FLOODFILLRECURSIVE(I, q, *new-color*)	

instead. Either way the output is the same, so a stack-based frontier is recommended because its memory management is simpler. An example of the algorithm applied to a binary image is illustrated in Figure 4.22.

Variations on the algorithm are easy to obtain by making minor modifications to this basic pseudocode. One common variation is to leave the original input image intact and instead to change an output image. This variation is implemented in Algorithm 4.6, where each pixel $O(x, y)$ in the output is set if the pixel $I(x, y)$ in the input would have been changed by the previous algorithm. This version of the algorithm will be used in the next section for connected components, as well as later for segmentation.[†] It does not return a value, since it operates on the output image that is passed into the procedure.

An interesting connection exists between floodfill and dilation, namely that floodfill is equivalent to repeated conditional dilations. The **conditional dilation** of an image I with respect to another image C is defined as dilation followed by intersection:[‡]

$$I \oplus_C B \equiv (I \oplus B) \cap C \qquad (4.75)$$

ALGORITHM 4.5 Perform floodfill on an image (fast version using frontier)

FLOODFILL(I, p, *new-color*)

Input: image I, seed pixel p, and new color
Output: all pixels in I connected to p are colored *new-color*

1	*orig-color* $\leftarrow$ $I(p)$	If seed pixel is already new color, then terminate.
2	**if** *orig-color* $\neq$ *new-color* **then**	
3	*frontier*.PUSH(p)	Otherwise, set frontier to seed pixel, and set pixel to new color.
4	$I(p) \leftarrow$ *new-color*	
5	**while** *frontier*.SIZE > 0 **do**	As long as frontier is not empty, examine a pixel from frontier; if any neighbor has the same color as the original seed pixel color, then set neighbor to new color, and add neighbor to frontier.
6	$p \leftarrow$ *frontier*.POP()	
7	**for** $q \in \mathcal{N}(p)$ **do**	
8	**if** $I(q) ==$ *orig-color* **then**	
9	*frontier*.PUSH(q)	
10	$I(q) \leftarrow$ *new-color*	

[†] Section 10.1.3 (p. 450).
[‡] *Conditional erosion* is defined similarly, by replacing the intersection operator with union and the dilation with erosion.

Figure 4.22 Step-by-step illustration of the 4-neighbor FLOODFILL algorithm on a small image. The frontier is shown below the image. Starting from the seed pixel labeled 1, the interior region of white pixels is changed to yellow by the algorithm, while orange is used to indicate the pixels being considered in the current expansion. The labels are artificially introduced to aid in associating pixels in the image with those in the frontier.

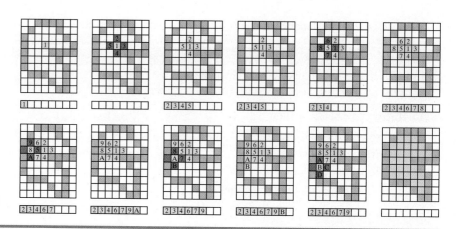

The image I is called the *marker image*, while C is called the *mask image*. If B is an *elementary* structuring element (either B_4 or B_8) containing 1s only in pixels that are a distance of 1 from the center, then conditional dilation is also known as **geodesic dilation** of size 1, denoted $I \oplus_C^1 B$. The geodesic dilation of size n is therefore the repetition of n conditional dilations, each with the same elementary structuring element:

$$I \oplus_C^n B \equiv \underbrace{\left(\left(\left(I \oplus_C B \right) \oplus_C B \right) \cdots \oplus_C B \right)}_{n \text{ iterations}} \tag{4.76}$$

We are usually interested in repeating conditional (or geodesic) dilation until convergence, a process known as **morphological reconstruction by dilation**:

$$I \oplus_C^\infty B \equiv \underbrace{\left(\left(\left(I \oplus_C B \right) \oplus_C B \right) \cdots \oplus_C B \right)}_{k \text{ iterations}} \tag{4.77}$$

where k is the smallest number such that $I \oplus_C^{k+1} = I \oplus_C^k$. In other words, the dilations are repeated until they have no effect, $I \oplus_C^\infty B = \left(\left(I \oplus_C^\infty B \right) \oplus_C B \right)$.

It is easy to see that the 4-neighbor floodfill of a 0-valued region in a binary image with seed pixel $\mathbf{p}$ is equivalent to $P \oplus_{\bar{I}}^\infty B_4$, where P is an image with a 1 at $\mathbf{p}$ and 0s everywhere

---

ALGORITHM 4.6 Perform floodfill, saving the output in a separate image

FLOODFILLSEPARATEOUTPUT(I, O, p, *new-color*)

Input: image I, seed pixel p, and new color preallocated output image O (same size as I)
Output: all pixels in O connected to p in I are colored *new-color*

```
1   orig-color ← I(p)
2   frontier.PUSH(p)
3   O(p) ← new-color
4   while frontier.SIZE > 0 do
5       p ← frontier.POP()
6       for q ∈ N(p) do
7           if I(p) == orig-color AND O(q) ≠ new-color then
8               frontier.PUSH(q)
9               O(q) ← new-color
```

else. The first iteration computes $(P \oplus B_4) \cap \neg I$, which includes the seed pixel and its 4-neighbors. The next iteration grows the region to include the 4-neighbors of all of these pixels, but the intersection with the complement of the original image retains only those pixels whose value in the original image is 0. This process continues until the entire region is filled, at which time repeated iterations do not change the output.

Viewing floodfill as conditional dilations is valuable due to the extremely compact mathematical notation that can be used to describe the computation, as seen in the $P \oplus_{\neg I}^{\infty} B_4$ above. Indeed, this ability to represent fairly complex operations using the basic building blocks of dilation or erosion is one of the strengths of mathematical morphology. On the other hand, conditional dilation is not used in practical implementations due to the substantial computational overhead incurred by touching every pixel at each iteration.

Closely related concepts are **opening by reconstruction** and **closing by reconstruction**, defined as

$$(I \oplus^n B) \check{\ominus}_I^{\infty} B' \qquad (\text{closing by reconstruction}) \qquad (4.78)$$

$$(I \check{\ominus}^n B) \oplus_I^{\infty} B' \qquad (\text{opening by reconstruction}) \qquad (4.79)$$

which are the opening and closing, respectively, by reconstruction of I of size n. In other words, the image I is used as the mask image, whereas an eroded or dilated version of I is used as the marker image. The two SEs B and B' may be the same or they may be different. Recall that morphological opening removes small objects, and the subsequent dilation restores the remaining objects. Similarly, opening by reconstruction (which is not a morphological opening but rather an algebraic opening) also removes small objects but then restores the remaining objects *exactly*. In other words, all image features that do not contain the SE are removed, while the others are not changed.

4.2.3 Connected Components

Recall that two pixels are said to be connected if there is a path between them consisting of pixels all having the same value. A **connected component** is defined as a maximal set of pixels that are all connected with one another. The connected components of an image are the *equivalence classes* of the image with respect to the *equivalence relation* "is connected to," where an equivalence relation is a reflexive, symmetric, and transitive relation that partitions a set into disjoint subsets (which are the equivalence classes).

Connected component labeling is the process of assigning a unique identifier to every pixel in the image indicating to which connected component it belongs. For example, this process is used to separate the various foreground objects in a binary image. Given a binary image with ON pixels signifying foreground and OFF pixels indicating background, the result of a connected component labeling algorithm is a two-dimensional array (the same size as the image) in which each element has been assigned an integer label indicating the region to which its pixel belongs. That is, all the pixels in one contiguous foreground region are assigned one label, while all the pixels in a different contiguous foreground region are assigned a different label, all the pixels in a contiguous background region are assigned yet another label, and so forth. Thus, connected components is a partitioning problem, because it assigns the image pixels to a relatively small number of discrete groups.

One way to implement connected components is by repeated applications of floodfill, starting each iteration with a new unlabeled pixel as the seed point. This is shown in Algorithm 4.7. Initially the output label array L is created to be the same size as the input image, all elements in this output array are unlabeled, and a global label is set to zero. Then the image is scanned, and whenever a pixel is encountered that has not yet been labeled, floodfill is applied to the image with that pixel as the seed pixel, filling the elements in the output array with the global label. The global label is then incremented, and the scan is

ALGORITHM 4.7 Perform connected components by repeated applications of floodfill

CONNECTEDCOMPONENTSBYREPEATEDFLOODFILL(I)

Input: image I
Output: integer image L containing a label for each pixel

1   **for** $(x, y) \in L$ **do**
2       $L(x, y) \leftarrow$ UNLABELED
3   *next-label* $\leftarrow 0$
4   **for** $(x, y) \in L$ **do**
5   **if** $L(x, y) ==$ UNLABELED **then**
6   FLOODFILLSEPARATEOUTPUT($I, L, (x, y),$ *next-label*)
7   *next-label* $\leftarrow_+ 1$
8   **return** L

continued. This relatively simple procedure labels each pixel with the value of its contiguous region. One advantage of this algorithm is that the regions are labeled with consecutive labels of 0, 1, 2, ... so that the number of regions found is given by the global label.

A more common approach, sometimes known as the **classic connected components algorithm**, involves scanning the image twice, as shown in Algorithm 4.8. In the first pass, the image is sequentially scanned from left to right and from top to bottom, and

ALGORITHM 4.8 Perform the classic union-find connected components algorithm

CONNECTEDCOMPONENTSBYUNIONFIND(I)

Input: image I
Output: label image L containing a label for each pixel

1   ▶ first pass
2   **for** $y \leftarrow 0$ **to** *height* $- 1$ **do**   ▶ Scan the image from top to bottom,
3   **for** $x \leftarrow 0$ **to** *width* $- 1$ **do**   from left to right.
4           $v \leftarrow I(x, y)$ ▶ If pixel value
5   **if** $v == I(x - 1, y)$ AND $v == I(x, y - 1)$ **then**   ▶ is the same as both neighbors,
6               $L(x, y) \leftarrow L(x - 1, y)$ then set the label to either label arbitrarily,
7   SETEQUIVALENCE($L(x - 1, y), L(x, y - 1)$)   and declare the labels equivalent.
8   **elseif** $v == I(x - 1, y)$ **then**   ▶ Otherwise, if pixel value is
9               $L(x, y) \leftarrow L(x - 1, y)$ the same as either neighbor,
10   **elseif** $v == I(x, y - 1)$ **then**   then set label to the label
11              $L(x, y) \leftarrow L(x, y - 1)$ of the appropriate pixel.
12   **else**   ▶ If pixel value is different from both neighbors,
13              $L(x, y) \leftarrow$ *next-label*   then declare a new region,
14   *next-label* $\leftarrow_+ 1$ and increment the global label.
15 ▶ second pass
16 **for** $(x, y) \in L$ **do**   ▶ For each pixel,
17      $L(x, y) \leftarrow$ GETEQUIVALENTLABEL($L(x, y)$)   traverse the equivalence table.
18 **return** L

SETEQUIVALENCE(a, b)

1   $a' \leftarrow$ GETEQUIVALENTLABEL(a)
2   $b' \leftarrow$ GETEQUIVALENTLABEL(b)
3   **if** $a' > b'$ **then**
4       $equiv[a'] \leftarrow b'$
5   **else**
6       $equiv[b'] \leftarrow a'$

GETEQUIVALENTLABEL(a)

1   **if** $a == equiv[a]$ **then**
2   **return** a
3   **else**
4       $equiv[a] \leftarrow$ GETEQUIVALENTLABEL($equiv[a]$)
5   **return** $equiv[a]$

all the pixels are labeled with preliminary labels based on a subset of their neighbors. For 4-neighbor connectedness, the algorithm compares a pixel with its two neighbors above and to the left; for 8-neighbor connectedness, the pixel is also compared with the two neighbors diagonally above-left and above-right, as shown in Figure 4.23. While performing the preliminary labeling, an equivalence table is built to keep track of which preliminary labels need to be merged. In the second pass, the label of each pixel is set to the equivalence of its preliminary label, using the equivalence table. This approach is also known as a **union-find** algorithm because it performs the two operations of finding regions and merging them. It is the first algorithm we have considered where the order in which the pixels are processed matters.

An example of this 4-connected version of the algorithm at work can be seen in Figure 4.24, while the output on a real image is displayed in Figure 4.25. To extend the code to 8-neighbors, simply insert two additional tests comparing the pixel with its neighbors $I(x - 1, y - 1)$ and $I(x + 1, y - 1)$; and set equivalences between any of the four neighboring pixels (left, above, above-left, and above-right) with the same value as the pixel. Note that out-of-bounds accessing has been ignored; to turn this pseudocode into executable code, bounds checking must be added in the **if** and **elseif** clauses, so that the top-left pixel $(0, 0)$ in the image falls through to the **else** clause, and all remaining pixels along the top row and left column are only compared with existing pixels.

Algorithm 4.8 relies on two helper functions. The first function, SETEQUIVALENCE, sets the equivalence between two labels, storing the equivalence in a one-dimensional array of integers, $equiv$. The array is initialized with its own indices, i.e., $equiv[i] \leftarrow i$ for all i. The convention is adopted that $equiv[i] \leq i$, to avoid creating cycles in the data structure. In other words, the smallest label in each set is taken to be the representative label of the set. (It is assumed that the array grows dynamically in size or is created large enough to hold the total number of labels encountered.) The second helper function, GETEQUIVALENTLABEL, returns an equivalent label by simply accessing the array, using recursion to ensure that

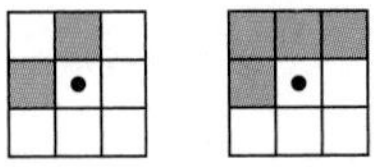

Figure 4.23 Masks for the 4-neighbor and 8-neighbor versions of the classic union-find connected components algorithm. The colored pixels are neighbors of the central pixel that are examined by the algorithm.

Figure 4.24 Classic union-find connected components algorithm on an example binary image. From left to right: The input image, the labels after the first pass, and the labels after the second pass. Below the image is the equivalence table, with green arrows pointing from a label to its equivalent label. Notice that the final image contains gaps; for example, no pixel is labeled 2.

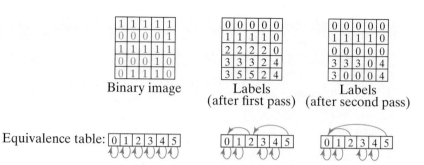

Binary image   Labels (after first pass)   Labels (after second pass)

Equivalence table: 0 1 2 3 4 5   0 1 2 3 4 5   0 1 2 3 4 5

the smallest possible label has been found. While getting an equivalent label, the array is updated with the smallest possible equivalent label. An alternative is to traverse the equivalence table once between the two passes, after which GETEQUIVALENTLABEL(a) can simply call *equiv*[a] without having to resort to recursion.

Both algorithms for connected components are linear in the number of pixels. To be more precise, the union-find algorithm applied to an image with n pixels is $O(n\alpha(n))$, where $\alpha(n)$ is the inverse Ackerman function that grows so extremely slowly that $\alpha(n) \leq 4$ for any conceivable image.[†] The four-neighbor floodfill version requires touching most pixels seven times (to set the output to UNLABELED in the initialization, to check whether the pixel has been labeled, and five times during the floodfill to set the pixel and check its label from the four directions); pixels along the border of two regions may require slightly more. The union-find algorithm involves touching each pixel just four times (the first pass, the second pass, and the check from the pixels to its right and below). Thus, in practice the union-find algorithm is slightly more efficient in run time despite the additional computation required by the equivalence table. However, one drawback of union-find is that it leaves gaps in the labels. That is, the final result might have (as in the example of Figure 4.24) a region 1 and a region 3, but no region of pixels labeled 2. This inconvenience can be removed by another pass through the equivalence table to produce a new equivalence table in which the base labels are sequential.

With either algorithm, it is easy to compute region properties such as area, moments (discussed later in the chapter), and bounding box. All of these quantities can be updated during the connected components algorithm with appropriate calculations each time an output pixel is set, with minimal overhead. The extension is left as an exercise.[‡]

Figure 4.25 Because the connected components algorithm assumes neighboring pixels have the exact same value, it works best on images with a small number of values. Shown here are an input image quantized to four gray levels (left) and the result of connected components (right), pseudocolored for display.

Stan Birchfield

[†] For example, $\alpha(n) = 4$ for an image with a googol $n = (10^{100})$ of pixels.
[‡] Problem 4.28.

4.2.4 Boundary Tracing

Given a contiguous region of pixels found by a floodfill or connected components algorithm, it is oftentimes useful to find its **boundary**. We distinguish between the **region boundary**, which is the smallest set of pixels that encloses (in some sense) all the pixels in the region, and the **hole boundary**, which is the smallest set of pixels that encloses all the holes (if any) inside the region. If the region contains no holes, then the hole boundary is the empty set. The union of the region and hole boundaries is the **complete boundary**. In a discrete image, it is not trivial to define what is meant by "enclosing" a set of pixels. This difficulty leads to two alternatives: the **inner boundary**, which consists of all pixels in the region that are next to some pixel not in the region, and the **outer boundary**, which consists of all pixels not in the region that are next to some pixel in the region. If we let R be a region represented as a binary image, that is $R(\mathbf{p}) = $ ON if $\mathbf{p}$ is in the region, and $R(\mathbf{p}) = $ OFF otherwise, where $\mathbf{p} = (x, y)$ is a pixel, then these definitions are given by

$$\text{inner boundary} = \{\mathbf{p} : \mathbf{p} \in R, \exists \mathbf{q} \in \mathcal{N}(\mathbf{p}), \mathbf{q} \notin R\} \tag{4.80}$$

$$\text{outer boundary} = \{\mathbf{p} : \mathbf{p} \notin R, \exists \mathbf{q} \in \mathcal{N}(\mathbf{p}), \mathbf{q} \in R\} \tag{4.81}$$

Technically, these definitions are for the inner *complete* boundary (with the inner *region* boundary and inner *hole* boundary as subsets) and the outer *complete* boundary (with the outer *region* boundary and outer *hole* boundary as subsets), respectively. Figure 4.26 illustrates these definitions.

The inner complete boundary of the region can be computed easily enough via the difference between the region itself and an eroded version of the region:

$$\text{inner boundary} = R \text{ AND } \neg(R \ominus B) \tag{4.82}$$

where B is either B_4 or B_8. That is, each pixel in the output is ON if and only if the corresponding pixel in R is ON, but in the eroded version it is OFF. Similarly, the outer complete boundary can be computed using dilation:

$$\text{outer boundary} = (R \oplus B) \text{ AND } \neg R \tag{4.83}$$

However, outer boundaries should be computed with care because they cannot easily be represented for any region that touches the image border. In either case, the choice of

Figure 4.26 A binary region with one hole, and different definitions of the boundary of the region. All results are shown using 4-neighbors. Colored cells are ON, white cells are OFF.

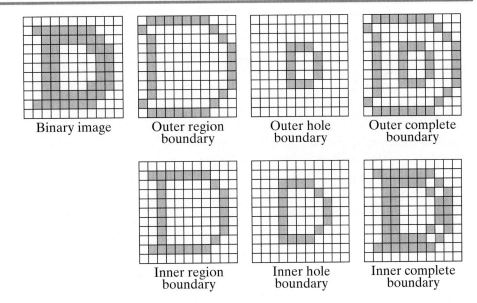

Binary image

Outer region boundary

Outer hole boundary

Outer complete boundary

Inner region boundary

Inner hole boundary

Inner complete boundary

the structuring element B will affect the result in a nonintuitive way: B_4 will produce an 8-connected boundary (see Figure 4.26), whereas B_8 will produce a 4-connected boundary.

While mathematical morphology provides a conveniently compact description of the boundary, procedures based on morphology—such as those in Equations (4.82)–(4.83)—only return the (unordered) *set* of pixels on the boundary. However, for some applications it is necessary to compute the boundary as a path, i.e., as an (ordered) *sequence* of pixels. A simple procedure for computing the boundary of a region as a path is the **wall-following algorithm**, also known as **Moore's boundary tracing algorithm**.[†] The wall-following algorithm derives its name from the analogy of a blindfolded person desiring to traverse the edges of a room. By holding out his left arm stiff to the side and his right arm stiff in front, the person continually walks straight until either contact with the left wall has been lost (in which case the person turns left) or a wall is detected in front (in which case the person turns right).

In a similar manner, the wall-following algorithm traverses the boundary of a region by examining pixels in front and to the left, turning appropriately based upon the values of the pixels. The algorithm, shown in Algorithm 4.9, computes the clockwise inner boundary, but other variations are easily obtained with slight modifications. We adopt the convention of the **Freeman chain code** directions, shown in Figure 4.27, in which the

ALGORITHM 4.9 Perform wall following

WALLFOLLOW(I)

Input: binary image I containing a single ON region
Output: clockwise sequence of pixels on the inner boundary of the region

1	$p \leftarrow p_0 \leftarrow$ FINDBOUNDARYPIXEL(I)	▷ Find first ON pixel from top-right corner.
2	$dir \leftarrow 0$	▷ Set initial direction to the right.
3	**repeat**	
4	$\quad$ *boundary-path*.PUSH(p)	
5	$\quad$ **if** LEFT (I, p, dir) $==$ ON **then**	▷ Turn left and move forward.
6	$\quad\quad$ $dir \leftarrow$ TURNLEFT(dir)	
7	$\quad\quad$ $p \leftarrow$ MOVEFORWARD(p, dir)	
8	$\quad$ **elseif** FRONT(I,p,dir) $==$ OFF **then**	▷ Turn right.
9	$\quad\quad$ $dir \leftarrow$ TURNRIGHT(dir)	
10	$\quad$ **else**	
11	$\quad\quad$ $p \leftarrow$ MOVEFORWARD(p, dir)	▷ Move forward.
12	$\quad$ **until** $p == p_0$ AND $dir == dir_0$	
13	$\quad$ **return** *boundary-path*	

FINDBOUNDARYPIXEL(I)

Input: binary image I containing a single ON region
Output: a pixel on the inner boundary of the region

1	**for** $y \leftarrow 0$ **to** *height* $- 1$ **do**	▷ Scan from top to bottom
2	$\quad$ **for** $x \leftarrow$ *width* $- 1$ **to** 0 **step** -1 **do**	and from right to left
3	$\quad\quad$ **if** $I(x,y) ==$ ON **then**	returning the first ON pixel encountered.
4	$\quad\quad\quad$ **return** (x, y)	▷ (Note: The order of scanning affects the starting direction.)

[†] After G. A. Moore, unrelated to E. F. Moore of the Moore neighborhood.

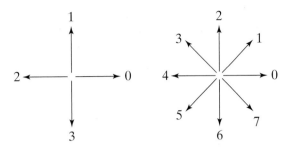

list of directions proceeds counterclockwise starting from the right (positive x axis). For 4-neighbor connectedness, *dir* takes on values in the set $\{0, 1, 2, 3\}$, while for 8-neighbor connectedness, *dir* takes on values of $\{0, 1, \ldots, 7\}$. With this convention, TURNLEFT and TURNRIGHT return the next and previous direction in the list, respectively, using modulo arithmetic: TURNLEFT(*dir*) returns $(dir - 1, z)$, while TURNRIGHT(*dir*) returns $(dir + 1, z)$, assuming z-neighbor connectedness. MOVEFORWARD computes the pixel attained after moving forward according to the current direction and the current pixel. Recalling that the y axis points down, this means MOVEFORWARD$(x, y, 0)$ returns $(x + 1, y)$, and MOVEFORWARD$(x, y, 1)$ returns $(x, y - 1)$ if $z = 4$, or $(x + 1, y - 1)$ if $z = 8$. An example of the wall-following algorithm is shown in Figure 4.28.

While the algorithm is straightforward, careful attention must be paid to the starting position. For example, if we select a random pixel inside the region and scan its neighbors iteratively until a boundary pixel is found, there is danger of finding a pixel on the hole boundary rather than the region boundary, in which case the wall-following algorithm will simply trace around the hole. Therefore, to ensure that FINDBOUNDARYPIXEL returns a pixel on the region boundary, it is best to start from a pixel known to be outside the region (at the image border, for example), scanning until a boundary pixel is found. The starting direction must be facing in the general direction of the exterior. For the ending condition, it is necessary (as shown in the code) to test for both pixel location *and* direction to handle the case of a single-pixel-thick isthmus or cape (see Figure 4.9).

Wall following is useful for several tasks. To compute the perimeter of a region, for example, apply the 8-neighbor version of WALLFOLLOW, then compute the distance along the resulting path using the techniques described in the next section. The 8-neighbor version

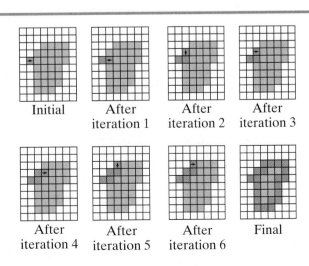

Initial   After iteration 1   After iteration 2   After iteration 3

After iteration 4   After iteration 5   After iteration 6   Final

will trace exactly the same pixels that result from subtracting the B_4-eroded image from the original region. As a result, the path computed by the 8-neighbor version contains pixels that are 4-neighbors of the background. It can be shown that consecutive pixels in the 8-neighbor path are m-adjacent.

The 4-neighbor version of wall following can be used as well, but in the resulting sequence of pixels the distance between any two consecutive pixels will be 1. As a result, the perimeter will be simply the number of pixels remaining after subtracting the eroded image from the original region, using B_8 as the structuring element. The path computed by the 4-neighbor version contains pixels that are 8-neighbors of the background.

4.2.5 Hole Filling

Sometimes a region contains small holes, and it is desired to fill the holes. A conceptually simple way to do this is to use morphological reconstruction by dilation:

$$\text{region without holes} = F \oplus^{\infty}_R B \tag{4.84}$$

where F is a marker image with at least one ON-pixel somewhere along the border of the image (but outside the region), and R is a binary image with ON pixels inside the region and OFF pixels outside the region as well as inside the holes. A practical implementation of this approach uses floodfill to fill the background, after which the complement of the filled region yields the desired hole-free region. This approach is especially useful if it is desired to fill in all the holes of all the regions of the image. In such a case the marker image F should be set to $F = I_{border} \cap \neg I$, where I_{border} contains ON pixels along the border of the image and OFF pixels everywhere else.

If the region is small compared to the size of the image, then the approach just described wastes much computation in filling in the entire background. An alternative is to use wall following to find the region boundary (as distinguished from the hole boundary), then perform floodfill on the binary image consisting of just the region boundary. This same approach can also be used to fill in all the regions of an image by applying the procedure repeatedly.

4.3 Computing Distance in a Digital Image

For many applications, such as measuring the perimeter of a region or the length of an object, it is necessary to compute the distance between two pixels in an image. In this section we discuss various techniques for efficiently estimating such a quantity.

4.3.1 Distance Functions

Let $\mathbf{p} = (p_x, p_y)$ and $\mathbf{q} = (q_x, q_y)$ be the coordinates of two pixels in an image. A function $d(\mathbf{p}, \mathbf{q})$ of two vectors is a distance function (or *metric*) if it satisfies three properties:

- $d(\mathbf{p}, \mathbf{q}) \geq 0$ and $d(\mathbf{p}, \mathbf{q}) = 0$ iff $\mathbf{p} = \mathbf{q}$ (non-negativity and reflexivity)
- $d(\mathbf{p}, \mathbf{q}) = d(\mathbf{q}, \mathbf{p})$ (symmetry)
- $d(\mathbf{p}, \mathbf{q}) \leq d(\mathbf{p}, \mathbf{r}) + d(\mathbf{r}, \mathbf{q})$ (triangle inequality)

for all possible coordinates $\mathbf{p}, \mathbf{q}, \mathbf{r} \in \mathbb{R}^2$, where iff means "if and only if". A function satisfying only the first two conditions is called a *semi-metric*, an example being the quadratic function $\|\mathbf{p} - \mathbf{q}\|^2 = (p_x - q_x)^2 + (p_y - q_y)^2$.[†]

[†] A function satisfying, in addition to all three conditions above, a particular fourth condition that we shall describe later is called an *ultrametric*, as covered in Section 10.3.4 (p. 481).

EXAMPLE 4.4	Show that the quadratic function does not obey the triangle inequality.
Solution	As a counter-example, let $\mathbf{p} = (-1, 0)$, $\mathbf{q} = (1, 0)$, and $\mathbf{r} = (0, 0)$. Then $d(\mathbf{p}, \mathbf{q}) = 4$, $d(\mathbf{p}, \mathbf{r}) = 1$, and $d(\mathbf{r}, \mathbf{q}) = 1$. Since $4 > 1 + 1$, the triangle inequality does not always hold for the quadratic function, and therefore it is not a metric.

Three common distance functions (metrics) for pixels are the **Euclidean distance**, which is the square root of the quadratic function, and two other functions that are approximations to Euclidean:

$$d_E(\mathbf{p}, \mathbf{q}) = \sqrt{(p_x - q_x)^2 + (p_y - q_y)^2} \quad \text{(Euclidean)}$$

$$d_4(\mathbf{p}, \mathbf{q}) = |p_x - q_x| + |p_y - q_y| \qquad \text{(Manhattan, or city-block)}$$

$$d_8(\mathbf{p}, \mathbf{q}) = \max(|p_x - q_x|, |p_y - q_y|) \qquad \text{(chessboard)}$$

The **Manhattan distance** is known as d_4 because the pixels that are one unit of distance away are the 4-neighbors of the pixel, while the **chessboard distance** is known as d_8 because the pixels that are one unit of distance away are the 8-neighbors of the pixel. It is worth noting that Manhattan always overestimates Euclidean, while chessboard always underestimates it: $d_8(\mathbf{p}, \mathbf{q}) \leq d_E(\mathbf{p}, \mathbf{q}) \leq d_4(\mathbf{p}, \mathbf{q})$. Moreover, the chessboard distance is never more than 30% away from the Euclidean distance, and the Manhattan distance is never more than 42% away:

$$0.7 d_E(\mathbf{p}, \mathbf{q}) < d_8(\mathbf{p}, \mathbf{q}) \leq d_E(\mathbf{p}, \mathbf{q}) \leq d_4(\mathbf{p}, \mathbf{q}) < 1.42 d_E(\mathbf{p}, \mathbf{q}) \tag{4.85}$$

The proof of these inequalities is left to the reader.[†]

These distance metrics are related to the **vector norm**. The L^p-norm of a vector $\mathbf{v} \in \mathbb{R}^n$ is defined as

$$\left\| \mathbf{v} \right\|_p \equiv \left(\sum_{i=1}^{n} |v_i|^p \right)^{\frac{1}{p}} \tag{4.86}$$

where v_i is the ith element of $\mathbf{v}$. The most common values of p are 1, 2, and ∞:

$$\|\mathbf{v}\|_1 = |v_1| + \cdots + |v_n| \qquad (\text{absolute value-, or, } L^1\text{-norm}) \tag{4.87}$$

$$\|\mathbf{v}\|_2 = \sqrt{v_1^2 + \cdots + v_n^2} \qquad (\text{Euclidean-, or } L^2\text{-norm}) \tag{4.88}$$

$$\|\mathbf{v}\|_\infty = \max\{|v_1|, \ldots, |v_n|\} \qquad (\text{maximum-, or } L^\infty\text{-norm}) \tag{4.89}$$

Where there is no subscript, the Euclidean-norm can safely be assumed: $\|\mathbf{v}\| = \|\mathbf{v}\|_2$. It is easy to see that the L^1-norm of $\mathbf{v} \equiv \mathbf{p} - \mathbf{q}$ is the Manhattan distance, the L^2-norm is the Euclidean distance, and the L^∞-norm is the chessboard distance.

Another metric that should be mentioned in this context is the **Mahalanobis distance**. As we have just seen, if $\mathbf{v}$ is the vector from one point to another, then the Euclidean distance between those points is given by $\|\mathbf{v}\|_2 = \sqrt{\mathbf{v}^\top \mathbf{v}}$, which is the square root of the inner product of the vector with itself. The Mahalanobis distance introduces a covariance matrix within the inner product, $\sqrt{\mathbf{v}^\top \mathbf{C}^{-1} \mathbf{v}}$, which has the effect of scaling the different axes differently:

$$d_{Mahal}(\mathbf{p}, \mathbf{q}; \mathbf{C}) = \sqrt{(\mathbf{p} - \mathbf{q})^\top \mathbf{C}^{-1} (\mathbf{p} - \mathbf{q})} \qquad (\text{Mahalanobis}) \tag{4.90}$$

where $\mathbf{C}$ is the covariance matrix. Obviously, if the covariance matrix is the identity matrix, then the Mahalanobis distance reduces to the Euclidean distance.

[†] Problems 4.31 and 4.32.

4.3.2 Path Length

Now suppose that we wish to calculate the length of a specific path ϕ between pixels $\mathbf{p}$ and $\mathbf{q}$, which as we saw earlier is defined as a sequence of pixels beginning with $\mathbf{p}$ and ending with $\mathbf{q}$ such that each successive pixel in the path is adjacent to the previous pixel. Let n_o be the number of **isothetic moves** in the path, where an isothetic move is one that is horizontal or vertical (i.e., the two pixels are 4-neighbors of each other). Similarly, let n_d be the number of **diagonal moves** in the path (i.e., the two pixels are D-neighbors of each other). Generally the path will use m-adjacency, but if 4-adjacency is used, then n_d is simply zero.

Even in a continuous space, the length of a path (or curve) is not always well-defined. Consider, for example, the well-known fractal question, "What is the length of the British coastline?" Depending upon the scale of interest, the resulting values can be significantly different from one another. Similarly, it is impossible to precisely define or solve the problem in a discrete image. Nevertheless, one reasonable approach is to sum the Euclidean distances between consecutive pixels along the path. Since the Euclidean distance between two pixels that are 4-neighbors of each other is 1, and the Euclidean distance between two pixels that are D-neighbors of each other is $\sqrt{2}$, this is equivalent to measuring the length of the path ϕ as

$$\text{length}(\phi) = n_o + n_d\sqrt{2} \qquad\qquad (\text{Freeman}) \qquad (4.91)$$

which is known as the **Freeman formula**. An alternate approach is to rearrange the node pairs and use the **Pythagorean theorem** to estimate the length of the curve as the hypotenuse of the resulting right triangle:

$$\text{length}(\phi) = \sqrt{n_d^2 + (n_o + n_d)^2} \qquad\qquad (\text{Pythagorean}) \qquad (4.92)$$

While the Freeman formula generally overestimates the length of a curve, the Pythagorean theorem usually underestimates it. Insight into the problem is obtained by noticing that the previous two equations can be written as special cases of the more general formula:

$$\text{length}(\phi) = \sqrt{n_d^2 + (n_d + cn_o)^2} + (1 - c)n_o \qquad (\text{Kimura}) \qquad (4.93)$$

where $c = 0$ for the Freeman formula and $c = 1$ for the Pythagorean theorem. By setting c to $\frac{1}{2}$, a compromise is achieved between overestimation and underestimation known as the **Kimura** method. In practice, this method works well, as shown in Figure 4.29.

4.3.3 Chamfering

We have seen how to compute the distance between two pixels, as well as the distance between two pixels along a specific path. But what if we want to compute a large number of distances? Such a problem arises, for example, when performing template matching using intensity edges, in which we need to compute distances between all the intensity edges in a template and their closest match in the image. For reasons of computational efficiency, it is not feasible to compute all of these distances directly. Instead, it is better to precompute a **distance transform**, which is an array that stores the distance from each pixel in the image to its nearest element in the set of interest (e.g., the intensity edges). A computational trick allows such an array to be computed efficiently using only a small number of passes through the image (usually just 2).

Let us define the (a,b) **chamfer distance** between pixel $\mathbf{p} = (p_x, p_y)$ and pixel $\mathbf{q} = (q_x, q_y)$ as $d_{a,b}(\mathbf{p}, \mathbf{q}) = \min_\phi\{an_o + bn_d\}$, where a and b are nonnegative values, n_o and n_d are the number of isothetic and diagonal moves in the path ϕ, respectively, and the minimum is computed over all possible paths between the two pixels. We shall assume that $0 < a \le b$, in which case the (a,b) chamfer distance $d_{a,b}$ is a metric. We also assume that $b \le 2a$ (known as the **Montanari condition**), which ensures that diagonal moves are not ignored. Rather than searching over all possible paths, simple observation reveals that the

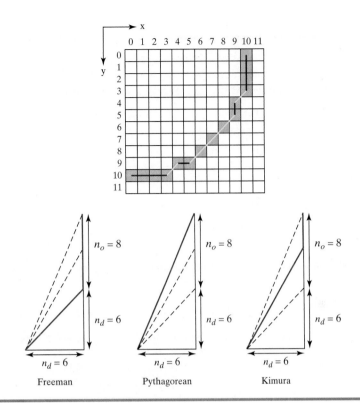

Figure 4.29 A discretized 90-degree sector of a circle with radius 10, where purple lines indicate isothetic moves, and yellow lines indicate diagonal moves. The number of isothetic and diagonal moves are $n_o = 8$ and $n_d = 6$, respectively. The true arc length is $10(\pi/2) = 15.7$. The estimated path length according to the three formulas is 16.5 (Freeman), 15.2 (Pythagorean), and 15.7 (Kimura).

shortest path always consists of a horizontal or vertical line segment, along with a diagonal line segment if the two pixels do not share the same column or row. That is, if we define $d_x = |p_x - q_x|$ and $d_y = |p_y - q_y|$, then the distance between $\mathbf{p}$ and $\mathbf{q}$ is simply given by $d_{a,b}(\mathbf{p}, \mathbf{q}) = an_o + bn_d$, where $n_d = \min(d_x, d_y)$ is the number of moves along the diagonal line segment and $n_o = \max(d_x, d_y) - n_d$ is the number of remaining isothetic moves. This single formula arises because the Montanari condition favors diagonal moves. If, on the other hand, the Montanari condition does not hold, then diagonal moves are ignored, in which case the distance is $d_{a,b}(\mathbf{p}, \mathbf{q}) = a(d_x + d_y)$, which is a scaled version of the Manhattan distance.

Assuming there is no obstruction between $\mathbf{p}$ and $\mathbf{q}$, the Euclidean distance is usually considered the "correct" distance. Other distance functions, known as **quasi-Euclidean**, are approximations to this Euclidean distance. It can be shown that the (a,b) chamfer distance that best approximates Euclidean is the one with $a = 1$ and $b = \frac{1}{\sqrt{2}} + \sqrt{\sqrt{2} - 1} \approx 1.351$, where the value of b is just slightly less than that used in the Freeman formula. A nearby integer ratio is 4/3, so if there is a need to avoid floating point computations, then $d_{3,4}$ yields a reasonable approximation to the Euclidean distance (scaled by the factor 3). It is easy to see that if $a = 1$ and $b = \infty$, the chamfer distance reduces to Manhattan because it ignores diagonal moves; or if $a = 1$ and $b = 1$, then it reduces to chessboard because it treats isothetic and diagonal moves equally. This relationship between Euclidean and quasi-Euclidean helps shed light on why this procedure is called *chamfering*. In woodworking, chamfering refers to the process of reducing the harsh 90-degree angles of a surface by introducing a beveled edge. In a similar manner, the chamfer distance in an image approximates the Euclidean distance by smoothing out the harsh corners of the Manhattan or chessboard distances by allowing appropriately weighted diagonal moves.

ALGORITHM 4.10 Compute the chamfer distance for all pixels in an image

CHAMFER(I, a, b)

Input: binary image I, chamfer parameters a and b
Output: image D containing the distance of each pixel to the nearest ON pixel

```
1  ▶ first pass
2  for y ← 0 to height − 1 do
3       for x ← 0 to width − 1 do
4            if I(x, y) = ON then
5                 D(x, y) ← 0
6            else
7                 D(x, y) ← MIN(∞, a + D(x − 1, y), a + D(x, y − 1),
                       b + D(x − 1, y − 1), b + D(x + 1, y − 1))
8  ▶ second pass
9  for y ← height − 1 to 0 step −1 do
10      for x ← width − 1 to 0 step −1 do
11           if I(x, y) ≠ ON then
12                D(x, y) ← MIN(D(x, y), a + D(x + 1, y), a + D(x, y + 1),
                       b + D(x + 1, y + 1), b + D(x − 1, y + 1))
13 return D
```

Computing the (a,b) chamfer distance is straightforward, as shown in Algorithm 4.10. It involves two passes through the image, with the first pass scanning the image from left-to-right and top-to-bottom, then the second pass scanning in the reverse direction. For each pixel, four of its 8-neighbors are examined in one direction, then the other four in the reverse direction. One can think of the algorithm as casting shadows from the foreground pixels in the two diagonal directions (southeast in the first pass, and northwest in the second). After the first pass, the distances from every pixel to all of the pixels above and to the left have been computed, and after the second pass, the distances to all of the pixels have been computed.

In the pseudocode, $\delta(x, y) = \infty$ if x or y is out of bounds, and ∞ is meant to represent a large number that is greater than any possible distance in the image. The chamfering algorithm is similar to connected components in that the manner in which boundary pixels should be handled is specified precisely: out-of-bounds pixels should simply be ignored, so that the minimum is computed over fewer than five values when the pixel is along the image border. Figure 4.30 shows an example of the chamfering algorithm applied to an image, while Figure 4.31 illustrates one use of the chamfer distance. For the special case of Manhattan distance ($a = 1$ and $b = \infty$), the pseudocode simplifies to only require examining two of the 4-neighbors in each pass, as shown in Algorithm 4.11.

Since Manhattan always overestimates Euclidean, while chessboard always underestimates it, a combination of the two can be used. One approach is to alternate the computations between the two distance metrics as the image is scanned; this requires two passes through the image, as usual. Another approach is to compute both d_4 and d_8 and then to combine the results: $\max(d_8(p, q), \frac{2}{3}d_4(p, q))$, which requires four passes through the image. Of course, as mentioned earlier, a good approximation can also be obtained by simply computing $d_{3,4}$ and then dividing by three.

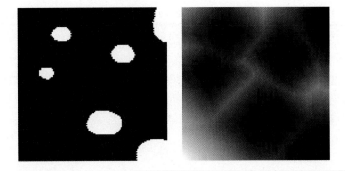

Figure 4.30 A binary image and its chamfer distance (brighter pixels indicate larger distances).

4.3.4 Exact Euclidean Distance

The chamfer distance approximations of the previous section are attractive due to their simplicity and computational efficiency. Nevertheless, exact Euclidean distances, if needed, can also be computed efficiently, with a running time that is linear in the number of pixels. Here we present a technique that, similar to computing the chamfer distance, requires just two passes through the image. Unlike the chamfer distance, however, this algorithm processes the image in one direction (rows or columns) in the first pass, then processes the orthogonal direction (columns or rows) in the second pass.

Consider a row of a binary image, which can be treated as a 1D function $I(x)$. To find the distance from every pixel in the row to the nearest ON pixel in the row is straightforward. The key insight is to recognize that the squared Euclidean distance is given by the lower envelope of vertical parabolas whose vertices are placed at each $(x, 0)$ for which $I(x)$ is ON, as shown in Figure 4.32. For example, suppose that the entire image is 0 except for one ON pixel at x_0. Then the Euclidean distance of any pixel in the image to the nearest ON pixel is given by $\delta(x) = \sqrt{(x - x_0)^2}$. Now, although in 1D it would be easy to cancel the square with the square root, such cancellation is not possible in 2D. Therefore, in anticipation of our later extension to 2D, let us consider the squared distance $\delta^2(x) = (x - x_0)^2$. The shape of this function is obviously a vertical parabola with vertex at $(x_0, 0)$. Now suppose the image consists of two ON pixels, one at x_0 and one at x_1. Then the squared distance from any pixel x to the nearest ON pixel would be the minimum of these two values, i.e., $\delta^2(x) = \min\{(x - x_0)^2, (x - x_1)^2\}$. Moreover, the two parabolas intersect halfway between the points at $(x_0 + x_1)/2$, leading to

$$\delta^2(x) = \begin{cases} (x - x_0)^2 & \text{if } x < \frac{1}{2}(x_0 + x_1) \\ (x - x_1)^2 & \text{otherwise} \end{cases} \tag{4.94}$$

Figure 4.31 For a region with concavities, its centroid may not even lie within the region. Therefore, the location with maximum chamfer distance (computed on the inverted image, so that the distance to the background is computed) is often a better estimate of the "center" of a region, because it yields the center of the largest part of the region.

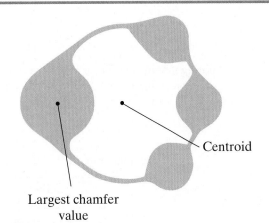

Centroid

Largest chamfer value

ALGORITHM 4.11 Compute the Manhattan distance for every pixel using chamfering

CHAMFERMANHATTAN(I)

Input: binary image I, chamfer parameters a and b
Output: image D containing the Manhattan distance of each pixel to the nearest ON pixel

```
1  ▶ first pass
2  for y ← 0 to height − 1 do
3        for x ← 0 to width − 1 do
4              if I(x, y) == ON then
5                    D(x, y) ← 0
6              else
7                    D(x, y) ← MIN(∞, 1 + D(x − 1, y), 1 + D(x, y − 1))
8  ▶ second pass
9  for y ← height − 1 to 0 step −1 do
10       for x ← width − 1 to 0 step −1 do
11             if I(x, y) ≠ ON then
12                   D(x, y) ← MIN(D(x, y), 1 + D(x + 1, y), 1 + D(x, y + 1))
13 return D
```

This argument is easily extended for any number of ON pixels.

The first pass of the algorithm, therefore, scans the 1D function and creates an array of the x coordinates of all the ON pixels. The elements of this array correspond to the parabolas, so the array stores everything we need in order to compute the distance function for all pixels in the row. To use the array, we simply consider the pixels in the row sequentially and compute the squared distance $(x - x_k)^2$, where x_k is the x-coordinate of the nearest parabola. The procedure for this first pass, shown in Algorithm 4.12, uses a dynamic array v to hold the x-coordinates of the parabolas and outputs a 1D array δ' that is the same length as the input. (We use *length* to refer to the number of pixels in the 1D horizontal or vertical slice, which is either the width or height, respectively, of the original 2D image.) Lines 8–9 advance to the next parabola when $x > \frac{1}{2}(x_k + x_{k+1})$ by updating the index k of the current parabola.

Now that we have seen how to compute the squared distance of a 1D signal (which is the first pass of the algorithm), let us see how it fits into the rest of the procedure. Given a 2D binary image I, let us construct a 2D function F so that

$$F(x, y) = \begin{cases} 0 & \text{if } I(x,y) = \text{ON} \\ \infty & \text{otherwise} \end{cases} \tag{4.95}$$

Omitting the square root for simplicity, the 2D exact squared Euclidean distance function is therefore computed by finding, for each pixel (x, y), the nearest pixel (x', y') for which $F(x', y') = 0$:

$$D^2(x, y) = \min_{x', y'} (x - x')^2 + (y - y')^2 + F(x', y') \tag{4.96}$$

$$= \min_{y'} \underbrace{(\min_{x'} (x - x')^2 + F(x', y'))}_{\text{first pass}} + (y - y')^2 \tag{4.97}$$

$$\underbrace{\phantom{= \min_{y'} (\min_{x'} (x - x')^2 + F(x', y')) + (y - y')^2}}_{\text{second pass}}$$

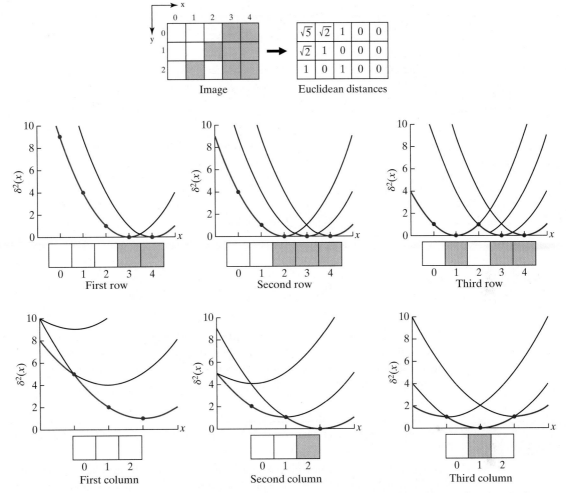

Figure 4.32 Computation of exact Euclidean distance from each pixel in a binary image to the nearest ON (blue) pixel. The first pass processes rows of the image to compute squared distances using the lower envelope of the parabolas. (The lower envelope is shown in red.) The second pass processes columns of the image using parabolas determined by the first pass. Based on P. F. Felzenszwalb and D. P. Huttenlocher. Pictorial structures for object recognition. *International Journal of Computer Vision*, 61(1): 55-79, Jan. 2005.

This separability of the horizontal and vertical processing is the trick that enables us to solve the problem efficiently. Separating the equation, we see that the two passes of the algorithm are similar:

$$D'^2(x, y) \equiv \min_{x'}(x - x')^2 + F(x', y) \qquad \text{(first pass)} \qquad (4.98)$$

$$D^2(x, y) = \min_{y'}(y - y')^2 + F'(x, y') \qquad \text{(second pass)} \qquad (4.99)$$

where $F' \equiv D'$. The first pass, whose procedure we have just examined, can be simplified because F is either 0 or ∞ everywhere, and values of ∞ can be ignored because we seek the minimum. As hinted in Equation (4.97), we shall describe the algorithm by processing the rows in the first pass, then the columns in the second pass; but the opposite order could just as easily have been adopted.

The second pass of the algorithm is slightly more complicated. The 1D function used as input in this case is not a row of the original binary image but rather a column of the output from the first pass. That is, after processing all the rows to compute the squared distances

ALGORITHM 4.12 Compute the exact Euclidean distance for all pixels in a 1D binary image

EXACTEUCLIDEANDISTANCE1DBINARY(I)

Input: 1D binary image I with length *length*
Output: 1D array δ' containing the squared Euclidean distance of each point to the nearest ON pixel

```
1  ▶ compute lower envelope
2  for x ← 0 to length − 1 do
3      if I[x] == ON then                              ▶ Store the x coordinates of all ON pixels
4          v.PUSH(x)                                   (this captures the lower envelope of parabolas).
5  ▶ fill in values
6  k ← 0
7  for x ← 0 to length − 1 do                          ▶ For each pixel in the row
8      while k ≤ v. SIZE − 2 AND x > (v[k] + v[k + 1])/2 do    advance to the next
9          k ←+ 1                                      parabola if necessary.
10     δ'[x] ← (x − v[k])²                             ▶ Evaluate the nearest parabola.
11 return δ'
```

along the rows, these values are then stored in a 2D array, whose columns are then processed in a similar manner to before. Let us imagine transposing this array, so that the variable x can continue to be used to index the 1D array even though it is a vertical slice of the original image. In the second pass, the squared Euclidean distance is again given by the lower

EXAMPLE 4.4

Compute the exact Euclidean distance of all pixels in Figure 4.32 to the nearest ON pixel.

Solution

Along the first row, the first pixel is at a distance of 3 from the nearest ON pixel, the second pixel is at a distance of 2, the third pixel is at a distance of 1, and the remaining two pixels are at a distance of 0. Similarly, along the second row, the distances are 2, 1, 0, 0, and 0. Along the third row, the distances are 1, 0, 1, 0, and 0. Putting these together, the squared distance from each pixel to the nearest ON pixel along the same row is given by

$$\begin{bmatrix} 9 & 4 & 1 & 0 & 0 \\ 4 & 2 & 0 & 0 & 0 \\ 1 & 0 & 1 & 0 & 0 \end{bmatrix}$$

Therefore, along the first column, we place a parabola at $x = 0$ with vertex at $\delta^2(x) = 9$, another one at $x = 1$ with vertex at $\delta^2(x) = 4$, and another one at $x = 2$ with vertex at $\delta^2(x) = 1$, where x is the row. The lower envelope of these parabolas yields the squared distances of 5, 2, and 1. Similarly, along the second column we place parabolas with vertices at $\delta^2(x) = 4$, 2, and 0, and the lower envelope yields the squared distances of 2, 1, and 0. Along the third column the parabolas are placed with vertices at 1, 0, and 1, and the lower envelope yields 1, 0, and 1. Putting these values together yields the squared distances given by

$$\begin{bmatrix} 5 & 2 & 1 & 0 & 0 \\ 2 & 1 & 0 & 0 & 0 \\ 1 & 0 & 1 & 0 & 0 \end{bmatrix}$$

Taking the square root yields the Euclidean distances shown in the figure.

envelope of the parabolas, but this time a parabola is placed not only at the ON pixels but rather at every pixel. Moreover, the parabolas do not in general touch the x axis but instead are offset vertically. That is, the parabolas are placed so that each center is at $(x, f'(x))$ for every x, where $f'(x)$ is the squared distance from the first pass.

The procedure is given in Algorithm 4.13. First the rows are processed, then the columns. We use slice notation, so $F(:, y)$ means the yth row of F, whereas $F(x, :)$ means the xth column of F. The 1D subroutine requires us to keep track of $k + 1$, which is the number of parabolas in the lower envelope, and two arrays. The array v has $k + 1$ values so that $v[k]$ is the

ALGORITHM 4.13 Compute the exact Euclidean distance for all pixels in an image

ExactEuclideanDistance(I)

Input: binary image I
Output: image containing the Euclidean distance of each pixel to the nearest ON pixel

1	▶ Compute lower envelope along rows	
2	$F \leftarrow (1 - I) * \infty$	▶ $F(x, y)$ is 0 for foreground, $+\infty$ for background.
3	**for** $y \leftarrow 0$ **to** $height - 1$ **do**	▶ For each row,
4	$\quad \delta' \leftarrow$ ExactEuclideanDistance1D($F(:, y)$)	process the row,
5	$\quad$ **for** $x \leftarrow 0$ **to** $width - 1$ **do**	and store the 1D result
6	$\quad\quad D'(x, y) \leftarrow \delta'[x]$	in a row of δ.
7	▶ Compute lower envelope along columns	
8	**for** $x \leftarrow 0$ **to** $width - 1$ **do**	▶ For each column,
9	$\quad \delta' \leftarrow$ ExactEuclideanDistance1D($D'(x, :)$)	process the column,
10	$\quad$ **for** $y \leftarrow 0$ **to** $height - 1$ **do**	and store the 1D result
11	$\quad\quad D'(x, y) \leftarrow \delta'[y]$	in a column of δ.
12	**return** SquareRoot(D')	

ExactEuclideanDistance1D(f)

Input: 1D function f with length $length$
Output: 1D array δ' containing the squared Euclidean distance of each point to the function f

1	$k \leftarrow 0$	▶ $k + 1$ is number of parabolas in lower envelope.
2	$v[0] \leftarrow 0$	▶ $v[k]$ is horizontal coordinate of kth parabola in lower envelope.
3	$z[0] \leftarrow -\infty$	▶ k^{th} parabola is in lower envelope from $z[k]$ to $z[k + 1]$.
4	$z[1] \leftarrow +\infty$	
5	▶ Compute lower envelope	
6	**for** $x \leftarrow 1$ **to** $length - 1$ **do**	
7	$\quad$ ▶ x' is horizontal position of intersection of parabola at x and parabola at $v[k]$	
8	$\quad x' \leftarrow ((f[x] + x^2) - (f[v[k]] + (v[k])^2)) / (2 * (x - v[k]))$	
9	$\quad$ **if** $x' > z[k]$ **then**	▶ Case 1: add parabola at x to lower envelope from x' to $+\infty$.
10	$\quad\quad k \leftarrow_+ 1$	
11	$\quad\quad v[k] \leftarrow x$	
12	$\quad\quad z[k] \leftarrow x'$	
13	$\quad\quad z[k + 1] \leftarrow +\infty$	
14	$\quad$ **else**	▶ Case 2: Remove parabola at $v[k]$ from lower envelope
15	$\quad\quad k \leftarrow_- 1$	(we will add parabola at x in subsequent iteration),

```
16              x ←_ 1                                    and advance to next.
17   ▶ Fill in values
18   k ← 0
19   for x ← 0 to length − 1 do            ▶ For each pixel in the row / column,
20       while z[k + 1] < x do                       advance to the next
21           k ←_+ 1                               parabola if necessary.
22       δ′[x] ← (x − v[k])² + f[v[k]]     ▶ Evaluate the nearest parabola.
23   return δ′
```

horizontal coordinate of the kth parabola in the lower envelope. The array z has $k + 1$ values so that the kth parabola is in the lower envelope from $z[k]$ to $z[k + 1]$. That is, z keeps track of which parabolas define the lower envelope, while v keeps track of where the parabolas are located. Straightforward algebra reveals that two vertical parabolas with vertices at $(x, f(x))$ and $(x_k, f(x_k))$ intersect at the horizontal position given by

$$x' = \frac{(f(x) + x^2) - (f(x_k) + x_k^2)}{2(x - x_k)}$$

(4.100)

When processing the parabolas sequentially in the second pass, two possible cases arise. In the first case, $x' > z[k]$, so the lower envelope must be modified to include the parabola from x starting at x'. In the second case, $x' \leq z[k]$, which indicates that the kth parabola is not part of the lower envelope and should therefore be removed. Note that Algorithm 4.12 is just a special case of EXACTEUCLIDEANDISTANCE1D when the input is binary, because the value of f is 0 for any parabola, which precludes case 2 (in Line 14) from occurring.

4.4 Region Properties

We now turn our attention to computing various properties of binary regions. Such properties are useful for classifying objects, detecting defects in manufactured parts, and pattern recognition, among other applications.

4.4.1 Moments

Many of the properties encountered in this section build on the foundational concept of **moments**. Let us represent an image region by a nonnegative **mass density function** $f(x, y) \geq 0$ defined over the image domain, where the function generally returns larger values inside the region than outside. We will focus our attention on the simple case of a binary region in which $f(x, y) = 1$ inside the region and $f(x, y) = 0$ outside the region—imagine a region found by thresholding, for example. Nevertheless, the formulas for moments apply to any nonnegative function, such as the result of some algorithm that generates a probability map (but note that none of the following analysis requires $\sum_{x,y} f(x, y) = 1$, as would be required by a probability density function).

Regular Moments

Given the discrete function f and nonnegative integers p and q, the pq^{th} **moment** of a 2D region is defined as:

$$m_{pq} \equiv \sum_x \sum_y x^p y^q f(x, y)$$

(4.101)

We say that the pq^{th} moment is of order $p + q$. Thus, the zeroth-order moment is m_{00}, the first-order moments are m_{10} and m_{01}, and the second-order moments are m_{20}, m_{02}, and m_{11}. Computing these moments is easy, requiring a single pass through the image, as shown in Algorithm 4.14.

The **centroid** $(\bar{x}, \bar{y})$ of the region is defined as the weighted average of the pixels and is easily computed from the zeroth and first moments:

$$(\bar{x}, \bar{y}) = \frac{1}{\sum_{x,y} f(x, y)} \left(\sum_{x,y} x f(x, y), \sum_{x,y} y f(x, y) \right) \tag{4.102}$$

$$= \left(\frac{m_{10}}{m_{00}}, \frac{m_{01}}{m_{00}} \right) \tag{4.103}$$

If f were a continuous function, it could be viewed as the mass density function of a solid planar body, the moments of which (by replacing the sums in Equation (4.101) with integrals) capture the inertial properties of the body. The zeroth moment, for example, yields the mass of the body, while the centroid captures its center of mass, and the second-order moments are related to its moments of inertia. In the case of a binary function f, a cardboard cutout (or any other flat material with uniform mass density) with the same shape as the region will remain horizontal when suspended by a string attached at the centroid.

Central Moments

The regular moments that we have just defined will differ depending on where in the image the region is located. To provide translation invariance, the pq^{th} **central moment** is defined as the pq^{th} regular moment about the centroid:

$$\mu_{pq} \equiv \sum_x \sum_y (x - \bar{x})^p (y - \bar{y})^q f(x, y)$$

It is easy to show that the central moments are functions of the regular moments:

$$\mu_{00} = m_{00} \tag{4.104}$$

$$\mu_{10} = 0 \tag{4.105}$$

$$\mu_{01} = 0 \tag{4.106}$$

ALGORITHM 4.14 Compute the zeroth, first, and second-order moments of an image

COMPUTEMOMENTS(f)

Input: image f
Output: zeroth-, first-, and second-order moments of the image

1    $m_{00} \leftarrow m_{10} \leftarrow m_{01} \leftarrow m_{20} \leftarrow m_{02} \leftarrow m_{11} \leftarrow 0$
2   **for** $(x, y) \in f$ **do**
3        $m_{00} \leftarrow m_{00} + f(x, y)$
4        $m_{10} \leftarrow m_{10} + x * f(x, y)$
5        $m_{01} \leftarrow m_{01} + y * f(x, y)$
6        $m_{20} \leftarrow m_{20} + x * x * f(x, y)$
7        $m_{02} \leftarrow m_{02} + y * y * f(x, y)$
8        $m_{11} \leftarrow m_{11} + x * y * f(x, y)$
9   **return** $m_{00}, m_{10}, m_{01}, m_{20}, m_{02}, m_{11}$

ALGORITHM 4.15 Compute the central moments of an image

COMPUTECENTRALMOMENTS(f)

Input: image f
Output: zeroth-, first-, and second-order central moments of the image

1   $(m_{00}, m_{10}, m_{01}, m_{20}, m_{02}, m_{11}) \leftarrow$ COMPUTEMOMENTS(f)
2   $\mu_{00} \leftarrow m_{00}$
3   $\bar{x} \leftarrow m_{10}/m_{00}$
4   $\bar{y} \leftarrow m_{01}/m_{00}$
5   $\mu_{20} \leftarrow m_{20} - \bar{x} * m_{10}$
6   $\mu_{02} \leftarrow m_{02} - \bar{y} * m_{01}$
7   $\mu_{11} \leftarrow m_{11} - \bar{y} * m_{10}$ ➤ (or equivalently, $\mu_{11} \leftarrow m_{11} - \bar{x} * m_{01}$)
8   **return** $\mu_{00}, \mu_{20}, \mu_{02}, \mu_{11}$

$$\mu_{20} = m_{20} - \bar{x}m_{10} \tag{4.107}$$

$$\mu_{02} = m_{02} - \bar{y}m_{01} \tag{4.108}$$

$$\mu_{11} = m_{11} - \bar{y}m_{10} = m_{11} - \bar{x}m_{01} \tag{4.109}$$

These equations allow us to compute the central moments with just a single pass through the image, shown in Algorithm 4.15, and there is no need to compute μ_{10} or μ_{01} because they are always zero. It is easy to verify that the equations above are special cases of the general formula for computing central moments from regular moments:

$$\mu_{pq} = \sum_{i=0}^{p} \sum_{j=0}^{q} (-1)^{i+j} \binom{p}{i}\binom{q}{j} \bar{x}^i \bar{y}^j m_{p-i, q-j} \tag{4.110}$$

where $\binom{n}{k} \equiv \dfrac{n!}{k!(n-k)!}$ is a binomial coefficient, and $n!$ is the factorial of n.

Normalized Central Moments

Under a uniform scale change $x' = \alpha x, y' = \alpha y, \alpha \neq 0$, the central moments change according to $\mu'_{pq} = \alpha^{p+q+2}\mu_{pq}$. This result is easily shown using the continuous formulation of moments:

$$\mu'_{pq} = \iint (x' - \bar{x}')^p (y' - \bar{y}')^q f\left(\frac{x'}{\alpha}, \frac{y'}{\alpha}\right) dx' \, dy' \tag{4.111}$$

$$= \iint (\alpha x - \alpha\bar{x})^p (\alpha y - \alpha\bar{y})^q f(x, y)\alpha^2 \, dx \, dy \tag{4.112}$$

$$= \alpha^{p+q+2} \iint (x - \bar{x})^p (y - \bar{y})^q f(x, y) \, dx \, dy \tag{4.113}$$

$$= \alpha^{p+q+2}\mu_{pq} \tag{4.114}$$

since, by change of variables, $dx' = \alpha \, dx$ and $dy' = \alpha \, dy$. For the zeroth moment $(p = q = 0)$, this yields $\mu'_{00} = \alpha^2\mu_{00}$. As a result, if we normalize a region's central moment by dividing by $\mu_{00}^{\frac{p+q+2}{2}}$, we obtain a quantity that does not change with scale, which can be seen as follows:

$$\frac{\mu'_{pq}}{(\mu'_{00})^{\frac{p+q+2}{2}}} = \frac{\alpha^{p+q+2}\mu_{pq}}{(\alpha^2\mu_{00})^{\frac{p+q+2}{2}}} = \frac{\alpha^{p+q+2}\mu_{pq}}{\alpha^{p+q+2}\mu_{00}^{\frac{p+q+2}{2}}} = \frac{\mu_{pq}}{(\mu_{00})^{\frac{p+q+2}{2}}} \tag{4.115}$$

This observation leads to the definition of the pq^{th} **normalized central moment** as

$$\eta_{pq} \equiv \frac{\mu_{pq}}{\mu_{00}^{\gamma}} \tag{4.116}$$

where $\gamma \equiv \frac{p+q+2}{2}$ for $p + q \geq 2$. The case $p + q = 1$ is not included simply because $\mu_{10} = \mu_{01} = 0$, as shown earlier. Other approaches to scale normalization are possible, such as $\mu_{pq}/(\mu_{20} + \mu_{02})^{\gamma/2}$, which is also scale invariant as can be seen by using the same reasoning as above for dividing by μ_{00}^{γ}.

Hu Moments

So far we have seen the central moments, which are invariant to translation, and the normalized central moments, which are invariant to translation and uniform scaling. The **Hu moments**, which are invariant to translation, uniform scaling, and rotation, are natural extensions. These are given by

$$\phi_1 = \eta_{20} + \eta_{02}$$

$$\phi_2 = (\eta_{20} - \eta_{02})^2 + 4\eta_{11}^2$$

$$\phi_3 = (\eta_{30} - 3\eta_{12})^2 + (3\eta_{21} - \eta_{03})^2$$

$$\phi_4 = (\eta_{30} + \eta_{12})^2 + (\eta_{21} + \eta_{03})^2$$

$$\phi_5 = (\eta_{30} - 3\eta_{12})(\eta_{30} + \eta_{12})[(\eta_{30} + \eta_{12})^2 - 3(\eta_{21} + \eta_{03})^2]$$
$$\quad + (3\eta_{21} - \eta_{03})(\eta_{21} + \eta_{03})[3(\eta_{30} + \eta_{12})^2 - (\eta_{21} + \eta_{03})^2]$$

$$\phi_6 = (\eta_{20} - \eta_{02})[(\eta_{30} + \eta_{12})^2 - (\eta_{21} + \eta_{03})^2] + 4\eta_{11}(\eta_{30} + \eta_{12})(\eta_{21} + \eta_{03})$$

$$\phi_7 = (3\eta_{21} - \eta_{03})(\eta_{30} + \eta_{12})[(\eta_{30} + \eta_{12})^2 - 3(\eta_{21} + \eta_{03})^2]$$
$$\quad + (3\eta_{12} - \eta_{30})(\eta_{21} + \eta_{03})[3(\eta_{30} + \eta_{12})^2 - (\eta_{21} + \eta_{03})^2]$$

The first six values are also invariant to reflection, while ϕ_7 changes sign upon reflection, allowing us to distinguish between mirror images. (We call the first six values *invariants*, while the last value is a *pseudoinvariant*.) If η_{pq} is replaced by μ_{pq} in these equations, the invariance to scale disappears while all other properties remain.

Zernike Moments

For rotational invariance, the **complex Zernike**[†] **moments**, which in practice exhibit reduced sensitivity to discretization noise, are even better than the Hu moments. Like the regular, central, and normalized central moments, but unlike the Hu moments, the complex Zernike moments can be computed for any order, thus enabling reconstruction of the region to arbitrary precision. To understand these moments, let us define a new coordinate system with its origin at the region centroid and scaled so that

$$\tilde{x} \equiv \frac{x - \bar{x}}{r} \qquad \tilde{y} \equiv \frac{y - \bar{y}}{r} \tag{4.117}$$

where r is the radius of a circle that entirely encloses the region of interest. It is easy to see that if x and y lie within a circle centered at $(\bar{x}, \bar{y})$ with radius r, then $\tilde{x}^2 + \tilde{y}^2 \leq 1$, that is,

---

[†] Pronounced ZERN-uh-kee.

$(\tilde{x}, \tilde{y})$ lies within the unit circle. This is important because the complex Zernike moments derive their rotational invariance from the orthogonality of the Zernike polynomials, but this property only holds within the unit circle.

The complex Zernike moments of order $p \geq 0$ are defined as a summation over the circle centered at $(\bar{x}, \bar{y})$ with radius r:

$$Z_{pq} \equiv \frac{p+1}{\pi} \sum_x \sum_y f(x, y) V_{pq}^*(\tilde{x}, \tilde{y}) \tag{4.118}$$

with the restriction that $p - |q| = 2n$ for some nonnegative integer n. As a result of this restriction, the moments are $Z_{00}, Z_{11}, Z_{20}, Z_{22}, Z_{31}, Z_{33}, Z_{40}, Z_{42}, Z_{44}$, and so on. Usually the first 9 to 12 moments are computed. Although q can be negative, the moments with negative q do not contribute new information, since it can be shown that $Z_{pq} = Z_{p,-q}^*$ for any p and q, where the asterisk (*) denotes the complex conjugate.

Since the difference between p and $|q|$ is even, their sum must be even as well, so let us define n' to be the nonnegative integer such that $p + |q| = 2n'$. The function V_{pq}^* is then the complex conjugate of a Zernike polynomial of degree p and angular dependence q:

$$V_{pq}(\tilde{x}, \tilde{y}) \equiv \sum_{m=0}^{n} (-1)^m \frac{(p-m)!}{m!(n-m)!(n'-m)!} (\tilde{x}^2 + \tilde{y}^2)^{\frac{p}{2}-m} e^{jq\theta} \tag{4.119}$$

$$= \sum_{\substack{k=|q| \\ p-k \text{ is even}}}^{p} B_{pqk} (\tilde{x}^2 + \tilde{y}^2)^{\frac{k}{2}} e^{jq\theta} \tag{4.120}$$

where $\tan \theta = \tilde{y}/\tilde{x}, j \equiv \sqrt{-1}$, and the second equality comes from substituting $m = (p-k)/2$, leading to

$$B_{pqk} = (-1)^{\frac{p-k}{2}} \frac{\left(\dfrac{p+k}{2}\right)!}{\left(\dfrac{p-k}{2}\right)! \left(\dfrac{k+|q|}{2}\right)! \left(\dfrac{k-|q|}{2}\right)!} \tag{4.121}$$

The Zernike moments are related to the regular moments according to

$$Z_{pq} = \frac{p+1}{\pi} \sum_{\substack{k=|q| \\ p-k \text{ is even}}}^{p} \sum_{a=0}^{k'} \sum_{b=0}^{|q|} (\mp j)^b \binom{k'}{a} \binom{|q|}{b} B_{pqk} m_{k-2a-b, 2a+b} \tag{4.122}$$

where $k' = \frac{1}{2}(k - |q|)$, and the regular moments are computed according to the normalized coordinate system defined by $\tilde{x}$ and $\tilde{y}$. (Note that in this equation the sign of j is opposite the sign of q.)

4.4.2 Area

The area of a region is given by its zeroth moment $m_{00} = \mu_{00}$. For a binary region, this is simply the number of pixels in the region. Alternatively, given the pixels defining the boundary (as in wall follow), the area of the polygon defined by these boundary pixels is given by

$$\text{area} = \frac{1}{2} \sum_{i=0}^{n-1} (x_{i+1} y_i - x_i y_{i+1}) \tag{4.123}$$

Note that, given two binary images I_1 and I_2, we have

$$\text{area}(I_1) + \text{area}(I_2) = \text{area}(I_1 \cap I_2) + \text{area}(I_1 \cup I_2) \tag{4.124}$$

4.4.3 Perimeter

The perimeter of a region is typically computed by applying the Kimura distance to the boundary found by wall following. Similar to area, we have

$$\text{perimeter}(I_1) + \text{perimeter}(I_2) \approx \text{perimeter}(I_1 \cap I_2) + \text{perimeter}(I_1 \cup I_2) \quad (4.125)$$

where the approximation is due to discretization effects.

4.4.4 Orientation

The orientation of a region is determined by the relationships between the x and y coordinates of the pixels in the region. These relationships are captured by the 2×2 **covariance matrix**, which is defined as the expected value of the outer products of the pixel coordinates after shifting the coordinate system to the region centroid, multiplied by $\rho(x, y) \equiv f(x, y) / \sum_{x,y} f(x, y)$, which is a normalized version of the mass density function so that $\sum_{x,y} \rho(x, y) = 1$. If we let $\mathbf{x} = [x \quad y]^{\mathsf{T}}$ and $\bar{\mathbf{x}} = [\bar{x} \quad \bar{y}]^{\mathsf{T}}$ be vector representations of the point (x, y) and centroid $(\bar{x}, \bar{y})$, respectively, and $E[\cdot]$ the expected value,[†] then the covariance matrix is written as

$$\mathbf{C}_{\{2 \times 2\}} = E[(\mathbf{x} - \bar{\mathbf{x}})(\mathbf{x} - \bar{\mathbf{x}})^{\mathsf{T}} \rho(\mathbf{x})] \quad (4.126)$$

$$= \frac{1}{\sum_{x,y} f(x, y)} \sum_{x,y} (\mathbf{x} - \bar{\mathbf{x}})(\mathbf{x} - \bar{\mathbf{x}})^{\mathsf{T}} f(x, y) \quad (4.127)$$

$$= \frac{1}{\sum_{x,y} f(x, y)} \sum_{x,y} \begin{bmatrix} x - \bar{x} \\ y - \bar{y} \end{bmatrix} \begin{bmatrix} x - \bar{x} & y - \bar{y} \end{bmatrix} f(x, y) \quad (4.128)$$

$$= \frac{1}{\sum_{x,y} f(x, y)} \sum_{x,y} \begin{bmatrix} (x - \bar{x})^2 & (x - \bar{x})(y - \bar{y}) \\ (x - \bar{x})(y - \bar{y}) & (y - \bar{y})^2 \end{bmatrix} f(x, y) \quad (4.129)$$

$$= \frac{1}{\sum_{x,y} f(x, y)}$$

$$\begin{bmatrix} \sum_{x,y}(x - \bar{x})^2 f(x, y) & \sum_{x,y}(x - \bar{x})(y - \bar{y}) f(x, y) \\ \sum_{x,y}(x - \bar{x})(y - \bar{y}) f(x, y) & \sum_{x,y}(y - \bar{y})^2 f(x, y) \end{bmatrix} \quad (4.130)$$

$$= \frac{1}{\mu_{00}} \begin{bmatrix} \mu_{20} & \mu_{11} \\ \mu_{11} & \mu_{02} \end{bmatrix} \quad (4.131)$$

where the equalities follow from straightforward substitution, and the (optional) subscript on $\mathbf{C}$ indicates its dimensions. As can be seen, the covariance matrix of an image region is a simple function of its central moments.

Returning to our earlier physical analogy, the covariance matrix is part of the **inertia moment tensor** of the body. Just as the area captures the body's resistance to linear forces, the values in this matrix capture the body's resistance to rotational forces about the axes, as illustrated in Figure 4.33. The diagonal elements μ_{20}/μ_{00} and μ_{02}/μ_{00}, known as the **moments of inertia**, capture the resistance of the body to rotation about the x and y axes, respectively, since $(x - \bar{x})$ and $(y - \bar{y})$ are distances to the axes. The off-diagonal element μ_{11}/μ_{00}, called the **product of inertia**, captures the twist that the body will undergo around

---

[†] The *expected value* of a random variable Z with probability density function $p_Z(z)$ is defined as the mean of the distribution: $\sum_z z \, p_Z(z)$.

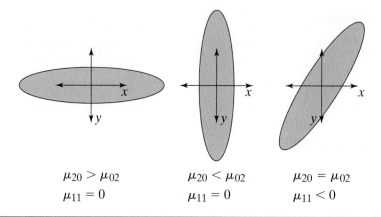

$$\mu_{20} > \mu_{02}$$
$$\mu_{11} = 0$$

$$\mu_{20} < \mu_{02}$$
$$\mu_{11} = 0$$

$$\mu_{20} = \mu_{02}$$
$$\mu_{11} < 0$$

one axis when it is rotated around the other axis, which is related to the asymmetry of the body. Note that μ_{11}, unlike the other quantities, can be negative.

Every rigid, physical body has a **figure axis**, which is defined as the axis about which the moment of inertia is minimized. This axis passes through the centroid of the body. For a circular body, the axis is arbitrarily defined, but for an elongated body the figure axis will be aligned with the direction of elongation. For example, much less energy is required to rotate a baseball bat about its axis of symmetry than to swing the bat, so the baseball bat's axis of symmetry is its figure axis.

The orientation of a 2D region is the angle θ of its figure axis, which is related in a rather simple way to the second-order central moments:

$$\tan 2\theta = \left(\frac{2\mu_{11}}{\mu_{20} - \mu_{02}} \right) \tag{4.132}$$

If the moments are computed using a standard image coordinate system with the positive x axis pointing right and the positive y axis pointing down, then the angle θ will be clockwise with respect to the positive x axis. It would be tempting to invert this equation, but this would be wrong:

$$\theta \neq \frac{1}{2} \arctan \left(\frac{2\mu_{11}}{\mu_{20} - \mu_{02}} \right) \tag{4.133}$$

Can you guess why? The inverse tangent function (arctan) returns an angle between $-\pi/2$ and $\pi/2$, so Equation (4.133) yields a value for θ between $-\pi/4$ to $\pi/4$, which of course does not represent the full range of possible line orientations. The solution to this problem is to keep the numerator and denominator separate, using their signs to compute an inverse tangent between $-\pi$ and π in the appropriate quadrant; multiplication by $\frac{1}{2}$ then yields an angle between $-\pi/2$ and $\pi/2$. Most programming languages have such a function, usually called ATAN2 (y,x). Since the right side of Equation (4.132) indicates the rise over the run, the numerator is the argument for y, while the denominator is the argument for x, as shown in Algorithm 4.16.

While proving Equation (4.132) is left as an exercise for the reader, some insight into the problem of orientation can be gained from our physical analogy. For any rigid body, it is possible to rotate the coordinate system so that the product of inertia goes to zero, i.e., $\mu_{11} = 0$. When this occurs, the new x and y axes are aligned with the **principal axes of inertia** of the body (also known as the principal axes of rotation), one of which is the figure

ALGORITHM 4.16 Compute the orientation of a region in an image

COMPUTEORIENTATION(I)

Input: binary image I containing a single ON region
Output: angle of major axis of binary region, clockwise from positive x axis

1 $\mu_{00}, \mu_{20}, \mu_{02}, \mu_{11} \leftarrow$ COMPUTECENTRALMOMENTS(I)
2 **return** $(1/2) *$ ATAN2$(2\mu_{11}, \mu_{20} - \mu_{02})$

axis. The moments of inertia about the principal axes of inertia are known as the **principal moments of inertia**. Rotating the coordinate system by θ is the same as rotating the body by $-\theta$, so let

$$\mathbf{R}_{\{2 \times 2\}} = \begin{bmatrix} \cos\theta & \sin\theta \\ -\sin\theta & \cos\theta \end{bmatrix} \tag{4.134}$$

be the rotation matrix that causes this to happen, i.e., it aligns the x axis with the figure axis. Then, the covariance matrix in the new, aligned coordinate system is

$$\mathbf{C}'_{\{2 \times 2\}} = E\left[(\mathbf{Rx} - \mathbf{R\bar{x}})(\mathbf{Rx} - \mathbf{R\bar{x}})^{\mathsf{T}}\rho(\mathbf{x})\right] \tag{4.135}$$

$$= \frac{1}{\sum_{x,y} f(x,y)} \sum_{x,y} \mathbf{R}(\mathbf{x} - \mathbf{\bar{x}})(\mathbf{x} - \mathbf{\bar{x}})^{\mathsf{T}}\mathbf{R}^{\mathsf{T}} f(x,y) \tag{4.136}$$

$$= \mathbf{R}\left(\frac{1}{\sum_{x,y} f(x,y)} \sum_{x,y} (\mathbf{x} - \mathbf{\bar{x}})(\mathbf{x} - \mathbf{\bar{x}})^{\mathsf{T}} f(x,y)\right)\mathbf{R}^{\mathsf{T}} \tag{4.137}$$

$$= \mathbf{RCR}^{\mathsf{T}} \tag{4.138}$$

where in the third line $\mathbf{R}$ and $\mathbf{R}^{\mathsf{T}}$ are pulled out of the summation because they do not depend on x or y, and the final equation results from comparing with Equation (4.127).

Since $\mu_{11} = 0$ in the aligned coordinate system, $\mathbf{C}'$ is a diagonal matrix. Thus, if we could determine the rotation matrix $\mathbf{R}$ that diagonalizes the covariance matrix, we would have an alternate way of determining the orientation θ of the region. It turns out that this is easily done by **eigendecomposition** of the matrix. Recall that a 2×2 matrix has two eigenvalues, λ_1 and λ_2, and two corresponding eigenvectors, $\mathbf{v}_1$ and $\mathbf{v}_2$. Eigenvalues and eigenvectors are paired together, so λ_1 goes with $\mathbf{v}_1$, and λ_2 goes with $\mathbf{v}_2$. The eigenvalues and eigenvectors of a matrix reveal its structure. In the case of an arbitrary input vector $\mathbf{x}$, the output vector $\mathbf{x}' = \mathbf{Cx}$ does not necessarily have any obvious relationship to the input. However, when the input vector is an eigenvector of the matrix, then the output is a scaled version of the input:

$$\mathbf{Cv}_1 = \lambda_1 \mathbf{v}_1 \qquad \mathbf{Cv}_2 = \lambda_2 \mathbf{v}_2 \tag{4.139}$$

where the eigenvalues specify the amount of scaling. We can think of the eigenvectors as the characteristic frequencies of the matrix, with the matrix resonating at those frequencies similar to the way a tuning fork resonates in response to a sound wave at a particular frequency.

If we stack the eigenvectors into the columns of a matrix $\mathbf{P} \equiv \begin{bmatrix} \mathbf{v}_1 & \mathbf{v}_2 \end{bmatrix}$ and the eigenvalues of $\mathbf{C}$ into a diagonal matrix $\mathbf{\Lambda} \equiv diag(\lambda_1, \lambda_2)$, then we obtain a compact representation of the diagonalized covariance matrix:

$$\mathbf{\Lambda} = \mathbf{P}^{\mathsf{T}}\mathbf{CP} \tag{4.140}$$

which is easy to show as follows. Since $\mathbf{C}$ is real and symmetric, its eigenvectors are ortho-normal (orthogonal and unit norm) as long as the eigenvalues are distinct.[†] Therefore, $\mathbf{P}^\mathsf{T} = \mathbf{P}^{-1}$. By substitution from Equation (4.139), we have

$$\mathbf{CP} = \mathbf{C}[\mathbf{v}_1 \quad \mathbf{v}_2] = [\lambda_1\mathbf{v}_1 \quad \lambda_2\mathbf{v}_2] = [\mathbf{v}_1 \quad \mathbf{v}_2]\begin{bmatrix} \lambda_1 & 0 \\ 0 & \lambda_2 \end{bmatrix} = \mathbf{P\Lambda} \qquad (4.141)$$

or $\mathbf{CP} = \mathbf{P\Lambda}$. Multiplying both sides on the left by $\mathbf{P}^\mathsf{T}$ yields Equation (4.140).

Comparing Equation (4.140) with Equation (4.134), we see that $\mathbf{P}^\mathsf{T}$ plays the role of $\mathbf{R}$ by rotating the covariance matrix so that it becomes diagonal, and therefore

$$\mathbf{P} = \mathbf{R}^\mathsf{T} = \begin{bmatrix} \cos\theta & -\sin\theta \\ \sin\theta & \cos\theta \end{bmatrix} \qquad (4.142)$$

so that $\mathbf{v}_1 = [\cos\theta \quad \sin\theta]^\mathsf{T}$ and $\mathbf{v}_2 = [-\sin\theta \quad \cos\theta]^\mathsf{T}$. Thus, the eigenvectors are the principal axes of the region, and, if we follow the common convention of ordering the eigenvalues so that $\lambda_1 \geq \lambda_2$, then $\mathbf{v}_1$ yields the figure axis.

One final detail needs to be considered. There is a sign ambiguity in the eigenvectors, which can be seen in Equation (4.139), because if $\mathbf{v}_i$ satisfies the equation, then so does $-\mathbf{v}_i$, for $i = 1, 2$. The ambiguity can also be seen in Equation (4.140), because $(\mathbf{PM})^\mathsf{T}\mathbf{C}(\mathbf{PM})$ is also equal to $\mathbf{\Lambda}$, where $\mathbf{M}$ is any of the following:

$$\begin{bmatrix} 1 & 0 \\ 0 & 1 \end{bmatrix} \quad \begin{bmatrix} -1 & 0 \\ 0 & 1 \end{bmatrix} \quad \begin{bmatrix} 1 & 0 \\ 0 & -1 \end{bmatrix} \quad \begin{bmatrix} -1 & 0 \\ 0 & -1 \end{bmatrix} \qquad (4.143)$$

Geometrically, we note that the axis defined by θ and the axis defined by $\theta + \pi$ are the same, which is why $[\cos\theta \quad \sin\theta]^\mathsf{T}$ and $[-\cos\theta \quad -\sin\theta]^\mathsf{T}$ refer to the same axis. As a result, Equation (4.142) is slightly misleading, because our convention says that $-\pi/2 < \theta \leq \pi/2$, in which case $\cos\theta \geq 0$. But when the eigenvectors of $\mathbf{C}$ are computed, there is no guarantee on the sign of any of the four values v_{11}, v_{21}, v_{12}, and v_{22}, where $\mathbf{v}_1 \equiv [v_{11} \quad v_{21}]^\mathsf{T}$ and $\mathbf{v}_2 \equiv [v_{12} \quad v_{22}]^\mathsf{T}$. To adhere to our convention, then, we should flip the sign on $\mathbf{v}_1$ if v_{11} is negative. The angle θ is therefore given by the arcsine (inverse sine) of v_{21}, subject to the sign of v_{11}, as shown in Algorithm 4.17.

Recall that an orthogonal matrix is one whose transpose is its inverse, e.g., $\mathbf{R}^{-1} = \mathbf{R}^\mathsf{T}$. A rotation matrix has an additional property; namely, its determinant is $+1$. Although $\mathbf{P}$ is guaranteed to be an orthogonal matrix,[‡] its determinant is not constrained to be $+1$ but may instead be -1. Therefore, the matrix $\mathbf{P}$ may cause not only a rotation but also an undesir-able mirror reflection, which yields a left-handed coordinate system—another reason that Equation (4.142) is misleading. As a result, even after correcting the sign of $\mathbf{v}_1$ according to our convention, care must be taken whether $\mathbf{v}_2$ or $-\mathbf{v}_2$ yields the orthogonal axis in the righthand sense. Because of our convention on θ, this means that we should use $-\mathbf{v}_2$ instead of $\mathbf{v}_2$ if v_{12} is positive, if we care about producing a right-handed coordinate system from the axes. Note, however, that this correction is irrelevant if orientation is all that is desired.

4.4.5 Best-Fitting Ellipse

Suppose we have a binary region defined as the set of pixels inside an ellipse centered at the origin:

$$\text{ellipse region} = \{(x, y) : ax^2 + bxy + cy^2 \leq 1\} \qquad (4.144)$$

[†] If the eigenvalues are not distinct, the eigenvectors can nevertheless be chosen to be orthogonal.

[‡] Somewhat confusingly, according to standard terminology a matrix is called *orthogonal* if its columns are *orthonormal*.

ALGORITHM 4.17 Compute the orientation of a region in an image (eigendecomposition method)

COMPUTEORIENTATIONBYEIGENDECOMPOSITION(I)

Input: binary image I containing a single ON region
Output: angle of major axis of binary region, clockwise from positive x axis

1   $\mu_{00}, \mu_{20}, \mu_{02}, \mu_{11} \leftarrow$ COMPUTECENTRALMOMENTS(I)

2   $\mathbf{C} \leftarrow \frac{1}{\mu_{00}} \begin{bmatrix} \mu_{20} & \mu_{11} \\ \mu_{11} & \mu_{02} \end{bmatrix}$

3   $\lambda_1, \lambda_2, \mathbf{v}_1, \mathbf{v}_2 \leftarrow$ EIGEN($\mathbf{C}$) $\blacktriangleright$ $\lambda_1 \geq \lambda_2, \mathbf{v}_1 = [v_{11} \quad v_{21}]^\mathsf{T}$, and $\mathbf{v}_2 = [v_{12} \quad v_{22}]^\mathsf{T}$

4   **if** $v_{11} \geq 0$ **then**

5   **return** ASIN(v_{21})

6   **else**

7   **return** ASIN($-v_{21}$)

with $4ac > b^2$ and $a, c > 0$ to ensure that the equation describes an ellipse (rather than a parabola or hyperbola). Conveniently, it can be shown that the second-order central moments of such a region are related to the coefficients of the ellipse in a simple way:

$$\mathbf{C} = \frac{1}{\mu_{00}} \begin{bmatrix} \mu_{20} & \mu_{11} \\ \mu_{11} & \mu_{02} \end{bmatrix} = \eta \begin{bmatrix} c & -\frac{b}{2} \\ -\frac{b}{2} & a \end{bmatrix} \tag{4.145}$$

where $\eta \equiv 4 \det(\mathbf{C})$, and $\det(\mathbf{C}) = (\mu_{20}\mu_{02} - \mu_{11}^2)/\mu_{00}^2 = \frac{\eta^2}{4}(4ac - b^2)$ is the determinant of the covariance matrix, so that $\eta = 1/(4ac - b^2)$. While proving the relationship in Equation (4.145) is omitted here, its truth can be partially verified by recalling that covariance matrices are always *positive semidefinite*, which implies that $\det(\mathbf{C}) \geq 0$, or $\mu_{20}\mu_{02} \geq \mu_{11}^2$, which implies $4ac \geq b^2$. It is also true that all the diagonal elements of a positive semidefinite matrix are nonnegative, so $a, c \geq 0$. Another useful exercise is to generate actual binary elliptical regions according to Equation (4.144), measure their moments, and verify that Equation (4.145) holds.

The eigenvalues of $\mathbf{C}$ can be computed by solving the characteristic equation $\det(\mathbf{C} - \lambda \mathbf{I}_{\{2 \times 2\}}) = 0$, where $\mathbf{I}_{\{2 \times 2\}}$ is the 2×2 identity matrix, leading to

$$\lambda_{1,2} = \frac{1}{2\mu_{00}} \left(\mu_{20} + \mu_{02} \pm \sqrt{(\mu_{20} - \mu_{02})^2 + 4\mu_{11}^2} \right) \tag{4.146}$$

where λ_1 takes the plus sign and λ_2 takes the minus sign, so that $\lambda_1 \geq \lambda_2$. It is easy to show that the sum of the eigenvalues is the trace of the covariance matrix, $\lambda_1 + \lambda_2 = \operatorname{tr}(\mathbf{C}) = (\mu_{20} + \mu_{02})/\mu_{00}$, and that the product of the eigenvalues is its determinant, $\lambda_1 \lambda_2 = \det(\mathbf{C}) = (\mu_{20}\mu_{02} - \mu_{11}^2)/\mu_{00}^2$. From the eigendecomposition of the previous subsection, we note that the eigenvalues are invariant to rotation, and their ratio is invariant to scale. Also note that, because $\mathbf{C}$ is positive semidefinite, both eigenvalues are real and nonnegative.

Every noncircular ellipse has a **major axis** and a **minor axis**, which are equivalent to the principal axes previously mentioned, with the major axis being identical to the figure axis. There is a simple relationship between the length of these axes and the eigenvalues of the covariance matrix:

$$\text{semimajor axis length} = 2\sqrt{\lambda_1} \tag{4.147}$$

$$\text{semiminor axis length} = 2\sqrt{\lambda_2} \tag{4.148}$$

where the major or minor axis length is defined as the distance between the two intersection points of the ellipse boundary with the major or minor axis, respectively, and the semi-major and semiminor axis lengths are one-half of these. (If the ellipse were a circle, then the axis length would be the diameter, and the semi-axis length would be the radius.) To verify that these equations are true, consider (without loss of generality) the simple case of an ellipse that is aligned with the coordinate axes, so that $b = 0$, and the boundary of the ellipse is given by $ax^2 + cy^2 = 1$. Such an ellipse crosses the x axis at $x = \pm\sqrt{\frac{1}{a}}$, and it crosses the y axis at $y = \pm\sqrt{\frac{1}{c}}$. Now suppose (also without loss of generality) the ellipse is wider than it is tall, so that $a \leq c$, which implies $\mu_{02} \leq \mu_{20}$. Then, by substitution of Equation (4.145),

$$\text{semimajor axis length} = \sqrt{\frac{1}{a}} = \sqrt{\frac{\eta\mu_{00}}{\mu_{02}}} = \sqrt{\frac{4\mu_{20}\,\mu_{02}\,\mu_{00}}{\mu_{00}^2\,\mu_{02}}} = \sqrt{\frac{4\mu_{20}}{\mu_{00}}} = 2\sqrt{\lambda_1} \quad (4.149)$$

$$\text{semiminor axis length} = \sqrt{\frac{1}{c}} = \sqrt{\frac{\eta\mu_{00}}{\mu_{20}}} = \sqrt{\frac{4\mu_{20}\,\mu_{02}\,\mu_{00}}{\mu_{00}^2\,\mu_{20}}} = \sqrt{\frac{4\mu_{02}}{\mu_{00}}} = 2\sqrt{\lambda_2} \quad (4.150)$$

Since the eigenvalues are not affected by rotation, this result holds no matter the orientation of the ellipse. If $a > c$, then the roles of a and c are swapped in Equations (4.149)–(4.150), but Equations (4.147)–(4.148) still hold.

Thus we see that given any arbitrary binary 2D region, the "best fitting ellipse" of the region, which is defined as the ellipse with the same second-order central moments as the region, is given by $ax^2 + 2bxy + cy^2 = 1$, where the coefficients are determined by Equation (4.145) after first computing the second-order central moments of the region. The major and minor axes of this ellipse are given by the eigenvectors of the covariance matrix formed from the second-order central moments, and the lengths of these axes are related to the square roots of the eigenvalues. In case it is not clear why a square root is needed, recall from Equation (4.141) that

$$\mathbf{C} = \mathbf{P} \begin{bmatrix} \lambda_1 & 0 \\ 0 & \lambda_2 \end{bmatrix} \mathbf{P}^\mathsf{T} \quad (4.151)$$

so that $diag(\lambda_1, \lambda_2)$ is the covariance matrix after the region has been rotated by $\mathbf{P}$ so that it is axis-aligned. Therefore, these eigenvalues are indeed the variances: $\lambda_1 = \sigma_1^2$ and $\lambda_2 = \sigma_2^2$, and their square roots are the standard deviations: $\sqrt{\lambda_1} = \sigma_1$ and $\sqrt{\lambda_2} = \sigma_2$. Since standard deviations are lengths, this explains why length is proportional to the square root of the eigenvalue.

We can go a step further by considering the 2D Gaussian (or bell) curve. We will study the Gaussian in more detail in the next chapter, but for now let us note that the **isotropic** (meaning the same in all directions) 2D Gaussian is in the shape of a bell. The **level set** of a 2D function $G(x, y)$ at level h is the set of points such that $G(x, y) = h$, for some constant h. Geometrically, the level set is the intersection of the function with the horizontal $z = h$ plane, and it can be thought of as a horizontal slice through the function. Because an isotropic Gaussian is rotationally symmetric, its level set will take the shape of a circle. An **anisotropic** (meaning not the same in all directions) Gaussian, on the other hand, is not rotationally symmetric, and therefore its level set is in the form of an ellipse. For every 2D elliptical region, there is an associated 2D Gaussian function with the same mean and covariance matrix, so that λ_1 and λ_2 are the variances of the two random variables of this 2D Gaussian. This analysis also gives meaning to the factor of 2 in Equations (4.147)–(4.148) , because it says that the ellipse captures $\pm 2\sigma_i$, $i = 1, 2$, in each direction, or 95.45% of the area under the Gaussian curve.

4.4.6 Compactness

Compactness is a measure of how close the pixels in the region are to the center of the region. Since the most compact shape is a circle, we define compactness as

$$\text{compactness} = \frac{4\pi(\text{area})}{(\text{perimeter})^2} \tag{4.152}$$

For a continuous region, compactness ranges from 0 to 1, with 1 indicating a circle, since the area of a circle with radius r is πr^2, and its perimeter is $2\pi r$. Note, however, that discretization effects can cause the resulting compactness to be slightly greater than 1 for discretized shapes that resemble circles.

The area of an ellipse is a natural generalization of the area of a circle, with the semimajor and semiminor axis lengths playing the role of the radius. If we let $\ell_1 \equiv 2\sqrt{\lambda_1}$ and $\ell_2 \equiv 2\sqrt{\lambda_2}$ be these lengths, according to Equations (4.147)–(4.148), then the area is $\pi\ell_1\ell_2$. The perimeter of an ellipse is more difficult to calculate, but one approximation that works reasonably well is $2\pi\sqrt{(\ell_1^2 + \ell_2^2)/2}$. Substituting into Equation (4.152) yields the compactness of an ellipse:

$$\text{compactness of ellipse} \approx \frac{2 \cdot 4\pi^2 \ell_1 \ell_2}{4\pi^2(\ell_1^2 + \ell_2^2)} = \frac{2\ell_1\ell_2}{\ell_1^2 + \ell_2^2} = \frac{2\alpha}{\alpha^2 + 1} \tag{4.153}$$

where $\alpha \equiv \frac{\lambda_1}{\lambda_2} \geq 1$ is a scale factor relating the lengths of the axes, so that $\lambda_1 = \alpha\lambda_2$. Note that the compactness is 1 for $\alpha = 1$ (circle), $\frac{4}{5} = 0.8$ for $\alpha = 2$, $\frac{8}{17} \approx 0.47$ for $\alpha = 4$, and so on.

4.4.7 Eccentricity

The eccentricity of a region measures its elongatedness—that is, how far it is from being rotationally symmetric around its centroid. Since the eigenvalues capture the variance in the two principal directions, the eccentricity of a region is defined as the difference between these variances, normalized by the larger variance. Similar to the axis lengths, we take the square root:

$$\text{eccentricity} = \sqrt{\frac{\lambda_1 - \lambda_2}{\lambda_1}} \tag{4.154}$$

which ranges from 0 (when the region is a circle) to 1 (when the region is a straight line). Substituting Equation (4.146) yields the eccentricity in terms of the moments:

$$\text{eccentricity} = \sqrt{\frac{2\sqrt{(\mu_{20} - \mu_{02})^2 + 4\mu_{11}^2}}{\mu_{20} + \mu_{02} + \sqrt{(\mu_{20} - \mu_{02})^2 + 4\mu_{11}^2}}} = \sqrt{\frac{2\beta}{\text{tr}(\mathbf{C}) + \beta}} \tag{4.155}$$

where $\beta \equiv \sqrt{\text{tr}^2(\mathbf{C}) - 4\det(\mathbf{C})}$, and $\text{tr}^2(\mathbf{C})$ is the square of the trace of the covariance matrix. In the case of an axis-aligned ellipse, $\mu_{11} = 0$, and Equation (4.155) simplifies to

$$\text{eccentricity (when axis-aligned)} = \sqrt{\frac{|\mu_{20} - \mu_{02}|}{\mu_{20} + \mu_{02} + |\mu_{20} - \mu_{02}|}} \tag{4.156}$$

Figure 4.34 shows an example of compactness and eccentricity.

Other definitions for eccentricity could be imagined, but there are several distinct advantages to the definition in Equation (4.154). First, it matches the standard mathematical

Figure 4.34 Left: A circle is the most compact shape, with a compactness of 1. Middle: A shape whose compactness is less than 1. Right: The eccentricity of the shape is computed as the eccentricity of the best-fitting ellipse.

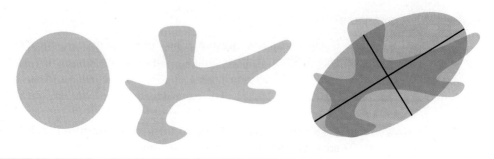

definition of the eccentricity of an ellipse, which is the ratio of the distance between the ellipse foci to the length of the major axis. Thus, the eccentricity of the region is equivalent to the eccentricity of the best-fitting ellipse. Secondly, due to the normalization by λ_1, it is invariant to scale, which is a desirable property because it ties the eccentricity to the shape of the object without regard to its size in the image.

For comparison, an alternative definition that has been proposed by some authors is the ratio of the principal axes of inertia:

$$\sqrt{\frac{\lambda_2}{\lambda_1}} = \sqrt{\frac{\mu_{20} + \mu_{02} - \sqrt{(\mu_{20} - \mu_{02})^2 + 4\mu_{11}^2}}{\mu_{20} + \mu_{02} + \sqrt{(\mu_{20} - \mu_{02})^2 + 4\mu_{11}^2}}} \qquad (4.157)$$

which, when axis-aligned, reduces to

$$\sqrt{\frac{\lambda_2}{\lambda_1}} = \sqrt{\frac{\mu_{02}}{\mu_{20}}} \qquad (4.158)$$

This definition shares many of the desirable properties of the definition above, except that it has to be interpreted in the opposite manner because it yields 1 for a circle and 0 for a line. To fix this problem, we could try to subtract the ratio from 1:

$$1 - \sqrt{\frac{\lambda_2}{\lambda_1}} \qquad (4.159)$$

which also ranges from 0 (when the region is a circle) to 1 (when the region is a straight line). However, this definition has a more serious drawback in that there is no straightforward relationship between a change in the axis lengths and the corresponding change in the value of the eccentricity. For example, a doubling of the ratio of the two axis lengths does not lead to a doubling of the eccentricity.

An even worse measure is the difference between the two eigenvalues:

$$\lambda_1 - \lambda_2 = \frac{1}{\mu_{00}} \sqrt{(\mu_{20} - \mu_{02})^2 + 4\mu_{11}^2} \qquad (4.160)$$

which ranges from 0 (when the region is a circle) to ∞ (when the region is a line). This definition suffers from two problems: a doubling of the difference between the two principal axes leads to a quadrupling of the eccentricity, and the eccentricity is dependent upon the scale of the region.

Finally, the following definition is sometimes proposed:

$$\frac{(\mu_{20} - \mu_{02})^2 + 4\mu_{11}}{\mu_{00}} \qquad (4.161)$$

but it is not clear what the justification for this measure is. Sometimes it is assumed to be the ratio of the principal axes of inertia, but this is not true, as we have just seen the correct formula in Equation (4.157). Moreover, this equation is fundamentally flawed because of the mismatch of units in the numerator, where one moment is added to the square of another. If we recall from Equation (4.114) that a scale change of α causes the second order moments $(\mu_{20}, \mu_{02}, \text{ and } \mu_{11})$ to increase by a factor of α^4, we notice that this same scale change causes the first term in the numerator to scale by a factor of α^8, while the second term scales by only α^4. This mismatch leads to unpredictable behavior in the overall quantity.

4.4.8 Convex Hull

A set of points in the plane is **convex** if any straight line between two points in the set lies entirely within the set. The **convex hull** of a set is the smallest convex set containing the set. The difference between the set and its convex hull is called the **convex deficiency**, which is sometimes used as a descriptor of the shape of the object. The convex deficiency is somewhat related to compactness, because compact shapes (i.e., those whose value is close to 1 according to Equation (4.152)) by their very nature have little to no convex deficiency; the converse, however, is not necessarily true because elongated convex shapes such as ellipses or rectangles can have arbitrarily small values for compactness without any convex deficiency.

Given a set of points in the plane, we can think of the convex hull as being the region defined by the shape of a rubber band placed around the set. Consider a polygon defined by a clockwise sequence of vertices $\mathbf{p}_0, \mathbf{p}_1, \mathbf{p}_2, \ldots, \mathbf{p}_n$. Each vertex is locally either convex or **concave** (the opposite of convex). If we sequentially remove each concave vertex until there are no more concave vertices, the resulting shape will be the convex hull, as shown in Figure 4.35. To test whether a vertex is locally concave, let $\mathbf{v}_i \equiv \mathbf{p}_i - \mathbf{p}_{i-1}$ be the vector joining two consecutive vertices, $\mathbf{p}_{i-1}$ and $\mathbf{p}_i$. Such a vector splits the entire plane into two half-planes, one to the right of the vector and one to the left. If the next vertex $(\mathbf{p}_{i+1})$ is in the left half-plane, then the middle vertex $(\mathbf{p}_i)$ is on a concavity, which means that it is in the interior of the convex hull and therefore *definitely not* on the boundary of the convex hull. On the other hand, if the next vertex is in the right half-plane, then the middle vertex is *possibly* on the boundary of the convex hull, depending upon additional information. In Figure 4.36, for example, $\mathbf{p}_2$ is to the right of the vector joining $\mathbf{p}_0$ and $\mathbf{p}_1$, and therefore the middle vertex $(\mathbf{p}_1)$ is, as far as we know based upon this single test, on the boundary of the convex hull. However, $\mathbf{p}_3$ is to the left of the vector joining $\mathbf{p}_1$ and $\mathbf{p}_2$, and therefore the middle vertex $(\mathbf{p}_2)$ is at a concavity and needs to be removed.

Mathematically, it is easy to test whether the third vertex is in the right or left half-plane defined by the vector joining the first two vertices. If we treat the vectors $\mathbf{v}_i$ and $\mathbf{v}_{i+1}$ as lying in a 3D space, then their cross product, $\mathbf{v}_i \times \mathbf{v}_{i+1}$, yields a vector perpendicular to the plane. According to the right-hand rule with our standard image coordinate system, if the z-coordinate of $\mathbf{v}_i \times \mathbf{v}_{i+1}$ is greater than 0, then $\mathbf{p}_{i+1}$ lies in the right half-plane defined by $\mathbf{v}_i$; whereas if the z-coordinate of $\mathbf{v}_i \times \mathbf{v}_{i+1}$ is less than 0, then $\mathbf{p}_{i+1}$ lies in the left half-plane

Figure 4.35 Lᴇꜰᴛ: An arbitrarily-shaped region in the plane. Rɪɢʜᴛ: The convex hull of the region is the shape that results from enveloping the region with a rubber band, which removes all concavities. All vertices are locally convex except for $\mathbf{p}_2$.

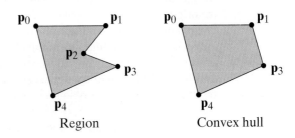

Region   Convex hull

Figure 4.36 LEFT: The vertex $\mathbf{p}_2$ is to the right of the vector joining $\mathbf{p}_0$ and $\mathbf{p}_1$, and therefore $\mathbf{p}_1$ is, as far as we know based upon this single test, on the boundary of the convex hull. RIGHT: The vertex $\mathbf{p}_3$ is to the left of the vector joining $\mathbf{p}_1$ and $\mathbf{p}_2$, and therefore $\mathbf{p}_2$ is at a concavity and needs to be removed.

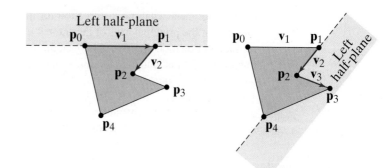

defined by $\mathbf{v}_i$. The sign of this z-coordinate is given by the sign of the determinant of a matrix containing the coordinates of the points, where $\mathbf{p}_i = (x_i, y_i)$, and so on:

$$\mathbf{p}_i \text{ is } \begin{cases} \text{possibly on boundary} & \text{if } d \geq 0 \\ \text{definitely in interior} & \text{if } d < 0, \end{cases} \quad \text{where } d \equiv \det\left(\begin{bmatrix} x_{i-1} & y_{i-1} & 1 \\ x_i & y_i & 1 \\ x_{i+1} & y_{i+1} & 1 \end{bmatrix}\right) \tag{4.162}$$

This test leads naturally to an algorithm for computing the convex hull of a contiguous binary region, shown in Algorithm 4.18. First wall following yields the boundary of the region as a sequence of pixels. Then we consider triplets of adjacent pixels along this boundary. For each triplet the test in Equation (4.162) is performed. If the third point is in the left half-plane, then as far as we know it might be one of the vertices of the convex hull, so the point is retained, and the algorithm continues. On the other hand, if it is in the right half-plane, then the previous point is at a concavity and is therefore definitely not a vertex of the convex hull because it lies in the interior of the convex hull. In this case, we remove the point and back up by one pixel in our traversal, then continue. By sequentially considering pixels along the boundary of the region and testing whether the next pixel is within or outside the right half-plane, a minimalistic sequence of half-planes can be generated so that all pixels in the boundary are within the convex hull defined by the intersection of these half planes. The points that remain at the end of this processing are the vertices of the polygon defining the convex hull. Note from the pseudocode that some additional logic is necessary to handle the special case of the initial pixel.

This consideration of half-planes also leads to an elegant definition of the convex hull in terms of mathematical morphology. More specifically, the convex hull is the intersection of the dilation of the region with the set of SEs containing half-planes at all orientations:

$$\text{convex hull} = \bigcap_{\text{all } \theta} I \oplus B_\theta \tag{4.163}$$

where B_θ is a half-plane at orientation θ, as illustrated in Figure 4.37. In this context, a half-plane SE has infinite extent with 1s on one side of a line and 0s on the other side. Suppose, for example, the region is dilated with a half plane defined by a vertical line passing through the origin, with 1s to the right of the line. Then, according to the definition of dilation, the dilation of the region with this SE yields a 1 anywhere to the left of the rightmost pixel in the region. That is, as the structuring element is translated to the left, there will always be overlap between the region and the structuring element. Only as the structuring element is translated to the right of the rightmost pixel is there no overlap, and hence a 0 output. The result is the same as if a rigid sheet were draped over the region with gravity pointing in the direction perpendicular to the line. As the line is rotated, different outputs result, and the convex hull is the intersection of all such dilations.

ALGORITHM 4.18 Compute the vertices of the convex hull of a binary image region

---

Comp.uteConvexHullVertices(I)

Input: binary image I with a single foreground (ON) region
Output: clockwise sequence of vertices defining the convex hull of the region as a polygon

1 $c \leftarrow$ WallFollow(I)
2 c. PushBack($c[0]$) ➤ Duplicate first vertex for wraparound
3 $i \leftarrow 1$ ($c[c$. Length $- 1] = c[0]$).
4 **while** $i < c$. Length -1 **do** ➤ Consider vertices in order, and
5   **if** IsClockwise($c[i-1], c[i], c[i+1])$] **then** if p_{i-1}, p_i, p_{i+1} are convex,
6        $i \leftarrow_+ 1$ then continue.
7   **else** ➤ Otherwise they are concave,
8   c. Remove(i) so remove vertex $c[i]$, and
9        $i \leftarrow_- 1$ back up to previous vertex.
10 c. Pop() ➤ Remove duplicate vertex, and
11 **if** NOT IsClockwise($c[c$. Length $-1], c[0], c[1])$ **then** remove first vertex
12   c. Remove(0) if concave.
13 **return** c

IsClockwise(p_1, p_2, p_3)

Input: three points in the image plane (with x axis pointing right, y axis pointing down)
Output: whether the line from p_2 to p_3 is a clockwise rotation of the line from p_1 to p_2 (returns false if rotation is counterclockwise or points are collinear)

1 $d \leftarrow \det\left(\begin{bmatrix} x_1 & y_1 & 1 \\ x_2 & y_2 & 1 \\ x_3 & y_3 & 1 \end{bmatrix}\right)$ ➤ Compute determinant of 3×3 matrix, where $p_1 = (x_1, y_1), p_2 = (x_2, y_2),$ and $p_3 = (x_3, y_3)$.

2 **return** $d > 0$ ➤ If determinant is positive, then clockwise rotation.

Dilating a region with an infinite number of infinite SEs is (obviously) computationally prohibitive. However, as the number of orientations is increased, such an approach yields an increasingly good approximation to the true answer. Therefore, a reasonable approach is to dilate the region with the four half planes oriented along the cardinal directions (up, down, right, and left), then dilate with four planes oriented at 45 degrees from these (which yields northwest, northeast, southwest, and southeast), and so on, with an appropriate stopping criterion. Even so, this approach is much more computationally intensive than the one presented above, and it results in only an approximation.

4.4.9 Euler Number

Topology is the study of properties of objects that are preserved under continuous deformations of the objects. Such deformations allow for bending, stretching, and compressing, but not tearing or sewing. The mathematical name for such a deformation is a **homotopy**,[†]

[†] If the homotopy has an inverse, it is known as a *homeomorphism*. For example, the mapping from a circle to an ellipse (or jelly bean, for that matter) is a homeomorphism, because the mapping is one-to-one. On the other hand, the mapping from a circle to a line is just a homotopy because multiple points on the circle map to the same point on the line.

ALGORITHM 4.19 Fill in the convex hull of a binary image region

FILLCONVEXHULL(I)

Input: binary image I with a single foreground (ON) region
Output: pixels inside convex hull of region are set to ON

1   $v \leftarrow$ COMPUTECONVEXHULLVERTICES(I)
2   $x_{min}, x_{max} \leftarrow$ COMPUTEPOLYGONCROSSINGS(v)
3   **for** $y \leftarrow 0$ **to** $height - 1$ **do**
4   **for** $x \leftarrow x_{min}[y]$ **to** $x_{max}[y]$ **do**
5           $I(x, y) \leftarrow$ ON

although they are more colloquially known as "rubber sheet deformations." As an example, consider the letters "A" and "P", as shown in Figure 4.38. By a continuous deformation (homotopy) of the points on the letter, one can be changed into another without affecting the connectivity (topology) of the region. That is, any path connecting any two points within one shape has a corresponding path in the other shape. In contrast, the letters "A" and "B" are not related by a homotopy, because the conversion from "A" to "B" requires tearing a hole (or joining two protrusions to create a hole). Similarly, the letters "A" and "C" are not related by a homotopy, because the conversion from "A" to "C" requires sewing up a hole. In both of these cases, there exist paths in one shape whose corresponding deformed paths in the other shape require passing through the hole of the other shape.

An important topological invariant is the **Euler number**[†] (or *Euler characteristic*), which is defined as the number of regions minus the number of holes:

$$\text{Euler number} \equiv \text{number of regions} - \text{number of holes} \tag{4.164}$$

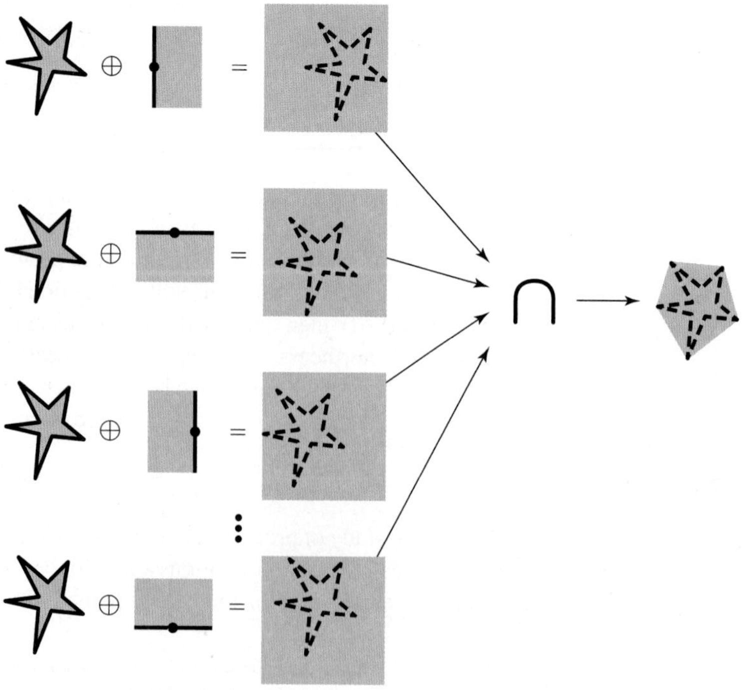

Figure 4.37 The convex hull of a region can be computed by the intersection of the dilation of the region by an infinite set of half planes of different orientations.

[†] Pronounced OIL-ur.

Figure 4.38 Top: The letters "A" and "P" are related by a homotopy, because there is a continuous deformation that relates the two shapes. Bottom: The letters "A" and "B" are not related by a homotopy, because there is not a continuous deformation that relates the two shapes. Rather, tearing the region to produce the extra hole is necessary (or sewing the hole in the case of the reverse transformation).

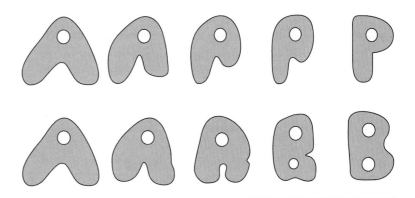

To get a sense of this concept, the Euler numbers of the capital and lowercase letters of the English alphabet are shown in Table 4.4.

Given a region in the plane, suppose *vertices* are added to the boundary of the region, and *edges* (which can be straight or curved lines) are drawn between these vertices, as shown in Figure 4.39; the edges divide the region into one or more *faces*. Let us enforce the following rules: Every vertex has at least one edge connected to it, the region has at least one (implicit if necessary) vertex, a vertex is declared wherever the edges intersect, and the outer and hole boundaries count as edges, too. According to the **Poincaré formula**, the Euler number is then equivalent to the number of vertices minus the number of edges plus the number of faces:

$$\text{Euler number} = \text{number of vertices} - \text{number of edges} + \text{number of faces} \quad (4.165)$$

where vertices and edges outside the region are not counted. This formula states that the Euler number is independent of how the region is divided by the vertices, edges, and faces. In the figure, for example, for the first region there is a single (implicit) vertex, a single edge (the outer contour), and a single face (the filled blue region), so the Euler number is $1 - 1 + 1 = 1$. If another vertex is added and an edge is drawn between the vertices, then there are 2 vertices, 3 edges (the outer contour has been split in two), and 2 faces, leading to $2 - 3 + 2 = 1$; and similarly for the other tesselations.

Equation (4.165) hints that it might be possible to compute the Euler number from local operations. Indeed, the Euler number obeys the **inclusion-exclusion principle**, similar to the operation on sets:

$$E(I_1 \cup I_2) = E(I_1) + E(I_2) - E(I_1 \cap I_2) \quad (4.166)$$

where I_1 and I_2 are two subsets of the plane, and $E(I)$ is the Euler number of I, as shown in Figure 4.40. Now suppose we already know the Euler number of a subset I, and we want to calculate how much the Euler number will change if it is merged with another region ΔI. By substitution, we have

$$\Delta E \equiv E(I \cup \Delta I) - E(I) = E(\Delta I) - E(I \cap \Delta I) \quad (4.167)$$

regions	holes	Euler	letters
1	2	−1	B g
1	1	0	A a b D d e O o P p q R
1	0	1	C c E F f G H h I J K k L l M m N n r S s T t U u V v W w X x Y y Z z
2	0	2	i j

TABLE 4.4 The Euler number of the lowercase and uppercase characters in the English alphabet.

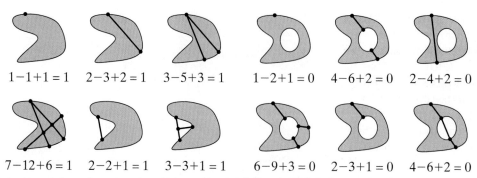

$$1-1+1=1 \quad 2-3+2=1 \quad 3-5+3=1 \quad 1-2+1=0 \quad 4-6+2=0 \quad 2-4+2=0$$

$$7-12+6=1 \quad 2-2+1=1 \quad 3-3+1=1 \quad 6-9+3=0 \quad 2-3+1=0 \quad 4-6+2=0$$

Figure 4.39 Various tesselations of a region whose Euler number is 1 (left) and 0 (right), showing that the Euler number is not dependent upon the particular tesselation chosen. Under each figure is the number of vertices minus edges plus faces according to Equation (4.165). Note there is implicitly at least one vertex, edges intersect at vertices, and external vertices or edges are not counted.

where ΔE is the amount that the Euler number changes when the ΔI region is unioned with I. In other words, the change in the Euler number is the Euler number of the added subset minus the overlap between the subsets.

As an application of this principle, the Euler number can be determined by dividing the plane into strips and counting the number of convexities and concavities within the strips:

$$\text{Euler number} = \text{number of upstream convexities} - \text{number of upstream concavities} \quad (4.168)$$

where "upstream" points opposite the sweeping direction. In Figure 4.41, for example, the sweeping direction is horizontal (to the right), and the first strip contains 1 convexity, the second strip 1 convexity, and the third strip 1 concavity. The Euler number is therefore $1 + 1 - 1 = 1$, which is what we expect since there is 1 region with 0 holes. The rationale behind Equation (4.168) is that whenever a convexity is encountered, a new region appears, thereby increasing the Euler number, but a concavity joins two regions, thereby decreasing the Euler number.

It is not hard to imagine how the locally countable property of the Euler number could apply to a square lattice such as a binary image. As the image is swept from left to right and top to bottom, any time a 2×2 array of pixels is encountered with a 1 in the lower-right corner and 0s in the other 3 pixels, there is a convexity. Similarly, any time there is a 0 in the lower right corner and there are 1s in the other pixels, there is a concavity, as seen from a sweep from the top-left corner to the bottom-right corner. Therefore, it is not too surprising that the Euler number is related to the number of the first occurrence minus the number of the second occurrence.

In reality, the square lattice complicates this simple analysis, requiring us to sweep from all four cardinal directions (from the four corners of the image) and average the results. This leads to the algorithm presented in Algorithm 4.20. Every 2×2 array of pixels in the zero-padded image, overlapping as needed, is examined. Each of these arrays is known as a *bit quad* since it contains 4 binary pixels. Since each pixel is either ON or OFF, there are

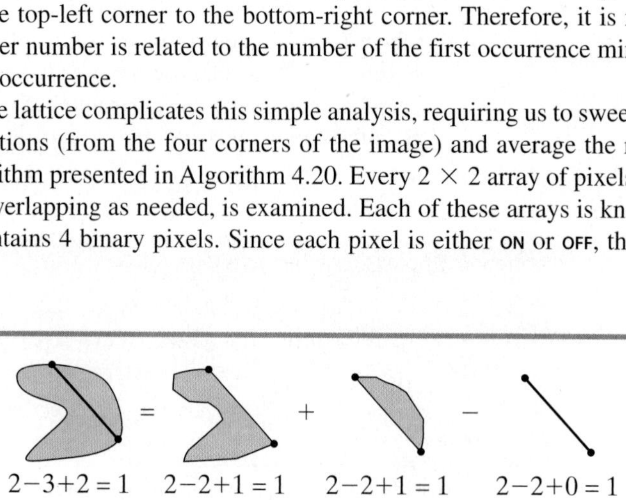

Figure 4.40 Euler number of the union of two regions is the sum of the Euler numbers of the individual regions minus the Euler number of the intersection.

$$2-3+2=1 \quad 2-2+1=1 \quad 2-2+1=1 \quad 2-2+0=1$$

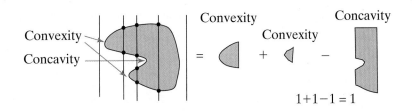

Figure 4.41 Euler number is the number of convexities minus the number of concavities.

$2^4 = 16$ possible bit quads. These are divided into 6 sets, depending upon the number and arrangement of the ON pixels:

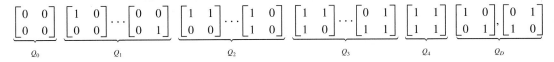

where $\mathcal{Q}_i$, $i = 0, \ldots, 4$ is the set of bit quads with i 4-connected ON pixels, and $\mathcal{Q}_D$ is the set of bit quads with ON pixels in the 2 corners. Note that $\mathcal{Q}_1, \mathcal{Q}_2,$ and $\mathcal{Q}_3$ each contain 4 elements, which vary from each other by 90-degree rotations.

Let n_i refer to the number of times that type $\mathcal{Q}_i$ appears. Then it can be shown that the Euler number of a binary image is given by:

$$E_4(I) = \frac{1}{4}(n_1 - n_3 + 2n_D) \qquad \text{(4-connectedness)} \qquad (4.169)$$

$$E_8(I) = \frac{1}{4}(n_1 - n_3 - 2n_D) \qquad \text{(8-connectedness)} \qquad (4.170)$$

where E_4 is the Euler number assuming 4-connectedness for the foreground (and 8-connectedness for the background), while E_8 is the Euler number assuming 8-connectedness for the foreground (and 4-connectedness for the background).

ALGORITHM 4.20 Compute the Euler number

COMPUTEEULERNUMBER(*I, connectedness*)

Input: binary image *I*, whether foreground is 4-connected or 8-connected
Output: the Euler number of *I*

1 $count \leftarrow 0$
2 $sign \leftarrow +1$ **if** $connectedness = 4$ **or** -1 **if** $connectedness = 8$
3 **for** $y \leftarrow 0$ **to** $height$ **do**
4   **for** $x \leftarrow 0$ **to** $width$ **do**   ▶ For each pixel,
5          $n_{\backslash} \leftarrow I(x, y) + I(x - 1, y - 1)$ examine ON pixels in 2 × 2 window.
6          $n_{/} \leftarrow I(x - 1, y) + I(x, y - 1)$ ▶ Note: For out-of-bounds pixels, $I(x, y) = 0$.
7   **if** $n_{\backslash} + n_{/} = 1$ **then**   ▶ If there is 1 ON pixel,
8              $count \leftarrow_+ 1$ then increment counter.
9   **elseif** $n_{\backslash} + n_{/} = 3$ **then**   ▶ If there are 3 ON pixels,
10             $count \leftarrow_- 1$ then decrement counter.
11   **elseif** $(n_{\backslash} = 2$ AND $n_{/} = 0)$ OR $(n_{\backslash} = 0$ AND $n_{/} = 2)$ **then**   ▶ If ON pixels
12             $count \leftarrow_+ 2 * sign$ are diagonal, then add or subtract 2.
13 **return** $count/4$ ▶ Divide by 4 to get Euler number.

Bit quads can also be used to compute the area and perimeter of a region:

$$\text{area} = \frac{1}{4}(n_1 + 2n_2 + 3n_3 + 4n_4 + 2n_D) \tag{4.171}$$

$$\text{perimeter} = n_1 + n_2 + n_3 + 2n_D \tag{4.172}$$

where the first equation is identical to the zeroth moment we showed before, i.e., counting all the ON pixels (with appropriate handling of the image borders). The second equation, however, computes the length of the perimeter along the edges between pixels, rather than by connecting the centers of the pixels. Thus a 2×2 square region of 4 ON pixels has a perimeter of 8, not 4. If it is desired to estimate properties of an underlying continuous object that has been discretized, somewhat more accurate formulas are as follows:

$$\text{area} = \frac{1}{4}(n_1 + 2n_2 + 3.5n_3 + 4n_4 + 3n_D) \tag{4.173}$$

$$\text{perimeter} = n_2 + \frac{1}{\sqrt{2}}(n_1 + n_3 + 2n_D) \tag{4.174}$$

Of course, it is also possible to compute the Euler number by simply running the connected components algorithm while keeping track of whether each region is ON or OFF, as well as whether each OFF region touches the image border. The Euler number is then the number of ON regions minus the number of OFF regions that do not touch the image border.

4.5 Skeletonization

Another characteristic of a binary region is its **skeleton**. For the moment, let us ignore discretization effects and consider only continuous regions. With this simplification, there are two alternate but equivalent definitions of the skeleton, illustrated in Figure 4.42. In the first, known as **Blum's medial axis transform**, we imagine a region of dry, flammable grass, surrounded by dirt that does not burn. If the region is set on fire along its boundary, then the wave front of the fire will propagate inward. Assuming the grass burns at a constant rate throughout the region, the **medial axis** is defined as the set of points where two or more wave fronts meet, and the skeleton is defined as the medial axis. Alternatively, the skeleton is defined as the locus (i.e., set of locations) of the centers of all the maximal disks, where a **maximal disk** is a circle that fits entirely within the region and touches the boundary in at least two places. It is easy to verify that these definitions are equivalent.

A related concept is the **quench function**, which is defined as the radius of the associated maximal disk for each point on the skeleton, or (equivalently) the distance traveled by the wave front before it was quenched at the medial axis. It is easy to see that the skeleton and quench function together contain enough information to uniquely reconstruct the region.

Figure 4.43 shows six example binary shapes and their associated skeletons. The skeleton of a filled square (not shown) is an X, and the skeleton of a filled rectangle is similar except that the X is stretched along the longer dimension. Similarly, the skeleton of a filled circle is a point, while the skeleton of an ellipse is a line segment whose endpoints are the centers of the circles contained with the ellipse and tangent to the ellipse ends. The skeleton of a filled plus or X depends upon the shape of the region at the extremities; with rounded extremities (as shown in the figure), the skeleton does not reach to the boundary of the region, whereas with pointed extremities it does. Finally, the skeleton of a hollow square is a thin square on the inside, plus straight line segments pointing to each corner.

Figure 4.42 The skeleton of a binary region is defined as the locus of points where the wave fronts of fires set to the boundary meet, or equivalently as the locus of the centers of the maximal balls.

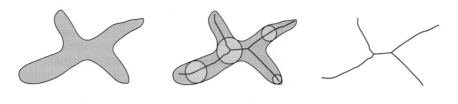

One of the first things you will notice about skeletons is that, even in their definition, they are extremely sensitive to noise. Even a tiny amount of noise can have a huge impact on the nature of the skeleton. Figure 4.44 shows two examples. In the first example, an otherwise perfectly noise-free filled rectangle is corrupted with a single pixel protruding from its side. This single pixel alters the skeleton (again, by definition, not dependent on any algorithm) to include a rather long protrusion from its middle segment to the boundary of the object. In addition, because the pixel is shaped like a square, this protrusion forks to touch the four corners of the pixel square as well. The fact that the skeleton, by definition, touches each corner of a region is even more problematic when discretization occurs. This phenomenon is illustrated in the second example, in which a circle is approximated by a set of pixels on a discrete lattice. Even ignoring all other sources of noise, the jagged edges from the pixel corners cause the skeleton to change shape beyond recognition from the simple point of the continuous circle. Notice that this phenomenon is due solely to the jagged corners; it occurs even when the skeleton itself is represented in the continuous domain.

To add to the problems already mentioned, it is not obvious how to translate the continuous definition of skeleton to the case in which the skeleton is represented discretely. In fact no definition has been found that works well in all cases, and instead we must settle for a reasonable approximation. That is, we seek a set of pixels (the discretized skeleton) that roughly corresponds (in some sense) to the true skeleton of a continuous shape obtained by interpolating the boundary points. As a result of this imprecision, there are two types of **skeletonization** algorithms. C-type (for "corner") algorithms seek to preserve the skeleton segments that touch the corners of the region, whereas S-type (for "smooth") algorithms seek to ignore the corners of the region and instead preserve only its overall shape. In the subsections that follow, we consider several approaches to skeletonization that fall into these two categories, all of which aim to compute a skeleton that is connected, maximally thin, and minimally eroded.

4.5.1 Skeletonization by Thinning

A common approach to skeletonization is to repeatedly thin the image until the result converges. We saw one version of this approach already, namely morphological thinning.[†] An ordered set of structuring elements (SEs) is applied repeatedly to the image until convergence, and the final result yields an approximation to the skeleton. At its core, each iteration of thinning identifies pixels that can be removed (that is, set to OFF) from the image without affecting the connectivity of the foreground regions, and while having minimal impact on their shape.

Figure 4.43 Six different continuous shapes (blue) and their skeletons (thin red lines).

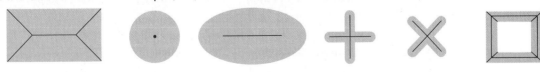

[†] Section 4.1.7 (p. 148).

Figure 4.44 The definition of the skeleton of a region is very sensitive to noise in the input. Left: Even a single pixel can drastically affect the skeleton, such as this small protrusion that causes an entirely new branch to be added. Right: Because a skeleton is required to touch each corner, the "true" skeleton of a discretized circle looks nothing like that of the continuous circle that it approximates.

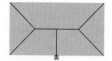

By changing the test that identifies the pixels to remove, as well as changing the order in which to remove them, alternate skeletonization-by-thinning algorithms are obtained. Two such algorithms are described in the next two subsections, but first we must lay the groundwork by attempting to understand under what circumstances it is advisable to remove a pixel based on the values of its 8-neighbors. To do this, let us consider some fundamental characteristics of 3×3 binary patterns.

There are $2^8 = 256$ possible 3×3 binary patterns with an ON pixel in the center, since there are 8 pixels around the center, and each pixel can take one of two values. Let σ represent the number of ON pixels in this outer ring of 8 pixels. The number of 3×3 binary patterns with a particular value of σ is given by "8 choose σ", represented mathematically as $\binom{8}{\sigma} \equiv \frac{8!}{\sigma!(8-\sigma)!}$, where $n! \equiv n \cdot (n-1) \cdot (n-2) \cdots 1$ is the factorial operator for any positive integer n, and $0! \equiv 1$. Substituting, we see that for $\sigma = 0$ there is just one pattern because $\binom{8}{0} = 1$; for $\sigma = 1$ there are $\binom{8}{1} = 8$ patterns; for $\sigma = 2$ there are $\binom{8}{2} = 28$ patterns; for $\sigma = 3$ there are $\binom{8}{3} = 56$ patterns; for $\sigma = 4$ there are $\binom{8}{4} = 70$ patterns; and so on. It is easy to verify that $1 + 8 + 28 + 56 + 70 + 56 + 28 + 8 + 1 = 256$.[†] If we discard patterns that are identical except for a rotation and/or reflection, we are left with just 50 unique patterns, shown in Figure 4.45.

Let us define the **connection number**, represented as ψ, as the number of regions that are 8-connected to the central pixel (minus 1 if there is a 4-connected cross in the center). In other words, the connection number is the number of 8-connected foreground (ON) regions in the 3×3 pattern that would remain if the central pixel were set to OFF, minus the number of holes that would be created.[‡] It turns out that ψ can be computed as the number of 0–1 (OFF-ON) transitions around the 4-neighbors of the central pixel, plus the number of isolated foreground pixels in the corners:

$$\psi = \underbrace{\psi_{01,4}}_{0-1 \text{ transitions}} + \underbrace{\psi_{010,c}}_{\text{isolated corners}} \tag{4.175}$$

If the pixels in a 3×3 neighborhood around a central pixel p_0 are labeled as follows:

$$\begin{bmatrix} p_8 & p_1 & p_2 \\ p_7 & p_0 & p_3 \\ p_6 & p_5 & p_4 \end{bmatrix}$$

then these quantities are computed as

$$\psi_{01,4} \equiv (\bar{p}_1 \cdot p_3) + (\bar{p}_3 \cdot p_5) + (\bar{p}_5 \cdot p_7) + (\bar{p}_7 \cdot p_1) \tag{4.176}$$

$$\psi_{010,c} \equiv (\bar{p}_1 \cdot p_2 \cdot \bar{p}_3) + (\bar{p}_3 \cdot p_4 \cdot \bar{p}_5) + (\bar{p}_5 \cdot p_6 \cdot \bar{p}_7) + (\bar{p}_7 \cdot p_8 \cdot \bar{p}_1) \tag{4.177}$$

where the overbar means binary complement. In this equation, $\bar{p}_i \cdot p_j = 1$ iff $p_i = 0$ (OFF) and $p_j = 1$ (ON), otherwise 0; and $\bar{p}_i \cdot p_j \cdot \bar{p}_k = 1$ iff $p_i = 0$ (OFF), $p_j = 1$ (ON), and $p_k = 0$ (OFF).

[†] As an aside, this analysis generalizes to an elegant, non-obvious formula: $2^n = \sum_{i=0}^{n} \binom{n}{i}$ for any integer $n \geq 0$.

[‡] Another term that is often used is the **crossing number**, represented by χ, which is defined as twice the connection number, that is, $\chi \equiv 2\psi$.

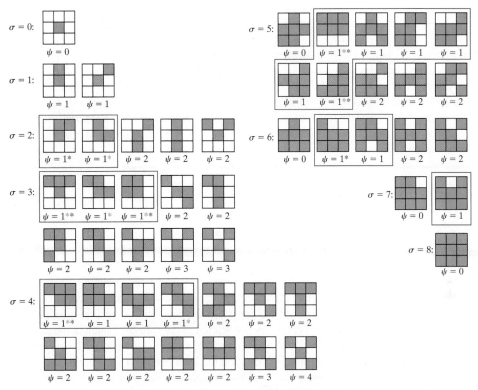

Figure 4.45 The 50 unique 3 × 3 binary patterns with ON in the center, along with their σ value (number of ONs in the 8-neighbors) and ψ value (connection number). The remaining 206 patterns are obtained by rotating and/or reflecting these patterns, which does not affect either σ or ψ. For all patterns with $\psi \neq 0$, the connection number is the number of 8-connected foreground regions that result if the central pixel is set to OFF. For patterns with $\sigma \neq 0$ and $\psi = 0$, setting the central pixel to OFF creates a hole. The purple boxes enclose the patterns whose central pixel is removed by the sigma-psi algorithm, while the green and brown asterisks indicate the patterns whose central pixel is removed by morphological thinning and Zhang-Suen, respectively. The former algorithm is more aggressive in removing pixels than the other two.

The values of σ and ψ provide valuable information about the role of the central pixel. Examining the 50 unique 3 × 3 patterns, it is evident that a pixel with $\psi > 1$ connects multiple regions, that is, ψ is either the number of 8-connected regions that would result if the central pixel were set to OFF (if $\Sigma \neq 0$ and $\psi \neq 0$), or the number of holes created (if $\psi = 0$). More specifically, a pixel with $\sigma = \psi = 2$ is in the middle of a line segment, a pixel with $\sigma = \psi = 3$ is at a T-junction, a pixel with $\sigma = \psi = 4$ is at an X-crossing, and a pixel with $\sigma = 1$ is at the end of a line segment. In all these cases, the pixel cannot be removed without disconnecting multiple regions, thus destroying the property that the thinning algorithm retains the connectivity of the original region. By inspection, a more general statement can be made: removing (that is, setting to OFF) the central pixel whenever $\psi = 1$ retains connectivity of the regions. However, removing the pixel when $\sigma = 1$ erodes the region more than is necessary, thus violating our requirement of minimal erosion.

4.5.2 Sigma-Psi Algorithm

This analysis leads naturally to the **sigma-psi (σ-ψ) algorithm**, which iterates through an image, examining pixels and deleting those for which $\sigma \neq 1$ and $\psi = 1$. The process continues until convergence. Figure 4.46 shows the result of this algorithm on the same

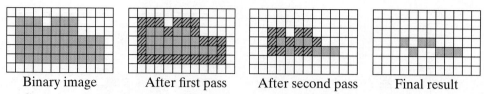

| Binary image | After first pass | After second pass | Final result |

Figure 4.46 Skeletonization using the σ-ψ algorithm. Each pixel is examined in turn, and if $\sigma \neq 1$ and $\psi = 1$, then it is deleted. In the first pass, all the pixels along the border of the region are deleted. In the second pass, a number of additional pixels are deleted, with the particular pixels chosen being dependent upon the order in which they are examined. The final result is indeed a thinned version of the input. This is an S-type algorithm, because the segments touching the corners are not preserved.

binary image as that of Figure 4.15. In this case the pixels were processed in order from top-to-bottom and from left-to-right. The result of this decision is that the skeleton tends to be shifted down in the image, because the pixels are removed in the order in which they are encountered. At first glance, it might appear that a solution to this problem of downward shifting would be to remove pixels in one-pixel-thick increments by simply flagging the pixels for deletion if they pass the $\sigma \neq 1$, $\psi = 1$ test, deleting them all at once after the entire image has been examined, and then repeating this process until convergence. Unfortunately, sigma-psi is too aggressive in the way that it flags pixels for deletion to adopt this approach. That is, if pixels are not deleted as they are encountered, the algorithm may lead to a disconnected skeleton. In the figure, for example, the pixel at location (7, 4), which is the third colored pixel from the right on the bottom row of the penultimate graphic, has $\sigma = 4$ and $\psi = 1$ and therefore would be flagged for deletion at the same time its neighbors are flagged. This conclusion is similar to our earlier statement that morphological thinning must process the corner SEs as a sequence rather than as a set to avoid disconnecting regions.

From the figure, it is clear that the sigma-psi algorithm is S-type, since it does not produce a skeleton that reaches to the corners of the regions. In contrast, the morphological thinning algorithm that we encountered previously is C-type, as can be seen by recalling the output in Figure 4.15, which does reach to the corners. Which type of skeleton is preferred depends upon the application.

4.5.3 Zhang-Suen Algorithm

A closely-related skeletonization-by-thinning method is the **Zhang-Suen algorithm**. Zhang-Suen repeatedly applies two subiterations to the image. In the first subiteration, pixels are flagged for removal if they meet the following 4 tests:

$$\text{a) } 2 \leq \sigma \leq 6 \tag{4.178}$$

$$\text{b) } \psi_{01,8} = 1 \tag{4.179}$$

$$\text{c) } p_1 \cdot p_3 \cdot p_5 = 0 \tag{4.180}$$

$$\text{d) } p_3 \cdot p_5 \cdot p_7 = 0 \tag{4.181}$$

where $\psi_{01,8}$ is the number of 0-1 (**OFF-ON**) transitions along the 8-neighbors of p_0, that is,

$$\psi_{01,8} \equiv (\bar{p}_1 \cdot p_2) + (\bar{p}_2 \cdot p_3) + (\bar{p}_3 \cdot p_4) + (\bar{p}_4 \cdot p_5) + (\bar{p}_5 \cdot p_6) + (\bar{p}_6 \cdot p_7) + (\bar{p}_7 \cdot p_8) + (\bar{p}_8 \cdot p_1) \tag{4.182}$$

Once the entire image has been examined, the pixels that have been flagged are removed. In the second subiteration, pixels are flagged for removal if they meet four tests, the first two of which are identical to those above, while the latter two are slightly changed:

$$\text{c') } p_1 \cdot p_3 \cdot p_7 = 0 \tag{4.183}$$

$$\text{d') } p_1 \cdot p_5 \cdot p_7 = 0 \tag{4.184}$$

As before, after the entire image has been examined, the pixels that have been flagged are removed. These two subiterations are repeated until convergence. Note that the two tests in Equations (4.180)–(4.181), and the two tests in Equations (4.183)–(4.184), are equivalent to the following two expressions, respectively:

$$c, d) \quad \bar{p}_3 \textbf{ or } \bar{p}_5 \textbf{ or } (\bar{p}_1 \cdot \bar{p}_7) \tag{4.185}$$

$$c', d') \quad \bar{p}_1 \textbf{ or } \bar{p}_7 \textbf{ or } (\bar{p}_3 \cdot \bar{p}_5) \tag{4.186}$$

That is, the first subiteration deletes points on the east or south boundaries, or on the northwest corner, while the second subiteration deletes points on the north or west boundaries, or on the southeast corner. The result of Zhang-Suen on the same binary image is shown in Figure 4.47, where the similarity with sigma-psi is obvious.

4.5.4 NF2 Algorithm

An alternative to using thinning to perform skeletonization is to use the distance transform of the complement of the image, which yields the distance from each foreground pixel to the nearest background pixel. Peaks and ridges of the resulting distance function then indicate the points on the skeleton. However, since there is no simple way to detect the ridges of a 2D function, algorithms based on such an approach tend to work poorly in practice, yielding instead disconnected skeletons or noisy results.

One of the more robust and effective techniques that is based on the distance function is known as **NF2**. This algorithm was developed in the robotics community for path planning and is therefore not widely known in the image processing community. The algorithm, presented in Algorithm 4.21, follows the wave front analogy, creating a frontier of pixels along the boundary of the region and allowing them to propagate inward at a constant rate. As the wave fronts propagate, the algorithm keeps track of, for each pixel on the frontier, the closest boundary pixel $b(\cdot)$ as well as the distance $d(\cdot)$ to that boundary pixel. When two wave fronts meet, the overlapping pixel is added to the skeleton S if their corresponding boundary pixels are at least a small distance apart. This distance threshold, represented as τ in the code, is typically set to a number between 2 and 6, depending on the expected noise level in the image. For simplicity, the pseudocode shows the version of the algorithm using the Manhattan distance, but other distance measures could be used if more accurate results are desired. Note that NF2 computes the distance transform on the fly rather than as a preprocessing step.

An example illustrating several steps of the execution of NF2 is shown in Figure 4.48. Initially, all pixels inside the region are set to a distance of infinity to their nearest boundary pixel, and all pixels on the boundary are set to a distance of 0. Then the frontier is created and allowed to propagate inward by a distance of 1 each iteration, thus computing the distance of each neighboring pixel, as well as determining when two wave fronts collide. Notice that NF2 tends to produce a fairly thin connected discrete skeleton. Another advantage to NF2 is its computational efficiency, because it does not require multiple passes through the image like the skeletonization-by-thinning algorithms do.

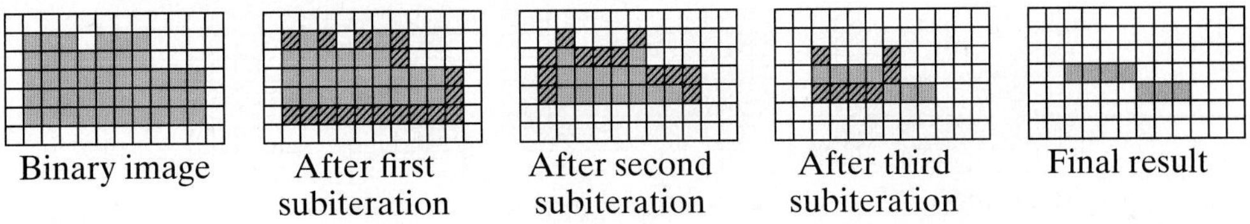

Figure 4.47
The Zhang-Suen algorithm applied to a binary image.

Binary image   After first subiteration   After second subiteration   After third subiteration   Final result

ALGORITHM 4.21 NF2 algorithm for skeletonization (using Manhattan distance)

SKELETONBYNF2(I)

Input: binary image I containing a single region R of ON pixels
Output: set S of pixels comprising skeleton of region

> Initialization

1	**for** each pixel $p \in R$ **do**	> For each pixel in region, initialize
2	$\quad d(p) \leftarrow \infty$	its distance to nearest background pixel to infinity.
3	**for** each pixel $p \notin R$ **do**	> For each pixel in background,
4	$\quad$ **if** there exists $q \in \mathcal{N}(p)$ s.t. $q \in R$ **then**	if it is a boundary pixel, then
5	$\quad\quad d(p) \leftarrow 0$	set its distance to nearest background pixel to zero,
6	$\quad\quad b(p) \leftarrow p$	set its nearest background pixel to itself,
7	$\quad\quad$ *next-frontier*. PUSH(p)	and push the pixel onto the zeroth frontier.

> Main loop

8	$\delta \leftarrow 0$	> Set distance to zero.
9	**repeat**	
10	$\quad$ *frontier* $\leftarrow$ *next-frontier*	> Copy the frontier generated in previous iteration.
11	$\quad$ **for** each $p \in$ *frontier* **do**	> For each pixel in the frontier,
12	$\quad\quad$ **for** each $q \in \mathcal{N}(p)$ s.t. $q \in R$ **do**	if it has a neighbor inside the region
13	$\quad\quad\quad$ **if** $d(q) == \infty$ **then**	that has not yet been visited, then
14	$\quad\quad\quad\quad d(q) \leftarrow \delta + 1$	set the neighbor's distance to one more than itself,
15	$\quad\quad\quad\quad b(q) \leftarrow b(p)$	set its nearest background pixel to be the same,
16	$\quad\quad\quad\quad$ *next-frontier*.PUSH(q)	and push the neighbor onto the next frontier.
17	$\quad\quad\quad$ **elseif** DIST$(b(q), b(p)) > \tau$ **then**	> Otherwise, if background pixels are far
18	$\quad\quad\quad\quad$ **if** $p \notin S$ **then**	enough apart, then the wave fronts meet; store only one
19	$\quad\quad\quad\quad\quad S$.PUSH($q$)	of the pixels to avoid generating double-thick skeleton.
20	$\quad \delta \leftarrow_+ 1$	> Increment the distance for the next iteration.
21	**until** *frontier*. SIZE $== 0$	> Repeat until frontier is empty.

Figure 4.48 Skeletonization using NF2. A binary input image, and the intermediate result using $\tau = 2$ after Lines 2, 7, and 19, respectively, for $\delta = 0$, followed by the intermediate results after Line 19 with $\delta = 1, 2, 3$. Green indicates pixels on the skeleton as the algorithm proceeds, while blue is used in the final graphic to depict the skeleton.

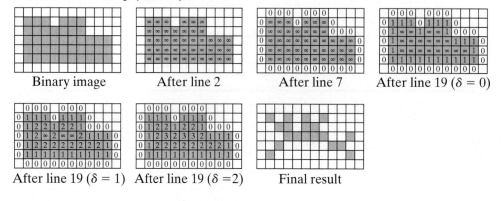

Binary image   After line 2   After line 7   After line 19 ($\delta = 0$)

After line 19 ($\delta = 1$)   After line 19 ($\delta = 2$)   Final result

A comparison of the various skeletonization algorithms on two different binary images is shown in Figure 4.49. Here it is obvious that NF2, like morphological thinning, is C-type, whereas the others are S-type. C-type algorithms produce actual skeletons when the corners in the region are meaningful, but they produce noisier outputs when discretization effects obscure the true shape of the underlying continuous region. The thinning algorithms are all similar in their approach, but they differ both in the test used to determine whether to delete pixels, as well as the order in which the pixels are processed and deleted. By examination it is easy to verify that morphological thinning identifies a subset of the patterns detected by the sigma-psi algorithm for deletion, since $\psi = 1$ and $\sigma \neq 1$ for all of the SEs in Figure 4.14. Similarly, Zhang-Suen is more conservative than sigma-psi, because several patterns that satisfy $\psi_{01,4} = 1$ have $\psi_{01,8} \neq 1$. These relationships are highlighted in Figure 4.45, where it is seen that only a subset of the pixels flagged for removal by sigma-psi are flagged for removal by either morphological thinning or Zhang-Suen. In general, it should be kept in mind that skeletonization is a delicate process, and the output can vary widely depending upon the details of the implementation, as well as upon any postprocessing used to clean up the result.

4.6 Boundary Representations

Earlier in the chapter[†] we discussed a variety of properties that can be computed of a binary region for the purpose of distinguishing the shape of the region from other shapes. In this section we continue that discussion, but here we focus on ways to represent the *boundary* of such a region. Since the interior of a region and its boundary are complementary, the representations discussed here can be thought of as a way to enrich those provided by the region properties considered earlier.

4.6.1 Chain Code

Given a binary region, the first step is to apply a boundary tracing algorithm (such as the one discussed earlier[‡]) to yield a sequence of pixels around the perimeter of the region.

Figure 4.49
Comparison of the various skeletonization algorithms on two different binary images.

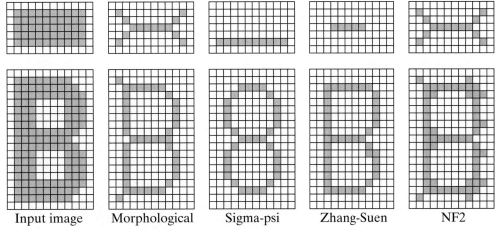

Input image   Morphological   Sigma-psi   Zhang-Suen   NF2

---

† Section 4.4 (p. 174).
‡ Section 4.2.4 (p. 161).

Treating each pair of consecutive pixels as vertices of a line segment, the sequence yields the simplest possible boundary representation, namely a polygon, which can represent the boundary with any arbitrary degree of precision by introducing or removing vertices with real-valued coordinates (perhaps interpolated between pixels). Nevertheless, the drawback of such a polygon representation is that it stores the absolute coordinates of the pixels, thus tying the representation to the actual location of the region in the image.

For distinguishing the shape of the boundary, generally we want to transform the sequence into a representation that is invariant to translation, rotation, and/or scale changes, as well as to the starting pixel. Perhaps the simplest, and certainly one of the oldest, techniques for doing so is the **chain code**. The most famous chain code is the **Freeman chain code**, which we saw in Figure 4.27. In a chain code, only the first point's absolute coordinates are stored, while all other points are represented by their position relative to the previous point in the chain. Thus, if the first point's coordinates are ignored, then the chain code provides a representation that is translation invariant. Figure 4.50 shows an example of a binary region and its 4-connected boundary, which is given by the following sequence of pixel coordinates: $(1,1)$, $(2,1)$, $(3,1)$, $(4,1)$, $(4,2)$, $(5,2)$, ..., $(1,2)$. Using the 4 cardinal directions of the compass (east, north, west, and south), the second pixel is east of the first pixel, the 5th pixel is south of the 4th pixel, and so on, leading to the representation E-E-E-S-E-E-S-E-W-S-W-W-S-W-W-N-W-N-N-N. By assigning the numerical values of 0, 1, 2, and 3 to these directions according to Figure 4.27, the Freeman chain code representation of this region is given by 00030030232232212111. Similarly, the 8-connected boundary is the sequence $(1,1)$, $(2,1)$, $(3,1)$, $(4,1)$, $(5,2)$, ..., $(1,2)$, which is represented by E-E-E-SE-E-SE-SW-W-SW-W-W-NW-N-N-N, or more compactly as 000707545443222. (Note that although 0 means east in both representations, the other numbers differ in their meaning between 4- and 8-connectedness.) Like other representations, the chain code converts the 2D representation of pixel coordinates into a 1D representation, thus simplifying the description for matching.

To make the representation rotation-invariant, the relative positions of consecutive pixels can be encoded as the number of left- and right-hand turns. This representation can be thought of as a derivative of the chain code, and it can be generated easily as a by-product of the wall-following algorithm. For example, in a 4-connected boundary it is possible to drive forward (F), turn right (R), turn left (L), or make a U-turn (U). The derivative of the 4-connected chain code above is therefore given by F-F-F-R-L-F-R-L-U-L-R-F-L-R-F-R-L-R-F-F, or 00031031213013031300. Similarly, the derivative of the 8-connected chain code is F-F-F-R_{45}-L_{45}-R_{45}-R_{90}-R_{45}-L_{45}-R_{45}-F-R_{45}-R_{45}-F-F, or 000717671707700, where the subscripts indicate the rotation angle in degrees.

4.6.2 Minimum-Perimeter Polygon

Another early approach to boundary representation is the **minimum-perimeter polygon (MPP)**. From the original discretized region, let us create a continuous boundary using some method (which could be as simple as connecting the centers of the boundary pixels

Figure 4.50 Simple binary region (left), with its 4-connected boundary (middle) and 8-connected boundary (right).

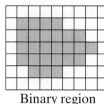
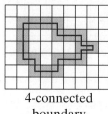
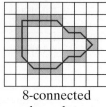

Binary region   4-connected   8-connected
                       boundary   boundary

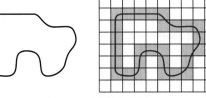 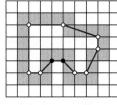

Figure 4.51
Continuous boundary (left), boundary overlaid on a discrete grid (middle-left), minimum-perimeter polygon (MPP) (middle-right), and MPP displayed without the grid (right).

Continuous boundary · Continuous boundary on discrete grid · Minimum-perimeter polygon (MPP) on discrete grid · Minimum-perimeter polygon (MPP)

to form a polygon). Then let us overlay a grid on the continuous boundary and mark the cells through which the boundary passes, where the grid resolution may or may not be the same as the original image resolution. If we imagine a rubber band that circumscribes the boundary but is required to remain inside the marked cells, the MPP is defined as the shape taken by the rubber band as it automatically stretches and compresses to minimize its internal energy.

From the example in Figure 4.51, it is easy to see that the rubber band will take the shape of a polygon whose vertices are one of two kinds: each vertex is either a convex corner of the 4-connected interior or the point opposite a concave corner of the 4-connected interior. In the figure the interior is white, and the convex corners are illustrated with white circles, while the vertices opposite the concave corners are illustrated with dark circles. It is easy to demonstrate that every convex vertex of the polygon is indeed a convex corner of the interior region, but the converse is not true. Similarly, every concave vertex of the polygon is a point opposite a concave corner of the interior, but the converse is not true. The actual algorithm for computing the MPP of a shape at a particular resolution is left as an exercise for the reader.

4.6.3 Signature

A common approach to representing the boundary of a region is via some type of **signature**. For all signature representations, the region is typically first rotated using the principal axis to ensure that the representation is independent of the starting pixel. Then the left-right ambiguity is resolved by projecting the region onto the principal axis and computing some property of the projection function, such as the side with the most mass.

The most basic type of signature is known as the **centroidal profile**, or r-θ curve. This approach captures the distance r from the center of the region as a function of the angle θ, as illustrated in Figure 4.52. Two drawbacks are obvious with this representation, illustrated in Figure 4.53. First, shapes with concavities might result in multivalued functions. Although

Figure 4.52 The centroidal profile (r-θ plot) of several shapes.

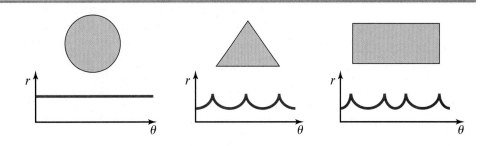

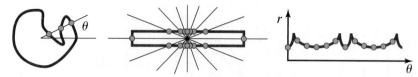

Figure 4.53 One drawback to the centroidal profile is the possibility of multiple values for a given angle (left). Another drawback is that uniform sampling of the angles does not necessarily lead to uniform sampling along the boundary, particularly for a region with a large aspect ratio (middle, right).

we could simply record the minimum value, this solution results in a loss of information, making it impossible to recover the original shape from the function. Secondly, for regions with a large aspect ratio, uniform sampling of the angle does not lead to uniform sampling of the region boundary, causing some parts of the boundary to be represented in greater detail than other parts.

To ensure uniform sampling of the shape around the boundary, the distance r from the center can be plotted as a function of the arclength s as the boundary is traversed. This is known as the **radial representation**, or r-s curve. The resulting curve is guaranteed to be single-valued (i.e., a true function), but this representation introduces an ambiguity whenever the tangent passes through the centroid. To resolve the ambiguity, it is necessary to store an extra bit to capture whether the boundary turns back on itself.

An alternative representation is to store the angle ψ between a fixed reference line (e.g., the positive x-axis) and the tangent to the boundary at the point at arclength s. This is known as the **tangential representation**, or ψ-s curve. This curve can be thought of as a continuous version of the chain code representation. Horizontal lines in the ψ-s curve correspond to straight lines on the boundary, since the tangent angle ψ is not changing. If the ψ-s curve is monotonically increasing, then the shape is convex.

4.6.4 Fourier Descriptor

None of the boundary representations mentioned above explicitly attempts to be robust when there is noise in the input. One important type of noise is occlusion, which causes missing features in the boundary. Another type of noise is the high-frequency noise due to sampling resolution that can interfere with the overall shape in which we are interested. We will revisit this issue later when we consider ways to fit polylines to curves,[†] but for now we mention that the Fourier transform is naturally suited to provide a multiscale representation in which low-frequency components capture the overall shape, while high-frequency components respond to data that may not be of interest. By retaining only a few of the components, a faithful approximation to the boundary with a much more compact representation can often be obtained.

In this approach, the sequence of pixel coordinates is treated as a vector-valued function of the arclength s. With a closed boundary, this is a periodic function, which is then expanded as a Fourier series (although in practice the discrete Fourier transform is always used[‡]). The Fourier series coefficients provide increasingly detailed representations of the boundary and therefore, depending upon the complexity of the boundary, it may often be faithfully

---

[†] Section 7.3 (p. 341).

[‡] For more details on the discrete Fourier transform, see Chapter 6 (p. 272).

represented with just a few of the coefficients. With proper parameterization, it is also possible to ensure that the Fourier coefficients are translation-, rotation-, and scale-invariant.

4.6.5 B-Spline

B-splines ("basis splines") were briefly mentioned in the previous chapter.[†] The term *spline* comes from building construction, where it refers to a thin, flexible strip of wood or metal; draftsmen later used splines for drawing curved lines, which led to its adoption as the name of a specific mathematical description of curved lines. In this section we cover the simplest type of B-spline, the uniform cubic B-spline, paying particular attention to some of the standard procedures for fitting and using this representation.

We should mention that B-splines are closely related to other popular types of curve representations. *Hermite splines*,[‡] for example, are used to interpolate data when the first derivatives are available. *Bézier curves* are useful for modeling curves where intuitive user control of the curve is needed. Bézier curves are useful for short curves with small number of control points, whereas B-splines are appropriate for long continuous curves. *NURBS (non-uniform rational B-splines)* are a generalization of both B-splines and Bézier curves that are popular in the graphics community because of their ability to handle both analytical shapes (conic sections) and freeform shapes in a consistent manner.

Computing a Point Along the Spline

A uniform cubic B-spline is represented as a sequence of **control points** $\mathbf{q}_i$, $i = 0, \ldots, n + 1$ in the plane. Let us collect these points in a matrix $\mathbf{Q}$ of size $(n + 2) \times 2$, where the ith row of $\mathbf{Q}$ contains the ith control point $\mathbf{q}_i$. The spline is parameterized by a real parameter s, which varies from 1 to n, that is, $1 \leq s \leq n$. A point on the spline is represented as $\mathbf{x}(s) = (x(s), y(s))$. Computing $\mathbf{x}(s)$ is fairly straightforward. First define the matrix $\mathbf{M}$ which contains the coefficients of the B-spline basis functions necessary to maintain C^2 continuity (that is, continuity in the function itself, as well as in its first and second derivatives):

$$\mathbf{M} = \frac{1}{6}\begin{bmatrix} -1 & 3 & -3 & 1 \\ 3 & -6 & 3 & 0 \\ -3 & 0 & 3 & 0 \\ 1 & 4 & 1 & 0 \end{bmatrix} \tag{4.187}$$

which we saw earlier in Equation (3.88). Next, let i be the largest integer that is no greater than s, that is, $i = \lfloor s \rfloor$, and let $\alpha = s - i$ be the fractional leftover value, so that $i = 1, \ldots, n$, and $0 \leq \alpha < 1$. Then,

$$\mathbf{x}(s) = [x(s) \ y(s)] = \mathbf{v}^{\mathsf{T}} \mathbf{M} \mathbf{Q}_i \tag{4.188}$$

where

$$\mathbf{v}^{\mathsf{T}} \equiv [\alpha^3 \ \alpha^2 \ \alpha \ 1] \quad \text{and} \quad \mathbf{Q}_i \equiv \begin{bmatrix} \mathbf{q}_{i-1} \\ \mathbf{q}_i \\ \mathbf{q}_{i+1} \\ \mathbf{q}_{i+2} \end{bmatrix} \tag{4.189}$$

[†] Section 3.8.4 (p. 115).
[‡] Section 3.8.3 (p. 110).

Note that $s = i = 1$ at the beginning of the spline, while $s = i = n$ at the end of the spline; in both cases $\alpha = 0$, so that

$$\mathbf{x}(1) = \frac{1}{6}(\mathbf{q}_0 + 4\mathbf{q}_1 + \mathbf{q}_2) \tag{4.190}$$

$$\mathbf{x}(n) = \frac{1}{6}(\mathbf{q}_{n-1} + 4\mathbf{q}_n + \mathbf{q}_{n+1}) \tag{4.191}$$

Computing the Slope of a Spline

To compute the slope of the tangent of the spline at a point, simply take partial derivatives:

$$\frac{\partial x(s)}{\partial s} = \mathbf{v'}^{\mathsf{T}}\mathbf{M}\mathbf{x}_i$$

$$\frac{\partial y(s)}{\partial s} = \mathbf{v'}^{\mathsf{T}}\mathbf{M}\mathbf{y}_i \tag{4.192}$$

where

$$\mathbf{v'}^{\mathsf{T}} = \begin{bmatrix} 3\alpha^2 & 2\alpha & 1 & 0 \end{bmatrix} \tag{4.193}$$

is the derivative of $\mathbf{v}$ with respect to α. The slope is then given by

$$\frac{\partial y(s)}{\partial x(s)} = \frac{\partial y(s)}{\partial s}\frac{\partial s}{\partial x(s)} = \frac{\mathbf{v'}^{\mathsf{T}}\mathbf{M}\mathbf{y}_i}{\mathbf{v'}^{\mathsf{T}}\mathbf{M}\mathbf{x}_i} \tag{4.194}$$

Constructing the Spline

Suppose we wish to interpolate a sequence of data points $\mathbf{x}_1, \mathbf{x}_2, \ldots, \mathbf{x}_n$ in the plane with a B-spline curve, where $\mathbf{x}_i \equiv (x_i, y_i)$ is the ith data point. Interpolation implies that we set $\mathbf{x}(i) = \mathbf{x}_i$ for $i = 1, \ldots, n$. Notice that, in the case of s being an integer, $\alpha = 0$, and therefore Equation (4.188) simplifies to the following:

$$\mathbf{x}(s) = \frac{1}{6}\begin{bmatrix} 1 & 4 & 1 \end{bmatrix}\begin{bmatrix} \mathbf{q}_{i-1} \\ \mathbf{q}_i \\ \mathbf{q}_{i+1} \end{bmatrix} \tag{4.195}$$

Collecting all the data points together yields

$$\begin{bmatrix} \mathbf{x}_1 \\ \mathbf{x}_2 \\ \vdots \\ \mathbf{x}_n \end{bmatrix} = \frac{1}{6}\begin{bmatrix} 1 & 4 & 1 & 0 & 0 & \cdots & 0 \\ 0 & 1 & 4 & 1 & 0 & \cdots & 0 \\ & & & \vdots & & & \\ 0 & \cdots & 0 & 0 & 1 & 4 & 1 \end{bmatrix}\begin{bmatrix} \mathbf{q}_0 \\ \mathbf{q}_1 \\ \vdots \\ \mathbf{q}_{n+1} \end{bmatrix} \tag{4.196}$$

If the spline represents a closed curve, then the constraints simply wrap around:

$$
\begin{bmatrix} \mathbf{x}_1 \\ \mathbf{x}_2 \\ \vdots \\ \mathbf{x}_n \end{bmatrix} = \frac{1}{6} \begin{bmatrix} 4 & 1 & 0 & 0 & 0 & \cdots & 1 \\ 1 & 4 & 1 & 0 & 0 & \cdots & 0 \\ 0 & 1 & 4 & 1 & 0 & \cdots & 0 \\ & & & \vdots & & & \\ 0 & \cdots & 0 & 0 & 1 & 4 & 1 \\ 1 & \cdots & 0 & 0 & 0 & 1 & 4 \end{bmatrix} \begin{bmatrix} \mathbf{q}_1 \\ \mathbf{q}_2 \\ \vdots \\ \mathbf{q}_n \end{bmatrix}
\tag{4.197}
$$

and we set $\mathbf{q}_0 = \mathbf{q}_n$ and $\mathbf{q}_{n+1} = \mathbf{q}_1$ to maintain $\mathcal{C}^2$ continuity. On the other hand, if the spline represents an open curve, then additional equations are needed to make the system invertible. One possible approach is to add two artificial data points:

$$
\mathbf{x}_0 \equiv 2\mathbf{x}_1 - \mathbf{x}_2
\tag{4.198}
$$

$$
\mathbf{x}_{n+1} \equiv 2\mathbf{x}_n - \mathbf{x}_{n-1}
\tag{4.199}
$$

which are chosen to linearly extend the curve at both ends. We also add two new control points, $\mathbf{q}_{-1}$ and $\mathbf{q}_{n+2}$, which disappear from the equations by setting $\mathbf{q}_{-1} \equiv \mathbf{q}_0$ and $\mathbf{q}_{n+2} \equiv \mathbf{q}_{n+1}$. The result is

$$
\begin{bmatrix} \mathbf{x}_0 \\ \mathbf{x}_1 \\ \vdots \\ \mathbf{x}_{n+1} \end{bmatrix} = \frac{1}{6} \begin{bmatrix} 5 & 1 & 0 & 0 & 0 & \cdots & 0 \\ 1 & 4 & 1 & 0 & 0 & \cdots & 0 \\ 0 & 1 & 4 & 1 & 0 & \cdots & 0 \\ 0 & 0 & 1 & 4 & 1 & \cdots & 0 \\ \vdots & \vdots & \vdots & \vdots & \vdots & \ddots & \vdots \\ 0 & \cdots & 0 & 0 & 1 & 4 & 1 \\ 0 & \cdots & 0 & 0 & 0 & 1 & 5 \end{bmatrix} \begin{bmatrix} \mathbf{q}_0 \\ \mathbf{q}_1 \\ \vdots \\ \mathbf{q}_{n+1} \end{bmatrix}
\tag{4.200}
$$

Thus, given $\mathbf{x}_1, \ldots, \mathbf{x}_n$, either Equation (4.197) or Equation (4.200) can be solved for $\mathbf{q}_0, \ldots, \mathbf{q}_{n+1}$.

4.7 Further Reading

Mathematical morphology and morphological processing techniques trace their roots to work done at the Centre de Morphologie Mathématique in the Ecole des Mines de Paris in Fontainebleau, France beginning in 1964. The pioneers in this field were Matheron and Serra, the latter of whom heads the Center for Mathematical Morphology to this day. The classic text in the field is that of Serra [1982], but it is a heavy mathematical read. For a more practical treatment, the work of Soille [2003] is recommended. Another well-known early paper in this field is that of Haralick et al. [1987]. For recent work on morphological processing (extended to 4D spatio-temporal volumes) see the paper by Luengo-Oroz et al. [2012]. Minkowski addition is due to Minkowski [1901], but Minkowski subtraction was introduced a half-century later not by Minkowski himself but by Hadwiger [1950]. Apparently, the original version of Minkowski subtraction included a reflection: $X \ominus B = \bigcap_{b \in B} X_{-b}$, but the general consensus for at least the past several decades has been to define it in the manner done in this chapter.

The classic connected components algorithm, with an extra step to resolve the labels between the first and second pass through the image, is due to Rosenfeld and Pfaltz [1966]. The union-find algorithm, described by Tarjan [1975], was first incorporated into the classic connected components algorithm by Dillencourt et al. [1992]. Nowadays when people refer to the "classic connected components algorithm" they typically are referring to this more recent version that takes advantage of union-find. For a recent paper on connected components and computing region attributes simultaneously, consult the paper by Gabbur et al. [2010]. There are numerous papers on connected components, such as that of Chang et al. [2004], which show how to compute connected components while simultaneously computing region contours (wall following). The wall-following algorithm, also known as boundary tracing, contour tracing, the Moore neighborhood algorithm, or the Moore-neighbor tracing algorithm, is due to Moore [1968].

The Manhattan chamfer distance algorithm was originally described by Rosenfeld and Pfaltz [1966] in the same paper that introduced the classic connected components algorithm. Other early work on distance functions can be found in the papers of Rosenfeld and Pfaltz [1968] and Montanari [1968], where the Montanari condition is first described. The classic work on distance transforms and chamfer metrics that approximate Euclidean distance is the well-known paper of Borgefors [1986]. The exact Euclidean distance algorithm presented here is from Felzenszwalb and Huttenlocher [2004]; alternative approaches can be found in papers by Breu et al. [1995] and Maurer et al. [2003]. The Kimura distance function is from Kimura et al. [1999].

Moments can be found in any image processing book, such as Gonzalez and Woods [2008] or Jain [1989]. Hu moments were introduced by Hu [1962], while Legendre and Zernike moments were introduced by Teague [1980]. An overview of different types of moments, including geometric, Legendre, Zernike, pseudo-Zernike, rotational, and complex moments can be found in the work of Teh and Chin [1988], along with an experimental analysis that shows the superiority of Zernike and pseudo-Zernike moments for image reconstruction. An example of an application using Zernike moments is that of Boland et al. [1998]. The formula for computing area can be found in Ballard and Brown [1982]. Some of the equations for the best-fitting ellipse are in Shapiro and Stockman [2001]. The equation for eccentricity, which is standard in the mathematical community, is almost entirely absent from the image processing literature, with one notable exception being the text by Burger and Burge [2008]; the alternate formula for eccentricity in Equation (4.161) is found in the books by [Ballard and Brown 1982, p. 255] and [Jain 1989, p. 392]. The approach of computing the convex hull by dilating with half planes of various orientations is described by Soille [2003].

The bit quad algorithm for computing the Euler number is due to Gray [1971]. Perhaps the most successful use of the Euler number to date has been for automatic thresholding, as described by Rosin [1998] and Snidaro and Foresti [2003]. Some researchers have found Euler number to be one of the most clinically useful parameters for discriminating cervical abnormalities, see Pogue et al. [2000], and it has been used to a limited extent in document image processing, see Srihari [1986]. Morphological thinning as presented here uses the SEs found in Sonka et al. [2008]. The sigma-psi algorithm is described by Davies [2005] using the crossing number $\chi = 2\psi$ instead of the connection number ψ. The Zhang-Suen skeleton algorithm is from Zhang and Suen [1984], whereas NF2 is described in Barraquand and Latombe [1991]. Blum's medial axis is from Blum [1967]. For a more recent application of skeletons, see the work on the shock graphs such as Giblin and Kimia [2003].

The Freeman chain code is from Freeman [1961]. The minimum-perimeter polygon (MPP) is first described by Sklansky et al. [1972]. The name B-spline was introduced in Schoenberg [1971], and there are many good resources on B-splines, such as the works of Bartels et al. [1987] and Mortenson [1997].

PROBLEMS

4.1 Define mathematical morphology.

4.2 Write the set representation of the binary image $\mathcal{A}$, and the array representation of the binary image $\mathcal{B}$.

$$\mathcal{A} = \begin{bmatrix} 1 & 1 & 1 & 0 \\ 0 & 1 & 0 & 0 \\ 1 & 1 & 1 & 0 \\ 0 & 0 & 0 & 0 \end{bmatrix} \qquad \mathcal{B} = \{(1, 1), (2, 1), (1, 2), (2, 2), (3, 2), (1, 3), (3, 3)\}$$

4.3 Apply the set operators of Figure 4.2 to the images $\mathcal{A}$ and $\mathcal{B}$ of the previous question, using $\mathbf{b} = (1, 1)$. That is, compute $\mathcal{A} \cup \mathcal{B}$, $\mathcal{A} \cap \mathcal{B}$, $\mathcal{A}_b$, $\check{\mathcal{B}}$, $\neg \mathcal{A}$, and $\mathcal{A} \backslash \mathcal{B}$. Write the results as arrays.

4.4 Compute Minkowski addition for sets $\mathcal{A}_1$ and $\mathcal{B}$, as well as Minkowski subtraction for sets $\mathcal{A}_2$ and $\mathcal{B}$, shown below. Ignore the out-of-bounds pixels.

(a) Use the center-in approach.

(b) Repeat, using the center-out approach.

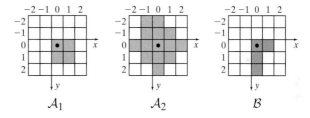

4.5 What is the difference between erosion and Minkowski subtraction?

4.6 Compute the dilation of the image $\mathcal{A}$ below using both center-in and center-out approaches. In both cases, *do not* reflect the structuring element $\mathcal{B}$. In which approach is reflection necessary to ensure that the output exhibits the same orientation as the input?

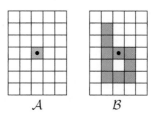

4.7 Prove Equation (4.15) from Equations (4.14), (4.28), and (4.29).

4.8 Recall the fruit image at the beginning of the chapter, which is reproduced below for convenience. On the two thresholded results shown, identify the name that best describes each of the labeled artifacts A–E: lake, bay, channel, cape, isthmus, or island. Which morphological operator (opening or closing) should be applied to the image on the left to remove noise? To the image on the right?

 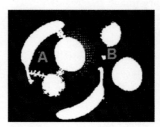 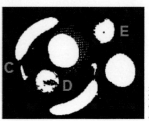

4.9 Thin the following binary image using the SEs shown in Figure 4.14 using the two variations described in the text.

(a) Apply all 8 SEs as a sequence until convergence.

(b) Apply the 4 edge SEs as a set until convergence, then apply the 4 corner SEs until convergence.

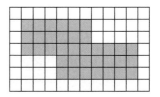

4.10 Determine the skeleton of the image shown in the previous question using

(a) the sigma-psi algorithm.

(b) the Zhang-Suen algorithm.

4.11 Implement the NF2 algorithm and run it on the image of the previous questions as well as on a binary image of your choice.

4.12 Which of the labeled pixels below are 4-neighbors of the central pixel c? 8-neighbors? diagonal neighbors?

	a	b
	c	d
e		

4.13 Implement the floodfill method of Algorithm 4.5 in your favorite programming language. Test your code on the synthetic image of Figure 4.22, along with another image you create. Now modify the code to use 8 neighbors.

4.14 Implement the connected components method of Algorithm 4.8. Test your code on a synthetic image.

4.15 Compute the Euclidean, Manhattan, and chessboard distances from each pixel in a 5×5 image to the central pixel. What shape do the isocontours take in each case?

4.16 Implement the Manhattan chamfer distance algorithm of Algorithm 4.11. Find an interesting binary image, then test your program on it.

4.17 Given the following binary image:

$$I = \begin{bmatrix} 0 & 0 & 1 \\ 1 & 0 & 1 \\ 0 & 1 & 1 \end{bmatrix}$$

(a) Compute the zeroth-, first-, and second-order regular moments.

(b) Compute the zeroth-, first-, and second-order central moments.

(c) Compute the covariance matrix, along with its eigenvalues and eigenvectors.

(d) Find the parameters of the best-fitting ellipse.

(e) Compute the eccentricity, orientation, and axis lengths of the best fitting ellipse.

4.18 Define the convex hull, then draw the convex hull for the following shape.

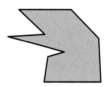

4.19 What is the Euler number of each of the following shapes? Verify your results using the Poincaré formula of Equation (4.165)?

4.20 Prove each of the following equations for sets $\mathcal{A}$, $\mathcal{B}$, and $\mathcal{C}$:

(a) $\mathcal{A} \ominus (\mathcal{B} \oplus \mathcal{C}) = (\mathcal{A} \ominus \mathcal{B}) \ominus \mathcal{C}$

(b) $\mathcal{A} \check{\ominus} (\mathcal{B} \oplus \mathcal{C}) = (\mathcal{A} \check{\ominus} \mathcal{B}) \check{\ominus} \mathcal{C}$

(c) $\mathcal{A} \oplus (\mathcal{B} \cup \mathcal{C}) = (\mathcal{A} \oplus \mathcal{B}) \cup (\mathcal{A} \oplus \mathcal{C})$

(d) $\mathcal{A} \ominus (\mathcal{B} \cup \mathcal{C}) = (\mathcal{A} \ominus \mathcal{B}) \cap (\mathcal{A} \ominus \mathcal{C})$

(e) $\mathcal{A} \check{\ominus} (\mathcal{B} \cup \mathcal{C}) = (\mathcal{A} \check{\ominus} \mathcal{B}) \cap (\mathcal{A} \check{\ominus} \mathcal{C})$

4.21 Show that each of following expressions is equivalent to the statement, "$\mathcal{A} \oplus \mathcal{B}$ is the set of points $\{\mathbf{z}\}$ such that for each $\mathbf{z}$ there is some $\mathbf{a}$ in $\mathcal{A}$ and some $\mathbf{b}$ in $\mathcal{B}$ whose sum is $\mathbf{z}$."

(a) $\mathcal{A} \oplus \mathcal{B} = \{\mathbf{z} : \mathbf{z} = \mathbf{a} + \mathbf{b}, \mathbf{a} \in \mathcal{A}, \mathbf{b} \in \mathcal{B}\}$

(b) $\mathcal{A} \oplus \mathcal{B} = \bigcup_{\mathbf{b} \in \mathcal{B}} \mathcal{A}_{\mathbf{b}}$

(c) $\mathcal{A} \oplus \mathcal{B} = \bigcup_{\mathbf{a} \in \mathcal{A}} \mathcal{B}_{\mathbf{a}}$

(d) $\mathcal{A} \oplus \mathcal{B} = \{\mathbf{z} : \mathbf{A} \cap (\check{\mathcal{B}})_{\mathbf{z}} \neq \emptyset\}$

4.22 Show that each of the following expressions is equivalent to the statement, "$\mathcal{A} \ominus \mathcal{B}$ is the set of points $\{\mathbf{z}\}$ such that for each $\mathbf{z}$ and for all $\mathbf{b} \in \mathcal{B}$, the point $\mathbf{z} - \mathbf{b}$ is in $\mathcal{A}$."

(a) $\mathcal{A} \ominus \mathcal{B} = \{\mathbf{z} : \mathbf{z} - \mathbf{b} \in \mathcal{A}, \forall \mathbf{b} \in \mathcal{B}\}$

(b) $\mathcal{A} \ominus \mathcal{B} = \{\mathbf{z} : \mathbf{z} - \mathbf{a} \notin \mathcal{B}, \forall \mathbf{a} \notin \mathcal{A}\}$

(c) $\mathcal{A} \ominus \mathcal{B} = \{\mathbf{z} : \mathbf{z} = \mathbf{a} + \mathbf{b}, \exists \mathbf{a} \in \mathcal{A}, \forall \mathbf{b} \in \mathcal{B}\}$

(d) $\mathcal{A} \ominus \mathcal{B} = \{\mathbf{z} : \mathbf{z} = \mathbf{a} + \mathbf{b}, \forall \mathbf{a} \notin \mathcal{A}, \exists \mathbf{b} \notin \mathcal{B}\}$

(e) $\mathcal{A} \ominus \mathcal{B} = \bigcap_{\mathbf{b} \in \mathcal{B}} \mathcal{A}_{\mathbf{b}}$

(f) $\mathcal{A} \ominus \mathcal{B} = \bigcap_{\mathbf{a} \notin \mathcal{A}} \neg \mathcal{B}_{\mathbf{a}}$

(g) $\mathcal{A} \ominus \mathcal{B} = \{\mathbf{z} : \check{\mathcal{B}}_{\mathbf{z}} \subseteq \mathcal{A}\}$

4.23 In robotics and motion planning the *Minkowski difference*, usually denoted $\mathcal{A} \ominus \mathcal{B}$, is not related to Minkowski subtraction at all but rather is equivalent to a reflected dilation, $\mathcal{A} \oplus \check{\mathcal{B}}$. Explain why this operation computes the set of locations at which a 2D robot $\mathcal{A}$ collides with object $\mathcal{B}$ assuming translation-only motion of the robot.

4.24 Demonstrate by a simple example that repeated applications of opening or closing do nothing.

4.25 Structuring elements are not always 3×3 arrays of 1s. *Annular opening* involves dilating an image by a donut-shaped structuring element (a ring of 1s with 0s in the middle). If $\mathcal{A}$ is an image of scattered tiny blobs, and $\mathcal{B}$ is an appropriately sized donut-shaped structuring element, then $(\mathcal{A} \oplus \mathcal{B}) \cap \mathcal{A}$ removes all isolated regions in the image. Sketch a simple example to demonstrate this.

4.26 The floodfill algorithm as presented in Algorithm 4.5 performs essentially a depth-first search of the space due to its use of a stack data structure. If a queue is used instead, then the algorithm will perform a breadth-first search. If the algorithm is modified to look not for *identically* colored pixels, but rather *similarly* colored pixels, then explain why a breadth-first search would be preferable.

4.27 If memory is limited, a fixed-memory floodfill algorithm can be designed using a variant of the wall-following algorithm. The boundary of the painted region is traced while painting new pixels adjacent to the boundary, until no such pixels exist. Such an algorithm is not used in practice because, although it is fairly efficient for nearly convex shapes, much time is wasted determining the next pixel to paint when it is applied to complex shapes. Nevertheless, write pseudocode for this algorithm.

4.28 Another alternative to the floodfill algorithm presented in this chapter processes scanlines rather than individual pixels. All the pixels reachable from the seed horizontally are painted, then the scanline above is examined and the reachable pixels painted, and so forth. When no more pixels can be painted, the process is repeated for the scanlines below the seed pixel. Thus, instead of pushing individual pixels onto the stack, this algorithm pushes the start (or end) coordinate of each disjointed set of horizontally connected pixels onto the stack. Write pseudocode for this algorithm.

4.29 The classic connected components algorithm of Algorithm 4.8 can easily be modified to calculate properties of the regions. Such quantities are updated during the algorithm by inserting appropriate calculations to update these quantities each time an output pixel is set, with minimal overhead. Write pseudocode to show how to calculate the area, moments, minimum and maximum gray level, and bounding box. Also show how to calculate the Euler number, i.e., to count the number of regions and holes.

4.30 Write pseudocode for other variations of the wall-following algorithm:

(a) counterclockwise interior boundary

(b) 8-connected interior boundary

(c) 4-connected exterior boundary

4.31 Manually apply the wall-following algorithm on the image of the letter "D" in Figure 4.26.

(a) Use the 4-neighbor version of Algorithm 4.9. Verify that the output is an 8-connected boundary.

(b) Use the modified 8-neighbor version you developed in Problem 4.30b. Verify that the output boundary is 4-connected.

4.32 Prove that the Manhattan distance always overestimates Euclidean, while the chessboard distance always underestimates it: $d_8(\mathbf{p}, \mathbf{q}) \leq d_E(\mathbf{p}, \mathbf{q}) \leq d_4(\mathbf{p}, \mathbf{q})$. *Hint*: Note that for any two nonnegative numbers $a, b \geq 0$,

$$\frac{1}{2}(a + b) \leq \frac{1}{\sqrt{2}}\sqrt{a^2 + b^2} \leq \max(a, b) \leq \sqrt{a^2 + b^2} \leq a + b.$$

4.33 Prove that the chessboard distance is never more than 30% away from the Euclidean distance, and the Manhattan distance is never more than 42% away: $0.7d_E(\mathbf{p}, \mathbf{q}) < d_8(\mathbf{p}, \mathbf{q}) \leq d_E(\mathbf{p}, \mathbf{q}) \leq d_4(\mathbf{p}, \mathbf{q}) < 1.42d_E(\mathbf{p}, \mathbf{q})$.

4.34 Apply the double angle formula, $\tan 2\theta = \frac{2\tan\theta}{1 - \tan^2\theta}$, to Equation (4.132) to obtain equivalent expressions for the orientation of a region (assuming $\mu_{11} \neq 0$):

$$\tan\theta = \frac{\mu_{02} - \mu_{20} + \sqrt{(\mu_{20} - \mu_{02})^2 + 4\mu_{11}^2}}{2\mu_{11}} = \frac{2\mu_{11}}{\mu_{20} - \mu_{02} + \sqrt{(\mu_{20} - \mu_{02})^2 + 4\mu_{11}^2}}.$$

4.35 Show that the curve $ax^2 + 2bxy + cy^2 = 1$ is an ellipse if and only if $ac > b^2$. *Hint*: Solve the equation for y, then examine what happens when x goes to infinity.

4.36 Prove that the orientation defined in Equation (4.132) describes the line about which the moment of inertia is minimized. (*Hint*: The moment of inertia of an object with mass density function $f(x, y)$ about a line with angle θ is given by $\sum\sum (x \sin\theta - y \cos\theta)^2 f(x, y)$. To minimize, differentiate and set to zero.)

4.37 Prove that the orientation defined in Equation (4.132) is the angle of the eigenvector corresponding to the largest eigenvalue of the covariance matrix. (*Hint*: Solve the system of equations $(\mathbf{C} - \lambda\mathbf{I})\mathbf{v} = 0$ for $\mathbf{v}_1$ and $\mathbf{v}_2$. Then use the definition $\mathbf{v}_1 = \begin{bmatrix} \cos\theta & \sin\theta \end{bmatrix}^\mathsf{T}$ and the double angle formula for tangent to recognize that $\tan 2\theta = \frac{2v_{11}v_{21}}{v_{11}^2 - v_{21}^2}$, where $\mathbf{v}_1 = \begin{bmatrix} v_{11} & v_{21} \end{bmatrix}^\mathsf{T}$.)

4.38 Prove that the orientation defined in Equation (4.132) describes the line that minimizes the sum of the squares of the perpendicular distances between the coordinates of the pixels in the region and the line.

4.39 Given a perfect square rotated at some arbitrary angle, what does Equation (4.132) yield for the orientation? Explain.

4.40 Design a vision system to detect a binary template in an image. Assume that the template is at the same scale as the image, so that only translation needs to be taken into account. Use the chamfer distance to efficiently compute a matching score associated with each location in the image. Write code to implement this procedure, and display the probability map showing the matching score for each location in the image (ignoring the borders). Also display a template-sized rectangle around the peak in the map.

4.41 Identify several different types of small, readily available objects, and gather several instances of each. Examples might be coins, buttons, pencils, keys, and so forth. Place the objects on a single-colored table or floor, and take a picture that looks down on the scene. Write code to threshold the image, clean up the noise, label the components, compute various properties of the foreground regions, and automatically classify the regions according to the appropriate category. Now rearrange the objects, take another picture, and run the same code. Note your observations on whether the algorithm performed robustly on the new image.

CHAPTER 5
Spatial-Domain Filtering

In Chapter 3 we explored a variety of simple point transformations in which each output pixel is a function of a single input pixel. An even more powerful class of transformations, known as spatial-domain filtering, occurs when each output pixel is a function of the input pixel and its neighbors. The two most common uses of spatial-domain filtering are noise removal and edge detection, which are accomplished using lowpass and highpass filters, respectively. In this chapter we study both types of filters in the spatial domain, reserving frequency analysis of such filters to a later chapter. Because of the importance of the Gaussian (bell-curve) function, we spent a great deal of effort describing the creation, analysis, and use of Gaussian convolution kernels and their derivatives.

5.1 Convolution

The most common way to filter an image in the spatial domain is convolution. We begin with the 1D case, then move to the 2D case, followed by a discussion about how convolution relates to other types of filters and operators.

5.1.1 1D Convolution

Let us begin with a simple example. Suppose we are given the 1D signal

$$[1 \quad 5 \quad 6 \quad 7]$$

and want to compute the average of each sample and its two neighbors. One way to view such a computation is to imagine another signal, called a **kernel**, consisting of three 1s that is slid across the original signal:

$$
\begin{array}{ccccccccccccccccccccc}
\cdot & 1 & 5 & 6 & 7 & \cdot & & \cdot & 1 & 5 & 6 & 7 & \cdot & & \cdot & 1 & 5 & 6 & 7 & \cdot & & \cdot & 1 & 5 & 6 & 7 & \cdot \\
1 & 1 & 1 & & & & & & 1 & 1 & 1 & & & & & & 1 & 1 & 1 & & & & & & 1 & 1 & 1 \\
& & \Downarrow & & & & & & & \Downarrow & & & & & & & & \Downarrow & & & & & & & \Downarrow & & \\
& & 2 & & & & & & & 4 & & & & & & & & 6 & & & & & & & 4.33 & &
\end{array}
$$

At each position, the values that are aligned vertically are multiplied by each other, then the products are summed and divided by 3. The output is therefore

$$
\begin{bmatrix} 2 & 4 & 6 & 4.33 \end{bmatrix}
$$

because $(1 + 5)/3 = 2$, $(1 + 5 + 6)/3 = 4$, and so on. Notice that when the kernel is near the beginning or end of the signal, the computation involves out-of-bounds pixels whose values are unknown (indicated by the dots). Some assumption must be made about these out-of-bounds samples in order to complete the computation. Any of the approaches discussed earlier[†] are applicable, but here we assume the values are 0 for simplicity.

The preceding example illustrates the important concept of **convolution**. More precisely, the discrete convolution of a 1D signal f with a kernel g is defined as[‡]

$$
f'(x) = f(x) \circledast g(x) \equiv \sum_{i=-\infty}^{\infty} f(x-i)g(i) \tag{5.1}
$$

$$
= \sum_{i=-\tilde{w}}^{w-\tilde{w}-1} f(x-i)g(i) \tag{5.2}
$$

where we use the prime $(')$ to denote the output (so f' should not be confused with the derivative), w is the width (or, equivalently, the length) of the kernel, and the second equality assumes that $g(x) = 0$ for all $x < -\tilde{w}$ or $x \geq w - \tilde{w}$. The **origin** $\tilde{w}$ of the kernel indicates the location where the result is stored, which is usually defined to be the index nearest the center. That is, $\tilde{w} \equiv \lfloor \frac{1}{2}(w-1) \rfloor$, so that $\tilde{w} = 0$ if the width is 1 or 2, $\tilde{w} = 1$ if the width is 3 or 4, and so on. For example, if we use underscore to indicate the origin,

$$
\begin{array}{llll}
w = 1 & \tilde{w} = 0 & \text{kernel: } & \begin{bmatrix} \underline{1} \end{bmatrix} \\
w = 2 & \tilde{w} = 0 & \text{kernel: } & \begin{bmatrix} \underline{1} & 1 \end{bmatrix} \\
w = 3 & \tilde{w} = 1 & \text{kernel: } & \begin{bmatrix} 1 & \underline{1} & 1 \end{bmatrix} \\
w = 4 & \tilde{w} = 1 & \text{kernel: } & \begin{bmatrix} 1 & \underline{1} & 1 & 1 \end{bmatrix}
\end{array}
$$

although we oftentimes omit the underscore when the central element of the kernel is the origin, so $\begin{bmatrix} 1 & 1 & 1 \end{bmatrix}$ means $\begin{bmatrix} 1 & \underline{1} & 1 \end{bmatrix}$. To ensure that the kernel has an unambiguous center, kernels are almost always created with an odd number of elements, in which case, $\tilde{w}$ is referred to as the **half-width** (because $w = 2\tilde{w} + 1$, so that $\tilde{w} = \frac{1}{2}(w-1)$ without any need for performing the floor operation), and the equation above simplifies to

$$
f'(x) = f(x) \circledast g(x) = \sum_{i=-\tilde{w}}^{\tilde{w}} f(x-i)g(i) \tag{5.3}
$$

[†] Section 4.1.3 (p. 138)

[‡] Although convolution is often denoted by the symbol *, we use the notation $\circledast$ to avoid confusion with multiplication and complex conjugate.

Convolution is closely related to **cross-correlation**, which is defined as

$$f'_{corr}(x) = f(x) \check{\circledast} g(x) \equiv \sum_{i=-\infty}^{\infty} f^*(x+i)g(i) = \sum_{i=-\tilde{w}}^{w-\tilde{w}-1} f^*(x+i)g(i) \tag{5.4}$$

where the superscript asterisk (*) indicates the complex conjugate. By comparing this equation with Equation (5.1), we see that if the signal is real, so that $f = f^*$, then the only difference between the two operations is that convolution reflects (flips) the kernel, whereas cross-correlation does not. Most of the kernels we will encounter are either *symmetric*, $g(k) = g(-k)$, in which case $f'(x) = f'_{corr}(x)$, or *antisymmetric*, $g(k) = -g(-k)$, in which case $f'(x) = -f'_{corr}(x)$. Therefore, it is usually okay to neglect to flip the kernel, as long as we remember to flip the sign of the result when the kernel is antisymmetric. Similarly, it is easy to see that convolution is always commutative, that is, $f \circledast g = g \circledast f$, whereas cross-correlation is commutative only when the signal is real and the kernel is symmetric.

A discrete signal is stored in memory as a 1D array of values with non-negative indices. To handle this detail, the kernel is typically shifted by $\tilde{w}$, defining $g[x] \equiv g(x - \tilde{w})$ as the 1D array that holds the samples of the function g, where $g[x]$ is valid for $x = 0, 1, \ldots, w - 1$, in which case convolution can be rewritten as

$$f'(x) = f(x) \circledast g(x) = \sum_{i=0}^{w-1} f[x + \tilde{w} - i] \cdot g[i] \tag{5.5}$$

For example, if $g(\cdot)$ is a three-element kernel defined for $x \in \{-1, 0, 1\}$, the array $g[\cdot]$ contains values for $i \in \{0, 1, 2\}$ such that $g[0] = g(-1)$, $g[1] = g(0)$, and $g[2] = g(1)$. Conveniently, note that $\tilde{w}$, which in this case is 1, is always the zero-based index of the central element of the kernel.

With this additional detail in mind, the pseudocode for 1D convolution is shown in Algorithm 5.1. As we saw in the preceding example, this code illustrates that convolution is a **shift-multiply-add** operation: viewing the signal f as a 1D image, the kernel is shifted (or slid) across the image, and at each pixel the elements of the kernel are individually multiplied (indicated by the asterisk) by the values in the image, followed by a summation of the resulting products. Brackets are used inside the pseudocode to emphasize that $g[x]$ is merely accessing an element of the discrete array. While f' is actually defined for $n + w - 1$ values according to Equation (5.5), where n is the length of the original signal, for simplicity we adopt the common image processing practice of setting the size of the output to be the same as that of the input. Also, near the two ends of the signal, the computation uses indices that exceed the size of f, as mentioned earlier; such out-of-bounds details have been omitted to avoid cluttering the code.

ALGORITHM 5.1 Convolve a 1D signal/image with a 1D kernel

CONVOLVE1D(f, g)

Input: 1D signal f with length n, 1D kernel g with length w
Output: the convolution of f and g

```
1   w̃ ← ⌊(w − 1)/2⌋
2   for x ← 0 to n − 1 do
3         val ← 0
4         for i ← 0 to w − 1 do
5               val ← val + g[i] * f[x + w̃ − i]
6         f'[x] ← val
7   return f'
```

EXAMPLE 5.1	Suppose we have an input signal $f = \begin{bmatrix} 8 & 24 & 48 & 32 & 16 \end{bmatrix}$ with five elements, and a kernel $g = \frac{1}{4}\begin{bmatrix} 1 & 2 & 1 \end{bmatrix}$ with three elements. That is, $n = 5$ and $w = 3$. What is $f \circledast g$, if we use replication to handle out-of-bounds values?
Solution	As shown in Figure 5.1, first we extend the signal past the borders using the method of replication (which, in the case of extending by just one sample, is equivalent to reflection). Then we slide the kernel across the signal and record, at each pixel, the sum of the element-wise multiplications. The output signal $f' = \begin{bmatrix} 12 & 26 & 38 & 32 & 20 \end{bmatrix}$ is the same width as the input. It is important to notice that convolution must never be done in place; otherwise, if f and f' are stored in the same place in memory, then the values computed for f' will corrupt those being read from f.

The relationship between the input f and the output f' depends upon the type of the kernel g. Two types of kernels are common. **Smoothing kernels** perform an averaging of the values in a local neighborhood and therefore reduce the effects of noise. Such kernels are often used as the first stage of preprocessing an image that has been corrupted by noise, in order to restore the original image. **Differentiating kernels**, on the other hand, accentuate the places where the signal is changing rapidly in value and are therefore used to extract useful information from images, such as the boundaries of objects, for purposes such as object detection. Smoothing kernels are lowpass filters, whereas differentiating kernels are highpass filters. To avoid changing the overall gray level of the output, smoothing kernels have the property that all the elements of the kernel sum to one, $\sum_i g(i) = 1$, whereas differentiating kernels have the property that all the elements of the kernel sum to zero, $\sum_i g(i) = 0$, since the derivative of a constant image is zero. Smoothing kernels are usually symmetric, whereas differentiating kernels are either symmetric or antisymmetric depending on whether the order of differentiation is even or odd, respectively. In the next few sections we will consider these two types of kernels in more detail.

Figure 5.1 An example of 1D convolution.

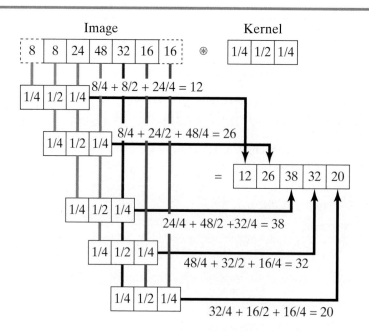

5.1.2 Convolution as Matrix Multiplication

Sometimes it is convenient to view discrete convolution as the multiplication of a matrix by a vector to produce another vector: the input vector is formed from the original signal, the matrix is formed from the convolution kernel, and the output vector is the result of the convolution. For example, consider the 1D input signal $f = \begin{bmatrix} 8 & 24 & 48 & 32 & 16 \end{bmatrix}$ and kernel $g = \frac{1}{4}\begin{bmatrix} 1 & 2 & 1 \end{bmatrix}$ that we saw earlier. It is easy to see that the following holds:

$$
\underbrace{\begin{bmatrix} 12 \\ 26 \\ 38 \\ 32 \\ 20 \end{bmatrix}}_{f'} = \frac{1}{4} \underbrace{\begin{bmatrix} 1 & 2 & 1 & 0 & 0 & 0 & 0 \\ 0 & 1 & 2 & 1 & 0 & 0 & 0 \\ 0 & 0 & 1 & 2 & 1 & 0 & 0 \\ 0 & 0 & 0 & 1 & 2 & 1 & 0 \\ 0 & 0 & 0 & 0 & 1 & 2 & 1 \end{bmatrix}}_{\substack{\textit{convolution matrix} \\ \mathbf{G}}} \underbrace{\begin{bmatrix} 8 \\ 8 \\ 24 \\ 48 \\ 32 \\ 16 \\ 16 \end{bmatrix}}_{f} \tag{5.6}
$$

where the left vector is simply the output $f' = f \circledast g$ shaped like a traditional column-wise vector, the right vector is the input f after extending the signal by replication, and the matrix is constructed by sliding the convolution kernel g horizontally by one position for each successive row. More generally, if the input signal f has n elements and the convolution kernel g has w elements, then convolution can always be represented as the following matrix multiplication:

$$
\mathbf{f}'_{\{n \times 1\}} = \mathbf{G}_{\{n \times n'\}} \mathbf{f}_{\{n' \times 1\}} \tag{5.7}
$$

where $\mathbf{G}$ is the convolution matrix constructed from the kernel g, $\mathbf{f}$ is the extended input signal stored in vector format, $\mathbf{f}'$ is the output as a vector, and $n' = n + w - 1$. Typically $n \gg w$, so $n \approx n'$. (If we ignore the border effects and set $n = n'$, then the convolution matrix is a *Toeplitz matrix*, meaning that every diagonal descending from the top left to the bottom right contains a constant value.)

5.1.3 Convolution as Fourier Multiplication

It is also worth noting that convolution in the spatial domain is equivalent to multiplication in the frequency domain. That is, if $f'(x) = f(x) \circledast g(x)$ is the convolution of two signals f and g, if $\mathcal{F}\{f\}$ and $\mathcal{F}\{g\}$ are the Fourier transforms of the two signals, respectively, and if $\mathcal{F}\{f'\}$ is the Fourier transform of the output, then the latter is the multiplication of the former two: $\mathcal{F}\{f'\} = \mathcal{F}\{f\} \cdot \mathcal{F}\{g\}$. An alternate way to compute the convolution of two signals, then, is the compute the inverse Fourier transform of the multiplication of the two Fourier transforms:

$$
f'(x) = f(x) \circledast g(x) = \mathcal{F}^{-1}\{\mathcal{F}\{f(x)\} \cdot \mathcal{F}\{g(x)\}\} \tag{5.8}
$$

Two important points must be noted, however. Since multiplication is less expensive than convolution, this trick can result in significant computational savings when the convolution kernel is large, because the overhead of computing the forward and inverse Fourier transforms is less than the amount of computation saved by not having to slide the kernel. Therefore, this trick is widely used in signal processing, where convolution kernels can be large, and where frequency-domain filters are commonly inserted into the pipeline. In image processing, however, convolution kernels tend to have only a few elements, so that a direct implementation of convolution is usually faster. A second point is that the equation above is true only in the case of circular convolution, that is convolution in which the signals are considered to be periodic. As a result, zero padding is necessary if linear convolution is desired. We will consider frequency-domain processing in more detail in the next chapter.

5.1.4 Linear Versus Nonlinear Systems

Before continuing, let us briefly consider filtering in general, to better appreciate how convolution fits into the larger context. In signal and image processing, as well as in related fields, a **system** is an operator that produces an output from an input. A system is said to be **linear** if both the *scaling* and *additivity* properties hold for all possible inputs:

$$\mathcal{L}(\alpha f) = \alpha \mathcal{L}(f) \qquad (\text{scaling}) \qquad (5.9)$$

$$\mathcal{L}(f_1 + f_2) = \mathcal{L}(f_1) + \mathcal{L}(f_2) \quad (\text{additivity}) \qquad (5.10)$$

where $\mathcal{L}$ represents the system, so that $\mathcal{L}(f)$ is the output of applying the system to the input f, and α is a scalar. According to the scaling property, a scaled version of the input causes the output to be scaled by the same amount, while the additivity property says that the output resulting from the sum of two inputs is simply the sum of the two individual outputs. Together, these properties are referred to as *superposition*:

$$\mathcal{L}(\alpha_1 f_1 + \alpha_2 f_2) = \alpha_1 \mathcal{L}(f_1) + \alpha_2 \mathcal{L}(f_2) \quad (\text{superposition}) \qquad (5.11)$$

The system $\mathcal{L}$ is linear if and only if this equation holds for all inputs and all scalars. If a system is not linear, then it is said to be **nonlinear**.

Another important property involves the response of the system to a shifted version of the input. A system is called **shift-invariant** if a shift in the input causes a shift in the output by the same amount. More precisely, if $f'(x) = \mathcal{L}(f(x))$ is the output of applying the system to the input signal $f(x)$, then the system is shift-invariant if and only if

$$f'(x - x_0) = \mathcal{L}(f(x - x_0)) \qquad (5.12)$$

Linear shift-invariant systems,[†] i.e., systems that are both linear and shift-invariant, are particularly important due to their convenient mathematical properties. Such systems are perfectly described by convolution with a (possibly infinite) kernel:

$$f'(x) = f(x) \circledast g(x) = \sum_{i=-\infty}^{\infty} f(x - i) g(i) \qquad (5.13)$$

A discrete linear shift-invariant system is perfectly described by its **impulse response**, which is defined as the output $\mathcal{L}(\delta)$ that results from applying the system to the function with a value of 1 at the origin and 0 everywhere else:

$$\delta(x) = \begin{cases} 1 & \text{if } x = 0 \\ 0 & \text{otherwise} \end{cases} \qquad (5.14)$$

also known as the **Kronecker delta function**. By setting f to δ in Equation (5.13), it should be easy to see that the impulse response of a linear shift-invariant system described by convolution is simply its convolution kernel.

Depending upon the impulse response, there are two types of linear shift-invariant systems. A **finite impulse response (FIR)** filter is a system for which the impulse

---

[†] Mathematically, linear shift-invariant systems are identical to the more well-known linear time-invariant (LTI) systems from signal processing. The difference in nomenclature arises because with images the independent variable is the pixel location rather than time.

response is finite in duration. FIR filters are perfectly described by convolution with a finite-length kernel. On the other hand, if the response continues forever, then we have an **infinite impulse response (IIR)** filter. Due to the impossibility of constructing a kernel with infinite duration, IIR filters are implemented using feedback from earlier computations. That is, unlike FIR filters, for which convolution must *not* be performed in place, IIR filters *do* perform their computation in place by storing the result into the same pixels that will be read in the next iteration, so that the output of the system is fed back as part of the input to the system. In practice, most filters are either linear FIR filters (and thus implemented using convolution) or nonlinear filters, although IIR filters can be used to perform fast filtering with large kernels. To develop some concrete appreciation for the differences between these types of filters, consider the following example.

EXAMPLE 5.2

Are the following systems linear or nonlinear, shift varying or shift-invariant, and FIR or IIR?

1. $f'(x) = \mathcal{L}(f(x)) = (f(x))^2$

2. $f'(x) = \mathcal{L}(f(x)) = \frac{1}{4}f(x-1) + \frac{1}{2}f(x) + \frac{1}{4}f(x+1)$

3. $f'(x) = \mathcal{L}(f(x)) = f'(x-1) + f(x)$

Solution

The three systems are analyzed as follows:

1. This system squares the input. To see that this is nonlinear, notice that $(\alpha f(x))^2 \neq \alpha^2(f(x))^2$ in general, thus violating the scaling principle. Nevertheless, it is shift-invariant, since the output at any given position is only dependent upon the input at that position. Since the concept of an impulse response is not meaningful in the case of nonlinear filters, we do not usually characterize them as being either FIR or IIR.

2. This system is just a convolution of the input f with the kernel $g = \frac{1}{4}\begin{bmatrix} 1 & 2 & 1 \end{bmatrix}$. Therefore, it is a linear shift-invariant system that is also an FIR filter.

3. This system is known as an accumulator because it sums all the values of f up to the present location: $f'(x) = \sum_{i=-\infty}^{x} g(x-i)f(i)$, where $g(x)$ is 1 for all $x \geq 0$ and 0 otherwise (the unit step function). The impulse response of this system is the unit step function, which extends forever. Therefore, this is a linear shift-invariant system that is also an IIR filter.

5.1.5 2D Convolution

Although 1D signals are easier to analyze, our goal is to perform filtering not on a 1D signal but rather on a 2D image. Thankfully, the extension of convolution to two dimensions is straightforward:

$$I'(x, y) = I(x, y) \circledast G(x, y) = \sum_{i=0}^{w-1}\sum_{j=0}^{h-1} I(x + \tilde{w} - i, y + \tilde{h} - j)G(i, j) \tag{5.15}$$

where w and h are the width and height of the kernel, respectively; $\tilde{h} \equiv \lfloor \frac{1}{2}(h-1) \rfloor$ is the half-height of the kernel, just like $\tilde{w}$ is the half-width; and the kernel G is assumed to be shifted by $\tilde{w}$ horizontally and $\tilde{h}$ vertically, so that all indices are nonnegative. To convolve an image with a 2D kernel, simply flip the kernel about both the horizontal and vertical axes, then slide the kernel along the image, computing the sum of the elementwise multiplication at each pixel between the kernel and the input image, as shown in Algorithm 5.2.

ALGORITHM 5.2 Convolve an image with a 2D kernel

CONVOLVE2D(I, G)

Input: 2D image $I_{\{width \times height\}}$, 2D kernel $G_{\{w \times h\}}$
Output: the 2D convolution of I and G

1 $\tilde{h} \leftarrow \lfloor (h-1)/2 \rfloor$
2 $\tilde{w} \leftarrow \lfloor (w-1)/2 \rfloor$
3 **for** $y \leftarrow 0$ **to** $height - 1$ **do**
4   **for** $x \leftarrow 0$ $width - 1$ **do**
5          $val \leftarrow 0$
6   **for** $j \leftarrow 0$ **to** $h - 1$ **do**
7   **for** $i \leftarrow 0$ **to** $w - 1$ **do**
8                  $val \leftarrow_+ G(i, j) * I(x + \tilde{w} - i, y + \tilde{h} - j)$
9          $I'(x, y) \leftarrow val$
10 **return** I'

5.2 Smoothing by Convolving with a Gaussian

The simplest smoothing kernel is the **box filter**, in which every element of the kernel has the same value. Since the elements must sum to one, the elements of a box filter of length w each have the value $1/w$. Some examples of box filters are $\frac{1}{3}\begin{bmatrix} 1 & 1 & 1 \end{bmatrix}$, and $\frac{1}{5}\begin{bmatrix} 1 & 1 & 1 & 1 & 1 \end{bmatrix}$. In practice, kernels are usually created to have an odd length to avoid undesired shifting of the output.

5.2.1 Gaussian Kernels

Convolving a box filter with itself yields an approximation to a **Gaussian kernel**. The continuous 1D Gaussian function is the familiar "bell curve", defined as

$$gauss_{\sigma^2}(x) \equiv \frac{1}{\sqrt{2\pi\sigma^2}} \exp\left(-\frac{x^2}{2\sigma^2}\right) \tag{5.16}$$

where σ^2 is the variance, and the normalization factor $1/\sqrt{2\pi\sigma^2}$ ensures that $\int_{-\infty}^{\infty} gauss_{\sigma^2}(x)\, dx = 1$. The 1D and 2D Gaussians are shown in Figure 5.3. Earlier we explained that if a signal of length n is convolved with a kernel of length w, the length of the result is $n + w - 1$. While we often choose to retain only the n values, sometimes it is necessary to retain all the values. For example, the simplest nontrivial box filter is $\frac{1}{2}\begin{bmatrix} 1 & 1 \end{bmatrix}$, which, when convolved with itself, leads to

$$\frac{1}{2}\begin{bmatrix} 1 & 1 \end{bmatrix} \circledast \frac{1}{2}\begin{bmatrix} 1 & 1 \end{bmatrix} = \frac{1}{4}\begin{bmatrix} 1 & 2 & 1 \end{bmatrix} \tag{5.17}$$

which is a kernel with $w = 3$, $\tilde{w} = 1$, $g[0] = \frac{1}{4}$, $g[1] = \frac{1}{2}$, and $g[2] = \frac{1}{4}$. This discrete kernel approximates a Gaussian with $\sigma^2 = \frac{1}{2}$. An additional iteration yields

$$\frac{1}{4}\begin{bmatrix} 1 & 2 & 1 \end{bmatrix} \circledast \frac{1}{4}\begin{bmatrix} 1 & 2 & 1 \end{bmatrix} = \frac{1}{16}\begin{bmatrix} 1 & 4 & 6 & 4 & 1 \end{bmatrix} \tag{5.18}$$

Figure 5.2 Binomial (left) and trinomial (right) triangles for constructing Gaussian kernels. The $(2k + 1)^{\text{th}}$ row of the binomial triangle approximates a Gaussian with $\sigma^2 = k/2$, while the $(k + 1)^{\text{th}}$ row of the trinomial triangle approximates a Gaussian with $\sigma^2 = \frac{2k}{3}$, where k is a non-negative integer.

```
              1                                 1
            1   1                           1   1   1
          1   2   1                     1   2   3   2   1
        1   3   3   1                1   3   6   7   6   3   1
      1   4   6   4   1          1   4  10  16  19  16  10   4   1
       Binomial triangle                 Trinomial triangle
```

which approximates a Gaussian with $\sigma^2 = 1$. Ignoring the normalization factor, these Gaussians can easily be remembered as the odd rows of the binomial triangle, also known as Pascal's triangle,[†] with the $(2k + 1)^{\text{th}}$ row approximating a Gaussian with $\sigma^2 = k/2$, as shown in Figure 5.2. Similarly, the $(k + 1)^{\text{th}}$ row of the trinomial triangle approximates a Gaussian with $\sigma^2 = \frac{2k}{3}$. For example, $\frac{1}{3}\begin{bmatrix} 1 & 1 & 1 \end{bmatrix}$ approximates a Gaussian with $\sigma^2 = \frac{2}{3}$, while

$$\frac{1}{3}\begin{bmatrix} 1 & 1 & 1 \end{bmatrix} \circledast \frac{1}{3}\begin{bmatrix} 1 & 1 & 1 \end{bmatrix} = \frac{1}{9}\begin{bmatrix} 1 & 2 & 3 & 2 & 1 \end{bmatrix} \tag{5.19}$$

approximates a Gaussian with $\sigma^2 = \frac{4}{3}$.

5.2.2 Computing the Variance of a Smoothing Kernel

To compute the variance of an arbitrary smoothing kernel g, one might be tempted to apply the formulas

$$\mu = \frac{1}{w}\sum_i g(i) \qquad (\text{wrong}) \tag{5.20}$$

$$\sigma^2 = \frac{1}{w}\sum_i (g(i) - \mu)^2 \quad (\text{wrong}) \tag{5.21}$$

Figure 5.3 A Gaussian is a bell curve. From left to right: The 2D isotropic Gaussian viewed as an image where the gray level of each pixel is proportional to the value of the Gaussian function at that point, the 2D isotropic Gaussian viewed as a surface in 3D, and the 1D Gaussian function (or, equivalently, a slice through the 2D Gaussian function, obtained by intersecting it with a vertical plane).

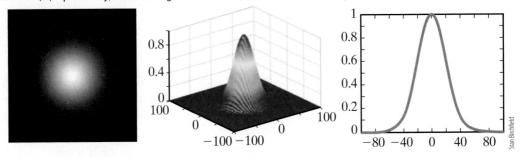

Stan Birchfield

[†] Blaise Pascal (1623–1662) was a French mathematician who developed probability theory and proved that light travels through a vacuum; he also invented the mechanical calculator, hydraulic press, and syringe; and he made important contributions to Christian philosophy, most notably the famous Pascal's Wager.

but these compute the mean and variance of the *values* of the kernel, rather than the mean and variance of the kernel along the *domain*. The correct way is to calculate the mean and variance of the coordinates of the elements of the kernel, with the values serving as weights:

$$\mu = \frac{\sum_i i g(i)}{\sum_i g(i)} \qquad (\text{correct}) \tag{5.22}$$

$$\sigma^2 = \frac{\sum_i (i - \mu)^2 g(i)}{\sum_i g(i)} \qquad (\text{correct}) \tag{5.23}$$

As an example, the mean and variance of $g = \frac{1}{4}\begin{bmatrix} 1 & 2 & 1 \end{bmatrix}$ are

$$\mu = \frac{1}{4}(0 \cdot 1 + 1 \cdot 2 + 2 \cdot 1) = 1 \tag{5.24}$$

$$\sigma^2 = \frac{1}{4}((0-1)^2 \cdot 1 + (1-1)^2 \cdot 2 + (2-1)^2 \cdot 1) = \frac{1}{2} \tag{5.25}$$

As mentioned earlier, with an odd-length kernel the center of the Gaussian is equal to the half-width, or $\mu = \tilde{w}$.

5.2.3 Separability

Given two 1D kernels, a 2D kernel can be constructed by convolving them with each other, with one oriented vertically and the other oriented horizontally. For example,

$$\frac{1}{3}\begin{bmatrix} 1 \\ 1 \\ 1 \end{bmatrix} \circledast \frac{1}{3}\begin{bmatrix} 1 & 1 & 1 \end{bmatrix} = \frac{1}{9}\begin{bmatrix} 1 & 1 & 1 \\ 1 & 1 & 1 \\ 1 & 1 & 1 \end{bmatrix} \tag{5.26}$$

Since convolution itself is commutative, we could also write this by reversing the order of the kernels, but writing the vertical kernel first provides a helpful visual aid, because the result is the same as the matrix multiplication of the two vectors (i.e., the outer product of the two vectors).

When a 2D kernel can be decomposed into the convolution of two 1D kernels, we say that the kernel is **separable**. Every 2D axis-aligned Gaussian kernel is separable, such as the 2D isotropic Gaussian:

$$Gauss_{\sigma^2}(x, y) \equiv \frac{1}{2\pi\sigma^2} \exp\left(-\frac{x^2 + y^2}{2\sigma^2}\right) \tag{5.27}$$

where the normalization $1/2\pi\sigma^2$ is again designed to ensure that $\iint Gauss_{\sigma^2}(x, y)\, dx\, dy = 1$. To show the separability of this function, apply the law of exponents to the convolution of an arbitrary 2D signal $I(x,y)$ and a 2D isotropic Gaussian, ignoring the normalization factor for simplicity:

$$I(x, y) \circledast Gauss_{\sigma^2}(x, y) = \sum_i \sum_j I(x - i, y - j) \exp\left(\frac{-(i^2 + j^2)}{2\sigma^2}\right) \tag{5.28}$$

$$= \sum_i \underbrace{\left[\sum_j I(x - i, y - j) \exp\left(\frac{-j^2}{2\sigma^2}\right)\right]}_{I(x,y)\,\circledast\,gauss_{\sigma^2}(y)} \exp\left(\frac{-i^2}{2\sigma^2}\right) \tag{5.29}$$

$$= \left(I(x, y) \circledast gauss_{\sigma^2}(y) \right) \circledast gauss_{\sigma^2}(x) \tag{5.30}$$

$$= \left(I(x, y) \circledast gauss_{\sigma^2}(x) \right) \circledast gauss_{\sigma^2}(y) \tag{5.31}$$

where one kernel is oriented vertically while the other is oriented horizontally. Thus we see that convolving I with $Gauss_{\sigma^2}(x, y)$ is the same as convolving I with a vertical 1D Gaussian kernel $gauss_{\sigma^2}(y)$, followed by a horizontal 1D Gaussian kernel $gauss_{\sigma^2}(x)$; or vice versa, since the order does not matter. As an example,

$$I(x, y) \circledast \left(\frac{1}{16} \begin{bmatrix} 1 & 2 & 1 \\ 2 & 4 & 2 \\ 1 & 2 & 1 \end{bmatrix} \right) = \left(I(x, y) \circledast \frac{1}{4} \begin{bmatrix} 1 \\ 2 \\ 1 \end{bmatrix} \right) \circledast \left(\frac{1}{4} \begin{bmatrix} 1 & 2 & 1 \end{bmatrix} \right) \tag{5.32}$$

because

$$\frac{1}{4} \begin{bmatrix} 1 \\ 2 \\ 1 \end{bmatrix} \circledast \frac{1}{4} \begin{bmatrix} 1 & 2 & 1 \end{bmatrix} = \frac{1}{16} \begin{bmatrix} 1 & 2 & 1 \\ 2 & 4 & 2 \\ 1 & 2 & 1 \end{bmatrix} \tag{5.33}$$

and because convolution is associative: $(f \circledast g_1) \circledast g_2 = f \circledast (g_1 \circledast g_2)$. A discrete 2D kernel is separable if and only if all of its rows and columns are linearly dependent (i.e., scalar multiples of one another), meaning that the kernel (viewed as a matrix) is rank 1.

Another way to derive the separability of the Gaussian is to notice from Equations (5.16) and (5.27) that the 2D Gaussian can also be viewed as simply the product of two 1D Gaussians:

$$Gauss_{\sigma^2}(x, y) \equiv \frac{1}{2\pi\sigma^2} \exp\left(-\frac{x^2 + y^2}{2\sigma^2} \right) \tag{5.34}$$

$$= \frac{1}{\sqrt{2\pi\sigma^2}} \exp\left(-\frac{x^2}{2\sigma^2} \right) \cdot \frac{1}{\sqrt{2\pi\sigma^2}} \exp\left(-\frac{y^2}{2\sigma^2} \right) \tag{5.35}$$

$$= gauss_{\sigma^2}(x) \cdot gauss_{\sigma^2}(y) \tag{5.36}$$

This equation is true for any point (x,y), which means that if we let $\mathbf{g}_{\sigma^2}$ be the 1D Gaussian kernel represented as a (vertically oriented) vector, then the 2D Gaussian is just the outer product of this vector with itself: $Gauss_{\sigma^2}(x, y) = \mathbf{g}_{\sigma^2}\mathbf{g}_{\sigma^2}^{\mathsf{T}}$. In fact, this observation holds for any separable kernel, because the convolution of two 1D kernels in orthogonal directions is equivalent to the outer product of the two kernels when they are viewed as vectors. For example, the outer product $\mathbf{g}\mathbf{g}^{\mathsf{T}} = \begin{bmatrix} 1 & 2 & 1 \end{bmatrix}^{\mathsf{T}} \begin{bmatrix} 1 & 2 & 1 \end{bmatrix}$ is given by

$$\begin{bmatrix} 1 \\ 2 \\ 1 \end{bmatrix} \begin{bmatrix} 1 & 2 & 1 \end{bmatrix} = \begin{bmatrix} 1 & 2 & 1 \\ 2 & 4 & 2 \\ 1 & 2 & 1 \end{bmatrix} \tag{5.37}$$

which (ignoring the normalization factor) is equivalent to Equation (5.33).

Separable convolution is shown in Algorithm 5.3 using two 1D kernels, one for the horizontal and one for the vertical operation. For simplicity this code assumes that the length of both kernels is the same, which is nearly always true in practice, although this assumption is not important. Note that convolution requires a temporary image to store the result of the first convolution, since convolution cannot be done in place. Note also that it is critical to convolve every row (that is, including the first and last) of the image for a horizontal kernel, and every column of the image for a vertical kernel. This is because the second convolution uses values computed in the first convolution. With a little extra work to handle out-of-bounds pixels, the values for all the pixels in the output image can be computed. If the 2D kernel

ALGORITHM 5.3 Convolve an image with a separable 2D kernel

$\textsc{ConvolveSeparable}(I, g_h, g_v)$

Input: 2D image $I_{\{width \times height\}}$, 1D kernels g_h and g_v each of length w
Output: the 2D convolution of I and $g_v \circledast g_h$

```
1   ▶ convolve horizontal
2   for y ← 0 to height − 1 do
3       for x ← w̃ to width − 1 − w̃ do
4           val ← 0
5           for i ← 0 to w − 1 do
6               val ← val + g_h[i] * I(x + w̃ − i, y)
7           I_tmp(x, y) ← val
8   ▶ convolve vertical
9   for y ← w̃ to height − 1 − w̃ do
10      for x ← 0 to width − 1 do
11          val ← 0
12          for i ← 0 to w − 1 do
13              val ← val + g_v[i] * I_tmp(x, y + w̃ − i)
14          I'(x, y) ← val
15  return I'
```

is of size $w \times w$, then the amount of computation in separable convolution is $O(2w)$ rather than $O(w^2)$, which can be a significant savings in computation over the full 2D convolution.

Just as 1D convolution can be viewed as matrix multiplication, so can 2D convolution. One way to achieve this is to stack all the pixels of the image row-by-row into a single vector, then form the convolution matrix $\mathbf{G}$ by stacking successively shifted versions of the different rows of the kernel to form an equation similar to Equation (5.7). However, when the kernel is separable there is a more compact, elegant notation:

$$\mathbf{I}' = \mathbf{G}_h \mathbf{I} \mathbf{G}_v^\mathsf{T} \tag{5.38}$$

where $\mathbf{G}_h$ is the convolution matrix constructed from the 1D kernel $\mathbf{g}_h$, $\mathbf{G}_v$ is the convolution matrix constructed from the 1D kernel $\mathbf{g}_v$, the 2D convolution kernel is $\mathbf{g}_v \mathbf{g}_h^\mathsf{T}$, and $\mathbf{I}$ is the image I viewed as a matrix and extended appropriately to handle border issues. To see this result, notice from Equation (5.7) that $\mathbf{I}_{tmp} \equiv \mathbf{G}_v \mathbf{I}^\mathsf{T}$ convolves along the columns of I, and applying the convolution again to the transposed result convolves along the rows of I, yielding the final 2D convolution: $\mathbf{I}' = \mathbf{G}_h (\mathbf{I}_{tmp})^\mathsf{T} = G_h \mathbf{I} G_v^\mathsf{T}$.

5.2.4 Constructing Gaussian Kernels

To construct a 1D Gaussian kernel with an arbitrary standard deviation σ, simply sample the continuous zero-mean Gaussian function $gauss_{\sigma^2}(x) \equiv \frac{1}{\sqrt{2\pi\sigma^2}} \exp\left(\frac{-x^2}{2\sigma^2}\right)$ and normalize by dividing each element by the sum of all the elements, $\sum_i gauss_{\sigma^2}[i]$, as shown in Algorithm 5.4. This sum is the zeroth moment of the signal. The property $\sum_i gauss_{\sigma^2}[i] = 1$ is important to ensure that the overall brightness of the image does not change as a result of the smoothing. Another way to look at this is that the smoothing of a constant image should not change the image. Note that the continuous normalization factor $\frac{1}{\sqrt{2\pi\sigma^2}}$, which ensures $\int_{-\infty}^{\infty} gauss_{\sigma^2}(x)\, dx = 1$ in the continuous domain, can be ignored since it disappears anyway when the discrete normalization step is performed. The discrete normalization step, however, cannot be ignored because the continuous normalization factor alone will not ensure that $\sum_i gauss_{\sigma^2}[i] = 1$, due to discretization effects. Also note in the pseudocode that we must subtract $\tilde{w}$ from the index while constructing the kernel, since $gauss_{\sigma^2}(x)$ has zero mean, but our discrete Gaussian kernel $gauss_{\sigma^2}[i]$ is centered around $i = \tilde{w}$.

Given a desired standard deviation, a reasonable approach to choosing an appropriate kernel half-width is the following:

$$\tilde{w} \approx 2.5\sigma - 0.5 \qquad (5.39)$$

where the approximation indicates that a rounding of the value on the right-hand side must occur, since $\tilde{w}$ is an integer. Line 1 of Algorithm 5.4 uses this expression, shown as pseudocode in Algorithm 5.5. To derive this expression, note that the central sample $gauss[\tilde{w}]$ in the discrete Gaussian approximates the region between $x = -0.5$ and $x = 0.5$, as shown in Figure 5.4. Similarly, the adjacent sample $gauss[\tilde{w} + 1]$ approximates the region between $x = 0.5$ and $x = 1.5$, and an arbitrary sample $gauss[\tilde{w} + k]$ approximates the region between $x = k - 0.5$ and $x = k + 0.5$. Since $w - 1 = 2\tilde{w} = \tilde{w} + \tilde{w}$, the final sample $gauss[w - 1]$ approximates the region between $x = \tilde{w} - 0.5$ and $x = \tilde{w} + 0.5$. Therefore, a kernel of width w approximately captures the area

$$\int_{-\tilde{w}-0.5}^{\tilde{w}+0.5} gauss(x)\, dx$$

under the original continuous Gaussian function. You may know from the definition of a Gaussian that 68.27% of the area under the Gaussian is captured in the region $\sigma \le x \le \sigma$, 95.45% in the region $2\sigma \le x \le 2\sigma$, as summarized in Table 5.1. By setting $\tilde{w} + 0.5 = 2.5\sigma$, 98.76% of the area under the Gaussian is captured, to ensure that the

---

ALGORITHM 5.4 Create a 1D Gaussian kernel

$\textsc{CreateGaussianKernel}(\sigma)$

Input: floating-point standard deviation σ
Output: 1D Gaussian kernel (as an array with w elements)

1 $\tilde{w} \leftarrow \textsc{GetKernelHalfWidth}(\sigma)$ ➤ Determine a reasonable halfwidth $\tilde{w}$, using, e.g., Algorithm 5.5.
2 $w \leftarrow 2\tilde{w} + 1$ ➤ Compute the (odd) width w from the halfwidth $\tilde{w}$.
3 $norm \leftarrow 0$ ➤ Initialize the normalization factor to zero.
4 **for** $i \leftarrow 0$ **to** $w - 1$ **do**   ➤ Construct the w-element kernel by sampling
5      $gauss[i] \leftarrow \exp(-(i - \tilde{w}) * (i - \tilde{w})/(2 * \sigma * \sigma))$ the continuous Gaussian function,
6      $norm \leftarrow_+ gauss[i]$ while keeping track of the normalization factor.
7 **for** $i \leftarrow 0$ **to** $w - 1$ **do**   ➤ Apply the normalization factor
8      $gauss[i] \leftarrow_/ norm$ to ensure that $\sum_{i=0}^{w-1} gauss[i] = 0$.
9 **return** $gauss$

ALGORITHM 5.5 Compute the appropriate halfwidth of a 1D Gaussian kernel with a given standard deviation

$\textsc{GetKernelHalfWidth}(\sigma)$

1 **return** $\textsc{Round}(2.5\sigma - 0.5)$

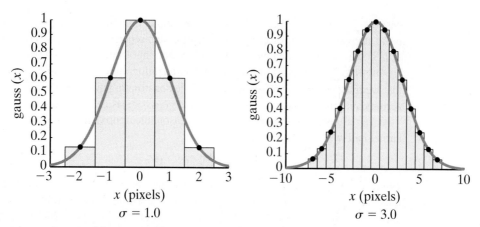

Figure 5.4 A continuous Gaussian (blue curve) and its discrete approximation (black circles) for two different values of σ. The shaded area under the rectangles approximates the area under the continuous curve between $x = -2.5\sigma$ and $x = +2.5\sigma$. The width and half-width of the kernels are $w = 5$, $\bar{w} = 2$ (for $\sigma = 1.0$), and $w = 15$, $\bar{w} = 7$ (for $\sigma = 3.0$), according to $\bar{w} + 0.5 = 2.5\sigma$ and $w = 2\bar{w} + 1$.

kernel well approximates a Gaussian. However, if a less accurate approximation is acceptable, then the 2.5 factor multiplying the standard deviation can be reduced accordingly.

If $gauss_{\sigma^2}$ refers to a 1D Gaussian kernel with variance σ^2, then some common 1D Gaussian kernels are as follows:

$$gauss_{0.25} = \frac{1}{8}\begin{bmatrix} 1 & 6 & 1 \end{bmatrix} \tag{5.40}$$

$$gauss_{0.333} = \frac{1}{6}\begin{bmatrix} 1 & 4 & 1 \end{bmatrix} \tag{5.41}$$

$$gauss_{0.375} = \frac{1}{16}\begin{bmatrix} 3 & 10 & 3 \end{bmatrix} \tag{5.42}$$

$$gauss_{0.5} = \frac{1}{4}\begin{bmatrix} 1 & 2 & 1 \end{bmatrix} \tag{5.43}$$

$$gauss_{1.0} = \frac{1}{16}\begin{bmatrix} 1 & 4 & 6 & 4 & 1 \end{bmatrix} \tag{5.44}$$

Note that Equation (5.39) returns the correct lengths for these variances, which are computed in the same manner described before. However, the variance of the discrete Gaussian kernel will in general be different from that of the underlying continuous Gaussian function from which it was sampled. From Table 5.2, we see that this difference in variance can be as high as 30%.

Now that we can create Gaussian kernels and convolve an image with horizontal and vertical kernels, we can smooth an image by convolving with a 2D Gaussian kernel, as shown in Algorithm 5.6. Although the code shows the simplest case of an isotropic Gaussian, because that is the most common case, it would be easy to extend the code to the anisotropic

TABLE 5.1 The area under the Gaussian curve within intervals defined by different factors of the standard deviation. This is sometimes called the $68 - 95 - 99.7$ rule.

domain	area under curve
$[-\sigma \leq x \leq \sigma]$	68.27%
$[-2\sigma \leq x \leq 2\sigma]$	95.45%
$[-2.5\sigma \leq x \leq 2.5\sigma]$	98.76%
$[-3\sigma \leq x \leq 3\sigma]$	99.73%

kernel	discrete σ^2	continuous σ^2	error
$gauss_{0.25} = \frac{1}{8}\begin{bmatrix} 1 & 6 & 1 \end{bmatrix}$	0.25	0.28	10.7%
$gauss_{0.333} = \frac{1}{6}\begin{bmatrix} 1 & 4 & 1 \end{bmatrix}$	0.333	0.36	7.4%
$gauss_{0.375} = \frac{1}{16}\begin{bmatrix} 3 & 10 & 3 \end{bmatrix}$	0.375	0.42	10.7%
$gauss_{0.5} = \frac{1}{4}\begin{bmatrix} 1 & 2 & 1 \end{bmatrix}$	0.5	0.72	30.6%
$gauss_{1.0} = \frac{1}{16}\begin{bmatrix} 1 & 4 & 6 & 4 & 1 \end{bmatrix}$	1.0	1.17	14.5%

TABLE 5.2 The discrete Gaussian kernel has, in general, a different variance from the underlying continuous function from which it was sampled. Shown are the differences for several different values of σ^2.

Gaussian case. Results of convolving an image with 2D isotropic Gaussians with different variances are shown in Figure 5.5.

5.2.5 Evaluating Gaussian Kernels

In this section we examine in more detail the relationship between the width of the Gaussian kernel and its standard deviation. The reader should feel free to skip this section upon first reading.

Some authors have argued that it is not possible to build a faithful Gaussian kernel with just three samples $(w = 3)$. The argument is based on recognizing that there are conflicting constraints: Only a narrow (small σ) Gaussian will be accurately represented by just three samples, but a narrow Gaussian in the spatial domain leads to a wide Gaussian (large σ') in the frequency domain, leading to aliasing.[†] This is because the Fourier transform of a Gaussian is $\mathcal{F}\{\exp(-\frac{x^2}{2\sigma^2})\} \propto \exp(-\frac{\omega^2}{2(1/\sigma)^2})$, where ω is the angular frequency and $\sigma' = 1/\sigma$ is the standard deviation of the Fourier transformed signal.[‡] To see this numerically, notice that capturing 98.76% of the Gaussian yields the constraint $w \geq 5\sigma$, where w is the width of the kernel, since the region from -2.5σ to 2.5σ has a width of 5σ. Now because the sampling frequency is 1 sample per pixel, Nyquist's sampling theorem

ALGORITHM 5.6 Smooth an image by convolving with a 2D Gaussian kernel

SMOOTH(I, σ)

Input: image I, standard deviation σ
Output: result of convolving I with a 2D isotropic Gaussian with σ

1   $gauss \leftarrow$ CREATEGAUSSIANKERNEL(σ)
2   $I' \leftarrow$ CONVOLVESEPARABLE$(I, gauss, gauss)$
3   **return** I'

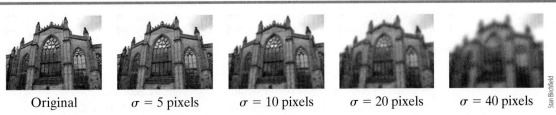

Original $\sigma = 5$ pixels $\sigma = 10$ pixels $\sigma = 20$ pixels $\sigma = 40$ pixels

Stan Birchfield

Figure 5.5 A 2304 × 1728 image, and the result of smoothing by convolving with an isotropic Gaussian with different standard deviations.

[†] Aliasing is discussed in more detail in Section 6.1.3 (p. 275).
[‡] The Fourier transform is covered in more detail in the next chapter.

says the cutoff frequency is 0.5, implying a cutoff *angular* frequency of $2\pi(0.5) = \pi$. To keep 98.76% of the area of the Fourier transform between $-\pi$ and π, we must therefore (assuming $w = 3$) set $2\pi \geq 5\sigma' = 5(1/\sigma)$, which leads to $\sigma \geq \frac{5}{2\pi} = 0.8$. Putting the spatial constraint $w \geq 5\sigma$ together with the frequency constraint $\sigma \geq 0.8$ implies $w \geq 5$, assuming w is odd. In other words, this reasoning leads to the conclusion that the kernel must contain at least 5 samples to faithfully represent the Gaussian.

However, such a conclusion is unwarranted. The 3-element kernel is widely used in practice, and for good reason. A more appropriate question to ask is: What is the best that a 3-element kernel can do? Instead of requiring that the kernel capture 98.76% of the area of the Gaussian, let us seek a value for σ that maximizes the area preserved in both the spatial and frequency domains. Let α represent the value such that this preserved area is in the region $-\alpha\sigma \leq \times \leq \alpha\sigma$, so that $w = 2\alpha\sigma$. The same area is captured in the frequency domain when $2\pi = 2\alpha\sigma'$, or $\sigma = 1/\sigma' = \alpha/\pi$. Since we are interested in the case $w = 3$, we can solve the equations $\alpha\sigma = 1.5$ and $\alpha/\sigma = \pi$ for the two unknowns to yield $\alpha = \sqrt{1.5\pi} \approx 2.17$ and $\sigma = \sqrt{1.5/\pi} \approx 0.69$. From the definition of the Gaussian, this value of $\alpha = 2.17$ implies that 97% of the Gaussian is captured in both the spatial and frequency domains, which is quite acceptable for many applications. Moreover, it is interesting to note that this particular value of $\sigma^2 = 0.48$ ($\sigma = 0.69$) is very close to the $\sigma^2 = 0.5$ ($\sigma = 0.71$) of the 3×1 kernel obtained by the binomial triangle. Therefore, according to the criterion of balancing the area under the spatial- and frequency-domain signals, $\sigma = 0.69$ is the optimal 3×1 Gaussian kernel, which is closely approximated by $gauss_{0.5} = \frac{1}{4}\begin{bmatrix} 1 & 2 & 1 \end{bmatrix}$.[†]

Now that we have seen that the best 3-element Gaussian kernel is very close to the one given by the binomial triangle, let us analyze the other kernels in the triangle. Table 5.3 shows the first few of these kernels, along with the value for α (which indicates the preserved region $-\alpha\sigma \leq x \leq \alpha\sigma$) and the area under the Gaussian curve (obtained using a Gaussian Q-function table). While the Gaussian is faithfully preserved in all cases, the binomial kernel is wider than it needs to be for increasing values of σ. For example, with $\sigma = 1.41$, a width of $w = 7$ would capture nearly 98.76% of the curve (since $5\sigma \approx 7$), but the binomial triangle uses $w = 9$ to represent this Gaussian. The trinomial triangle results in more compact Gaussians, as can be seen from the bottom portion of the table.

Now let us examine how faithfully Equation (5.39) captures the corresponding Gaussian. The formula in the procedure is $\tilde{w} = \text{ROUND}(2.5\sigma - 0.5)$. Therefore, this procedure will output half-width $\tilde{w}$ if and only if $\tilde{w} - 0.5 \leq 2.5\sigma - 0.5 < \tilde{w} + 0.5$, assuming values halfway between integers round up. Solving for σ yields $\tilde{w}/2.5 \leq \sigma < (\tilde{w} + 1)/2.5$. Since $w = 2\tilde{w} + 1$, we can solve for $\alpha = w/2\sigma$, which corresponds to the preserved region $x \in [-\alpha\sigma, \alpha\sigma]$. Table 5.4 shows the minimum and maximum values of α for each odd width w, along with the minimum and maximum values of the area under the curve. Here we see the Gaussian is faithfully represented and compact. In fact, the width computed is the same as that of the trinomial triangle in all examples shown. For values of σ very near the lower end of each range, the kernels are not as compact as they could be. If this is a concern, the formula could be replaced by reducing the multiplicative factor, for example, $\tilde{w} = \text{ROUND}(2.2\sigma - 0.5)$.

5.2.6 Smoothing with Large Gaussians

As the width of the kernel increases, the amount of computation required to convolve a signal with the kernel increases proportionally. Therefore, as the variance increases, convolving with a Gaussian becomes increasingly expensive. One way to solve this problem could be to replace the single large convolution with multiple smaller convolutions, taking advantage

---

[†] Nevertheless, perhaps it can be argued that 3 elements are not sufficient to accurately capture the derivative of a Gaussian, since the 3×1 first-and second-derivatives do not have a well-defined variance, Section 5.3.1 (p. 234) and Section 5.4 (p. 240).

a	$W = 2a + 1$	binomial kernel	$\sigma^2 = \frac{a}{2}$	σ	$\alpha = \frac{w}{2\sigma}$	area
1	3	$\frac{1}{4}\begin{bmatrix} 1 & 2 & 1 \end{bmatrix}$	0.5	0.71	2.12	96.60%
2	5	$\frac{1}{16}\begin{bmatrix} 1 & 4 & 6 & 4 & 1 \end{bmatrix}$	1.0	1.0	2.50	98.76%
3	7	$\frac{1}{64}\begin{bmatrix} 1 & 6 & 15 & 20 & 15 & 6 & 1 \end{bmatrix}$	1.5	1.22	2.86	99.58%
4	9	$\frac{1}{256}\begin{bmatrix} 1 & 8 & 28 & 56 & 70 & 56 & 28 & 8 & 1 \end{bmatrix}$	2.0	1.41	3.18	99.85%

a	$W = 2a + 1$	trinomial kernel	$\sigma^2 = \frac{2a}{3}$	σ	$\alpha = \frac{w}{2\sigma}$	area
1	3	$\frac{1}{3}\begin{bmatrix} 1 & 1 & 1 \end{bmatrix}$	0.67	0.82	1.84	93.42%
2	5	$\frac{1}{9}\begin{bmatrix} 1 & 2 & 3 & 2 & 1 \end{bmatrix}$	1.33	1.15	2.17	97.00%
3	7	$\frac{1}{27}\begin{bmatrix} 1 & 3 & 6 & 7 & 6 & 3 & 1 \end{bmatrix}$	2.00	1.41	2.47	98.65%
4	9	$\frac{1}{81}\begin{bmatrix} 1 & 4 & 10 & 16 & 19 & 16 & 10 & 4 & 1 \end{bmatrix}$	2.67	1.63	2.76	99.42%

TABLE 5.3 The area under the curve is well captured by Gaussian kernels given by the binomial and trinomial triangles. However, as σ increases, the kernels waste computation because they are much wider than necessary to faithfully capture the Gaussian to a reasonable amount.

of the property that the convolution of two Gaussian kernels with variances σ_1^2 and σ_2^2 is a Gaussian with variance $\sigma_1^2 + \sigma_2^2$;[†] therefore n repeated convolutions with $gauss_{\sigma^2}$ is equivalent to a single convolution with $gauss_{n\sigma^2}$.

Unfortunately this approach generally increases, rather than decreases, the amount of computation needed. For example, suppose we want to convolve with $gauss_{100}$ (that is, a Gaussian with $\sigma^2 = 100$). The straightforward approach would be to convolve with a kernel of halfwidth $\tilde{w} = \text{Round}(2.5(\sigma) - 0.5) = 25$, or width of 51. But according to the preceding analysis, an equivalent approach would be to first convolve with $gauss_{50}$, then convolve with $gauss_{50}$ again. Since each of these kernels has halfwidth $\tilde{w} = \text{Round}(2.5(\sqrt{50}) - 0.5) = 17$, or width of 35, this approach convolves with 71 elements instead of 51, thus actually requiring more computation than the straightforward approach— a conclusion that is consistent with what we said about Pascal's triangle, namely, that it leads to Gaussian kernels that are inefficient because they are too wide.

Nevertheless, a more computationally efficient approach *can* be achieved using the **central limit theorem**, which says that the repeated convolution of any nonnegative kernel with itself always converges (under very mild assumptions) to the shape of a Gaussian. More specifically, if we let b_w be the box filter of width w, then n convolutions of the filter with itself approximates $gauss_{\sigma^2}$, where $\sigma_2 = \frac{n+1}{12}(w^2 - 1)$, since the variance of b_w is $(w^2 - 1)/12$. Thus, the variance of $b_w \circledast b_w \circledast b_w$ is $(w^2 - 1)/4$, while the variance of

$\tilde{w}$	$w = 2\tilde{w} + 1$	σ range	$\alpha = w/2\sigma$ range	area max	area min
1	3	[0.4,0.8)	[3.75,1.88)	–	93.99%
2	5	[0.8,1.2)	[3.13,2.08)	99.83%	96.25%
3	7	[1.2,1.6)	[2.92,2.19)	99.65%	97.15%
4	9	[1.6,2.0)	[2.81,2.25)	99.50%	97.56%

TABLE 5.4 The area under the curve is well captured by Gaussian kernels given by GetKernelHalfWidth. For standard deviations very near the lower end of each range, the kernels could be reduced in size while maintaining acceptable accuracy, but for the most part the representation is compact. The third and fourth columns use interval notation, e.g., $0.4 \leq \sigma < 0.8$.

[†] See Problem 5.32.

$b_w \circledast b_w \circledast b_w \circledast b_w$ is $(w^2 - 1)/3$. As demonstrated in Figure 5.6, the convolution of a signal with a Gaussian can be well approximated as a series of convolutions with the box filter:

$$f \circledast gauss_{\sigma^2} \approx f \circledast (b_w \circledast b_w \circledast b_w \circledast b_w) = (((f \circledast b_w) \circledast b_w) \circledast b_w) \circledast b_w \qquad (5.45)$$

where generally only 2 to 4 convolutions are needed to yield a good approximation.

Convolution with a box filter is extremely fast because, ignoring the normalization factor, the box filter simply sums the elements of the signal overlapping the kernel. For example, if we let $f' = f \circledast b_7$, then at position x the computation is

$$f'(x) = f(x - 3) + f(x - 2) + f(x - 1) + f(x) + f(x + 1) + f(x + 2) + f(x + 3) \quad (5.46)$$

while at the next position $(x + 1)$, the computation is

$$f'(x + 1) = f(x - 2) + f(x - 1) + f(x) + f(x + 1) + f(x + 2) + f(x + 3) + f(x + 4) \quad (5.47)$$

$$= f'(x) - f(x - 3) + f(x + 4) \qquad (5.48)$$

Figure 5.6 Repeated convolutions of a box filter leads to a Gaussian. The vertical bars (in blue) show the samples of discrete kernels obtained by convolving a box filter with itself, for $w = 3$ (top), $w = 5$ (middle), and $w = 7$ (bottom); while the overlaid plots (in red) show the sampled Gaussians with the same variance as the kernels. After just a few convolutions, the approximation is extremely accurate, thus enabling efficient approximation of arbitrarily-sized Gaussians.

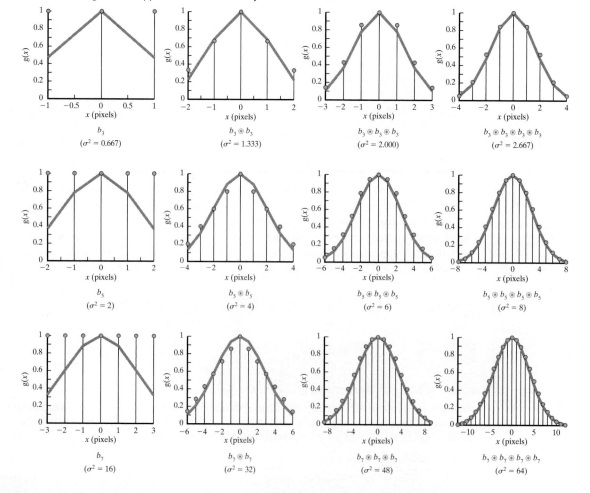

Thus, as the kernel slides across the signal, each output value is computed by simply adding the new value and subtracting the old value—without any multiplication, and independent of the length of the kernel. In this way, the box filter of any length can be implemented efficiently (in constant time) by just one subtraction and one addition per column, as shown in Algorithm 5.7.

An equivalent approach is to first compute a running sum[†] of the signal, then add and subtract values from this running sum. Such an approach reveals that convolving with a box filter is equivalent to performing an integration over a window, followed by a differentiation (subtracting two values from each other). More generally, it can be shown that for any continuous signal f and kernel g, the convolution of the signal and kernel is equivalent to the convolution of the signal integrated n times with the kernel differentiated n times:

$$f \circledast g = \left(\underbrace{\int_{-\infty}^{x} \cdots \int_{-\infty}^{x}}_{n \text{ integrations}} f(x)\,dx \right) \circledast \frac{d^n g(x)}{dx^n} \tag{5.49}$$

where n is an arbitrary nonnegative integer. This fact is easy to prove using the Fourier transform (which we shall consider in more detail in the next chapter). Since integration in the spatial domain is equivalent to division by $j\omega$ in the frequency domain, while differentiation is equivalent to multiplication by $j\omega$, where $j = \sqrt{-1}$ and ω is the angular frequency, the multiplication and division cancel each other, so the convolution of the integral with the derivative is, in the frequency domain, nothing but $\mathcal{F}\{f\} \cdot \mathcal{F}\{g\} \cdot j\omega/j\omega = \mathcal{F}\{f\} \cdot \mathcal{F}\{g\}$. Recalling that multiplication in the frequency domain is equivalent to convolution in the spatial domain, Equation (5.49) is established.

5.2.7 Integral Image

The running sum is easily extended to 2D, where it is known as the **integral image**.[‡] Given a single-channel 2D image I, its integral image S is computed by scanning the image from the top-left to the bottom-right corners, computing the running sum as

$$S(x, y) = I(x, y) - S(x - 1, y - 1) + S(x - 1, y) + S(x, y - 1) \tag{5.50}$$

ALGORITHM 5.7 Convolve a 1D signal/image with a 1D box kernel

CONVOLVEBOX(f, w)

Input: 1D signal f with length n, 1D box kernel with length w
Output: the convolution of f and g

1 $\tilde{w} \leftarrow \lfloor (w - 1)/2 \rfloor$
2 $val \leftarrow 0$
3 **for** $i \leftarrow -\tilde{w}$ **to** $\tilde{w}$ **do**
4      $val \leftarrow val + f[i]$
5 **for** $x \leftarrow 0$ **to** $n - 1$ **do**
6      $h[x] \leftarrow val$
7      $val \leftarrow val + f[x + \tilde{w} + 1] - f[x - \tilde{w}]$
8 **return** h

[†] Section 3.3.2 (p. 87).
[‡] In computer graphics, the integral image is known as a *summed area table (SAT)*.

It is easy to see that $S(x', y') = \sum_{x=0}^{x'} \sum_{y=0}^{y'} I(x, y)$. Once the integral image has been calculated (usually as a preprocessing step), then the sum of values inside any rectangle can be computed with simply one addition and two subtractions:

$$\sum_{x=x_0}^{x_1} \sum_{y=y_0}^{y_1} I(x, y) = S(x_0 - 1, y_0 - 1) + S(x_1, y_1) - S(x_0 - 1, y_1) - S(x_1, y_0 - 1) \quad (5.51)$$

where (x_0, y_0) and (x_1, y_1) are the top-left and bottom-right coordinates of the rectangle, respectively.

5.3 Computing the First Derivative

Now that we have seen how to apply a lowpass filter to an image to perform smoothing, we turn our attention in this section to highpass filters. A highpass filter is one that preserves the local differences in the input signal, which can be detected by computing the derivative of the signal. Large values in the derivative indicate important parts of the signal, which we shall consider in more detail in Chapter 7 when we consider edge detection.

5.3.1 Gaussian Derivative Kernels

The simplest approach to estimating the derivative is to compute **finite differences**, which means to subtract one value in the signal from another. If the values are adjacent, this is equivalent to convolving with the kernel $\begin{bmatrix} 1 & -1 \end{bmatrix}$. Since this kernel has an even number of elements, the center of the kernel can be placed on either element, leading to the so-called **forward difference kernel**, $\begin{bmatrix} 1 & -\underline{1} \end{bmatrix}$, and **backward difference kernel**, $\begin{bmatrix} \underline{1} & -1 \end{bmatrix}$. (If the convention of the origin appears reversed, remember that convolution flips the kernel before performing shift-multiply-add.)

In the real world, the input signal has typically been corrupted by some type of noise. That is, the input signal to which we have access is actually a combination of the underlying noise-free signal in which we are interested and noise that has unfortunately been mixed with the signal in some way. As a result, it is usually wise to perform at least some smoothing to the image before differentiating to help reduce the effects of such noise. This can be achieved by convolving the signal with a smoothing kernel before convolving with a differentiating kernel. The simplest smoothing kernel is $\frac{1}{2} \begin{bmatrix} 1 & 1 \end{bmatrix}$, leading to

$$\left(f(x) \circledast \frac{1}{2} \begin{bmatrix} 1 & 1 \end{bmatrix} \right) \circledast \begin{bmatrix} 1 & -1 \end{bmatrix} = f(x) \circledast \left(\frac{1}{2} \begin{bmatrix} 1 & 1 \end{bmatrix} \circledast \begin{bmatrix} 1 & -1 \end{bmatrix} \right) = f(x) \circledast \frac{1}{2} \begin{bmatrix} 1 & 0 & -1 \end{bmatrix} \quad (5.52)$$

where $\frac{1}{2} \begin{bmatrix} 1 & 0 & -1 \end{bmatrix}$ is the **central difference kernel**. In other words, convolving the image with a smoothing kernel, then convolving the result with a differentiating kernel, is equivalent to convolving with another kernel that is the combination of the two, due to the associativity of convolution. Note that the origins of the two kernels must be different to avoid undesirable shifting of the signal:

$$\frac{1}{2} \begin{bmatrix} 1 & \underline{1} \end{bmatrix} \circledast \begin{bmatrix} \underline{1} & -1 \end{bmatrix} = \frac{1}{2} \begin{bmatrix} \underline{1} & 1 \end{bmatrix} \circledast \begin{bmatrix} 1 & -\underline{1} \end{bmatrix} = \frac{1}{2} \begin{bmatrix} 1 & \underline{0} & -1 \end{bmatrix} \quad (5.53)$$

With larger smoothing kernels, computing finite differences between neighboring smoothed pixels does not yield favorable results. Instead, it is better to convolve the image with a smoothed differentiating kernel, which (since differentiation and convolution are associative) is equivalent to differentiating the smoothed signal:

$$\frac{d}{dx}\left(f(x) \circledast g(x)\right) = f(x) \circledast \left(\frac{d}{dx}g(x)\right)$$ (5.54)

where f is the input and g is the smoothing kernel. Since the Gaussian is the most common smoothing kernel, we focus our attention on the derivative of Gaussian, which is easily shown in the continuous domain to be just a scaled version of the x coordinate times the Gaussian:

$$\frac{d}{dx}gauss_{\sigma^2}(x) = \frac{d}{dx}\left(\frac{1}{\sqrt{2\pi\sigma^2}}\exp\left(\frac{-x^2}{2\sigma^2}\right)\right) = -\frac{x}{\sigma^2\sqrt{2\pi\sigma^2}}\exp\left(\frac{-x^2}{2\sigma^2}\right) = -\frac{x}{\sigma^2}gauss_{\sigma^2}(x)$$ (5.55)

For more compact notation, we denote this Gaussian derivative as $\dot{g}auss_{\sigma^2}(x)$, with a dot over the letter "g".

The overall shape of the Gaussian derivative kernel is evident in Figure 5.7. The function is antisymmetric, meaning $\dot{g}auss(-x) = -\dot{g}auss(x)$ for all x, and therefore it is zero at $x = 0$. Once the Gaussian derivative kernel has been constructed, the derivative of the signal is computed by simply convolving with the kernel. Examining the antisymmetric shape, it is evident that such a convolution is actually computing a weighted average of all the values to the right of the kernel center and subtracting a weighted average of all the values to the left. This equivalent view of the procedure is also illustrated in the figure and is sometimes helpful in understanding the connection between finite differences and Gaussian derivative kernels.

To construct the 1D Gaussian derivative kernel, simply sample the continuous Gaussian derivative, then normalize. Similar to Gaussian normalization, which imposes the constraint that convolution with a constant signal should not change the signal, Gaussian derivative normalization imposes the constraint that convolution with a ramp should yield the slope (i.e., derivative) of the ramp. For example, suppose $\begin{bmatrix} a & b & c \end{bmatrix}$ is an unnormalized differentiating kernel. Convolution of the kernel with $\begin{bmatrix} 0 & 1 & 2 \end{bmatrix}$, which is a ramp signal with a slope of 1, is given by $0{\cdot}c + 1{\cdot}b + 2{\cdot}a$ (since convolution flips the kernel), and this result should equal 1. Since the Gaussian derivative crosses the y-axis at $x = 0$, we know that the central element is zero, $b = 0$, and therefore $2a = 1$, or $a = \frac{1}{2}$. Since the Gaussian derivative is antisymmetric, the normalized 3-element central difference operator is thus given by

$$\dot{g}auss_{0.5} = \frac{1}{2}\begin{bmatrix} 1 & 0 & -1 \end{bmatrix}$$ (5.56)

More generally, it is easy to verify that normalization of a kernel of length w requires dividing each element of the kernel by $\sum_{i=0}^{w-1} i\,\dot{g}auss[w - 1 - i]$, where $\dot{g}auss[i]$ is the ith element of the unnormalized kernel. Because the Gaussian derivative is antisymmetric,

Figure 5.7 The 1D Gaussian derivative (left), along with an equivalent view of the operation (right) in which a weighted sum of the values on one side are subtracted from a weighted sum of the values on the other side.

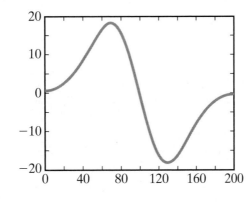

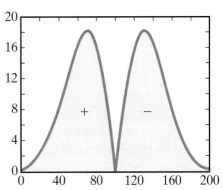

$\dot{gauss}[w - 1 - i] = -\dot{gauss}[i]$ for all i, assuming the kernel is odd width, so it may be easier to remember the normalization divisor as the negative of the first moment of the kernel: $-\sum_{i=0}^{w-1} i\, \dot{gauss}[i]$. Applying this reasoning to the unnormalized central difference operator $\begin{bmatrix} 1 & 0 & -1 \end{bmatrix}$, we find that the normalization divisor is given by $\sum_{i=0}^{w-1} i\, \dot{gauss}[i] = 0 \cdot 1 + 1 \cdot 0 + 2 \cdot (-1) = -2$, leading to Equation (5.56). The procedure for constructing a Gaussian derivative kernel, shown in Algorithm 5.8, is similar to the one used to construct the 1D Gaussian kernel except for the normalization factor and the factor of $i - \tilde{w}$ in Line 5, which is simply a shifted version of x. Note that an important step is missing from this code: after sampling, the value $(1/w)\sum_{i=0}^{w-1} i\, \dot{gauss}\text{-}deriv[i]$ should be subtracted from each element to ensure that the sum of all the elements is zero; this should be performed before computing the normalization factor.

Note that, although we use $\dot{gauss}_{\sigma^2}$ to denote the Gaussian derivative kernel with variance σ^2, there is no easy way to compute the variance of a derivative kernel, and in fact in the case of $w = 3$ all Gaussian derivative kernels are identical no matter the variance. That is, $\frac{1}{2}\begin{bmatrix} 1 & 0 & -1 \end{bmatrix}$ is the *only* 3×1 Gaussian derivative kernel, because no matter how wide the Gaussian, the sample in the middle is zero, the two remaining samples are opposite each other, and the normalization procedure always cancels the values to leave the same $\frac{1}{2}$ factor in front. As a result, the nominal variance of $\sigma^2 = 0.5$ was chosen somewhat arbitrarily. Also, keep in mind that although the normalization factor is needed in order to compute the slope of the function, in practice it can be ignored whenever the relative slope, rather than the actual slope, is all that is needed.

At first glance it may seem odd that the differentiating kernel ignores the central pixel. However, this is a natural consequence of the fact that the derivative of a Gaussian is zero at $x = 0$. Another way to see this is that, because the kernel sums to zero and is antisymmetric, its central pixel has to be zero. If neither of these explanations is entirely satisfactory, perhaps it will be helpful to note (as mentioned earlier) that the centralized difference operator actually averages the two slopes computed by the *forward difference*, $f(x + 1) - f(x)$, and the *backward difference*, $f(x) - f(x - 1)$. That is, if $f'(x) = f(x) \circledast \dot{gauss}_{0.5}$, then

$$f'(x) = \frac{1}{2}\left(\frac{f(x) - f(x - 1)}{1} + \frac{f(x + 1) - f(x)}{1} \right) = \frac{f(x + 1) - f(x - 1)}{2} \quad (5.57)$$

as shown in Figure 5.8.

ALGORITHM 5.8 Create a 1D derivative of a Gaussian kernel

CREATEGAUSSIANDERIVATIVEKERNEL (σ)

Input: floating-point standard deviation σ
Output: 1D Gaussian derivative kernel

1 $\tilde{w} \leftarrow$ GETKERNELHALFWIDTH (σ)
2 $w \leftarrow 2\tilde{w} + 1$
3 $norm \leftarrow 0$
4 **for** $i \leftarrow 0$ **to** $w - 1$ **do**
5      $gauss\text{-}deriv[i] \leftarrow (i - \tilde{w}) * \exp(-(i - \tilde{w}) * (i - \tilde{w})/(2 * \sigma * \sigma))$
6      $norm \leftarrow \_ i * gauss\text{-}deriv[i]$
7 **for** $i \leftarrow 0$ **to** $w - 1$ **do**
8      $gauss\text{-}deriv[i] \leftarrow_{/} norm$
9 **return** $gauss\text{-}deriv$

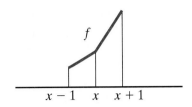

Figure 5.8 The central difference operator is the average of the forward and backward differences, i.e., the average of the two slopes.

5.3.2 Image Gradient

The derivative of a function of one variable is defined as

$$\frac{df}{dx} \equiv \lim_{\Delta x \to 0} \frac{f(x + \Delta x) - f(x)}{\Delta x} \tag{5.58}$$

The generalization of derivative to 2D is the **gradient**, the vector whose elements are the partial derivatives of the function along the two axes:

$$\nabla f(x, y) = \left[\frac{\partial f}{\partial x} \quad \frac{\partial f}{\partial y} \right]^\mathsf{T} \tag{5.59}$$

where the superscript $^\mathsf{T}$ denotes transpose. As shown in Figure 5.9, if the image is viewed as a surface $z = f(x, y)$, where z is the height of the surface at any point, the gradient is a vector pointing uphill. If we let $\mathbf{d} \equiv \nabla f(x_0, y_0)$ be the gradient evaluated at a point (x_0, y_0), and $\mathbf{e} \equiv [x - x_0 \quad y - y_0]^\mathsf{T}$ be the vector from (x_0, y_0) to an arbitrary point (x, y), then the equation of the tangent plane to the surface at the point (x_0, y_0) is given by the inner product of the two vectors plus the value at that point: $\hat{z} = \mathbf{d}^\mathsf{T} \mathbf{e} + f(x_0, y_0)$.

Computing the image gradient requires convolving the image with a Gaussian kernel to reduce the effects of noise, then computing the partial derivatives in the orthogonal directions. Due to the associative property of convolution, this is equivalent to convolving the image with the partial derivatives of a 2D Gaussian:

$$\frac{\partial I(x, y)}{\partial x} = \frac{\partial}{\partial x}(I(x, y) \circledast Gauss(x, y)) = I(x, y) \circledast \frac{\partial \, Gauss(x, y)}{\partial x} \tag{5.60}$$

$$\frac{\partial I(x, y)}{\partial y} = \frac{\partial}{\partial y}(I(x, y) \circledast Gauss(x, y)) = I(x, y) \circledast \frac{\partial \, Gauss(x, y)}{\partial y} \tag{5.61}$$

Figure 5.9 The derivative of a 1D function (left), and the gradient of a 2D function (right).

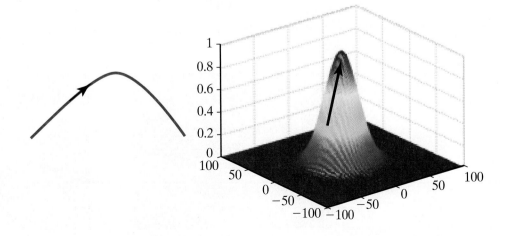

Continuous versions of these 2D convolution kernels are shown in Figure 5.10.

Thankfully, the partial derivative of a 2D Gaussian is separable, so that the convolution of an image I with the 2D kernel for the partial derivative is equivalent to two successive convolutions with 1D kernels:

$$I(x, y) \circledast \frac{\partial \, Gauss_{\sigma^2}(x, y)}{\partial x} = \iint - \frac{\xi}{\sigma^2} I(x - \xi, y - \eta) \exp\left\{ \frac{-(\xi^2 + \eta^2)}{2\sigma^2} \right\} d\xi d\eta \tag{5.62}$$

$$= \iint - \frac{\xi}{\sigma^2} I(x - \xi, y - \eta) \exp\left\{ \frac{-\eta^2}{2\sigma^2} \right\} \exp\left\{ \frac{-\xi^2}{2\sigma^2} \right\} d\xi d\eta \tag{5.63}$$

$$= \int - \frac{\xi}{\sigma^2} \left[\int I(x - \xi, y - \eta) \exp\left\{ \frac{-\eta^2}{2\sigma^2} \right\} d\eta \right] \exp\left\{ \frac{-\xi^2}{2\sigma^2} \right\} d\xi \tag{5.64}$$

$$= (I(x, y) \circledast gauss_{\sigma^2}(y)) \circledast \dot{g}auss_{\sigma^2}(x) \tag{5.65}$$

$$= (I(x, y) \circledast \dot{g}auss_{\sigma^2}(x)) \circledast gauss_{\sigma^2}(y) \tag{5.66}$$

where $\dot{g}auss_{\sigma^2}(x) = \frac{d}{dx} gauss_{\sigma^2}$ and where the final equation is due to the fact that the order of convolution does not matter. Similar analysis shows that the partial derivative in y is also separable:

$$I(x, y) \circledast \frac{\partial \, Gauss_{\sigma^2}(x, y)}{\partial y} = (I(x, y) \circledast gauss_{\sigma^2}(x)) \circledast \dot{g}auss_{\sigma^2}(y) \tag{5.67}$$

$$= (I(x, y) \circledast \dot{g}auss_{\sigma^2}(y)) \circledast gauss_{\sigma^2}(x) \tag{5.68}$$

Therefore, to compute the gradient of an image, we must differentiate along the x and y axes by convolving with a smoothing 1D kernel and a differentiating 1D kernel in the orthogonal direction. That is, to compute the partial derivative with respect to x, convolve with a horizontal derivative of a Gaussian, followed by a vertical Gaussian (or switch the order of

Figure 5.10 The 2D Gaussian partial derivatives in the x and y directions, shown as 3D plots (top) and images (bottom).

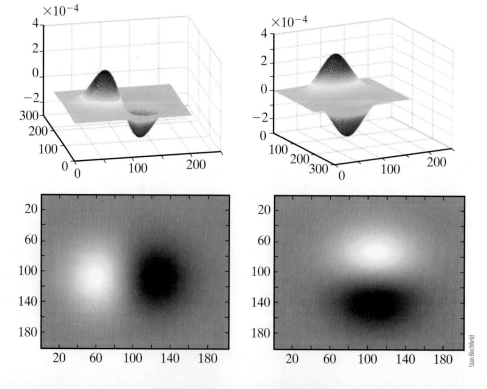

these two, since convolution is commutative); to compute the partial derivative with respect to y, convolve with a horizontal Gaussian, followed by a vertical derivative of a Gaussian.

The simplest 2D differentiating kernel is the **Prewitt operator**, which is obtained by convolving a 1D Gaussian derivative kernel with a 1D box filter in the orthogonal direction:

$$Prewitt_x = \frac{1}{3}\begin{bmatrix} 1 \\ 1 \\ 1 \end{bmatrix} \circledast \frac{1}{2}\begin{bmatrix} 1 & 0 & -1 \end{bmatrix} = \frac{1}{6}\begin{bmatrix} 1 & 0 & -1 \\ 1 & 0 & -1 \\ 1 & 0 & -1 \end{bmatrix}$$

$$Prewitt_y = \frac{1}{3}\begin{bmatrix} 1 & 1 & 1 \end{bmatrix} \circledast \frac{1}{2}\begin{bmatrix} 1 \\ 0 \\ -1 \end{bmatrix} = \frac{1}{6}\begin{bmatrix} 1 & 1 & 1 \\ 0 & 0 & 0 \\ -1 & -1 & -1 \end{bmatrix}$$

It can be shown that applying the Prewitt operator is equivalent to solving for the plane that minimizes the least-squares-error over the 3×3 window, where all the pixels are treated equally. Again, keep in mind that the normalization factor of $\frac{1}{6}$ can be ignored in most applications.

The **Sobel operator** is more robust, as it uses the Gaussian $(\sigma^2 = 0.5)$ for the smoothing kernel:

$$Sobel_x = gauss_{0.5}(y) \circledast \dot{g}auss_{0.5}(x) = \frac{1}{4}\begin{bmatrix} 1 \\ 2 \\ 1 \end{bmatrix} \circledast \frac{1}{2}\begin{bmatrix} 1 & 0 & -1 \end{bmatrix} = \frac{1}{8}\begin{bmatrix} 1 & 0 & -1 \\ 2 & 0 & -2 \\ 1 & 0 & 1 \end{bmatrix}$$

$$Sobel_y = gauss_{0.5}(x) \circledast \dot{g}auss_{0.5}(y) = \frac{1}{4}\begin{bmatrix} 1 & 2 & 1 \end{bmatrix} \circledast \frac{1}{2}\begin{bmatrix} 1 \\ 0 \\ -1 \end{bmatrix} = \frac{1}{8}\begin{bmatrix} 1 & 2 & 1 \\ 0 & 0 & 0 \\ -1 & -2 & -1 \end{bmatrix}$$

The **Scharr operator** is similar to Sobel, but with a smaller variance $(\sigma^2 = 0.375)$ in the smoothing kernel:

$$Scharr_x = gauss_{0.375}(y) \circledast \dot{g}auss_{0.5}(x) = \frac{1}{16}\begin{bmatrix} 3 \\ 10 \\ 3 \end{bmatrix} \circledast \frac{1}{2}\begin{bmatrix} 1 & 0 & -1 \end{bmatrix} = \frac{1}{32}\begin{bmatrix} 3 & 0 & -3 \\ 10 & 0 & -10 \\ 3 & 0 & -3 \end{bmatrix}$$

$$Scharr_y = gauss_{0.375}(x) \circledast \dot{g}auss_{0.5}(y) = \frac{1}{16}\begin{bmatrix} 3 & 10 & 3 \end{bmatrix} \circledast \frac{1}{2}\begin{bmatrix} 1 \\ 0 \\ -1 \end{bmatrix} = \frac{1}{32}\begin{bmatrix} 3 & 10 & 3 \\ 0 & 0 & 0 \\ -3 & -10 & -3 \end{bmatrix}$$

The advantage of the Scharr operator is that it is more rotationally invariant than other 3×3 Gaussian kernels.

For completeness, we mention the classic **Roberts cross operator**,[†] which is the oldest pair of gradient kernels:

$$Roberts_1 = \frac{1}{\sqrt{2}}\begin{bmatrix} 1 & 0 \\ 0 & -1 \end{bmatrix}$$

$$Roberts_2 = \frac{1}{\sqrt{2}}\begin{bmatrix} 0 & -1 \\ 1 & 0 \end{bmatrix}$$

The Roberts kernels suffer from two drawbacks. First, they compute derivatives along the diagonal directions rather than along the x and y axes, which is more of an inconvenience

---

[†] Lawrence G. Roberts (1937–), after writing the first Ph.D. dissertation in computer vision (in 1963), went on to become the architect who designed the original ARPANET, which became the Internet.

than any fundamental limitation. Secondly, because they have even dimensions, the kernels are not centered. That is, they do not compute the derivative at a pixel, but rather between pixels. Nevertheless, for some applications two-point differences are better than three-point central differences, because they use all pixels in the computation (as opposed to ignoring the central pixel) and because they are more compact.

Once we have computed the gradient of the image, it is often desirable to compute the magnitude of the gradient. A natural way to do this would be to compute the Euclidean norm of the gradient vector: $|\nabla f| = \sqrt{f_x^2 + f_y^2}$, where f_x and f_y are the two components of the gradient vector, i.e., $\nabla f = [f_x \quad f_y]^T$. A computationally efficient approximation can be obtained by computing the sum of the absolute values: $|\nabla f| \approx |f_x| + |f_y|$. A third option is to select the maximum of the two absolute values: $|\nabla f| \approx \max(|f_x|, |f_y|)$. Note that these three choices are, respectively, the Euclidean, Manhattan, and chessboard distances[†] between the origin and the point (f_x, f_y). The pseudocode to compute the image gradient is provided in Algorithm 5.9, using the Manhattan distance. For each pixel in the image, the partial derivatives in x and y are computed, then converted into a magnitude and phase representation. For other distance metrics, only Line 6 changes. Figure 5.11 shows the result of applying this computation to an image. Results of computing the gradient magnitude of an image, using different variances for the Gaussian, are shown in Figure 5.12.

5.4 Computing the Second Derivative

Just as the finite difference operator approximates the first derivative, the difference between differences approximates the second derivative. This can be seen in 1D by convolving the function with the noncentralized difference operator, then convolving the result again with the same operator:

$$(f(x) \circledast [1 \quad -1]) \circledast [1 \quad -1] = f(x) \circledast ([1 \quad -1] \circledast [1 \quad -1]) = f(x) \circledast [1 \quad -2 \quad 1] \quad (5.69)$$

which yields the second-derivative convolution kernel $[1 \quad -2 \quad 1]$. It turns out that, just as there is only one 3×1 kernel for computing the derivative of a Gaussian, this is the only 3×1 kernel for computing the second derivative of a Gaussian. Note that the normalization factor is 1, already included in the kernel. Another way to look at this is to express the first and second derivatives as $df/dx \approx f(x) - f(x - 1)$ and $d^2f/dx^2 \approx (f(x + 1) - f(x)) - (f(x) - f(x - 1)) = f(x + 1) - 2f(x) + f(x - 1)$, respectively.

ALGORITHM 5.9 Compute the gradient of an image

COMPUTEIMAGEGRADIENT(I, σ)

1 $gauss$ = CREATEGAUSSIANKERNEL(σ)
2 $gauss\text{-}deriv$ = CREATEGAUSSIANDERIVATIVEKERNEL(σ)
3 G_x = CONVOLVESEPARABLE$(I, gauss\text{-}deriv, gauss)$
4 G_y = CONVOLVESEPARABLE$(I, gauss, gauss\text{-}deriv)$
5 **for** $(x, y) \in I$ **do**
6      $G_{mag} = |G_x(x, y)| + |G_y(x, y)|$
7      $G_{phase} = $ATAN2$(G_y(x, y), G_x(x, y))$
8 **return** G_{mag}, G_{phase}

[†] Section 4.3.1 (p. 164).

Figure 5.11 TOP: An image. LEFT: The partial derivatives of the image in the x and y directions, which together form the two components of the gradient of the image. RIGHT: The magnitude and phase of the gradient.

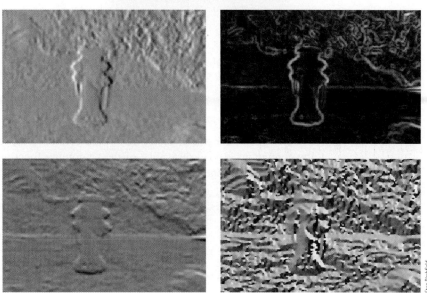

For a function (or image) of two variables, the second-derivative in the x and y directions can be obtained by convolving with the appropriately oriented second-derivative kernel:

$$\frac{\partial^2 I(x, y)}{\partial x^2} = I(x, y) \circledast \begin{bmatrix} 1 & -2 & 1 \end{bmatrix} \tag{5.70}$$

$$\frac{\partial^2 I(x, y)}{\partial y^2} = I(x, y) \circledast \begin{bmatrix} 1 \\ -2 \\ 1 \end{bmatrix} \tag{5.71}$$

while the cross-derivative is obtained as

$$\frac{\partial^2 I(x, y)}{\partial x\, \partial y} = \frac{\partial^2 I(x, y)}{\partial y\, \partial x} = I(x, y) \circledast \frac{1}{4} \begin{bmatrix} -1 & 0 & 1 \\ 0 & 0 & 0 \\ 1 & 0 & -1 \end{bmatrix} \tag{5.72}$$

| Original | $\sigma = 1$ pixel | $\sigma = 5$ pixels | $\sigma = 10$ pixels | $\sigma = 20$ pixels |

Figure 5.12 An 816 × 612 image, and the gradient magnitude computed using an isotropic Gaussian with different standard deviations.

where the latter kernel is obtained by convolving the centered first-derivative $\frac{1}{2}\begin{bmatrix} 1 & 0 & -1 \end{bmatrix}$ with an oriented version of itself:

$$\frac{1}{2}\begin{bmatrix} 1 & 0 & -1 \end{bmatrix} \circledast \frac{1}{2}\begin{bmatrix} 1 \\ 0 \\ -1 \end{bmatrix} = \frac{1}{2}\begin{bmatrix} 1 \\ 0 \\ -1 \end{bmatrix} \circledast \frac{1}{2}\begin{bmatrix} 1 & 0 & -1 \end{bmatrix} = \frac{1}{4}\begin{bmatrix} -1 & 0 & 1 \\ 0 & 0 & 0 \\ 1 & 0 & -1 \end{bmatrix} \quad (5.73)$$

Alternatively, using the non-centered first-derivative leads to a more compact but non-centered cross-derivative kernel:

$$\begin{bmatrix} 1 & -1 \end{bmatrix} \circledast \begin{bmatrix} 1 \\ -1 \end{bmatrix} = \begin{bmatrix} 1 \\ -1 \end{bmatrix} \circledast \begin{bmatrix} 1 & -1 \end{bmatrix} = \begin{bmatrix} -1 & 1 \\ 1 & -1 \end{bmatrix} \quad (5.74)$$

If larger kernels are desired, then the property that we saw with the first derivative, namely that differentiation of a smoothed signal is equivalent (in the continuous domain) to convolution with a differentiated Gaussian, can be used:

$$\frac{d^2}{dx^2}(f(x) \circledast g(x)) = f(x) \circledast \left(\frac{d^2 g(x)}{dx^2} \right) \quad (5.75)$$

In other words, a discrete, smoothed second-derivative kernel can be created by sampling the continuous second-derivative of the smoothing function. The second derivative of a Gaussian, in the continuous domain, is calculated to be

$$\ddot{g}auss_{\sigma^2}(x) \equiv \frac{d^2}{dx^2} gauss_{\sigma^2}(x) \quad (5.76)$$

$$= \frac{d}{dx}\left(-\frac{x}{\sigma^2} gauss_{\sigma^2}(x) \right) \quad (5.77)$$

$$= -\frac{1}{\sigma^2}\left(gauss_{\sigma^2}(x) + x\frac{d\, gauss_{\sigma^2}(x)}{dx} \right) \quad (5.78)$$

$$= -\left(\frac{1}{\sigma^2} - \frac{x^2}{\sigma^4} \right)gauss_{\sigma^2}(x) \quad (5.79)$$

where we use double dots to denote the second derivative. After sampling this continuous function, and shifting so that the sum of elements is zero normalization requires dividing each element by one-half the second centralized moment of the elements: $\frac{1}{2}\sum_{i=-\tilde{w}}^{\tilde{w}} i^2 \ddot{g}auss[i]$, where $\tilde{w}$ is the halfwidth of the kernel, $\ddot{g}auss$ is the unnormalized kernel, and $i = 0$ refers to the central element of the kernel. The justification for this formula is that convolution with a parabola, $y = x^2$, should yield the second derivative, which is 2. Note that with the second derivative it is important to use the centralized moment, whereas with the zeroth and first derivatives, either centralized or non-centralized moments produce the same result.

5.4.1 Laplacian of Gaussian (LoG)

Extending the previous discussion to 2D leads naturally to the **Laplacian operator** ∇^2, which is defined as the divergence of the gradient of a function. While the concept of divergence[†] is beyond our scope, in 2D Cartesian coordinates the Laplacian of an image is just the sum of the second derivatives along the two orthogonal axes:

$$\nabla^2 I = \nabla \cdot \nabla I = \begin{bmatrix} \frac{\partial}{\partial x} & \frac{\partial}{\partial y} \end{bmatrix} \begin{bmatrix} \frac{\partial I}{\partial x} & \frac{\partial I}{\partial y} \end{bmatrix}^{\mathsf{T}} = \frac{\partial^2 I}{\partial x^2} + \frac{\partial^2 I}{\partial y^2} \quad (5.80)$$

[†] Section 2.5.1 (p. 57).

Because of associativity, computing the Laplacian of a smoothed image is the same as convolving the image with the **Laplacian of Gaussian (LoG)**:

$$\frac{\partial^2(I \circledast Gauss(x,y))}{\partial x^2} + \frac{\partial^2(I \circledast Gauss(x,y))}{\partial y^2} = I \circledast \left(\frac{\partial^2 Gauss(x,y)}{\partial x^2} + \frac{\partial^2 Gauss(x,y)}{\partial y^2} \right) \quad (5.81)$$

which is also known as the inverted **"Mexican hat"** operator because of its shape, shown in Figure 5.13. The LoG is rotationally symmetric and is a **center-surround filter** because it consists of a central core of negative values surrounded by an annular ring of positive values. As a result, it maintains a connection with biology, because certain cells in the retina perform center-surround operations and therefore function like LoG filters.[†] The LoG is actually not a lowpass or highpass filter but rather a bandpass filter, though small 3×3 discrete LoG kernels operate essentially like highpass filters.

Not surprisingly, the 2D Gaussian second-derivative filter is separable:

$$I(x,y) \circledast \frac{\partial^2 Gauss(x,y)}{\partial x^2} = \iint \frac{1}{\sigma^2}\left(1 - \frac{\xi^2}{\sigma^2}\right) I(x - \xi, y - \eta) \exp\left\{\frac{-(\xi^2 + \eta^2)}{2\sigma^2}\right\} d\xi d\eta$$

$$= \iint \frac{1}{\sigma^2}\left(1 - \frac{\xi^2}{\sigma^2}\right) I(x - \xi, y - \eta) \exp\left\{\frac{-\xi^2}{2\sigma^2}\right\} \exp\left\{\frac{-\eta^2}{2\sigma^2}\right\} d\xi d\eta$$

$$= \int \left[\int \frac{1}{\sigma^2}\left(1 - \frac{\xi^2}{\sigma^2}\right) I(x - \xi, y - \eta) \exp\left\{\frac{-\xi^2}{2\sigma^2}\right\} d\xi \right] \exp\left\{\frac{-\eta^2}{2\sigma^2}\right\} d\eta$$

$$= (I(x) \circledast \ddot{g}auss(x)) \circledast gauss(y) \quad (5.82)$$

$$= (I(x) \circledast gauss(y)) \circledast \ddot{g}auss(x) \quad (5.83)$$

Therefore, since the LoG is the sum of two separable kernels, it is no more expensive to compute than the gradient. But the LoG itself is not separable but rather the sum of two separable computations:

$$\frac{\partial^2(I \circledast Gauss)}{\partial x^2} + \frac{\partial^2(I \circledast Gauss)}{\partial y^2} = I(x,y) \circledast \left(\frac{\partial^2 Gauss(x,y)}{\partial x^2} + \frac{\partial^2 Gauss(x,y)}{\partial y^2} \right) \quad (5.84)$$

$$\begin{aligned} = (I(x,y) \circledast \ddot{g}auss(x)) \circledast gauss(y) \\ + (I(x,y) \circledast \ddot{g}auss(y)) \circledast gauss(x) \end{aligned} \quad (5.85)$$

Figure 5.13 The Laplacian of Gaussian, presented as an image (left), a 2D plot (middle), and a 1D slice through the 2D plot (right). The center-surround nature of the operator is evident in the image, while the inverted Mexican hat shape is evident in the plots.

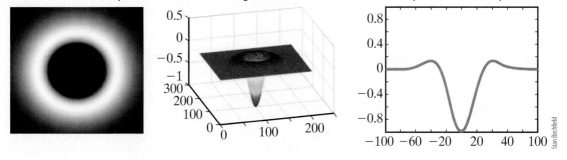

[†] Section 2.1.2 (p. 21).

A discrete LoG kernel is obtained by computing $\frac{\partial^2 \, Gauss}{\partial x^2} + \frac{\partial^2 \, Gauss}{\partial y^2}$ with the appropriate value for σ, then sampling. For 3×3 kernels, the differentiating kernel is fixed as $\begin{bmatrix} 1 & -2 & 1 \end{bmatrix}$ (because it is the only 3×1 second-derivative Gaussian kernel), and the smoothing kernel is determined by σ. The simplest Laplacian kernel involves no smoothing and is therefore simply the sum of the horizontal and vertical second-derivative kernels (considered as the middle row and column, respectively, of a square matrix whose remaining elements are zero):

$$LoG_{0.0} = \begin{bmatrix} 1 & -2 & 1 \end{bmatrix} + \begin{bmatrix} 1 \\ -2 \\ 1 \end{bmatrix} = \begin{bmatrix} 0 & 0 & 0 \\ 1 & -2 & 1 \\ 0 & 0 & 0 \end{bmatrix} + \begin{bmatrix} 0 & 1 & 0 \\ 0 & -2 & 0 \\ 0 & 1 & 0 \end{bmatrix} = \begin{bmatrix} 0 & 1 & 0 \\ 1 & -4 & 1 \\ 0 & 1 & 0 \end{bmatrix} \tag{5.86}$$

where we let LoG_{σ^2} signify the isotropic LoG kernel with variance σ^2. As another example, for $\sigma^2 = 0.25$, we have $\frac{1}{8}\begin{bmatrix} 1 & 6 & 1 \end{bmatrix}$ as the smoothing kernel, leading to

$$\frac{\partial^2 \, Gauss_{0.25}}{\partial x^2} = \begin{bmatrix} 1 & -2 & 1 \end{bmatrix} \circledast \frac{1}{8}\begin{bmatrix} 1 \\ 6 \\ 1 \end{bmatrix} = \frac{1}{8}\begin{bmatrix} 1 & -2 & 1 \\ 6 & -12 & 6 \\ 1 & -2 & 1 \end{bmatrix} \tag{5.87}$$

$$\frac{\partial^2 \, Gauss_{0.25}}{\partial y^2} = \frac{1}{8}\begin{bmatrix} 1 & 6 & 1 \end{bmatrix} \circledast \begin{bmatrix} 1 \\ -2 \\ 1 \end{bmatrix} = \frac{1}{8}\begin{bmatrix} 1 & 6 & 1 \\ -2 & -12 & -2 \\ 1 & 6 & 1 \end{bmatrix} \tag{5.88}$$

$$LoG_{0.25} = \frac{\partial^2 \, Gauss_{0.25}}{\partial x^2} + \frac{\partial^2 \, Gauss_{0.25}}{\partial y^2} = \frac{1}{4}\begin{bmatrix} 1 & 2 & 1 \\ 2 & -12 & 2 \\ 1 & 2 & 1 \end{bmatrix} \tag{5.89}$$

Repeating this procedure for other values of σ^2 yields alternative 3×3 LoG kernels, shown in Table 5.5.

The center-surround property is evident by noticing that in all of these 3×3 cases the kernel is equivalent to a scalar times the difference between the average value of the neighbors and the value of the central pixel: $\nabla^2 I = h(\bar{I} - I)$, where $\bar{I}$ is the average gray level of the 8-neighbors, and the scalar h is the negative of the central kernel weight. For example, convolving with $LoG_{0.0}$, computes

$$\nabla^2 I(x, y) = I(x + 1, y) + I(x - 1, y) + I(x, y - 1) + I(x, y + 1) - 4(I(x, y) \tag{5.90}$$

$$= 4(\bar{I}(x, y) - I(x, y)) \tag{5.91}$$

where

$$\bar{I}(x, y) = \frac{1}{4}\left(I(x + 1, y) + I(x - 1, y) + I(x, y - 1) + I(x, y + 1)\right) \tag{5.92}$$

so in this case $h = 4$. Results of computing the LoG of an image, using different variances for the Gaussian, are shown in Figure 5.14.

$\sigma^2 = 0.0$	$\sigma^2 = 0.167$	$\sigma^2 = 0.20$	$\sigma^2 = 0.25$	$\sigma^2 = 0.33$	$\sigma^2 = 0.5$
$\begin{bmatrix} 0 & 1 & 0 \end{bmatrix}$	$\frac{1}{12}\begin{bmatrix} 1 & 10 & 1 \end{bmatrix}$	$\frac{1}{10}\begin{bmatrix} 1 & 8 & 1 \end{bmatrix}$	$\frac{1}{8}\begin{bmatrix} 1 & 6 & 1 \end{bmatrix}$	$\frac{1}{6}\begin{bmatrix} 1 & 4 & 1 \end{bmatrix}$	$\frac{1}{4}\begin{bmatrix} 1 & 2 & 1 \end{bmatrix}$
$\begin{bmatrix} 0 & 1 & 0 \\ 1 & -4 & 1 \\ 0 & 1 & 0 \end{bmatrix}$	$\frac{1}{6}\begin{bmatrix} 1 & 4 & 1 \\ 4 & -20 & 4 \\ 1 & 4 & 1 \end{bmatrix}$	$\frac{1}{5}\begin{bmatrix} 1 & 3 & 1 \\ 3 & -16 & 3 \\ 1 & 3 & 1 \end{bmatrix}$	$\frac{1}{4}\begin{bmatrix} 1 & 2 & 1 \\ 2 & -12 & 2 \\ 1 & 2 & 1 \end{bmatrix}$	$\frac{1}{3}\begin{bmatrix} 1 & 1 & 1 \\ 1 & -8 & 1 \\ 1 & 1 & 1 \end{bmatrix}$	$\frac{1}{2}\begin{bmatrix} 1 & 0 & 1 \\ 0 & -4 & 0 \\ 1 & 0 & 1 \end{bmatrix}$

TABLE 5.5 Various discrete 3×3 LoG kernels. For each choice of variance, the middle row shows the 1D smoothing kernel, while the last row shows the resulting LoG kernel.

Original $\sigma = 1$ pixel $\sigma = 5$ pixels $\sigma = 10$ pixels $\sigma = 20$ pixels

Figure 5.14 A 2304 $\times$ 1728 image, and the result of convolving with an isotropic LoG with different standard deviations.

5.4.2 Difference of Gaussians (DoG)

The Laplacian of Gaussian (LoG) is closely connected to the **difference of Gaussians (DoG)**. More specifically, a close approximation to the LoG is obtained by subtracting one Gaussian from another, where the variances of the two Gaussians have a particular relationship to one another. To see this connection, simply differentiate the continuous *gauss(x)* with respect to the scale parameter σ:

$$\frac{d\, gauss_{\sigma^2}(x)}{d\sigma} = \frac{d}{d\sigma}\left(\frac{1}{\sqrt{2\pi}}\sigma^{-1}e^{-\frac{1}{2}x^2\sigma^{-2}}\right) \tag{5.93}$$

$$= \frac{1}{\sqrt{2\pi}}\left(-\sigma^{-2} + \sigma^{-1}(x^2\sigma^{-3})\right)e^{-\frac{x^2}{2\sigma^2}} \tag{5.94}$$

$$= \frac{1}{\sqrt{2\pi}}\left(-\frac{1}{\sigma^2} + \frac{x^2}{\sigma^4}\right)e^{-\frac{1}{2}x^2\sigma^{-2}} \tag{5.95}$$

$$= -\sigma\left(\frac{1}{\sigma^2} - \frac{x^2}{\sigma^4}\right)gauss_{\sigma^2}(x) \tag{5.96}$$

Comparing this result with the second derivative of the Gaussian in Equation (5.79) yields a surprising relationship:

$$\frac{d\, gauss_{\sigma^2}(x)}{d\sigma} = \sigma\frac{d^2\, gauss_{\sigma^2}(x)}{dx^2} \tag{5.97}$$

Since we cannot actually compute the derivative of the Gaussian with respect to σ, it is approximated using finite differences. From Equation (5.58) this yields

$$\frac{d\, gauss(x;\sigma)}{d\sigma} \approx \frac{gauss(x;\sigma+\delta) - gauss(x;\sigma)}{\delta} \tag{5.98}$$

where we introduce the notation $gauss(x;\sigma)$ to refer to $gauss_{\sigma^2}(x)$, and δ is a small change in standard deviation. Combining this expression with Equation (5.97) yields

$$gauss(x;\sigma+\delta) - gauss(x;\sigma) \approx \delta\sigma\frac{d^2\, gauss(x;\sigma)}{dx^2} \tag{5.99}$$

$$= (\rho - 1)\sigma^2\frac{d^2\, gauss(x;\sigma)}{dx^2} \tag{5.100}$$

where $\rho \equiv \frac{\sigma + \delta}{\sigma}$ is the ratio of the two standard deviations. As $\delta \to 0$ ($\rho \to 1$), the DoG approximates the scaled LoG. However, small values of δ lead to low sensitivity of the filter, because as the two Gaussians approach each other in width, their responses become identical. It has been found empirically that $\rho = 1.6$ yields a reasonable tradeoff, leading to

$$DoG(x; \sigma) \approx 0.6\sigma^2 \, LoG(x; \sigma) \tag{5.101}$$

That the scaled LoG is well approximated by the difference of two Gaussians is shown in Figure 5.15.

The relationship in 2D is the same as that in 1D:

$$\frac{d \, Gauss(x, y)}{d\sigma} = \sigma \left(\frac{\partial^2 \, Gauss(x, y)}{\partial x^2} + \frac{\partial^2 \, Gauss(x, y)}{\partial y^2} \right) \tag{5.102}$$

To see this, note that a 2D, zero-mean isotropic Gaussian is given by

$$Gauss(x, y) = \frac{1}{2\pi\sigma^2} \exp\left(-\frac{x^2 + y^2}{2\sigma^2} \right) \tag{5.103}$$

Differentiating twice with respect to x yields

$$\frac{\partial^2 \, Gauss(x, y)}{\partial x^2} = \frac{\partial^2}{\partial x^2} \left(\frac{1}{2\pi\sigma^2} \exp\left(-\frac{x^2 + y^2}{2\sigma^2} \right) \right) \tag{5.104}$$

$$= \frac{1}{\sigma^2} \left(1 - \frac{x^2}{\sigma^2} \right) Gauss(x, y) \tag{5.105}$$

and by symmetry,

$$\frac{\partial^2 \, Gauss(x, y)}{\partial y^2} = \frac{1}{\sigma^2} \left(1 - \frac{y^2}{\sigma^2} \right) Gauss(x, y) \tag{5.106}$$

Putting these together yields

$$LoG(x, y; \sigma) = \frac{\partial^2 \, Gauss(x, y)}{\partial x^2} + \frac{\partial^2 \, Gauss(x, y)}{\partial y^2} = \left(\frac{x^2 + y^2 - 2\sigma^2}{2\pi\sigma^6} \right) \exp\left(-\frac{x^2 + y^2}{2\sigma^2} \right) \tag{5.107}$$

$$= \left(\frac{x^2 + y^2}{\sigma^4} - \frac{2}{\sigma^2} \right) Gauss(x, y) \tag{5.108}$$

Differentiating with respect to σ yields

$$\frac{d \, Gauss(x, y)}{d\sigma} = \frac{d}{d\sigma} \left(\frac{1}{2\pi\sigma^2} \exp\left(-\frac{x^2 + y^2}{2\sigma^2} \right) \right) \tag{5.109}$$

$$= -\frac{2}{\sigma} \left(1 - \frac{x^2 + y^2}{2\sigma^2} \right) Gauss(x, y) \tag{5.110}$$

Comparing Equation (5.108) with Equation (5.110) yields Equation (5.102).

Figure 5.15 LEFT: Two Gaussians whose ratio of standard deviations is 1.6. RIGHT: The difference of Gaussians (solid blue) and 1D Laplacian of Gaussian (solid red). The scaled DoG (dashed blue) approximates the LoG.

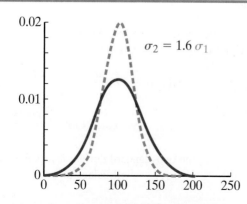

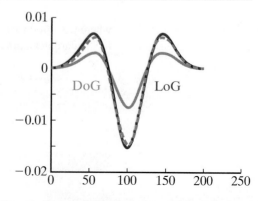

5.5 Nonlinear Filters

All the filters we have considered so far in this chapter rely upon convolution and are therefore *linear shift-invariant* filters, as we have already explained. Although there exist filters that are *linear* but not *shift-invariant*, we will not spend time discussing these, since they are not commonly used. Instead, we turn our attention in this section to *nonlinear* filters, which do not obey the superposition principle. In this section we consider several important, widely-used nonlinear filters.

5.5.1 Median Filter

In choosing what type of filter to use, it is important to identify the type of **noise** that one expects to be present in the image. For example, the lowpass filter obtained by convolving with a Gaussian is the optimal filter for minimizing the effects of **additive white Gaussian noise (AWGN)**. To understand what this means, imagine adding to each pixel of a noise-free image a random value drawn from a Gaussian distribution:

$$I'(x, y) = I(x, y) + \xi, \qquad \xi \sim \mathcal{N}(0, \sigma^2) \tag{5.111}$$

In this equation, I is the noise-free image, I' is the image corrupted by additive white Gaussian noise, and ξ is the random value drawn independently for each pixel from a Gaussian distribution (also known as the *normal distribution*, hence the $\mathcal{N}$) with a mean of zero and a variance of σ^2.[†] Pseudocode for adding white Gaussian noise to an image is given in Algorithm 5.10, where we assume that a procedure exists called RANDGAUSSIAN that generates a random value according to a Gaussian distribution, and we have made sure to clamp the value so that it stays within the image range (assuming a bit-depth of 8). Such a procedure is sometimes used to create synthetically corrupted images for measuring the robustness of algorithms.

Another type of noise is **salt-and-pepper noise**, in which each pixel is set to either the minimum ("pepper") or maximum ("salt") possible gray level, or it remains unchanged:

$$I'(x, y) = \begin{cases} 0 & \text{if } 0 \le \xi < p \\ 255 & \text{if } p \le \xi < p + q \qquad \xi \sim U(0,1) \\ I(x, y) & \text{otherwise} \end{cases} \tag{5.112}$$

where the random variable ξ is drawn from a uniform distribution between 0 and 1, inclusive. The probability of a pixel becoming salt is p; and the probability of a pixel becoming pepper is q. It is easy to see that $0 \le p \le 1$, $0 \le q \le 1$, and $0 \le p + q \le 1$.

ALGORITHM 5.10 Add independent Gaussian noise to an image

ADDGAUSSIANNOISE(I, σ)

Input: grayscale image I, standard deviation σ of the Gaussian distribution
Output: image corrupted by additive Gaussian noise

1   **for** $(x, y) \in I$ **do**
2        $\xi \leftarrow$ RANDGAUSSIAN(σ)
3        $I'(x, y) \leftarrow \max(0, \min(255, I(x, y) + \xi))$
4   **return** I'

[†] Considered as a random variable, we say that ξ is independent and identically distributed (i.i.d.).

For an image corrupted by salt-and-pepper noise, convolving with a Gaussian is a bad idea, because such an operation will change the values of *all* the pixels, even of those that are not corrupted. Moreover, such an operation will result in a blurry appearance due to the weighted averaging operation. A much better solution in such a case is the **median filter**, which replaces each pixel with the median of all the gray levels in a local neighborhood, generally defined by a square window. The median filter works on the assumption that a small percentage of the pixels has been corrupted. A comparison of the results of median filtering and Gaussian smoothing on an image corrupted by salt-and-pepper noise is shown in Figure 5.16.

Computing the median of a set of values involves sorting the values, then selecting the one in the middle. Therefore, the standard median filter algorithm on an image is computationally demanding, because for each pixel we must perform $O(w^2 \log w)$ operations to sort the gray levels in the $w \times w$ window centered at the pixel. With a slight trick, a much more efficient $O(w)$ algorithm is able to compute the exact same result as the standard algorithm. The key to this algorithm is to store the graylevel histogram of the pixels in the window, along with the median and the number of pixels whose gray level is less than or equal to the median. These two extra values of information can then be updated in (approximately) constant time for each pixel. This faster algorithm is shown in Algorithm 5.11. For simplicity, out-of-bounds details have been omitted from the pseudocode, and for compactness, the semicolons in lines 11, 13, 15, and 17 allow multiple lines of code to be displayed on a single line.

5.5.2 Non-Local Means

A particularly effective way to reduce the effects of noise in an image is to compute a weighted average over *all* pixels in the image, a technique known as **non-local means (NLM)**:

$$I'(x, y) = \frac{1}{\eta} \sum_{(x', y') \in I} w_{\mathbf{x}, \mathbf{x}'} I(x', y') \tag{5.113}$$

where $\eta = \sum_{(x', y') \in I} w_{\mathbf{x}, \mathbf{x}'}$ is a normalization factor and $w_{\mathbf{x}, \mathbf{x}'}$ is the weight associated with the pixels $\mathbf{x} = (x, y)$ and $\mathbf{x}' = (x', y')$, computed based on the similarity in appearance of the two graylevel patterns in the windows surrounding the pixels. As a result, similar-looking regions have more influence on the outcome than regions whose appearance is far from the window around the target pixel. A typical implementation of non-local means, shown in Algorithm 5.12, compares pixels using a distance function:

$$w_{\mathbf{x}, \mathbf{x}'} = e^{-\frac{d_I^2(\mathbf{x}, \mathbf{x}')}{2\sigma^2}} \tag{5.114}$$

where the distance function sums the difference in gray levels between corresponding pixels in some window $\mathcal{W}$:

$$d_I(\mathbf{x}, \mathbf{x}') = \sum_{\boldsymbol{\delta} \in \mathcal{W}} I(\mathbf{x} + \boldsymbol{\delta}) - I(\mathbf{x}' + \boldsymbol{\delta}) \tag{5.115}$$

Figure 5.16 **LEFT:** An 816×612 image, and its corruption by salt-and-pepper noise. Right: The result of applying 3×3 median and Gaussian filters, respectively, to the corrupted image. Notice that the median filter removes the noise much better than the Gaussian filter does.

Original image   Image corrupted by salt-and-pepper noise   Image restored using median filter   Image restored using Gaussian filter

ALGORITHM 5.11 Perform median filtering on an image

MEDIANFILTER (I, w)

Input: grayscale image I, width w of square window for computing median
Output: median-filtered image

1	$\tilde{w} \leftarrow \lfloor (w - 1)/2 \rfloor$	➤ Determine half-width.
2	**for** $y \leftarrow 0$ **to** $height - 1$ **do**	
3	$\quad \mathcal{W} \leftarrow \{(-\tilde{w}, -\tilde{w}), \ldots, (\tilde{w}, \tilde{w})\}$	➤ 2D window of pixel coordinates.
4	$\quad h \leftarrow$ HISTOGRAM $(I(w))$	➤ Compute graylevel histogram over window.
5	$\quad med \leftarrow$ MEDIAN(h)	➤ Compute the median of the pixels.
6	$\quad n_m \leftarrow \sum_{i=0}^{med} h[i]$	➤ Number of pixels whose gray level is less than or equal to med.
7	$\quad$ **for** $x \leftarrow 0$ **to** $width - 1$ **do**	
8	$\quad\quad I'(x, y) \leftarrow med$	➤ Set output pixel to the median value.
9	$\quad\quad$ **for** $y' \leftarrow -\tilde{w}$ **to** $\tilde{w}$ **do**	
10	$\quad\quad\quad v \leftarrow I(x - \tilde{w}, y + y')$	➤ Update histogram by removing pixels along left edge.
11	$\quad\quad\quad h[v] \leftarrow_{-} 1;$ **if** $v \leq med$ **then** $n_m \leftarrow_{-} 1$	
12	$\quad\quad\quad v \leftarrow I(x + \tilde{w} + 1, y + y')$	➤ Update histogram by adding pixels along right edge.
13	$\quad\quad\quad h[v] \leftarrow_{+} 1;$ **if** $v \leq med$ **then** $n_m \leftarrow_{+} 1$	
14	$\quad\quad$ **while** $n_m < \lfloor w * w/2 \rfloor$ **do**	➤ Update the median.
15	$\quad\quad\quad med \leftarrow_{+} 1; \quad n_m \leftarrow_{+} h[med]$	
16	$\quad\quad$ **while** $n_m > \lfloor w * w/2 \rfloor$ **do**	
17	$\quad\quad\quad n_m \leftarrow_{-} h[med]; \quad med \leftarrow_{-} 1$	
18	**return** I'	

Even more robust results can be achieved by applying a Gaussian weighting function to this window, so that pixels near the center achieve more influence over the computation of $d_I(\mathbf{x}, \mathbf{x}')$. This straightforward extension is left as an exercise for the reader.

5.5.3 Bilateral Filtering

Recall that the convolution of a signal $f(x)$ with a kernel $g(x)$ is given by sliding a flipped version of the kernel across the signal and computing the sum of the elementwise multiplications:

$$f'(x) = f(x) \circledast g(x) = \frac{1}{\eta} \sum_i f(i)g(x - i) \tag{5.116}$$

where $\eta = \sum_i g(i)$ is the normalization factor. Convolution is a linear operation because it operates the same on every pixel, regardless of the value of the pixel.

Many applications benefit from also taking into account the values of the pixels during the filtering. This concept leads to the **bilateral filter**, which contains two kernels, a *spatial kernel* and a *range kernel*. The spatial kernel g_s weights neighboring samples according to their proximity to the central sample, while the range kernel g_r weights neighboring samples according to their similarity in value to the central sample:

$$f'(x) = f(x) \odot \langle g_s(x), g_r(z) \rangle = \frac{1}{\eta(x)} \sum_i f(i)g_s(x - i)g_r(f(x) - f(i)) \tag{5.117}$$

ALGORITHM 5.12 Perform non-local means filtering on an image

NonLocalMeans(I, w)

Input: grayscale image I, set of pixels $\mathcal{W}$ specifying window
Output: smoothed image from applying non-local means filtering

1 **for** $(x, y) \in I'$ **do**   ▸ For each pixel in the output image (same size as input image),
2      $val \leftarrow 0$ initialize value to zero,
3      $norm \leftarrow 0$ and normalization factor to zero.
4   **for** $(x', y') \in I$ **do**   ▸ For each pixel in input image,
5          $d \leftarrow 0$ initialize distance to zero.
6   **for** $(\delta_x, \delta_y) \in \mathcal{W}$ **do**   ▸ Compute the dissimilarity between the two windows.
7              $d \leftarrow_+ I(x + \delta_x, y + \delta_y) - I(x' + \delta_x, y' + \delta_y)$
8          $w \leftarrow \exp(-d * d / (2 * \sigma * \sigma))$ ▸ Set the weight to the similarity.
9          $val \leftarrow_+ w * I(x', y')$ ▸ Update the sum of the weighted pixels,
10         $norm \leftarrow_+ w$ and the normalization factor.
11     $I'(x, y) \leftarrow val/norm$ ▸ Set the output pixel to the normalized value.
12 **return** I'

where $\odot$ indicates bilateral filtering, $\eta(x) = \sum_i g_s(x - i) g_r(f(x) - f(i))$ is the normalization factor, and the variable z is used to emphasize that the input values to g_r are not pixel coordinates but rather gray levels. If the range kernel is the identity function, that is, $g_r(f(x) - f(i)) = 1$, then bilateral filtering reduces to convolution.

The extension to 2D is straightforward:

$$I'(x, y) = I(x, y) \odot \langle G_s(x, y), g_r(z) \rangle = \frac{1}{\eta(x, y)} \sum_i \sum_j I(i, j) G_s(x - i, y - j)$$

$$g_r(I(x, y) - I(i, j)) \tag{5.118}$$

where $\eta(x, y) = \sum_i \sum_j G_s(x - i, y - j) g_r(I(x, y) - I(i, j))$ is the normalization factor. For the same reasons that Gaussian kernels are used in convolution, Gaussian kernels are also typically used in bilateral filtering:

$$G_s(x, y) = e^{-\frac{x^2 + y^2}{2\sigma_s^2}} \qquad g_r(z) = e^{-\frac{z^2}{2\sigma_r^2}} \tag{5.119}$$

Because the bilateral filter weights pixels according to their proximity, it produces a smooth blur, and because it weights pixels according to their values, it tends not to smooth across intensity edges. The filter thus achieves **edge-preserving smoothing**, as shown in Figure 5.17. Within a fairly homogeneous region of intensity, bilateral filtering is similar to convolution, but near an edge, it only takes into account the pixels on the near side of the edge, thus preserving the edge. Unlike convolution, which after repeated applications eventually converges to a flat image in which every pixel takes on the average value in the image, bilateral filtering typically converges to an image in which every pixel is assigned the average value of the similar-colored pixels nearby.[†] Repeated applications of bilateral filtering thus yield a cartoon-like image, as shown in Figure 5.18.

[†] Actually, such convergence occurs only if the range kernel is truncated (e.g., set to zero after several standard deviations), or if the iterations are stopped when the change in the image is small — almost always the case in practice. With an infinite-support range kernel, bilateral filtering eventually leads to a flat image just like convolution.

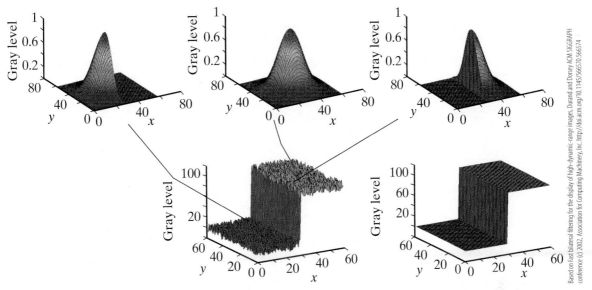

Based on Fast bilateral filtering for the display of high-dynamic-range images, Durand and Dorsey ACM SIGGRAPH conference (c) 2002, Association for Computing Machinery, Inc. http://doi.acm.org/10.1145/566570.566574 -

Figure 5.17 Bilateral filtering of a noisy step edge preserves the crisp edge as it smooths out the noise on either side of the edge. The top row shows the kernel at three locations: Far from the edge, the kernel approximates a Gaussian, whereas near the edge it approximates half a Gaussian.

There are three parameters in a Gaussian bilateral filter: the spatial standard deviation σ_s, the range standard deviation σ_r, and the number of iterations n_{iter}:

$$I'(x, y) = I(x, y) \circledcirc^{n_{iter}} \langle G_s(x, y), g_r(z) \rangle = \underbrace{(I(x, y) \circledcirc \langle G_s(x, y), g_r(z) \rangle) \cdots \circledcirc \langle G_s(x, y), g_r(z) \rangle}_{\text{repeated } n_{iter} \text{ times}} \quad (5.120)$$

The code to implement bilateral filtering is straightforward, as shown in Algorithm 5.13. Since the values of a Gaussian are nearly negligible beyond several standard deviations, we can limit the processing to a fixed-size window of radius $2.5\sigma_s$ in the spatial domain to limit computation. For relatively small values of σ_s, such a straightforward approach yields an algorithm that, while not necessary efficient, is nevertheless tolerable for many applications.

5.5.4 Bilateral Filtering for Large Windows

When the window size is large, bilateral filtering is extremely slow, requiring computation that is proportional to the size of the window for every pixel in the image. Just as we showed that Gaussian convolution can be achieved in constant $O(1)$ per-pixel time by approximating the

Figure 5.18 Repeated applications of bilateral filtering yield a cartoon-like image in which the colors are flattened in local regions. The result (right) was obtained by applying $n_{iter} = 5$ iterations of bilateral filtering on the input (left).

ALGORITHM 5.13 Perform bilateral filtering on an image

$\textsc{BilateralFilter}(I, \sigma_s, \sigma_r, n_{iter})$

Input: grayscale image I, standard deviations σ_s and σ_r of Gaussian spatial and range kernels, number n_{iter} of iterations
Output: bilateral-filtered image

1 **for** $k \leftarrow 1$ **to** n_{iter} **do**	For each iteration,
2 **for** $(x, y) \in I$ **do**	and for each pixel in the image,
3 $val \leftarrow 0$	initialize the value to zero,
4 $norm \leftarrow 0$	and the normalization factor to zero.
5 **for** $(\delta_x, \delta_y) \in \mathcal{W}$ **do**	For each pixel in a $\pm 2.5\sigma_s$ window,
6 $d_s^2 \leftarrow \delta_x * \delta_x + \delta_y * \delta_y$	compute squared spatial distance,
7 $d_r \leftarrow I(x, y) - I(x + \delta_x, y + \delta_y)$	and range difference
8 $w \leftarrow \exp(-d_s^2/(2 * \sigma_s^2)) * \exp(-(d_r * d_r)/(2 * \sigma_r^2))$	to compute weight.
9 $val \leftarrow_+ w * I(x + \delta_x, y + \delta_y)$	Accumulate weighted sum
10 $norm \leftarrow_+ w$	and update normalization factor.
11 $I'(x, y) \leftarrow val/norm$	Set output to normalized weighted sum.
12 $I \leftarrow I'$	Copy entire output image to input for next iteration.
13 **return** I'	

Gaussian kernel with the convolution of box kernels, bilateral filtering with Gaussian kernels can also be achieved in constant $O(1)$ per-pixel time by approximating the Gaussian range kernel. Several approaches have been proposed for achieving such an efficient approximation. To understand how this is done for one such approach, let us first describe the notion of *shiftability*, after which we will see that the Gaussian can be well approximated by a *raised cosine*.

A function $\phi(z)$ is said to be **shiftable** if for every translation τ, we have

$$\phi(z - \tau) = \sum_{i=1}^{n} c_i(\tau)\phi_i(z) \tag{5.121}$$

where c_i are the interpolating coefficients, ϕ_i are the global basis functions, and n is the order of shiftability. Using shiftability, a local kernel can be decomposed into weighted sums of the basis functions. It can be shown that the only smooth functions that are shiftable are composed of sums and products of the polynomials and exponentials. For example, consider the function

$$\phi(z) = \sum_{i=1}^{n} a_i e^{\alpha_i z} \tag{5.122}$$

It is easy to see that this function shifted by τ can be expressed as the weighted sum of terms depending only upon z, using weights depending only upon τ:

$$\phi(z - \tau) = \sum_{i=1}^{n} a_i e^{\alpha_i(z-\tau)} \tag{5.123}$$

$$= \sum_{i=1}^{n} \underbrace{e^{-\alpha_i \tau}}_{c_i(\tau)} \underbrace{a_i e^{\alpha_i z}}_{\phi_i(z)} \tag{5.124}$$

Another example of a shiftable function is the cosine: $\cos(z - \tau) = \cos(\tau)\cos(z) + \sin(\tau)\sin(z)$, and similarly the sine: $\sin(z - \tau) = \cos(\tau)\sin(z) - \sin(\tau)\cos(z)$. In other words, any sinusoid[†] can be expressed as the linear combination of two fixed sinusoids.

The key to making bilateral filtering efficient is to approximate the Gaussian. Among the many ways to approximate the Gaussian, one of the most effective for exploiting shiftability is that of the *raised cosine*, which is the cosine raised to some power. It can be shown that as the power increases, the raised cosine indeed converges to a Gaussian:

$$\lim_{n \to \infty} \left[\cos\left(\frac{\gamma z}{\sqrt{n}}\right) \right]^n = \exp\left(-\frac{\gamma^2 z^2}{2} \right) \tag{5.125}$$

where the scaling by $\sqrt{n}$ prevents the expression from degenerating to zero almost everywhere. The rate of convergence of raised Gaussians is much faster than that of other expressions such as the Taylor series polynomials, thus requiring fewer terms to achieve a good approximation. A reasonable choice is $n = 3$, as seen in Figure 5.19. Note that $\sigma_r^2 \approx 1/\gamma^2$, where σ_r^2 is the variance of the Gaussian, so that we can set $\gamma = \sigma_r^{-1}$ given some desired variance σ_r.

To see how to apply this technique, let us consider the case $n = 1$. Although this yields a rather crude approximation to the Gaussian, it simplifies the math considerably, enabling us to develop some intuition before tackling the more difficult case $n > 1$. In other words, let the range kernel be given by a windowed Gaussian:

$$g_r(z) \equiv \phi(z) = \begin{cases} \cos(\gamma z) & \text{if } -\alpha \le z \le \alpha \\ 0 & \text{otherwise} \end{cases} \tag{5.126}$$

where $\alpha \equiv \frac{\pi}{2\gamma}$ so that the function is nonnegative and unimodal with a peak at $z = 0$. Plugging this expression into the bilateral filter equation above yields

$$f'(x) = f(x) \odot \langle g_s(x), g_r(z) \rangle \tag{5.127}$$

$$\approx \frac{1}{\eta(x)} \sum_i f(i) g_s(x - i) \cos(\gamma(f(x) - f(i))) \tag{5.128}$$

Figure 5.19 The raised cosine approximates the Gaussian. The black line shows $\exp(-z^2/2\sigma^2)$ for $\sigma = 4$. The red lines show the raised cosine, $[\cos(\gamma z/\sqrt{n})]^n$, for $n = 1, 2,$ and 3, where $\gamma = 1/\sigma$. For $n = 2$ the raised cosine still exhibits considerable oscillation, but for $n \ge 3$ it well approximates the Gaussian shape. For $n \ge 20$ (not shown), it is nearly indistinguishable from the Gaussian.

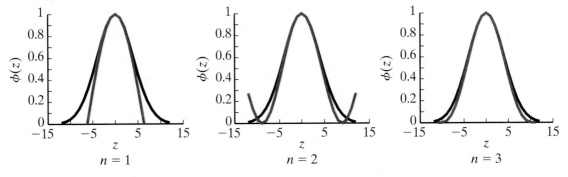

[†] Since the term *sinusoid* allows for arbitrary phase, it includes both sine and cosine.

$$= \frac{1}{\eta(x)} \sum_i f(i) g_s(x - i) [\cos(\gamma f(x)) \cos(\gamma f(i)) - \sin(\gamma f(x)) \sin(\gamma f(i))] \quad (5.129)$$

$$= \frac{1}{\eta(x)} (\cos(\gamma f(x)) \phi_1(x) - \sin(\gamma f(x)) \phi_2(x)) \quad (5.130)$$

where we notice from the last expression that the basis functions are simply convolutions of modulations of the original signal with a standard spatial convolution kernel:

$$\phi_1(x) = \sum_i f(i) \cos(\gamma f(i)) g_s(x - i) = f(x) \cos(\gamma f(x)) \circledast g_s(x) \quad (5.131)$$

$$\phi_2(x) = \sum_i f(i) \sin(\gamma f(i)) g_s(x - i) = f(x) \sin(\gamma f(x)) \circledast g_s(x) \quad (5.132)$$

Similarly, the normalization factor is given by

$$\eta(x) = \sum_i g_s(x - i) \cos(\gamma(f(x) - f(i))) \quad (5.133)$$

$$= \cos(\gamma f(x)) \sum_i \cos(\gamma f(i)) g_s(x - i) - \sin(\gamma f(x)) \sum_i \sin(\gamma f(i)) g_s(x - i) \quad (5.134)$$

$$= \cos(\gamma f(x)) [\cos(\gamma f(x)) \circledast g_s(x)] - \sin(\gamma f(x)) [\sin(\gamma f(x)) \circledast g_s(x)] \quad (5.135)$$

This is a truly remarkable result because it shows that the bilateral filter with an approximate Gaussian range kernel can be written as a linear combination of convolutions with the spatial kernel.

Implementing this approach to bilateral filtering is straightforward. The input signal $f(x)$ is multiplied pointwise by the function $\cos(\gamma f(x))$, as well as by the function $\sin(\gamma f(x))$. The results are then convolved with a Gaussian kernel with standard deviation σ_s to yield ϕ_1 and ϕ_2, as in Equations (5.131) and (5.132). The two signals ϕ_1 and ϕ_2 are then multiplied pointwise by $\cos(\gamma f(x))$ and $\sin(\gamma f(x))$ again, respectively, and the results are subtracted to form the numerator in Equation (5.130). A similar approach, but without the original input signal $f(x)$, yields the denominator $\eta(x)$, as in Equation (5.135). Dividing the numerator and denominator by each other yields the final result. Notice that every step of this algorithm is constant $O(1)$ per-pixel time, if we use the trick shown earlier for reducing Gaussian smoothing to constant per-pixel time. The pseudocode for 2D bilateral filtering using this rather crude approximation of the Gaussian range kernel is given in Algorithm 5.14.

Now let us consider $n > 1$, which yields a much better Gaussian approximation. Using the expression $\cos \theta = \frac{1}{2}(e^{j\theta} + e^{-j\theta})$ and the binomial theorem, it can be shown that

$$\phi(z) = \left[\cos\left(\frac{\gamma z}{\sqrt{n}} \right) \right]^n = \sum_{i=0}^n 2^{-n} \binom{n}{i} \exp\left(j \frac{(2i - n)\gamma z}{\sqrt{n}} \right) \quad (5.136)$$

where $j = \sqrt{-1}$. Note that we have $n + 1$ terms, leading to $2(n + 1)$ basis images. But if n is even, then the expression inside the summation is constant when $i = n/2$, thus reducing the number of basis images by 1.

Plugging this into the bilateral filter equation above yields

$$f'(x) = f(x) \odot \langle g_s(x), g_r(z) \rangle \quad (5.137)$$

$$\approx \frac{1}{\eta(x)} \sum_i f(i) g_s(x - i) \left(\cos\left(\frac{\gamma(f(x) - f(i))}{\sqrt{n}} \right) \right)^n$$

ALGORITHM 5.14 Perform fast bilateral filtering on an image using the crude approximation of a cosine for the range kernel

$\text{BILATERALFILTERFASTBUTCRUDE}(I, \sigma_s, \sigma_r, n_{iter})$

Input: grayscale image I, standard deviations σ_s and σ_r of Gaussian spatial and range kernels (range kernel is approximated as a cosine) number n_{iter} of iterations

Output: bilateral-filtered image

1    $\gamma \leftarrow 1/\sigma_r$ ▸ Determine γ from the standard deviation of the range kernel.

2   **for** $k \leftarrow 1$ **to** n_{iter} **do**   ▸ For each iteration of the bilateral filter,

3   **for** $(x, y) \in I$ **do**   compute the

4           $H_1(x, y) \leftarrow \cos(\gamma * I(x, y))$ coefficient images H_1 and H_2

5           $H_2(x, y) \leftarrow \sin(\gamma * I(x, y))$ (which are also used for normalization),

6           $G_1(x, y) \leftarrow I(x, y) * H_1(x, y)$ and the

7           $G_2(x, y) \leftarrow I(x, y) * H_2(x, y)$ basis images G_1 and G_2.

8        $G_1' \leftarrow \text{SMOOTH}(G_1, \sigma_s)$ ▸ Smooth the basis images

9        $G_2' \leftarrow \text{SMOOTH}(G_2, \sigma_s)$ by convolving with a spatial Gaussian, and

10       $H_1' \leftarrow \text{SMOOTH}(H_1, \sigma_s)$ smooth the normalization images

11       $H_2' \leftarrow \text{SMOOTH}(H_2, \sigma_s)$ by convolving with a spatial Gaussian.

12   **for** $(x, y) \in I'$ **do**   ▸ Set each pixel in the output (same size as input)

13          $num \leftarrow H_1(x, y) * G_1'(x, y) + H_2(x, y) * G_2'(x, y)$ to the weighted sum

14          $den \leftarrow H_1(x, y) * H_1'(x, y) + H_2(x, y) * H_2'(x, y)$ of basis images divided by

15          $I'(x, y) \leftarrow num/den$ weighted sum of normalization images.

16       $I \leftarrow I'$

17   **return** I'

$$= \frac{1}{\eta(x)} \sum_i f(i) g_s(x - i) \left(\sum_{i'=0}^{n} 2^{-n} \binom{n}{i'} \exp\left(j \frac{(2i' - n)\gamma(f(x) - f(i))}{\sqrt{n}} \right) \right)$$

$$= \frac{1}{\eta(x)} \sum_i f(i) g_s(x - i)$$

$$\left(\sum_{i'=0}^{n} 2^{-n} \binom{n}{i'} \underbrace{\exp\left(j \frac{(2i' - n)\gamma f(x)}{\sqrt{n}} \right)}_{c_{i'}(f(x))} \underbrace{\exp\left(-j \frac{(2i' - n)\gamma f(i)}{\sqrt{n}} \right)}_{\alpha_{i'}(f(i))} \right)$$

where again we notice that the basis functions are simply convolutions of functions of the original signal, after multiplying by a complex exponential, with a standard spatial smoothing kernel. Similarly, the normalization factor is given by

$$\eta(x) = \sum_i g_s(x - i) \left(\cos\left(\frac{\gamma(f(x) - f(i))}{\sqrt{n}} \right) \right)^n \tag{5.138}$$

$$= \sum_{i'=0}^{n} \left(2^{-n} \binom{n}{i'} \exp\left(j \frac{(2i' - n)\gamma f(x)}{\sqrt{n}} \right) \right) \left(\sum_i \exp\left(-j \frac{(2i' - n)\gamma f(i)}{\sqrt{n}} \right) g_s(x - i) \right)$$

This fast version of bilateral filtering, extended to 2D, is given in Algorithm 5.15, where $\text{FACT}(n) = n!$ is the factorial function, needed because $\binom{n}{i} = \frac{n!}{i!\,(n-i)!}$.

Bilateral filtering is used for a variety of applications. Besides general denoising purposes, it is widely used for tone mapping high-dynamic-range (HDR) images, as shown in Figure 5.20. The bilateral filter is applied to HDR intensity values in the log domain, followed by contrast reduction to yield low-dynamic-range (LDR) intensity values; glare reduction can be achieved in a similar manner. Bilateral filtering can also be used for low-light enhancement (LLE) using two images of a scene, one taken with a flash and the other taken without a flash, as shown in Figure 5.21. The flash image captures detail, while the no-flash image captures the unmodified viewing conditions (original scene ambience), so applying the bilateral filter to the latter to remove noise and then applying a cross-bilateral filter to the former using the latter allows the two images to be combined to preserve the important properties of each.

ALGORITHM 5.15 Perform fast bilateral filtering on an image using the raised cosine for the range kernel

$\text{BilateralFilterFast}(I, \sigma_s, \sigma_r, n_{iter})$

Input: grayscale image I, standard deviations σ_s and σ_r of Gaussian spatial and range kernels (range kernel is approximated as a raised cosine) number n_{iter} of iterations

Output: bilateral-filtered image

```
1   γ ← 1/σ_r
2   n ← 3                                          ▷ For a reasonable approximation, it is recommended that n ≥ 3.
3   for k ← 1 to n_iter do
4       for i ← 0 to n do
5                                                  ▷ Compute basis image G'_i and coefficient image D_i, along with H'_i for normalization
6           for (x, y) ∈ I do
7               v ← γ * (2 * i − n) * I(x, y)/Sqrt(n)                      ▷ scalar
8               β ← Fact(n)/(Fact(i) * Fact(n − i) * Pow(2, −n))          ▷ scalar
9               H_i(x, y) ← [cos(v) sin(v)]^T                             ▷ 2 × 1 vector represents complex number
10              G_i(x, y) ← I(x, y) * H_i(x, y)                          ▷ 2 × 1 vector represents complex number
11              D_i(x, y) ← [cos(v) sin(v)]^T * β                        ▷ 2 × 1 vector represents complex number
12          G'_i ← Smooth(G_i, σ_s)                     ▷ Smooth real and imaginary channels separately.
13          H'_i ← Smooth(H_i, σ_s)                     ▷ Smooth real and imaginary channels separately.
14
15                                                  ▷ For each pixel, divide the elementwise multiplication of the images.
16      for (x, y) ∈ I do
17          num ← 0
18          den ← 0
19          for i ← 0 to n do
20              num ←_+ D_i(x, y) * G'_i(x, y)            ▷ multiplication of two complex numbers
21              den ←_+ D_i(x, y) * H'_i(x, y)            ▷ multiplication of two complex numbers
22          I'(x, y) ← Real(num/den)                      ▷ Extract real component after complex division.
23      I ← I'
24  return I'
```

Figure 5.20 Tone mapping
of high dynamic range
images is often performed
using bilateral filtering.

Samot / Shutterstock.com

5.5.5 Mean-Shift Filter

Another edge-preserving filter closely related to the bilateral filter is the **mean-shift filter**. In the mean-shift filter, each pixel iteratively moves in such a way as to seek the nearest mode in the joint spatial-range space until convergence. More specifically, let $\bar{\mathbf{x}}_i \equiv (x_i, y_i, v_i)$ be the coordinates and value of the ith pixel, where $I(x_i, y_i) = v_i$. If we consider the set of points $\{(x_i, y_i, v_i)\}_{i=1}^{n}$, where n is the number of pixels in the image, as discrete samples of a probability distribution function (PDF), then an estimate $\hat{f}$ of the underlying continuous PDF can be constructed as

$$\hat{f}(\bar{\mathbf{x}}) = \frac{1}{n} \sum_{i=1}^{n} K\left(\frac{\bar{\mathbf{x}} - \bar{\mathbf{x}}_i}{h}\right) \propto \frac{1}{n} \sum_{i=1}^{n} k\left(\frac{\|\bar{\mathbf{x}} - \bar{\mathbf{x}}_i\|^2}{h^2}\right) \tag{5.139}$$

where $K(\cdot)$ is the **kernel function** with **bandwidth** h, and where the rightmost part assumes that $K(\cdot)$ is a radially symmetric kernel, so that $K(\mathbf{z}) \propto k(\|\mathbf{z}\|^2)$, where $k(\cdot)$ is called the **profile** of the kernel $K(\cdot)$, with $\mathbf{z}$ an arbitrary vector. Note that $k(\cdot)$ is only defined for nonnegative scalar values, and proportionality is used so that normalization constants can be ignored in the profile.

Figure 5.21 One of
the more interesting
applications of the
bilateral filter is to
combine a flash
image (left) with
a no-flash image
(middle) to preserve
both detail and the
original viewing
conditions of the
scene (right).

G. Petschnig, R. Szeliski, M. Agrawala, M. Cohen, H. Hoppe, and K. Toyama, "Digital photography with flash and no-flash image pairs," ACM Transactions on Graphics (SIGGRAPH), 23(3):664-672, August 2004. © 2004 Association for Computing Machinery, Inc. Reprinted by permission.

Two popular kernels for mean-shift filtering are the **Gaussian** (or **normal**) **kernel**, and the **Epanechnikov kernel**, which are given (along with their profiles) as follows:

$$K_N(\mathbf{z}) = \frac{1}{(2\pi)^{d/2}} \exp\left(-\tfrac{1}{2}\|\mathbf{z}\|^2\right) \qquad k_N(z) = \exp\left(-\tfrac{1}{2}z\right) \; z \geq 0$$

$$K_E(\mathbf{z}) = \begin{cases} \frac{d+2}{2c_d}(1 - \|\mathbf{z}\|^2) & \text{if } \|\mathbf{z}\| \leq 1 \\ 0 & \text{otherwise} \end{cases} \qquad k_E(z) = \begin{cases} 1 - z & \text{if } 0 \leq z \leq 1 \\ 0 & \text{if } z > 1 \end{cases} \qquad (5.140)$$

where c_d is the volume of the unit d-dimensional hypersphere and d is the dimensionality of the space ($d = 2$ for spatial coordinates, $d = 1$ for range values). For example, with Gaussian kernels, Equation (5.139) becomes

$$\hat{f}(\overline{\mathbf{x}}) \propto \frac{1}{n} \sum_{i=1}^{n} \exp\left(-\frac{\|\overline{\mathbf{x}} - \overline{\mathbf{x}}_i\|^2}{2h^2}\right) \qquad (5.141)$$

so that the bandwidth is the standard deviation. In the case of filtering images, $\overline{\mathbf{x}}$ is not in a Euclidean space, since the spatial and range dimensions have different units. As a result, it is necessary to use different bandwidth parameters for the different dimensions, so that $k(\|\overline{\mathbf{x}} - \overline{\mathbf{x}}_i\|^2/h^2)$ is replaced with $k(\|\mathbf{x} - \mathbf{x}_i\|^2/h_s^2 + (v - v_i)^2/h_r^2)$, where $\mathbf{x} \equiv (x, y)$ and $\mathbf{x}_i \equiv (x_i, y_i)$, and h_s and h_r are the spatial and range bandwidth parameters, respectively.

An extremum of the PDF can be found by differentiating Equation (5.139), which, if we let $g(z) \equiv -\partial k(z)/\partial \overline{\mathbf{x}}$, yields:

$$\frac{\partial \hat{f}(\overline{\mathbf{x}})}{\partial \overline{\mathbf{x}}} = 0 \qquad (5.142)$$

$$\propto \sum_{i=1}^{n} \frac{\partial}{\partial \overline{\mathbf{x}}}\left(k\left(\frac{\|\overline{\mathbf{x}} - \overline{\mathbf{x}}_i\|^2}{h^2}\right)\right)(\overline{\mathbf{x}} - \overline{\mathbf{x}}_i) \qquad (5.143)$$

$$= \sum_{i=1}^{n} g\left(\frac{\|\overline{\mathbf{x}} - \overline{\mathbf{x}}_i\|^2}{h^2}\right)(\overline{\mathbf{x}}_i - \overline{\mathbf{x}}) \qquad (5.144)$$

$$= \left[\sum_{i=1}^{n} g\left(\frac{\|\overline{\mathbf{x}} - \overline{\mathbf{x}}_i\|^2}{h^2}\right)\right]\left[\underbrace{\frac{\sum_{i=1}^{n} g\left(\frac{\|\overline{\mathbf{x}} - \overline{\mathbf{x}}_i\|^2}{h^2}\right)\overline{\mathbf{x}}_i}{\sum_{i=1}^{n} g\left(\frac{\|\overline{\mathbf{x}} - \overline{\mathbf{x}}_i\|^2}{h^2}\right)}}_{\text{weighted mean}} - \overline{\mathbf{x}}\right] \qquad (5.145)$$

$$\underbrace{\phantom{\left[\sum_{i=1}^{n} g\left(\frac{\|\overline{\mathbf{x}} - \overline{\mathbf{x}}_i\|^2}{h^2}\right)\right]\left[\frac{\sum_{i=1}^{n} g\left(\frac{\|\overline{\mathbf{x}} - \overline{\mathbf{x}}_i\|^2}{h^2}\right)\overline{\mathbf{x}}_i}{\sum_{i=1}^{n} g\left(\frac{\|\overline{\mathbf{x}} - \overline{\mathbf{x}}_i\|^2}{h^2}\right)} - \overline{\mathbf{x}}\right]}}_{\text{mean-shift}}$$

Since the derivative of the exponential is an exponential, $g_N(z) = \tfrac{1}{2}k_N(z)$ in the case of the Gaussian kernel, and $g_E(z) = 1$ in the case of the Epanechnikov kernel. Let us define the kernel $G(\cdot)$ so that $G(\mathbf{z}) \propto g(\|\mathbf{z}\|^2)$, then $K(\cdot)$ is called the **shadow** of $G(\cdot)$. It is not difficult to see that the Epanechnikov kernel is the shadow of the uniform kernel, while the Gaussian kernel is the shadow of itself.

In Equation (5.145) we have labeled the weighted mean of the points using the kernel centered at $\overline{\mathbf{x}}$, as well as the difference between the mean and the current estimate $\overline{\mathbf{x}}$, where the latter is known as the **mean-shift**. Since the extremum occurs when the mean-shift is zero, if we let $\bar{\mathbf{x}}^{(t)}$ be the estimate for the tth iteration, then the estimate in the next iteration can be set to $\bar{\mathbf{x}}^{(t+1)} = \bar{\mathbf{x}}^{(t)} + \Delta\bar{\mathbf{x}}^{(t)}$, where the difference between consecutive estimates $\Delta\bar{\mathbf{x}}^{(t)} \equiv \bar{\mathbf{x}}^{(t+1)} - \bar{\mathbf{x}}^{(t)}$ is just the mean-shift. This leads to the **mean-shift filtering algorithm**, which repeatedly sets the estimate to the weighted mean:

$$\bar{\mathbf{x}}^{(t+1)} = \frac{\sum_{i=1}^{n} g\left(\frac{\|\bar{\mathbf{x}}^{(t)} - \bar{\mathbf{x}}_i\|^2}{h^2}\right)\bar{\mathbf{x}}_i}{\sum_{i=1}^{n} g\left(\frac{\|\bar{\mathbf{x}}^{(t)} - \bar{\mathbf{x}}_i\|^2}{h^2}\right)} \tag{5.146}$$

for $t = 1, 2, \ldots$ until convergence, where $\bar{\mathbf{x}}^{(0)}$ is the initial estimate. The algorithm performs gradient ascent, which can be seen by noticing from Equations (5.142) and (5.145) that the mean-shift vector is always proportional to the density estimate, and therefore it points toward the maximum increase in the density.

The mean-shift algorithm is guaranteed to converge as long as the kernel profile is convex and monotonically decreasing, which are true for both the Gaussian and Epanechnikov kernels. For the latter, the algorithm converges in a finite number of steps, since the number of locations with unique mean values is finite. With a Gaussian, the algorithm requires an infinite number of steps, but it is easy in practice to terminate when the norm of the mean-shift vector is below a threshold. The convergence of the algorithm arises from the automatic adaptation of the step size in the gradient ascent. From the denominator in Equation (5.146) we notice that regions of low-density values yield large step sizes, whereas high-density regions lead to smaller step sizes. Thus, as the estimate approaches the nearest extremum, the step sizes decrease until they reach zero, either actually or asymptotically.

The mean-shift filtering procedure is shown in Algorithm 5.16. In Line 2, $\{\bar{\mathbf{x}}\}$ is initialized to the spatial coordinates and range value of a pixel in the image. Lines 4–11 contain the core procedure for updating $\bar{\mathbf{x}}$ by accumulating the vector numerator and scalar denominator

ALGORITHM 5.16 Apply the mean-shift filtering algorithm to an image

MeanShiftFilter(I, h_s, h_r)

Input: grayscale image I, bandwidth parameters h_s and h_r
Output: output image I' resulting from the mean-shift (edge-preserving) filter

```
1   for (x, y) ∈ I do                                                        ▶ For each pixel in the image,
2       (x', y', v') ← (x, y, I(x, y))                                       initialize x̄ = (x', y', v').
3       repeat
4           num ← (0, 0, 0)                                                  ▶ Loop through all
5           den ← 0                                                          the other pixels in the image,
6           for (xᵢ, yᵢ) ∈ I do                                             accumulating the vector
7               w ← g(((x' − xᵢ)² + (y' − yᵢ)²)/h²ₛ + (v' − vᵢ)²/h²ᵣ)       numerator Σg(·)x̄ᵢ,
8               num ←₊ w * (xᵢ, yᵢ, I(xᵢ, yᵢ))                              and scalar
9               den ←₊ w                                                     denominator Σg(·) of Equation (5.146),
10          mean-shift ← num/den − (x', y', v')                             and updating x̄ accordingly.
11          (x', y', v') ← num/den                                          ▶ This is repeated until the norm of the
12      until Norm(mean-shift) < τ                                          mean-shift vector is below a threshold.
13      I'(x, y) ← v'                                                        Update the pixel value using the value in x̄.
14  return I'
```

of Equation (5.146), looping through all the pixels in the image to compute the weighted average. After the algorithm has converged, the output gray level is set to the final gray level from the iterations, discarding the shift in spatial coordinates. Note that the algorithm requires four nested for loops, which explains why mean-shift filtering can be quite slow. In practice, the Gaussian kernel is typically truncated past $\pm 2.5 h_s$ or so, thus greatly reducing the amount of computation; the Epanechnikov kernel is truncated by definition. Note also that when the Epanechnikov kernel is used, the weight in Line 10 reduces to 1 for points in the hypersphere and 0 for points outside, since the Epanechnikov kernel is the shadow of the uniform kernel. An example of mean-shift filtering is shown in Figure 5.22.

The similarity between the bilateral filter and the mean-shift filter is obvious from studying Equations (5.118) and (5.146). In fact, the two equations are nearly identical in the case of the Gaussian kernel. The primary difference is that the bilateral filter updates the gray levels of the pixels without shifting the spatial coordinates of the pixels, whereas the mean-shift filter also shifts the spatial coordinates. Even though the shifted spatial coordinates are eventually discarded, this aspect of the computation is what enables the mean-shift algorithm to seek the nearest extremum of the PDF, and it is what guarantees that the algorithm converges to a cartoon-like image, whereas a bilateral filter that runs forever (without a truncated range kernel) will eventually yield a flat image in which all pixels have the same gray level.

5.5.6 Anisotropic Diffusion

In physics, **diffusion** refers to the movement of molecules from regions of high concentration to regions of low concentration. If we allow the graylevel pixel values to indicate the amount of concentration, then diffusion applied to an image would darken the bright regions and brighten the dark regions. And if run forever, diffusion would eventually lead to an image in which every pixel has the same value, namely the mean of all the pixel values.

Consider, for example, the following procedure. Let $I^{(t)}$ refer to the image after the tth iteration, with $I^{(0)} \equiv I$ being the initial image. Then repeatedly smooth the image by computing for $t = 1, 2, \ldots$, a weighted average between each pixel and its 4-neighbors:

$$
\begin{aligned}
I^{(t+1)}(x, y) = I^{(t)}(x, y) \\
+ \lambda(I^{(t)}(x - 1, y) - I^{(t)}(x, y)) \\
+ \lambda(I^{(t)}(x + 1, y) - I^{(t)}(x, y))
\end{aligned}
$$

Figure 5.22 An image (left) and the cartoon-like result of mean-shift filtering (right), using $h_s = 32$ and $h_r = 16$.

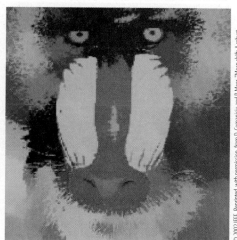

$$+\lambda(I^{(t)}(x, y - 1) - I^{(t)}(x, y))$$
$$+\lambda(I^{(t)}(x, y + 1) - I^{(t)}(x, y))$$
$$= (1 - 4\lambda)I^{(t)}(x, y) + 4\lambda\bar{I}^{(t)}(x, y) \qquad (5.147)$$

where the scalar λ governs the step size, and

$$\bar{I}^{(t)}(x, y) \equiv \frac{1}{4}(I^{(t)}(x - 1, y) + I^{(t)}(x + 1, y) + I^{(t)}(x, y - 1) + I^{(t)}(x, y + 1)) \qquad (5.148)$$

is the average of the 4-neighbors of the pixel at (x,y). To ensure convergence, $0 \le \lambda \le \frac{1}{4}$; if $\lambda = \frac{1}{4}$, then the computation simply replaces each pixel with the average of its 4-neighbors at each iteration; if $\lambda < \frac{1}{4}$, then the computation increases or decreases the pixel value to make it more like the average of its 4-neighbors.

Equation (5.147) is very similar to Gaussian smoothing, and it is easy to see that it will eventually lead to a flat image in which every pixel has the same value. This procedure is known as **isotropic diffusion**, because it propagates the pixel values in all directions (that is, all of the four cardinal directions, due to discretization effects) equally. In contrast, **anisotropic diffusion** means to smooth the image differently in the different directions. Typically, this involves smoothing the image everywhere except across intensity edges, like the edge-preserving smoothing behavior of the bilateral and mean-shift filters. In anisotropic diffusion, the image values are smoothed by repeatedly performing local averages, but doing so in a way that weights neighboring pixels less if they lie on an intensity edge. In this way, the anisotropic diffusion process blurs the image within regions but not across boundaries.

The implementation of anisotropic diffusion is straightforward. For any iteration t, only the latest 2D image $I^{(t)}$ is stored, from which the diffusion coefficients are computed for each pixel and direction. To preserve image boundaries caused by intensity edges, let us define

$$C^{(t)}(x, y, \Delta x, \Delta y) \equiv g(|I^{(t)}(x + \Delta x, y + \Delta y) - I^{(t)}(x, y)|) \qquad (5.149)$$

where $C^{(t)}(x, y, \Delta x, \Delta y)$ is the diffusion coefficient at pixel (x,y) in the direction $(\Delta x, \Delta y)$ at iteration t, and g is a monotonically decreasing function, such as $g(z) \equiv \exp(-z^2/s^2)$ or $g(z) \equiv (1 + z^2/s^2)^{-1}$, where s is a scale parameter. Similar to the isotropic procedure above, the new value of a pixel as

$$I^{(t+1)}(x, y) = I^{(t)}(x, y)$$
$$+\lambda C^{(t)}(x, y, -1, 0)(I^{(t)}(x - 1, y) - I^{(t)}(x, y))$$
$$+\lambda C^{(t)}(x, y, +1, 0)(I^{(t)}(x + 1, y) - I^{(t)}(x, y))$$
$$+\lambda C^{(t)}(x, y, 0, -1)(I^{(t)}(x, y - 1) - I^{(t)}(x, y))$$
$$+\lambda C^{(t)}(x, y, 0, +1)(I^{(t)}(x, y + 1) - I^{(t)}(x, y)) \qquad (5.150)$$

where $0 \le \lambda \le \frac{1}{4}$ to ensure convergence.

To gain an appreciation for the underlying math, consider the family of continuous-domain images obtained by convolving the original image $I(x, y)$ with Gaussian kernels having continuously increasing variance:

$$I(x, y, t) \equiv I(x, y) \circledast Gauss_t(x, y) \qquad (5.151)$$

where $t \ge 0$ is a *continuous* scale parameter that governs the amount of smoothing, $I(x, y, 0) \equiv I(x, y)$, and x and y are treated as continuous values as well. It can be shown that the resulting 3D volume defined in Equation (5.151) is the solution to the so-called **heat equation**, which is a physical equation describing the evolution of a heat distribution

$I(x, y, t)$ over time t in a homogeneous medium with isotropic conductivity, given an initial heat distribution $I(x, y, 0) \equiv I(x, y)$:

$$\frac{\partial I(x, y, t)}{\partial t} = \frac{1}{2}\nabla^2 I(x, y, t) \tag{5.152}$$

where ∇^2 is the Laplacian operator. This heat equation is a type of **diffusion equation**. Note that in the physical equation the variable t is actual time, whereas in the implementation above it is "pseudo-time" that corresponds to the iteration of the algorithm. The fact that the Gaussian kernel is the unique kernel for solving this equation also follows from the fact that it is the Green's function of this heat equation at an infinite domain.

Similarly, the continuous *anisotropic* diffusion equation states that the change over time is equal to the divergence of the weighted gradient:

$$\frac{\partial I(x, y, t)}{\partial t} = \nabla \cdot (C(x, y, t, \Delta x, \Delta y) \nabla I(x, y, t)) \tag{5.153}$$

$$= C(x, y, t, \Delta x, \Delta y) \nabla^2 I(x, y, t) + (\nabla C(x, y, t, \Delta x, \Delta y))^\mathsf{T} \nabla I(x, y, t) \tag{5.154}$$

where $C(x, y, t, \Delta x, \Delta y)$ is the diffusion coefficient at pixel (x, y, t) in the direction $(\Delta x, \Delta y)$, $\nabla \cdot$ is the divergence, ∇ is the spatial gradient, and the second equation is obtained from the first by the product rule of differentiation, noting that $\nabla \cdot \nabla I = \nabla^2 I$. It is easy to see that this anisotropic equation reduces to the isotropic equation of Equation (5.152) if $C(x, y, t, \Delta x, \Delta y) = \frac{1}{2}$, since $\nabla C = 0$ if C is constant. The implementations above arise from the approximation $\frac{\partial I(x, y, t)}{\partial t} \approx I^{(t+1)}(x, y) - I^{(t)}(x, y)$, and by using the 3×3 LoG kernel defined by $\sigma^2 = 0$.

5.5.7 Adaptive Smoothing

Closely related to anisotropic diffusion is the concept of **adaptive smoothing**. Adaptive smoothing simply computes a weighted average of the neighbors of a pixel, with weights that discourage smoothing across boundaries:

$$I^{(t+1)}(x, y) = \frac{\sum_i \sum_j I^{(t)}(x + i, y + j) w^{(t)}(x + i, y + j)}{\sum_i \sum_j w^{(t)}(x + i, y + j)} \tag{5.155}$$

where the summation is conducted over a local neighborhood of the pixel and the weights are defined to be inversely related to the magnitude of the image gradient:

$$w^{(t)}(x, y) \equiv \exp\left(-\frac{\|\nabla I(x, y)\|^2}{2s^2}\right) \tag{5.156}$$

where s is another scaling parameter. Comparing Equation (5.155) with Equation (5.118), it is clear that the bilateral filter is simply a special case of adaptive smoothing, with the particular choice of weights determined by Gaussian spatial and range kernels.

5.6 Grayscale Morphological Operators

Another important class of nonlinear image filters is that of morphological operators. In the previous chapter we discussed binary morphological operators, such as erosion, dilation, opening, and closing. These concepts can be naturally extended to grayscale images. Grayscale morphology is used for a variety of image processing tasks such as feature detection, segmentation, and sharpening.

5.6.1 Grayscale Dilation and Erosion

Recall from the previous chapter that a binary image can be viewed as a set of ON pixels, so that the dilation or erosion of a binary image is given by the union or intersection of translated versions of the structuring element placed within the image:

$$I \oplus B = \bigcup_{\delta \in B} I_\delta = \bigcup_{\delta \in I} B_\delta = \{\boldsymbol{\delta} : \check{B}_\delta \cap I \neq \emptyset\} \qquad (\text{binary or grayscale}) \qquad (5.157)$$

$$I \ominus B = \bigcap_{\delta \in \check{B}} I_\delta = \bigcap_{\delta \notin I} \neg \check{B}_\delta = \{\boldsymbol{\delta} : B_\delta \subseteq I\} \qquad (\text{binary or grayscale}) \qquad (5.158)$$

where $\boldsymbol{\delta} = (\delta_x, \delta_y)$, and we have shown both center-in and center-out formulations for completeness.

The extension to grayscale images is straightforward: just as a binary image can be viewed as a set of (x,y) coordinates such that $I(x, y) = $ ON, a grayscale image can be viewed as a set of (x,y,v) coordinates such that $I(x, y) \leq v$. In other words, if we view the grayscale image as a function of (x,y), then the image can be viewed as a set containing all the points at or below the function, as illustrated in Figure 5.23. Similarly, a grayscale structuring element (SE) is the set of points at or below the grayscale function defining the SE. With this expanded understanding of the sets I and B, the definition of grayscale dilation and erosion remains identical to Equations (5.157)–(5.158) above except with $\boldsymbol{\delta} = (\delta_x, \delta_y, \delta_v)$ and with $\check{B}$ flipping not only δ_x and δ_y but also δ_v.

As an alternate view, recall that binary erosion sets the central pixel to ON if *all* of the pixels overlapping the SE are ON; otherwise it sets the central pixel to OFF. Similarly, binary dilation sets the central pixel of a binary image to ON if *any* of the pixels overlapping the reflected SE are ON; otherwise it sets the central pixel to OFF. Equating OFF and ON with the binary values 0 and 1, respectively, it is easy to see that this computation is equivalent to

$$(I \oplus B)(x, y) = \max_{(\delta_x, \delta_y) \in \check{B}} I(x + \delta_x, y + \delta_y) \qquad (\text{binary dilation}) \qquad (5.159)$$

$$(I \ominus B)(x, y) = \min_{(\delta_x, \delta_y) \in B} I(x + \delta_x, y + \delta_y) \qquad (\text{binary erosion}) \qquad (5.160)$$

since 0 is the minimum binary value and 1 is the maximum binary value. Extending these definitions to grayscale images is straightforward:

$$(I \oplus B)(x, y) = \max_{(\delta_x, \delta_y, \delta_v) \in \check{B}} I(x + \delta_x, y + \delta_y) - \delta_v \qquad (\text{grayscale dilation}) \qquad (5.161)$$

Figure 5.23 A grayscale image can be viewed as the set of all 3D points (x, y, v) underneath the graylevel function.

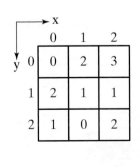

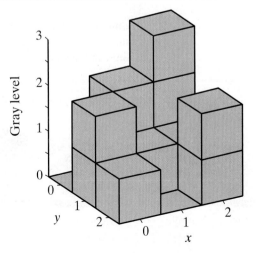

$$(I \,\check{\ominus}\, B)(x, y) = \min_{(\delta_x, \delta_y, \delta_v) \in B} I(x + \delta_x, y + \delta_y) - \delta_v. \quad \text{(grayscale erosion)} \qquad (5.162)$$

These expressions make it obvious what to expect from applying these operators to an image: dilation brightens an image, while erosion darkens it. (Note that in the case of dilation, δ_v in $\check{B}$ means $-\delta_v$ in B, so that Equation (5.161) actually adds the value in B to the image gray level rather than subtracting it.) Note that, for grayscale dilation/erosion to achieve the exact same result as binary dilation/erosion, the foreground and background pixels in the SE must be set to 0 and $-\infty$, respectively. For example, the grayscale versions of the B_4 and B_8 kernels of Equation (4.44) are, respectively

$$\begin{bmatrix} -\infty & 0 & -\infty \\ 0 & 0 & 0 \\ -\infty & 0 & -\infty \end{bmatrix} \quad \text{and} \quad \begin{bmatrix} 0 & 0 & 0 \\ 0 & 0 & 0 \\ 0 & 0 & 0 \end{bmatrix}$$

We distinguish between two kinds of grayscale structuring elements: a **flat SE** has the same value for all pixels in its domain, while a **non-flat SE** does not, where the domain is defined as the set of locations for which the SE has values greater than negative infinity. Thus, for example, the two SEs above are flat. Flat SEs are nearly always used in practice for reasons such as the following: it is difficult to select meaningful values for a non-flat SE, non-flat SEs can yield results outside of the valid range, and non-flat SEs incur significant additional computational burden.

5.6.2 Grayscale Opening and Closing

Grayscale opening and closing are defined in the same way as binary opening and closing:

$$I \bullet B = (I \oplus B) \,\check{\ominus}\, B \quad \text{(grayscale closing)} \qquad (5.163)$$

$$I \circ B = (I \,\check{\ominus}\, B) \oplus B \quad \text{(grayscale opening)} \qquad (5.164)$$

Just as with the binary versions, the grayscale operators are duals of each other with respect to complementation and reflection:

$$\neg(I \oplus B) = \neg I \,\check{\ominus}\, \check{B} \quad \text{(duality of grayscale dilation / erosion)} \qquad (5.165)$$

$$\neg(I \,\check{\ominus}\, B) = \neg I \oplus \check{B} \quad \text{(duality of grayscale dilation / erosion)} \qquad (5.166)$$

$$\neg(I \bullet B) = (\neg I \circ \check{B}) \quad \text{(duality of grayscale closing / opening)} \qquad (5.167)$$

$$\neg(I \circ B) = (\neg I \bullet \check{B}) \quad \text{(duality of grayscale closing / opening)} \qquad (5.168)$$

Opening a grayscale image with an SE removes light details smaller than the SE, while closing removes dark details smaller than the SE. Results of grayscale dilation, erosion, closing, and opening on an image are shown in Figure 5.24.

Figure 5.24 An image, and the result of applying grayscale dilation, erosion, closing, and opening, respectively, using a flat, circular structuring element.

Original image   Grayscale dilation   Grayscale erosion   Grayscale closing   Grayscale opening

5.6.3 Top-Hat Transform

A particularly useful transform that builds on grayscale opening and closing is the **top-hat transform**, illustrated in Figure 5.25. The **white top-hat (WTH)** transform, also known as "top-hat by opening", is the difference between the original image and the grayscale opening of the image and therefore preserves objects that are brighter than their surroundings, while the **black top-hat (BTH)** transform, also known as "top-hat by closing", is the difference between the grayscale closing of the original image and the image itself and therefore preserves objects that are darker than their surroundings:

$$I'_{WTH} = I - (I \circ B) \quad (\text{white top-hat}) \tag{5.169}$$

$$I'_{BTH} = (I \bullet B) - I \quad (\text{black top-hat}) \tag{5.170}$$

The **self-complementary top-hat** is defined as $I'_{WTH} + I'_{BTH}$, and it extracts all image structures that cannot contain the SE whatever their relative contrast. Because grayscale closing (like its binary counterpart) is extensive (assuming that the SE includes the origin) and because grayscale opening (also like its binary counterpart) is anti-extensive (also assuming that the SE includes the origin), the grayscale values in the output of the WTH or BTH are always nonnegative. It can be shown that the WTH is non-increasing and idempotent, while the BTH is neither idempotent nor increasing. The top-hat transform can be used to correct for uneven illumination, with WTH used for dark backgrounds and BTH for bright backgrounds. The top-hat transform with a large isotropic SE acts as a highpass filter, removing the low frequencies of the illumination gradient. Opening with a large SE, on the other hand, tends to remove relevant image structures but preserves the slowly varying illumination function. A simple neighborhood-based contrast operator is to take the image and add the WTH then subtract the BTH, i.e., $(I + I'_{WTH} - I'_{BTH})$. Results of the various top-hat transforms on an image are shown in Figure 5.26.

Figure 5.25 White top-hat (WTH) transform (left), and black top-hat (BTH) transform (right).

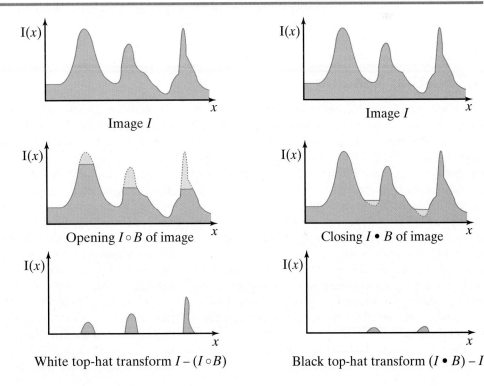

Image *I*

Opening $I \circ B$ of image

White top-hat transform $I - (I \circ B)$

Image *I*

Closing $I \bullet B$ of image

Black top-hat transform $(I \bullet B) - I$

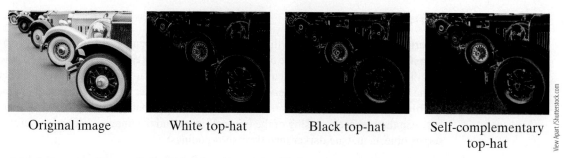

Original image   White top-hat   Black top-hat   Self-complementary
                                                                         top-hat

Figure 5.26 An image, and the result of applying the white top-hat, black top-hat, and self-complementary top-hat transforms, respectively, using a flat, circular structuring element.

5.6.4 Beucher Gradient

Another highpass filter is the **Beucher gradient**, which is defined as the difference between the grayscale dilation of the image and the grayscale erosion. Assuming an isotropic SE, this operator outputs the maximum variation within the disk rather than the slope. Nevertheless, the result of the computation approximates (and becomes equivalent to, as the radius tends toward zero) the norm of the traditional gradient vector. The Beucher gradient is the most common morphological gradient, but two other definitions of the morphological gradient are the half-gradient by erosion (or internal gradient), defined as the difference between the original image and the eroded image, and the half-gradient by dilation (or external gradient), defined as the difference between the dilated image and the original image.

To compute the directional gradient, one can use a line SE instead of an isotropic SE. One must be careful, however, to define the direction of the gradient as the perpendicular to the direction that outputs the minimum directional gradient, rather than the direction that outputs the maximum directional gradient. For example, a line SE applied to a line in the image outputs the same morphological gradient in all directions except for the direction of the line, where the output is zero.

5.7 Further Reading

The concepts of convolution, cross-correlation, and linear time- (or space-) invariant systems can be found in any good signal or image processing book (such as Oppenheim and Schafer [1999] or Jain [1989]), although the distinction between FIR and IIR filters is typically emphasized more in the signal, rather than image, processing literature. While viewing convolution as matrix multiplication is not particularly useful in everyday applications, it does provide a natural means to study the set of all linear transformations of an image. Seitz and Baker [2009] call this concept *filter flow* and present an intriguing approach to estimate various transformations that fall within this set, such as geometric transformations, vignetting, radial distortion, lighting changes, blur, optical flow, and even stereo.

The importance of the Gaussian is widely known and used. For a thorough treatment regarding the evaluation of Gaussian kernels, with some alternate conclusions than those presented here, see Trucco and Verri [1998]. The concept of *cascaded convolution*, in which convolution with a Gaussian kernel is approximated by repeated convolutions with box filters, is due to Wells [1986]. The $O(1)$ computation of Wells's approach relies on the moving average, which is related to the summed-area table of Crow [1984] and the space-variant filtering with arbitrarily sized polynomial kernels of Heckbert [1986]. For a discussion of the need for diagonal convolution to preserve 2D circular symmetry in cascaded convolution, see Rau and McClellan [1997].

An alternate approach to efficient Gaussian convolution, which is also independent of the width of the Gaussian, is to use IIR filters. This approach was first described by Deriche [1987, 1990] and later revisited to avoid having to recompute the coefficients of the filter in Deriche [1993]. A similar approach was adopted by Young and van Vliet [1995]. Yet another approach to efficient (though not constant-time) large-kernel convolution, called *hierarchical discrete correlation*, can be found in Burt [1981]. The integral image is due to Viola and Jones [2004], which is based on earlier work in computer graphics and fits within the more general boxlets framework of Simard et al. [1998].

The Roberts operator can be found in Roberts [1963], which many regard as the first publication in computer vision. Several years later, the Prewitt operator was developed by Prewitt [1970] in the context of biomedical image processing. The Sobel operator was first presented in a talk at the Stanford Artificial Intelligence Project in 1968 by I. Sobel and G. Feldman entitled, "A 3x3 Isotropic Gradient Operator for Image Processing," but the earliest known reference to the operator in print remains Pingle [1969], who attributed it to Sobel. Ironically, this "isotropic" operator is (like other discrete kernels) not really isotropic once discretization effects are considered, leading to the development of the Scharr operator, which is described (in German) by Scharr [2000].

The use of Laplacian of Gaussian in image processing is due to the pioneering work of Marr and Hildreth [1980], who also first proved the equivalency between the Laplacian of Gaussian and the Difference of Gaussians (in the limit). This theory influenced later work on scale space, including that of Witkin [1983] and Lindeberg [1990, 1994] which later influenced the popular SIFT feature detector of Lowe [2004]. The Gaussian pyramid was proposed almost simultaneously by the early papers of Burt [1981] and Crowley [1981], while the Laplacian pyramid is discussed in the works of Burt [1981], Burt and Adelson [1983], and Crowley [1981]. A more recent paper discussing efficient implementation details is that of Crowley et al. [2002].

The efficient $O(w)$ median filter using the graylevel histogram is due to Huang et al. [1979]. A faster $O(\log w)$ algorithm can be found in Weiss [2006], but it is difficult to implement. An even faster, constant-time $O(1)$ algorithm that also makes use of the graylevel histogram and is quite easy to understand can be found in Perreault and Hëert [2007], where additional implementation details are provided to vectorize the computation, enabling the method to achieve approximately the same speed as that of Weiss in practice. The non-local means (NLM) algorithm is due to Buades et al. [2005].

The origins of the bilateral filter can be traced to the nonlinear Gaussian filters of Aurich and Weule [1995] and the SUSAN framework of Smith and Brady [1997],

although it was independently rediscovered by Tomasi and Manduchi [1998], who gave the filter its present name. An overview of the bilateral filter is given by Paris et al. [2009]. To improve computational efficiency, several approaches have been proposed. Durand and Dorsey [2002] use the bilateral grid, whereas Weiss [2006] assumes that the spatial weight kernel is a box function. Porikli [2008] presents a variety of ways to achieve efficient $O(1)$ computation of various forms of the bilateral filter using either integral histograms or the Taylor series expansion to express the Gaussian using power terms of the image. Other approaches are described by Chen et al. [2007] and Yang et al. [2009]. The efficient algorithm presented in this chapter is from Chaudhury et al. [2011], who show how to approximate the Gaussian using raised cosines, which provide a better approximation; in follow-up work, Chaudhury [2011] explains the concept of shiftability. An alternate approach on fully connected graphs using polyhedral lattices is in Adams et al. [2010], and a real-time implementation of the bilateral filter for computational photography applications is described in Rithe et al. [2013].

Anisotropic diffusion is due to Perona and Malik [1990], for which a good reference is Weickert [1998]. Adaptive smoothing is found in Saint-Marc et al. [1991]. Barash [2002] draws the connection between anisotropic diffusion, adaptive smoothing, and bilateral filtering, and fixes adaptive smoothing to make it consistent with the anisotropic diffusion equation; and in follow-up work Barash and Comaniciu [2004] also connect these with mean-shift. The original mean-shift algorithm, which has been called Gaussian blurring mean-shift (GBMS) because the original data values are changed each iteration, can be found in Fukunaga and Hostetler [1975]. Interest in mean-shift was renewed by Cheng [1995], which introduced Gaussian mean-shift (GMS), the version explained in this chapter. The difference between GBMS and GMS, including the superior performance of the latter, is explained by Rao et al. [2009]. Further developments to mean-shift, along with practical applications such as filtering and segmentation, are due to Comaniciu and Meer [2002]. Mean-shift has also been used for tracking by a variety of researchers, such as Comaniciu et al. [2003], Avidan [2005], and Birchfield and Rangarajan [2005].

The top-hat transform is due to Meyer [1979], while the Beucher gradient was introduced in Rivest et al. [1993]. For further information, consult either the book by Serra [1982] or the one by Soille [2003].

PROBLEMS

5.1 Given the 1D kernel $g(x) = \begin{bmatrix} 8 & 4 & 1 & 4 & 9 & 3 & 2 & 4 & 2 \end{bmatrix}$, answer the following: a) What is its width? b) Half-width? c) Zero-based index of the central element?

5.2 Convolve the 1D input signal $f(x) = \begin{bmatrix} 5 & 4 & 0 & 3 & 8 & 2 \end{bmatrix}$ with the kernel $g(x) = \frac{1}{4}\begin{bmatrix} 1 & 2 & 1 \end{bmatrix}$. To properly handle the borders, extend the input by replicating the values, and set the output length to be the same as the input. Would the result of cross-correlation be the same as or different from that of convolution? Explain.

5.3 Convolve the 2D image I with the 2D kernel G, both given below. To properly handle the borders, extend the input by replicating the values, and set the output size to be the same as the input.

$$I = \begin{bmatrix} 5 & 4 & 0 & 3 \\ 6 & 2 & 1 & 8 \\ 7 & 9 & 4 & 2 \end{bmatrix} \quad G = \frac{1}{16}\begin{bmatrix} 1 & 2 & 1 \\ 2 & 4 & 2 \\ 1 & 2 & 1 \end{bmatrix}$$

5.4 Repeat the computation of the previous problem using the separable version of the kernel. First convolve with the horizontal $\frac{1}{4}\begin{bmatrix} 1 & 2 & 1 \end{bmatrix}$, then with the vertical $\frac{1}{4}\begin{bmatrix} 1 & 2 & 1 \end{bmatrix}$.

5.5 Convolve the following grayscale image with a 3×3 Gaussian (computed using Pascal's triangle), minimizing the number of computations used. Handle borders by extension. What is the normalizing constant?

$$\begin{bmatrix} 8 & 2 & 1 & 5 \\ 0 & 1 & 3 & 0 \\ 1 & 0 & 1 & 6 \\ 0 & 4 & 0 & 1 \end{bmatrix}$$

5.6 Write the convolution matrix associated with the convolution kernel $g(x) = \frac{1}{14}\begin{bmatrix} 1 & 3 & 6 & 3 & 1 \end{bmatrix}$. Assume an input of length 5, and that the input is extended by replicating the values.

5.7 For each of the kernels below, specify whether it is a smoothing or a differentiating kernel.

(a) $\frac{1}{32}\begin{bmatrix} 2 & 4 & 6 & 8 & 6 & 4 & 2 \end{bmatrix}$

(b) $\frac{1}{6}\begin{bmatrix} 1 & 2 & 3 & 2 & 1 & 0 & -1 & -2 & -3 & -2 & -1 \end{bmatrix}$

(c) $\frac{1}{9}\begin{bmatrix} 9 & 1 & -1 & -9 \end{bmatrix}$

(d) $\frac{1}{11}\begin{bmatrix} 1 & 9 & 1 \end{bmatrix}$

5.8 For each of the smoothing kernels below, specify the normalizing constant a:

(a) $\frac{1}{a}\begin{bmatrix} 1 & 3 & 6 & 3 & 1 \end{bmatrix}$

(b) $\frac{1}{a}\begin{bmatrix} 22 & 99 & 22 \end{bmatrix}$

(c) $\frac{1}{a}\begin{bmatrix} 3 & 16 & 109 & 16 & 3 \end{bmatrix}$

5.9 For each of the following filters, draw the impulse response, and specify whether it is FIR or IIR, where f is the input and f' is the output.

(a) $f'(x) = f(x - 1) + 2f(x) + f(x + 1)$

(b) $f'(x) = f(x - 1) - f(x + 1)$

(c) $f'(x) = f(x) + f'(x - 1)$

5.10 Is the system $f'(x) = af(x) + b$ linear (in the sense defined in Section 5.1.4), where a and b are scalars? Why or why not?

5.11 Compute the variance of the following kernel: $\begin{bmatrix} 1 & 2 & 3 & 4 & 5 & 4 & 3 & 2 & 1 \end{bmatrix}$.

5.12 Construct the 7×1 Gaussian kernel using the binomial triangle, including the normalization. Compute its variance.

5.13 Construct a 3×1 Gaussian kernel by sampling a continuous Gaussian with $\sigma^2 = 0.4$. What is the resulting variance of the discrete kernel?

5.14 Construct a 1D Gaussian kernel by sampling the Gaussian function with $\sigma = 1.8$. Use Equation (5.39) to determine the length of the kernel, and be sure to normalize properly. What is the standard deviation of the kernel after sampling?

5.15 Is the following kernel separable? If so, then separate into two 1D kernels. If not, then state why not.

$$\begin{bmatrix} 3 & 6 & 9 \\ 6 & 12 & 18 \\ 12 & 24 & 36 \end{bmatrix}$$

5.16 Write the matrices $\mathbf{G}_v$ and $\mathbf{G}_h$ for applying the horizontal kernel $g_h = \begin{bmatrix} 1 & 2 & 1 \end{bmatrix}^{\mathsf{T}}$ and vertical kernel $g_v = \begin{bmatrix} 3 & 5 & 3 \end{bmatrix}^{\mathsf{T}}$ as the matrix multiplication in Equation (5.38) to a 4×4 image with reflection.

5.17 Another way to compute the partial derivatives of a 2D array of pixel values is to fit a plane $I(x, y) \approx ax + by + c$ to the values (in the least squares sense), then compute the partial derivatives of the plane.

(a) Show that for a 3×3 array, the result of this procedure is identical to applying the Prewitt kernels.

(b) What are the results of applying this procedure to 2×2, 4×4, or 5×5 arrays?

(c) Show that the magnitude of the gradient computed using the Roberts' cross operator is the magnitude of the gradient computed using a 2×2 array multiplied by the factor $\sqrt{2}$.

5.18 Show that the Sobel kernels are equivalent to the convolution of 2×2 finite difference kernels along both axes with a 2×2 box filter.

5.19 We saw in Section 5.1.2 that a linear shift-invariant system can be represented as matrix multiplication. Does the same hold true for a linear shift-varying system? If so, then what property of such a matrix indicates whether a linear system is shift-invariant or not?

5.20 A scanline of a grayscale image has the following values: $\begin{bmatrix} 7 & 5 & 3 & 2 & 5 & 0 & 8 & 9 \end{bmatrix}$. Convolve this scanline with a) Gaussian, b) Gaussian derivative, and c) Gaussian second-derivative kernels, all with $\sigma^2 = 0.5$. Handle borders with reflection.

5.21 Explain why there is a) only one 3×1 Gaussian derivative kernel, and b) only one 3×1 Gaussian second-derivative kernel.

5.22 Smoothing a digital image is similar to defocusing the lens of the camera, because both approaches result in a blurred image. Answer the following:

(a) Can you think of any differences between the results of the two approaches? (*Hint:* Consider what happens to the image of a bright light.)

(b) What shape should the convolution kernel be to simulate the defocusing ability of a lens?

5.23 Show that magnitude of the gradient is not isotropic when implemented discretely, by comparing and contrasting the Euclidean, Manhattan, and chessboard versions on the two images below, using the Prewitt, Sobel, and Scharr operators. Which operator, and which metric, yields the most consistent behavior on these two inputs?

$$\begin{bmatrix} 1 & 1 & 0 \\ 1 & 1 & 0 \\ 1 & 1 & 0 \end{bmatrix} \quad \begin{bmatrix} 1 & 0 & 0 \\ 1 & 1 & 0 \\ 1 & 1 & 1 \end{bmatrix}$$

5.24 Prove that the output of a discrete linear shift-invariant system is the convolution of the input signal and the impulse response. *Hint:* Express the input as the sum of weighted Kronecker delta functions, then apply the additivity, scaling, and shift-invariant properties.

5.25 The isotropic Gaussian is the only rotationally symmetric 2D function that is separable. Prove that it is rotationally symmetric. (*Hint:* A 2D function is rotationally symmetric if and only if $(\partial f / \partial x) y = (\partial f / \partial y) x$.)

5.26 We have argued that in general, separable convolution is more efficient than non-separable convolution. Can you imagine a scenario in which the separable implementation might actually be slower?

5.27 A 3×3 Laplacian of Gaussian (LoG) kernel applied to the image below should equal 1, since the change in slope along the x direction is 1, and the change in slope along the y direction is 0. (That is, the slope between the first and second columns is zero, and the slope between the second and third columns is one, so the change in slope is $1 - 0 = 1$.) Select four of the LoG kernels introduced in this chapter, and show that they all satisfy this criterion.

$$\begin{bmatrix} 0 & 0 & 1 \\ 0 & 0 & 1 \\ 0 & 0 & 1 \end{bmatrix}$$

5.28 Is the following a LoG kernel? Why or why not?

$$\frac{1}{3} \begin{bmatrix} 2 & -1 & 2 \\ -1 & -4 & -1 \\ 2 & -1 & 2 \end{bmatrix}$$

5.29 Table 5.5 lists a number of different LoG kernels. Show that the following is also a LoG kernel, and compute its variance.

$$\frac{1}{8} \begin{bmatrix} 1 & 6 & 1 \\ 6 & -28 & 6 \\ 1 & 6 & 1 \end{bmatrix}$$

5.30 Derive the expression for the third-derivative Gaussian kernel, $\frac{d^3}{dx^3} gauss_{\sigma^2}(x)$.

5.31 Given the smoothing kernel $gauss_{0.125} = \frac{1}{16} \begin{bmatrix} 1 & 14 & 1 \end{bmatrix}$, calculate the associated 3×3 LoG kernel.

5.32 Prove that the convolution of two Gaussians is a Gaussian. For simplicity, assume continuous 1D signals.

5.33 Another way to verify the normalization of a LoG kernel is to convolve the kernel with the paraboloid $x^2 + y^2$ and ensure that the result equals 4, since $\nabla^2(x^2 + y^2) = \frac{\partial^2}{\partial x^2}(x^2 + y^2) + \frac{\partial^2}{\partial y^2}(x^2 + y^2) = 2 + 2 = 4$.

(a) What is the 3×3 image that results from sampling the central part of the paraboloid $x^2 + y^2$?

(b) Using this image, select four of the LoG kernels introduced in this chapter, and show that they all satisfy this criterion.

(c) Would the result be different if the image were obtained by sampling a non-central part of the paraboloid? Why or why not?

5.34 We mentioned that convolution must not be performed in place. Show an example where performing convolution in place erases nearly all image information.

5.35 Prove that convolution is commutative.

5.36 State the 3×3 convolution kernel whose effect is to shift the image to the right by one pixel.

5.37 Describe a scenario in which the normalization factors associated with Gaussian derivatives are important.

5.38 What filter would you use to remove salt-and-pepper noise? Is this a linear or non-linear filter?

5.39 Write pseudocode for implementing convolution with a box kernel of arbitrary size, computing the running sum to ensure that the procedure is computationally efficient.

5.40 In some applications, it is important to reverse the effects of convolution. If the kernel is known, how might this deconvolution be performed?

5.41 Show that the application of a 2D LoG kernel to a constant or ramp graylevel function yields zero.

5.42 Verify that convolving the signal $\begin{bmatrix} 6 & 8 & 3 \end{bmatrix}$ with the kernel $\frac{1}{4}\begin{bmatrix} 1 & 2 & 1 \end{bmatrix}$ twice is identical to convolving the signal with $\frac{1}{16}\begin{bmatrix} 1 & 4 & 6 & 4 & 1 \end{bmatrix}$.

5.43 Compute the integral image of the following grayscale image. Use the integral image to compute the sum of the inner 3×3 array of pixels.

$$
\begin{bmatrix}
218 & 87 & 246 & 63 & 175 \\
106 & 161 & 231 & 32 & 207 \\
16 & 141 & 136 & 140 & 202 \\
86 & 253 & 55 & 112 & 188 \\
73 & 85 & 165 & 209 & 99
\end{bmatrix}
$$

5.44 Compute the a) grayscale dilation and grayscale erosion of the 5×5 image shown in the previous question, using a 3×3 flat SE consisting of all zeros (the grayscale version of B_8). Then compute b) the grayscale closing and opening, and c) the white top-hat and black top-hat transforms. Handle borders with reflection.

5.45 In an attempt to remove noise, an image is convolved with a 3×3 Gaussian kernel composed from two 1D kernels, namely, the horizontal $gauss_{0.5}$ kernel and the vertical $gauss_{0.5}$ kernel. Then, to differentiate, the resulting smoothed image is convolved with horizontal $\frac{1}{2}\begin{bmatrix} 1 & 0 & -1 \end{bmatrix}$ and vertical $\frac{1}{2}\begin{bmatrix} 1 & 0 & -1 \end{bmatrix}$ kernels. Write the equivalent 2D kernels that, if the original image were convolved with them, would yield the same result. What do you notice about this kernel that is undesirable? What are the implications regarding smoothing before differentiating?

5.46 Write code in your favorite language to construct the Gaussian, first-derivative, and second-derivative kernels, all parameterized by the standard deviation σ. Then write code to apply these kernels to a grayscale image, computing the smoothed image, the gradient components in x and y, the gradient magnitude, and the LoG.

5.47 Implement both the bilateral and mean-shift filters and apply them to a grayscale image. Compare and contrast the two algorithms. Then apply them separately to the color channels of an RGB image. Describe the output that results.

CHAPTER 6
Frequency-Domain Processing

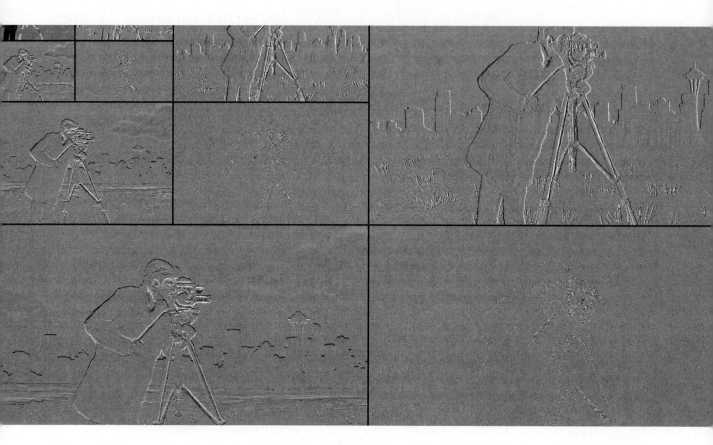

I n the previous chapter we considered ways to transform an image by filtering in the spatial domain. A complementary approach is to filter in the frequency domain using the well-known principle that convolution in the spatial domain is equivalent to multiplication in the frequency domain. In this chapter we discuss the Fourier transform in an effort to explore this concept of the frequency domain, particularly as it relates to discrete signals. We then examine frequency-domain approaches to filtering and their connection with spatial-domain filtering.

6.1 Fourier Transform

Suppose we have a one-dimensional continuous signal $g(t)$, such as an audio signal containing speech or music. Oftentimes we want to be able to analyze such a signal by determining which frequencies are present. Such information can be used in a variety of ways, for example to *classify* the signal (whether it contains primarily high or low frequencies) or to *filter* the signal (to remove, for example, high-frequency noise or a low-frequency hum).

6.1.1 Forward Transform

The standard technique for performing frequency analysis of a signal $g(t)$ is to compute its **Fourier transform**[†] $G(f)$, which is defined as the integration of the signal after first multiplying by a certain complex exponential:[‡]

$$G(f) \equiv \mathcal{F}\{g\} \equiv \int_{-\infty}^{\infty} g(t)e^{-j2\pi ft}\, dt \tag{6.1}$$

where $\mathcal{F}$ indicates the Fourier transform, t indicates time (the domain of the original signal), f indicates frequency (the domain of the transformed signal), and $j \equiv \sqrt{-1}$. If t is measured in seconds, then f is measured in inverse seconds, also known as hertz ($1\text{ Hz} = 1/\text{sec}$).

This equation, which at first glance may seem intimidating, can be made more understandable by applying **Euler's formula**, $e^{j\theta} = \cos\theta + j\sin\theta$:

$$G(f) = \underbrace{\int_{-\infty}^{\infty} g(t)\,\cos 2\pi ft\, dt}_{G_{even}} + j\underbrace{\int_{-\infty}^{\infty} -g(t)\,\sin 2\pi ft\, dt}_{G_{odd}} \tag{6.2}$$

or $G(f) = G_{even}(f) + jG_{odd}(f)$. Here we see that for any given frequency f, we obtain a measure indicating the presence of that frequency in the signal by multiplying the signal by a cosine and sine with frequency f and integrating them separately. Using complex numbers is just a convenient way of allowing us to express two separate quantities, G_{even} and G_{odd}, in a single equation. We could just as easily have defined the Fourier transform to be a pair of numbers for each frequency: $G(f) \equiv (G_{even}(f), G_{odd}(f))$, but this definition would lose some of the elegant mathematics that comes free when we use complex numbers.

The reason we call these two numbers G_{even} and G_{odd} is that the former captures the frequency information in a signal with even symmetry, while the latter captures the frequency information in a signal with odd symmetry. That is, $G_{odd} = 0$ for any signal with $g(t) = g(-t)$ for all t, and $G_{even} = 0$ for any signal with $g(t) = -g(-t)$ for all t. For a signal that has neither even nor odd symmetry, the two numbers together capture the frequency information.

We say that the function g is a **time-domain** signal if its domain is time (e.g., an audio signal), or a **spatial-domain** signal if its domain is some spatial coordinate (e.g., an image graylevel function along a row or column of a camera's imaging sensor). In either case the Fourier transform is the standard way to convert the original signal into the frequency domain, and the math is the same for both. The resulting **frequency-domain** representation of the signal is like a reverse phone book in which the entries are sorted by phone number rather than by name. This alternate representation makes it easy to discover information that is hidden in the original signal, such as which frequencies are present.

To see how the Fourier transform works, let us consider a simple example.

EXAMPLE 6.1

Compute the Fourier transform of $g(t) = \cos 2000\pi t$.

Solution

This continuous signal is a pure even sinusoid with frequency $f = 1000\text{ Hz} = 1\text{ kHz}$. We expect, therefore, that the Fourier transform will contain only real values ($G_{odd} = 0$ because the signal has even symmetry), and that the Fourier transform will somehow indicate that

---

[†] Joseph Fourier (1768–1830) was a French mathematician and physicist who also played a key, if indirect, role in deciphering the Rosetta Stone.

[‡] In this chapter we depart from our usual practice of using capital letters to indicate 2D functions, in order to follow the common notation of using capital letters to denote frequency-domain signals.

the signal contains only a single frequency. From Euler's formula it is not hard to show that $\cos\theta = \frac{1}{2}(e^{j\theta} + e^{-j\theta})$, which plugs into Equation (6.1) to yield

$$G(f) = \int_{-\infty}^{\infty} \frac{1}{2}(e^{j2000\pi t} + e^{-j2000\pi t})e^{-j2\pi ft}\,dt \tag{6.3}$$

$$= \frac{1}{2}\int_{-\infty}^{\infty}(e^{j2\pi(1000-f)t} + e^{-j2\pi(1000+f)t})\,dt \tag{6.4}$$

$$= \frac{1}{2}(\delta(f - 1000) + \delta(f + 1000)) \tag{6.5}$$

where $\delta(f)$ is the **Dirac delta function**, which is defined (informally at least) as an infinite spike at the origin with unit area. In other words,

$$\delta(f - f_0) = \begin{cases} \infty & \text{if } f = f_0 \\ 0 & \text{otherwise} \end{cases} \tag{6.6}$$

and $\int_{-\infty}^{\infty}\delta(f)\,df = 1$. This Fourier transform pair is illustrated in Figure 6.1. In case you are wondering why there is a spike at both the positive and negative frequencies (or, rather, what is the meaning of a negative frequency), consider a spinning wheel. The Fourier transform captures the frequency at which the wheel spins, but it cannot distinguish whether the wheel spins clockwise or counterclockwise; the positive and negative frequencies indicate these two possibilities.

Although deriving Equation (6.5) from Equation (6.4) is not trivial, when we consider the inverse Fourier transform below we will show that Equation (6.5) is indeed the correct answer. Additional intuition can be gained by simply substituting values: for example, it is not surprising that $G(f)$ blows up at $f = 1000$, since $e^{j2\pi(1000-f)t} = e^0 = 1$, leading to $\int_{-\infty}^{\infty}1\,dt$, which is unbounded; similarly, it is not surprising that $G(f) = 0$ if $f \neq 1000$, since $e^{j2\pi(1000-f)t}$ is just a complex exponential, and the oscillations of sine and cosine functions cause their positive and negative portions to cancel each other when integrated.

Figure 6.1 A continuous time-domain signal (left) and its Fourier transform (right). The latter reveals that the signal is a pure sinusoid with frequency 1000 Hz, since it contains two infinite spikes (Dirac deltas) at $f = 1$ kHz and $f = -1$ kHz. Note that the multiplicative factor $\frac{1}{2}$ has no effect on the display. See the text for an explanation of the negative frequency.

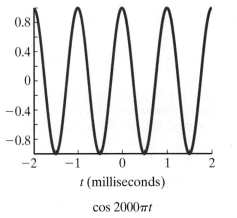

t (milliseconds)

$\cos 2000\pi t$

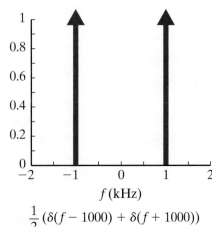

f (kHz)

$\frac{1}{2}(\delta(f - 1000) + \delta(f + 1000))$

6.1.2 Inverse Transform

One of the convenient properties of the Fourier transform is that it is reversible. That is, the original signal can be recovered from its frequency-domain representation by computing the **inverse Fourier transform**, which surprisingly is defined in exactly the same way as the forward Fourier transform except for the sign in the exponent, and the fact that the integral is computed over frequency rather than over time:

$$g(t) = \mathcal{F}^{-1}\{G\} \equiv \int_{-\infty}^{\infty} G(f)e^{j2\pi ft}\, df \tag{6.7}$$

For all practical purposes (that is, under rather mild mathematical assumptions), the forward and inverse transforms cancel each other, i.e., $\mathcal{F}^{-1}\{\mathcal{F}\{g\}\} = g$ and $\mathcal{F}\{\mathcal{F}^{-1}\{G\}\} = G$.

EXAMPLE 6.2

Compute the inverse Fourier transform of $G(f) = \frac{1}{2}(\delta(f - 1000) + \delta(f + 1000))$.

Solution

To solve this problem, we make use of the **sifting property** of the Dirac delta function, namely,

$$\int_{-\infty}^{\infty} h(f)\delta(f - f_0)\, df = h(f_0) \tag{6.8}$$

for any function $h(f)$. The sifting property is easy to see, since the Dirac delta function multiplies the entire function by zero except for the value at $f = f_0$. Plugging into Equation (6.7) yields

$$g(t) = \int_{-\infty}^{\infty} \frac{1}{2}(\delta(f - 1000) + \delta(f + 1000))e^{j2\pi ft}\, df \tag{6.9}$$

$$= \frac{1}{2}\int_{-\infty}^{\infty} \delta(f - 1000)e^{j2\pi ft}\, df + \frac{1}{2}\int_{-\infty}^{\infty} \delta(f + 1000)e^{j2\pi ft}\, df \tag{6.10}$$

$$= \frac{1}{2}e^{j2000\pi t} + \frac{1}{2}e^{-j2000\pi t} \tag{6.11}$$

$$= \cos 2000\pi t \tag{6.12}$$

where the last equality arises from Euler's formula. Thus we see that the forward transform of the previous example, which is difficult to derive analytically, is readily obtained by considering the problem in reverse.

6.1.3 Sampling and Aliasing

When a continuous signal is sampled, it becomes a discrete signal. It would be natural to assume that some information is lost in the process of sampling, thereby making it impossible to reconstruct the original signal from its samples. In fact, however, according to the **Nyquist-Shannon sampling theorem,**[†] it can be shown that if a certain condition holds true, then the discrete samples contain just as much information as the original signal, so that the original signal can be reconstructed *exactly* from the discrete samples. This condition is that the sampling rate must be greater than the **Nyquist rate**, which is twice the

---

[†] H. Nyquist (1889–1976) and C. Shannon (1916–2001) were pioneers in information theory at AT&T Bell Labs. The latter's Master's thesis is sometimes considered the most important such work ever produced.

highest frequency in the signal.[‡] A signal that contains such a maximum frequency is called a **band-limited signal**, so this theorem only applies to band-limited signals.

Figure 6.2 shows a continuous signal sampled at three different frequencies. When the sampling frequency is greater than the Nyquist rate, we say that the signal is **oversampled**, in which case perfect reconstruction is possible. When the sampling frequency is lower than the Nyquist rate, the signal is **undersampled**, and important information about the signal is irrecoverably lost. Similarly, when the sampling frequency is exactly the Nyquist rate (i.e., two samples for each period of the highest frequency), the signal is **critically sampled**, and the original signal is also (just barely) unrecoverable.

When a signal is undersampled, **aliasing** occurs. An *alias* is an assumed name, so it is as if the high frequency (which is higher than half the sampling rate) shows up as a different frequency. If f is the frequency of the signal being sampled, and f_s is the sampling rate, then the aliased frequency is given by $|f_s n - f|$, where $n = \text{Round}(f/f_s)$. (Note that if $f_s > 2f$, then $n = 0$ and the aliased frequency is identical to the actual frequency, i.e., there is no aliasing.) In the right side of Figure 6.2, for example, $f = 1000$ and $f_s = 1250$, so the aliased frequency is $|1250 \cdot 1 - 1000| = 250$, because $n = \text{Round}(1000/1250) = 1$. In other words, when the signal is sampled at $t = 0.8$ ms, it has undergone 0.8 periods. The value at this time is exactly what would have been obtained by sampling a 250 Hz signal, which would have undergone only 0.2 periods. At this sampling rate, therefore, it is impossible to tell a 1 kHz signal (the blue curve in the figure) from a 250 Hz signal (the red curve). This phenomenon is readily seen in old Western movies, where, as a wagon speeds up, its wheels appear at first to speed up, then slow down, then rotate in the opposite direction. For this reason, aliasing is also known as the "wagon wheel effect."

6.1.4 Four Versions of the Fourier Transform

The Fourier transform introduced in Equation (6.1) is actually one of several variations of the concept. As we have just seen, signals can be either continuous or discrete, and they can be defined everywhere (infinite duration) or over a limited domain (finite duration). These choices lead to four versions of the Fourier transform, as shown in Table 6.1. The version we have considered so far is applicable to continuous, infinite duration signals, in which both the time and frequency values are defined for all real numbers. The discrete-time Fourier transform

Figure 6.2 A continuous 1 kHz time-domain sinusoid sampled with 3 different sampling frequencies: 5000 Hz (left), 2000 Hz (middle), and 1250 Hz (right). The Nyquist rate, which is twice the frequency of the signal, is 2000 Hz. Sampling at higher than the Nyquist rate preserves the information in the signal, while sampling at lower than the Nyquist rate leads to aliasing. In this case the frequency of the aliased signal (red curve) is 250 Hz.

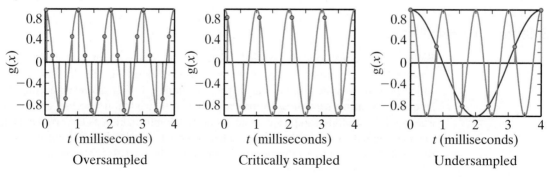

Oversampled   Critically sampled   Undersampled

[‡] Do not confuse the *Nyquist rate* (twice the highest frequency), which is a property of the signal, with the *Nyquist frequency* (half the sampling rate), which is a property of the sampling system.

	continuous	discrete
infinite duration	**Fourier transform** $$G(f) = \int_{-\infty}^{\infty} g(t)e^{-j2\pi ft}\,dt$$ $$g(t) = \int_{-\infty}^{\infty} G(f)e^{j2\pi ft}\,df$$ $$t \in \mathbb{R} \quad f \in \mathbb{R}$$	**Discrete-time Fourier transform (DTFT)** $$G(f) = \sum_{x=-\infty}^{\infty} g(x)e^{-j2\pi fx}$$ $$g(x) = \int_{-\frac{1}{2}}^{\frac{1}{2}} G(f)e^{j2\pi fx}\,df$$ $$x \in \mathbb{Z} \quad f \in \left[-\tfrac{1}{2},\tfrac{1}{2}\right]$$
finite duration (periodic)	**Fourier series** $$G(k) = \frac{1}{T}\int_{-\frac{T}{2}}^{\frac{T}{2}} g(t)e^{-j2\pi kt/T}\,dt$$ $$g(t) = \sum_{k=-\infty}^{\infty} G(k)e^{j2\pi kt/T}$$ $$t \in \left[-\tfrac{T}{2},\tfrac{T}{2}\right] \quad k \in \mathbb{Z}$$	**Discrete Fourier transform (DFT)** $$G(k) = \sum_{x=0}^{w-1} g(x)e^{-j2\pi kx/w}$$ $$g(x) = \frac{1}{w}\sum_{k=0}^{w-1} G(k)e^{j2\pi kx/w}$$ $$x \in \mathbb{Z}_{0:w-1} \quad k \in \mathbb{Z}_{0:w-1}$$

TABLE 6.1 The four versions of the Fourier transform.

(DTFT), on the other hand, applies to signals defined only for discrete values of the domain (but still extending forever), in which case the frequencies are defined only up to the value of $\frac{1}{2}$ due to the Nyquist-Shannon sampling theorem just mentioned. In the Fourier series, the roles are reversed, so that the continuous signal is represented as an infinite sum of weighted sinusoids. Finally, the discrete Fourier transform (DFT) applies to signals that have been sampled a finite number of times, so that both the samples and the frequencies are discrete and finite.

6.2 Discrete Fourier Transform (DFT)

In this chapter we focus our attention primarily on the last of the four versions, namely, the **discrete Fourier transform (DFT)**. The DFT is arguably the most practical of the versions, since to be stored in a digital computer a continuous signal must be sampled a finite number of times. As a result, all real-world signals are stored as discrete, finite-duration signals, so that if you ever run across the Fourier transform of a real-world signal, you are probably looking at a DFT. Moreover, the discrete mathematics behind the DFT is much simpler than that of the continuous Fourier transform, making it much easier to establish results and recognize connections between different aspects of the theory. One of the nice properties of the DFT is that, unlike some of the other versions, it always exists.[†]

6.2.1 Forward Transform

Let $g(x)$ be a 1D discrete signal with w samples. The DFT of g is defined as the summation of the signal after multiplying by a certain complex exponential:

$$G(k) = \mathcal{F}\{g(x)\} = \sum_{x=0}^{w-1} g(x)e^{-j2\pi kx/w} \tag{6.13}$$

where x and k are integers. Recognizing the similarity between this equation and the continuous version in Equation (6.1) , we see that the discrete version replaces the integral with a summation, and f with $\frac{k}{w}$, so that the latter plays the role of a discrete frequency. Similarly, in the discrete domain the sifting property is achieved with the **Kronecker delta function**, defined as

---

[†] The Fourier series, for example, only exists for signals that satisfy the *Dirichlet conditions*.

$$\delta(k - k_0) = \begin{cases} 1 & \text{if } k = k_0 \\ 0 & \text{otherwise} \end{cases} \tag{6.14}$$

The DFT takes a discrete signal consisting of w samples ($x = 0, 1, 2, \ldots, w - 1$) and produces an output also consisting of w samples ($k = 0, 1, 2, \ldots, w - 1$). Typically, the input signal is real-valued, whereas the output is complex-valued due to the use of complex exponentials. As with the continuous version, Equation (6.13) can be rewritten by noting that Euler's formula, $e^{j\theta} = \cos \theta + j \sin \theta$, allows us to express the complex exponentials in terms of sines and cosines:

$$G(k) = \mathcal{F}\{g(x)\} = \sum_{x=0}^{w-1} g(x)\left(\cos \frac{2\pi k}{w}x - j \sin \frac{2\pi k}{w}x\right) \tag{6.15}$$

where again it is obvious that $f = \frac{k}{w}$. This formula leads to a straightforward implementation for computing the DFT, presented in Algorithm 6.1. Although this pseudocode is perfectly valid, it is not widely used due to its inefficiency. A more efficient algorithm is the **fast Fourier transform (FFT)**, which by a clever trick reuses intermediate computations to reduce the running time from $O(w^2)$ to $O(w \log w)$ — a substantial improvement. All modern implementations of the DFT use some variation of the FFT algorithm, and most versions of the FFT algorithm require the length of the signal to be a power of 2, i.e., $w = 2^n$ where n is an integer. If this condition does not hold, then the signal is zero-padded to increase its length to the next power of 2.

6.2.2 Inverse Transform

Like the continuous Fourier transform, the DFT is reversible. That is, given the DFT of a signal, the original signal can be recovered by applying the **inverse DFT**:

$$g(x) = \mathcal{F}^{-1}\{G(k)\} = \frac{1}{w}\sum_{k=0}^{w-1} G(k)e^{j2\pi kx/w} \tag{6.16}$$

ALGORITHM 6.1 Compute the DFT of a 1D signal (slow version)

DISCRETEFOURIERTRANSFORM($g[0], \ldots, g[w - 1]$)

Input: real 1D signal g of length w
Output: real (G_{even}) and imaginary (G_{odd}) components of the DFT of g

```
1   for k ← 0 to w − 1 do
2       G_even[k] ← 0
3       G_odd[k] ← 0
4       f ← k/w
5       for x ← 0 to w − 1 do
6           G_even[k] ← G_even[k] + g[x] * cos(2 * π * f * x)
7           G_odd[k] ← G_odd[k] − g[x] * sin(2 * π * f * x)
8   return G_even, G_odd
```

Again, the inverse transform is identical to the forward transform except for the sign of the exponential and the summation variable. The scaling factor $\frac{1}{w}$ is needed to ensure that the two transforms are inverses of each other, that is, $\mathcal{F}^{-1}\{\mathcal{F}\{g\}\} = g$ and $\mathcal{F}\{\mathcal{F}^{-1}\{G\}\} = G$. But this factor may be placed in either transform, or alternatively $\frac{1}{\sqrt{w}}$ may be placed in front of both; what is important is that the multiplication of the two numbers is equal to $\frac{1}{w}$. The implementation of the inverse transform is made clear using Euler's formula:

$$g(x) = \mathcal{F}^{-1}\{G(k)\} = \frac{1}{w}\sum_{k=0}^{w-1}G(k)\left(\cos\frac{2\pi k}{w}x + j\sin\frac{2\pi k}{w}x\right) \tag{6.17}$$

$$= \frac{1}{w}\sum_{k=0}^{w-1}(G_{even}(k) + jG_{odd}(k))\left(\cos\frac{2\pi k}{w}x + j\sin\frac{2\pi k}{w}x\right) \tag{6.18}$$

$$= g_{real}(x) + jg_{imag}(x) \tag{6.19}$$

where

$$g_{real}(x) = \frac{1}{w}\sum_{k=0}^{w-1}G_{even}(k)\cos\frac{2\pi k}{w}x - G_{odd}(k)\sin\frac{2\pi k}{w}x \tag{6.20}$$

$$g_{imag}(x) = \frac{1}{w}\sum_{k=0}^{w-1}G_{odd}(k)\cos\frac{2\pi k}{w}x + G_{even}(k)\sin\frac{2\pi k}{w}x = 0 \tag{6.21}$$

Note that if $g(x)$ is real, then $g_{imag}(x) = 0$ for all x, so that the imaginary component can be discarded while computing the inverse DFT. This leads to Algorithm 6.2 as one way to compute the inverse DFT. In practice the inverse FFT algorithm should be used for efficiency.

6.2.3 Properties

Several important properties of the DFT are fairly straightforward to prove from the definition:

- The DFT is **linear**. That is, if $G(k) = \mathcal{F}\{g(x)\}$ and $H(k) = \mathcal{F}\{h(x)\}$ are the Fourier transforms of two signals, then the Fourier transform of a weighted combination of the signals is simply the weighted combination of their Fourier transforms, using the same weights:

$$\mathcal{F}\{ag(x) + bh(x)\} = a\mathcal{F}\{g(x)\} + b\mathcal{F}\{h(x)\} \tag{6.22}$$

which follows from the definition of the DFT:

$$\mathcal{F}\{ag(x) + bh(x)\} = \sum_{x=0}^{w-1}(ag(x) + bh(x))e^{-j2\pi kx/w} \tag{6.23}$$

$$= a\sum_{k=0}^{w-1}g(x)e^{-j2\pi kx/w} + b\sum_{k=0}^{w-1}h(x)e^{-j2\pi kx/w} \tag{6.24}$$

$$= a\mathcal{F}\{g(x)\} + b\mathcal{F}\{h(x)\} \tag{6.25}$$

ALGORITHM 6.2 Compute the inverse DFT of a 1D signal (slow version)

DISCRETEFOURIERTRANSFORMINVERSE $(G_{even}[0], \ldots, G_{even}[w-1], G_{odd}[0], \ldots, G_{odd}[w-1])$

Input: real (G_{even}) and imaginary (G_{odd}) components of the DFT of a real 1D signal
Output: real 1D signal g_{real} of length w whose DFT is $G_{even} + jG_{odd}$

```
1   for x ← 0 to w − 1 do
2       g_real[x] ← 0
3       for k ← 0 to w − 1 do
4           f ← k/w
5           g_real[x] ← g_real[x] + 1/w (G_even[x] * cos(2 * π * f * x) + G_odd[x] * sin(2 * π * f * x))
6   return g_real
```

A compact way to state the linearity property is

$$g(x) \overset{DFT}{\Longleftrightarrow} G(k) \tag{6.26}$$

$$h(x) \overset{DFT}{\Longleftrightarrow} H(k) \tag{6.27}$$

$$ag(x) + bh(x) \overset{DFT}{\Longleftrightarrow} aG(k) + bH(k) \tag{6.28}$$

- The DFT is **periodic**. Given an input discrete signal of w samples, the DFT also is composed of w (possibly complex) samples. By convention, this DFT is defined as $G(k), k = 0, \ldots, w - 1$. But what if we try to evaluate Equation (6.13) for some arbitrary value of k? It is easy to show that $G(k) = G(k + nw)$ for any integer n, since

$$e^{-j2\pi(k+nw)x/w} = e^{-j2\pi kx/w} \cdot e^{-j2\pi nx} = e^{-j2\pi kx/w} \tag{6.29}$$

where the last equality uses Euler's formula to deduce $e^{-j2\pi nx} = \cos 2\pi nx - j \sin 2\pi nx = 1$, since n and x are both integers. As a result, the DFT is periodic. This property of **periodicity** means that even though $G(k)$ can be evaluated for any value of k, it can be uniquely represented using the same number of samples as the original signal, since G is simply replicated forever in both directions. Similarly, the function $\mathcal{F}^{-1}\{G(k)\}$ is periodic with the same period w and can be evaluated for any value of x, even though the original signal is defined only for $x = 0, \ldots, w - 1$. Another way to look at this is that the DFT assumes that the original signal is defined over all possible integers but is periodic outside the values given. The periodicity of the original signal and the DFT are illustrated in Figure 6.3. In other words,

$$g(x + nw) = g(x) \overset{DFT}{\Longleftrightarrow} G(k) = G(k + nw), \quad x, k, n, w \in \mathbb{Z} \tag{6.30}$$

EXAMPLE 6.3	Is each of the following discrete signals symmetric about the origin: $\begin{bmatrix} 1 & \underline{2} & 1 \end{bmatrix}$, $\begin{bmatrix} \underline{4} & 3 & 3 \end{bmatrix}$, and $\begin{bmatrix} \underline{5} & 0 & 5 \end{bmatrix}$? (Recall that the underscore indicates the origin.)
Solution	The first signal, $\begin{bmatrix} 1 & \underline{2} & 1 \end{bmatrix}$, is obviously symmetric about the origin. Due to the periodicity property of the DFT, the second signal $\begin{bmatrix} \underline{4} & 3 & 3 \end{bmatrix}$ can be thought of as extending forever in all directions, i.e., $\begin{bmatrix} \cdots & 3 & 3 & \underline{4} & 3 & 3 & 4 & 3 & 3 & 4 & 3 & 3 & \cdots \end{bmatrix}$, which is the same as $\begin{bmatrix} 3 & \underline{4} & 3 \end{bmatrix}$, which also is symmetric about the origin. Applying the periodicity property to the third signal $\begin{bmatrix} \underline{5} & 0 & 5 \end{bmatrix}$, we see that it is equivalent to $\begin{bmatrix} \cdots & 5 & 0 & 5 & \underline{5} & 0 & 5 & \cdots \end{bmatrix}$, which is not symmetric about the origin.

Figure 6.3 Periodicity of the DFT. The discrete signal consisting of eight samples $x = 0, \ldots, 7$ (red, left) gives rise to the DFT consisting of eight samples $k = 0, \ldots, 7$ (red, right). If the DFT is evaluated for other values of k, or if the inverse DFT of the DFT is evaluated for other values of x, the signal repeats with period $w = 8$.

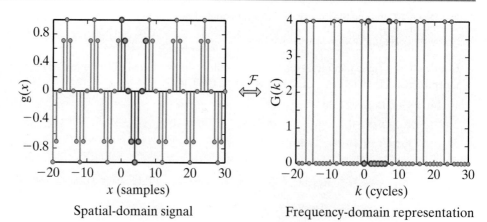

Spatial-domain signal

Frequency-domain representation

- Computing the DFT of a shifted signal is the same as multiplying the DFT of the original, unshifted signal by an appropriate complex exponential. Known as the **shift theorem**, this property is easy to prove:

$$\mathcal{F}\{g(x - x_0)\} = \sum_{x=0}^{w-1} g(x - x_0)e^{-j2\pi kx/w} \tag{6.31}$$

$$= \sum_{x'=-x_0}^{w-1-x_0} g(x')e^{-j2\pi k(x' + x_0)/w} \tag{6.32}$$

$$= \sum_{x'=0}^{w-1} g(x')e^{-j2\pi kx'/w}e^{-j2\pi kx_0/w} \tag{6.33}$$

$$= \mathcal{F}\{g(x)\}e^{-j2\pi kx_0/w} \tag{6.34}$$

where the second line uses a change of variables $x' = x - x_0$, and the third line follows from the periodicity of the DFT. In other words,

$$g(x) \quad \overset{DFT}{\Longleftrightarrow} \quad G(k) \tag{6.35}$$

$$g(x - x_0) \quad \overset{DFT}{\Longleftrightarrow} \quad G(k)e^{-j2\pi kx_0/w} \tag{6.36}$$

Keep in mind that with discrete sequences the shift theorem only holds when x_0 is an integer.

- **Modulation**, which is the dual of the shift theorem, states that multiplying a signal by a complex exponential causes a shift in the frequency domain:

$$\mathcal{F}\{g(x)e^{j2\pi k_0 x/w}\} = \sum_{x=0}^{w-1} g(x)e^{j2\pi k_0 x/w}e^{-j2\pi kx/w} \tag{6.37}$$

$$= \sum_{x=0}^{w-1} g(x)e^{-j2\pi(k - k_0)x/w} \tag{6.38}$$

$$= G(k - k_0) \tag{6.39}$$

where $G(k) = \mathcal{F}\{g(x)\}$. Substituting, we see that to shift by half the width of the signal, we must set $k_0 = \frac{w}{2}$, so that

$$G\left(k - \frac{w}{2}\right) = \mathcal{F}\{g(x)e^{j\pi x}\} = \mathcal{F}\{g(x)(-1)^x\} \tag{6.40}$$

where the last equality follows from Euler's formula and the fact that $\cos \pi x$ is 1 if x is even, or 0 if x is odd, assuming x is an integer. Therefore, if the original signal is multiplied by $(-1)^x$, the resulting DFT will be centered. In other words,

$$g(x) \quad \overset{DFT}{\Longleftrightarrow} \quad G(k) \tag{6.41}$$

$$g(x)e^{j2\pi k_0 x/w} \quad \overset{DFT}{\Longleftrightarrow} \quad G(k - k_0) \tag{6.42}$$

$$g(x)(-1)^x \quad \overset{DFT}{\Longleftrightarrow} \quad G\left(k - \frac{w}{2}\right) \tag{6.43}$$

Like the previous property, these formulas only apply to discrete sequences when k_0 is an integer. Therefore, when applying the latter formula be sure that the width of the signal is even, or the DFT will be shifted by a nonintegral amount, thus distorting the values.

- The **scaling property** says that if the signal is stretched in the spatial domain, then the Fourier transform is compressed in the frequency domain, and vice versa:

$$g(x) \quad \overset{\mathcal{F}}{\Longleftrightarrow} \quad G(k) \tag{6.44}$$

$$g(ax) \quad \overset{\mathcal{F}}{\Longleftrightarrow} \quad \frac{1}{a}G\left(\frac{k}{a}\right) \tag{6.45}$$

for $a \neq 0$. Note that this property strictly holds only for the continuous Fourier transform. For the DFT this is only an approximation, since no matter the value of a (apart from the trivial case $a = 1$), the scaling property involves noninteger indices.

- The complex exponentials are **orthogonal** to one another:

$$\sum_{k=0}^{w-1} e^{j2\pi kx/w} e^{-j2\pi kx'/w} = \begin{cases} w & \text{if } x = x' \\ 0 & \text{otherwise} \end{cases} \tag{6.46}$$

The first case $(x = x')$ is easy to show:

$$\sum_{k=0}^{w-1} e^{j2\pi k(x-x')/w} = \sum_{k=0}^{w-1} e^0 = \sum_{k=0}^{w-1} 1 = w \tag{6.47}$$

The second case $(x \neq x')$ uses the well-known formula for the sum of a geometric series, $\sum_{k=0}^{w-1} a^k = (1 - a^w)/(1 - a)$. Let us define $\delta \equiv x - x'$, then

$$\sum_{k=0}^{w-1} e^{j2\pi k\delta/w} = \frac{1 - e^{j2\pi\delta}}{1 - e^{j2\pi\delta/w}} = \frac{e^{j\pi\delta}(e^{-j\pi\delta} - e^{j\pi\delta})}{e^{-j\pi\delta/w}(e^{j\pi\delta/w} - e^{-j\pi\delta/w})} = e^{j\pi(w-1)\delta/w} \frac{\sin \pi\delta}{\sin \pi\delta/w} = 0 \tag{6.48}$$

where the last equality follows since $\sin \pi\delta = 0$ whenever δ is an integer, and $\sin \pi\delta/w \neq 0$ whenever δ is not a multiple of w.

- The DFT of a real-valued signal exhibits **Hermitian** symmetry, which means that its real component is even-symmetric, and its imaginary component is odd-symmetric. This follows naturally from the fact that the cosine function is even-symmetric, whereas the sine function is odd-symmetric:

$$G_{even}(-k) = \sum_{x=0}^{w-1} g(x) \cos(-j2\pi(-k)x/w) \tag{6.49}$$

$$= \sum_{x=0}^{w-1} g(x) \cos(-j2\pi kx/w) \tag{6.50}$$

$$= G_{even}(k) \tag{6.51}$$

$$G_{odd}(-k) = \sum_{x=0}^{w-1} g(x) \sin(-j2\pi(-k)x/w) \tag{6.52}$$

$$= \sum_{x=0}^{w-1} -g(x) \sin(-j2\pi kx/w) \tag{6.53}$$

$$= -G_{odd}(k) \tag{6.54}$$

That is, a real $g(x)$ leads to a Hermitian $G(k)$, and therefore $|G(-k)| = |G(k)|$ and $\angle G(-k) = -\angle G(k)$. The converse is also true: a Hermitian $G(k)$ leads to $g_{imag}(x) = 0$ for all x. When frequency-domain filters are introduced in Section 6.4, we will ensure that the filters also are Hermitian, so that the imaginary components remaining after the inverse DFT will be zero (at least to the level of machine precision).

- **Even and odd symmetry.** The DFT of a real-valued, even-symmetric signal is also real-valued and even-symmetric. The DFT of a real-valued, odd-symmetric signal is purely imaginary-valued and odd-symmetric. These properties arise because the sum of an odd function about the origin is zero, and the product of an even and odd function is odd. From Equation (6.15),

$$g(x) \text{ is even} \Rightarrow G(k) \text{ is real, even}: \quad G(k) = \sum_{x} g(x) \cos \frac{2\pi k}{w} x - j \sum_{x} \underbrace{g(x) \sin \frac{2\pi k}{w} x}_{\text{even·odd} = \text{odd}}{}^{0}$$

$$g \text{ is odd} \Rightarrow G(k) \text{ is imaginary, odd}: \quad G(k) = \sum_{x} \underbrace{g(x) \cos \frac{2\pi k}{w} x}_{\text{odd · even} = \text{odd}}{}^{0} - j \sum_{x} g(x) \sin \frac{2\pi k}{w} x$$

Thus, even-symmetric kernels, such as Gaussian or Laplacian of Gaussian, have frequency responses that are real and even, while odd-symmetric kernels, such as the derivative of Gaussian, have frequency responses that are imaginary and odd.

- **Parseval's theorem** states that the energy is preserved in the frequency domain, where the energy is defined as the sum of the squares of the magnitudes of the elements:

$$\sum_{x=0}^{w-1} |g(x)|^2 = \sum_{k=0}^{w-1} |G(k)|^2 \qquad (6.55)$$

This property is also known as the **unitarity** property of the DFT.

- The **DC component** of the signal is captured by $G(0)$, which is just the sum of the values in $g(x)$, i.e., $G(0) = \sum_{x=0}^{w-1} g(x)$, since $e^0 = 1$. For this reason, $G(0)$ is referred to as the DC component, where this term alludes to the *direct current* in an electrical circuit—that is, the amount of current flowing through the wire, ignoring oscillations.

- Convolution in the time (or spatial) domain is equivalent to multiplication in the frequency domain, and vice versa:

$$g_1(x) \circledast g_2(x) \quad \overset{DFT}{\Longleftrightarrow} \quad G_1(k)G_2(k) \qquad (6.56)$$

$$g_1(x)g_2(x) \quad \overset{DFT}{\Longleftrightarrow} \quad \frac{1}{w}G_1(k) \circledast G_2(k) \qquad (6.57)$$

It is important to note, however, that due to the periodicity of the DFT the convolution here is *circular convolution*, and hence this is known as the **circular convolution theorem**. If standard convolution is desired, the signals must be zero-padded with a sufficient number of values first, as explained next.

6.2.4 Zero Padding

It is often said that convolution in the time domain is equivalent to multiplication in the frequency domain. While this statement is true for continuous-time infinite-duration signals, special care must be taken in applying it to discrete signals. We cannot convolve two discrete signals by simply computing the DFT of each, multiplying the results, and computing the inverse DFT, for two reasons. First, as mentioned above, multiplication in the frequency domain is actually equivalent to *circular* convolution in the spatial domain, so if regular convolution is desired, we must first zero pad one of the signals. Secondly, if the two signals are of different lengths, then their Fourier transforms will have different lengths, thus precluding their multiplication; again zero-padding is the answer. The following example should make these concepts clear.

EXAMPLE 6.4

Suppose we want to convolve the input signal $g = \begin{bmatrix} \underline{6} & 4 & 1 & 0 & 1 & 4 \end{bmatrix}$ with the lowpass Gaussian filter $h(x) = \begin{bmatrix} 1 & \underline{2} & 1 \end{bmatrix}$, where the underscore indicates the value at the origin. Show how to zero pad the signals for a frequency-domain implementation of (a) circular convolution, and (b) linear convolution.

Solution

(a) Obviously we cannot simply multiply the DFTs because they are of different lengths. Instead, we must zero pad the kernel and use the periodicity property. Let us define $h_{zeropad}(x) \equiv \begin{bmatrix} 0 & 0 & 1 & \underline{2} & 1 & 0 \end{bmatrix} = \begin{bmatrix} \underline{2} & 1 & 0 & 0 & 0 & 1 \end{bmatrix}$, whose DFT is $\begin{bmatrix} \underline{4} & 3 & 1 & 0 & 1 & 3 \end{bmatrix}$. The DFT of the signal is $\begin{bmatrix} \underline{16} & 9 & 1 & 0 & 1 & 9 \end{bmatrix}$. The circular convolution is obtained from the inverse DFT of the multiplication of the two DFTs:

$$g'_{circ}(x) = g(x) \circledast h_{zeropad}(x) = \mathcal{F}^{-1}\{[16 \cdot 4 \quad 9 \cdot 3 \quad 1 \cdot 1 \quad 0 \cdot 0 \quad 1 \cdot 1 \quad 9 \cdot 3]\} \quad (6.58)$$

$$= \mathcal{F}^{-1}\{[\underline{64} \quad 27 \quad 1 \quad 0 \quad 1 \quad 27]\} \quad (6.59)$$

$$= \begin{bmatrix} \underline{20} & 15 & 6 & 2 & 6 & 15 \end{bmatrix} \quad (6.60)$$

(b) For linear convolution, we also need to zero pad the input signal to prevent the convolution kernel from wrapping past the signal boundary. To ensure both signals are of the same length, we also need to add additional zeros: By the periodicity property,

$$g_{zeropad}(x) = \begin{bmatrix} 0 & \underline{6} & 4 & 1 & 0 & 1 & 4 & 0 \end{bmatrix} = \begin{bmatrix} \underline{6} & 4 & 1 & 0 & 1 & 4 & 0 & 0 \end{bmatrix} \quad (6.61)$$

$$h_{zeropad}(x) = \begin{bmatrix} 0 & 0 & 0 & 1 & \underline{2} & 1 & 0 & 0 \end{bmatrix} = \begin{bmatrix} \underline{2} & 1 & 0 & 0 & 0 & 0 & 0 & 1 \end{bmatrix} \quad (6.62)$$

The linear convolution is then given by

$$g'(x) = g(x) \circledast h(x) = \mathcal{F}^{-1}\{\mathcal{F}\{g_{zeropad}\} \cdot \mathcal{F}\{h_{zeropad}\}\} \quad (6.63)$$

$$= \begin{bmatrix} \underline{15} & 15 & 6 & 2 & 6 & 9 \end{bmatrix} \quad (6.64)$$

While the forward and inverse DFTs are difficult to verify without a computer, the final results are easily computed using the standard spatial-domain techniques that we have already studied. The circular convolution of g and h allows the first and last values of the signal to wrap around when the convolution kernel is at the borders:

$$g'_{circ}(0) = 1 \cdot 4 + 2 \cdot 6 + 1 \cdot 4 = 20 \quad (6.65)$$

$$g'_{circ}(1) = 1 \cdot 6 + 2 \cdot 4 + 1 \cdot 1 = 15 \quad (6.66)$$

$$g'_{circ}(2) = 1 \cdot 4 + 2 \cdot 1 + 1 \cdot 0 = 6 \quad (6.67)$$

$$g'_{circ}(3) = 1 \cdot 1 + 2 \cdot 0 + 1 \cdot 1 = 2 \quad (6.68)$$

$$g'_{circ}(4) = 1 \cdot 0 + 2 \cdot 1 + 1 \cdot 4 = 6 \quad (6.69)$$

$$g'_{circ}(5) = 1 \cdot 1 + 2 \cdot 4 + 1 \cdot 6 = 15 \quad (6.70)$$

or $g'_{circ}(x) = \begin{bmatrix} \underline{20} & 15 & 6 & 2 & 6 & 15 \end{bmatrix}$. Linear convolution is obtain in the same manner, except that g is zero padded as necessary:

$$g'(0) = 1 \cdot 0 + 2 \cdot 6 + 1 \cdot 4 = 15 \quad (6.71)$$

$$g'(5) = 1 \cdot 1 + 2 \cdot 4 + 1 \cdot 0 = 9 \quad (6.72)$$

and so forth.

6.2.5 Magnitude and Phase

Often it is convenient to convert the real and imaginary components of the Fourier transform into **polar coordinates**:

$$G(k) = G_{even}(k) + jG_{odd}(k) = |G(k)|e^{j\angle G(k)} \tag{6.73}$$

where $G_{even}(k)$ and $G_{odd}(k)$ are the real and imaginary parts of G, respectively, and

$$|G(k)| = \sqrt{G_{even}^2(k) + G_{odd}^2(k)} \tag{6.74}$$

$$\angle G(k) = \tan^{-1}\left(\frac{G_{odd}(k)}{G_{even}(k)}\right) \tag{6.75}$$

are the magnitude and phase, respectively.

Filters are characterized by their phase, of which there are three types. By far, the most common type of digital filter is **zero-phase**, which means that $\angle G(k) = 0$ for all k. As we have just seen, a real-valued convolution kernel with even symmetry about the origin yields a real-valued Fourier transform with even symmetry about the origin, and is therefore a zero-phase filter. All Gaussian filters, such as $\begin{bmatrix} 1 & \underline{2} & 1 \end{bmatrix}$, and all Laplacian of Gaussian filters, such as $\begin{bmatrix} 1 & \underline{-2} & 1 \end{bmatrix}$, where the underscore indicates the origin, are zero-phase.

If the kernel is real-valued with even symmetry about an index other than the origin, then it is **linear-phase**. For example, $\begin{bmatrix} \underline{1} & 2 & 1 \end{bmatrix}$ is linear-phase. This is easy to see from the shift theorem, since a shift of the signal by x_0 causes the Fourier transform to be multiplied by $e^{-j2\pi kx_0/w}$, indicating that the phase is $-2\pi kx_0/w$, which is a linear function of k. Linear-phase filters are common in signal processing due to the need for causal processing, but they are uncommon in image processing. However, as we just saw, a real-valued convolution kernel with odd symmetry about the origin yields a purely imaginary Fourier transform with odd symmetry about the origin. More precisely, the magnitude of the Fourier transform is even, while the phase is either $\pm \pi/2$ everywhere. Such filters are considered *generalized linear phase* and can, for all practical purposes, be treated as linear-phase. All first-derivative Gaussian filters, such as $\begin{bmatrix} 1 & \underline{0} & -1 \end{bmatrix}$, fall into this category.

Finally, **nonlinear-phase** filters arise when dealing with old-fashioned analog circuitry or IIR filters. However, as long as we work with convolution (and hence digital FIR filters), we will not encounter these.

6.2.6 Interpreting Discrete Frequencies

One aspect of the DFT that may not be obvious at first is how to interpret discrete frequencies. To overcome this difficulty, let us consider the simple example of the continuous signal $g(x) = \cos\frac{2\pi}{8}x$, shown in the left side of Figure 6.4. If g is a spatial-domain signal and x is expressed in meters, then the frequency of the signal is $\frac{1}{8}$ *cycles per meter*, while the *period* of the signal (which is the inverse of the frequency) is 8 *meters per cycle*. Indeed, as can be seen from the figure, the signal repeats every 8 meters, so that $g(x + 8) = g(x)$, or more generally, $g(x + 8n) = g(x)$, where n is an arbitrary integer.

Now if the continuous signal is sampled at locations $x = 0, 1, \ldots, 7$, we obtain the discrete signal shown in the right side of the figure, where the units of x are now samples rather than meters. Therefore, the frequency of the discrete signal is $\frac{1}{8}$ *cycles per sample*, while the period of the signal is 8 *samples per cycle*. Like radians, "cycle" is a dimensionless unit that can be ignored whenever convenient, so it is equivalent to say that the frequency is $\frac{1}{8}$ inverse samples, while the period is 8 samples.

Although it is obvious from the shape of the plot that the signal is exactly one period of a cosine waveform, keep in mind that the DFT computation operates solely on the eight

Figure 6.4 LEFT: A continuous spatial-domain signal $\cos\frac{2\pi}{8}x$ is sampled at locations $x = 0, 1, \ldots, 7$. RIGHT: The discrete signal resulting from the sampling. Note that the units for the domain have changed from meters to samples.

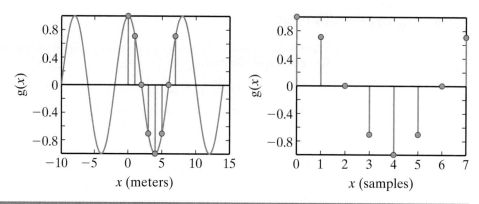

sampled values in the right side of the figure: $1, \frac{1}{\sqrt{2}}, 0, -\frac{1}{\sqrt{2}}, -1, -\frac{1}{\sqrt{2}}, 0, \frac{1}{\sqrt{2}}$. Applying Equation (6.13) yields

$$G(0) = 1e^0 + \frac{1}{\sqrt{2}}e^0 + 0e^0 + \cdots + \frac{1}{\sqrt{2}}e^0 = 0 \tag{6.76}$$

$$G(1) = 1e^0 + \frac{1}{\sqrt{2}}e^{-j2\pi/8} + 0e^{-j4\pi/8} + \cdots + \frac{1}{\sqrt{2}}e^{-j14\pi/8} = 4 \tag{6.77}$$

$$G(2) = 1e^0 + \frac{1}{\sqrt{2}}e^{-j4\pi/8} + 0e^{-j8\pi/8} + \cdots + \frac{1}{\sqrt{2}}e^{-j28\pi/8} = 0 \tag{6.78}$$

$$\vdots \tag{6.79}$$

$$G(7) = 1e^0 + \frac{1}{\sqrt{2}}e^{-j14\pi/8} + 0e^{-j28\pi/8} + \cdots + \frac{1}{\sqrt{2}}e^{-j98\pi/8} = 4 \tag{6.80}$$

which is summarized as

$$G(k) = \begin{cases} 4 & \text{if } k = 1 \text{ or } k = 7 \\ 0 & \text{otherwise} \end{cases} \tag{6.81}$$

and displayed in Figure 6.5. There is a spike at $k = 1$ and another spike at $k = 7$; or equivalently at $f = \frac{1}{8}$ and $f = \frac{7}{8}$, since $w = 8$. The first spike is what we expect, since the original continuous signal has a period of 8 meters, and therefore the discrete signal has a period of

Figure 6.5 LEFT: The DFT of the discrete signal shown in the right side of Figure 6.4, shown as a function of the discrete index k. MIDDLE: The DFT shown as a function of $f = k/w$, where $w = 8$ is the number of samples in the original discrete signal. RIGHT: The DFT of the discrete signal shifted to show positive and negative frequencies.

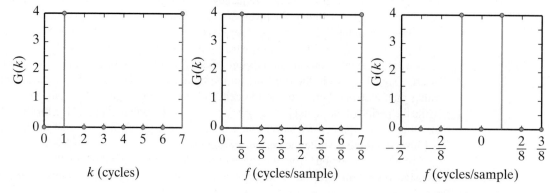

8 samples. To understand the spike at $k = 7$, recall from our discussion on periodicity that $f = \frac{w-1}{w}$ is the same as $f = -\frac{1}{w}$, so in this case $f = \frac{7}{8}$ is the same as $f = -\frac{1}{8}$. Therefore, this additional frequency of $f = \frac{7}{8}$, which at first glance appears extraneous, is actually none other than the negative of the frequency that we already know is present in the signal. In fact, the relationship between $f = \frac{1}{8}$ and $f = -\frac{1}{8}$ can be observed in the sum of complex exponentials:

$$\cos 2\pi fx = \frac{1}{2}\left(e^{2\pi fx} + e^{-2\pi fx}\right) \tag{6.82}$$

so that a sinusoid can be thought of as containing both the positive and negative frequencies. One way to visualize this relationship is to display the values for which $k \geq \frac{1}{2}$ to the left of the values for which $k < \frac{1}{2}$, as shown in the right side of the figure.

6.2.7 Basis Functions

The DFT illustrates a foundational concept in signal analysis, namely **basis functions**. A basis function is a scalar function defined over the same domain as the original signal that, when linearly combined with other basis functions, yields the signal:

$$g(x) = \sum_k \alpha_k \psi_k(x) \tag{6.83}$$

where g is the signal, α_k is the kth scalar weight, and $\psi_k(x)$ is the kth basis function, defined over the same domain as g. Rearranging Equation (6.83) into matrix form yields

$$\underbrace{\begin{bmatrix} g(0) \\ g(1) \\ \vdots \\ g(w) \end{bmatrix}}_{g} = \underbrace{\begin{bmatrix} \psi_0(0) & \psi_1(0) & \cdots & \psi_w(0) \\ \psi_0(1) & \psi_1(1) & \cdots & \psi_w(1) \\ \vdots & \vdots & \ddots & \vdots \\ \psi_0(w) & \psi_1(w) & \cdots & \psi_w(w) \end{bmatrix}}_{\Psi} \underbrace{\begin{bmatrix} \alpha_0 \\ \alpha_1 \\ \vdots \\ \alpha_w \end{bmatrix}}_{\alpha} \tag{6.84}$$

that is, $\mathbf{g} = \Psi\alpha$, or $\alpha = \Psi^{-1}\mathbf{g}$. It is easy to see that the basis functions are given by the columns of Ψ, i.e., $\Psi = \begin{bmatrix} \psi_0 & \psi_1 & \cdots & \psi_w \end{bmatrix}$, where $\psi_k \equiv \begin{bmatrix} \psi_k(0) & \psi_k(1) & \cdots & \psi_k(w) \end{bmatrix}^{\text{T}}$ is the kth basis function in vector form. If Ψ is orthogonal, then $\Psi^{-1} = \Psi^{\text{T}}$, in which case $\alpha = \Psi^{\text{T}}\mathbf{g}$, and the basis functions are equivalently given by the rows of Ψ^{T}.

The simplest set of basis functions is $\{\mathbf{e}_k\}_{k=1}^{w}$, where $\mathbf{e}_k \equiv \begin{bmatrix} 0 & \cdots & 0 & 1 & 0 & \cdots & 0 \end{bmatrix}^{\text{T}}$ is a vector of zeros with a one in the kth position:

$$\begin{aligned} g(x) &= \begin{bmatrix} g(0) & g(1) & g(2) & \cdots & g(w-1) \end{bmatrix} \\ &= g(0)\begin{bmatrix} 1 & 0 & 0 & \cdots & 0 \end{bmatrix} \\ &\quad + g(1)\begin{bmatrix} 0 & 1 & 0 & \cdots & 0 \end{bmatrix} \\ &\quad + g(2)\begin{bmatrix} 0 & 0 & 1 & \cdots & 0 \end{bmatrix} \\ &\quad \vdots \\ &\quad + g(w-1)\begin{bmatrix} 0 & 0 & 0 & \cdots & 1 \end{bmatrix} \end{aligned}$$

The basis functions define a transform such that

$$\alpha_k = \sum_{x=0}^{w-1} g(x)e_k(x) \tag{6.85}$$

$$g(x) = \sum_{k=0}^{w-1} \alpha_k e_k(x) \tag{6.86}$$

where $e_k(x)$ is the xth element of $\mathbf{e}_k$.

Notice that $\mathbf{e}_k$ is simply the unit vector along the kth axis in a Cartesian space. For example, suppose for simplicity that the signal consists of just two samples: $g(0)$ and $g(1)$. If we think of the signal as a vector $\mathbf{g} = \begin{bmatrix} g(0) & g(1) \end{bmatrix}^{\mathsf{T}}$, it is trivial to see that $\mathbf{g} = g(0)\begin{bmatrix} 1 & 0 \end{bmatrix}^{\mathsf{T}} + g(1)\begin{bmatrix} 0 & 1 \end{bmatrix}^{\mathsf{T}}$. In this example, $\mathbf{e}_0 = \begin{bmatrix} 1 & 0 \end{bmatrix}^{\mathsf{T}}$ and $\mathbf{e}_1 = \begin{bmatrix} 0 & 1 \end{bmatrix}^{\mathsf{T}}$ are the basis functions, and $g(0)$ and $g(1)$ are the weights that cause the linear combination of the basis functions to exactly represent the original signal g. As shown in Figure 6.6, the signal can be visualized as a point $(g(0), g(1))$ in the plane, $\mathbf{e}_0$ and $\mathbf{e}_1$ are unit vectors along the axes, and the weights $g(0)$ and $g(1)$ are obtained by *projecting* the signal onto the basis functions: $g(0) = \mathbf{g}^{\mathsf{T}}\mathbf{e}_0$ and $g(1) = \mathbf{g}^{\mathsf{T}}\mathbf{e}_1$.

The DFT operates in much the same way. The forward DFT performs an *analysis* of the signal by determining the contributions of the various frequencies in the signal, while the inverse DFT performs a *synthesis* of the signal as a weighted sum of sines and cosines. The sines and cosines at different frequencies are the basis functions of the DFT, and the weights for any particular signal are given by the output of the DFT applied to the signal. Basis functions, as in the case of simple unit axes or in the case of Fourier sines and cosines, are often orthogonal to one another, but we will see examples later in this chapter of non-orthogonal basis functions.

6.2.8 DFT as Matrix Multiplication

Sometimes it is helpful to consider the DFT as the multiplication of a matrix by a vector. That is, if we let g be the vector containing the input signal, and $\mathbf{g}_{\mathcal{F}}$ the vector containing the frequency-domain representation, then $\mathbf{g}_{\mathcal{F}} = \mathbf{F}_w\mathbf{g}$ is the forward DFT, while $\mathbf{g} = \mathbf{F}_w^{-1}\mathbf{g}_{\mathcal{F}}$ is the inverse DFT, where $\mathbf{F}_w$ is the $w \times w$ *normalized DFT matrix*, obtained by rearranging Equation (6.13) in matrix form:

$$
\underbrace{\begin{bmatrix} G(0) \\ G(1) \\ \vdots \\ G(w-1) \end{bmatrix}}_{\mathbf{g}_{\mathcal{F}}} = \frac{1}{\sqrt{w}} \underbrace{\begin{bmatrix} 1 & 1 & 1 & \cdots & 1 \\ 1 & e^{-j2\pi/w} & e^{-j4\pi/w} & \cdots & e^{-j2\pi(w-1)/w} \\ 1 & e^{-j4\pi/w} & e^{-j8\pi/w} & \cdots & e^{-j4\pi(w-1)/w} \\ \vdots & \vdots & \vdots & \ddots & \vdots \\ 1 & e^{-j2\pi(w-1)/w} & e^{-j4\pi(w-1)/w} & \cdots & e^{-j2\pi(w-1)^2/w} \end{bmatrix}}_{\mathbf{F}_w} \underbrace{\begin{bmatrix} g(0) \\ g(1) \\ \vdots \\ g(w-1) \end{bmatrix}}_{\mathbf{g}} \qquad (6.87)
$$

where the ik^{th} element is given by $(1/\sqrt{w})e^{-j2\pi ik/w}$, if i and k are zero-based indices. The basis functions are given by the columns of $\mathbf{F}_w^{-1}$. Since the matrix is both orthogonal and symmetric, $\mathbf{F}_w^{-1} = \mathbf{F}_w^{\mathsf{T}} = \mathbf{F}_w$, these are the same as the columns (or rows) of $\mathbf{F}_w$. These basis functions are orthogonal to one another, that is, $\mathbf{f}_i^{\mathsf{T}}\mathbf{f}_k^* = 0$ if $i \neq k$, where $\mathbf{f}_i$ is the ith column of $\mathbf{F}$, and * is the complex conjugate. Because of the normalization factor, all the basis functions have unit norm, that is, $\|\mathbf{f}_i\|^2 = \mathbf{f}_i^{\mathsf{T}}\mathbf{f}_i^* = 1$ for all i. Recall in our

Figure 6.6 In a standard Cartesian coordinate system, unit vectors along the coordinate axes act like basis functions, with the elements of a vector being equivalent to the projection of the vector onto these vectors. Shown are the basis functions $\mathbf{e}_0$ and $\mathbf{e}_1$ along the *x* and *y* axes, respectively.

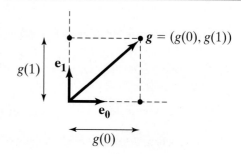

earlier definition of the DFT that the forward and inverse normalization factors can be set arbitrarily as long as their product is $1/w$. Here we distribute the normalization factor equally, with $1/\sqrt{w}$ in the forward transform and $1/\sqrt{w}$ in the inverse transform, which ensures that $\mathbf{F}_w$ is a unitary matrix (the complex version of an orthogonal matrix), so that $\mathbf{F}_w^\mathsf{T}\mathbf{F}_w^* = \mathbf{F}_w^*\mathbf{F}_w^\mathsf{T} = \mathbf{I}_{\{n \times n\}}$, where $\mathbf{I}_{\{n \times n\}}$ is the $n \times n$ identity matrix. Therefore, since $\mathbf{F}_w$ is symmetric, its inverse is its complex conjugate $(\mathbf{F}^{-1} = \mathbf{F}^*)$, which is what we expect because the inverse DFT is exactly the same as the forward DFT except for the sign of the complex exponential.

6.3 Two-Dimensional DFT

Now that we have established the foundation of the discrete Fourier transform, we are ready to discuss its application to image processing. The 2D DFT is a natural extension of the 1D case: simply replace the single frequency k with two frequencies in the two directions, k_x and k_y, so that kx/w becomes $k_x x/w + k_y y/h$. That is, if $g(x, y)$ is a 2D signal (such as an image) defined over the domain $x = 0, 1, \ldots, w - 1$ and $y = 0, 1, \ldots, h - 1$, then the forward and inverse DFTs are given by

$$G(k_x, k_y) = \sum_{x=0}^{w-1} \sum_{y=0}^{h-1} g(x, y) e^{-j2\pi \mathbf{x}^\mathsf{T}\mathbf{f}} \qquad (\text{forward DFT}) \qquad (6.88)$$

$$g(x, y) = \frac{1}{wh} \sum_{k_x=0}^{w-1} \sum_{k_y=0}^{h-1} G(k_x, k_y) e^{j2\pi \mathbf{x}^\mathsf{T}\mathbf{f}} \qquad (\text{inverse DFT}) \qquad (6.89)$$

where $\mathbf{x} = \begin{bmatrix} x & y \end{bmatrix}^\mathsf{T}$ and $\mathbf{f} = \begin{bmatrix} \frac{k_x}{w} & \frac{k_y}{h} \end{bmatrix}^\mathsf{T}$, so that $\mathbf{x}^\mathsf{T}\mathbf{f} = \frac{k_x x}{w} + \frac{k_y y}{h}$. As with the 1D transform, keep in mind that the placement of the scaling factor $1/wh$ is arbitrary.

6.3.1 Separability

A straightforward implementation of the 2D DFT is shown in Algorithm 6.3. The frequency representation G is stored in the same manner as a complex image would be, with two numbers per element (the real and imaginary components). Lines 1–2 loop over all the elements in this 2D array, computing the values by performing an elementwise sum in Lines 9–10 according to Equation (6.88), taking advantage of Euler's formula. The pseudocode is terribly inefficient, with an asymptotic running time of $O(w^4)$, assuming $w \approx h$.

The speed can be increased substantially by taking advantage of the fact that the 2D DFT is separable. To see that this is indeed the case, simply expand Equation (6.88), substitute the product of the exponents for the exponent of the sums, and recognize that the exponents in the product are themselves dependent upon only one of the two variables, either x or y:

$$G(k_x, k_y) = \sum_{x=0}^{w-1} \sum_{y=0}^{h-1} g(x, y) e^{-j2\pi(k_x x/w + k_y y/h)} \qquad (6.90)$$

$$= \sum_{x=0}^{w-1} \underbrace{\left(\sum_{y=0}^{h-1} g(x, y) e^{-j2\pi k_y y/h} \right)}_{G_y(x; k_y)} e^{-j2\pi k_x x/w} \qquad (6.91)$$

where $G_y(x; k_y)$ is the 1D DFT of column y of $g(x, y)$. This equation says that the 2D DFT can be computed by first computing the 1D DFT of each column independently, then

ALGORITHM 6.3 2D DFT (slow version)

DiscreteFourierTransform2D(I)

Input: grayscale image I of size $width \times height$
Output: real (G_{even}) and imaginary (G_{odd}) components of the 2D DFT of I

1   **for** $k_y \leftarrow 0$ **to** $height-1$ **do**
2   **for** $k_x \leftarrow 0$ **to** $width-1$ **do**
3           $G_{even}(k_x, k_y) \leftarrow 0$
4           $G_{odd}(k_x, k_y) \leftarrow 0$
5           $f_x \leftarrow k_x/width$
6           $f_y \leftarrow k_y/height$
7   **for** $y \leftarrow 0$ **to** $height-1$ **do**
8   **for** $x \leftarrow 0$ **to** $width-1$ **do**
9                   $G_{even}(k_x, k_y) \leftarrow_+ I(x, y) * \cos(2 * \pi * (f_x * x + f_y * y))$
10                  $G_{odd}(k_x, k_y) \leftarrow_- I(x, y) * \sin(2 * \pi * (f_x * x + f_y * y))$
11 **return** G_{even}, G_{odd}

computing the 1D DFT of the resulting rows. Of course due to symmetry this order can be reversed by processing the rows first, then the columns, without affecting the result. The separable algorithm is shown in Algorithm 6.4. First the DFT is computed of each row of the image, storing the result in the corresponding row of a temporary array. (The colon operator selects all the values in the row.) Then the DFT is computed of each column of the temporary array. Note that in the first iteration all the input values are real, whereas in the second iteration the values are complex, thus necessitating the call to the DiscreteFourier TransformComplex procedure we saw before, which is here abbreviated DFTC. The running time of this pseudocode for the 2D DFT is $O(w^3)$. To further increase speed a real implementation would use the FFT to compute the 1D DFTs of the separable algorithm, leading to a running time of $O(w^2 \log w)$.

The separability of the 2D DFT leads to an elegant, compact representation of the 2D DFT using matrix notation, similar to Equation (5.38):

$$\mathbf{G}_{\mathcal{F}} = \mathbf{F}_h \mathbf{G} \mathbf{F}_w \qquad (6.92)$$

where $\mathbf{G}$ is the original 2D signal (or image) treated as a $w \times w$ matrix; $\mathbf{F}_w$ is the $w \times w$ 1D DFT matrix given in Equation (6.87); $\mathbf{F}_h$ is the $h \times h$ 1D DFT matrix defined in the exact same

ALGORITHM 6.4 2D DFT (separable version, still slow)

DiscreteFourierTransform2DSeparable(g)

Input: grayscale image I of size $width \times height$
Output: real (G_{even}) and imaginary (G_{odd}) components of the 2D DFT of I

1   **for** $y \leftarrow 0$ **to** $height-1$ **do**
2       $Temp_{even}(:, y), Temp_{odd}(:, y) \leftarrow$ DiscreteFourierTransform$(g(0, y), \ldots, g(width-1, y))$
3   **for** $x \leftarrow 0$ **to** $width-1$ **do**
4       $G_{even}(x, :), G_{odd}(x, :) \leftarrow$ DFTC$(Temp_{even}(x, 0), \ldots, Temp_{even}(x, height-1),$
                                       $Temp_{odd}(x, 0), \ldots, Temp_{odd}(x, height-1))$
5   **return** G_{even}, G_{odd}

manner but substituting h for w in Equation (6.87); and $\mathbf{G}_{\mathcal{F}}$ is the 2D DFT of $\mathbf{G}$. This result is easy to derive and illustrated in Figure 6.7. Let $\mathbf{g}_i^\top$ be the ith row of $\mathbf{G}$ containing the values in the ith (zero-based index) row $y = i$ of the original 2D signal $g(x, y)$, oriented horizontally; that is, $\mathbf{G}^\top = [\mathbf{g}_0 \ \cdots \ \mathbf{g}_{h-1}]$. The vector $\mathbf{F}_w\mathbf{g}_i$ is therefore the 1D DFT of those values, oriented vertically. Let $\textbf{Temp} \equiv \mathbf{F}_w\mathbf{G}^\top$ be a temporary matrix (the same size as $\mathbf{G}^\top$) whose columns are the 1D DFTs of the columns of $\mathbf{G}^\top$. If we transpose this matrix and premultiply it by $\mathbf{F}_h$, then we will compute the 1D DFTs along the vertical direction of the original signal, yielding the desired result: $\mathbf{G}_{\mathcal{F}} = \mathbf{F}_h\textbf{Temp}^\top = \mathbf{F}_h(\mathbf{F}_w\mathbf{G}^\top)^\top = \mathbf{F}_h\mathbf{G}\mathbf{F}_w^\top = \mathbf{F}_h\mathbf{G}\mathbf{F}_w$, where the last equality follows from the symmetry of the 1D DFT matrix, as we saw earlier. Note that if the original signal $\mathbf{G}$ is square, then Equation (6.92) reduces simply to $\mathbf{G}_{\mathcal{F}} = \mathbf{F}_w\mathbf{G}\mathbf{F}_w$.

6.3.2 Projection-Slice Theorem

The **projection** of a continuous function $g(x, y)$ of two variables onto a line at some orientation θ is the 1D function that results from integrating the function along rays perpendicular to the line. Let us define a **slice** through a 2D continuous function $G(f_x, f_y)$ at θ as the 1D function obtained by ignoring all values except those along the line. The **projection-slice theorem**, also known as the **Fourier slice theorem**, says that the Fourier transform of the projection of g onto a line through the origin is the same as the 1D slice of G at the same orientation, where $G = \mathcal{F}\{g\}$. In other words, the Fourier transform of the projection is the slice of the Fourier transform, as shown in Figure 6.8.

This theorem is easily proved for the case of a horizontal slice along the x-axis:

$$G(f_x, 0) = \int_{-\infty}^{\infty} \int_{-\infty}^{\infty} g(x, y)e^{-j2\pi f_x x}\, dx\, dy \qquad (6.93)$$

$$= \int_{-\infty}^{\infty} \underbrace{\left[\int_{-\infty}^{\infty} g(x, y)dy\right]}_{g_p(x)} e^{-j2\pi f_x x}\, d \qquad (6.94)$$

$$= \mathcal{F}\{g_p(x)\} \qquad (6.95)$$

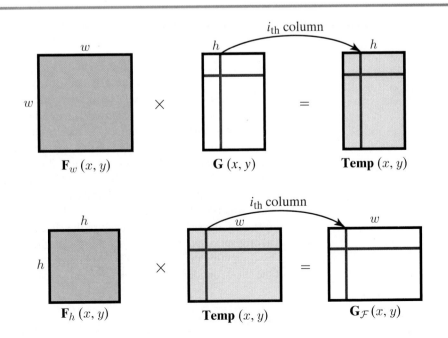

Figure 6.7 The 2D DFT as a pair of matrix multiplies, utilizing the principle of separability. The 1D DFT matrix is multiplied by the transpose of the original signal (treated as a matrix) to compute the 1D DFTs along the rows of **G** (columns of **G**ᵀ). Then this result is premultiplied by the 1D DFT matrix to compute the 1D DFTs along the columns of **G**, yielding the 2D DFT **G**_𝓕.

Figure 6.8 Projection-slice theorem.

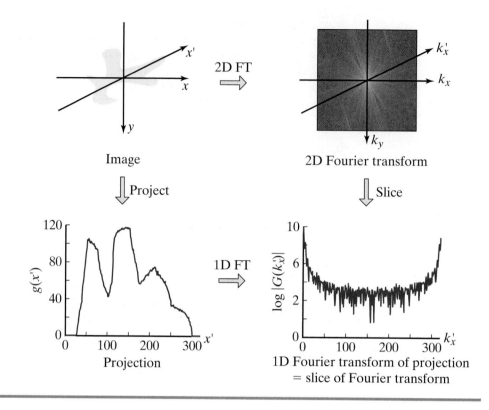

where $g_p(x) = \int g(x, y)\,dy$ is the projection of g onto the x-axis. In other words, a horizontal slice through the Fourier transform at $f_y = 0$ is identical to the Fourier transform of the projection of the original signal onto the x-axis. It will be obvious that the theorem applies along any orientation, after we show (later in this section) that a rotation of the image yields a rotation of the Fourier transform by the same amount in the same direction. This theorem also applies to the case of a discrete function, although when θ is not an integral multiple of 90 degrees, discretization effects cause the two functions to only approximate each other.

The projection-slice theorem is important in the reconstruction of an object from images of its slices. Imaging by slices is known as **tomography**, and the process of recovering an object from image slices is known as **tomographic reconstruction**. CAT scans (from *computed axial tomography*),[†] for example, operate by collecting cross-sectional slices through an object at various orientations. At each orientation, the sensor accumulates the light that passes through the object along a line, a process that is essentially a natural integration because the amount of radiation detected along a line is related to the sum of all the absorbances of the material along that line. Mathematically, the integration of a signal along all possible lines is known as the **Radon transform**, and it is widely used in tomographic reconstruction. By the projection-slice theorem, the 1D Fourier transform of the slices obtained by the sensor is equivalent to the slices of the 3D Fourier transform of the (unknown) original signal. It is easy to see that the original signal can be recovered (in theory at least) by computing the inverse 3D Fourier transform of the combined 1D Fourier transforms of the slices, although in practice more numerically stable approaches are often used.

[†] Section 2.4.2 (p. 54).

6.3.3 Displaying the 2D DFT

To display the 2D DFT of an image, the first step is to separate it into two 2D arrays containing the magnitude and phase, respectively. The phase is linearly scaled to the range of the display (typically 0 to 255) and shown as a grayscale image. If this same procedure were followed for the magnitude, however, the display would show a purely black image with a single bright pixel in the top-left corner. This top-left pixel, in a manner analogous to the 1D situation that we discussed earlier, captures the DC component, which is typically so much larger than the other values that they all appear to be zero, as shown in Figure 6.9.

To overcome this problem, it is common practice to display the logarithm of the magnitude instead, as shown in Figure 6.10. This display, however, still does not reveal the structure of the DFT magnitude very well because the DC component is in the top-left corner. To shift it to the center of the image, the four quadrants of the DFT should be cropped and pasted in a manner similar to the approach taken in the 1D case described earlier. In other words, if we imagine dividing the DFT into 4 quadrants, with the top-left labeled A, the top-right labeled B, the bottom-left labeled C, and the bottom-right labeled D, then the data must be shifted so that the order of display is (from top-left to bottom-right) D-C-B-A. A simple handy trick to do this, from Equation (6.43), is to simply multiply the value of each pixel in the signal by $(-1)^{x+y}$ prior to computing the DFT, since $\mathcal{F}\{g(x,y)(-1)^{x+y}\} = G(k_x - \frac{w}{2}, k_y - \frac{h}{2})$, where $G(k_x, k_y) = \mathcal{F}\{g(x,y)\}$, and w and h are the width and height, respectively, as shown in Figure 6.11.

Another way to think about this procedure is to remember from the periodicity property that the DFT treats the signal as if it were replicated forever in all directions, and therefore the DFT is replicated in all directions as well. To illustrate this, Figure 6.12 shows the image replicated four times. If the DFT were applied to each of these images separately, we would have 4 DFTs, each with quadrants A, B, C, and D, and each containing the DC component[†] at the top-left corner of the A quadrant. When the DFT is applied to the combined image, the resulting DFT is simply the concatenation of all 16 quadrants. The information in the four central quadrants, namely D-C-B-A, is identical to that in the DFT of the single image, but rearranged so that the DC component appears in the center, which is what we want. This manner of rearranging the display also reveals that the values along the middle of the image in both the horizontal and vertical directions are significantly larger than all the other values; there will be more about this in the next section.

Figure 6.9 Slice through the magnitude of the 2D DFT (first row). This first row includes the DC component, shown as a circle ('o'). Left: Without the log, the dynamic range is so great that nearly all frequencies appear to have zero contribution. Right: Applying the logarithm reduces the dynamic range to increase visibility of the components.

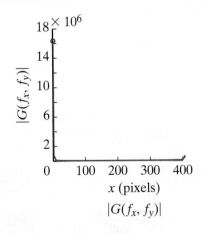

$|G(f_x, f_y)|$

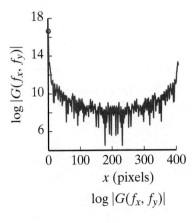

$\log |G(f_x, f_y)|$

[†] Please note that the two quadrants named C and D are unrelated to the *direct current (DC)* acronym.

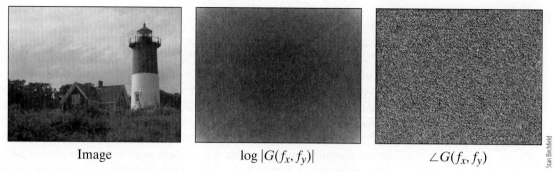

| Image | $\log |G(f_x, f_y)|$ | $\angle G(f_x, f_y)$ |

Figure 6.10 An image and its 2D DFT shown as magnitude and phase. (To increase the dynamic range of the display, the log of the magnitude is shown.) The DC component, which is the top-left corner of the magnitude, is difficult to see.

6.3.4 Linear Image Transforms

The properties of the 1D DFT outlined in Section 6.2.3 also apply to the 2D DFT, namely linearity, periodicity, shift theorem (an important application of which we just saw), modulation, orthogonality, Hermitian symmetry, unitarity, and so forth. Other properties generally hold as well, and the extension from 1D to higher dimensions is usually straightforward and obvious.

One such property is as follows. Suppose we have a continuous signal $g(\mathbf{x})$ with Fourier transform $G(\mathbf{f})$, and we want to find the Fourier transform of the related signal $g(\mathbf{x}')$, where $\mathbf{x}'$ and $\mathbf{x}$ are related by a **linear transform**:

$$\mathbf{x}' = \mathbf{A}\mathbf{x} \tag{6.96}$$

where $\mathbf{A}$ is a square matrix. It is easy to show that the Fourier transform of the transformed signal is given by

$$g(\mathbf{x}) \overset{\mathcal{F}}{\Longleftrightarrow} G(\mathbf{f}) \tag{6.97}$$

$$g(\mathbf{x}') \overset{\mathcal{F}}{\Longleftrightarrow} \frac{1}{\det(\mathbf{A})} G(\mathbf{f}') \tag{6.98}$$

where $\mathbf{f}' = \mathbf{A}^{-T}\mathbf{f}$. This expression is only approximately true for the DFT because of discretization issues but is nevertheless quite important in practice.

For example, suppose the coordinate system of the image is rotated by an angle θ so that $\mathbf{x}' = \mathbf{A}\mathbf{x}$, where $\mathbf{x} = \begin{bmatrix} x & y \end{bmatrix}^T$ and $\mathbf{f} = \begin{bmatrix} f_x & f_y \end{bmatrix}^T$. Then

$$\mathbf{A} = \mathbf{R}_\theta = \begin{bmatrix} \cos\theta & -\sin\theta \\ \sin\theta & \cos\theta \end{bmatrix} \tag{6.99}$$

Figure 6.11 Multiplying the image by $(-1)^{x+y}$ prior to taking the DFT causes the result to be shifted so that the DC component is in the center. On the right is shown the logarithm of the magnitude of the DFT of the post-multiplied image.

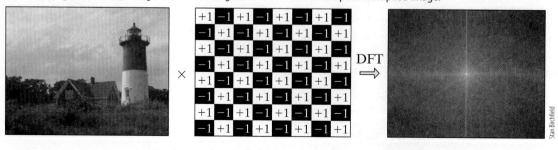

Figure 6.12 Top: The DFT treats the input as a replicated input, and produces a replicated output. Bottom: It is easier to visualize the DFT by shifting it so that the DC component is in the center, which causes no loss of information. The quadrants A, B, C, and D present in the original DFT output are also present in the shifted output, just in a different order.

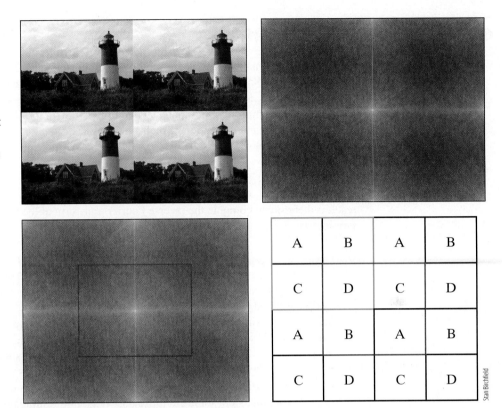

so that

$$g(x', y') \quad \overset{\mathcal{F}}{\Longleftrightarrow} \quad G(f_x', f_y') \tag{6.100}$$

where $\mathbf{f}' = \mathbf{A}\mathbf{f} = \mathbf{R}_\theta \mathbf{f}$, because for a rotation matrix, $\mathbf{R}_\theta^{-\mathsf{T}} = \mathbf{R}_\theta$, and $\det(\mathbf{R}_\theta) = 1$. In other words, the frequencies rotate in the same direction as the image. We now have an explanation for the bright horizontal and vertical lines in the middle of the shifted DFT in Figure 6.12. These lines arise from the sharp contrast between the top and bottom rows of the original image, as well as between the left and right columns, which are visible in the figure. If these wrapping effects of the DFT were not so overwhelming, then the DFT would more clearly reveal the dominant gradient directions in the image.

Another specific linear transform that appears often in practice is scaling. In 2D, scaling of the form $x' = ax, y' = by$ involves

$$\mathbf{A} = \begin{bmatrix} a & 0 \\ 0 & b \end{bmatrix} \tag{6.101}$$

and therefore

$$g(ax, by) \quad \overset{\mathcal{F}}{\Longleftrightarrow} \quad \frac{1}{|ab|} G\left(\frac{f_x}{a}, \frac{f_y}{b}\right) \tag{6.102}$$

That is, shrinking in one domain causes expansion in the other.

6.4 Frequency-Domain Filtering

One of the important applications of the DFT is to filter an image. In the previous chapter we looked at spatial-domain techniques to filter an image, whereas here we consider frequency-domain techniques. In reality these two approaches are often equivalent due to the circular convolution theorem, which says that the Fourier transform of the convolution of two signals is equivalent to the multiplication of their Fourier transforms. In other words, if $g(x,y)$ and $h(x,y)$ are two signals, and if $g' = g \circledast h$, then it is also true that

$$G'(k_x, k_y) = G(k_x, k_y)H(k_x, k_y) \tag{6.103}$$

where $G' = \mathcal{F}\{g'\}$, $G = \mathcal{F}\{g\}$, and $H = \mathcal{F}\{h\}$. An alternate way of obtaining the desired result is therefore to compute the inverse Fourier transform of the multiplication of the two Fourier transforms:

$$g'(x, y) = \mathcal{F}^{-1}\{\mathcal{F}\{g(x, y)\}\mathcal{F}\{h(x, y)\}\} \tag{6.104}$$

assuming that appropriate care has been taken in zero-padding. As discussed earlier, this is an FIR filter with h as the impulse response and H as the frequency response, assuming g is the original signal.

There are two primary reasons for considering frequency-domain approaches. First, it is often more intuitive to design and analyze filters in the frequency domain. That is, even if a filter is eventually implemented as a spatial convolution, it is usually much easier to understand the purpose of the filter by studying its frequency response than by using the weights of the convolution kernel. Secondly, multiplication is less computationally expensive than convolution, so in some circumstances (e.g., large kernels) the frequency-domain implementation can be faster than the spatial-domain implementation, despite the overhead required to compute forward and inverse Fourier transforms. However, be aware that this argument is more applicable in signal processing than it is in image processing because large kernels are rarely necessary in the latter, due to the prevalency of multiresolution analysis, which we consider in the next chapter.

Whether spatial- or frequency-domain, filtering is used primarily for two applications: namely, restoration and enhancement. In **restoration**, the goal is to remove the effects of noise that has, in some way or another, degraded the image quality from its original condition (or its potential condition, if the corruption occurred prior to capture). **Enhancement**, on the other hand, involves accentuating or sharpening features to make the image more useful, going beyond simply a pure, noise-free image. In previous chapters we saw techniques for restoration, such as Gaussian smoothing and median filtering, and we also saw techniques for enhancement, such as histogram equalization and level slicing. In this section we consider how to accomplish these goals via frequency-domain methods using lowpass, highpass, and bandpass filters.

6.4.1 Lowpass Filtering

A **lowpass filter** allows low frequencies to pass through while attenuating high frequencies. In image processing, pixels whose values are similar to their neighbors remain relatively unchanged, while sharp transitions are smoothed. Lowpass filtering is used primarily for restoration—that is, to remove noise that has corrupted the signal.

Ideal Lowpass Filter

The **ideal lowpass filter**, also known as the box filter, perfectly passes all frequencies below a certain cutoff, while perfectly attenuating all frequencies above the cutoff. The set of frequencies below the cutoff is called the **passband**, while the set of frequencies above the cutoff is called the **stopband**. The **frequency response** of a filter will be denoted as $H(f)$, which is a complex function of the frequency f, so that $H = |H|e^{j\angle H}$, where $|H|$ is the magnitude and $\angle H$ is the phase, as mentioned before. While the phase is an important consideration in the design of causal filters for time-domain signal processing, it is not very important for image processing because convolution kernels can be centered on the output pixel, thus incurring no shift in the spatial domain. Therefore, we will assume $\angle H = 0$ for the rest of our discussion, allowing us to focus solely upon the magnitude of the frequency response of the filters we encounter, since $H = |H|$. The magnitude of the ideal lowpass filter is

$$|H_{ilp}(f)| = \begin{cases} 1 & \text{if } f \leq f_c \\ 0 & \text{otherwise} \end{cases} \tag{6.105}$$

where f_c is the cutoff frequency. Equation (6.105) shows the 1D continuous case for simplicity, but the extension to the discrete 2D case is straightforward:

$$|H(k_x, k_y)| = \begin{cases} 1 & \text{if } d(k_x, k_y) \leq d_c \\ 0 & \text{otherwise} \end{cases} \tag{6.106}$$

where d_c is the discrete cutoff frequency, and

$$d(k_x, k_y) = \left\|\left[k_x - \tfrac{w}{2}, k_y - \tfrac{h}{2}\right]\right\| = \sqrt{\left(k_x - \frac{w}{2}\right)^2 + \left(k_y - \frac{h}{2}\right)^2} \tag{6.107}$$

is the distance from the origin in frequency space.

The convolution kernel associated with the ideal lowpass filter is given by its inverse Fourier transform:

$$h(x) = \int_{-\infty}^{\infty} H_{ilp}(f)e^{j2\pi fx}\, df = \int_{-f_c}^{f_c} e^{j2\pi fx}\, df \tag{6.108}$$

$$= \frac{1}{j2\pi x}\left(e^{j2\pi f_c x} - e^{-j2\pi f_c x}\right) = \frac{\sin 2\pi f_c x}{\pi x} = 2f_c \operatorname{sinc} 2f_c x \tag{6.109}$$

where $\operatorname{sinc} x \equiv \frac{\sin \pi x}{\pi x}$ is the normalized sinc function.[†] Equation (6.105) is called a **rect function** since, when plotted, it looks like a rectangle. Thus, the Fourier transform of a sinc function is a rect function, and vice versa. To apply the ideal lowpass filter in the frequency domain, simply compute the Fourier transform of the signal, then multiply all frequencies above the cutoff frequency by zero. To apply the same filter in the spatial domain, convolve the signal with the sinc kernel.

A fundamental principle in filter design is that there is no perfect filter, and therefore the best we can do is to strike a practical balance between the various trade-offs in order to achieve the desired performance. The reason for this limitation is that no filter can have a finite extent in both the spatial and frequency domains. In other words, every filter must extend either infinitely in space, infinitely in frequency, or infinitely in both. To say this another way, no filter can be both bandlimited and timelimited. A **bandlimited** filter is one whose values in the frequency domain are zero for all $f > f_{max}$, where f_{max} is some constant. A **timelimited** filter is one whose values are zero in the time (or spatial) domain for all $x > x_{max}$, where x_{max} is some constant.

Consider, for example, Figure 6.13, which shows the ideal lowpass filter being applied to a 1D signal. With the ideal lowpass filter, $H(f)$ has finite extent, but $h(x)$ extends forever. Since it is not possible to convolve a signal with a kernel (such as sinc) that has an infinite

---

[†] Section 3.8.5 (p. 118).

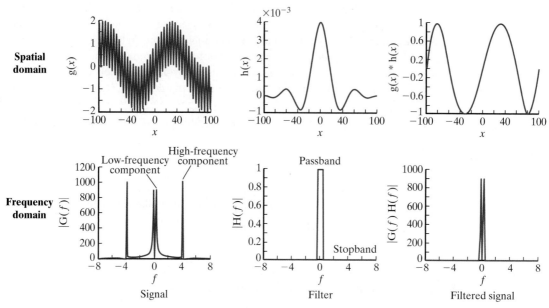

Figure 6.13 The function $\sin(f_1 x) + \sin(f_2 x)$ filtered by an ideal lowpass filter. In the frequency domain, the Fourier transform of the signal is multiplied by a box function. Equivalently, in the spatial domain, the signal is convolved with a sinc function. In this example the filter successfully removes the high-frequency component from the signal, leaving only the low-frequency component.

domain, the ideal lowpass filter is not realizable in the spatial domain. Another drawback of the sinc function is that it oscillates about the y-axis. Therefore, even the ideal lowpass filter is not a perfect filter because it gives rise to **ringing** in the output signal. Ringing, which occurs when the output signal contains oscillations that are not present in the input signal, is related to the **Gibbs phenomenon**, which occurs when a function with a jump discontinuity (such as the rect function) is approximated by a finite number of Fourier coefficients. Ringing is generally considered undesirable because it causes the signal to overshoot or undershoot, which can cause the signal to be clipped to the maximum or minimum value, thus further distorting the shape of the signal. In the example of Figure 6.14, which illustrates the process of applying a lowpass filter to an image, ringing is evident.

Figure 6.14 The process of frequency-domain filtering. From left to right: The DFT of the image is computed and multiplied by the frequency-domain filter, followed by the inverse DFT to yield the filtered image. Notice in this example that the ideal lowpass filter causes significant ringing in the output.

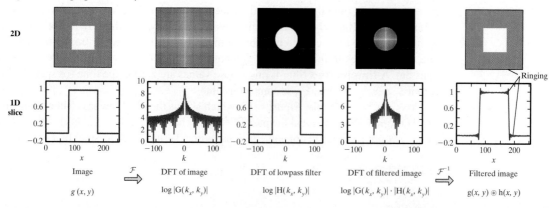

Windowing

Any filter can be implemented in either the spatial or frequency domains. That is, we can implement the filter either as a finite array of values in the frequency domain (in which case it will have infinite extent in the spatial domain), or as a finite array of values in the spatial domain (in which case it will have infinite extent in the frequency domain). Either way, it is necessary to multiply the filter by a **window** function, which is a nonnegative function that decreases monotonically from the center, such as the rect function or any of various bell-shaped curves.[†]

Let us consider what happens if we multiply (in the spatial domain) the sinc kernel by a rect window function. Since multiplication in the spatial domain is equivalent to convolution in the frequency domain, this is equivalent to (in the frequency domain) convolving the rect function (which is the Fourier transform of the sinc) with the sinc function (which is the Fourier transform of the rect). As can be seen from Figure 6.15, the windowing operation results in **ripples** in the frequency domain response of the filter.

Gaussian Lowpass Filter

This fundamental trade-off between the extent of the filter in the spatial and frequency domains leads naturally to the **Gaussian lowpass filter**, which is defined as

$$|H_{glp}(f)| = e^{-f/2f_c^2} \tag{6.110}$$

where the standard deviation f_c of the Gaussian plays the role of the cutoff frequency. It can be shown that the Gaussian filter is the perfect balance between the extent in the two domains, in the sense that it is the shape that minimizes the product of the spatial- and frequency-domain functions. Because the Fourier transform of the Gaussian is another Gaussian and the Gaussian is a monotonic function on either side of the mean, it is easy to see that filtering with a Gaussian does not yield any ripples (in the frequency domain) or produce any ringing (in the spatial domain). However, one of the drawbacks of the Gaussian is its very mild **roll-off** from the passband to the stopband, unlike the steep

Figure 6.15 When the ideal lowpass filter (left) is multiplied by a window function (middle), the resulting filter exhibits ripples (right). Note that the bottom middle plot shows the absolute value of the sinc function.

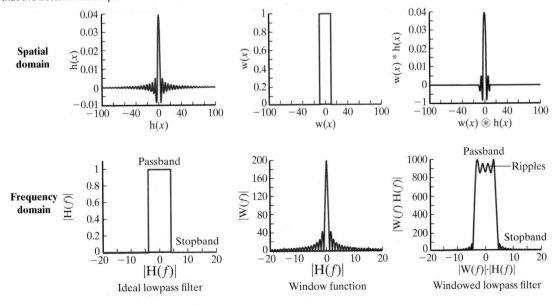

[†] Such as the Hann, Hamming, or Bartlett-Hann window functions.

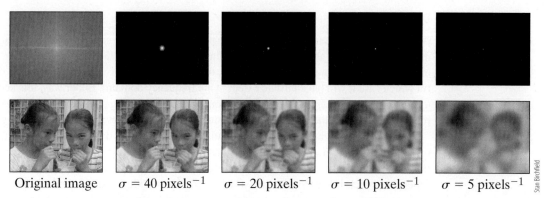

Original image $\sigma = 40$ pixels^{-1} $\sigma = 20$ pixels^{-1} $\sigma = 10$ pixels^{-1} $\sigma = 5$ pixels^{-1}

Figure 6.16 An image, and the result of Gaussian low-pass filtering in the frequency domain with different variances. The top row shows the DFT of the image and the magnitude of the frequency response of each filter. The smoothed images are the inverse DFT of the multiplication of the image DFT with the various filter frequency responses. Note that a large variance in the frequency domain yields less smoothing, whereas a small variance yields more smoothing.

roll-off of the windowed box function. These two alternatives are analogous to *underdamped* and *overdamped* control systems, where a quick response time goes hand-in-hand with overshooting and ringing. The result of Gaussian lowpass filtering on an image, with different variances, is shown in Figure 6.16.

Butterworth Lowpass Filter

Filter design is a delicate process that has been studied extensively by the signal processing community for decades. To overcome the undesired effects of the simplistic rect and Gaussian filters, filter designers have proposed a number of more sophisticated filters. One of the more common is the **elliptic filter**, which is a general type of filter that allows the designer to independently specify the amount of ripple in the passband and stopband. As the amount of ripple goes to zero in the passband or stopband, the elliptic filter is well approximated by a **Chebyshev filter**. If the ripple goes to zero in both the passband and stopband, the elliptic filter approximates a **Butterworth filter**. The Butterworth filter is also known as the *maximally flat* filter (since all derivatives exist and are zero at the origin), and it is generally considered a good compromise between various trade-offs and is therefore widely used in signal processing applications. The magnitude-squared of the Butterworth lowpass filter of order n is given by

$$|H_{blp}(f)|^2 = \frac{1}{1 + (f/f_c)^{2n}} \tag{6.111}$$

Taking the square root of this expression yields the magnitude of the filter, from which it is clear that the frequency response is $|H_{blp}(f)| = 1/\sqrt{2}$ at the cutoff frequency $f = f_c$, for any value of n. Unlike the elliptic and Chebyshev filters, the Butterworth filter is monotonic in both the passband and stopband. While the Butterworth roll-off for order $n = 2$ is notoriously slow, as n increases, the shape of the Butterworth approximates the ideal lowpass filter, as shown in Figure 6.17.

Sometimes you will see this equation without the square:

$$|H_{sblp}(f)| = \frac{1}{1 + (f/f_c)^{2n}} \tag{6.112}$$

which could be called the *"sloppy Butterworth"*. While the sloppy Butterworth does not possess any particularly interesting spectral properties (and is therefore not used in signal processing applications), its simplicity (i.e., lack of a square root) makes it a somewhat popular choice for the more forgiving area of image processing, where specific spectral properties are much less important than the overall shape of the function.

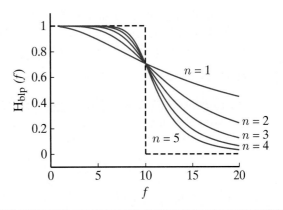

Figure 6.17 The magnitude of the Butterworth lowpass filter for $n = 1$ to $n = 5$ (solid lines). As n increases, the Butterworth response approaches the ideal lowpass filter (dashed line).

Lanczos Filter

When the sinc function is multiplied by the first lobe of another sinc function, the result is the **Lanczos filter**,[†] which is a high-quality filter that is widely used in image processing, particularly for smoothing an image before downsampling. Because of its high computational requirements compared to simple Gaussian kernels,[‡] however, it is not well suited to real-time applications. While the Lanczos filter can be implemented in the spatial domain as easily as the frequency domain, it belongs in this chapter because it is more easily explained in relation to the ideal lowpass filter.

To understand the details of the Lanczos filter, recall that the Fourier transform of a rect function is a sinc function, and vice versa. More specifically, if τ and β are scaling factors in the spatial and frequency domains, respectively, then we have the following Fourier pairs:

$$\text{rect}\left(\frac{x}{\tau}\right) = \begin{cases} 1 & \text{if } |x| \leq \dfrac{\tau}{2} \\ 0 & \text{otherwise} \end{cases} \quad \overset{\mathcal{F}}{\Longleftrightarrow} \quad \tau \, \text{sinc}(\tau f) = \frac{\sin \pi \tau f}{\pi \tau f} \tag{6.113}$$

$$\beta \, \text{sinc}(\beta x) = \frac{\sin \pi \beta x}{\pi \beta x} \quad \overset{\mathcal{F}}{\Longleftrightarrow} \quad \text{rect}\left(\frac{f}{\beta}\right) = \begin{cases} 1 & \text{if } |f| \leq \dfrac{\beta}{2} \\ 0 & \text{otherwise} \end{cases} \tag{6.114}$$

The Lanczos convolution kernel is the product of two kernels, one that performs the work of the ideal lowpass filter, and one that performs windowing. For an ideal lowpass filter with cutoff frequency f_c, we have $\beta = 2f_c$, leading to

$$w_1(x) = 2f_c \text{sinc}(2f_c x) = 2f_c \frac{\sin 2\pi f_c x}{2\pi f_c x} = \frac{\sin 2\pi f_c x}{\pi x} \tag{6.115}$$

For the window function, we apply the first lobe of another sinc:

$$w_2(x) = \begin{cases} \text{sinc}\left(\frac{x}{\tilde{w}}\right) = \dfrac{\sin \pi \frac{x}{\tilde{w}}}{\pi \frac{x}{\tilde{w}}} & \text{if } x = -\tilde{w}, \ldots, \tilde{w} \\ 0 & \text{otherwise} \end{cases} \tag{6.116}$$

where $w = 2\tilde{w} + 1$ is the width of the kernel. Multiplied together, these two yield the Lanczos convolution kernel:

$$h(x) = w_1(x)w_2(x) = \frac{\sin 2\pi f_c x}{\pi x} \cdot \frac{\sin \pi \frac{x}{\tilde{w}}}{\pi \frac{x}{\tilde{w}}} \qquad x = -\tilde{w}, \ldots, \tilde{w} \tag{6.117}$$

[†] Recall the closely-related concept of Lanczos interpolation in Section 3.8.5 (p. 118).
[‡] Section 5.2 (p. 222).

Smaller kernels exhibit more noticeable ringing, while larger kernels result in wasted computation because they do not yield noticeable improvement in the output quality. Typical values for w lie in the range of 19 to 31.

6.4.2 Highpass Filtering

In the frequency domain, the magnitude of a **highpass filter** is just the magnitude of an allpass filter minus the magnitude of the corresponding lowpass filter. Since an allpass filter does not attenuate any of the frequencies, it has the value 1 everywhere, leading to

$$|H_{highpass}(f)| = 1 - |H_{lowpass}(f)| \tag{6.118}$$

which is illustrated in Figure 6.18 for an ideal lowpass filter.

Applying this equation to the lowpass filters of the previous section yields the following:

$$|H_{ihp}(f)| = \begin{cases} 1 & \text{if } f \geq f_c \\ 0 & \text{otherwise} \end{cases} \qquad \text{(ideal highpass filter)} \tag{6.119}$$

$$|H_{bhp}(f)| = 1 - \sqrt{\frac{1}{1 + (f/f_c)^{2n}}} \qquad \text{(Butterworth highpass filter)} \tag{6.120}$$

$$|H_{sbhp}(f)| = \frac{1}{1 + (f_c/f)^{2n}} \qquad \text{(sloppy Butterworth highpass filter)} \tag{6.121}$$

$$|H_{ghp}(f)| = 1 - e^{-f/2f_c^2} \qquad \text{(Gaussian highpass filter)} \tag{6.122}$$

When the filter is zero-phase, the filter is equal to its magnitude, $H(f) = |H(f)|$, thus simplifying Equation (6.118) to $H_{highpass}(f) = 1 - H_{lowpass}(f)$. This leads to a simple relationship in the spatial domain between convolution with a highpass kernel h_{hp} and its corresponding lowpass kernel h_{lp}:

$$g'(x) = g(x) \circledast h_{hp}(f) \tag{6.123}$$
$$= \mathcal{F}^{-1}\{G(f)H_{hp}(f)\} \tag{6.124}$$
$$= \mathcal{F}^{-1}\{G(f)(1 - H_{lp}(f))\} \tag{6.125}$$
$$= \mathcal{F}^{-1}\{G(f)\} - \mathcal{F}^{-1}\{G(f)H_{lp}(f)\} \tag{6.126}$$
$$= g(x) \circledast (\delta(x) - h_{lp}(x)) \tag{6.127}$$
$$= g(x) - g(x) \circledast h_{lp}(x) \tag{6.128}$$

where Equation (6.127) follows from Equation (6.125), and Equation (6.128) follows from Equation (6.126). To derive Equation (6.127), recall that the Fourier transform of a Dirac delta function is

$$G(f) = \int \delta(x)e^{-j2\pi fx}\,dx = 1 \tag{6.129}$$

from the sifting property in Equation (6.8), so that the inverse Fourier transform of an allpass filter is a delta function. The result of Gaussian highpass filtering on an image, with different variances, is shown in Figure 6.19.

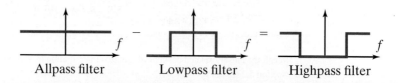

Figure 6.18 A highpass filter is the allpass filter minus a lowpass filter

Allpass filter   Lowpass filter   Highpass filter

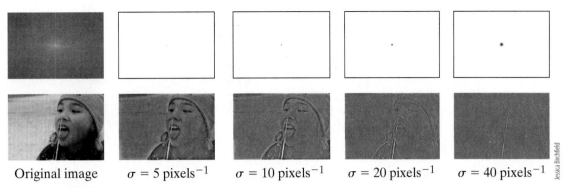

Original image $\sigma = 5$ pixels^{-1} $\sigma = 10$ pixels^{-1} $\sigma = 20$ pixels^{-1} $\sigma = 40$ pixels^{-1}

Figure 6.19 An image, and the result of Gaussian high-pass filtering in the frequency domain with different variances. The top row shows the DFT of the image and the magnitude of the frequency response of each filter. Bright values indicate frequencies that are passed, whereas dark values indicate frequencies that are attenuated.

6.4.3 Bandpass Filtering

A **bandpass filter** rejects both low and high frequencies, instead passing only frequencies in a certain band. The ideal bandpass filter is

$$|H(f)| = \begin{cases} 1 & \text{if } f_{lo} \leq f \leq f_{hi} \\ 0 & \text{otherwise} \end{cases} \tag{6.130}$$

where the passband is between f_{lo} and f_{hi}.

Laplacian of Gaussian (LoG) filter

By far the most common bandpass filter for image processing is the **Laplacian of Gaussian (LoG) filter**, which we saw in the previous chapter:

$$|H(f)| = -f^2 e^{-f/2f_c^2} \tag{6.131}$$

This expression follows from the well-known formula for the Fourier transform of the nth derivative of an arbitrary function g:

$$\mathcal{F}\left\{\frac{d^n g(x)}{dx^n}\right\} = (jf)^n G(f) \tag{6.132}$$

by letting g and G be the Gaussian, letting $n = 2$, and recognizing that $j^2 = -1$. The LoG filter is sometimes known as the **Laplacian filter**.

It is worth noting that 3 elements are not sufficient to capture the bandpass nature of the Laplacian. In the previous chapter we noted that the only 3×1 second-derivative Gaussian kernel is $\begin{bmatrix} 1 & -2 & 1 \end{bmatrix}$. The DFT of this kernel is $\begin{bmatrix} -3 & 0 & -3 \end{bmatrix}$, which removes the low-frequency DC component while passing the other high frequency. In other words, every 3×3 LoG kernel acts like a highpass filter. To keep the DFT nonnegative (and hence zero-phase) we typically use the negative LoG kernel $\begin{bmatrix} -1 & 2 & -1 \end{bmatrix}$, whose DFT is $\begin{bmatrix} 3 & 0 & 3 \end{bmatrix}$. It is easy to see that the negative LoG kernel is just the scaled difference between allpass and lowpass filters:

$$\underbrace{\begin{bmatrix} -1 & 2 & -1 \end{bmatrix}}_{\text{highpass}} = 3\left(\underbrace{\begin{bmatrix} 0 & 1 & 0 \end{bmatrix}}_{\text{allpass}} - \frac{1}{3}\underbrace{\begin{bmatrix} 1 & 1 & 1 \end{bmatrix}}_{\text{lowpass}}\right) \tag{6.133}$$

or in 2D,

$$
\underbrace{\begin{bmatrix} 0 & -1 & 0 \\ -1 & 4 & -1 \\ 0 & -1 & 0 \end{bmatrix}}_{\text{highpass}} = \eta \left(\underbrace{\begin{bmatrix} 0 & 0 & 0 \\ 0 & 1 & 0 \\ 0 & 0 & 0 \end{bmatrix}}_{\text{allpass}} - \frac{1}{6} \underbrace{\begin{bmatrix} 0 & 1 & 0 \\ 1 & 2 & 1 \\ 0 & 1 & 0 \end{bmatrix}}_{\text{lowpass}} \right)
\tag{6.134}
$$

where $\eta = 6$. Since the convolution of the image with an allpass filter is just the image itself, we have

$$
I \circledast (-LoG) = \eta(I - I \circledast h_{lowpass})
\tag{6.135}
$$

where $h_{lowpass}$ is the 3×3 lowpass kernel and η is the associated scaling factor. Table 6.2 provides the lowpass kernels for several 3×3 LoG kernels, along with their scaling factors.

EXAMPLE 6.5

Given the following lowpass kernel:

$$
h_{lowpass}(x, y) = \frac{1}{9} \begin{bmatrix} 1 & 1 & 1 \\ 1 & 1 & 1 \\ 1 & 1 & 1 \end{bmatrix} = \frac{1}{3} \begin{bmatrix} 1 \\ 1 \\ 1 \end{bmatrix} \circledast \frac{1}{3} \begin{bmatrix} 1 & 1 & 1 \end{bmatrix}
\tag{6.136}
$$

find the equivalent LoG kernel that satisfies Equation (6.6), and find the scaling factor η.

Solution

An image does not change when it is convolved with the allpass filter. Combined with the linearity property of convolution, this yields

$$
\begin{aligned}
I'(x, y) &= \eta(I(x, y) - I(x, y) \circledast h_{lowpass}(x, y)) \tag{6.137} \\
&= \eta \left(I(x, y) \circledast \begin{bmatrix} 0 & 0 & 0 \\ 0 & 1 & 0 \\ 0 & 0 & 0 \end{bmatrix} - I(x, y) \circledast \frac{1}{9} \begin{bmatrix} 1 & 1 & 1 \\ 1 & 1 & 1 \\ 1 & 1 & 1 \end{bmatrix} \right) \tag{6.138} \\
&= \eta \left(I(x, y) \circledast \left(\begin{bmatrix} 0 & 0 & 0 \\ 0 & 1 & 0 \\ 0 & 0 & 0 \end{bmatrix} - \frac{1}{9} \begin{bmatrix} 1 & 1 & 1 \\ 1 & 1 & 1 \\ 1 & 1 & 1 \end{bmatrix} \right) \right) \tag{6.139} \\
&= I(x, y) \circledast \frac{\eta}{9} \begin{bmatrix} -1 & -1 & -1 \\ -1 & 8 & -1 \\ -1 & -1 & -1 \end{bmatrix} \tag{6.140}
\end{aligned}
$$

The convolution kernel in the final line is recognized as the $LoG_{0.33}$ kernel of Table 5.5, multiplied by $-\eta$, where $\eta = 3$. The result is shown in the penultimate column of Table 6.2.

Unsharp Masking and Highboost Filtering

The Laplacian leads to a popular way to enhance an image known as **sharpening**. This approach takes advantage of a peculiarity of the human visual system, namely that neurons in the retina distort the intensity values based on neighboring intensities. Such a distortion is evidenced in the well-known **Mach bands illusion**, illustrated in Figure 6.20, which reveals that the human brain perceives exaggerated intensity changes near intensity edges. Capitalizing on this phenomenon, the sharpening trick to image enhancement introduces artificial Mach bands by exaggerating intensity edges.

Sharpening is almost always performed with a Laplacian kernel, and it can be done in either the spatial or frequency domain. In the spatial domain, simply subtract a blurred

$\begin{bmatrix} 0 & 1 & 0 \\ 1 & -4 & 1 \\ 0 & 1 & 0 \end{bmatrix}$	$\frac{1}{6}\begin{bmatrix} 1 & 4 & 1 \\ 4 & -20 & 4 \\ 1 & 4 & 1 \end{bmatrix}$	$\frac{1}{5}\begin{bmatrix} 1 & 3 & 1 \\ 3 & -16 & 3 \\ 1 & 3 & 1 \end{bmatrix}$	$\frac{1}{4}\begin{bmatrix} 1 & 2 & 1 \\ 2 & -12 & 2 \\ 1 & 2 & 1 \end{bmatrix}$	$\frac{1}{3}\begin{bmatrix} 1 & 1 & 1 \\ 1 & -8 & 1 \\ 1 & 1 & 1 \end{bmatrix}$	$\frac{1}{2}\begin{bmatrix} 1 & 0 & 1 \\ 0 & -4 & 0 \\ 1 & 0 & 1 \end{bmatrix}$
$\sigma^2 = 0.0$	$\sigma^2 = 0.167$	$\sigma^2 = 0.20$	$\sigma^2 = 0.25$	$\sigma^2 = 0.33$	$\sigma^2 = 0.5$
$\frac{1}{6}\begin{bmatrix} 0 & 1 & 0 \\ 1 & 2 & 1 \\ 0 & 1 & 0 \end{bmatrix}$	$\frac{1}{36}\begin{bmatrix} 1 & 4 & 1 \\ 4 & 16 & 4 \\ 1 & 4 & 1 \end{bmatrix}$	$\frac{1}{25}\begin{bmatrix} 1 & 3 & 1 \\ 3 & 9 & 3 \\ 1 & 3 & 1 \end{bmatrix}$	$\frac{1}{16}\begin{bmatrix} 1 & 2 & 1 \\ 2 & 4 & 2 \\ 1 & 2 & 1 \end{bmatrix}$	$\frac{1}{9}\begin{bmatrix} 1 & 1 & 1 \\ 1 & 1 & 1 \\ 1 & 1 & 1 \end{bmatrix}$	$\frac{1}{4}\begin{bmatrix} 1 & 0 & 1 \\ 0 & 0 & 0 \\ 1 & 0 & 1 \end{bmatrix}$
—	$\frac{1}{6}\begin{bmatrix} 1 & 4 & 1 \end{bmatrix}$	$\frac{1}{5}\begin{bmatrix} 1 & 3 & 1 \end{bmatrix}$	$\frac{1}{4}\begin{bmatrix} 1 & 2 & 1 \end{bmatrix}$	$\frac{1}{3}\begin{bmatrix} 1 & 1 & 1 \end{bmatrix}$	$\frac{1}{2}\begin{bmatrix} 1 & 0 & 1 \end{bmatrix}$
—	$\sigma_s^2 = 0.333$	$\sigma_s^2 = 0.40$	$\sigma_s^2 = 0.5$	$\sigma_s^2 = 0.67$	$\sigma_s^2 = 1.0$
$\eta = 6$	$\eta = 6$	$\eta = 5$	$\eta = 4$	$\eta = 3$	$\eta = 2$

TABLE 6.2 The lowpass kernels associated with various LoG kernels. The first two rows show the discrete 3×3 LoG kernels from Table 5.5, along with their variances σ^2. The next four rows show the corresponding 3×3 lowpass kernel $h_{1owpass}$ so that $I \circledast (-LoG) = \eta(I - I \circledast h_{lowpass})$, along with the 3×1 generator for the separable lowpass kernel and its variance σ_s^2, and the associated scaling factor η. (The 3×3 lowpass kernel in the first column is not separable.)

version of the image from the image, then add the result back to the original image. Inserting scaling factors a and b, this is represented mathematically as

$$I'(x, y) = I(x, y) + (aI(x, y) - b\bar{I}(x, y)) \tag{6.141}$$

$$= I(x, y) + (aI(x, y) - bI(x, y) \circledast h_{lowpss}) \tag{6.142}$$

$$= (1 + a - b)I(x, y) + bI(x, y) \circledast h_{highpass} \tag{6.143}$$

$$= (1 + a - b)I(x, y) + \frac{b}{\eta}I(x, y) \circledast (-LoG(x, y)) \tag{6.144}$$

where $\bar{I}(x, y)$ is a blurred version of the image, $h_{highpass} = h_{allpass} - h_{lowpass}$ is the 2D highpass kernel, $-LoG$ is the zero-phase negative-LoG kernel, and the last equality follows from Equation (6.135) when the kernel is 3×3. This technique is known as either **unsharp masking** or **highboost filtering**. Sometimes the former name is reserved for the case when $a = b = 1$, while the latter term is reserved for the case when $a = b > 1$, but such a distinction seems unnecessary since all cases involve boosting, or emphasizing, the high frequencies. The process of unsharp masking is illustrated in Figure 6.21.

The expression in Equation (6.144) is equivalent to the following convolution:

$$I'(x, y) = I(x, y) \circledast h_{USM}(x, y) \tag{6.145}$$

Figure 6.20 The Mach bands illusion. Left: Image consisting of dark and light regions, with a linear transition between them. The human visual system hallucinates a dark band left of the transition and a bright band right of the transition. Right: 1D slice through the image, showing the actual graylevel function and the perceived function.

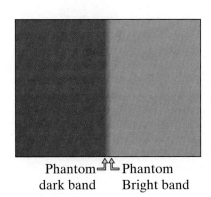

Phantom⤒⤓Phantom
dark band   Bright band

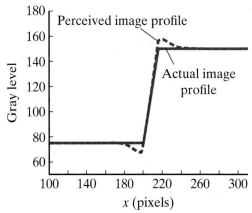

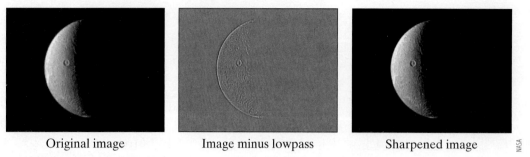

Original image   Image minus lowpass   Sharpened image

Figure 6.21 From left to right: An image of Saturn's moon Dione, the result of subtracting the low-pass filtered version of the image from itself, and the sharpened image resulting from adding this subtraction back to the original image.

where

$$h_{USM} \equiv (1 + a - b)h_{allpass} - \frac{b}{\eta}LoG \tag{6.146}$$

Different choices for the LoG kernel lead to different convolution kernels, such as

$$h_{USM,0.0}(x, y) = \begin{bmatrix} 0 & 0 & 0 \\ 0 & 1+a-b & 0 \\ 0 & 0 & 0 \end{bmatrix} - \frac{b}{6}\begin{bmatrix} 0 & 1 & 0 \\ 1 & -4 & 1 \\ 0 & 1 & 0 \end{bmatrix} = \frac{b}{6}\begin{bmatrix} 0 & -1 & 0 \\ -1 & \frac{6}{b}(1+a)-2 & -1 \\ 0 & -1 & 0 \end{bmatrix}, \text{ or} \tag{6.147}$$

$$h_{USM,0.33}(x, y) = \begin{bmatrix} 0 & 0 & 0 \\ 0 & 1+a-b & 0 \\ 0 & 0 & 0 \end{bmatrix} - \frac{b}{3}\cdot\frac{1}{3}\begin{bmatrix} 1 & 1 & 1 \\ 1 & -8 & 1 \\ 1 & 1 & 1 \end{bmatrix} = \frac{b}{9}\begin{bmatrix} -1 & -1 & -1 \\ -1 & \frac{9}{b}(1+a)-1 & -1 \\ -1 & -1 & -1 \end{bmatrix} \tag{6.148}$$

Figure 6.22 shows the results of sharpening an image using unsharp masking via Equation (6.148).

Due to the circular convolution theorem, the equivalent representation of unsharp masking in the frequency domain is

$$I'(x, y) = \mathcal{F}^{-1}\{\mathcal{F}\{I(x, y)\} \cdot H_{USM}(k_x, k_y)\} \tag{6.149}$$

where

$$H_{USM}(k_x, k_y) = \mathcal{F}\{h_{USM}\} = (1 + a - b) + bH_{highpass}(K_x, k_y) \tag{6.150}$$

Figure 6.22 The process of image sharpening: The image is convolved with the LoG, and the result is subtracted from the original image. The edge in the right column appears sharper than that in the left column.

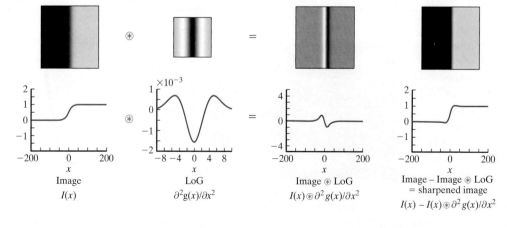

where the highpass filter $H_{highpass}$ is, in the 3×3 case, the DFT of the scaled negative-LoG kernel. The filter H_{USM} is sometimes known as a **high-frequency-emphasis** filter.

In case you are wondering about the name "unsharp masking"—which may seem odd since it is a technique used to sharpen an image—the term comes from the old days of photography (before the digital age). A picture taken by a camera would be stored on a piece of film called a "negative." In a darkroom a light would be shone through the negative to produce a positive image, which was then developed using special chemicals. To accomplish unsharp masking, the light would first be shone through a thin piece of glass to create a slightly out-of-focus positive image, which would then be aligned with the original in-focus negative image for another round of exposure. Light shining simultaneously through the in-focus negative and out-of-focus positive images would then create the unsharp masked version.

6.4.4 Homomorphic Filtering

In an earlier chapter[†] we looked at the notion of *intrinsic images*, in which a 2D image is decomposed into the individual causes of the image such as illumination and reflectance. **Homomorphic filtering** is a simple and classic technique to separate these two components—that is, to estimate a model that explains for each pixel the contribution due to light shining in the scene and the contribution due to the surface reflecting the light.

In Equation (2.11) we saw that the irradiance E at a point (x, y) is the product of the light Λ and the surface reflectance R:

$$E(x, y) = \Lambda(x, y)R(x, y) \tag{6.151}$$

where we have ignored the dependency on wavelength for simplicity. Taking the Fourier transform of both sides does not help, because there is no formula relating the product of two DFTs:

$$\mathcal{F}\{E(x, y)\} \neq \mathcal{F}\{\Lambda(x, y)\}\mathcal{F}\{R(x, y)\} \tag{6.152}$$

However, if we first take the logarithm of the image (which, after gamma expansion, we assume to be the irradiance), then a simple relationship emerges:

$$\log E(x, y) = \log \Lambda(x, y) + \log R(x, y) \tag{6.153}$$

$$\mathcal{F}\{\log E(x, y)\} = \mathcal{F}\{\log \Lambda(x, y)\} + \mathcal{F}\{\log R(x, y)\} \tag{6.154}$$

Typically the lighting function Λ contains primarily low frequencies, while the reflectance function R contains more high-frequency information. In homomorphic filtering, the DFT of the logarithm of the image is computed, then the result is filtered using a lowpass or highpass filter to process reflectance and lighting differently. Finally, the exponential of the inverse DFT of the result is computed. Figure 6.23 shows the results of homomorphic filtering on an image to reveal details in the shadows that are not visible in the original image.

Figure 6.23 Left: An image with severe shadows, and the result of homomorphic filtering using a high-frequency filter to reduce the influence of lighting. Right: the result of multiplying the image by a constant and adding a constant to the image, for comparison. Note the ability of homomorphic filtering to reveal details in the shadow of the canon that are not visible in any of the other images.

Original image   Homomorphic filtered   Increased gain   Increased bias

[†] Section 2.2.5 (p. 42).

6.5 Localizing Frequencies In Time

As we have seen throughout this chapter, the Fourier transform is a handy technique for extracting frequency information from a 1D or 2D signal, as well as to filter such information. A serious drawback of the Fourier transform, however, is that it only indicates the presence of a certain frequency in the signal without providing any information about the location of that frequency within the signal. In other words, although the Fourier transform indicates *which* frequencies are present in the signal, it does not tell us anything about *where* they are located.

Musical notation, such as that shown in Figure 6.24, provides a helpful way to look at this problem. Such notation captures not only *which* notes are played, but also *when* they are played. The original 1D time-domain signal is, in some sense, a projection of this notation onto the time axis, making it easy to determine when the notes are played but not so easy to determine which notes are played. The Fourier transform, on the other hand, can be thought of as a projection of this notation onto the frequency axis, making it easy to determine which notes are played but not when they are played. What we desire is a way to transform the original signal that captures both types of information.

6.5.1 Gabor Limit

To better understand this trade-off, suppose you want to play a short pulse of a note on a musical instrument, like a flute. You place your fingers carefully over the flute and blow into the mouthpiece for a brief period of time. The question is, How long should you blow? If the pulse is too short, then it will not be easy to tell *which* note was played, but if your pulse is too long, then it will not be easy to tell *when* the note was played (because it will not appear as a pulse anymore). In the former case the pulse cannot be localized in *frequency*, whereas in the latter case it cannot be localized in *time*. This fundamental trade-off is captured in the **Gabor limit**,[†] which says that a signal cannot be localized simultaneously in both frequency and time.

The Gabor limit forces us into a fundamental trade-off between localizing in frequency and localizing in time. If our goal is to balance this trade-off the best we can, then it is clear that the duration of the pulse should be related to the frequency that we are trying to localize. That is, a high-frequency tone should receive a shorter pulse, while a low-frequency tone should receive a longer pulse, as illustrated in Figure 6.25. The pulse should be long enough to capture at least one period (or cycle) of the tone, but not so long as to capture an unnecessarily large number of periods. Since the period is inversely proportional to the frequency, this tells us that the desired pulse duration should also be inversely proportional to the frequency.

Figure 6.24 Musical notation is a convenient way of expressing frequencies as they occur in time. The 1D time-domain signal can be thought of as the projection of this musical notation onto the time axis, while the 1D frequency-domain representation is the projection onto the frequency axis.

[†] Dennis Gabor (1900–1979), in addition to his work on wavelets, is best known for inventing holography, for which he received the Nobel Prize in Physics.

Figure 6.25 The duration of a pulse of a pure frequency should be determined by the frequency. Shown here are two signals (high-frequency on the left, low-frequency on the right) modulated by two different window functions (short-duration window on top, long-duration window on bottom). In the top-right the pulse is too short, so that the frequency is not discernible, whereas in the bottom-left the pulse is too long, making it difficult to precisely locate the pulse. In the top-left and bottom-right, the width of the pulse is appropriately chosen.

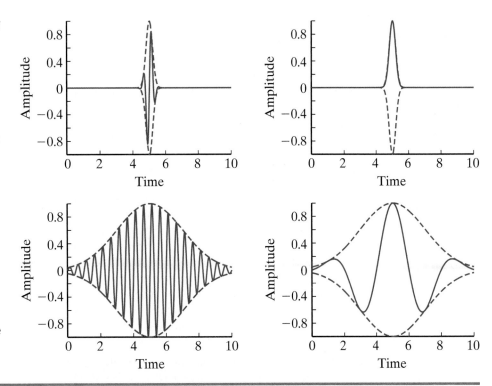

6.5.2 Short-Time Fourier Transform (STFT)

One approach to capturing time-localized frequencies of a signal is to slide a window function (e.g., a Gaussian) across the signal and at each location in time compute the Fourier transform of the windowed signal. This approach, known as the **short-time Fourier transform (STFT)** or the **windowed Fourier transform**, is computationally expensive because it requires computing many Fourier transforms for a single signal. More importantly, it violates the principle of the Gabor limit, which says that we should multiply the signal by a windowing function whose width is inversely related to the frequency we are trying to capture. In other words, high frequencies need narrow windows, while low frequencies require wide windows; but the STFT multiplies the signal by the windowing function *before* extracting the frequencies, and thus it subjects all frequencies to the same windowing function. It is unavoidable, then, that the windowing function will be too wide for high frequencies and too narrow for low frequencies. What we need is a way to adaptively adjust the windowing function applied *to* the signal based on the frequencies that we are seeking to extract *from* the signal. While this goal might at first glance appear to be enormously difficult (or even impossible), there is in fact a simple, computationally efficient way to achieve it: namely, the wavelet transform.

6.6 Discrete Wavelet Transform (DWT)

The key idea of the wavelet transform is to determine the locations of frequencies in a signal in such a way that the frequencies are taken into account when determining their location. Like the Fourier transform, the wavelet transform can be either discrete or continuous, and it can be applied to either infinite-duration or finite-duration signals. We will focus our attention primarily upon the **discrete wavelet transform (DWT)** because it is both easier to understand and more practical. But for the most part, the discussion below will apply equally well to any of the versions.

Like the Fourier transform, whose basis functions are sines and cosines, the wavelet transform also projects the signal onto basis functions. Starting with a **mother wavelet**, denoted by $\psi(x)$, the wavelet transform basis functions are obtained by scaling and shifting the mother wavelet:

$$\psi_{a,b}(x) \equiv \frac{1}{\sqrt{a}} \psi\left(\left\lfloor \frac{x-b}{a} \right\rfloor \right) \tag{6.155}$$

where the translation b determines the location of the wavelet, and the scaling (or dilation) $a > 0$ governs its frequency. Note that $\psi = \psi_{1,0}$. The normalization $\frac{1}{\sqrt{a}}$ ensures that $\|\psi_{a,b}(x)\|$ is unaffected by a or b. For a discrete signal, both a and b are integers, and the floor operator is necessary to ensure that the argument to the mother wavelet is an integer; for a continuous signal, the floor operator can be removed, and a and b allowed to be real.

The **discrete wavelet transform (DWT)** of a 1D discrete signal $g(x)$ is a 2D array of values $G(a,b)$, where each element in the array is the sum of the elementwise product of the signal with the appropriate wavelet function:

$$G(a,b) \equiv \sum_x g(x)\psi_{a,b}(x) \tag{6.156}$$

This equation clearly reveals that the wavelet transform is by its very nature massively redundant, because it replaces a 1D function $g(x)$ with a 2D function $G(a,b)$. If a and b are allowed to take on any integer values, then the transform is **overcomplete**, because it contains more information than is necessary to represent the original signal faithfully. The beauty of the wavelet transform is that this redundancy can be removed and still retain all the essential information by spacing a and b appropriately. For a signal with length w, we typically set $a = 2^j$ and $b = 2ak$, where $j = 0, \ldots, \log_2 w - 1$ and $k = 0, \ldots, \frac{w}{2a} - 1$, so that successive frequencies are separated by an octave, and the translation keeps neighboring wavelets well-separated. This is called **critical sampling**, and it yields sparse basis functions that balance the competing design goals of accurately representing the signal while not being too redundant, just as the Gabor limit tells us to do. A wavelet transform that is critically sampled is called **complete**, because it retains exactly the information of the original signal, enabling the inverse to be computed.

As an example, when $w = 8$, we have $j \in \{0, 1, 2\}$, with the values of k depending upon j as shown in Table 6.3. Each of the entries in the table yields one of the wavelet basis functions for $w = 8$, so that there are just 7 basis functions (4 in the first row, 2 in the second row, and 1 in the bottom row). As a result, when an 8-element signal is projected onto the basis functions using Equation (6.156), only 7 values will be obtained as outputs, meaning that information has been lost. Thus, in our attempt to remove the redundant information, we removed too much information. This problem is easily corrected by augmenting with an additional basis function, which in general is needed to preserve all the information in the original signal, whenever critical sampling is performed.

j	a	k	b
0	1	0,1,2,3	0,2,4,6
1	2	0,1	0,4
2	4	0	0

TABLE 6.3 The values of a and b needed to generate a complete wavelet basis for an 8-element input.

This additional basis function, often called the **father wavelet** (or *scaling function*), is denoted $\phi(x)$, analogous to the mother wavelet $\psi(x)$. Oftentimes, the wavelet transform is **orthogonal**, meaning that the mother and father wavelets form an orthonormal basis for the space when translated by even multiples:

$$\sum_x \psi(x - 2i)\psi(x - 2j) = \sum_x \phi(x - 2i)\phi(x - 2j) = \begin{cases} 1 & \text{if } i = j \\ 0 & \text{otherwise} \end{cases} \tag{6.157}$$

$$\sum_x \psi(x - 2i)\phi(x - 2j) = 0 \tag{6.158}$$

where i and j are arbitrary integers. To create a father wavelet orthogonal to the mother wavelet, simply reverse the order of the values and negate every other element:

$$\phi(x) = (-1)^{x+1}\psi(w - 1 - x) \tag{6.159}$$

where w is the length of the kernel, which is even.[†]

6.6.1 Haar Wavelets

The simplest and oldest type of wavelet is the **Haar wavelet**.[‡] In the continuous domain the Haar mother wavelet is simply two adjacent *boxcar functions*[§] of opposite sign:

$$\psi(x) = \begin{cases} 1 & \text{if } 0 < x < \frac{1}{2} \\ -1 & \text{if } \frac{1}{2} < x < 1 \\ 0 & \text{otherwise} \end{cases} \tag{6.160}$$

and the wavelet functions are critically sampled, as shown in Figure 6.26. In the discrete domain the Haar mother wavelet is a sequence of 1s followed by a sequence of -1s, along with some scaling. Not only are Haar wavelets the easiest way to learn about the wavelet transform, they also form the basis of widely used techniques, such as the features used in commercially available face detectors. This popularity stems from the speed at which they can be computed due to their reliance upon simple sums and differences.

The discrete Haar mother wavelet is just a 1 at $x = 0$ and a -1 at $x = 1$, appropriately scaled so the norm is 1:

$$\psi(x) \equiv \frac{1}{\sqrt{2}} \begin{cases} 1 & \text{if } x = 0 \\ -1 & \text{if } x = 1 \\ 0 & \text{otherwise} \end{cases} \tag{6.161}$$

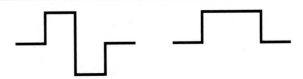

Figure 6.26 Haar basis functions are based on boxcar functions. Shown are the mother (left) and father (right) wavelets.

[†] If necessary, simply append a zero element to the end of the discrete wavelet to make w even.

[‡] Alfréd Haar (1885–1933) was a Hungarian mathematician.

[§] A boxcar function is constant over some particular interval but zero everywhere else.

To represent the Haar wavelet as a discrete array, simply ignore the zeros (which implicitly extend the array in both directions):

$$\phi(x) = \frac{1}{\sqrt{2}} \begin{bmatrix} \underline{1} & 1 \end{bmatrix} \quad \text{(father wavelet)} \tag{6.162}$$

$$\psi(x) = \frac{1}{\sqrt{2}} \begin{bmatrix} \underline{1} & -1 \end{bmatrix} \quad \text{(mother wavelet)} \tag{6.163}$$

where the father wavelet is obtained by the QMF relation in Equation (6.159). Note that ψ acts like a highpass filter, and indeed $\sum_x \psi(x) = 0$. Similarly, ϕ acts like a lowpass filter, but unlike a lowpass convolution kernel, the sum of the elements is not 1, but rather the energy of the signal is 1. That is, $\sum_x \phi(x) \neq 1$ but $\sum_x \phi^2(x) = 1$; as a result, the father wavelet has the effect of increasing the overall level of the signal, which, in the case of an image, brightens the image. Because of the quadrature mirror relation, the energy in the highpass filter is also 1: $\sum_x \psi^2(x) = 1$.

EXAMPLE 6.6

Show that the discrete Haar wavelet satisfies (6.159) and is orthogonal.

Solution

The length of the kernels is $w = 2$. From Equation (6.163) we have $\psi(0) = -\psi(1) = \frac{1}{\sqrt{2}}$. Plugging into Equation (6.159) yields

$$\phi(0) = (-1)^1 \psi(2 - 1 - 0) = -\psi(1) = \frac{1}{\sqrt{2}} \tag{6.164}$$

$$\phi(1) = (-1)^2 \psi(2 - 1 - 1) = \psi(0) = \frac{1}{\sqrt{2}} \tag{6.165}$$

which indeed matches Equation (6.162). It is easy to show that Equations (6.157)–(6.158) are satisfied:

$$\left(\frac{1}{\sqrt{2}}\right)^2 + \left(\frac{1}{\sqrt{2}}\right)^2 = \left(\frac{1}{\sqrt{2}}\right)^2 + \left(-\frac{1}{\sqrt{2}}\right)^2 = 1 \tag{6.166}$$

$$\left(\frac{1}{\sqrt{2}}\right)^2 - \left(-\frac{1}{\sqrt{2}}\right)^2 = 0 \tag{6.167}$$

EXAMPLE 6.7

Write the translated and scaled versions $\psi_{1,1}$ and $\psi_{2,0}$ of the Haar mother wavelet in Equation (6.161).

Solution

Setting $b = 1$ causes the wavelet to shift to the right by 1:

$$\psi_{1,1}(x) = \frac{1}{\sqrt{2}} \begin{cases} 1 & \text{if } x = 1 \\ -1 & \text{if } x = 2 \\ 0 & \text{otherwise} \end{cases} \tag{6.168}$$

since $\psi_{1,1}(x) = \psi(x - 1)$. Similarly, setting $a = 2$ causes the width of the wavelet to expand by 2:

$$\psi_{2,0}(x) = \frac{1}{2} \begin{cases} 1 & \text{if } x = 0 \\ 1 & \text{if } x = 1 \\ -1 & \text{if } x = 2 \\ -1 & \text{if } x = 3 \\ 0 & \text{otherwise} \end{cases} \tag{6.169}$$

since $\psi_{2,0}(x) = \frac{1}{\sqrt{2}} \psi\left(\lfloor \frac{x}{2} \rfloor\right)$. Note that the normalization factor ensures that the energy in the signal remains constant: $\|\psi_{1,0}(x)\| = \|\psi_{1,1}(x)\| = \|\psi_{2,0}(x)\| = 1$. The results are plotted in Figure 6.27.

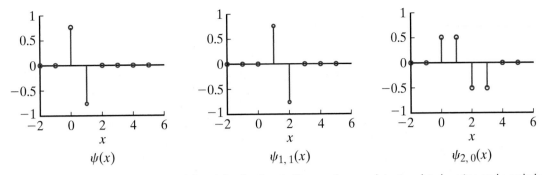

Figure 6.27 The wavelets of Example 6.7. From left to right: the discrete Haar mother wavelet, a translated version, and a scaled version.

The basis functions for the discrete Haar wavelet transform are the scaled and translated versions of the mother wavelet, along with the appropriately scaled father wavelet. For a signal with $w = 8$, for example, these scales and translations are given by the values in Table 6.3. Ignoring normalization for simplicity, the scaled father wavelet $(j = 2, a = 4)$ consists of eight 1s, and the correspondingly scaled mother wavelet consists of four 1s followed by four -1s. At the next scale $(j = 1, a = 2)$, the width of the latter wavelet is reduced by a factor of 2, so it consists of two 1s followed by two -1s, with 0s filling in the remaining elements. At this scale the translation is either $b = 0$ or $b = 4$, to avoid overlap of the nonzero elements of the wavelet functions within a scale. At the finest scale $(j = 0, a = 1)$, the width of the nonzero elements of the wavelet function is only 2, with translations of 0, 2, 4, and 6. These basis functions can be summarized as shown in Table 6.4, where the positive sign $(+)$ indicates the number $+1$, while the negative sign $(-)$ indicates the number -1. These basis functions are displayed in Figure 6.28.

| EXAMPLE 6.8 | Compute the discrete wavelet transform, using the Haar wavelet, of the 1D signal $g(x) = \begin{bmatrix} 1 & 7 & 3 & 0 & 5 & 4 & 2 & 9 \end{bmatrix}$, ignoring normalization for simplicity. |

Solution

To compute the DWT, simply sum the elementwise products of the signal with the basis functions. Because the basis functions consist of only $+1$s or -1s, this procedure requires simply adding or subtracting values. If we let $G(0, 0)$ indicate the DC component obtained from the scaled father wavelet, then this yields:

$$G(0,0) = 1 + 7 + 3 + 0 + 5 + 4 + 2 + 9 = \boxed{31} \tag{6.170}$$

$$G(4,0) = 1 + 7 + 3 + 0 - 5 - 4 - 2 - 9 = \boxed{-9} \tag{6.171}$$

$$G(2,0) = 1 + 7 - 3 - 0 = \boxed{5} \tag{6.172}$$

$$G(2,4) = 5 + 4 - 2 - 9 = \boxed{-2} \tag{6.173}$$

$$G(1,0) = 1 - 7 = \boxed{-6} \tag{6.174}$$

$$G(1,2) = 3 - 0 = \boxed{3} \tag{6.175}$$

$$G(1,4) = 5 - 4 = \boxed{1} \tag{6.176}$$

$$G(1,6) = 2 - 9 = \boxed{-7} \tag{6.177}$$

The final wavelet transform then consists of these 8 values, which we arrange by convention as the low frequency component followed by the components of increasingly higher frequencies: $\mathcal{W}\{g\} = [31, -9, 5, -2, -6, 3, 1, -7]$. As we shall see later, this transform is invertible, so the original signal can easily be recovered from these 8 values.

j	k	a	b	basis function
2	0	4	0	$\phi_{4,0} = \eta_j [+ \; + \; + \; + \; + \; + \; + \; +]$
2	0	4	0	$\psi_{4,0} = \eta_j [+ \; + \; + \; + \; - \; - \; - \; -]$
1	0	2	0	$\psi_{2,0} = \eta_j [+ \; + \; - \; - \; 0 \; 0 \; 0 \; 0]$
1	1	2	4	$\psi_{2,4} = \eta_j [0 \; 0 \; 0 \; 0 \; + \; + \; - \; -]$
0	0	1	0	$\psi_{1,0} = \eta_j [+ \; - \; 0 \; 0 \; 0 \; 0 \; 0 \; 0]$
0	1	1	2	$\psi_{1,2} = \eta_j [0 \; 0 \; + \; - \; 0 \; 0 \; 0 \; 0]$
0	2	1	4	$\psi_{1,4} = \eta_j [0 \; 0 \; 0 \; 0 \; + \; - \; 0 \; 0]$
0	3	1	6	$\psi_{1,6} = \eta_j [0 \; 0 \; 0 \; 0 \; 0 \; 0 \; + \; -]$

TABLE 6.4 Basis functions for an 8-element discrete Haar wavelet, where $+$ means $+1$, and $-$ means -1. The scaling factors are given by $\eta_j = 2^{-(j+1)/2} = \frac{1}{\sqrt{2a}}$, which leads to $\frac{1}{2\sqrt{2}}$ for the first 2 rows, $\frac{1}{2}$ for the next 2 rows, and $\frac{1}{\sqrt{2}}$ for the bottom 4 rows.

6.6.2 DWT as Matrix Multiplication

Just as we showed that the discrete Fourier transform (DFT) can be viewed as matrix multiplication,[†] so can the discrete wavelet transform (DWT). For an 8-element signal, for example, the Haar wavelet matrix is

$$\begin{bmatrix} G(0) \\ G(1) \\ G(2) \\ G(3) \\ G(4) \\ G(5) \\ G(6) \\ G(7) \end{bmatrix} = \frac{1}{\sqrt{2}} \underbrace{\begin{bmatrix} 1/2 & 1/2 & 1/2 & 1/2 & 1/2 & 1/2 & 1/2 & 1/2 \\ 1/2 & 1/2 & 1/2 & 1/2 & -1/2 & -1/2 & -1/2 & -1/2 \\ 1/\sqrt{2} & 1/\sqrt{2} & -1/\sqrt{2} & -1/\sqrt{2} & 0 & 0 & 0 & 0 \\ 0 & 0 & 0 & 0 & 1/\sqrt{2} & 1/\sqrt{2} & -1/\sqrt{2} & -1/\sqrt{2} \\ 1 & -1 & 0 & 0 & 0 & 0 & 0 & 0 \\ 0 & 0 & 1 & -1 & 0 & 0 & 0 & 0 \\ 0 & 0 & 0 & 0 & 1 & -1 & 0 & 0 \\ 0 & 0 & 0 & 0 & 0 & 0 & 1 & -1 \end{bmatrix}}_{\mathbf{H}_8} \begin{bmatrix} g(0) \\ g(1) \\ g(2) \\ g(3) \\ g(4) \\ g(5) \\ g(6) \\ g(7) \end{bmatrix} \qquad (6.178)$$

where we define $G(i)$ to be the ith element in the output, as in Equations (6.170)–(6.177). It is easy to verify that this matrix is orthogonal, that is, $\mathbf{H}_8 \mathbf{H}_8^{\mathsf{T}} = \mathbf{H}_8^{\mathsf{T}} \mathbf{H}_8 = \mathbf{I}_{\{8 \times 8\}}$.

6.6.3 Fast Wavelet Transform (FWT)

In the previous example, scaled versions of the wavelet function were repeatedly applied. Examining Equations (6.170)–(6.177), we notice that this approach involves a fair amount of redundant computation. A much faster approach, known as the **fast wavelet transform (FWT)**, simply computes the high- and low-frequency components first, then downsamples the low-passed signal and repeats until the length of the signal is too small to continue.

More specifically, if g is the signal, ϕ is the lowpass kernel, and ψ is the highpass kernel, then the computation at a single resolution is given by

$$g_{low}(x) = (g \circledast \phi) \downarrow 2 = \sum_k g(2x + k)\phi(k) \qquad (6.179)$$

[†] Section 6.2.8 (p. 288).

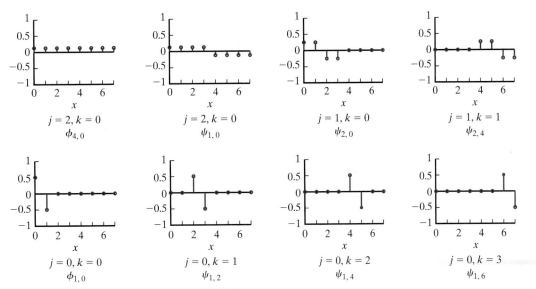

Figure 6.28 The 8 basis functions for a complete 8-element discrete Haar wavelet.

$$g_{high}(x) = (g \,\check{\circledast}\, \psi) \downarrow 2 = \sum_k g(2x + k)\psi(k),\qquad(6.180)$$

where the subsampling operator is defined as $(g \downarrow k)(x) = g(kx)$. In other words, we can conceptually think about convolving (or, rather, cross-correlating, since the kernel is not flipped, hence the symbol $\check{\circledast}$) the signal g with the kernels ϕ and ψ, then downsampling both results by 2 to get the lowpass and highpass signals—which is shown in the middle of the equations. But in reality the wasted computation is avoided by shifting the kernels by 2 samples each time—shown on the right of the equations.

After the low- and high-frequency components have been computed for one resolution, the low-frequency component is used as the signal for the next resolution. That is, at resolution i, Equations (6.179)–(6.180) become

$$g_{low}^{(i+1)}(x) = (g_{low}^{(i)} \,\check{\circledast}\, \phi) \downarrow 2 = \sum_k g_{low}^{(i)}(2x + k)\phi(k)\qquad(6.181)$$

$$g_{high}^{(i+1)}(x) = (g_{low}^{(i)} \,\check{\circledast}\, \psi) \downarrow 2 = \sum_k g_{low}^{(i)}(2x + k)\psi(k)\qquad(6.182)$$

where $g_{low}^{(0)} \equiv g$ is the original signal. Since the signal is downsampled by 2 each iteration, the method is straightforward if the length w of the signal is a power of 2; if not, then zero-padding is necessary. The final wavelet transform is given by stacking the high-frequency components at all levels, along with the low-frequency component at the lowest level, into a vector.

Pseudocode for the FWT is provided in Algorithm 6.5, where Lines 3 and 5 indicate a concatenation of the lowpass and highpass components. Unlike the FFT, whose description was omitted from the text due to its rather extensive bookkeeping, the FWT is easy to understand and implement, particularly in the case of Haar wavelets. In fact, the wavelet transform is one of those rare cases in which no engineering trade-off has to be made: Not only does the wavelet transform provide more information than the Fourier transform, but the algorithm to compute it is much simpler to implement and faster to run. If w is the length of the signal, the FWT can be computed in linear $O(w)$ time, compared with the $O(w \log w)$ asymptotic running time for the FFT.

ALGORITHM 6.5 Compute the discrete wavelet transform of a 1D signal using Haar basis functions

$\text{FastWaveletTransformHaar}(g[0], \ldots, g[w-1])$

Input: real 1D signal g of length w, which is a power of 2
Output: DWT of g (low frequency followed by increasingly higher frequencies)

1 $G_{lowpass}, G_{highpass} \leftarrow \text{HaarOneLevel}(g[0], \ldots, g[w-1])$
2 **if** $w == 2$ **then**
3   **return** $[G_{lowpass} \quad G_{highpass}]$
4 **else**
5   **return** $[\text{FastWaveletTransformHaar}(G_{lowpass}) \quad G_{highpass}]$

$\text{HaarOneLevel}(g[0], \ldots, g[w-1])$

1 **for** $k \leftarrow 0$ **to** $\frac{w}{2} - 1$ **do**
2        $v_0 \leftarrow g[2 * k]$
3        $v_1 \leftarrow g[2 * k + 1]$
4        $G_{lowpass}[k] \leftarrow v_0 + v_1$
5        $G_{highpass}[k] \leftarrow v_0 - v_1$
6 **return** $G_{lowpass}, G_{highpass}$

EXAMPLE 6.9

Apply the fast wavelet transform (FWT) to the signal $g^{(0)} = g(x) = [1 \ 7 \ 3 \ 0 \ 5 \ 4 \ 2 \ 9]$ using Haar wavelets. Ignore normalization for simplicity.

Solution

Ignoring the normalization factors, simply add and subtract pairs of adjacent elements of the signal:

$$1 + 7 = \boxed{8} \qquad 3 + 0 = 3 \qquad 5 + 4 = 9 \qquad 2 + 9 = 11$$
$$1 - 7 = \boxed{-6} \qquad 3 - 0 = \boxed{3} \qquad 5 - 4 = \boxed{1} \qquad 2 - 9 = \boxed{-7}$$

Then repeat the same procedure on the sums:

$$8 + 3 = 11 \qquad 9 + 11 = 20$$
$$8 - 3 = \boxed{5} \qquad 9 - 11 = \boxed{-2}$$

and again:

$$11 + 20 = \boxed{31}$$
$$11 - 20 = \boxed{-9}$$

Using the convention of Algorithm 6.5, the outlined results are concatenated in reverse order, so that the low-frequency component comes first. The resulting wavelet transform is therefore $\mathcal{W}\{g\} = [31 \quad -9 \quad 5 \quad -2 \quad -6 \quad 3 \quad 1 \quad -7]$, which is exactly the same result obtained in Example 6.8 .

Alternatively, Equations (6.181) and (6.182) can be applied:

$$g^{(1)}(x) = \frac{1}{\sqrt{2}}[1 + 7 \quad 3 + 0 \quad 5 + 4 \quad 2 + 9] = \frac{1}{\sqrt{2}}[8 \quad 3 \quad 9 \quad 11] \qquad (6.183)$$

$$g^{(1)}_{high}(x) = \frac{1}{\sqrt{2}}[1 - 7 \quad 3 - 0 \quad 5 - 4 \quad 2 - 9] = \frac{1}{\sqrt{2}}\left[\boxed{-6} \quad \boxed{3} \quad \boxed{1} \quad \boxed{-7}\right] \qquad (6.184)$$

At the next resolution,

$$g^{(2)}(x) = \frac{1}{2}[8 + 3 \quad 9 + 11] = \frac{1}{2}[11 \quad 20] \qquad (6.185)$$

$$g^{(2)}_{high}(x) = \frac{1}{2}[8 - 3 \quad 9 - 11] = \frac{1}{2}\left[\boxed{5} \quad \boxed{-2}\right] \qquad (6.186)$$

and at the final resolution,

$$g^{(3)}(x) = \frac{1}{2\sqrt{2}}[11 + 20] = \frac{1}{2\sqrt{2}}\left[\boxed{31}\right] \qquad (6.187)$$

$$g^{(3)}_{high}(x) = \frac{1}{2\sqrt{2}}[11 - 20] = \frac{1}{2\sqrt{2}}\left[\boxed{-9}\right] \qquad (6.188)$$

which is the same result as before except for the scaling factors. To see that the normalization factors preserve the energy in the original signal, note that

$$1^2 + 7^2 + 3^2 + 0^2 + 5^2 + 4^2 + 2^2 + 9^2 = 185$$

$$\left(\frac{31}{2\sqrt{2}}\right)^2 + \left(\frac{-9}{2\sqrt{2}}\right)^2 + \left(\frac{5}{2}\right)^2 + \left(\frac{-2}{2}\right)^2 + \left(\frac{-6}{\sqrt{2}}\right)^2 + \left(\frac{3}{\sqrt{2}}\right)^2 + \left(\frac{1}{\sqrt{2}}\right)^2 + \left(\frac{-7}{\sqrt{2}}\right)^2 = 185.$$

6.6.4 Inverse Wavelet Transform

Just as the Fourier transform admits an inverse, so does the wavelet transform. In the case of the DWT, the matrix formulation just introduced makes it easy to discover the inverse. For example, the forward Haar transform is

$$\begin{bmatrix} v'_0 \\ v'_1 \end{bmatrix} = \frac{1}{\sqrt{2}}\begin{bmatrix} 1 & 1 \\ 1 & -1 \end{bmatrix}\begin{bmatrix} v_0 \\ v_1 \end{bmatrix} \qquad (6.189)$$

where $v_0 \equiv g(2x)$, $v_1 \equiv g(2x + 1)$, $v'_0 \equiv g_{low}(x)$, and $v'_1 \equiv g_{high}(x)$. It is easy to see that the matrix is its own inverse, so that

$$\begin{bmatrix} v_0 \\ v_1 \end{bmatrix} = \frac{1}{\sqrt{2}}\begin{bmatrix} 1 & 1 \\ 1 & -1 \end{bmatrix}\begin{bmatrix} v'_0 \\ v'_1 \end{bmatrix} \qquad (6.190)$$

As with the Fourier transform, the specific values of the forward and inverse normalization factors are not important, as long as their product is (in the case of Haar) $\frac{1}{\sqrt{2}} \cdot \frac{1}{\sqrt{2}} = \frac{1}{2}$. Therefore, an alternative is to place the normalization factor entirely in the inverse transform, which is oftentimes more convenient:

$$\begin{bmatrix} v'_0 \\ v'_1 \end{bmatrix} = \begin{bmatrix} 1 & 1 \\ 1 & -1 \end{bmatrix}\begin{bmatrix} v_0 \\ v_1 \end{bmatrix} \qquad (6.191)$$

$$\begin{bmatrix} v_0 \\ v_1 \end{bmatrix} = \frac{1}{2}\begin{bmatrix} 1 & 1 \\ 1 & -1 \end{bmatrix}\begin{bmatrix} v'_0 \\ v'_1 \end{bmatrix} \qquad (6.192)$$

Recalling the definitions of v_0, v_1, v'_0, and v'_1, and carefully noting the resolution, the inverse Haar wavelet transform in Equation (6.192) can be rewritten as

$$g^{(i-1)}(2x) = \frac{1}{2}\left(g^{(i)}_{low}(x) + g^{(i)}_{high}(x)\right) \qquad (6.193)$$

$$g^{(i-1)}(2x + 1) = \frac{1}{2}\left(g^{(i)}_{low}(x) - g^{(i)}_{high}(x)\right) \qquad (6.194)$$

The pseudocode is provided in Algorithm 6.6.

ALGORITHM 6.6 Compute the inverse DWT of a 1D signal using Haar basis functions

FastInverseWaveletTransformHaar$(G[0], \ldots, G[w-1])$

Input: DWT G of a real 1D signal (low frequency followed by increasingly higher frequencies)
Output: real 1D signal g of length w whose DWT is G

1   **if** $w = 2$ **then**
2   **return** HaarOneLevelInverse$(G[0], \ldots, G[w-1])$
3   **else**
4   *first-half* $\leftarrow G[0], \ldots, G[\frac{w}{2}-1]$
5   *second-half* $\leftarrow G[\frac{w}{2}], \ldots, G[w]$
6   **return** [FastInverseWaveletTransformHaar(*first-half*)   *second-half*]

HaarOneLevelInverse$(G[0], \ldots, G[w-1])$

1   **for** $k \leftarrow 0$ **to** $\frac{w}{2} - 1$ **do**
2         $v_0 \leftarrow G[k]$
3         $v_1 \leftarrow G[k + \frac{w}{2}]$
4         $g[2*k] \leftarrow (v_0 + v_1)/2$
5         $g[2*k+1] \leftarrow (v_0 - v_1)/2$
6   **return** g

EXAMPLE 6.10

From the previous example, we have that $\mathcal{W}\{[1 \quad 7 \quad 3 \quad 0 \quad 5 \quad 4 \quad 2 \quad 9]\} = [31 \quad -9 \quad 5 \quad -2 \quad -6 \quad 3 \quad 1 \quad -7]$, using the simplified normalization of Equation (6.191). Apply the inverse fast wavelet transform to the result to verify this result.

Solution

Applying Equation (6.192) to the first two values of the inverse wavelet transform yields:

$$f^{(2)}[2n] = \frac{1}{2}[31 + (-9)] = [11] \tag{6.195}$$

$$f^{(2)}[2n+1] = \frac{1}{2}[31 - (-9)] = [20] \tag{6.196}$$

Concatenating the values yields $f^{(2)}[n] = [11 \quad 20]$. At the next resolution, these two values are combined with the next two in the sequence:

$$f^{(1)}[2n] = \frac{1}{2}[11 + 5 \quad 20 + (-2)] = [8 \quad 9] \tag{6.197}$$

$$f^{(1)}[2n+1] = \frac{1}{2}[11 - 5 \quad 20 - (-2)] = [3 \quad 11] \tag{6.198}$$

Interleaving these values for the even and odd indices yields $f^{(1)}[n] = [8 \quad 3 \quad 9 \quad 11]$. These four values are then combined with the final four:

$$f^{(0)}[2n] = \frac{1}{2}[8 + (-6) \quad 3 + 3 \quad 9 + 1 \quad 11 + (-7)] = [1 \quad 3 \quad 5 \quad 2] \tag{6.199}$$

$$f^{(0)}[2n+1] = \frac{1}{2}[8 - (-6) \quad 3 - 3 \quad 9 - 1 \quad 11 - (-7)] = [7 \quad 0 \quad 4 \quad 9] \tag{6.200}$$

Again interleaving these values yields the final result: $f^{(0)}[n] = [1 \quad 7 \quad 3 \quad 0 \quad 5 \quad 4 \quad 2 \quad 9]$, which is what we expect.

Note that the fast wavelet transform (FWT) operates on the signal at multiple resolutions. First the sums and differences are computed at the original resolution. Then the sums are used to effectively subsample the signal, and the sums and differences are computed on a signal that contains half as many samples as the original. This process of computing sums and differences and downsampling is repeated until the signal is too small to continue. As we shall see in the next chapter, there is a close connection between the FWT and the well-known pyramidal decomposition of an image; an operation is applied to repeatedly subsampled versions of the signal to reveal frequency information at each scale. The underlying theory behind these algorithms is known as **multiresolution analysis (MRA)**, and it applies to critically sampled wavelet transforms.

6.6.5 Daubechies Wavelets

Despite the simplicity and popularity of Haar wavelets, the abrupt transitions of the box-car functions yield objectionable artifacts that are unacceptable for some applications. A generalization of Haar wavelets are **Daubechies wavelets**—in fact, Haar is a special case of Daubechies. Recall that a wavelet family is defined entirely by the mother wavelet, but that the father wavelet (or scaling function) is needed for computational efficiency by the DWT algorithm. Like Haar wavelets, Daubechies wavelets are orthogonal, so that the father wavelet is determined from Equation (6.159).

The key idea behind the Daubechies wavelet is to achieve the highest number of vanishing moments for a defined support width. A smooth signal can be locally approximated by a polynomial, and the moments of the signal are a measure of how similar the signal is to the powers of x in the polynomial. A *vanishing moment* occurs when the moment is zero, in which case the signal bears no resemblance, and therefore the low-order polynomial features of the signal are removed by the wavelet transform, leaving only higher-order features. Thus, the number of vanishing moments are related to the compression ability of the wavelet, so that the Daubechies wavelets are designed to attain the maximum compression (in some sense) for a given amount of computation (support width).

Daubechies wavelets are defined given a certain kernel width, which is an even number ranging from 2 to 20. The Daubechies D2 wavelet is identical to the Haar wavelet, D4 is the simplest non-Haar Daubechies wavelet, and so forth. The number of vanishing moments is given by half the number, so D2 has 1 vanishing moment (constant signals transform to zero), D4 has 2 vanishing moments (linearly sloped signals transform to zero), and so forth. It is beyond our scope to discuss how to construct these kernels, but it is important to note that Equation (6.159) makes it easy to determine the scaling kernel from the wavelet kernel, and vice versa. Specifically, we simply reverse the order of the values and negate every other element. For example, the scaling and wavelet kernels for D2 are given by

$$\phi(x) = \frac{1}{\sqrt{2}} \begin{bmatrix} \underline{1} & 1 \end{bmatrix} = \begin{bmatrix} \underline{0.70710678} & 0.70710678 \end{bmatrix} \quad (\text{D2 scaling}) \quad (6.201)$$

$$\psi(x) = \frac{1}{\sqrt{2}} \begin{bmatrix} \underline{1} & -1 \end{bmatrix} = \begin{bmatrix} \underline{0.70710678} & -0.70710678 \end{bmatrix} \quad (\text{D2 wavelet}) \quad (6.202)$$

while the scaling and wavelet kernels for D4 are given by

$$\phi(x) = \frac{1}{4\sqrt{2}} \begin{bmatrix} \underline{1 + \sqrt{3}} & 3 + \sqrt{3} & 3 - \sqrt{3} & 1 - \sqrt{3} \end{bmatrix}$$

$$\approx \begin{bmatrix} \underline{0.48296291} & 0.83651630 & 0.22414387 & -0.12940952 \end{bmatrix} \quad (\text{D4 scaling}) \quad (6.203)$$

$$\psi(x) = \frac{1}{4\sqrt{2}} \begin{bmatrix} \underline{1 - \sqrt{3}} & -3 + \sqrt{3} & 3 + \sqrt{3} & -1 - \sqrt{3} \end{bmatrix}$$

$$\approx \begin{bmatrix} \underline{-0.12940952} & -0.22414387 & 0.83651630 & -0.48296291 \end{bmatrix} \quad (\text{D4 wavelet}) \quad (6.204)$$

Note that the sum of the elements of each wavelet kernel is 0, the sum of the elements of each scaling function is $\sqrt{2}$, and the norm of each kernel is 1. Verifying the vanishing moments is left as an exercise.[†]

| EXAMPLE 6.11 | Apply the fast wavelet transform (FWT) to the signal $g^{(0)} = g(x) = \begin{bmatrix} 1 & 7 & 3 & 0 & 5 & 4 & 2 & 9 \end{bmatrix}$ that we saw in Example 6.9 using D4 Daubechies wavelets. For simplicity, assume periodicity. |

Solution

If we let $\phi(x) = \begin{bmatrix} h_0 & h_1 & h_2 & h_3 \end{bmatrix}$, so that $h_0 \approx 0.48$, $h_1 \approx 0.84$, $h_2 \approx 0.22$, and $h_3 \approx -0.13$, then the lowpass and highpass values from the finest resolution are computed as follows:

$$
\begin{bmatrix}
g_{low}^{(1)}(0) \\
g_{low}^{(1)}(1) \\
g_{low}^{(1)}(2) \\
g_{low}^{(1)}(3) \\
g_{high}^{(1)}(0) \\
g_{high}^{(1)}(1) \\
g_{high}^{(1)}(2) \\
g_{high}^{(1)}(3)
\end{bmatrix}
=
\begin{bmatrix}
h_0 & h_1 & h_2 & h_3 & 0 & 0 & 0 & 0 \\
0 & 0 & h_0 & h_1 & h_2 & h_3 & 0 & 0 \\
0 & 0 & 0 & 0 & h_0 & h_1 & h_2 & h_3 \\
h_2 & h_3 & 0 & 0 & 0 & 0 & h_0 & h_1 \\
h_3 & -h_2 & h_1 & -h_0 & 0 & 0 & 0 & 0 \\
0 & 0 & h_3 & -h_2 & h_1 & -h_0 & 0 & 0 \\
0 & 0 & 0 & 0 & h_3 & -h_2 & h_1 & -h_0 \\
h_1 & -h_0 & 0 & 0 & 0 & 0 & h_3 & -h_2
\end{bmatrix}
\begin{bmatrix}
1 \\ 7 \\ 3 \\ 0 \\ 5 \\ 4 \\ 2 \\ 9
\end{bmatrix}
\approx
\begin{bmatrix}
7.011 \\
2.0520 \\
5.0445 \\
7.8128 \\
0.8111 \\
1.8625 \\
-4.2173 \\
-4.8203
\end{bmatrix}
$$

At the next level we have

$$
\begin{bmatrix}
g_{low}^{(2)}(0) \\
g_{low}^{(2)}(1) \\
g_{high}^{(2)}(0) \\
g_{high}^{(2)}(1)
\end{bmatrix}
=
\begin{bmatrix}
h_0 & h_1 & h_2 & h_3 \\
h_2 & h_3 & h_0 & h_1 \\
h_3 & -h_2 & h_1 & -h_0 \\
h_1 & -h_0 & h_3 & -h_2
\end{bmatrix}
\begin{bmatrix}
7.011 \\
2.0520 \\
5.0445 \\
7.8128
\end{bmatrix}
\approx
\begin{bmatrix}
5.2222 \\
10.2778 \\
-0.9208 \\
2.4698
\end{bmatrix}
$$

Concatenating these values yields the result, in order of increasing frequency, as $\mathcal{W}\{g\} \approx \begin{bmatrix} 5.2 & 10.3 & -0.9 & 2.5 & 0.8 & 1.9 & -4.2 & -4.8 \end{bmatrix}$. It is easy to verify that both the 8×8 and 4×4 matrices above are orthogonal.

Notice that in this example we allowed the coefficients to "wrap around" the edge of the matrix, which assumes periodicity in the signal (like circular convolution); this assumption can be removed by extending the signal past the borders and increasing the number of columns in the matrix to accommodate the shifting wavelet kernel.

6.6.6 2D Wavelet Transform

The wavelet transform is easily generalized to 2D. Usually the 2D wavelets are separable so that they can be expressed as the multiplication of two 1D wavelets, in which case Equations (6.181)–(6.182) can be extended as follows:

$$
g_{LL}^{(i+1)}(x, y) = \sum_{k_x} \sum_{k_y} g_{LL}^{(i)}(2x + k_x, 2y + k_y) \phi(k_x) \phi(k_y) \tag{6.205}
$$

$$
g_{HL}^{(i+1)}(x,y) = \sum_{k_x} \sum_{k_y} g_{LL}^{(i)}(2x + k_x, 2y + k_y) \psi(k_x) \phi(k_y) \tag{6.206}
$$

$$
g_{LH}^{(i+1)}(x, y) = \sum_{k_x} \sum_{k_y} g_{LL}^{(i)}(2x + k_x, 2y + k_y) \phi(k_x) \psi(k_y) \tag{6.207}
$$

[†] Problem 6.35.

$$g_{HH}^{(i+1)}(x, y) = \sum_{k_x} \sum_{k_y} g_{LL}^{(i)}(2x + k_x, 2y + k_y)\psi(k_x)\psi(k_y) \tag{6.208}$$

where $g_{LL}^{(0)}(x, y) \equiv g(x, y)$ and $g(x,y)$ is the original image. At each iteration the signal is downsampled by two in each direction after applying the wavelet and/or scaling functions. In this notation, g_{LL} contains the low-frequency components in both directions, g_{HL} contains the high-frequency components in the x direction but the low-frequency components in the y direction, g_{LH} is the opposite, and g_{HH} contains the high-frequency components in the diagonal direction. The downsampled low-frequency version is always fed as input to the next level.

Due to separability, Equations (6.205)–(6.208) are easy to compute. Let g be a $w \times h$ image. First, ϕ and ψ are applied to the image in the horizontal direction to yield

$$g_L^{(1)}(x, y) \equiv \sum_k g(2x + k, y)\phi(k) \tag{6.209}$$

$$g_H^{(1)}(x, y) \equiv \sum_k g(2x + k, y)\psi(k) \tag{6.210}$$

both of which are of size $\frac{w}{2} \times h$. Then ϕ and ψ are applied to the image in the vertical direction to yield

$$g_{LL}^{(1)}(x, y) = \sum_k g_L^{(i)}(x, 2y + k)\phi(k) \tag{6.211}$$

$$g_{HL}^{(1)}(x, y) = \sum_k g_H^{(i)}(x, 2y + k)\phi(k) \tag{6.212}$$

$$g_{LH}^{(1)}(x, y) = \sum_k g_L^{(i)}(x, 2y + k)\psi(k) \tag{6.213}$$

$$g_{HH}^{(1)}(x, y) = \sum_k g_H^{(i)}(x, 2y + k)\psi(k) \tag{6.214}$$

all four of which are of size $\frac{w}{2} \times \frac{h}{2}$. In other words, after one iteration the $w \times h$ image has been replaced by four images, each of which is one-fourth the size of the original. (Of course, the order is arbitrary and can therefore be reversed, i.e., vertical before horizontal.) Repeating this procedure using $g_{LL}^{(i)}$ for subsequent iterations then yields the 2D wavelet transform of the image. See Figure 6.29 for an example, where the normalization factor from a lowpass convolution kernel was used instead of the wavelet normalization factor to prevent the father wavelet from brightening the image.

6.6.7 Gabor Wavelets

The final wavelet that we will consider is the **Gabor wavelet**, which is a complex sinusoid multiplied by a Gaussian window. In 1D, the wavelet is given by

$$\psi(x) = \underbrace{e^{-\alpha x^2}}_{\text{Gaussian}} \cdot \underbrace{e^{j\omega x}}_{\text{sinusoid}} \tag{6.215}$$

As with the Fourier transform, using complex numbers is just a mathematical convenience. Conceptually, the Gabor wavelet consists of even and odd components: $\psi(x) = \psi_{even}(x) + j\psi_{odd}(x)$, where

$$\psi_{even}(x) = e^{-\alpha x^2} \cos(\omega x) \tag{6.216}$$

Figure 6.29 An image (top-left) and the $g_L^{(1)}$ and $g_H^{(1)}$ (top-right) that result from convolving and downsampling the image with the scaling and wavelet functions. Note that together $g_L^{(1)}$ and $g_H^{(1)}$ have the same number of pixels as the original image and therefore contain the same amount of information. Continuing the process yields the four downsampled signals for the first level of the wavelet transform (bottom-right). The final discrete wavelet transform (DWT) of the image after three levels of processing (bottom-left). Again, note that the number of pixels is the same, and therefore the transform is invertible.

$$\psi_{odd}(x) = e^{-\alpha x^2}\sin(\omega x) \tag{6.217}$$

where $\alpha = 1/(2\sigma^2)$, σ is the width of the Gaussian envelope, $\omega = 2\pi f = 2\pi/\tau$ is the angular frequency, f is the frequency, and τ is the period of the sinusoid. The Gabor wavelet has the property that it minimizes the uncertainty in the spatial and frequency domains.

A 1D Gabor wavelet is shown in Figure 6.30. The primary parameter is the frequency f of the sinusoid (or equivalently, the period $\tau = 1/f$ or angular frequency $\omega = 2\pi f$). The secondary parameter is $\kappa \equiv \sigma/\tau$, which governs the width of the Gaussian relative to the period of the sinusoid. Recall from our discussion of the Gabor limit (see Figure 6.25) that the Gaussian envelope should not be too wide or too narrow. Since 95.45% of the Gaussian is captured within $\pm 2\sigma$ and the Gaussian should capture approximately 1 period, it is recommended to set $\kappa \approx \frac{1}{2}$, so that $\sigma = \frac{\tau}{2} = \frac{1}{2f}$, or $\alpha = 2f^2$. This ratio works well when the frequencies are spaced by 1 octave; for higher spacing, the ratio should be decreased accordingly, e.g., $\kappa \approx 0.4$ for a spacing of 1.5 octaves.

To apply the DWT to a 1D signal using Gabor wavelets, discrete kernels are first created of the even and odd components at the highest frequency of interest which, according to the Nyquist frequency mentioned earlier, restricts $\tau > 2$. Typically $\tau \approx 3$ at this finest level. Discrete kernels are then constructed for the remaining frequencies of interest, reducing the frequency by some factor each time, until the minimum desired frequency. (If the factor is 2, then the frequencies are spaced an octave apart.) The Gabor wavelets are then applied to the signal by performing a convolution-like operation (computing the sum of an elementwise product) at each frequency, except that the spatial shift is frequency-dependent, typically approximately $\tau/2$. This computes $\psi((x-a)/b)$, where the parameter a governs the spatial shift, and b governs the frequency shift. The result is an overcomplete (assuming appropriate spatial and frequency spacing) representation of the signal. Gabor wavelets

Figure 6.30 1D Gabor wavelet, showing even (left) and odd (right) components, using $\sigma = 1$ and $\tau = 2$.

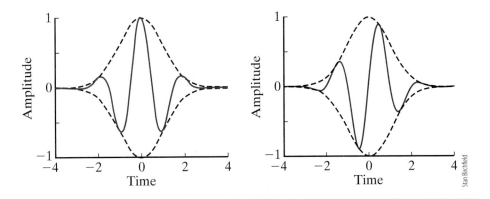

are not orthogonal, so it is not possible to easily construct an invertible transform, making them not well suited for use in compression. Instead, Gabor wavelets are useful for extracting frequency information at different locations in the signal for the purpose of detecting localized features in the signal.

In 2D, the complex sinusoid is a plane wave, and the 2D Gaussian envelope is aligned with the direction of the wave, given by the angle θ. If we let the coordinate transformation be $x' = x \cos \theta - y \sin \theta$, $y' = x \sin \theta + y \cos \theta$, then the 2D Gabor wavelet is

$$\psi(x, y) = e^{-(\alpha x'^2 + \beta y'^2)} e^{j\omega x'}$$

where, as before, $\alpha = 1/(2\sigma^2)$, σ is the width of the Gaussian envelope along the direction of the wave, $\omega = 2\pi f = 2\pi/\tau$ is the angular frequency, f is the frequency, and τ is the period of the sinusoid. Typically we set $\beta = \alpha/4$, so that the Gaussian is twice as wide in the direction orthogonal to the direction of the wave. A 2D Gabor wavelet is shown in Figure 6.31.

2D Gabor wavelets are characterized not only by their frequency but also by their orientation in the plane. Thus, an appropriate spacing must be chosen not only in space and frequency but also in orientation. Otherwise, 2D Gabor wavelets are applied in a manner similar to 1D Gabor wavelets.

Figure 6.31 Gabor 2D wavelets are achieved by multiplying a plane wave sinusoid with a Gaussian window function aligned with the direction of the wave propagation. Shown are the even (top) and odd (bottom) components, both as a 3D plot and as an image, using $\sigma = 1, \tau = 2, \theta = 30°$, and $\beta = \alpha/4$.

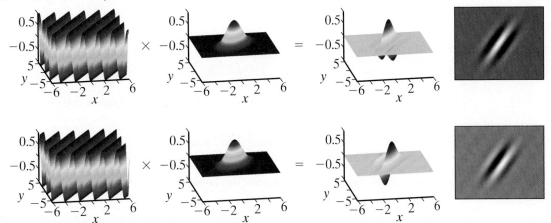

Studies have shown that mammalian visual systems apply filters similar to 2D Gabor wavelets in order to extract spatially-localized frequency-dependent features from the retinal image. The continuous Fourier transform of Equation (6.214) is

$$\Psi_{\text{Gabor}}(\omega') = \mathcal{F}\{\psi(x)\} = e^{-(\omega'-\omega)^2/\alpha} \tag{6.218}$$

which is a Gaussian centered at the angular frequency ω. In fact, mammalian visual systems encode information in a way that is more like a Gaussian shape on a logarithmic frequency axis rather than a linear frequency axis. As a result, the log-Gabor wavelet has been proposed, which has the following frequency response:

$$\Psi_{\text{log-Gabor}}(\omega') = e^{-\frac{(\log\frac{\omega'}{\omega})^2}{2(\log\frac{\zeta}{\omega})^2}} \tag{6.219}$$

where ζ/ω is chosen to ensure that the width of the Gaussian in the frequency domain is approximately one octave. If viewed on a linear frequency scale, log-Gabor wavelets have an extended tail in the high frequencies. As a result, the log-Gabor wavelet spreads information equally across the different frequencies, as opposed to the standard Gabor wavelet, which overrepresents the low frequencies. Since the inverse of the Fourier transform in Equation (6.219) cannot be obtained numerically, log-Gabor kernels in the spatial domain must be obtained numerically.

6.7 Further Reading

The Fourier transform is due to the work of Fourier, Lagrange, and others in the 18th and early 19th centuries. The FFT algorithm is primarily due to Cooley and Tukey [1965], although it was later realized that Gauss had discovered a very similar algorithm around 1805. The Lanczos filter is due to Duchon [1979], where the kernel has between 19 and 51 elements sampled at integer positions. Detailed explanations of the classic unsharp masking technique for photographers (in a darkroom) is provided in a series of articles by Bond [1996, 1997]; also see the article by Wainwright [2004]. Wavelets were first invented by Haar [1910, 1911]. Daubechies wavelets were first presented by Daubechies [1988]. Gabor's contribution to time-frequency analysis can be found in Gabor [1946], and an analysis of Gabor wavelets for image representation is due to Lee [1996]. Multi-resolution analysis, and its connection with the pyramidal algorithm, is described by Mallat [1989]; see also Mallat and Zhong [1992]. Log-Gabor wavelets are described by Field [1987]. For the use of wavelets for FBI fingerprint compression, see the work of Bradley et al. [1993] and Bradley and Brislawn [1994]. Space has not permitted us to discuss the closely related concept of watermarking, for which a seminal paper is that of Cox et al. [1997].

PROBLEMS

6.1. For each of the following continuous signals, state the frequency, and write the Fourier transform: (a) $\cos 8\pi t$, (b) $\cos(16t + 8)$, (c) $\sin 44t$.

6.2. Prove that (a) for a continuous signal with even symmetry, its Fourier transform is real and (b) for a continuous signal with odd symmetry, its Fourier transform is imaginary.

6.3. Explain the difference between the Dirac delta function and the Kronecker delta function.

6.4. Explain the difference between the Nyquist rate and the Nyquist frequency.

6.5. Explain why 44.1 kHz is a common audio sampling rate.

6.6. Suppose a signal is sampled at 1000 Hz. For each of the following frequencies, state whether aliasing will occur and, if so, what is the aliased frequency: (a) 300 Hz, (b) 600 Hz, (c) 1200 Hz.

6.7. A finite-duration, discrete signal is composed of 5 samples. List the discrete frequencies that are captured by the DFT of the signal.

6.8. Show that Equation (6.16) is indeed the inverse of Equation (6.13), that is, that the two expressions form a Fourier transform pair.

6.9. Explain why it is not recommended to directly implement Equation (6.1) in practice.

6.10. Compute the DFT of the discrete signal $\begin{bmatrix} 3 & 8 & 1 & 7 \end{bmatrix}$ both by hand and by using an existing software package. Also compute the magnitude and phase of the DFT. What can you infer about where the software package places the scaling factor?

6.11. What is the DFT of $\begin{bmatrix} 1 & 0 & \cdots & 0 \end{bmatrix}$?

6.12. Compute the DFT of the following signals. In each case explain the meaning of the frequencies captured by the DFT.

(a) $\cos \frac{2\pi}{8} x$, sampled at $x = 0, \ldots, 7$

(b) $\sin \frac{2\pi}{8} x$, sampled at $x = 0, \ldots, 7$

(c) $\cos \frac{2\pi}{8} x$, sampled at $x = 0, \ldots, 15$

(d) $\cos \frac{2\pi}{16} x$, sampled at $x = 0, \ldots, 15$

6.13. Find the cutoff frequency of the simple lowpass filter with kernel $\begin{bmatrix} 1 & 1 & 1 \end{bmatrix}$ in inverse samples. Assume the cutoff frequency is where the Fourier transform reaches half its maximum value.

6.14. Plot the basis functions for the 4-point DFT.

6.15. For a 32-element sequence, what is the equivalent negative frequency corresponding to the following positive frequencies: (a) $\frac{7\pi}{32}$, (b) $\frac{15\pi}{32}$, (c) $\frac{19\pi}{32}$?

6.16. Suppose we wish to convolve a 5-element discrete signal with a 3-element discrete kernel, using the circular convolution theorem. How much zero-padding is required?

6.17. (a) Compute the 2D DFT of the following grayscale image. (b) Multiply the image by $(-1)^{x+y}$ and repeat the computation. (c) How are the two results related?

$$\begin{bmatrix} 151 & 222 & 160 & 88 \\ 79 & 24 & 23 & 197 \\ 143 & 78 & 152 & 92 \\ 84 & 123 & 71 & 209 \end{bmatrix}$$

6.18. Explain the Radon transform.

6.19. Explain why the shifted 2D DFT usually has large values along the horizontal and vertical axes, forming the shape of a plus sign?

6.20. Suppose an image is rotated clockwise by 30 degrees. How does this change the 2D DFT?

6.21. List the two primary applications for filtering.

6.22. Explain the following terms: (a) Gibbs phenomenon, (b) ringing, (c), overshoot, (d) undershoot, (e) clipping.

6.23. Explain why it is not possible in practice to apply an ideal low pass filter in the time domain.

6.24. Given the following low pass convolution kernel, compute its corresponding high pass convolution kernel by subtracting from an allpass filter (in the frequency domain): $[2 \quad 4 \quad \underline{6} \quad 4 \quad 2]$.

6.25. In Example 6.5 we showed that a particular 2D LoG kernel can be obtained by the subtraction of a lowpass kernel from an allpass kernel:

$$\underbrace{\frac{1}{3}\begin{bmatrix} 1 & 1 & 1 \\ 1 & -8 & 1 \\ 1 & 1 & 1 \end{bmatrix}}_{\text{highpass}} = -3\left(\underbrace{\begin{bmatrix} 0 & 0 & 0 \\ 0 & 1 & 0 \\ 0 & 0 & 0 \end{bmatrix}}_{\text{allpass}} - \underbrace{\frac{1}{9}\begin{bmatrix} 1 & 1 & 1 \\ 1 & 1 & 1 \\ 1 & 1 & 1 \end{bmatrix}}_{\text{lowpass}}\right)$$

Show that this is also true for the other LoG kernels in Table 5.5, as well as for the LoG kernel in Problem 5.29. Indicate which, if any, of these lowpass kernels is not Gaussian.

6.26. Explain the subtle differences between the terms unsharp masking, highboost filtering, and high-frequency emphasis, as they are sometimes used.

6.27. Construct the two unsharp masking kernels for the case $a = b = 2$. How would you decide which of the two kernels to use?

6.28. Explain the principle behind homomorphic filtering.

6.29. What is the primary drawback of the short-time Fourier transform?

6.30. Given the mother wavelet $\psi(x) = \frac{1}{\eta}[2 \quad 4 \quad 6 \quad -1 \quad -3 \quad -5]$, use Equation (6.159) to compute the corresponding father wavelet $\phi(x)$. What is the value of η to ensure that the norm of the wavelet is 1?

6.31. What is the result in Example 6.8 if normalization is considered?

6.32. Show that the discrete Haar wavelet preserves the energy in a constant signal.

6.33. Suppose a mother wavelet is given by $\frac{1}{\sqrt{10}}[1 \quad -3]$. Adopting the approach of critical sampling to generate a complete wavelet transform, write all the basis functions for an 8-element input, using Table 6.2 and Equation (6.159).

6.34. Apply the FWT using D4 Daubechies wavelets to the input signal $g(x) = [5 \quad 7 \quad 6 \quad 4 \quad 3 \quad 1 \quad 8 \quad 9]$.

6.35. Verify that D2 Daubechies wavelets have 1 vanishing moment, and D4 Daubechies wavelets have 2 vanishing moments. That is, show that D2 or D4 applied to a constant signal $g(x) = a$ yields 0, and that D4 applied to a linearly sloped signal $g(x) = mx + b$ also yields 0.

6.36. How well does the discrete Haar wavelet preserve the energy in the signal $f(x) = [1 \quad 7 \quad 3 \quad 0 \quad 5 \quad 4 \quad 2 \quad 9]$ at each level?

6.37. This question is about representing the DWT as a matrix transform.

(a) If we let $\gamma \equiv 1/\sqrt{2}$, write the 8×8 Haar DWT matrix in Equation (6.178) in terms of γ. Show how this matrix can be derived from repeated applications of Equation (6.189).

(b) Apply a similar procedure to compute the 8×8 matrix for the D4 Daubechies wavelet from Equations (6.203)–(6.204). Verify that the D4 Daubechies basis functions (rows of the matrix) are orthogonal.

6.38. Show that if the 2D signal $f(x, y)$ is separable, then so is its Fourier transform $F(k_x, k_y)$.

6.39. Use the continuous Fourier transform to prove that Equation (6.98) is true for continuous signals.

6.40. Implement the 2D fast wavelet transform using Haar wavelets, and apply to a gray-scale image.

6.41. As mentioned in the text, the fast wavelet transform is based upon multiresolution analysis (MRA). Not all wavelets admit multiresolution analysis, but for those that do, their father ϕ and mother ψ wavelets satisfy the following pair of equations for some sequence of coefficients g and h:

$$\phi(t) = \sqrt{2} \sum_k h(k) \phi(2t - k) \tag{6.220}$$

$$\psi(t) = \sqrt{2} \sum_k g(k) \phi(2t - k) \tag{6.221}$$

The first equation, known as a *refinement equation* for the father wavelet, captures its self-similarity at multiple resolutions, while the second equation captures the relationship between the mother and father wavelets. Show that the continuous Haar wavelet satisfies these two equations, and find the corresponding sequences g and h.

CHAPTER 7
Edges and Features

For many applications it is important to be able to detect intensity edges in an image, which indicate locations in the image where the graylevel function changes drastically. Similarly, detecting and matching feature points (also known as interest operators or corners) either within an image or between images is useful for a variety of purposes. In this chapter we discuss these closely related problems.

7.1 Multiresolution Processing

One of the curious properties of images is that objects can appear at any size. For example, an upright person with a real-world height of 2 meters can occupy a region in the image whose height is 4 or 400 pixels (or any number between or beyond), all depending on the distance from the camera to the object, the focal length, and the sensor resolution. Conversely, an image region whose height is 10 pixels could be a person, a skyscraper, or a tiny bug. This loss of scale information is a property unique to the imaging process, arising from the projection of the 3D world onto a 2D image.

7.1.1 Gaussian Pyramid

One of the most common ways to deal with this loss of scale information is to process the image at multiple resolutions. For example, suppose we want to search an image to find all the human faces, so we look for faces at a certain scale in the original image, then downsample the image by a factor of 2 in each direction and look for faces at the same scale in the downsampled version, then downsample again, and so on. A face that appears to occupy, say, a 40×40 region in the original image will occupy only a 20×20 region in the downsampled image, a 10×10 region in the twice downsampled image, and so forth. As a result, faces at a variety of scales (and hence people at different distances to the camera) can be detected using a relatively straightforward procedure that interleaves detection and downsampling.

Because each successive image is smaller than its predecessor, stacking the images on top of one another yields the shape of a pyramid. For this reason the sequence of images is known as an **image pyramid**. It is usually a bad idea to downsample an image directly, because of the undesirable effect of aliasing;[†] instead, the image should first be smoothed to remove the high frequencies.[‡] Among the many ways to smooth an image, the most popular is to convolve with a Gaussian kernel, leading to a **Gaussian pyramid**. Given an image $I(x, y)$, let us define the zeroth level of the Gaussian pyramid as the image itself:

$$I^{(0)}(x, y) \equiv I(x, y) \tag{7.1}$$

then let us define each successive image in the Gaussian pyramid as the downsampled, smoothed version of the previous image:

$$I^{(i+1)}(x, y) \equiv \left(I^{(i)}(x, y) \circledast Gauss_{\sigma^2}(x, y)\right) \downarrow 2 \tag{7.2}$$

where $i = 0, 1, \ldots$ is a nonnegative integer, and $I \downarrow 2$ means to downsample I by a factor of 2 in both horizontal and vertical directions. Notice that, in the absence of aliasing, downsampling does not lose any information between the smoothed version of the image and the downsampled, smoothed version. An example Gaussian pyramid with a downsampling factor of 2 is shown in Figure 7.1.

Any of the standard 3×3 Gaussian kernels works well in practice. One advantage of the kernel $\frac{1}{4}\begin{bmatrix} 1 & 2 & 1 \end{bmatrix}$, for which $\sigma^2 = 0.5$, is that it satisfies the **equal contribution property**, meaning that each pixel in the image contributes an equal amount to the downsampled version. To see this, suppose we have a 1D signal f and a Gaussian kernel given

Figure 7.1 Four levels of a Gaussian pyramid, obtained with $\sigma^2 = 0.5$ and a downsampling factor of 2.

$I^{(0)}$ $\qquad\qquad\qquad$ $I^{(1)}$ $\qquad$ $I^{(2)}$ $\quad$ $I^{(3)}$

Stan Birchfield

[†] Aliasing is covered in Section 6.1.3 (p. 275).

[‡] Although aliasing, strictly speaking, refers to sampling a *continuous* signal, the effect of downsampling a *discrete* signal is similar, because the original continuous signal can no longer be exactly reconstructed if the downsampling is excessive.

by $\begin{bmatrix} a & b & a \end{bmatrix}$, where $b \geq a$. Then the elements of the smoothed signal f' are given by $f'(1) = af(0) + bf(1) + af(2)$, $f'(2) = af(1) + bf(2) + af(3)$, $f'(3) = af(2) + bf(3) + af(4)$, and so forth. When f' is downsampled by a factor of 2, every other element is discarded. Suppose, for example, we keep the even elements and discard the odd ones; then $f'(2) = af(1) + bf(2) + af(3)$, $f'(4) = af(3) + bf(4) + af(5)$, $f'(6) = af(5) + bf(6) + af(7)$, and so forth. Ignoring border effects, it is easy to see that each sample $f(i)$, where i is odd, contributes $2a$, while each sample $f(i)$, where i is even, contributes b. Therefore, equal contribution among all pixels implies $b = 2a$. Since $b + 2a = 1$ for normalization, this yields $b = \frac{1}{2}$ and $a = \frac{1}{4}$, which is the Gaussian kernel just mentioned.

Although a downsampling factor of 2 is convenient, sometimes it is desirable to downsample the image by finer amounts to provide a less noticeable transition between levels of the pyramid. For example, if we downsample by $\sqrt[4]{2} \approx 1.19$ each time, then successive images are given by

$$I^{(i+1)}(x, y) \equiv (I^{(i)}(x, y) \circledast Gauss_{\sigma'^2}(x, y)) \downarrow \sqrt[4]{2} \tag{7.3}$$

where some form of interpolation should be used to facilitate downsampling by a non-integral amount. In this case $I^{(1)}$ will be $\frac{1}{1.19} \approx 0.84$ as large as $I^{(0)}$ in each direction, $I^{(2)}$ will be $\frac{1}{(1.19)^2} \approx 0.71$ as large, $I^{(3)}$ will be $\frac{1}{(1.19)^3} \approx 0.59$ as large, and so forth. Conveniently, $I^{(4)}$ will be half as large as $I^{(0)}$ in each direction, $I^{(8)}$ will be half as large as $I^{(4)}$, and $I^{(12)}$ will be half as large as $I^{(8)}$. Each reduction by a factor of two is known as an **octave**. Images $I^{(0)}$ through $I^{(3)}$ are in the first octave, while $I^{(4)}$ through $I^{(7)}$ are in the second octave. More generally, if the downsampling factor is $\sqrt[n]{2} = 2^{\frac{1}{n}}$, then there are n images per octave. Since repeated convolutions with a Gaussian are equivalent to a single convolution with a Gaussian whose variance is the sum of the individual variances, we define $\sigma'^2 \equiv \frac{1}{n}\sigma^2$ to ensure that the overall smoothing between octaves is the same as between consecutive levels of Equation (7.2). An example Gaussian pyramid with a downsampling factor of $\sqrt[4]{2}$ is shown in Figure 7.2.

Figure 7.2 Twelve levels of a Gaussian pyramid, obtained with $\sigma'^2 = \frac{1}{4}(0.5) = 0.125$ and a downsampling factor of $\sqrt[4]{2}$. Note that $I^{(4)}$ is half as large as $I^{(0)}$ in each direction, and that $I^{(8)}$ is half as large as $I^{(4)}$.

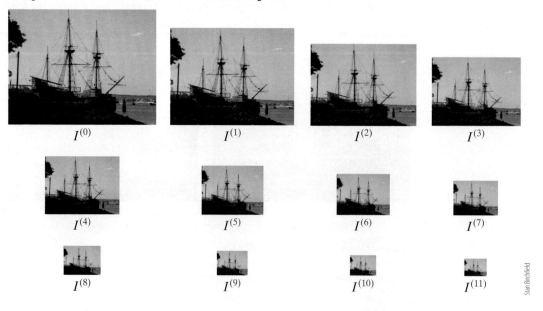

$I^{(0)}$ $I^{(1)}$ $I^{(2)}$ $I^{(3)}$

$I^{(4)}$ $I^{(5)}$ $I^{(6)}$ $I^{(7)}$

$I^{(8)}$ $I^{(9)}$ $I^{(10)}$ $I^{(11)}$

7.1.2 Laplacian Pyramid

A Gaussian pyramid is a lowpass pyramid. For some applications a bandpass pyramid is preferable, particularly for extracting image features such as interest points. Since the Laplacian is a bandpass operator, convolving the image with a Laplacian of Gaussian (LoG) kernel with increasing variance yields the **Laplacian pyramid**, which is the most popular bandpass pyramid:

$$L^{(i+1)}(x, y) \equiv (I^{(0)}(x, y) \circledast LoG_{(i+1)\sigma^2}(x, y)) \downarrow (i + 1)d \tag{7.4}$$

where d is the amount of downsampling between consecutive levels, so that $(i + 1)d$ is the total amount of downsampling in the $(i + 1)^{\text{th}}$ level. Note the difference between this equation and Equation (7.2): Here the images are not processed successively by feeding the previous output from level i as input to level $i + 1$. Instead, the original image is convolved each time with successively wider LoG kernels, with successively larger downsampling factors—an extremely inefficient computation.

Thankfully there is a better way. Recall that the LoG is well approximated by the difference of two Gaussians (DoG).[†] As a result, the Laplacian pyramid can be computed by successively smoothing and downsampling the image as in Equations (7.2) or (7.3), yielding a Gaussian pyramid; then the consecutive slices in the Gaussian pyramid are subtracted from one another to yield the Laplacian pyramid. However, there are two details that should be kept in mind. First, in order to subtract consecutive levels of the Gaussian pyramid, the previous image must be subtracted from its smoothed version *before* downsampling (so that the two images being subtracted are the same size):

$$I_{temp}^{(i+1)}(x, y) \equiv I^{(i)}(x, y) \circledast Gauss_{\sigma_i^2}(x, y) \tag{7.5}$$

$$L^{(i+1)}(x, y) \equiv I_{temp}^{(i+1)}(x, y) - I^{(i)}(x, y) \tag{7.6}$$

$$I^{(i+1)}(x, y) \equiv (I_{temp}^{(i+1)}(x, y)) \downarrow d \tag{7.7}$$

where the levels of the Laplacian pyramid are given by $L^{(1)}, L^{(2)}, L^{(3)}$, and so forth.

Secondly, as hinted by the subscript on the variance in Equation (7.5), the variance of the Gaussian might not be the same in each iteration. Recall from Equation (5.101) that the LoG is well approximated by the difference of two Gaussians whose variances are related by a constant ratio ρ^2:

$$Gauss_{\rho^2\sigma^2} - I \circledast Gauss_{\sigma^2} \approx (\rho - 1)\sigma^2 LoG_{\sigma^2} \tag{7.8}$$

As a result, to ensure that the ratio remains constant, the standard deviation of the ith Gaussian-smoothed image should be set as

$$\sigma_i \equiv \rho^i \sigma_0 \tag{7.9}$$

so that $\sigma_{i+1}/\sigma_i = \rho$ for all i. Alternatively, if the Gaussian-smoothed images that are used in constructing the Laplacian pyramid are from a traditional Gaussian pyramid as in Equations (7.2) or (7.3), then the ratio of the variances of successive levels in the Laplacian pyramid will not be constant.

To avoid the messiness of downsampling by a nonintegral amount, oftentimes downsampling is performed only when the factor is a power of two. As a result, all images within each octave are of the same size, whereas images in the next octave are half as big (in each direction) as those in the previous octave. As before, let n be the number of images per

---

[†] Section 5.4.2 (p. 245).

octave. Then $d = \sqrt[n]{2}$, and each image $I^{(i)}_{temp}$ is downsampled by a factor of 2 only if $i + 1$ is divisible by n. For example, if $n = 2$, then $I^{(0)}$ and $I^{(1)}$ are the same size as each other, $I^{(2)}$ and $I^{(3)}$ are half as large, $I^{(4)}$ and $I^{(5)}$ are one quarter as large, and so forth. To implement this procedure, change Equation (7.7) to

$$I^{(i+1)}(x, y) \equiv \begin{cases} (I^{(i+1)}_{temp}(x, y)) \downarrow 2 & \text{if } \mathrm{mod}(i + 1, n) = 0 \\ (I^{(i+1)}_{temp}(x, y)) & \text{otherwise} \end{cases} \qquad (7.10)$$

An example Laplacian pyramid is shown in Figure 7.3. To ensure that each octave is convolved with the same sequence of variances relative to the image size, the variance ratio should be set to $\rho = 2^{\frac{1}{n}}$. (To see this number, note that downsampling occurs after image $I^{(n)}$, which implies that $\sigma_n = 2\sigma_0$; combining with $\sigma_n = \rho^n \sigma_0$ from Equation (7.9) reveals $\rho^2 = 2^{\frac{2}{n}}$.) A reasonable choice for the initial variance is $\sigma_0^2 = \frac{1}{n}(0.5)$, although other choices are possible.

7.1.3 Scale Space

Returning to the Gaussian pyramid, if the downsampling step is omitted, then the procedure of successive smoothing yields a stack of images, all the same size, which are increasingly blurred. Since the sole purpose of downsampling is to avoid unnecessary processing of redundant information, the stack of Gaussian-blurred images contains essentially the same information as that of the Gaussian pyramid.

Now let us take this idea one step further. Consider the family of images obtained by convolving the original image $I(x,y)$ with Gaussian kernels having *continuously increasing* variance:

$$I(x, y, t) \equiv I(x, y) \circledast Gauss_t(x, y) \qquad (7.11)$$

Figure 7.3 Laplacian pyramid with $n = 2$ images per octave. The images are successively convolved with a Gaussian, then downsampled at the end of each octave to produce something that closely resembles a Gaussian pyramid. Differences between successive Gaussian-smoothed images yield DoGs, which approximate LoGs. The initial variance is $\frac{1}{2}(0.5) = 0.25$, and the ratio between successive standard deviations is $\rho = \sqrt{2}$.

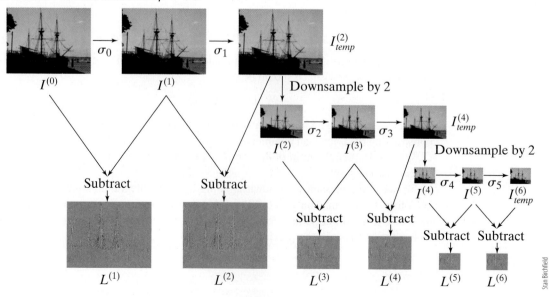

Stan Birchfield

where $t \geq 0$ is a *continuous* scale parameter that governs the amount of smoothing, and $I(x, y, 0) \equiv I(x, y)$. Treating x and y as continuous values as well, the resulting 3D continuous volume is known as the **scale space** of the image, and $t = \sigma^2$ is the **scale-space parameter**. Equivalently, the scale space can be seen as an embedding of the original image into a one-parameter family of derived images constructed by convolving the original image with a one-parameter family of Gaussian kernels of increasing variance, where t is the parameter. Recall that the concept of scale space was briefly mentioned in the context of the heat equation,[†] since Equation (7.11) is identical to Equation (5.151). The Gaussian pyramid, or rather the stack of Gaussian-blurred images, is simply a sampled version of the scale space.

The family of derived images in the scale space represents the original image at various levels of scale, as shown in Figure 7.4. As t increases, the amount of blurring of the image increases, and the amount of preserved detail from the original image decreases. For the scale space to be a meaningful representation of the image, it is important that several basic properties, called the **scale-space axioms**, should be satisfied. Among these axioms is the **causality criterion**, which ensures that the number of local extrema does not increase as we proceed to coarser levels of scale. In other words, the maxima are flattened while the minima are raised. It can be shown that the Gaussian kernel is unique in that it is the only convolution kernel that guarantees this result. The scale space, therefore, is usually constructed using the Gaussian kernel, and known as the **Gaussian scale space**.

In the case of a one-dimensional image, an extremum in the first derivative corresponds to a zero crossing in the second derivative. Since maxima in $\partial I / \partial x$ usually indicate an interesting location in the image — perhaps the boundary of an object — the zero crossings of $\partial^2 I / \partial x^2$ indicate potentially interesting locations. Since no new structure is introduced by the Gaussian blurring, these zero crossings form curves in scale space that always start and terminate at the original image, as shown in Figure 7.5. Empirically, there is a close relationship between the length of the curves and the perceptual saliency of the regions corresponding to them. Therefore, such an approach can be used to detect significant image structures from the scale-space representation by looking for the cusp of the longer second-derivative zero-crossing curves. The extension to a 2D image is straightforward, since an extremum in $\|\nabla I\|$ corresponds to a zero crossing in $\nabla^2 I$. We shall revisit this concept later in the chapter.[‡]

Figure 7.4 The Gaussian scale space of an image consists of a continuous 3D volume in which each slice is an increasingly blurred version of the original image. Shown here are ten sample images from the scale space.

[†] Section 5.5.6 (p. 260).
[‡] Section 7.4.6 (p. 347).

Figure 7.5 An example of 1D scale space. Top: 1D signal (bottom) smoothed by convolving with Gaussian kernels with increasing variance (toward the top). Bottom: Zero-crossings of the second derivative form curves that always start and end at the bottom. Cusps indicate the location and scale of significant image structures.

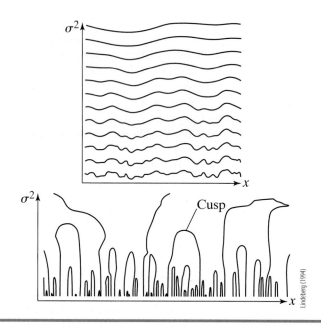

7.2 Edge Detection

Intensity edges are pixels in the image where the intensity (or, more precisely, graylevel) function changes rapidly. Sometimes these intensity edge pixels are known as **edgels** ("edge elements"), analogous to pixels ("picture elements"). From an information-theoretic point of view, these are the locations that carry the most information because the gray levels at these pixels are the least predictable from the values of their neighbors. Intensity edges retain a surprisingly large amount of information about the scene, as seen by the line drawings in Figure 7.6, as compared with the original images in Figure 7.7. Viewing only the line drawings, most human viewers can recognize these scenes effortlessly. Such demonstrations suggest that intensity edges are important for natural and, hence, artificial vision; indeed, the early days of computer vision were focused on line drawing images of polyhedral objects, where it was shown that robust interaction with the world was possible—even with limited algorithmic and computational complexity—using just the edges of the objects in the scene.

As shown in Figure 7.8, there are four types of intensity edges. The simplest and most important type is the **step edge**, which occurs when a light region is adjacent to a dark region. **Line edges** occur when a thin light (or dark) object, such as a wire, is in front of a dark (or light) background. At a **roof edge**, the change is not in the lightness[†] itself but rather the derivative of the lightness. And, finally, a **ramp edge** occurs when the lightness changes slowly across a region.

Figure 7.6 Intensity edges capture a rich representation of the scene. The scenes and objects in these line drawings are, with little difficulty, recognizable by the average human viewer. For the original images, turn to Figure 7.7.

D. B. Walther, B. Chai, E. Caddigan, D. M. Beck, and L. Fei-Fei, "Simple line drawings sffice for functional MRI decoding of natural scene categories," Proceedings of the National Academy of Sciences (PNAS), 108(23):9661-9666, 2011.

[†] Lightness is defined in Section 2.3.2 (p. 43).

D. B. Walther, B. Chai, E. Caddigan, D. M. Beck, and L. Fei-Fei, "Simple line drawings su ce for functional MRI decoding of natural scene categories," Proceedings of the National Academy of Sciences (PNAS), 108(23):9661-9666, 2011.

Figure 7.7 The original images from which the line drawings shown in Figure 7. 6 were obtained.

We shall focus our attention primarily upon step edges. In 1D, a step edge is accompanied by a large value in the first derivative, which is either positive if the pixels get brighter as one goes from left to right, or negative if the pixels get darker from left to right. Often the orientation of the edge is not important, in which case only the absolute value of the first derivative (in 1D) or the magnitude of the gradient (in 2D) is important. Therefore, the simplest way to find intensity edges in an image is to compute the gradient magnitude, as illustrated in Figure 7.9, where the gradient can be computed using any of the standard derivative kernels that we studied earlier,[†] such as Prewitt, Sobel, Scharr, Roberts, or the sampled Gaussian derivative. For some applications a binary decision is needed, in which case the gradient magnitude is thresholded, while for other applications it is better to retain the non-thresholded values.

7.2.1 Canny Edge Detector

The **Canny edge detector** is a classic algorithm for detecting intensity edges in a gray-scale image that, like the simple approach just described, relies on the gradient magnitude. Even today, Canny remains a popular algorithm due to its good performance, computational efficiency, and ease of implementation. The algorithm involves three steps, as shown in Algorithm 7.1. First the gradient of the image is computed, including the magnitude and phase. In the next step, called **non-maximum suppression**, any pixel is set to zero whose gradient magnitude is not a local maximum in the direction (as indicated by the phase) of the gradient. Finally, edge linking (also known as hysteresis thresholding or double thresholding[‡]) is performed to discard pixels without much support. The result is a binary image whose edge pixels along one-pixel-thick boundaries are ON, while all other pixels are OFF.

Non-maximum suppression is illustrated in Figure 7.10. The gradient magnitude of each pixel is compared with the gradient magnitude of the two pixels along the direction of the gradient vector. As shown in the figure, the gradient direction (also known as the phase) is quantized into one of four different directions, and the pixel is compared with either its neighbors to the left and right, its neighbors above and below, or its neighbors along one of the diagonals, depending upon the phase. For example, if θ is the clockwise angle from the positive x axis, then $\frac{3\pi}{8} \leq \theta < \frac{5\pi}{8}$ or $\frac{3\pi}{8} \leq \theta + \pi < \frac{5\pi}{8}$, where θ is the phase, causes the pixel to be compared with its neighbors above and below, where the latter test is needed

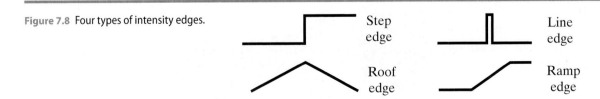

Figure 7.8 Four types of intensity edges.

Step edge

Line edge

Roof edge

Ramp edge

[†] Section 5.3.2 (p. 237).
[‡] Section 10.1.3 (p. 450).

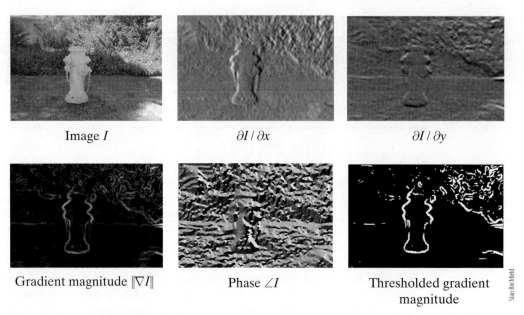

Image I $\partial I / \partial x$ $\partial I / \partial y$

Gradient magnitude $\lVert \nabla I \rVert$ Phase $\angle I$ Thresholded gradient
                                                                                                          magnitude

Figure 7.9 Top: An image and its partial derivatives in the x and y directions. Bottom: The gradient magnitude and phase of the image, along with the thresholded gradient magnitude.

to handle all possible values between $-\pi$ and π. If the gradient magnitude of the pixel is not at least as great as that of the two appropriate neighbors, it is artificially set to zero, as presented in Algorithm 7.2.

As shown in Algorithm 7.3, the process of edge linking is very similar to the floodfill procedure of Algorithm 4.5.[†] Any pixel whose non-maximum-suppressed gradient magnitude value is greater than a high threshold, τ_{high}, is a potential seed pixel. Floodfill is performed from these seed pixels, with the expansion continuing as long as the pixel value is greater than a low threshold, τ_{low}. The thresholds can be set manually or automatically. A common way to set them automatically is to sort the gradient magnitude values in the image, then set τ_{high} to the value that forces at least $100\alpha\%$ of the pixels to be edge pixels, then set $\tau_{high} = \beta\tau_{low}$, where reasonable values are $\alpha = 0.1$ and $\beta = 0.2$. Note that this is an

ALGORITHM 7.1 Detect intensity edges in an image using the Canny algorithm

CANNY(I, σ)

Input: grayscale image I, standard deviation σ
Output: set of pixels constituting one-pixel-thick intensity edges

1 $G_{mag}, G_{phase} \leftarrow$ COMPUTEIMAGEGRADIENT(I, σ)
2 $G_{localmax} \leftarrow$ NONMAXSUPPRESSION(G_{mag}, G_{phase})
3 $\tau_{low}, \tau_{high} \leftarrow$ COMPUTETHRESHOLDS$(G_{localmax})$
4 $I'_{edges} \leftarrow$ EDGELINKING$(G_{localmax}, \tau_{low}, \tau_{high})$
5 **return** I'_{edges}

[†] Section 4.2.2 (p. 154).

Figure 7.10 Non-maximum suppression. The gradient direction (or phase) θ is quantized into one of four values, shown by the colored wedges of the circle. The quantized phase governs which of the two neighbors to compare with the pixel. If the gradient magnitude of the pixel is not at least as great as both neighbors, then it is set to zero. This has the effect of thinning the edges, as shown in the inset.

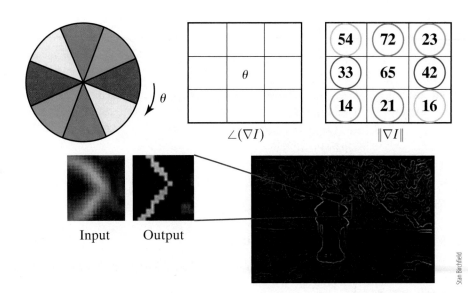

$\angle(\nabla I)$ $\|\nabla I\|$

Input   Output

extremely simple computation, because if there are n pixels in the image and if the sorting is performed in descending order, then τ_{high} is the gradient magnitude of the jth pixel in the list, where $j = \text{Round}(\alpha n)$. An example of edge linking with hysteresis, which yields the final Canny output, is shown in Figure 7.11.

Despite the ease with which the Canny edge detector is implemented, the algorithm is undergirded by a rich theoretical foundation. For example, one question that arises from the above discussion is how to select the filter for computing the gradient or, if the Gaussian derivative is used, what value to choose for the standard deviation. As it turns out, a large sigma yields a better signal-to-noise ratio (SNR), but a smaller sigma yields a more

ALGORITHM 7.2 Perform non-maximal suppression

$\textsc{NonMaxSuppression}(G_{mag}, G_{phase})$

Input: gradient magnitude and phase
Output: gradient magnitude with all nonlocal maxima set to zero

```
1   for (x, y) ∈ G_mag do                                                    ▶ For each pixel,
2        θ ← G_phase(x, y)                                                      adjust the phase
3        if θ ≥ 7π/8 then θ ← θ − π                                             to ensure that
4        if θ < −π/8 then θ ← θ + π                                             −π/8 ≤ θ < 7π/8.
5        if   −π/8 ≤ θ < π/8 then neigh₁ ← G_mag(x − 1, y), neigh₂ ← G_mag(x + 1, y)
6        elseif π/8 ≤ θ < 3π/8 then neigh₁ ← G_mag(x − 1, y − 1), neigh₂ ← G_mag(x + 1, y + 1)
7        elseif 3π/8 ≤ θ < 5π/8 then neigh₁ ← G_mag(x, y − 1), neigh₂ ← G_mag(x, y + 1)
8        elseif 5π/8 ≤ θ < 7π/8 then neigh₁ ← G_mag(x − 1, y + 1), neigh₂ ← G_mag(x + 1, y − 1)
9        if v ≥ neigh₁ AND v ≥ neigh₂ then                                   ▶ If the pixel is a local maximum
10            G_localmax(x, y) ← G_mag(x, y)                                    in the direction of the gradient,
11       else                                                                   then retain the value;
12            G_localmax(x, y) ← 0                                              otherwise set it to zero.
13  return G_localmax
```

$$1 \quad \textbf{for } (x, y) \in G_{mag} \textbf{ do}$$
$$2 \quad\quad \theta \leftarrow G_{phase}(x, y)$$
$$3 \quad\quad \textbf{if } \theta \geq \tfrac{7\pi}{8} \textbf{ then } \theta \leftarrow \theta - \pi$$
$$4 \quad\quad \textbf{if } \theta < -\tfrac{\pi}{8} \textbf{ then } \theta \leftarrow \theta + \pi$$
$$5 \quad\quad \textbf{if } \; -\tfrac{\pi}{8} \leq \theta < \tfrac{\pi}{8} \textbf{ then } neigh_1 \leftarrow G_{mag}(x - 1, y), neigh_2 \leftarrow G_{mag}(x + 1, y)$$
$$6 \quad\quad \textbf{elseif } \tfrac{\pi}{8} \leq \theta < \tfrac{3\pi}{8} \textbf{ then } neigh_1 \leftarrow G_{mag}(x - 1, y - 1), neigh_2 \leftarrow G_{mag}(x + 1, y + 1)$$
$$7 \quad\quad \textbf{elseif } \tfrac{3\pi}{8} \leq \theta < \tfrac{5\pi}{8} \textbf{ then } neigh_1 \leftarrow G_{mag}(x, y - 1), neigh_2 \leftarrow G_{mag}(x, y + 1)$$
$$8 \quad\quad \textbf{elseif } \tfrac{5\pi}{8} \leq \theta < \tfrac{7\pi}{8} \textbf{ then } neigh_1 \leftarrow G_{mag}(x - 1, y + 1), neigh_2 \leftarrow G_{mag}(x + 1, y - 1)$$
$$9 \quad\quad \textbf{if } v \geq neigh_1 \textbf{ AND } v \geq neigh_2 \textbf{ then}$$
$$10 \quad\quad\quad G_{localmax}(x, y) \leftarrow G_{mag}(x, y)$$
$$11 \quad\quad \textbf{else}$$
$$12 \quad\quad\quad G_{localmax}(x, y) \leftarrow 0$$
$$13 \quad \textbf{return } G_{localmax}$$

ALGORITHM 7.3 Perform edge linking

$\textsc{EdgeLinking}\left(G_{localmax}, \tau_{low}, \tau_{high}\right)$

Input: local gradient magnitude maxima $G_{localmax}$, along with low and high thresholds
Output: binary image I'_{edges} indicating which pixels are along linked edges

1   **for** $(x, y) \in G_{localmax}$ **do**
2   **if** $G_{localmax}(x, y) > \tau_{high}$ **then**
3   *frontier*.$\textsc{Push}(x, y)$
4          $I'_{edges}(q) \leftarrow$ ON
5   **while** *frontier*. $\textsc{Size} > 0$ **do**
6       $p \leftarrow$ *frontier*. $\textsc{Pop}()$
7   **for** $q \in \mathcal{N}(p)$ **do**
8   **if** $G_{localmax}(q) > \tau_{low}$ **then**
9   *frontier*.$\textsc{Push}(q)$
10         $I'_{edges}(q) \leftarrow$ ON
11   **return** I'_{edges}

accurate location for the edge—a dilemma known as the **localization-detection tradeoff**. To derive the optimal step detector, two criteria are specified: the detector should yield low false positive and false negative rates (that is, good *detection*), and the detected edge should be close to the true edge (that is, good *localization*). To quantify these two criteria, let the true edge be given by an ideal step:

$$g(x) = \begin{cases} 0 & \text{if } x < 0 \\ a & \text{if } x \geq 0 \end{cases} \tag{7.12}$$

where a is the height of the edge. Let the actual edge in the image be the true edge plus noise: $g(x) + \xi(x)$, where $\xi(x) \sim \mathcal{N}(0, n_0^2)$ is zero-mean Gaussian noise with variance n_0^2. Let $f(x)$ be the impulse response of the filter we are trying to find. The good detection criterion seeks to minimize the signal-to-noise ratio (SNR), that is, the ratio of the response of the filter to the true edge and the root-mean-square response of the filter to the noise:

$$SNR = \frac{a}{n_0} \Sigma(f), \quad \text{where} \quad \Sigma(f) \equiv \frac{\left| \int_{-\tilde{w}}^{0} f(x)\,dx \right|}{\sqrt{\int_{-\tilde{w}}^{\tilde{w}} f^2(x)\,dx}} \tag{7.13}$$

Figure 7.11 Edge linking with hysteresis, also known as double thresholding or hysteresis thresholding. Thresholding the gradient magnitude with the low threshold produces too many edge pixels (left), while thresholding with the high threshold produces too few edge pixels (middle). Edge linking with hysteresis combines the benefits of both (right), to produce the final Canny edge detector output.

Stan Birchfield

For the localization criterion, the reciprocal of the RMS distance of the marked edge from the center of the true edge is used. Skipping the mathematical derivation, which is too involved to cite here, this criterion is given by

$$LOC = \frac{1}{\sqrt{E[x_0^2]}} = \frac{a}{n_0}\Lambda(\dot{f}), \quad \text{where} \quad \Lambda(\dot{f}) \equiv \frac{|\dot{f}(0)|}{\sqrt{\int_{-\tilde{w}}^{\tilde{w}} \dot{f}^2(x)\,dx}} \tag{7.14}$$

where x_0 is the distance of the detected edge from the true edge, $E[\cdot]$ is the expectation, and $\dot{f} = df/dx$ is the derivative of f. Notice that $\Sigma(f)$ and $\Lambda(\dot{f})$ are two measures of the performance of the filter, and they depend only on the filter, not on the noise n_0 or the magnitude a of the true edge.

The localization-detection tradeoff is now easily shown by scaling the domain of the function: $f'(x) \equiv f(\frac{x}{\gamma})$, where γ is a scaling factor. Substituting into Equations (7.13) and (7.14),

$$\Sigma(f') = \sqrt{\gamma}\Sigma(f) \quad \text{and} \quad \Lambda(\dot{f}') = \frac{1}{\sqrt{\gamma}}\Lambda(\dot{f}) \tag{7.15}$$

In other words, if the filter is stretched to make it larger $(\gamma > 1)$, then the detection response increases, because a broader impulse response will have a larger SNR. At the same time, a larger filter reduces the localization performance, because it integrates information far from the edge. Multiplying the two criteria together achieves performance that is independent of scale, $\Sigma(f)\Lambda(\dot{f})$, revealing that the optimal 1D step edge detector is a simple difference operator or box filter:

$$f(x) = \begin{cases} -1 & \text{if } x < 0 \\ 1 & \text{if } x \geq 0. \end{cases} \tag{7.16}$$

However, the problem with a box filter is that it causes many local maxima to be detected. To solve this problem, a third criterion is introduced, called the **single response constraint**, which says that the detector should return only one point for each true edge point. In other words, it must minimize the number of local maxima around the true edge created by noise. Solving this numerical optimization problem yields a signal whose shape is nearly the same as the derivative of a Gaussian. Extending this reasoning to 2D, the gradient computed from the partial derivatives of a 2D Gaussian are used and steered to the appropriate 1D direction across the edge.

7.2.2 Marr-Hildreth Operator

There is a close relationship between the first and second derivatives, because first-derivative extrema are accompanied by second-derivative zero crossings. In other words, if $f(x)$ is a 1D function of x and if df/dx is a maximum at $x = x_0$, then $d^2f/dx^2 = 0$ at $x = x_0$. Since the function is discretely sampled, instead of looking for coordinates where the second derivative is exactly zero (an extremely rare phenomenon), we instead look for coordinates where the second derivative changes sign, the so-called **zero crossings**. Extending this logic to 2D, the maxima of the image gradient are accompanied by zero crossings in the LoG. Therefore, instead of using the gradient magnitude to compute intensity edges, we can use the zero crossings of the Laplacian of Gaussin (LoG), as shown in Figure 7.12. This approach is known as the **Marr-Hildreth operator**, and it predates the Canny edge detector by several years. Although the LoG still has other applications, such as the sign of the LoG as a texture pattern, its use as an edge detector is primarily of historical interest,

Figure 7.12 An image, with the result of applying the Laplacian of Gaussian (LoG) and the sign of the Laplacian of Gaussian (sLoG). The zero crossings of the LoG are an approach to edge detection that is not widely used due to the drawback of isotropic smoothing.

Image LoG

Zero crossings Sign of the LoG (sLoG)

Stan Birchfield

since it has all but been replaced by the gradient magnitude for that purpose. The primary drawback of the LoG is that it is isotropic, meaning that it smooths *across* as well as *along* edges, as opposed to the gradient vector, which can be used to treat pixels differently across and along the edge, as is done by the Canny algorithm.

7.2.3 Frei-Chen Edge Detection

Another approach to edge detection that is largely of historical interest is that of **Frei-Chen**. This approach uses a set of nine 3×3 kernels to provide an orthogonal basis for the 9-dimensional space of 3×3 subimages:

$$W_1 = \frac{1}{\sqrt{8}} \begin{bmatrix} -1 & -\sqrt{2} & -1 \\ 0 & 0 & 0 \\ 1 & \sqrt{2} & 1 \end{bmatrix} \quad W_4 = \frac{1}{\sqrt{8}} \begin{bmatrix} \sqrt{2} & -1 & 0 \\ -1 & 0 & 1 \\ 0 & 1 & -\sqrt{2} \end{bmatrix} \quad W_7 = \frac{1}{6} \begin{bmatrix} 1 & -2 & 1 \\ -2 & 4 & -2 \\ 1 & -2 & 1 \end{bmatrix} \quad (7.17)$$

$$W_2 = \frac{1}{\sqrt{8}} \begin{bmatrix} 1 & 0 & -1 \\ \sqrt{2} & 0 & -\sqrt{2} \\ 1 & 0 & -1 \end{bmatrix} \quad W_5 = \frac{1}{2} \begin{bmatrix} 0 & 1 & 0 \\ -1 & 0 & -1 \\ 0 & 1 & 0 \end{bmatrix} \quad W_8 = \frac{1}{6} \begin{bmatrix} -2 & 1 & -2 \\ 1 & 4 & 1 \\ -2 & 1 & -2 \end{bmatrix} \quad (7.18)$$

$$W_3 = \frac{1}{\sqrt{8}} \begin{bmatrix} 0 & -1 & \sqrt{2} \\ 1 & 0 & -1 \\ -\sqrt{2} & 1 & 0 \end{bmatrix} \quad W_6 = \frac{1}{2} \begin{bmatrix} -1 & 0 & 1 \\ 0 & 0 & 0 \\ 1 & 0 & -1 \end{bmatrix} \quad W_9 = \frac{1}{3} \begin{bmatrix} 1 & 1 & 1 \\ 1 & 1 & 1 \\ 1 & 1 & 1 \end{bmatrix} \quad (7.19)$$

Let **b** be the 9-element vector obtained by reshaping the 3×3 subimage surrounding a pixel, and let $\mathbf{w}_i$ be the 9-element vector obtained by reshaping the kernel W_i in the same manner. The first four vectors ($\mathbf{w}_1, \mathbf{w}_2, \mathbf{w}_3, \mathbf{w}_4$) form the basis for the edge subspace, while the next four ($\mathbf{w}_5$, $\mathbf{w}_6$, $\mathbf{w}_7$, $\mathbf{w}_8$) form the basis for the line subspace. Edges are therefore detected by projecting vector $\mathbf{b}$ onto the edge subspace:

$$\theta_{edge} = \arccos \sqrt{\frac{\sum_{i=1}^{4} (\mathbf{w}_i^\mathsf{T}\mathbf{b})^2}{\sum_{i=1}^{9} (\mathbf{w}_i^\mathsf{T}\mathbf{b})^2}} \tag{7.20}$$

with smaller values indicating the likelihood that the pixel is an edge point. Alternatively, large values of

$$\frac{\sum_{i=1}^{4} (\mathbf{w}_i^\mathsf{T}\mathbf{b})^2}{\mathbf{b}^\mathsf{T}\mathbf{b}} \tag{7.21}$$

indicate an edge point.

7.3 Approximating Intensity Edges with Polylines

Once the intensity edges have been found, as described in the previous section, it is often desirable to approximate the edges with a more abstract representation such as line segments or parametric curves. In this section we focus on polylines, which are sequences of line segments. The first step is to store the edgels as an ordered *sequence* of points, rather than as an unordered *set* of points, which is easily accomplished with rather minor modifications to the Canny algorithm. Next, the polylines are fit to the point sequence using any of several techniques.

7.3.1 Douglas-Peucker Algorithm

The classic algorithm for fitting a polyline to a sequence of points is the **Douglas-Peucker algorithm** (also known as the Ramer-Douglas-Peucker algorithm), illustrated in Figure 7.13. First, a straight line called the *anchor-floater line* is drawn from the first to the last point. For each intermediate point, its distance to the line is computed. If all such distances are below a threshold, then the intermediate points are discarded. Otherwise the point with the maximum distance to the anchor-floater line is retained, called the *critical point* for that anchor-floater line. The sequence is then subdivided at the critical point, and the process is repeated for two new anchor-floater lines, one from the start to the critical point, and the other from the critical point to the end. This process continues recursively until all points are within the specified tolerance of the anchor-floater lines.

7.3.2 Repeated Elimination of the Smallest Area

One drawback of the Douglas-Peucker algorithm is that it is suboptimal, because it treats the points associated with each anchor-floater line separately. A holistic approach that considers all points in each iteration involves repeatedly eliminating the point with the least **effective area**, where the effective area is the area of the triangle formed by the point, its predecessor, and its successor. Points with a small effective area have little influence on the perception of the complete polyline and can thus be eliminated without significantly affecting the overall shape. This approach thus progressively eliminates geometric features from the smallest to the largest, repeating the process until the effective area of all points is above some threshold. Note that this approach sequentially determines which point to eliminate,

Sequence of points   Iteration 1   Iteration 2   Iteration 3   Final polyline

Figure 7.13 The Douglas-Peucker algorithm recursively subdivides a polyline by computing the largest distance (orange line) from the points in the polyline to the anchor-floater lines (blue lines).

as opposed to Douglas-Peucker which sequentially determines which point to retain. One minor extension to the algorithm is to store the eliminated points in a list sorted by their effective area, which allows the original curve to be fitted to arbitrary resolution on the fly.

7.4 Feature Detectors

While an edge detector finds pixels where the magnitude of the gradient is large, a **feature point detector** (or *interest operator*) seeks pixels where the graylevel values vary locally in more than one direction, as shown in Figure 7.14. Such pixels are interesting because they are unique and distinguishable from other pixels using only the local information in the immediate neighborhood. Such pixels are called **feature points** (or *interest points*). Since these feature points lie, among other places, at the corners formed by two perpendicular lines, they are sometimes called *corner points*.

7.4.1 Moravec Interest Operator

One of the earliest feature detectors was the **Moravec interest operator**. Given a pixel $\mathbf{x} = (x, y)$ in an image I, let $\mathcal{R}$ be the set of pixels in a small neighborhood around the pixel. Although the image patch $\mathcal{R}$ can be any set of pixels, it is often a square window, say 3×3, centered at the pixel. The Moravec operator shifts the pixels horizontally by a small amount and compares the difference between the original graylevel pattern and the shifted version. Then the process is repeated by shifting the pixels vertically by a small amount, comparing the difference in the same way. More precisely, let

$$\epsilon_{\mathcal{R}}(\Delta \mathbf{x}) \equiv \sum_{\mathbf{x} \in \mathcal{R}} \left(I(\mathbf{x}) - I(\mathbf{x} + \Delta \mathbf{x}) \right)^2 \tag{7.22}$$

Figure 7.14 An image (left) with feature points overlaid in red (right). These feature points were detected with the Tomasi-Kanade operator, but similar results are obtained with other feature detectors. Note that feature points do not occur in untextured areas or along intensity edges, but rather where the graylevel values vary in multiple directions.

be the **sum of squared differences (SSD)** between the image patch and its shifted version, where $\Delta\mathbf{x} = (\Delta x, \Delta y)$ is the shift. If the difference is large in both directions, then the patch is interesting because it contains graylevel variation in multiple directions. Moravec defines the **cornerness** of a pixel as the minimum SSD as the window is shifted left, right, up, and down:

$$\text{cornerness} \equiv \min\{\epsilon_{\mathcal{R}}(-1, 0), \epsilon_{\mathcal{R}}(1, 0), \epsilon_{\mathcal{R}}(0, -1), \epsilon_{\mathcal{R}}(0, 1)\}, \quad (\text{Moravec}) \quad (7.23)$$

where the cornerness of a pixel indicates how locally interesting it is.

An example of the Moravec operator applied to a synthetic image of an aligned square, a rotated square, and a circle is shown in Figure 7.15. Examining the aligned square, it is clear that the pixels fall into one of three categories: 1) Where the scene is untextured, none of the SSD values is high; 2) where there is an intensity edge, some of the SSD values are low while others are high; and 3) where there is a corner, all of the SSD values are high. Therefore, the Moravec interest operator correctly detects the corners of the aligned square. However, it is obvious from the diagonal edges of the rotated square and along the sides of the circle that the operator suffers from being anisotropic (i.e., not rotationally invariant).

7.4.2 Harris Corner Detector

A more robust interest operator can be derived by expanding $I(\mathbf{x} + \Delta\mathbf{x})$ using the first-order Taylor series:

$$I(\mathbf{x} + \Delta\mathbf{x}) \approx I(\mathbf{x}) + (\Delta\mathbf{x})^{\mathsf{T}}\frac{\partial I}{\partial \mathbf{x}} \quad (7.24)$$

Substituting into (7.22) and introducing an optional weighting function $0 \le w(\mathbf{x}) \le 1$, to allow some pixels to count more than others, yields

$$\epsilon_{\mathcal{R}}(\Delta\mathbf{x}) \approx \sum_{\mathbf{x}\in\mathcal{R}} w(\mathbf{x})\left((\Delta\mathbf{x})^{\mathsf{T}}\frac{\partial I}{\partial \mathbf{x}}\right)^2 \quad (7.25)$$

$$= \sum_{\mathbf{x}\in\mathcal{R}} w(\mathbf{x})(\Delta\mathbf{x})^{\mathsf{T}}\left(\frac{\partial I}{\partial \mathbf{x}}\right)\left(\frac{\partial I}{\partial \mathbf{x}}\right)^{\mathsf{T}}(\Delta\mathbf{x}) \quad (7.26)$$

$$= (\Delta\mathbf{x})^{\mathsf{T}}\underbrace{\sum_{\mathbf{x}\in\mathcal{R}} w(\mathbf{x})\left(\frac{\partial I}{\partial \mathbf{x}}\right)\left(\frac{\partial I}{\partial \mathbf{x}}\right)^{\mathsf{T}}}_{\mathbf{Z}}(\Delta\mathbf{x}) \quad (7.27)$$

$$= (\Delta\mathbf{x})^{\mathsf{T}}\mathbf{Z}(\Delta\mathbf{x}) \quad (7.28)$$

Figure 7.15 From left to right: A synthetic image, the SSD of each pixel as the image is shifted left and up, and the cornerness as measured by the Moravec interest operator. High cornerness values occur at the corners of the aligned square, as well as along all diagonal lines.

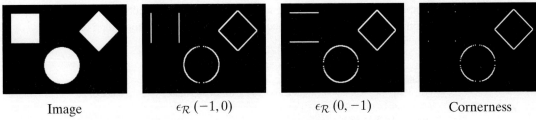

| Image | $\epsilon_{\mathcal{R}}(-1, 0)$ | $\epsilon_{\mathcal{R}}(0, -1)$ | Cornerness |

where $\mathbf{Z}$ is a symmetric 2×2 matrix composed of the outer products of the gradient vectors of the pixels in the window:

$$\mathbf{Z} \equiv \sum_{\mathbf{x} \in \mathcal{R}} w(\mathbf{x}) \begin{bmatrix} I_x^2(\mathbf{x}) & I_x(\mathbf{x})I_y(\mathbf{x}) \\ I_x(\mathbf{x})I_y(\mathbf{x}) & I_y^2(\mathbf{x}) \end{bmatrix} = \begin{bmatrix} z_x & z_{xy} \\ z_{xy} & z_y \end{bmatrix} \tag{7.29}$$

where $I_x \equiv \partial I/\partial x$ and $I_y \equiv \partial I/\partial y$ denote the partial derivatives of the image along the coordinate axes, so that the gradient is given by $\nabla I = \partial I/\partial x = [I_x \quad I_y]^{\mathsf{T}}$; and where $z_x \equiv \sum_{\mathbf{x}} w(\mathbf{x})I_x^2(\mathbf{x})$, $z_y \equiv \sum_{\mathbf{x}} w(\mathbf{x})I_y^2(\mathbf{x})$, and $z_{xy} \equiv \sum_{\mathbf{x}} w(\mathbf{x})I_x(\mathbf{x})I_y(\mathbf{x})$ are the elements of $\mathbf{Z}$. Typically the weighting function is either uniform to treat all pixels equally, or a Gaussian to weight pixels near the center of the window more than those far away.

The matrix $\mathbf{Z}$, which is determined solely by first derivatives, goes by various names. Some authors call it the **gradient covariance matrix**, because if $w(\mathbf{x}) = \frac{1}{|\mathcal{R}|}$ and the mean is zero, then Equation (7.29) is the covariance matrix of the gradient vectors of the pixels inside the window. Other authors refer to it as the **autocorrelation matrix**, the *structure tensor*, or the *second moment matrix*, due to the connection between the covariance and second moments exhibited in Equation (4.145). It is also known as the Hessian matrix but, as we shall see later, this term should be avoided due to the confusion it can cause.

Since, as we saw earlier,[†] the covariance matrix fits an ellipse to the data so that the principal axes of the ellipse are captured by the eigenvalues of the matrix, these eigenvalues are crucial to understanding the structure of the covariance matrix. Three cases are possible: if both eigenvalues are large, then the pixel values vary in multiple directions, making the pixel region uniquely distinguishable from its local surroundings; if one eigenvalue is large while the other is small, then the region straddles an intensity edge; finally, if both eigenvalues are small, then the region is untextured because the gradient magnitude is small for all pixels in the region.

To find distinguishable features, then, a search is conducted for pixels whose gradient covariance matrix contains two large eigenvalues. One of the most popular approaches to feature detection is the **Harris corner detector**,[‡] which measures cornerness using the trace and determinant of the matrix:

$$\text{cornerness} \equiv \det(\mathbf{Z}) - k\,(\text{trace}(\mathbf{Z}))^2 \qquad (\text{Harris}) \tag{7.30}$$

$$= z_x z_y - z_{xy}^2 - k(z_x + z_y)^2 \tag{7.31}$$

$$= \lambda_1 \lambda_2 - k(\lambda_1 + \lambda_2)^2 \tag{7.32}$$

where λ_1 and λ_2 are the eigenvalues of $\mathbf{Z}$, and the second equality comes from the fact that the determinant of a square matrix is the product of its eigenvalues, while its trace is the sum of its eigenvalues, which we showed earlier.[§] The constant k is a small factor whose value is typically recommended to be in the vicinity of 0.04. The second term is used to reduce the chance of selecting a point with a single very large eigenvalue. Note that since the eigenvalues are inherently invariant to rotation, Harris features are largely invariant to rotation as well. In fact, repeated studies have shown variations of Harris to be some of the most reliable detectors in the presence of image rotations, illumination transformations, and perspective deformations. Another advantage of Harris is that the determinant / trace trick simplifies the computation by eliminating the need to compute square root, and it also eliminates the possibility of dividing by zero, since the eigenvalues are not computed explicitly.

[†] Section 4.4.5 (p. 182).

[‡] The Harris detector is sometimes known as the *Plessey operator*, after the name of the company employing the inventor at the time of discovery.

[§] Section 4.4.5 (p. 182).

7.4.3 Tomasi-Kanade Feature Detector

An alternative to Harris is the **Tomasi-Kanade operator** (also known as the **Shi-Tomasi operator**), which measures cornerness using the minimum eigenvalue:

$$\text{cornerness} \equiv \min(\{\lambda_1, \lambda_2\}) = \lambda_2 \qquad (\text{Tomasi-Kanade}) \qquad (7.33)$$

where the final equality assumes that the eigenvalues have been sorted in decreasing order, so that $\lambda_1 \geq \lambda_2$. Recall that the eigenvalues are obtained by solving the characteristic equation $\det(\mathbf{Z} - \lambda \mathbf{I}_{\{2 \times 2\}}) = 0$, leading to

$$\lambda_2 = \frac{1}{2}\left((z_x + z_y) - \sqrt{(z_x - z_y)^2 + 4z_{xy}^2}\right) \qquad (7.34)$$

which is the same as Equation (4.146). Tomasi-Kanade requires more computation than Harris and has generally been found to be slightly less robust, though the performances of the two are quite comparable.

A less well-known alternative is to divide the determinant by the trace:

$$\frac{1}{\text{cornerness}} \equiv \frac{\text{trace}(\mathbf{Z})}{\det(\mathbf{Z})} = \frac{\lambda_1 + \lambda_2}{\lambda_1 \lambda_2} = \frac{1}{\lambda_1} + \frac{1}{\lambda_2} \qquad (\text{parallel resistors}) \qquad (7.35)$$

which treats the eigenvalues like parallel resistors in an electrical circuit. If two resistors r_1 and r_2 are connected in parallel, then the combined resistance is r, where $\frac{1}{r} = \frac{1}{r_1} + \frac{1}{r_2}$. It is easy to show that the combined resistance is never greater than the smallest resistor, and likewise, the cornerness is never greater than the smallest eigenvalue:

$$\frac{\lambda_2}{2} \leq \frac{\det(\mathbf{Z})}{\text{trace}(\mathbf{Z})} \leq \lambda_2 \qquad (7.36)$$

To avoid divide-by-zero errors, simply add a small number to the denominator or avoid the division whenever both eigenvalues are smaller than a threshold.

Contour plots of Harris, Tomasi-Kanade, and parallel resistors, along with the determinant of the matrix, versus λ_1 and λ_2 are given in Figure 7.16. All the measures have similar shapes because they all attempt to maximize the two eigenvalues. Note that the isocontours of the determinant are the lines $\lambda_2 = 1/\lambda_1$, with asymptotes $\lambda_1 = 0$ and $\lambda_2 = 0$. As a result, one large eigenvalue can overcome an arbitrarily small eigenvalue. Harris avoids this problem by subtracting the square of the trace, thus penalizing situations where one of the eigenvalues is small. The asymptotes of Tomasi-Kanade are $\lambda_1 = \lambda_2 = c$, where c is the cornerness value, whereas the asymptotes of both Harris and parallel resistors are angled in slightly to decrease the effects of one small eigenvalue.

7.4.4 Beaudet Detector

The **Hessian** of a function is a matrix containing the second-order partial derivatives of the function. Therefore, the Hessian of the region surrounding a pixel is given by

$$H \equiv \sum_{\mathbf{x} \in \mathcal{R}} w(\mathbf{x}) \begin{bmatrix} I_{xx}(\mathbf{x}) & I_{xy}(\mathbf{x}) \\ I_{xy}(\mathbf{x}) & I_{yy}(\mathbf{x}) \end{bmatrix} \qquad (7.37)$$

where $I_{xx} \equiv \partial^2 I / \partial x^2$, $I_{yy} \equiv \partial^2 I / \partial y^2$, and $I_{xy} \equiv \partial^2 I / \partial x \partial y$ are the second derivatives. The **Beaudet detector**[†] uses these second derivatives and is defined as the determinant of the Hessian:

$$\text{cornerness} \equiv I_{xx} I_{yy} - I_{xy}^2 \qquad (\text{Beaudet}) \qquad (7.38)$$

[†] Also known as the DET operator.

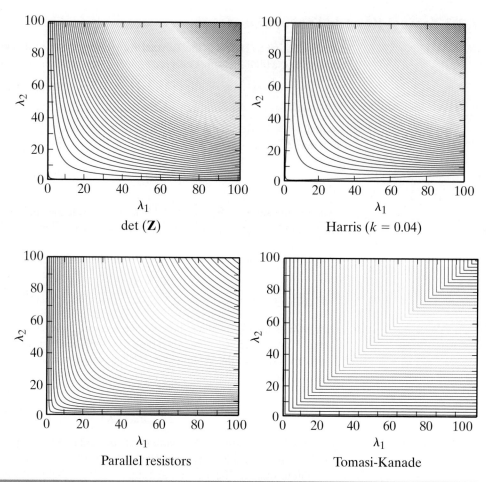

The Beaudet measure is related to the Gaussian curvature of the signal, and it is rotation invariant.

As mentioned earlier, the second moment matrix $\mathbf{Z}$ in Equation (7.29) is sometimes referred to as the Hessian, because it is an approximation to the Hessian of $\frac{1}{2}I^2(\mathbf{x})$:

$$\frac{\partial^2}{\partial x^2}\left(\frac{1}{2}I^2(\mathbf{x})\right) = \frac{\partial}{\partial x}\left(I(\mathbf{x})\frac{\partial I(\mathbf{x})}{\partial x}\right) = \left(\frac{\partial I(\mathbf{x})}{\partial x}\right)^2 + \left(I(\mathbf{x})\frac{\partial^2 I(\mathbf{x})}{\partial x^2}\right) \approx I_x^2 \tag{7.39}$$

$$\frac{\partial^2}{\partial y^2}\left(\frac{1}{2}I^2(\mathbf{x})\right) = \frac{\partial}{\partial y}\left(I(\mathbf{x})\frac{\partial I(\mathbf{x})}{\partial y}\right) = \left(\frac{\partial I(\mathbf{x})}{\partial y}\right)^2 + \left(I(\mathbf{x})\frac{\partial^2 I(\mathbf{x})}{\partial y^2}\right) \approx I_y^2 \tag{7.40}$$

$$\frac{\partial^2}{\partial x \partial y}\left(\frac{1}{2}I^2(\mathbf{x})\right) = \frac{\partial}{\partial x}\left(I(\mathbf{x})\frac{\partial I(\mathbf{x})}{\partial y}\right) = \left(\frac{\partial I(\mathbf{x})}{\partial x}\frac{\partial I(\mathbf{x})}{\partial y}\right) + \left(I(\mathbf{x})\frac{\partial^2 I(\mathbf{x})}{\partial x \partial y}\right) \approx I_x I_y \tag{7.41}$$

where the approximations ignore the second-order terms. We avoid using this name because of the confusion that can result, since the Hessian is a matrix of second derivatives, whereas $\mathbf{Z}$ is a matrix of first derivatives.

7.4.5 Kitchen-Rosenfeld

The classic **Kitchen-Rosenfeld interest point detector** combines first- and second-order derivatives to find the maximum curvature along an intensity isocontour, weighted by the nonmaximum-suppressed gradient magnitude:

$$\text{cornerness} \equiv \frac{I_{xx}I_y^2 + I_{yy}I_x^2 - 2I_{xy}I_xI_y}{I_x^2 + I_y^2} \qquad (\text{Kitchen-Rosenfeld}) \qquad (7.42)$$

The Kitchen-Rosenfeld detector, like Beaudet, is no longer widely used.

7.4.6 SIFT Feature Detection

Despite the success of Harris and the other variants just mentioned, an even more popular approach is the **SIFT feature detector** (for *Scale Invariant Feature Transform*), which is illustrated in Figure 7.17. The first step of SIFT is to build a Laplacian pyramid of the image. If s is the number of scales per octave at which the features are detected, then the pyramid uses $s + 3$ Gaussians per octave; in the figure $s = 2$, so the difference between $2 + 3 = 5$ Gaussians yields 4 LoG-approximated images, from which the two scales are processed. The second step is to determine, for every pixel and for every scale, whether the pixel is a local maximum among its 26 neighbors (8 in the same image, 9 at the next smallest scale, and 9 at the next largest scale). A final step discards pixels in untextured areas or along intensity edges.

For each local maximum in scale space just detected, its position can be refined by repeatedly solving a linear system to yield floating-point coordinates. More specifically, a second-order Taylor series expansion is computed of the scale space image:

$$I(\mathbf{x} + \Delta\mathbf{x}) \approx I(\mathbf{x}) + \frac{\partial I(\mathbf{x})^{\mathsf{T}}}{\partial\mathbf{x}}(\Delta\mathbf{x}) + \frac{1}{2}(\Delta\mathbf{x})^{\mathsf{T}}\frac{\partial^2 I(\mathbf{x})}{\partial\mathbf{x}^2}(\Delta\mathbf{x}) \qquad (7.43)$$

Figure 7.17 SIFT features are detected by computing a Laplacian pyramid, then looking for local maxima among the 26 neighbors of a pixel. For each octave in this drawing there are 5 Gaussians, 4 LoGs (approximated by DoGs), and two scales (indicated by red arrows); the remaining two LoGs are used only in the neighborhood computation.

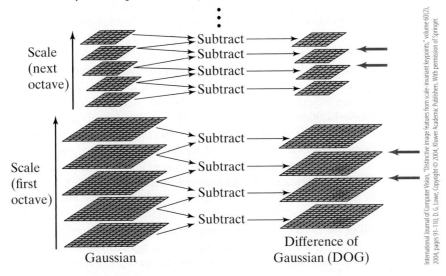

Scale (next octave)

Scale (first octave)

Gaussian

Subtract

Difference of Gaussian (DOG)

International Journal of Computer Vision, "Distinctive image features from scale-invariant keypoints," volume 60(2), 2004, pages 91–110, D. G. Lowe, Copyright © 2004, Kluwer Academic Publishers. With permission of Springer.

where $\mathbf{x} \equiv \begin{bmatrix} x & y & \sigma \end{bmatrix}^{\mathsf{T}}$. Taking the derivative of this function and setting it to zero yields

$$\frac{\partial I(\mathbf{x})^{\mathsf{T}}}{\partial \mathbf{x}} + \frac{\partial^2 I(\mathbf{x})}{\partial \mathbf{x}^2}(\Delta \mathbf{x}) = 0 \qquad (7.44)$$

or

$$\Delta \mathbf{x} = -\left(\frac{\partial^2 I(\mathbf{x})}{\partial \mathbf{x}^2}\right)^{-1} \frac{\partial I(\mathbf{x})}{\partial \mathbf{x}} \qquad (7.45)$$

Once this minimization has stabilized, the value at that point can be computed by plugging the computed value of $\Delta \mathbf{x}$ into Equation (7.43) using only the linear term for simplicity:

$$I(\mathbf{x} + \Delta \mathbf{x}) \approx I(\mathbf{x}) + \frac{\partial I(\mathbf{x})^{\mathsf{T}}}{\partial \mathbf{x}}(\Delta \mathbf{x}) \qquad (7.46)$$

The popularity of the SIFT feature detector is due to its leveraging of scale space to find features regardless of their scale in the image.

7.5 Feature Descriptors

Once features have been detected in multiple images using one of the techniques described above, the features are often matched across the images. For example, suppose we have an image of a known object from a database, along with a query image that contains the object at some arbitrary position, orientation, and scale. By matching features between the two images, we can infer whether the object is present and, if so, the pose at which the object is located. In order for such an approach to work, it is necessary to compute and match **feature descriptors** that are invariant to changes in pose and illumination.

7.5.1 SIFT Feature Descriptor

Among the many features descriptors that have been proposed, one of the most widely used is the **SIFT feature descriptor**. Although the SIFT feature descriptor typically goes hand-in-hand with the SIFT feature detector, this is not necessary, since the descriptor can be applied anywhere in the image. The algorithm works as follows. Once a feature has been detected using the SIFT feature detector or some other means, the first step is to sample the image gradient magnitudes and orientations in the neighborhood surrounding the feature, using the scale at which the feature was detected to specify the size of the neighborhood and amount of gradient smoothing. The dominant gradient orientation is computed in a manner similar to the one described earlier, and all gradient orientations are rotated relative to this orientation, to make the computation invariant to image rotation. The gradient magnitudes are then weighted by a single Gaussian (whose width is determined by the scale of the detected feature) in order to increase the weight of pixels near the center.

The gradient vectors are quantized into one of several possible orientations and then accumulated over discrete spatial regions into a 3D histogram over space and scale. For example, Figure 7.18 shows the gradients of all the pixels in an 8×8 neighborhood surrounding the feature; there are 8 possible orientations, and $8 \cdot 8 = 64$ gradient vectors which are accumulated into a 4×4 grid. The orientations of the gradient vectors of the 16 pixels in the top-left subarray are accumulated in the histogram bins associated with the top-left cell of the keypoint descriptor array, the values in the top-right subarray are accumulated in the histogram bins associated with the top-right cell of the keypoint descriptor array, and so forth. Each gradient vector votes for the appropriate bin in the histogram with a weight

International Journal of Computer Vision features from scale-invariant keypoints," volume 60(2), 2004, pages 91–110, D. G. Lowe, Copyright © 2004, Kluwer Academic Publishers. With permission of Springer.

Figure 7.18 The SIFT feature descriptor is computed by accumulating the orientations of the gradient vectors in a neighborhood of the feature point into a 3D array over position and orientation.

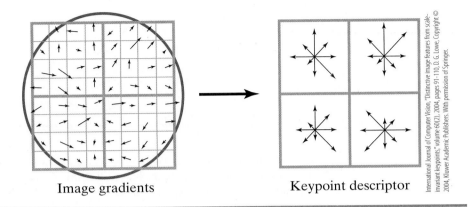

Image gradients   Keypoint descriptor

proportional to the gradient magnitude, using trilinear interpolation to distribute values to neighboring bins in a robust manner. The values in the 3D histogram are then concatenated to form a vector that describes the feature. Because the figure shows 8 orientations and $2 \cdot 2 = 4$ subarrays, this example yields a 32-dimensional vector, but in practice there are usually 8 orientations and $4 \cdot 4 = 16$ positions, leading to a 128-dimensional vector. To achieve illumination invariance, the vector is normalized to unit length by dividing by its L^2-norm.

Figure 7.19 shows an application of SIFT feature detections and descriptors. On the left are images of a toy frog and toy train from a database. In the middle is the query image. On the right are the detected feature points that match features in the database, along with the detected objects that were obtained by aggregating the results of the individual features. Notice that the objects are detected despite significant difference in pose, as well as occlusion.

7.5.2 Gradient Location and Orientation Histogram (GLOH)

An extension of the SIFT descriptor is the **gradient location and orientation histogram (GLOH)**. As with SIFT, the gradient of the image is computed, and the gradient orientations are accumulated in a histogram. However, instead of using a rectangular grid of pixels, a log-polar grid is used to specify 17 spatial bins from 2 annuli and 8 orientations, in addition to one bin in the center, as shown in Figure 7.20. With 16 quantized gradient orientations, the histogram contains $17 \cdot 16 = 272$ bins, which are then reduced to a 128-element vector

Figure 7.19 SIFT feature matching results. SIFT feature descriptors from the query image (middle) are matched against descriptors from the database (left) to detect objects at various poses and lighting conditions, and even with severe occlusion (right).

International Journal of Computer Vision, "Distinctive image features from scale-invariant keypoints," volume 60(2), 2004, pages 91–110, D. G. Lowe, Copyright © 2004, Kluwer Academic Publishers. With permission of Springer.

Figure 7.20 The GLOH feature descriptor involves sampling gradient orientations in a log-polar grid.

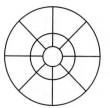

using PCA[†] applied to a large database of image patches. GLOH feature descriptors have been shown to be slightly more distinctive than SIFT descriptors when matching images with rotation, scale, and viewpoint changes.

7.5.3 Shape Context

A closely related descriptor designed specifically for binary images is the **shape context**. As shown in Figure 7.21, the shape context of a point on the boundary of an object is computed as a 2D histogram over spatial locations arranged in a log-polar grid similar to that of GLOH, except that the center is also divided into wedges. With 5 radii and 12 angles, the resulting histogram contains 60 bins. Each bin of the histogram contains the sum of edge points within the region defined by the bin. Note that gradient orientation is not used, and all edge points contribute equally to the histogram. The shape context has been used successfully in matching binary shapes.

7.5.4 Histogram of Oriented Gradients (HOG)

Another popular image descriptor is the **histogram of oriented gradients (HOG)**, which is a vector of concatenated histograms of gradient orientations. Since the SIFT feature descriptor is also a vector of concatenated histograms of gradient orientations, it can in some sense be thought of as a HOG. However, the term HOG is usually reserved for a descriptor computed over a dense rectangular region of the image rather than just at a feature point.

The HOG descriptor was developed in the context of pedestrian detection. A 64×128 rectangular window is slid across the image, and at each location the HOG descriptor of the window is computed and then evaluated[‡] to determine whether a pedestrian is in the window. The window is divided into a dense array of non-overlapping *cells* consisting of $8 \cdot 8 = 64$ pixels. Within each cell a histogram of gradient orientations is computed by allowing each pixel

Figure 7.21 The shape context captures the shape of a binary region by counting the number of edge pixels in a log-polar grid. From left to right: A binary shape, the log-polar grid, and the resulting histogram at a particular point.

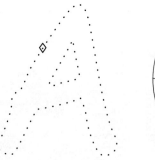
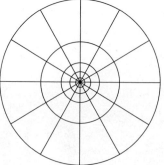
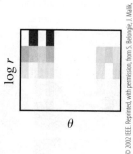

[†] Section 12.3.5 (p. 589).
[‡] Using the techniques of Chapter 12 (p. 560)

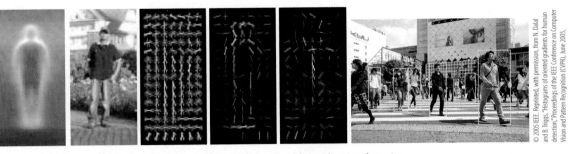

© 2005 IEEE. Reprinted, with permission, from N. Dalal and B. Triggs, "Histograms of oriented gradients for human detection," Proceedings of the IEEE Conference on Computer Vision and Pattern Recognition (CVPR), June 2005, Tupungato / Shutterstock.com.

Figure 7.22 Histograms of oriented gradients (HOGs) are widely used for pedestrian detection.

to cast a weighted vote for its orientation, where the weight is given by its gradient magnitude, and the orientations are quantized into either 9 possibilities (if $0°$ to $180°$ orientations are used, that is, the sign of the gradient is ignored) or 18 possibilities (if $0°$ to $360°$ orientations are used). The cells are grouped into overlapping *blocks*, where each block contains $2 \cdot 2 = 4$ neighboring cells. To provide some amount of illumination invariance, the histograms of the cells within a block are concatenated to form a vector, and the vector is then normalized by dividing by its L^2-norm. Note that, because blocks are overlapping, each cell's histogram is used multiple times to create the vectors for the blocks within which it lies. As with any computer vision algorithm, many variations of this approach are possible by changing the shape of the cells, the manner of normalization, and so forth, but the description presented here is of one of the more common variations. Due to their dense nature, HOG descriptors are able to capture subtle variations in the window, making them enormously successful in the task of detecting pedestrians and other shape-based object classes.

7.6 Further Reading

The Laplacian pyramid is due to Burt and Adelson [1983], where a description of the equal contribution property can also be found. Mallat [1989] is the classic paper that links wavelets, multiresolution analysis, and pyramid algorithms. Scale space is introduced in the paper by Witkin [1983]. A thorough discussion of the Gaussian and Laplacian pyramids and their relationship to scale space can be found in Lindeberg [1994]. An even more in-depth treatment of scale space can be found in Lindeberg [1993]. The causality criterion and the concept of the deep structure of the image are due to Koenderink [1984], where the Hessian matrix is also discussed. Babaud et al. [1986] show the uniqueness of the Gaussian kernel for constructing a scale space. Applications of scale space to feature detection can be found in Canny [1986], Mallat and Zhong [1992], Lindeberg [1998a], and Lindeberg [1998b].

Everyone should read the delightful early paper of Attneave [1954], which connects intensity edges with the notions of predictability and redundancy. The early work on interpreting line drawing images of polyhedral objects is due to Roberts [1963], Huffman [1971], and Clowes [1971]. The Marr-Hildreth edge detector was proposed

by Marr and Hildreth [1980], which was replaced by the classic work of Canny [1986]. Canny's paper is a dense read but contains some real gems for anyone patient enough to read it carefully. The sign of the Laplacian of Gaussian is used by Nishihara [1984]. Although space has not permitted a thorough discussion, edge detectors have been compared empirically using Pratt's figure of merit, see Abdou and Pratt [1979], and by the approach of Bowyer and Phillips [1998]. For a more recent approach to edge detection, see the probability of boundary (Pb) detector by Martin et al. [2004].

The Douglas-Peucker line-fitting algorithm is due to Douglas and Peucker [1973], which was slightly preceded by the independent work of Ramer [1972] — hence the name Ramer-Douglas-Peucker. Hershberger and Snoeyink [1992] propose a speedup to Douglas-Peucker with a worst-case running time of $O(n \log n)$, whereas Douglas-Peucker is $O(n^2)$. The algorithm to repeatedly eliminate the smallest area is due to Visvalingam and Whyatt [1992].

The Moravec interest operator is from Moravec [1977]. The classic operators of Beaudet [1978] and

Kitchen and Rosenfeld [1982] are primarily of historic interest only, having been replaced by more recent approaches. The Harris feature detector, which is still widely used, is presented in Harris and Stephens [1988]. The Tomasi-Kanade detector was first described by Tomasi and Kanade [1991], although it is more widely known from the paper by Shi and Tomasi [1994], which explains the alternate name of Shi-Tomasi. Although the second-moment matrix is called the Hessian in Baker and Matthews [2004] in the context of Gauss-Newton minimization for point feature tracking, the term Hessian is usually reserved for the matrix of second-derivatives, as in Bay et al. [2008]. Several widely cited studies have been conducted to compare different feature detectors, such as that of Schmid et al. [2000] and Mikolajczyk and Schmid [2005], which have largely concluded that Harris (or some variation of it) is the most repeatable. Another interesting study is that of Kenney et al. [2005], which concluded that Tomasi-Kanade is the best feature detector according to a particular set of axioms.

The SIFT feature detector and descriptor were introduced by Lowe [2004], the GLOH descriptor is from Mikolajczyk and Schmid [2005], the shape context is due to Belongie et al. [2002], and HOG is presented in Dalal and Triggs [2005]. A number of other feature detectors and/or descriptors have emerged over the years, such as SURF from Bay et al. [2008], FAST from Rosten and Drummond [2006], and DAISY from Tola et al. [2010]. Other work of historical interest is the discovery of receptive fields in the human visual system by Hubel and Wiesel [1962] and Olshausen and Field [1996] and the local jets of Koenderink and van Doorn [1987]. Another relevant piece of work is that of Ozuysal et al. [2007] on fast keypoint recognition.

We did not have space to discuss texture in detail, but the classic work of Julesz [1981] and Julesz and Bergen [1983] on textons should be consulted for historical context. Another classic work on texture is that of Laws [1980]. Steerable filters were introduced by Freeman and Adelson [1991]. A remarkably simple and effective algorithm for texture synthesis can be found in the well-known work of Efros and Leung [1999]. Additional information on visual texture can be found in the overview of Tuceryan and Jain [1993].

PROBLEMS

7.1 Given that the area (i.e., the number of pixels) of the original image is a, and the downsampling factor is $\sqrt[4]{2}$,

(a) Compute the area of the following levels of a Gaussian pyramid: $I^{(1)}, I^{(2)}, I^{(5)}, I^{(6)}$.

(b) Verify that the following relation holds:

$$\frac{\text{area of } I^{(1)}}{\text{area of } I^{(2)}} = \frac{\text{area of } I^{(5)}}{\text{area of } I^{(6)}}$$

7.2 Assuming a downsampling factor of 2, calculate the family of 5-element symmetric kernels that satisfies the equal contribution property. Do any of these kernels look familiar from Pascal's triangle?

7.3 Suppose we wish to construct a Gaussian pyramid with $n = 3$ images per octave.

(a) What is the downsampling factor?

(b) How should σ'^2 be chosen to ensure that the overall smoothing between octaves is $\sigma^2 = 1.2$?

7.4 Suppose we wish to construct a Laplacian pyramid with $n = 5$ images per octave.

(a) What should be the variance ratio ρ in order to ensure that each octave is convolved with the same sequence of variances relative to the image size?

(b) What variance should be applied for pyramid levels 1, 2, and 3 (i.e., what are $\sigma_0^2, \sigma_1^2, \sigma_2^2$)?

7.5 Explain why the causality criterion is important in computing the scale space.

7.6 List the four types of intensity edges.

7.7 Explain why the Canny edge detector fails at the intersection of two lines.

7.8 Perform non-maximal suppression on the following gradient magnitude and phase images. (Compute results only for the inner 3 × 3 array.)

3	3	3	3	3
3	10	9	5	3
3	20	8	7	3
3	5	30	10	3
3	3	3	3	3

magnitude

0	0	0	0	0
0	0	$-\frac{4\pi}{8}$	0	0
0	$\frac{10\pi}{8}$	$\frac{6\pi}{8}$	0	0
0	$\frac{4\pi}{8}$	0	$\frac{2\pi}{8}$	0
0	0	0	0	0

phase

7.9 Explain the localization-detection tradeoff.

7.10 Why is the Marr-Hildreth operator a bad edge detector?

7.11 Stacking the Frei-Chen kernels into 1D vectors, verify that they form an orthonormal set.

7.12 We looked at ways to detect step edges. Describe how you might go about detecting ridge edges.

7.13 Use the Douglas-Peucker algorithm to fit a polyline to the following sequence of points. Set the threshold to 1.5, where 1 is the length of a square in the grid. Show the output after each step of the algorithm.

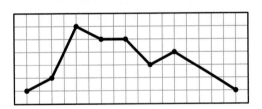

7.14 How is the Harris corner detector better than the Moravec interest operator?

7.15 Explain why you would want to perform non-maximum suppression after computing the Harris cornerness measure on the pixels of an image. How would you modify the non-maximum suppression procedure of Algorithm 7.2 to apply to Harris?

7.16 Derive the closed-form expression for the eigenvalues of a 2 × 2 gradient covariance matrix.

7.17 Prove that the trace of a 2 × 2 covariance matrix is the sum of its eigenvalues, and that the determinant of the matrix is their product. Is this true for any 2 × 2 matrix?

7.18 Given the following directional derivatives, compute Harris and Tomasi-Kanade cornerness measures of the central pixel, assuming a 3 × 3 window and uniform weighting for all the pixels in the window.

$$I_x = \begin{bmatrix} -5 & -9 & 5 \\ 7 & 3 & -8 \\ -6 & 9 & 3 \end{bmatrix} \quad I_y = \begin{bmatrix} 2 & -7 & -6 \\ -1 & 8 & 9 \\ -5 & 2 & 3 \end{bmatrix}$$

7.19 What is the Hessian of a function? Why is it potentially confusing to use the term Hessian to refer to the gradient covariance matrix?

7.20 True or false: the Scale Invariant Feature Transform (SIFT) is invariant to translation, rotation, and scale.

7.21 Explain the difference between a feature detector and a feature descriptor.

7.22 Implement the detection of sparse features points in an image. Use either the Harris corner detector or the Tomasi-Kanade method of thresholding the minimum eigenvalue to compute a measure of "cornerness" for every pixel in the image, using a small 3×3 window for constructing the gradient covariance matrix. Then perform non-maximal suppression to set the "cornerness" to zero for every pixel that is not a local maximum in a 3×3 neighborhood (using either 4- or 8-neighbors). Note that, unlike Canny, this non-maximum suppression will not care about the direction in which neighbors lie relative to the pixel, but instead will consider all the pixels in the neighborhood at once. You may also want to either enforce a minimum distance between features, or to simply allow no more than 1 feature in each 8×8 image block.

7.23 Implement the Canny edge detector. Your code should accept a single scale parameter (σ) as input. There should be three steps to your code: gradient estimation, non-maximum suppression, and thresholding with hysteresis (i.e., double-thresholding). For the gradient estimation, convolve the image with the derivative of a Gaussian (i.e., convolve with a 2D Gaussian derivative, implemented using the separable property), rather than computing finite differences in the smoothed image. Do not worry about image borders; the simplest solution is to simply set the border pixels in the convolution result to zero rather than extending the image. Automatically compute the threshold values based upon image statistics. Display intermediate results (e.g., the two $x-$ and $y-$ gradient components, the gradient magnitude and angle, and the edges before thresholding), in addition to the final result.

CHAPTER 8
Compression

I n this chapter we consider the problem of image compression, whose purpose is to reduce both storage size and transmission time by representing an image using fewer bits than would be required otherwise. The two primary approaches to compression are *lossless* compression, in which absolutely no information about the original image is lost, and *lossy* compression, in which the image is approximated in such a way that the distortions are not objectionable. While there is some truth to the statement that compression is all about impatience (if we simply wait a few years, then storage and transmission rates will increase, due to Moore's Law), the topic of compression is here to stay thanks to the huge size of images and videos, the large compression ratios that can be achieved, and the insatiable appetite of consumers for more and more data.

8.1 Basics

It should come as no surprise that images and videos take up a huge amount of storage space. For example, a raw, uncompressed image taken by a 5-megapixel consumer-level camera requires $5 \cdot 10^6$ pixels $\cdot$ 24 bits per pixel/8 bits per byte = 15 megabytes. megabytes. So if you were to take an average of 10 photographs per day, then over a period of ten years you would accumulate

15 megabytes per picture $\cdot$ 10 photographs per day $\cdot$ 3652.5 days per decade = 548 gigabytes

of image data. Similarly, a raw, uncompressed 2-hour HDTV movie requires

$$120 \text{ minutes} \cdot 60 \, \frac{\text{seconds}}{\text{minute}} \cdot 30 \, \frac{\text{frames}}{\text{second}} \cdot 1920 \cdot 1280 \, \frac{\text{pixels}}{\text{frame}} \cdot 3 \, \frac{\text{bytes}}{\text{pixel}} = 1.6 \cdot 10^{12} \text{ bytes}$$

or almost 2 terabytes; a collection of 1000 movies would therefore require 2 petabytes. On a typical image-sharing website, hundreds of millions of photographs are uploaded every day, amounting to several exabytes per year of images.[†] These numbers are staggering, and although we are starting to reach the point where memory is cheap enough that we can begin to think about storing large collections of raw images and videos at home or on a server, limited transmission speeds and the desire to store these data on mobile devices, not to mention rapidly increasing rates of content creation, continue to motivate the need for compressing and decompressing the data.

An overview of a compression/decompression system is provided in Figure 8.1. A stream of bits (in our case an image) is fed to a **compressor**, which converts the stream to a smaller stream of bits. This new stream is then either stored as a file on disk or transmitted across a network, where on the other end a **decompressor** restores the original image. Sometimes the compressor and decompressor are known as a *coder* and *decoder*, respectively, so that the software part of the system is collectively known as a **codec**.

When we say that the decompressor restores the original image, we must make an important distinction because there are two types of compression. In **lossless compression**, the restored image is *exactly* the same as the original image, so that no information has been lost. Lossless compression techniques are applicable to any type of data, such as text, an image, a database of addresses, or a file containing an executable. On the other hand, the image restored by **lossy compression** is only *similar* to the original image. Lossy compression techniques are applicable to data arising from real-world measurements, such as an audio signal, a photographic image, or a signal captured by some other type of sensor. In such data, certain aspects of the data are more important than others, both because of noise in the signal and the inability of the human perceptual system to distinguish certain subtle characteristics of the data. By taking into account these perceptual inequities, the information that is lost by a well-designed lossy compression algorithm will not be noticeable to the person viewing the restored image.

The key idea behind compression is the distinction between **data** and **information**. Data are the bits, stored in the computer and, taken in and of themselves, carry no inherent meaning. Information, on the other hand, can be thought of as the message being conveyed by the data, or rather the meaning that can be inferred from the data. This distinction is made clear in the following example.

EXAMPLE 8.1	Suppose you want to share with someone the number 3.14159265358979323846264433832... What is the data, and what is the information?
Solution	If you were to send this number as a series of digits, it would literally take forever. On the other hand, if you were to agree with the receiver beforehand that the Greek letter π represents the number, then you could simply send only a small number of bits. In this context, the *information* is the ratio of a circle's circumference to its diameter, while the data is either the infinite string of digits or the single Greek letter. Obviously, the latter is a much more efficient encoding of the information than the former.

[†] A megabyte is a million (10^6) bytes; a gigabyte is a billion (10^9) bytes; a terabyte is a trillion (10^{12}) bytes; a petabyte is a million billion (10^{15}) bytes; and an exabyte is a billion billion (10^{18}) bytes.

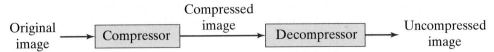

Figure 8.1 In a typical compression / decompression system, an image is first compressed, then the compressed image is either stored or transmitted, after which it is decompressed to yield either the original image (in the case of lossless compression) or an approximation to it (in the case of lossy compression).

An important characteristic of the efficiency of the compression process is the **compression ratio**, which is defined as the ratio of the number of bits used to store the original image to the number of bits used to store the compressed image:

$$\text{compression ratio} \equiv \frac{\text{number of bits in uncompressed image}}{\text{number of bits in compressed image}} \qquad (8.1)$$

In most cases the compression ratio will be greater than 1, and it is typically on the order of 5:1 (read "5 to 1"), 10:1, or even 100:1, depending upon the type of data.

Compression takes into account the fact that there is often **redundancy** in the data being used to carry or store the information. As a result, the same amount of information can be represented with less data. Redundancy—which can be thought of as the "wasted space" in the data—refers to the unnecessary repetition in the data used to store or transmit the information. Without redundancy, data compression is impossible. The compression ratio is bound by the **relative redundancy**, which is defined as the difference between the number of bits used in the uncompressed signal and the number of bits needed to represent the information, normalized by the former:

$$\text{relative redundancy} \equiv \frac{\text{number of bits used } - \text{ number of bits needed}}{\text{number of bits used}} \qquad (8.2)$$

leading to the following bound:

$$\text{relative redundancy} \geq 1 - \frac{1}{\text{compression ratio}} \qquad (8.3)$$

For example, if 90% of the data is extraneous, then the relative redundancy is 90%, or 0.90, in which case the compression ratio can be no more than 10:1 without losing information.

8.1.1 Redundancy in an Image

There are three causes of redundancy. **Coding redundancy** refers to the fact that not all bit patterns (usually called *symbols*) are equally likely. For example, in an English sentence the letter "E" is much more likely than the letter "Q". So if the same number of bits is used to represent all letters, the bits will be used inefficiently. A more efficient coding scheme, therefore, assigns fewer bits to the letters that appear more frequently and more bits to the letters that appear less frequently. Similarly, if the graylevel histogram of an image is not flat, then not all pixel values are equally likely; in the case of a dark image, for example, the pixels with smaller values are more likely than the ones with larger values. By representing the more likely values with fewer bits and the less common values with more bits, a more efficient image representation can be achieved.

A second type of redundancy is called **interpixel redundancy**, which refers to correlation between pixels. If the pixels are in the same image, then this is known as **spatial redundancy**; whereas if they are in adjacent frames of a video sequence, it is known as

temporal redundancy. To understand spatial redundancy, the correlation between pairs of pixels separated by δ along a row of the image is computed as

$$\text{corr}(\delta) = \frac{1}{width'} \sum_{x=0}^{width'-1} I(x,y)I(x+\delta,y) \qquad (8.4)$$

where $width' \equiv width - \delta$. For a typical photograph the normalized value $\frac{\text{corr}(1)}{\text{corr}(0)}$ will be between 0.90 and 0.98, which means that neighboring pixels are usually highly correlated. In other words, it is easy to *predict* the next pixel's value (or at least a good approximation) from the current pixel's value. For example, in the following image:

$$\begin{bmatrix} 33 & 33 & 34 & 33 \\ 33 & 62 & 62 & 62 \\ 62 & 62 & 145 & 145 \\ 145 & 146 & 145 & 145 \end{bmatrix} \qquad (8.5)$$

if we were to simply predict that each pixel's value is the same as the previous pixel's value (considered in row-major order), we would be correct more than 50% of the time. Moreover, the difference between the predicted and actual values would almost always be less than 32, so that fewer than 5 bits per pixel (on average) would be needed instead of the full 8 bits.

Finally there is **psychovisual redundancy**, which refers to the fact that not all errors are noticeable to a human observer. While the first two forms of redundancy are used by lossless compression techniques, this last one is required by lossy compression. For example, consider the following grayscale image:

$$\begin{bmatrix} 32 & 33 & 254 & 255 \\ 33 & 32 & 255 & 254 \\ 33 & 33 & 255 & 254 \\ 32 & 33 & 255 & 255 \end{bmatrix} \qquad (8.6)$$

If these values were displayed as an image, your eyes would be drawn to the sharp intensity edge between the smaller values on the left and the larger values on the right; and the relatively minor differences between 32 and 33 on the left, and between 254 and 255 on the right, would likely not be noticeable at all. Therefore, if all the pixels with value 32, say, were changed to 33, and all the pixels with value 254 were set to 255, the resulting image would require fewer bits to encode (since then only two gray levels would be used), but the difference would not even be noticed at all when the values were viewed as an image. Lossy compression takes advantage of such imperceptible differences to yield even greater compression ratios than would be possible using lossless compression alone.

8.1.2 Graphic Drawings Versus Photographs

In considering the difference between lossy and lossless compression, it is important to keep in mind that there are, roughly speaking, two types of images. Photographs captured by a camera contain a continuous range of colors (or gray levels), so that although neighboring pixels generally have similar values, it is rare for them to have the exact same value. Graphic drawings, on the other hand, often use a limited palette of colors, and they often contain large single-colored regions in which neighboring pixels have the exact same value.

Lossless compression relies on both coding and interpixel redundancy. Because photographs contain very little of these, lossless compression does not yield impressive results on this type of image. (Compression ratios rarely exceed 2:1.) Applied to a graphic image, however, lossless compression will often yield compression ratios of 20:1 or even 100:1, a huge savings. Therefore, as a general rule, graphic images should be compressed using lossless techniques.

Lossy compression, on the other hand, takes advantage of the psychovisual redundancy present in photographs in order to compress more than would be possible with lossless techniques alone. Depending upon the desired quality of the output, lossy compression ratios are typically on the order of 10:1 for an image or 100:1 for a video. Lossy compression should therefore, as a general rule, be applied to photographs and videos. While lossless compression can also be applied to photographs if it is required to ensure that no information has been lost (e.g., medical or space images), one should *almost never* apply lossy compression to a graphic image, because not only will the compression ratio actually be worse with lossy than with lossless compression, but the lossy compression will also result in undesirable visible artifacts in the output. It is not that graphic images do not contain any psychovisual redundancy, but rather that the human visual system is quite capable of detecting the sharp edges that occur in graphic drawings, and lossy compression tends to blur these edges as an unintended artifact. Examples of these two types of images and the effects of lossy compression are shown in Figure 8.2.

8.1.3 Information Theory

The branch of mathematics underlying lossless compression is known as **information theory**, which provides the tools necessary to answer questions such as how to determine the theoretical lower limit of the amount of data needed to encode a certain amount

Figure 8.2 Grayscale photograph (top) versus binary graphic drawing (middle). Uncompressed, both 8-bit-per-pixel images contain 400 kB of data; with lossless compression they require 235 kB and 3 kB, respectively. As the quality of the lossy compression (in this case JPEG) is decreased, the size decreases. However, even a large reduction in quality for a graphic image does not reduce the file size as low as lossless compression. The bottom row displays the horizontal graylevel profile across one of the bright lines in the graphic, showing the degradation due to lossy compression. Even though the artifacts of lossy compression are not noticeable in the image at the low resolution displayed here, the signal has been significantly degraded, and for no benefit.

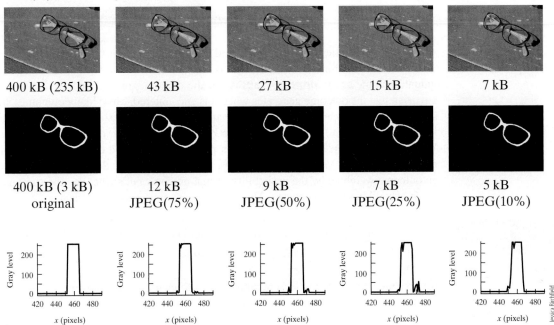

of information. A key quantity for answering such questions is **entropy**. You may have encountered this term before as a measure of disorder in a thermodynamic system: When the system is highly ordered it has low entropy, but when it is disordered it has high entropy—and nature always tends toward the latter. Similarly, in information theory, entropy is used to describe the amount of unpredictability in a data stream: a data stream that is highly ordered (predictable) has low entropy, while an unorganized stream has high entropy. And just as maximum entropy is obtained by a random thermodynamic system, so in information theory maximum entropy is obtained by a random data stream.

It is important to note that entropy does not address the notion of "meaning" or "useful content" at all. As a result, when we say that entropy captures the amount of information in a stream, we are using the term "information" in a very narrow sense; in its broadest sense, the term is difficult to define and elusive to quantify. For example, when listening to a high quality lecture from a brilliant luminary, we might say that the presentation contains a large quantity of information, whereas a poorly given talk by a politician who simply repeats key phrases that the constituents want to hear may contain little useful information. If both the lecture and the political speech are encoded by digitizing the audio waveforms captured by the microphones into a sequence of bytes, both sequences might very well contain the same, or similar, amounts of entropy, even though the amount of useful content is vastly different between them. Nevertheless, despite this limitation, entropy is a simple measure that captures many of the properties that we intuitively expect in a measure of information, and therefore the entropy of a stream is widely used as a measure of the amount of information in the stream.

Entropy of a Single Random Variable

Entropy is formally defined in the following way. Suppose a **random variable** X can take on any one of several possible **outcomes** $x \in \mathcal{X}$, with probabilities given by $p(x)$. For example, a coin has two possible outcomes, so that the set $\mathcal{X}$ contains two elements, $\mathcal{X} = \{\text{HEADS}, \text{TAILS}\}$; the two possibilities for the random variable are $X = \text{HEADS}$ or $X = \text{TAILS}$; and the probabilities of these assignments are $p(\text{Heads})$ and $p(\text{Tails})$, respectively. The entropy of X is given by the negative of the weighted average of the logarithm of the probabilities:

$$H(X) = -\sum_{x \in \mathcal{X}} p(x) \log p(x) \tag{8.7}$$

where the probabilities are the weights, and the letter H is the standard symbol for entropy. To understand this formula, it is helpful to think of the **information content** in an individual outcome as the **unpredictability** of the outcome, or equivalently the amount of **surprise** that the outcome generates when it is encountered. More likely outcomes are *less* surprising, whereas less likely outcomes are *more* surprising, so the information content $\log \frac{1}{p(x)} = -\log p(x)$ in a single outcome x increases when $p(x)$ decreases, and vice versa. When an outcome is certain to occur, its probability is 1, and therefore its information content is zero. The value $-p(x) \log p(x)$ in Equation (8.7) is the information content in an outcome weighted by the probability of that outcome occurring, and the summation is taken over all possible outcomes. The following example makes this clear.

EXAMPLE 8.2

What is the entropy of each of the following random variables:

(a) Flipping an unfair coin that is heads on both sides?
(b) Flipping a standard, fair coin?
(c) Rolling a fair 4-sided die (a pyramid with a triangular-shaped base)?
(d) Rolling a standard, fair 6-sided die?

<table>
<tr><td>**Solution**</td><td>

First, let us propose answers based on intuition, without using the formula in Equation (8.7):

(a) The probabilities are $p(\text{HEADS}) = 1$ and $p(\text{TAILS}) = 0$. Whenever the coin is flipped, it will always show heads. As a result, there is no unpredictability in the coin flip, and hence no surprise. Therefore, the information content of heads is zero. Although the information content of tails is infinite (we would definitely be surprised to see tails show up as the result of the flip!), its probability is zero and therefore does not change the overall information content, which is zero.

(b) When the fair coin is flipped, it is difficult to predict which outcome will occur, since $p(\text{HEADS}) = p(\text{TAILS}) = \frac{1}{2}$. But the outcome will be either heads or tails, and these two states can be stored in a single bit ("binary digit") of data. The information in the fair coin toss is therefore 1 bit.

(c) The probability of any particular side is $\frac{1}{4}$, and therefore it requires 2 bits to store the outcome of a single roll.

(d) The probability of any particular side is $\frac{1}{6}$. Intuitively, because there is more uncertainty in the outcome, we expect there to be slightly more information content than in the 4-sided die.

</td></tr>
</table>

Not surprisingly, applying Equation (8.7) yields the same answers:

(a) entropy of 2-headed coin flip: $H(X) = -0 \log_2 0 - 1 \log_2 1 = 0$ bits $\qquad$ (8.8)

(b) $\qquad$ entropy of fair coin flip: $H(X) = -0.5 \log_2 \dfrac{1}{2} - 0.5 \log_2 \dfrac{1}{2} = 1$ bit $\qquad$ (8.9)

(c) entropy of fair 4-sided die roll: $H(X) = 4 \left(-0.25 \log_2 \dfrac{1}{4} \right) = 2$ bits $\qquad$ (8.10)

(d) entropy of fair 6-sided die roll: $H(X) = 6 \left(-\dfrac{1}{6} \log_2 \dfrac{1}{6} \right) \approx 2.585$ bits $\qquad$ (8.11)

where the convention $0 \log 0 = 0$ has been used. Note that the base of the logarithm changes only the units in which the answer is expressed; here base-2 is used, so these results are in bits.

In three of the four cases above, all outcomes are equally likely, and therefore the *probability mass function* of the random variable is uniform. This is the case of *maximum entropy*. When the probability mass function is less uniform, the entropy (and hence the amount of information) decreases.

<table>
<tr><td>**EXAMPLE 8.3**</td><td>

Compute the entropy of the random variable associated with the toss of an unfairly weighted coin, with $p(\text{HEADS}) = \frac{1}{4}$, and $p(\text{TAILS}) = \frac{3}{4}$.

</td></tr>
<tr><td>**Solution**</td><td>

The information content in each of the 2 possible outcomes is

$$\text{information content in HEADS outcome} = -log_2 \frac{1}{4} = 2 \text{ bits} \qquad (8.12)$$

$$\text{information content in TAILS outcome} = -log_2 \frac{3}{4} \approx 0.415 \text{ bits} \qquad (8.13)$$

These results tell us that the TAILS outcome, which is much more predictable than the HEADS outcome, has less information content. This is because we would be correct 75% of the time

</td></tr>
</table>

if we were to simply guess tails each time. However, whenever the coin flip results in heads, we are surprised, because heads events are more rare.

The entropy of the coin flip is given by the weighted average of the information content of the individual outcomes, or

$$H(X) = -\frac{1}{4}\log_2\frac{1}{4} - \frac{3}{4}\log_2\frac{3}{4} \approx 0.25(2) + 0.75(0.415) \approx 0.811 \text{ bits} \quad (8.14)$$

Notice that the information content in the coin flip has reduced as a result of the bias, because there is more predictability in the outcome.

Entropy of a Sequence of i.i.d. Random Variables

What if we are interested in a sequence of experiments, such as a sequence of coin tosses or die rolls? Two scenarios are common, covered in this and the following subsection. In the first scenario, the individual experiments are assumed to be **independent and identically distributed (i.i.d.)**, meaning that the outcome of each experiment is independent of all the other experiments and the probability of the different outcomes remains the same throughout all the experiments (that is, the probability mass function does not change). A simple example of i.i.d. variables would be the outcomes of a series of coin flips: the outcome of each flip is independent of all the other flips, and the properties of the coin do not change. The entropy of a sequence of n i.i.d. experiments is simply $nH(X)$, where $H(X)$ is the entropy of an individual experiment.

EXAMPLE 8.4	Compute the entropy of a sequence of ten flips of the unfairly weighted coin above, with the distribution of $p(\text{HEADS}) = \frac{1}{4}$ and $p(\text{TAILS}) = \frac{3}{4}$ for each flip. Assume the outcomes of the flips are independent of one another.
Solution	Since the experiments are i.i.d., the entropy is same for each flip: $H(X_i) \approx 0.811$ from the example above. Therefore simply multiply the number of experiments by the entropy of each experiment to yield the entropy of the sequence:

$$H(X_{1:10}) = 10H(X_i) = 10(0.811) = 8.11 \text{ bits} \quad (8.15)$$

A fundamental result of information theory is **Shannon's source coding theorem**,[†] which says that the optimal encoding of a sequence is bounded by its entropy. In other words, given a sequence of symbols assumed to be i.i.d. outcomes of a random variable generated according to some probability mass function, the average codeword length per symbol can be no less than the entropy of a single experiment. This theorem is important because it gives us a theoretical limit on how much we can expect to losslessly compress any given data stream, under this assumption.

Entropy of a Stationary Sequence of Random Variables

In the second scenario, the sequence of experiments is assumed to be the result of a **stationary** random process, which means that the joint probability distribution of any subset of the sequence of random variables does not change with respect to shifts in the time index. For example, in a stationary process, $p(X_i, X_{i+1})$ is the same no matter the value of i. *Stationarity* can be thought of as a generalization of *independence*, because a process

---

[†] Claude Shannon (1916–2001) is known as the "father of information theory." He wrote what is widely regarded as the most important master's thesis (1937) of all time, because it proposed the basic concept underlying all modern digital computers, namely applying Boolean algebra to digital circuit design.

of i.i.d. random variables is always stationary, but the random variables in a stationary sequence are not necessarily independent of each other.

A first-order **Markov chain** is a sequence of random variables in which the conditional probability given the immediately previous variable is the same as the conditional probability given all the previous variables:

$$p(X_i|X_{i-1}) = p(X_i|X_1, X_2, \ldots, X_{i-1}) \tag{8.16}$$

In other words, if the value of the immediately preceding outcome (in this case X_{i-1}) is known, then everything that we need to know in order to predict the current outcome (in this case X_i) is known as well. The Markov chain assumption is particularly useful when studying stationary processes, because it simplifies the computation of entropy. Furthermore, in the case of a stationary process, $p(X_i|X_{i-1})$ is the same no matter the value of i.

The *entropy rate* of a stochastic process captures how the entropy of the sequence grows with the length of the sequence. It can be shown that the entropy rate of a stationary Markov stochastic process is simply the *conditional entropy* $H(X_i|H_{i-1})$, where conditional entropy is related to entropy in the same way that conditional probability is related to probability. If the Markov chain has stationary distribution μ and transition matrix $\mathbf{P}$, the entropy rate is given by

$$H(X_{1:n}) = H(X_i|H_{i-1}) = -\sum_{i,j} \mu_i p(j|i) \log p(j|i) \tag{8.17}$$

where $X_{1:n}$ refers to the random process with n timesteps, μ_i is an element of μ, and $p(j|i)$ is an element of $\mathbf{P}$. It can be shown that the minimum expected codeword length per symbol of a stationary process is given by the entropy rate of the process. The key word here is "expected," which says that while there is no theoretical bound on the compression of a stationary process as there is for an i.i.d. process; we can expect that on average the minimum codeword length is given by the entropy rate. Therefore, the entropy rate of a stationary process and the entropy of each variable in an i.i.d. process can be considered as being more or less equivalent from a practical point of view, because both give us a way to quantify the amount of compression that can be realistically achieved.

EXAMPLE 8.5

Suppose a binary random process outputs a stream of bits such that 90% of the time the next bit is identical to the previous bit. What is the entropy rate of this process?

Solution

This is a stationary Markov random process. A typical sequence might look something like this:

000000000001111111110000000000001111111111 . . .

where the length of each run of 0s or 1s will be, on average, 10. Clearly this is not an i.i.d. process, since each bit is heavily dependent upon the previous bit. Overall, however, 0 is just as likely as 1, so the stationary distribution is uniform: $\mu_0 = \mu_1 = \frac{1}{2}$. The transition matrix is given by $p(0|0) = p(1|1) = 0.9$ and $p(0|1) = p(1|0) = 0.1$. Therefore, from the equation above we have

$$H(X_{1:n}) = -\frac{1}{2}(0.9) \log_2 0.9 - \frac{1}{2}(0.9) \log_2 0.9 - \frac{1}{2}(0.1) \log_2 0.1$$

$$- \frac{1}{2}(0.1) \log_2 0.1 \tag{8.18}$$

$$= -0.9 \log_2 0.9 - 0.1 \log_2 0.1 \approx 0.47 \text{ bits} \tag{8.19}$$

Therefore, we can expect to compress this stream by about a factor of 2.

8.2 Lossless Compression

Now that we have considered the basic principles of information theory, let us examine several algorithms for lossless compression that are based on these principles.

8.2.1 Huffman Coding

One way to view an image is as the result of an i.i.d. process. In other words, the sequence of pixel values is considered as the outcome of a large number of experiments, each of which draws a pixel value according to some probability distribution. In the case of a grayscale image, for example, each byte is treated as the output of an unfairly weighted 256-sided die, with the probability of each gray level given by a probability mass function.

EXAMPLE 8.6

What is the probability mass function of the following 4×4 image? For simplicity, assume only 3 bits per pixel, so *ngray* = 8.

$$\begin{bmatrix} 7 & 3 & 2 & 0 \\ 0 & 2 & 3 & 1 \\ 7 & 2 & 3 & 3 \\ 2 & 3 & 5 & 4 \end{bmatrix}$$

Solution

Without additional information, the best approximation to the probability mass function is obtained by computing the normalized graylevel histogram, i.e., by counting the percentage of each gray level's occurrence. The gray level 0 occurs 2 times, the gray level 1 occurs 1 time, the gray level 2 occurs 4 times, and so forth. This leads to the following probability mass function:

$$p(0) = \frac{2}{16} \quad p(1) = \frac{1}{16} \quad p(2) = \frac{4}{16} \quad p(3) = \frac{5}{16} \quad p(4) = \frac{1}{16} \quad p(5) = \frac{1}{16} \quad p(6) = \frac{0}{16} \quad p(7) = \frac{2}{16}$$

A **source code** φ for a random variable is a mapping from values (or symbols) to **code words**, which are bit strings. For example, the uncompressed image above would typically be represented by the mapping

$$\varphi(0) = 000$$
$$\varphi(1) = 001$$
$$\varphi(2) = 010$$
$$\varphi(3) = 011$$
$$\varphi(4) = 100$$
$$\varphi(5) = 101$$
$$\varphi(6) = 110$$
$$\varphi(7) = 111$$

which requires 3 bits per pixel to store the image. With 16 pixels total, the uncompressed image above would be stored as a sequence of 48 bits:

1 1 1 0 1 1 0 1 0 0 0 0 0 0 0 1 0 0 1 1 0 0 1 1 1 1 0 1 0 0 1 1 0 1 1 0 1 0 0 1 1 1 0 1 1 0 0
  7   3   2   0   0   2   3   1   7   2   3   3   2   3   5   4

where we have assumed row-major ordering of the pixels.

EXAMPLE 8.7	Compute the entropy of the image in Example 8.6.
Solution	The entropy is given by Equation (8.7), using the probabilities already computed:

$$\text{entropy} = -\frac{2}{16}\log_2\frac{2}{16} - \frac{1}{16}\log_2\frac{1}{16} - \frac{4}{16}\log_2\frac{4}{16} - \frac{5}{16}\log_2\frac{5}{16}$$

$$-\frac{1}{16}\log_2\frac{1}{16} - \frac{1}{16}\log_2\frac{1}{16} - \frac{0}{16}\log_2\frac{0}{16} - \frac{2}{16}\log_2\frac{2}{16}$$

$$\approx 2.52 \text{ bits per pixel.}$$

According to Shannon's source coding theorem, this result gives us the smallest possible expected codeword length of any algorithm based on the i.i.d. assumption.

Instead of assigning the same number of bits to each pixel regardless of its symbol, data compression can be achieved by assigning shorter codewords to the more frequent symbols and longer codewords to the less frequent symbols. This is called a **variable-length code**. In the example above, the symbol 3 should be represented by the shortest codeword, since it is the most common; while the symbol 6 should be represented by the longest codeword, since it is the least common. The codeword length is the number of bits in the codeword, and the expected codeword length is the weighted average of the codeword lengths, with weights given by the probability mass function:

$$\text{expected codeword length} = \sum_{x \in \mathcal{X}} |\varphi(x)| p(x) \tag{8.20}$$

where $|\varphi(x)|$ is the length of the codeword $\varphi(x)$.

Huffman coding is a simple algorithm that generates a set of codes with the minimum expected codeword length. In other words, under the assumption that the symbols were generated independently according to the probability mass function and under the constraint that each symbol is coded separately from the other symbols, Huffman coding is optimal. The Huffman algorithm is straightforward, forming a binary tree by starting with each symbol as a leaf node, then iteratively combining the two least likely nodes, where the probability of a node is the sum of its own probability with all the probabilities of its children.

Figure 8.3 illustrates the algorithm applied to the example image above. In the first iteration, the two least likely symbols have probabilities $\frac{0}{16}$ and $\frac{1}{16}$. Notice that there is a tie here, and we are free to break the tie however we decide. A Huffman code is not guaranteed, therefore, to be unique. In this case we choose (somewhat arbitrarily) to combine the symbols 5 and 6, for a combined probability of $\frac{1}{16}$. In the next iteration, the two least likely symbols both have probabilities $\frac{1}{16}$, but again the tie is broken arbitrarily by merging symbol 4 with symbol 5–6. This procedure is repeated until all the nodes are merged under a single root. The binary labels 0 and 1 are then assigned to the two branches proceeding from each non-leaf node, and the codeword for each symbol is given by the concatenation of these labels. Notice that the most likely symbols receive the shortest codewords, while the rarest symbols receive the longest codewords. Keep in mind that this is simple example; in a real implementation the nodes would be stored in a priority queue to make it easy to determine the two least likely symbols for any given iteration. Also, the branches in the tree usually cross one another, because the combined nodes are not guaranteed to be contiguous ranges of the leaves as they are here.

Figure 8.3 An illustration of the Huffman algorithm on the image of Example 8.6. The 8 symbols are the leaf nodes, whose values are in green. By successively combining the two least likely symbols, we arrive at a tree structure that yields the optimal code. The final codewords are at the bottom in red.

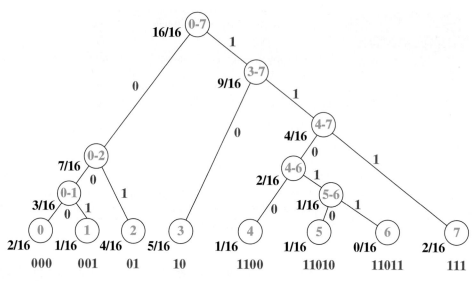

Using this Huffman code, the compressed image would be stored as a sequence of 42 bits:

$$\underbrace{1\,1\,1}_{7}\ \underbrace{1\,0}_{3}\ \underbrace{0\,1}_{2}\ \underbrace{0\,0\,0}_{0}\,\underbrace{0\,0\,0}_{0}\ \underbrace{0\,1}_{2}\ \underbrace{1\,0}_{3}\ \underbrace{0\,0\,1}_{1}\,\underbrace{1\,1\,1}_{7}\ \underbrace{0\,1}_{2}\ \underbrace{1\,0}_{3}\ \underbrace{1\,0}_{3}\ \underbrace{0\,1}_{2}\ \underbrace{1\,0}_{3}\ \underbrace{1\,1\,0\,1\,0}_{5}\ \underbrace{1\,1\,0\,0}_{4}$$

In this particular example, the small amount of savings is hardly worth it, even though only $42/16 \approx 2.62$ bits are used per pixel on average, which is close to the theoretical limit. In practice, however, Huffman coding can yield a significant amount of compression when the probability mass function is less uniform than this one.

Huffman coding yields what is known as a **prefix code**. In a prefix code, no codeword is a prefix of any other codeword. To decompress a compressed image, therefore, we simply need to scan the bitstream one bit at a time and, whenever a sequence of bits has been accumulated that matches a codeword in the codebook, the corresponding symbol is output. For example, suppose the first bit is a 1. If the next bit is 0, we know that the codeword is 10, which represents the symbol 3. But if the next bit is another 1, then the sequence is 11, which is not a codeword, indicating that we must continue grabbing bits.

Huffman coding is not typically applied to the raw image pixel values but is nevertheless widely used as a means of compressing values resulting from another type of compression. For example, both JPEG compression and the CCITT encoding used by fax machines apply modified Huffman algorithms to the results of run-length encoding. Huffman coding is also used by the Deflate algorithm, which is used by the file compression programs PKZIP, gzip, and ZIP, as well as by PNG image files, where Huffman coding is applied to the output of a precompression predictive filter, described in more detail later in the chapter.

8.2.2 Lempel-Ziv Encoding

We have just seen that, if the source is i.i.d. and the probability distribution is known, then Huffman coding is optimal. Why, then, is there a need for anything else? Well, first of all, Huffman coding requires the distribution (or equivalently the codebook or binary tree) to be transmitted along with the compressed bitstream—an additional overhead that detracts from the savings obtained by compression. Second, it is often the case that the probability distribution is unknown, or the source is not i.i.d. Third, Huffman coding must process the source twice: once to construct the probability distribution (from which the codebook is produced), and another time to perform encoding.

A final drawback of Huffman coding is that it assumes a fixed block size. For example, in Example 8.6 each triplet of bits was treated as a pixel and encoded, which leads to an entropy of 2.52 bits per pixel. But different block sizes lead to different values for the entropy, with increasingly larger block sizes typically leading to lower entropy (and hence higher theoretical compression), but more overhead due to a larger codebook. This is illustrated in the following example.

EXAMPLE 8.8

Compute the entropy of the image in Example 8.6 with block sizes of 1, 2, 3, and 4 bits.

Solution

In row-major order, the uncompressed image is represented by the following bitstream:

$$111011010000000010011001111010011011010011101100$$

This bitsream contains 24 ones and 24 zeros. Using a block size of 1 bit, the probability mass function is given by $p(0) = p(1) = 24/48 = 0.5$. The entropy is 1 bit per block, or 3 bits per input pixel. No compression is possible.

Considered as a bitstream of 2-bit blocks, the image is

$$11\ 10\ 11\ 01\ 00\ 00\ 00\ 00\ 10\ 01\ 10\ 01\ 11\ 10\ 10\ 01\ 10\ 11\ 01\ 00\ 11\ 10\ 11\ 00$$

leading to a probability mass function of $p(00) = 6/24, p(01) = 5/24, p(10) = 7/24, p(11) = 6/24$. From Equation (8.7) the entropy is therefore

$$\text{entropy} = -\frac{1}{4}\log_2\frac{1}{4} - \frac{5}{24}\log_2\frac{5}{24} - \frac{7}{24}\log_2\frac{7}{24} - \frac{1}{4}\log_2\frac{1}{4} \approx 1.99 \text{ bits per block} \qquad (8.21)$$

or 2.98 bits per pixel, which is just slightly better than what was achieved with 1-bit blocks. 3-bit blocks were covered in Example 8.7, leading to an entropy of 2.52 bits per pixel, a considerable improvement.

Considered as a bitstream of 4-bit blocks, the image is

$$1110\ 1101\ 0000\ 0000\ 1001\ 1001\ 1110\ 1001\ 1011\ 0100\ 1110\ 1100$$

leading to a probability mass function of

$p(0000) = 2/12$	$p(0100) = 1/12$	$p(1000) = 0/12$	$p(1100) = 1/12$
$p(0001) = 0/12$	$p(0101) = 0/12$	$p(1001) = 3/12$	$p(1101) = 1/12$
$p(0010) = 0/12$	$p(0110) = 0/12$	$p(1010) = 0/12$	$p(1110) = 3/12$
$p(0011) = 0/12$	$p(0111) = 0/12$	$p(1011) = 1/12$	$p(1111) = 0/12$

From Equation (8.7) we have

$$\text{entropy} = -4\left(\frac{1}{12}\log_2\frac{1}{12}\right) - \frac{1}{6}\log_2\frac{1}{6} - 2\left(\frac{3}{12}\log_2\frac{3}{12}\right) \approx 2.64 \text{ bits per block} \qquad (8.22)$$

or 1.32 bits per pixel, allowing even more compression.

Summarizing, for this example the entropy varies significantly depending upon the block size:

number of bits per block	entropy per pixel
1	3.00 bits
2	2.98 bits
3	2.52 bits
4	1.32 bits

To overcome these limitations, we turn our attention to **universal source coding**, which refers to a coding procedure that does not depend on the probability mass distribution of the source. One popular procedure for universal data compression, known as **Lempel-Ziv**, is easy to implement and achieves a rate of compression that asymptotically (as the length of the sequence tends to infinity) approaches the source entropy. Instead of assuming that the input process is i.i.d., as does Huffman coding, Lempel-Ziv only assumes that the input is stationary. In fact, Lempel-Ziv leverages inter-pixel redundancy in order to compress the sequence. Like Huffman coding, a variant of Lempel-Ziv is used by the Deflate algorithm, which is used by the popular PNG image file format, as explained later.

The key idea behind Lempel-Ziv is to look for subsequences that have appeared before. For example, suppose the input is the bitstream from Example 8.8:

$$111011010000000010011001111010011011010011101100$$

Now suppose the bits are scanned from left to right, and a comma is inserted every time we encounter a subsequence that we have not encountered before. Thus, the first 1 is new to us, so we place a comma after it. But the second 1 has been encountered before, so we refrain adding a comma until after the third 1, since the subsequence 11 is new to us. Continuing this procedure leads to the following:

$$1, 11, 0, 110, 10, 00, 000, 001, 0011, 00111, 101, 00110, 1101, 001110, 1100$$

Unlike Huffman coding, which builds the codebook only after processing the entire stream, Lempel-Ziv builds the codebook, also known as the **dictionary**, on the fly as the data are compressed. Dictionary entries consist of (*index, bit*) pairs, where *index* is the index of the dictionary entry containing the prefix, and *bit* is the bit appended to the prefix to obtain the subsequence corresponding to this entry. If we let $dict[i]$ refer to the ith dictionary entry, and assuming that $dict[0]$ refers to the empty subsequence, the input can thus be represented as

$$\underbrace{(0,1)}_{dict[1]}, \underbrace{(1,1)}_{dict[2]}, \underbrace{(0,0)}_{dict[3]}, \underbrace{(2,0)}_{dict[4]}, \underbrace{(1,0)}_{dict[5]}, \underbrace{(3,0)}_{dict[6]}, \underbrace{(6,0)}_{dict[7]}, \underbrace{(6,1)}_{dict[8]}, \underbrace{(8,1)}_{dict[9]}, \underbrace{(9,1)}_{dict[10]}, \underbrace{(5,1)}_{dict[11]}, \underbrace{(9,0)}_{dict[12]}, \underbrace{(4,1)}_{dict[13]}, \underbrace{(10,0)}_{dict[14]}, \underbrace{(4,0)}_{dict[15]}$$

where the dictionary is illustrated in Figure 8.4. The first subsequence of 1 is represented by the dictionary entry (0,1), because it is 1 appended to the empty sequence. The next subsequence of 11 is represented by (1,1), because it is 1 appended to the sequence given by $dict[1]$. The process continues to the final subsequence of 1100, which is represented by (4,0) since it is 0 appended to $dict[4]$, which itself is 0 appended to $dict[2]$, which is the subsequence 11.

If we assume that the dictionary has 16 entries, then each entry requires 5 bits (4 for the index, and 1 for the appended bit). Therefore, the original sequence of 48 bits has been transformed into a sequence of $5 \cdot 15 = 75$ bits, which might make it appear that Lempel-Ziv is not very good at compression. In practice, however, on longer input sequences the compression ability of Lempel-Ziv is significant and—as mentioned above—approaches the entropy of the source. It is also important to keep in mind that there are many implementation details that are omitted from this simple example that can be used to further improve performance.

8.2.3 Lempel-Ziv-Welch Algorithm

One variant of Lempel-Ziv is the **Lempel-Ziv-Welch (LZW)** algorithm, which also builds the dictionary on the fly as the data are compressed. The trick behind LZW is to notice that the bit appended to each dictionary entry is not needed. Therefore, a dictionary entry in LZW consists only of the index of the previous dictionary entry. The patented LZW algorithm is used by the GIF image file format but is not widely used anymore since GIF has largely been replaced by PNG. In fact, PNG stands for "PNG is not GIF" and was developed intentionally to avoid the patent restrictions[†] of LZW by reverting to the simpler version of Lempel-Ziv.

[†] The patent has since expired.

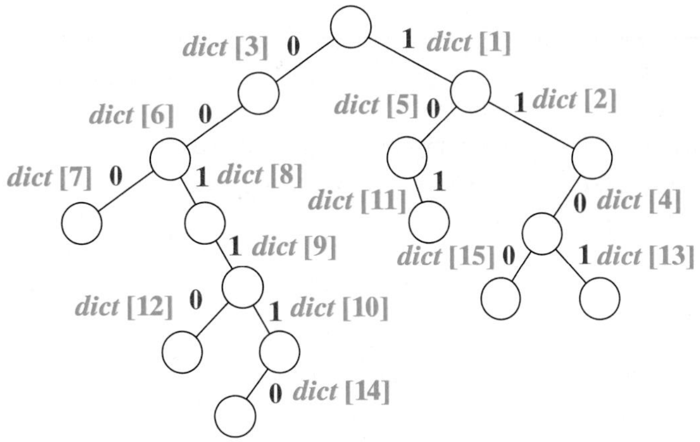

Figure 8.4 An illustration of the dictionary created by the Lempel-Ziv algorithm on the bitstream of Example 8.8. Each dictionary entry captures a subsequence composed of a bit (in red) appended to the subsequence of its parent, with successively longer subsequences farther down in the tree. For example, the subsequence of $dict[14]$ is 001110, which is 0 appended to $dict[10]$.

LZW encoding works as follows. An initial alphabet size is determined, and the dictionary is initialized for each symbol of the alphabet. We will assume without loss of generality that the input is scanned a single bit at a time, as in the Lempel-Ziv example of the previous section, so that the alphabet has two entries: $\mathcal{X} = \{0, 1\}$, and the dictionary is initialized with $dict[0] = 0$ and $dict[1] = 1$. To avoid confusing the input bits with the dictionary indices, let us introduce the letters $A \equiv 0$ and $B \equiv 1$ to represent the input bits, so that $\mathcal{X} = \{A, B\}$, and the dictionary is initialized with $dict[0] = A$ and $dict[1] = B$. As the input is scanned, the dictionary is searched to find the longest subsequence in the dictionary that matches the input. When such a match cannot be found, the index of the previous dictionary entry (that is, the one corresponding to the subsequence without the most recent bit) is output, and the sequence (with the most recent bit) is added to the dictionary. For example, suppose AAB is represented in the dictionary by $dict[3]$, but AABA is not in the dictionary yet. If AABA is encountered, then 3 will be output, and AABA will be added to the dictionary after the last entry. The process continues until the entire input has been scanned, as presented in Algorithm 8.1.

EXAMPLE 8.9

Suppose we have a 3-bit sequence beginning with A. What is the output of LZW on all four possible input sequences?

Solution

Since the dictionary is initialized with only two entries, when the second bit is encountered, there is no match in the dictionary. At this point LZW will output 0 (since the first bit is A), and AX will be added to the dictionary, where X is the second bit. The four possible input sequences are the following, with a stopcode (represented by #) appended to each input:

- A,AA,#. $dict[2] == AA$ from the first 2 bits. When the third A is encountered, the subsequence AA is in the dictionary, so nothing is output until the stop code is encountered. At this point the subsequence is AA#, which is obviously not in the dictionary, causing 2 to be output for all but the last symbol. The output is therefore 02.
- A,A,B,#. $dict[2] == AA$ from the first 2 bits. When the third bit of B is encountered, the subsequence AB is not in the dictionary, so 0 is output. When the stop code is encountered, B# is not in the dictionary, so 1 is output for B. The output is therefore 001.
- A,B,A,#. $dict[2] == AB$ from the first 2 bits. When the third bit of A is encountered, the subsequence BA is not in the dictionary, so 1 is output. When the stop code is encountered, 0 is output for the final A. The output is therefore 010.
- A,B,B,#. $dict[2] == AB$ from the first 2 bits. When the third bit of B is encountered, the subsequence BB is not in the dictionary, so 1 is output. When the stop code is encountered, 1 is output for the final B. The output is therefore 011.

ALGORITHM 8.1 Lempel-Ziv-Welch compress / encode

LEMPELZIVWELCHENCODE$(f, \mathcal{A})$

Input: 1D array f of symbols from an alphabet $\mathcal{A}$

Output: Bitstream representing the compressed array

1	**for** $i \leftarrow 0$ **to** $|\mathcal{A}| - 1$ **do**	➤ Initialize dictionary to contain all symbols.
2	$dict[i] \leftarrow a_i$	➤ a_i is i^{th} symbol of $\mathcal{A}$.
3	$next\text{-}index \leftarrow |\mathcal{A}|$	
4	$match \leftarrow \emptyset$	➤ Initialize longest matching string.
5	**repeat**	
6	$a \leftarrow$ read next value from input array	
7	$match' \leftarrow$ CONCATENATE$(match, a)$	
8	**if** $dict[i] == match'$ for some i **then**	➤ $match'$ is in dictionary,
9	$match \leftarrow match'$	so continue to build longest matching string.
10	**else**	➤ Longest matching string has been found,
11	output i for which $dict[i] == match$	so output index of $match$, and
12	$dict[next\text{-}index] \leftarrow match'$	add $match'$ to dictionary.
13	$next\text{-}index \leftarrow_+ 1$	➤ Increment $next\text{-}index$, and
14	$match \leftarrow a$	reset longest matching string.

To summarize

$$AAA \Rightarrow 02$$

$$AAB \Rightarrow 001$$

$$ABA \Rightarrow 010$$

$$ABB \Rightarrow 011$$

Not that the outputs are unique, even though LZW discarded the appended bit used by the more basic Lempel-Ziv algorithm.

LZW decoding, whose pseudocode is given in Algorithm 8.2, works as follows. The encoder and decoder must agree on the alphabet size beforehand, and the dictionary is initialized for each symbol of the alphabet in the exact same way as the encoder. The input stream consists only of dictionary indices, so the decoder must build up the dictionary on the fly in the exact same way as was done by the encoder. The first value is guaranteed to be a valid index into the dictionary, so the corresponding symbol is immediately output, and the decoder begins to build the largest matching subsequence as that symbol. As the input is scanned, beginning with the second value, two situations are possible. If the index encountered is already in the dictionary, then the subsequence corresponding to that entry is output, and an entry is added to the dictionary corresponding to the longest matching string so far with the first bit of the subsequence *from the dictionary* appended. On the other hand, if the index is not yet in the dictionary, then an entry is added to the dictionary corresponding to the longest matching subsequence so far with the first bit of the *longest matching subsequence* appended; after which this dictionary entry is output. The reason behind this decision is that the only situation in which the encoder could have output an index not yet in the dictionary is if the current subsequence ends with the same symbol with which it begins. The next example will help to illustrate this perhaps nonintuitive principle.

EXAMPLE 8.10	Decode the streams from the previous example.

Solution
The dictionary is initialized as before with two entries. The first index encountered is 0, so A is output, and A also becomes the longest matching substring. The four streams are as follows:

- 02. When 2 is encountered, it is noticed that $dict[2]$ does not exist, so we cannot look in the dictionary to find the corresponding subsequence. However, we know that the encoder added an entry $dict[2] = AX$, where X is unknown, and we know that if the second symbol had been a B, then the third symbol would have caused a subsequence like BX, which would definitely not be in the dictionary, and therefore 1 would have been output as the second index. Therefore, the second symbol must have been an A, and therefore $dict[2] = AA$.
- 001. All of these indices are in the initial directory, so the original sequence is trivially recovered by looking in the dictionary to find AAB.
- 010. Similarly, the original sequence is easily found from the dictionary as ABA.
- 011. Similarly, the original sequence is easily found from the dictionary as ABB.

To summarize

$$02 \Rightarrow AAA$$
$$001 \Rightarrow AAB$$
$$010 \Rightarrow ABA$$
$$011 \Rightarrow ABB$$

To better understand the mechanics of the LZW algorithm, let us consider a more thorough example.

ALGORITHM 8.2 Lempel-Ziv-Welch decompress / decode

LEMPELZIVWELCHDECODE$(f', \mathcal{A})$

Input: Compressed bitstream f' containing the indices of dictionary entries
Output: Original uncompressed 1D array of symbols

```
1   for i ← 0 to |A| − 1 do                              ▶ Initialize dictionary to contain all symbols.
2       dict[i] ← aᵢ                                      ▶ aᵢ is iᵗʰ symbol of A.
3   next-index ← |A|
4   k ← read next value from bitstream
5   output dict[k]
6   match ← val ← dict[k]
7   repeat
8       k ← read next value from bitstream
9       if k < next-index then                           ▶ dict[k] exists, so
10          val ← dict[k]                                 get val from dictionary, and
11          dict[next-index] ← CONCATENATE(match, val[0]) add new match to dictionary.
12      else                                             ▶ dict[k] does not yet exist, so
13          dict[next-index] ← CONCATENATE(match, match[0]) add new match to dictionary, and
14          val ← dict[k]                                 get val; note k = next-index.
15      next-index ←₊ 1                                  ▶ Increment next-index.
16      output val
17      match ← val
```

EXAMPLE 8.11

Encode the following 4×4 image using Lempel-Ziv-Welch. For simplicity, assume only 1 bit per pixel, so *ngray* = 2, and use a 3-bit dictionary.

$$\begin{bmatrix} 0 & 0 & 0 & 1 \\ 0 & 0 & 1 & 0 \\ 0 & 1 & 0 & 1 \\ 0 & 1 & 0 & 1 \end{bmatrix}$$

Solution

As before, to avoid confusing the pixel values with the dictionary indices, let us assign $A \equiv 0$ and $B \equiv 1$, so that the image in row-major order is given by *AAABAABAABABABAB*. Initially, the dictionary is as follows:

$dict[0] = A$ $dict[1] = B$ $dict[2] = ?$ $dict[3] = ?$ $dict[4] = ?$ $dict[5] = ?$ $dict[6] = ?$ $dict[7] = ?$

Processing the data sequentially yields the following:

match	a	match'	in dictionary?	dictionary	output	comment
				$dict[0] \leftarrow A$ $dict[1] \leftarrow B$		initialization
$\emptyset$	A	A	Yes			
A	A	AA	No	$dict[2] \leftarrow AA$	0	$dict[0] == A$
A	A	AA	Yes			
AA	B	AAB	No	$dict[3] \leftarrow AAB$	2	$dict[2] == AA$
B	A	BA	No	$dict[4] \leftarrow BA$	1	$dict[1] == B$
A	A	AA	Yes			
AA	B	AAB	Yes			
AAB	A	AABA	No	$dict[5] \leftarrow AABA$	3	$dict[3] == AAB$
			Yes			
A	A	AA	Yes			
AA	B	AAB	Yes			
AAB	A	AABA	No			
AABA	B	AABAB	No	$dict[6] \leftarrow AABAB$	5	$dict[5] == AABA$
B	A	BA	Yes			
BA	B	BAB	No	$dict[7] \leftarrow BAB$	4	$dict[4] == BA$
B	A	BA	Yes			
BA	B	BAB	Yes			
BAB	#	BAB#	No		7	$dict[7] == BAB$

In other words, the output of LZW compression is 0213547, which encodes the input in the following way:

$$\underbrace{0}_{A} \ \underbrace{2}_{AA} \ \underbrace{1}_{B} \ \underbrace{3}_{AAB} \ \underbrace{5}_{AABA} \ \underbrace{4}_{BA} \ \underbrace{7}_{BAB}$$

Since we have allowed only 8 dictionary entries, the simplest approach would be to use 3 bits per index, leading to the following bitstream:

$$\underbrace{000}_{A}\ \underbrace{010}_{AA}\ \underbrace{001}_{B}\ \underbrace{011}_{AAB}\ \underbrace{101}_{AABA}\ \underbrace{100}_{BA}\ \underbrace{111}_{BAB}$$

which requires 21 bits. Since the uncompressed image required only 16 bits, this is hardly a savings. With a longer input, however, the dictionary eventually adapts to the structure of the input, and LZW can be shown to asymptotically approach the entropy of the input.

EXAMPLE 8.12

Use Lempel-Ziv-Welch to decode the bitstream 0213547 of the previous example.

Solution

We assume that the encoder and decoder have agreed beforehand that there is 1 bit per symbol and 3 bits per dictionary entry.

The dictionary is initialized in the same manner. Processing the data sequentially leads to the following:

match	k	$dict[k]$ exists?	val	dictionary	output	comment
				$dict[0] \leftarrow A$		initialization
				$dict[1] \leftarrow B$		
	0	Yes	A		A	$dict[0] == A$
A	2	No	AA	$dict[2] \leftarrow AA$	AA	
AA	1	Yes	B	$dict[3] \leftarrow AAB$	B	$dict[1] == B$
B	3	Yes	AAB	$dict[4] \leftarrow BA$	AAB	$dict[3] == AAB$
AAB	5	No	AABA	$dict[5] \leftarrow AABA$	AABA	
AABA	4	Yes	BA	$dict[6] \leftarrow AABAB$	BA	$dict[4] == BA$
BA	7	No	BAB	$dict[7] \leftarrow BAB$	BAB	

In Example 8.11 the 16-bit input has been transformed to a 21-bit output, since there are 7 indices and 3 bits per index. Again, it is difficult to devise a simple example that showcases LZW's ability to actually compress data. Nevertheless, as was mentioned in the discussion above regarding Lempel-Ziv, LZW does achieve asymptotically optimal compression as the length of the input tends toward infinity, and for typically sized inputs LZW is quite competitive with other algorithms. Although not discussed here, an important way to reduce the wasted bits in the compressed sequence is to progressively increase the number of bits for transmitting each index as the number of indices grows. That is, at first only 1 bit is output for the index, then once the dictionary entry $dict[2]$ is added, 2 bits are output for all indices, since 2 bits are required to represent the number 2 in binary, and so on.

Another important issue required in an actual implementation regards what to do when the dictionary is full. One approach is to flush the dictionary when that occurs and begin building it again. Another approach is to replace the entries that are used least, in order to allow the dictionary to adapt to the input over time. In fact, an adaptive approach is useful for all of the compression algorithms that we have studied thus far, because image statistics can vary significantly for different parts of an image.

8.2.4 Arithmetic Coding

Another compression technique is known as **arithmetic coding**. Instead of attempting to code the individual symbols of the input, arithmetic coding encodes the entire sequence at once by transforming it into a single real number between 0 and 1. The encoder computes the probability mass function of the input symbols, which it then converts into a cumulative mass function by successively adding the values. Since the cumulative mass function is a monotonic function that begins at 0 and ends at 1, it can be used to divide the interval between 0 and 1 into subintervals corresponding to the symbols. The encoder reads the first symbol and determines to which subinterval it corresponds. It then divides this subinterval in the same manner as it divided the original interval, and the process is repeated with the next symbol of the sequence. By continuing to subdivide the interval in this manner, eventually the encoder reaches a subinterval associated with the entire sequence. It then proceeds to represent this real number as a binary number with enough digits to guarantee that it can be uniquely decoded. This binary number is then the code for the sequence. Although it may not be obvious at first glance, arithmetic coding can be seen as a generalization of Huffman coding that is better able to handle probabilities that deviate significantly from powers of 2, because it takes the entire sequence into account during coding. In fact, Hufmman and arithmetic coding correspond closely to one another when the probabilities in the probability mass function are a power of 2 (e.g., $\frac{1}{2}$ or $\frac{1}{4}$). Nevertheless, arithmetic coding is not as widely used as the other methods, despite being included in the obscure parts of the specifications of several image file formats.

8.2.5 Run-Length Encoding

One of the simplest and oldest lossless compression algorithms is **run-length encoding (RLE)**. The idea is to encode n consecutive identical symbols x as the pair (n,x). An image is then encoded as a sequence of these length-value pairs. Run-length encoding does not perform well on photographs, where adjacent pixels are almost never exactly equal, but it can be used quite effectively on line drawings or graphic images which contain large sections of identically colored pixels. RLE is also applied to JPEG coefficients after first transforming and quantizing image blocks, as we shall see later in the chapter.

Run-length encoding is particularly effective on binary (black-and-white) images, and is therefore used in the CCITT compression standard for fax machines. With only two possible colors per pixel, there is a high probability that neighboring pixels will be identical. Moreover, with only two possible values, it is sufficient to store only the length, rather than the length and the value. As long as there is an agreed-upon convention about which value occurs first, all the other values can be determined by assuming that they alternate between adjacent runs (otherwise they would have been combined into a single run). One convention is to specify the first value of each row, which allows rows to begin with different values (a form of adaptation mentioned earlier). Another convention is to assume that every row begins with a certain color (white, for example) and then to allow the length of the first run of the row to be zero.

EXAMPLE 8.13	Compute the run-length encoding of the following row of bits from a binary image:

00000111110000000011111110000000110000000111111111

Solution

The row consists of 5 0s, followed by 5 1s, followed by 7 0s, followed by 7 1s, and so forth. The resulting RLE is therefore the following sequence of length-value pairs:

$$(5, 0), (5, 1), (7, 0), (7, 1), (6, 0), (2, 1), (7, 0), (9, 1)$$

Adopting the convention that the row begins with the value 1, this leads to the following sequence of lengths:

$$0, 5, 5, 7, 7, 6, 2, 6, 9$$

If we assume that we need 4 bits per length, then the original sequence of 48 bits has been compressed to just 36 bits.

It is common practice to achieve even greater compression by applying a variable-length coding scheme like Huffman coding to the results of run-length encoding.

8.2.6 Predictive Coding

Another simple but effective method of compression is **predictive coding**, in which both the encoder and decoder utilize an identical filter to predict the next value based on previous values. If the filter accurately models the imaging process, then the prediction error (i.e., the difference between the predicted and actual values) will be small. Predictive coding is typically used as a prefilter to produce differences that are then encoded using an entropy-based scheme such as Huffman coding. This differencing process reduces the statistical dependencies between adjacent pixels and typically results in a signal with lower entropy than the original values.

If f_i is the ith actual value of a 1D input signal (e.g., an image scanline), and $\hat{f}_i$ is the ith predicted value, then the simplest approach is to predict the value of the pixel to be the same as that of the previous pixel:

$$\hat{f}_i = f_{i-1} \tag{8.23}$$

so that the prediction error is the difference between adjacent pixel values:

$$\epsilon_i \equiv f_i - \hat{f}_i = f_i - f_{i-1} \tag{8.24}$$

We call this the **copycat predictor**, and its prediction error is identical to the result of convolving the image with the kernel $\begin{bmatrix} 1 & -1 \end{bmatrix}$. It is an amazing fact of nature that essentially any photograph of any scene will yield a copycat prediction error whose histogram is shaped like the one in Figure 8.5. Mathematically, this function can be modeled as a *Laplace distribution*, also known as the double exponential distribution, given by

$$\psi(x) = \frac{1}{x\sqrt{2}} e^{-\frac{|x|\sqrt{2}}{\sigma}} \tag{8.25}$$

Compared with the rounded shape of the Gaussian distribution, the Laplace distribution has high *kurtosis*, meaning that the distribution is sharply peaked, with small tails and rapid drop-off away from the mean. As a result, the standard deviation and entropy of the copycat prediction error will be significantly smaller than those of the original signal. Even a highly textured scene will result in some reduction, but with less textured scenes one can expect more dramatic reduction, oftentimes on the order of 2:1.

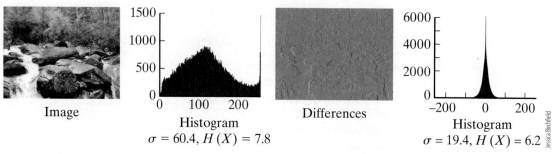

Figure 8.5 Due to the orderly nature of the world, neighboring pixels in photographs are highly correlated. On the left is an image and its graylevel histogram. On the right are the differences computed by convolving the image with a $\begin{bmatrix} 1 & -1 \end{bmatrix}$ kernel and its histogram. The latter histogram, which follows the shape of the Laplace distribution, has a much smaller standard deviation and entropy.

EXAMPLE 8.14

Consider an 8-bit grayscale image with a row consisting of a 1D ramp of graylevel values from the minimum to maximum gray levels:

$$0, 1, 2, 3, 4, 5, 6, 7, 8, 9, \ldots, 255$$

Compute the entropy of this row of the original image, as well as the entropy of the result of the simple predictive filter described in Equation (8.23).

Solution

With an alphabet of 256 symbols and a flat probability mass function (since each value appears exactly once), the entropy is maximal at 8 bits per pixel. The simple predictive filter transforms the image into the sequence of prediction errors:

$$1, 1, 1, 1, 1, 1, 1, 1, 1, 1, 1, \ldots, 1$$

which has an entropy of 0 bits per pixel. Obviously this is an extreme example, and there will be some overhead associated with transmitting the first value of the row. Nevertheless, it illustrates the power of using a predictive filter in reducing entropy.

In 2D, the prediction of a pixel is usually based not only on previous pixels on the same row but also on pixels of previous rows. Let $\ell \equiv I(x - 1, y)$, $u \equiv I(x, y - 1)$, and $d \equiv I(x - 1, y - 1)$ be the pixel values left, up, and diagonally left-and-up, respectively, from the pixel $p \equiv I(x, y)$ to be predicted. As the image is scanned left-to-right and top-to-bottom, all three of these neighboring pixels will have been encountered prior to the pixel p. A particularly useful predictor, called the **planar predictor**, is given by treating the ℓ, u, and d values as "heights" of the pixels, and fitting a plane through the three 3D points $(x - 1, y, \ell)$, $(x, y - 1, u)$, and $(x - 1, y - 1, d)$, leading to $\hat{p} = \ell + u - d$, or

$$\underbrace{\hat{I}_{planar}(x, y)}_{\hat{p}} \equiv \underbrace{I(x - 1, y)}_{\ell} + \underbrace{I(x, y - 1)}_{u} - \underbrace{I(x - 1, y - 1)}_{d} \tag{8.26}$$

To see this result, note that the 3D point halfway between the first two points is $(x - \frac{1}{2}, y - \frac{1}{2}, \frac{1}{2}(\ell + u))$, so that $\hat{p} = d + 2(\frac{1}{2}(\ell + u) - d) = \ell + u - d$. Alternatively, note that the horizontal gradient is $u - d$ and the vertical gradient is $\ell - d$, so that the planar predictor yields the value that assumes that the gradient is the same for both columns and rows: $\ell + (u - d) = u + (\ell - d)$. Either way, the prediction error for the planar predictor is

$$\epsilon(x, y) \equiv \underbrace{I(x, y)}_{p} - \underbrace{\hat{I}_{planar}(x, y)}_{\hat{p}} = p - \ell - u + d = (p - \ell) - (u - d) \quad (8.27)$$

which is the difference between the prediction errors for copycat predictors for the two adjacent rows (or columns).

A popular variation of the planar predictor is the **Paeth predictor**, which outputs the value of the pixel (among the three neighbors) that is closest to the planar prediction. That is, if we let $\hat{p}$ be the output of planar prediction, then

$$\hat{I}_{Paeth}(x, y) \equiv \begin{cases} \ell & \text{if } |\hat{p} - \ell| \leq |\hat{p} - u| \text{ and } |\hat{p} - \ell| \leq |\hat{p} - d| \\ u & \text{if } |\hat{p} - u| \leq |\hat{p} - d| \\ d & \text{otherwise} \end{cases} \quad (8.28)$$

where the order here matters, so that ties are broken in the same way by the encoder and decoder, with ℓ preferred first, then u, and finally d. The Paeth predictor can be thought of as a combination of the copycat and planar predictors, because its output is always the value of one of the neighbors (like copycat), but the choice of neighbor is determined by fitting a plane (like planar). The rationale behind the Paeth predictor is that, in some cases at least, it requires a smaller vocabulary than the planar predictor, thus leading to less entropy. For example, with an 8-bit-per-pixel image the output of the Paeth predictor will always be a number between 0 and 255, whereas the planar predictor can produce any value between -255 and 510 (note: $0 + 0 - 255 = -255$ and $255 + 255 - 0 = 510$), which is nearly three times as many possible outputs. On the other hand, the planar predictor should typically produce errors that are closer to zero, so whether the planar or Paeth predictor is better overall depends upon the input.

Another variation is the **LOCO-I** (Low complexity lossless compression for images) algorithm, defined as follows:

$$\hat{I}_{Loco}(x, y) \equiv \begin{cases} \min(\ell, u) & \text{if } d \geq \max(\ell, u) \\ \max(\ell, u) & \text{if } d \leq \min(\ell, u) \\ \hat{p} & \text{otherwise} \end{cases} \quad (8.29)$$

LOCO-I can be thought of as a primitive edge detector that tends to select u if there is a vertical intensity edge to the left of the pixel or ℓ if there is a horizontal intensity edge above the pixel. If there is no such edge, then it selects the value from the planar predictor. The LOCO-I predictor requires only simple operations, making it fast to execute; it has been shown to exhibit good compression performance despite the crude nature of the edge detection.

It should be mentioned that predictive coding can also be used for lossy compression. That is, if the difference between the predicted and actual values is too small to be distinguished by the human visual system, then we can take advantage of the psychovisual redundancy by quantizing the difference. One such technique is known as **differential pulse-code modulation (DPCM)**, which has two flavors: one type of DPCM computes the difference between adjacent quantized samples, while the other type of DPCM computes the quantized difference with respect to the output of a local model of the decoder.

8.2.7 Example: PNG Compression

Let us now consider how these principles are put to work in an actual lossless compression scheme, namely the one behind the popular PNG image file format. This discussion focuses primarily upon the compression implementation, with only brief mention of the specifics of the file format itself. A PNG file contains a *PNG signature* followed by series of *chunks*.

Each chunk contains either image data or metainformation about the image such as the bit depth, the type of image (grayscale, truecolor RGB, or indexed-color), the image dimensions, whether an alpha channel is present, or whether the rows are interlaced. Conceptually, the compressed data stream is the concatenation of all the image data chunks into a single byte stream, without regard to the chunk boundaries.

The encoder applies two operations to the image pixel data as the image is scanned left-to-right, top-to-bottom. First, the raw pixel values are *filtered* by transmitting the difference between the raw pixel value and one of five possible prediction filters: 1) the pixel value to the left, ℓ; 2) the pixel value above, u; 3) the average of these two, rounded down, $\lfloor (\ell + u)/2 \rfloor$; 4) the Paeth predictor; or 5) the value 0, meaning that no prediction occurs, and the raw pixel value is transmitted unchanged. These filters are applied not to pixels themselves but rather to bytes, so that different color channels (as well as the alpha channel) are processed separately, and both bytes in a 16-bit-per-pixel grayscale image are also processed separately. If an image has fewer than 8 bits per pixel, then the filtering step operates on multiple pixels, which of course reduces its effectiveness. Bytes to the left of the first pixel for any given row, or bytes above the first row, are treated as if they were zero.

The filtering is applied adaptively. That is, although all the pixels in a row of the image use the same filter, each row is allowed to use a different filter, independent of the other rows; a byte indicating the filter type is prepended to each row. The encoder is free to select the filters in any way it chooses, but typically some sort of search is performed to identify the best filter for each row.

In the second operation, the filtered pixel values are compressed by the Deflate algorithm, which treats the data as a 1D stream of bytes. Deflate breaks the stream into a series of *blocks*, where each block contains either uncompressed data (i.e., filtered pixel values) or compressed data. Uncompressed blocks are limited to a maximum size of 65,535 bytes, but compressed blocks have no limit in size. Compression involves a combination of the Lempel-Ziv algorithm and Huffman coding. Lempel-Ziv is used to scan the bytestream looking for duplicated byte sequences within a certain number of the most recent previously encountered bytes; the maximum length of a duplicated sequence is 258 bytes, and the maximum distance between the duplicated sequences is 32,768 bytes. (Duplications are allowed to span block boundaries, which, by the way, have no relationship to chunk boundaries.) These (length, distance) pairs, along with the "literal" values of the unduplicated sequences, are stored in two Huffman trees, one that holds the codes for both literals and lengths, and another that holds the codes for distances. Although it is possible to use fixed Huffman codes, most blocks use dynamic Huffman coding, in which the Huffman trees are determined for each block and sent along with the block. The result of the Deflate algorithm is called a "zlib datastream."

Thus we see that a popular image compression scheme, namely, PNG, combines aspects of several algorithms that we have discussed, including filtering, the Paeth predictor, Huffman coding, and Lempel-Ziv. By taking advantage of the strengths of these techniques, higher compression ratios can be obtained than would be possible when using any single technique alone.

8.3 Lossy Compression

We now turn our attention to **lossy compression** techniques. Lossy compression takes advantage of psychovisual redundancy in the image data to reduce the amount of information in the image beyond what is possible with lossless compression alone. As mentioned earlier, however, one should keep in mind that lossy compression is best suited for photographs of natural scenes containing smooth variations of color. For line drawings or graphic images with sharp boundaries between regions, lossy compression produces noticeable artifacts while actually achieving worse (smaller) compression ratios than loss-less compression.

8.3.1 Measuring the Quality of Lossy Compression

With lossy compression, an image $I(x, y)$ is transformed into a smaller number of bits that, when decompressed, yields not the original image but an approximation $\hat{I}(x, y)$ to the original image. Perhaps the simplest way of measuring the error in the approximation is given by the mean squared error (MSE):

$$\text{mean squared error} = \frac{1}{width \cdot height} \sum_{x=0}^{width-1} \sum_{y=0}^{height-1} \left(I(x, y) - \hat{I}(x, y) \right)^2 \qquad (8.30)$$

known as the **distortion function** (or *distortion measure*). Obviously the mean squared error does not accurately measure distortion as perceived by the human visual system, since it does not take perception into account.

The branch of information theory dealing with the quality of lossy compression is known as **rate distortion theory**. Rate distortion theory answers two questions: 1) For a given compression ratio, what is the best approximation that can be achieved, and conversely, 2) For a given approximation, what is the best compression ratio that can be achieved? We will not cover rate distortion theory in any detail, but as a general rule, the greater the compression, the worse the approximation. A typical lossy compression algorithm can achieve compression ratios on the order of 10 to 1 for a still image, or 100 to 1 for a video, without any noticeable loss in quality. That is, in such cases the original image and the approximation look identical to a human observer. Nevertheless, artifacts do exist, and these artifacts are visible when zooming in on the image. When the image is repeatedly compressed and decompressed (possibly with some editing in between), these artifacts lead to what is known as **generation loss**, which is why lossy compression should not be used for images or videos that will be edited.

8.3.2 Transform Coding

The human visual system is more sensitive to small variations in intensity over large areas than to large variations over small areas (e.g., at an intensity edge). In other words, distortion caused by reducing accuracy in the high-frequency components of an image is much less noticeable to a human observer than distortion caused by reducing accuracy in the low-frequency components. This observation motivates **transform coding**, in which the image is transformed to a different domain that better captures the information of the signal as it will be perceived. For this reason transform coding is also known as **perceptual coding**. Typically the other domain is some version of the frequency domain, so that more bits can be devoted to low-frequency components, while fewer bits are devoted to high-frequency components.

Transform coding can be applied to the entire image at once, but usually the image is first divided into non-overlapping *blocks*, and transform coding is applied to each block separately. This approach is sometimes known as **block transform coding**, and it is by far the most common way to perform lossy image compression. For computational reasons the blocks are usually all the same size, and their dimensions are powers of 2, with the most common sizes being 8×8 and 16×16. Larger block sizes provide more opportunities for compression but require more computation and allow for less adaptation.

As illustrated in Figure 8.6, transform coding of a block involves three steps. First, an orthogonal transform is applied to the image to transform the original pixel values into

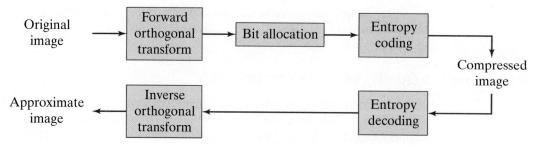

Figure 8.6 Transform coding typically involves three steps to compress an image: an orthogonal transform, bit allocation, and entropy coding. To decompress the image the steps are reversed, but since information is lost during bit allocation, the result is an approximation of the original image.

coefficients in a different domain (e.g., the frequency domain). After transforming each block to the new domain, many of the coefficients will be small and hence can be quantized or set to zero without much loss in perceptual quality of the image when transformed back into the original domain. Therefore the second step, known as **bit allocation**, applies scalar quantization to these coefficients to reduce the number of bits needed to represent the signal. (This is the step in which information is actually lost.) Finally, the quantized coefficients are compressed even further using a lossless entropy coding scheme such as Huffman coding. We will focus primarily upon the orthogonal transform itself, since the second step is fairly straightforward, and entropy coding was discussed in detail earlier.

Suppose we have a 1D array $g(x)$ of n values, and let $\mathbf{F}$ be an $n \times n$ matrix with orthonormal columns that transforms these values into a frequency-like domain, so that $\mathbf{g}' = \mathbf{F}\mathbf{g}$ is the **orthogonal linear transform** of $\mathbf{g}$ according to $\mathbf{F}$, where $\mathbf{g}$ contains the values of g stacked into a vector. We can recover the original signal exactly by applying the inverse procedure: $\mathbf{g} = \mathbf{F}^{-1}\mathbf{g}'$, where, since $\mathbf{F}$ is unitary,[†] the inverse is just the transpose of its complex conjugate: $\mathbf{F}^{-1} = (\mathbf{F}^*)^{\mathsf{T}}$. It is easy to see that $\mathbf{g}$ is just a linear combination of the basis functions given by the columns of $\mathbf{F}^{-1}$:

$$\mathbf{g} = \mathbf{F}^{-1}\mathbf{g}' = \begin{bmatrix} \mathbf{b}_0 & \cdots & \mathbf{b}_{n-1} \end{bmatrix} \begin{bmatrix} g_0' \\ \vdots \\ g_{n-1}' \end{bmatrix} = \sum_{i=0}^{n-1} g_i' \mathbf{b}_i \tag{8.31}$$

where $\mathbf{b}_i$ is the ith column of $\mathbf{F}^{-1}$, and g_i' is the ith element of $\mathbf{g}'$. In other words, the values g_i' can be interpreted as coefficients which, when multiplied by the basis vectors $\mathbf{b}_i$, $i = 0, \ldots, n-1$, allow us to reconstruct the original signal, as we saw earlier.[‡] Since $\mathbf{F}$ is unitary, each basis vector has unit norm: $\|\mathbf{b}_i\| = 1$, so that the amount of energy contributed to the reconstruction by each coefficient is given by that coefficient. As a result, small changes to the coefficients result in small changes to the reconstruction. Therefore, the original signal can be approximated by setting the small coefficients to zero while truncating the remaining coefficients to a limited number of digits:

$$\mathbf{g} \approx \sum_{i=0}^{n-1} \varphi_\tau(g_i') \mathbf{b}_i \tag{8.32}$$

[†] Recall from Section 6.2.8 (p. 288) that a matrix with orthonormal columns is called *unitary* if its elements are complex, or (as a special case) *orthogonal* if all its elements are real. The transform, on the other hand, is called orthogonal whether or not its values are complex.

[‡] Section 6.2.7 (p. 287).

where φ_τ is a function that sets all values below a threshold τ to zero, while truncating the representation of the remaining values:

$$\varphi_\tau(v) = \begin{cases} \text{TRUNCATE}(v) & \text{if } |v| \geq \tau \\ 0 & \text{otherwise} \end{cases} \tag{8.33}$$

As we saw earlier,[†] this approach extends naturally to 2D. Suppose we have an $n \times n$ image block $G(x,y)$ of values, and let $\mathbf{F}$ be the same unitary $n \times n$ matrix seen above. Then $\mathbf{G}' = \mathbf{F}\mathbf{G}\mathbf{F}^\mathsf{T}$ is the orthogonal linear transform, where $\mathbf{G}$ is the $n \times n$ matrix form of $\mathbf{G}$, and $\mathbf{G}'$ is an $n \times n$ array of (possibly complex) coefficients. The approximate reconstruction is given by

$$\mathbf{G} \approx \sum_{x=0}^{n-1} \sum_{y=0}^{n-1} \varphi_\tau(G'_{x,y}) \mathbf{B}_{x,y} \tag{8.34}$$

where $G'_{x,y}$ is the $(x,y)^{\text{th}}$ element of $\mathbf{G}'$, and the basis functions $\mathbf{B}_{x,y} = \mathbf{b}_x \mathbf{b}_y^\mathsf{T}$ are given by the outer products of the basis functions of $\mathbf{F}$.

The procedure is best illustrated by an example.

EXAMPLE 8.15

Perform the first two steps of transform coding on the following 1D array of pixel values $\mathbf{g} = \begin{bmatrix} 94 & 95 & 96 & 97 \end{bmatrix}^\mathsf{T}$ using the following simple orthogonal transform:

$$\mathbf{F} = \begin{bmatrix} 1 & 0 & 0 & 0 \\ 0 & 1 & 0 & 0 \\ 0 & 0 & \frac{1}{\sqrt{2}} & \frac{1}{\sqrt{2}} \\ 0 & 0 & -\frac{1}{\sqrt{2}} & \frac{1}{\sqrt{2}} \end{bmatrix}$$

Solution

It is easy to verify that $\mathbf{F}$ is orthogonal, that is, when multiplied by its transpose it yields the 4×4 identity matrix. To transform the values to the new domain, simply multiply $\mathbf{F}$ by $\mathbf{g}$:

$$\mathbf{g}' = \mathbf{F}\mathbf{g} = \begin{bmatrix} 1 & 0 & 0 & 0 \\ 0 & 1 & 0 & 0 \\ 0 & 0 & 0.7071 & 0.7071 \\ 0 & 0 & -0.7071 & 0.7071 \end{bmatrix} \begin{bmatrix} 94 \\ 95 \\ 96 \\ 97 \end{bmatrix} = \begin{bmatrix} 94.0 \\ 95.0 \\ 136.4716 \\ 0.7071 \end{bmatrix}$$

Perfect reconstruction is obtained by applying the inverse transform:

$$\mathbf{g} = \mathbf{F}^{-1} \mathbf{g}' = \begin{bmatrix} 1 & 0 & 0 & 0 \\ 0 & 1 & 0 & 0 \\ 0 & 0 & 0.7071 & -0.7071 \\ 0 & 0 & 0.7071 & 0.7071 \end{bmatrix} \begin{bmatrix} 94.0 \\ 95.0 \\ 136.4716 \\ 0.7071 \end{bmatrix}$$

$$= 94.0 \begin{bmatrix} 1 \\ 0 \\ 0 \\ 0 \end{bmatrix} + 95.0 \begin{bmatrix} 0 \\ 1 \\ 0 \\ 0 \end{bmatrix} + 136.4716 \begin{bmatrix} 0 \\ 0 \\ 0.7071 \\ 0.7071 \end{bmatrix} + 0.7071 \begin{bmatrix} 0 \\ 0 \\ -0.7071 \\ 0.7071 \end{bmatrix} = \begin{bmatrix} 94.0 \\ 95.0 \\ 96.0 \\ 97.0 \end{bmatrix}$$

Of the four coefficients, 94.0, 95.0, 136.4716, and 0.7071, the last one is much smaller than the others, so it does not contribute much to the reconstruction. Similarly, by a

---

[†] Section 6.3.1 (p. 289).

straightforward argument regarding significant digits, the trailing digits of the third coefficient are of decreasing importance. As a result, we can set the fourth coefficient to zero entirely, and we can truncate the third coefficient to the approximate value of 136, both without significantly affecting the reconstruction. With these changes, the approximate reconstruction is given by

$$
\mathbf{g} \approx \mathbf{F}^{-1}\varphi_\tau(\mathbf{g}') \;=\; 94.0\begin{bmatrix}1\\0\\0\\0\end{bmatrix} + 95.0\begin{bmatrix}0\\1\\0\\0\end{bmatrix} + 136.0\begin{bmatrix}0\\0\\0.7071\\0.7071\end{bmatrix} = \begin{bmatrix}94.0\\95.0\\96.1665\\96.1665\end{bmatrix}
$$

which when rounded to the nearest integer yields the approximate signal $[94 \quad 95 \quad 96 \quad 96]^\mathsf{T}$.

The reconstruction has a mean squared error of only $\frac{1}{4}(1)^2 = 0.25$.

This example has illustrated how a 1D image with 4 pixel values can be represented in a different domain with only 3 values (the truncated coefficients), yielding a reconstruction with a small mean squared error. To simplify the math, we used a simple matrix $\mathbf{F}$, but in the real world one of the most important decisions in designing a transform coding system is which orthogonal transform to use; various alternatives are presented in the next several sections.

8.3.3 Discrete Fourier Transform (DFT)

Perhaps the most familiar orthogonal transform is the discrete Fourier transform (DFT), which we studied extensively in Chapter 6. For an $n \times n$ image block, recall the forward and inverse DFTs are given by

$$
G'_{\mathcal{F}}(k_x, k_y) = \frac{1}{n}\sum_{x=0}^{n-1}\sum_{y=0}^{n-1} G(x, y)e^{-j2\pi \mathbf{x}^\mathsf{T}\mathbf{f}} \tag{8.35}
$$

$$
G(x, y) = \frac{1}{n}\sum_{k_x=0}^{n-1}\sum_{k_y=0}^{n-1} G'_{\mathcal{F}}(k_x, k_y)e^{j2\pi \mathbf{x}^\mathsf{T}\mathbf{f}} \tag{8.36}
$$

where $\mathbf{x} = [x \quad y]^\mathsf{T}$ and $\mathbf{f} = \frac{1}{n}[k_x \quad k_y]^\mathsf{T}$, so that $\mathbf{x}^\mathsf{T}\mathbf{f} = \frac{1}{n}(k_x x + k_y y)$, and we have chosen here to distribute the normalization equally between the forward and inverse transforms. In matrix notation, from Equation (6.92) we have

$$
\mathbf{G}'_{\mathcal{F}} = \mathbf{F}_n \mathbf{G}\mathbf{F}_n \tag{8.37}
$$

where $\mathbf{F}_n$ is the normalized $n \times n$ 1D DFT matrix, and the basis functions are the 2D complex exponentials given by the outer products of the columns of $\mathbf{F}_n$.

EXAMPLE 8.16

What are the basis functions for the DFT of a 2×2 image block?

Solution

From Equation (6.87), it is easy to see that the 1D DFT matrix for a signal with 2 elements is given by

$$
\mathbf{F}_2 = \frac{1}{\sqrt{2}}\begin{bmatrix}1 & 1\\ 1 & -1\end{bmatrix} \tag{8.38}
$$

Since this matrix is its own inverse, that is, $\mathbf{F}_2^{-1} = \mathbf{F}_2$, the 2D basis functions are given by the outer products of the columns of the matrix:

$$\mathbf{B}_{0,0} = \mathbf{b}_0\mathbf{b}_0^{\mathsf{T}} = \frac{1}{2}\begin{bmatrix} 1 \\ 1 \end{bmatrix}\begin{bmatrix} 1 & 1 \end{bmatrix} = \frac{1}{2}\begin{bmatrix} 1 & 1 \\ 1 & 1 \end{bmatrix} \quad \mathbf{B}_{0,1} = \mathbf{b}_0\mathbf{b}_1^{\mathsf{T}} = \frac{1}{2}\begin{bmatrix} 1 \\ 1 \end{bmatrix}\begin{bmatrix} 1 & -1 \end{bmatrix} = \frac{1}{2}\begin{bmatrix} 1 & -1 \\ 1 & -1 \end{bmatrix}$$

$$\mathbf{B}_{1,0} = \mathbf{b}_1\mathbf{b}_0^{\mathsf{T}} = \frac{1}{2}\begin{bmatrix} 1 \\ -1 \end{bmatrix}\begin{bmatrix} 1 & 1 \end{bmatrix} = \frac{1}{2}\begin{bmatrix} 1 & 1 \\ -1 & -1 \end{bmatrix} \quad \mathbf{B}_{1,1} = \mathbf{b}_1\mathbf{b}_1^{\mathsf{T}} = \frac{1}{2}\begin{bmatrix} 1 \\ -1 \end{bmatrix}\begin{bmatrix} 1 & -1 \end{bmatrix} = \frac{1}{2}\begin{bmatrix} 1 & -1 \\ -1 & 1 \end{bmatrix}$$

The DFT has the nice property that it actually captures discrete frequencies. However, the DFT is wasteful because every real value in the input signal is transformed into a complex value in the output. As we shall see below, the DCT overcomes this inherent drawback of the DFT.

8.3.4 Walsh-Hadamard Transform (WHT)

An alternative to the DFT is the **Walsh-Hadamard transform (WHT)**, whose forward and inverse transforms of an $n \times n$ image block are given by

$$G'_{\mathcal{H}}(k_x, k_y) = \frac{1}{n}\sum_{x=0}^{n-1}\sum_{y=0}^{n-1} G(x, y)(-1)^{\phi(\mathbf{x},\mathbf{f})} \tag{8.39}$$

$$G(x, y) = \frac{1}{n}\sum_{u=0}^{n-1}\sum_{v=0}^{n-1} G'_{\mathcal{H}}(k_x, k_y)(-1)^{\phi(\mathbf{x},\mathbf{f})} \tag{8.40}$$

where $\mathbf{x}$ and $\mathbf{f}$ are defined as before, and

$$\phi(\mathbf{x}, \mathbf{f}) \equiv \sum_{i=0}^{m-1} b_i(x)b_i(k_x) + b_i(y)b_i(k_y) \tag{8.41}$$

where $n = 2^m$, and $b_i(z) \in \{0, 1\}$ is the ith bit in the binary representation of z for any integer z:

$$z = \sum_{i=0}^{m-1} 2^i b_i(z) = 2^{m-1}b_{m-1}(z) + \cdots + 2b_1(z) + b_0(z) \tag{8.42}$$

In matrix form, the Walsh-Hadamard transform is represented as

$$\mathbf{G}'_{\mathcal{H}} = \mathbf{H}_m\mathbf{G}\mathbf{H}_m \tag{8.43}$$

where $\mathbf{H}_m$ is the $n \times n$ **Hadamard matrix** used in the 1D transform, defined recursively as follows:

$$\mathbf{H}_m \equiv \frac{1}{\sqrt{2}}\begin{bmatrix} \mathbf{H}_{m-1} & \mathbf{H}_{m-1} \\ \mathbf{H}_{m-1} & -\mathbf{H}_{m-1} \end{bmatrix} \tag{8.44}$$

with $\mathbf{H}_0 \equiv 1$. Note that the Hadamard matrix has a particular simple form, being composed (apart from the scaling factor) of elements which are either $+1$ or -1. Moreover, the matrix is symmetric and is its own inverse, that is, $\mathbf{H}_m = \mathbf{H}_m^{\mathsf{T}}$ and $\mathbf{H}_m\mathbf{H}_m = \mathbf{I}_n$, where $\mathbf{I}_{\{n\times n\}}$ is the $n \times n$ identity matrix, so that the forward and inverse transforms are identical. The Walsh-Hadamard transform can be thought of as a simpler version of the Fourier transform since it involves (apart from the scaling factor) only sums and differences.

EXAMPLE 8.17

Compute the 1D Hadamard matrices for $m = 0, 1, 2,$ and 3.

Solution

For brevity let us represent the elements of the Hadamard matrix by the symbols $+$ or $-$ to indicate $+1$ or -1, respectively. The 1×1 matrix is then given by $\mathbf{H}_0 = +$. Applying the recursive definition yields the 2×2 matrix as

$$\mathbf{H}_1 = \frac{1}{\sqrt{2}} \begin{bmatrix} + & + \\ + & - \end{bmatrix}$$

which is exactly the same as the normalized DFT matrix. The 4×4 matrix is

$$\mathbf{H}_2 = \frac{1}{2} \begin{bmatrix} + & + & + & + \\ + & - & + & - \\ + & + & - & - \\ + & - & - & + \end{bmatrix}$$

while the 8×8 matrix is

$$\mathbf{H}_3 = \frac{1}{2\sqrt{2}} \begin{bmatrix} + & + & + & + & + & + & + & + \\ + & - & + & - & + & - & + & - \\ + & + & - & - & + & + & - & - \\ + & - & - & + & + & - & - & + \\ + & + & + & + & - & - & - & - \\ + & - & + & - & - & + & - & + \\ + & + & - & - & - & - & + & + \\ + & - & - & + & - & + & + & - \end{bmatrix}$$

Examining the matrices above, it is easy to verify that an alternate definition for the Hadamard matrix assigns the $(x, y)^{\text{th}}$ element (with zero-based indexing starting at the top-left element) as follows:

$$\mathbf{H}_m(x, y) = \frac{1}{2^{(m/2)}} (-1)^{\mathbf{b}_x^{\mathsf{T}} \mathbf{b}_y} \tag{8.45}$$

where the exponent is the inner product of the binary representations of x and y:

$$\mathbf{b}_z \equiv [b_{m-1}(z) \quad \cdots \quad b_2(z) \quad b_1(z) \quad b_0(z)]^{\mathsf{T}} \tag{8.46}$$

where $z \in \{x, y\}$.

The basis functions for the 1D Walsh-Hadamard transform are the rows (or equivalently the columns, since the matrix is symmetric) of the Hadamard matrix. These basis functions, known as the **Walsh functions**, have a peculiar property. Examining the matrices in Example 8.17, notice that for each row the number of sign transitions, that is, the transitions from $+$ to $-$ or from $-$ to $+$, is unique to that row. In other words, the number of transitions in all rows comprise the set $\{0, 1, \ldots, n - 1\}$. For example, the first row of $\mathbf{H}_1$ has 0 transitions, while the second row has 1 transition; together these are the numbers in $\{0, 1\}$. Similarly, for $\mathbf{H}_2$ the transitions are (from top to bottom) 0, 3, 1, and 2; these are the numbers in $\{0, 1, 2, 3\}$. For $\mathbf{H}_3$ they are (from top to bottom) 0, 7, 3, 4, 1, 6, 2, 5; these are the numbers in the set $\{0, \ldots, 7\}$; and so forth.

The number of transitions is known as **sequency**. If the rows are reordered so that the number of sign changes increases with the row, then we have what is known as a *sequency-ordered* Hadamard matrix, as opposed to the recursively defined matrix of the previous example, which is known as the *naturally ordered* Hadamard matrix. Sequency-ordering is sometimes known as *Walsh ordering*, and the sequency-ordered Hadamard matrix is sometimes known as the *Walsh matrix*. Notice that the sequency-ordered matrix is also symmetric.

EXAMPLE 8.18

Compute the sequency-ordered 1D Hadamard matrices for $m = 3$.

Solution

Rearranging the rows of $\mathbf{H}_3$ from the previous example yields:

$$(\text{Sequency ordered}) \ \mathbf{H}_3 = \frac{1}{2\sqrt{2}} \begin{bmatrix} + & + & + & + & + & + & + & + \\ + & + & + & + & - & - & - & - \\ + & + & - & - & - & - & + & + \\ + & + & - & - & + & + & - & - \\ + & - & - & + & + & - & - & + \\ + & - & - & + & - & + & + & - \\ + & - & + & - & - & + & - & + \\ + & - & + & - & + & - & + & - \end{bmatrix}$$

The 2D basis functions are given by the outer products of the Walsh functions. As with the Fourier transform, for an $n \times n$ image block there are n 1D basis functions, and therefore n^2 2D basis functions. For example, with a 4×4 image block the 1D transform matrix is of size 4×4, so that there are 4 1D basis functions and 16 2D basis functions.

EXAMPLE 8.19

Compute the 16 2D Walsh-Hadamard basis functions for a 4×4 image. Ignore the scaling factor for simplicity.

Solution

The basis functions are given by the outer products of the columns of $\mathbf{H}_2$ and illustrated in Figure 8.7:

$$\mathbf{B}_{0,0} = \begin{bmatrix} + & + & + & + \\ + & + & + & + \\ + & + & + & + \\ + & + & + & + \end{bmatrix} \quad \mathbf{B}_{0,1} = \begin{bmatrix} + & - & + & - \\ + & - & + & - \\ + & - & + & - \\ + & - & + & - \end{bmatrix} \quad \mathbf{B}_{0,2} = \begin{bmatrix} + & + & - & - \\ + & + & - & - \\ + & + & - & - \\ + & + & - & - \end{bmatrix} \quad \mathbf{B}_{0,3} = \begin{bmatrix} + & - & - & + \\ + & - & - & + \\ + & - & - & + \\ + & - & - & + \end{bmatrix}$$

$$\mathbf{B}_{1,0} = \begin{bmatrix} + & + & + & + \\ - & - & - & - \\ + & + & + & + \\ - & - & - & - \end{bmatrix} \quad \mathbf{B}_{1,1} = \begin{bmatrix} + & - & + & - \\ - & + & - & + \\ + & - & + & - \\ - & + & - & + \end{bmatrix} \quad \mathbf{B}_{1,2} = \begin{bmatrix} + & + & - & - \\ - & - & + & + \\ + & + & - & - \\ - & - & + & + \end{bmatrix} \quad \mathbf{B}_{1,3} = \begin{bmatrix} + & - & - & + \\ - & + & + & - \\ + & - & - & + \\ - & + & + & - \end{bmatrix}$$

$$\mathbf{B}_{2,0} = \begin{bmatrix} + & + & + & + \\ + & + & + & + \\ - & - & - & - \\ - & - & - & - \end{bmatrix} \quad \mathbf{B}_{2,1} = \begin{bmatrix} + & - & + & - \\ + & - & + & - \\ - & + & - & + \\ - & + & - & + \end{bmatrix} \quad \mathbf{B}_{2,2} = \begin{bmatrix} + & + & - & - \\ + & + & - & - \\ - & - & + & + \\ - & - & + & + \end{bmatrix} \quad \mathbf{B}_{2,3} = \begin{bmatrix} + & - & - & + \\ + & - & - & + \\ - & + & + & - \\ - & + & + & - \end{bmatrix}$$

$$\mathbf{B}_{3,0} = \begin{bmatrix} + & + & + & + \\ - & - & - & - \\ - & - & - & - \\ + & + & + & + \end{bmatrix} \quad \mathbf{B}_{3,1} = \begin{bmatrix} + & - & + & - \\ - & + & - & + \\ - & + & - & + \\ + & - & + & - \end{bmatrix} \quad \mathbf{B}_{3,2} = \begin{bmatrix} + & + & - & - \\ - & - & + & + \\ - & - & + & + \\ + & + & - & - \end{bmatrix} \quad \mathbf{B}_{3,3} = \begin{bmatrix} + & - & - & + \\ - & + & + & - \\ - & + & + & - \\ + & - & - & + \end{bmatrix}$$

Figure 8.7 The 16 2D Wash-Hadamard basis functions for a 4×4 image.

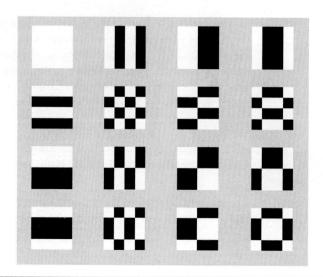

Although the WHT typically achieves better compression than the DFT (at least for block sizes less than about 32×32), it is not nearly as effective as the DCT, which is described next. Moreover, it does not compute coefficients that correspond to actual discrete frequencies.

8.3.5 Discrete Cosine Transform (DCT)

By far the most popular orthogonal transform is the **discrete cosine transform (DCT)**, whose forward and inverse transforms of an $n \times n$ image block are given by

$$G'_{DC}(k_x, k_y) = \sum_{x=0}^{n-1} \sum_{y=0}^{n-1} \eta(k_x)\eta(k_y)G(x, y) \cos\left(\frac{(2x+1)\pi k_x}{2n}\right) \cos\left(\frac{(2y+1)\pi k_y}{2n}\right) \qquad (8.47)$$

$$G(x, y) = \sum_{k_x=0}^{n-1} \sum_{k_y=0}^{n-1} \eta(k_x)\eta(k_y)G'_{DC}(k_x, k_y) \cos\left(\frac{(2x+1)\pi k_x}{2n}\right) \cos\left(\frac{(2y+1)\pi k_y}{2n}\right) \qquad (8.48)$$

where

$$\eta(k) \equiv \begin{cases} \frac{1}{\sqrt{n}} & \text{if } k = 0 \\ \sqrt{\frac{2}{n}} & \text{otherwise} \end{cases} \qquad (8.49)$$

In matrix form, the 2D DFT of an $n \times n$ image block G is given by $\mathbf{G}'_{DC} = \mathbf{C}_n\mathbf{G}\mathbf{C}_n^T$, where $\mathbf{C}_n$ is the $n \times n$ 1D normalized DCT matrix. Notice that this is the first transform we have considered that is not necessarily symmetric ($\mathbf{C}_n \neq \mathbf{C}_n^T$ for $n > 2$), so that the transpose is important. The fact that the forward and inverse transforms have identical form, as can be seen by examining Equations (8.47)–(8.48), means that the inverse of the matrix is equal to its transpose: $\mathbf{C}_n^{-1} = \mathbf{C}_n^T$.

The popularity of the DCT lies in its ability to overcome the fundamental limitation of the DFT, namely that the DFT requires twice as much data to be stored because it converts real values into complex values. There are two ways to look at this limitation, either of which explains why the DCT is able to store the same information with fewer coefficients. First, recall

| EXAMPLE 8.20 | Compute the 1D normalized DCT matrix for $n = 2$ and $n = 4$. |

| Solution | The 1D forward DCT is given by |

$$g'_{DC}(k) = \eta(k) \sum_{x=0}^{n-1} g(x) \cos\left(\frac{(2x+1)\pi k}{2n}\right) \tag{8.50}$$

or in matrix form, $\mathbf{g}'_{DC} = \mathbf{C}_n \mathbf{g}$. Note that the columns of $\mathbf{C}_n$ are multiplied by the elements of the spatial-domain signal $\mathbf{g}$ to produce the rows of the frequency-domain signal $\mathbf{g}'_{DC}$, as in Equation (6.87). Therefore, for $n = 2$ we substitute $x = 0, 1$ along the columns and $k = 0, 1$ along the rows into Equation (8.50) to yield the *unnormalized* 1D DCT matrix:

$$\begin{bmatrix} 1 & 1 \\ \cos\frac{\pi}{4} & \cos\frac{3\pi}{4} \end{bmatrix} \approx \begin{bmatrix} 1 & 1 \\ 0.7071 & -0.7071 \end{bmatrix}$$

Note that the Euclidean norm of the first row $\begin{bmatrix} 1 & 1 \end{bmatrix}$ is $\sqrt{2} = \sqrt{n}$, and the norm of the second row $\begin{bmatrix} 0.7071 & -0.7071 \end{bmatrix}$ is $1 = \sqrt{n/2}$, which is what we expect since $\eta(0) = 1/\sqrt{n}$ and $\eta(1) = \sqrt{2/n}$. With the proper normalization, then, we have

$$\mathbf{C}_2 = \frac{1}{\sqrt{2}} \begin{bmatrix} 1 & 1 \\ 1 & -1 \end{bmatrix}$$

which is identical to the 2×2 normalized DFT and Hadamard matrices.
For $n = 4$ we substitute $x = 0, 1, 2, 3$ along the columns and $k = 0, 1, 2, 3$ along the rows to yield the unnormalized 1D DCT matrix:

$$\begin{bmatrix} 1 & 1 & 1 & 1 \\ \cos\frac{\pi}{8} & \cos\frac{3\pi}{8} & \cos\frac{5\pi}{8} & \cos\frac{7\pi}{8} \\ \cos\frac{\pi}{4} & \cos\frac{3\pi}{4} & \cos\frac{5\pi}{4} & \cos\frac{7\pi}{4} \\ \cos\frac{3\pi}{8} & \cos\frac{9\pi}{8} & \cos\frac{15\pi}{8} & \cos\frac{21\pi}{8} \end{bmatrix} \approx \begin{bmatrix} 1 & 1 & 1 & 1 \\ 0.9239 & 0.3827 & -0.3827 & -0.9239 \\ 0.7071 & -0.7071 & -0.7071 & 0.7071 \\ 0.3827 & -0.9239 & 0.9239 & -0.3827 \end{bmatrix}$$

Multiplying the first row by $\eta(0) = 1/\sqrt{n} = 1/2$ and the other rows by $\eta(k) = \sqrt{2/n} = 1/\sqrt{2}, k \neq 0$ yields

$$\mathbf{C}_4 \approx \frac{1}{2} \begin{bmatrix} 1 & 1 & 1 & 1 \\ 1.3066 & 0.5412 & -0.5412 & -1.3066 \\ 1 & -1 & -1 & 1 \\ 0.5412 & -1.3066 & 1.3066 & -0.5412 \end{bmatrix}$$

which is a matrix whose rows are orthonormal. It is easy to verify that $\mathbf{C}_2 \mathbf{C}_2^\mathsf{T} = \mathbf{I}_{\{2 \times 2\}}$ and $\mathbf{C}_4 \mathbf{C}_4^\mathsf{T} = \mathbf{I}_{\{4 \times 4\}}$ as mentioned above, where $\mathbf{I}_{\{n \times n\}}$ is the $n \times n$ identity matrix.

that a signal with n real elements in the spatial domain yields a DFT with n complex elements in the frequency domain. That is, the DFT transforms n scalars into $2n$ scalars, thus doubling the amount of representation needed. Obviously the $2n$ scalars contain the same information as the original n scalars, so there is tremendous redundancy in the representation. The reason for this inefficiency is that the DFT is designed to handle complex signals in the spatial domain as well, so that it does not take advantage of the special case when the signal is real.

Another way to view the limitation is to notice the boundary effects of the DFT due to its inherent periodicity. Recall that the DFT treats the incoming signal as if it were periodic, in effect gluing the beginning of the signal to the end, and vice versa, forever. But the beginning of a signal is not typically related to its end, since they are spatially separated by such a wide amount. Therefore a discontinuity is usually formed when the signal is glued in this way, and such a discontinuity leads to the Gibbs phenomenon discussed earlier,[†] which causes the DFT to devote unusually large coefficients to the high frequencies necessary to precisely preserve this shape, even though the shape is unimportant because it arises simply as an artifact of our having fed the raw signal blindly to the DFT. Thus the discontinuity leads to great inefficiency in the DFT coefficients' ability to capture the original signal.

Since the DFT of a real, even-symmetric signal is real and even-symmetric, if we could somehow transform the signal, which is already real, into an even-symmetric signal, then there would be no discontinuity, and we could discard the imaginary components altogether. But how can we make a signal even-symmetric? Suppose for example that we have a signal with four values, $g(0) = a$, $g(1) = b$, $g(2) = c$, and $g(3) = d$, that is, the signal is $[\underline{a} \quad b \quad c \quad d]$, where the underline indicates the origin at $x = 0$. Two approaches come to mind:

1. The first choice is to replicate all but the first and last values by reflecting about the last value. In this example, this approach leads to the 6 values $[\underline{a} \quad b \quad c \quad d \quad c \quad b]$, where for clarity we have colored the new values red. To see the symmetry, it may be helpful to explicitly replicate the symmetric signal:

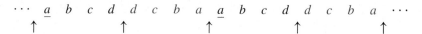

where the arrows indicate the values about which the signal is even-symmetric. Starting at any arrow, you will encounter the same values whether you travel left or right.

2. The second choice is to replicate *all* the values by reflecting about the last value. Continuing our example, this approach leads to 8 values: $[\underline{a} \quad b \quad c \quad d \quad d \quad c \quad b \quad a]$, and the replicated signal is

$$\cdots \quad \underline{a} \quad b \quad c \quad d \quad d \quad c \quad b \quad a \quad \underline{a} \quad b \quad c \quad d \quad d \quad c \quad b \quad a \quad \cdots$$

Notice here that the symmetry is about the point halfway between two values.

The first choice, known as the first version of the discrete cosine transform, or **DCT-I**, has two drawbacks. First, it does not treat all the values equally, because some values appear twice in the replicated signal, while others appear only once. Secondly, the length of the symmetric signal is no longer a power of 2, even if the original signal has a length that is a power of 2. All modern algorithms for efficiently computing transforms, such as the fast Fourier transform (FFT), fast Walsh-Hadamard transform, and fast cosine transform (FCT), rely on the signal having a length that is a power of 2, so breaking that constraint causes significant difficulty in efficient implementation.

Even worse is the second choice, which will not work at all because the location of the symmetry is wrong. That is, for the DFT to yield a real-valued frequency-domain representation, the symmetry must be about the origin, in order to allow $g(1)$ to be paired with $g(n-1)$, $g(2)$ to be paired with $g(n-2)$, and so forth, so that the complex exponentials can be paired to form cosines according to Euler's formula. Although the symmetric signal above is of no use, with a slight modification it yields the second version of the discrete cosine transform, known as **DCT-II**, which is what is represented in Equation (8.50) as well as in Equation

---

[†] Section 6.4.1 (p. 296).

(8.47) . The modification is to insert 0s in between each value, which for the example above leads to 16 values: $[0\ \ a\ \ 0\ \ b\ \ 0\ \ c\ \ 0\ \ d\ \ 0\ \ d\ \ 0\ \ c\ \ 0\ \ b\ \ 0\ \ a]$. We refer to this process as **zero-zipping**. For clarity the replicated zero-zipped symmetric signal is

$$\cdots\ 0\ \ b\ \ 0\ \ a\ \ \underset{\uparrow}{0}\ \ a\ \ 0\ \ b\ \ 0\ \ c\ \ 0\ \ d\ \ 0\ \ \underset{\uparrow}{d}\ \ 0\ \ c\ \ 0\ \ b\ \ 0\ \ a\ \ \underset{\uparrow}{0}\ \ a\ \ 0\ \ b\ \ \cdots$$

from which it is apparent that the symmetry is about the origin and the zero-zipped symmetric signal is four times as long as the original. The DFT of this zero-zipped symmetric signal yields the same result as the DCT given in Equation (8.50) up to a scale factor, along with some redundant values that we are free to discard. To see this, let us define $\varepsilon_n(z) \equiv e^{-j2\pi z/n}$ for brevity, and note that $\varepsilon_n(n-z) = \varepsilon_n(-z)$. The original signal is of length $n = 4$, while the replicated signal is of length $\breve{n} = 4n = 16$. If we let $\breve{g}$ be the replicated signal, then applying the DFT leads to

$$g'_{\mathcal{F}}(k) = \sum_{x=0}^{\breve{n}-1} \breve{g}(x)\varepsilon_{\breve{n}}(kx)$$

$$= \breve{g}(1)\varepsilon_{\breve{n}}(k) + \breve{g}(3)\varepsilon_{\breve{n}}(3k) + \breve{g}(5)\varepsilon_{\breve{n}}(5k) + \cdots + \breve{g}(15)\varepsilon_{\breve{n}}(15k)$$

$$= a\varepsilon_{\breve{n}}(k) + b\varepsilon_{\breve{n}}(3k) + c\varepsilon_{\breve{n}}(5k) + d\varepsilon_{\breve{n}}(7k) + d\varepsilon_{\breve{n}}(9k) + c\varepsilon_{\breve{n}}(11k) + b\varepsilon_{\breve{n}}(13k) + a\varepsilon_{\breve{n}}(15k)$$

$$= 2a\cos 2\pi\frac{k}{4n} + 2b\cos 2\pi\frac{3k}{4n} + 2c\cos 2\pi\frac{5k}{4n} + 2d\cos 2\pi\frac{7k}{4n}$$

$$= 2\sum_{x=0}^{n-1} g(x) \cos\frac{(2x+1)\pi k}{2n}$$

where we have used Euler's formula to simplify $\varepsilon_{\breve{n}}(z) + \varepsilon_{\breve{n}}(\breve{n}-z) = 2\cos\frac{2\pi z}{\breve{n}} = 2\cos\frac{\pi z}{2n}$. Except for the normalization factor, this last equation is identical to the DCT formula in Equation (8.50). Note that since the replicated signal has 16 values, its DFT also has 16 values. However, it is easy to show that only the first 4 values need to be stored, since the remaining values can be derived from them.[†] To summarize, what we have shown is that computing the DCT-II of a signal of length n is equivalent to computing the DFT of the zero-zipped symmetric signal, then retaining only the first n values.

8.3.6 Karhunen-Loève Transform (KLT)

All of the previous transforms are image-independent, so no matter what image is being compressed, the basis functions are the same. Not surprisingly, even better results can be achieved if the basis functions are selected in a way that is dependent upon the image. The most straightforward way to do this is to use the **Karhunen-Loève transform (KLT)**,[‡] which requires that we have a statistical model of the image, namely, the mean value of each pixel and a covariance matrix capturing the linear relationships between the pixels. This mean image and covariance matrix are usually computed from an ensemble of images taken of either the same scene or of similar scenes, and the eigenvectors of the covariance matrix yield the basis functions.

[†] Problem 8.22.

[‡] Also known as the **Hotelling transform** or **principal components analysis (PCA)**, this technique has a variety of uses outside image compression. We will explore PCA in more detail in Section 12.3.5 (p. 589).

Mathematically, for a 1D signal $\mathbf{g}$ of length n, the transform is defined as $\mathbf{g}' \equiv \mathbf{P}^{\mathsf{T}}(\mathbf{g} - \mu)$, where μ is the mean vector and $\mathbf{P}$ is an $n \times n$ orthogonal matrix whose columns are the eigenvectors of the covariance matrix $\mathbf{C}$. Since this is an orthogonal transform, it is invertible, so that the original vector $\mathbf{g}$ can be recovered from the transformed vector $\mathbf{g}'$ by $\mathbf{g} = \mathbf{P}\mathbf{g}' + \mu$. In other words, just as we saw with previous transforms, $\mathbf{g}'$ can be thought of as the coefficients that are multiplied by the basis vectors, where the basis vectors are the columns of $\mathbf{P}$, that is, the eigenvectors of $\mathbf{C}$.

Unlike other transforms, KLT comes with a principled way to truncate the coefficients. Since the eigenvalues capture the variance in the decorrelated data, the most important eigenvectors are the ones associated with the largest eigenvalues. Therefore, if we assume that the columns of $\mathbf{P}$ are sorted in decreasing order according to their associated eigenvalue, lossy compression can be achieved by discarding all but the first k columns of $\mathbf{P}$, resulting in an $n \times k$ matrix $\mathbf{P}_k$, where $k < n$. Then the original point $\mathbf{g}$ can be approximated by a k-dimensional vector $\mathbf{a}'_{\mathbf{g}} \equiv \mathbf{P}_k^{\mathsf{T}}(\mathbf{g} - \mu)$ using the formula $\mathbf{g} \approx \mathbf{P}_k \mathbf{a}'_{\mathbf{g}} + \mu$. It can be shown that this approach yields the *optimal transform*, in the sense that it is the linear transform that achieves the lowest mean squared error for a given number of coefficients, or equivalently minimizes the number of coefficients for a certain desired mean squared error. Keep in mind, though, that KLT is not necessarily optimal with respect to any single image but rather the optimality is achieved *on average* when the images are drawn according to the distribution defined by μ and $\mathbf{C}$. Unfortunately, in practice, there are few situations where such a statistical model is known.

Nevertheless, there is an unexpected connection between KLT and DCT which helps to justify the widespread adoption of the latter. Suppose we have a 1D image whose pixel values are drawn from a zero-mean, unit-variance distribution, so that $\mu = 0$ and $E[g(x)g(x)] = \sigma^2 = 1$ for each pixel, where $E[\cdot]$ is the expected value. If we assume that the image was generated according to a first-order Markov process with constant correlation coefficient ρ, where $0 \leq \rho < 1$, then $E[g(x)g(x+1)] = \rho$, $E[g(x)g(x+2)] = \rho^2$, and so forth. In that case, the covariance matrix is the following Toeplitz matrix:

$$
\mathbf{C}_{Markov} = \begin{bmatrix}
1 & \rho & \rho^2 & \cdots & \rho^{n-1} \\
\rho & 1 & \rho & \cdots & \rho^{n-2} \\
\rho^2 & \rho & 1 & \cdots & \rho^{n-3} \\
\vdots & & & \ddots & \vdots \\
\rho^{n-1} & \rho^{n-2} & \rho^{n-3} & \cdots & 1
\end{bmatrix}
\tag{8.51}
$$

For a wide range of values for ρ, it can be shown that the eigenvectors of this matrix look very similar to the basis functions of the DCT. Since the KLT is the optimal linear transform, the DCT also performs (according to the assumptions presented here) in a near-optimal manner as well.

8.3.7 Example: JPEG Compression

By far the most widely used lossy image compression format is **JPEG** (named for the Joint Photographic Experts Group), which is based on entropy encoding of quantized DCT coefficients. JPEG operates on a single image with 1 to 4 components. A grayscale image contains 1 component, whereas an RGB image contains 3 components. Although the standard itself says nothing about color spaces, typically an RGB image is first transformed to a different color space that separates the luminance from chrominance information, yielding one luminance component and two chrominance components.[†] The components in a JPEG

[†] Technically, these are the nonlinear quantities *luma* and *chroma*, as we shall see in the next chapter; however, in this section we retain the terms used in the official JPEG standard.

image are either all the same size, or they are subsampled by integer amounts. For example, it is common for the chrominance components to be downsampled by a factor of 2 both horizontally and vertically, since the human eye is less sensitive to spatial variations in chrominance than in luminance.

In this section we focus our attention primarily upon the **baseline process**, which is the most popular variation of the JPEG specification, illustrated in Figure 8.8. In the baseline process the image is non-overlapping 8×8 blocks of samples are processed independently. (A *sample* is a pixel value per component, and if the dimensions of any component are not divisible by 8, then the last row and/or column of samples is replicated as necessary.) The DCT is computed of each 8×8 block (also called a *data unit*), thus transforming the original data into 64 frequency domain coefficients. But to reduce the dynamic range of the DCT computation, the samples are first **level shifted** to a signed representation by subtracting the value $\frac{ngray}{2}$ from each sample, where *ngray* is the number of possible values for each sample. For the baseline process, the value of a sample is limited to 8 bits per component, so *ngray* = 256, and therefore 128 is subtracted from each sample to yield a level-shifted value in the range from -128 to $+127$ before the DCT computation. Substituting $n = 8$ into Equation (8.47) and letting u and v be the frequency indices yields the following equation for the forward level-shifted DCT:

$$G'_{DC}(u, v) = \frac{1}{4}\eta'(u, v) \sum_{x=0}^{7} \sum_{y=0}^{7} (G(x, y) - 128) \cos\left(\frac{(2x + 1)\pi u}{16}\right) \cos\left(\frac{(2y + 1)\pi v}{16}\right) \quad (8.52)$$

where

$$\eta'(u, v) \equiv \begin{cases} \frac{1}{2} & \text{if } u = 0 \text{ and } v = 0 \\ 1 & \text{otherwise} \end{cases} \quad (8.53)$$

Lossy compression is achieved by dividing the 64 DCT coefficients by the corresponding values in a **quantization table**, then rounding to the nearest integer:

$$G'_{quantized}(u, v) = \text{Round}\left(\frac{G'_{DC}(u, v)}{Q(u, v)}\right) \quad (8.54)$$

where $Q(u,v)$ is the entry in the quantization table. (In the baseline process the quantized values are stored with 11 bits of precision.) The encoder is free to use any quantization table, but the reference quantization tables for luminance and chrominance shown in Figure 8.9 are widely used. Typically the quality of the compressed image is controlled by a quality factor q, where $1 \le q \le 100$, in which case

$$Q(u, v) = \text{Clamp}(s(q) \cdot Q_{ref}(u, v), 1, 255) \quad (8.55)$$

Figure 8.8 JPEG compression overview.

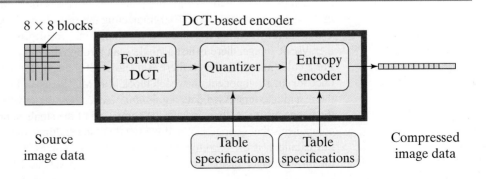

Figure 8.9 Reference
JPEG quantization tables
for luminance and
chrominance.

$$\begin{bmatrix} 16 & 11 & 10 & 16 & 24 & 40 & 51 & 61 \\ 12 & 12 & 14 & 19 & 26 & 58 & 60 & 55 \\ 14 & 13 & 16 & 24 & 40 & 57 & 69 & 56 \\ 14 & 17 & 22 & 29 & 51 & 87 & 80 & 62 \\ 18 & 22 & 37 & 56 & 68 & 109 & 103 & 77 \\ 24 & 35 & 55 & 64 & 81 & 104 & 113 & 92 \\ 49 & 64 & 78 & 87 & 103 & 121 & 120 & 101 \\ 72 & 92 & 95 & 98 & 112 & 100 & 103 & 99 \end{bmatrix} \qquad \begin{bmatrix} 17 & 18 & 24 & 47 & 99 & 99 & 99 & 99 \\ 18 & 21 & 26 & 66 & 99 & 99 & 99 & 99 \\ 24 & 26 & 56 & 99 & 99 & 99 & 99 & 99 \\ 47 & 66 & 99 & 99 & 99 & 99 & 99 & 99 \\ 99 & 99 & 99 & 99 & 99 & 99 & 99 & 99 \\ 99 & 99 & 99 & 99 & 99 & 99 & 99 & 99 \\ 99 & 99 & 99 & 99 & 99 & 99 & 99 & 99 \\ 99 & 99 & 99 & 99 & 99 & 99 & 99 & 99 \end{bmatrix}$$

Luminance table, $Q^{\ell}_{ref}(u,v)$ Chrominance table, $Q^{c}_{ref}(u,v)$

where $\text{CLAMP}(z,a,b) \equiv \min(\max(z,a),b)$, Q_{ref} is either the luminance Q^{ℓ}_{ref} or chrominance Q^{c}_{ref} as appropriate, and the most widely used scaling definition is given by

$$s(q) \equiv \begin{cases} 50/q & \text{if } q \le 50 \\ 2 - q/50 & \text{if } q > 50 \end{cases} \tag{8.56}$$

Note that smaller values in the quantization table cause more information to be retained, while larger values cause more information to be lost. For example, if the DCT coefficient is 128, then a quantization divisor of 5 yields $\text{ROUND}(128/5) = 26$, while a divisor of 200 yields $\text{ROUND}(128/200) = 0$. It is easy to see that $q = 50$ uses the reference quantization table, $q = 100$ effectively sets the quantization table to all 1s, and $q = 1$ leads to very large values in the quantization table. The reference quantization tables are designed such that if their values are divided by two, that is, if $s = \frac{1}{2}$, then the compression incurs essentially no perceptual loss, that is, the compressed image and the original image look identical. This occurs when $q = 75$, and there is rarely any legitimate reason to increase q beyond this value. If the implementation uses integer-only arithmetic operations, then the equations for $s(q)$ are multiplied by 100, that is, $s(q) = 5000/q$ and $s(q) = 200 - 2q$, then $s(q) \cdot Q_{ref}$ is divided by 1000 and rounded before clamping.

Each block is then scanned in a **zigzag** fashion, as shown in Figure 8.10, to order the 64 quantized DCT coefficients approximately from the lowest frequency to the highest frequency. As we saw in the previous chapter, the top-left value, $G'_{quantized}(0,0)$, is the DC ("direct current") coefficient and captures the absolute lowest frequency. Since this lowest of all frequency components changes slowly throughout an image, there is high spatial correlation between the DC components of neighboring blocks, so to reduce the amount of data further a DPCM predictive coding procedure is applied to subtract the DC coefficient of the previous block from that of the current block. The remaining 63 coefficients, which are sometimes known as AC ("alternating current"), contain many zeros, so they are run-length encoded (RLE) to yield (*skip,value*) pairs, where *skip* is the number of zeros, and *value* is the next non-zero value, with (0,0) indicating the end-of-block (EOB).

For entropy coding, the blocks are grouped into **minimum coded units (MCUs)**. For a grayscale image, there is just a single scan through the image, so each block is an MCU. For a color image, however, there is a choice. If the encoder scans the image three times, once for each component, then each block is an MCU (just as in grayscale), and we say that the resulting compressed data are *noninterleaved*. On the other hand, if the encoder scans the image just once, then an MCU is composed of the smallest number of blocks in all the components that go together, and we say that the resulting compressed data is *interleaved*. For example, if the chrominance components are downsampled by 2 in each direction, then 4 luminance blocks correspond to the same pixels as 1 block from each of the two

Figure 8.10 Zigzag scanning of JPEG components proceeds from low- to high-frequency components.

DC

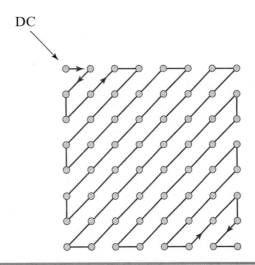

chrominance components, so that each MCU contains information from a 16×16 array in the original image. These three possibilities are illustrated as follows:

$$\underbrace{Y_{00}}_{MCU_1} \; \underbrace{Y_{01}}_{MCU_2} \; \underbrace{Y_{02}}_{MCU_3} \; \underbrace{Y_{03}}_{MCU_4} \; \underbrace{Y_{04}}_{MCU_5} \; \underbrace{Y_{05}}_{MCU_6} \; \cdots \quad \text{(grayscale)}$$

$$\underbrace{\underbrace{Y_{00}}_{MCU_1} \; \underbrace{Y_{01}}_{MCU_2} \; \underbrace{Y_{02}}_{MCU_3} \; \cdots}_{scan_1} \; \underbrace{\underbrace{C^{(1)}_{00}}_{MCU_1} \; \underbrace{C^{(1)}_{01}}_{MCU_2} \; \underbrace{C^{(1)}_{02}}_{MCU_3} \; \cdots}_{scan_2} \; \underbrace{\underbrace{C^{(2)}_{00}}_{MCU_1} \; \underbrace{C^{(2)}_{01}}_{MCU_2} \; \underbrace{C^{(2)}_{02}}_{MCU_3} \; \cdots}_{scan_3} \quad \text{(color noninterleaved)}$$

$$\underbrace{\underbrace{Y_{00}Y_{01}Y_{10}Y_{11}C^{(1)}_{00}C^{(2)}_{00}}_{MCU_1} \; \underbrace{Y_{02}Y_{03}Y_{12}Y_{13}C^{(1)}_{01}C^{(2)}_{01}}_{MCU_2} \; \underbrace{Y_{04}Y_{05}Y_{14}Y_{15}C^{(1)}_{02}C^{(2)}_{02}}_{MCU_3} \; \cdots,}_{scan} \quad \text{(color interleaved)}$$

where Y_{ij} contains the DPCM value and RLE pairs for the ij^{th} luminance block, and similarly $C^{(1)}$ and $C^{(2)}$ capture the DPCM value and RLE pairs for the chrominance blocks. The MCUs are entropy coded using Huffman encoding.

The JPEG bitstream is organized according to the *interchange format*, which contains *marker segments* and *entropy-coded data segments* separated by *markers*, which are two-byte codes where the first byte is $\mathtt{FF}$ and the second byte is nonzero. These markers identify the structural parts of the bitstream and make it possible to discover information without having to decode the entire image. The frame begins with up to 4 quantization tables, followed by the *frame header* and the scans. The frame header specifies the dimensions of the image, the subsampling factors of the scans, and which quantization table is used by each scan, among other information. Each scan begins with one or more Huffman tables, followed by the *scan header* and the entropy-coded data segment(s) for that scan. The scan header specifies the components contained in the scan, the Huffman table(s) used by the different components in the scan, and other information. Typically each scan uses two Huffman tables per component, that is, a DC Huffman table for the DPCM values and an AC Huffman table for the RLE pairs. Each entropy-coded data segment contains the Huffman-coded MCUs, with any byte $\mathtt{FF}$ immediately followed by $\mathtt{00}$ (a process known as *byte stuffing*) to avoid ambiguity with the markers.

Surprisingly, the JPEG interchange format, which specifies the format of the compressed bitstream, does not specify the file format itself. The most common file formats are JFIF (JPEG File Interchange Format) and the more recent Exif (Exchangeable image file format). Although technically speaking these are two separate formats, in practice most digital cameras output a JFIF file with Exif metadata appended, so that a user typically need not distinguish between the two. The metadata contains additional information about the conditions in which the picture was taken, such as the camera manufacturer and model; the aperture, focal length, and exposure time of the camera; and the time and GPS (global positioning system) coordinates of the camera when the picture was taken. A JFIF/Exif file begins with a "start of image" (SOI) marker (bytes FFD8), followed by an application-specific APP0 marker (bytes FFE0) to indicate the JFIF information, later followed by the APP1 marker (bytes FFE1) to indicate the Exif metadata. Note that although JPEG itself does not specify the color space, JFIF/Exif files typically store the ICC color profile of the original RGB color space, which is converted to YCbCr using CCIR Rec. 601 to separate the luminance from chrominance components.[†] Results of applying JPEG compression to an image, with various levels of quality, are shown in Figure 8.11.

In addition to the baseline process described in this section, the JPEG standard contains a number of extensions and variations, such as 12 bits per pixel per component, arithmetic instead of Huffman coding, progressive instead of sequential scanning, and even a lossless mode (which was for the most part replaced by JPEG-LS which is based on LOCO-I predictive coding). These extensions and variations, however, are not as commonly used as the baseline process.

8.3.8 Wavelet-Based Compression

Eight years after the JPEG compression format was finalized, the same committee proposed a replacement called JPEG 2000. Instead of using the DCT, JPEG 2000 is based upon the discrete wavelet transform (DWT).[‡] The main advantage of JPEG 2000 over JPEG is that the DWT naturally provides a multiresolution representation of the image, which is much more

Figure 8.11 An image,[§] along with the results of JPEG compression with different levels of quality. Underneath each image is shown the average number of bits per pixel, and the compression ratio. With $q = 75$ there is no visible difference between the compressed image and the original; with $q = 40$ only minor changes to background pixels can be noticed; with $q = 10$ and $q = 5$ compression artifacts are apparent.

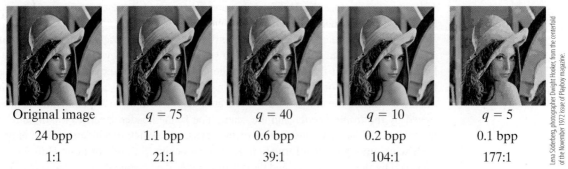

Original image	$q = 75$	$q = 40$	$q = 10$	$q = 5$
24 bpp	1.1 bpp	0.6 bpp	0.2 bpp	0.1 bpp
1:1	21:1	39:1	104:1	177:1

Lena Söderberg, photographer Dwight Hooker, from the centerfold of the November 1972 issue of Playboy magazine.

[†] See Chapter 9 for more details regarding color.

[‡] Section 6.6 (p. 309). Like the original JPEG format, JPEG 2000 also contains a lossless variation, but it is almost never used.

[§] This is the well-known but somewhat controversial Lena (pronounced "LENN-eh") image. It is included here for its historical importance in being the standard image used to evaluate image compression algorithms for nearly half a century.

flexible than the DCT. In particular, JPEG 2000 supports **progressive decoding**, whereby a decoder produces an approximate image before the entire bitstream has been received, then incrementally improves the fidelity of the image as more bits are received—a feature that is important in Internet applications. Unfortunately, JPEG 2000 was never widely adopted, primarily due to the increased complexity of the encoder/decoder, only modest increases in the compression ratio, and the fact that JPEG had already attained wide-spread popularity.

JPEG 2000 uses the **CDF 9/7 wavelet**,[†] which has 9 coefficients and 7 vanishing moments. Compared with the traditional Daubechies D10 wavelet, which has 10 coefficients and $\frac{10}{2} = 5$ vanishing moments, CDF 9/7 achieves more vanishing points with fewer coefficients. The CDF 9/7 *analysis* lowpass and highpass kernels, respectively, are as follows:

$$\phi \approx [0.037828 \quad -0.023849 \quad -0.110624 \quad 0.377403 \quad \underline{0.852699} \quad 0.377403 \quad -0.110624 \quad -0.023849 \quad 0.037828]$$

$$\psi \approx [0 \quad 0.064539 \quad -0.040689 \quad -0.418092 \quad \underline{0.788486} \quad -0.418092 \quad -0.040689 \quad 0.064539 \quad 0 \quad]$$

Not only is the CDF 9/7 wavelet part of the JPEG 2000 specification, it is also widely used by the Federal Bureau of Investigation (FBI) for storing fingerprint images.

Note that $|\phi|^2 = 1.0404$ in the kernel above, which is not 1. This means that CDF 9/7 is not energy-preserving, and hence not orthogonal; instead it is **biorthogonal**. Compared with an orthogonal wavelet, which satisfies Equation (6.158),[‡] a biorthogonal wavelet satisfies

$$\sum_x \phi(x - 2i)\tilde{\psi}(x - 2j) = \sum_x \tilde{\phi}(x - 2i)\psi(x - 2j) = 0 \tag{8.57}$$

where $\tilde{\psi}$ and $\tilde{\phi}$ are the *synthesis* lowpass highpass kernels, which in this case are given by changing the sign of every other element (and shifting in origin to ensure Equation (8.57) is satisfied)

$$\tilde{\phi} \approx [0 \quad -0.064539 \quad -0.040689 \quad 0.418092 \quad 0.788486 \quad \underline{0.418092} \quad -0.040689 \quad -0.064539 \quad 0 \quad]$$

$$\tilde{\psi} \approx [0.037828 \quad 0.023849 \quad -0.110624 \quad -0.377403 \quad 0.852699 \quad \underline{-0.377403} \quad -0.110624 \quad 0.023849 \quad 0.037828]$$

A comparison of orthogonality and biorthogonality is shown in Figure 8.12. Note that a biorthogonal wavelet approximates an orthogonal wavelet in the autocorrelation of its lowpass and highpass kernels, but the analysis and synthesis kernels are different. As a result, biorthogonality retains several of the advantages of orthogonality but relaxes the energy-preserving property.

Why is biorthogonality so important? Well, from an earlier discussion[‡] we know that Daubechies D2–20 wavelets are particularly good at compression, because they are designed to maximize the number of vanishing moments. However, when a wavelet transform is applied to a signal, the number of resulting coefficients is larger than the original signal. For example, if a 1D signal of length n is convolved with a wavelet kernel of length w, the result is an array of $n + w - 1$ values, which is larger than the original n. Although this drawback can be mitigated by using circular convolution, doing so introduces its own problem; namely, if the signal is not periodic (which it typically is not), then circular convolution causes undesirable artifacts at the border. To fix this problem, we must replicate the signal to make it symmetric, as we did with the DCT. Recall, however, that with an n-element input, the DCT produces only n unique coefficients, with the remaining $3n$ coefficients being redundant—all because the DCT itself is symmetric. In contrast, the Daubechies D2–20 wavelets are not symmetric, so there is no redundancy in the output. As a result, the number

---

[†] Also referred to as Daubechies 9/7, a variant of Cohen-Daubechies-Feauveau (CDF) 9/7.
[‡] Section 6.6 (p. 309).

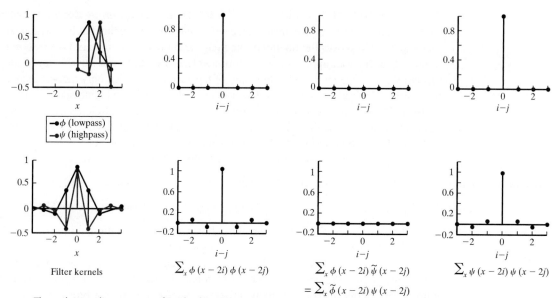

Figure 8.12 The orthogonality property of Daubechies D4 (top) and the biorthogonality property of CDF 9/7 (bottom) wavelets. With D4, the analysis and synthesis filters are the same, whereas with CDF 9/7 they are different. In addition, CDF 9/7 is not energy preserving, as shown by the fact that the inner product of the lowpass with itself, and the highpass with itself, is only approximately the Kronecker delta.

of coefficients is even greater than would be with linear convolution, thus detracting from any compression benefits from using the filter in the first place.

To restate the problem, the ideal wavelet function for compression would be orthogonal and symmetric. An orthogonal transform preserves energy, while a symmetric filter properly handles border effects without introducing unnecessary coefficients (because a symmetric filter has linear phase). As it turns out, however, the only compactly supported orthogonal linear phase wavelet is the Haar wavelet, which as noted earlier has its own drawbacks due to its harsh edges. Therefore, one of these two requirements must be compromised, and since symmetry is absolutely essential to achieve compression, the orthogonality property must be compromised. Biorthogonality is, in some sense, the best we can do to achieve a symmetric filter with approximate energy conservation, while at the same time maintaining separation between the axes of the basis functions.

8.4 Compression of Videos

Up to now we have concentrated on the compression of images, for which three types of redundancy have been exploited in various ways. For video, a fourth type of redundancy is present. *Temporal redundancy* refers to the high correlation between corresponding pixels of adjacent image frames, which enables pixel values in one image to be predicted not only from neighboring pixels in the same image frame but also from pixels in nearby frames. Below we briefly consider two common approaches to video compression.

8.4.1 M-JPEG Compression

The simplest video compression approach is that of **Motion JPEG (M-JPEG)**, which is literally nothing more than JPEG compression (using the exact same approach and format just discussed) of each image frame in the video sequence. Because M-JPEG performs compression only *within* a frame (*intraframe compression*), it is not able to achieve compression ratios any higher than JPEG itself, typically on the order of 10:1. The main advantage of M-JPEG, besides its simplicity, is that it does not suffer from generation loss in the temporal dimension. As a result, M-JPEG is the de facto standard for video editing.

8.4.2 MPEG Compression

Another popular video standard is **MPEG** (named for the Moving Picture Experts Group), which not only performs intraframe compression like M-JPEG but also compression between frames (*interframe compression*). As a result, MPEG is able to exploit temporal redundancy to achieve compression ratios greater than 100:1. Like JPEG and M-JPEG, MPEG divides the image into 8×8 blocks and operates on subsampled chrominance components, leading to 16×16 *macroblocks* that contain 4 blocks of luminance data and 2 blocks of subsampled chrominance data. The image frames of the video sequence are treated in one of three ways. Some image frames, called **I-frames** because only intraframe compression is applied, are compressed using JPEG. These frames can be decoded independently of all other frames and therefore play the role of *key frames*. Typically I-frames occur approximately twice per second. Other frames, called **P-frames**, are allowed to store the difference between themselves and a preceding frame (either an I- or another P-frame), as well as intraframe compression. This difference relies on *motion estimation*, that is, the calculation of motion vectors for each macroblock to determine the correspondence between consecutive frames to aid the prediction that exploits temporal redundancy. Since P-frames can be decoded based only on previous frames, causal processing of the video sequence is sufficient. The third type of frame, known as **B-frames** because they support backward prediction, are allowed to store the difference between themselves and the immediately preceding or following frame, as well as intraframe compression. Obviously a B-frame cannot be decoded until the next *anchor frame* (either an I- or P-frame) is decoded, thus making the decoding process non-causal and complicating the buffering procedure for the decoder.

Several MPEG formats have been developed over the years, each building on the previous one. **MPEG-1** (also known as Rec. H.261) is typically applied to relatively low-resolution video of approximately 320×240 images, achieving bit rates less than 2 megabits per second. **MPEG-2** (Rec. H.262) provides support for interlaced video, thus effectively doubling the image resolution. Most DVD (originally "digital video disc," now "digital versatile disc") videos are stored in MPEG-2 format. **MPEG-4** (Rec. H.264) improves the coding efficiency over MPEG-2 but is an evolving, complex standard with a variety of "parts" and "levels" with support for advanced concepts such as multiview (e.g., stereo) video. Blu-ray discs originally supported only MPEG-2 but now also support some versions of MPEG-4. Keep in mind that these standards specify not only the video encoding but also the audio encoding, which is beyond our present scope. For example, the widely popular MP3 audio format is not MPEG-3 but rather Audio Layer III of either MPEG-1 or MPEG-2.

8.5 Further Reading

The principles of information theory were laid down by Shannon in a series of extremely influential papers beginning with his seminal publication, Shannon [1948]. A good introduction to the field can be found in any textbook such as that by Cover and Thomas [1991]. Huffman coding was introduced by Huffman [1952]. The Lempel-Ziv algorithm, also known as LZ77, was introduced by Ziv and Lempel [1977]. The same authors extended their work the next year in Ziv and Lempel [1978], which describes an algorithm sometimes known as LZ78, which was further expanded by Welch [1984] to form the Lempel-Ziv-Welch (LZW) algorithm. The GIF (graphics interchange format) image file format, created in 1987, is based on LZW. The PNG (portable network graphics) image file format was created in 1996 and is based on LZ77 instead to avoid the patent restrictions of GIF, which have since expired.

The Paeth algorithm is presented by Paeth [1991]. The LOCO-I algorithm is due to Weinberger et al. [2000]. A good overview of transform coding can be found in Goyal [2001]. The DCT was originally described by Ahmed et al. [1974], and its history was related years later by its lead author, Ahmed [1991]. For recent proposals to improve upon MSE for evaluating image quality, see the structural similarity (SSIM) index of Wang et al. [2004], as well as the work of Sheikh et al. [2005].

The JPEG standard was released as ITU-T Rec. T.81 in 1992. The lossless JPEG-LS (ITU-T Rec. T.87), based on the LOCO-I algorithm, was released in 1998. JPEG 2000 was published as ISO/IEC 15444 in 2000 (and republished as ITU-T Rec. T.800 in 2002). The Cohen-Daubechies-Feauveau (CDF) wavelets are described in Cohen et al. [1992], while the Daubechies 9/7 wavelet used in JPEG 2000 is described in Usevitch [2001] and Unser and Blu [2003]. MPEG-1 (ITU-T Rec. H.261) was released in 1988 and modified in 1993, MPEG-2 (ITU-T Rec. H.262) was released in 1995 and modified as recently as 2013, while MPEG-4 (ITU-T Rec. H.264) was released in 2003 and modified as recently as 2014.

PROBLEMS

8.1 Explain the difference between data and information.

8.2 List the three causes of redundancy.

8.3 What are some differences between lossless and lossy compression? On what type of image is lossless compression more appropriate? (In fact, lossy compression should never be used on this type of image.) On what type of image is lossy compression more appropriate?

8.4 Suppose a 500×500 8-bit-per-pixel grayscale image is compressed to 50,000 bits. Compute the compression ratio.

8.5 What is the entropy of the random variable associated with rolling an unfair 6-sided die with the following distribution: $p(1) = \frac{1}{16}, p(2) = \frac{1}{8}, p(3) = \frac{3}{16}, p(4) = \frac{1}{4}, p(5) = \frac{1}{16}, p(6) = \frac{5}{16}$?

8.6 Does Shannon's source coding theorem apply to an i.i.d. sequence of random variables, or to a stationary sequence?

8.7 Consider the image below.

$$\begin{bmatrix} 5 & 1 & 3 & 2 \\ 4 & 1 & 3 & 4 \\ 0 & 3 & 3 & 6 \\ 7 & 6 & 1 & 7 \end{bmatrix}$$

(a) What is the probability mass function, assuming 3 bits per pixel?

(b) Write the sequence of bits representing the uncompressed image. How many bits are needed?

(c) Generate the Huffman codebook.

(d) Apply the Huffman codebook to compress the image.

(e) Ignoring the size of the codebook itself, what compression ratio is achieved by this approach?

8.8 Compute the entropy of the image in the previous problem using block sizes of 1, 2, 3, and 4 bits.

8.9 Explain what is meant by a prefix code.

8.10 Use the Lempel-Ziv algorithm to encode the following data sequence:

ABAABABABBABBABBABABAABBABBAAABBA

8.11 Encode the following 4×4 image using Lempel-Ziv-Welch assuming 1 bit per pixel and a 3-bit dictionary, processing the pixels in row-major order. To avoid confusion, use $A = 0$ and $B = 1$.

$$\begin{bmatrix} 0 & 1 & 0 & 1 \\ 1 & 0 & 1 & 1 \\ 0 & 0 & 1 & 1 \\ 0 & 1 & 0 & 0 \end{bmatrix}$$

8.12 Encode the following sequences using RLE encoding. For each sequence, explain whether RLE should be used.

(a)   *AAAABBBBBBAAACCCCCCCBBB*

(b)   *ABCBDAABCDADCA*

8.13 How does the Laplace distribution differ from the Gaussian distribution?

8.14 Use (a) the planar predictor, (b) the Paeth predictor, (c) the LOCO-I algorithm to compute lower-right pixel in the following image.

$$\begin{bmatrix} 25 & 26 \\ 32 & ? \end{bmatrix}$$

8.15 The LOCO-I predictor is also known as the *median edge detector (MED)* because it computes the median of the three values ℓ, u, and $\hat{p}$. Show that this is true.

8.16 Why was the PNG file format invented? Does it perform lossless or lossy compression? What compression techniques are used in PNG?

8.17 Golomb coding and the related Rice coding are compression methods not mentioned in this chapter. Search online to discover when these approaches are appropriate and where they are used. Explain your findings.

8.18 Assuming that the second image is an approximation of the first image, compute the mean squared error.

$$\begin{bmatrix} 0 & 2 & 7 & 3 \\ 3 & 5 & 4 & 1 \\ 1 & 8 & 0 & 6 \\ 9 & 7 & 5 & 3 \end{bmatrix} \qquad \begin{bmatrix} 4 & 0 & 5 & 1 \\ 2 & 8 & 3 & 7 \\ 8 & 9 & 2 & 6 \\ 3 & 0 & 8 & 4 \end{bmatrix}$$

8.19 Write the 8×8 1D orthogonal linear transform for the following:

(a)   Discrete Fourier transform (DFT)

(b)   Walsh-Hadamard transform (WHT)

(c)   Discrete cosine transform (DCT)

8.20 Apply each of the transforms from the previous problem to the following grayscale image.

$$\begin{bmatrix} 6 & 5 & 9 & 6 & 1 & 0 & 3 & 2 \\ 2 & 1 & 3 & 5 & 1 & 3 & 7 & 8 \\ 8 & 1 & 2 & 9 & 5 & 2 & 7 & 5 \\ 3 & 3 & 3 & 3 & 8 & 8 & 7 & 10 \\ 5 & 8 & 6 & 8 & 9 & 3 & 5 & 1 \\ 7 & 3 & 5 & 8 & 1 & 5 & 1 & 4 \\ 9 & 8 & 4 & 4 & 6 & 2 & 2 & 1 \\ 10 & 2 & 8 & 6 & 5 & 6 & 9 & 10 \end{bmatrix}$$

8.21 The DCT of $\begin{bmatrix} 0 & 8 & 6 & 9 & 1 & 4 & 3 & 8 \end{bmatrix}$ is equivalent to the first 8 values of the DFT of what sequence?

8.22 Show that $\check{n} - n$ of the values in the DFT of the replicated signal in DCT-II can be determined from the first n values, where $\check{n} = 4n$ and n is the length of the original signal.

8.23 Compute the eigenvectors of Equation (8.51) for $\rho = 0, 0.5$, and 0.75, and compare with the DCT basis functions, using $n = 4$. For which value of ρ are the eigenvectors most similar to the basis functions?

8.24 Can the Karhunen-Loève transform (KLT) be used to compress a single image? Why or why not?

8.25 How is JPEG compression different from JPEG 2000 compression?

8.26 Apply the reference JPEG quantization table for luminance in Figure 8.9, with a quality of $q = 50$, to the DCT coefficients computed in Problem 8.20. Show the result in zigzag order.

8.27 Explain the difference between I-, P-, and B-frames in MPEG compression.

8.28 How are MPEG and M-JPEG different? List some pros and cons of each.

8.29 What is steganography? (Search online if necessary.) Explain a simple steganographic method for images. How feasible do you think it is to develop a method of steganography that works even in the case of lossy compression? Explain.

8.30 Implement transform coding for both the DCT and WHT. Then compress a photograph to compare the performance of truncating $x\%$ of coefficients, where x is a value you select.

8.31 Compress a photograph of your choosing using JPEG for quality values of $q = 5$, 10, 20, 40, 60, 75, and 100. For what values can you not tell any difference between the compressed image and the original? Calculate the MSE for each and plot.

8.32 Repeat the previous exercise with a line drawing image. Zoom in on the results to examine the degradation in more detail.

CHAPTER 9
Color

In previous chapters we devoted most of our attention to processing grayscale images, an approach justified by the fact that the grayscale (or luminance) portion of an image contains most of the information. In fact, more than 75% of a typical video stream is devoted solely to luminance, and we have no difficulty in viewing an old black-and-white movie or functioning in a dark environment. Nevertheless, it is plainly evident that color enhances our visual experience significantly, and it provides crucial information to facilitate tasks that would otherwise be much more difficult. Color cameras are now commonplace, typically providing three values per pixel from the sensor itself. Although for many applications these different color channels can be processed simply as three separate streams without further analysis, other applications require a more thorough understanding of the subject in order to properly utilize these values. Therefore, to understand color more deeply, in this chapter we turn to two disparate and sometimes conflicting disciplines, namely, color science and video engineering. The subtle interplay between the rigor of the former and the practical trade-offs made by the latter sometimes leads to nonobvious implications when they are combined, as we shall see.

9.1 Physics and Psychology of Color

We begin our exploration by considering both the physics and psychology of color.

9.1.1 The Rainbow and the Color Wheel

As explained earlier,[†] the light traveling from the sun or some other light source is an electromagnetic wave whose oscillations usually contain a mixture of different wavelengths. When Isaac Newton published his work on optics at the dawn of the 18th century, he demonstrated to the world for the first time that the light that we usually consider to be white is actually composed of a range of different wavelengths corresponding to the colors of the rainbow, from red at one end to violet at the other end. This range of colors is referred to as the visible **spectrum**, and the pure colors in the rainbow are therefore known as the **spectral colors**.

Newton was also the first person to arrange the colors into a circle, known as the **color wheel** (or **color circle**), by wrapping the colors of the rainbow so that red and violet, which are the farthest from each other in the rainbow, are adjacent in the color wheel. Here we see the difference between color as it exists in the **physical** world, and color as it exists in the **psychological** world of our mind, as shown in Figure 9.1.[‡] Physically, the spectral colors are *linear* in the sense that they lie on a straight line parameterized by wavelength. Psychologically, however, the spectral colors are *circular*, in the sense that they are perceived in ways that are best modeled by connecting the two ends of the spectrum. For example, just as green is a mixture of the two adjacent colors blue and yellow (■ = ■ + ■), and orange is a mixture of the two adjacent colors red and yellow (■ = ■ + ■), so too purple[§] is a mixture of the two adjacent colors red and blue (■ = ■ + ■), although red and blue should, from a physical point of view, have little in common with each other since they are at nearly opposite ends of the spectrum. The color wheel is widely used in art to identify complementary colors on opposite sides

Figure 9.1 The spectral colors of the rainbow are modeled physically as a straight line (left), or psychologically as a circle (right).

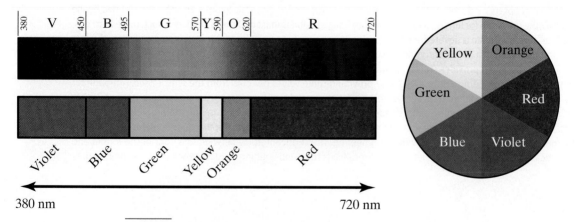

[†] Section 2.2.1 (p. 33)

[‡] Newton is also the reason schoolchildren memorize the colors of the rainbow by the 7-letter acronym ROYGBIV (or VIBGYOR, depending on whether the colors are ordered by increasing frequency or increasing wavelength). Newton divided the rainbow into 7 colors because he considered 7 to be the perfect number and therefore somewhat arbitrarily included the color indigo. Most people would agree, however, that indigo is difficult to discern as a separate color, leaving us with the 6 colors shown in the figure; of course, other ways of discretizing the rainbow are possible.

[§] Note that while violet is a pure wavelength, purple is a mixture.

of the wheel and so forth; our focus in this chapter, however, will be on the underlying science rather than aesthetics, in order to *quantitatively* understand the relationships between various colors.

9.1.2 Spectral Power Distributions (SPDs)

Returning to the physical world, the **spectral power distribution (SPD)** represents the amount of optical power in each wavelength of light, as shown in Figure 9.2. Although the SPD is a continuous function, it is generally measured at discrete intervals (using an instrument called a **spectroradiometer**) to yield a vector of values. The length of this vector is determined by both the quantization spacing and the definition of the visible spectrum. For our purposes, we will assume that the SPD can be adequately represented by a 35-dimensional vector capturing the values in 10 nm bands from 380 nm to 720 nm, which roughly corresponds to the visible spectrum.

SPDs obey the principle of **superposition**. That is, the SPD that results from shining two lights together is equal to the sum of the SPDs of the two individual lights. If we let $t_1(\lambda)$, $t_2(\lambda)$, and $t_{combined}(\lambda)$ be the corresponding elements of the SPD at wavelength λ for the first, second, and combined lights, respectively, then

$$t_{combined}(\lambda) = t_1(\lambda) + t_2(\lambda) \tag{9.1}$$

The principle of superposition allows us to predict the SPD that will occur when spectral components are mixed together by measuring the power at each wavelength separately for each component.

When the light hits an opaque surface, some amount of the light is absorbed by the surface, while the rest is reflected. For example, a red object will tend to reflect longer wavelengths, while a blue object will reflect shorter wavelengths. The SPD of the light that is reflected from the surface is equal to the incident SPD minus the spectrum that is absorbed, or approximately the elementwise multiplication of the SPD of the incident light and the surface reflectance spectrum. That is, if we let $t_{incident}(\lambda)$, $t_{reflected}(\lambda)$, and $t_{absorbed}(\lambda)$ be corresponding elements of the incident, reflected, and absorbed SPD vectors, respectively, then

$$t_{reflected}(\lambda) \quad = \quad t_{incident}(\lambda) - t_{absorbed}(\lambda) \tag{9.2}$$

$$\approx \quad t_{incident}(\lambda) \cdot \rho_{surface}(\lambda) \tag{9.3}$$

where $\rho_{surface}(\lambda) \approx 1 - t_{absorbed}(\lambda)/t_{incident}(\lambda)$. The SPD measured by a sensor, then, captures the interplay of the SPD of the light source and the reflectance spectrum of the surface in the scene.

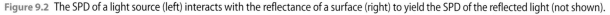

Figure 9.2 The SPD of a light source (left) interacts with the reflectance of a surface (right) to yield the SPD of the reflected light (not shown).

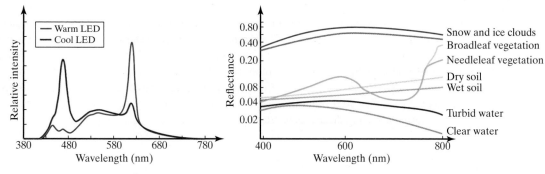

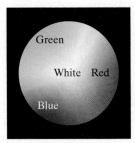 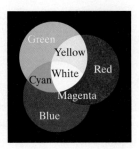 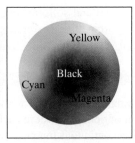 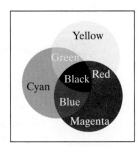

Figure 9.3 In the additive color model (left), mixing all colors yields white, whereas in a subtractive color model (right), mixing all colors yields black. Both color models contain the same colors in the same order, although the additive model is linear, while subtractive models are highly nonlinear.

9.1.3 Additive and Subtractive Colors

From Equation (9.1), light sources *add* light to a scene, so that the more lights that are shone, the brighter the scene becomes, and all the colors mixed together yield white. For this reason, the color model that explains the color that results from mixing lights is known as the **additive color model**. In contrast, Equation (9.2) says that surfaces *subtract* light from the scene, so that the more ink or paint pigments are mixed on the surface, the darker the scene becomes, and all the colors mixed together yield black. A color model that explains the color resulting from the mixing of ink or paint is therefore known as a **subtractive color model**. As shown in Figure 9.3, the color wheels for both additive and subtractive color models contain the same colors in the same order. For additive colors, the wheel contains the primary colors of red (■), green (■), and blue (■), with the mixed colors of yellow, cyan, and magenta in between; for subtractive colors, the wheel contains the primary colors of cyan (■), magenta (■), and yellow (■), with the mixed colors of blue, red, and green in between.

An important difference between lights and pigments is that mixing lights is easy to model mathematically using the principle of superposition, as in Equation (9.1). That is, light in the physical world is *linear*, not only in the sense that the spectral colors lie on a line (as mentioned above) but also in the mathematical sense of a linear system obeying the principle of superposition.[†] Pigments and inks, on the other hand, interact with each other in ways that are much more difficult to model mathematically. Subtractive color models, therefore, are highly *nonlinear*, which is why we have used an approximation sign in Equation (9.3) to deter the thought that simply adding the reflectance of one paint to the reflectance of another paint yields the reflectance of the mixture of the two. In this chapter we will focus our attention primarily upon the additive color model of linear light.

9.2 Trichromacy

The human visual system perceives colors in three dimensions, a phenomenon known as **trichromacy** ("three colors"). These dimensions do not refer to the 3D world in which we live, but rather to the three axes that are required of any color space modeling human color perception. In this section we explore both the **physiological** point of view, in which trichromacy is due to the human retina containing three types of cone photoreceptors, as well as the **psychovisual** point of view, where trichromacy is discoverable by color matching experiments using three independent primary colors.

[†] Section 5.1.4 (p. 220)

9.2.1 Spectral Sensitivity Functions (SSFs)

The process of vision begins when light hits photoreceptors in the retina that measure the amount of light incident upon their surfaces. Photoreceptors are not equally sensitive to all wavelengths but rather are more efficient at measuring some wavelengths of light than others. The **spectral sensitivity function (SSF)**, which we shall denote by $s(\lambda)$, captures the relative efficiency of a photoreceptor in detecting light for each visible wavelength λ. Assuming a linear system, the output of the photoreceptor can be viewed as the integration $\int s(\lambda)t(\lambda)\,d\lambda$ over all wavelengths of the incident SPD, denoted by $t(\lambda)$, weighted by the SSF, denoted $s(\lambda)$. Using a discrete approximation, this output becomes an inner product between the vector $\mathbf{s}$ capturing the SSF of the photoreceptor and the vector $\mathbf{t}$ capturing the SPD of the incident light: $\mathbf{s}^\mathsf{T}\mathbf{t}$. Assuming that these are represented by 35-dimensional vectors, as mentioned earlier, then $\mathbf{s}, \mathbf{t} \in \mathbb{R}^{35}$.

Although the human retina contains two types of photoreceptors (rods and cones), color vision refers to the sensory experience that occurs at normal light levels (photopic vision),[†] where the response of the cones dominates, and the response of the rods can thus be safely ignored. The human retina has three types of cones known appropriately as the S-, M-, and L-cones because they respond, respectively, to short-, medium-, and long-wavelength light. The SSFs of these three types of cones have been carefully measured by scientists for the human retina, resulting in the plots shown in Figure 2.2. These SSFs, which are also known as the **cone fundamentals**, are represented as $\mathbf{s}_S$, $\mathbf{s}_M$, and $\mathbf{s}_L$. From the figure, it is clear why it is not entirely appropriate to call these the red-, green-, and blue-cones, since the L-cone SSF peak is very close to that of the M-cone and, in fact, is not near red at all.

Stacking these vectors into a matrix $\mathbf{S}^\mathsf{T} \equiv \begin{bmatrix} \mathbf{s}_L & \mathbf{s}_M & \mathbf{s}_S \end{bmatrix}$ so that $\mathbf{S}$ is 3×35,[‡] the photopic imaging process of the human visual system can be represented mathematically as

$$\mathbf{v} = \mathbf{St} \tag{9.4}$$

where $\mathbf{s}_S, \mathbf{s}_M, \mathbf{s}_L, \mathbf{t} \in \mathbb{R}^{35}$, $\mathbf{t}$ is the SPD incident upon the sensor, and the output $\mathbf{v} \in \mathbb{R}^3$ contains a measure of the number of photons sensed by the three cone types. Thus, the color imaging process of a single location on the retina can be viewed as the projection from a point in an approximately 35-dimensional space to a point in a 3D space. Keep in mind that this equation assumes that the human imaging process is linear, which is true only when the nonlinearities in the photoreceptors and lenses are taken into account.

9.2.2 Color Matching Functions (CMFs)

While the SSFs of the cones explain the low-level *physiological* basis for trichromacy, additional insight can be gained by conducting *psychovisual* experiments. In a **color matching experiment**, a person (called the *observer*) is placed in a dark room facing a wall, as shown in Figure 9.4. A test light with an SPD is shone onto one side of the wall, while onto the other side of the wall is shone a mixture of several different primary lights. The observer views the wall through a small circular hole cut in an opaque barrier, so that the hole subtends a small portion (say 2°) of the observer's field of view; a thin opaque divider extends toward the observer from the wall to separate the left and right sides. As a result, the observer sees a circle with two halves (one side showing the test light, and the other side showing the mixture of primary lights). The observer is asked to control the intensities of the primary lights until both sides of the circle appear identical.

[†] Section 2.1.2 (p. 21)

[‡] Alternatively, $\mathbf{S}$ (as well as $\mathbf{C}$ in the next section) could be defined as 35×3 matrices, but the convention used here simplifies the notation considerably by reducing the number of transposes required.

Based on B. A. Wandell. Foundations of Vision. Sunderland, Mass.: Sinauer Associates, Inc, 1995.

Figure 9.4 In a color matching experiment, a person is asked to specify the intensities of three primary colors such that their combination visually matches a test light, when viewed through a small aperture in a dark room.

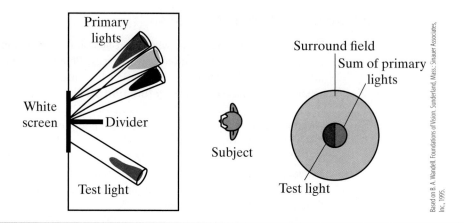

It turns out that for a normal (non-color-blind) human observer, exactly three primary lights (usually red, green, and blue) are needed to match any test light. The intensities of the three lights, known as the **tristimulus values**, control the power emitted by them. (It may be helpful to imagine the observer controlling the intensities of the primary lights by turning three knobs, and the tristimulus values are the positions of the knobs.) Mathematically, the relationship is given by

$$\mathbf{t} \cong u_r\mathbf{p}_r + u_g\mathbf{p}_g + u_b\mathbf{p}_b \tag{9.5}$$

where $\mathbf{p}_i \in \mathbb{R}^{35}$ is the SPD of the ith primary light, $i \in \{r, g, b\}$, u_i is the ith intensity, and the symbol $\cong$ means that the two SPDs on either side look identical. In other words, due to the loss of dimensionality in the sensation of color, a given triplet of tristimulus values could have arisen from an infinite number of possible SPDs; two SPDs $\mathbf{t}$ and $\mathbf{t}'$ are called **metamers** if they themselves are not identical but nevertheless look identical (i.e., their tristimulus values are identical), meaning $\mathbf{t} \neq \mathbf{t}'$ but $\mathbf{t} \cong \mathbf{t}'$.

Two caveats are necessary for Equation (9.5) to be true. First, the primary lights must be independent of each other, meaning that none of the lights can be matched by a mixture of the other primary lights. That is,

$$\mathbf{p}_i \not\cong \alpha\mathbf{p}_j + \beta\mathbf{p}_k \tag{9.6}$$

for any combination of i,j,k and for all scalars α and β. Secondly, negative weights must be allowed, which is achieved in the experiment by moving the corresponding primary light(s) to the same side as the test light, for example,

$$\mathbf{t} + u_r\mathbf{p}_r \cong u_g\mathbf{p}_g + u_b\mathbf{p}_b \tag{9.7}$$

For any physically realizable set of primaries, negative values must be allowed in order to match all possible test lights. This is an important fact whose implications will be explored later.

If the color matching experiment is conducted using a series of monochromatic test lights, then the **color matching functions (CMFs)** of a given set of primary lights can be measured. A *monochromatic light* is a light that shines a single wavelength, that is, a pure spectral color. The tristimulus values from the series of experiments can be stacked into a 3×35 matrix $\mathbf{C}$, with each column of the matrix holding the tristimulus values for a different wavelength, and with the rows of the matrix containing the CMFs for the primary lights. Notice from the CMFs shown in Figure 9.5 using typical red, green, and blue primary lights that some values are negative: red is negative from 445 to 525 nm, green is negative from 390 to 440 nm, and blue is negative from 530 to 640 nm. Also shown in the figure are the cone SSFs after applying a 3×3 linear transform, to show the close agreement of the two concepts. Such linear transforms will be explored in more detail in the next section.

Figure 9.5 Color matching functions (solid lines) using a particular set of RGB primaries. The overlaid circles are obtained by a linear 3×3 transform from the cone SSFs shown in Figure 2.2. The transformed cone SSFs closely agree with the CMFs.

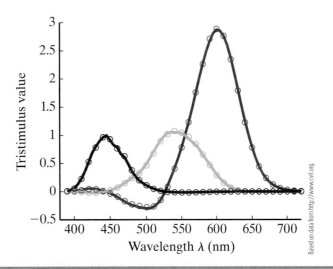

Based on data from http://www.cvrl.org.

Even though SSFs and CMFs are both represented by 3×35 matrices, it is important to keep the distinction clearly in mind: an SSF is obtained by *physiological* measurements of the cones (or *physical* measurements of a camera), while a CMF is obtained by *psychovisual* measurements of a human observer. All the values in an SSF are nonnegative, whereas any set of physically realizable CMFs must contain some negative values. Nevertheless, despite their distinct origins and properties, SSFs and CMFs serve a similar purpose in that they both convert an SPD incident on the sensor to three values that numerically represent the color being viewed. The overlay in Figure 9.5 emphasizes this connection.

9.2.3 Grassmann's Law

Just as physical light obeys the principle of superposition, so do the tristimulus values. More specifically, if $\mathbf{u} = \begin{bmatrix} u_r & u_g & u_b \end{bmatrix}^{\mathsf{T}}$ are the weights of the three primaries needed to match a test light with SPD $\mathbf{t}$ and if the weights $\mathbf{u}' = \begin{bmatrix} u_r' & u_g' & u_b' \end{bmatrix}^{\mathsf{T}}$ match another test light $\mathbf{t}'$, then the test light $\alpha \mathbf{t} + \beta \mathbf{t}'$ formed by a weighted combination of the two individual test lights is matched by the same weighted combination of the tristimulus values: $\alpha \mathbf{u} + \beta \mathbf{u}'$, where α and β are scalars. This result, known as **Grassmann's Law**,[†] can be restated as

$$\alpha \mathbf{t} + \beta \mathbf{t}' \;\cong\; (\alpha u_r + \beta u_r')\mathbf{p}_r + (\alpha u_g + \beta u_g')\mathbf{p}_g + (\alpha u_b + \beta u_b')\mathbf{p}_b \tag{9.8}$$

whenever

$$\mathbf{t} \;\cong\; u_r \mathbf{p}_r + u_g \mathbf{p}_g + u_b \mathbf{p}_b \tag{9.9}$$

$$\mathbf{t}' \;\cong\; u_r' \mathbf{p}_r + u_g' \mathbf{p}_g + u_b' \mathbf{p}_b \tag{9.10}$$

The implication of Grassmann's Law is that color matching is linear:

$$\mathbf{u} = \mathbf{C}\mathbf{t} \tag{9.11}$$

where $\mathbf{t} \in \mathbb{R}^{35}$ is the SPD of the test light, $\mathbf{C}$ is the same 3×35 matrix whose rows are the CMFs of the three primaries, and $\mathbf{u} \in \mathbb{R}^{3}$ are the tristimulus values. This equation is easy to see by substituting a monochromatic light so that $\mathbf{t}$ contains zeros everywhere except for a single element, which causes $\mathbf{u}$ to select a column from $\mathbf{C}$; by linearity any test light can

---

[†] Hermann Grassmann (1809–1877) was a German linguist and mathematician.

be written as a linear combination of scaled monochromatic lights. As with the SSFs, keep in mind that the linearity of the CMFs assumes that the viewing conditions are carefully controlled so that we can safely ignore the overall luminance level, chromatic adaptation, viewing angle, or surrounding background. Also note that Grassmann's Law does not hold in very bright or very dim conditions, in which color matching is no longer linear.

For a given set of primaries with CMFs $\mathbf{C}$, along with two test lights $\mathbf{t}$ and $\mathbf{t}'$, the tristimulus values of the test lights are given by $\mathbf{u} = \mathbf{Ct}$ and $\mathbf{u}' = \mathbf{Ct}'$. Further, if $\mathbf{t}$ and $\mathbf{t}'$ are metamers, then their tristimulus values are identical, leading to

$$\mathbf{u} = \mathbf{Ct} = \mathbf{Ct}' = \mathbf{u}' \qquad (9.12)$$

As we shall see, the linearity of color matching implied by Grassmann's Law means that two SPDs that are metamers for any set of independent primaries are metamers for *all* sets of independent primaries, since any pair of CMFs is related by a 3×3 linear transform. That is, the concept of metamer transcends the choice of primaries, and if $\mathbf{t}$ and $\mathbf{t}'$ are metamers then Equation (9.12) holds for any set of CMFs $\mathbf{C}$.

We now have the mathematical machinery necessary to compute the 3×3 linear approximation of the CMFs overlaid in Figure 9.5. Combining Equations (9.4) and (9.11) yields an expression for the tristimulus values as a linear transformation of the sensor values: $\mathbf{u} \approx \mathbf{CS}^+\mathbf{v}$, where $\mathbf{S}^+$ is the Moore-Penrose pseudoinverse[†] of $\mathbf{S}$. Applying this result to all spectral colors then yields $\mathbf{C} \approx (\mathbf{CS}^+)\mathbf{S}$, which reveals the 3×3 transformation as $\mathbf{CS}^+$. Note that the close agreement between the measured CMFs and the estimated CMFs is by no means guaranteed but arises only because of the way in which photons absorbed by the cones are translated into perceptual signals in the brain. In addition, the close agreement is because the SSFs shown in Figure 2.2 were carefully constructed to ensure that important details were taken into account, such as the spectral absorption of the cornea and the inert pigments in the eye.

9.2.4 Luminous Efficiency Function (LEF)

A variation of the color matching experiment is to ask the observer whether one color is brighter or darker than another. Such an experiment is difficult to perform with arbitrary colors—imagine trying to determine whether red is darker than blue. Therefore, the experiment is conducted by using a pair of spectral colors with nearby wavelengths and recording the relative power of the two lights. By repeating this experiment for the full range of visible wavelengths, a **luminous efficiency function (LEF)** is obtained. The **photopic LEF** corresponds to normal light levels where the cones dominate due to the saturation of the rods, whereas the **scotopic LEF** corresponds to low light levels where the rods dominate due to the lack of sensitivity of the cones.

The two functions are plotted in Figure 9.6. Not surprisingly, the scotopic LEF closely matches the rod SSF (not shown), and the photopic LEF can be well approximated as a weighted combination of the cone SSFs: $\mathbf{s} \approx \mathbf{S}^\mathsf{T}\mathbf{w}$, where $\mathbf{s} \in \mathbb{R}^{35}$ is the photopic LEF, the columns of $\mathbf{S}^\mathsf{T}$ are the cone SSFs, and $\mathbf{w} \in \mathbb{R}^3$ are three weights determined empirically. Also shown in the figure for comparison is the second tristimulus value of the CIE 1931 *XYZ* space, which is described in more detail in the next section.

9.2.5 Psychological Primaries

Before continuing further, it might be helpful to first settle a potentially nagging doubt. That is, we have seen from the principles of trichromacy that there are three primary colors, but what are these colors, and why? In grade school, most of us learn from mixing paints that the three (subtractive) primary colors are red (■), blue (■), and yellow ( ), and that green

---

[†] The pseudoinverse is covered in Section 11.1.5 (p. 520).

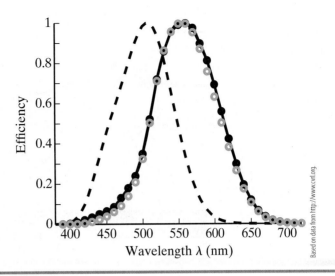

Figure 9.6 Photopic (solid black) and scotopic (dashed black) luminous efficiency functions (LEFs), from Figure 2.9. Close approximations to the photopic LEF are given by a linearly weighted combination of the cone SSFs (black circles), and the *Y* channel from the CIE 1931 *XYZ* space (green circles).

Based on data from http://www.cvrl.org.

is a mixture of blue and yellow (■ = ■ + ■). Then, as we grow older, we learn that the three (additive) primary colors, which are used in cameras, televisions, and computer monitors, are actually red (■), green (■), and blue (■), and that yellow is a mixture of green and red (■ = ■ + ■). This seems quite odd indeed, since red and blue show up in both cases, but yellow and green are different. Moreover, although green *almost* looks like a combination of blue and yellow, yellow in no way looks like a mixture of green and red. Some comfort is supposed to be gained, perhaps, by noticing that the three subtractive primary colors used in printer inks are cyan (■), magenta (■), and yellow (■), which appear something like our grade-school primaries. But cyan and magenta do not appear to be pure colors at all, but rather as diluted versions of the pure colors blue and red.

How do we make sense of these apparent discrepancies? Well, although there is no disputing that the **trichromatic theory** of color vision (also known as the **Young-Helmholtz theory**[†]) is enormously powerful—most of the material in this chapter is based on it—nevertheless, it does not explain all the known phenomena of color vision. For example, there are colors that appear to be both red and blue (the purples ■), both red and yellow (the oranges ■), both blue and green (the variations of teal ■), or both yellow and green (the variations of chartreuse ■). But psychologically it is impossible to perceive a yellowish blue or a bluish yellow (■ ■)—such a color inevitably adopts a greenish tint, since blue and yellow make green (subtractively). In the same way, we cannot perceive a combination of red and green (■ ■), which inevitably adopts a yellowish tint, since red and green make yellow (additively). In other words, it makes sense to talk about a "reddish blue" (purple ■) or a "greenish yellow" (chartreuse ■) or a "reddish yellow" (orange ■) or a "greenish blue" (teal ■) but not a "reddish green" nor a "bluish yellow"—the latter are known as **impossible colors**.

To explain these phenomena, a scientist named Hering[‡] proposed the **opponent process theory**, which states that the human visual system senses the differences between colors along three axes illustrated in Figure 9.7. For more than half a century a heated debate raged among the advocates of the trichromatic theory and the proponents of the opponent process

---

[†] Thomas Young (1773–1829) proposed the existence of three types of photoreceptors in 1802, while Hermann von Helmholtz (1821–1894) further developed the theory in 1850.

[‡] Ewald Hering (1834–1918) proposed the opponent process theory in 1892.

Figure 9.7 The opponent color process computes differences between L-, M-, and S-cone outputs, producing values along red-green, blue-yellow, and black/white axes.

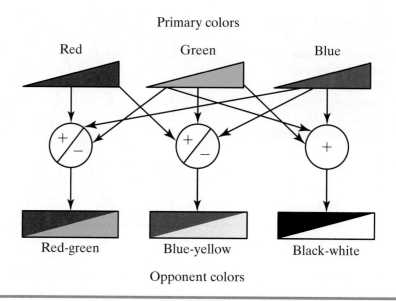

theory. In the end the debate was resolved by experimentally verifying that, in fact, both theories are true: photoreceptors in the retina measure light according to three components, while cells later in the visual pipeline transform the component color signals into opponent color signals that capture the relative amounts of red/green, blue/yellow, and black/white.[†] This explains why both theories are successful at explaining various aspects of our psychological color experience.

In other words, while there are three primary colors at the lowest level (either red, green, and blue for additive colors or cyan, magenta, and yellow for subtractive colors), there are in fact six **psychological primaries**, as shown in Figure 9.8. These colors are red (■), green (■), blue (■), yellow (□), black (■), and white (□) In this context we are not referring to primaries as SPDs that are mixed in a physical imaging system, but rather the colors that appear to human viewers to be pure. For example, even if we are intellectually convinced that white light is obtained by mixing all the colors of the rainbow, the color white still appears to be pure. Similarly, even though we know that (in theory at least) black can

Figure 9.8 The six psychological primaries: red, green, blue, yellow, black, and white.

[†] Gunnar Svaetichin (1915–1981) confirmed the existence of the three photoreceptors in the retina in 1956, while the husband-and-wife-team of Leo Hurvich (1910–2009) and Dorothea Jameson (1920–1998) confirmed the existence of opponent cells later in the visual pipeline in 1957.

be obtained by mixing all the different colored paint pigments, the color black still appears to be pure. In fact, this is the reason that printers include four inks (adding black to the three subtractive primaries), because it is very difficult to mix the primaries in such a way that pure black is achieved. And even when we know from science that green (■) can be obtained by mixing blue (■) and yellow ( ) paints, green still *appears* to us to be pure in some sense; and no matter how much evidence is presented that yellow ( ) light results from combining red (■) and green (■) lights, it is impossible to see it as anything other than a pure color with no relationship to the two colors of which it is composed.

9.3 Designating Colors

Over the years, color scientists have proposed a number of different ways to precisely designate colors to facilitate unambiguous communication. The goal of these models is to provide, as much as is possible in the face of color blindness and chromatic adaptation, a precise mapping from the limited number of color names to the millions of perceptible colors. In this section we consider several of these models, culminating in the ubiquitous CIE *XYZ* color space.

9.3.1 Hue, Chroma, and Lightness

Although red, green, and blue values are the natural representation for cameras and displays, such values are awkward for describing colors. Not only is it nonintuitive to describe a color by the hex value FFFF00 rather than the name "yellow", such values are also device-dependent because they are affected by the primaries being used as well as the gamma correction involved. A more natural way to describe colors is by their hue, chroma, and lightness. The **hue** is the dominant perceived color, that is, the similarity of the color to one of the so-called **unique hues**, for example, the non-gray psychological primaries (red, green, blue, and yellow). The **chroma** is the purity of the color, that is, the degree of difference between the color and neutral gray; it is closely related to **saturation**.[†] The **lightness** (or **value**) is the nonlinearly transformed luminance which captures, in some sense, the brightness of the color. In Figure 9.3, for any given color in the circle, hue can be thought of as the angle around the circle, chroma is the distance from the circle center, and lightness is the height out of the page (with the black-white axis perpendicular to the circle). For example, a bright pure blue would have a blue hue (obviously), a high chroma, and a high lightness. A dark orange would have a hue somewhere between yellow and red, a high chroma, and a low lightness, and a bright pink would have the hue of red, a low chroma (because it is mixed with white), and a high lightness. In the following subsections we consider various ways to quantify these phenomena.

9.3.2 Munsell Color System

The earliest attempt to quantify colors was the **Munsell color system**,[‡] first published in 1905 and still in use today, although it has largely been superseded by the CIE system described later. The color of an object is established by comparing it to a large number of samples in the *Munsell Book of Color*, taking care to ensure that the proper illuminant is

---

[†] *Saturation* is the chroma relative to the maximum possible chroma for the given lightness. Saturation always ranges from 0 to 1 (0% to 100%), whereas the range of chroma is dependent upon the particular lightness and hue. The subjective perception of color, analogous to *brightness*, is known as **colorfulness**.

[‡] Alfred H. Munsell (1858–1918) was an American painter and art teacher whose motivation was to teach colors to children. He established the Munsell Color Company to continue the work after his death.

used and the surrounding conditions are controlled. Colors are named according to their hue, value, and chroma. Hue is divided into ten steps, and each step is subdivided into ten sub-steps, for a total of 100 integer hues. The ten steps include the five principal hues (red, yellow, green, blue, and purple) and five intermediate hues between them. The hue is given as the letter(s) identifying the step, and the number identifying the sub-step, with 5 indicating the middle of the step. After the hue comes the value, which ranges from 0 (darkest) to 10 (lightest), and the chroma, which ranges from 0 to about 20 although there is no theoretical upper limit. A color is then specified by listing the hue, value, and chroma in that order. For example, a bright orange used for safety vests is 5YR 6/15, where 5YR means the middle (5) of the yellow-red (YR) hue (that is, halfway between the yellow and red hues), 6/ means medium lightness, and 15 means highly saturated. The enduring nature of the Munsell color system is due to the careful experimentation conducted in devising it.

9.3.3 Natural Color System (NCS)

An alternative approach, the **Natural Color System (NCS)**, was developed in Sweden in the 1930s and is still used in several European nations. Unlike the Munsell and CIE systems, the NCS emphasizes the logical description of color experience and is based on Hering's color opponency model. The NCS system visualizes colors as lying within a double square-shaped pyramid whose vertices are the six psychological primaries: the top and bottom of the pyramid represent white and black, while the four cardinal directions represent red, green, blue, and yellow. A color is specified by its **blackness** (how dark the color is), **chromaticness** (how chromatically strong the color is), and a percentage between two adjacent color hues. For example, a certain yellow is NCS 0580-Y10R, meaning 5% darkness, 80% chromaticness, and a mixture that is 90% yellow and 10% red. A color's **whiteness** is defined so that the whiteness, blackness, and chromaticness add to 100%, which in this case leads to a whiteness of $100 - 5 - 80 = 15\%$.

9.3.4 ISCC-NBS System

Also developed in the 1930s and also emphasizing the logical names of colors is the **ISCC-NBS System of Color Designation**. The ISCC-NBS system is based on 13 basic color categories, which include 10 hue names (such as pink, red, orange, and brown) and three neutral colors (black, white, and gray). Intermediate categories are obtained by combining these names, such as "reddish orange" and "purplish pink". Categories are further subdivided through the use of modifiers, such as "vivid", "brilliant", or "dark". The publication *Color: Universal Language and Dictionary of Names* contains both the language and the dictionary for describing hundreds of colors referenced to other systems such as Munsell's.

9.3.5 CIE Chromaticity Diagram

Despite the benefits and uses of the previously described systems, by far the most influential and widely used model for quantifying color is the **CIE 1931 *XYZ* color space**, along with its companion the **CIE 1964 *XYZ* color space**. These models were adopted by the International Commission on Illumination, or **Commission Internationale de l'Éclairage (CIE)**, in the years 1931 and 1964, respectively. The CIE 1931 *XYZ* space is based upon color matching experiments conducted in the 1920s by Wright and Guild in which the observer's field of view was blocked so that only the inner 2° of the retina was used. The CIE 1964 *XYZ* space, on the other hand, is based upon experiments conducted in the 1950s by Stiles and Burch, as well as Speranskaya, using a 10° field of view. Therefore the resulting CMFs are sometimes known as the **CIE 1931 2° Standard Colorimetric Observer** and **CIE 1964**

10° Supplementary Standard Colorimetric Observer, respectively. Although the 1964 standard is recommended in situations involving a larger field of view, we will focus primarily upon the 1931 standard since it is still the most widely used. In fact, if you ever come across the term CIE *XYZ* without any qualifier, it is probably referring to CIE 1931 *XYZ*.

As we shall see in the next section, because of Grassmann's Law all CMFs are related to one another via 3 × 3 linear transforms. Each CIE *XYZ* system is therefore completely specified by a 3 × 3 transform matrix from *RGB* values to *XYZ* values, along with the primaries used to define *RGB* in each case.[†] These matrices, which we shall present later, allow us to transform the CMFs obtained in *RGB* space to the CMFs corresponding to *XYZ* space, the results of which are shown in Figure 9.9. Two characteristics of these CMFs are worth noting. First, each space was designed so that the middle coordinate *Y* approximates the photopic luminous efficiency function (LEF)—recall Figure 9.6. Secondly, all values are non-negative everywhere. Of course, this non-negativity means that the primaries corresponding to the *XYZ* space are not physically realizable, but it makes the *XYZ* space particularly convenient to work with mathematically.

To separate the luminance from the chrominance, the **normalized coordinates** are obtained by dividing by the sum of all three *XYZ* values:

$$x \equiv \frac{X}{X + Y + Z} \tag{9.13}$$

$$y \equiv \frac{Y}{X + Y + Z} \tag{9.14}$$

$$z \equiv \frac{Z}{X + Y + Z} \tag{9.15}$$

Since $x + y + z = 1$, these three normalized values are not independent, and therefore only two are needed. Discarding z yields the **CIE chromaticity coordinates** (x,y).

Plotting the CIE chromaticity coordinates for all visible wavelengths yields the tongue-shaped curve in Figure 9.10. This plot is known as the **CIE chromaticity diagram**, and it is probably the single most important visualization tool for understanding the quantification of human color sensation. Given any color in a scene, the tristimulus values can be measured using the CIE 1931 RGB primaries, then the appropriate 3 × 3 transformation can be

Figure 9.9 Color matching functions (CMFs) for CIE 1931 *XYZ* (left) and CIE 1964 *XYZ* (right). Shown are the CMFs for *X* (red), *Y* (green), and *Z* (blue). The black outline for the *Y* CMF emphasizes that it is designed to approximate the luminous efficiency function (LEF).

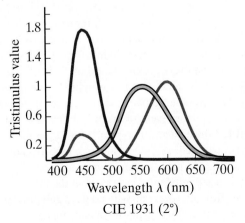

CIE 1931 (2°)

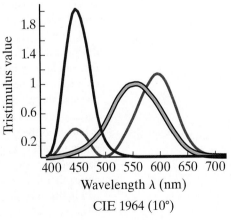

CIE 1964 (10°)

[†] The primaries used in the experiments for the 1931 standard consist of the wavelengths 700 nm (red), 546.1 nm (green), and 435.8 nm (blue); for the 1964 standard they are 645.2 nm (red), 526.3 nm (green), and 444.4 nm (blue).

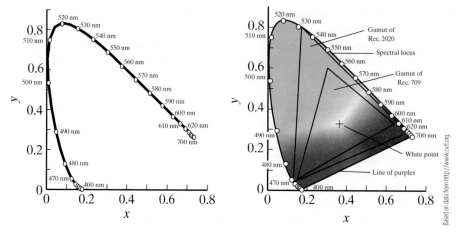

Based on data from http://www.cvrl.org.

Figure 9.10 CIE 1931 chromaticity diagram. LEFT: The strict diagram showing only the (*x,y*) chromaticity coordinates of the pure wavelengths. RIGHT: The diagram with approximate colors filled in, and with different parts of the diagram labeled, for better visualization.

applied to yield the *XYZ* coordinates, which are then normalized to obtain the (*x,y*) chromaticity coordinates, which are then plotted on the diagram. In this way, every possible color can be mapped to a point on the diagram. Note that we do not have to run an actual color matching experiment to plot a color on the diagram, because all we need is the SPD of the color. That is, if we let $\tilde{C}$ be the CMFs for CIE *XYZ* space, which are readily available, then the *XYZ* coordinates are obtained simply as $\begin{bmatrix} X & Y & Z \end{bmatrix}^\mathsf{T} = \tilde{C}\mathbf{t}$, where **t** is the SPD of the color; normalization yields the coordinates (*x,y*).

The chromaticity coordinates (*x,y*) are the basis for all modern approaches to **colorimetry**, the scientific measurement of color. Strictly speaking, the CIE chromaticity diagram consists only of the (*x,y*) coordinates of colors, not the colors themselves, and it is therefore misleading to draw the diagram with colors filled in. Nevertheless, for visualization purposes it is often helpful to color the diagram with approximate colors, as is done in the figure. The diagram clearly reveals several facts about color perception and reproduction. First, the diagram closely resembles the color wheel, where the **spectral locus** traces the chromaticity coordinates of the pure wavelengths, and the **line of purples** connects red and violet: all visible colors lie within this tongue-shaped region. Note that colors along the line of purples are nonspectral colors—magenta, for example, is not in the rainbow.

Secondly, the diagram clearly reveals the colors that can be displayed by any given device. Most displays contain three primary lights (some flavor of red, green, and blue), and these lights can be plotted on the diagram—based on their SPDs—to form a triangle. Any color that can be displayed by the device is a weighted combination of these three lights; because the weights must be nonnegative, such colors must lie within the triangle whose vertices are the chromaticity coordinates of the three lights. The set of colors in this triangle is known as the **gamut** of the device, and typically the larger the gamut the better. It is easy to see from the diagram why red, green, and blue are usually chosen as the primary colors, because they yield the triangle with the largest area and therefore the largest gamut. Also notice that Rec. 2020 has a significantly larger gamut than that of Rec. 709.

Finally, it is obvious from this diagram why no physically realizable device can produce all colors, because some colors will always lie outside this triangle (or outside the polygon defined by a finite number of primary lights, whose gamut is given by the convex hull of their coordinates). Keep in mind that, although the diagram is applicable to subtractive as well as to additive color spaces, the gamut of any subtractive device (e.g., an ink-based printer) will not be a triangle but rather an arbitrarily-shaped region due to the nonlinearities involved with paints and pigments.

The **white point** is the chromaticity of the color used as the reference white. The color of a light source, or **illuminant**, is characterized by its relative SPD—that is, by its SPD after normalizing by some arbitrary amount since the overall intensity of the light is irrelevant. The white point of an illuminant is given by the chromaticity coordinates of a white object viewed under the illuminant. Light sources are usually modeled as blackbody radiators, in which case their chromaticity coordinates are uniquely determined by their absolute temperature.[†] The **blackbody locus** is the path traced in (x, y) coordinates as the temperature of the blackbody source is raised. In 1963 the CIE adopted the "D" series of standard illuminants, in which the subscript indicates the first two digits of the illuminant's **correlated color temperature (CCT)**. While the choice of white point is application specific, by far the most common is D_{65}, which represents average daylight with a CCT of 6504 K and has chromaticity coordinates $(x, y) = (0.3127, 0.3290)$. For indoor lighting, sometimes a white point corresponding to tungsten incandescent bulbs or fluorescent lamps, or increasingly LEDs, is used. The graphic arts industry uses D_{50}, which has a CCT of 5003 K and chromaticity coordinates $(x, y) = (0.3457, 0.3585)$. Another definition of the white point uses the illuminant that contains equal energy among all wavelengths, called **CIE Standard Illuminant E**, or **equal energy white**. Since the CIE CMFs are designed to have equal area under each curve, the chromaticity coordinates of Illuminant E are $(x, y) = (1/3, 1/3)$. Alternatively, for a given set of primaries we could define the white point determined by setting all three primaries to output equal power, which is found at the centroid of the triangle defining the gamut.

The chromaticity diagram also reveals why hue, chroma, and lightness are a natural way to verbally describe a color. Given any color's (x, y) chromaticity coordinates, the hue of the color can be determined by drawing a line from the color through the white point, then finding the intersection of this line with the spectral locus. For this reason, the hue is sometimes called the "dominant wavelength", although for some colors the line does not intersect the spectral locus but rather the line of purples, in which case there is not an actual wavelength. The chroma of a color is related to the distance to the white point, so that colors on the spectral locus (or line of purples) are fully saturated, while the white point is fully unsaturated. The lightness is a third coordinate coming out of the page. Keep in mind, however, that the mapping described here should be considered as informal, since the xy space is not perceptually uniform, a point we consider in more detail later.

9.4 Linear Color Transformations

In this section we show how Grassmann's Law leads to a 3×3 linear transformation between any two CMFs or between any two linear color spaces defined by tristimulus values for some set of primaries, whether the primaries are real or or non-physically realizable. Using this result, we describe the procedure for calibrating between cameras and displays, including the important step of white balancing.

9.4.1 Transforming Between CMFs

We have seen that color matching functions (CMFs) depend upon the primaries chosen. That is, two different sets of primaries will lead to two different sets of CMFs. Because CMFs are linear functions (due to Grassmann's Law), it should not be surprising that the CMFs

---

[†] Section 2.5.3 (p. 60)

for two sets of primaries are related by a 3×3 linear transformation. More specifically, the transformation between color matching functions $\mathbf{C}_{\{3 \times 35\}}$ and $\mathbf{C}'_{\{3 \times 35\}}$ corresponding to primaries $\mathbf{P}_{\{35 \times 3\}} = \begin{bmatrix} \mathbf{p}_r & \mathbf{p}_g & \mathbf{p}_b \end{bmatrix}$ and $\mathbf{P}'_{\{35 \times 3\}} = \begin{bmatrix} \mathbf{p}'_r & \mathbf{p}'_g & \mathbf{p}'_b \end{bmatrix}$, respectively, is given by the 3×3 matrix $\mathbf{M} \equiv \mathbf{C}'\mathbf{P}$, which is composed of one set of CMFs and the other set of primaries:

$$\mathbf{C}' = \underbrace{\mathbf{C}'\mathbf{P}}_{\mathbf{M}}\, \mathbf{C} \tag{9.16}$$

To see this result, let $\mathbf{t} \in \mathbb{R}^{35}$ be the SPD of a test light. Suppose the tristimulus values for matching $\mathbf{t}$ using the primaries $\mathbf{P}$ are given by $\mathbf{u} \in \mathbb{R}^3$. Similarly, if we let $\mathbf{u}'$ be the tristimulus values for $\mathbf{P}'$ to match the same test light, then from Equation (9.11) we have

$$\mathbf{u} = \mathbf{Ct} \tag{9.17}$$
$$\mathbf{u}' = \mathbf{C}'\mathbf{t} \tag{9.18}$$

When the intensities of the first primaries are set to the tristimulus values specified by $\mathbf{u}$, the resulting SPD is $\mathbf{Pu}$; and when the intensities of the second primaries are set to the tristimulus values specified by $\mathbf{u}'$, the resulting SPD is $\mathbf{P}'\mathbf{u}'$. In both cases, the resulting SPD looks identical to the test light. In other words, $\mathbf{t}$, $\mathbf{Pu}$, and $\mathbf{P}'\mathbf{u}'$ are metamers, which from Equation (9.12) implies

$$\mathbf{Ct} = \mathbf{CPu} = \mathbf{CP}'\mathbf{u}' \tag{9.19}$$
$$\mathbf{C}'\mathbf{t} = \mathbf{C}'\mathbf{Pu} = \mathbf{C}'\mathbf{P}'\mathbf{u}' \tag{9.20}$$

where we note that metamers are metamers no matter which CMF is used. By substituting Equation (9.18) into Equation (9.20), we get

$$\mathbf{u}' = \mathbf{C}'\mathbf{t} = \underbrace{\mathbf{C}'\mathbf{P}}_{\mathbf{M}}\, \mathbf{u} \tag{9.21}$$

which reveals that $\mathbf{M} \equiv \mathbf{C}'\mathbf{P}$ is the 3×3 matrix relating the tristimulus values in the two color spaces. Similarly, substituting Equation (9.17) into Equation (9.20) yields

$$\mathbf{C}'\mathbf{t} = \mathbf{C}'\mathbf{Pu} = \underbrace{\mathbf{C}'\mathbf{P}}_{\mathbf{M}}\, \mathbf{Ct} \tag{9.22}$$

Since this equation is true for any test light $\mathbf{t}$, we have the desired result: $\mathbf{C}' = (\mathbf{C}'\mathbf{P})\mathbf{C} = \mathbf{MC}$. Note that although we did not use Equation (9.19) in the derivation, the result reveals that Equations (9.19) and (9.20) are related by a simple multiplication by $\mathbf{M}$.

9.4.2 Transforming between Cameras and Displays

When a color camera senses the incoming light, each pixel yields values for the red, green, and blue channels. Each of these values can be thought of as tristimulus values in some color space. When a display device, such as a computer monitor, produces light to be seen, it does so by producing for each pixel a linear combination of the red, green, and blue primaries according to three input values. If the three values captured by the camera were the same three values needed to drive the display, then there would be no need for further discussion. However, cameras and displays typically operate in different color spaces, so to ensure faithful reproduction it is necessary to transform the tristimulus values in the camera color space to three values in the display color space.

More precisely, suppose we have a camera that captures RGB values according to

$$\mathbf{v}_c = \mathbf{St} \tag{9.23}$$

where $\mathbf{S}_{\{3\times 35\}}$ contains the SSFs of the pixels, analogous to the sensing of the human cones considered earlier, and $\mathbf{t} \in \mathbb{R}^{35}$ is a test light. Now suppose we have a display with primaries $\mathbf{P}_{\{35\times 3\}}$ so that when the display is driven by the values $\mathbf{v}_d$ the resulting SPD is $\mathbf{P}\mathbf{v}_d$. We seek, if possible, the 3×3 linear transform $\mathbf{M}_{cd}$ that maps incoming tristimulus values to output values: $\mathbf{v}_d = \mathbf{M}_{cd}\mathbf{v}_c$.

For accurate color reproduction, we want the output $\mathbf{P}\mathbf{v}_d$ of the display and the original SPD $\mathbf{t}$ to be metamers:

$$\mathbf{Ct} = \mathbf{CPv}_d = \mathbf{CPM}_{cd}\mathbf{v}_c = \mathbf{CPM}_{cd}\mathbf{St} \tag{9.24}$$

where $\mathbf{C}_{\{3\times 35\}}$ contains the CMFs corresponding to the display primaries. Since this relationship must be true for all test lights $\mathbf{t}$, it means that

$$\mathbf{C} = \underbrace{\mathbf{CP}}_{3\times 3} \mathbf{M}_{cd}\mathbf{S} \tag{9.25}$$

or, since $\mathbf{CP}$ is invertible,

$$\mathbf{M}_{cd}\mathbf{S} = (\mathbf{CP})^{-1}\mathbf{C} \tag{9.26}$$

This is a standard linear equation in the unknown transform $\mathbf{M}_{cd}$ given the SSFs $\mathbf{S}$ of the camera, the primaries $\mathbf{P}$ of the display, and the CMFs $\mathbf{C}$ corresponding to the primaries.

One important point to note about Equation (9.26) is that there is no guarantee that for an arbitrary 3×35 matrix $\mathbf{S}$ and an arbitrary 3×35 matrix $(\mathbf{CP})^{-1}\mathbf{C}$ on the right-hand side, there exists a 3×3 linear transform between them. That is, there is no inherent reason why the three values sensed by an arbitrary camera should bear any relationship whatsoever to the tristimulus values of any color space. Nevertheless, cameras are generally designed so that the SSFs of the three color filters in front of the pixels are similar enough to the SSFs of the human cones that the three sensed values can approximately be related to the tristimulus values using a linear transform. This requirement that $\mathbf{S} = \mathbf{MC}$ for some $\mathbf{M}_{\{3\times 3\}}$ and some color space $\mathbf{C}$ is known as the **Luther condition**.[†] Note that $\mathbf{C}$ here could be any color space, since all CMFs are known to be related by a linear transform according to Grassmann's Law; or equivalently we could require $\mathbf{S} = \mathbf{M\tilde{S}}$ (yielding a different $\mathbf{M}$, of course), where $\mathbf{\tilde{S}}$ are the cone SSFs, since if measured properly the cone SSFs are also related to the CMFs via a linear transform as we saw in Figure 9.5.

In practice, it is quite tedious and error-prone to measure $\mathbf{S}$, $\mathbf{P}$, and $\mathbf{C}$, so that solving Equation (9.26) is not the usual way of obtaining $\mathbf{M}_{cd}$. Instead, the camera and display are separately calibrated to find the linear transforms to CIE *XYZ*. To calibrate a color camera, pictures are taken of objects with known CIE *XYZ* coordinates under a standard illuminant. The *RGB* values are captured by the camera, and a linear fit is performed to the data to find the best 3×3 camera transform $\mathbf{M}_c$:

$$\begin{bmatrix} X_1 & X_2 & & X_n \\ Y_1 & Y_2 & \cdots & Y_n \\ Z_1 & Z_2 & & Z_n \end{bmatrix} = \mathbf{M}_c \begin{bmatrix} R_1 & R_2 & & R_n \\ G_1 & G_2 & \cdots & G_n \\ B_1 & B_2 & & B_n \end{bmatrix} \tag{9.27}$$

where n is the number of measurements. In industry, a widely used calibration target is the **Macbeth ColorChecker** color-rendition chart, which consists of 24 square patches laid out in a 4×6 array, shown in Figure 9.11. Each patch has a surface reflectance function similar

---

[†] Robert Luther (1868–1945) was a German chemist who died in Dresden just months after the controversial Allied bombing of that city at the end of World War II. The Luther condition is also known as the *Maxwell-Ives criterion*.

Figure 9.11 Macbeth ColorChecker
color-rendition chart.

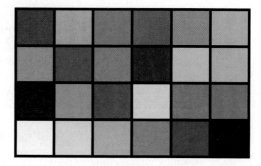

to a naturally occurring surface, such as human skin, foliage, blue sky, chicory flower, and so forth, as well as the additive and subtractive primary colors and a range of grays. Once the camera has been calibrated, the sensed *RGB* values can be converted to *XYZ* using

$$\begin{bmatrix} X \\ Y \\ Z \end{bmatrix} = \mathbf{M}_c \begin{bmatrix} R_c \\ G_c \\ B_c \end{bmatrix} \tag{9.28}$$

where R_c, G_c, and B_c are the three raw color values obtained by the camera at a given pixel. Recall from what we saw before that the matrix is composed as $\mathbf{M}_c = \tilde{\mathbf{C}}\mathbf{S}^+$, where $\tilde{\mathbf{C}}$ contains the CMFs for the *XYZ* space (Figure 9.5), and $\mathbf{S}$ contains the SSFs of the three color filters of the camera.

Similarly, if we let R_d, G_d, and B_d be the three values sent to the color channels of the display (e.g., a computer monitor), then the *XYZ* coordinates being displayed are given by a different 3×3 display transform $\mathbf{M}_d$:

$$\begin{bmatrix} X \\ Y \\ Z \end{bmatrix} = \mathbf{M}_d \begin{bmatrix} R_d \\ G_d \\ B_d \end{bmatrix} \tag{9.29}$$

To calibrate the display (e.g., a computer monitor), typically a **colorimeter** is used to measure the output for a certain number of known color values. A colorimeter is similar to a spectroradiometer except that the former measures wideband spectral energy filtered by 3 filters whose transmittance spectra match the CIE CMFs; in this way the colorimeter outputs the *XYZ* tristimulus values for any light. For example, if each of the three primaries is turned on by itself, the colorimeter will reveal the columns of $\mathbf{M}_d$. Also recall from before that $\mathbf{M}_d = \tilde{\mathbf{C}}\mathbf{P}$.

Once the camera and display have been calibrated, the 3×3 transformation relating them can be expressed as $\mathbf{M}_{cd} \equiv \mathbf{M}_d^{-1}\mathbf{M}_c$, since we want the *XYZ* tristimulus values to match:

$$\begin{bmatrix} R_d \\ G_d \\ B_d \end{bmatrix} = \underbrace{\mathbf{M}_d^{-1}\mathbf{M}_c}_{\mathbf{M}_{cd}} \begin{bmatrix} R_c \\ G_c \\ B_c \end{bmatrix} \tag{9.30}$$

Note that this result is identical to that of Equation (9.26), since $\mathbf{M}_{cd} = \underbrace{(\tilde{\mathbf{C}}\mathbf{P})}_{\mathbf{M}_d}^{-1}\underbrace{\tilde{\mathbf{C}}\mathbf{S}^+}_{\mathbf{M}_c}$.

9.4.3 White Balancing

The transformation of the previous subsection is based on matching a test light with the light emitted by the display. In practice, however, the camera is usually viewing opaque objects reflecting light emitted by a source (illuminant). Under a different illuminant, the values produced by the three sensors in the camera will be different, and yet we want the perceived color to be the same. The human visual system has a remarkable, and as yet not fully understood, ability to adjust to different viewing conditions, a phenomenon known as **chromatic adaptation** or **color constancy**. The same object when viewed under bright sunlight, overcast daylight, an incandescent bulb, or a fluorescent light is generally perceived the same, even when the sensations produced in the three cones are not the same.[†]

The goal of **white balancing** is to adjust the values produced by the camera according to the illuminant to ensure that neutral colors (grays) remain neutral, and that all other colors maintain their appearance even as the illuminant changes. While there is no simple mathematical relationship that guarantees this result, in practice a diagonal scaling in some color space works well. This approach requires a 3×3 transformation into the particular color space, followed by a diagonal scaling, and finally the inverse transformation. That is, if we redefine $\begin{bmatrix} R_c & G_c & B_c \end{bmatrix}^{\mathsf{T}}$ as the output of the camera after this transform, while $\begin{bmatrix} R'_c & G'_c & B'_c \end{bmatrix}^{\mathsf{T}}$ is the raw input before the transformation, then we have

$$
\begin{bmatrix} R_c \\ G_c \\ B_c \end{bmatrix} = \mathbf{M}_w^{-1} \begin{bmatrix} \alpha_r & 0 & 0 \\ 0 & \alpha_g & 0 \\ 0 & 0 & \alpha_b \end{bmatrix} \mathbf{M}_w \begin{bmatrix} R'_c \\ G'_c \\ B'_c \end{bmatrix} \tag{9.31}
$$

where the diagonal scaling is captured by α_r, α_g, and α_b, and the choice of $\mathbf{M}_w$ determines the color space. One approach that works reasonably well in practice, known as the **von Kries's method**, is to choose $\mathbf{M}_w$ to transform to the color space define by the cone SSFs. An approach that does not work as well is to select $\mathbf{M}_w$ to transform to CIE *XYZ*, which is sometimes known as a "wrong von Kries method". In practice, the best transforms have been found to correspond to narrow cone spaces, that is, color spaces in which the sensors have their sensitivity more narrowly concentrated than the cone SSFs. Such spectrally sharpened methods include the widely used **Bradford transform** and **Sharp transform**.

9.4.4 ICC Profiles

In practice the problem of transforming between the color spaces of various devices is made more complicated by the fact that printers operate in a highly nonlinear, subtractive space, and that the viewing conditions for all devices are not necessarily the same. For example, computer monitors tend to be viewed in relatively bright environments, but movies are usually watched in dark environments. Therefore, when displaying a white object on a monitor it might be necessary to maximize the output of all three primaries, but this leaves no additional intensity with which to display specular highlights. When displaying a movie, on the other hand, it may be possible to display white with less than full intensity of the primaries, since the human eye will have adapted to the dark environment, so that specular highlights can be displayed with greater intensity. **Rendering intent** refers to the specification that governs how an output device handles colors that are mapped to out-of-gamut values, that is, values that are not capable of being displayed by the device.

[†] A perfect example highlighting the imperfections of chromatic adaptation is the photograph of a certain black-and-blue (or white-and-gold) dress, which caused explosive interest on the Internet in February 2015. (See https://en.wikipedia.org/wiki/The_dress_(viral_phenomenon)).

The current standard, since 2004, for specifying the color properties of a device is the **ICC profile** established by the **International Color Consortium**. Three different types of devices can have ICC profiles: input devices (e.g., cameras or scanners), display devices (e.g., computer monitors), and output devices (e.g., printers). Manufacturers carefully calibrate their devices so that they are shipped with ICC profiles; in this way, various input and output devices can be connected without knowing about each other, thus facilitating accurate color reproduction. An ICC profile specifies the transformation from the values in the color space of the device to one of two standard color spaces, either CIE 1931 *XYZ* or CIELAB (defined later), where the transformation takes the form of either a 3×3 matrix or a lookup table (LUT) with values for different wavelengths. The profile corresponds to a reference viewing condition and the standard illuminant D_{50}. Each profile also includes one of four rendering intents: absolute colorimetric, relative colorimetric, perceptual, or saturation. The first two preserve, respectively, either the absolute colorimetry of in-gamut colors, or the colorimetry of in-gamut colors relative to the white point. The latter two compromise the colorimetry of in-gamut colors in order to better handle out-of-gamut colors, with perceptual intent preserving the full range of colors, and saturation intent preserving saturated colors for displaying computer graphics.

9.4.5 CAMs and CATs

The conclusion that color perception can be adequately described using three dimensions is supported by color matching experiments conducted in a carefully controlled viewing environment with a limited field of view, no context, and a particular state of adaptation. In real-world viewing, with different illumination levels and wider fields of view, however, more complex models are necessary. Such models are known as **color appearance models (CAMs)**, with the most recent standard being **CIECAM02**. In a CAM, the visual perception of color appearance is defined by 6 dimensions, including 3 that describe the object properties (lightness, chroma, and hue) and 3 that describe the illumination environment (brightness, colorfulness, and saturation). In practice, colorfulness, chroma, and saturation are all interrelated, so some have suggested that 5 dimensions may be sufficient for describing color appearance. A CAM includes a **chromatic adaptation transform (CAT)** whose goal is to achieve some measure of color constancy by adapting the white point, and it also includes equations for calculating the 6 values just mentioned.

9.5 Color Spaces

Now that we have considered the fundamentals of color sensation and representation, in this section we examine several of the more commonly used color spaces, along with the transformations between them.

9.5.1 *RGB* and *R'G'B'*

The color space most natural for cameras and displays is *RGB*. For cameras, the term *RGB* refers to the fact that the peaks in the SSFs of the filters placed over the pixels' photodetectors correspond roughly to red, green, and blue; for displays the term refers to the fact that the three primaries used look like red, green, and blue. To carefully distinguish between the values in the linear light space in which Grassmann's Law applies, and the nonlinear values resulting from gamma compression, we use *RGB* to refer to the former and *R'G'B'* to refer to the latter. If you load pixel values from a file containing a color image, you are most likely accessing nonlinear *R'G'B'*, not linear *RGB*, unless the pixel values were stored in raw format.

It is important to note that *RGB* values do not uniquely define a color, because the values depend upon which color space is being used. An *RGB* space is uniquely defined by its primaries and its white point (along with the transfer function used for gamma compression in the case of nonlinear $R'G'B'$). Given the linear *XYZ* tristimulus values of the three primaries and the whitepoint, the transformation from linear *RGB* to *XYZ* is

$$
\begin{bmatrix} X \\ Y \\ Z \end{bmatrix} = \underbrace{\begin{bmatrix} X_r & X_g & X_b \\ Y_r & Y_g & Y_b \\ Z_r & Z_g & Z_b \end{bmatrix}}_{\text{primaries}} \begin{bmatrix} A_r & 0 & 0 \\ 0 & A_g & 0 \\ 0 & 0 & A_b \end{bmatrix} \begin{bmatrix} R \\ G \\ B \end{bmatrix}
\tag{9.32}
$$

where the scaling coefficients A_r, A_g, and A_b are given by solving

$$
\underbrace{\begin{bmatrix} X_r & X_g & X_b \\ Y_r & Y_g & Y_b \\ Z_r & Z_g & Z_b \end{bmatrix}}_{\text{primaries}} \begin{bmatrix} A_r \\ A_g \\ A_b \end{bmatrix} = \underbrace{\begin{bmatrix} X_w \\ Y_w \\ Z_w \end{bmatrix}}_{\text{white point}}
\tag{9.33}
$$

In practice the overall intensities of the primaries are not available. Instead we have access only to their chromaticity coordinates (x_r, y_r), (x_g, y_g), and (x_b, y_b). Inserting $\mathbf{F}^{-1}\mathbf{F}$ just after the matrix containing the primaries in Equations (9.32) and (9.33), where $\mathbf{F} \equiv diag(Y_r, Y_g, Y_b)$, yields

$$
\begin{bmatrix} X \\ Y \\ Z \end{bmatrix} = \begin{bmatrix} X_r & X_g & X_b \\ Y_r & Y_g & Y_b \\ Z_r & Z_g & Z_b \end{bmatrix} \begin{bmatrix} Y_r^{-1} & 0 & 0 \\ 0 & Y_g^{-1} & 0 \\ 0 & 0 & Y_b^{-1} \end{bmatrix} \begin{bmatrix} Y_r & 0 & 0 \\ 0 & Y_g & 0 \\ 0 & 0 & Y_b \end{bmatrix} \begin{bmatrix} A_r & 0 & 0 \\ 0 & A_g & 0 \\ 0 & 0 & A_b \end{bmatrix} \begin{bmatrix} R \\ G \\ B \end{bmatrix}
\tag{9.34}
$$

$$
\begin{bmatrix} X_r & X_g & X_b \\ Y_r & Y_g & Y_b \\ Z_r & Z_g & Z_b \end{bmatrix} \begin{bmatrix} Y_r^{-1} & 0 & 0 \\ 0 & Y_g^{-1} & 0 \\ 0 & 0 & Y_b^{-1} \end{bmatrix} \begin{bmatrix} Y_r & 0 & 0 \\ 0 & Y_g & 0 \\ 0 & 0 & Y_b \end{bmatrix} \begin{bmatrix} A_r \\ A_g \\ A_b \end{bmatrix} = \begin{bmatrix} X_w \\ Y_w \\ Z_w \end{bmatrix}
\tag{9.35}
$$

which lead to equations that we can use:

$$
\begin{bmatrix} X \\ Y \\ Z \end{bmatrix} = \begin{bmatrix} x_r/y_r & x_g/y_g & x_b/y_b \\ 1 & 1 & 1 \\ z_r/y_r & z_g/y_g & z_b/y_b \end{bmatrix} \begin{bmatrix} \alpha_r & 0 & 0 \\ 0 & \alpha_g & 0 \\ 0 & 0 & \alpha_b \end{bmatrix} \begin{bmatrix} R \\ G \\ B \end{bmatrix}
\tag{9.36}
$$

$$
\begin{bmatrix} x_r/y_r & x_g/y_g & x_b/y_b \\ 1 & 1 & 1 \\ z_r/y_r & z_g/y_g & z_b/y_b \end{bmatrix} \begin{bmatrix} \alpha_r \\ \alpha_g \\ \alpha_b \end{bmatrix} = \begin{bmatrix} X_w \\ Y_w \\ Z_w \end{bmatrix}
\tag{9.37}
$$

since $x_r \equiv X_r/S_r$, $y_r \equiv Y_r/S_r$, and $\alpha_r \equiv A_r Y_r$, where $S_r \equiv X_r + Y_r + Z_r$, so that $X_r/Y_r = x_r S_r/y_r S_r = x_r/y_r$, and similarly for the other color channels. If we further assume that the values for R, G, and B range from 0 to 1, and similarly for Y, and if we assume that the maximum intensity of all three primaries ($R = G = B = 1$) leads to the maximum luminance ($Y = 1$), then we have the additional constraint that $\alpha_r + \alpha_g + \alpha_b = 1$.

Putting this all together, if we are given the chromaticities of the primaries and the whitepoint, Equation (9.37) can be solved for α_r, α_g, and α_b. Once these values have been found, we then impose the constraint that $\alpha_r + \alpha_g + \alpha_b = 1$ to remove the effects of any scaling in (X_w, Y_w, Z_w). Then these normalized values are plugged back into Equation (9.36) to yield the transformation from (R, G, B) to (X, Y, Z).

The chromaticities and white points of several well-known video and image standards are listed in Table 9.1. (Their nonlinear transfer functions were covered in a previous chapter.[†]) The analog color television signal broadcast in the United States and Japan, among other countries, from its adoption in 1953 to its replacement by digital video in 2009, is known as **NTSC** (for the National Television System Committee). The NTSC primaries correspond to the phosphor technology used at the time, and the illuminant is the now-obsolete CIE Standard Illuminant C. Phosphor technology improved considerably over the next decade, so that when the European equivalent **PAL** (for Phase Alternating Line) was standardized in 1966, the primaries were changed to something very close to those in use today. The chromaticity coordinates and white points of these NTSC and PAL standards were essentially inherited by the two versions of **Rec. 601**,[‡] which was adopted in 1982 for encoding interlaced analog video signals as digital video.

| EXAMPLE 9.1 | Compute the 3×3 matrix conversion between RGB and CIE XYZ using the NTSC primaries and white point shown in Table 9.1. |

Solution:

From the first row of the table we have $x_r/y_r = 0.67/0.33 = 2.03$, and $z_r/y_r = (1 - x_r - y_r)/y_r = (1 - 0.67 - 0.33)/0.33 = 0$ for the red channel. Repeating for the other color channels, Equation (9.37) is used to solve for the coefficients:

$$\begin{bmatrix} \alpha_r \\ \alpha_g \\ \alpha_b \end{bmatrix} = \begin{bmatrix} 2.0303 & 0.2958 & 1.75 \\ 1 & 1 & 1 \\ 0 & 0.1127 & 9.75 \end{bmatrix}^{-1} \begin{bmatrix} 0.310 \\ 0.316 \\ 0.374 \end{bmatrix} = \begin{bmatrix} 0.0945 \\ 0.1853 \\ 0.0362 \end{bmatrix} \Rightarrow \begin{bmatrix} 0.2990 \\ 0.5864 \\ 0.1146 \end{bmatrix} \qquad (9.38)$$

where the final values are obtained by normalizing so that $\alpha_r + \alpha_g + \alpha_b = 1$. Plugging these values back into Equation (9.36) yields the desired result:

$$\begin{bmatrix} X^C \\ Y^C \\ Z^C \end{bmatrix} = \begin{bmatrix} 2.0303 & 0.2958 & 1.75 \\ 1 & 1 & 1 \\ 0 & 0.1127 & 9.75 \end{bmatrix} \begin{bmatrix} 0.2990 & 0 & 0 \\ 0 & 0.5864 & 0 \\ 0 & 0 & 0.1146 \end{bmatrix} \begin{bmatrix} R^C_{NTSC} \\ G^C_{NTSC} \\ B^C_{NTSC} \end{bmatrix} \qquad (9.39)$$

$$= \underbrace{\begin{bmatrix} 0.6070 & 0.1734 & 0.2006 \\ 0.2990 & 0.5864 & 0.1146 \\ -0.0000 & 0.0661 & 1.1175 \end{bmatrix}}_{\mathbf{M}^C_{NTSC}} \begin{bmatrix} R^C_{NTSC} \\ G^C_{NTSC} \\ B^C_{NTSC} \end{bmatrix} \qquad (9.40)$$

where we have introduced the superscript to indicate the white point being used (in this case CIE Standard Illuminant C) and the subscript to indicate the choice of RGB primaries (in this case those specified by NTSC), and the notation $\mathbf{M}^a_b$ indicates a transform from the primaries in b using the white point a. What is particularly interesting about this result is the middle row, which says that the luminance is given by

$$Y^C = 0.299 R^C_{601} + 0.587 G^C_{601} + 0.114 B^C_{601} \qquad (9.41)$$

where we have changed the subscript to 601 because these are the numbers adopted by Rec. 601. This equation is the basis behind the well-known **3-6-1 rule** used to convert RGB to grayscale, which we shall examine in more detail later.

[†] Section 2.3.2 (p. 43).
[‡] Officially ITU-R Recommendation BT.601, formerly known as CCIR 601.

	red	green	blue	white point	
NTSC	(0.67, 0.33)	(0.21, 0.71)	(0.14, 0.08)	(0.310, 0.316)	CIE C
PAL	(0.64, 0.33)	(0.29, 0.60)	(0.15, 0.06)	(0.3127, 0.3290)	CIE D_{65}
Rec. 709, sRGB	(0.640, 0.330)	(0.300, 0.600)	(0.150, 0.060)	(0.3127, 0.3290)	CIE D_{65}
Rec. 2020	(0.708, 0.292)	(0.170, 0.797)	(0.131, 0.046)	(0.3127, 0.3290)	CIE D_{65}

TABLE 9.1 The (x, y) chromaticity coordinates of the primaries and white point of several video and image standards. NTSC and PAL are the original North American and European analog video signals, respectively, no longer in widespread use; Rec. 709 is the modern standard used in HDTV, sRGB is the standard for desktop computing, and Rec. 2020 is the latest standard used in UHDTV.

The modern video and image standards used by nearly all devices today share the same chromaticities and white points as each other, differing only in their transfer function. **Rec. 709**[†] was adopted in 1990 as the standard for high-definition television (HDTV), as well as for studio video and broadcast television. The **sRGB** standard was developed in 1996 by leading personal computer manufacturers as a standard for personal desktop computers, monitors, printers, and the Internet, and it is also used by LCD monitors, digital cameras, and scanners. **Rec. 2020** was adopted in 2012 as the standard for ultra-high-definition television (UHDTV). As a general rule one should assume, in the absence of an embedded color profile or additional information, that any 8-bit-per-channel still image is in the sRGB color space, while any video is in that of Rec. 709.

Repeating the procedure above for the Rec. 709 standard (or equivalently, sRGB, since the linear color spaces are the same) yields

$$\begin{bmatrix} X^{D65} \\ Y^{D65} \\ Z^{D65} \end{bmatrix} = \underbrace{\begin{bmatrix} 0.4124 & 0.3576 & 0.1805 \\ 0.2126 & 0.7152 & 0.0722 \\ 0.0193 & 0.1192 & 0.9505 \end{bmatrix}}_{\mathbf{M}_{709}^{D65}} \begin{bmatrix} R_{709}^{D65} \\ G_{709}^{D65} \\ B_{709}^{D65} \end{bmatrix} \qquad (9.42)$$

Thus the more modern color space leads to a **2-7-1** rule for computing luminance:

$$Y^{D65} = 0.2126 R_{709}^{D65} + 0.7152 G_{709}^{D65} + 0.0722 B_{709}^{D65} \qquad (9.43)$$

Note that in both Equations (9.40) and (9.42) the middle row sums to 1, and each column of the matrix contains the tristimulus values of the corresponding primary. Thus, $0.607/(0.607 + 0.299) = 0.67, 0.1734/(0.1734 + 0.5864 + 0.0661) = 0.21,$ $0.2006/(0.2006 + 0.1146 + 1.1175) = 0.14,$ and so on. And of course the conversion from XYZ back to RGB is simply the inverse of the matrix.

When converting from one RGB space to another, it is common to transform the RGB values to a standard reference coordinate system (e.g., CIE XYZ), then convert from the reference coordinate system to the other RGB space. That is, the overall transform involves multiplying one 3×3 matrix by the inverse of the other matrix, but care must be taken because an additional matrix needs to be introduced if the white points are different. For example, although the matrices above transform both $NTSC$ and Rec. 709 to CIE XYZ, they use different white points, so the transform between them is actually $(\mathbf{M}_{709}^{D65})^{-1} \mathbf{N}_C^{D65} \mathbf{M}_{NTSC}^C$, where $\mathbf{N}_b^a$ transforms between the two white points a and b. This is very important because, although Rec. 709 and sRGB use D_{65}, the ICC profiles used for transforming between color spaces use D_{50} instead.

[†] Officially ITU-R Recommendation BT.709.

EXAMPLE 9.2	Compute the 3×3 matrix conversion between D_{65} and D_{50}.
Solution:	Since we are free to use any primaries we like, let us use the NTSC primaries. Similar to Example 9.1, we have the following for D_{50}:

$$\begin{bmatrix} \alpha_r \\ \alpha_g \\ \alpha_b \end{bmatrix} = \begin{bmatrix} 2.0303 & 0.2958 & 1.75 \\ 1 & 1 & 1 \\ 0 & 0.1127 & 9.75 \end{bmatrix}^{-1} \begin{bmatrix} 0.3457 \\ 0.3585 \\ 0.2958 \end{bmatrix} = \begin{bmatrix} 0.1148 \\ 0.2158 \\ 0.0278 \end{bmatrix} \Rightarrow \begin{bmatrix} 0.3203 \\ 0.6020 \\ 0.0777 \end{bmatrix} \quad (9.44)$$

so that

$$\begin{bmatrix} X^{D50} \\ Y^{D50} \\ Z^{D50} \end{bmatrix} = \begin{bmatrix} 2.0303 & 0.2958 & 1.75 \\ 1 & 1 & 1 \\ 0 & 0.1127 & 9.75 \end{bmatrix} \begin{bmatrix} 0.3203 & 0 & 0 \\ 0 & 0.6020 & 0 \\ 0 & 0 & 0.0777 \end{bmatrix} \begin{bmatrix} R_{NTSC}^{D50} \\ G_{NTSC}^{D50} \\ B_{NTSC}^{D50} \end{bmatrix} \quad (9.45)$$

$$= \underbrace{\begin{bmatrix} 0.6503 & 0.1781 & 0.1359 \\ 0.3203 & 0.6020 & 0.0777 \\ -0.0000 & 0.0678 & 0.7573 \end{bmatrix}}_{\mathbf{M}_{NTSC}^{D50}} \begin{bmatrix} R_{NTSC}^{D50} \\ G_{NTSC}^{D50} \\ B_{NTSC}^{D50} \end{bmatrix} \quad (9.46)$$

Similarly, for D_{65} we have

$$\begin{bmatrix} \alpha_r \\ \alpha_g \\ \alpha_b \end{bmatrix} = \begin{bmatrix} 2.0303 & 0.2958 & 1.75 \\ 1 & 1 & 1 \\ 0 & 0.1127 & 9.75 \end{bmatrix}^{-1} \begin{bmatrix} 0.3127 \\ 0.3290 \\ 0.3583 \end{bmatrix} = \begin{bmatrix} 0.0953 \\ 0.1993 \\ 0.0344 \end{bmatrix} \Rightarrow \begin{bmatrix} 0.2897 \\ 0.6056 \\ 0.1047 \end{bmatrix} \quad (9.47)$$

so that

$$\begin{bmatrix} X^{D65} \\ Y^{D65} \\ Z^{D65} \end{bmatrix} = \begin{bmatrix} 2.0303 & 0.2958 & 1.75 \\ 1 & 1 & 1 \\ 0 & 0.1127 & 9.75 \end{bmatrix} \begin{bmatrix} 0.2897 & 0 & 0 \\ 0 & 0.6056 & 0 \\ 0 & 0 & 0.1047 \end{bmatrix} \begin{bmatrix} R_{NTSC}^{D65} \\ G_{NTSC}^{D65} \\ B_{NTSC}^{D65} \end{bmatrix} \quad (9.48)$$

$$= \underbrace{\begin{bmatrix} 0.5881 & 0.1791 & 0.1832 \\ 0.2897 & 0.6056 & 0.1047 \\ -0.0000 & 0.0682 & 1.0208 \end{bmatrix}}_{\mathbf{M}_{NTSC}^{D65}} \begin{bmatrix} R_{NTSC}^{D65} \\ G_{NTSC}^{D65} \\ B_{NTSC}^{D65} \end{bmatrix} \quad (9.49)$$

The conversion from D_{65} to D_{50} is then given by

$$\begin{bmatrix} X^{D50} \\ Y^{D50} \\ Z^{D50} \end{bmatrix} = \underbrace{\mathbf{M}_{NTSC}^{D50} (\mathbf{M}_{NTSC}^{D65})^{-1}}_{\mathbf{N}_{D65}^{D50}} \begin{bmatrix} X^{D65} \\ Y^{D65} \\ Z^{D65} \end{bmatrix} = \underbrace{\begin{bmatrix} 1.1206 & -0.0301 & -0.0649 \\ 0.0621 & 0.9797 & -0.0355 \\ -0.0164 & 0.0333 & 0.7414 \end{bmatrix}}_{\mathbf{N}_{D65}^{D50}} \begin{bmatrix} X^{D65} \\ Y^{D65} \\ Z^{D65} \end{bmatrix}$$

$$(9.50)$$

9.5.2 $Y'P_B P_R$

It is often desirable to separate the **luminance**, which is related to the perception of brightness, from the **chrominance**, which captures the parameters that we normally associate with the notion of color (e.g., hue and chroma). This separation is performed by transforming the three RGB sensor values into a color space in which the luminance is described by one axis, while the chrominance is described by the remaining two axes. As we shall see, nearly all color spaces except for $RGB / R'G'B'$ are based upon such a separation. (The XYZ space, while not explicitly maintaining this separation, can nevertheless be transformed by simply concatenating the luminance value Y with the chromaticity coordinates x and y to yield the xyY color space.)

The first such color space that we shall consider is $Y'P_B P_R$, which is sometimes used for transmitting analog video signals. The prime ($'$) indicates that the value Y' is not the luminance used in color science. In fact, it is not even gamma-corrected luminance but rather the nonlinear quantity known to video engineers as **luma**, which is computed as the weighted average of the three gamma-corrected $R'G'B'$ values: $Y' = \alpha_r R' + \alpha_g G' + \alpha_b B'$. For Rec. 601 and Rec. 709, this leads to the following definitions:

$$^{601}Y' \equiv 0.299R' + 0.587G' + 0.114B' \tag{9.51}$$

$$^{709}Y' \equiv 0.2126R' + 0.7152G' + 0.0722B' \tag{9.52}$$

where the leading superscript indicates the coefficients used in computing luma, and the trailing super/subscripts have been dropped because the exact choice of primaries and white point is not important.

Compared with the middle row of the colorimetric expressions in Equations (9.41) and (9.43), the most important difference is that Equations (9.51)–(9.52) above operate in the gamma-corrected $R'G'B'$ space and, as a result, they are no longer correct in any colorimetric sense. To understand the reason behind defining luma in a non-colorimetric way, we must consider the historical context in which the decision was made. As shown in Figure 9.12, the correct way of transmitting RGB values back in the days of analog video transmission would have been, in the encoder, to convert to luminance Y (and chrominance) using a 3×3 transform, then apply gamma compression to reduce the perceptual effects of noise by transmitting the luma Y' instead of the luminance Y. The decoder would then have applied gamma expansion to recover the luminance, then applied the inverse 3×3 transform to yield linear RGB values. So far so good. However, because cathode ray tube (CRT) displays, which were the prevailing technology in the 1950s, inherently apply a nonlinear transform function to the input voltages, a nonlinear function resembling gamma compression had to be inserted to cancel this effect. Therefore, to simplify the decoding process, video engineers reversed the order of the gamma and linear transform blocks, thus rendering two of the blocks in the diagram unnecessary since they cancel each other. This engineering tradeoff sacrificed colorimetric accuracy for implementation simplicity. Although this tradeoff would not be necessary with today's technology, the effects of this decision are still with us due to their influence upon widely-used video standards.

The nonlinear chrominance components,[†] P_B and P_R, are the color differences computed with respect to the blue and red channels, since green is the most similar to luma:

$$P_B \equiv \frac{1}{2} \cdot \frac{B' - Y'}{1 - \alpha_b} = \frac{-\alpha_r R' - \alpha_g G' + (1 - \alpha_b)B'}{2(1 - \alpha_b)} \tag{9.53}$$

[†] Just as nonlinear luminance is called *luma*, so nonlinear chrominance is called *chroma*, Section 2.3.4 (p. 52). However, to avoid confusion with the other meaning of the term "chroma", we will avoid the use of the term here.

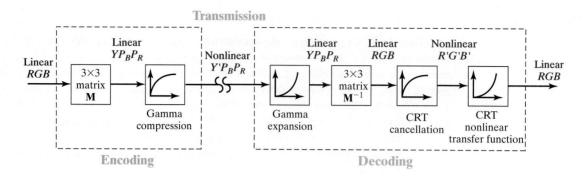

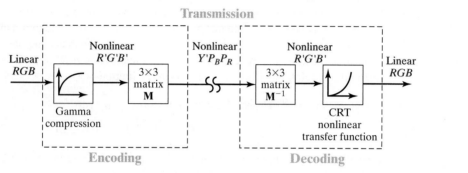

Based on Poynton (2003)

Figure 9.12 The theoretically correct way of transmitting analog *RGB* (top), and the engineering approximation developed in the 1950s (bottom). The approximation swaps the order of the linear 3 × 3 transform and the nonlinear gamma compression or expansion, to simplify decoding by canceling two of the blocks in the diagram.

$$P_R \equiv \frac{1}{2} \cdot \frac{R' - Y'}{1 - \alpha_r} = \frac{(1 - \alpha_r)R' - \alpha_g G' - \alpha_b B'}{2(1 - \alpha_r)} \tag{9.54}$$

The scaling factors ensure that P_B and P_R range from -0.5 to $+0.5$, whereas Y' ranges from 0 to 1. Plugging in the Rec. 601 coefficients leads to the transformation from $R'G'B'$ to $^{601}Y'P_BP_R$:

$$\begin{bmatrix} ^{601}Y' \\ P_B \\ P_R \end{bmatrix} = \begin{bmatrix} 1 & 0 & 0 \\ 0 & 1/(2 \cdot 0.886) & 0 \\ 0 & 0 & 1/(2 \cdot 0.701) \end{bmatrix} \begin{bmatrix} 0.299 & 0.587 & 0.114 \\ -0.299 & -0.587 & 0.886 \\ 0.701 & -0.587 & -0.114 \end{bmatrix} \begin{bmatrix} R' \\ G' \\ B' \end{bmatrix} \tag{9.55}$$

$$= \begin{bmatrix} 0.299 & 0.587 & 0.114 \\ -0.1687 & -0.3313 & 0.5 \\ 0.5 & -0.4187 & -0.0813 \end{bmatrix} \begin{bmatrix} R' \\ G' \\ B' \end{bmatrix} \tag{9.56}$$

and similarly using the Rec. 709 coefficients:

$$\begin{bmatrix} ^{709}Y' \\ P_B \\ P_R \end{bmatrix} = \begin{bmatrix} 0.2126 & 0.7152 & 0.0722 \\ -0.1146 & -0.3854 & 0.5 \\ 0.5 & -0.4542 & -0.0458 \end{bmatrix} \begin{bmatrix} R' \\ G' \\ B' \end{bmatrix} \tag{9.57}$$

Note that in both cases the first row sums to 1, while the second and third rows sum to zero. Conversion back to $R'G'B'$ is obtained by simply inverting the matrix in each case.

9.5.3 $Y'C_BC_R$

More common than $Y'P_BP_R$, which is used for analog signals, is $Y'C_BC_R$, which is used for digital signals. The latter is a scaled and offset version of the former:

$$\begin{bmatrix} _{219}Y' \\ C_B \\ C_R \end{bmatrix} \equiv \begin{bmatrix} 16 \\ 128 \\ 128 \end{bmatrix} + \begin{bmatrix} 219Y' \\ 224P_B \\ 224P_R \end{bmatrix} \tag{9.58}$$

where the leading subscript 219 indicates the scaling of luma. Note that $_{219}Y'$, C_B, and C_R on the left side each fit within 8 bits: $16 \leq {}_{219}Y' \leq 235$ and $16 \leq C_B, C_R \leq 240$. At first it may seem strange that the full range of 256 values is not used, but this extra *footroom* (below 16) and *headroom* (above 235 for luma or 240 for nonlinear chrominance) is necessary to allow for some undershoot or overshoot when filtering, without incurring clipping. Combining Equations (9.56) and (9.58) yields the transform from 8-bit $R'G'B'$ to 8-bit $_{219}^{601}Y'C_BC_R$ using the Rec. 601 coefficients:

$$\begin{bmatrix} _{219}^{601}Y' \\ C_B \\ C_R \end{bmatrix} = \begin{bmatrix} 16 \\ 128 \\ 128 \end{bmatrix} + \frac{1}{255} \begin{bmatrix} 65.4810 & 128.5530 & 24.9660 \\ -37.7968 & -74.2032 & 112.0000 \\ 112.0000 & -93.7860 & -18.2140 \end{bmatrix} \begin{bmatrix} _{255}R' \\ _{255}G' \\ _{255}B' \end{bmatrix} \tag{9.59}$$

where we have added a scaling factor of 1/255 so that the equation applies when $_{255}R'$, $_{255}G'$, and $_{255}B'$ are also stored as 8 bits each, i.e., $0 \leq {}_{255}R', {}_{255}G', {}_{255}B' \leq 255$.

The JFIF format for JPEG uses Rec. 601 and no headroom or footroom, so that Y', C_B, and C_R encompass the full range from 0 to 255. With R', G', and B' also ranging from 0 to 255, the conversion is given by adding offsets to Equation (9.56):

$$\begin{bmatrix} _{255}^{601}Y' \\ C_B \\ C_R \end{bmatrix} \equiv \begin{bmatrix} 0 \\ 128 \\ 128 \end{bmatrix} + \begin{bmatrix} 0.299 & 0.587 & 0.114 \\ -0.1687 & -0.3313 & 0.5 \\ 0.5 & -0.4187 & -0.0813 \end{bmatrix} \begin{bmatrix} _{255}R' \\ _{255}G' \\ _{255}B' \end{bmatrix} \tag{9.60}$$

Conversion back to $R'G'B'$ is obtained by simply inverting these equations. But when headroom and footroom exist, care must be taken in the inversion because if any processing of the image has occurred, some values might be in the invalid range below 16 or above 235 (in the case of Y') or above 240 (in the case of C_B or C_R). Clamping such values can lead to undesirable visual distortions in the result.

As with $Y'P_BP_R$, $Y'C_BC_R$ is not an absolute color space but rather a way of encoding *RGB* information. The actual color displayed depends on the *RGB* primaries used, the white point, and the nonlinear transfer function. A value expressed as $Y'C_BC_R$ is only predictable if these quantities are known, or if an ICC profile is used.

9.5.4 Y'UV

Originally the term $Y'UV$ referred to the color space used by the transmission of PAL analog video signals. The original scaling factors set the maximum U and V values to 0.436 and 0.615, respectively, instead of 0.5. From Equations (9.53) and (9.54) this yields

$$^{601}Y' \equiv \alpha_r R' + \alpha_g G' + \alpha_b B' = 0.299R' + 0.587G' + 0.114B \tag{9.61}$$

$$U \equiv 0.436 \cdot \frac{B' - Y'}{1 - \alpha_b} = 0.4921(B' - Y') \tag{9.62}$$

$$V \equiv 0.615 \cdot \frac{R' - Y'}{1 - \alpha_r} = 0.8773(R' - Y') \tag{9.63}$$

Surprisingly, PAL uses the Rec. 601 coefficients to convert to luma, as in Equation (9.51), even though the PAL primaries are different from those of NTSC.

Since the advent of digital video, the term $Y'UV$ has adopted a more ambiguous meaning and should generally be avoided in favor of the more specific $Y'C_BC_R$. Often, $Y'UV$ simply means some scaled version of $Y'C_BC_R$, and it is essential to know the scaling factors in order to properly interpret the values. One commonly used definition is the headroom/footroom version of Equation (9.59), which can be approximated by integer-only computations as:

$$\begin{bmatrix} {}^{601}_{219}Y' \\ U \\ V \end{bmatrix} \approx \begin{bmatrix} 16 \\ 128 \\ 128 \end{bmatrix} + \frac{1}{256}\begin{bmatrix} 66 & 129 & 25 \\ -38 & -74 & 112 \\ 112 & -94 & -18 \end{bmatrix}\begin{bmatrix} {}_{255}R' \\ {}_{255}G' \\ {}_{255}B' \end{bmatrix} \tag{9.64}$$

with the inverse transform given by

$$\begin{bmatrix} {}_{255}R' \\ {}_{255}G' \\ {}_{255}B' \end{bmatrix} \approx \frac{1}{256}\begin{bmatrix} 298 & 0 & 409 \\ 298 & -100 & -208 \\ 298 & 516 & 0 \end{bmatrix}\left(\begin{bmatrix} {}^{601}_{219}Y' \\ U \\ V \end{bmatrix} - \begin{bmatrix} 16 \\ 128 \\ 128 \end{bmatrix}\right) \tag{9.65}$$

9.5.5 Y'IQ

As $Y'UV$ was the color space used by analog PAL, the color space used originally by analog NTSC was $Y'IQ$, which is $Y'UV$ with the chrominance axes rotated by 33 degrees and exchanged:

$$\begin{bmatrix} {}^{601}Y' \\ I \\ Q \end{bmatrix} \equiv \begin{bmatrix} 1 & 0 & 0 \\ 0 & -\sin(33°) & \cos(33°) \\ 0 & \cos(33°) & \sin(33°) \end{bmatrix}\begin{bmatrix} {}^{601}Y' \\ U \\ V \end{bmatrix} \tag{9.66}$$

However, even NTSC switched from using $Y'IQ$ to $Y'UV$ in the early 1970s, so that $Y'IQ$ has been obsolete for a long time now.

9.5.6 Converting from R'G'B' to Grayscale

We are now in a position to consider a problem that arises frequently in practice, namely how to properly convert an $R'G'B'$ image to grayscale. Conventional wisdom suggests that one should always use the 3-6-1 rule of Equation (9.51):

$$gray = 0.299R' + 0.587G' + 0.114B' \tag{9.67}$$

This is the transformation you will most often find in image processing and computer vision software. However, as we saw in our analysis above, this transformation has never been correct in a colorimetric sense because it applies coefficients derived from linear RGB space to a nonlinear $R'G'B'$ space; moreover, it is based upon obsolete primary chromaticities and an obsolete white point. In contrast, the proper way to convert $R'G'B'$ to grayscale is to apply the appropriate inverse nonlinear transfer function (gamma expansion) to yield linear RGB values, then apply the Rec. 709 coefficients of Equation (9.52), then apply the appropriate nonlinear transfer function (gamma compression). An easy way to convince yourself of the potential error that can occur by ignoring gamma is to apply Equation (9.67) to a purely blue image ($R' = G' = 0$, $B' = 255$ for all pixels); the result will be $gray = 29$ everywhere, which will be noticeably darker than the original.

Having said that, it is comforting to know that, unless the particular application at hand requires colorimetric fidelity, there is no need to go to all this trouble. One of the reasons the 3-6-1 rule is still widely used is that it works well in practice, but in fact just about any reasonably weighted average of the three components yields a visually acceptable result, as we saw earlier.[†] For many applications, therefore, simpler methods of conversion are feasible, such as the one we introduced in Equation (3.37),

$$gray = \frac{1}{4}(R' + 2G' + B')$$

(9.68)

which has the advantage that it can be implemented with only bitshifts and without any floating-point multiplication or division. As an aside, note that (as we discussed previously) the inverse conversion from grayscale to $R'G'B'$ is easy: simply replicate the value three times.

9.5.7 Opponent Colors

Because of their effective use by the human visual system, opponent color spaces are often useful for image processing applications. One natural approach is to define the two color dimensions c_1 and c_2 as blue minus yellow and red minus green, respectively, as in the human visual system:

$$c_1 \equiv \text{blue} - \text{yellow} = B' - \frac{1}{2}(R' + G')$$

(9.69)

$$c_2 \equiv \text{red} - \text{green} = R' - G'$$

(9.70)

leading to

$$\begin{bmatrix} Y' \\ c_1 \\ c_2 \end{bmatrix} \equiv \begin{bmatrix} \alpha_r & \alpha_g & \alpha_b \\ -\frac{1}{2} & -\frac{1}{2} & 1 \\ 1 & -1 & 0 \end{bmatrix} \begin{bmatrix} R' \\ G' \\ B' \end{bmatrix}$$

(9.71)

where Y' is the luma, and we can set $\alpha_r = \alpha_g = \alpha_b = \frac{1}{3}$ for simplicity or use one of the standard sets of luma coefficients.

An alternative, which we shall call the **Hanbury transformation**, is to define c_1 and c_2 as follows:

$$\begin{bmatrix} Y' \\ c_1 \\ c_2 \end{bmatrix} \equiv \begin{bmatrix} \alpha_r & \alpha_g & \alpha_b \\ 1 & -\frac{1}{2} & -\frac{1}{2} \\ 0 & \frac{\sqrt{3}}{2} & -\frac{\sqrt{3}}{2} \end{bmatrix} \begin{bmatrix} R' \\ G' \\ B' \end{bmatrix}$$

(9.72)

which has the advantage that it maps the corners of the $R'G'B'$ color cube to a hexagon in the c_1-c_2 chromaticity plane, a concept explored in more detail next.

9.5.8 *HSV* and *HSL*

Consider the $R'G'B'$ color cube shown in Figure 9.13, whose eight vertices correspond to the three additive and subtractive primaries (red, green, blue, cyan, magenta, and yellow), along with black and white. By convention black is at $(0, 0, 0)$, white is at $(1, 1, 1)$, and red, green, and blue are at $(1, 0, 0)$, $(0, 1, 0)$, and $(0, 0, 1)$, respectively. Note that the R', G', and B' axes form a righthand coordinate system. Also shown in the figure is the line of grays connecting the white and black vertices, along with an arbitrary point at $\mathbf{p} = (0.8, 0.6, 0.4)$

[†] Section 3.4.1 (p. 93).

for illustrative purposes. While one could also define the *RGB* color cube using linear light intensities, it rarely (if ever) makes sense to do so, since it is usually desirable to preserve perceptual uniformity; hence the choice of $R'G'B'$ here.

A curious property of the $R'G'B'$ color cube is that, for each additive primary (red, green, or blue), there is exactly one face that connects it with three other vertices, each of which is not an additive primary. Thus, the red face connects red with magenta, yellow, and white; the green face connects green with cyan, yellow, and white; and the blue face connects blue with cyan, magenta, and white. If the cube is rotated by 45 degrees about the R' axis, then rotated again by $\hat{\theta}$ about the original G' axis, where $\hat{\theta} \approx -35.26$ degrees,[†] then the cube will be balanced on the black vertex, with the white vertex directly above it. Viewed from above with orthographic projection, the cube adopts the shape of a regular hexagon whose six vertices are the three additive and three subtractive primaries, with the gray axis coming out of the page and piercing the center of the hexagon. Compared with Figure 9.3, we see that this hexagon is simply our old friend the color wheel, with a slightly different shape. Mathematically the transformation from $R'G'B'$ to color difference axes c_1 and c_2 is obtained by concatenating the rotation matrices, along with appropriate scaling factors for the nonlinear chromaticity and luma axes:

$$
\begin{bmatrix} c_1 \\ c_2 \\ Y' \end{bmatrix} = \begin{bmatrix} \sqrt{\frac{3}{2}} & 0 & 0 \\ 0 & \sqrt{\frac{3}{2}} & 0 \\ 0 & 0 & \frac{1}{3}\sqrt{3} \end{bmatrix} \begin{bmatrix} \cos\hat{\theta} & 0 & \sin\hat{\theta} \\ 0 & 1 & 0 \\ -\sin\hat{\theta} & 0 & \cos\hat{\theta} \end{bmatrix} \begin{bmatrix} 1 & 0 & 0 \\ 0 & \cos 45° & -\sin 45° \\ 0 & \sin 45° & \cos 45° \end{bmatrix} \begin{bmatrix} R' \\ G' \\ B' \end{bmatrix} \quad (9.73)
$$

$$
= \begin{bmatrix} 1 & -\frac{1}{2} & -\frac{1}{2} \\ 0 & \frac{\sqrt{3}}{2} & -\frac{\sqrt{3}}{2} \\ \frac{1}{3} & \frac{1}{3} & \frac{1}{3} \end{bmatrix} \begin{bmatrix} R' \\ G' \\ B' \end{bmatrix} \quad (9.74)
$$

which is simply the Hanbury transformation with $\alpha_r = \alpha_g = \alpha_b = \frac{1}{3}$, and a reordering of the rows. Notice that this transformation maps red $(R' = 1, G' = B' = 0)$ to $(c_1, c_2) = (1, 0)$, green $(G' = 1, R' = B' = 0)$ to $(-\frac{1}{2}, \frac{\sqrt{3}}{2})$, and blue $(B' = 1, R' = G' = 0)$ to $(-\frac{1}{2}, -\frac{\sqrt{3}}{2})$.

The hexagon is naturally divided into six equally-sized sectors, and points fall into one or the other sector depending upon the relative values of R', G', and B'. By convention the hue is defined as the counterclockwise angle with respect to the red axis, so that sector 0 contains all the points such that $R' \geq G' \geq B'$ (hue from 0 to 60 degrees), sector 1 contains all the points such that $G' \geq R' \geq B'$ (hue from 60 to 120 degrees), and so forth. R' is the maximum value in sectors 0 and 5, G' is the maximum value in sectors 1 and 2, and B' is the maximum value in sectors 3 and 4. These relationships can be seen from the colored faces of the 3D color cube, or from the lines that bisect the primary colors in the 2D hexagon.

For any given color (R', G', B') located at a certain point within the cube, the point lies within a **subcube** anchored at black whose length along each side is $V_{max} \equiv \max(R', G', B')$, shown in the left side of Figure 9.13. This subcube projects orthographically onto the page as a smaller hexagon within the complete hexagon already mentioned. For clarity, we shall refer to the smaller hexagon as a **sub-hexagon**. The size of the sub-hexagon depends on V_{max}, with $V_{max} = 1$ yielding the complete hexagon, and $V_{max} = 0$ yielding a point. If we stack all these sub-hexagons on top of one another we get a hexagon-based pyramid called a **hexcone**, which is illustrated in Figure 9.14.

With these preliminaries in place, we are now in a position to begin to quantify the concepts of hue, chroma, saturation, and value that we mentioned earlier in the context of

---

[†] The exact angle is given by $\cos\hat{\theta} = \sqrt{2/3}$, $\sin\hat{\theta} = -1/\sqrt{3}$.

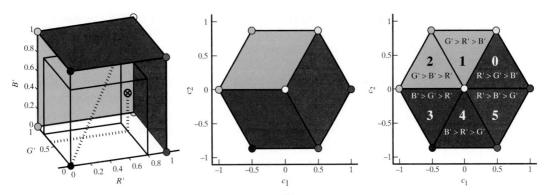

Figure 9.13 LEFT: The $R'G'B'$ color cube, with the red, green, and blue faces colored. Also shown is the line of grays and an arbitrary point at $\mathbf{p} = (0.8, 0.6, 0.4)$ along with its subcube. Middle: Top-down view of the cube after rotating so that white is above black. RIGHT: The 2D hexagon within which all (R', G', B') colors project orthographically parallel to the line of grays. The hexagon is divided into six sectors (numbered 0 through 5), and each point falls into a particular sector depending on the relative ordering of R', G', and B'.

the Munsell color system. The *HSV* and *HSL* color spaces were developed in the 1970s for computer graphics applications. Specifically, they are used widely in "color picker" user interfaces that allow a person to select a color in a more intuitive way than by specifying $R'G'B'$ values directly. They are also commonly used in image processing and computer vision applications because they separate luma from nonlinear chrominance, placing the latter values along axes that facilitate, to some extent, meaningful color differences.

A confusing set of choices exists in defining a color space based on these principles, which over time has led to a mishmash of related techniques all going by different but similar-sounding names. Thus, we have *HSV* (hue-saturation-value), *HSL* (hue-saturation-lightness), *HSI* (hue-saturation-intensity), *HSB* (hue-saturation-brightness), and so forth. In many of these cases the names are misleading because the terms are not used precisely, and the

Figure 9.14 LEFT: The hexcone obtained by stacking the sub-hexagons associated with each value of V_{max}. RIGHT: The point $\mathbf{p} = (0.8, 0.6, 0.4)$ projects onto the nonlinear chromaticity plane at $(c_1, c_2) \approx (0.3, 0.2)$, which lies within sector 0. Its sub-hexagon with spokes of length 0.8 is shown within the complete hexagon, and the intersection of the ray toward $\mathbf{p}$ with the sub-hexagon is shown at $\approx (0.6, 0.35)$. For any point, the hue is the counterclockwise angle with respect to the red axis, the chroma is the distance from the center to the point, and the saturation is the chroma normalized by the distance from the center to the sub-hexagon along the ray passing through $\mathbf{p}$.

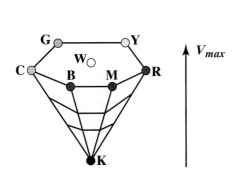

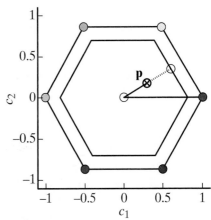

differences between the spaces are not necessarily captured by the names. For example, the term "value" in *HSV* is not used in the standard way, *HSL* does not use true "lightness", *HSI* usually does not denote true linear "intensity", and *HSB* does not capture actual "brightness", which is a subjective concept. Similarly, the "hue" in *HSV* slightly differs from that of *HSL*.

To avoid such confusion, we shall describe the original formulations of *HSV* and *HSL*, treating these as the official definitions, being careful to maintain the proper distinction between the two. In truth, they are both quite similar, with the most significant difference being that *HSV* is based on orthographic projection, whereas *HSL* is based on perspective projection. This choice primarily affects the saturation, which, as we will show, is not really the best quantity to be measuring anyway.

HSV

The *HSV* color space is based on the **hexcone model**, which involves orthographic projection along a direction parallel to the line of grays. The value V is simply the height within the cone, namely, V_{max}, while the saturation and hue are defined within the sub-hexagon specified by V_{max}. The saturation S is the distance from the center to the point, relative to the distance to the outer edge of the sub-hexagon along the same ray. That is, $S = 0$ is a shade of gray, while $S = 1$ means that the color is on the edge of the sub-hexagon and therefore is purely saturated. If we define $V_{min} \equiv \min(R', G', B')$, then any color can be split into the sum of a term involving two colors and a term representing a shade of gray that dilutes the color. For example, if $R' \geq G' \geq B'$, then we have $(R', G', B') = (V_\delta, G' - V_{min}, 0) + (V_{min}, V_{min}, V_{min})$, where $V_\delta \equiv V_{max} - V_{min}$. The saturation is then given by V_δ/V_{max}. The hue is approximated by H, the proportional length within the given sector along the direction parallel to the outer rim of the hexagon, added to the sector number. This definition yields an approximate hue between 0 and 6, which can then be converted to the range 0 to 1 or $0°$ and $360°$, as desired, by a simple conversion factor.

Conversion from $R'G'B'$ to *HSV* involves computing the corresponding coordinates within the hexcone given a point in the color cube, then normalizing to compute saturation. Assuming $0 \leq R', G', B' \leq 1$, the conversion is given by

$$V \equiv V_{max} \tag{9.75}$$

$$S \equiv V_\delta/V_{max} \tag{9.76}$$

$$H \equiv 60° \cdot \begin{cases} (G' - B')/V_\delta + 0 & \text{if } R' = V_{max} & (\text{red sectors 0 and 5}) \\ (B' - R')/V_\delta + 2 & \text{if } G' = V_{max} & (\text{green sectors 1 and 2}) \\ (R' - G')/V_\delta + 4 & \text{if } B' = V_{max}, & (\text{blue sectors 3 and 4}) \end{cases} \tag{9.77}$$

where $0 \leq S, V \leq 1$. In the definition of H, the fractions range from -1 to 1, so that when added to the central values of 0, 2, and 4 they yield a result that ranges from -1 to 5. Therefore, if the pixel is in sector 5, the resulting angle will be between $-60°$ and $0°$, so to ensure that $0° \leq H < 360°$, simply add $360°$ to the result above if $H < 0°$. Note that if $S = 0$, hue is not defined due to divide by zero.

If we define $\varsigma \equiv H/60°$, then the inverse transformation back to $R'G'B'$ is given by

$$(R', G', B') = \begin{cases} (V, V_3, V_1) & \text{if } \lfloor \varsigma \rfloor = 0 & (\text{sector 0}) \\ (V_2, V, V_1) & \text{if } \lfloor \varsigma \rfloor = 1 & (\text{sector 1}) \\ (V_1, V, V_3) & \text{if } \lfloor \varsigma \rfloor = 2 & (\text{sector 2}) \\ (V_1, V_2, V) & \text{if } \lfloor \varsigma \rfloor = 3 & (\text{sector 3}) \\ (V_3, V_1, V) & \text{if } \lfloor \varsigma \rfloor = 4 & (\text{sector 4}) \\ (V, V_1, V_2) & \text{if } \lfloor \varsigma \rfloor = 5, & (\text{sector 5}) \end{cases} \tag{9.78}$$

where

$$V_1 \equiv V(1 - S) \tag{9.79}$$

$$V_2 \equiv V(1 - \alpha S) \tag{9.80}$$

$$V_3 \equiv V(1 - (1 - \alpha)S) \tag{9.81}$$

and $\alpha \equiv \mathsf{s} - \lfloor \mathsf{s} \rfloor$.

The saturation $S = V_\delta / V_{max}$ can be interpreted geometrically as the chroma divided by the maximum chroma, where the chroma is defined as the distance from the hexagon center to the point, calculated as

$$C \equiv \sqrt{c_1^2 + c_2^2} = \sqrt{R'^2 + G'^2 + B'^2 - R'G' - R'B' - G'B'} \tag{9.82}$$

and the maximum chroma for the particular value V_{max} and the approximate hue H is given by the distance from the hexagon center to the sub-hexagon along the ray passing through the point, which can be shown to be[†]

$$C_{max} = CV_{max}/V_\delta \tag{9.83}$$

leading to saturation as the normalized chroma: $S = C/C_{max} = V_\delta/V_{max}$, which is the same as Equation (9.76).

If the exact hue angle within the hexagon is desired, it is given by

$$\tan \overline{H} = \frac{c_2}{c_1} = \frac{\sqrt{3}(G' - B')}{2R' - G' - B'} \tag{9.84}$$

where ATAN2 should be used to ensure that the resulting angle is in the correct quadrant,[‡] and the overbar indicates that this is not always identical to H, although they do happen to agree whenever the hue is a multiple of $30°$. Over the entire $R'G'B'$ cube, the maximum discrepancy between H in Equation (9.77) and $\overline{H}$ in Equation (9.84) is less than $1.2°$.

HSL

The *HSL* color space is based on the **triangle model**, in which the point (R', G', B') is perspectively projected onto the $R' + G' + B' = 1$ plane which passes through the red, green, and blue vertices of the color cube. This projection yields the intersection of the plane with the line passing through the black vertex and the point, as shown in Figure 9.15. Given any point in the cube, its projection onto this plane lands within the triangle defined by the red, green, and blue vertices. The line of grays intersects the triangle at its centroid. The hue is the counterclockwise angle with respect to the red axis, and the saturation is the distance from the centroid to the perspectively projected point normalized by the distance from the centroid to the edge of the triangle along the ray passing through the perspectively projected point. Assuming $0 \leq R', G', B' \leq 1$, the conversion from $R'G'B'$ to *HSL* is given by

$$\overline{L} \equiv \frac{1}{3}(R' + G' + B') \tag{9.85}$$

$$\overline{S} \equiv 1 - \frac{V_{min}}{\overline{L}} \tag{9.86}$$

$$\overline{H} \equiv \tan^{-1} \frac{\sqrt{3}(G' - B')}{2R' - G' - B'} \tag{9.87}$$

[†] The rather tedious derivation is omitted.
[‡] Section 4.4.4 (p. 179).

Figure 9.15 LEFT: In the *HSL* color space, a point in the $R'G'B'$ color cube is perspectively projected onto the plane defined by $R' + G' + B' = 1$, with all projections passing through the origin. RIGHT: Top-down view (white is directly above black) of the triangle defined by the intersection of the plane and the cube. The complete hexagon is shown for reference.

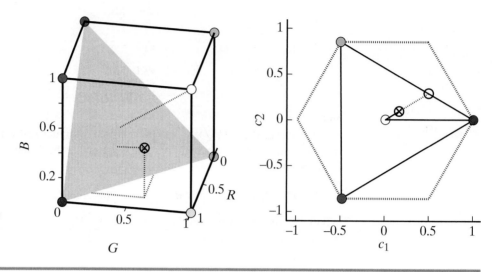

where the overbars are used for clarity. $\overline{H}$ is the exact angle, identical to Equation (9.84); $\overline{S}$ is the saturation in the triangle model, which is quite different from the saturation in the hexcone model; and the overbar in $\overline{L}$ reminds us that this is neither true lightness nor luminance but rather luma with coefficients $\alpha_r = \alpha_g = \alpha_b = \frac{1}{3}$. Conversion back to $R'G'B'$ depends on the sector:

$$(R', G', B') = \begin{cases} (\overline{V}_2, \overline{V}_3, V_{min}) & \text{if } 0° \le \overline{H} < 120° \quad (\text{sectors 0 and 1}) \\ (V_{min}, \overline{V}_2, \overline{V}_3) & \text{if } 120° \le \overline{H} < 240° \quad (\text{sectors 2 and 3}) \\ (\overline{V}_3, V_{min}, \overline{V}_2) & \text{if } 240° \le \overline{H} < 360° \quad (\text{sectors 4 and 5}) \end{cases} \tag{9.88}$$

where

$$V_{min} = \overline{L}(1 - \overline{S}) \tag{9.89}$$

$$\overline{V}_2 \equiv \overline{L}\left(1 + \overline{S}\frac{\cos(\overline{H} - h)}{\cos(60° + h - \overline{H})}\right) \tag{9.90}$$

$$\overline{V}_3 \equiv 3\overline{L} - (V_{min} + \overline{V}_2) \tag{9.91}$$

and $h \equiv 120° \cdot \lfloor \overline{H}/120° \rfloor$.

To understand the definition of saturation, note that the point (R', G', B') perspectively projects onto the plane at $\frac{1}{D}(R', G', B')$, where $D \equiv R' + G' + B' = 3\overline{L}$, so that $(R' + G' + B')/D = 1$. If we let (c_1, c_2) refer to the coordinates of the projection of (R', G', B') onto the c_1-c_2 plane by the Hanbury transformation in Equation (9.74), and $(\overline{c}_1, \overline{c}_2)$ be the coordinates of the projection of $\frac{1}{D}(R', G', B')$, then it is easy to see that

$$\begin{bmatrix} \overline{c}_1 \\ \overline{c}_2 \end{bmatrix} = \frac{1}{D}\begin{bmatrix} c_1 \\ c_2 \end{bmatrix} \tag{9.92}$$

which reveals that the two points are related to each other by a simple scaling along the ray from the origin, so that both points share the same hue, that is, $\tan^{-1}\frac{c_1}{c_2} = \tan^{-1}\frac{\overline{c}_1}{\overline{c}_2}$. The chroma resulting from this perspective projection is

$$\overline{C} \equiv \sqrt{\overline{c}_1^2 + \overline{c}_2^2} = \frac{1}{D}\sqrt{c_1^2 + c_2^2} = \frac{C}{D} \tag{9.93}$$

where C is the chroma from orthographic projection defined in Equation (9.82).

As before, saturation is the normalized chroma: $\overline{S} = \frac{\overline{C}}{\overline{C}_{max}}$, where $\overline{C}_{max}$ is the distance from the triangle centroid to the intersection of the ray with the triangle, calculated as[†]

$$\overline{C}_{max} = \frac{\overline{C}}{3(\overline{L} - V_{min})} \tag{9.94}$$

which leads to $\overline{S} = \frac{3(\overline{L} - V_{min})}{3\overline{L}} = 1 - \frac{V_{min}}{\overline{L}}$, as in Equation (9.86).

Sometimes you will see the following definition for hue:

$$\overline{H} \equiv \begin{cases} \theta & \text{if } B' \leq G' \\ 360° - \theta & \text{otherwise} \end{cases} \tag{9.95}$$

where

$$\theta \equiv \cos^{-1}\left(\frac{\frac{1}{2}[(R' - G') + (R' - B')]}{\sqrt{(R' - G')^2 + (R' - B')(G' - B')}}\right) \tag{9.96}$$

or its equivalent using the trigonometry identity $\cos^{-1}\theta = 90° - \tan^{-1}(\theta/\sqrt{1 - \theta^2})$. Simple algebraic manipulation reveals that both Equation (9.95) and its inverse tangent equivalent are identical to the much more compact Equation (9.87).

For completeness we should mention that the original formulation of *HSL* allows arbitrary coefficients α_r, α_g, and α_b in Equation (9.85), as long as $\alpha_r + \alpha_g + \alpha_b = 1$. However, relaxing the constraint that $\alpha_r = \alpha_g = \alpha_b = \frac{1}{3}$ means that the "line of grays" is no longer a line but a curve, which greatly complicates the math with no real practical benefit for most applications.

Using Chroma Instead of Saturation

In reality, saturation does not do a good job of capturing the purity of a color. Recall that both *HSV* and *HSL* were devised for the context of computer graphics and user interfaces, in which it makes sense for the quantity presented to the user (saturation) to always range from 0 to 1. For image processing and analysis, however, it is more important that the quantity capture some notion of the color purity. To see that saturation does not do this, consider the color $(R', G', B') = (0.01, 0, 0)$, assuming that the values range from 0 to 1. On the screen such a color would be considered black and colorless, but its saturation is 1 in both *HSV* and *HSL*, as if it were completely saturated. To overcome this problem, it is often preferable to use the chroma, which is obtained by removing the normalization, resulting in color spaces that could be called *HCV* or *HCL*. There are four choices for chroma, any of which may be used: the actual distances C and $\overline{C}$ in Equations (9.82) and (9.93), respectively, or their approximations $C' \equiv V_{\delta}$ in the case of *HSV* or $\overline{C}' \equiv 1 - V_{min}$ in the case of *HSL*. Figure 9.16 shows an example of chroma versus saturation for a real image.

9.5.9 CIE *XYZ*, *L* u* v**, and *L* a* b**

In many applications we would like to be able to measure the difference between two colors. Clearly, such differences only make sense when they are small, because (for example) it makes no sense to ask whether red looks more like green than blue looks like yellow. Nevertheless, even with small differences, the choice of color space is important. In particular, one cannot expect to obtain a meaningful result by performing such a difference in an *RGB* or *R'G'B'* space. To see that CIE *XYZ* has the same limitation, note that in Figure 9.10 green occupies a much larger area than yellow, so that the distance between two colors

---

[†] As before, the derivation is omitted.

Original   HSV   HSL   Max − min

Reprinted from Pattern Recognition Letters, 29(4), A. Hanbury, "Constructing cylindrical coordinate colour spaces," Pages 494–500, Copyright 2008, with permission from Elsevier.

Figure 9.16 Saturation versus chroma. Chroma more consistently captures the intuitive notion of the pureness of the color than does saturation.

that are very far from each other (e.g., red and yellow) can be the same as the distance between two colors that look quite similar (e.g., two types of green). Quantitatively, such a result can be obtained by performing a variation of the color matching experiment in which two color patches are presented to the observer under the same lighting, and the observer is asked whether the two patches are the same color. The threshold for being able to distinguish between two colors is known as the just-noticeable difference (JND),[†] and if the contours of the JND regions are plotted on the CIE chromaticity diagram, they appear as ellipses which are known as **MacAdam ellipses**.

In a perceptually uniform color space the JND contours are circles, so that Euclidean distances in the color space capture the perceptual dissimilarity between colors. One way to achieve an approximation to such an ideal is to apply a projective transform to the *XYZ* coordinates:

$$u' \equiv \frac{4X}{X + 15Y + 3Z} = \frac{4x}{3 - 2x + 12y} \tag{9.97}$$

$$v' \equiv \frac{9Y}{X + 15Y + 3Z} = \frac{9y}{3 - 2x + 12y} \tag{9.98}$$

where (u', v') define the **CIE 1976 Uniform color space (UCS)**. The quantities u' and v' denote the CIE 1976 successors to the obsolete 1960 CIE u and v quantities, related by $u = u'$, $v = \frac{2}{3}v'$.

The CIE 1976 UCS is used by the **CIE 1976 L*u*v* color space (CIELUV)**.[‡] The transformation from *XYZ* to $L*u*v*$ is given by applying a nonlinear transfer function to the relative luminance, then multiplying by shifted versions of u' and v'.

$$L* \equiv 116 f(Y/Y_n) - 16 \tag{9.99}$$

$$u* \equiv 13L*(u' - u'_n) \tag{9.100}$$

$$v* \equiv 13L*(v' - v'_n) \tag{9.101}$$

[†] Section 2.1.2 (p. 21).

[‡] CIELUV is written without any space and is pronounced "SEA-love". Similarly, CIELAB is pronounced "SEA-lab".

where the relative luminance is obtained by dividing the CIE luminance Y by the luminance Y_n of the reference white, and where the nonlinear transfer function f is defined as the cube root for most values, with a linear slope at low intensities chosen so that the function and its derivative are continuous at the junction:

$$f(t) \equiv \begin{cases} t^{1/3} & \text{if } 0.008856 \leq t \leq 1 \\ 7.787t + 16/116 & \text{if } 0 \leq t < 0.008856 \end{cases} \tag{9.102}$$

The values u'_n and v'_n are the u' and v' of the reference white, whose tristimulus values are given by X_n, Y_n, and Z_n, respectively. We should note that the shifting of u' and v' in $L^*u^*v^*$ has no basis in psychovisual experiments, so there is no reason to expect them to perform well in scenarios far from the white point used to derive them. The quantity L^*, known as CIE "lightness" and ranging from 0 to 100, is the standard approximation to the human perceptual response to luminance, designed by taking into account Stevens' Power Law.[†] It closely models the Munsell value, except that the range is different.

An alternative to CIELUV is the **CIE 1976 L*a*b* color space (CIELAB)**, which is also designed to be perceptually uniform. Rather than using a projective transform, as in CIELUV, CIELAB relies upon color differences similar to opponent colors:

$$L^* \equiv 116f(Y/Y_n) - 16 \tag{9.103}$$

$$a^* \equiv 500 \left(f(X/X_n) - f(Y/Y_n) \right) \tag{9.104}$$

$$b^* \equiv 200 \left(f(Y/Y_n) - f(Z/Z_n) \right) \tag{9.105}$$

where f is defined in Equation (9.102). The a^* axis roughly corresponds to the red-green opponent colors, while the b^* axis corresponds to the blue-yellow opponent colors. Although the committee at the time (in 1976) could not agree between the two alternatives for chroma (hence resulting in two standards), over the years it has generally been agreed upon that $L^*a^*b^*$ is the more accurate model, so that $L^*a^*b^*$ is the standard approach today when a perceptually uniform color space is needed. Even so, note by comparing with Equation (9.31) that $L^*a^*b^*$ itself is flawed because it is based upon a "wrong von Kries" method, with scaling performed (incorrectly) in XYZ space rather than in the space of the L-, M-, and S-cones.

The total color difference $\sqrt{(L_1^* - L_2^*)^2 + (u_1^* - u_2^*)^2 + (v_1^* - v_2^*)^2}$ in $L^*u^*v^*$ space is approximately perceptually uniform. That is, two colors whose difference is small are not perceptually distinct, whereas two colors whose difference is large are noticeably different. Similarly, in $L^*a^*b^*$ the total color difference is given by $\sqrt{(L_1^* - L_2^*)^2 + (a_1^* - a_2^*)^2 + (b_1^* - b_2^*)^2}$. More recent formulations, such as **CIE94** and **CIEDE2000**, include additional normalization and terms to improve perceptual normalization.

9.5.10 CMYK

The final color space we shall consider is *CMYK* (cyan-magenta-yellow-black). Unlike all the other color spaces we have mentioned, CMYK is a subtractive color space used for printing. In theory, it is simply the additive inverse of *RGB*:

$$\begin{bmatrix} C \\ M \\ Y \end{bmatrix} = 1 - \begin{bmatrix} R \\ G \\ B \end{bmatrix} \tag{9.106}$$

[†] Section 2.3.2 (p. 43).

so that pure red $(R = 1, G = B = 0)$ is a mixture of magenta and yellow, and so forth, a relationship that is obvious from the color wheel and chromaticity diagram. In practice, however, as we have mentioned, paint pigments and inks are highly nonlinear, and therefore subtractive color spaces are much more difficult to characterize mathematically. Although we often say that cyan, magenta, and yellow are the primary colors of printing, this is really just an approximation necessitated by economics. Inexpensive printers can faithfully reproduce a wide range of colors using just mixtures of these three colors, but the color gamut of such printers (which are not triangles but rather irregular shapes) tend to be much smaller than the gamut of displays. As a result, high-end printers use more than three primary colors (e.g., orange and green), and paint mixers in paint stores typically combine 12 or more colors.

9.6 Further Reading

The material in this chapter lies at the intersection of color science (colorimetry), video engineering, and image processing. There are many good books on color science, such as the classic by Wyszecki and Stiles [1982] or the more recent book by Lee [2005]. An excellent short introduction to color appearance in general has been written by Harold [2001] as well as Brainard and Stockman [2009], and another easily accessible presentation can be found in the book by Fairchild [2011]. Such works cover the CIE chromaticity diagram, CIE color spaces, CAMs and CATs, the difference between SSFs and CMFs, and the difference between saturation, chroma, and colorfulness. Helpful descriptions can also be found in Wandell [1995] and Palmer [1999], which are written from a psychological point of view. For an introductory presentation aimed at the prepress and printing industries, see Green [1999]. And, of course, the book that started it all is the classic work of Newton [1704].

Early work on color matching is due to Grassmann [1854]. Grassmann influenced Maxwell [1860], who performed detailed color matching experiments of his own, providing an early description of metamers. The CIE 1931 CMFs are based on the 2° experiments conducted independently by Wright [1929] and Guild [1931]. However, these experiments did not measure the CMFs directly, so the CIE 1931 CMFs had to be obtained by assuming that the CIE 1924 photopic LEF was a linear combination of the CMFs, which led to errors in the estimated CMFs due to the fact that the CIE 1924 LEF curve has a noticeable error in the small-wavelength (blue) region. As explained by Stockman and Sharpe [1999], corrections later made by Judd [1951] and Vos [1978] alleviate this problem. The CIE 1964 CMFs are based on the 10° experimental results presented by Stiles and Burch [1959] and, to a lesser extent, on those by Speranskaya [1959]. The cone SSFs

were first measured by Svaetichin [1956]. The CMFs shown in Figure 9.5 are from the 2° experiments of Stiles and Burch [1955]; while the cone SSFs in Figure 2.2 are from Stockman and Sharpe [2000] and are based on the 10° data of Stiles and Burch [1959], among other sources. The data for the CMFs, SSFs, and CIE chromaticity diagram can be found online.[†]

Some of the material in this chapter relies upon the video engineering insights of Poynton [2003]. In that work the reader will find descriptions on transforming between cameras and displays, the mathematics of white balancing, the distinction between luminance and luma, and so forth. The diagram in Figure 9.12 explaining the origin of luma is adapted from that work, and the notation $^{601}_{255}Y'$ can also be found there. The distinction between the two terms luma and luminance were clarified in Engineering Guideline EG 28, which was adopted by SMPTE in 1993. The official Rec. 601, Rec. 709, and Rec. 2020 specifications can be found online.[‡] The term "wrong von Kries" is due to Terstiege [1972].

The Munsell Color System was first described in the classic books of Munsell [1905, 1915]. The so-called Hanbury transformation is from Hanbury [2008]. The Bradford transform is due to Lam [1985], while the Sharp transform can be found in Finlayson and Süsstrunk [2000]. The Macbeth ColorChecker is the work of McCamy et al. [1976]. Alvy Ray Smith is the pioneer behind *HSV* and *HSL*, and the equations presented here are from his original publication of Smith [1978]. In particular, our HSL model is identical to Smith's original HSL model (except for our restriction on the luma coefficients); it is also identical to the model called HSI in Gonzalez and Woods [2008]. For measuring the difference between color distributions, see the earth mover's distance as described by Rubner et al. [1998].

[†] Colour and Vision Research Laboratory, http://www.cvrl.org

[‡] http://www.itu.int

PROBLEMS

9.1 Define each of the following terms, and explain how they are related to each other:

(a) spectral power distribution (SPD)

(b) spectral sensitivity function (SSF)

(c) color matching function (CMF)

(d) luminous efficiency function (LEF)

9.2 The spectral power distributions (SPDs) of three monochromatic light sources A, B, and C are shown in the plot below.

(a) What color does light A look like? B? C?

(b) If lights B and C are shone together, what color will result?

(c) The table below shows the reflected power measured when the lights were shone in different combinations on an unknown surface. What color is the surface? Explain your answer.

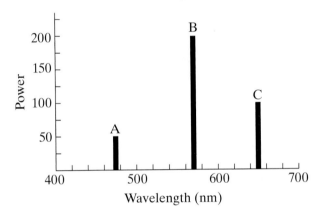

incident light	A	B	C	A + B	A + C	B + C
reflected power	40	80	10	100	30	70

9.3 Explain the relationship between the rainbow and the color wheel.

9.4 Can an SSF be estimated for the rods? Can an LEF be estimated for the cones? Why or why not?

9.5 Compare and contrast the additive and the subtractive color models.

9.6 What is a metamer?

9.7 If a plot has some negative values, list all of the possibilities that it could be: SPD, SSF, LEF, or CMF. Repeat for a plot without any negative values.

9.8 By combining adjacent spectral bands, the 3×35 cone fundamental matrix can be represented in simplified form by the following 3×7 matrix:

$$\mathbf{S} = \begin{bmatrix} \mathbf{s}_L^T \\ \mathbf{s}_M^T \\ \mathbf{s}_S^T \end{bmatrix} = \begin{bmatrix} 0.0134 & 0.0858 & 0.4985 & 0.9598 & 0.6523 & 0.1059 & 0.0037 \\ 0.0174 & 0.1472 & 0.6548 & 0.8497 & 0.2173 & 0.0100 & 0.0002 \\ 0.3751 & 0.7165 & 0.0758 & 0.0013 & 0.0000 & 0 & 0 \end{bmatrix}$$

where the first column captures the multiband 380–430 nm, the second column captures 430–480 nm, and so forth, to the last column, which captures 670–720 nm.

(a) What are the outputs of the three cones for the test light $\mathbf{t} = \begin{bmatrix} 8 & 9 & 1 & 9 & 6 & 2 & 3 \end{bmatrix}^T$?

(b) Find a metamer of this test light.

9.9 Suppose that, for a given set of primaries, a particular test light **t** is matched with the tristimulus values $u_r = 1.0$, $u_g = 0.5$, and $u_b = 0.7$. Assuming that Grassman's Law applies, what are the tristimulus values when the intensity of the test light is doubled?

9.10 If a set of CMFs has no negative values, what does this tell you about the primaries used?

9.11 By combining adjacent spectral bands, the CMFs for CIE 1931 XYZ can be represented in simplified form by the following 3×7 matrix:

$$\tilde{C} = \begin{bmatrix} 0.0523 & 0.2843 & 0.0419 & 0.4864 & 0.9633 & 0.2932 & 0.0150 \\ 0.0016 & 0.0505 & 0.4120 & 0.9547 & 0.5982 & 0.1145 & 0.0055 \\ 0.2515 & 1.5581 & 0.3137 & 0.0131 & 0.0007 & 0.0000 & 0 \end{bmatrix}$$

Using the cone fundamentals from Prob. 9.8, calculate the 3×3 matrix that transforms cone outputs to tristimulus values.

9.12 Explain how the Y in the CIE XYZ space relates to the luminance.

9.13 Name the rival to trichromatic theory.

9.14 Explain the difference between chroma and saturation.

9.15 For each of the names below, provide an everyday name to indicate what color is represented; also specify which color system is being used.

(a) 10R 3/6

(b) 2Y 8/16

(c) NCS 4055-R95B

(d) NCS 0580-Y10R

9.16 Search online for the CIE 1931 CMFs from Wright and Guild.[†] Plot the data and verify that they match Figure 9.9. Repeat for the CIE 1964 CMFs from Stiles and Burch.

9.17 Can any color be matched as the linear combination of just two monochromatic primaries, if (a) the primaries are fixed or (b) the primaries can be selected according to the color being matched? Explain why or why not using the CIE chromaticity diagram.

9.18 Answer the following:

(a) What is the approximate correlated color temperature (CCT) of standard illuminant D_{75}?

(b) As the CCT increases, does the dominant wavelength increase or decrease? (*Hint:* Recall Planck's Law from Figure 2.39.)

(c) Use Wien's displacement law[‡] to compute the dominant wavelength of D_{75}.

9.19 Three of the colors on the Macbeth ColorChecker color-rendition chart are "blue flower", "moderate red", and "foliage", with CIE 1931 xyY coordinates of (0.2651,0.2400,24.27), (0.4533,0.3058,19.77), and (0.3372,0.4220,13.29), respectively. Suppose these colors produce RGB values of (0.35,0.40,0.92), (0.85,0.42,0.31), and (0.27,0.81,0.33), respectively. Compute the 3×3 transformation matrix $\mathbf{M}_c^{-1}$ to transform an RGB triplet sensed by the camera into the XYZ color space. Are the RGB values gamma-compressed? Why or why not?

---

† For example, see http://www.cvrl.org
‡ Section 2.5.3 (p. 60).

9.20 Explain why ICC profiles are useful.

9.21 Are the CMFs that result from a color matching experiment dependent upon the primary lights used? Explain your answer.

9.22 Give some reasons why photographs taken of the same scene by cameras calibrated using the Macbeth ColorChecker color-rendition chart may still not look identical.

9.23 What is white balancing, and why is it important?

9.24 Using the parameters of Rec. 2020 provided in Table 9.1:

(a) Compute the 3×3 matrix conversion between RGB and CIE XYZ.

(b) What are the luma coefficients?

(c) Compute the 3×3 matrix that transforms from $R'G'B'$ to $Y'P_BP_R$ using 10 bits per sample, for which the black level is defined as code 64, and the nominal peak level is code 940. (10-bit chroma values are similarly scaled by a factor of 4 from their 8-bit values.)

(d) What is the transform from $R'G'B'$ to $Y'C_BC_R$?

9.25 Convert the gamma-corrected color $(R', G', B') = (20, 40, 200)$ to (a) HSV color space and (b) HSL color space. Also, (c) compute the chroma and max chroma.

9.26 Convert the color with HSV values $H = 120°$, $S = 0.7$, $V = 0.7$ to $R'G'B'$ space.

9.27 Convert the color with luminance $Y = 0.4$ and chromaticity coordinates $(x, y) = (0.25, 0.4)$ into

(a) CIE 1976 $L^* u^* v^*$ color space, given $Y_n = 0.54$, $u'_n = 0.2009$, and $v'_n = 0.4610$

(b) CIE 1976 $L^* a^* b^*$ color space, given $X_n = 0.3$, $Z_n = 0.66$

9.28 In order to convert from $R'G'B'$ to CIE XYZ, what must be known about the $R'G'B'$ space?

9.29 Several prominent high-tech corporations (one rhymes with "loft" and the other rhymes with "frugal") use red, green, blue, and yellow in their logos. Based on what you know about primary colors, explain why you think this combination has been found to be appealing.

9.30 Explain the difference between purple, violet, magenta, and pink.

9.31 Suppose your friend wants to build a computer monitor that uses cyan, magenta, and yellow primary lights. Is this possible? What will be the result?

9.32 Is color matching always linear? Why or why not?

9.33 Explain from the CIE chromaticity diagram why CMFs always have some negative values.

9.34 What is the difference between the photopic and scotopic LEFs?

9.35 Explain why chroma is better than saturation at capturing the purity of a color.

9.36 What standard illuminant is used in an ICC profile?

9.37 Convert $(R', G', B') = (0, 0, 255)$ to (a) luma $^{601}Y'$, (b) luma $^{709}Y'$, (c) grayscale using Equation (9.68), and (d) gamma expansion, followed by conversion to luminance, followed by gamma compression, using Rec. 709. What do you conclude from these numbers?

9.38 For computing distances between colors, what color space is recommended?

9.39 Show that Equation (9.77) is identical to the expression found in Smith [1978]:

$$H \equiv 60° \cdot \begin{cases} 5 + b & \text{if } G' \le B' \le R' \quad \triangleright \text{ sector 0} \\ 1 - g & \text{if } B' \le G' \le R' \quad \triangleright \text{ sector 1} \\ 1 + r & \text{if } B' \le R' \le G' \quad \triangleright \text{ sector 2} \\ 3 - b & \text{if } R' \le B' \le G' \quad \triangleright \text{ sector 3} \\ 3 + g & \text{if } R' \le G' \le B' \quad \triangleright \text{ sector 4} \\ 5 - r & \text{if } G' \le R' \le B' \quad \triangleright \text{ sector 5} \end{cases} \qquad (9.107)$$

where

$$r \equiv \frac{V - R'}{V - V_{min}} \qquad g \equiv \frac{V - G'}{V - V_{min}} \qquad b \equiv \frac{V - B'}{V - V_{min}}$$

9.40 In the $R'G'B'$ cube, points (R', G', B') and $(R' + x, G' + x, B' + x)$ lie along a line parallel to the line of grays, for any offset x. Due to the orthographic projection, in *HSV* both points project onto the same location in the hexagon and therefore have the same hue, chroma, and saturation. Verify this statement for chroma by substituting these values into (9.82).

9.41 Show that Equation (9.87) is equivalent to Equations (9.95)–(9.96).

CHAPTER 10
Segmentation

Now that we have covered the primary problems of image processing in the preceding chapters, the remainder of this book is focused upon the primary problems of image analysis. The first of these, segmentation, is a *bottom-up process* that groups pixels in an image based on their low-level, local properties such as color or texture, so that each region contains pixels that look similar to one another. Segmentation involves two subproblems, namely, determining a model of the pixels in each region and assigning each pixel to the best fitting model. Since the models are not known beforehand and since there are no labeled training data, segmentation is a problem of *unsupervised learning*. Generally, the goal is to group pixels that are projections of the same object in the scene, although it must be kept in mind that there is no objective definition of "object." For example, whether a person should be segmented from the background (keeping the head and body together as one object), or whether the head should be segmented separately from the body (two objects), can only be decided based on the particular task at hand.

There are two flavors of segmentation problems. In some cases it is desired to segment dense data, for example, pixels in an image, in which case the models tend to be fairly simple, capturing some notion of homogeneity in color or texture. In other situations the goal is to cluster sparse data, such as feature points detected in an image, in which case the models tend to be more complex, such as the equation for a line or plane. Although there is nothing fundamentally different between these two types of problems, in practice the various techniques tend to be more applicable for one type of problem than another. In this chapter we focus on segmenting images, whereas the next chapter addresses fitting more complex models to sparse data.

10.1 Thresholding

The simplest variant of the segmentation problem is **foreground/background segmentation**. Earlier we saw that thresholding produces a binary image in which foreground pixels have a value of 1, whereas background pixels have a value of 0.[†] Logically, these values can be considered as ON or OFF, respectively. The most difficult part of thresholding is to determine an appropriate threshold value, whether manually or automatically. In this section we discuss several well-known methods for automatically determining the threshold value.

10.1.1 Global Thresholding

Let τ be a single **global threshold**. Once τ has been determined, it is applied to every pixel in a straightforward manner:

$$I'(x, y) = \begin{cases} \text{ON} & \text{if } I(x, y) > \tau \\ \text{OFF} & \text{otherwise} \end{cases} \tag{10.1}$$

Most algorithms for automatically determining the threshold rely on examining the gray-level histogram. Typically there is one peak corresponding to the foreground pixels and another peak corresponding to the background pixels, with a valley in between. The goal is to find the gray level near the minimum of the valley in order to best separate the foreground from the background.

Two simple, widely used global thresholding techniques are known as the Ridler-Calvard algorithm and Otsu's method, both of which are described in this section. In a real implementation it may be advantageous to preprocess the image before applying such algorithms, such as first smoothing the image by convolving with a Gaussian, in order to make the histogram peaks more distinctive; or first computing the intensity edges and applying the algorithms only to the pixels near the edges, to reduce the asymmetry that occurs in the size of the histogram peaks when the foreground objects are small relative to the large background. Nevertheless, to simplify the presentation, we describe the basic algorithms without any such preprocessing.

Ridler-Calvard Algorithm

Figure 10.1 reproduces the image of fruit on a background that we saw before, along with its graylevel histogram. The histogram has one large well-defined mode on the left due primarily to pixels in the background, and another mode on the right, much broader and noisier, which is due primarily to foreground pixels. Therefore, a good threshold is one which separates these two modes of the histogram or, equivalently, lies in the valley between the two hills (at approximately gray level 130, in this case). A simple iterative algorithm to achieve this is the **Ridler-Calvard algorithm**. Let τ be a threshold, and let $\mu_{\blacktriangleleft}$ be the mean gray level of all the pixels whose gray level is less than or equal to τ, while $\mu_{\triangleright}$ is the mean gray level of all the pixels whose gray level is greater than τ. If we assume that the background is darker than the foreground, then $\mu_{\blacktriangleleft}$ is the mean of the background pixels, whereas $\mu_{\triangleright}$ is the mean of the foreground pixels.

[†] Section 3.2.4 (p. 83)

Figure 10.1 LEFT: A grayscale image of several types of objects (fruit) on a dark background (conveyor belt). RIGHT: The graylevel histogram of the image.

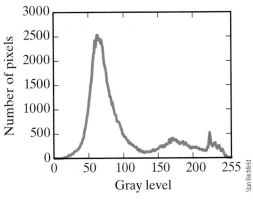

Let $h[\ell]$, $\ell = 0, \ldots, \zeta - 1$ be the value of the graylevel histogram for gray level ℓ, where $\zeta \equiv ngray = 256$ is the number of gray levels. Recalling our earlier discussion of moments,[†] it is easy to show that

$$\mu_{\blacktriangleleft} = \frac{m_1[\tau]}{m_0[\tau]} \quad \text{and} \quad \mu_{\triangleright} = \frac{m_1[\zeta - 1] - m_1[\tau]}{m_0[\zeta - 1] - m_0[\tau]} \tag{10.2}$$

where

$$m_0[\tau] = \sum_{\ell=0}^{\tau} h[\ell] \quad \text{and} \quad m_1[\tau] = \sum_{\ell=0}^{\tau} \ell h[\ell] \tag{10.3}$$

are the zeroth and first moments, respectively, of the histogram h from gray level 0 to τ. Note that m_0 is just the cumulative normalized histogram $\bar{c}$, scaled so that $m_0[\zeta - 1]$ is the number of pixels in the image.

The Ridler-Calvard algorithm iteratively computes the two means based on the current estimate of the threshold, then sets the threshold to the average of the two means. This classic algorithm, shown in Algorithm 10.1, requires one pass through the image to compute the histogram, then one pass through the histogram to compute the arrays m_0 and m_1, followed by a small number of iterations consisting only of constant-time operations. (Note that in Line 1 the normalized histogram could be used instead of the regular histogram since the divisions in Lines 9 and 10 cancel the scaling factor.)

As we shall see in the next chapter, Ridler-Calvard is simply the k-means algorithm applied to a graylevel histogram. Although iterative algorithms usually require a good starting point—and k-means is no exception—Ridler-Calvard is powerful because in practice any initial value will converge to the same solution, because the computation is one-dimensional. An example of the algorithm achieving a good solution despite a terrible initial value for τ is illustrated in Figure 10.2.

The Ridler-Calvard algorithm is based on the assumption that the foreground and background gray levels are distributed as Gaussians with equivalent standard deviations. In such a case, the optimal decision boundary occurs where the two distributions intersect:

$$\frac{1}{\sqrt{2\pi\sigma^2}} \exp\left\{-\frac{(\tau - \mu_{\blacktriangleleft})^2}{2\sigma^2}\right\} = \frac{1}{\sqrt{2\pi\sigma^2}} \exp\left\{-\frac{(\tau - \mu_{\triangleright})^2}{2\sigma^2}\right\} \tag{10.4}$$

[†] Section 4.4.1 (p. 174).

ALGORITHM 10.1 Compute an image threshold using the Ridler-Calvard algorithm

RIDLER-CALVARD(I)

Input: grayscale image I
Output: threshold value τ

1   $h \leftarrow$ COMPUTEHISTOGRAM(I)
2   $m_0[0] \leftarrow h[0]$
3   $m_1[0] \leftarrow 0$
4   **for** $k \leftarrow 1$ **to** $\zeta - 1$ **do**
5        $m_0[k] \leftarrow m_0[k-1] + h[k]$
6        $m_1[k] \leftarrow m_1[k-1] + k * h[k]$
7   $\tau \leftarrow \zeta/2$ ▶ reasonable initial value, but not important
8   **repeat**
9        $\mu_\blacktriangleleft \leftarrow m_1[\tau]/m_0[\tau]$
10      $\mu_\triangleright \leftarrow (m_1[\zeta - 1] - m_1[\tau])/(m_0[\zeta - 1] - m_0[\tau])$
11      $\tau \leftarrow$ ROUND $\left(\frac{1}{2}(\mu_\blacktriangleleft + \mu_\triangleright)\right)$
12 **until** τ does not change
13 **return** τ

where $\mu_\blacktriangleleft$ and $\mu_\triangleright$ are the mean gray levels of the two groups of pixels. Solving this equation for τ yields

$$\tau = \frac{(\mu_\blacktriangleleft + \mu_\triangleright)}{2} \tag{10.5}$$

which appears in Line 11 of the algorithm. This derivation illustrates an important point in the design and analysis of algorithms that process noisy, real-world data: namely, such algorithms often make statistical assumptions about the data they are processing whether these assumptions are explicitly acknowledged or not, and specifying such assumptions explicitly enables the algorithm to be separated from the model. In other words, understanding the underlying statistical assumptions allows us to separate the specific steps involved in

Figure 10.2 Step-by-step example of the Ridler-Calvard algorithm applied to the image of Figure 10.1. Note that even with an initial threshold far from the true solution, the algorithm converges in only five iterations. The top row shows the histogram. The green arrow pointing down indicates the threshold at each iteration, while the gold arrows pointing up indicate the two means. The bottom row shows the result of thresholding the image using the threshold for that iteration.

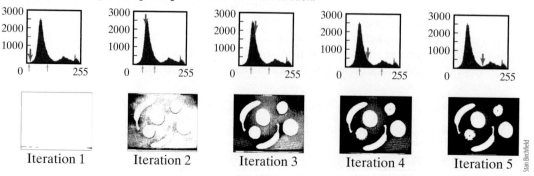

Iteration 1   Iteration 2   Iteration 3   Iteration 4   Iteration 5

the computation from the goal that the computation achieves. Making this distinction clear often yields new insights into the problem, making it easier to evaluate the algorithm and suggest potential improvements. In the case of Ridler-Calvard, the analysis reveals that the algorithm assumes that the foreground and background variances are identical.

Otsu's Method

If we relax this assumption and instead allow the two regions to have different variances, we arrive at **Otsu's method**. The goal of Otsu's method is to find the threshold τ that minimizes the *within-class variance*, which is defined as the weighted sum of the variances of the two groups of pixels:

$$\sigma_w^2(\tau) \equiv p_\blacktriangleleft(\tau)\sigma_\blacktriangleleft^2(\tau) + p_\triangleright(\tau)\sigma_\triangleright^2(\tau) \tag{10.6}$$

where τ is the unknown threshold, $p_\blacktriangleleft(\tau)$ is the proportion of pixels whose gray level is less than or equal to τ, and $\mu_\blacktriangleleft(\tau)$ and $\sigma_\blacktriangleleft^2(\tau)$ are their mean and variance in gray level, respectively. These quantities, as well as the analogous values for the foreground region, are given by

$$p_\blacktriangleleft(\tau) = \sum_{\ell=0}^{\tau}\overline{h}[\ell] = \frac{m_0[\tau]}{m_0[\zeta-1]} \qquad p_\triangleright(\tau) = \sum_{\ell=\tau+1}^{\zeta-1}\overline{h}[\ell] = \frac{m_0[\zeta-1]-m_0[\tau]}{m_0[\zeta-1]} \tag{10.7}$$

$$\mu_\blacktriangleleft(\tau) = \frac{\sum_{\ell=0}^{\tau}\ell\overline{h}[\ell]}{\sum_{\ell=0}^{\tau}\overline{h}[\ell]} = \frac{m_1[\tau]}{m_0[\tau]} \qquad \mu_\triangleright(\tau) = \frac{\sum_{\ell=\tau+1}^{\zeta-1}\ell\overline{h}[\ell]}{\sum_{\ell=\tau+1}^{\zeta-1}\overline{h}[\ell]} = \frac{m_1[\zeta-1]-m_1[\tau]}{m_0[\zeta-1]-m_0[\tau]} \tag{10.8}$$

$$\sigma_\blacktriangleleft^2(\tau) = \frac{\sum_{\ell=0}^{\tau}\overline{h}[\ell](\ell-\mu_\blacktriangleleft(\tau))^2}{\sum_{\ell=0}^{\tau}\overline{h}[\ell]} \qquad \sigma_\triangleright^2(\tau) = \frac{\sum_{\ell=\tau+1}^{\zeta-1}\overline{h}[\ell](\ell-\mu_\triangleright(\tau))^2}{\sum_{\ell=\tau+1}^{\zeta-1}\overline{h}[\ell]} \tag{10.9}$$

where $\overline{h}$ is the normalized graylevel histogram. Note that $p_\blacktriangleleft(\tau) + p_\triangleright(\tau) = 1$ for all values of τ, and the mean gray level of all the pixels is given by $\mu = m_1[\zeta-1]/m_0[\zeta-1]$.

If we define the *between-class variance* as

$$\sigma_b^2(\tau) \equiv p_\blacktriangleleft(\tau)(\mu_\blacktriangleleft(\tau)-\mu)^2 + p_\triangleright(\tau)(\mu_\triangleright(\tau)-\mu)^2 \tag{10.10}$$

it can be shown that the sum of the within- and between-class variances is the total variance of all the pixel values, by simply expanding the definition of σ^2:

$$\sigma^2 = \sum_{\ell=0}^{\zeta-1}\overline{h}[\ell](\ell-\mu)^2 \tag{10.11}$$

$$= \sum_{\ell=0}^{\tau}\overline{h}[\ell](\ell-\mu_\blacktriangleleft+\mu_\blacktriangleleft-\mu)^2 + \sum_{\ell=\tau+1}^{\zeta-1}\overline{h}[\ell](\ell-\mu_\triangleright+\mu_\triangleright-\mu)^2 \tag{10.12}$$

$$= \sum_{\ell=0}^{\tau}\overline{h}[\ell]\left((\ell-\mu_\blacktriangleleft)^2 + 2(\ell-\mu_\blacktriangleleft)\underbrace{(\mu_\blacktriangleleft-\mu)}_{0} + (\mu_\blacktriangleleft-\mu)^2\right)$$

$$+ \sum_{\ell=\tau+1}^{\zeta-1}\overline{h}[\ell]\left((\ell-\mu_\triangleright)^2 + 2(\ell-\mu_\triangleright)\underbrace{(\mu_\triangleright-\mu)}_{0} + (\mu_\triangleright-\mu)^2\right) \tag{10.13}$$

$$= p_{\blacktriangleleft}(\tau)\sigma_{\blacktriangleleft}^2(\tau) + p_{\blacktriangleleft}(\tau)(\mu_{\blacktriangleleft} - \mu)^2 + p_{\triangleright}(\tau)\sigma_{\triangleright}^2(\tau) + p_{\triangleright}(\tau)(\mu_{\triangleright} - \mu)^2 \quad (10.14)$$

$$= \sigma_w^2(\tau) + \sigma_b^2(\tau) \tag{10.15}$$

where the dependence of $\mu_{\blacktriangleleft}$ and $\mu_{\triangleright}$ on τ has been omitted to avoid clutter, and the proof that the middle terms go to zero is left as an exercise.[†]

Since the total variance σ^2 does not depend on the threshold τ, minimizing σ_w^2 is the same as maximizing σ_b^2. The advantage of the latter is that it is dependent only upon first-order properties (means) rather than second-order properties (variances), thus making it easier to compute. Substituting the expressions for $p_{\blacktriangleleft}(\tau), \mu_{\blacktriangleleft}(\tau), p_{\triangleright}(\tau)$, and $\mu_{\triangleright}(\tau)$ into Equation (10.10) and simplifying yields

$$\sigma_b^2(\tau) = \frac{(m_1[\tau] - \mu m_0[\tau])^2}{m_0[\tau](m_0[\zeta - 1] - m_0[\tau])} \tag{10.16}$$

Since there is a small number of possible thresholds (usually just 256), Otsu's method iterates through all these possible values for τ to find the one that maximizes the quantity $\sigma_b^2(\tau)$. The quality of the result yielded by the threshold τ is given by the following measure of separability:

$$\eta(\tau) = \frac{\sigma_b^2(\tau)}{\sigma^2} \tag{10.17}$$

Otsu's method, shown in Algorithm 10.2, begins with the same precomputation as Ridler-Calvard, and it also can be performed with either the standard histogram or the normalized histogram, since the division in Lines 7 and 10 cancel the normalization.

ALGORITHM 10.2 Compute an image threshold using Otsu's method

OTSU(I)

Input: grayscale image I
Output: threshold value τ

1   $h \leftarrow$ COMPUTEHISTOGRAM(I)
2   $m_0[0] \leftarrow h[\ell]$
3   $m_1[0] \leftarrow \ell * h[\ell]$
4   **for** $\ell \leftarrow 1$ **to** $\zeta - 1$ **do**
5         $m_0[\ell] \leftarrow m_0[\ell - 1] + h[\ell]$
6         $m_1[\ell] \leftarrow m_1[\ell - 1] + \ell * h[\ell]$
7   $\mu \leftarrow m_1[\zeta - 1]/m_0[\zeta - 1]$
8   $\hat{\sigma}_b^2 \leftarrow 0$
9   **for** $\ell \leftarrow 0$ **to** $\zeta - 1$ **do**
10        $\sigma_b^2 \leftarrow (m_1[\ell] - \mu m_0[\ell])^2/(m_0[\ell] * (m_0[\zeta - 1] - m_0[\ell]))$
11   **if** $\sigma_b^2 > \hat{\sigma}_b^2$ **then**
12              $\hat{\sigma}_b^2 \leftarrow \sigma_b^2$
13              $\tau \leftarrow \ell$
14 **return** τ

[†] Problem 10.3.

Similar to Ridler-Calvard, Otsu's method requires one pass through the image, then one pass through the histogram. The key difference between the two algorithms is that Otsu's method performs an exhaustive search over all possible $\zeta = 256$ thresholds via another pass through the histogram, rather than iteratively converging on a solution from a starting point. Figure 10.3 shows a comparison of the outputs of Ridler-Calvard and Otsu's method on the fruit image.

10.1.2 Adaptive Thresholding

Global thresholding techniques such as the Ridler-Calvard algorithm and Otsu's method do not perform well when the image noise characteristics vary across the image. For example, strong lighting conditions can produce highlights in one part of the image and shadows in other parts. To overcome such difficulties, **adaptive thresholding** techniques are needed, in which the threshold used at any given pixel in the image is based upon local statistical properties in the neighborhood of the pixel:

$$I'(x, y) = \begin{cases} \text{ON} & \text{if } I(x, y) > \tau(x, y) \\ \text{OFF} & \text{otherwise} \end{cases} \qquad (10.18)$$

where $\tau(x, y)$ indicates that the threshold varies as a function of the pixel. Figure 10.4 shows an image with strong lighting variations and the result of adaptive thresholding.

One way to perform adaptive thresholding is to divide the image into blocks and to run a global thresholding algorithm, such as Ridler-Calvard or Otsu's method, on each block to determine the threshold for that block. Indeed, one of the classic techniques, known as **Chow-Kaneko**, does exactly that: the image is divided into overlapping blocks, and the histogram is examined for each block to determine a threshold value for the block. Interpolation between these threshold values then yields a threshold function defined over the entire image. Alternatively, a preprocessing step can be applied, such as the white top-hat (WTH) transform,[†] followed by a global threshold.

An alternate family of approaches compares each pixel with the statistics of its surrounding neighborhood. The simplest way to do this is to convolve the image with a smoothing kernel (e.g., a Gaussian), then set the threshold as

$$\tau(x, y) = t \cdot \mu(x, y) \qquad (10.19)$$

where $0 \leq t \leq 1$ is a scalar and $\mu(x, y)$ is the mean of the neighborhood, so that each pixel is set to ON if it is greater than $100t$ percent of the mean. For example, if t is set to 0.8, then

Figure 10.3 From left to right: Input image, output of the Ridler-Calvard algorithm, and output of Otsu's method. On this particular image, the outputs are almost indistinguishable.

Stan Birchfield

[†] Section 5.6.3 (p. 265)

Stan Birchfield

Figure 10.4 Example
of adaptive thresholding.

a pixel whose neighborhood mean is 128 will be set to OFF only if it is less than or equal to $0.8 \cdot 128 = 102$. When processing a text document (such as the one in Figure 10.4), the algorithm should be conservative in declaring pixels OFF but generous in declaring pixels ON, which means that t should be small. Note that the mean can be computed using either a Gaussian, $\mu(x, y) \equiv I(x, y) \circledast G(x, y)$, or a box filter, $\mu(x, y) \equiv I(x, y) \circledast B(x, y)$, and that the latter can be efficiently computed using the integral image. Alternatively, the median or some other operation on the neighborhood can be used.

Niblack's method takes this idea to the next level by considering both the mean and standard deviation of the pixel values in the neighborhood. The threshold is then given by

$$\tau(x, y) = \mu(x, y) - k \cdot \sigma(x, y) \qquad (10.20)$$

where $\sigma(x, y)$ is the standard deviation and k is a scalar with a recommended value of $k = 0.2$. In a region containing significant variation (e.g., a region with text on a background page) the threshold is lowered to reduce the number of spurious off pixels, but in a homogeneous region (e.g., just the background page with no text), the threshold is close to $\mu(x, y)$, making it susceptible to noise. **Sauvola's method** overcomes this drawback by weighting the standard deviation by its dynamic range:

$$\tau(x, y) = \mu(x, y) \cdot \left[1 - k \cdot \left(1 - \frac{\sigma(x, y)}{r} \right) \right] \qquad (10.21)$$

where k is between 0.2 and 0.5 and $r \equiv \frac{\zeta}{2} = 128$ is the maximum standard deviation for an 8-bit image. Note that this equation is the same as Equation (10.19) with an adaptive t based on the local standard deviation. In a high-contrast area $\sigma(x, y)$ is close to r, in which case the threshold is just the mean. In a low-contrast area, the threshold is reduced by the amount based on the decrease in contrast (since $\sigma(x, y) \leq r$, the factor multiplied by k is always nonnegative), down to a minimum of $\mu(x, y) \cdot (1 - k)$. For example, if $k = 0.2$ then the threshold varies between $\mu(x, y)$ and $0.8\mu(x, y)$.

10.1.3 Hysteresis Thresholding

The concept of **hysteresis thresholding** uses a low threshold τ_{low} and a high threshold τ_{high}. Any pixel is labeled ON if it is either above the high threshold or above the low threshold and connected to another pixel that is above the high threshold, as shown in Figure 10.5.

Figure 10.5 An illustration
of hysteresis thresholding,
also known as double
thresholding.

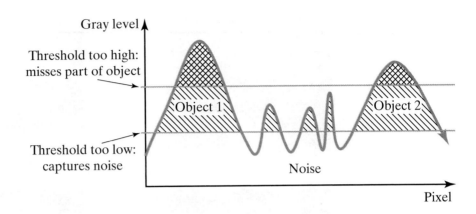

Figure 10.5 An illustration of hysteresis thresholding, also known as double thresholding.

In many situations hysteresis thresholding is able to achieve more robust results than are possible with a single threshold. The term *hysteresis* indicates that the output of a pixel is not known immediately, but rather that an earlier computation (namely, the high threshold) must be performed first. Hysteresis thresholding is also known as **double thresholding**.[†]

If we let $T_{low}(x, y)$ and $T_{high}(x, y)$ be binary images with ON pixels wherever the original pixel value is above the low or high threshold, respectively, then hysteresis thresholding can be represented compactly using our notation for morphological reconstruction by dilation[‡] as

$$I' = T_{high} \oplus^{*}_{T_{low}} B \tag{10.22}$$

where B is either B_4 or B_8, and I' is the output hysteresis-thresholded image. The pseudocode is given in Algorithm 10.3, where floodfill is used instead of dilation for computational efficiency. As before, this algorithm assumes that the low and high thresholds have already been determined through some other means. An example of hysteresis thresholding is shown in Figure 10.6.

ALGORITHM 10.3 Perform hysteresis thresholding on an image

HYSTERESISTHRESHOLD$(I, \tau_{low}, \tau_{high})$

Input: grayscale image I, along with low and high thresholds
Output: binary image I' from hysteresis thresholding

1   **for** $(x, y) \in I$ **do**
2       $T_{low}(x, y) \leftarrow 1$ **if** $I(x, y) > \tau_{low}$ **else** 0
3       $I'(x, y) \leftarrow$ OFF
4   **for** $(x, y) \in I$ **do**
5   **if** $I(x, y) > \tau_{high}$ **then**
6   FLOODFILLSEPARATEOUTPUT$(T_{low}, I', (x, y),$ ON$)$
7   **return** I'

[†] Hysteresis thresholding is also a step in the Canny edge detector, Section 7.2.1 (p. 335).
[‡] Section 4.2.2 (p. 154)

Stan Birchfield

Figure 10.6 An example of hysteresis thresholding.

Image

Low threshold retains some background

High threshold removes some foreground

Combined using floodfill on low threshold with seeds from high threshold

10.1.4 Multilevel Thresholding

All the techniques considered so far output an image with just two labels. Sometimes, however, the graylevel histogram has multiple peaks, with distinct valleys between the peaks, and it is desired to assign a different output value to each peak. This is called **multilevel thresholding**. For example, one might wish to apply two thresholds to yield three different output values:

$$I'(x, y) = \begin{cases} v_1 & \text{if } I(x, y) \leq \tau_1 \\ v_3 & \text{if } I(x, y) > \tau_2 \\ v_2 & \text{otherwise} \end{cases} \tag{10.23}$$

where $\tau_1 < \tau_2$. It is straightforward to extend Otsu's method to the case of multiple levels, leading to the **multilevel Otsu method**. Extending Equation (10.10) to the case of three levels, for example, yields

$$\sigma_b^2 = p_1 (\mu_1 - \mu)^2 + p_2 (\mu_2 - \mu)^2 + p_3 (\mu_3 - \mu)^2 \tag{10.24}$$

where the dependence of the variables upon τ_1 and τ_2 is omitted for clarity. Including this dependence, these variables are defined as

$$p_1(\tau_1) = \sum_{\ell=0}^{\tau_1} \overline{h}[\ell] \qquad \mu_1(\tau_1) = \frac{1}{p_1(\tau_1)} \sum_{\ell=0}^{\tau_1} \ell \overline{h}[\ell]$$

$$p_2(\tau_2) = \sum_{\ell=\tau_1+1}^{\tau_2} \overline{h}[\ell] \qquad \mu_2(\tau_2) = \frac{1}{p_2(\tau_2)} \sum_{\ell=\tau_1+1}^{\tau_2} \ell \overline{h}[\ell] \tag{10.25}$$

$$p_3(\tau_1, \tau_2) = 1 - p_1(\tau_1) - p_2(\tau_2) \qquad \mu_3(\tau_1, \tau_2) = \frac{1}{p_3(\tau_1, \tau_2)} \sum_{\ell=\tau_2+1}^{\zeta-1} \ell \overline{h}[\ell]$$

where we note that $\mu = \sum_{\ell=0}^{\xi-1} \ell \bar{h}[\ell]$ is not dependent on either threshold. The most straightforward approach to implementing the multilevel Otsu method is to exhaustively search over all possible threshold combinations. In practice, such an implementation is not unreasonable given modern processing power, since it is rare that multilevel thresholding would ever be used for more than three values (two thresholds). It is left as an exercise for the reader[†] to describe a more efficient implementation that eliminates redundant computations.

10.2 Deformable Models

Thresholding, as we have just seen, segments the foreground from the background on a per-pixel basis, and the outcome of each pixel is somewhat independent of the other pixels. A more powerful approach is to enforce similar outcomes among neighboring pixels to ensure that the output is smooth. Although this approach can be applied to multiple foreground objects, to simplify the presentation we will consider the case of a single foreground object on the background, and a **deformable model** is fit to the foreground object using the image data. In this section we consider two types of approaches to using deformable models: namely, active contours and level sets.

10.2.1 Active Contours (Snakes)

An **active contour**, also known as a **snake**, is a 2D deformable model that represents some structure in the image as a parametric curve. If we let $0 \leq s \leq 1$ be the scalar parameter that governs the location along the curve, then a continuous contour is represented as $\mathbf{c}(s) = (x(s), y(s))$. The curve may be *open*, $\mathbf{c}(0) \neq \mathbf{c}(1)$, or *closed*, $\mathbf{c}(0) = \mathbf{c}(1)$, as shown in Figure 10.7. The mechanics are largely the same in both cases, but we will concentrate primarily on the latter case, since a closed contour performs the work of segmentation by enclosing a region in the image that (hopefully) corresponds to some object in the world.

Given some initial conditions, the contour automatically deforms to minimize the energy of the system, where the energy consists of an internal energy term, E_{int}, and an external energy term, E_{ext}. With these two terms, the energy of a continuous snake is given by

$$E_{\mathbf{c}} = \int_0^1 E(\mathbf{c}(s)) \, ds = \int_0^1 \left(E_{ext}(\mathbf{c}(s)) + E_{int}(\mathbf{c}(s)) \right) ds \tag{10.26}$$

The external energy attempts to align the contour with the image data, and it is evaluated along the contour as a function of the gray level at each location:

$$E_{ext}(x, y) = g(I(x, y)) \tag{10.27}$$

Figure 10.7 Contours can be either open (left) or closed (right). An active contour, also known as a snake, evolves in such a way as to minimize an energy functional that takes into account the image data as well as the internal smoothness of the curve.

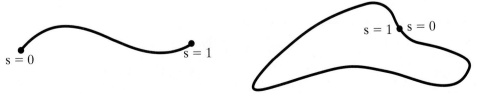

where $g(\cdot)$ is a monotonically nonincreasing function of the image gradient magnitude, such as one of the following:

$$g(I(x,y)) \equiv \begin{cases} -\|\nabla I(x,y)\|^2, \text{ or} \\ 1/(\|\nabla I(x,y)\|^2 + \epsilon), \text{ or} \\ \exp\left(-\|\nabla I(x,y)\|^2\right) \end{cases} \qquad (10.28)$$

where ϵ is a small constant to avoid dividing by zero. Any such function will cause the snake to snap to the nearest step edge surrounding the object. An alternative is to use a line edge model such as $g(x,y) \equiv I(x,y)$ that seeks dark locations in the image or $g(x,y) \equiv -I(x,y)$ that seeks bright locations in the image. This external energy term may also incorporate additional information supplied by the user, such as mouse clicks indicating either attraction or repulsion points.

The internal energy (or smoothness) seeks to preserve the internal consistency of the contour. Computer vision problems are often underconstrained due to ambiguity in the data, thus requiring one or more regularizing smoothness terms to overcome this ambiguity. Two terms usually govern the smoothness of a snake, namely the **elasticity** and **stiffness** terms:

$$E_{int}(\mathbf{c}(s)) = \underbrace{\frac{1}{2}\alpha \left\|\frac{d\mathbf{c}(s)}{ds}\right\|^2}_{\text{elasticity}} + \underbrace{\frac{1}{2}\beta \left\|\frac{d^2\mathbf{c}(s)}{ds^2}\right\|^2}_{\text{stiffness}} \qquad (10.29)$$

where α and β are shown here as constants but could depend upon s, and the derivative is given by

$$\frac{d\mathbf{c}}{ds} \equiv \lim_{\Delta s \to 0} \frac{\mathbf{c}(s + \Delta s) - \mathbf{c}(s)}{\Delta s} = \lim_{\Delta s \to 0} \frac{\Delta \mathbf{c}}{\Delta s} \qquad (10.30)$$

where $\Delta \mathbf{c}$ is the tangential vector along the contour, as shown in Figure 10.8. The elasticity is a first-order term, and it causes the snake to act like a membrane (stretching balloon or elastic band), with α controlling the tension along the contour. The stiffness is a second-order term, and it causes the snake to act like a thin metal plate (metal wire), with β controlling the rigidity of the contour. The elasticity term discourages stretching, while the stiffness term discourages bending.

To better understand the behavior caused by the elasticity and stiffness terms, some examples are shown in Figure 10.9. The first-order elasticity term pulls the points along the contour closer to each other, so that the elastic energy penalty is related to the length of the curve: stretching the curve increases the elastic energy, breaking the band yields infinite elastic energy, and minimizing the elastic energy shrinks the curve to a point. Similarly, the second-order stiffness term resists bending, so that forming a crease yields infinite stiffness energy. In the case of an open contour, minimizing the stiffness energy causes the curve

Figure 10.8 Definition of $\Delta \mathbf{c}$ as the vector between nearby points on the contour of the snake.

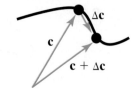

Figure 10.9 The effects on snake shape caused by the elasticity and stiffness terms.

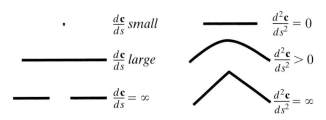

to assume the shape of a straight line, whereas in the case of a closed contour, minimizing the stiffness energy molds the curve into the shape of a circle.

Oftentimes when you see an energy functional like the one in Equation (10.29), it is perfectly valid to replace the square of the quantity with its absolute value, since the primary purpose of the square is simply to ensure that the result is nonnegative. However, in the case of snakes, the square in $\|d\mathbf{c}/ds\|^2$ is intentional and important; it cannot be replaced by the absolute value. The reason for this is that the square causes the term to provide a force that pushes the points away from each other, thus yielding approximately equal spacing between them; with the absolute value there is no such force. To see this, note that $\delta_{12}^2 + \delta_{23}^2 \geq \delta_{13}^2$ for any $|\delta_{12}| + |\delta_{23}| = |\delta_{13}|$, where $|\delta_{ij}|$ is the distance between the ith and j^{th} points. For example, suppose in the 1D case we have three points located at $x = 0, 1, 2$ so that they are equally spaced. Then $\delta_{12} = \delta_{23} = 1$, $\|d\mathbf{c}/ds\|^2$ yields $1^2 + 1^2 = 2$, and $\|d\mathbf{c}/ds\|$ yields $1 + 1 = 2$. But now if we move the points to $x = 0, 1.5, 2$ so that they are no longer equally spaced, then $\delta_{12} = 1.5, \delta_{23} = 0.5$, so that $\|d\mathbf{c}/ds\|^2$ yields $(1.5)^2 + (0.5)^2 = 2.5$, but $\|d\mathbf{c}/ds\|$ yields $1.5 + 0.5 = 2$. Thus, we see that the square leads to a higher energy when the points are not equally spaced, but the absolute value makes no such distinction.

For computational reasons, the contour is usually represented discretely as a sequence of n points: $\mathbf{v} = \langle \mathbf{v}_0, \mathbf{v}_1, \ldots, \mathbf{v}_{n-1} \rangle$, where $\mathbf{v}_i = (x_i, y_i), i = 0, \ldots, n - 1$. Although these points could represent control points of a B-spline or some other parametric curve representation, the simplest approach is to treat the contour as a polygon whose vertices are the points. This is known as the **buoy-rope** or **marker-string** representation, where the points are the buoys (or markers), and the connections between adjacent points are the ropes (or strings) connecting them. With this interpretation, the data term is simply evaluated at each point, and the derivatives are replaced by their discrete approximations:

$$E_{ext}(\mathbf{v}) = \sum_{i=0}^{n-1} g(I(x_i, y_i)) \tag{10.31}$$

$$E_{int}(\mathbf{v}) = \frac{1}{2}\sum_{i=0}^{n-1} \alpha \|\mathbf{v}_{i+1} - \mathbf{v}_i\|^2 + \beta \|\mathbf{v}_{i+1} - 2\mathbf{v}_i + \mathbf{v}_{i-1}\|^2 \tag{10.32}$$

where the indices include an implicit modulo n, so that $\mathbf{v}_n = \mathbf{v}_0$, and so forth. The total energy is given by $E_\mathbf{v} = E_{ext}(\mathbf{v}) + E_{int}(\mathbf{v})$.

Conceptually, the snake energy is minimized in an iterative manner, where each iteration performs an exhaustive local search around the current contour estimate, as shown in Algorithm 10.4. That is, for each of the n points, we search among all 8-neighbors of the point to see if any of them yield less energy. Because the points are connected with each other, we cannot treat them independently but instead must consider all 9^n possibilities, as shown in Figure 10.10. This conceptual algorithm, of course, is cost prohibitive because it requires computation time that is exponential in the number of points.

ALGORITHM 10.4 Active contour minimization (cost-prohibitive version)

SNAKE-CONCEPTUAL(I, $\mathbf{v}$)

Input: grayscale image I
             initial sequence of vertices $\mathbf{v}_0, \ldots, \mathbf{v}_{n-1}$ outlining the foreground region
Output: refined sequence of vertices

1 $\hat{\mathbf{v}} \leftarrow \langle \mathbf{v}_0, \ldots, \mathbf{v}_{n-1} \rangle$ ▷ Initialize foreground contour
2 **repeat**
3        $diff \leftarrow 0$
4   **for** $\mathbf{v} \in \mathcal{N}(\hat{\mathbf{v}})$ **do**   ▷ For each contour near the current one,
5   **if** $E(\mathbf{v}) < E(\hat{\mathbf{v}})$ **then**   if its energy is less,
6                    $diff \leftarrow \text{MAX}(diff, E(\hat{\mathbf{v}}) - E(v))$ then store the difference,
7                    $\hat{\mathbf{v}} \leftarrow \mathbf{v}$ and store the contour.
8   **until** $diff = 0$ ▷ Terminate when the minimization converges.
9   **return** $\hat{\mathbf{v}}$

To enable efficient computation, we note that the discrete energy functional formulates the contour as a Markov chain, because the probability of a point being at one of
the 9 locations is conditionally independent of all the other points given its immediate
neighbors. If $\beta = 0$, then the contour is a first-order Markov chain, so that if we know the
locations of $\mathbf{v}_{i-1}$ and $\mathbf{v}_{i+1}$ then the change in energy caused by moving $\mathbf{v}_i$ is independent
of all the other points, because its data term depends only on itself and its smoothness terms
depend only upon its immediately previous and following neighbors, which are fixed. If
$\beta \neq 0$, then the contour is a second-order Markov chain, and the effects of moving $\mathbf{v}_i$ are
independent of all other points given $\mathbf{v}_{i-2}$, $\mathbf{v}_{i-1}$, $\mathbf{v}_{i+1}$, and $\mathbf{v}_{i+2}$.

As a result of the Markov property, there is an efficient algorithm to exhaustively search
all 9^n possibilities, taking advantage of the redundancy in computation. The algorithm falls
under the general paradigm known as **dynamic programming**, and the specific algorithm
is known as the **Viterbi algorithm**. We describe the Viterbi algorithm in detail for the simpler case where $\beta = 0$, then briefly mention how to extend it to the case where $\beta \neq 0$. While
the brute-force exponential search requires an exorbitant $O(9^n)$ calculations, the Viterbi

Figure 10.10 One iteration of snake minimization involves searching over 9^n possible local variations of the current estimate. Shown is
a simple example with $n = 4$, the current estimate as a solid black line, and 2 of the 9^4 possibilities shown as dashed colored lines.

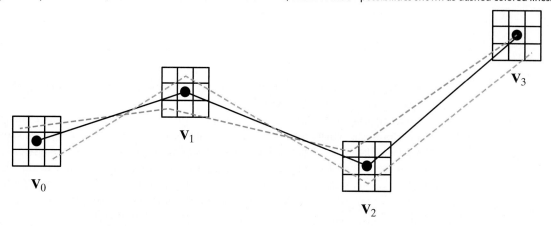

algorithm yields the exact same solution in just $O(9^2n)$ time, which is a considerable savings since it is linear in the number of points.

Minimizing a First-Order, Open Contour

For simplicity, let us assume that $\beta = 0$ and that the contour is open, so that $\mathbf{v}_0$ is the first point. If we were to conduct an exhaustive search, we would consider all 9 possibilities for the location of $\mathbf{v}_0$, then all 9 possibilities for $\mathbf{v}_1$ for each of the 9 possibilities of $\mathbf{v}_0$, and so on. It is easy to see that this approach leads to redundant computations, with the same value being computed multiple times. The Viterbi algorithm avoids this waste by storing intermediate results in a 2D array that we shall call Φ.

The Φ array has 9 rows and n columns, shown in Figure 10.11. Each entry $\phi_{i,j}$, where i is the column and j is the row, stores the total energy of the first $i + 1$ points in the snake, under the assumption that the last (i^{th}) point has been shifted by j from the current location. For example, $\phi_{1,4}$ stores the total energy of the snake consisting of $\mathbf{v}_0$ and $\mathbf{v}_1$, assuming that $\mathbf{v}_1$ is shifted by $j = 4$. The value j, which ranges from 0 to 8, inclusive, is simply a 1D encoding of the 2D shift (δ_x, δ_y):

$$\delta_x \equiv \lfloor j/3 \rfloor - 1 \tag{10.33}$$

$$\delta_y \equiv (j \bmod 3) - 1 \tag{10.34}$$

For example $j = 0$ indicates $(\delta_x, \delta_y) = (-1, -1)$, $j = 4$ indicates $(\delta_x, \delta_y) = (0, 0)$, and $j = 8$ indicates $(\delta_x, \delta_y) = (1, 1)$. We will let $\mathbf{v}_i^{(j)} \equiv \mathbf{v}_i + (\delta_x, \delta_y)$ denote the jth shift of the ith point, where $\mathbf{v}_i$ is the current location of the ith point.

The Viterbi algorithm proceeds in two steps. In the first step the array is filled up by columns, from left to right, using the values in the previous column. The first column involves just the data term, so that

$$\phi_{0,j} = E_{ext}(\mathbf{v}_0^{(j)}) \tag{10.35}$$

for all j. Starting with the second column, the energy stored in each row is the data term for that row plus the minimum of the energy for all 9 possibilities of the previous point:

$$\phi_{i,j} = E_{ext}(\mathbf{v}_i^{(j)}) + \min_{j'=0,\,\ldots,\,8} \left\{ \phi_{i-1,j'} + \frac{1}{2}\alpha \left\| \mathbf{v}_{i-1}^{(j')} - \mathbf{v}_i^{(j)} \right\|^2 \right\} \tag{10.36}$$

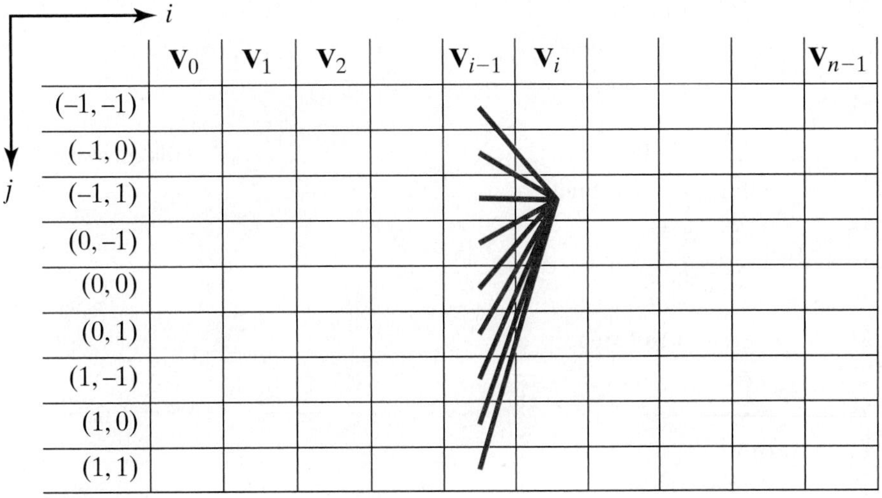

Figure 10.11 Array used by the Viterbi algorithm for efficient snake minimization. Each entry holds $\phi_{i,j}$, the energy of the snake ending with $\mathbf{v}_i$ at location j.

Once the array has been filled, the second step of the algorithm is to traverse the array from right to left, starting with the minimum entry in the last column. Given an entry $\phi_{i,j}$ in the ith column, the entry in the preceding column is given by

$$j_{prev} = \arg\min_{j'=0,\ldots,8} \left\{ \phi_{i-1,j'} + \frac{1}{2}\alpha \left\| \mathbf{v}_{i-1}^{(j')} - \mathbf{v}_i^{(j)} \right\|^2 \right\} \tag{10.37}$$

For computational reasons, it is helpful to store these values of j_{prev} in a second array, which we call Π, during the first step of the algorithm, so that the second step involves simply selecting $j_{prev} = \pi_{i,j}$.

The pseudocode is provided in Algorithm 10.5. Lines 2–13 implement the first step of the algorithm, namely to fill the Φ array (as well as the Π array for efficiency). Lines 15–23 implement the second step of the algorithm, namely to traverse the array from right to left.

ALGORITHM 10.5 Minimization of first-order, open active contour (one iteration)

SNAKE-OPENITER(I, v)

Input: grayscale image I

initial (open) sequence of vertices defining the foreground $\mathbf{v}_0, \ldots, \mathbf{v}_{n-1}$

Output: refined sequence of vertices after one iteration

1 ➤ Fill 2D array
2 **for** $j \leftarrow 0$ **to** 8 **do**   ➤ Fill first column using only data terms.
3      $\phi_{0,j} \leftarrow E_{ext}(\mathbf{v}_0^{(j)})$
4 **for** $i \leftarrow 1$ **to** $n-1$ **do**   ➤ For each remaining column,
5   **for** $j \leftarrow 0$ **to** 8 **do**   look at all rows of the previous column,
6          $\hat{\varphi} \leftarrow \infty$ and find the row j_{prev}
7   **for** $j' \leftarrow 0$ **to** 8 **do**   that yields the minimum energy, $\hat{\phi}$.
8              $\varphi \leftarrow \phi_{i-1,j'} + \frac{1}{2}\alpha \left\| \mathbf{v}_{i-1}^{(j')} - \mathbf{v}_i^{(j)} \right\|^2$
9   **if** $\varphi < \hat{\varphi}$ **then**
10                 $\hat{\varphi} \leftarrow \varphi$
11                 $j_{prev} \leftarrow j'$
12         $\phi_{i,j} \leftarrow E_{ext}(\mathbf{v}_i^{(j)}) + \hat{\varphi}$ ➤ Store the results in the Φ and Π arrays.
13         $\pi_{i,j} \leftarrow j_{prev}$
14 ➤ Traverse 2D array
15 $\hat{\varphi} \leftarrow \infty$
16 **for** $j' \leftarrow 0$ **to** 8 **do**   ➤ In the last column, find the row j with the minimum value.
17   **if** $\phi_{n-1,j'} < \hat{\varphi}$ **then**
18         $\hat{\varphi} \leftarrow \phi_{n-1,j'}$
19         $j \leftarrow j'$
20 $\mathbf{v} \leftarrow \langle \mathbf{v}_{n-1}^{(j)} \rangle$ ➤ Store the new coordinates of the final vertex.
21 **for** $i \leftarrow n-1$ **to** 1 **step** -1 **do**   ➤ For each previous vertex,
22     $j \leftarrow \pi_{i,j}$ find the row j from which it came, and
23     $\mathbf{v} \leftarrow \langle \mathbf{v}_{i-1}^{(j)} \mathbf{v} \rangle$ store the new coordinates.
24 **return** $\mathbf{v}$

Minimizing a First-Order, Closed Contour

With an open contour, the preceding procedure is guaranteed to find the global minimum of all 9^n local perturbations of the current snake. With a closed contour, however, it is not so obvious how to perform the minimization. For starters, the previous computation is missing a data term, namely the connection between the first and last point in the snake, $\frac{1}{2}\alpha\|\mathbf{v}_{n-1} - \mathbf{v}_0\|$. This term can easily be added by inserting a small amount of code in Algorithm 10.5 just after Line 13 to treat the last column differently, just as the first column is treated differently. However, the resulting procedure would not be guaranteed to find the global minimum, and in fact it would nearly always produce suboptimal results, particularly near the beginning and end of the curve.

The simplest solution to this problem is to run the open-curve algorithm twice, as shown in Algorithm 10.6. Although neither run is guaranteed to yield optimal results, in practice the points far from the beginning and end are not likely to be affected by the approximation caused by adding the extra computation to the last column. As a result, it is reasonable to assume that after the first run the middle point is in the optimal position. The second run, therefore, cements the middle point at the location determined by the first run, then runs the minimization again on the remaining points. Note the primes ($'$) after the procedure calls in Lines 2 and 4, which indicate that minor changes must be made to the procedure in these two calls. The specific changes are left as an exercise for the reader.[†]

Minimizing a Second-Order Contour

When $\beta \neq 0$, the system is second-order, because each point is conditionally dependent upon all other points given not only its immediate neighbors, but also the immediate neighbors of the immediate neighbors. The second-order system is minimized using the same Viterbi algorithm but using a 2D array with $9^2 = 81$ rows and n columns. Each column corresponds to the 81 possible locations for a pair of adjacent points. Since column i represents the locations for $\mathbf{v}_{i-1}$ and $\mathbf{v}_i$, and column $i - 1$ represents the locations for $\mathbf{v}_{i-2}$ and $\mathbf{v}_{i-1}$, there is redundancy in the array since $\mathbf{v}_{i-1}$ is represented in both places. As a result, for any given entry $\phi_{i,k}$, only 9 of the 81 possibilities of the previous column are consistent with that entry. That is, $\pi_{i,k}$, is restricted to the set $\{k': k' \Diamond k\}$, where $k' \Diamond k$ means that k and k' are consistent with each other. (The cardinality of the set is 9.) The computation involves

$$\phi_{i,k} = E_{ext}(\mathbf{v}_i^{(j)}) + \min_{\substack{k'=0,\ldots,80 \\ \text{s.t. } k' \Diamond k}}\left\{\phi_{i-1,k'} + \frac{1}{2}\alpha\left\|\mathbf{v}_{i-1}^{(j')} - \mathbf{v}_i^{(j)}\right\|^2 + \frac{1}{2}\beta\left\|\mathbf{v}_{i-2}^{(j'')} - 2\mathbf{v}_{i-1}^{(j')} + \mathbf{v}_i^{(j)}\right\|^2\right\} \quad (10.38)$$

ALGORITHM 10.6 Minimization of first-order, closed active contour

SNAKE-CLOSED(I, v)

Input: grayscale image I

        initial (closed) sequence of vertices defining the foreground $\mathbf{v}_0, \ldots, \mathbf{v}_{n-1}$

Output: refined sequence of vertices

1   **while** not converged **do**
2       $\langle\mathbf{v}_0, \ldots, \mathbf{v}_{n-1}\rangle \leftarrow$ SNAKE-OPENITER$'(I, \langle\mathbf{v}_0 \ldots, \mathbf{v}_{n-1}\rangle)$
3   **while** not converged **do**
4       $\langle\mathbf{v}_{\frac{n}{2}}, \ldots, \mathbf{v}_{n-1}, \mathbf{v}_0, \ldots, \mathbf{v}_{\frac{n}{2}-1}\rangle \leftarrow$ SNAKE-OPENITER$'(I, \langle\mathbf{v}_{\frac{n}{2}}, \ldots, \mathbf{v}_{n-1}, \mathbf{v}_0, \ldots, \mathbf{v}_{\frac{n}{2}-1}\rangle)$
5   **return** $\langle\mathbf{v}_0, \ldots, \mathbf{v}_{n-1}\rangle$

[†] Problem 10.17. Also see Problem 10.18 for a solution that *is* guaranteed to find the optimal solution, albeit at considerable computational expense.

where the positions j, j', and j'' can be obtained from the row numbers k and k'. The computation is otherwise the same as in the first-order case and, for a closed contour, is run twice, with the second run cementing the location of the middle point.

Minimizing Using Calculus of Variations

An alternative approach to updating the snake is to employ the **calculus of variations**. Recall that in calculus, if x is a local minimum of a function $y = f(x)$, then $\frac{df}{dx} = 0$ at x. In other words, to find a local minimum of a function, compute the derivative of the function, set it to zero, and solve for x. Similarly, in the calculus of variations, we take derivatives, set them to zero, and solve for the unknown. The difference is that instead of the function $f(x)$, the calculus of variations seeks to minimize a **functional**[†] $\int_a^b f(x, q(x), \dot{q}(x))\, dx$, where x is the *independent variable* (usually either time or space), q is the *dependent variable* whose value depends upon x, $\dot{q} \equiv \frac{dq}{dx}$ is the derivative of the function $q(x)$ with respect to the independent variable, and a and b are constants. The expression of derivative that is set to zero is known as the **Euler-Lagrange equation**:

$$\frac{\partial f}{\partial q} - \frac{d}{dx}\left(\frac{\partial f}{\partial \dot{q}}\right) = 0 \tag{10.39}$$

In other words, when this equation is solved for x, it yields the values of x for which the functional $\int_a^b f(x, q(x), \dot{q}(x))dx$ is minimum.[‡]

More generally, suppose we have m independent and n dependent variables. If we let $x_1, \ldots, x_m$ be the independent variables and $q_1, \ldots, q_n$ be the dependent variables, then the functional

$$\int_{a_1}^{b_1} \cdots \int_{a_m}^{b_m} f\left(x_1, \ldots, x_m, q_1, \ldots, q_n, \frac{\partial q_1}{\partial x_1}, \ldots, \frac{\partial q_n}{\partial x_1}, \ldots, \frac{\partial q_1}{\partial x_m}, \frac{\partial q_n}{\partial x_m}\right) dx_1 \cdots dx_m \tag{10.40}$$

is at a local minimum where the following Euler-Lagrange equations are satisfied:

$$\frac{\partial f}{\partial q_k} - \sum_{i=1}^m \frac{\partial}{\partial x_i}\left(\frac{\partial f}{\partial \left(\frac{\partial q_k}{\partial x_i}\right)}\right) = 0, \qquad k = 1, \ldots, n \tag{10.41}$$

Note that there are n equations, and each equation contains $m + 1$ terms. If $m = n = 1$, then this equation reduces to Equation (10.39). Similarly, if the functional contains higher derivatives, such as $\int_a^b f(x, q, \dot{q}, \ddot{q}, \ldots)\, dx$, then a local minimum occurs where

$$\frac{\partial f}{\partial q} - \frac{d}{dx}\left(\frac{\partial f}{\partial \dot{q}}\right) + \frac{d^2}{dx^2}\left(\frac{\partial f}{\partial \ddot{q}}\right) - \cdots = 0 \tag{10.42}$$

Applying the calculus of variables to the problem at hand, we note that $\mathbf{c}$ is a function of s, that $\mathbf{c}$ depends upon the first and second derivatives of s, and our goal is to search over all possible functions $\mathbf{c}(s)$ to find the one that minimizes the energy functional $E_{\mathbf{c}}$ in Equation (10.26) . To solve this problem, then, we must solve the following Euler-Lagrange equation:

$$\frac{\partial E}{\partial \mathbf{c}} - \frac{d}{ds}\left(\frac{\partial E}{\partial \dot{\mathbf{c}}}\right) + \frac{d^2}{ds^2}\left(\frac{\partial E}{\partial \ddot{\mathbf{c}}}\right) = 0 \tag{10.43}$$

[†] Whereas a *function* takes a variable as input, a *functional* takes a function as input, which in this case is q, a function of the variable x.

[‡] Technically, the point is either a local minimum, local maximum, or saddle point (collectively known as *stationary points*), but we shall use the technique exclusively to search for a local minimum.

where s is the independent variable, $\mathbf{c}$ contains the dependent variables (because x and y depend upon s), E stands for $E(c(\mathbf{s})) = E_{ext}(\mathbf{c}) + E_{int}(\mathbf{c})$, $\dot{\mathbf{c}} \equiv d\mathbf{c}/ds$, and $\ddot{\mathbf{c}} \equiv d^2\mathbf{c}/ds^2$. Since E_{int} does not directly depend upon $\mathbf{c}$, we have $\partial E/\partial \mathbf{c} = \partial E_{ext}/\partial \mathbf{c} = \nabla E_{ext}$. Similarly, E_{ext} does not directly depend upon $\dot{\mathbf{c}}$ or $\ddot{\mathbf{c}}$, so $\partial E/\partial \dot{\mathbf{c}} = \partial \mathbf{E}_{int}/\partial \dot{\mathbf{c}} = \alpha \dot{\mathbf{c}}$, and $\partial E/\partial \ddot{\mathbf{c}} = \partial \mathbf{E}_{int}/\partial \ddot{\mathbf{c}} = \beta \ddot{\mathbf{c}}$. Substituting into Equation (10.43) yields the final equation to be solved:

$$\nabla E_{ext}(\mathbf{c}) - \alpha \frac{d^2\mathbf{c}}{ds^2} + \beta \frac{d^4\mathbf{c}}{ds^4} = 0 \qquad (10.44)$$

Approximating the derivatives with finite differences yields

$$\nabla E_{ext}(\mathbf{v}_i) + \alpha(\mathbf{v}_i - \mathbf{v}_{i-1}) - \alpha(\mathbf{v}_{i+1} - \mathbf{v}_i) + \beta(\mathbf{v}_{i-2} - 2\mathbf{v}_{i-1} + \mathbf{v}_i)$$
$$- 2\beta(\mathbf{v}_{i-1} - 2\mathbf{v}_i + \mathbf{v}_{i+1}) + \beta(\mathbf{v}_i - 2\mathbf{v}_{i+1} + \mathbf{v}_{i+2}) = 0 \qquad (10.45)$$

Since this equation equals zero if the forces are perfectly balanced, the deviation of the left-hand side from zero indicates the amount of imbalance, and the sign of the left-hand side indicates the direction of imbalance. As a result, we set the right-hand side to $-\lambda \Delta \mathbf{v}_i$, where λ is the step size, and $\Delta \mathbf{v}_i$ is the change in $\mathbf{v}_i$ between iterations.

Stacking the n equations from Equation (10.45), one per vertex, leads to the full system to be solved, which can be rewritten in matrix form as

$$\mathbf{A}\mathbf{x}^{(k)} - \mathbf{f}_x(\mathbf{x}^{(k-1)}, \mathbf{y}^{(k-1)}) = -\lambda(\mathbf{x}^{(k)} - \mathbf{x}^{(k-1)})$$
$$\mathbf{A}\mathbf{y}^{(k)} - \mathbf{f}_y(\mathbf{x}^{(k-1)}, \mathbf{y}^{(k-1)}) = -\lambda(\mathbf{y}^{(k)} - \mathbf{y}^{(k-1)}) \qquad (10.46)$$

where $\mathbf{x} \equiv [x_0 \ \cdots \ x_{n-1}]^T$, $\mathbf{y} \equiv [y_0 \ \cdots \ y_{n-1}]^T$, $\mathbf{v}_i = [x_i \ \ y_i]^T$, and the superscript indicates the iteration number. Here $\mathbf{A}$ is the $n \times n$ pentadiagonal banded matrix consisting of the α and β terms of Equation (10.45), while $\mathbf{f}_x(\mathbf{x}, \mathbf{y})$ and $\mathbf{f}_y(\mathbf{x}, \mathbf{y})$ are the x and y components of the force $-\nabla E_{ext}(\mathbf{v})$ evaluated at all the points, stacked into vectors. Note that $\mathbf{f}_x$ and $\mathbf{f}_y$ are evaluated using the estimated coordinates from the previous iteration, because that is the only choice available to us. (It is not possible to use the estimate from the current iteration until it has completed.) As a result, we say that the method is *explicit* with regard to the external forces. However, since $\mathbf{A}$ is a constant matrix, we are free to multiply it by either the estimates from the previous or current iterations; the latter is chosen because it yields better convergence properties. For this reason, the method is said to be *implicit* with regard to the internal forces. Because the entire system of Equation (10.46) combines implicit and explicit approaches, it is said to be *semi-implicit*. The system is solved iteratively until convergence.

10.2.2 Gradient Vector Flow

The active contours, or snakes, of the previous section have a number of drawbacks. One obvious shortcoming is that the elasticity term exerts a contraction force that tends to shrink the contour. As a result, care must be taken to ensure that (in the case of a closed contour) the snake is initialized to a curve that is larger than the image region on which it should settle. To counter this effect, one approach is to introduce an additional term that exerts an outward force that tends to expand the snake. The resulting formulation is called a **balloon**, and various attempts have been made to combine contraction and expansion forces to achieve a hybrid snake-balloon model. Although these approaches have achieved moderate success, snakes (with or without balloons) tend to be quite sensitive to the initial conditions.

An even better solution is to replace the force $\nabla F \equiv -\nabla E_{ext}$ with the **gradient vector flow (GVF)** field $\mathbf{g}(x, y) \equiv (U(x, y), V(x, y))$, where (x, y) are coordinates in the image plane. Recall that two fundamental operators for vector fields are the divergence and curl.[†] According to the Helmholtz theorem, a static vector field can be decomposed into two components, namely, a *solenoidal* field for which the divergence is zero everywhere, and an *irrotational* field for which the curl is zero everywhere. It can be shown that the force field arising from $-\nabla E_{ext}$ (used in a traditional snake) is a static irrotational field.[‡] The GVF field generalizes this concept by introducing a solenoidal component, which not only enables GVF snakes to overcome the initialization problem of standard snakes by being much less sensitive to the initial position but also enables them to fill concavities, a capability that eludes traditional snakes.

To make these concepts more concrete, let us consider how GVF snakes work in practice. A GVF snake is solved in two steps. The first step, namely, estimating $\mathbf{g}$, involves solving the following energy functional:

$$\mathcal{E} = \iint \mu(U_x^2 + U_y^2 + V_x^2 + V_y^2) + \|\nabla F\|^2 \|\mathbf{g} - \nabla F\|^2 \, dx \, dy \qquad (10.47)$$

where μ is a scalar that governs the relative importance of the smoothness and data terms, and the coordinates (x,y) have been omitted for simplicity. The smoothness term contains $U_x \equiv \partial U / \partial x$ and so forth, while the data term seeks to match $\mathbf{g}$ to the force ∇F, weighted by the amount of image information near the pixel, $\|\nabla F\|^2$. In other words, where there is information in the image (e.g., near an intensity edge), then the data term is weighted more, whereas the smoothness term dominates in regions where there is less information (e.g., in an untextured area).

The Euler-Lagrange equations of Equation (10.47) lead to

$$\mu \nabla^2 U - (U - f_x)(f_x^2 + f_y^2) = 0 \qquad (10.48)$$

$$\mu \nabla^2 V - (V - f_y)(f_x^2 + f_y^2) = 0 \qquad (10.49)$$

where $f_x = \partial F / \partial x$ and $f_y = \partial F / \partial y$, similar to the vector quantities in Equation (10.46), except here they are scalars because they are evaluated at a particular pixel. Setting the right-hand side to $-\lambda \Delta U$ and $-\lambda \Delta V$ and using the standard approximation for the Laplacian, yields

$$\mu(\bar{U}^{(k-1)} - U^{(k-1)}) - (U^{(k-1)} - f_x)b = -\lambda(U^{(k)} - U^{(k-1)}) \qquad (10.50)$$

$$\mu(\bar{V}^{(k-1)} - V^{(k-1)}) - (V^{(k-1)} - f_x)b = -\lambda(V^{(k)} - V^{(k-1)}) \qquad (10.51)$$

where

$$\bar{U}(x, y) \equiv \frac{1}{4}(U(x-1, y) + U(x+1, y) + U(x, y-1) + U(x, y+1)) \qquad (10.52)$$

$$\bar{V}(x, y) \equiv \frac{1}{4}(V(x-1, y) + V(x+1, y) + V(x, y-1) + V(x, y+1)) \qquad (10.53)$$

$b \equiv f_x^2 + f_y^2$, and the superscripts indicate the iteration number. Rearranging terms yields

---

[†] Section 2.5.1 (p. 57).

[‡] In fact, any vector field that can be expressed as the gradient of a scalar field is irrotational.

$$U^{(k)} = -\frac{1}{\lambda}\left(\mu\overline{U}^{(k-1)} + (-b - \lambda - \mu)U^{(k-1)} + f_x b\right) \tag{10.54}$$

$$V^{(k)} = -\frac{1}{\lambda}\left(\mu\overline{V}^{(k-1)} + (-b - \lambda - \mu)V^{(k-1)} + f_y b\right) \tag{10.55}$$

Recognizing that these equations are applicable for each pixel (x,y), they can be solved iteratively to determine U and V, and hence $\mathbf{g}$, for every pixel.

In the second step, $\mathbf{g}$ is used as the external force in solving for the snake. That is, once $\mathbf{g}$ has been determined, it can be substituted for $-\nabla E_{ext}$ in Equation (10.44) to yield

$$-\mathbf{g}(\mathbf{v}) - \alpha\frac{d^2\mathbf{v}}{d\mathbf{s}^2} + \beta\frac{d^4\mathbf{v}}{d\mathbf{s}^4} = 0 \tag{10.56}$$

By similar reasoning to that above, the snake can then be found by iteratively solving

$$\mathbf{A}\mathbf{x}^{(k)} - \mathbf{g}_x(\mathbf{x}^{(k-1)}, \mathbf{y}^{(k-1)}) = -\lambda(\mathbf{x}^{(k)} - \mathbf{x}^{(k-1)})$$

$$\mathbf{A}\mathbf{y}^{(k)} - \mathbf{g}_y(\mathbf{x}^{(k-1)}, \mathbf{y}^{(k-1)}) = -\lambda(\mathbf{y}^{(k)} - \mathbf{y}^{(k-1)}) \tag{10.57}$$

where $\mathbf{g}_x \equiv \partial\mathbf{g}/\partial x$, and $\mathbf{g}_y \equiv \partial\mathbf{g}/\partial y$. An example of a GVF snake is shown in Figure 10.12.

10.2.3 Level Set Method

Despite their influence upon the field of computer vision, snakes have several fundamental limitations. First, because snakes are parameterized by a sequence of points, snake minimization has the undesirable property that the outcome depends upon the parameterization chosen. Secondly, snakes do not handle topological changes, such as when one contour splits into two, or when two contours merge into one. Although various attempts have been made over the years to address this shortcoming, the resulting behavior is usually quite brittle due to the heuristic nature of the methods. Similarly, care must be taken when minimizing a snake to prevent the contour from crossing itself, which would result in a curve that no longer encloses a well-defined region. Finally, it is not easy to extend snakes to higher dimensions beyond 2D.

The **level set representation** (also known as a *geometric active contour*) overcomes these limitations by representing the curve *implicitly* rather than *explicitly*. Instead of modifying the curve directly, the level set method evolves an implicit function according to a partial differential equation (PDE), which then modifies the curve as a side effect. The level set method naturally handles topological changes, allowing the curve to split or merge as needed; it is *parameterless* since it does not store the curve explicitly, and it naturally

Figure 10.12 An example of a GVF snake. Unlike traditional snakes, GVF snakes can be initialized either inside or outside the object (or a combination of the two, as shown here), and they handle concavities.

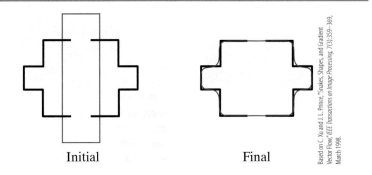

Initial   Final

Based on C. Xu and J. L. Prince, "Snakes, Shapes, and Gradient Vector Flow," IEEE Transactions on Image Processing, 7(3):359–369, March 1998.

extends to higher dimensions. In addition to these advantages, the level set method is easy to implement.

Let $\mathbf{c}(s) = (x(s), y(s))$, $0 \leq s \leq 1$, be a closed curve in the image plane, as before. Now let us embed the curve in a dimension that is one higher than itself. That is, we define an **implicit function** $\Phi(x, y)$ that is a function of two variables such that the zeroth level set of Φ is the curve $\mathbf{c}$. (Recall that the *level set* of a function is the set of points such that the function evaluated at those points is constant.[†]) In other words, $\Phi(x, y) = 0$ if and only if $\mathbf{c}(s) = (x, y)$ for some value of s. By convention, we let $\Phi(\cdot) > 0$ inside the curve and $\Phi(\cdot) < 0$ outside the curve.[‡]

Now suppose the curve evolves over time, starting from an initial curve. Let us define $\mathbf{c}(s,t)$ as a family of curves so that at any time $t \geq 0$, $\mathbf{c}(s, t)$ is a curve $\mathbf{c}(s)$. Since the representation is nonparametric, we are only concerned with motion perpendicular to the curve at any point; motion tangential to the curve simply affects the parameterization, such as the locations of the points along the contour in the case of a snake. As a result, the change in the curve over time (that is, as the algorithm progresses), is proportional to the normal vector:

$$\frac{\partial \mathbf{c}(s, t)}{\partial t} = g(I(s)) \cdot \mathbf{n}(s) \tag{10.58}$$

where $\mathbf{n}(s)$ is the inward normal vector to the curve at s, and $g(I(s))$ is a function of the image that specifies the speed with which the contour moves. Usually g is a nonincreasing function of the image gradient magnitude, as in Equation (10.27), so that when the contour is far from an intensity edge the contour moves at a higher speed and when the contour is

Figure 10.13 The curve is the zeroth level set of the implicit function. As the implicit function evolves, so does the curve. Note that the level set method naturally handles topology changes, as shown here.

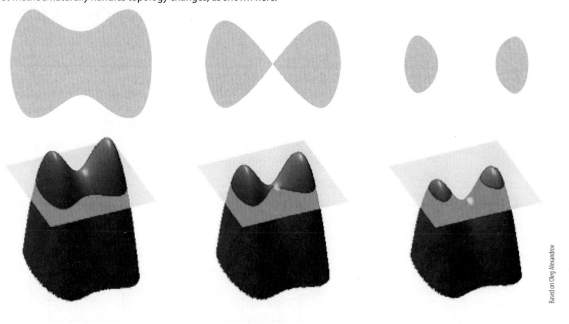

[†] Section 4.4.5 (p. 182)

[‡] Some authors follow the opposite convention.

aligned with an intensity edge, it moves at a slower speed. Note that if $g(\cdot) > 0$ then the contour shrinks, whereas $g(\cdot) < 0$ causes the contour to expand.[†]

Although it may not be obvious how to evolve the *curve* according to the explicit formulation in Equation (10.58), it is equivalent (as shown below) to evolve the *implicit function* according to

$$-\frac{\partial \Phi(x, y, t)}{\partial t} = g(I(x, y)) \cdot \|\nabla \Phi(x, y, t)\| \tag{10.59}$$

which resembles a **Hamilton-Jacobi equation**. In an implementation, Φ is defined over the same rectangular lattice as the image, so that for any iteration specified by t (which can be thought of as "pseudo-time"), $\Phi(x, y)$ is defined for all pixel coordinates. By moving the value of $\Phi(x, y)$ down or up, the curve is moved inward or outward, respectively, as illustrated in Figure 10.14.

It is straightforward to show that Equations (10.58) and (10.59) are mathematically equivalent. The first-order Taylor series approximation of $\frac{d\Phi}{dt}$ yields

$$\frac{d\Phi}{dt} \approx \frac{\partial \Phi}{\partial x}\frac{dx}{dt} + \frac{\partial \Phi}{\partial y}\frac{dy}{dt} + \frac{\partial \Phi}{\partial t} \tag{10.60}$$

Now, since $\Phi(x, y, t) = 0$ along the curve, it is obviously the case that $\frac{d\Phi}{dt} = 0$ along the curve, thus yielding a relation between the change in Φ over time and the motion of the contour points:

$$-\frac{\partial \Phi(x, y, t)}{\partial t} = (\nabla \Phi(x, y, t))^{\mathsf{T}}\left(\frac{\partial \mathbf{c}(s, t)}{\partial t}\right) \tag{10.61}$$

since $\mathbf{c} = \begin{bmatrix} x & y \end{bmatrix}^{\mathsf{T}}$, and $\nabla \Phi = \begin{bmatrix} \frac{\partial \Phi}{\partial x} & \frac{\partial \Phi}{\partial y} \end{bmatrix}^{\mathsf{T}}$. Substituting Equation (10.58) into Equation (10.61) yields

$$-\frac{\partial \Phi(x, y, t)}{\partial t} = g(I(x, y)) \cdot (\nabla \Phi(x, y, t))^{\mathsf{T}} n(s) \tag{10.62}$$

Figure 10.14 The curve is the zeroth level set of the implicit function. As the implicit function moves down or up, the curve moves inward or outward. Here the implicit function is moving down, causing the curve to move inward (shrink). The horizontal (green) arrows indicate the motion induced on the curve by the vertical changes (blue arrows) in the implicit function.

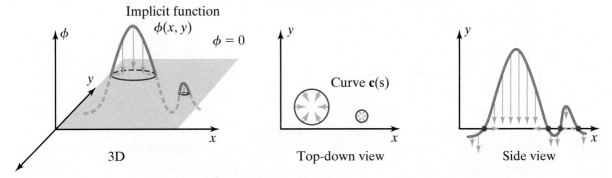

[†] Note that if the first choice in (10.28) is selected, then a large constant should be added to ensure that $g(\cdot)$ is positive.

which is closer to what we want, since only motion normal to the curve is important. Note that $I(s)$ and $I(x,y)$ are identical, as long as (x,y) are the coordinates corresponding to the point along the contour at s. To compute the inward normal, note that $\Phi(\cdot) = 0$ along the curve, so that $\Phi_s \equiv \frac{\partial \Phi}{\partial s} = 0$ along the curve as well. By the chain rule of differentiation, this leads to

$$\frac{\partial \Phi}{\partial s} = \frac{\partial \Phi}{\partial x}\frac{\partial x}{\partial s} + \frac{\partial \Phi}{\partial y}\frac{\partial y}{\partial s} = (\nabla \Phi)^{\mathsf{T}} \mathbf{c}_s = 0 \qquad (10.63)$$

where $\mathbf{c}_s \equiv \frac{\partial \mathbf{c}}{\partial s}$. This equation tells us that the gradient vector $\nabla \Phi(x, y.t)$ is perpendicular to the tangential vector along the curve. As a result, the unit-length inward normal vector to the curve is given by $\mathbf{n} = \frac{\nabla \Phi}{\|\nabla \Phi\|}$, where the convention that $\Phi(\cdot) > 0$ inside the curve leads to the absence of a negative sign. Substituting this definition of $\mathbf{n}$ back into Equation (10.62) yields

$$-\frac{\partial \Phi(x, y, t)}{\partial t} = g(I(x, y)) \cdot (\nabla \Phi)^{\mathsf{T}}\left(\frac{\nabla \Phi}{\|\nabla \Phi\|}\right) = g(I(x, y)) \cdot \frac{\|\nabla \Phi\|^2}{\|\nabla \Phi\|} = g(I(x, y)) \cdot \|\nabla \Phi\| \qquad (10.64)$$

which is identical to Equation (10.59).

Implementing the level set method is straightforward, as shown in Algorithm 10.7. In the simplest approach, the implicit function is initialized to -1 along the outer border of the image and $+1$ everywhere else, so that the zero-level contour encloses almost the entire image. The gradient magnitude of the image is computed, and Φ is then repeatedly updated using Equation (10.59) until the motion of the curve slows down below a threshold. The only trick in the implementation is the need to *reinitialize* Φ every now and then to be a signed distance function to the contour. The reason for this reinitialization is that Equation (10.59) is mathematically valid only along the contour, as can be seen by its derivation. As

---

ALGORITHM 10.7 Evolve a contour using the level set method

LevelSetMethod(I)

Input: grayscale image I
Output: 2D array Φ in which foreground object pixels are indicated by $\Phi(x, y) > 0$

```
1   G_I ← GradientMagnitude(I)                          ➤ Compute gradient magnitude of image.
2   for (x, y) ∈ I do                                   ➤ Initialize Φ to −1 along image border, +1 everywhere else.
3       val ← −1 if (x = 0 or x = width −1 or y = 0 or y = height −1) else +1
4       Φ_prev(x, y) ← Φ(x, y) ← val
5   speed ← ∞
6   while speed > τ do                                  ➤ While contour is moving at sufficient speed,
7       G_Φ ← GradientMagnitude(Φ)                      compute gradient magnitude of Φ,
8       for (x, y) ∈ I do                               then for each pixel,
9           Φ(x, y) ←_+ G_I(x, y) * G_Φ(x, y)           update Φ using Equation (10.59),
10          C(x, y) ← TRUE if Φ(x, y) > 0 or FALSE otherwise    and threshold Φ.
11      Φ ← SignedChamferDistance(C)                    ➤ Reinitialize Φ using signed chamfer distance.
12      speed ← max(|Φ − Φ_prev|)                       ➤ Set the speed to the maximum distance moved by any pixel,
13      Φ_prev ← Φ                                      and store a copy of Φ for the next iteration.
14  return Φ
```

Figure 10.15 Level set reinitialization using the signed distance function. The contour remains unchanged, but the implicit function is more well-behaved after reinitialization.

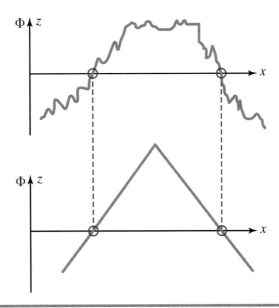

a result, far from the contour the implicit function can develop kinks and other undesirable characteristics. In the pseudocode, Φ is reinitialized every iteration, but this is not usually necessary in an actual implementation. An alternate approach is to avoid reinitialization altogether by incorporating an additional term in the energy functional that penalizes the deviation of Φ from the signed distance function.

Reinitialization spends a great deal of computation updating the value of Φ for every pixel in the image. But pixels far from the contour are not important, since Equation (10.59) is not valid for them anyway, so their values are not likely to influence the computation of the contour. This observation leads to the **narrow band method** in which Φ is updated and maintained only for those pixels that are within a narrow band around the contour. A related observation is that, if the sign of the speed never changes, then the contour always shrinks (or always expands), in which case the contour crosses each pixel exactly once during the course of the algorithm execution. This leads to the **fast marching method**, which computes for each pixel the time that the contour crosses that pixel.

For many natural phenomena, the speed at which the curve moves is proportional to the curvature at that point. The analogy is the propagation of the front of a fire. As the fire burns the grass, any long convex extension of the fire into the grass will be surrounded by ambient temperature and will therefore burn more slowly, whereas a concave extension of the grass into the fire will be surrounded by a tremendous amount of heat and will therefore burn more quickly. Similarly, a rubber sheet or balloon will exert more force where its curvature is greatest. By including curvature, Equations (10.58) and (10.59) become equations describing **mean curvature motion**:

$$\frac{\partial \mathbf{c}(s,t)}{\partial t} = g(I(s)) \cdot \kappa(s) \cdot \mathbf{n}(s) \tag{10.65}$$

$$-\frac{\partial \Phi(x,y,t)}{\partial t} = g(I(x,y)) \cdot \kappa(x,y) \cdot \|\nabla \Phi(x,y,t)\| \tag{10.66}$$

where $\kappa(\cdot) = \text{div}(\nabla \Phi / \|\nabla \Phi\|) + \epsilon$ is the curvature, defined as the divergence of the normalized gradient of the implicit function, plus a small constant to ensure that the speed always remains nonzero. In Cartesian coordinates the divergence[†] of a vector field $A(x, y) \equiv [a(x, y) \quad b(x, y)]^{\mathsf{T}}$ is defined as

$$\text{div}(A) = \nabla \cdot A = \frac{\partial a}{\partial x} + \frac{\partial b}{\partial y} \tag{10.67}$$

so that the divergence of the gradient, $\nabla A \equiv \left[\frac{\partial A}{\partial x} \quad \frac{\partial A}{\partial y}\right]^{\mathsf{T}}$, is the Laplacian:

$$\nabla \cdot \nabla A = \frac{\partial^2 A}{\partial x^2} + \frac{\partial^2 A}{\partial y^2} = A_{xx} + A_{yy} = \nabla^2 A \tag{10.68}$$

where $A_{xx} \equiv \partial^2 A/\partial x^2$ and $A_{yy} \equiv \partial^2 A/\partial y^2$. When we include the normalization of the vector, the math is a bit more complicated, but nevertheless we arrive at an expression that is more suitable to implementation:

$$
\begin{aligned}
\kappa(x, y) &= \text{div}\left(\frac{\nabla \Phi}{\|\nabla \Phi\|}\right) = \nabla \cdot \left(\frac{\nabla \Phi}{\|\nabla \Phi\|}\right) = \nabla \cdot \frac{[\Phi_x \quad \Phi_y]^{\mathsf{T}}}{\sqrt{\Phi_x^2 + \Phi_y^2}} = \frac{\partial}{\partial x}\left(\frac{\Phi_x}{\sqrt{\Phi_x^2 + \Phi_y^2}}\right) + \frac{\partial}{\partial y}\left(\frac{\Phi_y}{\sqrt{\Phi_x^2 + \Phi_y^2}}\right) \\
&= \frac{\Phi_{xx}}{(\Phi_x^2 + \Phi_y^2)^{\frac{1}{2}}} - \frac{\Phi_x(\Phi_x\Phi_{xx} + \Phi_y\Phi_{xy})}{(\Phi_x^2 + \Phi_y^2)^{\frac{3}{2}}} + \frac{\Phi_{yy}}{(\Phi_x^2 + \Phi_y^2)^{\frac{1}{2}}} - \frac{\Phi_y(\Phi_x\Phi_{xy} + \Phi_y\Phi_{yy})}{(\Phi_x^2 + \Phi_y^2)^{\frac{3}{2}}} \\
&= \frac{\Phi_{xx}\Phi_y^2 - 2\Phi_x\Phi_y\Phi_{xy} + \Phi_{yy}\Phi_x^2}{(\Phi_x^2 + \Phi_y^2)^{\frac{3}{2}}}
\end{aligned}
\tag{10.69}
$$

where $\Phi_x \equiv \partial\Phi/\partial x$, $\Phi_y \equiv \partial\Phi/\partial y$, $\Phi_{xx} \equiv \partial^2\Phi/\partial x^2$, $\Phi_{yy} \equiv \partial^2\Phi/\partial y^2$, $\Phi_{xy} \equiv \partial^2\Phi/\partial x\partial y$, and we have ignored ϵ for simplicity.

10.2.4 Geodesic Active Contours

The level set method, as we have just seen, is based upon curve evolution rather than energy minimization. The speed of the curve evolution is related to the magnitude of the image gradient, so that ideally the speed becomes zero when the image gradient is zero. In practice, however, the gradient magnitude never actually reaches zero because of noise, and therefore the level set method does not actually converge. Instead, the curve slows as it reaches the intensity edge, then speeds back up again after it passes the intensity edge—that is, if the algorithm is allowed to continue. To prevent this from happening, heuristic stopping conditions are often introduced. These heuristics detract from the mathematical purity of the approach and require careful fine tuning.

A popular solution to this drawback is the **geodesic active contour**, which connects energy minimization with curve evolution. Combining Equations (10.26), (10.27), and (10.29), the classic snake energy is given by

$$E(\mathbf{c}) = \frac{1}{2}\alpha \int_0^1 \left\|\frac{d\mathbf{c}(s)}{ds}\right\|^2 ds + \int_0^1 g(I(s)) ds \tag{10.70}$$

[†] Section 2.5.1 (p. 57).

where we have assumed $\beta = 0$ for simplicity, and where $g(\cdot)$ depends only upon the magnitude of the gradient of I. It can be shown[†] that the contour $\mathbf{c}$ that minimizes this energy functional also minimizes

$$E_{geod}(\mathbf{c}) = \int_0^1 g(I(s)) \left\| \frac{d\mathbf{c}(s)}{ds} \right\| ds \tag{10.71}$$

under the assumption of conservation of energy: $\frac{1}{2}\alpha\|d\mathbf{c}(s)/ds\|^2 + g(I(s)) = 0$.

In a Euclidean space the shortest distance between two points is along a straight line connecting them. The Euclidean length of the contour is therefore given by $\int \|d\mathbf{c}/ds\| ds$, which simply integrates the derivative. Comparing this quantity with Equation (10.71), we see that the latter applies the weight $g(\cdot)$ in calculating the length of the curve. This weight causes the computation to yield the length of the **geodesic**, where a geodesic is defined as a minimal-length curve between two points in a Riemannian manifold.[‡] To gain some intuition behind the idea of a geodesic, consider two points on the surface of the Earth. The shortest distance between them is not a straight line but rather the arc along the great circle connecting them. In a similar way, the geodesic active contour framework equates boundary detection with finding a curve of minimal weighted length.

Minimizing Equation (10.71) involves solving either of the following equivalent PDEs:

$$\frac{\partial \mathbf{c}(s,t)}{\partial t} = g(I(s)) \cdot \kappa(s) \cdot \mathbf{n}(s) - (\nabla g(I(s)))^\mathsf{T} \mathbf{n}(s) \cdot \mathbf{n}(s) \tag{10.72}$$

$$\frac{-\partial \Phi(x,y,t)}{\partial t} = g(I(x,y)) \cdot \kappa(x,y) \cdot \|\nabla \Phi(x,y,t)\| - (\nabla g(I(x,y)))^\mathsf{T} \nabla \Phi(x,y,t) \tag{10.73}$$

Comparing Equation (10.72) with Equation (10.65), notice the extra term $-((\nabla g)^\mathsf{T} \mathbf{n})\mathbf{n}$. Similarly, comparing Equation (10.73) with Equation (10.66), there is an extra term $(\nabla g)^\mathsf{T} \nabla \Phi$. In both cases the extra term attracts the curve to the intensity edge no matter the direction from which the curve approaches, as shown in Figure 10.16. As a result, it should be clear that the geodesic active contour approach indeed solves the problem of convergence mentioned above.

10.2.5 Chan-Vese Algorithm

We come now to the **Chan-Vese level set algorithm**. All the previous active contour and geometric contour methods presented in this chapter have relied upon intensity edges to guide the contour to the boundary of the object. Intensity edges, however, make up a small percentage of any given image, making their basin of attraction small. As a result, it is difficult for a technique that relies solely upon intensity edges to work properly if the initial contour is far from the edge. Regions, on the other hand, are much larger than intensity edges. Therefore, techniques based upon image regions have much larger basins of attraction. The Chan-Vese algorithm in particular, like geodesic active contours, works when the contour is initialized inside or outside the boundary (or partly inside and outside), and it has a wide basin of attraction.

[†] The derivation relies upon *Maupertuis' principle*, a special case of the principle of least action, which states that the path followed by a physical system is the one with the shortest length; and *Fermat's principle*, which states that the path taken by a ray of light is the one that is traversed in the least time.

[‡] A *Riemannian manifold* is a generalization of Euclidean space in which the distance between two points is given by their inner product.

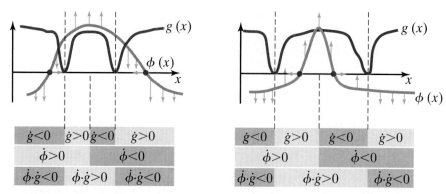

Figure 10.16 The extra term $(\nabla g)^{\mathsf{T}} \nabla \Phi$ in Equation (10.73) causes the curve evolution to stop at the intensity edge. Shown here is a 1D slice. $g(x)$ is the inverse of the magnitude of the gradient of the image, so that $g(x) = 0$ occurs where the intensity edges lie (at the boundary of the object). Left: Φ encloses the object and therefore needs to shrink. From the diagram it is clear that $\frac{\partial \Phi}{\partial x} \cdot \frac{\partial g}{\partial x} < 0$ at the contour, which indeed causes the contour to shrink. Right: Φ is enclosed by the object and therefore needs to expand. From the diagram it is clear that $\frac{\partial \Phi}{\partial x} \cdot \frac{\partial g}{\partial x} > 0$ at the contour, which indeed causes the contour to expand. The argument for the extra term in Equation (10.72) is similar.

Instead of using intensity edges, Chan-Vese balances the probabilities of foreground and background pixels, where foreground pixels lie inside the contour, while background pixels lie outside the contour. (Chan-Vese assumes a closed contour.) Any model for the appearance of foreground and background can be used, but the most basic formulation assumes the foreground region consists of pixels with approximate gray level v_i, while the background region consists of pixels with approximate gray level v_o. In this case the Chan-Vese energy functional to be minimized is given by

$$E(\mathbf{c}) = \lambda_\ell E_\ell(\mathbf{c}) + \lambda_a E_a(\omega) + \lambda_i E_i(\mathbf{c}) + \lambda_o E_o(\mathbf{c}) \tag{10.74}$$

which is the weighted sum of four terms governing, respectively, the length of the contour, the area enclosed by the contour, the similarity of the inside region to the value v_i, and the similarity of the outside region to the value v_o:

$$E_\ell(\mathbf{c}) = \int_\Omega \|\nabla h(\Phi(x, y))\| \, dx \, dy \qquad \text{(length)} \tag{10.75}$$

$$E_a(\mathbf{c}) = \int_\Omega h(\Phi(x, y)) \, dx \, dy \qquad \text{(area)} \tag{10.76}$$

$$E_i(\mathbf{c}) = \int_\Omega |I(x, y) - v_i|^2 h(\Phi(x, y)) \, dx \, dy \qquad \text{(inside)} \tag{10.77}$$

$$E_o(\mathbf{c}) = \int_\Omega |I(x, y) - v_o|^2 (1 - h(\Phi(x, y))) \, dx \, dy \qquad \text{(outside)} \tag{10.78}$$

where Ω is the domain of the image, and $h(\cdot)$ is the thresholding operator known as the **Heaviside function** defined as $h(z) = 1$ if $z \geq 0$, and $h(z) = 0$ otherwise. As a result, $h(\Phi(x, y))$ is 1 inside the contour and 0 outside the contour. If we define H as the 2D function such that $H(x, y) = h(\Phi(x, y))$, then $\|\nabla H(\mathbf{x})\|$ is large along the contour and 0 everywhere else.

As an aside, the Chan-Vese energy functional is closely related to the classic **Mumford-Shah** energy functional, which is designed to partition an image into multiple regions according to a minimal length curve while reconstructing the noisy signal in each region:

$$E(\hat{I}, \mathbf{c}) = \iint_{\Omega} (\hat{I}(x, y) - I(x, y))^2 \, dx \, dy + \lambda_s \iint_{\Omega \backslash \mathbf{c}} \|\nabla \hat{I}(x, y)\|^2 \, dx \, dy + \lambda_\ell \int_{\mathbf{c}} \left\| \frac{d\mathbf{c}(s)}{ds} \right\|^2 ds \quad (10.79)$$

where I is the original noisy signal, $\hat{I}$ is the reconstructed signal, the second term enforces smoothness of the reconstruction, and the third term minimizes the length of the curve. A simplified version of this problem, known as the **minimal partition problem**, occurs when the reconstructed signal is piecewise-constant, that is, constant within each region. Chan-Vese can therefore be considered a solution to the special case of the minimal partition problem in which there is just a single contour separating the inside region from the outside region.

The Chan-Vese energy in Equation (10.74) is minimized using iterations of the form

$$\Delta\Phi(\mathbf{x}) = -h_\Phi(\Phi(\mathbf{x})) \left[\lambda_a + \lambda_i |I(\mathbf{x}) - v_i|^2 - \lambda_o |I(\mathbf{x}) - v_o|^2 - \lambda_\ell \cdot \text{div}\left(\frac{\nabla\Phi(\mathbf{x})}{\|\nabla\Phi(\mathbf{x})\|} \right) \right] \quad (10.80)$$

where $\Delta\Phi(\mathbf{x})$ is the change in the value of Φ for the pixel at $\mathbf{x} \equiv (x, y)$ between iterations, and v_i and v_o are computed as the average of the gray levels in the inner and outer regions, respectively:

$$v_i = \frac{\iint_{\Omega} I(x, y) h(\Phi(x, y)) \, dx \, dy}{\iint_{\Omega} h(\Phi(x, y)) \, dx \, dy} \quad (10.81)$$

$$v_o = \frac{\iint_{\Omega} I(x, y) (1 - h(\Phi(x, y))) \, dx \, dy}{\iint_{\Omega} (1 - h(\Phi(x, y))) \, dx \, dy} \quad (10.82)$$

The factor $h_\Phi(\Phi(\mathbf{x})) \equiv \partial h(\Phi(\mathbf{x}))/\partial\Phi(\mathbf{x})$ is, in theory, the delta function, because it is the derivative of a step function. In practice, however, we approximate $h(\cdot)$ with a **regularized Heaviside function** (or soft threshold operator), such as

$$h(z) \approx \frac{1}{2}\left(1 + \frac{2}{\pi} \arctan\left(\frac{z}{a} \right) \right) \quad (10.83)$$

whose derivative is given by

$$h_z(z) = \frac{1}{\pi} \cdot \frac{a}{a^2 + z^2} \quad (10.84)$$

where a governs the width of the transition. For convenience, we usually set $a = 1$, leading to

$$h_\Phi(\Phi(\mathbf{x})) = \frac{1}{\pi} \cdot \frac{1}{1 + (\Phi(\mathbf{x}))^2} \quad (10.85)$$

The pseudocode, shown in Algorithm 10.8 , reveals that the technique follows the same basic pattern as the standard level set method except that the image pixels are compared

ALGORITHM 10.8 Evolve a contour using the Chan-Vese level set method.

CHANVESELEVELSETMETHOD(I)

Input: grayscale image I
Output: 2D array Φ in which foreground object pixels are indicated by $\Phi(x, y) > 0$

1    $\Phi_{prev} \leftarrow \Phi \leftarrow$ INITIALIZEIMPLICITFUNCTION()
2    $speed \leftarrow \infty$
3   **while** $speed > \tau$ **do**   ▶ While contour is moving at sufficient speed,
4       $v_i \leftarrow$ MEANGRAYLEVELINSIDECONTOUR(I, Φ) compute the mean gray levels
5       $v_o \leftarrow$ MEANGRAYLEVELINSIDECONTOUR$(I, -\Phi)$ of the inner and outer regions.
6   **for** $(x, y) \in I$ **do**   ▶ For each pixel,
7         $h_z \leftarrow 1/(\pi * (1 + \Phi(x, y) * \Phi(x, y)))$ compute
8         $\phi_x \leftarrow \Phi(x, y) - \Phi(x - 1, y)$ the first
9         $\phi_y \leftarrow \Phi(x, y) - \Phi(x, y - 1)$ and
10        $\phi_{xx} \leftarrow \Phi(x + 1, y) - 2\Phi(x, y) + \Phi(x - 1, y)$ second
11        $\phi_{yy} \leftarrow \Phi(x, y + 1) - 2\Phi(x, y) + \Phi(x, y - 1)$ derivatives
12        $\phi_{xy} \leftarrow \Phi(x, y - 1) + \Phi(x - 1, y) - \Phi(x - 1, y - 1) - \Phi(x, y)$ in order to
13        $num \leftarrow \phi_{xx} * \phi_y * \phi_y - 2 * \phi_x * \phi_y * \phi_{xy} + \phi_{yy} * \phi_x * \phi_x$ compute the
14        $den \leftarrow \text{Pow}(\phi_x * \phi_x + \phi_y * \phi_y, 1.5)$ curvature
15        $\kappa \leftarrow num/den$ using Equation (10.69).
16        $d_i \leftarrow (I(x, y) - v_i) * (I(x, y) - v_i)$ ▶ Compute the differences between
17        $d_o \leftarrow (I(x, y) - v_o) * (I(x, y) - v_o)$ the actual and expected gray levels.
18        $\Phi(x, y) \leftarrow_- h_z * (\lambda_a + \lambda_i * d_i - \lambda_o * d_o - \lambda_\ell * \kappa)$ ▶ Update Φ using Equation (10.80),
19        $C(x, y) \leftarrow$ TRUE **if** $\Phi(x, y) > 0$ **or** FALSE **otherwise**   and threshold Φ.
20      $\Phi \leftarrow$ SIGNEDCHAMFERDISTANCE(C) ▶ Reinitialize Φ using signed chamfer distance.
21      $speed \leftarrow \max(|\Phi - \Phi_{prev}|)$ ▶ Set the speed to the maximum distance moved by any pixel.
22      $\Phi_{prev} \leftarrow \Phi$ ▶ Store a copy of Φ for the next iteration.
23   **return** Φ

MEANGRAYLEVELINSIDECONTOUR(I, Φ)

Input: grayscale image I, implicit function Φ
Output: mean gray level inside contour specified by Φ

1    $sum \leftarrow 0$
2    $n \leftarrow 0$
3   **for** $(x, y) \in \Phi$ **do**
4   **if** $\Phi(x, y) > 0$ **then**
5         $sum \leftarrow_+ I(x, y)$
6         $n \leftarrow_+ 1$
7   **return** sum/n

with the mean gray levels inside and outside the contour, rather than using the intensity gradient of the image. Note that Lines 8–15 are used to compute the curvature using Equation (10.69), applying the noncentered kernels that we considered earlier.[†] Since Chan-Vese is insensitive to the initial contour, the initialization is performed by INITIALIZEIMPLICITFUNCTION (whose implementation will depend upon the problem at hand). Note that Chan-Vese can be extended easily to incorporate any appearance model of the inner and outer regions, such as using color histograms, with only slight modifications to the code.

To derive Equation (10.80), let us rewrite the energy in Equation (10.74) as

$$E(\mathbf{c}) = E(\Phi) = \int_{\Omega} F(\mathbf{x}, \Phi, \Phi_x, \Phi_y) \, d\mathbf{x} \tag{10.86}$$

where

$$F(\mathbf{x}, \Phi, \Phi_x, \Phi_y) \equiv \lambda_\ell \|\nabla H(\mathbf{x})\| + \lambda_a h(\Phi(\mathbf{x})) + \lambda_i |I(\mathbf{x}) - \mathbf{v_i}|^2 \mathbf{h}(\Phi(\mathbf{x}))$$

$$+ \lambda_o |I(\mathbf{x}) - v_o|^2 [1 - h(\Phi(\mathbf{x}))] \tag{10.87}$$

Noting that x and y are the independent variables and Φ is the dependent variable, the Euler-Lagrange equations are given by

$$\frac{\partial F}{\partial \Phi} - \frac{\partial}{\partial x}\left(\frac{\partial F}{\partial \Phi_x}\right) + \frac{\partial}{\partial y}\left(\frac{\partial F}{\partial \Phi_y}\right) = 0 \tag{10.88}$$

where the partial derivative with respect to Φ is

$$\frac{\partial F}{\partial \Phi} = \lambda_a h_\Phi(\Phi(\mathbf{x})) + \lambda_i |I(\mathbf{x}) - v_i|^2 h_\Phi(\Phi(\mathbf{x})) - \lambda_o |I(\mathbf{x}) - v_o|^2 h_\Phi(\Phi(\mathbf{x})) \tag{10.89}$$

$$= h_\Phi(\Phi(\mathbf{x}))(\lambda_a + \lambda_i |I(\mathbf{x}) - v_i|^2 - \lambda_o |I(\mathbf{x}) - v_o|^2) \tag{10.90}$$

where $h_\Phi \equiv \partial h(\Phi)/\partial \Phi$ is the delta function. The other derivatives are as follows:

$$\frac{\partial F}{\partial \Phi_x} = \frac{\partial}{\partial \Phi_x}(\lambda_\ell \|\nabla H(\mathbf{x})\|) = \lambda_\ell h_\Phi(\Phi(\mathbf{x}))\frac{\partial}{\partial \Phi_x}(\|\nabla\Phi(\mathbf{x})\|) = \lambda_\ell h_\Phi(\Phi(\mathbf{x}))\frac{\partial}{\partial \Phi_x}(\Phi_x^2(\mathbf{x}) + \Phi_y^2(\mathbf{x}))^{\frac{1}{2}}$$

$$= \lambda_\ell h_\Phi(\Phi(\mathbf{x}))\frac{1}{2}(\Phi_x^2(\mathbf{x}) + \Phi_y^2(\mathbf{x}))^{-\frac{1}{2}}(2\Phi_x(\mathbf{x})) = \lambda_\ell h_\Phi(\Phi(\mathbf{x}))\frac{\Phi_x(\mathbf{x})}{\|\nabla\Phi(\mathbf{x})\|} \tag{10.91}$$

$$\frac{\partial F}{\partial \Phi_y} = \lambda_\ell h_\Phi(\Phi(\mathbf{x}))\frac{\Phi_y(\mathbf{x})}{\|\nabla\Phi(\mathbf{x})\|} \tag{10.92}$$

where we have used the fact that the gradient of $H(\mathbf{x})$ can be expressed as

$$\nabla H(\mathbf{x}) = \left[\frac{\partial h(\Phi)}{\partial x} \quad \frac{\partial h(\Phi)}{\partial y}\right]^{\mathsf{T}} = \left[\frac{\partial h(\Phi)}{\partial \Phi}\frac{\partial \Phi}{\partial x} \quad \frac{\partial h(\Phi)}{\partial \Phi}\frac{\partial \Phi}{\partial y}\right]^{\mathsf{T}} = h_\Phi(\Phi)[\Phi_x \quad \Phi_y]^{\mathsf{T}} = h_\Phi(\Phi) \cdot \nabla\Phi \tag{10.93}$$

so that $\|\nabla H(\mathbf{x})\| = h_\Phi(\Phi) \cdot \|\nabla\Phi\|$.

[†] Section 5.4 (p. 240)

Plugging these derivatives into Equation (10.88) yields

$$
\begin{aligned}
0 &= h_\Phi(\Phi(\mathbf{x}))[\lambda_a + \lambda_i|I(\mathbf{x}) - v_i|^2 - \lambda_o|I(\mathbf{x}) - v_o|^2] - \frac{\partial}{\partial x}\left(\lambda_\ell h_\Phi(\Phi(\mathbf{x}))\frac{\Phi_x(\mathbf{x})}{\|\nabla\Phi(\mathbf{x})\|}\right) \\
&\quad - \frac{\partial}{\partial y}\left(\lambda_\ell h_\Phi(\Phi(\mathbf{x}))\frac{\Phi_y(\mathbf{x})}{\|\nabla\Phi(\mathbf{x})\|}\right) \\
&= h_\Phi(\Phi(\mathbf{x}))\left[\lambda_a + \lambda_i|I(\mathbf{x}) - v_i|^2 - \lambda_o|I(\mathbf{x}) - v_o|^2] - \lambda_\ell \cdot \mathrm{div}\left(\frac{\nabla\Phi(\mathbf{x})}{\|\nabla\Phi(\mathbf{x})\|}\right)\right] \quad (10.94)
\end{aligned}
$$

If we could solve this equation directly for Φ, we would have an answer for the Φ that minimizes the energy functional above. However, because this is a nonlinear equation, we instead adopt a gradient descent approach to solve for Φ, treating the value on the right-hand side as the deviation from the true solution. Given our convention that $\Phi(\cdot) > 0$ inside the contour, note that a positive value in the term involving v_i indicates that the zero level set of implicit function is outside the contour, in which case Φ needs to be reduced, thus leading to the negative sign in Equation (10.80) . A flowchart of the Chan-Vese algorithm is shown in Figure 10.17, and the result of the algorithm in Figure 10.18.

10.3 Image Segmentation

So far we have considered the simplified case of segmenting just two regions, namely the foreground and background. Now we turn our attention to the general case of image segmentation, in which the goal is to carve an image into an arbitrary number of disjoint regions such that the pixels within each region share some visual property such as similarity in color, texture, or motion, as shown in Figure 10.19. Before we consider a number of popular algorithms for automatic segmentation, let us first consider an influential movement in psychology that attempts to explain how the human visual system performs segmentation.

Figure 10.17 Flowchart of the Chan-Vese algorithm. First derivatives are taken to compute the divergence of the normalized gradient of Φ. The heart of the algorithm is to compute $\Delta\Phi$ to update Φ. Then $\Delta\Phi$ is added to Φ of the previous iteration to yield Φ of the current iteration, followed by reinitialization of Φ.

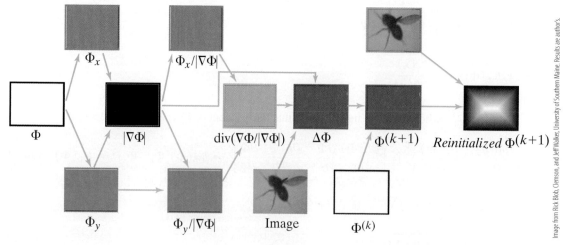

Figure 10.18 The result of the Chan-Vese algorithm on a grayscale image.

10.3.1 Gestalt Psychology

Consider the well-known Müller-Lyer illusion in Figure 10.20. Even though the two horizontal lines are exactly the same length, one appears longer than the other. The reason for this is that the human visual system groups the horizontal and diagonal lines so that they appear as a single unit, and the perception is performed at such a low level in the system that it is impossible to "undo" the merge. No matter how hard a viewer may try, it is not possible to separate the lines in order to see that the two horizontal lines are, in fact, identical. Similarly, in the famous Kanisza triangle, a white triangle is perceived, even though the figure does not contain a triangle at all but rather some thin lines and circles with missing pieces; because these pieces are aligned in a certain way, our visual system perceives a white triangle occluding three circles and a thin triangle. Similarly, the specific arrangement of the cones in the figure causes us to perceive an underlying sphere.

In the 1920s, a group of German psychologists led by Max Wertheimer studied phenomena such as these. Their question was, "How does the visual system organize the lightness

Figure 10.19 Image segmentation involves carving an image into an arbitrary number of regions, each containing pixels that share some property of appearance (in this case color). Shown below are four different ways of displaying the output of a segmentation algorithm: the boundaries between regions, the label of each region as an integer (only some labels are shown), a random color for each region, or the mean RGB color of each region.

 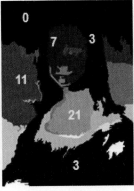  

Boundaries   Labels   Pseudocolors   Mean colors

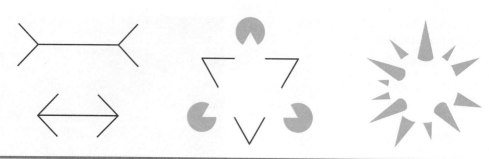

patterns on the retina into coherent objects in order to give us a visual experience of the world?" Their answer was that the perceptual system does not simply collect and combine incoming sensory information to give a picture of the world, but instead actively organizes it. The fundamental principle that they articulated is known as the **Law of Prägnanz**,[†] which says that every stimulus pattern is seen in such a way that the resulting structure is as simple as possible. In other words, this law can be thought of as Occam's Razor[‡] applied to perception, because it favors the simplest and most stable interpretations. By reacting against the prevailing atomistic view of the time, these scientists founded the **gestalt school of psychology**,[§] which emphasized grouping as the key to understanding visual perception. They used the term *Gestalt Qualität* to refer to the set of internal relationships that makes the individual light sensations perceived as a whole. Some of the more well-known gestalt factors are shown in Figure 10.21.

Despite the fact that gestalt psychology provides a compelling description of our experience, it is too imprecise to be easily translated into a specific algorithm. That is, saying that the visual system groups pixels together (through some mysterious process) rather than treating them independently is not of much help to a computer vision practitioner unless one is also informed *how* such grouping takes place. And if we know how to group the pixels, then there does not seem to be much difference in saying that the pixels are treated as a unit

Figure 10.21 Various gestalt relationships.

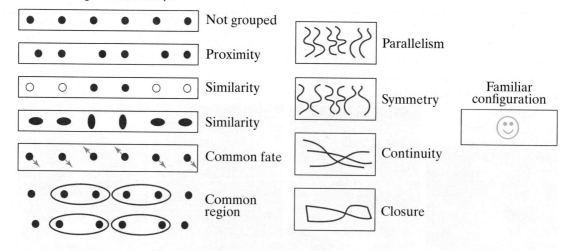

---

versus saying that they are treated as individual pixels, because the grouping process itself has no other option than to treat them that way. So, while the gestalt approach may provide inspiration for segmentation, we must look elsewhere for specific algorithms.

10.3.2 Splitting and Merging

Let us define the segmentation of an image I as a set of regions $\mathcal{R}_1, \ldots, \mathcal{R}_n$ such that every pixel is in exactly one region:

$$I = \bigcup_{i=1}^{n} \mathcal{R}_i \qquad \text{(covers entire image)} \qquad (10.95)$$

$$\mathcal{R}_i \cap \mathcal{R}_j = \emptyset \quad \text{for all} \quad i \neq j \qquad \text{(non-overlapping)} \qquad (10.96)$$

where each region is also a set of pixels, and additionally we assume that the pixels in each region are connected.

In the classic view of the problem of segmentation, a predicate $h(\mathcal{R}_i)$ measures the **homogeneity** of a region. That is, $h(\mathcal{R}_i)$ returns true if all the pixels in region $\mathcal{R}_i$ satisfy the homogeneity criterion, and false otherwise. A proper segmentation of the image therefore satisfies two criteria:

$$h(\mathcal{R}_i) = \text{TRUE for all } i \qquad \text{(each region is homogeneous)} \qquad (10.97)$$

$$h(\mathcal{R}_i \cup \mathcal{R}_j) = \text{FALSE for all adjacent } \mathcal{R}_i, \mathcal{R}_j, i \neq j \quad \text{(adjacent regions are different)} \qquad (10.98)$$

Image segmentation algorithms typically fall into one of two categories. Some algorithms begin with each pixel as a separate region, then recursively merge adjacent regions whenever they are similar to each other. Such algorithms are known as **merging** algorithms. At each step of the computation, Equation (10.97) is satisfied, and the goal of the algorithm is to continue merging until Equation (10.98) is satisfied as well. Other algorithms are known as **splitting** algorithms because they begin with the entire image as a single region, then recursively split regions whenever they are found to be nonhomogeneous. With such algorithms, at each step of the computation Equation (10.98) is satisfied, and the goal of the algorithm is to continue splitting until Equation (10.97) is satisfied as well. Merging is also known as *agglomerative clustering*, while splitting is known as *divisive clustering*.

The classic **split-and-merge** algorithm combines these two ideas. First the image is recursively split until each region passes the homogeneity test. To facilitate this splitting, a **quad-tree** data structure is used. A quad-tree is a tree where each nonleaf node has exactly four children. The splitting computation works as follows. First the entire image (which forms the root of the tree) is tested for homogeneity. If it fails the test (which it will certainly do for all but the most boring of images), then it is split down the middle vertically and horizontally into four equally-sized regions, and the process is repeated recursively for each of the four regions, until all the regions are homogeneous.

Once the splitting is complete, then the merging process begins. There are various ways to perform merging, but one approach involves a variation on the classic connected components algorithm.[†] The image is scanned from left-to-right and top-to-bottom, and the homogeneity test is performed on the regions corresponding to adjacent regions to decide whether to merge them. Unlike the classic connected components algorithm, however, where the order of scanning is arbitrary (as long as the algorithm is modified accordingly), the order in which the image is scanned for merging will affect the results, since the properties of the regions change when they are merged. An example of the split-and-merge algorithm applied to a grayscale image is shown in Figure 10.22.

[†] Section 4.2.3 (p. 157).

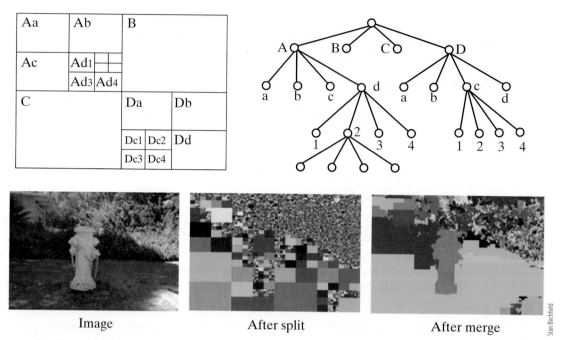

Figure 10.22 Top: The quad-tree data structure used in splitting. Bottom: The split-and-merge algorithm applied to a grayscale image. The algorithm is able to find the fire hydrant and most of the ground, though it oversegments the textured background.

10.3.3 Region Growing

Even though split-and-merge sounds like a good idea, in practice a more effective technique is simply to apply agglomerative clustering, which, in the context of image segmentation, is known as **region growing**. In region growing, each pixel is initialized as a separate region, and adjacent regions that look similar are successively merged. The pseudocode for growing a single region is displayed in Algorithm 10.9. The reader should immediately recognize the similarity of this procedure with the floodfill algorithm of Algorithm 4.6.[†] Floodfill merges pixels only when they have *identical* values, whereas region growing merges pixels when they have *similar* values. Just as with floodfill, region growing starts with an initial seed pixel $\mathbf{p} = (x, y)$. A stack called a frontier is used to keep track of the pixels on the boundary of the region as it is being grown. At each iteration, a pixel is popped off the frontier, and its neighbors are examined. For each neighbor, if the appearance is similar to the model of the region, then that neighbor is added to the region, and the model is updated; the process continues until the frontier is empty.

Three lines distinguish the procedure from the basic floodfill: Line 1 initializes the model, Line 7 compares whether the pixel is *similar* to the model (rather than *identical* to the seed pixel), and Line 10 updates the model. The implementation of these lines depends upon the model chosen. One simple, natural, and effective choice is the Gaussian model. By maintaining the mean and variance of the pixel values in the region, we can measure the similarity of a pixel to the model by evaluating the Gaussian at that value. More specifically, for a grayscale image the mean and variance are given by

$$\mu = \frac{1}{n} \sum_{i=1}^{n} I(\mathbf{p}_i) \tag{10.99}$$

[†] Section 4.2.2 (p. 154)

ALGORITHM 10.9 Region growing for a single region

GROWSINGLEREGION(*I*,*O*,*p*,*new-label*)

Input: Image *I*, output image *O*, seed pixel *p*, and new label *new-label*
Output: Pixels in *O* are set

1   *model*.INITIALIZE(*I*(*p*))
2   *frontier*.PUSH(*p*)
3   *O*(*p*) ← *new-label*
4   **while** *frontier*.SIZE > 0 **do**
5   *p* ← *frontier*. POP( )
6   **for** *q* ∈ 𝒩(*p*) **do**
7   **if** *model*.ISSIMILAR(*I*(*q*)) AND *O*(*q*) ≠ *new-label* **then**
8   *frontier*.PUSH(*q*)
9   *O*(*q*) ← *new-label*
10   *model*.UPDATE(*I*(*q*))

MODEL.INITIALIZE(*VAL*)

1   *model*.*s* ← *val*
2   *model*.*s̃* ← *val* * *val*
3   *model*.*n* ← 1
4   *model*.τ^2 ← 2.5 * 2.5   ➤ example threshold to capture $\pm 2.5\sigma$

MODEL.ISSIMILAR(*VAL*)

1   μ ← *model*.*s*/*model*.*n*
2   σ^2 ← *model*.*s̃*/*model*.*n* − *model*.μ * *model*.μ
3   d^2 ← (*val*−*model*.μ) * (*val*−*model*.μ)
4   **return** $d^2 \leq$ *model*.τ^2 * σ^2

MODEL.UPDATE(*VAL*)

1   *model*.*s* ←₊ *val*
2   *model*.*s̃* ←₊ *val* * *val*
3   *model*.*n* ←₊ 1

$$\sigma^2 \;=\; \frac{1}{n}\sum_{i=1}^{n}\left(I(\mathbf{p}_i) - \mu\right)^2 \tag{10.100}$$

where $\mathbf{p}_1, \mathbf{p}_2, \ldots, \mathbf{p}_n$ are the pixels in the region. The dissimilarity of a pixel $\mathbf{q}$ is then measured by its distance from the mean, relative to the standard deviation: $d(\mathbf{q}; I, \mu, \sigma) = |I(\mathbf{q}) - \mu|/\sigma$. The reader may notice that this is simply the Mahalanobis distance[†] in one dimension. The pixel is considered to be similar to the model if $d(\mathbf{q}; I, \mu, \sigma) < \tau$, where τ is a constant threshold. One advantageous property of the Gaussian is that there is a natural and intuitive way to select the threshold. Usually

[†] Section 4.3.1 (p. 164).

$2.0 \leq \tau \leq 3.0$, where $\tau = 2.0$ captures $\pm 2\sigma$ of the Gaussian, or 95% of the area under the curve, while $\tau = 3.0$ captures $\pm 3\sigma$ of the Gaussian, or 99.7% of the area under the curve. Another advantage of using the Gaussian is that it is easy to update the mean and variance by simply maintaining the running sum of values $s_n \equiv \sum_{i=1}^{n} I(\mathbf{p}_i)$ and running sum of squares $\tilde{s}_n \equiv \sum_{i=1}^{n} I^2(\mathbf{p}_i)$. It is easy to show that $\mu = s_n/n$ and $\sigma^2 = \tilde{s}_n/n - \mu^2$. To avoid having to compute the square root, the function IsSIMILAR takes advantage of the fact that $d < \tau$ is equivalent to $d^2 < \tau^2$, or $(I(\mathbf{q}) - \mu)^2 < \tau^2 \sigma^2$. Note that IsSIMILAR, as it is written here, will always set σ^2 to 0 whenever $n = 1$, because the variance is impossible to estimate from a single value. This will prevent merging from ever taking place. Moreover, the variance estimate is not accurate for small sets. As a result, in a real implementation this simple pseudocode should be slightly modified to encourage merging when n is small.

It is easy to extend this basic procedure to color images, or to any other image with a vector of values per pixel. For vector-valued images, the mean vector is computed by stacking the means of the different channels, and the covariance matrix is computed similarly. The similarity function uses the Mahalanobis distance with the multivariate Gaussian.[§] Oftentimes the full covariance matrix is not needed, and good results can be achieved by assuming the covariance matrix is diagonal, thus ignoring the covariances and using only the variances along the dimensions.

Just as the connected components of an image can be computed by repeatedly applying floodfill, so the entire image can be segmented, applying a label to every pixel, by repeatedly applying the region growing procedure. The resulting algorithm is shown in Algorithm 10.10, where the function GROWSINGLEREGION is repeatedly called until all pixels have been labeled. At each iteration, the next seed pixel can be selected either by sequentially scanning the image or by selecting an unlabeled pixel whose value is similar to its neighbors (to avoid starting the growing procedure on an intensity edge). Despite its simplicity, region growing is an effective segmentation technique. Figure 10.23 shows the results of region growing with a Gaussian RGB model for each region on an example image. As was done here, it is generally recommended to enforce a minimum region size to reduce the effects of noise.

ALGORITHM 10.10 Image segmentation by region growing

REGIONGROW(I)

Input: image I
Output: label image L

1   $label \leftarrow 0$
2   **for** $(x, y) \in I$ **do**
3          $L(x, y) \leftarrow$ UNLABELED
4   **for** $(x, y) \in I$ **do**
5   **if** $L(x, y) =$ UNLABELED **then**
6   GROWSINGLEREGION($I, L, (x, y), label$)
7                 $label \leftarrow_+ 1$
8   **return** L

[§] Section 12.2.4 (p. 580).

Figure 10.23 Left: An RGB image. Right: Regions found by region growing, pseudocolored.

P. Chockalingam, N. Pradeep, and S. T. Birchfield. Adaptive fragments-based tracking of non-rigid objects using level sets. In Proceedings of the International Conference on Computer Vision, Oct. 2009.

10.3.4 Hierarchical Clustering Scheme (HCS)

Image segmentation can be viewed as a specific application of data clustering, which is a general problem that arises in many empirical domains. One popular approach to data clustering is the **hierarchical clustering scheme (HCS)**. An HCS operates on a fully connected graph containing a vertex for each data point. The weights of the edges are, not surprisingly, the distances between the data points as computed in some feature space. The procedure follows the agglomerative clustering approach, beginning by assigning each data point to its own cluster, then iteratively applying the following two steps: first, the two closest clusters are merged, and secondly, the weights from the remaining clusters to the new cluster are updated. This simple procedure successively merges clusters until all clusters have been merged. Although the result is not exactly a segmentation per se, the procedure bears close resemblance to several popular segmentation algorithms, making HCS a helpful framework for considering various related issues.

Figure 10.24 shows a simple example of the HCS that results from applying this procedure to a set of five data points. Since the minimum distance is between vertices **a** and **b**, these two are merged first. Then, the distance between vertices **c** and **d** is minimum, so these are merged next. Of the remaining distances, the minimum is between **e** and the cluster {**c**, **d**}. Finally, the clusters {**a**, **b**} and {**c**, **d**, **e**} are merged. The result is a hierarchical clustering scheme that can be viewed as a **dendrogram**, which looks like a binary tree with a leaf node for each vertex in the original graph. Dendrograms are usually drawn

Figure 10.24 Left: An example of five data points labeled a through e, viewed as a graph with the edge weights indicating the distances (in some feature space) between the points. Middle columns: Initially considering each data point as a separate cluster, sequential iterations of the HCS procedure merge the two closest clusters until all clusters have been merged. Because the weights satisfy the ultrametric inequality, no updating of the weights is needed. Right: The dendrogram is a way to visualize the resulting hierarchical clustering, with the original data points along the horizontal axis and the distances used for merging along the vertical axis.

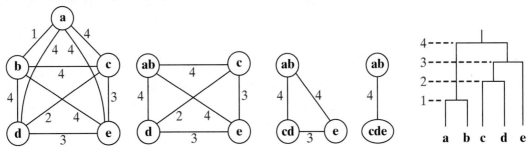

so that the distance encountered during the merge of each pair of clusters is shown along the vertical axis.

The reader may be wondering why only one step of the HCS procedure was used in this example when we said there were two steps involved. The reason we did not have to update the weights is that the weights in this particular graph satisfy what is known as the **ultrametric inequality**. Let $\mathcal{A}$ and $\mathcal{B}$ be the two clusters in the graph that will be merged in a particular iteration of the procedure, and let $\psi(\mathcal{A}, \mathcal{B})$ denote the distance between them. Then, for any other cluster $\mathcal{C}$ in the graph, the distance between $\mathcal{C}$ and the new cluster $\{\mathcal{A}, \mathcal{B}\}$ needs to be defined once $\mathcal{A}$ and $\mathcal{B}$ are merged so that the procedure can continue. It turns out that if $\psi(\mathcal{A}, \mathcal{C}) = \psi(\mathcal{B}, \mathcal{C})$, as in the simple example above, then the choice is obvious, namely, $\psi(\{\mathcal{A}, \mathcal{B}\}, \mathcal{C}) = \psi(\mathcal{A}, \mathcal{C}) = \psi(\mathcal{B}, \mathcal{C})$. This will happen when for each cluster triplet the distance function satisfies the ultrametric inequality, namely,

$$\psi(\mathcal{A}, \mathcal{B}) \leq \max(\psi(\mathcal{A}, \mathcal{C}), \psi(\mathcal{B}, \mathcal{C})) \quad \text{for all } \mathcal{A}, \mathcal{B}, \text{ and } \mathcal{C}. \quad (\text{ultrametric inequality}) \quad (10.101)$$

Recall that a distance function is a *metric* if it satisfies the properties of nonnegativity, symmetry, and the triangle inequality.[†] A distance function is an **ultrametric** if it satisfies not only these three properties but also the ultrametric inequality.

It is easy to see that the ultrametric inequality implies the triangle inequality, but not vice versa. It is also easy to see that if three distances satisfy the ultrametric inequality, then two of the distances are equal, and the third distance is no greater than those two. For example, the set of distances $\{3, 4, 4\}$ satisfies the ultrametric property, but the set $\{3, 4, 5\}$ does not. (That is, $3 \leq \max(4, 4)$ and $4 \leq \max(3, 4)$ but $5 \nleq \max(3, 4)$.) Note that with $\{3, 4, 4\}$, the two clusters separated by a distance of 3 will be merged, so that $\psi(\mathcal{A}, \mathcal{C}) = \psi(\mathcal{B}, \mathcal{C}) = 4$. More generally, if we have a set of clusters with distances defined between all pairs using a distance function that satisfies the ultrametric inequality, then if we let $\mathcal{A}$ and $\mathcal{B}$ be the closest clusters, that is, $\psi(\mathcal{A}, \mathcal{B}) \leq \psi(\mathcal{X}, \mathcal{Y})$ for any clusters $\mathcal{X}$ and $\mathcal{Y}$, then we know that $\psi(\mathcal{A}, \mathcal{Z}) = \psi(\mathcal{B}, \mathcal{Z})$ for any cluster $\mathcal{Z}$. In the same way, all the weights in the example graph above satisfy the ultrametric inequality, because in each case the distance from all vertices in the new cluster to all other clusters is the same. For example, when merging $\{\mathbf{c}, \mathbf{d}\}$ and $\mathbf{e}$, the distance between $\{\mathbf{c}, \mathbf{d}\}$ and $\{\mathbf{a}, \mathbf{b}\}$ is the same as the distance between $\mathbf{e}$ and $\{\mathbf{a}, \mathbf{b}\}$, making it easy to determine the distance between $\{\mathbf{c}, \mathbf{d}, \mathbf{e}\}$ and $\{\mathbf{a}, \mathbf{b}\}$.

In real-world situations, however, the distance function will not be an ultrametric, and therefore we must decide how to define the distance between two clusters $\mathcal{A}$ and $\mathcal{B}$. The three most common approaches are to compare the closest points from the two sets, the farthest points from the two sets, or the centroid of the two sets, as listed here:

$$\psi(\mathcal{A}, \mathcal{B}) \;=\; \min\{\psi(a, b), a \in \mathcal{A}, b \in \mathcal{B}\} \quad (\text{single-link clustering}) \quad (10.102)$$

$$\psi(\mathcal{A}, \mathcal{B}) \;=\; \max\{\psi(a, b), a \in \mathcal{A}, b \in \mathcal{B}\} \quad (\text{complete-link clustering}) \quad (10.103)$$

$$\psi(\mathcal{A}, \mathcal{B}) \;=\; \frac{1}{|\mathcal{A}||\mathcal{B}|} \sum_{a \in \mathcal{A}, b \in \mathcal{B}} \psi(a, b) \quad (\text{group-average clustering}) \quad (10.104)$$

These clustering choices are illustrated in Figure 10.25. Note that the region-growing approach of the previous subsection uses group-average clustering, with a slight modification to compute the average distance relative to the standard deviation. Figure 10.26 shows a more realistic example of an HCS, using both single-link and complete-link clustering. In the case of single-link clustering, the edges that are used for merging form a minimum

---

[†] Section 4.3.1 (p. 164).

Figure 10.25 Three common clustering choices between two sets of points are the minimum distance between points in the two sets (single-link clustering), the maximum distance (complete-link), and the distance between the centroids (group-average).

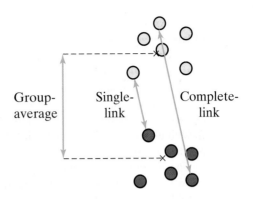

spanning tree for the graph. The importance of this observation will be clear later in the chapter when we consider the Felzenszwalb-Huttenlocher algorithm.[†]

10.3.5 Watershed Method

When rain falls on the side of a hill or mountain, the water flows downhill to the valley below. Unless the water is absorbed by the ground or is evaporated, it flows into the nearest creek, which joins with a nearby river, when eventually (in the typical case) empties into the

Figure 10.26 Top: A more realistic example of five data points, with weights that do not satisfy the ultrametric inequality. Shown are the iterations for single-link (middle) and complete-link (bottom) clustering approaches.

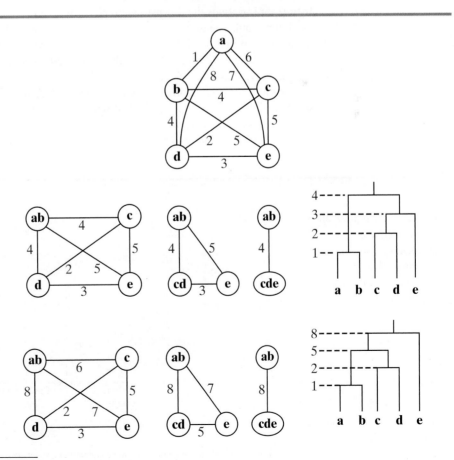

† Section 10.4.1 (p. 490).

ocean. As illustrated in Figure 10.27, a **catchment basin**, or **drainage basin**, is an area of land where all the surface water in the area eventually flows to a single point at a lower elevation. The ridge that separates adjacent drainage basins is known as a **ridgeline**, or **drainage divide**. In hilly country, this ridgeline connects the peaks of the mountains in the range, while in relatively flat areas the ridgeline is more difficult to locate. A **continental divide** is a ridgeline that separates two catchment basins, each of which flows into a different ocean. For example, rain that falls in the Rocky Mountains of the United States might flow to either the Atlantic or Pacific Ocean, depending on which side of the continental divide it lands.

These topographic concepts lead naturally to an approach of image segmentation known as **watershed**[†] **segmentation** which combines ideas from region growing and edge detection. The watershed approach to segmentation determines the segmented regions of the original image by computing the catchment basins of the terrain indicated by the **segmentation function**, which is a function (defined over the same domain as the image) that yields small values in the interior of regions but large values along the boundaries between regions. This segmentation function is viewed as the surface of a terrain (or **topographic surface**) in which the mountains (large values) indicate boundaries between regions, and the valleys (small values) indicate the region interiors. Typical choices for the segmentation function include the gradient magnitude of the image, the output of a boundary detector, a distance function from approximate region centers, or any other representation that emphasizes the boundaries of regions.

There are two basic flavors of the watershed approach. One flavor, known as **tobogganing**, finds the downstream path from each pixel (imagined as a raindrop) to the local minimum. The drawback with this technique is that discretization effects in a digital image make it impossible to determine the downward direction using only local information when there is a plateau. As a result, if the pixel (raindrop) follows the path of gradient descent, it will get stuck in the plateau.

The more common approach is known as **immersion**. In this approach, we imagine the segmentation function as a rigid 3D structure which is punctured at each local minimum and slowly submersed in water. As the segmentation function is pushed down, the water flows through the punctured holes to fill up the catchment basins, each of which is assigned

Figure 10.27 In watershed segmentation, the segmentation function is interpreted as a topographical surface. The most common choice for the segmentation function is the gradient magnitude image.

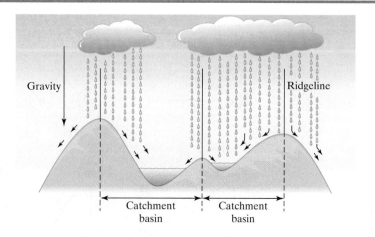

[†] The term *watershed* itself is ambiguous; in some communities a watershed is a catchment basin, whereas in other communities a watershed is a ridgeline.

a unique label according to which hole the water flowed through. When the water levels from adjacent catchment basins reach the ridgeline between them, we imagine building a dam of infinite height at the ridgeline to prevent the two bodies of water from mixing (which would confuse the unique labels). Once the entire segmentation function has been submerged, every point on the surface has received a unique label, which then indicates the identity of the region containing the point.

To gain some intuition behind the immersion approach to watershed segmentation, Figure 10.28 shows the contour lines of a segmentation function containing just 6 levels (that is, each pixel in the segmentation function takes a value 0 through 5), along with the output after each step of the algorithm. In the first step, Level 0 is considered (that is, all pixels whose segmentation function value is 0). There are two such regions, and the floodfill procedure is applied to both with separate labels (indicated by different colors). In the second step, Level 1 is considered. One of the regions is disconnected, so the floodfill procedure is applied with a third label (indicated by yet another color). At the same time, the two existing regions are grown to include the Level 1 pixels that are connected to them. In the third and fourth steps, the existing regions continue to grow by assimilating Levels 2 and 3. In the fifth step the regions continue to grow using Level 4, but they bump into each other, causing early termination. In the sixth step, regions in the corners are grown without hindrance. From this example we see that the immersion approach involves considering each level in sequential order. At each level, three possibilities exist for any contiguous group of pixels: the group is either disconnected from all existing regions, connected to exactly one existing region, or connected to more than one region. The first case is handled by the floodfill procedure, while the latter two cases are handled by growing the existing regions in a breadth-first manner to ensure that regions meet halfway when necessary. Note that at each level, the **geodesic influence zone** of each existing catchment basin is determined, which is the set of unlabeled pixels that are contiguous with the basin and closer to that basin than to any other.

Figure 10.28 Step-by-step results of immersion-based watershed segmentation on a segmentation function with 6 levels (0 through 5). The different colors indicate the unique labels of the three different regions. The contour lines of the segmentation function are shown, with numbers indicating the levels of the pixels.

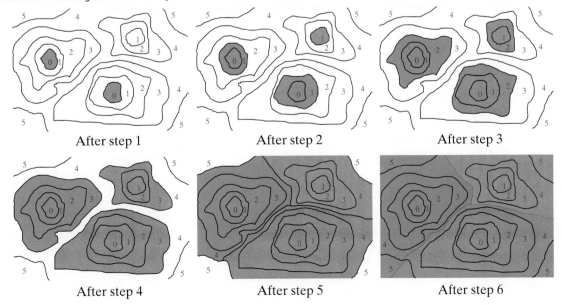

After step 1   After step 2   After step 3

After step 4   After step 5   After step 6

One of the problems with the standard watershed approach is that it is highly susceptible to noise in the segmentation function. That is, any tiny wrinkle in the function will create a local minimum, which then will generate a unique label for the (possibly very small) catchment basin containing the local minimum. This leads to **oversegmentation**, and the errors from the standard algorithm are usually quite severe. The most common solution to this problem, known as **marker-based watershed segmentation**, is to first use an independent procedure to generate a binary image where ON pixels indicate locations (called **markers**) where the local minimum is significant. The segmentation function is then only allowed to be punctured at the markers, so that the number of regions in the final segmentation (i.e., the number of unique labels) is equal to the number of markers.

ALGORITHM 10.11 Marker-based immersion watershed algorithm

WATERSHEDBYIMMERSION(I, M)

Input: grayscale image I, binary image M of the same size as I indicating markers
Output: label image L

➤ Initialize data structures

1   $G \leftarrow$ ROUND(GRADIENTMAGNITUDE(I))   ➤ Compute gradient magnitude.
2   **for** each value $k \leftarrow 0$ to $n_{grad} - 1$ **do**   ➤ Clear the pixel list associated with each
3       $pixellist$ [k].CLEAR()   possible gradient magnitude value.
4   **for** each pixel $p \in G$ **do**   ➤ Precompute pixel lists by storing coordinates of all
5       $pixellist$ [$G(p)$].PUSHBACK(p)   pixels, arranged by gradient magnitude value.
6       $L(p) \leftarrow$ UNLABELED   ➤ All pixels are initially unlabeled.
7   $next\text{-}label \leftarrow 0$
8   $frontier$.CLEAR()

➤ Flood topological surface one value at a time

9   **for** each value $k \leftarrow 0$ to $n_{grad} - 1$ **do**

➤ Grow existing catchment basins by one pixel, creating initial frontier

10   **for** each pixel p in $pixellist[k]$ **do**
11   **if** $L(p) ==$ UNLABELED and there exists a neighbor q of p
            such that $G(q) < k$ and $L(q) \neq$ UNLABELED **then**
12              $L(p) \leftarrow L(q)$ ➤ p is in an existing catchment basin, so copy label
13              $frontier$.PUSHBACK(p)   from neighbor q, and push p onto frontier.

➤ Continue to grow existing basins one pixel thick each iteration by expanding frontier

14   **while** $frontier$.SIZE > 0 **do**
15          $p \leftarrow frontier$.POPFRONT()
16   **if** there exists a neighbor q of p such that $G(q) \leq k$ and $L(q) ==$ UNLABELED **then**
17              $L(q) \leftarrow L(p)$ ➤ q is not labeled, so copy label
18              $frontier$.PUSHBACK(q)   from neighbor p, and push q onto frontier.

➤ Create new catchment basins

19   **for** each pixel p in $pixellist[k]$ **do**
20   **if** $L(p) ==$ UNLABELED and $M(p) ==$ ON **then**   ➤ p is still unlabeled, and a marker
21   FLOODFILL(L, p, $next\text{-}label$)   exists at p, so floodfill starting at p,
22              $next\text{-}label \leftarrow_+ 1$ and increment $next\text{-}label$.
23 **return** L

The final marker-based immersion watershed algorithm is detailed in Algorithm 10.11. Line 1 computes the gradient magnitude of the image as the segmentation function, but any other suitable computation may be substituted here. Note that the segmentation function is quantized (in this case using ROUND) so that there is a discrete set of possible values. For efficiency, the approach uses a data structure called *pixellist*, which is a fixed-size array of dynamically-sized arrays. That is, *pixellist[k]* is a list of the coordinates of all the pixels for which the segmentation function has a value of k. Lines 2–8 initialize this data structure, along with the output and the frontier. After the initialization, each possible value of the segmentation function is considered in turn, and three steps are performed. First, the existing catchment basins are grown by one pixel, initializing the frontier. Secondly, the frontier is expanded to grow the existing catchment basins until all adjacent pixels of the current level have been assimilated. Finally, any pixels of the current level that have not yet been labeled are floodfilled with a new label if there is a marker. Note that the frontier is a FIFO (first-in first-out) queue, which causes a breadth-first search to take place, thus enabling the different regions to grow at the same rate and therefore meet in the middle when necessary.

The careful reader may have noticed that the creation of new catchment basins in Lines 19–22 can be pulled out of the loop and placed before Line 9, since the markers are known beforehand. This observation is indeed true, and a real marker-based implementation should do so. Nevertheless, the code has been structured in the manner shown in Algorithm 10.11 so that it also implements the non-marker-based watershed algorithm if $M(p) = $ ON for all pixels p. One other subtle minor distinction between the two cases is that the test $G(q) \leq k$ in Line 16 may be written as $G(q) == k$ in the non-marker-based case, since all pixels with $G(q) < k$ will already have been labeled. In the marker-based case, however, the less-than sign is necessary to ensure that local minima without accompanying markers are filled in, and in the non-marker-based case, the less-than sign does not hurt.

The algorithm given above is a simplified version of what is usually presented. The most common algorithm, known as **Vincent-Soille**, involves the building of dams between the catchment basins. That is, when two catchment basins grow to the point that they touch each other, a dam is built between them to prevent water from flowing into the other catchment basin. By simply checking whether a pixel has been labeled, we have avoided the need for building dams and thereby simplified the algorithm considerably. As a result, Algorithm 10.11 may be considered as a **dam-less Vincent-Soille algorithm**. Note, however, that our algorithm does not find the boundaries between regions explicitly; it only labels the pixels. If boundaries are desired, they can be estimated by differentiating the resulting label image. In the worst case, which arises when the true boundary pixel is equidistant from two catchment basins, this approach introduces a one-pixel error in the location of the boundary; a simple modification removes this error.

An example of the immersion algorithm, without markers, on a 5×5 image with 10 levels (0 through 9) is shown in Figure 10.29. First, the pixel list is created, and each pixel in the label image is set to UNLABELED. Considering Level 0, since there are no catchment basins there is nothing to do but to floodfill the pixel with value 0. In Level 1, two of the pixels are adjacent to a labeled pixel and are therefore placed on the frontier, while new catchment basins are created for the other two pixels. In Level 2, all four pixels are placed on the frontier, and the process continues until all pixels have been labeled.

When implementing the marker-based watershed algorithm, it is important to create not only a marker for every foreground object but also a marker for the background. The typical way to find the foreground objects is thresholding. To find the background object, simply run non-marker-based watershed on the distance image to the foreground objects, then compute the edges of the result. (To compute the edges, simply look for pixels whose value

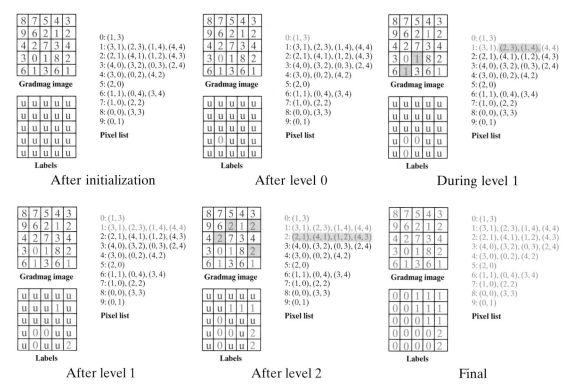

Figure 10.29 An example of non-marker-based immersion watershed on a simple 5 × 5 image with 10 levels (0 through 9). Shown are the results after several steps of the algorithm, followed by the final result. (Note that ties are broken arbitrarily, so other solutions are possible.) At each step, pixel coordinates in the pixel list with a value less than or equal to the current value are colored, and pixels on the frontier are shaded.

is different from their neighbors.) These edges, logically ORed with the foreground blobs, constitute the markers that are fed to the marker-based algorithm. The pseudocode for this end-to-end procedure is provided in Algorithm 10.12, and the corresponding flowchart is illustrated in Figure 10.30. One detail that should be mentioned is that a real implementation should be sure to prevent the possibility of the logical OR in Line 5 from causing the background marker to touch a foreground marker. A straightforward way to overcome this potential problem is to make the marker image a tertiary image, with each pixel having a

ALGORITHM 10.12 End-to-end watershed segmentation of an image

WATERSHEDWITHTHRESHOLDING(I, τ)

Input: grayscale image I, threshold τ
Output: label image L

1   $T \leftarrow$ NOT THRESHOLD(I, τ) ▸ Invert the thresholded image.
2   $D \leftarrow$ CHAMFER(T) ▸ Compute distance image to dark regions.
3   $W \leftarrow$ WATERSHEDBYIMMERSION(D, ON) ▸ Run non-marker-based watershed.
4   $E \leftarrow$ EDGEDETECTION(W) ▸ Detect region boundaries in watershed output.
5   $M \leftarrow T$ OR E ▸ Construct marker image by combining foreground and background.
6   **return** WATERSHEDBYIMMERSION(I, M) ▸ Run marker-based watershed.

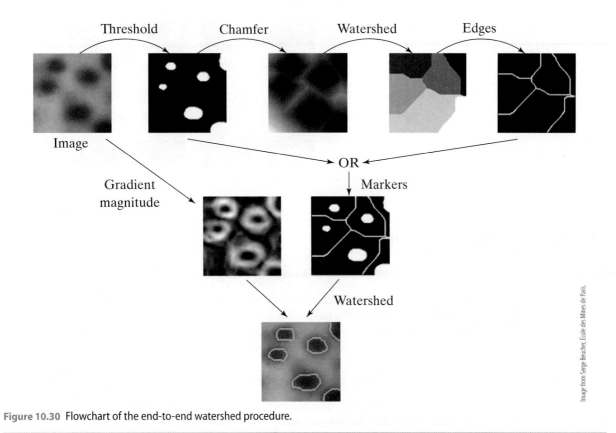

Figure 10.30 Flowchart of the end-to-end watershed procedure.

value indicating either no marker, foreground marker, or background marker, so that when the floodfill occurs, there is no bleeding between the foreground and background. This extension is left as an exercise for the reader.

10.3.6 Mean-Shift Segmentation

Recall that the mean-shift filter[†] is an approach to edge-preserving smoothing that results in a cartoon-like image by smoothing within regions but not across boundaries. The resulting cartoon-like image looks almost the same as the results of a segmentation algorithm. This leads to the **mean-shift segmentation algorithm**, which involves first running the mean-shift filter, then applying some postprocessing steps to the result. The algorithm achieves excellent results but is computationally expensive. Usually the algorithm is applied to a color image, using a color space like $L^*u^*v^*$ or $L^*a^*b^*$ in which distances are approximately perceptually correct. Typically, two postprocessing steps are applied: any two pixels that are nearby in both spatial coordinates and range are merged, and finally small regions are merged with nearby larger regions. The mean-shift segmentation algorithm reveals a close connection between segmentation and edge-preserving smoothing, and therefore any other edge-preserving smoothing algorithm can be used in the same manner. The result of mean-shift segmentation on an image is shown in Figure 10.31.

[†] Section 5.5.5 (p. 257).

Figure 10.31 Result of the mean-shift segmentation algorithm on an image of a clown.

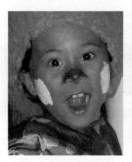 

10.4 Graph-Based Methods

We now turn our attention to graph-based methods of image segmentation. These algorithms all share the property that they view the image as a graph, with the vertices representing the pixels, the edges representing the connections between neighboring pixels, and the weights of the edges representing the dissimilarity (or similarity, depending upon the algorithm) between neighboring pixels. We begin with the popular Felzenszwalb-Huttenlocher algorithm, followed by normalized cuts, then *s-t* cuts and semantic segmentation.

10.4.1 Felzenszwalb-Huttenlocher (FH) Algorithm

One of the most popular image segmentation algorithms today is the **Felzenszwalb-Huttenlocher (FH) algorithm**, due to its computational efficiency, ease of implementation, and good results. The FH algorithm is often known as **graph-based segmentation**, but this name is not very descriptive, since many other graph-based segmentation algorithms exist, and they often have little in common with one another. The FH algorithm is closely related to computing the **minimum spanning tree (MST)** of a graph, which is the set of edges with minimum weight (meaning that the sum of the weights of its edges is minimum) that connects all the vertices in the graph such that the set contains no cycles (hence it is a tree). One of the two most commonly used methods for finding the MST of a graph is **Kruskal's algorithm**.[†] Therefore, we first describe Kruskal's algorithm itself, then we explain the slight modification to Kruskal's algorithm that turns it into the FH algorithm.

The mechanics of Kruskal's algorithm are simple. First the edges are sorted in non-decreasing order according to their weight. Then the edges are considered one at a time, in order, starting with the smallest-weight edge. For each edge, if the two vertices on either end of the edge are in different regions, then the two regions are merged. The procedure continues until all the vertices are in the same region. Why does this approach work? Because, in order to achieve a spanning tree, we know that two vertices will eventually have to merge, and since the edges are already sorted, we can do no better than to merge them using the current edge. Kruskal's algorithm is therefore an example of a greedy algorithm (meaning it makes decisions locally) that is nevertheless optimal.

The pseudocode for Kruskal's MST algorithm is provided in Algorithm 10.13. As with the classic connected components algorithm,[‡] this method uses a disjoint-set data structure

---

[†] The other is Prim's algorithm, which, for various reasons, is less applicable to the problem of image segmentation than Kruskal's.

[‡] Section 4.2.3 (p. 157).

ALGORITHM 10.13 Kruskal's minimum spanning tree (MST) algorithm

---

KRUSKALMST(I)

Input: image I
Output: minimum spanning tree T

1 INITIALIZE($width * height$)
2 $E \leftarrow$ CONSTRUCTWEIGHTEDEDGES(I)
3 $\langle e_1, \ldots, e_n \rangle \leftarrow$ SORTASCENDINGBYWEIGHT(E)
4 $T \leftarrow \phi$
5 **for** $(u, v) \leftarrow e_1$ **to** e_n **do**
6   **if** FINDSET$(u) \neq$ FINDSET(v) **then**
7          $T \leftarrow T \cup \{(u,v)\}$
8   MERGE(u,v)
9 **return** T

INITIALIZE(n)

1 **for** $i \leftarrow 0$ **to** $n - 1$ **do**
2      $equiv[i] \leftarrow i$

FINDSET(u)

1 **if** $u == equiv[u]$ **then**
2   **return** u
3 **else**
4      $equiv[u] \leftarrow$ FINDSET$(equiv[u])$
5   **return** $equiv[u]$

MERGE(u,v)

1 $a \leftarrow$ MIN(FINDSET(u),FINDSET(v))
2 $b \leftarrow$ MAX(FINDSET(u),FINDSET(v))
3 $equiv[b] \leftarrow a$

(called *equiv* here, for "equivalence table"). After initializing the disjoint-set data structure, the procedure CONSTRUCTWEIGHTEDEDGES generates a list of weighted edges for the graph by measuring the dissimilarity between neighboring pixels, typically using 4-neighbor connectedness and comparing pixels using the Euclidean distance in some color space. These edges are then sorted in ascending order by their weight, so that the pixels that are the most similar in appearance occur near the front of the list. Each edge $e_i = (u, v)$ is considered in turn, and the two vertices (pixels) of the edge are examined. If they are in separate regions, then they are merged; otherwise they are already in the same region, so there is nothing to do. The auxiliary procedures are as follows: INITIALIZE places each node into a separate set, FINDSET returns the "root node" of a set, and MERGE merges two sets. These latter two are exactly the same as the GETEQUIVALENTLABEL and SETEQUIVALENCE procedures, respectively, of connected components.

An example of Kruskal's algorithm applied to a simple 3×3 image is given in Figure 10.32, where the pixels are labeled a through i for convenience. The edge weights are sorted in nondecreasing order, then considered one at a time. The first edge is (h,i),

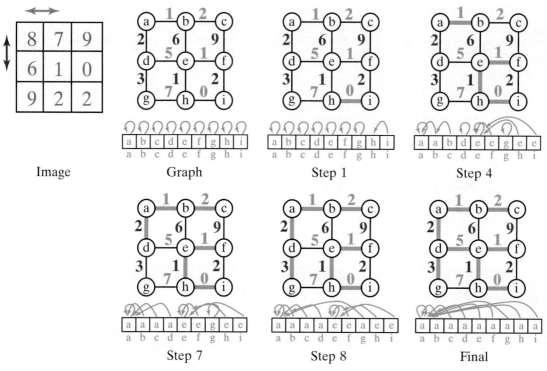

Image   Graph   Step 1   Step 4

Step 7   Step 8   Final

Figure 10.32 Kruskal's algorithm applied to a 3 × 3 image, whose gray levels are shown in brown. Next to the image is shown the graph whose edge weights are the absolute difference in gray level between neighboring pixels, using 4-neighbor connectedness. The remaining graphs show the edges (in orange) added to compute the minimum spanning tree of the graph, as the algorithm proceeds step by step. Note that the penultimate step (step 8) results in a compelling segmentation of the image.

which has weight 0. Since h and i are in different regions, the edge is added to the tree, and the pixels are merged. There is a three-way tie for second place between (a, b), (e, f), and (e, h), all with weight 1. In this case the tie is not important, since all three are added to the tree. Similarly, for weight 2 the tie can be broken arbitrarily without affecting the result. Note, however, that the edge (f, i) cannot be added to the tree since it would create a cycle, which is known because $\text{FINDSET}(f) = \text{FINDSET}(i) = e$. The final MST is found after edge (d, e) is added to the tree, since all pixels are spanned by the tree at that point. It is easy to see that Kruskal's algorithm finds the exact same hierarchical clustering found by the HCS procedure[†] with single-link clustering, assuming that the distance between non-adjacent pixels is set to infinity.

In this example, it is interesting to consider what happens just before the final result. That is, after the penultimate edge of (d, g) is added, there is a forest with two trees, one tree spanning pixels in the lower-right corner of the image with small gray levels, and another tree spanning the remaining pixels with large gray levels. Thus we see that, although Kruskal's algorithm eventually merges all the pixels into a single region, if it were terminated early it would segment the image.

Inspired by this observation, Felzenszwalb-Huttenlocher (FH) achieves the goal of segmentation by a simple modification to Kruskal's algorithm: instead of merging automatically whenever two pixels are in different regions, two pixels are merged only if the

[†] Section 10.3.4 (p. 481).

pixels are in different regions *and* if they look similar. As a result, this modified algorithm does not find a single MST but rather a forest of MSTs, one tree per region. The pseudocode is provided in Algorithm 10.14, where it is clear that the basic algorithmic structure is identical to that of Kruskal's MST algorithm, except that, since we do not care about the MSTs

ALGORITHM 10.14 Felzenszwalb-Huttenlocher minimum spanning tree-based segmentation

FELZENSZWALBHUTTENLOCHERSEGMENTATION(I, k)

Input: image I, scalar parameter k
Output: label image L

1   INITIALIZE($width * height$)
2   $I' \leftarrow$ SMOOTH(I)
3   $E \leftarrow$ CONSTRUCT EDGES(I')
4   $\langle e_1, \ldots, e_n \rangle \leftarrow$ SORTASCENDINGBYWEIGHT(E)
5   **for** $(u, v) \leftarrow e_1$ **to** e_n **do**
6       $u' \leftarrow$ FINDSET(u)
7       $v' \leftarrow$ FINDSET(v)
8   **if** $u' \neq v'$ and ISSIMILAR($w(u, v), u', v'; k$) **then**
9   MERGE($u', v', w(u, v)$)
10 **for** $(x, y) \in I$ **do**
11      $L(x, y) \leftarrow$ FINDSET($y * width + x$)
12 **return** L

INITIALIZE(n)

1   **for** $i \leftarrow 0$ **to** $n - 1$ **do**
2       $equiv[i] \leftarrow i$
3       $max\text{-}edge\text{-}weight[i] \leftarrow 0$
4       $num\text{-}pixels[i] \leftarrow 1$

FINDSET(u)

1   **if** $u == equiv[u]$ **then**
2   **return** u
3   **else**
4       $equiv[u] \leftarrow$ FINDSET($equiv[u]$)
5   **return** $equiv[u]$

MERGE(u, v, w)

1   $a \leftarrow$ MIN(FINDSET(u), FINDSET(v))
2   $b \leftarrow$ MAX(FINDSET(u), FINDSET(v))
3   $equiv[b] \leftarrow a$
4   $max\text{-}edge\text{-}weight[a] \leftarrow$ MAX(w, $max\text{-}edge\text{-}weight[a]$, $max\text{-}edge\text{-}weight[b]$)
5   $num\text{-}pixels[a] \leftarrow num\text{-}pixels[a] + num\text{-}pixels[b]$

ISSIMILAR($w, u, v; k$)

1   **return** $w <$ MIN($max\text{-}edge\text{-}weight[u] + k/ num\text{-}pixels[u]$, $max\text{-}edge\text{-}weight[v] + k/ num\text{-}pixels[v]$)

themselves, we do not bother to keep track of the variable T anymore. The disjoint-set data structure contains not only the equivalence table but also the maximum edge weight in the region, as well as the number of pixels in the region. Note that when merging, the weight w will always be greater than any weight considered so far (because of the ascending sort), and therefore the MAX of MERGE in line 4 is unnecessary (but it does no harm either).

The procedure IsSIMILAR compares the weight $w(u,v)$ of an edge (u,v) with the regions containing the pixels u' and v', using a single parameter k. With this parameter, the threshold of a region $\mathcal{R}$ is given by

$$\tau(\mathcal{R}) = \frac{k}{|\mathcal{R}|} \tag{10.105}$$

where $|\mathcal{R}|$ is the size of the region, meaning the number of pixels in the region. The internal variation of the region is given by the maximum weight of the edges in the MST:

$$v(\mathcal{R}) = \max_{e \in MST(\mathcal{R},E)} w(e) \tag{10.106}$$

Therefore, two regions $\mathcal{R}_1$ and $\mathcal{R}_2$ are merged if the weight w between them satisfies:

$$w < \min(v(\mathcal{R}_1) + \tau(\mathcal{R}_1), v(\mathcal{R}_2) + \tau(\mathcal{R}_2)) \tag{10.107}$$

which is line 1 of IsSIMILAR. We call the clustering scheme used by IsSIMILAR *smallest-neighbor clustering*, which is the same as single-link clustering when the distances between nonadjacent pixels are set to infinity. Like single-link clustering, this choice allows for large amounts of drift in appearance within a single region, and therefore it tends to undersegment images. It is important to note that, for this procedure to work, the division must be floating-point division; otherwise, once the region is bigger than k, no more merging will occur because $k/num\text{-}pixels < 1$, which truncates to zero, and k effectively becomes the maximum region size. As a result, the image must first be smoothed in order to convert the pixel values to floating point and to introduce slight differences between identical neighboring pixels. Therefore, the smoothing of the image in Line 2, which produces a floating-point value for each pixel, is not an optional step of the algorithm.

The FH algorithm can be seen as an elegant version of region growing that overcomes some of the limitations of the standard approach. While region growing is an effective technique, it leaves several important questions unanswered, such as 1) What merge criterion should be used? 2) How should we select the starting pixels? and 3) Among the several pixels adjacent to the region, which one should be considered next? The merge criterion, which defines what we mean when we say pixels "look similar," involves three aspects: the value used (e.g., grayscale, color, texture, stereo, motion, and so forth); the distance metric used (e.g., Euclidean or Manhattan); and the clustering type (single-link vs. complete-link vs. group-average). Such decisions are largely application-specific, and there is not much we can do to answer them at the algorithmic level. However, the other two questions are fundamentally important, because the outcome of the algorithm will depend heavily on the choice of starting pixels and on the selection of the adjacent pixel to consider next. Moreover, a serious drawback of the region-growing technique is that only one region is grown at a time, which introduces a bias toward earlier over later regions. Thus, the FH algorithm can be considered an improved version of region growing that, although it merges regions in a greedy manner, effectively grows the regions simultaneously in a fair and balanced way. The result of the FH algorithm on an image is shown in Figure 10.33.

The FH algorithm is limited to smallest-neighbor clustering, with other clustering schemes introducing prohibitive computational cost. In contrast, region growing can use

Figure 10.33 Result of the Felzenszwalb-Huttenlocher segmentation algorithm on an image of a knitted butterfly, shown as boundaries (red).

a variety of different options. Recall that our version of region growing in Algorithm 10.9 uses group-average clustering, where one cluster is the region being grown, and the second cluster consists of the single pixel being added to the region. It would be easy to modify that algorithm to use single-link clustering, where the distance between two nonadjacent pixels is defined as infinity, but complete-link clustering is cost prohibitive for most applications because it requires the pixel to be compared with every pixel in the region. Thankfully, the best results are usually achieved using either single-link or group-average clustering, the former allowing an arbitrary drift in appearance within regions while the latter enforces data-driven bounds on the compactness of the similarity of any given region.

10.4.2 Normalized Cuts

Suppose a graph is constructed such that each vertex represents a pixel, as we just saw, but the weights assigned to the edges are based on the *similarities* between neighboring pixels rather than their *dissimilarities*. An easy way to turn dissimilarities into similarities is to use the exponential function: $s(\mathbf{p}, \mathbf{q}) = \exp\{-d(\mathbf{p}, \mathbf{q})^2\}$, where $s(\mathbf{p},\mathbf{q})$ is the similarity between pixels $\mathbf{p}$ and $\mathbf{q}$, and $d(\mathbf{p},\mathbf{q})$ is their dissimilarity. The algorithms that we have been considering so far (including region growing, watershed, mean shift, and Felzenszwalb-Huttenlocher) are all examples of agglomerative clustering because they repeatedly merge regions. We now describe an approach based on divisive clustering because it repeatedly splits the image.

Any weighted graph can be represented by its **weighted adjacency matrix**, which is an $n \times n$ matrix containing the weights of the edges, where n is the number of vertices in the graph (i.e., pixels in the image). That is, the ij^{th} entry of the matrix is the weight of the edge between the ith and j^{th} vertices. Let $w(i, j)$ be the weight between pixels i and j. Define $\mathbf{W}$ as the $n \times n$ weighted adjacency matrix so that $w_{ij} = w(i, j)$ for all the elements of the matrix. The weights capture the similarity between the pixels, usually by combining spatial distance with distance in some feature space such as color or texture. For example,

$$w(i, j) \equiv \exp\left(-\frac{\|\mathbf{x}_i - \mathbf{x}_j\|^2}{2\sigma_s^2}\right) \exp\left(-\frac{\|I(\mathbf{x}_i) - I(\mathbf{x}_j)\|^2}{2\sigma_v^2}\right) \tag{10.108}$$

where $\mathbf{x}_i$ contains the (x,y) coordinates of the ith pixel, $I(\mathbf{x}_i)$ is the image value of the pixel, and σ_s and σ_v govern the expected spatial and value differences, respectively. Because these weights are related to the similarities between pixels, $\mathbf{W}$ is also known as the **affinity matrix**. Note that, unlike the region growing-based methods, in this approach the pixel neighbors can be defined in any way desired, such as 4- or 8- neighbors, or by taking all (or a random subset) of the pixels within a specified radius.

In **spectral graph theory**, various properties of a graph are revealed by studying the eigenvectors and eigenvalues of its affinity matrix, or of matrices derived from it. One such derived matrix is the **weighted degree matrix D**, where the ij^{th} element is defined as

$$d_{ij} = \begin{cases} d_i & \text{if } i = j \\ 0 & \text{otherwise} \end{cases} \tag{10.109}$$

where $d_i \equiv \sum_j w(i, j)$. Obviously, **D** is diagonal, and each element on the diagonal is the sum of all the weights of that row (or column, since **W** is symmetric). Other derived matrices are as follows:

$$\mathbf{L} \equiv \mathbf{D} - \mathbf{W} \qquad \qquad (\text{Laplacian matrix}) \tag{10.110}$$

$$\tilde{\mathbf{W}} \equiv \mathbf{D}^{-\frac{1}{2}}\mathbf{W}\mathbf{D}^{-\frac{1}{2}} \qquad (\text{normalized affinity matrix}) \tag{10.111}$$

$$\tilde{\mathbf{L}} \equiv \mathbf{D}^{-\frac{1}{2}}\mathbf{L}\mathbf{D}^{-\frac{1}{2}} = \mathbf{I}_{\{n \times n\}} - \tilde{\mathbf{W}} \qquad (\text{normalized Laplacian matrix}) \tag{10.112}$$

where $\mathbf{I}_{\{n \times n\}}$ is the $n \times n$ identity matrix. The normalized Laplacian matrix is also known as the *symmetric normalized Laplacian matrix*.

It can be shown that the Laplacian matrix **L** is always positive semi-definite, which means that all its eigenvalues are nonnegative. It is also the case that the smallest eigenvalue of **L** is always zero, and its corresponding eigenvector is a vector of all ones (denoted by **1**), which can easily be shown:

$$\mathbf{L}\mathbf{1} = (\mathbf{D} - \mathbf{W})\mathbf{1} = \mathbf{D}\mathbf{1} - \mathbf{W}\mathbf{1} = \mathbf{0} = 0\mathbf{1} \tag{10.113}$$

since **W1** simply sums the rows of **W**, thus **D1** and **W1** are identical. Let us denote this smallest eigenvalue as $\lambda_0 = 0$. The second-smallest eigenvalue (or, equivalently, the smallest nonzero eigenvalue), which we shall denote by λ_1, is known as the **spectral gap** or **Fiedler value**. It is also known as the **algebraic connectivity**, because the number of zero eigenvalues is equal to the number of components in the graph, so when a graph has noisy weights, λ_1 is an indication of how well connected the graph is. When the graph has multiple components, the Laplacian matrix (after possibly reordering the rows and columns) is block diagonal, with each block indicating the Laplacian of the individual component. As a result, when the graph is noisy, the eigenvector associated with λ_1 indicates which vertices are likely to belong to the most dominant component.

This analysis leads naturally to an image segmentation algorithm known as the **normalized cuts algorithm**. In this algorithm, the affinity matrix of the image is computed, and the eigenvector associated with the second-smallest eigenvalue of the normalized Laplacian matrix is computed and thresholded. Pixels whose value in the corresponding element of the eigenvector is above the threshold are considered part of the dominant component and labeled as such. The affinity matrix of the remaining pixels in the image is then computed, and the eigenvector associated with the second-smallest eigenvalue of the new normalized Laplacian matrix is computed and thresholded to yield the next dominant component. This process is repeated until the desired number of regions has been found. At each iteration, the threshold is determined either by selecting the value 0, or by selecting the median of all the values in the eigenvector, or by checking a number of equally spaced values and selecting the one that minimizes the normalized cut between the two regions resulting from the split; the latter approach is recommended.

To better understand the normalized cuts algorithm, consider the diagram and example in Figure 10.34. Let Ω be the set of pixels in the image (equivalently the set of vertices in the graph). Our goal is to partition the graph into two disjoint subgraphs containing vertex sets

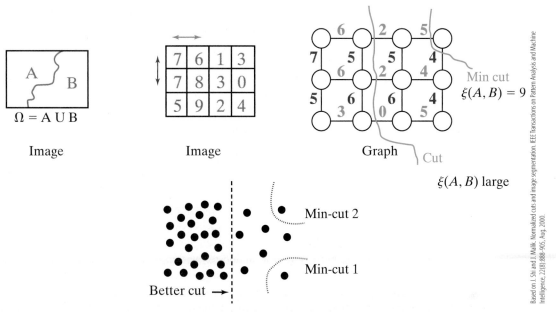

Based on J. Shi and J. Malik. Normalized cuts and image segmentation. IEEE Transactions on Pattern Analysis and Machine Intelligence, 22(8):888–905, Aug. 2000.

Figure 10.34 To divide an image into two regions, the normalized cuts algorithm formulates a graph with each pixel as a vertex, and each pair of adjacent pixels as an edge (4-neighbor connectedness is shown here for simplicity). The edge weights represent the affinity between adjacent pixels (computed here as a large constant number minus the absolute difference in intensity). The desired cut is in the middle of the image, but the cost of this cut will grow as the image gets larger, whereas the corner cut will remain a relatively small value. This illustrates the need for normalization.

$\mathcal{A}$ and $\mathcal{B}$, with $\mathcal{A} \cup \mathcal{B} = \Omega$ and $\mathcal{A} \cap \mathcal{B} = \emptyset$, by removing any edges connecting vertices in $\mathcal{A}$ with vertices in $\mathcal{B}$. A *cut* of the graph is a set of edges separating it into two subgraphs, and the *cost* of the cut is the sum of the weights of the edges in the cut. For any two arbitrary subsets $\mathcal{U}$ and $\mathcal{V}$ (not necessarily disjoint), let us define

$$\xi(\mathcal{U}, \mathcal{V}) = \sum_{u \in \mathcal{U}, v \in \mathcal{V}} w(u, v) \qquad (10.114)$$

as the sum of the weights between vertices in the two sets. Obviously, $\xi(\mathcal{A}, \mathcal{B})$ is the cost of the cut between $\mathcal{A}$ and $\mathcal{B}$. Simply finding the cut with the minimum cost leads to trivial segmentations, as illustrated in Figure 10.34, because it favors isolating small subsets of vertices. As a result, the *normalized cut* between $\mathcal{A}$ and $\mathcal{B}$ is defined as

$$\text{Ncut}(\mathcal{A}, \mathcal{B}) = \frac{\xi(\mathcal{A}, \mathcal{B})}{\xi(\mathcal{A}, \Omega)} + \frac{\xi(\mathcal{A}, \mathcal{B})}{\xi(\mathcal{B}, \Omega)} \qquad (10.115)$$

where the matrix is assumed to be symmetric so that $\xi(\mathcal{A}, \mathcal{B}) = \xi(\mathcal{B}, \mathcal{A})$.

It is easy to see that, for any subgraphs $\mathcal{U}$, $\mathcal{V}$, and $\mathcal{W}$, the following relation holds:

$$\xi(\mathcal{U}, \mathcal{V} \cup \mathcal{W}) = \xi(\mathcal{U}, \mathcal{V}) + \xi(\mathcal{U}, \mathcal{W}) \qquad (10.116)$$

from which it follows that

$$\xi(\mathcal{A}, \Omega) = \xi(\mathcal{A}, \mathcal{A} \cup \mathcal{B}) = \xi(\mathcal{A}, \mathcal{A}) + \xi(\mathcal{A}, \mathcal{B}) \qquad (10.117)$$

Thus, $\xi(\mathcal{A}, \Omega) + \xi(\mathcal{B}, \Omega) = \xi(\Omega, \Omega)$, which means that the sum of the two denominators is constant. The situation in Equation (10.115) is therefore analogous to minimizing the sum $\frac{1}{x} + \frac{1}{1-x}$, which is minimized when $x = \frac{1}{2}$. As a result, finding the cut that yields the minimum normalized cut results in a good segmentation because the denominators prevent $|\mathcal{A}|$ and $|\mathcal{B}|$ from being too small, thus balancing the size of the two subgraphs. Using Equation (10.117) and the similar equation for $\xi(\mathcal{B}, \Omega)$, the normalized cut can be rewritten as

$$\text{Ncut}(\mathcal{A}, \mathcal{B}) = \frac{\xi(\mathcal{A}, \Omega) - \xi(\mathcal{A}, \mathcal{A})}{\xi(\mathcal{A}, \Omega)} + \frac{\xi(\mathcal{B}, \Omega) - \xi(\mathcal{B}, \mathcal{B})}{\xi(\mathcal{B}, \Omega)} \tag{10.118}$$

$$= 2 - \left[\frac{\xi(\mathcal{A}, \mathcal{A})}{\xi(\mathcal{A}, \Omega)} + \frac{\xi(\mathcal{B}, \mathcal{B})}{\xi(\mathcal{B}, \Omega)} \right] \tag{10.119}$$

Finding the cut that minimizes the normalized cut is a discrete optimization problem. It can be shown that this particular problem is NP-hard, which means that we cannot hope to solve it for an exact solution for any realistically sized input in any reasonable amount of time; instead, we must settle for an approximation. The algorithm outlined above, which computes the eigenvector associated with the second-smallest eigenvalue, is the solution to a continuous version of the discrete problem. Although there is no guarantee that the continuous solution obtained by ignoring the discrete constraints bears any relationship to the discrete solution, in practice they tend to be closely related.

To see the connection between the discrete and continuous problems, let us define $\mathbf{a} \equiv [a_0 \quad a_1 \quad \cdots \quad a_{|\Omega|-1}]^\mathsf{T}$ as an indicator vector with a binary value for each pixel in the image, so that $a_i = 1$ if and only if pixel i is in $\mathcal{A}$, and equivalently $a_i = 0$ if and only if pixel i is in $\mathcal{B}$. Then it follows that

$$\text{Ncut}(\mathcal{A}, \mathcal{B}) = \frac{\xi(\mathcal{A}, \mathcal{B})}{\xi(\mathcal{A}, \Omega)} + \frac{\xi(\mathcal{B}, \mathcal{A})}{\xi(\mathcal{B}, \Omega)} \tag{10.120}$$

$$= \frac{\sum_{i \in A, j \notin A} w_{ij}}{\sum_{i \in A} \sum_{j} w_{ij}} + \frac{\sum_{i \notin A, j \in A} w_{ij}}{\sum_{i} \sum_{\substack{j \notin A}} w_{ij}} \tag{10.121}$$

$$= \frac{\mathbf{a}^\mathsf{T}(\mathbf{D} - \mathbf{W})\mathbf{a}}{\mathbf{a}^\mathsf{T}\mathbf{D}\mathbf{a}} + \frac{(1-\mathbf{a})^\mathsf{T}(\mathbf{D} - \mathbf{W})(1-\mathbf{a})}{(1-\mathbf{a})^\mathsf{T}\mathbf{D}(1-\mathbf{a})} \tag{10.122}$$

$$= \frac{\mathbf{a}^\mathsf{T}(\mathbf{D} - \mathbf{W})\mathbf{a}}{\mathbf{a}^\mathsf{T}\mathbf{D}\mathbf{a}} + \frac{\mathbf{a}^\mathsf{T}(\mathbf{D} - \mathbf{W})\mathbf{a}}{(1-\mathbf{a})^\mathsf{T}\mathbf{D}(1-\mathbf{a})} \tag{10.123}$$

The connection from the second to third lines is easy to see by noting that $\mathbf{a}^\mathsf{T}\mathbf{W}\mathbf{a} = \sum_{i \in A} \sum_{j \in A} w_{ij}$ and $\mathbf{a}^\mathsf{T}\mathbf{D}\mathbf{a} = \sum_{i \in A} \sum_{j} w_{ij}$. In other words, $\mathbf{a}^\mathsf{T}\mathbf{W}\mathbf{a}$ is a compact way of writing the sum of the weights between pixels in $\mathcal{A}$, and $\mathbf{a}^\mathsf{T}\mathbf{D}\mathbf{a}$ is the sum of the weights from pixels in $\mathcal{A}$ to any other pixel. Putting these together reveals a simple formula for the sum of the weights from pixels in $\mathcal{A}$ to pixels not in $\mathcal{A}$:

$$\mathbf{a}^\mathsf{T}(\mathbf{D} - \mathbf{W})\mathbf{a} = \mathbf{a}^\mathsf{T}(\mathbf{D} - \mathbf{W})\mathbf{a} = (1-\mathbf{a})^\mathsf{T}(\mathbf{D} - \mathbf{W})(1-\mathbf{a}) = \sum_{i \in A} \sum_{j \notin A} w_{ij}, \tag{10.124}$$

where the middle expression follows from $\sum_{i \notin A} \sum_{j \in A} w_{ij} = \sum_{i \in A} \sum_{j \notin A} w_{ij}$, since $\mathbf{W}$ is symmetric.

Now let us define the vector $\mathbf{y} \equiv \mathbf{a} - b(\mathbf{1} - \mathbf{a})$, where

$$b \equiv \frac{\sum_{i \in \mathcal{A}} d_i}{\sum_{i \notin \mathcal{A}} d_i} = \frac{\mathbf{a}^{\mathsf{T}} \mathbf{D} \mathbf{a}}{(\mathbf{1} - \mathbf{a})^{\mathsf{T}} \mathbf{D} (\mathbf{1} - \mathbf{a})} \tag{10.125}$$

so that $y_i = 1$ if and only if pixel i is in $\mathcal{A}$, and $y_i = -b$ if and only if pixel i is in $\mathcal{B}$. The vector $\mathbf{y}$ is obviously constrained so that each element is either 1 or $-b$. With some effort, it can be shown that if $\mathbf{a}$ is the indicator vector that minimizes the normalized cut in Equation (10.123), then the vector $\mathbf{y}$ obtained by $\mathbf{y} = \mathbf{a} - b(\mathbf{1} - \mathbf{a})$ is identical to the vector $\mathbf{y}$ that minimizes the *generalized Rayleigh quotient*:

$$\min_{\mathbf{a}} \mathrm{Ncut}(\mathcal{A}, \mathcal{B}) = \min_{\mathbf{y}} \frac{\mathbf{y}^{\mathsf{T}}(\mathbf{D} - \mathbf{W})\mathbf{y}}{\mathbf{y}^{\mathsf{T}} \mathbf{D} \mathbf{y}} \tag{10.126}$$

$$= \min_{\mathbf{z}} \frac{\mathbf{z}^{\mathsf{T}} \mathbf{D}^{-\frac{1}{2}}(\mathbf{D} - \mathbf{W}) \mathbf{D}^{-\frac{1}{2}} \mathbf{z}}{\mathbf{z}^{\mathsf{T}} \mathbf{z},} \tag{10.127}$$

where we have defined $\mathbf{z} \equiv \mathbf{D}^{\frac{1}{2}} \mathbf{y}$. Two constraints must be satisfied on the solution. First, $y_i \in \{1, -b\}$ for all i, because we seek a discrete solution that assigns each pixel to exactly one of the two regions. Secondly, $\mathbf{y}^{\mathsf{T}} \mathbf{D} \mathbf{1} = 0$, which arises from the definition of $\mathbf{y}$:

$$\mathbf{y}^{\mathsf{T}} \mathbf{D} \mathbf{1} = (\mathbf{a} - b(\mathbf{1} - \mathbf{a}))^{\mathsf{T}} \mathbf{D} \mathbf{1} = \mathbf{a}^{\mathsf{T}} \mathbf{D} \mathbf{1} - b(\mathbf{1} - \mathbf{a})^{\mathsf{T}} \mathbf{D} \mathbf{1} = \sum_{i \in \mathcal{A}} d_i - b \sum_{i \notin \mathcal{A}} d_i = \sum_{i \in \mathcal{A}} d_i - \sum_{i \in \mathcal{A}} d_i = 0 \tag{10.128}$$

Although this discrete problem is NP-hard, if we relax the first constraint and instead allow $\mathbf{y}$ to take on real values, then the minimized quotient is found by solving either the generalized eigenvalue problem involving $\mathbf{y}$ or, equivalently, the standard eigenvalue problem involving $\mathbf{z}$:

$$(\mathbf{D} - \mathbf{W})\mathbf{y} = \lambda \mathbf{D} \mathbf{y} \tag{10.129}$$

$$\mathbf{D}^{-\frac{1}{2}}(\mathbf{D} - \mathbf{W}) \mathbf{D}^{-\frac{1}{2}} \mathbf{z} = \lambda \mathbf{z} \tag{10.130}$$

It is easy to see that $\mathbf{z}_0 = \mathbf{D}^{\frac{1}{2}} \mathbf{1}$ is an eigenvector of Equation (10.130) with an eigenvalue of $\lambda_0 = 0$, because $\mathbf{D}^{-\frac{1}{2}}(\mathbf{D} - \mathbf{W}) \mathbf{D}^{-\frac{1}{2}} \mathbf{z}_0 = \mathbf{D}^{-\frac{1}{2}}(\mathbf{D} - \mathbf{W}) \mathbf{1} = \mathbf{0} = 0 \mathbf{1}$, as we saw before. Now we know that this is the smallest eigenvalue because the matrix $\mathbf{D}^{-\frac{1}{2}}(\mathbf{D} - \mathbf{W}) \mathbf{D}^{-\frac{1}{2}}$ is positive semidefinite, so it has no negative eigenvalues. Thus, the eigenvector associated with the smallest eigenvalue is not very interesting. However, because the matrix is real and symmetric, all of its eigenvectors are mutually orthogonal, and therefore $\mathbf{z}_1^{\mathsf{T}} \mathbf{z}_0 = \mathbf{y}_1^{\mathsf{T}} \mathbf{D} \mathbf{1} = 0$, where $\mathbf{z}_1$ is the eigenvector associated with the second smallest eigenvalue, λ_1. Thus, the second constraint above is automatically satisfied, yielding the solution that we have been seeking:

$$\mathbf{y}_1 = \arg \min_{\mathbf{y}:\mathbf{y}^{\mathsf{T}} \mathbf{D} \mathbf{1}} \frac{\mathbf{y}^{\mathsf{T}}(\mathbf{D} - \mathbf{W})\mathbf{y}}{\mathbf{y}^{\mathsf{T}} \mathbf{D} \mathbf{y}} \tag{10.131}$$

In summary, one iteration of the normalized cuts algorithm involves constructing $\mathbf{W}$ and $\mathbf{D}$ from the data, then solving the eigenvalue problem in Equation (10.130) for the eigenvector $\mathbf{z}_1$ associated with the second-smallest eigenvalue λ_1. This result is then transformed into $\mathbf{y}_1 = \mathbf{D}^{-\frac{1}{2}} \mathbf{z}_1$. If the resulting vector were to contain only the values 1 and $-b$, we would be done. In practice, however, $\mathbf{y}_1$ must be thresholded, with large values indicating pixels that belong to $\mathcal{A}$, and small values indicating pixels that belong to $\mathcal{B}$. This process is then repeated to find additional splits, until all the desired regions have been found.

The procedure for one iteration of the normalized cuts algorithm is provided in Algorithm 10.15. First the affinity matrix $\mathbf{W}$ is computed using some neighborhood $\mathcal{N}$, along with spatial and value Gaussians with standard deviations σ_s and σ_v, respectively. As mentioned above, one interesting characteristic of the normalized cuts algorithm is that any neighborhood function can be used, including nonadjacent pixels. From $\mathbf{W}$ we compute $\mathbf{D}, \mathbf{D}^{-\frac{1}{2}}$, and the normalized Laplacian, where we note that Line 3 performs matrix multiplication. The most difficult part of the algorithm is Line 4, which computes the eigenvector associated with the second-smallest eigenvalue. Since the normalized Laplacian is typically a large matrix, it is computationally expensive to compute all of the eigenvectors and eigenvalues. Since most of these are not needed, a faster way is to compute only a subset using, for example, the **Lanczos method**. The result is transformed to the vector $\mathbf{y}_1$, which is then thresholded to yield the final result. The threshold is determined in any of a number of ways. Here we search through all possible thresholds, for example, by trying all values between the minimum and maximum elements of $\mathbf{y}_1$ at some level of quantization (usually ten thresholds are sufficient). Once the procedure shown here has been performed, the two regions resulting from the segmentation are recursively processed to further divide the image, until the normalized cut exceeds a certain limit.

ALGORITHM 10.15 Normalized cuts segmentation (one iteration)

NORMALIZEDCUTSSEGMENTATION-ONEITERATION(I)

Input: image I
Output: binary image L

1   $\mathbf{W} \leftarrow$ CONSTRUCTAFFINITYMATRIX(I)
2   $\mathbf{D}, \mathbf{D}_{isr} \leftarrow$ CONSTRUCTDIAGONALSUMMATRIXANDSQUAREROOT($\mathbf{W}$)
3   $\mathbf{L}_{norm} \leftarrow \mathbf{D}_{isr} * (\mathbf{D} - \mathbf{W}) * \mathbf{D}_{isr}$
4   $\mathbf{z}_1 \leftarrow$ COMPUTESECONDSMALLESTEIGENVECTOR($\mathbf{L}_{norm}$)
5   $\mathbf{y}_1 \leftarrow \mathbf{D}_{isr} * \mathbf{z}_1$
6   $\tau \leftarrow$ FINDBESTTHRESHOLD($\mathbf{y}_1, \mathbf{W}$)
7   **for** $(x, y) \in I$ **do**
8       $i \leftarrow y * width + x$
9       $L(x, y) \leftarrow (\mathbf{y}_1(i) > \tau)$
10 **return** L

CONSTRUCTAFFINITYMATRIX(I)

1   $\mathbf{W} \leftarrow$ ZEROS(n,n)   ▶ Let n be the number of pixels, i.e., *width * height*.
2   **for** $(x, y) \in I$ **do**   ▶ Construct the affinity matrix by computing, for each pixel,
3   **for** $(x', y') \in \mathcal{N}(x, y)$ **do**   the affinity between it and every
4           $i \leftarrow y * width + x$ other pixel in some neighborhood $\mathcal{N}$.
5           $i' \leftarrow y' * width + x'$
6           $affinity \leftarrow$ EXP$\left(-((x - x')^2 + (y - y')^2)/(2 * \sigma_s^2)\right) *$ EXP$\left(-\|I(x, y) - I(x',y')\|^2)/(2 * \sigma_v^2)\right)$
7           $\mathbf{W}(i, i') \leftarrow_+ affinity$ ▶ Add the affinity to both sides of the diagonal
8           $\mathbf{W}(i', i) \leftarrow_+ affinity$ to ensure that the matrix remains symmetric.
9   **return** $\mathbf{W}$

ConstructDiagonalSumMatrixAndSquareRoot(W)

1 $\mathbf{D}_{isr} \leftarrow \mathbf{D} \leftarrow$ Zeros(n, n)
2 **for** $i \leftarrow 0$ **to** $n - 1$ **do**
3      $\mathbf{D}(i, i) \leftarrow \mathbf{W}(i, 0) + \mathbf{W}(i, 1) + \cdots + \mathbf{W}(i, n - 1)$
4      $\mathbf{D}_{isr}(i, i) \leftarrow 1/$ Sqrt$(\mathbf{D}(i,i))$
5 **return** $\mathbf{D}, \mathbf{D}_{isr}$

FindBestThreshold$(\mathbf{y}_1, \mathbf{W})$

1 *min-cut-cost* $\leftarrow \infty$
2 **for** $\tau \in \mathcal{T}$ **do**
3      $\mathbf{a} \leftarrow$ Threshold$(\mathbf{y}_1, \tau)$
4   *cut-cost* $\leftarrow$ ComputeCutCost$(\mathbf{W}, \mathbf{a})$
5   **if** *cut-cost* $<$ *min-cut-cost* **then**
6          $\tau_{best} \leftarrow \tau$
7   *min-cut-cost* $\leftarrow$ *cut-cost*
8 **return** τ_{best}

▶ Find the best threshold by iterating through all the possible thresholds in some set $\mathcal{T}$, computing the cost of the cut defined by each thresholded vector, and retaining the threshold that yields the minimum cost.

ComputeCutCost$(\mathbf{W}, \mathbf{a})$

1 *cost* $\leftarrow 0$
2 **for** $(i, j) \in \mathbf{W}$ **do**
3   **if** $\mathbf{a}(i) \neq \mathbf{a}(j)$ **then**
4   *cost* $\leftarrow_+ \mathbf{W}(i, j)$
5 **return** *cost*

▶ Compute the cost of the cut of the graph by summing the weights of all the edges that connect two pixels in different regions.

10.4.3 Minimum *s-t* Cut

The last several algorithms we have considered are applicable to general image segmentation, in which the image is carved into multiple regions according to some criterion of homogeneity. In this section we revisit the problem of foreground/background segmentation, in which the goal is to divide the image into exactly two regions whose appearance models are known (at least approximately) beforehand. One powerful and popular approach to this problem follows the paradigm of **interactive segmentation**, in which the user manually specifies some information to the system. For example, the user clicks and drags the mouse over the image in a graphical user interface to specify some pixels that belong to the foreground, as well as some other pixels that belong to the background. Usually just a few strokes on the image for each of the two categories are necessary to build a representative appearance model for each category that is rich enough to facilitate a clean, effective segmentation of the foreground from the background. Although any model can be used, a color histogram has proved to be sufficient for a variety of scenes.

Perhaps the most popular and powerful technique for interactive foreground/background segmentation is known as **graph cuts segmentation**, although we shall refer to it as **minimum *s-t* cut segmentation** in order to properly distinguish it from other methods involving other types of graph cuts (e.g., the minimum spanning tree-based approach or the normalized cuts algorithm). Let $\mathcal{G} = (\mathcal{V}, \mathcal{E}, w)$ be a weighted graph, where $\mathcal{V}$ is the set of vertices, $\mathcal{E}$ is the set of edges, and w is a weight function that returns a nonnegative real value for every edge. Suppose two of the vertices are special, with one being the "source" (called s) and the other being the "sink" (called t). A cut that separates the graph into two subgraphs $\mathcal{S}$ and $\mathcal{T}$, with $s \in \mathcal{S}$ and $t \in \mathcal{T}$, is known as an *s-t* cut. As with any graph cut, the cost of an *s-t* cut is given by the sum of the edges between the two sets:

$$c(\mathcal{S}, \mathcal{T}) \equiv \sum_{u \in \mathcal{S}, v \in \mathcal{T}, (u, v) \in \mathcal{E}} w(u, v) \tag{10.132}$$

The goal is to find the **minimum *s-t* cut** (which may not be unique, because multiple cuts can have the same cost). Unlike normalized cuts, which is a bottom-up procedure for clustering pixels without a previously known model, minimum *s-t* cut segmentation is applicable when two models for the two different regions (typically foreground and background) are provided.

We imagine the edges as hollow pipes, with their cross section proportional to the weight of the edge, so that the weight of an edge is also known as its **capacity**. A hose is connected to the source, and the water is turned on. The water flows from the source, through the pipes, and out through the sink. For every vertex except the source and sink, the amount of water flowing into the vertex is equal to the amount of water flowing out of it. This property is known as **flow conservation**. Let us assume that the rate of flow of the water throughout the graph is kept constant. Then as we increase the amount of water flowing into the source from the hose, the amount of water emerging from the sink will increase as well, until the capacity of the graph is reached. Once the graph has reached its capacity, no increase in the amount of water that can flow through it is possible.

Determining the maximum amount of water that can flow through a weighted graph is known as the **maximum flow problem**. According to the **max-flow min-cut theorem**, the maximum flow through the graph from the source to the sink is equal to the minimum cost among all possible cuts separating the source from the sink. That is, the maximum flow is the cost of the minimum *s-t* cut. Returning to our analogy of water flowing through pipes, it should be obvious that, once the graph has reached its capacity, the source can be separated from the sink entirely by removing all the **saturated edges**, that is, all edges that have reached their capacity. Removing all such edges divides the graph into two subgraphs, one containing s and the other containing t. This set of saturated edges is the minimum *s-t* cut (assuming it is unique), or rather it contains all the minimum *s-t* cuts (if there is more than one).

The application to image segmentation is as follows. Let I be the image, and let L be the binary labeling that results from the segmentation algorithm, with $L(\mathbf{x}) = $ ON for foreground pixels, and $L(\mathbf{x}) = $ OFF for background pixels, where $\mathbf{x} \equiv (x, y)$ is a pixel. Then, similar to what we saw earlier with active contours (snakes), our goal is to find the labeling L that minimizes an energy functional consisting of a data term and a smoothness term:

$$E(L) = \psi_{\text{const}} + \underbrace{\sum_{\mathbf{x} \in \mathcal{V}} \psi_{\mathbf{x}}(L(\mathbf{x}))}_{\text{data}} + \lambda \underbrace{\sum_{(\mathbf{x}, \mathbf{x}') \in \mathcal{E}} \psi_{\mathbf{x}, \mathbf{x}'}(L(\mathbf{x}), L(\mathbf{x}'))}_{\text{smoothness}} \tag{10.133}$$

where λ is a scaling factor capturing the relative importance of the two terms, and ψ_{const} is a constant that could be used to ensure that the energy remains nonnegative (although it is typically not needed).

The function $\psi_{\mathbf{x}}(L(\mathbf{x}))$ returns a value indicating how well $I(\mathbf{x})$ matches the model associated with $L(\mathbf{x})$, and $\psi_{\mathbf{x}, \mathbf{x}'}(L(\mathbf{x}), L(\mathbf{x}'))$ is a penalty function for neighboring pixels $\mathbf{x}$ and $\mathbf{x}'$ having different labels. For example we might model the foreground and background as Gaussians around a certain value:

$$\psi_{\mathbf{x}}(\ell) = \begin{cases} \exp(-(I(\mathbf{x}) - \mu_F)^2 / 2\sigma_F^2) & \text{if } \ell = \text{ ON} \\ \exp(-(I(x) - \mu_B)^2 / 2\sigma_B^2) & \text{if } \ell = \text{ OFF} \end{cases} \tag{10.134}$$

where μ_F and μ_B are the nominal values of the foreground and background distributions, respectively. The simplest penalty function, known as the **Potts model**, is to assign a constant penalty to any two neighboring pixels whose labels differ, regardless of the actual labels themselves:

$$\psi_{\mathbf{x},\mathbf{x}'}(\ell, \ell') = \begin{cases} 1 & \text{if } \ell \neq \ell' \text{ and } \mathbf{x}' \in \mathcal{N}(\mathbf{x}) \\ 0 & \text{otherwise} \end{cases} \tag{10.135}$$

We say that Equation (10.133) treats the image as a (first-order) **Markov random field (MRF)** because, viewed probabilistically, the value of any given pixel is conditionally independent of all the other pixels, given its neighbors. In other words, if we let $\ell_i \equiv L(\mathbf{x}_i)$, then the joint probability density function (PDF) of all the pixel labels is given by $p(L) = p(\ell_0, \ell_1, \ldots, \ell_{n-1})$. If we assume that the label of each pixel is conditionally independent of all the other pixels, given its neighbors, that is, $p(\ell_i|\ell_{n-1}, \ell_{n-2}, \ldots, \ell_0) = p(\ell_i|\mathcal{N}(\ell_i))$, then we can factorize the PDF as the product of factors depending on a single pixel or on pairs of pixels:

$$p(L) = \prod_i p(\ell_i) \prod_{i,i':\, i' \in \mathcal{N}(i)} p(\ell_i|\ell_{i'}) \tag{10.136}$$

Since the logarithm of a product is the sum of the logarithms, and since the logarithm function is monotonically increasing, therefore maximizing the probability $p(L)$ is the same as minimizing the following negative log probability:

$$-\log p(L) = -\sum_i \log p(\ell_i) - \sum_{i,i':\, i' \in \mathcal{N}(i)} \log p(\ell_i|\ell_{i'}) \tag{10.137}$$

If we assume that the probability functions take the form $p(\ell) = \exp(-\psi_{\mathbf{x}}(\ell))$ and $p(\ell, \ell') = \exp(-\psi_{\mathbf{x},\mathbf{x}'}(\ell, \ell'))$, then the connection between Equations (10.137) and (10.133) is obvious. A probability function that factorizes over positive functions defined on *cliques* of nodes and edges in a graph is called a **Gibbs distribution**. The connection between Gibbs distributions and MRFs, namely that a Gibbs distribution can be represented exactly by an MRF, is explained by the **Hammersley-Clifford theorem**.

To minimize Equation (10.133), we construct a graph with $n + 2$ vertices, where n is the number of pixels in the image. Each pixel is represented by a vertex, in addition to the two vertices that are created for the source and sink. Each pixel is connected to all its neighbors with edges whose weights are proportional to the difference in value between the two pixels. These edges are called *n*-links (for "neighbor"). Each pixel is also connected to both the source and the sink with edges that are related to the difference between the pixel and the opposite model (that is, the weight of the edge connecting the pixel with the source is related to the dissimilarity between the pixel and the model associated with the sink, and vice versa). These edges are called *t*-links (for "terminal"). Once the graph has been constructed, any standard algorithm to compute the maximum flow can be used. There are two basic flavors of such algorithms: the Ford-Fulkerson methods based on *augmenting paths*, and the Goldberg-Tarjan style *push-relabel* methods. In computer vision there is also the **Boykov-Kolmogorov algorithm**, based on augmenting paths, which is highly efficient because it takes advantage of the rectangular lattice structure of the graph when the graph represents the pixels in an image. No matter which algorithm is used, once the maximum flow of the graph has been found, the saturated edges divide the graph into the sets $\mathcal{S}$ and $\mathcal{T}$, and each pixel is then assigned the foreground or background depending upon whether its vertex belongs to $\mathcal{S}$ or $\mathcal{T}$.

Because $E(\cdot)$ maps binary vectors into real numbers (that is, treating L as a vector by scanning in row-major order, for example), $E(\cdot)$ is called a **pseudo-boolean function**. We should not take it for granted that the minimum s-t cut of the graph will yield the minimum of Equation (10.133) for any possible pseudo-boolean function. In fact, this approach only works in the case that the function is **submodular**, which means that every pairwise term $\psi_{\mathbf{x},\mathbf{x}'}$ satisfies the following inequality:

$$\psi_{\mathbf{x},\mathbf{x}'}\left(\text{OFF},\text{OFF}\right) + \psi_{\mathbf{x},\mathbf{x}'}\left(\text{ON},\text{ON}\right) \leq \psi_{\mathbf{x},\mathbf{x}'}\left(\text{OFF},\text{ON}\right) + \psi_{\mathbf{x},\mathbf{x}'}\left(\text{ON},\text{OFF}\right) \tag{10.138}$$

It can be shown that the global minimum of any submodular function can be computed in polynomial time as the minimum s-t cut in an appropriately constructed graph. It is easy to see that the Potts model always leads to a submodular function, since

$$0 + 0 \leq 1 + 1 \tag{10.139}$$

In fact, for the problem of image segmentation, nearly any penalty function leads to a submodular function, although non-submodular functions arise in other problems in computer vision, such as superresolution. It should be mentioned that, although we have explained the technique of graph cuts in the case of an energy functional that contains only unary and binary terms, the concept can be naturally extended to included higher-order terms, with the definition of submodular adjusted accordingly.

The pseudocode for performing binary segmentation by computing the minimum s-t cut is provided in Algorithm 10.16. A weighted graph is constructed with $n + 2$ vertices, where n is the number of pixels in the image. Each edge is either a t-link or an n-link. The t-links connect pixels to the source and sink vertices, while the n-links connect pixels to their neighbors. The weights of t-links are determined by GETDISTANCETOMODEL, which

ALGORITHM 10.16 Compute binary labeling using s-t graph cut

BINARYGRAPHCUT(I, *MODEL0*, *MODEL1*)

Input: image I, and two appearance models, *model0* and *model1*
Output: binary image L assigning each pixel in I to one of the two models

1   $n \leftarrow width * height$
2   $s \leftarrow n$
3   $t \leftarrow n + 1$
4   $\mathcal{V} \leftarrow \{0, \ldots, n - 1, s, t\}$
5   $\mathcal{E}_t \leftarrow$ CREATETLINKS(I, *model0*, *model1*, s, t)
6   $\mathcal{E}_n \leftarrow$ CREATENLINKS(I)
7   $\mathcal{E} = \mathcal{E}_n \bigcup \mathcal{E}_t$
8   $\mathcal{S}, \mathcal{T} \leftarrow$ COMPUTEMINSTCUT($\mathcal{V}, \mathcal{E}, s, t$)
9   **return** ASSIGNBINARYLABELING($\mathcal{S}, \mathcal{T}, s, t$)

CREATETLINKS(I, *MODEL0*, *MODEL1*, s, t)

1   $\mathcal{E}_t \leftarrow \emptyset$
2   **for** $(x, y) \in I$ **do**
3       $d_0 \leftarrow$ GETDISTANCETOMODEL($I(x,y)$, *model0*)
4       $d_1 \leftarrow$ GETDISTANCETOMODEL($I(x,y)$, *model1*)
5       $i \leftarrow y * width + x$
6       $\mathcal{E}_t \leftarrow \mathcal{E}_t \bigcup (i, s, d_0) \bigcup (i, t, d_1)$
7   **return** $\mathcal{E}_t$

```
CREATENLINKS(I)

1   E_n ← Ø
2   for (x, y) ∈ I do
3           for (x', y') ∈ N(x, y) do
4                   d ← exp(−‖I(x, y) − I(x', y')‖²/2σ²)
5                   i ← y * width + x
6                   i' ← y' * width + x'
7                   E_n ← E ∪ (i, i', λd)
8   return E_n
```

```
ASSIGNBINARYLABELING(S, T, s, t)

1    for i ∈ S do
2            if i ≠ s then
3                    y ← ⌊i/width⌋
4                    x ← i − y * width
5                    L(x, y) ← ON
6    for i ∈ T do
7            if i ≠ t then
8                    y ← ⌊i/width⌋
9                    x ← i − y * width
10                   L(x, y) ← OFF
11   return L
```

we assume exists for whatever appearance model has been chosen. Here, instead of the Potts model, the weights of n-links are assigned to be large when the pixels have similar values (thus discouraging cuts between similar-looking pixels), and small otherwise. Note the slight abuse of notation here in that we have included the edge weights with the edges themselves. Once the graph has been constructed, we call COMPUTEMINSTCUT, which can be any existing procedure for computing the minimum s-t cut of a graph, based on either augmenting paths or push-relabel. Once the cut has been found, the pixels connected to s are assigned the label ON, while the pixels connected to t are assigned the label OFF. This labeling may seem reversed, but note that if the distance to *model0* is small, then the minimum cut is likely to separate the pixel from *model0*, and vice versa. If we changed the code so that the t-link weights contained the affinity with the model (rather than the distance to the model), then the pixels connected to s would be assigned the label ON, while the pixels connected to t would be assigned the label OFF.

A popular technique for foreground/background segmentation based on this procedure is known as **GrabCut**. In GrabCut, the user selects a region of the image by drawing a bounding box. Two color histogram models are then created, one using the pixels inside the box and another using the pixels outside the box. With these two models, the minimum s-t cut of the graph constructed from the image pixels is found. New color histograms are then created using the assignment resulting from the minimum cut, and the process is repeated until convergence. Once the foreground has been separated from the background, it can be cut out of the original background and composited onto a new background. This technique, known as **rotoscoping**, can be seen as a more sophisticated version of green-screening.

10.4.4 Semantic Segmentation

A natural extension to binary segmentation using the minimum s-t cut is multilabel segmentation using the multiway cut. In multiway cut there are m labels, $m \geq 2$, and the goal is to assign each pixel to one of the labels. To solve this problem, a graph is constructed containing a vertex for each pixel in the image and a vertex for each possible label. Just as with the minimum s-t cut graph, in this expanded graph, t-links connect each pixel with each of the possible labels, and n-links connect pixels with their neighbors. The multiway cut is then found by an iterative procedure that is based on the minimum s-t cut algorithm.

Two algorithms for solving the multiway cut problem are common. In $\alpha-\beta$ **swap**, each pair of labels is considered in turn, and an appropriate graph is constructed using only those pixels that are currently assigned to either of the two labels. The minimum s-t cut algorithm is applied to determine which, if any, of these pixels should switch labels. This process is repeated until convergence. In α-**expansion**, each label is considered in turn, and an appropriate graph is constructed using all the pixels in the image. The minimum s-t cut algorithm is applied to determine which, if any, pixels should abandon their current label in favor of the label being considered. Comparing the two approaches, α-expansion is more computationally efficient, since it is linear in the number of labels rather than quadratic. It also comes with guarantees that the minimum that it finds will be within some constant factor of the global minimum, whereas no such guarantees are provided by the α-β swap algorithm. Nevertheless, both approaches yield excellent results in practice, often yielding results that are visually indistinguishable from the global minimum.

A popular application of multiway cut is the problem of **semantic segmentation**. In semantic segmentation, the goal is to assign a label to each pixel in the image indicating to what category of object it belongs, such as tree, car, road, building, or sky. Classifiers are trained for each of the categories beforehand, using techniques such as those discussed later,[†] and the multiway cut of the image graph is found using a technique such as α-expansion. Impressive results using such a technique have been achieved on fairly complex imagery. Typically such approaches do not yield explicit models, so that the probability of a pixel belonging to a particular category cannot be computed; instead the (dis)similarity, or distance, between the pixel and the various categories is determined, leading to a **conditional random field (CRF)**. Computationally MRFs and CRFs are for

Figure 10.35 In semantic segmentation, the pixels in an image are grouped and labeled according to a predetermined set of categories.

P. Kohli, L. Ladický, and P. H. S. Torr. Robust higher order potentials for enforcing label consistency, International Journal of Computer Vision, 82(3):302–324, May 2009.

[†] Chapter 12 (p. 560).

the most part indistinguishable, the primary difference being whether an explicit model is provided or merely a decision boundary; the latter also allows the prior smoothness term to be based upon the data. The distinction between MRFs and CRFs is analogous to that between a generative model and a discriminative model, explored further in Chapter 12.

10.5 Further Reading

There is no shortage of image thresholding algorithms. The iterative algorithm for determining the global threshold is due to Ridler and Calvard [1978], while Otsu [1979] developed the approach allowing for different variances for the background and foreground. An efficient extension of Otsu's method to multilevel thresholding can be found in Reddi et al. [1984] and Liao et al. [2001]. The Chow-Kaneko method for adaptive thresholding is described by Chow and Kaneko [1972]. The original Niblack approach can be found in Niblack [1986, pp. 115–116], and Sauvola's extension is in the paper by Sauvola and Pietikainen [2000]. An efficient implementation of Sauvola's method using integral images is described by Shafait et al. [2008]. Bradley and Roth [2007] present an alternate approach to adaptive thresholding using the integral image, and Blayvas et al. [2006] present a more sophisticated technique to determine the threshold surface. Another popular approach uses the local entropy surrounding the pixel, such as the algorithm by Pal and Pal [1989]. Several authors have conducted systematic studies to evaluate various thresholding algorithms—see Sahoo et al. [1988], Rosin and Ioannidis [2003], Sezgin and Sankur [2004], and Stathis et al. [2008]—but no clear winner has yet emerged. Hysteresis, or double thresholding, originated with the work of Canny [1986].

Snakes, or active contours, were first developed by Kass et al. [1988], where the snake energy is minimized using the calculus of variations. The alternative approach of using dynamic programming was proposed by Amini et al. [1990]. The gradient vector flow (GVF) snake is due to Xu and Prince [1997], Xu and Prince [1998], and Xu et al. [1999]. The level set method originated in the work of Osher and Sethian [1988] and is treated in detail by Sethian [1999]. The narrow band method was first described by Adalsteinsson and Sethian [1995]. Geodesic active contours were independently introduced by Caselles et al. [1993], Caselles et al. [1997], and Malladi et al. [1995]. A popular approach to avoid level set reinitialization is described in Li et al. [2005]. The Chan-Vese level set method is due to Chan and Vese [2001], and it is related to the Mumford-Shah functional described in Mumford and Shah [1985, 1989]. For the application of active contours to object tracking, see the work of

Paragios and Deriche [2000] and Yilmaz et al. [2004]. Intelligent scissors, which we did not have space to discuss, were introduced by Mortensen and Barrett [1995].

The classic split-and-merge algorithm is due to Horowitz and Pavlidis [1976], for which we have presented a variation. Hierarchical clustering schemes (HCS) and ultrametrics can be traced to the classic work of Johnson [1967] and have recently seen a resurgence of interest in Arbeláez [2006]. The classic work on gestalt psychology is that of Wertheimer [1938], where perceptual grouping and vision are argued to be the key to visual perception. The mean shift algorithm was introduced by Fukunaga and Hostetler [1975]. Interest in mean shift was revived by the work of Cheng [1995], and again by Comaniciu and Meer [2002], the latter of which addresses both segmentation and discontinuity-preserving smoothing.

The watershed approach was introduced by Beucher and Lantuàoul [1979]. Follow-up work can be found in the papers by Meyer and Beucher [1990] and Beucher and Meyer [1992]. The watershed algorithm presented here is a simplified dam-less version of the original algorithm by Vincent and Soille [1991], for which a description can also be found in Soille [2003]. The image foresting transform and its connection to the watershed approach, due to Falcão et al. [2004], is worth reading. An influential approach to region growing is described by Adams and Bischof [1994], and region growing combined with variable-order surface fitting can be found in Besl and Jain [1988]. Another influential approach that combines region growing with snakes is that of Zhu and Yuille [1996].

Image segmentation based on the minimum cut of a graph goes back to Wu and Leahy [1993], who also noticed the tendency of the minimum cut (without normalization) to favor small sets of pixels. The normalized cuts algorithm was developed by Shi and Malik [2000], and further analysis of the use of eigenvectors for segmentation is found in Weiss [1999], where the importance of normalization is shown. Ng et al. [2001] present a more advanced algorithm that, instead of iteratively bipartitioning the graph using the eigenvector associated with the second-smallest eigenvalue, partitions the graph into multiple regions simultaneously using multiple eigenvectors. The equivalency between a general weighted kernel k-means objective and various graph clustering objectives, such as normalized cut, ratio cut, and ratio association, are shown in Dhillon et al. [2007], where a fast algorithm that does not require eigenvectors is presented. The Fiedler value is due to Fiedler [1973].

Early work on solving image labeling problems is that of Geman and Geman [1984], where simulated annealing (a very slow algorithm) is used to minimize the cost of a Markov random field (MRF). The application of graph cuts, namely minimum s-t cuts, to solving binary image labeling problems goes back to the work of Greig et al. [1989]. Roy and Cox [1998] showed that the global energy of some multilabel MRFs could be minimized by finding the minimum s-t cut of a certain graph. Unfortunately, the graphs that are amenable to such a technique are not very useful in practice, because they are not discontinuity-preserving. Not long after, Boykov et al. [2001] proposed the α-β swap and α-expansion algorithms, which have proved enormously useful because they preserve discontinuities between regions. Conditional random fields (CRFs) were introduced into computer vision by Kumar and Hebert [2006], based on the work by Lafferty et al. [2001].

An analysis of what energy functions can be minimized by a binary graph cut, with the conclusion that such functions must be submodular, can be found in Kolmogorov and Zabih [2004]; further description of pseudo-boolean functions can be found in Boros and Hammer [2002]. Standard graph cut algorithms can be found in Cormen et al. [1990], while the efficient Boykov-Kolmogorov

algorithm that takes advantage of the lattice structure of image graphs is described by Boykov and Kolmogorov [2004]. Interactive segmentation and segmentation in higher dimensions using graph cuts is presented in Boykov and Jolly [2001] and Boykov and Funka-Lea [2006]. The interactive GrabCut segmentation algorithm is described by Rother et al. [2004]. For applications such as superresolution where non-submodular functions arise, see, for example, Rother et al. [2007].

An interesting recent development is to use clusters of pixels that are output from one or more segmentation algorithms, termed *superpixels*, introduced by Ren and Malik [2003], for various downstream algorithms. For an influential approach to texture segmentation using Gabor filters, see Jain and Farrokhnia [1991]. We have not had the space to discuss ways of evaluating the performance of segmentation algorithms. One influential work in this regard is that of Davies and Bouldin [1979], which describes the Davies-Bouldin index (DBI) for measuring the performance of a clustering scheme. The basic idea is to compute the mean over all clusters of the maximum ratio to other clusters, where the ratio is the sum of the standard deviations over the distance between the clusters. More recent work is the Berkeley segmentation database described by Martin et al. [2001], the Blobworld representation of Carson et al. [2002], and the hierarchical segmentation approach of Arbeláez et al. [2011]. For a representative approach to semantic segmentation, see the work of Kohli et al. [2009].

PROBLEMS

10.1 Segmentation corresponds to which branch of machine learning?

10.2 Derive Equation (10.5) from Equation (10.4).

10.3 Show that the middle terms in Equation (10.13) go to zero.

10.4 Derive the expression for between-class variance in Equation (10.16).

10.5 Sometimes Otsu's method is presented using the following recursive relations:

$$\mu_{\blacktriangleleft}(\tau+1) = \frac{p_{\blacktriangleleft}(\tau)\mu_{\blacktriangleleft}(\tau) + (\tau+1)\overline{h}[\tau+1]}{p_{\blacktriangleleft}(\tau+1)} \qquad \mu_{\triangleright}(\tau+1) = \frac{p_{\blacktriangleleft}(\tau)\mu_{\blacktriangleleft}(\tau) + (\tau+1)\overline{h}[\tau+1]}{p_{\blacktriangleleft}(\tau+1)}$$

so that the gray levels are scanned from $\tau = 0$ to $\tau = \zeta - 1$, and for each gray level these equations are used to compute the means of the two populations based on the previously calculated means.

(a) Derive these recursive relations from Equations (10.7) and (10.8).

(b) Explain why these recursive relations are no faster to evaluate than the original algorithm presented in this chapter.

10.6 Compute the optimal threshold for the following grayscale image using (a) the Ridler-Calvard algorithm and (b) Otsu's method:

$$\begin{bmatrix} 6 & 5 & 8 & 7 \\ 4 & 2 & 3 & 8 \\ 1 & 8 & 6 & 1 \end{bmatrix}$$

10.7 On the same image of the previous problem, compute the threshold using the following adaptive thresholding algorithms with a 3×3 sliding window: (a) Equation (10.19), (b) Niblack's method, and (c) Sauvola's method. For simplicity, show results only for the middle two pixels, use a box filter, and set t=0.8, k=0.2, and r=4.

10.8 Eliminate redundant computations from the multilevel Otsu method.

10.9 Perform hysteresis thresholding on the following image with a low threshold of 3 and a high threshold of 7 (that is, $I(x,y)>3$ and $I(x,y)>7$) to reveal an important concept in segmentation:

$$\begin{bmatrix} 4\ 6\ 0\ 3\ 2\ 6\ 2\ 2\ 0\ 4\ 4\ 1\ 0\ 3\ 6\ 1\ 2\ 1\ 4\ 4\ 0\ 0\ 6\ 5\ 2\ 5\ 7\ 5\ 1\ 2\ 2\ 6 \\ 4\ 2\ 6\ 9\ 7\ 2\ 8\ 7\ 9\ 3\ 2\ 8\ 8\ 7\ 0\ 7\ 5\ 5\ 1\ 1\ 8\ 7\ 1\ 3\ 8\ 1\ 5\ 1\ 7\ 6\ 7\ 2 \\ 2\ 8\ 2\ 1\ 1\ 1\ 7\ 1\ 2\ 0\ 9\ 2\ 0\ 2\ 2\ 2\ 6\ 1\ 0\ 6\ 0\ 3\ 6\ 2\ 6\ 2\ 7\ 5\ 1\ 6\ 1\ 6 \\ 3\ 8\ 1\ 8\ 5\ 2\ 8\ 8\ 1\ 6\ 0\ 9\ 7\ 1\ 6\ 1\ 6\ 3\ 1\ 7\ 7\ 7\ 9\ 0\ 5\ 3\ 4\ 5\ 0\ 7\ 1\ 4 \\ 3\ 6\ 2\ 2\ 8\ 2\ 5\ 3\ 0\ 6\ 0\ 3\ 0\ 8\ 1\ 3\ 8\ 1\ 1\ 7\ 0\ 1\ 9\ 2\ 8\ 3\ 2\ 6\ 3\ 9\ 2\ 6 \\ 6\ 0\ 6\ 9\ 7\ 0\ 7\ 6\ 7\ 1\ 5\ 8\ 6\ 2\ 7\ 2\ 8\ 1\ 1\ 6\ 1\ 1\ 8\ 2\ 8\ 6\ 7\ 0\ 3\ 6\ 2\ 5 \\ 7\ 7\ 3\ 2\ 2\ 6\ 1\ 2\ 3\ 6\ 0\ 3\ 3\ 6\ 6\ 6\ 0\ 5\ 5\ 1\ 5\ 5\ 3\ 6\ 3\ 1\ 1\ 6\ 6\ 2\ 6\ 4 \end{bmatrix}$$

10.10 Apply morphological reconstruction by dilation, Equation (10.22), to the image of the previous question and compare the results.

10.11 Both Algorithm 7.3 and Algorithm 10.3 are based on the floodfill algorithm, and they look quite similar. (a) Is there any difference in output between the two algorithms? (b) Can you think of any other reason the algorithms are written differently?

10.12 Both Ridler-Calvard and Otsu's method ignore the spatial relationships between pixels. A more sophisticated algorithm that uses this information is local entropy thresholding (LET). Draw a simple example image where such information would be important for determining the proper threshold.

10.13 Usually when the sum of a data term and smoothness term are presented, there is a scalar to govern the relative weight between them. Explain why there is no such scalar in Equation (10.26).

10.14 The gray levels of a bright blob in an image follow a Gaussian distribution with mean 204 and variance 92.6. Find the thresholds for ensuring that 95.45% of the pixels are segmented.

10.15 Suppose three consecutive points of a discrete snake have coordinates given by $v_{i-1} = (100,48)$, $v_i = (107,39)$, $v_{i+1} = (111,44)$. Compute the discrete approximation of (a) $d\mathbf{c}/ds$ and (b) $d^2\mathbf{c}/ds^2$ at the ith point.

10.16 Implement the minimization routine for a first-order closed snake. Run your code on an image of a single bright foreground object on a fairly dark background. How sensitive is the output of the algorithm to the initial conditions? Now modify your code to compute the norm for the smoothness term, instead of the squared norm; what difference do you notice in the output?

10.17 As mentioned in the text, in order for Algorithm 10.6 to work, Algorithm 10.5 must be modified to include a smoothness term between the first and last vertices, and the call from Line 4 of Algorithm 10.6 must cement the middle vertex. Make the appropriate modifications to the pseudocode so that Algorithm 10.6 actually works.

10.18 An alternative to Algorithm 10.6 is to instead fold the snake in half, pairing up points according to their order from the crease points. For example, a snake with 6 points could be folded so that v_0 and v_3 are the crease points, and the pairs are v_1 and v_5, as well as v_2 and v_4. For a first-order snake, a table is then created with 81 rows and (approximately) $n/2$ columns, where each column refers to each pair of points. The table is filled from left to right, then traversed from right to left, as before. Such an approach is guaranteed to find the global minimum, at the expense of additional computation. Write the pseudocode for this procedure, making the simplifying assumption that n is even.

10.19 Suppose the goal is to perform double thresholding on an image, followed by connected components. Show that this can be done in a single step by modifying the double thresholding procedure of Algorithm 10.3 to send the global counter as the last parameter to FLOODFILLSEPARATEOUTPUT. Demonstrate your pseudocode on a simple example.

10.20 Write the pseudocode for second-order snake minimization.

10.21 Explain the difference between a function and a functional.

10.22 Suppose we wish to use the Euler-Lagrange equations to solve for the equation of motion of a simple swinging pendulum. How many independent variables and how many dependent variables do you suppose there are? Based on your answer, state how many Euler-Lagrange equations there are, and how many terms in each equation.

10.23 Write out the **A** matrix of Equation (10.46) .

10.24 List several advantages of GVF snakes over traditional snakes.

10.25 Derive Equations (10.48)–(10.49) by applying the Euler-Lagrange equations to Equation (10.47) .

10.26 Sketch the 1D function $e^{-\frac{x^2}{100}} - \frac{1}{2}$, and indicate on your plot the zeroth level set.

10.27 Explain in your own words how Equation (10.59) can be used to evolve a contour.

10.28 Implement the Chan-Vese level set method, and run your code on an image containing a bright foreground on a dark background (or vice versa).

10.29 Explain the narrow band and fast marching methods.

10.30 Describe 4 ways to visualize the output of an image segmentation algorithm.

10.31 Briefly explain gestalt psychology, and list at least 5 gestalt factors.

10.32 What is the difference between agglomerative clustering and divisive clustering? For each of the following algorithms, specify whether it is an agglomerative or divisive approach: (a) region growing, (b) watershed, (c) mean shift segmentation, (d) Felzenszwalb-Huttenlocher, (e) normalized cuts, and (f) minimum s-t cut.

10.33 Implement the region-growing algorithm using a Gaussian model in RGB color space. Run your code on an image with brightly colored regions and notice where it succeeds and where it fails.

10.34 Show that the ultrametric inequality implies the triangle inequality.

10.35 Suppose that we have four points, which (in keeping with the notation of Section 10.3.4) we label as clusters $\mathcal{A}$, $\mathcal{B}$, $\mathcal{C}$, and $\mathcal{D}$. Let the distance between the first two points be $\psi(\mathcal{A}, \mathcal{B}) = 0.4$, and the distance between the remaining two points be $\psi(\mathcal{C}, \mathcal{D}) = 0.8$. In order for the distance function to satisfy the ultrametric inequality, what do we know about the other distances, that is, $\psi(\mathcal{A}, \mathcal{C})$, $\psi(\mathcal{A}, \mathcal{D})$, $\psi(\mathcal{B}, \mathcal{C})$, and $\psi(\mathcal{B}, \mathcal{D})$?

10.36 Compute the Mahalanobis distance between a 1D point at $x = 10$ and a Gaussian distribution with mean $\mu = 7$ and variance $\sigma^2 = 2.5$.

10.37 Draw the dendrogram corresponding to the graph in Figure 10.32, assuming single-link clustering.

10.38 Explain the primary drawback of tobogganing.

10.39 Modify Algorithm 10.11 and Example 10.29 with a tertiary marker image to prevent the background marker from accidentally merging with any of the foreground markers.

10.40 Implement the simplified Vincent-Soille algorithm and run it on a grayscale image of your choice.

10.41 Explain why markers are needed to obtain good results with the watershed algorithm.

10.42 We have seen that, despite the fact that the watershed algorithm is usually presented with dams, they are not necessary for the algorithm to perform well. Nevertheless, describe the one potential benefit of using dams that could be of interest to an application that requires the highest fidelity possible.

10.43 Describe the relationship between mean-shift filtering and mean-shift segmentation.

10.44 Given the following graph, draw the minimum spanning tree.

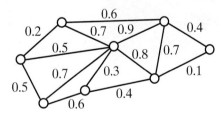

10.45 Implement the Felzenswalb-Huttenlocher segmentation algorithm and run it on a color image of your choice.

10.46 Explain the difference between an adjacency matrix and an affinity matrix.

10.47 Using the graph in Problem 10.44, construct the (a) weighted adjacency matrix, (b) weighted degree matrix, (c) Laplacian matrix, (d) normalized affinity matrix, and (e) normalized Laplacian matrix. Also compute the (f) Fiedler value. Order the nodes clockwise starting with the top node, ending with the node in the center.

10.48 Apply the first iteration of the normalized cuts method to segment the graph in the previous question, using zero as the threshold. What is the cost of this cut? Evaluate the quality of the result by seeing if you can find a better segmentation on your own.

10.49 What is the (a) max-flow min-cut theorem, and (b) Hammersley-Clifford theorem? What type of additional information must be available in order for these theorems to be applicable to image segmentation?

10.50 Construct a simple graph whose edge weights lead to a non-submodular function.

10.51 Name the two graph-based algorithms for solving the multiway cut problem.

CHAPTER 11
Model Fitting

I n the previous chapter we looked at various approaches to the problem of image segmentation. These algorithms were, for the most part, applied to data consisting of a dense 2D array of pixels arranged in a discrete lattice. More generally, however, the problem of segmentation, also known as clustering, can be applied to any set of data, whether dense or sparse. For example, we may wish to cluster a given set of 2D or 3D points. Such problems are proverbial chicken-and-egg problems because if we knew which points belonged together, we could fit a model to those points. Alternatively, if we knew the parameters of all the models, we could determine which model best explains each data point. Solving both of the subproblems simultaneously, however, is a real challenge. Moreover, we often do not know how many models there are, and some points may not belong to any model (i.e., they are outliers). In this chapter we focus primarily upon the subproblem of fitting various models to data, while the entire problem of simultaneously fitting models and clustering data points is considered toward the end of the chapter.

11.1 Fitting Lines and Planes

We begin our discussion with a problem that arises frequently in practice—namely, fitting a line or plane to a set of data points.

11.1.1 Ordinary Least Squares

Suppose we have a set of points in a 2D plane, and we wish to find the parameters of the line that best fits the points. Let $\{(x_i, y_i)\}_{i=1}^n$ be the set of coordinates of the n points. For now, let us assume the slope-intercept representation for a line: $y = mx + k$, where m is the slope of the line, and k is the y-intercept—that is, the value of y where the line intersects the y axis. Our goal is to find the parameters m and k that satisfy the following system of n linear equations in two unknowns:

$$mx_1 + k = y_1 \tag{11.1}$$

$$mx_2 + k = y_2 \tag{11.2}$$

$$\vdots \tag{11.3}$$

$$mx_n + k = y_n \tag{11.4}$$

If $n = 2$, then there are two equations and two unknowns, and we can solve for the unknowns exactly by finding the parameters of the line that passes through the two points. But usually $n > 2$, leading to an overdetermined system of linear equations, in which case we will not be able to find a line that passes through all the points exactly, due to noise in the measurements from which the point coordinates arise. Instead, we shall content ourselves with finding the line that "best" fits the points (in some sense).

To solve a system of equations like this, let us stack the unknowns into a vector $\mathbf{x} = \begin{bmatrix} m & k \end{bmatrix}^\mathsf{T}$, then rearrange the other values into a matrix $\mathbf{A}$ and vector $\mathbf{b}$:

$$\underbrace{\begin{bmatrix} x_1 & 1 \\ x_2 & 1 \\ \vdots \\ x_n & 1 \end{bmatrix}}_{\mathbf{A}_{\{n \times 2\}}} \underbrace{\begin{bmatrix} m \\ k \end{bmatrix}}_{\mathbf{x}_{\{2 \times 1\}}} = \underbrace{\begin{bmatrix} y_1 \\ y_2 \\ \vdots \\ y_n \end{bmatrix}}_{\mathbf{b}_{\{n \times 1\}}} \tag{11.5}$$

or $\mathbf{A}\mathbf{x} = \mathbf{b}$. It should be easy to see that this matrix equation is equivalent to the n scalar equations above.

Of course, since $\mathbf{A}$ is not a square matrix, we cannot simply take its inverse. Instead, we multiply both sides of the equation by its transpose, $\mathbf{A}^\mathsf{T}$, to yield the so-called **normal equations**:

$$\mathbf{A}^\mathsf{T}\mathbf{A}\mathbf{x} = \mathbf{A}^\mathsf{T}\mathbf{b} \tag{11.6}$$

In a real-world application, the 2×2 matrix $\mathbf{A}^\mathsf{T}\mathbf{A}$ is almost certainly invertible. (The only situation in which it is singular is when all the points are identical.) Therefore, both sides of the equation can be multiplied by the inverse of $\mathbf{A}^\mathsf{T}\mathbf{A}$ in order to solve for the desired solution:

$$\mathbf{x} = (\mathbf{A}^\mathsf{T}\mathbf{A})^{-1}\mathbf{A}^\mathsf{T}\mathbf{b} \tag{11.7}$$

This approach is known as **linear least squares**, or **ordinary least squares**. When it is applicable, ordinary least squares is an easy technique to use because it finds the solution in a finite number of algorithmic steps, with no initial guess needed. You may be wondering, in what sense is the solution found the "best" solution? Well, the ith **residual** is the difference between the left and right sides of the ith equation: $r_i \equiv mx_i + k - y_i$. If all the points fit the line exactly, then all the residuals would be zero. It can be shown that the value $\mathbf{x}$ that results from solving Equation (11.7) is the one that minimizes the sum of squared residuals: $\sum_{i=1}^n r_i^2 = \sum_{i=1}^n (mx_i + k - y_i)^2 = \|\mathbf{A}\mathbf{x} - \mathbf{b}\|^2$. In other words,

$$(\mathbf{A}^\mathsf{T}\mathbf{A})^{-1}\mathbf{A}^\mathsf{T}\mathbf{b} = \arg\min_{\mathbf{x}}\|\mathbf{A}\mathbf{x} - \mathbf{b}\|^2 \tag{11.8}$$

If we define $\mathbf{a} \equiv [x_1 \quad \cdots \quad x_n]^\mathsf{T}$ and $\mathbf{1} \equiv [1 \quad \cdots \quad 1]^\mathsf{T}$, then $\mathbf{A} = [\mathbf{a} \quad \mathbf{1}]$, from which it is easy to see that the matrix $\mathbf{A}^\mathsf{T}\mathbf{A}$ and vector $\mathbf{A}^\mathsf{T}\mathbf{b}$ can be computed by

$$\mathbf{A}^\mathsf{T}\mathbf{A} = \begin{bmatrix} \mathbf{a}^\mathsf{T}\mathbf{a} & \mathbf{a}^\mathsf{T}\mathbf{1} \\ \mathbf{a}^\mathsf{T}\mathbf{1} & n \end{bmatrix} \qquad \mathbf{A}^\mathsf{T}\mathbf{b} = \begin{bmatrix} \mathbf{a}^\mathsf{T}\mathbf{b} \\ \mathbf{b}^\mathsf{T}\mathbf{1} \end{bmatrix} \tag{11.9}$$

because $\mathbf{a}^\mathsf{T}\mathbf{a} = \sum_{i=1}^{n} x_i^2$, $\mathbf{a}^\mathsf{T}\mathbf{1} = \sum_{i=1}^{n} x_i$, $\mathbf{a}^\mathsf{T}\mathbf{b} = \sum_{i=1}^{n} x_i y_i$, $\mathbf{b}^\mathsf{T}\mathbf{1} = \sum_{i=1}^{n} y_i$, and $\mathbf{1}^\mathsf{T}\mathbf{1} = n$. Since the inverse is given by

$$(\mathbf{A}^\mathsf{T}\mathbf{A})^{-1} = \frac{1}{n(\mathbf{a}^\mathsf{T}\mathbf{a}) - (\mathbf{a}^\mathsf{T}\mathbf{1})^2} \begin{bmatrix} n & -\mathbf{a}^\mathsf{T}\mathbf{1} \\ -\mathbf{a}^\mathsf{T}\mathbf{1} & \mathbf{a}^\mathsf{T}\mathbf{a} \end{bmatrix} \tag{11.10}$$

we have from Equation (11.7) that

$$\begin{aligned}
\mathbf{x} &= \frac{1}{n(\mathbf{a}^\mathsf{T}\mathbf{a}) - (\mathbf{a}^\mathsf{T}\mathbf{1})^2} \begin{bmatrix} n & -\mathbf{a}^\mathsf{T}\mathbf{1} \\ -\mathbf{a}^\mathsf{T}\mathbf{1} & \mathbf{a}^\mathsf{T}\mathbf{a} \end{bmatrix} \begin{bmatrix} \mathbf{a}^\mathsf{T}\mathbf{b} \\ \mathbf{b}^\mathsf{T}\mathbf{1} \end{bmatrix} \\
&= \frac{1}{n(\mathbf{a}^\mathsf{T}\mathbf{a}) - (\mathbf{a}^\mathsf{T}\mathbf{1})^2} \begin{bmatrix} n(\mathbf{a}^\mathsf{T}\mathbf{b}) - (\mathbf{a}^\mathsf{T}\mathbf{1})(\mathbf{b}^\mathsf{T}\mathbf{1}) \\ -(\mathbf{a}^\mathsf{T}\mathbf{1})(\mathbf{a}^\mathsf{T}\mathbf{b}) + (\mathbf{a}^\mathsf{T}\mathbf{a})(\mathbf{b}^\mathsf{T}\mathbf{1}) \end{bmatrix}
\end{aligned} \tag{11.11}$$

The pseudocode for ordinary least squares for the simple case of fitting a line to a set of 2D points is given in Algorithm 11.1. With a 2×2 matrix, there is very little computational overhead to computing its inverse, and the solution provided in the pseudocode is acceptable. For larger matrices, however, it is best to avoid the expense of computing the matrix inverse. A more computationally efficient approach is to perform **Gauss-Jordan elimination** on both sides of Equation (11.6). Or, if the equation is to be solved multiple times with different values of $\mathbf{b}$, then it may be more appropriate to first factor the matrix $\mathbf{A}^\mathsf{T}\mathbf{A}$ using Cholesky decomposition, QR factorization, or SVD factorization, the latter of which is explained in more detail later. Two examples of ordinary least squares applied to sets of points are shown in Figure 11.1.

ALGORITHM 11.1 Ordinary Least Squares for 2D line fitting

ORDINARYLEASTSQUARES $\left(\{(x_i, y_i)\}_{i=1}^{n} \right)$

Input: set of n points $\{(x_i, y_i)\}_{i=1}^{n}$ in the plane
Output: slope m and y-intercept k of the line $y = mx + k$ that minimizes the sum-of-squared vertical residuals

$\sum_{i=1}^{n} (mx_i + k - y_i)^2$

1   $s_a \leftarrow s_b \leftarrow s_{aa} \leftarrow s_{bb} \leftarrow s_{ab} \leftarrow 0$
2   **for** $i \leftarrow 1$ **to** n **do**
3       $s_a \leftarrow_+ x_i$ ▷ Compute $\mathbf{a}^\mathsf{T}\mathbf{1}$.
4       $s_b \leftarrow_+ y_i$ ▷ Compute $\mathbf{b}^\mathsf{T}\mathbf{1}$.
5       $s_{aa} \leftarrow_+ x_i * x_i$ ▷ Compute $\mathbf{a}^\mathsf{T}\mathbf{a}$.
6       $s_{bb} \leftarrow_+ y_i * y_i$ ▷ Compute $\mathbf{b}^\mathsf{T}\mathbf{b}$.
7       $s_{ab} \leftarrow_+ x_i * y_i$ ▷ Compute $\mathbf{a}^\mathsf{T}\mathbf{b}$.
8   $den \leftarrow n * s_{aa} - s_a * s_a$ ▷ Compute denominator $den = n(\mathbf{a}^\mathsf{T}\mathbf{a}) - (\mathbf{a}^\mathsf{T}\mathbf{1})^2$.
9   $m \leftarrow (n * s_{ab} - s_a * s_b) / den$ ▷ Compute $(n(\mathbf{a}^\mathsf{T}\mathbf{b}) - (\mathbf{a}^\mathsf{T}\mathbf{1})(\mathbf{b}^\mathsf{T}\mathbf{1}))/den$.
10 $k \leftarrow (s_b * s_{aa} - s_a * s_{ab}) / den$ ▷ Compute $(-(\mathbf{a}^\mathsf{T}\mathbf{1})(\mathbf{a}^\mathsf{T}\mathbf{b}) + (\mathbf{a}^\mathsf{T}\mathbf{a})(\mathbf{b}^\mathsf{T}\mathbf{1}))/den$.
11 **return** m, k

Figure 11.1 Results of fitting a line to a set of points using ordinary least squares. This approach minimizes the sum of squared *vertical* distances. For nonvertical lines (left), the technique works reasonably well, but for vertical lines (right), it does not.

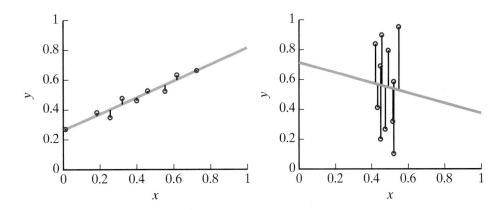

11.1.2 Normalization

One problem associated with the solution outlined above arises when the values for x_i are large. For example, suppose the values of x_i are on the order of, say, a hundred, and that there are ten points. Then the matrix $\mathbf{A}^\mathsf{T}\mathbf{A}$ would look something like this:

$$\mathbf{A}^\mathsf{T}\mathbf{A} \approx \begin{bmatrix} 100{,}000 & 1000 \\ 1000 & 1 \end{bmatrix} \tag{11.12}$$

Such a matrix is said to be **ill-conditioned** because, informally at least, it contains both very large values and very small values. An ill-conditioned matrix is bad because it is numerically unstable, meaning that a small amount of error in the input can cause a large error in the output. An ill-conditioned matrix is also out of balance, treating the different unknowns unequally, which is easily understood by imagining the units being used. Suppose, for example, that we were to calculate m in kilometers but k in micrometers; then an error on the order of ± 1 in k might be insignificant, while an error of ± 1 in m would have an enormous effect on the validity of the solution, because $1\ \mathrm{km} = 10^9\ \mu\mathrm{m}$.

The problem of ill-conditioning is overcome by **normalizing** the coordinates, which is typically achieved by first scaling the axes so that the errors of the different unknowns are balanced. In the particular problem at hand, however, a simpler solution to normalization is possible—namely, to shift all the input points so that their centroid is the origin. After the shift, the y-intercept k is guaranteed to be zero, thus eliminating one of the variables altogether. More specifically, first compute the mean of the points

$$\bar{\mathbf{x}} \equiv \begin{bmatrix} \bar{x} \\ \bar{y} \end{bmatrix} = \frac{1}{n} \sum_{i=1}^{n} \begin{bmatrix} x_i \\ y_i \end{bmatrix} \tag{11.13}$$

Then subtract the mean to obtain centered coordinates for each point:

$$\begin{bmatrix} \tilde{x}_i & \tilde{y}_i \end{bmatrix}^\mathsf{T} \equiv \begin{bmatrix} x_i - \bar{x} & y_i - \bar{y} \end{bmatrix}^\mathsf{T} \tag{11.14}$$

The best-fitting line through the centered coordinates passes through the origin, leading to

$$\underbrace{\begin{bmatrix} \tilde{x}_1 \\ \tilde{x}_2 \\ \vdots \\ \tilde{x}_n \end{bmatrix}}_{\tilde{\mathbf{a}}} m = \underbrace{\begin{bmatrix} \tilde{y}_1 \\ \tilde{y}_2 \\ \vdots \\ \tilde{y}_n \end{bmatrix}}_{\tilde{\mathbf{b}}} \tag{11.15}$$

whose solution is given by

$$m = \frac{\tilde{\mathbf{a}}^\mathsf{T}\tilde{\mathbf{b}}}{\tilde{\mathbf{a}}^\mathsf{T}\tilde{\mathbf{a}}} = \frac{\tilde{\mathbf{a}}^\mathsf{T}\tilde{\mathbf{b}}}{\|\tilde{\mathbf{a}}\|^2} \tag{11.16}$$

Once m has been computed, k can be found using the point-slope form, since the centroid is guaranteed to be on the line:

$$y - \bar{y} = m(x - \bar{x}) \tag{11.17}$$

$$y = mx + \bar{y} - m\bar{x} \tag{11.18}$$

leading to $k = \bar{y} - m\bar{x}$. The pseudocode for this normalized version, shown in Algorithm 11.2, is straightforward.

11.1.3 Total Least Squares

One drawback with ordinary least squares is that the error that it minimizes is the vertical distance from each data point to the line, which becomes less robust to noise as the absolute value of the slope of the line increases. And, of course, when the line is vertical, the slope-intercept representation does not work at all.

A more robust approach, called **total least squares**, is to minimize the perpendicular distance from each point to the line. For this method, the line is represented using three parameters a, b, and c, so that $ax + by + c = 0$. Since any solution to the equation can be multiplied by any nonzero scalar to yield another solution, this is called a **homogeneous equation**. That is, if a, b, c are the coefficients representing a line, then αa, αb, αc represent the same line for any $\alpha \neq 0$.

Because the representation uses 3 parameters, but the line has only 2 degrees of freedom, we impose the constraint that $\|\mathbf{n}\| = 1$, where $\mathbf{n} \equiv \begin{bmatrix} a & b \end{bmatrix}^\mathsf{T}$ is the normal vector to the line.

ALGORITHM 11.2 Ordinary least squares for 2D line fitting, with normalization

ORDINARYLEASTSQUARES-NORMALIZED($\{(x_i, y_i)\}_{i=1}^n$)

Input: set of n points $\{(x_i, y_i)\}_{i=1}^n$ in the plane
Output: slope m and y-intercept k of the line $y = mx + k$ that minimizes the sum-of-squared vertical residuals $\sum_{i=1}^n (mx_i + k - y_i)^2$

1 $\bar{x} \leftarrow \bar{y} \leftarrow 0$
2 **for** $i \leftarrow 1$ **to** n **do** ▷ Compute means.
3      $\bar{x} \leftarrow_+ x_i$
4      $\bar{y} \leftarrow_+ y_i$
5 $\bar{x} \leftarrow \bar{x} / n$
6 $\bar{y} \leftarrow \bar{y} / n$
7 $s_{aa} \leftarrow s_{ab} \leftarrow 0$
8 **for** $i \leftarrow 1$ **to** n **do**
9      $s_{aa} \leftarrow_+ (x_i - \bar{x}) * (x_i - \bar{x})$ ▷ Compute $\tilde{\mathbf{a}}^\mathsf{T}\tilde{\mathbf{a}}$.
10     $s_{ab} \leftarrow_+ (x_i - \bar{x}) * (y_i - \bar{y})$ ▷ Compute $\tilde{\mathbf{a}}^\mathsf{T}\tilde{\mathbf{b}}$.
11 $m \leftarrow s_{ab} / s_{aa}$ ▷ Compute $m = (\tilde{\mathbf{a}}^\mathsf{T}\tilde{\mathbf{b}})/(\tilde{\mathbf{a}}^\mathsf{T}\tilde{\mathbf{a}})$.
12 $k \leftarrow \bar{y} - m * \bar{x}$ ▷ Compute $k = \bar{y} - m\bar{x}$.
13 **return** m, k

With this constraint, the distance from the origin to the line along this normal vector is given by $|c|$, and the normal either points from the line to the origin (if $c > 0$) or from the origin to the line (if $c < 0$). The distance from a point (x, y) and the line is given by $|ax + by + c|$, so our goal is to minimize the sum of squared distances, $\sum_{i=1}^{n} (ax_i + by_i + c)^2$. As before, let us define the centroid of the points as $\bar{\mathbf{x}} \equiv \frac{1}{n} \sum_{i=1}^{n} [x_i \quad y_i]^\mathsf{T}$. If we define the covariance matrix as $\mathbf{C} \equiv \frac{1}{n} \sum_{i=1}^{n} [x_i - \bar{x} \quad y_i - \bar{y}]^\mathsf{T} [x_i - \bar{x} \quad y_i - \bar{y}]$, then it is easy to show that $\mathbf{n}$ is the eigenvector of $\mathbf{C}$ associated with the smallest eigenvalue, and that $c = -\bar{\mathbf{x}}^\mathsf{T} \mathbf{n}$.[†] The pseudocode is shown in Algorithm 11.3.

To derive the solution for $\mathbf{n}$, let us first write the n equations:

$$ax_1 + by_1 + c = 0 \tag{11.19}$$

$$ax_2 + by_2 + c = 0 \tag{11.20}$$

$$\vdots \tag{11.21}$$

$$ax_n + by_n + c = 0 \tag{11.22}$$

Stacking these equations into a matrix and pulling out the unknowns yields

$$\underbrace{\begin{bmatrix} x_1 & y_1 & 1 \\ x_2 & y_2 & 1 \\ \vdots & & \\ x_n & y_n & 1 \end{bmatrix}}_{\mathbf{A}_{\{n \times 3\}}} \underbrace{\begin{bmatrix} a \\ b \\ c \end{bmatrix}}_{\mathbf{x}_{\{3 \times 1\}}} = \underbrace{\begin{bmatrix} 0 \\ 0 \\ \vdots \\ 0 \end{bmatrix}}_{\mathbf{b}_{\{n \times 1\}}} \tag{11.23}$$

where the matrix $\mathbf{A}$ is different from before. Multiplying both sides of the equation by $\mathbf{A}^\mathsf{T}$ yields

$$\mathbf{A}^\mathsf{T} \mathbf{A} \mathbf{x} = \mathbf{0} \tag{11.24}$$

ALGORITHM 11.3 Total least squares for 2D line fitting

$\textsc{TotalLeastSquares}\left(\{(x_i, y_i)\}_{i=1}^{n}\right)$

Input: set of n points $\{(x_i, y_i)\}_{i=1}^{n}$ in the plane
Output: parameters a, b, and c of the line $ax + by + c$, where $\sqrt{a^2 + b^2} = 1$ and $c < 0$, that minimizes the sum-of-squared perpendicular residuals $\sum_{i=1}^{n} (ax_i + by_i + c)^2$

1 $[\bar{x} \quad \bar{y}] \leftarrow \frac{1}{n} \sum_{i=1}^{n} [x_i \quad y_i]$ ➤ Compute the mean of the points.
2 $\mathbf{C} \leftarrow \frac{1}{n} \sum_{i=1}^{n} [x_i - \bar{x} \quad y_i - \bar{y}]^\mathsf{T} [x_i - \bar{x} \quad y_i - \bar{y}]$ ➤ Construct covariance matrix.
3 $\mathbf{v}_1, \mathbf{v}_2, \lambda_1, \lambda_2 \leftarrow \textsc{Eigen}(\mathbf{C})$ ➤ Compute eigenvectors and associated eigenvalues, where $\lambda_1 \geq \lambda_2$.
4 $[a \quad b]^\mathsf{T} \leftarrow \mathbf{v}_2$ ➤ The eigenvector associated with the smallest eigenvalue yields the normal.
5 Set $c \leftarrow -\bar{\mathbf{x}}^\mathsf{T} \mathbf{n}$, where $\bar{\mathbf{x}}^\mathsf{T} = [\bar{x} \quad \bar{y}]$ and $\mathbf{n} = [a \quad b]^\mathsf{T}$ ➤ Compute the perpendicular
6 **if** $c > 0$ **then**   distance to the line.
7      $[a \quad b \quad c]^\mathsf{T} \leftarrow -[a \quad b \quad c]^\mathsf{T}$ ➤ Enforce convention (not necessary).
8 **return** a, b, c

[†] Note that n, the number of points, and $\mathbf{n}$, the normal vector, are unrelated.

Here we run into a problem. Because this equation is homogeneous (all zeros on the right-hand side), we cannot use the same approach we used before, which would simply yield $\mathbf{x} = (\mathbf{A}^\mathsf{T}\mathbf{A})^{-1}\mathbf{0} = \mathbf{0}$, the vector of all zeros, which is not very interesting. Instead, in the case of a homogeneous equation, we must impose an additional constraint, namely that the norm of the unknown vector is 1, or $\|\mathbf{x}\| = 1$.

As before, the variable c is eliminated by subtracting the mean to get

$$a\tilde{x}_1 + b\tilde{y}_1 = 0 \tag{11.25}$$

$$a\tilde{x}_2 + b\tilde{y}_2 = 0 \tag{11.26}$$

$$\vdots \tag{11.27}$$

$$a\tilde{x}_n + b\tilde{y}_n = 0 \tag{11.28}$$

Stacking these equations into a matrix and pulling out the unknowns yields

$$\underbrace{\begin{bmatrix} \tilde{x}_1 & \tilde{y}_1 \\ \tilde{x}_2 & \tilde{y}_2 \\ \vdots & \\ \tilde{x}_n & \tilde{y}_n \end{bmatrix}}_{\tilde{\mathbf{A}}_{\{n \times 2\}}} \underbrace{\begin{bmatrix} a \\ b \end{bmatrix}}_{\mathbf{n}_{\{2 \times 1\}}} = \underbrace{\begin{bmatrix} 0 \\ 0 \\ \vdots \\ 0 \end{bmatrix}}_{\mathbf{0}_{\{n \times 1\}}} \tag{11.29}$$

Multiplying both sides of the equation by $\tilde{\mathbf{A}}^\mathsf{T}$ yields

$$\tilde{\mathbf{A}}^\mathsf{T}\tilde{\mathbf{A}}\mathbf{n} = \mathbf{0} \tag{11.30}$$

In other words, our goal is to perform the following constrained minimization:

$$\min_{\|\mathbf{n}\|=1} \|\tilde{\mathbf{A}}^\mathsf{T}\tilde{\mathbf{A}}\mathbf{n}\| \tag{11.31}$$

It is easy to see that the covariance matrix is equivalent to $\mathbf{C} = \frac{1}{n}\tilde{\mathbf{A}}^\mathsf{T}\tilde{\mathbf{A}}$. As we have already seen,[†] because $\mathbf{C}$ is real and symmetric, its eigenvalues are real, and its eigenvectors are mutually orthogonal. If the eigenvectors $\mathbf{v}_1, \mathbf{v}_2$ and eigenvalues λ_1, λ_2 of $\mathbf{C}$ are computed, then $\mathbf{C}\mathbf{v}_i = \lambda_i\mathbf{v}_i$ for all $i = 1, 2$. Stacking the eigenvectors into a matrix $\mathbf{P} = \begin{bmatrix} \mathbf{v}_1 & \mathbf{v}_2 \end{bmatrix}$ and the eigenvalues into a diagonal matrix $\Lambda = \mathrm{diag}(\lambda_1, \lambda_2)$ yields the equation $\mathbf{CP} = \mathbf{P}\Lambda$. Rearranging, the matrix is factored into the product of a diagonal matrix and two orthogonal matrices: $\mathbf{C} = \mathbf{P}\Lambda\mathbf{P}^\mathsf{T}$, since $\mathbf{P}$ is an orthogonal matrix which implies that $\mathbf{P}^\mathsf{T}\mathbf{P} = \mathbf{I}_{\{n \times n\}}$.

Substituting, Equation (11.31) is revealed to be the same as

$$\min_{\|\mathbf{n}\|=1} \|\mathbf{P}\Lambda\mathbf{P}^\mathsf{T}\mathbf{n}\| = \min_{\|\mathbf{n}\|=1} \|\Lambda\mathbf{P}^\mathsf{T}\mathbf{n}\| \tag{11.32}$$

where we have ignored the scalar factor of $\frac{1}{n}$ since the equation is homogeneous, and the second expression arises from left-multiplying the first expression by $\mathbf{P}^\mathsf{T}$. Now, since $\mathbf{P}$ is orthogonal, it simply performs a rotation (with possible reflection) on the unit vector $\mathbf{n}$, so that the result is also unit norm, $\|\mathbf{P}^\mathsf{T}\mathbf{n}\| = 1$. It should be rather obvious that the smallest vector that will result from multiplying a diagonal matrix by a unit vector occurs when the input vector has all its energy concentrated in the row corresponding to the smallest value in the diagonal matrix. In other words, if we let $\mathbf{n}' \equiv \begin{bmatrix} n'_x & n'_y \end{bmatrix}^\mathsf{T} \equiv \mathbf{P}^\mathsf{T}\mathbf{n}$,

---

[†] Section 4.4.5 (p. 182)

then minimizing Equation (11.32) is the same as minimizing $(\lambda_1 n_x')^2 + (\lambda_2 n_y')^2$ subject to the constraint that $(n_x')^2 + (n_y')^2 = 1$. If we adopt the convention that the eigenvalues are sorted in decreasing order, $\lambda_1 \geq \lambda_2$, then this minimum occurs when $n_x' = 0$ and $n_y' = 1$, or $\mathbf{P}^\mathsf{T}\mathbf{n} = \begin{bmatrix} 0 & 1 \end{bmatrix}^\mathsf{T}$. Therefore $\mathbf{n} = \mathbf{P}\begin{bmatrix} 0 & 1 \end{bmatrix}^\mathsf{T}$, which is simply the right column of $\mathbf{P}$, or $\mathbf{v}_2$.

Another way to look at this problem is to notice that it is a constrained optimization problem. A common way of solving constrained optimization problems is through the use of **Lagrange multipliers**.[†] We want to minimize $\|\tilde{\mathbf{A}}\mathbf{n}\|^2 = \mathbf{n}^\mathsf{T}\tilde{\mathbf{A}}^\mathsf{T}\tilde{\mathbf{A}}\mathbf{n}$ subject to the constraint $\mathbf{n}^\mathsf{T}\mathbf{n} = 1$. To apply the technique of Lagrange multipliers, the constraint is rewritten so that it equals zero, which in our case yields $\mathbf{n}^\mathsf{T}\mathbf{n} - 1 = 0$. Then the **Lagrangian** is defined as the quantity being minimized minus a scalar multiple of the left-hand side of the constraint $\mathbf{n}^\mathsf{T}\tilde{\mathbf{A}}^\mathsf{T}\tilde{\mathbf{A}}\mathbf{n} - \lambda(\mathbf{n}^\mathsf{T}\mathbf{n} - 1)$, where the (unknown) scalar factor λ is called the **Lagrange multiplier**.[‡] We can solve for λ by setting this expression to zero and taking the derivative with respect to $\mathbf{n}$, yielding $\tilde{\mathbf{A}}^\mathsf{T}\tilde{\mathbf{A}}\mathbf{n} - \lambda\mathbf{n} = \mathbf{0}$, or $\tilde{\mathbf{A}}^\mathsf{T}\tilde{\mathbf{A}}\mathbf{n} = \lambda\mathbf{n}$. This equation shows that λ is an eigenvalue of $\tilde{\mathbf{A}}^\mathsf{T}\tilde{\mathbf{A}}$, and that $\mathbf{n}$ is the corresponding eigenvector.

11.1.4 Fitting a Plane

The procedure for fitting a plane to a set of 3D points is identical to that of fitting a line to a set of 2D points. A plane is represented by the four parameters in the homogeneous equation $ax + by + cz + d = 0$. The normal to the plane is given by $\mathbf{n} \equiv \begin{bmatrix} a & b & c \end{bmatrix}^\mathsf{T}$, where we impose $\|\mathbf{n}\| = 1$, and $|d|$ is the distance from the origin to the plane. The distance from a point (x, y, z) to the plane is given by $|ax + by + cz + d|$, so our goal is to minimize the sum of squared distances $\sum_{i=1}^{n} (ax_i + by_i + cz_i + d)^2$. As before, let us define the centroid of the points as $\bar{\mathbf{x}} \equiv \frac{1}{n}\sum_{i=1}^{n} \begin{bmatrix} x_i & y_i & z_i \end{bmatrix}^\mathsf{T}$. If all the points are shifted by their centroid, $\begin{bmatrix} \tilde{x}_i & \tilde{y}_i & \tilde{z}_i \end{bmatrix} = \begin{bmatrix} x_i - \bar{x} & y_i - \bar{y} & z_i - \bar{z} \end{bmatrix}$, then all the coordinates $\begin{bmatrix} \tilde{x}_i & \tilde{y}_i & \tilde{z}_i \end{bmatrix}$ are stacked as rows of the matrix $\tilde{\mathbf{A}}$, and the covariance matrix is given by $\mathbf{C} = \frac{1}{n}\tilde{\mathbf{A}}^\mathsf{T}\tilde{\mathbf{A}}$. The result is the same as before; $\mathbf{n}$ is the eigenvector of $\mathbf{C}$ associated with the smallest eigenvalue, normalized to unit norm, and $d = -\bar{\mathbf{x}}^\mathsf{T}\mathbf{n}$.

Figure 11.2 Results of fitting a line to the same set of points as Figure 11.1 using total least squares. This approach minimizes the sum of squared *perpendicular* distances. The technique works well for both nonvertical lines (left) and vertical lines (right).

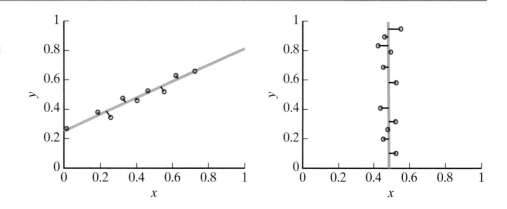

[†] Section 12.4.5 (p. 602).

[‡] Despite the use of the letter λ, the Lagrange multipler is in general not necessarily related to any eigenvalue (although in this particular case it is).

11.1.5 Singular Value Decomposition (SVD)

From the definition of eigenvalues and eigenvectors, it is easy to see that any square matrix $\mathbf{A}$ can be decomposed, or factorized, into the product of three matrices:

$$\mathbf{A} = \mathbf{P}\Lambda\mathbf{P}^{-1} \tag{11.33}$$

where Λ is a diagonal matrix consisting of the eigenvalues of $\mathbf{A}$, and the columns of $\mathbf{P}$ are the eigenvectors. If $\mathbf{A}$ is real and symmetric (or complex and Hermitian),[†] then all of its eigenvalues are real, and all of its eigenvectors corresponding to unique eigenvalues are orthogonal. This means that $\mathbf{P}$ is either orthogonal or can be constructed to be orthogonal, yielding $\mathbf{A} = \mathbf{P}\Lambda\mathbf{P}^{\mathsf{T}}$, as we saw earlier.[‡] In this section this idea is generalized by introducing the **singular value decomposition (SVD)**, which is a factorization technique applicable to *any* matrix. Because of its widespread importance and applicability, the SVD deserves a detailed discussion.

The SVD is defined as follows. Any real[§] $m \times n$ matrix $\mathbf{A}$ may be decomposed as the product of an orthogonal matrix $\mathbf{U}$, a diagonal matrix Σ, and another orthogonal matrix $\mathbf{V}$:

$$\mathbf{A}_{\{m \times n\}} = \mathbf{U}_{\{m \times m\}} \, \Sigma_{\{m \times n\}} \, \mathbf{V}^{\mathsf{T}}_{\{n \times n\}} \tag{11.34}$$

Since $\mathbf{U}$ and $\mathbf{V}$ are orthogonal matrices, $\mathbf{U}\mathbf{U}^{\mathsf{T}} = \mathbf{U}^{\mathsf{T}}\mathbf{U} = \mathbf{I}_{\{m \times m\}}$ and $\mathbf{V}\mathbf{V}^{\mathsf{T}} = \mathbf{V}^{\mathsf{T}}\mathbf{V} = \mathbf{I}_{\{n \times n\}}$. The matrix $\Sigma = diag(\sigma_1, \sigma_2, \ldots, \sigma_p)$ is a diagonal matrix whose entries are sorted in non-increasing order: $\sigma_1 \geq \sigma_2 \geq \cdots \geq \sigma_p \geq 0$, where $p \equiv \min(m, n)$. The columns of $\mathbf{U}$ are called the **left singular vectors**, the columns of $\mathbf{V}$ are the **right singular vectors**, and the values σ_i are the **singular values** of $\mathbf{A}$. It is not hard to see that Equation (11.34) is mathematically equivalent to the sum of the singular values multiplied by the outer product of the left and right singular vectors:

$$\mathbf{A} = \sum_{i=1}^{p} \sigma_i \mathbf{u}_i \mathbf{v}_i^{\mathsf{T}} \tag{11.35}$$

which is called the **SVD expansion**, where $\mathbf{u}_i$ is the ith column of $\mathbf{U}$, and $\mathbf{v}_i$ is the ith column of $\mathbf{V}$.

If the singular values are unique, then the singular vectors are also unique if we simply follow the convention of setting the first nonzero entry of every left singular vector to be positive. Otherwise, there is a sign ambiguity because a sign flip of a left singular vector $\mathbf{u}_i$ and its corresponding right singular vector $\mathbf{v}_i$ will leave the product $\mathbf{U}\Sigma\mathbf{V}^{\mathsf{T}}$ unchanged. In the case of repeated singular values, the singular vectors are no longer unique; any orthogonal vectors that span the subspace corresponding to the repeated singular values will do.

The singular vectors can be thought of as generalized eigenvectors, while the singular values are generalized eigenvalues. To see that the singular values act like eigenvalues while the singular vectors act like eigenvectors, note the following:

$$\mathbf{A}\mathbf{v}_i = \mathbf{U}\Sigma(\mathbf{V}^{\mathsf{T}}\mathbf{v}_i) = \mathbf{U}\Sigma\mathbf{e}_i = \sigma_i\mathbf{U}\mathbf{e}_i = \sigma_i\mathbf{u}_i \tag{11.36}$$

$$\mathbf{A}^{\mathsf{T}}\mathbf{u}_i = \mathbf{V}\Sigma^{\mathsf{T}}(\mathbf{U}^{\mathsf{T}}\mathbf{u}_i) = \mathbf{V}\Sigma^{\mathsf{T}}\mathbf{e}_i = \sigma_i\mathbf{V}\mathbf{e}_i = \sigma_i\mathbf{v}_i \tag{11.37}$$

where $\mathbf{e}_i \equiv \begin{bmatrix} 0 & \cdots & 0 & 1 & 0 & \cdots & 0 \end{bmatrix}^{\mathsf{T}}$ is the standard basis vector with a 1 in the ith element. These equations show the ease with which properties can be proved using the

---

[†] A symmetric matrix is equal to its transpose, while a Hermitian matrix is equal to its conjugate transpose. All real symmetric matrices are therefore Hermitian.

[‡] Section 4.4.4 (p.179)

[§] If the matrix is complex, then simply replace *transpose* with *conjugate transpose* everywhere.

SVD, due to its simplicity; in this case, the trick is that $\mathbf{u}_i$ is orthogonal to every column of $\mathbf{U}$ except for the ith column, and similarly for $\mathbf{v}_i$ and $\mathbf{V}$.

Applying similar principles to the $m \times m$ matrix $\mathbf{AA}^\mathsf{T}$ and the $n \times n$ **Gramian** $\mathbf{A}^\mathsf{T}\mathbf{A}$, we see that the left and right singular vectors are the eigenvectors of these two matrices, respectively:

$$\mathbf{AA}^\mathsf{T}\mathbf{u}_i = (\mathbf{U\Sigma V}^\mathsf{T})(\mathbf{V\Sigma}^\mathsf{T}\mathbf{U}^\mathsf{T})\mathbf{u}_i = \mathbf{U\Sigma\Sigma}^\mathsf{T}\mathbf{e}_i = \sigma_i^2\mathbf{U}\mathbf{e}_i = \sigma_i^2\mathbf{u}_i \qquad (11.38)$$

$$\mathbf{A}^\mathsf{T}\mathbf{A}\,\mathbf{v}_i = (\mathbf{V\Sigma}^\mathsf{T}\mathbf{U}^\mathsf{T})(\mathbf{U\Sigma V}^\mathsf{T})\mathbf{v}_i = \mathbf{V\Sigma}^\mathsf{T}\mathbf{\Sigma}\mathbf{e}_i = \sigma_i^2\mathbf{V}\mathbf{e}_i = \sigma_i^2\mathbf{v}_i \qquad (11.39)$$

while the first p eigenvalues are the squares of the singular values. In the special case that the matrix $\mathbf{A}$ is symmetric—that is, $\mathbf{A} = \mathbf{A}^\mathsf{T}$—then it is easy to show that the left and right singular vectors are identical to each other and to the eigenvectors of $\mathbf{A}$, and the singular values are the eigenvalues of $\mathbf{A}$.

The SVD makes it easy to reveal the structure of a matrix. It is not difficult to show that the rank, null space, range, and **Frobenius norm** of $\mathbf{A}$ are related to the SVD in the following way, where the Frobenius norm is the Euclidean norm of the vector composed of the elements of the matrix:

$$\text{rank}(\mathbf{A}) = r$$

$$\text{null}(\mathbf{A}) = \text{span}\,\{\mathbf{v}_{r+1}, \ldots, \mathbf{v}_n\}$$

$$\text{range}(\mathbf{A}) = \text{span}\,\{\mathbf{u}_1, \ldots, \mathbf{u}_r\}$$

$$\|\mathbf{A}\|_F \equiv \sqrt{\sum_{i=1}^{m}\sum_{j=1}^{n}a_{ij}^2} = \sqrt{\sum_{i=1}^{r}\sigma_i^2}$$

where r is the number of nonzero singular values, so that $\Sigma = diag(\sigma_1, \sigma_2, \cdots, \sigma_r, 0, \cdots, 0)$, and a_{ij} is the ij^{th} element of $\mathbf{A}$. If all of the singular values are nonzero then $r = p$; otherwise $\sigma_r > 0$, but $\sigma_{r+1} = 0$.

Clearly, for a matrix with rank r, the final $m - r$ columns of $\mathbf{U}$, as well as the final $n - r$ columns of $\mathbf{V}$, are not needed, since these values are multiplied by singular values which are set to zero, $\sigma_i = 0$. This observation leads naturally to the **compact version** of the SVD in which the unnecessary elements are removed:

$$\mathbf{A} = \underbrace{\begin{bmatrix} \mathbf{u}_1 & \mathbf{u}_2 & \cdots & \mathbf{u}_r \end{bmatrix}}_{\mathbf{U}_r} \underbrace{\begin{bmatrix} \sigma_1 & & & \\ & \sigma_2 & & \\ & & \ddots & \\ & & & \sigma_r \end{bmatrix}}_{\Sigma_r} \underbrace{\begin{bmatrix} \mathbf{v}_1^\mathsf{T} \\ \mathbf{v}_2^\mathsf{T} \\ \vdots \\ \mathbf{v}_r^\mathsf{T} \end{bmatrix}}_{\mathbf{V}_r^\mathsf{T}} \qquad (11.40)$$

In other words, in the compact version of the SVD, $\mathbf{A} = \mathbf{U}_r\Sigma_r\mathbf{V}_r^\mathsf{T}$, where $\mathbf{U}_r$ contains the first r columns of $\mathbf{U}$, Σ_r contains the first r singular values, and $\mathbf{V}_r$ contains the first r columns of $\mathbf{V}$. If $\mathbf{A}$ is $m \times n$, then $\mathbf{U}_r$ is $m \times r$, Σ_r is $r \times r$, and $\mathbf{V}_r$ is $n \times r$ (so that $\mathbf{V}_r^\mathsf{T}$ is $r \times n$). If $\mathbf{A}$ is a tall matrix $(m > n)$, then $\mathbf{V}_r = \mathbf{V}$, whereas if $\mathbf{A}$ is a short matrix $(m < n)$, then $\mathbf{U}_r = \mathbf{U}$, as illustrated in Figure 11.3. The compact version is more computationally efficient to compute and store, and it is often the version returned by software packages; for square, full-rank matrices, the two versions are identical.

We briefly mention several additional uses for the SVD. First, we all know that a square matrix $\mathbf{A}_{\{n \times n\}}$ is nonsingular if and only if it is full rank, i.e., $\sigma_i > 0$, $i = 1, \ldots, n$. In practice, however, a matrix can be full rank mathematically, while at the same time the

Tall matrix $m > n, p = n$

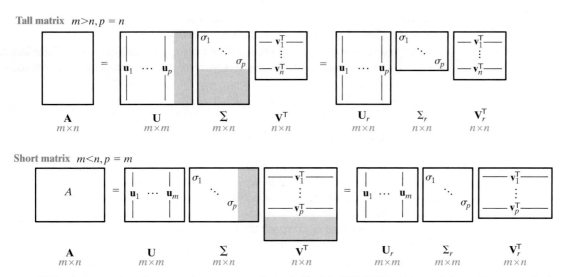

Short matrix $m < n, p = m$

Figure 11.3 If **A** is not square, then there are columns or rows of zeros in Σ of the SVD (indicated in gray), depending upon whether **A** is tall or short. Because these zeros are multiplied by either columns of **U** or rows of **V**, respectively, those columns or rows are not important, leading to the compact version of the SVD.

computation of its inverse leads to numerically unstable results. For example, although the ill-conditioned matrix in Equation (11.12) is, technically speaking, invertible, it is "close" to being singular, and therefore its inverse should not be computed. The SVD provides a simple and effective way to quantify this intuition. For any given matrix **A**, the ratio of the first singular value to the last nonzero singular value, $c \equiv \frac{\sigma_1}{\sigma_p} \geq 1$, is called the **condition number**.[†] If $c \approx 1$, then **A** is well-conditioned, but if c is large, then **A** is ill-conditioned. More specifically, if $\frac{1}{c}$ is comparable to the machine's floating-point precision, then the matrix should not be inverted because attempting to do so will lead to numerically unstable results.

Secondly, the **Moore-Penrose pseudoinverse** of a matrix **A** is defined as $\mathbf{A}^+ = \mathbf{V}\Sigma^+\mathbf{U}^\mathsf{T}$, where $\Sigma^+ = diag\left(\frac{1}{\sigma_1}, \frac{1}{\sigma_2}, \ldots, \frac{1}{\sigma_r}, 0, \ldots, 0\right)$. In other words, each nonzero singular value in Σ is inverted to obtain Σ^+, while zero singular values are left alone. It is easy to see that $\mathbf{A}\mathbf{A}^+ = \mathbf{U}\Sigma\mathbf{V}^\mathsf{T}\mathbf{V}\Sigma^+\mathbf{U}^\mathsf{T} = \mathbf{U}\Sigma\Sigma^+\mathbf{U}^\mathsf{T} = \mathbf{U}\mathbf{U}^\mathsf{T} = \mathbf{I}$, since $\mathbf{V}^\mathsf{T}\mathbf{V} = \mathbf{I}$, $\mathbf{U}\mathbf{U}^\mathsf{T} = \mathbf{I}$, and $\Sigma\Sigma^+ = \mathbf{I}$; and if the matrix **A** is square and full rank (and hence invertible), then $\mathbf{A}^+ = \mathbf{A}^{-1}$ because $\Sigma^+ = \Sigma^{-1}$. It can be shown that the least squares solution $(\mathbf{A}^\mathsf{T}\mathbf{A})^{-1}\mathbf{A}^\mathsf{T}\mathbf{b}$ to the problem $\mathbf{A}\mathbf{x} = \mathbf{b}$ is equivalent to $\mathbf{A}^+\mathbf{b}$, which is a generalization of $\mathbf{A}^{-1}\mathbf{b}$ to the case of a nonsquare matrix. In the case of a tie, the computation $\mathbf{A}^+\mathbf{b}$ yields, among all vectors that minimize $\|\mathbf{A}\mathbf{x} - \mathbf{b}\|$, the one with minimum norm.

Thirdly, the least squares solution to the homogeneous equation $\mathbf{A}\mathbf{x} = \mathbf{0}$ is given by the rightmost right singular vector of **A**, that is, $\mathbf{v}_n$; or equivalently, from Equation (11.39), the eigenvector of $\mathbf{A}^\mathsf{T}\mathbf{A}$ associated with the smallest eigenvalue. This is easily shown:

$$\arg\min_{\|\mathbf{x}\|=1} \|\mathbf{A}\mathbf{x}\|^2 = \arg\min_{\|\mathbf{x}\|=1} \|\mathbf{U}\Sigma\mathbf{V}^\mathsf{T}\mathbf{x}\|^2 = \mathbf{V}\left(\arg\min_{\|\mathbf{y}\|=1} \|\Sigma\mathbf{y}\|^2\right) = \mathbf{V}\mathbf{e}_n = \mathbf{v}_n \quad (11.41)$$

[†] If the matrix is real, symmetric, and positive definite, then the condition number is also the ratio of the largest to the smallest eigenvalues, since the eigenvalues are equal to the singular values and the eigenvalues are positive. More generally, if the matrix is real and normal, that is, $\mathbf{A}^\mathsf{T}\mathbf{A} = \mathbf{A}\mathbf{A}^\mathsf{T}$, then the condition number is $\max(\mathcal{L})/\min(\mathcal{L})$, where $\mathcal{L} \equiv \{|\lambda|\}_{i=1}^p$, and the absolute value is needed to handle the possibility of negative eigenvalues.

where $\mathbf{e}_n \equiv \begin{bmatrix} 0 & 0 & \cdots & 0 & 1 \end{bmatrix}^\mathsf{T}$. The first equality uses the definition of the SVD, the third equality uses the fact that the singular values are sorted in non-increasing order, and the final equality simply selects the final column of $\mathbf{V}$. To see the second equality, note that $\mathbf{U}$ and $\mathbf{V}$ perform a rotation (and possible reflection) without changing the length of the input: $\|\mathbf{Vx}\|^2 = (\mathbf{Vx})^\mathsf{T}(\mathbf{Vx}) = \mathbf{x}^\mathsf{T}\mathbf{V}^\mathsf{T}\mathbf{Vx} = \mathbf{x}^\mathsf{T}\mathbf{x} = \|\mathbf{x}\|^2$, since $\mathbf{V}^\mathsf{T}\mathbf{V} = \mathbf{I}$, and similarly for $\|\mathbf{Ux}\|^2$. Therefore, if we define $\mathbf{y} \equiv \mathbf{V}^\mathsf{T}\mathbf{x}$, then the problem reduces to minimizing $\|\mathbf{U\Sigma y}\|^2 = (\mathbf{y}^\mathsf{T}\mathbf{\Sigma}^\mathsf{T}\mathbf{U}^\mathsf{T})(\mathbf{U\Sigma y}) = \mathbf{y}^\mathsf{T}\mathbf{\Sigma}^\mathsf{T}\mathbf{\Sigma y} = \|\mathbf{\Sigma y}\|^2$ subject to the constraint $\|\mathbf{y}\| = 1$, where we note that $\mathbf{x} = \mathbf{Vy}$.

Finally, the SVD provides an easy way to find the closest matrix (in the Frobenius norm sense) to $\mathbf{A} = \mathbf{U\Sigma V}^\mathsf{T}$ with a particular rank. That is, if $\text{rank}(\mathbf{A}) = r$, then the matrix $\mathbf{A}'$ such that $\text{rank}(\mathbf{A}') = r' < r$ and $\|\mathbf{A} - \mathbf{A}'\|_F$ is minimized, is given by $\mathbf{A}' = \mathbf{U\Sigma}'\mathbf{V}^\mathsf{T}$, where $\mathbf{\Sigma}' = \text{diag}(\sigma_1, \sigma_2, \ldots, \sigma_{r'}, 0, \ldots, 0)$ is the diagonal matrix retaining the first r' singular values of $\mathbf{\Sigma}$. This provides a simple procedure for enforcing a **rank constraint** on a matrix by simply setting the smaller singular values to zero.

11.2 Fitting Curves

The closed-form least squares algorithm of the previous section is only applicable when the error to be minimized is quadratic in the unknowns. Such an error function is shaped like a parabola, which means that it is a convex function with a single global minimum and no local minima to confuse the algorithm. As a result, the problem can be solved in a single step. For example, the sum of squared perpendicular distances $\sum_{i=1}^{n}(ax_i + by_i + c)^2$ to the line is quadratic in the unknowns. Differentiating this error with respect to the unknowns yields equations that are linear in the unknowns, and such equations can be solved using closed-form methods.

Since curves are higher-order functions than lines, the sum of perpendicular distances to a curve is usually not quadratic. As a result, such *geometric error* functions cannot be minimized except by iterative non-linear methods. Nevertheless, good results can usually be obtained by approximating the geometric error with an *algebraic error*, which often is quadratic. In this section we cover several curves that arise in practice: namely, circles and ellipses.

11.2.1 Fitting a Circle

Consider the equation of a circle:

$$(x - h)^2 + (y - k)^2 = r^2 \tag{11.42}$$

where *(h, k)* is the center of the circle, and *r* is the radius. The perpendicular (or radial) distance from a point (x_i, y_i) to the circle is given by the absolute difference between its distance to the center of the circle and the radius: $d_i \equiv |r_i - r|$, where $r_i \equiv \sqrt{(x_i - h)^2 + (y_i - k)^2}$. Let us define the **geometric error** of a solution as the sum of squared perpendicular distances:

$$\varepsilon_{geom}(h, k, r) \equiv \sum_{i=1}^{n} d_i^2 = \sum_{i=1}^{n}(r_i - r)^2$$

$$= \sum_{i=1}^{n}(\sqrt{(x_i - h)^2 + (y_i - k)^2} - r)^2 \qquad (\text{geometric error}) \quad \textbf{(11.43)}$$

It is not possible to apply least squares to minimize this nonquadratic equation, because there is no closed-form solution. The only way to minimize this geometric error is by iterative nonlinear minimization.

A closed-form solution does exist, however, if instead of minimizing the geometric error we minimize the **algebraic error**:

$$\varepsilon_{alg}(h, k, r) \equiv \sum_{i=1}^{n} \delta_i^2 = \sum_{i=1}^{n} (r_i^2 - r^2)^2$$

$$= \sum_{i=1}^{n} ((x_i - h)^2 + (y_i - k)^2 - r^2)^2 \qquad \text{(algebraic error)} \qquad (11.44)$$

where $\delta_i \equiv (x_i - h)^2 + (y_i - k)^2 - r^2$ is the amount by which the two sides of Equation (11.42) differ. Usually the circle obtained by minimizing the algebraic error is a close approximation to the circle obtained by minimizing the geometric error, and even if the latter is desired, the former yields a starting point for the iterative minimization process of the latter.

To minimize the algebraic error, let us rewrite Equation (11.42) as

$$x^2 + y^2 + ax + by + c = 0 \qquad (11.45)$$

where $a \equiv -2h$, $b \equiv -2k$, and $c \equiv h^2 + k^2 - r^2$. Then, stacking the measurement equations yields

$$\underbrace{\begin{bmatrix} x_1 & y_1 & 1 \\ x_2 & y_2 & 1 \\ & \vdots & \\ x_n & y_n & 1 \end{bmatrix}}_{\check{\mathbf{A}}} \begin{bmatrix} a \\ b \\ c \end{bmatrix} = -\underbrace{\begin{bmatrix} x_1^2 + y_1^2 \\ x_2^2 + y_2^2 \\ \vdots \\ x_n^2 + y_n^2 \end{bmatrix}}_{\check{\rho}} \qquad (11.46)$$

or

$$\begin{bmatrix} a \\ b \\ c \end{bmatrix} = -(\check{\mathbf{A}}^{\mathsf{T}}\check{\mathbf{A}})^{-1}\check{\mathbf{A}}^{\mathsf{T}}\check{\rho} \qquad (11.47)$$

Once a, b, and c have been computed, we simply set $h = -\frac{a}{2}$, $k = -\frac{b}{2}$, and $r = \sqrt{h^2 + k^2 - c}$. This procedure, known as the **Kåsa method**, is straightforward, as shown in Algorithm 11.4.

A quick glance at Equation (11.46) reveals that the matrix $\check{\mathbf{A}}$ will be ill-conditioned if the values for x_i and y_i are large, which is typically the case with images. To normalize

ALGORITHM 11.4 Least squares fit of circle to set of points (algebraic error)—Kåsa method

FitCircle-Kåsa($\{(x_i, y_i)\}_{i=1}^{n}$)

Input: set of n points $\{(x_i, y_i)\}_{i=1}^{n}$ in the plane
Output: center (h, k) and radius r of circle that minimizes the algebraic error $\sum_{i=1}^{n} ((x_i - h)^2 + (y_i - k)^2 - r^2)^2$

1   Construct $\check{\mathbf{A}}$ and $\check{\rho}$ as shown in Equation (11.46)
2   Solve Equation (11.47) for a, b, and c
3   $h \leftarrow -a/2$
4   $k \leftarrow -b/2$
5   $r \leftarrow$ Sqrt $(h * h + k * k - c)$
6   **return** h, k, r

the coordinates, first shift all the points by subtracting the centroid $(\bar{x}, \bar{y})$ to define $(\tilde{x}, \tilde{y}) \equiv (x - \bar{x}, y - \bar{y})$. The equation of the circle in the shifted coordinate system is

$$\tilde{x}^2 + \tilde{y}^2 + \tilde{a}\tilde{x} + \tilde{b}\tilde{y} + \tilde{c} = 0 \tag{11.48}$$

where $\tilde{a} \equiv -2\tilde{h}$, $\tilde{b} \equiv -2\tilde{k}$, and $\tilde{c} \equiv \tilde{h}^2 + \tilde{k}^2 - r^2$, and where $(\tilde{h}, \tilde{k})$ is the center of the circle in the shifted coordinate system. (Note that the radius does not change.) What is interesting about this solution is that when the measurement equations are stacked, then multiplied on the left by the transpose of the matrix, they yield

$$\begin{bmatrix} \tilde{x}_1 & \tilde{x}_2 & \cdots & \tilde{x}_n \\ \tilde{y}_1 & \tilde{y}_2 & \cdots & \tilde{y}_n \\ 1 & 1 & \cdots & 1 \end{bmatrix} \begin{bmatrix} \tilde{x}_1 & \tilde{y}_1 & 1 \\ \tilde{x}_2 & \tilde{y}_2 & 1 \\ & \vdots & \\ \tilde{x}_n & \tilde{y}_n & 1 \end{bmatrix} \begin{bmatrix} \tilde{a} \\ \tilde{b} \\ \tilde{c} \end{bmatrix} = - \begin{bmatrix} \tilde{x}_1 & \tilde{x}_2 & \cdots & \tilde{x}_n \\ \tilde{y}_1 & \tilde{y}_2 & \cdots & \tilde{y}_n \\ 1 & 1 & \cdots & 1 \end{bmatrix} \begin{bmatrix} \tilde{x}_1^2 + \tilde{y}_1^2 \\ \tilde{x}_2^2 + \tilde{y}_2^2 \\ \vdots \\ \tilde{x}_n^2 + \tilde{y}_n^2 \end{bmatrix} \tag{11.49}$$

which simplifies to

$$\sum_{i=1}^{n} \begin{bmatrix} \tilde{x}_i^2 & \tilde{x}_i\tilde{y}_i & 0 \\ \tilde{x}_i\tilde{y}_i & \tilde{y}_i^2 & 0 \\ 0 & 0 & 1 \end{bmatrix} \begin{bmatrix} \tilde{a} \\ \tilde{b} \\ \tilde{c} \end{bmatrix} = -\sum_{i=1}^{n} \begin{bmatrix} \tilde{x}_i^3 + \tilde{x}_i\tilde{y}_i^2 \\ \tilde{x}_i^2\tilde{y}_i + \tilde{y}_i^3 \\ \tilde{x}_i^2 + \tilde{y}_i^2 \end{bmatrix} \tag{11.50}$$

where the zero elements in the final row and column of the matrix arise because $\sum_{i=1}^{n} \tilde{x}_i = \sum_{i=1}^{n} \tilde{y}_i = 0$, which follows from the definition of centroid. Because of these zeros, the parameter $\tilde{c}$ is decoupled from $\tilde{a}$ and $\tilde{b}$ and can therefore be solved separately. With a few minor substitutions, it can be seen that the shifted center is computed by solving the following linear system:

$$\underbrace{\begin{bmatrix} \tilde{x}_1 & \tilde{y}_1 \\ \tilde{x}_2 & \tilde{y}_2 \\ \vdots & \vdots \\ \tilde{x}_n & \tilde{y}_n \end{bmatrix}}_{\tilde{\mathbf{A}}} \begin{bmatrix} \tilde{h} \\ \tilde{k} \end{bmatrix} = \frac{1}{2} \underbrace{\begin{bmatrix} \tilde{x}_1^2 + \tilde{y}_1^2 \\ \tilde{x}_2^2 + \tilde{y}_2^2 \\ \vdots \\ \tilde{x}_n^2 + \tilde{y}_n^2 \end{bmatrix}}_{\tilde{\boldsymbol{\rho}}} \tag{11.51}$$

or

$$\begin{bmatrix} \tilde{h} \\ \tilde{k} \end{bmatrix} = \frac{1}{2} (\tilde{\mathbf{A}}^\top \tilde{\mathbf{A}})^{-1} \tilde{\mathbf{A}}^\top \tilde{\boldsymbol{\rho}} \tag{11.52}$$

Once the shifted center has been computed, the center in the original coordinate system is found by setting $h = \tilde{h} + \bar{x}$ and $k = \tilde{k} + \bar{y}$. From Equation (11.50), we see that $\tilde{c} = -\frac{1}{n} \sum_{i=1}^{n} (\tilde{x}_i^2 + \tilde{y}_i^2)$, and therefore the radius is given by

$$r = \sqrt{\tilde{h}^2 + \tilde{k}^2 - \tilde{c}} = \sqrt{\frac{1}{n} \sum_{i=1}^{n} (x_i - h)^2 + (y_i - k)^2} \tag{11.53}$$

whose derivation is left as an exercise.[†]

The pseudocode for this technique, which we call the **normalized Kåsa method**, is provided in Algorithm 11.5. Note that this approach involves solving a 2×2, rather than a 3×3, linear system, which has advantages numerically. Examples of circle fitting are shown in Figure 11.4.

[†] Problem 11.18.

ALGORITHM 11.5 Least squares fit of circle to set of points (algebraic error) – Normalized Kåsa method

FITCIRCLE-NORMALIZEDKÅSA $\left(\{(x_i, y_i)\}_{i=1}^{n}\right)$

Input: set of n points $\{(x_i, y_i)\}_{i=1}^{n}$ in the plane
Output: center (h, k) and radius r of circle that minimizes the algebraic error $\sum_{i=1}^{n}\left((x_i - h)^2 + (y_i - k)^2 - r^2\right)^2$

1   Compute centroid $(\bar{x}, \bar{y}) \leftarrow \frac{1}{n}\sum_{i=1}^{n}(x_i, y_i)$
2   Compute shifted points $(\tilde{x}_i, \tilde{y}_i) \leftarrow (x_i - \bar{x}, y_i - \bar{y})$ for all i
3   Construct $\tilde{\mathbf{A}}$ and $\tilde{\boldsymbol{\rho}}$ as shown in Equation (11.51)
4   Solve Equation (11.52) for $\tilde{h}$ and $\tilde{k}$
5   $(h, k) \leftarrow (\tilde{h} + \bar{x}, \tilde{k} + \bar{y})$
6   Compute r using Equation (11.53)
7   **return** h, k, r

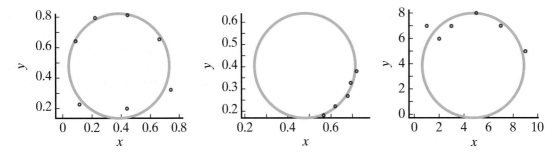

Figure 11.4 Three examples of results of fitting a circle to a set of points, using Algorithm 11.5. (Algorithm 11.4 produces indistinguishable results on these data.)

11.2.2 Fitting a Conic Section

The equation for a general **2D conic section** is given by the following homogeneous equation:

$$ax^2 + bxy + cy^2 + dx + ey + f = 0 \qquad (11.54)$$

where $a, b, c, d, e,$ and f are the parameters describing the conic section; or, in matrix form,

$$\begin{bmatrix} x & y & 1 \end{bmatrix} \underbrace{\begin{bmatrix} 2a & b & d \\ b & 2c & e \\ d & e & 2f \end{bmatrix}}_{\mathbf{C}} \begin{bmatrix} x \\ y \\ 1 \end{bmatrix} = 0 \qquad (11.55)$$

As before, note that these equations are homogeneous, so all the parameters can be multiplied by the same nonzero scalar without changing the equation. That is, $\mathbf{C}$ and $\alpha\mathbf{C}$ represent the same conic section, where $\alpha \neq 0$ is an arbitrary scaling factor.[†]

You may recall from geometry that a conic section arises as the intersection of an infinite double cylindrical cone and a plane. If $\det(\mathbf{C}) = 0$, then the conic section is degenerate

---

[†] Do not confuse the 3×3 conic section matrix with the covariance matrix; we use $\mathbf{C}$ for both.

and takes the shape of either a point, line, or pair of lines. Otherwise, the conic section takes one of three possible shapes, determined by the **discriminant** $b^2 - 4ac$:

$$\text{Conic section is} \begin{cases} \text{an ellipse} & \text{if } b^2 - 4ac < 0 \\ \text{a parabola} & \text{if } b^2 - 4ac = 0 \\ \text{a hyperbola} & \text{if } b^2 - 4ac > 0 \end{cases} \qquad (11.56)$$

Moreover, for the ellipse to be real, c and $\det(\mathbf{C})$ must have opposite signs, that is, $\text{sgn}(c) \neq \text{sgn}(\det(\mathbf{C}))$, otherwise the ellipse is imaginary. Note that the ellipse is a circle if $a = c$ and $b = 0$.

Minimizing the geometric error of a conic section is much harder than it is for a circle because we do not even know the shape of the curve until after we have solved for it. Nevertheless, minimizing the algebraic error, $\sum_{i=1}^{n}(ax_i^2 + bx_iy_i + cy_i^2 + dx_i + ey_i + f)^2$ is straightforward. By stacking the equations together, the algebraic error is given by $\|\mathbf{Da}\|^2 = \mathbf{a}^\mathsf{T}\mathbf{D}^\mathsf{T}\mathbf{Da}$, where $\mathbf{D}$ and $\mathbf{a}$ are defined as follows:

$$\underbrace{\begin{bmatrix} x_1^2 & x_1y_1 & y_1^2 & x_1 & y_1 & 1 \\ x_2^2 & x_2y_2 & y_2^2 & x_2 & y_2 & 1 \\ & & \vdots & & & \\ x_n^2 & x_ny_n & y_n^2 & x_n & y_n & 1 \end{bmatrix}}_{\mathbf{D}} \underbrace{\begin{bmatrix} a \\ b \\ c \\ d \\ e \\ f \end{bmatrix}}_{\mathbf{a}} = \mathbf{0}_{\{n \times 6\}} \qquad (11.57)$$

As we saw in the case of total least squares, the quantity $\|\mathbf{Da}\|^2$ is minimized by selecting rightmost right singular vector of $\mathbf{D}$. Equivalently, as explained above, we can define $\mathbf{S} \equiv \mathbf{D}^\mathsf{T}\mathbf{D}$ and minimize $\|\mathbf{Sa}\|^2 = \mathbf{A}^\mathsf{T}\mathbf{S}^\mathsf{T}\mathbf{Sa}$ by selecting the eigenvector of $\mathbf{S}$ associated with its smallest eigenvalue. The pseudocode is shown in Algorithm 11.6, but remember to first normalize the coordinates for more robust results.

11.2.3 Fitting an Ellipse

While fitting a conic section is easy, such a procedure is not of much practical benefit, because the shape of a conic section is too general. By far the most useful conic section is the ellipse, which arises commonly in practice as the projection of a circle onto an image plane. To minimize the algebraic error of an ellipse, we start with the same $\mathbf{D}$ and $\mathbf{S} = \mathbf{D}^\mathsf{T}\mathbf{D}$ matrices that we just saw. The constraint that describes an ellipse is given by $b^2 - 4ac < 0$ which, since scale does not matter, can be changed from an inequality constraint into an equality constraint: $b^2 - 4ac = -1$, where the -1 is chosen somewhat arbitrarily (any negative number will do). In matrix form, this equality constraint can be expressed as the constraint $\mathbf{a}^\mathsf{T}\mathbf{Ea} = 1$, where

$$\mathbf{E} \equiv \begin{bmatrix} 0 & 0 & 2 & 0 & 0 & 0 \\ 0 & -1 & 0 & 0 & 0 & 0 \\ 2 & 0 & 0 & 0 & 0 & 0 \\ 0 & 0 & 0 & 0 & 0 & 0 \\ 0 & 0 & 0 & 0 & 0 & 0 \\ 0 & 0 & 0 & 0 & 0 & 0 \end{bmatrix} \qquad (11.58)$$

ALGORITHM 11.6 Least squares fit of conic section to set of points (algebraic error)

FitConicSection $(\{(x_i, y_i)\}_{i=1}^n)$

Input: set of n points $\{(x_i, y_i)\}_{i=1}^n$ in the plane
Output: parameters a, b, c, d, e, and f describing conic section that minimizes the algebraic error
$$\sum_{i=1}^n (ax_i^2 + bx_iy_i + cy_i^2 + dx_i + ey_i + f)^2$$

1   Construct the $n \times 6$ measurement matrix $\mathbf{D}$ as shown in Equation (11.57)
2   Compute the 6×6 scatter matrix $\mathbf{S} \leftarrow \mathbf{D}^T\mathbf{D}$
3   Set $[a \quad b \quad c \quad d \quad e \quad f]^T$ to the eigenvector associated with the smallest eigenvalue of $\mathbf{S}$
4   **return** a, b, c, d, e, and f

Our problem is therefore to minimize $\|\mathbf{Da}\|^2 = \mathbf{a}^T\mathbf{D}^T\mathbf{Da}$ subject to the constraint $\mathbf{a}^T\mathbf{Ea} = 1$. Using Lagrange multipliers, we take the derivative of the quantity to be minimized and subtract the derivative of the constraint, multiplied by the unknown Lagrange multiplier λ, and set the result to zero. This yields $2\mathbf{D}^T\mathbf{Da} - 2\lambda\mathbf{Ea} = 0$, or $\mathbf{Sa} = \lambda\mathbf{Ea}$. The problem can thus be stated as follows:

$$\text{Solve} \quad \mathbf{Sa} \quad = \quad \lambda\mathbf{Ea} \tag{11.59}$$

$$\text{such that} \quad \mathbf{a}^T\mathbf{Ea} \quad = \quad 1 \tag{11.60}$$

Solving the generalized eigensystem in Equation (11.59) yields 6 generalized eigenvalue-eigenvector pairs, $(\lambda_i, \mathbf{a}_i)$, where $i = 1, \ldots, 6$. The eigenvector associated with the positive eigenvalue yields the desired result $\mathbf{a}$, as shown in the pseudocode of Algorithm 11.7. To understand the solution, let λ_i and $\mathbf{a}_i$ be one of the generalized eigenvalue-eigenvector pairs satisfying Equation (11.59). Plugging $\mathbf{Ea} = \frac{1}{\lambda}\mathbf{Sa}$ from Equation (11.59) into Equation (11.60), we have $1 = \mathbf{a}_i^T\mathbf{Ea}_i = \frac{1}{\lambda}\mathbf{a}_i^T\mathbf{Sa}$. Since $\mathbf{S}$ is positive definite, $\mathbf{a}_i^T\mathbf{Sa}_i > 0$. Therefore, this expression can be true only if $\lambda > 0$. It can be shown that exactly one of the generalized eigenvalues is guaranteed to be positive, thus leading to an unambiguous solution. This latter truth arises because the eigenvalues of $\mathbf{E}$ are $\{2, -1, -2, 0, 0, 0\}$, so exactly one is positive; and the generalized eigenvalues of Equation (11.59) have the same signs as the eigenvalues of $\mathbf{E}$, since $\mathbf{S}$ is positive definite and $\mathbf{E}$ is symmetric. Examples of fitting an ellipse to a set of points are provided in Figure 11.5.

ALGORITHM 11.7 Least squares fit of ellipse to set of points (algebraic error)

FitEllipse$(\{(x_i, y_i)\}_{i=1}^n)$

Input: set of n points $\{(x_i, y_i)\}_{i=1}^n$ in the plane
Output: parameters a, b, c, d, e, and f describing ellipse that minimizes the algebraic error
$$\sum_{i=1}^n (ax_i^2 + bx_iy_i + cy_i^2 + dx_i + ey_i + f)^2 \text{ subject to the constraint that } 4ac - b^2 = 1$$

1   Construct the $n \times 6$ measurement matrix $\mathbf{D}$ as shown in Equation (11.57)
2   Compute the 6×6 scatter matrix $\mathbf{S} \leftarrow \mathbf{D}^T\mathbf{D}$
3   Construct the ellipse constraint matrix $\mathbf{E}$ as shown in Equation (11.58)
4   Solve the generalized eigensystem $\mathbf{Sa} = \lambda\mathbf{Ea}$ in Equation (11.59) for λ_i and $\mathbf{a}_i$, $i = 1, \ldots, 6$
5   Find $k \in \{1, \ldots, 6\}$ such that $\lambda_k > 0$ ➤ Note: There is only one positive eigenvalue.
6   Set $[a \quad b \quad c \quad d \quad e \quad f]^T \leftarrow \mathbf{a}_k$
7   **return** a, b, c, d, e, and f

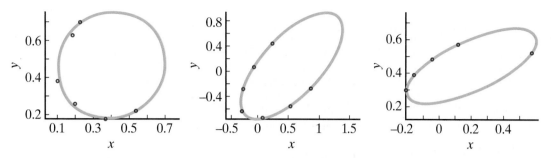

Figure 11.5 Three examples of fitting an ellipse to a set of points, using Algorithm 11.7.

11.2.4 Fitting a Filled Ellipse

Suppose that, instead of a set of points along the perimeter of an ellipse, we have a binary image region shaped like an ellipse. One approach would be to compute the boundary pixels, then apply the previous procedure and hope that the discretization effects are insignificant. Although such an approach will be satisfactory for many applications, more robust results can be achieved using *all* the pixels, including those in the interior. First calculate the regular and central moments of the region. Then, from our discussion of the best-fitting ellipse,[†] and in particular from Equation (4.144), the following relationships are established:

$$a = \frac{\mu_{02}}{\eta\mu_{00}} \tag{11.61}$$

$$b = \frac{-2\mu_{11}}{\eta\mu_{00}} \tag{11.62}$$

$$c = \frac{\mu_{20}}{\eta\mu_{00}} \tag{11.63}$$

where $\eta = 4(\mu_{20}\mu_{02} - \mu_{11}^2)/\mu_{00}^2$; and the centroid is $(\bar{x},\bar{y}) = \left(\frac{m_{10}}{\mu_{00}}, \frac{m_{01}}{\mu_{00}}\right)$, where $\mu_{00} = m_{00}$.

These parameters a, b, and c capture an ellipse that is centered at the origin whose shape is described by the equation $ax^2 + bxy + cy^2 = 1$. Shifting the entire ellipse by some amount (h,k) yields the general expression for an ellipse:

$$a(x - h)^2 + b(x - h)(y - k) + c(y - k)^2 - 1 = 0 \tag{11.64}$$

$$ax^2 + bxy + cy^2 + \underbrace{(-2ah - bk)}_{d}x + \underbrace{(-2ck - bh)}_{e}y + \underbrace{(ah^2 + bhk + ck^2 - 1)}_{f} = 0 \tag{11.65}$$

Setting (h, k) to the centroid $\left(\frac{m_{10}}{\mu_{00}}, \frac{m_{01}}{\mu_{00}}\right)$ yields expressions for the other parameters:

$$d = -2ah - bk = \frac{-\mu_{02}m_{10} + \mu_{11}m_{01}}{2(\mu_{20}\mu_{02} - \mu_{11}^2)} \tag{11.66}$$

$$e = -2ck - bh = \frac{-\mu_{20}m_{01} + \mu_{11}m_{10}}{2(\mu_{20}\mu_{02} - \mu_{11}^2)} \tag{11.67}$$

$$f = ah^2 + bhk + ck^2 - 1 = \frac{\mu_{02}m_{10}^2 - 2\mu_{11}m_{10}m_{01} + \mu_{20}m_{01}^2 - 4\mu_{00}(\mu_{20}\mu_{02} - \mu_{11}^2)}{4\mu_{00}(\mu_{20}\mu_{02} - \mu_{11}^2)} \tag{11.68}$$

[†] Section 4.4.5 (p.182).

ALGORITHM 11.8 Fit ellipse to binary image region

FITELLIPSETOFILLEDREGION(*I*)

Input: binary image *I* with $I(x, y) = 1$ inside the region and 0 outside
Output: parameters *a*, *b*, *c*, *d*, *e*, and *f* describing the ellipse

1 Compute the 0th-, 1st-, and 2nd-order regular moments of *I*
2 Compute the 0th-, 1st-, and 2nd-order central moments of *I*
3 Compute *a*, *b*, and *c* using Equations (11.61)–(11.63)
4 Compute *d*, *e*, and *f* using Equations (11.66)–(11.68)
5 **return** *a*, *b*, *c*, *d*, *e*, and *f*

This yields a direct approach for fitting an ellipse to a binary region using the regular and central moments, as shown in Algorithm 11.8. Examples demonstrating the results of this algorithm are provided in Figure 11.6.

11.2.5 Fitting a Filled Square

Consider a square of width *w*, and assume a binary image so that $I(x, y) = 1$ inside the square and 0 outside. If we further assume that the square is oriented with the axes, then in the continuous domain the second-order central moment along either axis is given by

$$\mu_{20} = \mu_{02} = \int_{-\frac{w}{2}}^{\frac{w}{2}} \int_{-\frac{w}{2}}^{\frac{w}{2}} x^2 \, dx \, dy = \left. \frac{x^3 y}{3} \right|_{-\frac{w}{2}}^{\frac{w}{2}} \left. \right|_{-\frac{w}{2}}^{\frac{w}{2}} = \frac{2}{3}\left(\frac{w}{2}\right)^3 w = \frac{w^4}{12} \qquad (11.69)$$

Therefore, the width can be computed using either moment:

$$w = (12\mu_{20})^{\frac{1}{4}} = (12\mu_{02})^{\frac{1}{4}} \qquad (11.70)$$

or, with noise, the average of the two. But this only works when the square is aligned with the axes. If it is at an angle, then we must first find the angle of orientation. The moments are of no use for this purpose because the best fitting ellipse for a square is a circle, which has no orientation.[†] Instead, a histogram of the orientations in the gradient image can be

Figure 11.6 Three examples of fitting an ellipse to a binary region, using Algorithm 11.8.

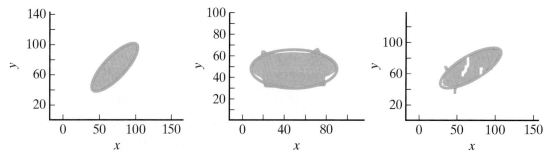

[†] That is, for a square of any orientation, $\mu_{20} = \mu_{02}$ and $\mu_{11} = 0$, so that Equation (4.131) involves $\frac{0}{0}$, which is undefined.

computed, from which the two peaks yield the orientation. Once the orientation θ has been determined, the second-order central moment along the axis defined by θ is computed by

$$\mu_\theta = \int ((x - \bar{x}) \cos \theta - (y - \bar{y}) \sin \theta)^2 \, dx \, dy \tag{11.71}$$

Note that if $\theta = 0$, then $\mu_\theta = \mu_{20}$, and if $\theta = \frac{\pi}{2}$, then $\mu_\theta = \mu_{02}$. The width is computed as the average of the values from the two orthogonal directions:

$$w = \frac{1}{2} \left((12\mu_\theta)^{\frac{1}{4}} + (12\mu_{\theta + \frac{\pi}{2}})^{\frac{1}{4}} \right) \tag{11.72}$$

Pseudocode of this procedure is provided in Algorithm 11.9.

11.2.6 Fitting a Filled Rectangle

Suppose we have a filled region shaped like a rectangle oriented with the x and y axes. Let us define the length ℓ to be the length along the x axis and the width w to be the width along the y axis. Then, using a similar procedure to that of a square,

$$\mu_{20} = \int_{-\frac{w}{2}}^{\frac{w}{2}} \int_{-\frac{\ell}{2}}^{\frac{\ell}{2}} x^2 \, dx \, dy = \frac{x^3 y}{3} \Big|_{-\frac{\ell}{2}}^{\frac{\ell}{2}} \Big|_{-\frac{w}{2}}^{\frac{w}{2}} = \frac{2}{3} \left(\frac{\ell}{2} \right)^3 w = \frac{w\ell^3}{12} \tag{11.73}$$

$$\mu_{02} = \int_{-\frac{w}{2}}^{\frac{w}{2}} \int_{-\frac{\ell}{2}}^{\frac{\ell}{2}} y^2 \, dx \, dy = \frac{x y^3}{3} \Big|_{-\frac{\ell}{2}}^{\frac{\ell}{2}} \Big|_{-\frac{w}{2}}^{\frac{w}{2}} = \frac{2}{3} \left(\frac{w}{2} \right)^3 \ell = \frac{w^3 \ell}{12} \tag{11.74}$$

Combining these two equations by multiplying and dividing yields

$$\frac{\mu_{20}}{\mu_{02}} = \frac{\ell^2}{w^2} \qquad \mu_{20}\mu_{02} = \frac{w^4 \ell^4}{144} \tag{11.75}$$

Although these are nonlinear equations, we can solve for the unknowns by straightforward substitution, yielding the desired result:

$$w = \left(\frac{144 \, \mu_{02}^3}{\mu_{20}} \right)^{\frac{1}{8}} \qquad \ell = \left(\frac{144 \mu_{20}^3}{\mu_{02}} \right)^{\frac{1}{8}} \tag{11.76}$$

ALGORITHM 11.9 Fit square to binary image region

FitSquareToFilledRegion(I)

Input: binary image I with $I(x, y) = 1$ inside the region and 0 outside
Output: width w and orientation θ describing square

1 Compute gradient of I
2 Compute histogram of the orientation of the pixels in the gradient image
3 Find the two dominant peaks separated by 90° to get the orientation θ
4 Compute the centroid of the region using the 0th- and 1st-order regular moments
5 Compute the 2nd-order central moment along the axis specified by θ using Equation (11.71)
6 Compute the 2nd-order central moment along the axis specified by $\theta + \frac{\pi}{2}$ using Equation (11.71)
7 Compute the width w using Equation (11.72)
8 **return** w, θ

ALGORITHM 11.10 Fit rectangle to binary image region

FITRECTANGLETOFILLEDREGION(I)

Input: binary image I with $I(x, y) = 1$ inside the region and 0 outside
Output: length ℓ, width w, and orientation θ describing rectangle

1 Compute the 0th-, 1st-, and 2nd-order regular moments of I
2 Compute the 0th-, 1st-, and 2nd-order central moments of I
3 Compute the orientation θ using the approach of the best-fitting ellipse
4 Compute the 2nd-order central moment along the axis specified by θ using Equation (11.71)
5 Compute the 2nd-order central moment along the axis specified by $\theta + \frac{\pi}{2}$ using Equation (11.71)
6 Compute the length ℓ and width w using Equation (11.77)
7 **return** ℓ, w, θ

A rectangle is easier to fit than a square because, if it has a significant aspect ratio, then its moments can be used to find the principal axes of the best-fitting ellipse. Once the orientation θ of the principal axis is known, its length and width are determined in a similar manner to that of a square, except that the values along the orthogonal axes are kept distinct. Assuming $\ell > w$, then the length ℓ is along the principal axis, and therefore the length and width are given by

$$\ell = \left(\frac{144\mu_\theta^3}{\mu_{\theta + \frac{\pi}{2}}} \right)^{\frac{1}{8}} \qquad w = \left(\frac{144\mu_{\theta + \frac{\pi}{2}}^3}{\mu_\theta} \right)^{\frac{1}{8}} \tag{11.77}$$

The pseudocode is provided in Algorithm 11.10.

11.2.7 Fitting a 3D Geometric Model

Fitting a 3D model is a natural extension to the procedures we have just seen for 2D models. Given a set of points $\{(x_i, y_i, z_i)\}_{i=1}^n$, the centroid is given by $\bar{\mathbf{x}} \equiv \frac{1}{n} \sum_{i=1}^n [x_i \quad y_i \quad z_i]^\mathsf{T}$. Then the 3×3 covariance matrix is $\mathbf{C} \equiv \frac{1}{n} \sum_{i=1}^n [x_i - \bar{x} \quad y_i - \bar{y} \quad z_i - \bar{z}]^\mathsf{T} [x_i - \bar{x} \quad y_i - \bar{y} \quad z_i - \bar{z}]$. The principal axes of the data are given by the eigenvectors of $\mathbf{C}$, and the square roots of the eigenvalues are related to the extent of the data along the axes. The procedure for fitting a sphere is a simple extension to that for fitting a circle, and similarly for the procedure for fitting an ellipsoid. A cube is like a square, and a cuboid is like a rectangle. For a cylinder, the axis of rotation is found as the eigenvector associated with the eigenvalue that is most unlike the other two eigenvalues, after which the length is determined by fitting a rectangle to the data projected to a plane containing the axis of rotation, and the diameter is determined by fitting a circle to the data projected to the plane perpendicular to the axis of rotation.

11.3 Fitting Point Cloud Models

Sometimes instead of a parametric model such as an ellipse or rectangle, the shape of the object is modeled as a discrete set of points, often called a **point cloud**. In this case our job is to find the best alignment between the noisy input point set and the model point set. In this section we cover two well-known techniques for solving this problem. We will describe the methods in 3D, but they work equally well in 2D.

11.3.1 Procrustes Analysis

Suppose we have a set of 3D points $\{(x_i, y_i, z_i)\}_{i=1}^n$ and another set of 3D points $\{(x_i', y_i', z_i')\}_{i=1}^n$, and let us assume that the correspondence between the two sets is known, so that point $\mathbf{x}_i = (x_i, y_i, z_i)$ corresponds to the point $\mathbf{x}_i' = (x_i', y_i', z_i')$, denoted by $(x_i, y_i, z_i) \leftrightarrow (x_i', y_i', z_i')$. One of these sets will be considered to be the model that is known beforehand, whereas the other set arises from current sensor data. Let us assume that the point sets are related by an unknown translation, rotation, and scaling. Therefore, we want to find the transformation so that when we stretch, shift, and rotate one point set, the result matches as closely as possible to the other point set. That is, our goal is to find s, $\mathbf{t}$, and $\mathbf{R}$ that minimizes the following sum of squared errors:

$$\min_{s,\mathbf{t},\mathbf{R}} \sum_{i=1}^n \| (s\mathbf{R}\mathbf{x}_i + \mathbf{t}) - \mathbf{x}_i' \|^2 \tag{11.78}$$

Finding this transformation is known as **Procrustes analysis**.[‡]

It can be shown that the solution to the least-squares problem above can be calculated as follows. First define the centroids $\bar{\mathbf{x}}$ and $\bar{\mathbf{x}}'$ as we have done before. Then define the variance $\sigma^2 \equiv \frac{1}{n} \sum_{i=1}^n \| \mathbf{x}_i - \bar{\mathbf{x}} \|^2$ of the first point set, and the covariance $\mathbf{C} \equiv \frac{1}{n} \sum_{i=1}^n (\mathbf{x}_i' - \bar{\mathbf{x}}')(\mathbf{x}_i - \bar{\mathbf{x}})^{\mathsf{T}} = \frac{1}{n}\tilde{\mathbf{B}}^{\mathsf{T}}\tilde{\mathbf{A}}$ between the sets, where

$$\tilde{\mathbf{A}} \equiv \begin{bmatrix} \tilde{x}_1 & \tilde{y}_1 & \tilde{z}_1 \\ \tilde{x}_2 & \tilde{y}_2 & \tilde{z}_2 \\ & \vdots & \\ \tilde{x}_n & \tilde{y}_n & \tilde{z}_n \end{bmatrix} \qquad \tilde{\mathbf{B}} \equiv \begin{bmatrix} \tilde{x}_1' & \tilde{y}_1' & \tilde{z}_1' \\ \tilde{x}_2' & \tilde{y}_2' & \tilde{z}_2' \\ & \vdots & \\ \tilde{x}_n' & \tilde{y}_n' & \tilde{z}_n' \end{bmatrix} \tag{11.79}$$

where $\tilde{x}_i \equiv x_i - \bar{x}_i$, $\tilde{x}_i' \equiv x_i' - \bar{x}_i'$, and so forth. Let ρ and ρ' be the root-mean-squared distance of the points to the corresponding centroid:

$$\rho = \sqrt{\frac{1}{n} \sum_{i=1}^n \| \mathbf{x}_i - \bar{\mathbf{x}} \|^2} \qquad \rho' = \sqrt{\frac{1}{n} \sum_{i=1}^n \| \mathbf{x}_i' - \bar{\mathbf{x}}' \|^2} \tag{11.80}$$

Then the transformation parameters are given by

$$\mathbf{R} = \mathbf{U}\mathbf{S}\mathbf{V}^{\mathsf{T}} \tag{11.81}$$

$$s = \frac{1}{\sigma^2}\mathrm{tr}(\mathbf{\Sigma}\mathbf{S}) = \frac{\rho'}{\rho} \tag{11.82}$$

$$\mathbf{t} = \bar{\mathbf{x}}' - s\mathbf{R}\bar{\mathbf{x}} \tag{11.83}$$

where $\mathbf{U}\mathbf{\Sigma}\mathbf{V}^{\mathsf{T}} = \mathbf{C}$ is the SVD of the covariance matrix, $\mathrm{tr}(\cdot)$ is the trace of a matrix (i.e., the sum of its diagonal elements), and the matrix $\mathbf{S}$ is defined as

$$\mathbf{S} = \begin{bmatrix} 1 & 0 & 0 \\ 0 & 1 & 0 \\ 0 & 0 & 1 \end{bmatrix} \text{ if } \det(\mathbf{C}) \geq 0, \quad \text{or} \quad \mathbf{S} = \begin{bmatrix} 1 & 0 & 0 \\ 0 & 1 & 0 \\ 0 & 0 & -1 \end{bmatrix} \text{ otherwise} \tag{11.84}$$

[†] When a traveling stranger would spend the night at his house, Procrustes, a character in Greek mythology, would either stretch the stranger's limbs or simply cut them off, in order to ensure that the stranger fit the bed.

so that $\mathbf{R}$ is a valid rotation matrix, i.e., $\det(\mathbf{R}) = \det(\mathbf{USV}^\mathsf{T}) = \mathbf{det}(\mathbf{S})\cdot\det(\mathbf{UV}^\mathsf{T}) = 1$. Note that since the singular values are all non-negative, $\det(\mathbf{S}) \geq 0$, so $\mathrm{sgn}(\det(\mathbf{C})) = \mathrm{sgn}(\det(\mathbf{UV}^\mathsf{T}))$. (If instead it is desired to allow for reflection, then simply set $\mathbf{R} = \mathbf{UV}^\mathsf{T}$ without considering the determinant of $\mathbf{C}$, in which case it is possible for $\det(\mathbf{R}) = -1$.) The procedure is provided in Algorithm 11.11 and can be easily adapted to 2D or any other dimensionality.

Through a change of variables, it is not difficult to see that Equation (11.78) is the same as

$$\min_{\mathbf{R}} \sum_{i=1}^{n} \left\| \mathbf{R}\left(\frac{\mathbf{x}_i - \bar{\mathbf{x}}}{\rho}\right) - \left(\frac{\mathbf{x}_i' - \bar{\mathbf{x}}'}{\rho'}\right) \right\|^2 \tag{11.85}$$

where the rotation matrix is computed by defining the shifted, scaled data:

$$\check{\mathbf{x}}_i \equiv \begin{bmatrix} \check{x}_i \\ \check{y}_i \\ \check{z}_i \end{bmatrix} \equiv \frac{(\mathbf{x}_i - \mu)}{\rho} \qquad \check{\mathbf{x}}_i' \equiv \begin{bmatrix} \check{x}_i' \\ \check{y}_i' \\ \check{z}_i' \end{bmatrix} \equiv \frac{(\mathbf{x}_i' - \mu')}{\rho'} \tag{11.86}$$

The matrices $\check{\mathbf{A}}$ and $\check{\mathbf{B}}$ are defined as

$$\check{\mathbf{A}} \equiv \begin{bmatrix} \check{x}_1 & \check{y}_1 & \check{z}_1 \\ \check{x}_2 & \check{y}_2 & \check{z}_2 \\ & \vdots & \\ \check{x}_n & \check{y}_n & \check{z}_n \end{bmatrix} = \frac{1}{\rho}\tilde{\mathbf{A}} \quad \check{B} \equiv \begin{bmatrix} \check{x}_1' & \check{y}_1' & \check{z}_1' \\ \check{x}_2' & \check{y}_2' & \check{z}_2' \\ & \vdots & \\ x_n' & y_n' & z_n' \end{bmatrix} = \frac{1}{\rho'}\tilde{\mathbf{B}} \tag{11.87}$$

from which the 3×3 scaled covariance matrix is computed as $\check{\mathbf{C}} \equiv \frac{1}{n}\check{\mathbf{B}}^\mathsf{T}\check{\mathbf{A}} = \frac{1}{\rho\rho'}\mathbf{C}$. The singular value decomposition of this matrix, $\check{\mathbf{C}} = \mathbf{U\Sigma V}^\mathsf{T}$, yields the desired rotation matrix

---

ALGORITHM 11.11 Procrustes analysis to align two point sets with known correspondence

PROCRUSTESANALYSIS $(\{(x_i, y_i, z_i)\}_{i=1}^{n}, \{(x_i', y_i', z_i')\}_{i=1}^{n})$

Input: two point sets with correspondence $(x_i, y_i, z_i) \leftrightarrow (x_i', y_i', z_i')$
Output: transformation parameters s, $\mathbf{R}$, and $\mathbf{t}$ that minimizes $\sum_{i=1}^{n} \|(s\mathbf{R}\mathbf{x}_i + \mathbf{t}) - \mathbf{x}_i'\|^2$

1   $\bar{\mathbf{x}} \equiv \begin{bmatrix} \bar{x} & \bar{y} & \bar{z} \end{bmatrix}^\mathsf{T} \leftarrow \frac{1}{n}\sum_{i=1}^{n} \begin{bmatrix} x_i & y_i & z_i \end{bmatrix}^\mathsf{T}$ ➤ Compute mean of first point set.

2   $\bar{\mathbf{x}}' \equiv \begin{bmatrix} \bar{x}' & \bar{y}' & \bar{z}' \end{bmatrix}^\mathsf{T} \leftarrow \frac{1}{n}\sum_{i=1}^{n} \begin{bmatrix} x_i' & y_i' & z_i' \end{bmatrix}^\mathsf{T}$ ➤ Compute mean of second point set.

3   $\sigma^2 \leftarrow \frac{1}{n}\sum_{i=1}^{n} \|\mathbf{x}_i - \bar{\mathbf{x}}\|^2$ ➤ Compute variance of first point set.

4   $\mathbf{C} \leftarrow \frac{1}{n}\sum_{i=1}^{n} (\mathbf{x}_i' - \bar{\mathbf{x}}')(\mathbf{x}_i - \bar{\mathbf{x}})^\mathsf{T}$ ➤ Compute covariance matrix.

5   $\mathbf{U}, \mathbf{\Sigma}, \mathbf{V}^\mathsf{T} = \mathrm{SVD}(\mathbf{C})$ ➤ Compute the singular value decomposition.

6   **if** *allow-reflection* **or** $\det(\mathbf{C}) \geq 0$ **then**   ➤ If reflection is allowed or $\mathbf{R}$ is already a rotation, then use the identity matrix;

7        $\mathbf{S} \leftarrow \mathrm{diag}(1, 1, 1)$

8   **else**   otherwise restrict $\mathbf{R}$ to be a rotation matrix by ensuring that $det(\mathbf{R}) = 1$.

9        $\mathbf{S} \leftarrow \mathrm{diag}(1, 1, -1)$

10   $\mathbf{R} \leftarrow \mathbf{USV}^\mathsf{T}$ ➤ Compute rotation matrix.

11   $s \leftarrow \frac{1}{\sigma^2}\mathrm{tr}(\mathbf{\Sigma S})$ ➤ Compute scaling.

12   $\mathbf{t} \leftarrow \bar{\mathbf{x}}' - s\mathbf{R}\bar{\mathbf{x}}$ ➤ Compute translation vector.

13   **return** s, $\mathbf{R}$, and $\mathbf{t}$

$\mathbf{R} = \mathbf{U}\mathbf{S}\mathbf{V}^\mathsf{T}$, where $\mathbf{S}$ is defined as before. This leads to an alternate procedure in which the points are shifted and scaled first, then the rotation matrix is computed, as shown in Algorithm 11.12.

11.3.2 Iterative Closest Point (ICP)

Oftentimes we are given two points sets but do not know the correspondence between them and, in fact, the two sets may contain different numbers of points. To estimate the transformation in such a case, the **iterative closest point (ICP)** algorithm is used. At each iteration of the algorithm, correspondence is established automatically by finding, for every point in one set, the closest point in the other set. With this temporary correspondence, Procrustes analysis is applied to find the optimal transformation parameters between the sets. The transformation is then applied to one point set to bring the two sets closer into alignment, and the process is repeated until the solution converges.

The procedure is presented in Algorithm 11.13. First, the initial transformation parameters are determined, either as the identity transformation (as shown in the code) or using some additional information. Then, within the loop that is repeated until convergence, each point in the first set is transformed according to the parameters, and the closest point in the other set is found. Procrustes is then applied to find the local transformation between the two point sets, and the parameters are then updated. If $s^{(j)}$, $\mathbf{R}^{(j)}$, $\mathbf{t}^{(j)}$ are the parameters from

ALGORITHM 11.12 Procrustes analysis to align two point sets with correspondence. (Alternate version)

PROCRUSTESANALYSIS($\{(x_i, y_i, z_i)\}_{i=1}^n, \{(x_i', y_i', z_i')\}_{i=1}^n$)

Input: two point sets with correspondence $(x_i, y_i, z_i) \leftrightarrow (x_i', y_i', z_i')$

Output: transformation parameters s, $\mathbf{R}$, and $\mathbf{t}$ that minimizes $\sum_{i=1}^n \|(s\mathbf{R}\mathbf{x}_i + \mathbf{t}) - \mathbf{x}_i'\|^2$

1 $\bar{\mathbf{x}} \equiv [\bar{x} \quad \bar{y} \quad \bar{z}]^\mathsf{T} \leftarrow \frac{1}{n}\sum_{i=1}^n [x_i \quad y_i \quad z_i]^\mathsf{T}$ ▷ Compute mean of first point set.

2 $\bar{\mathbf{x}}' \equiv [\bar{x}' \quad \bar{y}' \quad \bar{z}']^\mathsf{T} \leftarrow \frac{1}{n}\sum_{i=1}^n [x_i' \quad y_i' \quad z_i']^\mathsf{T}$ ▷ Compute mean of second point set.

3 $\rho \leftarrow \sqrt{\frac{1}{n}\sum_{i=1}^n \|\mathbf{x}_i - \bar{\mathbf{x}}\|^2}$ ▷ Compute scale of first point set.

4 $\rho' \leftarrow \sqrt{\frac{1}{n}\sum_{i=1}^n \|\mathbf{x}_i' - \bar{\mathbf{x}}'\|^2}$ ▷ Compute scale of second point set.

5 $\check{C} \leftarrow \frac{1}{n}\sum_{i=1}^n \left(\frac{\mathbf{x}_i' - \bar{\mathbf{x}}'}{\rho'}\right) * \left(\frac{\mathbf{x}_i - \bar{\mathbf{x}}}{\rho}\right)^\mathsf{T}$ ▷ Compute covariance matrix of shifted, scaled data.

6 $\mathbf{U}, \check{\Sigma}, \mathbf{V}^\mathsf{T} = \text{SVD}(\check{C})$ ▷ Compute the singular value decomposition.

7 **if** *allow-reflection* **or** $\det(\check{C}) \geq 0$ **then**   ▷ If reflection is allowed or $\mathbf{R}$ is already a rotation, then use the identity matrix;

8      $\mathbf{S} \leftarrow \text{diag}(1,1,1)$

9 **else**   otherwise restrict $\mathbf{R}$ to be a rotation matrix by ensuring that $\det(\mathbf{R}) = 1$.

10      $\mathbf{S} \leftarrow \text{diag}(1, 1, -1)$

11 $\mathbf{R} \leftarrow \mathbf{U}\mathbf{S}\mathbf{V}^\mathsf{T}$ ▷ Compute rotation matrix.

12 $s \leftarrow \dfrac{1}{\rho\rho'}$ ▷ Compute scaling.

13 $\mathbf{t} \leftarrow \bar{\mathbf{x}}' - s\mathbf{R}\bar{\mathbf{x}}$ ▷ Compute translation vector.

14 **return** s, $\mathbf{R}$, and $\mathbf{t}$

ALGORITHM 11.13 Iterative closest point (ICP) to align two point sets without correspondence

ITERATIVECLOSESTPOINT($\{\mathbf{x}_i\}_{i=1}^n, \{\mathbf{x}_i'\}_{i=1}^m$)

Input: two point sets without correspondence
Output: transformation parameters s, $\mathbf{R}$, and $\mathbf{t}$

1 $s \leftarrow 1, \mathbf{t} \leftarrow \mathbf{0}, \mathbf{R} \leftarrow \mathbf{I}$ ➤ Initialize transformation.
2 **repeat** until convergence
3   **for** $i \leftarrow 1$ **to** n **do**   ➤ For each point (assuming $n \leq m$), apply transformation and find closest point in other set.
4          $\mathbf{x}_i'' \leftarrow s\,\mathbf{R}\,\mathbf{x}_i + \mathbf{t}$
5          $c[i] \leftarrow \arg\min_j \|\mathbf{x}_i'' - \mathbf{x}_j'\|$
6      $s', \mathbf{R}', \mathbf{t}' \leftarrow$ PROCRUSTES($\{\mathbf{x}_i''\}_{i=1}^n, \{\mathbf{x}_{c[i]}'\}_{i=1}^n$)   ➤ Compute local transformation
7      $s \leftarrow s's$ and update parameters by composing them
8      $\mathbf{R} \leftarrow \mathbf{R}'\mathbf{R}$ with the local transformation parameters.
9      $\mathbf{t} \leftarrow s'\mathbf{R}'\mathbf{t} + \mathbf{t}'$
10 **return** s, $\mathbf{R}$, and $\mathbf{t}$

one iteration, and if $s^{(j-1)}$, $\mathbf{R}^{(j-1)}$, $\mathbf{t}^{(j-1)}$ are the parameters from the previous iteration, then the two sets of parameters are composed as

$$\mathbf{x}^{(j+1)} = s^{(j)}\mathbf{R}^{(j)}\mathbf{x}^{(j)} + \mathbf{t}^{(j)} \tag{11.88}$$

$$= s^{(j)}\mathbf{R}^{(j)}\left(s^{(j-1)}\mathbf{R}^{(j-1)}\mathbf{x}^{(j-1)} + \mathbf{t}^{(j-1)}\right) + \mathbf{t}^{(j)} \tag{11.89}$$

$$= \underbrace{s^{(j)}s^{(j-1)}}_{s}\underbrace{\mathbf{R}^{(j)}\mathbf{R}^{(j-1)}}_{\mathbf{R}}\mathbf{x}^{(j-1)} + \underbrace{s^{(j)}\mathbf{R}^{(j)}\mathbf{t}^{(j-1)} + \mathbf{t}^{(j)}}_{\mathbf{t}} \tag{11.90}$$

which is shown in Lines 7–9.

Although we have described the algorithm in the context of aligning point sets, ICP is a general-purpose algorithm that is applicable to aligning any 2D or 3D curves or surfaces as well. The algorithm is guaranteed to converge to the nearest local minimum of a mean-squared distance metric. The pseudocode just presented uses the **point-to-point error metric** because it computes the distance between points. A variation of the algorithm that is popular for aligning point sets arising from cameras is to use the **point-to-plane error metric**, which computes the distance between a point and the other plane. While point-to-point can be minimized in closed form using Procrustes, minimizing point-to-plane requires a nonlinear minimization approach, such as Levenberg-Marquardt. However, although each iteration of point-to-plane is therefore more expensive, it tends to converge faster in practice.

11.4 Robustness to Noise

The least squares approaches described so far work fine when the data are all corrupted by the same amount of noise, and that noise is well-behaved. More specifically, a least-squares algorithm produces the maximum likelihood estimate (MLE) when the data are corrupted by independent, identically distributed (i.i.d.) Gaussian noise. With i.i.d. noise, each data point deviates from its true value by an amount that is drawn from a Gaussian distribution with constant variance. In practice, however, it is often the case that some points are corrupted more than others. First we will examine the case in which, although most of the data are good, a small percentage are outliers; then we will consider what happens when most of the data are outliers, and only a small percentage are good. Before we consider these cases, however, we need to establish some basic concepts regarding estimation.

11.4.1 Maximum Likelihood Estimators

Suppose we want to estimate a quantity given a number of samples of the quantity. For simplicity, we shall consider the case in 1D, but these findings will easily generalize to higher dimensions. Let x be the quantity to estimate, and let $\{x_i\}_{i=1}^n$ be n samples that have been obtained somehow. From Bayes' rule,[‡] the conditional probability is

$$p(x|x_1, \ldots, x_n) = \frac{p(x_1, \ldots, x_n|x)p(x)}{p(x_1, \ldots, x_n)} \tag{11.91}$$

The goal is to maximize the left-hand side, which is called the **posterior** (or, *a posteriori* probability), which is the probability of the quantity being a certain value given all the data. That is, our goal is to find the **maximum a posteriori (MAP) estimate**:

$$\hat{x} = \arg\max_x p(x|x_1, \ldots, x_n) = \arg\max_x p(x_1, \ldots, x_n|x)p(x) \tag{11.92}$$

where we are free to ignore the denominator, $p(x_1, \ldots, x_n)$ because it is a fixed scaling constant that does not affect the result. The right-most factor $p(x)$ is called the **prior** (or, *a priori* probability). It is the probability that the quantity is a certain value, without considering the data at all. Oftentimes the prior is uniform, $p(x) = 1$, in which case the MAP estimate leads to the **maximum likelihood estimate (MLE)**:

$$\hat{x} = \arg\max_x p(x_1, \ldots, x_n|x) \tag{11.93}$$

If the samples are independent, then this is equivalent to the product of the individual conditional probabilities:

$$\hat{x} = \arg\max_x \prod_{i=1}^n p(x_i|x) \tag{11.94}$$

If we further assume that the noise is Gaussian, i.e., $x_i = x + \xi$, where $\xi \sim \mathcal{N}(0,\sigma^2)$, then we have

$$\hat{x} = \arg\max_x \prod_{i=1}^n e^{-\frac{(x-x_i)^2}{2\sigma^2}} \tag{11.95}$$

Since the logarithm is monotonically increasing, this is the same as maximizing the log, or equivalently, minimizing the **negative log likelihood**:

$$\hat{x} = \arg\min_x \sum_{i=1}^n \frac{(x-x_i)^2}{2\sigma^2} \tag{11.96}$$

If all the variances are the same, then this is the same as

$$\hat{x} = \arg\min_x \sum_{i=1}^n (x-x_i)^2 \tag{11.97}$$

[†] Bayes' rule is described in more detail in Section 12.2.1 (p. 572).

We now have arrived at a very interesting result. Equation (11.97) says that minimizing the **sum of squared differences (SSD)** (or, equivalently, the mean squared error, MSE) is the MLE of a variable whose samples are corrupted by i.i.d. additive Gaussian noise with constant variance. In 1D, this estimate is just the *mean*

$$x = \frac{1}{n} \sum_{i=1}^{n} x_i \tag{11.98}$$

which can be seen by differentiating Equation (11.97), setting the result to zero:

$$\sum_{i=1}^{n} (x - x_i) = 0 \tag{11.99}$$

and solving for *x*. In other words, all of the least squares methods that we have considered in this chapter have assumed that the data were corrupted by i.i.d. additive Gaussian noise. In some applications this is a valid assumption, but oftentimes when working with image data the situation is more complicated than that.

By a similar derivation, we can easily show that if the noise follows the double exponential, then the MLE estimator is given by

$$x = \arg \max_{x} \prod_{i=1}^{n} e^{-\frac{|x - x_i|}{2\sigma^2}} \tag{11.100}$$

$$= \arg \min_{x} \sum_{i=1}^{n} |x - x_i| \tag{11.101}$$

Thus, minimizing the **sum of absolute differences (SAD)** is the MLE of a variable whose samples are corrupted by independent, additive double-exponential noise. In 1D, this estimate is just the *median*:

$$x = \text{med}\{x_1, \ldots, x_n\} \tag{11.102}$$

11.4.2 Generalized Least Squares

We have just seen that if we know that the noise is Gaussian, then the best estimator (that is, the one that maximizes the likelihood) is the one that minimizes the sum of squared residuals. Now let us turn the problem around. Suppose we do not know what type of noise is added to the system, but we want to restrict ourselves to *linear estimators*. From now on, to emphasize that these techniques are applicable to any dimensionality, we shall use $\mathbf{x}_i$ to refer to the ith data point, and therefore to avoid confusion we shall use $\boldsymbol{\theta}$ to refer to the unknown parameters of the model. As it turns out, the **best linear unbiased estimator (BLUE)** for $\mathbf{A}\boldsymbol{\theta} = \mathbf{b}$ is the solution to

$$\mathbf{C}^{-1}\mathbf{A}\boldsymbol{\theta} = \mathbf{C}^{-1}\mathbf{b} \tag{11.103}$$

or $\boldsymbol{\theta} = (\mathbf{A}^{\mathsf{T}}\mathbf{C}^{-1}\mathbf{A})^{-1}\mathbf{A}^{\mathsf{T}}\mathbf{C}^{-1}\mathbf{b}$, where $\mathbf{C}$ is the covariance matrix of the data. In other words, if we do not know the specific noise distribution, but we do have access to the covariance matrix, then Equation (11.103) can be used to find the BLUE of the unknown vector $\boldsymbol{\theta}$. This technique is known as **generalized least squares**.

11.4.3 Iteratively Reweighted Least Squares (IRLS)

A special case of generalized least squares arises when $\mathbf{C}$ is the identity matrix, in which case Equation (11.103) reduces to ordinary least squares: $\boldsymbol{\theta} = (\mathbf{A}^{\mathsf{T}}\mathbf{A})^{-1}\mathbf{A}^{\mathsf{T}}\mathbf{b}$. In fact, generalized least squares provides us with a new way of looking at least squares. Not only is the least squares result the MLE under the assumption of i.i.d. Gaussian noise, but according to the **Gauss-Markov theorem**, it is also the BLUE under the assumption of uncorrelated (not necessarily independent) zero-mean (not necessarily Gaussian) noise with constant variance.

Another special case of generalized least squares arises when $\mathbf{C}$ is diagonal (and therefore the noise is uncorrelated) but not necessarily the identity matrix. When the noise for all the samples has equal variance, we say that the noise is *homoskedastic*. Obviously, ordinary least squares assumes homoskedastic noise. On the other hand, if the variances of the samples are not equal, then the noise is *heteroskedastic*. When the noise is heteroskedastic, then some values along the diagonal will be larger than others. Large values indicate large variances and hence unreliable data samples, whereas small values indicate small variances and hence reliable data samples. To give reliable samples more weight, a standard approach is to use a weight matrix $\mathbf{W} \equiv \mathbf{C}^{-1}$, leading to

$$\mathbf{WA}\boldsymbol{\theta} = \mathbf{Wb} \tag{11.104}$$

where

$$\mathbf{W} = \begin{bmatrix} \frac{1}{\sigma_1^2} & 0 & 0 \\ 0 & \ddots & 0 \\ 0 & 0 & \frac{1}{\sigma_n^2} \end{bmatrix} \tag{11.105}$$

where $\sigma_1, \ldots, \sigma_n$ are the variances. This problem is known as **weighted least squares**, whose solution is $\boldsymbol{\theta} = (\mathbf{A}^\mathsf{T}\mathbf{WA})^{-1}\mathbf{A}^\mathsf{T}\mathbf{Wb}$. The pseudocode is provided in Algorithm 11.14.

In practice, we rarely know the variances of the different data points beforehand. In fact, it is often the case that much of the data come from legitimate sources with low variance, while other data are outliers with huge variance. A single outlier can easily corrupt the minimization of least squares. As a result, we must use weighted least squares, but we must estimate the weights on the fly. This leads to the method of **iteratively reweighted least squares (IRLS)**. In IRLS, all the weights are initialized to 1, unless there is some prior information. Then weighted least squares is performed to determine the parameters of the model being fit. Once the model is fit, the distance from each data point to the model is computed. In the next iteration, points with small distances receive higher weight than points with larger distances. The process is repeated until convergence. If the noise is not so severe as to ruin the initial estimate, then this approach leads to quite robust results. The pseudocode is provided in Algorithm 11.15.

ALGORITHM 11.14 Weighted least squares

WEIGHTEDLEASTSQUARES $\left(\{(\mathbf{x}_i, w_i)\}_{i=1}^n\right)$

Input: set of n points $\{\mathbf{x}_i\}_{i=1}^n$ in the plane, with associated weights w_i, $i = 1, \ldots n$
Output: parameters $\boldsymbol{\theta}$ of the model that best fits the data

1 Construct diagonal matrix $\mathbf{W} = diag(w_1, \ldots, w_n)$
2 Construct matrix $\mathbf{A}$ from the points
3 Construct vector $\mathbf{b}$ from the points
4 Solve $\boldsymbol{\theta} = (\mathbf{A}^\mathsf{T}\mathbf{WA})^{-1}\mathbf{A}^\mathsf{T}\mathbf{Wb}$
5 **return** $\boldsymbol{\theta}$

ALGORITHM 11.15 Iteratively reweighted least squares (IRLS)

IterativelyReweightedLeastSquares($\{\mathbf{x}_i\}_{i=1}^n$)

Input: set of n points $\{\mathbf{x}_i\}_{i=1}^n$
Output: parameters $\boldsymbol{\theta}$ of the model that best fits the data

1   Initialize weights to uniform, $w_i \leftarrow 1, i = 1, \ldots, n$
2   **while** not convergence **do**
3   Construct diagonal matrix $\mathbf{W} = diag(w_1, \ldots, w_n)$
4         $\boldsymbol{\theta} \leftarrow$ WeightedLeastSquares($\{(\mathbf{x}_i, w_i)\}_{i=1}^n$)
5   Compute residuals $r_i, i = 1, \ldots, n$
6   Recompute weights using residuals $w_i \leftarrow e^{-r_i^2}, i = 1, \ldots, n$
7   **return** $\boldsymbol{\theta}$

11.4.4 M-Estimators

We have just seen that IRLS is able to handle the presence of outliers in the data, which would otherwise corrupt standard least squares. Let r_i be the residual of the ith data point. Least squares minimizes $\sum_{i=1}^n r_i^2$, whereas weighted least squares minimizes $\sum_{i=1}^n w_i r_i^2$, where $w_i = \frac{1}{\sigma_i^2}$ is the weight of the ith data point. One question remains: namely, how to select the weights in a principled way when the variances are unknown. One approach, known as **truncation** (or trimming), involves discarding outliers completely. That is, once a data sample has been determined to be too far from the fitting function, it is not considered at all in the minimization process. An alternate approach, known as **Winsorizing** (or clamping[†]), limits the influence of the outliers. As we shall see in a moment, the latter produces superior results in practice because, while we want to limit the effects of outliers, ignoring them altogether can cause the procedure to fall into a local minimum.

One of the more popular ways to Winsorize data for determining IRLS weights is to use an **M-estimator**.[‡] An M-estimator reduces the effect of outliers by minimizing some function of the residuals, $\sum_{i=1}^n \rho(r_i)$, where ρ is a symmetric function with a single global minimum at zero and no local minima. To reduce the effect of outliers, the function is

Figure 11.7 A single outlier can severely corrupt the result of least squares, in which case it is necessary to reweight the data samples to reduce the influence of outliers. Shown is the result from total least squares (left) and iteratively reweighted least squares (right).

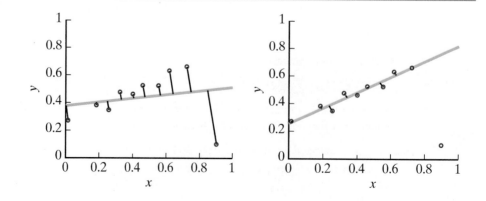

[†] The term "clamping" in computing is known as "clipping" in signal processing.
[‡] The "M" is from "maximum," as in "maximum likelihood."

chosen to always be less than or equal to the function used by least squares: $\rho(x) \le x^2$ for all x.

Let $\boldsymbol{\theta} \equiv [\theta_1 \quad \cdots \quad \theta_m]^T$ be the vector of parameters to be estimated. Then the minimum of $\sum_{i=1}^{n} \rho(r_i)$ will be achieved when

$$\sum_{i=1}^{n} \frac{\partial \rho(r_i)}{\partial \theta_j} = \sum_{i=1}^{n} \underbrace{\frac{\partial \rho(r_i)}{\partial r_i}}_{\psi(r_i)} \frac{\partial r_i}{\partial \theta_j} = 0, \qquad j = 1, \ldots, m \tag{11.106}$$

where $\psi(z) \equiv d\rho(z)/dz$ is called the **influence function**. If we define the **weight function** as

$$w(z) \equiv \frac{\psi(z)}{z} \tag{11.107}$$

then substituting into Equation (11.106) yields

$$\sum_{i=1}^{n} w(r_i) r_i \frac{\partial r_i}{\partial \theta_j} = 0. \qquad j = 1, \ldots, m \tag{11.108}$$

Note that this result is exactly what we obtain when we differentiate

$$\sum_{i=1}^{n} w_i r_i^2 = 0 \tag{11.109}$$

with respect to the parameter vector, assuming that $w_i = w(r_i)$ is constant. In other words, the M-estimator of θ that minimizes $\sum_{i=1}^{n} \rho(r_i)$ is identical to the result of applying IRLS to Equation (11.109), where $w(r_i)$ is evaluated on the residual from the previous iteration. This leads to a new robust technique for model fitting, namely to apply IRLS using an M-estimator function at each iteration to reduce the effects of outliers.

For this approach to work, it is important to get the scale correct. If the scale is too big, then nothing will be considered an outlier. If the scale is too small, then good data will be rejected. A common way to estimate the scale is to use the median of the residuals. The **interquartile range** of the Gaussian distribution is the middle 50%, and it lies between $\pm 0.6745\sigma$. Therefore, the number 0.6745σ is one-half of the interquartile range, which is the same as the median absolute deviation (MAD). If we assume that half of the data are inliers, this leads to an estimate of the scale s as

$$s = \frac{\text{med}_i |r_i|}{0.6745} = 1.48258 \, \text{med}_i |r_i| \tag{11.110}$$

The ith weight is then set to

$$w_i = \frac{\psi\left(\frac{r_i}{s}\right)}{\left(\frac{r_i}{s}\right)} = \frac{s}{r_i} \psi\left(\frac{r_i}{s}\right) \tag{11.111}$$

A number of different M-estimators have been proposed over the years. One of the most popular is the **Geman-McClure error function**:

$$\rho(z) \equiv \frac{1}{2} \cdot \frac{z^2}{z^2 + \tau^2} \tag{11.112}$$

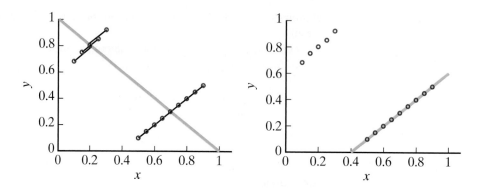

Figure 11.8 Iterative least squares while discarding outliers (left) may not always converge to the true solution because it ignores outliers completely. An M-estimator (right), on the other hand, is able to settle onto the majority consensus, since outliers incur a (roughly) constant penalty.

where τ is a fixed parameter and the $\frac{1}{2}$ scaling factor is optional but helps to simplify the derivative. When $z \ll \tau$, then $\rho(z)$ is shaped like z^2 and the data point is treated as it would be in a standard least squares estimator (i.e., as an inlier). When $z \gg \tau$, then $\rho(z)$ flattens so that the penalty becomes essentially constant (i.e., the data value is treated as an outlier). One advantage of the Geman-McClure error function, besides its simplicity, is that it is twice differentiable.

Other well-known M-estimator functions are shown in Table 11.1. These include the **Huber loss function** and **Tukey's biweight function**, which are provided primarily for historic reasons, although sometimes they are encountered. Huber is also known as **minimax** because $\psi(z) = \min(\max(z, -\tau), \tau)$. Note that the **Cauchy-Lorentz estimator** is sometimes called Cauchy, or Lorentz, or Lorentzian; it receives its name from the fact that its weight function is a scaled version of the Cauchy-Lorentz distribution.

Plots of these M-estimators and their influence functions are shown in Figure 11.9. An estimator is known as **redescending** if the influence function approaches zero as the error increases—that is, $\psi(z) \to 0$ as $z \to \infty$. Note that Geman-McClure, Tukey, the truncated quadratic, and Cauchy-Lorentz are all redescending. Redescending estimators have a saturating property because the influence of outliers is essentially constant no matter how bad

name	$\rho(z)$	$\psi(z)$	$w(z)$	
L^2	$\frac{1}{2}z^2$	z	1	
L^1	$\lvert z \rvert$	$\operatorname{sgn}(z)$	$\frac{1}{\lvert z \rvert}$	
Geman-McClure	$\frac{1}{2}\dfrac{z^2}{z^2+\tau^2}$	$\dfrac{z\tau^2}{(z^2+\tau^2)^2}$	$\dfrac{\tau^2}{(z^2+\tau^2)^2}$	
Huber	$\begin{cases}\frac{1}{2}z^2 \\ \tau(\lvert z\rvert - \frac{\tau}{2})\end{cases}$	$\begin{cases}z \\ \tau\operatorname{sgn}(z)\end{cases}$	$\begin{cases}1 \\ \frac{\tau}{\lvert z\rvert}\end{cases}$	if $\lvert z\rvert \le \tau$ otherwise
Tukey	$\begin{cases}\frac{\tau^2}{6}\left(1-(1-\frac{z^2}{\tau^2})^3\right) \\ \frac{\tau^2}{6}\end{cases}$	$\begin{cases}z(1-\frac{z^2}{\tau^2})^2 \\ 0\end{cases}$	$\begin{cases}(1-\frac{z^2}{\tau^2})^2 \\ 0\end{cases}$	if $\lvert z\rvert \le \tau$ otherwise
Truncated quadratic	$\begin{cases}\frac{1}{2}z^2 \\ \frac{1}{2}\tau^2\end{cases}$	$\begin{cases}z \\ 0\end{cases}$	$\begin{cases}1 \\ 0\end{cases}$	if $\lvert z\rvert \le \tau$ otherwise
Cauchy-Lorentz	$\log\left(1+\frac{1}{2}\frac{z^2}{\tau^2}\right)$	$\dfrac{z}{\tau^2+\frac{1}{2}z^2}$	$\dfrac{1}{\tau^2+\frac{1}{2}z^2}$	

TABLE 11.1 Some well-known M-estimators. The influence function $\psi(z)$ is obtained by differentiating $\rho(z)$, while the weight function $w(z)$ is the influence function divided by z.

Figure 11.9 Top: Popular
M-estimators and influence
functions, including L^2, L^1,
and Geman-McClure with
different parameter values
$\tau \in \{0.25, 0.5, 1, 2\}$.
Bottom: Other M-estimators,
with L^2 shown as a dashed
black line for comparison.
The parameter values
are $\tau = 1$ (Huber and
truncated quadratic), $\tau = 2$
(Tukey), and $\tau = 0.5$
(Cauchy-Lorentz). Note that
Geman-McClure, Tukey,
the truncated quadratic,
and Cauchy-Lorentz are
all redescending, because
$\psi(z) \to 0$ as $z \to \infty$.

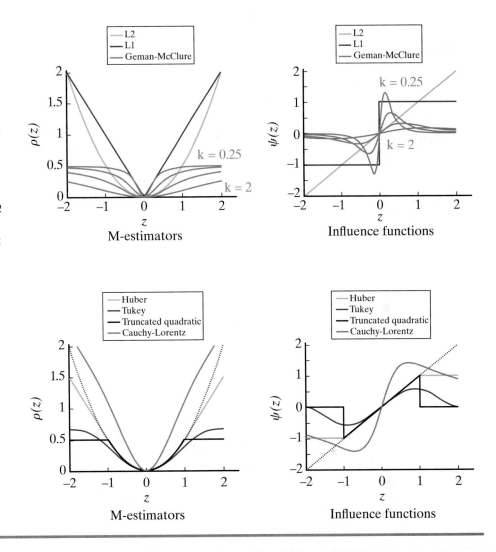

the outlier is. The influence function captures how much influence data points have on the solution. With the L^2-norm,[†] the influence of outliers increases linearly and without bound, which is why L^2 is so sensitive to outliers. With the L^1-norm, $\psi(r_i) = \mathrm{sgn}(r_i)$, so that (in the 1D case) all data to the right of the origin have positive weight, while all data to the left of the origin have negative weight. In such a case the data essentially cast votes for the direction in which to move the estimate, and each data point gets to cast exactly one vote, no matter how far away it is. This explains why L^1 leads (in 1D) to the median. Note that the term "influence" here is slightly misleading because a data point with zero influence still affects the solution; however, its effect is not dependent upon the particular value of the data point. For a data point to have no influence whatsoever, we would have to discard the data point entirely, as in truncation mentioned above.

Two ways to characterize an estimator are its breakdown point and its efficiency. The **breakdown point** is the fraction of data that can be arbitrarily bad without causing the solution to be arbitrarily bad. Least squares has a breakdown point of 0%, because even a single bad outlier can disrupt the solution in an unbounded way. The median has a

---

[†] Section 4.3.1 (p. 164).

ALGORITHM 11.16 Iteratively reweighted least squares (IRLS) with an M-estimator

IRLS-WithMEstimator $\left(\{\mathbf{x}_i\}_{i=1}^n\right)$

Input: set of n points $\{\mathbf{x}_i\}_{i=1}^n$
Output: parameters $\boldsymbol{\theta}$ of the model that best fits the data

1   Initialize weights to uniform, $w_i \leftarrow 1, i = 1, \ldots, n$
2   **while** not convergence **do**
3   Construct diagonal matrix $\mathbf{W} = diag(w_1, \ldots, w_n)$
4       $\boldsymbol{\theta} \leftarrow$ WeightedLeastSquares$\left(\{(\mathbf{x}_i, w_i)\}_{i=1}^n\right)$
5   Compute residuals $r_i, i = 1, \ldots, n$
6   Estimate scale $s \leftarrow 1.5 \operatorname{med}_i |r_i|$
7   Recompute weights $w_i \leftarrow \frac{s}{r_i} \psi\left(\frac{r_i}{s}\right), i = 1, \ldots, n$
8   **return** $\boldsymbol{\theta}$

breakdown point of 50%. The **efficiency** is the ratio of the theoretically lowest variance achievable to the actual variance achieved. It is not uncommon for robust estimators to achieve an efficiency $> 90\%$.

The pseudocode for IRLS with an M-estimator is shown in Algorithm 11.16. It is important to keep in mind that, although we are presenting M-estimators in the context of IRLS, IRLS is just one of several ways to use M-estimators. Another popular way is to apply Equation (11.108) directly using nonlinear minimization such as gradient descent.

11.4.5 Random Sample Consensus (RANSAC)

Traditional techniques, including the robust estimators of the previous section, rely on the smoothing assumption, which says that the estimate should improve as more data are used, because the errors average out. This assumption is valid for dealing with *measurement errors*, which are usually Gaussian, or at least well-behaved. However, in many computer vision problems, there is another type of error, called *classification errors*, that give rise to outliers which, as we have already seen, have the potential to dramatically corrupt the result of an estimator. Although the robust estimators just considered can handle some outliers, they have a breakdown point of no more than 50%. As a result, if more than half of the data are corrupted, they are not expected to find a good solution.

Recall from the previous chapter that segmentation algorithms can be divided into two general categories depending upon whether they split or merge the data. In the same way, all the estimators presented so far can be considered splitters, because they begin with all the data, then reclassify data into inliers and outliers. An alternate approach is for the estimator to begin with the smallest amount of data, then grow the result by merging. We now discuss two estimators that follow this paradigm of merging to achieve breakdown points much greater than 50% because they are designed to work with data overwhelmed by outliers. The first of these is known as **Random sample consensus (RANSAC)**. RANSAC is an iterative, nondeterministic algorithm that follows a hypothesize-and-test paradigm. First, a minimal sample set is randomly selected from the original set to form a hypothesized set of inliers. A minimal sample set contains the minimum number of points necessary to fit the particular model under consideration — two points for a line, three points for a plane, and so forth. Then, under the assumption that this minimal set contains only inliers, a model is fit. Finally, the model is tested for its quality. These three steps are repeated until a good model is found, or the maximum number of iterations has been reached. The pseudocode

ALGORITHM 11.17 Random sample consensus (RANSAC)

$\textsc{Ransac}(\{\mathbf{x}_i\}_{i=1}^n)$

Input: set of n points $\mathcal{P} \equiv \{\mathbf{x}_i\}_{i=1}^n$
Output: parameters $\boldsymbol{\theta}$ of the model that best fits the data

1 **repeat**
2   Randomly select minimal subset $\mathcal{S} \subset \mathcal{P}$ of points
3   Fit model $\boldsymbol{\theta}'$ to subset $\mathcal{S}$
4   Find all points in $\mathcal{P}$ that fit $\boldsymbol{\theta}'$ with residue less than threshold (inliers)
5   Compute quality of model $\boldsymbol{\theta}'$
6   If this is the best model seen so far, then store $\boldsymbol{\theta} \leftarrow \boldsymbol{\theta}'$
7 **until** a good model has been found, or the maximum number of iterations has been reached
8 **return** $\boldsymbol{\theta}$

for this general procedure of finding a single model is shown in Algorithm 11.17. It is easy to see how to repeatedly call the procedure to fit multiple models to data, excluding data that have already been fit in previous iterations.

In the original version of RANSAC, three parameters are needed by the algorithm. First, a fixed threshold is used to determine which points are inliers, and the quality of a model is simply the number of these inliers. This threshold can often be determined by knowing something about the system generating the data, or by trial and error. The second parameter, the minimum number of inliers needed to declare a model to be good, is usually determined by knowing something about the specific problem being solved.

The third parameter is the maximum number of iterations to try. To derive a reasonable value for this parameter, let $0 \le \epsilon \le 1$ be the estimated fraction of points that are outliers, and let m be the size of a minimal sample. (For fitting a line, $m = 2$.) To find a set with only inliers with probability p, we must process at least k sets. To compute k, consider the following: The probability of a point being an inlier is given by $1 - \epsilon$, so $(1 - \epsilon)^m$ is the probability of a random set of m points all being inliers, and $1 - (1 - \epsilon)^m$ is the probability of a random set of m points containing at least one outlier. Therefore, $(1 - (1 - \epsilon)^m)^k$ is the probability of k such random sets all containing at least one outlier, and $1 - (1 - (1 - \epsilon)^m)^k$ is the probability that at least one of the k random sets is outlier-free. Setting this to p yields $1 - (1 - (1 - \epsilon)^m)^k = p$. Rearranging yields

$$k = \frac{\log(1 - p)}{\log(1 - (1 - \epsilon)^m)} \approx 2(1 - \epsilon)^{-m} \tag{11.113}$$

where the approximation (corresponding to $p = 95\%$) is left as an exercise for the reader. Table 11.2 shows some approximate numbers for different values of ϵ and m (with $p \approx 95\%$).

There are several variations of the original RANSAC algorithm. One approach, called **Locally optimized RANSAC (Lo-RANSAC)**, fits the model parameters again using all the inliers, once the inliers have been found. Another variation is **M-estimator SAC (MSAC)**, in which an M-estimator is used to evaluate model quality in order to reduce the dependency upon the threshold. In other words, the original RANSAC implementation measures quality as $\sum_{i=1}^n \rho(r_i)$, where

$$\rho(z) = \begin{cases} 0 & \text{if } z \le \tau & \text{(inlier)} \\ 1 & \text{otherwise} & \text{(outlier)} \end{cases} \tag{11.114}$$

ϵ \ m	2	3	5	8
0.2	3	4	8	16
0.5	10	20	100	800
0.8	70	400	9000	10^6

TABLE 11.2 The approximate number k of RANSAC iterations needed for different values of m (the number of parameters needed to estimate the model) and ϵ (the proportion of outliers).

where τ is the threshold for determining whether a data point is an inlier. MSAC replaces this error function with one of the M-estimators we considered above. Alternatively, the dependency upon the threshold can be reduced by using the maximum likelihood estimate, as in **Maximum-likelihood estimation SAC (MLESAC)**. Finally, since the most expensive part of the computation is to test all the data points with the current model, some variations propose to test only a subset of the data. To simplify the confusing panoply of acronyms, we suggest using the term RANSAC to refer to any algorithm that fits the pattern of Algorithm 11.17, so that Lo-RANSAC would be known as RANSAC with local optimization, MSAC would be RANSAC with an M-estimator, and so forth.

11.4.6 Hough Transform

An alternative to RANSAC is the **Hough transform**,[†] which involves not only a transform but also an algorithm for finding parameterized shapes. The core principle behind the Hough transform is to allow the data to vote for discrete possibilities within the space of parameters; the choice with the maximum number of votes then yields the parameters of the shape that best explain the data. Like RANSAC, this procedure can be repeated multiple times to detect multiple instances of the shape.

To see how this works, let us consider the simplest case of detecting lines in an image. For the moment, let us parameterize a line by its slope m and y-intercept k, as we did earlier. Then the equation for a line is given by

$$y = \underbrace{m}_{\text{slope}} x + \underbrace{k}_{y-\text{intercept}} \tag{11.115}$$

The traditional way of interpreting this equation is that m and k are constants, so that the value of y can be computed for any given value of x. Alternatively, however, if we view x and y as constants arising from some data point, then the equation can be used to compute k for a given value of m (or vice versa). This is best seen by rearranging the equation:

$$k = \underbrace{-x}_{\text{slope}} m + \underbrace{y}_{y-\text{intercept}} \tag{11.116}$$

where $-x$ now plays the role of slope, and y plays the role of y-intercept. Suppose we are given a set of points $\{(x_i, y_i)\}_{i=1}^n$. Each point (x_i, y_i) defines a line in the **parameter space** defined by Equation (11.116) by substituting x_i for x and y_i for y. The line joining two points is therefore given by the intersection of their two respective lines in parameter space. If there is no noise in the system, then all the points lying on the line will also give rise to lines in parameter space that intersect at the same point. When noise is present, then the lines will not intersect exactly but rather will pass near one another.

[†] Pronounced HUFF.

To detect the parameters of the line given noisy data, the parameter space is discretized into a rectangular lattice of cells, and each data point votes for all the cells through which the corresponding line passes. The data structure that holds the votes is called the **accumulator**, which is just a 2D array of values. After all the data have voted, the entire accumulator is searched for the cell with the largest value, and the parameters corresponding to that cell yield the desired line.

As we have already seen, the slope-intercept form for a line has serious drawbacks. A better parameterization for a line is the Hessian normal form:

$$x \cos \theta + y \sin \theta + \rho = 0 \tag{11.117}$$

where the normal to the line is $[\cos \theta \quad \sin \theta]^{\mathsf{T}}$, the slope of the line is given by $-\cot \theta$, and $|\rho|$ is the perpendicular distance from the line to the origin. Each point (x,y) yields a sinusoidal curve in the parameter space described by θ and ρ. Two points yield two curves, and the line determined by joining the two points is given by the intersection of the two curves, as shown in Figure 11.10 for the case of multiple points. The procedure is then modified as follows. For each point (x,y), consider each possible angle $0 \leq \theta < \pi$, quantized into reasonably sized bins (e.g., 10 degrees each). For each angle, ρ is computed to yield a point (θ, ρ) in parameter space, and the corresponding accumulator cell is incremented. The peak is found in the same manner as before. Note that ρ can be negative, and there is no need to consider $\theta \geq \pi$, since (θ, ρ) and $(\theta + \pi, -\rho)$ correspond to the same line.

One difficulty of the Hough transform is selecting the proper discretization size for the accumulator. If the discretization is too coarse, then the maximum cell in the accumulator will bear little resemblance to the true solution, and multiple shapes will be accidentally merged. On the other hand, if the discretization is too fine, then the accumulator cells will be too small, and any noise in the data will cause the information to spread to neighboring cells, making it less likely that there will be a single peak at the desired solution. One approach is to use a coarse discretization, but to let each data point vote $2d$ times, where d is the number of dimensions. Along each dimension, the data point votes for the two neighboring bins using linear interpolation to determine the relative weighting for the two votes. Once all the votes have been counted, a geometric verification procedure can be run on each bin with sufficient votes to further refine the estimate.

In a real implementation, there are a few details to keep in mind. First, better results will be obtained if all the (x,y) coordinates are first translated to a coordinate system whose origin is at the center of the image. Thus,

$$\underbrace{\left(x - \frac{width}{2}\right)}_{x'} \cos \theta + \underbrace{\left(y - \frac{height}{2}\right)}_{y'} \sin \theta + \underbrace{\rho + \frac{width}{2} \cos \theta + \frac{height}{2} \sin \theta}_{\rho'} = 0 \tag{11.118}$$

Figure 11.10 The Hough transform represents a model using votes in parameter space. Shown are 5 points in the plane (left) and the resulting curves in parameter space (right). The intersection of these curves yields the parameters of the line: $\theta = \frac{3\pi}{4}$ and $\rho = \sqrt{2}$.

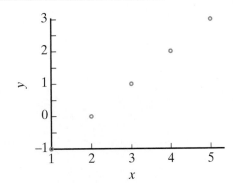

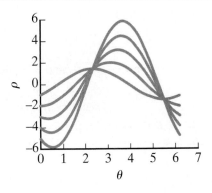

Secondly, the magnitude and orientation of the gradient vector—information that comes for free in the computation that finds the points—can be used to improve results. Instead of each pixel having an equal vote, the vote is made proportional to the gradient magnitude at that pixel, while ignoring pixels whose magnitude is below a threshold. Also, the angle can be used to determine θ directly or, assuming that the angle is within some amount of the true angle (say within $\pm 20°$), the accumulator cells within a small window around the gradient angle are incremented, rather than all the cells along the entire sinusoid. One final step that is sometimes desired is to detect the endpoints of the lines to yield line segments, which is left as an exercise for the reader. The pseudocode for detecting a single line, provided in Algorithm 11.18, uses A as the accumulator and a function g that maps gradient magnitude values to votes; the simplest such function would be a threshold that allows all pixels whose gradient magnitude exceeds a certain value to cast a single vote. Of course, the procedure can be repeated to detect multiple lines by applying nonmaximal suppression to the array once a peak is found.

The Hough transform is easily modified to detect other parameterized shapes, such as circles. If the radius is already known, then the accumulator is just a 2D array over quantized (x_c, y_c) positions for the center of the circle. Using $(x - x_c)^2 + (y - y_c)^2 = r^2$ as the equation of a circle centered at (x_c, y_c) with radius r, each (x,y) data point defines a circle in parameter space centered at (x,y) and having radius r. By incrementing the accumulator cells along this circle, the data are allowed to vote for the location of the circle. As before, the gradient orientation can be used to concentrate the votes near the detected orientation, rather than being spread out along the entire circle in parameter space. This procedure is easily extended to the case when the radius is unknown by using a 3D accumulator.

We could just as easily detect a circle by sliding a template of a circle across the image, summing gradient magnitude values along the perimeter of the circle in the template, and selecting the location with the highest score. In fact, such a computation would be identical to computing the Hough transform. Given this close connection between the Hough transform and template-based approaches to detection, it is not surprising that the Hough transform can be extended to handle arbitrary shapes. The resulting algorithm, known as the **generalized Hough transform**, represents the shape as a 2D binary template and utilizes a table whose entries are lists of offsets to the reference position given the orientation of an edge.

ALGORITHM 11.18 Hough transform for detecting a line

HoughLine(I)

Input: grayscale image I of size $width$ by $height$
Output: angle θ and distance ρ describing the most prominent line in the image

```
1   G_mag, G_phase ← ComputeImageGradient(I)
2   for (x, y) ∈ I do
3       x' ← x − width/2
4       y' ← y − height/2
5       θ̂ ← G_phase(x,y)
6       for θ ← θ̂ − θ_δ to θ̂ + θ_δ do
7           ρ' ← − x' cos θ − y' sin θ
8           A(Quantize(ρ'), Quantize(θ)) ←_+ g(G_mag(x, y))
9   ρ', θ ← FindMax(A)
10  ρ ← ρ − (width/2) cos θ − (height/2) sin θ
11  return ρ, θ
```

Finally, a standard implementation of the Hough transform uses a multidimensional array to represent the accumulator. The drawback with such an approach is that most of the cells remain empty, thus wasting space, which in turn leads to inefficient running times due to memory swapping. Moreover, it is often difficult to decide beforehand the range of the bins. One common way to overcome these limitations is to use a pseudo-random hash function to insert votes into a one-dimensional hash table, where collisions are easily detected.

11.5 Fitting Multiple Models

Although any of the methods considered so far in this chapter can be repeated to find multiple models, some algorithms by their very nature assume the existence of multiple models. We will now consider two such algorithms.

11.5.1 *K*-Means Clustering

In the previous chapter we looked at various clustering methods. By far the most popular clustering algorithm used today in scientific applications is **Lloyd's algorithm**, which pursues a greedy hill-climbing strategy to find a partition of the data that minimizes a squared-error criterion. Lloyd's algorithm is designed to solve the *k*-means clustering problem: Given an integer k and a set of n data points in $\mathbb{R}^d$, find k points (called *centers*) in $\mathbb{R}^d$ that minimize the mean squared distance from each data point to the nearest center. Although *k*-means clustering is NP-hard, Lloyd's algorithm is very easy to implement and executes quickly, and in practice it often converges to a good local minimum if initialized properly. Lloyd's algorithm is so popular that it has become synonymous with *k*-means, and it is oftentimes referred to as the *k*-means algorithm.

For each center $\mathbf{c}_j \in \mathbb{R}^d$, denote its *cluster* as the set of all data points for which $\mathbf{c}_j$ is the closest center. This set lies in the Voronoi cell[§] of $\mathbf{c}_j$. It is not hard to see that the optimal placement of a center is at the centroid of the associated cluster. As a result, Lloyd's algorithm alternates between two steps until convergence: First, all the centers are moved to the centroids of their clusters, and secondly, all points are assigned to the cluster associated with the nearest center. The pseudocode is provided in Algorithm 11.19.

ALGORITHM 11.19 *K*-means algorithm (i.e., Lloyd's algorithm) for finding clusters in a set of data

K-MEANSCLUSTERING $\left(\{\mathbf{x}_i\}_{i=1}^n\right)$

Input: set of n points $\{\mathbf{x}_i\}_{i=1}^n$
Output: cluster centers $c_j, j = 1, \ldots, k$ fitting the data

1   Initialize $c_j, j = 1, \ldots, k$
2   **repeat**
3   **for** $i \leftarrow 1$ **to** n **do**
4           $\delta_i \leftarrow \arg\min_j \|\mathbf{x}_i - \mathbf{c}_j\|$ ➤ Assign point to closest cluster.
5   **for** $j \leftarrow 1$ **to** k **do**
6           $c_j \leftarrow$ average of all $\mathbf{x}_i$ whose $\delta_i = j$ ➤ Move cluster center to centroid.
7   **until** convergence
8   **return** $c_j, j = 1, \ldots, k$

[§] Voronoi tesselation is illustrated in Section 12.3.3 (p. 587).

ALGORITHM 11.20 *K*-means algorithm (i.e., Lloyd's algorithm) for fitting models to a set of data

K-MEANSFITTING $\left(\{\mathbf{x}_i\}_{i=1}^n\right)$

Input: set of n points $\{\mathbf{x}_i\}_{i=1}^n$
Output: parameters $\theta_j, j = 1, \ldots, k$ of the models that best fit the data

1   Initialize $\theta_j, j = 1, \ldots, k$
2   **repeat**
3   **for** $i \leftarrow 1$ **to** n **do**
4           $\delta_i \leftarrow \arg\min_j d(\mathbf{x}_i, \theta_j)$ ➤ Assign point to cluster.
5   **for** $j \leftarrow 1$ **to** k **do**
6           $\theta_j \leftarrow \arg\max_\theta p(\theta|\{\mathbf{x}_i\})$ for all $\mathbf{x}_i$ whose $\delta_i = j$ ➤ Fit cluster model.
7   **until** convergence
8   **return** $\theta_j, j = 1, \ldots, k$

Lloyd's algorithm is simple and flexible and can be easily extended to fit more complex models to the clusters. Algorithm 11.20 shows the algorithm extended to fit multiple models to a set of data. Only two small changes must be made to the pseudocode: In Line 4, each point is assigned to the closest model using a distance function d, which generalizes the Euclidean distance used in the previous algorithm; and in Line 6, the model parameters are updated by maximizing the probability of the parameters given the points currently assigned to that particular model. For example, this pseudocode could be used to fit a certain number of lines to a set of data: Line 4 assigns each point to the closest line, and Line 6 performs least-squares fitting for each line using the points currently assigned to it.

Assuming all points are in general position (that is, no point is equidistant from two centers), Lloyd's algorithm is guaranteed to converge to a local minimum. Because it is a greedy algorithm, however, this local minimum may not be (and often is not) the global minimum. To avoid being trapped in a local minimum, careful attention must be paid to the initialization of the cluster centers. The standard approach is to initialize all the clusters randomly. A better approach, known as ***k*-means++**, is to select the first center at random, then to randomly select subsequent centers from the data points with probability proportional to the squared distance to the nearest center. In other words, first a point is selected at random to be $\mathbf{c}_1$. Then a PDF is created with the probability of selecting $\mathbf{x}_i$ as $\|\mathbf{x}_i - \mathbf{c}_1\|^2$, and a point is selected at random according to this PDF to be $\mathbf{c}_2$. Then a PDF with value $\min(\|\mathbf{x}_i - \mathbf{c}_1\|^2, \|\mathbf{x}_i - \mathbf{c}_2\|^2)$ is created for selecting a point to be $\mathbf{c}_3$, and so on.

If computational cost is not an issue, then a deterministic method called **global *k*-means** can be used to overcome the initialization problem. In this approach the centers are optimized by incrementally adding one center at a time as follows. Let $m = 1, \ldots, k$, and for each value of m run Lloyd's algorithm n times with the initial $m - 1$ cluster centers at the same locations found in the previous iteration, and the mth cluster center at position $\mathbf{x}_i, i = 1, \ldots, n$. For example, with $m = 1$, $\mathbf{c}_1^{(1)}$ is selected as the centroid of all the data. Then with $m = 2$, Lloyd's algorithm is run n times, starting with the initial conditions $\mathbf{c}_1 = \mathbf{c}_1^{(1)}$ and $\mathbf{c}_2 = \mathbf{x}_i, i = 1, \ldots, n$. The best of these n results yields the best pair of centers $\mathbf{c}_1^{(2)}$ and $\mathbf{c}_2^{(2)}$ to be used in the next iteration. Then with $m = 3$, Lloyd's algorithm is run n times, starting with the initial conditions $\mathbf{c}_1 = \mathbf{c}_1^{(2)}$, $\mathbf{c}_2 = \mathbf{c}_2^{(2)}$, and $\mathbf{c}_3 = \mathbf{x}_i, i = 1, \ldots, n$. The best of these n results yields the best triplet of centers $\mathbf{c}_1^{(3)}$, $\mathbf{c}_2^{(3)}$, and $\mathbf{c}_3^{(3)}$ to be used in the next iteration, and so on.

11.5.2 Expectation-Maximization (EM)

Lloyd's k-means algorithm makes use of **hidden variables** that indicate, for each data point, to which cluster (or model) it belongs. These hidden variables, also known as **latent variables**, lead to a sort of *chicken-and-egg problem*: If we knew the values of the hidden variables, then it would be easy to determine the parameters of the models, and if we knew the parameters of the models, then it would be easy to determine the most likely assignment of the data to those models (the hidden variables). This state of affairs leads naturally to the two-step algorithm of Algorithm 11.5.1, which alternates between computing the assignments, assuming that the parameters are known, and updating the parameters given the assignments. These two steps can be summarized as follows:

$$\delta_i \leftarrow \arg \max_{j=1,\,\ldots,\,k} p(\delta_i = j | \boldsymbol{\theta}_1, \ldots, \boldsymbol{\theta}_k), \quad i = 1, \ldots, n \qquad (11.119)$$

$$\boldsymbol{\theta}_j \leftarrow \arg \max_{\boldsymbol{\theta}} p_j(\boldsymbol{\theta} | \mathbf{x}_1, \ldots, \mathbf{x}_n, \delta_1, \ldots, \delta_n), \quad j = 1, \ldots, k \qquad (11.120)$$

where in the first line $p(\cdot)$ is the probability that the ith data point belongs to the jth model, and in the second line $p_j(\cdot)$ is the probability that the parameters for the jth model are θ, given the assignments.

Notice that k-means uses **hard assignments** for the data points. That is, each data point belongs to exactly one model during any iteration of the algorithm. A natural generalization is to allow so-called **soft assignments**, where each data point belongs to each of the models with a certain probability. This soft-assignment generalization of k-means is known as the **expectation-maximization (EM) algorithm**. Similar to k-means, the two steps of EM are to compute the soft assignments given the model parameters (known as the "E-step"), and update the model parameters by maximizing their likelihood given the hidden variables (known as the "M-step"):

E-step:

$$\pi_{ij} \leftarrow p(\delta_i = j | \boldsymbol{\theta}_1, \ldots, \boldsymbol{\theta}_k), \quad i = 1, \ldots, n, \quad j = 1, \ldots, k \qquad (11.121)$$

M-step:

$$\boldsymbol{\theta}_j \leftarrow \arg \max_{\boldsymbol{\theta}} p_j(\boldsymbol{\theta} | \mathbf{x}_1, \ldots, \mathbf{x}_n, \pi_{11}, \ldots, \pi_{nk}), \quad i = 1, \ldots, n, \quad j = 1, \ldots, k \qquad (11.122)$$

where the similarity with Equations (11.119)–(11.120) is obvious.

The EM algorithm is most commonly associated with **mixture models**, where each model specifies a probability density function (PDF) describing how likely it is to draw a certain data point. The overall PDF is the sum (or mixture) of the individual PDFs, weighted by the *a priori* probability of selecting each one:

$$p(\mathbf{x}_i) = \sum_{i=1}^{k} p(\mathbf{x}_i | \boldsymbol{\theta}_j) \pi_j \qquad (11.123)$$

where π_j is the *a priori* probability of selecting the jth model, and $p(\mathbf{x}_i | \boldsymbol{\theta}_j)$ is the probability of generating the data point $\mathbf{x}_i$ from the jth model. In other words, to generate a random data point we first randomly select a number from 1 to k using the PDF specified by $\pi_1, \ldots, \pi_k$ over the models. Then, given that the jth model was selected, a random data point is generated from that model.

The most common mixture model is the **Gaussian mixture model (GMM)** in which the individual models follow a Gaussian distribution. In that case, the model contains the mean and covariance matrix $\boldsymbol{\theta}_j = (\boldsymbol{\mu}_j, \boldsymbol{\Sigma}_j)$, and the conditional probability of $\mathbf{x}_i$ given $\boldsymbol{\theta}_j$ assumes the form of a multivariate Gaussian:[§]

$$p(\mathbf{x}_i | \boldsymbol{\theta}_j) = \frac{1}{(2\pi)^{\frac{d}{2}} |\boldsymbol{\Sigma}_j|^{\frac{1}{2}}} \exp\left(-\frac{1}{2}(\mathbf{x}_i - \boldsymbol{\mu}_j)^{\mathsf{T}} \boldsymbol{\Sigma}_j^{-1}(\mathbf{x}_i - \boldsymbol{\mu}_j)\right) \tag{11.124}$$

For a GMM, the E- and M-steps are

E-step:

$$\pi_{ij} \leftarrow \frac{\pi_j \exp\left(-\frac{1}{2}(\mathbf{x}_i - \boldsymbol{\mu}_j)^{\mathsf{T}} \boldsymbol{\Sigma}_j^{-1}(\mathbf{x}_i - \boldsymbol{\mu}_j)\right)}{\sum_{j'=1}^{k} \pi_{j'} \exp\left(-\frac{1}{2}(\mathbf{x}_i - \boldsymbol{\mu}_{j'})^{\mathsf{T}} \boldsymbol{\Sigma}_{j'}^{-1}(\mathbf{x}_i - \boldsymbol{\mu}_{j'})\right)}, \quad \begin{array}{l} i = 1, \ldots, n \\ j = 1, \ldots, k \end{array} \tag{11.125}$$

M-step:

$$\pi_j \leftarrow \frac{1}{n}\sum_{i=1}^{n} \pi_{ij}, \quad j = 1, \ldots, k \tag{11.126}$$

$$\boldsymbol{\mu}_j \leftarrow \frac{\sum_{i=1}^{n} \pi_{ij}\mathbf{x}_i}{\sum_{i=1}^{n} \pi_{ij}}, \quad j = 1, \ldots, k \tag{11.127}$$

$$\boldsymbol{\Sigma}_j \leftarrow \frac{\sum_{i=1}^{n} \pi_{ij}(\mathbf{x}_i - \boldsymbol{\mu}_j)(\mathbf{x}_i - \boldsymbol{\mu}_j)^{\mathsf{T}}}{\sum_{i=1}^{n} \pi_{ij}}, \quad j = 1, \ldots, k \tag{11.128}$$

Although deriving these equations from Equations (11.121) and (11.122) is not trivial, they are fairly intuitive. In Equation (11.125), the probability π_{ij} that $\mathbf{x}_i$ was drawn from the jth model is given by the probability of generating the value $\mathbf{x}_i$ using the mean and covariance of the jth Gaussian. In Equation (11.126), the *a priori* probability of selecting the jth model is the sum of the individual probabilities of all the data points, normalized to ensure that $\sum_j \pi_j = 1$. In Equations (11.127)–(11.128), mean and covariance are computed in the standard way, with the contribution of each data point proportional to the probability that the data point matches the particular model. Let us now apply these equations to a simple example.

EXAMPLE 11.1

Consider the clustering problem of a simple 1D dataset with two clusters centered at c_1 and c_2, and three points: $x_1 = 3$, $x_2 = 5$, $x_3 = 8$. Let the initial centers be $c_1 = 0$ and $c_2 = 12$. Apply the k-means and EM algorithms to these data.

Solution

First, let us apply k-means, using Equations (11.119)–(11.120). In the first step we assign $\delta_1 \leftarrow 1$, $\delta_2 \leftarrow 1$, $\delta_3 \leftarrow 2$, since 3 and 5 are closer to 0, while 8 is closer to 12; that is, $|3 - 0| < |3 - 12|$, $|5 - 0| < |5 - 12|$, and $|8 - 0| > |8 - 12|$. In the second step, since x_1 and x_2 are assigned to c_1, while x_3 is assigned to c_2, we update the centers as $c_1 = \frac{1}{2}(3 + 5) = 4$ and $c_2 = 8$. Convergence has been reached in just one iteration.

[§] Section 12.2.4 (p. 580).

Now let us apply EM, using Equations (11.125)–(11.128). Let us initialize the means and covariances to $\mu_1 = c_1 = 0$ and $\mu_2 = c_2 = 12$, and $\sigma_1^2 = \sigma_2^2 = 16$. In the first step (E-step), assign the probabilities based on the numerator of Equation (11.125):

$$\pi_{11} \leftarrow \exp\left(-\frac{1}{2\sigma_1^2}|3 - 0|^2\right) \approx 0.755, \quad \pi_{12} \leftarrow \exp\left(-\frac{1}{2\sigma_2^2}|3 - 12|^2\right) \approx 0.080 \quad \text{(11.129)}$$

$$\pi_{21} \leftarrow \exp\left(-\frac{1}{2\sigma_1^2}|5 - 0|^2\right) \approx 0.458, \quad \pi_{22} \leftarrow \exp\left(-\frac{1}{2\sigma_2^2}|5 - 12|^2\right) \approx 0.216 \quad \text{(11.130)}$$

$$\pi_{31} \leftarrow \exp\left(-\frac{1}{2\sigma_1^2}|8 - 0|^2\right) \approx 0.135, \quad \pi_{32} \leftarrow \exp\left(-\frac{1}{2\sigma_2^2}|8 - 12|^2\right) \approx 0.607 \quad \text{(11.131)}$$

then normalize by dividing by the denominator:

$$\pi_{11} \leftarrow \frac{0.755}{0.755 + 0.080} = 0.904, \quad \pi_{12} \leftarrow \frac{0.080}{0.755 + 0.080} = 0.096 \quad \text{(11.132)}$$

$$\pi_{21} \leftarrow \frac{0.458}{0.458 + 0.216} = 0.680, \quad \pi_{22} \leftarrow \frac{0.216}{0.458 + 0.216} = 0.320 \quad \text{(11.133)}$$

$$\pi_{31} \leftarrow \frac{0.135}{0.135 + 0.607} = 0.182, \quad \pi_{32} \leftarrow \frac{0.607}{0.135 + 0.607} = 0.818 \quad \text{(11.134)}$$

From these numbers we notice that x_1 and x_2 are much more likely to belong to the first model, whereas x_3 is more likely to belong to the second model, as we would expect. In the second step (M-step), assign

$$\pi_1 \leftarrow \frac{1}{3}(0.904 + 0.680 + 0.182) = 0.589 \quad \text{(11.135)}$$

$$\pi_2 \leftarrow \frac{1}{3}(0.096 + 0.320 + 0.818) = 0.411 \quad \text{(11.136)}$$

and

$$\mu_1 \leftarrow \frac{0.904(3) + 0.680(5) + 0.182(8)}{0.904 + 0.680 + 0.182} = 4.3 \quad \text{(11.137)}$$

$$\mu_2 \leftarrow \frac{0.096(3) + 0.320(5) + 0.818(8)}{0.096 + 0.320 + 0.818} = 6.8 \quad \text{(11.138)}$$

and similarly for the variances:

$$\sigma_1^2 \leftarrow \frac{0.904(3 - 4.3)^2 + 0.680(5 - 4.3)^2 + 0.182(8 - 4.3)^2}{0.904 + 0.680 + 0.182} = 2.46 \quad \text{(11.139)}$$

$$\sigma_2^2 \leftarrow \frac{0.096(3 - 4.3)^2 + 0.320(5 - 4.3)^2 + 0.818(8 - 4.3)^2}{0.096 + 0.320 + 0.818} = 2.91 \quad \text{(11.140)}$$

This ends the first iteration. The result of this iteration, as well as the next two, are shown in Figure 11.11. Because the initial variance is so high, the middle data point becomes associated almost equally between the two models, rather than being associated only with the first model, as in k-means.

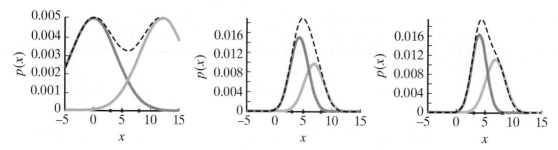

Figure 11.11 The first three iterations of the EM algorithm on Example 11.5.2, showing the data points (red circles), the first Gaussian model (blue curve), the second Gaussian model (green curve), and the combined PDF (dashed black curve).

The EM algorithm just presented is actually a simplified form of it. Nevertheless, the form just presented is, by far, the most widely used variation. A more detailed discussion of EM would require quite a bit of mathematical sophistication to show that the coordinate ascent approach described above is actually a form of lower-bound maximization, and that EM is always guaranteed to arrive at a local maximum of the likelihood function. This additional math would also reveal that the E-step actually constructs a tight lower-bound approximation to the true likelihood function, and the M-step finds the local maximum of that tight lower-bound approximation. The lower-bound approximation is the expectation of the complete-data log likelihood with respect to the PDF of the assignment probabilities—which explains the name of the E-step. An alternative to the EM algorithm is the generalized EM algorithm, which does not maximize the function each iteration, but rather only increases the likelihood.

11.6 Further Reading

The straightforward algorithm for fitting a circle by minimizing the algebraic distance is due to Kåsa [1976]. A number of methods for minimizing the geometric distance have been proposed over the years, such as Crawford [1983], Gander et al. [1994], and Chernov and Lesort [2005]. Note that some closed-form solutions for minimizing the algebraic distance use 4 parameters (making these approaches more sensitive to noise), rather than the 3 used by Kåsa (which is quite robust). For example, Figure 11.4c on page 490 of this book shows the Kåsa method successfully working on the same data shown in Figure 2.1 of Gander et al. [1994], for which the 4-parameter version yields noisy results. The generalized eigenvalue approach to fitting an ellipse is due to Fitzgibbon et al. [1999], which is related to the methods of Bookstein [1979] and Taubin [1991]. For a unifying view of the more sophisticated algebraic techniques known as the Pratt method, the Taubin method, and the hyperaccurate fit method, see the work of Al-Sharadqah and Chernov [2009].

The SVD approach to Procrustes analysis is originally due to Schönemann [1966] and Kabsch [1976],

and its popularity in computer vision is due to Umeyama [1991], which is a refinement of the work of Arun et al. [1987]. An alternate approach to aligning point clouds based on quaternions, which can be found in Horn [1987], was quite popular for several years before being largely replaced by the SVD method. For a comparison of four different algorithms, which finds them all to achieve essentially the same results, see Eggert et al. [1997]. The classic paper describing ICP is that of Besl and McKay [1992], which uses a point-to-point error metric. Chen and Medioni [1992] simultaneously proposed a nearly identical algorithm using a point-to-plane error metric. Another variant of ICP that was independently discovered is described in Zhang [1994], which handles outlier rejection and uses a k-d tree for finding the closest point. Rusinkiewicz and Levoy [2001] describe variants of ICP.

M-estimators were introduced by Huber [1964] and explained in great detail in his book, Huber [1981]. Another well-known book discussing M-estimators, with particular emphasis upon influence functions, is that of Hampel et al. [1986]. Tukey's influence in the field of robust estimation goes back to Tukey [1960]. The Geman-McClure estimator is due to Geman and McClure [1987]. We did not have space to discuss least median of squares (LMedS), which is another popular robust estimation technique that also achieves a breakdown of 50% but has relatively low efficiency; for more details see Rousseeuw [1984] and Rousseeuw and Leroy [1987].

RANSAC was introduced in the classic paper by Fischler and Bolles [1981], MSAC and MLE-SAC are described in Torr and Zisserman [2000], and Lo-RANSAC is from Chum et al. [2003]. The Hough transform was first described by Duda and Hart [1972], who called it the "generalized Hough transform," based on the early work of Hough [1959]. The Hough transform was later expanded to use the gradient direction in O'Gorman and Clowes [1976], and it was further generalized by Ballard [1981]. The history of the Hough transform is presented by Hart [2009].

Lloyd's algorithm for k-means was first presented at the Institute of Mathematical Sciences Meeting in 1957 but was not published until many years later by Lloyd [1982]. Other early work that presents essentially the same algorithm as Lloyd's is that of Forgy [1965]. MacQueen [1967] popularized the problem and coined the name k-means. Global k-means was proposed by Likas et al. [2003], and k-means++ is described by Arthur and Vassilvitskii [2007]. For an efficient variant of Lloyd's algorithm, see the filtering algorithm by Kanungo et al. [2002]. A well-cited survey on clustering techniques is that of Xu and Wunsch [2005]. The EM algorithm is presented in the classic work of Dempster et al. [1977], which is a challenging read.

PROBLEMS

11-1 Suppose we have the following points in the x-y plane: (0.08,1.16), (1.46,2.38), (2.67,3.91), (3.19,4.80), and (4.49,5.25).

(a) Write the matrix $\mathbf{A}$ and vector $\mathbf{b}$, as defined in Equation (11.5).

(b) Write the normal equations.

(c) Compute the condition number of $\mathbf{A}$.

(d) Compute the slope and y-intercept of the best-fitting line, in the ordinary least squares sense, using Algorithm 11.1.

(e) Repeat (d) using Algorithm 11.2.

(f) Compute the residual for (d) and (e).

11-2 Repeat the previous problem after first multiplying the $\times$ coordinate by 10^{20}. That is, the first point is $\left(0.08 \cdot 10^{20}, 1.16\right)$, and so forth. How do your results compare with those of the previous problem?

11-3 Consider the least squares problem $\mathbf{A}\mathbf{x} = \mathbf{b}$, where

$$\mathbf{A} = \begin{bmatrix} 0.7094 & 0.2760 \\ 0.7547 & 0.6797 \end{bmatrix} \quad \text{and} \quad \mathbf{b} = \begin{bmatrix} 0.6551 \\ 0.1626 \end{bmatrix}$$

(a) Write the two scalar equations represented by this matrix equation.

(b) Solve this problem using Gauss-Jordan elimination, i.e., multiply one of the equations by a scalar, subtract the equations to eliminate one of the variables, then back-substitute.

(c) Solve the problem by inverting the matrix $\mathbf{A}$, then multiplying the result by $\mathbf{b}$.

(d) Of (b) and (c), which one involves fewer computations?

(e) Generalize your result in (d) to larger systems of equations. That is, what is the computational complexity of the two approaches?

11-4 Suppose we have the following points in the *x-y* plane: (2.64, 0.33), (2.94, 1.31), (3.27, 2.09), (3.47, 3.30), and (3.43, 4.36).

    (a) Compute the slope and *y*-intercept of the best fitting line, in the ordinary least squares sense, using Algorithm 11.2.

    (b) Compute the parameters of the best fitting line using Algorithm 11.3.

    (c) Compare your results in (a) and (b). Which algorithm yields better results, and why?

11-5 Suppose you have a problem in which, according to the theory, $x_i^T B x_i = 0$ for every data point x_i. Now suppose you collect the following noisy data points (0.53, 0.22), (0.64, 0.21), (0.64, 0.79), (0.94, 0.26), and (0.77, 0.42). Formulate this as a least-squares problem, and solve for the elements of the 2×2 matrix **B**.

11-6 Compute the condition number of each of the following matrices. Assuming your machine has floating-point precision of 0.01, indicate whether each matrix is ill-conditioned.

(a) $\begin{bmatrix} 11.9 & 29.7 \\ 29.7 & 90.8 \end{bmatrix}$

(b) $\begin{bmatrix} 50.6 & -0.4 \\ -0.4 & 50.5 \end{bmatrix}$

(c) $\begin{bmatrix} 508.1 & 499.3 \\ 499.3 & 494.2 \end{bmatrix}$

(d) $\begin{bmatrix} 7.4 & 3.8 \\ 3.8 & 4.5 \end{bmatrix}$

11-7 For each of the following, state whether it is a homogeneous equation, assuming *x* and *y* are the variables:

    (a) $5x + 4y = 6$

    (b) $2x + 13y - 8 = 0$

    (c) $-22y = 44$

    (d) $6.3x + 14.6y = -8.1y$

11-8 A matrix has eigenvectors $[0.6483\ 0.6424\ 0.4088]^T$, $[0.4177\ 0.3520\ -0.8377]^T$, and $[0.2166\ -0.7404\ 0.6363]^T$ and eigenvalues 1.6213, 0.0614, and -0.2756. What is the matrix?

11-9 Show that

    (a) $\|x\|^2 = x^T x$ for any vector **x**, and

    (b) $\|Ax\|^2 = x^T A^T A x$ for any matrix **A** and vector **x**.

11-10 Given a diagonal matrix $\Lambda = diag(\lambda_1, \ldots, \lambda_n)$, show that $\|\Lambda x\|$ is minimized (with the constraint that $\|x\| = 1$) when $x = e_i$, where e_i is a vector of all zeros except a 1 in the ith element, and $|\lambda_i| \le |\lambda_j|$ for all *j*.

11-11 Fit a plane to the following points:

$$\{(1.16, 2.07, 3.23), (2.20, 2.80, 3.43), (2.81, 3.89, 3.23), (4.21, 5.02, 3.84)\}$$

11-12 Given a matrix with 8 rows and 6 columns,

    (a) Is the matrix tall or short?

    (b) Specify the dimensions of $\mathbf{U}$, $\mathbf{\Sigma}$, and $\mathbf{V}$ using the standard version of the SVD.

    (c) Specify the dimensions of $\mathbf{U}_r$, $\mathbf{\Sigma}_r$, and $\mathbf{V}_r$ using the compact version of the SVD.

    (d) How many singular values does the matrix have?

    (e) How would your answers in (a)-(d) change if the matrix had, instead, 6 rows and 8 columns?

11-13 Suppose the SVD of a matrix is given by the following:

$$\mathbf{U} = \begin{bmatrix} 0.862 & -0.506 & 0.018 \\ 0.331 & 0.537 & -0.776 \\ 0.383 & 0.675 & 0.630 \end{bmatrix} \quad \mathbf{\Sigma} = \begin{bmatrix} 1.034 & 0 \\ 0 & 0.457 \\ 0 & 0 \end{bmatrix} \quad \mathbf{V} = \begin{bmatrix} 0.942 & -0.335 \\ 0.335 & 0.942 \end{bmatrix}$$

    (a) Write the compact version of the SVD.

    (b) What is the matrix? (Hint: Do not forget to transpose the matrix $\mathbf{V}$.) Verify that you get the same result with the standard version and the compact version.

11-14 Write the Moore-Penrose pseudoinverse of the matrix in the previous problem.

11-15 Compute the Frobenius norm of the matrix $\mathbf{U}$ in Problem 11.13.

11-16 Explain the difference between geometric and algebraic error. What are the advantages of each?

11-17 Fit a circle to the following set of points, using (a) Algorithm 11.4, and (b) Algorithm 11.5:

$$\{(351.4, 916.7), (199.3, 750.1), (144.0, 431.3), (307.6, 124.3), (667.1, 162.3), (856.0, 384.5)\}$$

Compare the results.

11-18 At first glance it may not be obvious that the left and right sides of Equation (11.53) are equivalent.

    (a) Derive the right side of Equation (11.53) from the left side.

    (b) Show that the right side can also be obtained by differentiating the algebraic error in Equation (11.44) with respect to the squared radius, and setting the result to zero.

11-19 An alternate way to formulate the problem of circle fitting is to find the parameters a, b, c, and d that minimize $\sum_{i=1}^{n} (ax_i^2 + ay_i^2 + bx_i + cy_i + d)^2$, corresponding to the equation $a(x_i^2 + y_i^2) + bx_i + cy_i + d = 0$.

    (a) What constraints that must be met to ensure that the result is a circle? (*Hint:* Ensure that the radius is greater than zero, and avoid degeneracy.)

    (b) Does the (non-normalized) Kåsa method guarantee that these constraints are met?

    (c) How about the normalized Kåsa method?

11-20 Take the derivative of the shifted algebraic error of a circle:

$$\sum_{i=1}^{n} ((\tilde{x}_i - \tilde{h})^2 + (\tilde{y}_i - \tilde{k})^2 - r^2)^2 \qquad (11.143)$$

with respect to $\tilde{h}$ and $\tilde{k}$, and set the result to zero, to provide an alternate derivation of Equation (11.51). (*Hint:* Recall that $\sum_{i=1}^{n} \tilde{x}_i = \sum_{i=1}^{n} \tilde{y}_i = 0$.)

11-21 For each of the following conic sections, specify whether it is an ellipse, parabola, or hyperbola, and if it is an ellipse, specify whether it is real or imaginary, and whether it is a circle.

(a) $3x^2 + 5y^2 + 24x - 10y + 45 = 0$

(b) $3x^2 + 6xy + 3y^2 + 5x + 8y - 22 = 0$

(c) $4x^2 + 4y^2 - 32x - 82y + 17 = 0$

(d) $5x^2 + 8xy + 3y^2 - 9x - 21y - 44 = 0$

(e) $2x^2 + 1xy + 4y^2 - 4x - 24y + 40 = 0$

11-22 Fit an ellipse to the following set of points, using Algorithm 11.7:

$$\{(611.8, 606.7), (418.2, 764.6), (252.3, 826.0), (160.1, 761.7), (349.1, 317.3), (634.8, 206.1)\}$$

11-23 Suppose a binary image region with a centroid at (246.8, 217.0) has the following second-order central moments: $\mu_{00} = 60536.0$, $\mu_{20} = 886013794.4$, $\mu_{02} = 586810479.4$, $\mu_{11} = -632045701.7$. Fit an ellipse to this region using Algorithm 11.2.4, and sketch the ellipse using the principles of Section 4.4.5.

11-24 Compute the principal axes of the (a) aligned square and (b) oriented square below, using Equation (4.132):

$$\text{(a)} \begin{bmatrix} 0 & 0 & 0 & 0 & 0 \\ 0 & 1 & 1 & 1 & 0 \\ 0 & 1 & 1 & 1 & 0 \\ 0 & 1 & 1 & 1 & 0 \\ 0 & 0 & 0 & 0 & 0 \end{bmatrix} \qquad \text{(b)} \begin{bmatrix} 0 & 0 & 1 & 0 & 0 \\ 0 & 1 & 1 & 1 & 0 \\ 1 & 1 & 1 & 1 & 1 \\ 0 & 1 & 1 & 1 & 0 \\ 0 & 0 & 1 & 0 & 0 \end{bmatrix}$$

What do you notice from your results?

11-25 Align the following point sets using Procrustes analysis, both (a) with and (b) without scaling:

first set: $\quad \{(56.5, 940.1), (107.1, 867.0), (171.7, 767.5), (65.7, 793.9), (157.8, 913.7)\}$

second set: $\quad \{(729.3, 530.7), (743.1, 314.3), (743.2, 118.4), (591.0, 343.6), (869.8, 346.5)\}$

Does it make any difference if you allow reflection?

11-26 Show that the rotation matrix found by Procrustes analysis in 2D is

$$\mathbf{R} \equiv \begin{bmatrix} \cos\theta & -\sin\theta \\ \sin\theta & \cos\theta \end{bmatrix} \tag{11.161}$$

where the angle θ is the one that satisfies

$$\tan\theta = \frac{\sum_{i=1}^{n} x_i y_i' - x_i' y_i}{\sum_{i=1}^{n} x_i' x_i + y_i' y_i} \tag{11.162}$$

(*Hint:* Formulate the error as the sum of squared distances between the second set of points and the transformed points from the first set, then take the derivative with respect to θ, and set to zero.)

11-27 Explain how ICP relates to Procrustes analysis.

11-28 When does a MAP estimator become an MLE estimator?

11-29 Suppose you have a set of scalars visualized as points on the real number line.

(a) What is the value that minimizes the sum of squared differences between itself and all the scalars? Explain how you arrive at the answer.

(b) What is the value that minimizes the sum of absolute differences between itself and all the scalars? Explain, paying particular attention to the situation in which the number of scalars is even.

11-30 Explain the difference between homoskedastic and heteroskedastic noise.

11-31 How does truncation relate to Winsorizing?

11-32 (a) Is the M-estimator $\rho(z) = (1 - e^{-z}) \sin z$ redescending? (b) What about $\rho(z) = (1 - e^{-z} \sin z)$? Explain your answer in each case.

11-33 Implement RANSAC for 2D line fitting and show that the technique successfully finds a line even when more than 50% of the data points are corrupted.

11-34 Modify the pseudocode in Algorithm 11.18 to detect a circle of known radius. Explain how you would extend the approach to handle an unknown radius.

11-35 Explain how the Ridler-Calvard algorithm is a 1D version of k-means clustering.

11-36 What is the relationship between k-means and the most commonly used form of EM?

CHAPTER 12
Classification

Classification is the process of assigning a label to either an image or a region of an image. The label could indicate, for example, whether an image region contains a face, or whether it contains a car. The label could indicate the identity of the person or the type of car. Classification involves learning decision boundaries from manually labeled training data in order to correctly assign labels to unlabeled data. In this chapter we cover various algorithms and principles related to classification, which is also known as **supervised learning** or **pattern recognition**. First we describe some fundamental principles, such as the difference between training error and test error, and the various ways to evaluate classification systems. Then we look at statistical pattern recognition in general, and Bayesian decision theory in particular. Finally, we describe a number of widely used generative and discriminative methods for performing classification.

12.1 Fundamentals

We begin by discussing some fundamental concepts of the classification problem, such as the overall process of using training data to learn a decision boundary between categories, how to evaluate the performance of a classifier using test data, and some of the basic trade-offs between the complexity of the model and number of features on the one hand and generalization capability on the other hand.

12.1.1 Detection, Recognition, and Verification

Classification problems come in three basic forms. In a **detection** problem, the goal is to locate all instances of a particular class in an image. Face detection, for example, involves searching an image to find all the locations at which a human face is visible. In a **recognition** problem, it is known that an image region contains an item of a certain class, but the item's identity is unknown; a popular example of this type of problem is face recognition, where the goal is to label the identity of a detected human face. Finally, **verification** is a special case of recognition in which a hypothesized identity is provided, and the goal is to either verify or reject the identity. An example of a face verification system is one that determines whether an employee presenting an identification card at an entrance is indeed who he or she claims to be, using a picture taken by a camera mounted at the entrance.

All classification systems, whether designed for detection, recognition, or verification, involve the same basic steps and rely on the same underlying machinery. In all cases a finite number of discrete categories represent the possibilities. We shall represent these as $\omega_1, \ldots, \omega_N$. With detection and verification, there are just two categories $(N = 2)$, because the item of interest is either present or not, and the identity is either correct or not. Recognition, on the other hand, requires a category for each of the different instances of the class, as well as an additional category to represent an "unknown" or unrecognized instance.

An overview of a classification system is shown in Figure 12.1. Data are collected and manually labeled by a human teacher, and the data are separated into a training set and a test set. The first step is to transform the image pixels into a feature vector $\mathbf{x}$. The simplest such transformation is to stack the raw pixel values of the image region into a vector. However, to achieve better invariance to changes in lighting, pose, and structure, most real-world systems first preprocess the image to transform the raw pixel values into a separate feature space. In the case of face detection, for example, we could use features that estimate the symmetry, the bright spots on the cheeks, the dark spots near the eyes, and so forth. Such features are usually computed from the intensities in a rectangular window of pixels, but if segmentation has already been performed, then features related to the shape of the object can be used as well.

Once the transformation has been determined, the training data are used to learn a mapping $f : \mathbf{x} \mapsto \omega$, where $\mathbf{x}$ is the feature vector, and ω is the category. During the training process, the parameters of f are determined in order to satisfy two objectives. First, the classification error on the training set should be low, that is, $f(\mathbf{x})$ should for the most part return the same category as the label manually assigned to $\mathbf{x}$ for any feature vector in the training set. Secondly, the classification error should be low on data that has never been seen by the algorithm, that is, the mapping f should **generalize** well to the test set. Finally, once the classifier has been trained and evaluated, it can then be applied to unlabeled images for any particular application where the classifier is needed.

Figure 12.1 An overview of a classification system.

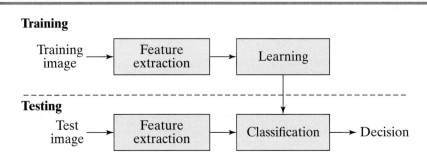

12.1.2 Classifiers, Discriminant Functions, and Decision Boundaries

Suppose a video camera is looking down on a conveyor belt, and we want to develop a system that can distinguish between the objects traveling down the belt. For simplicity, suppose there are only two types of objects, say shoes and hats, and that they are well separated so that only one object is seen at a time, making segmentation trivial. When an object enters the field of view of the camera, the camera takes a picture, and the system analyzes the image to determine whether the object is a shoe or a hat. To do this, the system first extracts features from the image, then feeds these features to a previously trained classifier, which then makes the decision.

Let ω_1 represent the category "shoe," and let ω_2 represent "hat." One of the most natural ways of representing a classifier is through the use of **discriminant functions**. For each category ω_i, its discriminant function, represented as $g_i(\mathbf{x})$, captures the likelihood that feature vector $\mathbf{x}$ belongs to it. A **classifier** then assigns feature $\mathbf{x}$ to category ω_i if the ith discriminant function yields a larger value than any other discriminant function:

$$f(\mathbf{x}) = \omega_i \quad \text{if } g_i(\mathbf{x}) > g_j(\mathbf{x}) \text{ for all } j \neq i \tag{12.1}$$

When there are just two categories, as in our example above, this reduces to

$$f(\mathbf{x}) = \begin{cases} \omega_1 & \text{if } g_1(\mathbf{x}) > g_2(\mathbf{x}) \\ \omega_2 & \text{otherwise} \end{cases} \tag{12.2}$$

in which case the classifier is known as a **dichotomizer**. For a dichotomizer it is more common to define a single discriminant function $g(\mathbf{x}) \equiv g_1(\mathbf{x}) - g_2(\mathbf{x})$, so that Equation (12.2) is simplified to

$$f(\mathbf{x}) = \begin{cases} \omega_1 & \text{if } g(\mathbf{x}) > 0 \\ \omega_2 & \text{otherwise} \end{cases} \tag{12.3}$$

The scenario is illustrated in Figure 12.2, where for simplicity we assume that the feature vectors contain just two dimensions so that $\mathbf{x} = \begin{bmatrix} x_1 & x_2 \end{bmatrix}^{\mathsf{T}}$. The feature vector dimensions could be any property derived from the image, such as the length and width of the region, the intensity and eccentricity of the region, and so forth. The feature vectors are plotted onto a **feature space**, and the learning algorithm then learns a **decision boundary** (or *decision surface*) to separate the data points from the two categories. The two regions on

Figure 12.2 Classification problem with two categories and a two-dimensional feature vector. Left: Training data points from one category plotted onto the feature space. Middle: Training data points from the other category plotted onto the same feature space. Right: Training data points plotted together, along with the learned decision boundary and the decision regions.

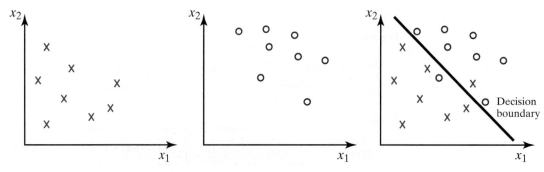

either side of the decision boundary in the feature space are known as **decision regions**. In this case $g_1(\mathbf{x}) = -g_2(\mathbf{x})$ for any $\mathbf{x}$, and the discriminant function g($\mathbf{x}$) is simply the signed distance to the decision boundary, so that the classifier f maps any point below the decision boundary to one category, while any point above the decision boundary is mapped to the other category.

12.1.3 Error, Loss, and Risk

Any real-world classifier will make mistakes, or **errors**. In the example above, an error occurs when a hat is labeled "shoe," or when a shoe is labeled "hat." In Figure 12.2, for example, there is a red "x" above the line and a blue "o" below the line that are in the wrong decision regions. A **loss function** indicates the cost of the classifier making a particular decision on a particular input. More precisely, a loss function L maps the actual output and desired output to a real number. The most commonly used loss function is the **zero-one (0–1) loss function**, which assigns a cost of 1 to every error, and a cost of 0 to every successful classification:

$$L(\omega, \hat{\omega}) = \begin{cases} 0 & \text{if } \omega = \hat{\omega} \\ 1 & \text{if } \omega \neq \hat{\omega} \end{cases} \tag{12.4}$$

where $\hat{\omega}$ is the correct label. With the zero-one loss function, the performance of a classifier can be determined by focusing solely on the errors.

The **risk** associated with the classifier and a particular decision is the product of the probability of making the decision and the cost associated with the decision. For example, the risk of assigning ω_1 to ω_2-type items, for a given classifier, is the probability of mistakenly assigning ω_1 to an ω_2-type item, times the loss associated with mistakenly assigning ω_1 to an ω_2-type item. The **total risk** of a classifier is the expectation of the loss function, which in our example would be the risk of assigning ω_1 to ω_2-type items, plus the risk of assigning ω_2 to ω_1-type items. However, the total risk is often unmeasurable because the joint distribution of the inputs and outputs is unknown. For example, we do not know the exact probability of obtaining a particular length measurement over the set of all possible hats and all possible shoes. As a result, we often settle for measuring the **empirical risk**, which is the average loss on the training set:

$$\frac{1}{n} \sum_{i=1}^{n} L(\omega_i, \hat{\omega}_i) \tag{12.5}$$

where n is the number of samples in the training set. The **structural risk** is the empirical risk plus an additional regularization term to penalize the complexity of the decision boundary, considered in more detail below.

12.1.4 Training Error, Test Error, and True Error

The classifier is trained on a training set, then evaluated on a test set. The error on the training set is the **training error,** and the error on the test set is the **test error.** A good classifier is one that not only performs well on the training set but also generalizes well to the test set. That is, we want not only low training error, but also (and, in fact, more importantly) low test error.

Both the training and test sets can be thought of as containing samples from the same underlying probability distribution. Hats, for example, might be more likely than shoes, and large hats might be more likely than small hats. Although such characteristics will be manifested in both sets, they will not necessarily have the exact same proportions. That is, the ratio of hats to shoes in the training set will typically be different from the ratio of hats to shoes in the test set, and both of these ratios will differ from the actual ratio in the seemingly infinite world of hats and shoes.

If we had access to large amounts of data—hundreds of millions of hats and shoes—then such discrepancies would be small indeed. In reality, however, it is often difficult to collect data sets that are large enough to accurately represent the underlying distribution. For example, to accurately capture the set of human faces for a face detector, one would need to capture not only images of billions of actual faces (based on the number of people in the world), but also images at a virtually infinite number of poses, lighting conditions, distances, ages, facial expressions, hair styles, backgrounds, camera exposures and gains, and so forth. As a result, for such problems, there will always remain a discrepancy between the sample errors (obtained using the finite training or test sets) and the **true error,** which is the probability of misclassifying a sample drawn at random from the underlying distribution.

Therefore, given the error, ϵ_{train}, of a classifier on a finite number of samples, two questions need to be answered in order to evaluate the accuracy of the classifier.[†] First, what is the probable error of the classifier on future samples that have not yet been seen? The answer to this question is easy: In the absence of additional information, the best estimate for the true error, ϵ_{true}, is simply the training error:

$$\epsilon_{true} \approx \epsilon_{train} \tag{12.6}$$

Secondly, how accurate is this estimate of the probable error? The answer to this question is given by

$$\epsilon_{true} = \epsilon_{train} \pm \eta \sqrt{\frac{\epsilon_{train}(1 - \epsilon_{train})}{n}} \tag{12.7}$$

where n is the number of samples in the training set, and the scalar η depends upon the desired confidence interval for the uncertainty estimate. For example, $\eta = 1.64$ for 90% confidence, $\eta = 1.96$ for 95% confidence, and $\eta = 2.58$ for 99% confidence. Equation (12.7) is statistically valid only when a large enough number of samples have been gathered, that is, whenever $n \geq \frac{5}{\epsilon_{train}(1 - \epsilon_{train})}$. For example, when $\epsilon_{train} = 0.1$, at least 50 samples are needed in order to use this equation.

EXAMPLE 12.1	Suppose we have a binary classification problem in 1D. Training data are collected for ten objects, 6 of which are hats and 4 of which are shoes. The lengths of the hats are 21, 24, 26, 27, 29, and 31; the lengths of the shoes are 30, 33, 34, and 35. Suppose a classifier is trained that outputs HAT if the length is less than 30 and SHOE otherwise. What is the true error of this classifier?
Solution	Since only one of the hats and none of the shoes are misclassified, the training error is $\epsilon_{train} = 0.1$, or 10%. Without additional information, the best estimate for the true error is therefore 0.1. However, the true error is not guaranteed to be exactly this number. Rather, if the number of samples were at least 50, then with 95% confidence we could say from Equation (12.7) that the true error is between 0 and 0.286, since $1.96\sqrt{\frac{(0.1)(0.9)}{10}} = 0.186$, and since the error cannot be negative.

[†] The relationship between accuracy and error is straightforward. For example, a classifier that yields 10% error is 90% accurate.

12.1.5 Bias-Variance Tradeoff, Overfitting, and Occam's Razor

There is an inherent tradeoff between improving performance on the training set and improving performance on the test set. If this statement is not intuitive at first glance, consider the example shown in Figure 12.3. A simple model (in this case a line) does a reasonable job of separating the two categories, but it leads to some errors. If the complexity of the model is increased, this error on the training data can be reduced arbitrarily, even to zero error. However, the danger of doing so is the possibility of **overfitting** to the specific training set, which will cause the algorithm to perform poorly on the test set, as well as on future input samples.

Two important concepts in this regard are **bias** (the expected discrepancy between the sample error and the true error), and **variance** (the expected variation in the error of the learned model as the training set is resampled from the underlying distribution). The training error exhibits nonzero bias, because it generally presents an overly optimistic measure of the accuracy of the classifier. The test error, on the other hand, usually exhibits zero bias, since the training and test sets are sampled independently. Even in the case of zero bias, however, there will be nonzero variance, because the measured error differs from the true error due to the finite number of samples.

Mathematically, the bias-variance decomposition of the squared error can be expressed as follows:

$$E[(f(x) - \omega)^2] = \underbrace{(E[f(x) - \tilde{f}(x)])^2}_{bias} + \underbrace{E[(f(x) - E[f(x)])^2]}_{variance} + E[\omega^2] \quad (12.8)$$

where $\tilde{f}$ is the true classifier (given infinite data), $E[\omega^2]$ is the **irreducible error** (which arises because the data are not perfectly separable), and $E[\cdot]$ is the expected value. Deriving this equation is straightforward and left as an exercise for the reader.[†] In this equation, the bias can be thought of as the error caused by the simplifying assumptions of the model, whereas the variance is the flexibility of the model to adapt to different training sets. In general, more complex models yield less bias but higher variance, whereas simpler models yield more bias but less variance—an observation known as the **bias-variance tradeoff.** For example, although the complex model in Figure 12.3 yields less bias (which is good), the stability of the linear model over various training sets (because a linear model does not have much flexibility to adapt to the data) is probably preferred in this case. That is, although high variance is desirable for fitting the training data, it often leads to overfitting. Philosophically, this observation is expressed in the principle known as **Occam's Razor,**[‡] which states that simpler models should be preferred over more complex models.

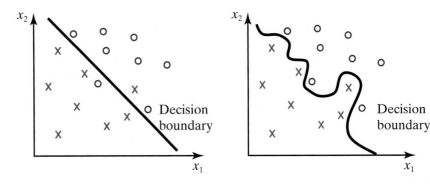

Figure 12.3 A low-order model (left) leads to some misclassifications on the training set, while a high-order model (right) leads to a reduced number of misclassifications. However, high-order models tend to overfit the training set, thus leading to poor generalization performance.

[†] Problem 12-9.

[‡] William of Ockham (c. AD 1287–1347) was a medieval friar, philosopher, and theologian in England who died just before the Black Death reached that country. Occam's Razor says that, all other things being equal, a more complicated explanation should not be chosen over a simpler explanation—in other words, the simplest explanation is the best.

EXAMPLE 12.2	Using the same training data as the previous example, suppose we train a classifier that outputs HAT if the length is either less than 30 or between 31 and 32, inclusive, or SHOE otherwise. Is this classifier better or worse than the previous one?
Solution	By allowing a more complex model, we have been able to reduce our training error to zero. However, the problem with this approach is that the training data is not a perfect sampling of the underlying distribution: If we were to collect another training set, it would likely be different. For example, suppose on another day that we measure 10 objects, and 5 happen to be hats while 5 happen to be shoes. The lengths of the hats are 22, 23, 25, 27, 28, and 29; while the lengths of the shoes are 30, 31, 32, 34, and 36. On this set of data the first classifier yields zero error, but the second classifier yields 0.2, or 20% error, because it has overfit to the first training data set. In other words, the model with the simple threshold is more likely to generalize to other samplings of the underlying distribution than is the complex model.

12.1.6 Holdout Method and Cross-Validation

The most straightforward way to evaluate the performance of a classifier is to collect some data, manually label the data, then partition the data into two nonoverlapping sets. One set (the training set) is used to train the classifier, while the other set (the test set) is used to evaluate it. This is known as the **holdout method,** and a good rule of thumb is to use two-thirds of the data for training and one-third for testing.

While the holdout method is a reasonable approach, it suffers from several drawbacks. First, since manually labeling data is so painful and time-consuming, we would like to be able to take advantage of *all* the labeled data in the training process. Otherwise the learned model is not as good as it could be because it did not have access to all the data. Secondly, the results of the evaluation will be heavily dependent upon the particular partitioning of the data; if for some reason the split happens to be a bad one, then the results could be skewed accordingly.

One approach that fixes these drawbacks is known as **random subsampling.** In random subsampling, the holdout method is repeated, with the data sampled each time at random to form the training and test sets. In each iteration a new model is learned from the training set and evaluated on the test set. The errors over all iterations are then averaged to yield the final evaluation of the classifier. With random subsampling, the size of the test set and the number of iterations can be chosen independently.

A more systematic approach to accomplish the same objective is **k-fold cross validation.** In this approach, all the labeled data are divided into k nonoverlapping sets such that every data sample is in exactly one of the sets.[†] One of the sets is chosen as the test set, with the other $k-1$ sets used as the training set, to compute the error of the classifier. This constitutes one iteration. The process is repeated k times, with a different set selected in turn for each iteration. The performance of the classifier is given by the average of the k results.

Larger values of k lead to less bias in the result at the expense of more computation, with $k = 5$ and $k = 10$ being common values (5-fold cross validation or 10-fold cross validation, respectively). A special case occurs when $k = n$ (that is, when k is the number of data samples), known as **leave-one-out cross validation (LOOCV).**[‡] In LOOCV, all the data except for one sample are used for training, with the isolated sample used for testing.

[†] If the relative proportions of the different labels are approximately the same in all the sets, then the cross validation is said to be *stratified.*

[‡] Another special case occurs when $k = 2$; it is similar to the holdout method, but it requires both sets to have equal size, and it treats them symmetrically by averaging the two results.

This process is then repeated for all samples in the data set, and the n results are averaged. An advantage of LOOCV is that it treats all data samples exactly the same, and therefore the result is deterministic. (In contrast, the result of k-fold cross validation depends upon the particular partitions chosen.) A drawback of LOOCV is the increased training time, because it requires n models to be learned, although for some models (such as linear regression), this additional expense can be mitigated.

In addition to evaluating a classifier, it is oftentimes necessary to tune its parameters. For example, it might be necessary to choose the number of nearest neighbors or the number of nodes in a decision tree, or the stopping point for backpropagation. In such a case, the data are split three ways. One set of the data is used for training, a second set is used for parameter tuning, and a third set is used for testing. First the training set is used to train the classifier, then the validation set is used to tune the parameters, and finally the test set is used to evaluate the classifier. A typical rule of thumb for these three sets is 50%, 25%, and 25%, respectively. It is straightforward to combine this idea with cross validation.

12.1.7 Model Selection and Regularization

As a result of the bias-variance tradeoff, an important decision is to determine the appropriate complexity for the model. If the model has too few parameters, then the model will not be able to adjust to the training data, and the error will be high. On the other hand, if the model has too many parameters, then although the model will fit the training data well, it will not generalize well to the test data. This problem is known as **model selection**.

The most common solution to the model selection problem is known as **regularization**, which minimizes not just the training error, but rather the sum of the training error and a regularizing term that is proportional to the complexity of the model, where the complexity can be measured in any of a variety of ways. For example, the **minimum description length (MDL) principle** seeks a model to explain the data such that the data and model together can be represented with the minimum number of bits. Alternatively, the **Akaike information criterion (AIC)**, is defined as

$$AIC \equiv -2 \ln \alpha + 2m \tag{12.9}$$

where α is the maximum likelihood achieved by the model on the data, and m is the number of parameters in the model. Closely related is the **Bayesian information criterion (BIC)**:[†]

$$BIC \equiv -2 \ln \alpha + m \ln n \tag{12.10}$$

where n is the number of samples in the data set. Note that for any reasonably-sized data set, $m \ln n > 2m$, and therefore BIC favors simpler models than AIC because it weights the model complexity more.

An alternate approach to regularization is to estimate the true error of the classifier using only the training error and the **Vapnik-Chervonenkis (VC) dimension** of the classifier space. As mentioned earlier, the true error can be thought of as the test error on an infinite test set containing all possible inputs. It can be shown that, with probability $1 - \eta$, the true error of a particular classifier is no more than the training error of the classifier plus the **VC confidence**:

$$\text{true error} \leq \text{training error} + \underbrace{\sqrt{\frac{\left(\log \frac{2n}{h} + 1\right)h - \log \frac{\eta}{4}}{n}}}_{\text{VC confidence}} \tag{12.11}$$

[†] Also known as the Schwarz criterion.

where n is the size of training data set, and h is the VC dimension of the **classifier space**,[†] which contains all possible classifiers of a particular type. For example, the classifier space could be the set of all nonvertical lines in the plane characterized by their slope and y-intercept, and a particular classifier within this space might be $y = 2x + 3$. The VC dimension is related to the complexity of the models within the space: a larger value of h indicates a space containing more complex models. As a result, Equation (12.11) can be used to select the right balance between model complexity and the bound on the generalization error, in a process known as **structural risk minimization**.

To understand the concept of VC dimension, it is imperative to grasp the concept of **shattering**. A classifier space is said to *shatter* a set of data points if, for every possible labeling of the points, there exists a classifier that assigns correct labels to all the points; the VC dimension of the classifier space is then the largest number of points that can be arranged so that they are shattered by the space. For example, the VC dimension of a linear classifier in a d-dimensional space is $d + 1$. This is easy to show for the case of $d = 2$ because a line can shatter 3 points in the plane, but it cannot shatter 4 points. That is, there exists an arrangement of 3 data points such that, no matter which of the $2^3 = 8$ possible binary labelings are assigned, there exists a line in the plane that will correctly separate these points from each other. But there is no arrangement of 4 points in the plane (assuming they are in general position, meaning no three are collinear) for which all labelings can be separated by a line.

12.1.8 Curse of Dimensionality and the Peaking Phenomenon

Let d be the dimensionality of the feature space, then $\mathbf{x} \in \mathbb{R}^d$. As the size of the feature vector increases, the space in which the learning algorithm must operate increases exponentially. For example, assuming that the classifier is a simple lookup table, and assuming that the space is divided into c equally-sized intervals for each dimension, the table will require c^d elements (i.e., c elements for $d = 1$, c^2 for $d = 2$, c^3 for $d = 3$, and so on). Thus, as d grows to any reasonably-sized number, it will be nearly impossible to collect enough training samples to accurately represent the space, leading to many elements in the table with zero entries. This is known as the **curse of dimensionality**, and it is not restricted to lookup tables because any type of classifier has to learn in a space whose size is exponential in the number of dimensions, making learning particularly challenging for real-world problems.

One consequence of the curse of dimensionality is the fact that adding more features to a classifier oftentimes decreases accuracy. Known as the **peaking phenomenon**, this behavior is counterintuitive in that one would expect that more information would yield better accuracy. Indeed, more information does yield better accuracy, *if* sufficient training data can be collected. However, increasing the dimension of the feature vector makes it that much harder to gather sufficient training data, thus making the data available to the learning algorithm less representative of the actual underlying distribution, and thus reducing accuracy. One approach to mitigate the peaking phenomenon is to carefully select the features according to their effect on accuracy, a process known as **feature selection**.

12.1.9 Evaluating Classification Results

With many binary classification problems, there is an inherent imbalance between the two categories. For example, we may wish to determine whether patients visiting a doctor have a particular disease. Since the number of healthy people far exceeds the number of sick people, the two categories are not interchangeable. (And even during an epidemic, when

---

[†] A classifier is sometimes known as a *hypothesis*, so that a classifier space is also known as a *hypothesis space*. Equivalent terms are *hypothesis class* or *concept class*.

the number of sick people is greatly increased, there still remains an important semantic difference between the two categories, so there is still a need to treat them differently.)

In such cases, one category is considered **positive**, whereas the other category is considered **negative**. These names do not indicate whether a particular attribute is desirable, but rather whether a particular attribute is present. Thus, for example, a patient who has the disease is considered a positive example, while a patient who is healthy is considered a negative example. This distinction gives rise to four possibilities for the output of a binary classifier:

- A **true positive (TP)** occurs when the classifier assigns the positive label to a positive example (e.g., it correctly says that a sick person is sick).
- A **true negative (TN)** occurs when the classifier assigns the negative label to a negative example (e.g., it correctly says that a healthy person is healthy.)
- A **false positive (FP)** occurs when the classifier assigns the positive label to a negative example (e.g., it mistakenly says that a healthy person is sick). Also known as a *false alarm*, *false detection*, or *Type I error*.
- A **false negative (FN)** occurs when the classifier assigns the negative label to a positive example (e.g., it mistakenly says that a sick person is healthy). Also known as a *false dismissal*, or *Type II error*.

A natural way to capture the performance of a classifier on a data set is to use a **confusion matrix**, whose rows correspond to the actual categories and whose columns correspond to the outputs of the classifier.[†] For example, suppose we have a data set of 100 people, 90 of whom are healthy and 10 of whom are sick. Suppose the classifier labels 18 of the healthy people as having the disease, and also 9 of the sick people as having the disease. If we let *TP* refer to the number of true positives, and so forth, then we have $TP = 9$, $TN = 72$, $FP = 18$, and $FN = 1$. Note that $TP + TN + FP + FN = 100$, which is the total number of people in the data set. The resulting confusion matrix is shown in Table 12.1 where, by convention, the entries are normalized by the size of the data set to yield probabilities.

Various performance measures based on the entries of the confusion matrix are listed in Table 12.2. The **true positive rate (TPR)** is the proportion of positive examples labeled positive, $TP/(TP + FN)$; the **true negative rate (TNR)** is the proportion of negative examples labeled negative, $TN/(FP + TN)$; and so on. These two measures are also known as the **sensitivity** and **specificity**, respectively, and an ideal classifier would achieve 100% for each. Another way to measure performance, applied primarily to image retrieval systems, is through **precision** and **recall**, which are the fraction of detected items that are relevant, and the fraction of relevant items that are detected, respectively. These, too, should be 100%. Note that TPR, sensitivity, and recall are synonyms.

A related measure is the **F-measure** (also known as the F_1 score), which is defined as the harmonic mean of precision and recall:

$$\text{F-measure} \equiv \frac{2}{\frac{1}{\text{precision}} + \frac{1}{\text{recall}}} = \frac{2 \cdot \text{precision} \cdot \text{recall}}{\text{precision} + \text{recall}} = \frac{TP}{2\,TP + FP + FN} \tag{12.12}$$

	test positive (disease detected)	test negative (not detected)
actual positive (sick person)	0.09 (*TP*)	0.01 (*FN*)
actual negative (healthy person)	0.18 (*FP*)	0.72 (*TN*)

TABLE 12.1 An example confusion matrix

[§] Some authors define the confusion matrix as the transpose of the one presented here.

Name	Formula	Alternate names
true positive rate (*TPR*)	$\frac{TP}{TP + FN}$	recall, sensitivity, hit rate
false negative rate (*FNR*)	$\frac{FN}{TP + FN}$	miss rate
true negative rate (*TNR*)	$\frac{TN}{FP + TN}$	specificity
false positive rate (*FPR*)	$\frac{FP}{FP + TN}$	false alarm rate, fallout
precision	$\frac{TP}{TP + FP}$	positive predictive value
accuracy	$\frac{TP + TN}{TP + TN + FP + FN}$	

TABLE 12.2 Performance measures of a binary classifier. Note that TPR and FNR are complementary: $TPR + FNR = 1$, and the same for TNR and FPR: $TNR + FPR = 1$.

Closely related is the **Jaccard coefficient**, defined as the intersection over union of the set $\mathcal{A}$ of actual positives (the samples in the first row of the confusion matrix) and the set $\mathcal{B}$ of detected positives (the samples in the first column of the confusion matrix).

$$\text{Jaccard coefficient} \equiv \frac{\mathcal{A} \cap \mathcal{B}}{\mathcal{A} \cup \mathcal{B}} = \frac{TP}{TP + FP + FN} \tag{12.13}$$

Many classifiers contain a parameter that governs the tradeoff between sensitivity and specificity. Imagine a threshold, for example, so that any number above the threshold is labeled positive, while any number below the threshold is labeled negative. As the threshold is increased, the classifier generates more true positives but also more false positives; and as the threshold is decreased, the classifier generates fewer true positives but also fewer false positives. To evaluate the performance of the classifier, then, it is not enough to measure its sensitivity and specificity for a particular choice of threshold. Rather, it is necessary to generate the curve through sensitivity-specificity space, where each point along the curve is a pair of numbers indicating the sensitivity and specificity for a particular value of the threshold.

Two types of curves are common, as shown in Figure 12.4. A **receiver operating characteristic (ROC) curve** plots the TPR versus the FPR, and the ideal curve is one that passes through the upper-left corner where $FPR = 0$ and $TPR = 1$. A **precision-recall (PR) curve** plots the precision versus the recall (or TPR), and the ideal curve passes through the upper-right corner where precision and recall are both 1.[†] ROC curves and PR curves are closely related, but the latter are more appropriate for skewed, or imbalanced, datasets. A **skewed dataset** is one in which the number of items in one category far exceeds the number of items in the other category, just like the imbalance in binary classification problems mentioned above. In the case of a skewed dataset, a PR curve can reveal details about a classifier's performance that are hidden by an ROC curve. To understand this phenomenon, imagine a dataset of people, only a tiny fraction of whom are sick. In this case the number of true negatives far exceeds the number of true positives, or $TN \gg TP$. Because the normalization factor of FPR contains $FP + TN$, a large change in FP can be masked by the extremely large value of TN, thus hiding the effects of false positives in an ROC curve. In

---

[‡] Other choices include the **detection error tradeoff (DET) curve**, which plots the FNR versus the FPR with nonlinear scaling along the axes to straighten the curve and thereby accentuate small probabilities; and the **Cumulative Match Characteristic (CMC) curve**, which plots the probability that the correct matching identity is contained within the top k results, where k is the horizontal axis. The latter is used primarily for evaluating biometric systems.

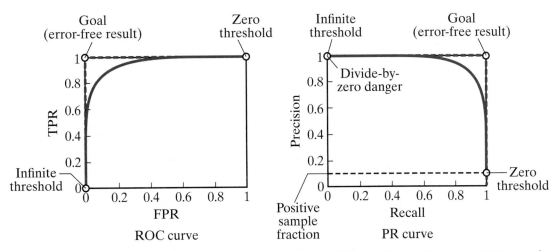

Figure 12.4 The performance of a classifier is often visualized through either an ROC curve (left) or a PR curve (right). An error-free classifier would generate a curve along the dashed blue lines.

contrast, because a PR curve utilizes precision, which normalizes by the total number of elements tested positive, such effects remain apparent.

Another problem with ROC curves is that, in computer vision problems, it is often extremely difficult to define the number of true negatives. For example, suppose there is a classifier to detect faces in an image. In most cases it is fairly straightforward to define *TP*, which is the number of actual faces in the image that are detected. But what is the value of *TN*? Most detection algorithms adopt a **sliding window** approach, in which a window is slid across the image and, at each location, the classifier is run. If *TN* is defined as the number of locations at which there is no face and the detector did not fire, then *TN* will be a very large number indeed, and the normalizing factor will then yield an extremely small value for *FPR*, even if the number of false positives rises to the level that appears unacceptable to a user. For this reason, oftentimes the FPR is reported in terms of the number of false positives per image, or something similar, even though such measures are not guaranteed to be between 0 and 1.

Classifiers are compared by comparing their curves. Typically this is done in one of two ways. The **equal error rate (EER)** is defined as the point on the ROC curve where *FPR = FNR*, or equivalently, *TPR = TNR*. This point is the intersection of the ROC curve with the diagonal line drawn from the upper-left to the lower-right corner. Two ROC curves can then be compared by comparing their EERs. The other, and more robust, way is to compute the **area under the curve (AUC)**,[†] with the larger number indicating a better classifier.

12.2 Statistical Pattern Recognition

Now that we have considered the fundamental concepts of classification, let us turn our attention to various techniques for solving the problem. There are two basic approaches to pattern recognition. In **syntactic (or structural) pattern recognition**, objects are represented by strings of symbols related by a formal grammar, or by nodes in a graph; for

---

[†] For those familiar with statistics, the AUC is equivalent to the Mann-Whitney U statistic.

example, two images could be matched by comparing the graphs of the objects detected in each. In **statistical pattern recognition**, on the other hand, objects are represented by feature vectors, and statistical analysis is applied to learn decision boundaries between categories. We focus our attention upon the latter, which is by far the more popular approach.

12.2.1 Bayes' Rule

We begin by recalling a foundational result from probability and statistics. Suppose a person throws darts at a board, and every dart is guaranteed to land somewhere on the board. Let Ω be the set of all locations on the board, and let $\mathbf{r} \in \Omega$ be a particular dart location. Now suppose two overlapping regions labeled $\mathcal{A}$ and $\mathcal{B}$ have been drawn on the board, as shown in Figure 12.5. Let $p(\mathbf{r} \in \mathcal{A})$ be the probability that a dart lands in region $\mathcal{A}$, and $p(\mathbf{r} \in \mathcal{B})$ the probability that it lands in region $\mathcal{B}$. With a slight abuse of notation, we shall refer to these simply as $p(\mathcal{A})$ and $p(\mathcal{B})$, respectively. If we know that a particular dart landed in region $\mathcal{B}$, the probability that it also landed in region $\mathcal{A}$ is called the *conditional probability* $p(\mathcal{A}|\mathcal{B})$. It is easy to see from the figure that $p(\mathcal{A}|\mathcal{B}) = p(\mathcal{A} \cap \mathcal{B})/p(\mathcal{B})$, and similarly, that $p(\mathcal{B}|\mathcal{A}) = p(\mathcal{A} \cap \mathcal{B})/p(\mathcal{A})$. Putting these two equations together yields the well-known **Bayes' rule**:

$$p(\mathcal{A}|\mathcal{B}) = \frac{p(\mathcal{B}|\mathcal{A})p(\mathcal{A})}{p(\mathcal{B})} \tag{12.14}$$

Bayes' rule is important because it allows us to convert an impossible problem into one that is solvable. To see what is meant by this, let us change the problem slightly. Let us suppose that Ω is divided into N disjoint regions $\Omega_1, \ldots, \Omega_N$, so that there is no intersection between any two regions: $\Omega_i \cap \Omega_j = \emptyset$ for all $i \neq j$. We shall also assume that each point in Ω is in some region, so that their union covers the entire space: $\bigcup_{i=1}^{N} \Omega_i = \Omega$. We will call the points *instances*, and the regions *categories*, so that Ω_i is a set containing all the instances in category ω_i. Since each instance belongs to exactly one of the categories, we have

$$\sum_{i=1}^{N} p(\mathbf{r} \in \Omega_i) = 1 \quad \forall \mathbf{r} \in \Omega \tag{12.15}$$

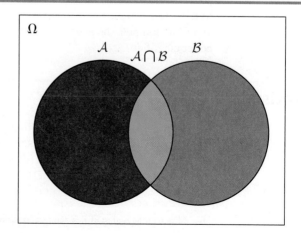

Figure 12.5 Bayes' rule, which relates one conditional probability $p(\mathcal{A}|\mathcal{B})$ to its reverse conditional probability $p(\mathcal{B}|\mathcal{A})$, can be understood by considering darts thrown at a dartboard Ω containing two possibly overlapping regions $\mathcal{A}$ and $\mathcal{B}$.

Figure 12.6 Another view of Bayes rule, in which the dartboard has been divided into N disjoint regions $\Omega_1, \ldots, \Omega_N$. ($N = 3$ in this illustration.) The region $\mathcal{X}$, shown in yellow, overlaps one or more regions and contains all the instances for which a particular feature measurement was obtained. The goal is to determine to which region an instance x belongs, given that the instance is in $\mathcal{X}$.

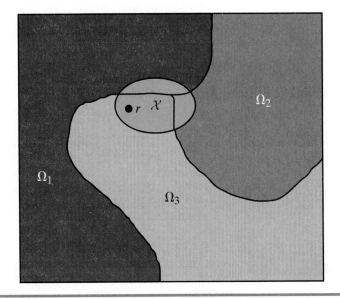

Now suppose that we have access to some particular instance, and we want to know to which category it belongs. Suppose we can measure some property, or *feature*, associated with the instance. Let $\mathcal{X}$ be the set of all instances that yield the particular measurement **x** that we obtained, as illustrated in Figure 12.6. By letting Ω_i play the role of region $\mathcal{A}$, and $\mathcal{X}$ play the role of region $\mathcal{B}$, Bayes' rule yields

$$p(\mathbf{r} \in \Omega_i | \mathbf{r} \in \mathcal{X}) = \frac{p(\mathbf{r} \in \mathcal{X} | \mathbf{r} \in \Omega_i) p(\mathbf{r} \in \Omega_i)}{p(\mathbf{r} \in \mathcal{X})} \qquad (12.16)$$

As before, we can rewrite the equation by simplifying the notation, letting ω_i refer to the category associated with region Ω_i, and letting **x** refer to the feature associated with the elements of $\mathcal{X}$:

$$p(\omega_i | \mathbf{x}) = \frac{p(\mathbf{x} | \omega_i) p(\omega_i)}{p(\mathbf{x})} \qquad (12.17)$$

or, in words,

$$\text{posterior} = \frac{\text{likelihood} \cdot \text{prior}}{\text{evidence}} \qquad (12.18)$$

The power of Bayes' rule lies in its ability to transform the difficult problem of computing the **posterior**, $p(\omega_i | \mathbf{x})$, into the easier problem of computing the **likelihood**, $p(\mathbf{x} | \omega_i)$. It is important to understand the distinction between these two terms. The posterior captures the problem that we are trying to solve—namely, to find the most likely category of an instance given information about the instance. In contrast, the likelihood specifies the probability of obtaining a particular feature **x** from an instance in a known category. In this manner, Bayes' rule allows us to perform *diagnostic* reasoning (computing the posterior) by instead performing *causal* reasoning (computing the likelihood).

The **prior** captures the probability of a particular result without access to any data. Sometimes instead of the term "posterior," you will see the term "a posteriori;" similarly you may see "a priori" instead of "prior." These mean the same thing: The posterior is indeed the *a posteriori* probability, because it is the probability after taking the data into account, whereas the prior is the *a priori* probability, because it does not depend on the data. The last term to consider, the **evidence**, is the probability of obtaining a particular measurement. But since the measurement has been obtained already, the evidence is the same for all possible categories, and therefore it is usually treated as an unimportant scaling factor, leading to a simpler expression for Bayes' rule:

$$\underbrace{p(\omega_i|\mathbf{x})}_{\text{posterior}} \propto \underbrace{p(\mathbf{x}|\omega_i)}_{\text{likelihood}} \underbrace{p(\omega_i)}_{\text{prior}} \tag{12.19}$$

where the equal sign has been replaced by the proportional sign.

Since our goal is to infer the proper category given some data, the most obvious solution is to select the category that yields the maximum value for the posterior. Known as **maximum a posteriori (MAP)** estimation, this approach selects the category $\hat{\omega}_i$ such that

$$\hat{\omega}_i = \arg\max_{\omega_i} p(\mathbf{x}|\omega_i)p(\omega_i) \tag{12.20}$$

where arg max returns the argument that maximizes the expression. In the case of a uniform prior, $p(\omega_i)$ is the same for all i, and MAP estimation therefore reduces to **maximum likelihood (ML)** estimation:

$$\hat{\omega}_i = \arg\max_{\omega_i} p(\mathbf{x}|\omega_i) \tag{12.21}$$

12.2.2 Bayesian Decision Theory

Bayesian decision theory is a statistical approach to classification based upon Bayes' rule. In this approach, we use the training set to estimate the **class-conditional probability density** $p(\mathbf{x}|\omega_i)$ for all i. Later we will discuss the various ways to do this, but for now note that since all the training data are manually labeled, we can take all the samples in the training set with label ω_i, compute the feature for each sample, then aggregate those features in some way to estimate the density. In other words, the class-conditional density captures how often feature $\mathbf{x}$ occurs in category ω_i.

What we want, of course, is the **posterior density**, $p(\omega_i|\mathbf{x})$. That is, given a measured feature $\mathbf{x}$, the posterior density yields the probability that the feature was obtained from a sample in category ω_i. It should be obvious that the sum of the posterior densities is 1:

$$\sum_{j=1}^{N} p(\omega_j|\mathbf{x}) = 1 \tag{12.22}$$

where N is the number of categories. Of course, Bayes' rule relates the two quantities:

$$p(\omega_i|\mathbf{x}) = \frac{p(\mathbf{x}|\omega_i)p(\omega_i)}{p(\mathbf{x})} \tag{12.23}$$

Substituting Equation (12.23) into Equation (12.22) leads to a relation on the class-conditional densities:

$$\sum_{j=1}^{N} p(\mathbf{x}|\omega_j)p(\omega_j) = p(\mathbf{x}) \tag{12.24}$$

so that the posterior in Equation (12.23) can be written as

$$p(\omega_i|\mathbf{x}) = \frac{p(\mathbf{x}|\omega_i)p(\omega_i)}{\sum_{j=1}^{N} p(\mathbf{x}|\omega_j)p(\omega_j)} \tag{12.25}$$

In the binary case, Equations (12.22) and (12.24) are simply

$$p(\omega_1|\mathbf{x}) + p(\omega_2|\mathbf{x}) = 1 \tag{12.26}$$

$$p(\mathbf{x}|\omega_1)p(\omega_1) + p(\mathbf{x}|\omega_2)p(\omega_2) = p(\mathbf{x}) \tag{12.27}$$

These two types of densities are illustrated in Figure 12.7 for a simple example with a one-dimensional feature.

If we let $L(\omega_i|\omega_j)$ be the loss associated with choosing label ω_i when the actual label is ω_j, then the **conditional risk** is defined as

$$R(\omega_i|\mathbf{x}) = \sum_{j=1}^{N} L(\omega_i|\omega_j)p(\omega_j|\mathbf{x}) \tag{12.28}$$

If we let $f: \mathbf{x} \mapsto \omega$ be the decision rule (or classifier), then the **overall risk**, which is the expected loss associated with the decision rule, is given by

$$R = \int R(f(\mathbf{x})|\mathbf{x})p(\mathbf{x}) \, d\mathbf{x} \tag{12.29}$$

where the integral is computed over the entire feature space. Obviously, if f is chosen so that it yields the smallest possible value for $R(f(\mathbf{x})|\mathbf{x})$ for every possible $\mathbf{x}$, then the overall risk is minimized. This leads to the following **Bayes decision rule**:

$$f(\mathbf{x}) = \arg\min_{\omega_i} R(\omega_i|\mathbf{x}) \tag{12.30}$$

In other words, for each input $\mathbf{x}$, f selects the category that minimizes the conditional risk. This minimum overall risk is known as the **Bayes risk**, and it is the best result that can be attained.

In the binary case, the Bayes decision rule in Equation (12.30) simplifies to

$$f(\mathbf{x}) = \begin{cases} \omega_1 & \text{if } R(\omega_1|\mathbf{x}) \le R(\omega_2|\mathbf{x}) \\ \omega_2 & \text{otherwise} \end{cases} \tag{12.31}$$

which, from Equation (12.28), is equivalent to

$$f(\mathbf{x}) = \begin{cases} \omega_1 & \text{if } \frac{p(\omega_1|\mathbf{x})}{p(\omega_2|\mathbf{x})} \ge \frac{L(\omega_1|\omega_2)}{L(\omega_2|\omega_1)} \\ \omega_2 & \text{otherwise} \end{cases} \tag{12.32}$$

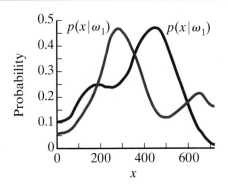 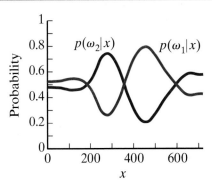

Figure 12.7 Example of class-conditional densities (left) and posterior densities (right) for a binary classification problem. Note that the former always integrate to 1 over all inputs, and the latter always sum to 1 for each input.

where we have assumed that $L(\omega_1|\omega_1) = L(\omega_2|\omega_2) = 0$, since it is not desirable to penalize correct responses. By applying Bayes' rule, this is equivalent to

$$f(\mathbf{x}) = \begin{cases} \omega_1 & \text{if } \frac{p(\mathbf{x}|\omega_1)}{p(\mathbf{x}|\omega_2)} \geq \tau \\ \omega_2 & \text{otherwise} \end{cases} \tag{12.33}$$

where $\frac{p(\mathbf{x}|\omega_1)}{p(\mathbf{x}|\omega_2)}$ is the **likelihood ratio**, and

$$\tau \equiv \frac{L(\omega_1|\omega_2)p(\omega_2)}{L(\omega_2|\omega_1)p(\omega_1)} \tag{12.34}$$

is a threshold. If ω_1 is treated as the positive category, then as the threshold is increased, the classifier will yield fewer false positives but also fewer true positives; whereas as the threshold is decreased, the classifier will yield fewer false negatives but also fewer true negatives. This is exactly the tradeoff captured by the ROC and PR curves mentioned earlier.

In computer vision, the zero-one loss function is nearly always used, in which case Equation (12.28) reduces to

$$R(\omega_i|\mathbf{x}) = \sum_{j \neq i} p(\omega_j|\mathbf{x}) = 1 - p(\omega_i|\mathbf{x}) \tag{12.35}$$

so that the Bayes decision rule minimizes the probability of error, or, rather, maximizes the *a posteriori* probability:

$$f(\mathbf{x}) = \arg \max_{\omega_i} p(\omega_i|\mathbf{x}) \tag{12.36}$$

which is known as **minimum error rate classification** or, equivalently, maximum *a posteriori* (MAP) classification. In this case the decision boundaries occur where the posteriors intersect, and the Bayes risk is the area under the curve generated by the minimum value among all posteriors for any value of the input feature, as shown in Figure 12.8. In the binary case, $L(\omega_2|\omega_1) = L(\omega_1|\omega_2) = 1$, and the Bayes decision rule reduces to selecting the category with the maximum posterior:

$$f(\mathbf{x}) = \begin{cases} \omega_1 & \text{if } p(\omega_1|\mathbf{x}) \geq p(\omega_2|\mathbf{x}) \\ \omega_2 & \text{otherwise} \end{cases} \tag{12.37}$$

Figure 12.8 With the zero-one loss function, the decision boundaries (vertical dashed lines) occur where the posteriors intersect, and the Bayes risk is the shaded area, or area under the curve generated by the minimum value among all posteriors for any value of x.

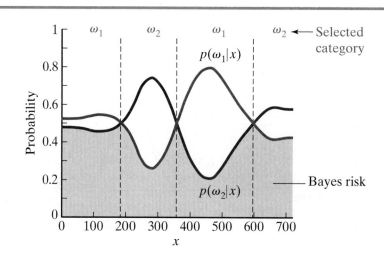

In addition, when the priors are equal, $p(\omega_1) = p(\omega_2)$, thus further simplifying the rule to

$$f(\mathbf{x}) = \begin{cases} \omega_1 & \text{if } p(\mathbf{x}|\omega_1) \geq p(\mathbf{x}|\omega_2) \\ \omega_2 & \text{otherwise} \end{cases} \qquad (12.38)$$

so that ω_1 is chosen if the likelihood ratio is greater than 1, or ω_2 otherwise.

To summarize, the general approach to learning a Bayesian classifier is to collect a training set, label the training set, compute features on the samples in the training set, and use these features to estimate the class-conditional densities. Then, the priors are estimated using either the distribution of the labels within the training set (if these are believed to be representative of the general population) or through some other means; if no such information is available, then the priors are set to 1. Since a zero-one loss function is nearly always used, once the Bayesian classifier has been learned, it can be applied to a new sample by extracting the feature vector from the sample and using Bayes' rule to estimate the posterior for each category, then choosing the category that yields the maximum posterior.

12.2.3 Parametric vs. Nonparametric Representations

There are two basic ways to represent probability distributions. A **parametric** representation uses an analytic expression with a small number of parameters, such as the coefficients of a polynomial, or the mean vector and covariance matrix of a Gaussian. A **nonparametric** representation, on the other hand, uses a data-driven approach, such as a histogram over the data, or simply the raw data themselves. In this section we present an example of a simple classification problem, for the purpose of illustrating the difference between parametric and nonparametric representations, as well as for providing some working knowledge of the use of class-conditional densities, posterior distributions, and likelihood ratios. First, a nonparametric approach is shown using graylevel histograms, then the problem is revisited using parametric Gaussian densities.

Consider the image shown at the top-left of Figure 12.9, which we have used before. In general, the foreground objects are lighter than the background, and therefore we expect the pixels on the foreground objects to have higher gray levels than those on the background. Suppose we wish to develop a classifier that will determine, based solely on a pixel's gray level, whether it is likely to be part of the foreground or background.

The first step is to manually label the pixels as belonging to one of the two categories, with the result shown in the top-right of the figure. Treating the image as a training image, we then construct the class-conditional densities, where ω_1 refers to the foreground and ω_2 refers to the background. For each of the two categories, a normalized histogram is constructed of the gray levels of all the pixels with that particular label in the ground truth. The resulting two normalized histograms $\bar{h}_1$ and $\bar{h}_2$, which are the class-conditional densities, are shown in the middle of Figure 12.9.

To compute the priors, simply count the number of pixels in each category, followed by an appropriate normalization. If we let h_1 and h_2 be the two non-normalized histograms, then the priors are computed as

$$p(\omega_1) = \frac{\sum_i h_1[i]}{\sum_i h_1[i] + \sum_i h_2[i]} = 0.26 \quad (\text{foreground prior}) \qquad (12.39)$$

$$p(\omega_2) = \frac{\sum_i h_2[i]}{\sum_i h_1[i] + \sum_i h_2[i]} = 0.74 \quad (\text{background prior}) \qquad (12.40)$$

Figure 12.9 Top: An image with ground truth binary labels. Middle: Class-conditional densities for the two categories represented nonparametrically as graylevel histograms. Bottom: Posteriors and log-likelihood ratio. The two categories have equal probability near gray level 110.

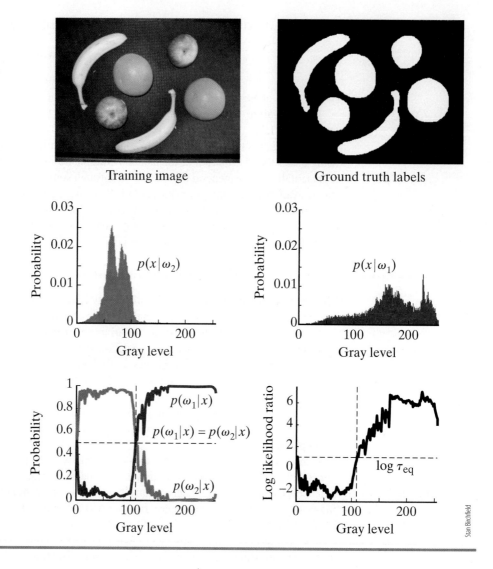

Training image   Ground truth labels

where $i = 0, \ldots, 255$. Note that $p(\omega_1) + p(\omega_2) = 1$. Keep in mind that this is just one way to compute priors. If the training data set cannot be trusted to be sufficiently representative of the underlying distribution, then both priors can simply be set to 0.5. For example, if we later want to run the algorithm on an image that contains just one piece of fruit, the prior from this particular procedure will skew the classifier into labeling more pixels as foreground than it would otherwise.

The posterior probability for any particular gray level i is computed, as in Equation (12.25), by multiplying the class-conditional density evaluated at i by the prior, and normalizing appropriately:

$$post_1[i] = \frac{\bar{h}_1[i]\, p(\omega_1)}{\bar{h}_1[i]\, p(\omega_1) + \bar{h}_2[i]\, p(\omega_2)} \quad \text{(foreground posterior)} \qquad (12.41)$$

$$post_2[i] = \frac{\bar{h}_2[i]\, p(\omega_2)}{\bar{h}_1[i]\, p(\omega_1) + \bar{h}_2[i]\, p(\omega_2)} \quad \text{(background posterior)} \qquad (12.42)$$

Again, note that $post_1[i] + post_2[i] = 1$ for all i, which is evident from the bottom-left of the figure.

The likelihood ratio is computed by taking the ratio of the class-conditional densities. By applying the logarithm, the ratio is converted into a difference, yielding the **log-likelihood ratio**:

$$log\text{-}like[i] \;=\; \log\frac{\overline{h}_1[i]}{\overline{h}_2[i]} = \log \overline{h}_1[i] - \log \overline{h}_2[i] \quad (\text{log-likelihood ratio}) \qquad (12.43)$$

resulting in a more natural dynamic range for display purposes. The log-likelihood is displayed in the bottom-right of the figure.

The posteriors cross at $p(\omega_1|\mathbf{x}) = p(\omega_2|\mathbf{x}) = 0.5$, or in terms of likelihood ratio:

$$\frac{p(\mathbf{x}|\omega_1)}{p(\mathbf{x}|\omega_2)} = \frac{p(\omega_2)}{p(\omega_1)} = \frac{0.74}{0.26} = 2.8 \equiv \tau_{eq} \qquad (12.44)$$

The horizontal line in the plot is at $\log(2.8) = 1.0$ (using the natural logarithm), while the vertical line is near gray level 110. Notice that if the distributions were different there could be more vertical lines, but there will be only one horizontal line.

As the threshold τ is varied, as in Equation (12.33), the classifier yields different results, five of which are shown in Figure 12.10. If we let ω_1 be the positive category, and ω_2 be the negative category, then as τ is decreased there are more true positives and also more false positives. The performance of the classifier is depicted in the ROC and PR curves shown in the figure. Keep in mind that in a real application, we would want to use a separate test

Figure 12.10 Top: Classification results using five different values for the likelihood ratio τ. As τ is decreased, the classifier yields more positives (both false and true). Bottom: Receiver operating characteristic (ROC) and precision-recall (PR) curves for the classifier, with the equal error rate (EER) shown for the former.

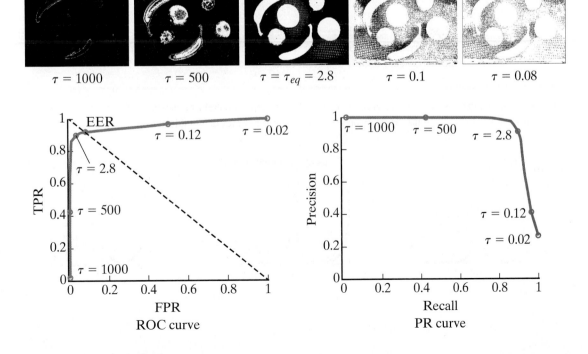

image for generating curves that were used to depict the performance of the classifier, but for illustrative purposes, the training image is sufficient.

So far we have use the nonparametric approach of graylevel histograms for modeling class-conditional densities. An alternate approach is to approximate these densities with Gaussian distributions by computing the mean and variance of the histograms:

$$\mu_1 = \sum_i \bar{h}_1[i]i \qquad \sigma_1^2 = \sum_i \bar{h}_1[i](i - \mu_1)^2 \tag{12.45}$$

$$\mu_2 = \sum_i \bar{h}_2[i]i \qquad \sigma_2^2 = \sum_i \bar{h}_2[i](i - \mu_2)^2 \tag{12.46}$$

The class-conditional densities, posteriors, and log-likelihood ratio using Gaussian distributions are all shown in Figure 12.11. The class-conditional densities cross near gray level 107, whereas the posteriors and log-likelihood cross near gray level 116. Notice that the fact that $p(\omega_1) < p(\omega_2)$ causes the threshold to move slightly toward the right (i.e., increases the threshold) to make it more likely that pixels are labeled background (ω_2) by the classifier.

12.2.4 Gaussian Densities

Let us continue our discussion of parametric approaches by considering the most popular parametric density, namely, the Gaussian. When it is a function of a single variable, the Gaussian has the following form:

$$p(x|\omega_i) = \frac{1}{\sqrt{2\pi}\sigma}e^{-\frac{(x - \mu_i)^2}{2\sigma_i^2}} \tag{12.47}$$

where μ_i and σ_i are the mean and variance, respectively, and where x is a scalar. When it is a function of multiple variables, it is the multivariate Gaussian:[†]

$$p(\mathbf{x}|\omega_i) = \frac{1}{(2\pi)^{\frac{d}{2}}|\mathbf{\Sigma}|^{\frac{1}{2}}}e^{-\frac{1}{2}(\mathbf{x} - \boldsymbol{\mu}_i)^{\mathsf{T}}\mathbf{\Sigma}_i^{-1}(\mathbf{x} - \boldsymbol{\mu}_i)} \tag{12.48}$$

where $\boldsymbol{\mu}_i$ and $\mathbf{\Sigma}_i$ are the mean vector and covariance matrix, respectively, and where $\mathbf{x} \in \mathbb{R}^d$. For the rest of this section, we will concern ourselves with the multivariate Gaussian, since the scalar Gaussian is a special case.

Figure 12.11 Corresponding plots for the classifier when the densities are approximated with parametric Gaussians. The posteriors are equally probable near gray level 116, while the class-conditional densities (since they do not take the priors into account) cross near gray level 107.

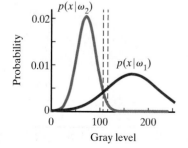
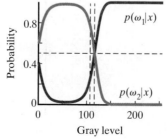
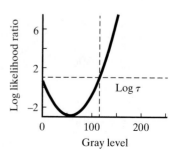

[†] Section 11.5.2 (p. 551).

The discriminant function for a posterior distribution can be chosen as any monotonically increasing function of that distribution, since it is only the order of the values that matters. For convenience, apply the natural logarithm and invoke Bayes' rule:

$$g_i(\mathbf{x}) \equiv \log p(\omega_i|\mathbf{x}) = \log p(\mathbf{x}|\omega_i)p(\omega_i) = \log p(\mathbf{x}|\omega_i) + \log p(\omega_i) \quad (12.49)$$

Substituting Equation (12.48) and ignoring the leading constant $(2\pi)^{-d/2}$ yields

$$g_i(\mathbf{x}) \equiv -\frac{1}{2}\log|\mathbf{\Sigma}_i| - \frac{1}{2}(\mathbf{x} - \boldsymbol{\mu}_i)^\mathsf{T}\mathbf{\Sigma}_i^{-1}(\mathbf{x} - \boldsymbol{\mu}_i) + \log p(\omega_i) \quad (12.50)$$

Classification using Gaussians is therefore straightforward: Simply select the category that yields the largest value for the discriminant function in Equation (12.50). However, analytically describing the decision boundaries is not easy, because the equation describes, in general, a hyperquadric, which can assume any of a variety of forms.

A special case occurs when all the Gaussian densities share the same covariance matrix, that is, $\mathbf{\Sigma}_i = \mathbf{\Sigma}$ for all i. In that case the leading term in Equation (12.50) can be dropped, leading to

$$g_i(\mathbf{x}) \equiv -\frac{1}{2}(\mathbf{x} - \boldsymbol{\mu}_i)^\mathsf{T}\mathbf{\Sigma}^{-1}(\mathbf{x} - \boldsymbol{\mu}_i) + \log p(\omega_i) \quad (12.51)$$

$$= -\frac{1}{2}(\mathbf{x}^\mathsf{T}\mathbf{\Sigma}^{-1}\mathbf{x} - 2\boldsymbol{\mu}_i^\mathsf{T}\mathbf{\Sigma}^{-1}\mathbf{x} + \boldsymbol{\mu}_i^\mathsf{T}\mathbf{\Sigma}^{-1}\boldsymbol{\mu}_i) + \log p(\omega_i) \quad (12.52)$$

Since $\mathbf{x}^\mathsf{T}\mathbf{\Sigma}^{-1}\mathbf{x}$ does not depend upon the category i, this term can be dropped, thus revealing that the discriminant function is linear:

$$g_i(\mathbf{x}) \equiv \mathbf{w}_i^\mathsf{T}\mathbf{x} + \beta_i \quad (12.53)$$

where

$$\mathbf{w}_i \equiv \mathbf{\Sigma}^{-1}\boldsymbol{\mu}_i \quad (12.54)$$

$$\beta_i \equiv -\frac{1}{2}\boldsymbol{\mu}_i^\mathsf{T}\mathbf{\Sigma}^{-1}\boldsymbol{\mu}_i + \log p(\omega_i) \quad (12.55)$$

With a linear discriminant function, the decision boundary between two adjacent regions is a *hyperplane*[†] defined by

$$\mathbf{w}^\mathsf{T}(\mathbf{x} - \mathbf{c}) = 0 \quad (12.56)$$

where

$$\mathbf{w} \equiv \mathbf{\Sigma}^{-1}(\boldsymbol{\mu}_i - \boldsymbol{\mu}_j) \quad (12.57)$$

$$\mathbf{c} \equiv \frac{1}{2}(\boldsymbol{\mu}_i + \boldsymbol{\mu}_j) + \frac{\boldsymbol{\mu}_i - \boldsymbol{\mu}_j}{(\boldsymbol{\mu}_i - \boldsymbol{\mu}_j)^\mathsf{T}\mathbf{\Sigma}^{-1}(\boldsymbol{\mu}_i - \boldsymbol{\mu}_j)}\log\frac{p(\omega_i)}{p(\omega_j)} \quad (12.58)$$

It is clear from Equation (12.58) that the hyperplane intersects the line joining the means at the point $\mathbf{c}$, and furthermore that the point $\mathbf{c}$ is exactly halfway between the means when the priors are equal. (If the priors are not equal, then $\mathbf{c}$ is shifted away from the mean of the more likely distribution.) In addition, when the priors are equal, the decision

---

[†] A hyperplane is a generalization of a plane to any number of dimensions, for example a plane when $d = 3$ or line when $d = 2$.

can be made by simply selecting the category whose mean is the closest using the Mahalanobis distance.[†]

A further special case occurs when all the variances of all the Gaussian densities are equal, that is, $\mathbf{\Sigma}_i = \sigma^2 \mathbf{I}_{\{d \times d\}}$ for all i. In that case, Equation (12.51) simplifies to

$$g_i(\mathbf{x}) \equiv -\frac{1}{2\sigma^2} \|\mathbf{x} - \boldsymbol{\mu}_i\|^2 + \log p(\omega_i) \tag{12.59}$$

As before, the discriminant function is linear, $g_i(\mathbf{x}) \equiv \mathbf{w}_i^\mathsf{T} \mathbf{x} + \beta_i$, where

$$\mathbf{w}_i \equiv \frac{1}{\sigma^2} \boldsymbol{\mu}_i \tag{12.60}$$

$$\beta_i \equiv -\frac{1}{2\sigma^2} \|\boldsymbol{\mu}_i\|^2 + \log p(\omega_i) \tag{12.61}$$

With a linear discriminant function, the decision boundary between two adjacent regions is a hyperplane defined by $\mathbf{w}^\mathsf{T}(\mathbf{x} - \mathbf{c}) = 0$, where

$$\mathbf{w} \equiv \boldsymbol{\mu}_i - \boldsymbol{\mu}_j \tag{12.62}$$

$$\mathbf{c} \equiv \frac{1}{2}(\boldsymbol{\mu}_i + \boldsymbol{\mu}_j) + \frac{\sigma^2(\boldsymbol{\mu}_i - \boldsymbol{\mu}_j)}{\|\boldsymbol{\mu}_i - \boldsymbol{\mu}_j\|^2} \log \frac{p(\omega_i)}{p(\omega_j)} \tag{12.63}$$

This hyperplane is perpendicular to the line joining the means at the point $\mathbf{c}$. As before, if the priors are equal, then $\mathbf{c}$ is exactly halfway between the means, and the decision can be made by simply selecting the category whose mean is the closest using the Euclidean distance.

12.3 Generative Methods

Classification methods come in one of two flavors:

- **Generative methods** construct a *generative model* for each category by learning the class-conditional densities, then determine how well the measured data fit into these models. By providing an explicit model for the probability density of each category, this type of approach is more flexible and enables synthetic data to be generated by sampling the distributions.
- **Discriminative methods** construct a *discriminative model* that directly captures the decision boundaries explicitly, without attempting to model the underlying densities.

We continue to explore generative methods in this section, followed by discriminative methods in the next section.

12.3.1 Histograms

One of the most common, and simplest, nonparametric generative models is the histogram. One of the primary advantages of histograms is that, as the histogram is being computed, the data can be discarded. Therefore, the amount of memory needed is related solely to the size of the histogram rather than to the amount of data. Nevertheless, a particularly acute drawback of histograms is that the amount of memory needed is exponential in the

---

[†] Section 4.3.1 (p. 164).

number of dimensions, so that histograms are not practical when d is much bigger than 3, and certainly infeasible when $d > 5$, which is an effect of the curse of dimensionality. (Recall that histograms can be computed on any type of data, not just gray levels.)

To see how this works in practice, let us consider a real-world application of histograms, namely the detection of skin-colored pixels in an RGB color image—that is, automatically determining which pixels in the image contain human skin. Let us simplify the problem so that the decision is made independently for each pixel based solely on the color of that pixel, so that the feature vector $\mathbf{x}$ is the RGB value[†] of the pixel. To train the classifier, a large number of images of a variety of scenes is collected, and the pixels in these images are then manually labeled according to whether they belong to the category of human skin color or not.

From these labeled pixels, both the prior $p(\omega_i)$ and the likelihood $p(\mathbf{x}|\omega_i)$ are estimated for the two categories using two normalized histograms, each containing the class-conditional density of one of the two classes. The likelihoods are obtained by looking up the values in the appropriate bins of the normalized histograms:

$$p((r, g, b)|\omega_1) \approx \bar{h}_{skin}(r, g, b) = \frac{n_{skin}^{(r,g,b)}}{n_{skin}} \tag{12.64}$$

$$p((r, g, b)|\omega_2) \approx \bar{h}_{nonskin}(r, g, b) = \frac{n_{nonskin}^{(r,g,b)}}{n_{nonskin}} \tag{12.65}$$

where (r, g, b) is the RGB triplet, $\bar{h}_i(r, g, b)$ is the ith normalized histogram evaluated at the triplet, $n_i^{(r,g,b)}$ is the number of pixels in the ith category with the particular RGB values, and $n_i = \sum_{(r,g,b)} n_i^{(r,g,b)}$ is the total number of pixels in the ith category, where $i = 1$ refers to "skin" and $i = 2$ refers to "nonskin." Assuming that the pixels selected for inclusion in the training set are representative of the pixels likely to be encountered in unlabeled data, the priors can be estimated using the fraction of the instances of the two categories:

$$p(\omega_1) \approx \frac{n_{skin}}{n_{skin} + n_{nonskin}} \quad \text{and} \quad p(\omega_2) \approx \frac{n_{nonskin}}{n_{skin} + n_{nonskin}} \tag{12.66}$$

For 24-bit color images, there are 256 possible values for red, 256 possible values for green, and 256 possible values for blue. As a result, there are $256^3 = 16{,}777{,}216$ possible colors for any given pixel. In practice, however, a histogram with such a large number of entries has two drawbacks. First, the amount of memory required is large, which negatively impacts computational performance. Secondly, with a limited amount of training data many of the bins will not contain any pixels. Such empty bins lead to the mistaken belief that the corresponding color is unlikely even if there is a large number of pixels in the adjacent bin, whose color is almost indistinguishable from that of the bin with no pixels. Therefore, the values are typically quantized into a smaller number of bins. For example, a histogram with $8^3 = 512$ bins can be created by dividing each of the red, green, and blue dimensions into eight equally-spaced intervals. Pixels with the same green and blue values will then be placed into different bins depending upon whether their red value falls within the range 0–31, 32–63, 64–95, . . . or 224–255.

Figure 12.12 shows the output of a histogram-based skin color detector on several images. Also shown are visualizations of the skin and nonskin histograms, obtained by projecting the histograms along the green-magenta axis and displaying the isocontours

---

[†] This is actually the nonlinear $R'G'B'$ value, see Section 9.5.1 (p. 420); other color spaces could be used as well.

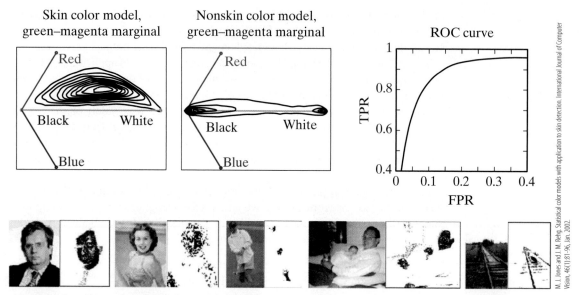

M. J. Jones and J. M. Rehg. Statistical color models with application to skin detection. International Journal of Computer Vision, 46(1):81–96, Jan. 2002.

Figure 12.12 Top-Left: Skin and non-skin color models, plotted as isocontours after projecting along the green-magenta axis. Note that skin color tends to feature more red than green or blue, while nonskin colors are fairly evenly distributed around the line of grays. Top-Right: ROC curve for the histogram-based skin color detector using 16^3 bins. Bottom: Results of the skin color detector on several images.

(that is, the contours of equal probability). Notice that skin-colored pixels contain more red than blue, and non-skin-colored pixels are fairly evenly distributed along the line of grays. Also shown in the figure is the ROC curve for a histogram with 16^3 cells. Not only does the smaller histogram have the advantage of requiring less memory, but it actually outperforms the 256^3 histogram (not shown) due to its better smoothing of the data.

12.3.2 Kernel Density Estimation (KDE)

In contrast to histograms, which discard all the data as the histogram is being constructed, **memory-based learning methods** (also known as *lazy learning* methods) do not process the training data at all but rather store them until a query comes in. The two most common forms of memory-based learning are **kernel density estimation (KDE)**, which we consider in this section, and nearest neighbors, which is covered in the next section. Like histograms, both of these techniques are nonparametric.

Suppose we have a set of n_j training data samples $\mathbf{x}_1, \ldots, \mathbf{x}_{n_j} \in \mathbb{R}^d$ with a particular label ω_j. The approach of kernel density estimation places a kernel function at the center of each data sample, and the probability density of that label is then the normalized sum of all these kernel functions:

$$p(\mathbf{x}|\omega_j) \approx \frac{1}{n_j} \sum_{i=1}^{n_j} \frac{1}{h^d} \varphi\left(\frac{\mathbf{x} - \mathbf{x}_i}{h}\right) \tag{12.67}$$

where h is the bandwidth of the kernel $\varphi(\cdot)$, which is usually related to the number of data samples, e.g., $h = h_0/\sqrt{n_j}$, where h_0 is a constant. The kernel function itself can be any symmetric, monotonically decreasing function from the origin, such as the box window function or the isotropic Gaussian, with the latter being the most common: $\varphi(\mathbf{x}) = \frac{1}{(\sqrt{2\pi})^d} \exp(-\|\mathbf{x}\|^2/2)$, leading to

$$p(\mathbf{x}|\omega_j) \approx \frac{1}{n_j} \sum_{i=1}^{n_j} \frac{1}{(h\sqrt{2\pi})^d} \exp\left(-\frac{\|\mathbf{x} - \mathbf{x}_i\|^2}{2h^2}\right) \tag{12.68}$$

Consider, for example, the 1D training set $x_1 = 2$, $x_2 = 5$, $x_3 = 8$, $x_4 = 9$, as shown in Figure 12.13. The probability density obtained by adding the individual **kernel densities** for the training samples, then normalizing to ensure that the density integrates to 1. When the variance of the kernel functions (or kernel densities) is small, then the resulting density closely follows each individual data sample, whereas when the variance is large, the resulting density smooths over individual data samples.

Kernel density estimation is also known as the **Parzen window** method. To derive the approach, let $p_v(\mathbf{x}|\omega_j)$ be the probability that a particular data sample $\mathbf{x}$ lands within a particular region $\mathcal{R}$ whose volume is v. If the true probability density $p(\mathbf{x}|\omega_j)$ varies smoothly, and the region is small enough that the density can be considered approximately constant within the region, then we have

$$p_v(\mathbf{x}|\omega_j) = \int_{\mathcal{R}} p(\mathbf{x}'|\omega_j)d\mathbf{x}' \approx p(\mathbf{x}|\omega_j)v \tag{12.69}$$

If n_j is the number of data samples with label ω_j, then the expected number $E[k_j]$ of samples within the volume is given by

$$E[k_j] = p_v(\mathbf{x}|\omega_j)n_j \approx p(\mathbf{x}|\omega_j)vn_j \tag{12.70}$$

Therefore, the probability peaks at $p_v(\mathbf{x}|\omega_j) = k_j/n_j$, so that if k_j samples fall within the volume, then the probability density can be estimated by

$$p(\mathbf{x}|\omega_j) \approx \frac{k_j}{vn_j} \tag{12.71}$$

If we let φ be the box kernel function that returns 1 if $\mathbf{x}$ is inside the unit hypercube centered at the origin:

$$\varphi(\mathbf{x}) \equiv \begin{cases} 1 & \text{if } \|\mathbf{x}\|_\infty \leq \frac{1}{2} \\ 0 & \text{otherwise} \end{cases} \tag{12.72}$$

Figure 12.13 Gaussian kernel density estimation of a 1D training set consisting of 4 samples. The kernel densities (red curves) are centered at each data sample, while the overall density (blue curve) is obtained by summing the kernel densities, then normalizing. The width of the kernels has a significant effect on the result, with $\sigma^2 = 0.5$ (left) and $\sigma^2 = 2$ (right) shown.

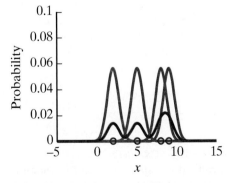
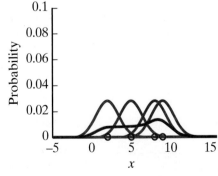

where $\|\mathbf{x}\|_\infty$ is the maximum element in the vector, then the number of samples falling within a hypercube with length h is

$$k_j = \sum_{i=1}^{n_j} \varphi\left(\frac{\mathbf{x} - \mathbf{x}_i}{h}\right) \tag{12.73}$$

and the volume of the hypercube is $v = h^d$. Substituting into Equation (12.71) yields

$$p(\mathbf{x}|\omega_j) \approx \frac{1}{n_j} \sum_{i=1}^{n_j} \frac{1}{h^d} \varphi\left(\frac{\mathbf{x} - \mathbf{x}_i}{h}\right) \tag{12.74}$$

Equation (12.67) is then obtained from this equation by simply relaxing the constraint that φ is the hypercube and instead allowing it to be any symmetric, monotonically decreasing function.

There is a close connection between Parzen windows and histograms. Let h refer to the width of a histogram bin along any particular dimension, and let $\mathbf{b_x}$ refer to the center of the histogram bin into which $\mathbf{x}$ falls. Then the value of the bin associated with $\mathbf{b_x}$ is the number of samples that fall within the bin: $\sum_{i=1}^{n_j} \varphi((\mathbf{x} - \mathbf{b_x})/h)$, where φ is the same hypercube as above. The normalized histogram approximates the density for $\mathbf{x}$ as follows:

$$p(\mathbf{x}|\omega_j) \approx \frac{1}{n_j} \sum_{i=1}^{n_j} \varphi\left(\frac{\mathbf{x} - \mathbf{b_x}}{h}\right) \tag{12.75}$$

The similarity between the two approaches should be apparent. In both cases, the samples that fall within scaled unit boxes are counted. The difference is that the centers of the boxes are fixed in a histogram, whereas the Parzen window boxes are centered at the data samples. Histograms perform some computation up front to quantize the data into fixed bins, after which evaluating the density requires a simple lookup. Parzen windows, on the other hand, simply store the raw data, but evaluating the density requires summing over all the stored data. The size of a histogram bin is related to the dimensionality of the space, whereas the size of the Parzen window representation is related to the number of data samples. Also, while the curse of dimensionality directly affects the size of the histogram, it only indirectly affects the Parzen window approach, because the number of samples needed grows exponentially with the number of dimensions.

The normalization by h^d in Equation (12.74) is necessary to ensure that the density integrates to 1. Such normalization is not needed for the histogram because the total number of samples contained in all the histogram bins is n, no matter the value of h. But for Parzen windows, increasing h means increasing the number of samples counted.

Kernel density estimation, or Parzen windows, also bears a close relationship to **locally weighted averaging (LWA)**. Let us redefine $\mathbf{x}_i$ as a training data sample with label $\ell(\mathbf{x}_i)$, let $n = n_1 + \cdots + n_N$ be the total number of training data samples of all categories, and let $\delta(a, b) = 1$ if $a = b$ or 0 if $a \neq b$. Then LWA is computed as

$$p(\omega_j|\mathbf{x}) = \frac{\sum_{i=1}^{n} \exp\left(-\dfrac{\|\mathbf{x} - \mathbf{x}_i\|^2}{2h^2}\right) \delta(\ell(\mathbf{x}_i), \omega_j)}{\sum_{i=1}^{n} \exp\left(-\dfrac{\|\mathbf{x} - \mathbf{x}_i\|^2}{2h^2}\right)} \tag{12.76}$$

which is nearly identical to Equation (12.68), since the numerator is the same as the $\exp(\cdot)$ expression in Equation (12.68), and the denominator simply ensures the proper normalization. Note, however, that Equation (12.68) is the class-conditional density, whereas Equation (12.76) is the posterior.

12.3.3 Nearest Neighbors

One of the simplest, and yet surprisingly powerful, classifiers is the **nearest-neighbor (NN) classifier**, which selects the category corresponding to the closest training sample:

$$f(\mathbf{x}) = \ell(\mathbf{x}_{\hat{i}}), \quad \text{where} \quad \hat{i} = \arg\min_i \|\mathbf{x} - \mathbf{x}_i\| \tag{12.77}$$

where $\ell(\mathbf{x}_{\hat{i}})$ is the label assigned to data sample $\mathbf{x}_{\hat{i}}$, and $\mathbf{x}_{\hat{i}}$ is the **nearest neighbor** of the query $\mathbf{x}$. Equation (12.77) is known as the **nearest-neighbor rule**. Like kernel density estimation, the nearest-neighbor classifier is a memory-based learning method because it involves storing all the training data, without any processing beforehand. It is also called a **1-nearest-neighbor** classifier, since it relies upon a single training sample.

More generally, **k-nearest neighbors (kNN)** implements the **k-nearest-neighbor rule**, which selects the category associated with the majority of the k closest neighbors to the query, thus achieving more robust results. Usually k is set so that $k = k_0\sqrt{n}$, where k_0 is a constant. This rule can be seen by applying Equation (12.71) and Bayes' rule to compute the posterior:

$$p(\omega_j|\mathbf{x}) = \frac{p(\mathbf{x}|\omega_j)p(\omega_j)}{p(\mathbf{x})} = \frac{p(\mathbf{x}|\omega_j)p(\omega_j)}{\sum_{j'=1}^N p(\mathbf{x}|\omega_{j'})p(\omega_{j'})} \approx \frac{\frac{k_j}{vn_j} \cdot \frac{n_j}{n}}{(k_1 + \cdots + k_N) \cdot \frac{1}{vn}} = \frac{k_j \cdot \frac{1}{vn}}{k \cdot \frac{1}{vn}} = \frac{k_j}{k} \tag{12.78}$$

where k is the number of samples of any label within the volume v. From this equation we see that the *a posteriori* probability of ω_j is simply the fraction of samples with that label. Therefore the minimum error rate is achieved by selecting the category that appears most frequently. A natural extension is to assign weights to the samples, for example using the inverse distance to the query, to allow some samples to vote more than others.

There is a close connection between k-nearest neighbors and the Parzen window method, both of which are nonparametric approaches. With Parzen windows, the bandwidth of the kernel (and therefore the volume v) is fixed for a given n, so that the number of data samples used to estimate the density varies throughout the feature space depending on the local density of the data samples. In contrast, with kNN, the number of data samples is fixed for a given n, and the kernel bandwidth (and therefore the volume v) varies throughout the feature space depending on the local density of the data samples. In either case, as n goes to infinity, the volume becomes infinitesimally small and yet contains an infinite number of samples. Therefore, kNN is able to approach the Bayes error rate. In fact, as n approaches infinity, it can be shown that the error rate of even 1-NN is no worse than twice that of the Bayesian classifier.

The training samples are known as **prototypes**. In effect, NN and kNN compare the query $\mathbf{x}$ with all the prototypes, using the nearest prototype(s) to determine the category of $\mathbf{x}$. Prototypes have been a popular approach to classification since the early days of computer vision. This is done by performing **template matching** (such as cross-correlation) to a set of 2D templates of the objects of interest in order to determine the proper category. Studies have shown that the human visual system uses a similar approach—for example, to classify colors by comparing with a small set of prototype colors called **focal colors**, which are largely language- and culture-independent. Such prototypes are far more effective than colors along the decision boundaries. The prototypes cause a **Voronoi tesselation** of the feature space, where all the feature values that are closest to a particular prototype are within the cell carved by that prototype, as shown in Figure 12.14. As the query moves across boundaries in the Voronoi tesselation, the set of nearest neighbors changes, thus causing discontinuities in the slope of the density.

Figure 12.14 Nearest neighbors assigns each possible feature to its nearest prototype, which can be represented by a Voronoi tesselation of the feature space. In this 2D example, the prototypes are the small red dots, while the Voronoi tesselation is the set of blue lines.

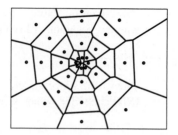

Drawbacks to nearest neighbor approaches include the large amount of memory needed to store the training data, as well as the computational complexity of comparing a query to all the training data. The basic algorithm is to simply perform a linear search through all the samples in the training data, comparing the query with each one. Three general approaches have been invented to reduce the burden of nearest neighbors. In the first approach, called **partial distance**, the difference between the query and each training sample is computed using only some of the dimensions, and if the distance exceeds a certain value, it is discarded. In the second approach, called **space partitioning**, a search tree such as a k-d tree is used to speed up the search. Finally, the training set can be **pruned** by removing samples whose nearest neighbors (i.e., adjacent cells in the Voronoi tesselation) are all of the same category, since such samples are far from the decision boundaries.

12.3.4 Naive Bayes

As mentioned earlier, the curse of dimensionality makes it impossible to use histograms for higher dimensional feature spaces. As a rule of thumb, anything beyond 3D is difficult to model with histograms, and anything beyond 5D is impossible. Such a large histogram would not only require too much memory, but also the table lookup would be extremely slow due to caching issues. Moreover, with high dimensionality, it is difficult to get enough training data, so that most of the bins in a large histogram will be zero, thus making a poor representation of the underlying distribution. In one study,[†] over a billion pixels were examined from images obtained via a web search crawl, and over 77% of the possible RGB colors were never encountered. That is a lot of zeros.

One trick for using histograms to model distributions in high dimensional feature spaces is to assume independence among the different elements of the feature vector. That is, if we let $x^{(i)}$ be the ith element in the feature vector, so that $\mathbf{x} = [x^{(1)} \quad x^{(2)} \quad \cdots \quad x^{(d)}]^{\mathsf{T}}$, then this assumption can be written as

$$p(\mathbf{x}|\omega_j) = p(x^{(1)}, x^{(2)}, \cdots, x^{(d)}|\omega_j) = \prod_{i=1}^{d} p(x^{(i)}|\omega_j) \tag{12.79}$$

This approach, known as **Naive Bayes**, in effect projects the d-dimensional distribution onto d separate axes, yielding d 1D distributions. As a result, the original d-dimensional histogram is replaced by d 1D histograms. Since multiplying small probabilities can lead to numerical instabilities, it is usually better to work with log likelihoods, which converts the multiplications into summations:

$$\log p(\mathbf{x}|\omega_j) = \log \prod_{i=1}^{d} p(x^{(i)}|\omega_j) = \sum_{i=1}^{d} \log p(x^{(i)}|\omega_j) \tag{12.80}$$

Figure 12.15 shows the results of a well-known Naive Bayes classifier for detecting faces despite their orientation (that is, whether they are facing the camera or sideways).

[†] The results of the study are shown in Figure 12.12.

H. Schneiderman and T. Kanade. A statistical method for 3D object detection applied to faces and cars. In Proceedings of the IEEE Conference on Computer Vision and Pattern Recognition (CVPR), June 2000.

Figure 12.15 Results of a view-invariant face detector using a Naive Bayes classifier with wavelet features.

12.3.5 Principal Components Analysis (PCA)

We come now to an important principle known as **dimensionality reduction**. According to the curse of dimensionality, it is undesirable to have too many dimensions in the feature space because of the additional memory required, the extra computation needed, the large number of training samples required, and the decrease in performance. It is therefore important to be able to distinguish the important dimensions from the less important dimensions, and then to discard the latter if possible.

A popular approach to dimensionality reduction is **principal components analysis (PCA)**. The basic idea behind PCA is to approximate the data with a multivariate Gaussian, then use the variances of the Gaussian along the different dimensions to determine the importance of those dimensions, discarding the dimensions with low variance. Recall from our earlier discussion of the best-fitting ellipse[†] that the covariance matrix rotates the feature space so as to best align the data with the new axes, and that the square root of the eigenvalues of the covariance matrix are proportional to the length of the axes of the Gaussian. Therefore, the eigenvalues indicate which dimensions should be kept and which can be discarded. PCA uses this same procedure of orthogonalization that we saw earlier, combined with a dimensionality reduction step based on this observation regarding the eigenvalues.

The details of PCA are as follows. Suppose we have a set of n points in a d-dimensional space: $\mathbf{x}_i \in \mathbb{R}^d$, $i = 1, \ldots, n$. The mean vector and covariance matrix are computed, respectively, as the average of the points and the normalized outer product of the centered points:[‡]

$$\mu = E[\mathbf{x}] = \frac{1}{n}\sum_{i=1}^{n}\mathbf{x}_i \tag{12.81}$$

$$\mathbf{C} = E[(\mathbf{x} - \mu)(\mathbf{x} - \mu)^\mathsf{T}] = \frac{1}{n}\sum_{i=1}^{n}(\mathbf{x}_i - \mu)(\mathbf{x}_i - \mu)^\mathsf{T} = \frac{1}{n}\mathbf{A}\mathbf{A}^\mathsf{T} \tag{12.82}$$

where $E[\cdot]$ is the expectation, and $\mathbf{A}$ is defined as

$$\mathbf{A} \equiv [\mathbf{x}_1 \quad \mathbf{x}_2 \quad \cdots \quad \mathbf{x}_n] - \mu\mathbf{1}^\mathsf{T}_{\{n \times 1\}} \tag{12.83}$$

where $\mathbf{1}_{\{n \times 1\}}$ is a vector of n 1s.

[†] Section 4.4.4 (p. 179) and Section 4.4.5 (p. 182)

[‡] Sometimes you will see the normalization of the covariance matrix to be $\frac{1}{n-1}$ instead of $\frac{1}{n}$. Called Bessel's correction, this approach is technically required to yield an unbiased estimate of the underlying distribution from which the sample points (x_i, y_i) are drawn. However, understanding this distinction is beyond the scope of this book, and thankfully, for any reasonably sized data set (e.g., $n \geq 100$), the difference between the two definitions is negligible.

Since $\mathbf{C}$ is real and symmetric, its eigenvalues are real, and its eigenvectors are mutually orthonormal. Let $\mathbf{v}_1, \ldots, \mathbf{v}_d$ be the eigenvectors, and $\lambda_1, \ldots, \lambda_d$ the corresponding eigenvalues, which for convenience are sorted in decreasing order: $\lambda_1 \geq \lambda_2 \geq \cdots \geq \lambda_d$. Recall from the definition of eigenvector that $\mathbf{C}\mathbf{v}_i = \lambda_i \mathbf{v}_i$ for all $i = 1, \ldots, d$. Stacking the eigenvectors into a matrix $\mathbf{P} = \begin{bmatrix} \mathbf{v}_1 & \cdots & \mathbf{v}_d \end{bmatrix}$ and the eigenvalues into a diagonal matrix $\mathbf{\Lambda} = diag(\lambda_1, \ldots, \lambda_d)$ thus yields the equation $\mathbf{C}\mathbf{P} = \mathbf{P}\mathbf{\Lambda}$, as we saw in Equation (4.141). Since $\mathbf{P}$ is an orthogonal matrix, $\mathbf{P}^{-1} = \mathbf{P}^\mathsf{T}$, and therefore we can factor the covariance matrix into the product of a diagonal matrix and two orthogonal matrices: $\mathbf{C} = \mathbf{P}\mathbf{\Lambda}\mathbf{P}^\mathsf{T}$.

Now let us define the following orthogonal transformation of the data:

$$\mathbf{x}' \equiv \mathbf{P}^\mathsf{T}(\mathbf{x} - \boldsymbol{\mu}) \tag{12.84}$$

with the inverse transform:

$$\mathbf{x} = \mathbf{P}\mathbf{x}' + \boldsymbol{\mu} \tag{12.85}$$

where both $\mathbf{x} \in \mathbb{R}^d$ and $\mathbf{x}' \in \mathbb{R}^d$. It is easy to show that the mean of $\mathbf{x}'$ is zero and the covariance matrix of $\mathbf{x}'$ is diagonal:

$$\boldsymbol{\mu}_{\mathbf{x}'} = \frac{1}{n}\sum_{i=1}^n \mathbf{y}_i = \frac{1}{n}\sum_{i=1}^n \mathbf{P}^\mathsf{T}(\mathbf{x}_i - \boldsymbol{\mu}) = \mathbf{P}^\mathsf{T}\left(\frac{1}{n}\sum_{i=1}^n \mathbf{x}_i - \frac{1}{n}\sum_{i=1}^n \boldsymbol{\mu}\right) = \mathbf{P}^\mathsf{T}(\boldsymbol{\mu} - \boldsymbol{\mu}) = \mathbf{0}_{\{d\times 1\}} \tag{12.86}$$

$$\mathbf{C}_{\mathbf{x}'} = \frac{1}{n}\sum_{i=1}^n \mathbf{y}_i\mathbf{y}_i^\mathsf{T} = \mathbf{P}^\mathsf{T}\left[\frac{1}{n}\sum_{i=1}^n (\mathbf{x}_i - \boldsymbol{\mu})(\mathbf{x}_i - \boldsymbol{\mu})^\mathsf{T}\right]\mathbf{P} = \mathbf{P}^\mathsf{T}\mathbf{C}\mathbf{P} = \mathbf{\Lambda} \tag{12.87}$$

where $\mathbf{0}_{\{d\times 1\}}$ is a vector of d zeros.

Since the covariance matrix of $\mathbf{x}'$ is diagonal, the transformed data are uncorrelated along the new axes. In other words, the orthogonalization step of PCA above approximates the data with a multivariate Gaussian so that the eigenvectors are aligned with the axes of the hyperellipsoid captured by the Gaussian. Meanwhile, the eigenvalues capture the variances of the Gaussian along the different axes. Therefore, if λ_i is small, the data do not deviate significantly from the ith axis, in which case the ith element in $\mathbf{x}'$ is not very important for the purpose of faithfully recreating the original data. The total variance in the data is given by the sum of the variances of $\mathbf{C}$, which is equivalent to the sum of the eigenvalues: $\sum_{i=1}^d \lambda_i$. The fraction of the total variance captured by the first k elements of $\mathbf{x}'$, then, is given by

$$\text{captured variance}(k) = \frac{\sum_{i=1}^k \lambda_i}{\sum_{i=1}^d \lambda_i} \tag{12.88}$$

To determine the number of dimensions to keep, a number of approaches can be used, such as looking for the "knee" in the plot of eigenvalues (the so-called **scree test**), or the **Kaiser criterion** (which retains only dimensions whose eigenvalue is greater than 1).

Let $\mathbf{P}_k$ be the first k columns of $\mathbf{P}$, so that $\mathbf{P}_k$ contains the first k eigenvectors of $\mathbf{C}$, where $k < d$. Then the following transformation can be defined:

$$\mathbf{a} \equiv \mathbf{P}_k^\mathsf{T}(\mathbf{x} - \boldsymbol{\mu}) \tag{12.89}$$

and the inverse transform:

$$\mathbf{x} \approx \mathbf{P}_k\mathbf{a} + \boldsymbol{\mu} \tag{12.90}$$

where $\mathbf{a} \in \mathbb{R}^k$ is a k-dimensional vector. These equations are similar to Equations (12.84) and (12.85), except that information has been lost due to the data reduction from $\mathbb{R}^d$ to $\mathbb{R}^k$,

Figure 12.16 LEFT: The first step of principal components analysis (PCA) fits a multivariate Gaussian to data, which effectively aligns the coordinate axes with the principal axes of the data. RIGHT: The second step of PCA reduces the dimensionality of the space by discarding the dimensions with the least variance. In this toy example, all the points lie in a 2D subspace (flat disk) of the 3D space, and thus the third dimension is not needed once the axes are aligned.

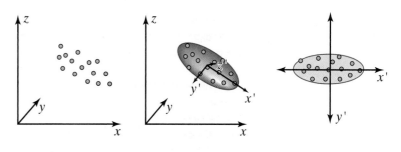

so that the inverse transformation only approximates the original point. As long as k is chosen so that the fraction of the variance captured—given by Equation (12.88)—is sufficiently large, the information in the d-dimensional point **x** is well approximated by the k-dimensional point **a**. For typical data sets, 95% or so of the variance can be captured with just a few eigenvalues, thus saving orders of magnitude storage and computation, as well as facilitating less sensitivity to noise.

One classic use of PCA is the **point distribution model (PDM)**, which captures the statistical variation in shape among the training set. Consider a 2D shape described by m points. Stacking the (x, y) coordinates together into a vector, the shape can be described by a d-dimensional vector, where $d = 2m$. Each instance of the shape is then a point in a d-dimensional space, and all the n training samples can be compactly stored in the $d \times n$ centered matrix **A** defined in Equation (12.83). Figure 12.17 shows the PDM of a database of 18 hand shapes, each of which contains $m = 72$ points. As a preliminary step, all the shapes are first aligned so that their mean

Figure 12.17 Point distribution model of a hand shape database. TOP: 10 of 18 training shapes. MIDDLE: From left to right: The mean shape, a scatter plot of the point coordinates from the training images overlaid on the mean shape (only every 3rd point is shown, pseudocolored), and a plot of the sorted eigenvalues (blue) and captured variance (red). Note that most of the variance is capture by the first few eigenvectors. BOTTOM: The effects of the first four eigenvectors. Shown is $\mu \pm 2\sigma_i \mathbf{v}_i$ (in black, red, and green, respectively) for $i = 1, \ldots, 4$, where $\sigma_i = \sqrt{\lambda_i}$.

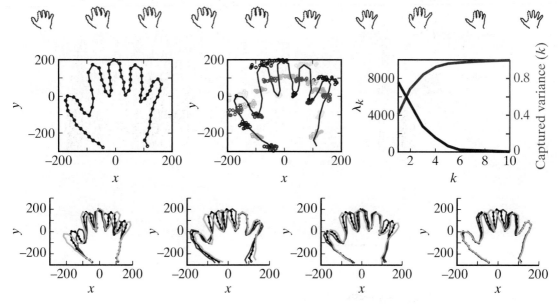

is at the origin, that is, $\sum_{k=1}^{m} \mathbf{x}_i^{(k)} = \mathbf{0}_{\{2 \times 1\}}$ for all $i = 1, \ldots, n$, where $\mathbf{x}_i^{(k)}$ is the kth point in the ith shape. Then the mean shape is computed according to Equation (12.81), and the covariance matrix is computed using Equation (12.82). On this dataset the first 5 eigenvalues capture 96% of the variance of the data, computed using Equation (12.88). The different eigenvectors capture the correlated wiggling of the different fingers.

Another classic application of PCA is to operate on the gray levels of images of faces, where the reshaped eigenvectors are known as **eigenfaces**. Although this technique is not effective enough to be used in practice, it has been widely influential in the field and therefore warrants our attention. Recall from an earlier discussion[†] that an image can be viewed as a vector. That is, the columns of the image can be stacked on top of one another to form a vertical vector, so that a $w \times h$ image is viewed as a point in a wh-dimensional space. Figure 12.18 shows 10 training face images, each of size 160×200, which have been shifted and scaled so as to align the major facial features as well as possible. Each image is therefore a point in a $160 \cdot 200 = 32000$-dimensional space, which is quite large indeed. Applying the procedure above to this dataset, the reshaped eigenvectors show the principal directions of variations of human faces.

The columns of $\mathbf{P}$, which is $d \times d$, form an orthonormal basis that spans the original d-dimensional space. Similarly, the columns of $\mathbf{P}_k$, which is $d \times k$, form an orthonormal basis that spans the k-dimensional subspace. Combining Equations (12.89) and (12.90) yields an expression for the closest point in the subspace to any query point $\mathbf{x}$:

$$\mathbf{x}_{closest} \approx \mathbf{P}_k(\mathbf{P}_k^\mathsf{T}(\mathbf{x} - \boldsymbol{\mu})) + \boldsymbol{\mu} \tag{12.91}$$

where the extra parentheses ensure that the enormous multiplication $\mathbf{P}_k\mathbf{P}_k^\mathsf{T}$ (which in our case would be 32000×32000) never occurs. Figure 12.19 shows the closest image in the subspace to a masked face image that simulates occlusion (this face was not part of the training database).

Similarly, given current data $\mathbf{x}$, Equation (12.89) can be used to compute the best model parameters to explain the data; then given an estimated change $\boldsymbol{\delta}_\mathbf{x}$ (determined by measuring image properties) in the data, the best change $\boldsymbol{\delta}_\mathbf{a}$ in parameters that explains the data is

$$\boldsymbol{\delta}_\mathbf{a} = \mathbf{P}_k^\mathsf{T}\boldsymbol{\delta}_\mathbf{x} \tag{12.92}$$

Figure 12.18 Top: 10 training images of faces. Bottom: Mean face, and top 4 eigenfaces, which together capture 75% of the total variance.

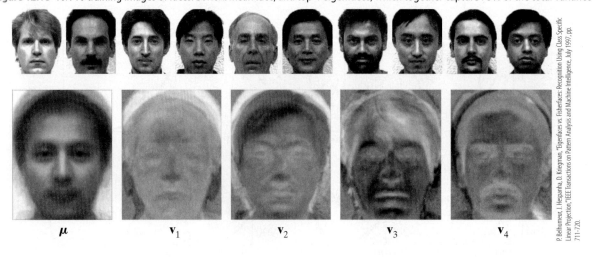

$\boldsymbol{\mu}$ $\mathbf{v}_1$ $\mathbf{v}_2$ $\mathbf{v}_3$ $\mathbf{v}_4$

P. Belhumeur, J. Hespanha, D. Kriegman, "Eigenfaces vs. Fisherfaces: Recognition Using Class Specific Linear Projection," IEEE Transactions on Pattern Analysis and Machine Intelligence, July 1997, pp. 711-720.

[†] Section 1.4.3 (p. 13)

Figure 12.19 Left: A face image not in the training database, with some pixels altered to simulate occlusion. Right: Projection of the image onto the subspace spanned by the 4 most significant eigenfaces.

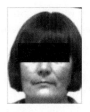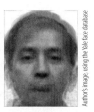

which can be shown from Equation (12.90):

$$\mathbf{x} + \boldsymbol{\delta_x} \approx \mathbf{P}_k(\mathbf{a} + \boldsymbol{\delta_a}) + \boldsymbol{\mu} \tag{12.93}$$

$$\boldsymbol{\delta_x} \approx \mathbf{P}_k \boldsymbol{\delta_a} \tag{12.94}$$

where Equation (12.92) is the least-squares result of solving Equation (12.94). The parameters should always remain within a hyperellipsoid bounded by $\pm 2.5\sigma_i$ (for 98.76% confidence) in each direction, where $\sigma_i \sqrt{\lambda_i}$. Therefore, after computing the change in parameters using Equation (12.92), any element of the parameter that is outside this hyperellipsoid should be adjusted accordingly.

This latter procedure is used by **active shape models (ASMs)**, which can be thought of as an extension to the active contour model we saw earlier.[†] During training, a PDM is used to construct a flexible model of the shape of an object. Then, given an initial estimate for the shape and pose parameters describing the object's appearance in an image, those parameters are updated in a simple iterative procedure in which the pose (usually translation, rotation, and scale) is updated using Procrustes analysis[‡] to best align the currently estimated shape with the model shape. Afterward, the shift in point coordinates $(\boldsymbol{\delta_x})$ is computed by searching for strong intensity gradients along rays perpendicular to the contour, then this shift is converted into a change in shape parameters using Equation (12.92). An extension is to learn a combined shape and graylevel appearance model (which effectively combines the ideas of PDMs and eigenfaces) to jointly search for pose, shape variation, and graylevel variation to fit the model to an image. This technique, known as **active appearance models (AAMs)**, has been used to model faces across variations in identity, expression, and pose, as shown in Figure 12.20.

Figure 12.20 An active appearance model (AAM) uses PCA to capture the variations in a joint shape-appearance space. On the left is the result of fitting an AAM to a face image. In the middle and right are the effects of the top four eigenvectors, with the mean displayed in the middle of each triplet.

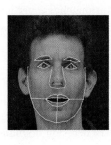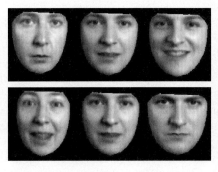

[†] Section 10.2.1 (p. 453).
[‡] Section 11.3.1 (p. 533).

One detail remains. Sometimes the covariance matrix $\mathbf{C}$ is too large for the eigenvalues and eigenvectors to be computed in any reasonable amount of time, or even at all. (In the eigenface example above, it is 32000×32000.) From linear algebra we know that the rank of the sum of two matrices is no greater than the sum of their ranks. That is, if $\mathbf{A}$ and $\mathbf{B}$ are matrices, then $\text{rank}(\mathbf{A} + \mathbf{B}) \leq \text{rank}(\mathbf{A}) + \text{rank}(\mathbf{B})$. As a result, since each outer product in Equation (12.82) has rank 1, the rank of $\mathbf{C}$ is bounded by $\text{rank}(\mathbf{C}) \leq \max(n, d)$. If $n < d$, as in our example here (where $n = 10$ and $d = 32000$), then the rank of $\mathbf{C}$ is no greater than n. Therefore, instead of computing the eigenvectors of the $d \times d$ matrix $\mathbf{C} = \frac{1}{n}\mathbf{A}\mathbf{A}^\mathsf{T}$, we can compute the eigenvectors of the much smaller $n \times n$ **normalized Gramian** $\tilde{\mathbf{C}} \equiv \frac{1}{n}\mathbf{A}^\mathsf{T}\mathbf{A}$. Then, each of the n eigenvectors of the original matrix $\mathbf{C}$ is given by $\mathbf{v}_i = \mathbf{A}\tilde{\mathbf{v}}_i$, where $\tilde{\mathbf{v}}_i$ is the ith eigenvector of $\tilde{\mathbf{C}}$, which is easily seen by starting with the eigenvector definition:

$$\underbrace{\frac{1}{n}\mathbf{A}^\mathsf{T}\mathbf{A}}_{\tilde{\mathbf{C}}}\ \tilde{\mathbf{v}}_i \;=\; \tilde{\lambda}_i\tilde{\mathbf{v}}_i \tag{12.95}$$

and premultiplying by $\mathbf{A}$:

$$\underbrace{\frac{1}{n}\mathbf{A}\mathbf{A}^\mathsf{T}}_{\mathbf{C}}\ \underbrace{\mathbf{A}\tilde{\mathbf{v}}_i}_{\mathbf{v}_i} \;=\; \tilde{\lambda}_i\ \underbrace{\mathbf{A}\tilde{\mathbf{v}}_i}_{\mathbf{v}_i} \tag{12.96}$$

Note that the resulting vectors need to be normalized to ensure that they are proper unit-norm eigenvectors.

12.4 Discriminative Methods

An alternative to explicitly modeling the probability distributions (as in the generative methods described in the previous section) is to learn the decision boundaries directly. After all, since the decision boundaries are what govern the accuracy of the classifier, it makes sense to focus our attention primarily upon these decision boundaries. As discussed earlier, algorithms that do this are known as *discriminative methods*. In this section we describe several of the more popular discriminative approaches to supervised learning.

12.4.1 Linear Discriminant Functions

Discriminant functions are used to indicate how well a data point matches a particular category. An important special case of the discriminative approach occurs when the discriminant function is linear, in which case the function is given by the equation

$$g_i(\mathbf{x}) \;=\; \mathbf{w}_i^\mathsf{T}\mathbf{x} + b_i \tag{12.97}$$

where $\mathbf{x} \in \mathbb{R}^d$ is a point in the feature space. Although at first glance linear discriminant functions may appear too restrictive for real applications, in fact they are worth studying in detail because they are capable of modeling arbitrarily complex decision boundaries through one of two clever extensions. One of these extensions (known as the *kernel trick*) leads to support vector machines (SVMs), while the other (combining multiple discriminant functions), leads to artificial neural networks (ANNs)—both of which are some of the most powerful classifiers on the market today.

To simplify the discussion, we shall focus on the case when the classifier is a dichoto-mizer (i.e., there are just two categories), so that the discriminant function is given by

$$g(\mathbf{x}) \equiv g_1(\mathbf{x}) - g_2(\mathbf{x}) = \mathbf{w}^\mathsf{T}\mathbf{x} + b \tag{12.98}$$

which yields the following decision boundary:

$$\mathbf{w}^\mathsf{T}\mathbf{x} + b = 0 \tag{12.99}$$

which describes a hyperplane in $\mathbb{R}^d$ parameterized by the vector $\mathbf{w} \equiv \mathbf{w}_1 - \mathbf{w}_2 \in \mathbb{R}^d$ and scalar $b \equiv b_1 - b_2 \in \mathbb{R}$. The value b is called the *bias*.

It is easy to see that the vector $\mathbf{w}$ is orthogonal (perpendicular) to the hyperplane. If we let $\mathbf{x}_1$ and $\mathbf{x}_2$ be two points on the hyperplane, then $\mathbf{w}^\mathsf{T}\mathbf{x}_1 + b = \mathbf{w}^\mathsf{T}\mathbf{x}_2 + b = 0$, which leads to $\mathbf{w}^\mathsf{T}(\mathbf{x}_1 - \mathbf{x}_2) = 0$. This means that $\mathbf{w}$ is orthogonal to the vector $\mathbf{x}_1 - \mathbf{x}_2$, and since the latter is in the hyperplane, it proves that $\mathbf{w}$ is orthogonal to the hyperplane. Similarly, it is easy to see that if $\|\mathbf{w}\| = 1$, then $|b|$ is the distance from the origin to the hyperplane. If we let a be the unknown distance, and if $\|\mathbf{w}\| = 1$, then $\mathbf{x} = a\mathbf{w}$ or $\mathbf{x} = -a\mathbf{w}$ (depending upon the orientation of $\mathbf{w}$) is the intersection of the ray along $\mathbf{w}$ with the hyperplane. This point must lie on the hyperplane, so from Equation (12.99) we have $\pm a\mathbf{w}^\mathsf{T}\mathbf{w} + b = 0$, or $\pm a\|\mathbf{w}\|^2 + b = 0$, or $a = |b|$.

To better understand these parameters graphically, consider Figure 12.21, which illus-trates a 2D linear discriminant function $(d = 2)$. Since Equation (12.99) is homogeneous, the hyperplane does not change when $\mathbf{w}$ and b are multiplied by the same nonzero scalar constant. Nevertheless, although the scale of $\mathbf{w}$ is not important for defining the decision boundary (hyperplane), it *is* important for defining the discriminant function. We adopt the convention that $\mathbf{w}$ and b are both multiplied by -1 if necessary so that $\mathbf{w}$ always points toward the positive training samples. When $\mathbf{w}$ is oriented in this manner (and $\|\mathbf{w}\| = 1$), it can be shown that the distance traveled along $\mathbf{w}$ from the origin to reach the hyperplane is $-b$. As a result, if $\mathbf{w}$ points toward the hyperplane (as in the left side of the figure), then $b < 0$, whereas if $\mathbf{w}$ points away from the hyperplane (as in the right side of the figure), then $b > 0$.

Because scaling is arbitrary, it might seem natural to adopt the convention that $\|\mathbf{w}\| = 1$, which we could do without loss of generality. However, we will not impose this constraint, because in the discussion below $\mathbf{w}$ is manipulated in various calculations, so that oftentimes the vector will not necessarily have unit norm. Instead, we will explicitly write $\mathbf{w}/\|\mathbf{w}\|$ when we want to indicate the unit norm vector. If we let $\delta(\mathbf{x})$ be the signed distance from the point $\mathbf{x}$ to the hyperplane, it is easy to see that

$$\delta(\mathbf{x}) \equiv \frac{\mathbf{w}^\mathsf{T}\mathbf{x} + b}{\|\mathbf{w}\|} = \frac{g(\mathbf{x})}{\|\mathbf{w}\|} \qquad (\text{signed distance}) \tag{12.100}$$

Figure 12.21 With our convention for a linear discriminant function, the vector **w** points from the origin toward the positive side of the space. Left: The vector **w** points toward the decision boundary, so $b < 0$. Right: The vector **w** points away from the decision boundary, so $b > 0$.

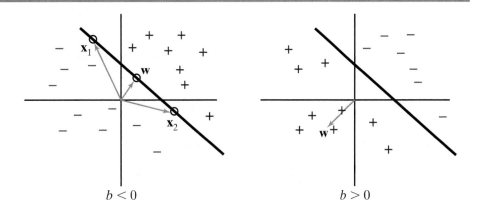

To derive this equation, let $\mathbf{x}'$ be an arbitrary point in the hyperplane. Then the vector $\mathbf{x} - \mathbf{x}'$ points from $\mathbf{x}$ to the hyperplane, and therefore the projection of this vector onto the unit vector $\mathbf{w}/\|\mathbf{w}\|$ yields the signed distance: $\mathbf{w}^\mathsf{T}(\mathbf{x} - \mathbf{x}')/\|\mathbf{w}\|$. Now, for any point $\mathbf{x}'$ in the hyperplane, we have $\mathbf{w}^\mathsf{T}\mathbf{x}' = -b$ from Equation (12.99). Therefore, the signed distance is $(\mathbf{w}^\mathsf{T}\mathbf{x} - \mathbf{w}^\mathsf{T}\mathbf{x}')/\|\mathbf{w}\| = (\mathbf{w}^\mathsf{T}\mathbf{x} + b)/\|\mathbf{w}\|$.

12.4.2 Fisher's Linear Discriminant (FLD)

Discriminant function analysis (DFA) is concerned with determining which continuous variables discriminate between discrete categories. One of the most widely used approaches to DFA is known as **Fisher's linear discriminant (FLD)**, which is a technique for learning the best linear discriminant function as in Equation (12.98). Suppose we have n data samples, $\mathbf{x}_1, \ldots, \mathbf{x}_n \in \mathcal{D} \subset \mathbb{R}^d$, and two categories ω_1 and ω_2. Let the subset $\mathcal{D}_1 \subset \mathcal{D}$ contain n_1 samples labeled ω_1, while the subset $\mathcal{D}_2 \subset \mathcal{D}$ contains n_2 samples labeled ω_2, where $n_1 + n_2 = n$, and $\mathcal{D} = \mathcal{D}_1 \cup \mathcal{D}_2$ and $\mathcal{D}_1 \cap \mathcal{D}_2 = \emptyset$. Our goal is to find the vector $\mathbf{w}$ so that the projection of $\mathbf{x}$ onto $\mathbf{w}$ separates the categories well. Let us define this projection as $y = \mathbf{w}^\mathsf{T}\mathbf{x}$, where $y \in \mathcal{Y} \subset \mathbb{R}$, and $\mathcal{Y} = \mathcal{Y}_1 \cup \mathcal{Y}_2$ and $\mathcal{Y}_1 \cap \mathcal{Y}_2 = \emptyset$. After $\mathbf{w}$ is determined, a value for the bias b is selected so as to yield an effective classifier.

Let the **sample mean** of the jth category, $j = 1, 2$, be defined as

$$\boldsymbol{\mu}_j \equiv \frac{1}{n_j} \sum_{\mathbf{x} \in \mathcal{D}_j} \mathbf{x} \tag{12.101}$$

and let the **projected sample mean** be

$$\tilde{\mu}_j \equiv \frac{1}{n_j} \sum_{y \in \mathcal{Y}_j} y = \frac{1}{n_j} \sum_{y \in \mathcal{Y}_j} \mathbf{w}^\mathsf{T}\mathbf{x} = \mathbf{w}^\mathsf{T}\boldsymbol{\mu}_j \tag{12.102}$$

which is simply a projection of the sample mean. Also define the **scatter** to be

$$\tilde{s}_j^2 \equiv \sum_{y \in \mathcal{Y}_j} (y - \tilde{\mu}_j)^2 \tag{12.103}$$

which is just n_j times the variance, and define the **total within-class scatter** as $\tilde{s} \equiv \tilde{s}_1^2 + \tilde{s}_2^2$.

Now the distance between the projected sample means is given by

$$|\tilde{\mu}_1 - \tilde{\mu}_2| = \|\mathbf{w}^\mathsf{T}(\boldsymbol{\mu}_1 - \boldsymbol{\mu}_2)\| \tag{12.104}$$

We want this distance to be large relative to the standard deviation of each class. In other words, Fisher's linear discriminant chooses $\mathbf{w}$ so as to maximize

$$\xi(\mathbf{w}) \equiv \frac{|\tilde{\mu}_1 - \tilde{\mu}_2|}{\tilde{s}_1^2 + \tilde{s}_2^2} \tag{12.105}$$

To derive the computation needed to maximize this quantity, let us define the **scatter matrix** of the jth category as

$$\mathbf{S}_j \equiv \sum_{\mathbf{x} \in \mathcal{D}_j} (\mathbf{x} - \boldsymbol{\mu}_j)(\mathbf{x} - \boldsymbol{\mu}_j)^\mathsf{T} \tag{12.106}$$

and define the **within-class scatter matrix** as $\mathbf{S}_w \equiv \mathbf{S}_1 + \mathbf{S}_2$. Then, from Equation (12.103),

$$\tilde{s}_j^2 = \sum_{\mathbf{x} \in \mathcal{D}_j} (\mathbf{w}^\mathsf{T}\mathbf{x} - \mathbf{w}^\mathsf{T}\boldsymbol{\mu}_j)^2 = \sum_{\mathbf{x} \in \mathcal{D}_j} \mathbf{w}^\mathsf{T}(\mathbf{x} - \boldsymbol{\mu}_j)(\mathbf{x} - \boldsymbol{\mu}_j)^\mathsf{T}\mathbf{w} = \mathbf{w}^\mathsf{T}\mathbf{S}_j\mathbf{w} \tag{12.107}$$

and therefore, $\tilde{s}^2 = \tilde{s}_1^2 + \tilde{s}_2^2 = \mathbf{w}^T \mathbf{S}_w \mathbf{w}$. Similarly,

$$(\tilde{\mu}_1 - \tilde{\mu}_2)^2 = (\mathbf{w}^T \boldsymbol{\mu}_1 - \mathbf{w}^T \boldsymbol{\mu}_2)^2 = \mathbf{w}^T (\boldsymbol{\mu}_1 - \boldsymbol{\mu}_2)(\boldsymbol{\mu}_1 - \boldsymbol{\mu}_2)^T \mathbf{w} = \mathbf{w}^T \mathbf{S}_B \mathbf{w} \quad (12.108)$$

where $\mathbf{S}_B \equiv (\boldsymbol{\mu}_1 - \boldsymbol{\mu}_2)(\boldsymbol{\mu}_1 - \boldsymbol{\mu}_2)^T$ is the **between-class scatter matrix**. Substituting Equations (12.107) and (12.108) into Equation (12.105) yields

$$\xi(\mathbf{w}) = \frac{\mathbf{w}^T \mathbf{S}_B \mathbf{w}}{\mathbf{w}^T \mathbf{S}_w \mathbf{w}} \quad (12.109)$$

As we saw earlier,[†] this is a generalized Rayleigh quotient, and the vector $\mathbf{w}$ that maximizes $\xi(\mathbf{w})$ satisfies the following generalized eigenvalue problem:

$$\mathbf{S}_B \mathbf{w} = \lambda \mathbf{S}_w \mathbf{w} \quad (12.110)$$

which has just one solution since the rank of $\mathbf{S}_B$ is 1. Assuming there are more training samples than there are dimensions in the feature space, that is, $n > d$, and assuming the data are not degenerate, then $\mathbf{S}_w$ is nonsingular, in which case we have

$$\mathbf{S}_w^{-1} \mathbf{S}_B \mathbf{w} = \lambda \mathbf{w} \quad (12.111)$$

Substituting the definition of $\mathbf{S}_B$ yields

$$\mathbf{S}_w^{-1} (\boldsymbol{\mu}_1 - \boldsymbol{\mu}_2) \underbrace{(\boldsymbol{\mu}_1 - \boldsymbol{\mu}_2)^T \mathbf{w}}_{\text{scalar}} = \lambda \mathbf{w} \quad (12.112)$$

Recognizing that the inner product is a scalar leads to the conclusion that $\mathbf{w}$ is parallel to $\mathbf{S}_w^{-1}(\boldsymbol{\mu}_1 - \boldsymbol{\mu}_2)$:

$$\mathbf{w} \propto \mathbf{S}_w^{-1}(\boldsymbol{\mu}_1 - \boldsymbol{\mu}_2) \quad (12.113)$$

Therefore, to get a unit vector, set

$$\mathbf{w} = \frac{\mathbf{S}_w^{-1}(\boldsymbol{\mu}_1 - \boldsymbol{\mu}_2)}{\|\mathbf{S}_w^{-1}(\boldsymbol{\mu}_1 - \boldsymbol{\mu}_2)\|} \quad (12.114)$$

Selecting an appropriate bias depends on the variances of the projected data, as well as on the loss function. If the variances are equal, then with the zero-one loss function simply set b to the point halfway between the projected sample means:

$$b = -\frac{1}{2}\mathbf{w}^T(\boldsymbol{\mu}_1 + \boldsymbol{\mu}_2) \quad (12.115)$$

where the negative sign arises from the convention mentioned above. The pseudocode for Fisher's linear discriminant is provided in Algorithm 12.1.

12.4.3 Perceptrons

Historically, a linear classifier is also called a **perceptron**. Combining the linear discriminant function $g(\mathbf{x}) = \mathbf{w}^T \mathbf{x} + b$ in Equation (12.98) with the classifier $f(x)$ in Equation (12.3), we see that a perceptron applies a linear combiner followed by a hard limiter (or threshold):

$$y = f(\mathbf{x}) = h(\mathbf{w}^T \mathbf{x} + b) \quad (12.116)$$

[†] Section 10.4.2 (p. 495).

ALGORITHM 12.1 Compute Fisher's linear discriminant on a set of binary labeled data

---

FISHERLINEARDISCRIMINANT $(\{(\mathbf{x}_i, y_i)\}_{i=1}^n)$

Input: set of n labeled data points $\{(\mathbf{x}_i, y_i)\}_{i=1}^n$, where
  $\mathbf{x}_i \in \mathbb{R}^d$ is the ith data point
  $y_i \in \{-1, +1\}$ is the binary label of the ith data point, so that
  $\mathbf{x}_i \in \mathcal{D}_1$ if $y_i = +1$, or $\mathbf{x}_i \in \mathcal{D}_2$ if $y_i = -1$.
Output: parameters $\mathbf{w}$ and b of a linear discriminant function to separate the data

1 $\boldsymbol{\mu}_1 \leftarrow \left(\sum_{\mathbf{x} \in D_1} \mathbf{x} \right) / |\mathcal{D}_1|$ ▷ Compute sample means using Equation (12.101),

2 $\boldsymbol{\mu}_2 \leftarrow \left(\sum_{\mathbf{x} \in D_2} \mathbf{x} \right) / |\mathcal{D}_2|$ where $|\mathcal{D}_i|$ is the cardinality of the ith set.

3 $\mathbf{S}_1 \leftarrow \sum_{\mathbf{x} \in D_1} (\mathbf{x} - \boldsymbol{\mu}_1)(\mathbf{x} - \boldsymbol{\mu}_1)^{\mathsf{T}}$ ▷ Compute the scatter matrices using Equation (12.106).

4 $\mathbf{S}_2 \leftarrow \sum_{\mathbf{x} \in D_2} (\mathbf{x} - \boldsymbol{\mu}_2)(\mathbf{x} - \boldsymbol{\mu}_2)^{\mathsf{T}}$

5 $\mathbf{S}_w \leftarrow \mathbf{S}_1 + \mathbf{S}_2$ ▷ Compute the within-class scatter matrix.

6 $\mathbf{w} \leftarrow \mathbf{S}_w^{-1}(\boldsymbol{\mu}_1 - \boldsymbol{\mu}_2)$ ▷ Compute the vector $\mathbf{w}$ using Equation (12.109).

7 $\mathbf{w} \leftarrow \mathbf{w}/\|\mathbf{w}\|$

8 $b \leftarrow -\frac{1}{2}\mathbf{w}^{\mathsf{T}}(\boldsymbol{\mu}_1 + \boldsymbol{\mu}_2)$ ▷ Set b to be halfway between the projected means (or some other suitable value).

9 **return** $\mathbf{w}, b$

---

where y is the output of the perceptron, and the threshold (Heaviside) function is defined as $h(z) = 1$ if $z \geq 0$, or 0 otherwise. Expanding the inner product, this equation can alternatively be written as

$$y = h\left(\sum_{k=1}^d w^{(k)} x^{(k)} + b \right) \tag{12.117}$$

where $\mathbf{x} \equiv \begin{bmatrix} x^{(1)} & x^{(2)} & \cdots & x^{(d)} \end{bmatrix}^{\mathsf{T}}$ and $\mathbf{w} \equiv \begin{bmatrix} w^{(1)} & w^{(2)} & \cdots & w^{(d)} \end{bmatrix}^{\mathsf{T}}$. A single perceptron is illustrated in Figure 12.22.

The **perceptron learning algorithm** is straightforward. Let y_i be the ground truth binary label of the ith data point in the training set. If the ith point is a positive training sample, then $y_i = +1$, and the signed distance from the point to the decision boundary

---

Figure 12.22 A single perceptron multiplies the individual elements of the input vector with the individual weights, adds the bias, then applies the threshold function: $y = h(w^{(1)}x^{(1)} + w^{(2)}x^{(2)} + \cdots + w^{(d)}x^{(d)} + b)$.

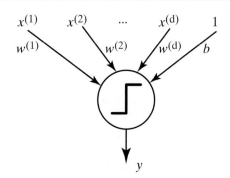

should be positive; whereas if the point is a negative training sample, then $y_i = -1$, and the signed distance from the point to the decision boundary should be negative. Therefore, if we let δ_i be this signed distance, then the ith point is misclassified if y_i and δ_i have different signs, which is captured mathematically by $y_i \delta_i < 0$. Applying the formula for signed distance in Equation (12.100), the ith point is misclassified if

$$\frac{y_i(\mathbf{w}^\mathsf{T}\mathbf{x}_i + b)}{\|\mathbf{w}\|} < 0 \tag{12.118}$$

which, since the denominator is guaranteed to be positive, is equivalent to the test

$$y_i(\mathbf{w}^\mathsf{T}\mathbf{x}_i + b) < 0 \tag{12.119}$$

The perceptron learning algorithm iterates through the training samples, and for each misclassified sample $\mathbf{x}_i$, the hyperplane parameters are updated as follows:

$$\mathbf{w}^{(k+1)} = \mathbf{w}^{(k)} + y_i\mathbf{x}_i \tag{12.120}$$

$$b^{(k+1)} = b^{(k)} + y_i \tag{12.121}$$

where k is the iteration number. In other words, in the case of a positive training sample $(y_i = +1)$, then $\mathbf{x}$ is added to $\mathbf{w}$, and b is incremented by 1; in the case of a negative training sample $(y_i = -1)$, then $\mathbf{x}$ is subtracted from $\mathbf{w}$, and b is decremented by 1. It can be shown that, if the data are linearly separable, this simple algorithm is guaranteed to converge in a finite number of steps to a decision boundary that separates the two categories. Each iteration effectively updates the *extended vector* $\tilde{\mathbf{w}} \equiv [\mathbf{w}^\mathsf{T} \; b]^\mathsf{T}$ by rotating it either toward or away from the extended input sample $\tilde{\mathbf{x}} \equiv [\mathbf{x}_i^\mathsf{T} \; 1]^\mathsf{T}$ to increase the value of $y_i\tilde{\mathbf{w}}^\mathsf{T}\tilde{\mathbf{x}}_i$ in Equation (12.119). Somewhat miraculously, the amount by which the vector $\mathbf{w}$ changes each time (the **learning rate**) decreases automatically with each iteration, thus causing convergence.

Although the perceptron learning algorithm is guaranteed to converge (if the data are linearly separable), there is no bound on the number of iterations required. One way to accelerate convergence is to weight each update by the amount of the error:

$$\mathbf{w}^{(k+1)} = \mathbf{w}^{(k)} + \alpha_i y_i\mathbf{x}_i \tag{12.122}$$

$$b^{(k+1)} = b^{(k)} + \alpha_i y_i \tag{12.123}$$

where $\alpha_i \equiv \frac{1}{\|\tilde{\mathbf{x}}_i\|^2}\left(-y_i\left((\mathbf{w}^{(k)})^\mathsf{T}\mathbf{x}_i + b\right) + \epsilon\right)$ is the learning rate, and $\epsilon > 0$ is a small positive number. Note that $\alpha_i > 0$. With this accelerated learning, it is easy to show that the ith sample is correctly classified after just one iteration, using the extended vector $\tilde{\mathbf{w}}^{(k+1)} = \tilde{\mathbf{w}}^{(k)} + \alpha_i y_i\tilde{\mathbf{x}}_i$. To simplify the notation let $\tilde{\mathbf{w}}$ stand for $\tilde{\mathbf{w}}^{(k)}$:

$$y_i(\tilde{\mathbf{w}}^{(k+1)})^\mathsf{T}\tilde{\mathbf{x}}_i = y_i(\tilde{\mathbf{w}} + \alpha_i y_i\tilde{\mathbf{x}}_i)^\mathsf{T}\tilde{\mathbf{x}}_i \tag{12.124}$$

$$= y_i\tilde{\mathbf{w}}^\mathsf{T}\tilde{\mathbf{x}}_i + y_i\left(\frac{1}{\|\tilde{\mathbf{x}}_i\|^2}(-y_i\tilde{\mathbf{w}}^\mathsf{T}\tilde{\mathbf{x}}_i + \epsilon)\right)y_i\tilde{\mathbf{x}}_i^\mathsf{T}\tilde{\mathbf{x}}_i \tag{12.125}$$

$$= y_i\tilde{\mathbf{w}}^\mathsf{T}\tilde{\mathbf{x}}_i - y_i\tilde{\mathbf{w}}^\mathsf{T}\tilde{\mathbf{x}}_i + \epsilon \tag{12.126}$$

$$= \epsilon \tag{12.127}$$

so that $y_i(\tilde{\mathbf{w}}^{(k+1)})^\mathsf{T}\tilde{\mathbf{x}}_i > 0$, which means that the point is classified correctly after the update.

The pseudocode of the accelerated perceptron learning algorithm is shown in Algorithm 12.2. Although $\mathbf{w}$ and b can be initialized to anything, a reasonable choice is to

ALGORITHM 12.2 Perceptron learning algorithm (with accelerated learning rate)

PERCEPTRONLEARNINGALGORITHM$\left(\{(\mathbf{x}_i, y_i)\}_{i=1}^{n}\right)$

Input: set of n labeled data points $\{(\mathbf{x}_i, y_i)\}_{i=1}^{n}$
   $\mathbf{x}_i \in \mathbb{R}^d$ is the ith data point
   $y_i \in \{-1, +1\}$ is the binary label of the ith data point
   $\mathbf{x}_i \in \mathcal{D}_1$ if $y_i = +1$, or $\mathbf{x}_i \in \mathcal{D}_2$ if $y_i = -1$.
Output: parameters $\mathbf{w}$ and b of a perceptron to separate the data (assuming the data are linearly separable)

1   $\mathbf{w} \leftarrow \left(\sum_{\mathbf{x} \in \mathcal{D}_1} \mathbf{x}\right)/|\mathcal{D}_1| - \left(\sum_{\mathbf{x} \in \mathcal{D}_2} \mathbf{x}\right)/|\mathcal{D}_2|$ ▶ Initialize the parameters.

2   $b \leftarrow |\mathcal{D}_1| - |\mathcal{D}_2|$
3   **while** not converged **do**   ▶ Until all points are correctly classified,
4   Randomly shuffle data points $\{(\mathbf{x}_i, y_i)\}_{i=1}^{n}$ shuffle the points (optional).
5   **for** $i \leftarrow 1$ **to** n **do**   ▶ For each data sample,
6   **if** $y_i(\mathbf{w}^\mathsf{T}\mathbf{x}_i + b) < 0$ **then**   if the sample is misclassified, then
7            $\alpha_i \leftarrow (-y_i(\mathbf{w}^\mathsf{T}\mathbf{x}_i + b) + \epsilon)/(\|\mathbf{x}_i\|^2 + 1)$ determine the learning rate,
8            $\mathbf{w} \leftarrow \mathbf{w} + \alpha_i y_i \mathbf{x}_i$ and update the hyperplane
9            $b \leftarrow b + \alpha_i y_i$ parameters.
10   **return** $\mathbf{w}, b$

compute the mean of the two sample subsets $\mathcal{D}_1$ and $\mathcal{D}_2$, then set $\mathbf{w}$ to the difference between the means and b to the difference between the sizes of the subsets:

$$\mathbf{w}_{init} = \frac{1}{|\mathcal{D}_1|}\sum_{\mathbf{x} \in \mathcal{D}_1} \mathbf{x} - \frac{1}{|\mathcal{D}_2|}\sum_{\mathbf{x} \in \mathcal{D}_2} \mathbf{x} \tag{12.128}$$

$$b_{init} = |\mathcal{D}_1| - |\mathcal{D}_2| \tag{12.129}$$

The algorithm converges when all input points have been classified correctly.

   One drawback of the perceptron learning algorithm is that if the data are not linearly separable, then the algorithm is not guaranteed to converge to a meaningful solution. This problem is solved with the so-called **pocket algorithm**, which runs the perceptron learning algorithm but retains the best hyperplane parameters (as if kept in its pocket) so far. Eventually, when convergence has been determined (e.g., a specified number of iterations has not yet caused significant change in the parameters), the best parameters are returned. There are different ways to determine which parameters are best, but a simple approach is to count the number of consecutive iterations that have not encountered a misclassified sample. The pseudocode for the pocket algorithm is shown in Algorithm 12.3 where, for simplicity, we have not included the accelerated learning rate. Note that none of these algorithms is guaranteed to return the best classifier in any quantitative sense of the term because they only train on the misclassified samples. In the next section we will see a better way to find the best separating hyperplane.

12.4.4 Maximum-Margin Classifiers

Returning to the case of linearly separable data, the perceptron learning algorithm of the previous subsection is able to find a hyperplane to separate two categories of data, when such a separating hyperplane exists. However, it is not able to find the best hyperplane among all

ALGORITHM 12.3 Pocket perceptron learning algorithm

PocketPerceptronLearningAlgorithm$\left(\{(\mathbf{x}_i, y_i)\}_{i=1}^n\right)$

Input: set of n labeled data points
Output: parameters of perceptron that tries to separate data (even if not linearly separable)

1 Initialize $\mathbf{w}$ and b similar to before
2 $h \leftarrow 0$　　　　　　　　　　　　　　　　　　　▶ number of consecutive successfully tested vectors
3 **while** not converged **do**
4 　　Randomly shuffle data points $\{(\mathbf{x}_i, y_i)\}_{i=1}^n$　　　　　　　▶ (optional)
5 　　**for** $i \leftarrow 1$ **to** n **do**
6 　　　　**if** $y_i(\mathbf{w}^\mathsf{T}\mathbf{x}_i + b) < 0$ **then** $\mathbf{w} \leftarrow \mathbf{w} + y_i\mathbf{x}_i, b \leftarrow b + y_i, h \leftarrow 0$
7 　　　　**else** $h \leftarrow_+ 1$
8 　　　　　　**if** $h > h^*$ **then** $\mathbf{w}^* \leftarrow \mathbf{w}, b^* \leftarrow b$
9 　　**return** $\mathbf{w}^*, b^*$

the possible separating hyperplanes. In this subsection we discuss the problem of finding the *optimal* separating hyperplane—that is, the one that maximizes the separation between the data samples in the two categories. This separation is known as the **margin**, and such an approach is therefore known as a **maximum-margin classifier**. The material in this subsection lays the groundwork for the support vector machines (SVMs) of the next subsection.

If a point $\mathbf{x}_i$ is classified correctly, then similar to Equation (12.118) its distance (not signed distance) to the separating hyperplane is given by

$$\gamma_i \equiv y_i\delta_i = \frac{y_i(\mathbf{w}^\mathsf{T}\mathbf{x}_i + b)}{\|\mathbf{w}\|} > 0 \tag{12.130}$$

which is known as the **geometric margin** of the classifier with respect to the point $\mathbf{x}_i$ labeled y_i. The geometric margin of the classifier with respect to the set of labeled points $\{(\mathbf{x}_i, y_i)\}_{i=1}^n$ is the minimum of the geometric margins of all the points:

$$\gamma \equiv \min_i \gamma_i = \min_i \frac{y_i(\mathbf{w}^\mathsf{T}\mathbf{x}_i + b)}{\|\mathbf{w}\|} \tag{12.131}$$

However, we are not content to find just any separating hyperplane but instead want to find the optimal separating hyperplane. That is, we want to find the largest value $\gamma \geq 0$ such that, for each point $\mathbf{x}_i$, $y_i\delta_i > \gamma$. Obviously, the larger the value of γ, the more separation between positive and negative data samples, and therefore the better the classifier because, assuming the training data is representative of the test data, there is reason to believe that the classifier will perform well on the test data. The value γ is the geometric margin of the optimal classifier.

Finding the optimal separating hyperplane, then, involves finding $\mathbf{w}$ and b that maximize the geometric margin:

$$\max_{\mathbf{w},b} \quad \gamma \tag{12.132}$$

$$\text{subject to} \quad \frac{y_i(\mathbf{w}^\mathsf{T}\mathbf{x}_i + b)}{\|\mathbf{w}\|} \geq \gamma \quad \text{for all} \ i = 1, \ldots, n \tag{12.133}$$

Recall that $\mathbf{w}$ and b are unique only up to a positive scaling factor. That is, if we multiply both $\mathbf{w}$ and b by the same positive number, the separating hyperplane and geometric margin are not changed. As a result, we are free to set $\|\mathbf{w}\|$ to whatever we want. For convenience let us set $\|\mathbf{w}\| = 1/\gamma$, so that maximizing γ is the same as minimizing $\|\mathbf{w}\|$, or equivalently minimizing $\frac{1}{2}\|\mathbf{w}\|^2$. The problem above is then changed into something more readily solvable:

$$\min_{\mathbf{w},b} \quad \frac{1}{2}\|\mathbf{w}\|^2 \tag{12.134}$$

$$\text{subject to} \quad y_i(\mathbf{w}^\mathsf{T}\mathbf{x}_i + b) \geq 1 \quad \text{for all } i = 1, \ldots, n \tag{12.135}$$

which is an optimization problem with a convex quadratic objective function and linear constraints, which can be solved using off-the-shelf quadratic programming software.

12.4.5 Support Vector Machine (SVM)

The **support vector machine (SVM)** is closely related to the maximum-margin classifier. Three extensions are necessary to derive the SVM from the approach in the previous subsection. First, the primal problem is transformed into a dual problem. Secondly, slackness variables are introduced to handle the case when the data are not linearly separable. Finally, the so-called kernel trick is used to transform the input data points to a new space where separability is more likely. Let us now consider these three extensions in turn.

Dual Approach to the Optimal Separating Hyperplane

According to the duality principle, optimization problems can be viewed from one of two perspectives. One of these is known as the **primal problem**, whereas the other is known as the **dual problem**. As shown below, solving the primal problem in Equations (12.134) and (12.136) is equivalent to solving the following dual problem:

$$\max_{\alpha_i} \quad \sum_{i=1}^{n} \alpha_i - \frac{1}{2}\sum_{i=1}^{n}\sum_{j=1}^{n} \alpha_i\alpha_j y_i y_j \mathbf{x}_i^\mathsf{T}\mathbf{x}_j \tag{12.136}$$

$$\text{subject to} \quad \alpha_i \geq 0 \quad \text{for all } i = 1, \ldots, n \tag{12.137}$$

$$\text{and} \quad \sum_{i=1}^{n} \alpha_i y_i = 0 \tag{12.138}$$

Therefore, to find the maximum-margin separating hyperplane for training data that are linearly separable, simply solve Equations (12.136)–(12.138) for $\alpha_1, \ldots, \alpha_n$, then compute the normal vector as a weighted combination of the input vectors:

$$\mathbf{w} = \sum_{i=1}^{n} \alpha_i y_i \mathbf{x}_i \tag{12.139}$$

and the scalar bias as

$$b = \frac{1}{2}\left(\min_{\mathbf{x}\in\mathcal{D}_1} \mathbf{w}^\mathsf{T}\mathbf{x} + \max_{\mathbf{x}\in\mathcal{D}_2} \mathbf{w}^\mathsf{T}\mathbf{x}\right) \tag{12.140}$$

Note from Equation (12.139) that input samples for which $\alpha_i = 0$ have no influence over the hyperplane parameters; these are points that are so far from the hyperplane that they are easily classified correctly. The hyperplane normal is thus a linear combination only of the **support vectors**, which are the input samples for which $\alpha_i > 0$. If we let $\mathcal{S}$ be the set of support vectors, then, after training the SVM, a query point $\mathbf{x}$ is classified as positive or negative according to the sign of its projection onto the hyperplane normal:

$$y = \text{sgn}(\mathbf{w}^\mathsf{T}\mathbf{x} + b) = \text{sgn}\left(\sum_{i=1}^{n} \alpha_i y_i \mathbf{x}^\mathsf{T}\mathbf{x}_i + b\right) = \text{sgn}\left(\sum_{\mathbf{x}' \in \mathcal{S}} \alpha_i y_i \mathbf{x}^\mathsf{T}\mathbf{x}' + b\right) \quad (12.141)$$

which is computationally efficient, since the number of support vectors is usually a small fraction of the training samples.

To derive the dual problem from the primal problem, we turn to the technique of **Lagrange multipliers**.[†] Suppose we want to minimize (or maximize) a function $f(\mathbf{x})$ subject to the equality constraint $g(\mathbf{x}) = 0$.[‡] The key insight of the Lagrange multiplier approach is that, at the solution $\mathbf{x}^*$, the gradient vectors of the two functions are parallel: $\nabla f(\mathbf{x}^*) \propto \nabla g(\mathbf{x}^*)$, or $\nabla f(\mathbf{x}^*) + \lambda \nabla g(\mathbf{x}^*) = 0$ for some scalar λ. While the proof of this statement can be found in any standard calculus text, we shall content ourselves here with an intuitive illustration. Suppose you are hiking on a hill, and your goal is to reach the highest elevation possible, but you are constrained to remain along a particular path carved out of the hillside. So you continue walking as long as your elevation is increasing, but once it stops increasing, you stop hiking as well. Let $f(\mathbf{x})$ be the elevation function, and let $g(\mathbf{x})$ be the implicit function whose zero level set defines the path. At the place where you stopped, the gradient ∇f of the elevation function, which always points uphill, is orthogonal to the path; it does not have a component tangent to the path, otherwise you would have continued to hike along the path. Since the gradient ∇g of the path is always orthogonal to the path, the two vectors ∇f and ∇g must be parallel to each other at the place you stopped hiking. Therefore, if the vectors are parallel, you are guaranteed to be at a local stationary point (either a maximum, minimum, or saddle point). As a result, the vectors being parallel is a necessary (but not sufficient) condition that the place where you are standing is the global extremum sought; whether it is actually the desired solution depends upon the shape of f.

Let us apply the technique of Lagrange multipliers to the following optimization problem:

$$\min \quad f(\mathbf{x}) \tag{12.142}$$

$$\text{subject to} \quad g_i(\mathbf{x}) = 0 \quad \text{for all } i = 1, \ldots, n \tag{12.143}$$

where the min could be replaced by max, depending upon the shape of f. By similar reasoning to that above, in the case of multiple constraints, $\nabla f(\mathbf{x})$ is parallel to a linear combination of $\nabla g_1(\mathbf{x}), \ldots, \nabla g_n(\mathbf{x})$. Therefore,

$$\nabla f(\mathbf{x}) + \lambda_1 \nabla g_1(\mathbf{x}) + \cdots + \lambda_n \nabla g_n(\mathbf{x}) = \nabla f(\mathbf{x}) + \sum_{i=1}^{n} \lambda_i \nabla g_i(\mathbf{x}) = \mathbf{0} \quad (12.144)$$

for some scalars $\lambda_1, \ldots, \lambda_n$. To solve the optimization problem, then, we construct the **Lagrangian** (or Lagrange function):

$$\psi(\mathbf{x}) \equiv f(\mathbf{x}) + \sum_i \lambda_i g_i(\mathbf{x}) \tag{12.145}$$

take derivatives, and set the result to zero:

$$\nabla \psi(\mathbf{x}) = \nabla f(\mathbf{x}) + \sum_i \lambda_i \nabla g_i(\mathbf{x}) = \mathbf{0} \tag{12.146}$$

[†] We encountered Lagrange multipliers in Section 11.1.3 (p. 516).

[‡] The functions f and g in this subsection are not related to the classifier or discriminant functions.

If $\mathbf{x} \in \mathbb{R}^n$, then this equation together with Equation (12.143) provide $2n$ equations for the $2n$ unknowns of $\lambda_1, \ldots, \lambda_n$ and the n elements of $\mathbf{x}$.

Lagrange multipliers can be generalized to also handle inequality constraints. Suppose we wish to optimize

$$\min \quad f(\mathbf{x}) \tag{12.147}$$

$$\text{subject to} \quad g_i(\mathbf{x}) = 0 \quad \text{for all } i = 1, \ldots, n \tag{12.148}$$

$$\text{and} \quad h_j(\mathbf{x}) \le 0 \quad \text{for all } j = 1, \ldots, n \tag{12.149}$$

Similar to above, we define the Lagrangian as

$$\psi(x) \equiv f(\mathbf{x}) + \sum_i \lambda_i g_i(\mathbf{x}) + \sum_j \alpha_j h_j(\mathbf{x}) \tag{12.150}$$

where $\alpha_1, \ldots, \alpha_n$ are scalars, and we set the derivatives to zero, as before, which yields $3n$ equations in $3n$ unknowns. For this approach to work, the **Karush-Kuhn-Tucker (KKT) conditions** must be satisfied. While these conditions are too technical for our purposes, we will draw attention to two of the KKT conditions in particular:

$$\alpha_j \ge 0 \quad (\text{non-negativity, or dual feasibility}) \tag{12.151}$$

$$\alpha_j h_j(\mathbf{x}*) = 0 \quad (\text{complementary slackness}) \tag{12.152}$$

where the first condition is obvious, and the importance of the second condition will become apparent below.

Now let us apply this approach to the optimization problem in Equations (12.134)–(12.135), which we rewrite as follows:

$$\min_{\mathbf{w}, b} \quad \frac{1}{2}\|\mathbf{w}\|^2 \tag{12.153}$$

$$\text{subject to} \quad 1 - y_i(\mathbf{w}^{\mathsf{T}}\mathbf{x}_i + b) \le 0 \quad \text{for all } i = 1, \ldots, n \tag{12.154}$$

so that the objective function to be differentiated is given by

$$\psi(\mathbf{w}, b) \equiv \frac{1}{2}\|\mathbf{w}\|^2 + \sum_{i=1}^{n} \alpha_i(1 - y_i(\mathbf{w}^{\mathsf{T}}\mathbf{x}_i + b)) \tag{12.155}$$

Taking the derivatives yields

$$\frac{\partial \psi(\mathbf{w}, b)}{\partial \mathbf{w}} = \mathbf{w} - \sum_{i=1}^{n} \alpha_i y_i \mathbf{x}_i = \mathbf{0} \tag{12.156}$$

$$\frac{\partial \psi(\mathbf{w}, b)}{\partial b} = -\sum_{i=1}^{n} \alpha_i y_i = 0 \tag{12.157}$$

Rearranging yields two important equations:

$$\mathbf{w} = \sum_{i=1}^{n} \alpha_i y_i \mathbf{x}_i \tag{12.158}$$

$$0 = \sum_{i=1}^{n} \alpha_i y_i \tag{12.159}$$

The first equation says that any solution to the optimization problem is a weighted sum of the input vectors, just as we saw regarding perceptrons in Equation (12.120), whereas the second equation says that the sum of the multipliers weighted by the ground truth labels is zero. Plugging Equation (12.158) into Equation (12.155) yields

$$\psi(\mathbf{w}, b) = \frac{1}{2}\mathbf{w}^\mathsf{T}\mathbf{w} + \sum_{i=1}^{n}\alpha_i - \sum_{i=1}^{n}\alpha_i y_i \mathbf{w}^\mathsf{T}\mathbf{x}_i - \sum_{i=1}^{n}\alpha_i y_i b \tag{12.160}$$

$$= \frac{1}{2}\sum_{i=1}^{n}\sum_{j=1}^{n}\alpha_i\alpha_j y_i y_j \mathbf{x}_i^\mathsf{T}\mathbf{x}_j + \sum_{i=1}^{n}\alpha_i - \sum_{i=1}^{n}\sum_{j=1}^{n}\alpha_i\alpha_j y_i y_j \mathbf{x}_i^\mathsf{T}\mathbf{x}_j - \sum_{i=1}^{n}\alpha_i y_i b \tag{12.161}$$

$$= -\frac{1}{2}\sum_{i=1}^{n}\sum_{j=1}^{n}\alpha_i\alpha_j y_i y_j \mathbf{x}_i^\mathsf{T}\mathbf{x}_j + \sum_{i=1}^{n}\alpha_i - b\sum_{i=1}^{n}\alpha_i y_i \tag{12.162}$$

$$= -\frac{1}{2}\sum_{i=1}^{n}\sum_{j=1}^{n}\alpha_i\alpha_j y_i y_j \mathbf{x}_i^\mathsf{T}\mathbf{x}_j + \sum_{i=1}^{n}\alpha_i \tag{12.163}$$

where Equation (12.159) is used in the final equation. Combining Equations (12.163), (12.151), and (12.159) yields the dual problem in Equations (12.136)–(12.138).

Ignoring the constraints in Equations (12.137)–(12.138), note that the dual problem in Equation (12.136) is convex in the unknown values for α_i, $i = 1, \ldots, n$. As a result, the function $\psi(\mathbf{w}, b)$ is a quadratic performance surface, which is easily seen by stacking these values into a vector $\boldsymbol{\alpha} \equiv \begin{bmatrix} \alpha_1 & \alpha_2 & \cdots & \alpha_n \end{bmatrix}^\mathsf{T}$:

$$\psi(\mathbf{w}, b) = \boldsymbol{\alpha}^\mathsf{T}\underbrace{\begin{bmatrix} y_1^2\mathbf{x}_1^\mathsf{T}\mathbf{x}_1 & y_1 y_2\mathbf{x}_1^\mathsf{T}\mathbf{x}_2 & \cdots & y_1 y_n\mathbf{x}_1^\mathsf{T}\mathbf{x}_n \\ \vdots & & \ddots & \vdots \\ y_1 y_n\mathbf{x}_1^\mathsf{T}\mathbf{x}_n & y_2 y_n\mathbf{x}_2^\mathsf{T}\mathbf{x}_n & \cdots & y_n^2\mathbf{x}_n^\mathsf{T}\mathbf{x}_n \end{bmatrix}}_{\mathbf{H}}\boldsymbol{\alpha} - \mathbf{1}_{\{n \times 1\}}^\mathsf{T}\boldsymbol{\alpha} \tag{12.164}$$

where $\mathbf{1}_{\{n \times 1\}}$ is a vector of n 1s. Taking the partial derivatives with respect to $\boldsymbol{\alpha}$, and setting the result to zero, yields $\mathbf{H}\boldsymbol{\alpha} = \mathbf{1}_{\{n \times 1\}}$, or $\boldsymbol{\alpha} = \mathbf{H}^{-1}\mathbf{1}_{\{n \times 1\}}$, where $\mathbf{H}$ is the symmetric positive semidefinite matrix in the equation above. If there were no constraints, then we could simply solve the optimization problem in one step in this manner. With the constraints, however, it is more efficient to perform stochastic gradient descent, which is made easier by the fact that there are no local minima due to the convex shape of the function being minimized.

Introducing Slack Variables

For all interesting real-world situations, the data are not linearly separable, and therefore some data will be misclassified no matter where the hyperplane is located. As a result, Equation (12.154) will not hold for all data samples. To obtain a meaningful result in such a case, we introduce the **slack variables** ξ_i, $i = 1, \ldots, n$, where $\xi_i = 0$ if the data sample $\mathbf{x}_i$ is classified correctly, or $\xi_i > 0$ if the sample is misclassified. Equations (12.153)–(12.154) are then modified to penalize the misclassifications, yielding the following primal problem:

$$\min_{\mathbf{w}, b, \xi} \quad \frac{1}{2}\|\mathbf{w}\|^2 + c\sum_{i=1}^{n}\xi_i \tag{12.165}$$

$$\text{subject to} \quad 1 - y_i(\mathbf{w}^\mathsf{T}\mathbf{x}_i + b) \leq \xi_i \quad \text{for all } i = 1, \ldots, n \tag{12.166}$$

$$\text{and} \quad \xi_i \geq 0 \quad \text{for all } i = 1, \ldots, n \tag{12.167}$$

where c is a constant that governs the misclassification penalty. Applying the Lagrange / KKT approach, this primal problem can be converted to the dual problem

$$\max_{\alpha_i} \quad \sum_{i=1}^{n} \alpha_i - \frac{1}{2} \sum_{i=1}^{n} \sum_{j=1}^{n} \alpha_i \alpha_j y_i y_j \mathbf{x}_i^\mathsf{T} \mathbf{x}_j \tag{12.168}$$

$$\text{subject to} \quad 0 \le \alpha_i \le c \quad \text{for all } i = 1, \ldots, n \tag{12.169}$$

$$\text{and} \quad \sum_{i=1}^{n} \alpha_i y_i = 0 \tag{12.170}$$

which is identical to Equations (12.136)–(12.138) except for the maximum constraint on the multipliers. As before, inputs for which $\alpha_i = 0$ are classified correctly and have no influence over the outcome. But inputs for which $\alpha_i = c$ are misclassified so badly that their influence on the outcome is bounded. From the complementary slackness condition of Equation (12.152), note that for each input either $\alpha_i = 0$ (indicating that it is classified correctly) or the inequality in Equation (12.166) is an equality (so that the slackness variable indicates the amount of misclassification).

Kernel Trick

Given two vectors $\mathbf{x}_i$ and $\mathbf{x}_j$, the Euclidean inner product between them is $\mathbf{x}_i^\mathsf{T} \mathbf{x}_j$, which is a property of vectors in a Euclidean space, because the inner product measures angles between vectors. A curious fact of the SVM is that the data are never accessed by themselves but only by their inner products with each other. That is, if we were to compute all the inner products between the training samples and each other, as well as between the training samples and any future query, then the data themselves could be discarded completely, and the classifier would run unhindered. This fact can be seen for training in Equations (12.164) and (12.168), and it can be seen for runtime in Equation (12.141).

As a result, the Euclidean inner product can be replaced with any other suitable mapping without changing the SVM training or runtime algorithm at all. To define "suitable," we turn to **Mercer's theorem**, which requires the mapping $K : \mathbb{R}^d \times \mathbb{R}^d \to \mathbb{R}$ to be continuous, symmetric, and positive definite. Such a mapping is known as a **Mercer kernel**.[†] In practice, however, any mapping that captures the intuitive notion of similarity tends to work, and such a mapping is simply called a *kernel*. The **kernel trick**, therefore, refers to the fact that we can replace the Euclidean inner product $\mathbf{x}_i^\mathsf{T} \mathbf{x}_j$ between input vectors with a kernel $K(\mathbf{x}_i, \mathbf{x}_j)$ applied to them. One of the more common kernels is the Gaussian kernel:

$$K(\mathbf{x}_i, \mathbf{x}_j) = \exp\left(-\frac{1}{2} \mathbf{x}_i^\mathsf{T} \mathbf{\Sigma}^{-1} \mathbf{x}_j \right) \tag{12.171}$$

where $\mathbf{\Sigma}$ is the covariance matrix of the multidimensional Gaussian, and the normalization factor is omitted because overall scaling is unimportant.

The power of the kernel trick is that a hyperplane in a high-dimensional space can represent arbitrary decision boundaries in a low-dimensional space if the dimension of the hyperplane is high enough. As a result, a kernel can be used to project the data into a high-dimensional space, then the SVM applies a linear decision boundary in this high-dimensional space, with the result being identical to a complex decision boundary in the original input space. The Gaussian kernel, as it turns out, projects the input data to an

---

[†] Mercer kernels generalize the notion of inner product and lead to a *reproducing kernel Hilbert space*. Similarly, any vector space with a suitable inner product defined is known as a *Hilbert space*, of which Euclidean is a special case.

infinite-dimensional space, which makes it clear why SVMs are such powerful tools for classification.

One approach to training an SVM is to use stochastic gradient ascent, which updates one multiplier at a time. Another approach that has been widely adopted is the **sequential minimal optimization (SMO) algorithm**, which updates pairs of weights at a time. A simplified version of SMO is presented in Algorithm 12.4. Although this code is not sufficiently detailed for a real implementation, it should convey the basic flavor of the approach, thus making it easier to understand the full algorithm.

ALGORITHM 12.4 Simplified SMO algorithm for training an SVM

SIMPLIFIEDSMO $(\{(\mathbf{x}_i, y_i)\}_{i=1}^{n}, c)$

Input: set of n labeled data points $\{(\mathbf{x}_i, y_i)\}_{i=1}^{n}$
$\qquad$ $\mathbf{x}_i \in \mathbb{R}^d$ is the ith data point
$\qquad$ $y_i \in \{-1, +1\}$ is the binary label of the ith data point
$\qquad$ weight c governing the misclassification penalty
Output: parameters $\mathbf{w}$ and b of the maximum-margin classifier

1 $\quad \alpha_i \leftarrow 0$ for $i = 1, \ldots, n$

2 $\quad b \leftarrow 0$

3 $\quad$ **while** not converged **do**

4 $\qquad$ **for** $i \leftarrow 1$ **to** n **do**

5 $\qquad\qquad \delta_i \leftarrow \left(\sum_{k=1}^{n} \alpha_k y_k K(\mathbf{x}_i, \mathbf{x}_k) + b \right) - y_i$

6 $\qquad\qquad$ **if** $(\delta_i y_i < -\epsilon$ and $\alpha_i < c)$ or $(\delta_i y_i > \epsilon$ and $\alpha_i > 0)$ **then**

7 $\qquad\qquad\qquad$ Select $j \neq i$ at random

8 $\qquad\qquad\qquad \delta_j = \left(\sum_{k=1}^{n} \alpha_k y_k K(\mathbf{x}_j, \mathbf{x}_k) + b \right) - y_j$

9 $\qquad\qquad\qquad$ **if** $y_i == y_j$ **then**

10 $\qquad\qquad\qquad\qquad v_{low} \leftarrow \max(0, \alpha_i + \alpha_j - c), \quad v_{high} \leftarrow \min(c, \alpha_i + \alpha_j)$

11 $\qquad\qquad\qquad$ **else**

12 $\qquad\qquad\qquad\qquad v_{low} \leftarrow \max(0, \alpha_j - \alpha_i), \quad v_{high} \leftarrow \min(c, c + \alpha_j - \alpha_i)$

13 $\qquad\qquad\qquad \eta \leftarrow 2K(\mathbf{x}_i, \mathbf{x}_j) - K(\mathbf{x}_i, \mathbf{x}_i) - K(\mathbf{x}_j, \mathbf{x}_j)$

14 $\qquad\qquad\qquad \alpha_j' \leftarrow \max(v_{low}, \min(v_{high}, \alpha_j - y_j(\delta_i - \delta_j)/\eta))$

15 $\qquad\qquad\qquad \alpha_i' \leftarrow \alpha_i + y_i y_j(\alpha_j - \alpha_j')$

16 $\qquad\qquad\qquad b_1 \leftarrow b - \delta_i - y_i(\alpha_i' - \alpha_i)K(\mathbf{x}_i, \mathbf{x}_i) - y_j(\alpha_j' - \alpha_j)K(\mathbf{x}_i, \mathbf{x}_j)$

17 $\qquad\qquad\qquad b_2 \leftarrow b - \delta_j - y_i(\alpha_i' - \alpha_i)K(\mathbf{x}_i, \mathbf{x}_j) - y_j(\alpha_j' - \alpha_j)K(\mathbf{x}_j, \mathbf{x}_j)$

18 $\qquad\qquad\qquad$ **if** $v_{low} \neq v_{high}$ and $\eta < 0$ and $|\alpha_j - \alpha_j'| > \epsilon$ **then**

19 $\qquad\qquad\qquad\qquad \alpha_i \leftarrow \alpha_i' \quad \alpha_j \leftarrow \alpha_j'$

20 $\qquad\qquad\qquad\qquad b \leftarrow \begin{cases} b_1 & \text{if } 0 < \alpha_i < c \\ b_2 & \text{if } 0 < \alpha_j < c \\ (b_1 + b_2)/2 & \text{otherwise} \end{cases}$

21 $\quad \mathbf{w} \leftarrow \sum_{i=1}^{n} \alpha_i y_i \mathbf{x}_i$

22 $\quad$ **return** $\mathbf{w}$, b

SVMs, as with any learning technique, can take any type of feature vector as input. One approach that has achieved some popularity, known as **bag of visual words**, involves treating the image like a document, and treating image features like words. To classify a document, a common approach is to build a sparse histogram (*bag of words*) over a predetermined vocabulary of prototype words, or codewords, by letting the words in the document vote for the codewords. (The vocabulary is usually created by applying a technique like k-means[†] to a large training set to cluster the features into prototypes.) This histogram is then fed to a learning algorithm to determine the type of document. In a similar way, visual features can be extracted from the image, then mapped to a previously learned set of visual codewords, forming a sparse histogram over codewords. This histogram is then fed to an SVM or other learning algorithm.

12.4.6 Neural Networks

As we saw earlier, the decision boundary for a single perceptron is a hyperplane. Although this shape is sufficient when the data are linearly separable, in many real-world situations a more complex decision boundary is needed. For example, Figure 12.23 shows the classic example of the XOR problem in which the data are not linearly separable, and therefore a perceptron will perform poorly. In the previous subsection we saw one way to overcome this limitation, namely to map the data to a higher-dimensional space in which the data *are* linearly separable, an approach known as the kernel trick. In this subsection we consider an alternate approach, namely to combine multiple perceptrons together into what is called an **artificial neural network (ANN)**.

An artificial neural network, oftentimes just called a neural network, can be thought of as a biologically-inspired computational model, since it simulates the network of neurons in the brain. However, keep in mind that there are many details of how the brain works that are not included in such a model. For example, traditional neural networks ignore the temporal aspects of the biological signals.

Several different types of neural networks are possible, as shown in Figure 12.24, depending upon the types of connections between the perceptrons. The classic, traditional architecture (and the most common) is the **feedforward neural network** in which the information flows in a single direction from the input to the output. That is, if the perceptrons are considered as nodes in a graph, with edges connecting the inputs of some perceptrons to the outputs of others, there are no cycles in such a network. The alternative to the feedforward network is the **recurrent neural network**, in which the output of some

Figure 12.23 In contrast to the AND and OR problems, which are linearly separable, the classic XOR problem is not linearly separable. Although hyperplanes (in this case lines) can be drawn to separate the data in the first two cases, no hyperplane performs well in the last case.

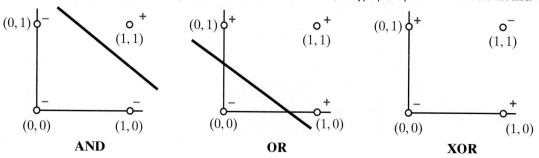

[†] Section 11.5.1 (p. 549).

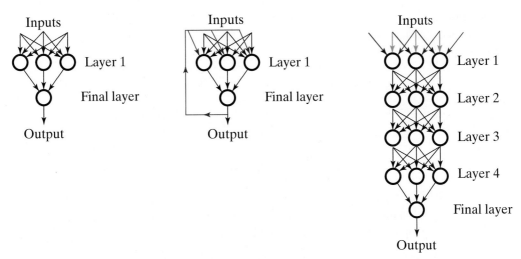

Figure 12.24 Different neural network architectures. From left to right: 2-layer feedforward network, 2-layer recurrent network, 4-layer convolutional network. In the latter, the edges in the first layer that share colors have the same weights.

perceptrons feeds back into the input of earlier perceptrons in the pipeline, thus creating cycles in the graph. **Deep neural networks**, which generally contain anywhere from 10 to 30 layers or more, are undergoing a resurgence of interest in the research community, as they have been shown to be capable of achieving leading-edge results on many challenging problems. Finally, **convolutional neural networks** tie the weights of early perceptrons together, so that the early layers essentially learn to extract convolution-like features from the image. In practice, both deep and convolutional networks are typically feedforward networks, though this is not required.

The most common arrangement for a feedfoward network is the **multilayer perceptron (MLP)**, in which the perceptrons are organized in a sequence of layers, as shown in the figure. The **input layer**, which is not shown, simply passes the inputs through, while the **output layer** yields the final result; the remaining layers are known as **hidden layers**. It can be proved that, just as a support vector machine with the kernel trick is capable in theory of separating any data set, a fully-connected two-layer feedforward neural network is capable of separating any data set. For example, Figure 12.25 shows such a network that solves the XOR problem by computing the following output:

$$y = h\big(h(x^{(1)} - 1) - 2h(x^{(1)} + x^{(2)} - 2) + h(x^{(2)} - 1) - 1\big) \qquad (12.172)$$

where $h(\cdot)$ is the threshold function. It is easy to verify that this output indeed yields the desired behavior.

The most common approach to training a feedforward neural network is the **backpropagation algorithm**. In this algorithm, errors in the output (the difference between the predicted value and the ground truth value) are propagated backward through the network, updating the weights of the output and hidden layers in a gradient descent fashion. In order for this algorithm to work, the threshold function must be differentiable. As a result, neural networks typically do not use a hard threshold but rather a soft threshold called a **sigmoid function** (because it is in the shape of an "S"). A common sigmoid function is the logistic function:

$$h(z) = \frac{1}{1 + e^{-z}} \qquad (12.173)$$

Figure 12.25 A simple two-layer feedforward neural network (multilayer perceptron) that solves the XOR problem.

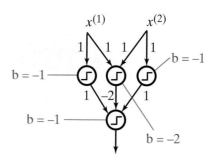

which approaches $h(z) \to 0$ as $z \to -\infty$ and $h(z) \to 1$ as $z \to +\infty$. More recently, it has been found that improved results are obtained using a **rectified linear unit (ReLU)**, which employs the biologically inspired rectifier:

$$h(z) = \max(z, 0) \tag{12.174}$$

Unlike the sigmoid function, ReLUs do not suffer from the *vanishing gradient problem*, because the gradient of this function remains constant even for large values of z, which is important for decreasing training time. Other advantages include the fact that, for positive numbers, the output is a scaled version of the input, thus introducing some amount of invariance to the gain of an image, and many units output the value zero, which introduces sparsity in the model.

A diagram of a historically important neural network is shown in Figure 12.26. In this system, which was the first widely-used face detector introduced two decades ago, the image is scanned at all locations and scales, and at each of these places a 20×20 window of image grayscale values is fed to the face detector. Preprocessing normalizes the grayscale values to correct for lighting variations and low contrast, then the window is fed to a multi-layer feedforward neural network, whose hidden layers have been handcrafted to take advantage of the peculiar structure of the human face. Training is performed using backpropagation, using a **bootstrap algorithm** to feed false positives back into the training process to reduce the occurrence of such errors. The algorithm detects faces at multiple scales by processing an image pyramid obtained by successive downsampling.[†] An arbitration function produces a

Figure 12.26 The Rowley-Baluja-Kanade neural network face detector applies a feedforward neural network to lighting-corrected 20×20 graylevel windows to detect faces in an image.

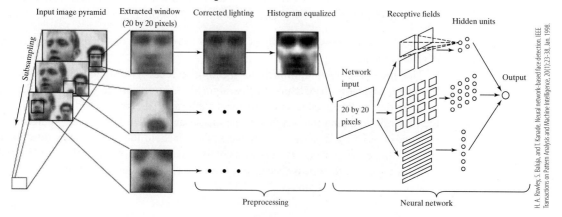

[†] Section 7.1.1 (p. 329).

final result from the overlapping rectangles output by the neural network on nearby locations and scales. It is important to note that there is little difference in the accuracy of this historic face detector and modern face detectors, apart from progress that has been made on computational efficiency and handling out-of-plane rotation. The lesson learned is therefore that the most important aspects of a detector are the training data and the features extracted rather than the specific algorithm used for classification and/or training.

12.4.7 Random Forests

It is inevitable that, on real-world data, a classifier will make mistakes. One way to improve accuracy is to train multiple classifiers, then let the classifiers run in parallel and vote, with the majority vote determining the outcome. Since an "ensemble" is a collection or group, this approach of training a group of classifiers is known as **ensemble learning**.

There are two primary approaches to ensemble learning. **Bagging** (for "bootstrap aggregating") refers to the process of training multiple classifiers by repeatedly resampling the training data (with replacement), and feeding each sampled set to a different classifier during training. Note that samples in the original training data may be sampled multiple times, or not at all, but all the sampled sets are approximations to the same underlying distribution. Such sampling with replacement is known as **bootstrapping**, hence the name *bootstrap aggregating*.[†]

The most popular use of bagging are **random forests**, which are collections of decision trees. A **decision tree** is a sequence of decisions, with each branch in the tree pivoting on a single element of the remaining input vector, and each leaf assigning either a single outcome or a probability distribution over possible outcomes. During training, at each stage of the tree the input vector is searched to find the element that does the best job of separating the data according to the training labels. This recursive training process is known as **CART** (for "classification and regression trees"). Decision trees are an example of an **unstable classifier**, which means that a small change in the training data set can have a large effect on the performance of the classifier. Unstable classifiers are known to benefit significantly from aggregating the results of multiple instances trained on slightly different data sets, as long as the different classifiers are not highly correlated. To achieve this behavior, the trees are trained together to ensure that they do not learn to split the data according to the same features. Decision trees and random forests are alternatives to SVMs and neural networks, with the added advantage that they tend to be easier to interpret than the latter approaches, which are more opaque due to their black-box nature.

The other popular approach to ensemble learning is known as **boosting**. Boosting seeks to answer two questions during the training process: namely, how important are the different elements of the feature vector, and which training samples are the hardest to classify. The most popular boosting algorithm is known as **AdaBoost**, which initially weights all the training samples equally by imposing a uniform distribution. After one round of training, the classifier has learned to classify the data with some level of accuracy, resulting in what is known as a **weak classifier**. In the second round of training, the classifier is again fed all the training data, but this time with the misclassified samples weighted more highly, resulting in another weak classifier. The process continues some number of iterations, each time reweighting the training samples and producing another weak classifier. Finally, all the weak classifiers are combined into a strong classifier using a weighted average of the results. The efficacy of boosting results from the fact that if each weak classifier produces a result that is at least a little bit better than random, then the linear combination of the weak classifiers will produce an even better result.

[†] The term *bootstrapping* is also used to refer to the somewhat unrelated idea of feeding errors back into the training process to improve the classifier, as seen earlier.

12.4.8 Attentional Cascade

Applying a decision tree to a new input involves performing a sequence of tests, with the output of each test determining which branch of the tree to follow for subsequent tests. Using a decision tree, a test is performed. Depending upon the outcome of the test, either the left or the right branch of the tree is taken, and the process is repeated until a decision is made. In this approach the final outcome is not known until the entire depth of the tree has been traversed, and, moreover, each test typically requires approximately the same amount of computation.

A slight variation of this approach is a **degenerate decision tree**, in which at each level of the tree only one branch leads to further tests, whereas the other branch terminates in a decision. If the goal is **rare event detection**, which is often the case, then the degenerate decision tree takes the form of an **attentional cascade**, in which only positive responses lead to further consideration, while negative responses immediately terminate. In other words, in an attentional cascade each test asks the question, "Is this input the object?" If the response is negative, then the answer can be considered a definitive "no," and the detector terminates with a negative response; if the response is positive, then the answer can be considered a tentative "maybe," and further tests are performed until either a negative response is encountered or the entire sequence of tests pass, in which case a positive response is generated.

The most successful face detector to date is the **Viola-Jones face detector**, shown in Figure 12.27, and it is fairly safe to assume that any face detector you see in practice is probably based on this algorithm—not because it produces higher accuracy than other detectors, but solely because of its superior computational performance. Viola-Jones achieves real-time performance through the use of two tricks. First, it uses Haar-like image features that are computed quickly using the integral image.[†] After the initial preprocessing to create the integral image, any feature can be computed at any location and any scale with just a few lookups, and therefore no time is wasted downsampling the image to search at multiple scales. Secondly, it uses an attentional cascade architecture that quickly discards unpromising image locations. In the original implementation, 32 stages of the cascade are used, with only a small fraction of the tens of thousands of possible image features needed to perform the sequence of tests. During training, discriminative features are found and thresholds are set so that tests are allowed to yield a large number of false positives, as long as the number of false negatives is low. In this way, when a test generates a negative response, it can be trusted with a high degree of confidence. A similar approach has also been used to detect other types of objects, such as pedestrians.

Figure 12.27 The Viola-Jones face detector combines Haar-like features computed using an integral image (left) with an attentional cascade architecture that quickly discards a large percentage of the input windows (middle) in order to find faces in an image in real time (right).

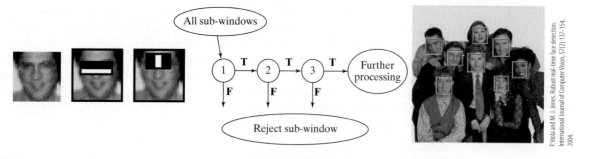

P. Viola and M. J. Jones. Robust real-time face detection. International Journal of Computer Vision, 57(2):137-154, 2004.

[†] Section 5.2.7 (p. 233) and Section 6.6.1 (p. 311).

12.4.9 Deformable Part-Based Model (DPM)

Rather than representing the object as a single rigid structure, an alternative is to represent it as a collection of rigid parts whose locations relative to one another are allowed to move somewhat. For example, a person's face can be described as a pair of eyes, pair of ears, nose, and mouth, along with some expected spatial relationships between these parts that may vary slightly from one individual to another, as shown in Figure 12.28. Similarly, a person's body can be described as a head, torso, a pair of arms, and a pair of legs.

The classic **pictorial structure** representation of an object consists of a collection of parts in a deformable configuration with spring-like connections between the parts. The representation utilizes an undirected graph in which the vertices are the parts, and the edges are the connections between the parts. For computational reasons, the graph is typically prohibited from containing cycles and therefore is in fact a tree. Efficient algorithms based on the Viterbi algorithm[†] can be used to search through the space of part locations to minimize a global energy function that seeks to match part models to image data while at the same time minimizing the deformations between the part locations compared with their ideal locations.

Pictorial structures are a type of **deformable part model (DPM)**. In a DPM, the graph captures a probabilistic model of the object, where the unary energy terms of the vertices measure the likelihoods of the part locations, and the binary energy terms of the springs provide a geometric prior on their relative locations. Finding an object in an image then becomes equivalent to finding the maximum likelihood configuration of the object according to the probabilistic graphical model. With a certain constraint on the allowed deformation costs, such costs can be efficiently computed using a generalized distance transform, leading to an algorithm that can compute the global minimum of the energy function in a time that is linear in the number of parts, as well as linear in the number of possible locations of those parts.

DPMs have been successfully applied to a variety of challenging object recognition problems. As an example, results of a well-known DPM system are shown in Figure 12.29. In this system the object is represented by a star configuration of parts located with respect to a root node corresponding to a lower resolution representation of the entire object. A fixed

Figure 12.28 A pictorial structure, or deformable parts model, represents an object as a collection of parts with spring-like connections between them. Shown here is a classic model of a face composed of two eyes, two ears, a nose, mouth, and some hair.

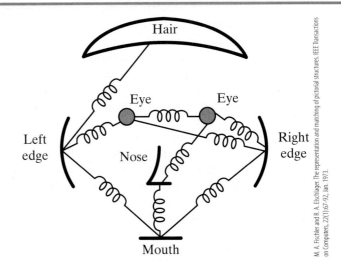

M. A. Fischler and R. A. Elschlager. The representation and matching of pictorial structures, IEEE Transactions on Computers, 22(1):67-92, Jan. 1973.

[†] Section 10.2.1 (p. 453).

P. Felzenszwalb, R. Girshick, D. McAllester, and D. Ramanan. Object detection with discriminatively trained part based models. IEEE Transactions on Pattern Analysis and Machine Intelligence, 32(9): 1627–1645, Sept. 2010, http://arxiv.org/pdf/1311.2524v5.pdf.

Person   Car   Horse   Sofa   Bicycle

Figure 12.29 Results of a well-known deformable part model-based system on the challenging problem of multicategory object classification.

number of parts are greedily placed at locations of high energy content around the root location, and the appearance of each part, along with the appearance of the root, are captured using dimensionally-reduced HOG features.[†] To handle large intracategory variations in appearance, such as images of a bicycle from a front view versus a side view, a collection of DPMs in a mixture model are used for each object category. It should be noted that, while systems based on DPMs have achieved impressive results, according to some studies their success lies primarily in their use of subcategories to model intracategory variations rather than in the deformable parts themselves, thus leaving open the question as to how important DPMs are for good performance in object recognition tasks.

12.4.10 Deep Learning

The traditional approach to classification, as we have seen, is to extract handcrafted features from images, then feed these features to a machine learning algorithm. For example, SIFT features,[‡] HOG features, or a bag of visual words can be extracted from an image and sent to an SVM or a traditional two-layer neural network. Such an approach, for lack of a better term, can be referred to as **shallow learning**. For almost two decades, beginning in the mid-1990s, this approach of shallow learning was state-of-the-art, and it has enabled successful commercial systems such as real-time face and pedestrian detection. However, very soon after the initial successes, the accuracy of such systems plateaued, so that for many years the research community struggled to obtain any significant improvement in accuracy on a variety of challenging problems using this paradigm. Handcrafted features capture low-level properties like intensity edges and blobs, but it is very difficult to construct mid-level features like curves and parts by hand.

The situation changed when, at the 2012 ImageNet competition, significantly improved results were obtained using a **deep learning** approach. Since then, deep learning has taken over the computer vision community, showing superior results on a wide variety of problems when compared with traditional approaches. Some example results on multicategory classification are shown in Figure 12.30. There is no agreed-upon definition of deep learning, but it generally refers to a machine learning model consisting of a large number of information processing stages. Although deep learning is a more generic term that encompasses a variety of techniques, for computer vision nearly all successes have been achieved with deep neural networks (DNNs), and in particular deep convolutional neural networks (CNNs), which were mentioned earlier in the chapter.

As it is usually constructed, a deep neural network consists of a large number of alternating convolutional and pooling layers, followed by a number of fully connected layers. As in a convolutional neural network, the weights in each convolutional layer are tied together to

---

[†] Section 7.5.4 (p. 350).

[‡] Section 7.5.1 (p. 348).

Figure 12.30 Deep learning is currently the leading approach for multicategory object detection, as well as several other important computer vision problems.

facilitate translational invariance. Each pooling layer then works like a local receptive field to aggregate information over a spatially nearby area. Since the number of outputs of the pooling layer is smaller than the number of inputs, the pooling layers perform downsampling as well as aggregation, thus facilitating processing at multiple resolutions. The most common form of pooling is to compute the maximum of all the nearby inputs, known as **max pooling**. The results of the final pooling layer are then fed to a traditional fully connected feedforward neural network with a number of layers.

One of the challenges of a deep neural network is how to train such a highly-dimensioned model. A deep network can easily contain tens or hundreds of millions of parameters, which leads to two problems. The first problem is overfitting. Without sufficient labeled training data (which is difficult to obtain), it is easy for the training process to yield a model that overfits to the data that it has seen. A common solution to this problem, known as **dropout**, is to randomly and temporarily remove a hidden unit from the network during training. Dropout effectively allows the training process to search through a variety of thinned network architectures with extensive weight sharing between them, leading to a resulting network that favors sparsity since many of the weights will be close to zero. The second problem is the training time required for searching for all those parameters, which is reduced by using modern hardware like GPUs (graphical processing units) and by using activation functions whose responses do not saturate (such as ReLUs). Nevertheless, training is typically performed using either standard backpropagation and/or stochastic gradient descent, along with (in some circumstances at least) pretraining of the network weights using large amounts of unlabeled data to speed up the process and avoid local minima.

12.5 Further Reading

A more in-depth treatment of the foundational material in this chapter can be found in the classic book on pattern recognition by Duda et al. [2001]. The excellent work by Bishop [2006] is also helpful, as is the survey paper by Jain et al. [2000]. The bias-variance tradeoff is well-explained by Mitchell [1997]; a good resource for cross validation is the paper by Kohavi [1995]. Precision-recall curves were first used in evaluating edge detectors by Abdou and Pratt [1979], whereas ROC curves gained popularity in computer vision through the work of Bowyer and Phillips [1998] and Rowley et al. [1998]. A comparison of PR and ROC curves can be found in Davis and Goadrich [2006]. The curse of dimensionality and peaking phenomenon are described in an early paper by Trunk [1979]. The Akaike Information Criterion is due to Akaike [1974], while the Bayesian Information Criterion is due to Schwarz [1978]. Of the few papers that cover syntactic pattern recognition, one of the more well-known is that of Luo and Hancock [2001].

The VC dimension was proposed by Vapnik and Chervonenkis [1971], whereas support vector machines were introduced by Cortes and Vapnik [1995]. Other good descriptions of SVMs can be found in the works of Vapnik [1995], Burges [1998], and Boyd and Vandenberghe [2004]. The SMO algorithm is due to Platt [1998]. Classification trees and the CART algorithm, bagging, and random forests are found in the work of Breiman et al. [1984], Breiman [1996], and Breiman [2001], respectively. The Adaboost algorithm is explained by Freund and Schapire [1999]. Geometric hashing, which is an efficient way to index a database of models, can be found in the work of Lamdan and Wolfson [1988].

Color histograms built using billions of pixels were used for skin color detection by Jones and Rehg [2002)], and earlier work on using color for detecting objects can be found in Swain and Ballard [1991]. Locally weighted learning is due to Atkeson et al. [1997]. Parzen windows and kernel density estimation can be traced to the work of Parzen [1962]. A recent defense of nearest neighbors was provided by Boiman et al. [2008], and an improvement to nearest neighbors called neighborhood component analysis can be found in Goldberger et al. [2004]. A psychological basis for nearest neighbors can be found in focal colors, which are described by Berlin and Kay [1969] and discussed in the book by Palmer [1999]. Naive Bayes is used for view-invariant face detection, as well as for car detection, by Schneiderman and Kanade [2000]. A well-cited work showing the benefit of summing (rather than multiplying) classifier outputs is that of Kittler et al. [1998]. The scree test is due to Cattell [1966], while the Kaiser criterion is due to Kaiser [1960].

Eigenfaces were introduced by Turk and Pentland [1991], which was motivated by the work of Kirby and Sirovich [1990] to represent faces using PCA. Later extensions include Fisherfaces by Belhumeur et al. [1997] and Laplacianfaces by He et al. [2005], the former being based upon Fisher's linear discriminant, which is introduced in the work of Fisher [1936]. Eigenvectors are used for representing 3D objects in the work of Murase and Nayar [1995]. Point distribution models and active shape models are due to Cootes et al. [1995], whereas active appearance models are due to Cootes et al. [2001]. The perceptron learning algorithm can be traced to the work of Rosenblatt [1958], while Gallant [1990] proposed the pocket algorithm. After a promising early start in the 1960s, neural network research came to a halt as the result of the publication of the book by Minsky and Papert [1969], which caused the so-called "AI winter." The research was revived with the work of Rumelhart et al. [1986], which introduced the backpropagation algorithm.

The first successful face detectors were the Rowley-Baluja-Kanade algorithm described in Rowley et al. [1998] and the simultaneous but less well-known approach of Sung and Poggio [1998]. The former is a pivotal paper in the history of computer vision, marking the introduction of machine learning to the field in a way that showed its power to solve challenging, real-world problems. The Viola-Jones face detector is presented in Viola and Jones [2004], with follow-up work on multiview face detection in Jones and Viola [2003] and pedestrian detection in Viola et al. [2005]. Other work on multiview face detection is that of Wu et al. [2004], and other work on pedestrian detection is that of Dalal and Triggs [2005] and Enzweiler and Gavrila [2009], the latter of which has found its way into commercial use in automobiles. Impressive work combining detection and tracking of pedestrians is that of Wu and Nevatia [2007]. A thorough, though somewhat outdated, survey of face detection can be found in Yang et al. [2002]. Space limitations have not permitted us to discuss face recognition, but several historically important papers are those of Brunelli and Poggio [1993], Wiskott et al. [1997], and Phillips et al. [2000], as well as the more recent work of Taigman et al. [2014].

The paper that sparked the recent explosion of interest in deep learning, and deep convolutional neural networks in particular, is that of Krizhevsky et al. [2012]. A good overview of deep learning can be found in the work of Deng and Yu [2013], which is focused primarily upon speech applications. Rectified linear units (ReLUs) are due to the work of Nair and Hinton [2010], and the purpose of dropout is explained by Srivastava et al. [2014]. Those interested in part-based classification should consult the early work of Fischler and Elschlager [1973] as well as the more recent papers by Felzenszwalb and Huttenlocher [2005] and Felzenszwalb et al. [2010]. Divvala et al. [2012] analyze the relative importance of deformable parts, latent discriminative learning, and subcategories in the accuracy of DPMs.

PROBLEMS

12-1 Classification corresponds to which branch of machine learning?

12-2 List the three types of classification problems, and briefly describe each.

12-3 Explain the importance of generalization.

12-4 Define a dichotomizer.

12-5 Suppose an inspection system detects 99.99% of the good parts correctly, but only 98% of the defective parts. Suppose the cost of incorrectly labeling a good part bad is 5 minutes of extra time for a person to manually inspect the part, but the cost of incorrectly labeling a bad part good is an average of 3 hours of extra time to correct the problem downstream in the assembly line. What is the total risk of the system? Which type of error dominates the total risk?

12-6 You are part of a team whose job is to develop a classifier of some kind, so you collect some data and manually label them. What are your two options regarding how to separate the training data from the test data? What are the pros and cons of each? What are you absolutely not allowed to do with these two datasets in any circumstance?

12-7 What are the two most popular approaches to cross-validation?

12-8 Briefly explain the bias-variance tradeoff.

12-9 Derive Equation (12.8).

12-10 Suppose you have a reasonably-sized dataset, and you compute both the Akaike and Bayesian information criteria (AIC and BIC). Which one do you expect to be greater?

12-11 Briefly explain the concept of structural risk minimization.

12-12 What is the VC dimension of a classifier with a parabola-shaped decision boundary (at any orientation) in 2D?

12-13 Explain how the curse of dimensionality, the peaking phenomenon, and feature selection are all related.

12-14 Suppose we have a binary classification problem with the following training data, using 2D features:

category	set of 2D features (x_1, x_2)
ω_1	$\{(-12, 3)\}$
ω_2	$\{(10, 5), (-5, 5), (9, 6), (13, 0)\}$

Assume that the discrimination functions returned by the training procedure are the following:

$$g_1(x) = 0.4x_1 + 0.5x_2 - 10$$
$$g_2(x) = 0.5x_1 + 0.3x_2 - 9$$

Calculate the accuracy of classification.

12-15 Suppose 145 test objects arrive in the detection zone of a conveyor belt, of which 95 are bananas. When a classifier is applied to these objects, 16 of the bananas are mislabeled, and 3 of the other objects are mislabeled as bananas.

(a) Compute TPR, TNR, FPR, and FNR, and build the confusion matrix.

(b) Calculate the sensitivity, specificity, and accuracy.

(c) Calculate the F-measure and the Jaccard coefficient.

12-16 Given the Receiver Operating Characteristic (ROC) curves of two classifiers below, find the better classifier using (a) the equal error rate (EER), and (b) the area under the curve (AUC).

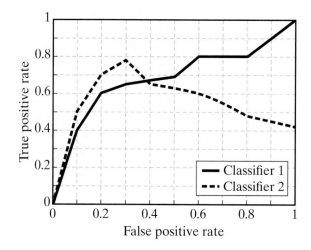

12-17 When is a precision-recall (PR) curve preferred to a receiver operating characteristic (ROC) curve?

12-18 Suppose that the population within a certain region is 51% male and that 42% of the males wear eyeglasses, while 52% of the females wear eyeglasses. One person is randomly selected for a survey, and after the person is selected, it is later learned that the person wears eyeglasses. Use Bayes' rule to calculate the probability that the selected person is male.

12-19 Given the means and covariance matrices of the Gaussian densities of three categories ω_1, ω_2, and ω_3 as

$$\boldsymbol{\mu}_1 = \begin{bmatrix} -1 \\ 1 \end{bmatrix} \quad \boldsymbol{\mu}_2 = \begin{bmatrix} 3 \\ 4 \end{bmatrix} \quad \boldsymbol{\mu}_3 = \begin{bmatrix} 6 \\ 9 \end{bmatrix} \quad \boldsymbol{\Sigma}_1 = \boldsymbol{\Sigma}_2 = \boldsymbol{\Sigma}_3 = \begin{bmatrix} 10 & 0 \\ 0 & 10 \end{bmatrix}$$

and the prior probabilities as $p(\omega_1) = 0.5$, $p(\omega_2) = p(\omega_3) = 0.25$, find the discriminant functions for each category, then calculate the decision boundary between the first two categories.

12-20 Is Bayes' rule limited to Gaussian distributions? Why or why not?

12-21 When is the maximum a posteriori (MAP) estimate the same as the maximum likelihood (ML) estimate?

12-22 True or false: For any value in the domain, the sum of all the class-conditional densities evaluated at that value equals 1.

12-23 Name one parametric and one nonparametric approach to representing a probability distribution.

12-24 When are two Gaussian densities separated by a hyperplane?

12-25 Explain the difference between a generative method and a discriminative method for classification.

12-26 Suppose you are given the following set of 1D training data along the x axis: $\{7, 12, 13, 15, 16\}$. Find the Parzen probability density function (pdf) estimate at $x = 15$, using the Gaussian function with variance 1 as the window function.

12-27 Explain how Parzen windows are related to locally weighted averaging (LWA).

12-28 Following Section 12.3.1 (p. 582), collect a dataset of images and label most of the pixels as either red or not red using a paint program. (It is best to leave pixels that are ambiguous as unlabeled, and to not use them for training.) Write a program to construct positive and negative histograms of the class-conditional densities, then to find all the red pixels in a query image using this model.

12-29 List some strengths and weaknesses of nearest-neighbor classification.

12-30 Construct a Naive Bayes classifier, using a Gaussian assumption, to classify whether a piece of fruit is an apple or plum based on the measured features, including weight and perimeter. The training data set is provided below.

category	weight (g)	perimeter (cm)
apple	450	10.5
apple	332	9.6
apple	289	8.2
apple	265	8.3
apple	306	8.5
plum	320	9.0
plum	235	8.1
plum	226	8.1
plum	308	8.7
plum	266	8.3

Apply the classifier to a test sample whose weight is 220 g and perimeter is 8.2 cm.

12-31 Given the following set of 10 samples in a 3D space, follow the steps of PCA to calculate the eigenvalues of the orthogonal transformed data. Suppose the threshold of the fraction of the captured variance is set as 95%, show whether it is possible to reduce the dimensionality of the data.

$$\{(7, 4, 5), (6, 5, 4), (8, 4, 1), (2, 6, 9), (3, 6, 6), (5, 7, 3), (3, 5, 9), (2, 8, 6), (1, 7, 5), (8, 5, 2)\}$$

12-32 What is the scree test?

12-33 Explain how active shape models (ASMs) and active appearance models (AAMs) are related.

12-34 Calculate the linear discriminant function using Fisher's linear discriminant (FLD) for the following 2D data sets:

$$\mathcal{D}_1 = \{(5, 3), (2, 6), (3, 5), (3, 6), (4, 7)\}$$
$$\mathcal{D}_2 = \{(7, 9), (6, 7), (9, 5), (8, 8), (10, 8)\}$$

For simplicity, set the bias to the point halfway between the projected means.

12-35 Implement (a) the perceptron learning algorithm with accelerated learning rate in Algorithm 12.2 and (b) the pocket perceptron learning algorithm in Algorithm 12.3. Apply each algorithm to the datasets of the previous problem.

12-36 Use Lagrange multipliers to find the maximum and minimum values of the function $f(x, y) = 8x^2 + 200y^4$ with the constraint $x^2 + 2y^2 = 8$.

12-37 (a) What are the key ideas that turn a maximum margin classifier into a support vector machine (SVM)? (b) Explain the kernel trick.

12-38 What logic operation does the following neural network perform? The weights of each branch are marked near the arrows, and the thresholds are as shown. The neurons output true if the input values pass the threshold, otherwise false.

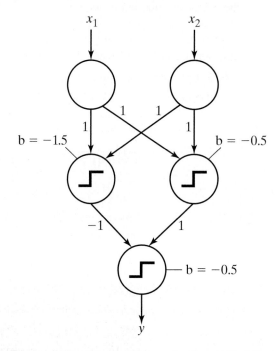

12-39 Plot the response of a logistic function and of a rectified linear unit (ReLU). List some of the advantages of the latter.

12-40 Briefly explain the concepts of bagging and boosting, and their relationship to one another.

12-41 Draw a degenerate decision tree, labeling the leaf nodes with "yes," "no," or "maybe."

12-42 What is a model containing parts with spring-like connections called?

12-43 Sketch the main components of a deep neural network. Explain, at a high level, how this approach works, including the principles of max pooling and dropout.

12-44 Implement your own face detector. Collect several hundred images of faces from the Internet and label them by hand by clicking on the images to define rectangles around the faces. (Be sure that all rectangles have the same aspect ratio.) Then collect several hundred images that contain no faces. Write code that performs preprocessing to generate a feature vector for a rectangle, and perform a sliding window search that computes a score for each hypothesis by passing the feature vector to a machine learning algorithm of your choice (using either your own implementation or code you find online). Report the accuracy of the detector on both the training set and on a separate test data, plotting results on an ROC or PR curve. Then write what you learned from this exercise.

CHAPTER 13
Stereo and Motion

I n previous chapters we have considered various ways to process a single image. In this chapter we consider what happens when multiple images of a scene are available from either an array of cameras that capture their images simultaneously, or from a single video camera that captures a sequence of images of a static or moving scene. In both cases the aim is to find corresponding pixels between images, which yields either *stereo correspondence* or *optical flow*, respectively. After discussing these two foundational problems, the rest of the chapter covers the geometrical principles for using such correspondences to yield 3D information, including projective geometry, camera calibration, and the geometry of multiple views. Together, these concepts are central to some of the most common techniques for understanding motion and estimating depth—and hence 3D geometry—from multiple images.

13.1 Human Stereopsis

It has been known at least since the time of Euclid[†] that depth perception in the human visual system is aided by the simultaneous viewing of two dissimilar images of the same object. Depth is perceived by the **retinal disparity** in the two images, which is the horizontal difference in the retinal locations of two projections of the same scene point.[‡] Stereo vision, or **stereopsis**, refers to the process of recovering 3D information about the world from

---

[†] Euclid (c. 300 BC) was a Greek mathematician known as the "Father of Geometry." His *Elements* remained the standard geometry textbook for more than 2000 years, even until the early 20th century AD.
[‡] See Figure 2.5 (p. 24).

multiple images of a scene taken at the same time by different imaging devices. By using additional sensory inputs, stereo overcomes the loss of dimensionality that is fundamental to the imaging process. Stereo comes from a Greek word meaning "solid." Like a stereo sound system, in which the brain fuses slightly different aural inputs to produce a fuller representation of the acoustic space, stereo vision involves fusing slightly different visual inputs to produce a solid representation of the nearby scene.

As an example, consider the pair of stereo images in Figure 13.1. To fuse these images, relax your eyes until the two images overlap, then concentrate on bringing the two copies of the statue head (or any other feature) into alignment, and wait patiently until a single 3D percept emerges.[†] The resulting 3D percept is sometimes called a **Cyclopean image** (after the famous one-eyed monster), because the fused image almost seems to result from an additional sensor in the center of your head. With relaxed-eye viewing, the separation between the images must not be greater than the **interpupillary distance (IPD)**, which is the distance between your two eyes (approximately 63 mm). This process of fusing stereo images has been a popular pastime of children and adults since *stereoscopes* were introduced in the 1830s and reappeared in various forms, most recently with the advent of *autostereograms* in the 1990s.

Many people find it easier to fuse a pair of stereo images whose positions have been reversed, as shown in Figure 13.2. In such a case, rather than relaxing the eyes so that they see beyond the page, the eyes must be crossed so that the fixation point is in front of the page. In this way the right eye sees the image on the left, while the left eye sees the image on the right. With cross-eyed display, the separation between the images is not limited by the interpupillary distance, so that the images may be placed much farther apart, and at much larger resolution, and still be fused.

Figure 13.1 A pair of stereo images. To fuse, relax your eyes so that you are seeing through the paper, and try to bring the two images together so that they come into alignment. Be patient, as it may take some time initially. If you experience difficulty fusing the images, it may help to place a vertical divider (such as a piece of cardboard) between the images so that each eye can see only one image. (Note that fusion is impossible if the distance between the two left edges of the images exceeds the interpupillary distance, which may occur if this page was enlarged by photocopying or viewed on screen at a large zoom setting.)

Left image   Right image

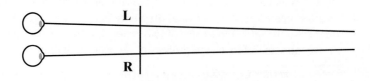

[†] Note, however, that some people are unable to fuse such images due to some form of stereoblindness.

Figure 13.2 The same pair of stereo images displayed for cross-eyed viewing.

Right image   Left image

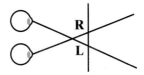

Stereo is not the only cue that contributes to depth perception. In fact, beyond a few meters the retinal disparity is too small to be detectable, and most of our depth perception arises from other cues, such as relative size, perspective, object overlap, contrast, lighting, shading, texture, and so forth. These factors explain why we are able to infer 3D information even from a single image. Nevertheless, the **random dot stereogram**, an example of which is shown in Figure 13.3, demonstrates conclusively that binocular fusion occurs even in the absence of these other cues. A random dot stereogram is formed by shifting a random pattern of black-and-white dots in some predefined shape to form the two images, then surrounding the patterns with identical patterns of dots. While other visual cues are no doubt used when they are available, the ability of the human visual system to fuse such random dot stereograms in the absence of any other visual cues proves that binocular fusion is a separate process in the visual system.

13.2 Matching Stereo Images

The success of the human visual system at inferring depth from a random dot stereogram leads naturally to the question of how to automatically infer depth by matching the pixels in two images. This is known as the **correspondence problem**—that is, to determine for each point in one image, its corresponding point in the other image. Several approaches to solving correspondence are covered in this section.

Figure 13.3 A random dot stereogram. If you relax your eyes, a small square should appear behind the large square. If you cross fuse, it should appear in front. Some isolated dots may appear to be at the wrong disparity, due to the inability of the visual system to properly segment all the pixels.

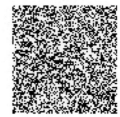

13.2.1 Correspondence

Two pixels are said to **correspond** if both pixels are projections along lines of sight of the same physical scene element. Once the correspondence between two points has been established, the coordinates of the two points can be compared to arrive at the **disparity** between them. The disparity between two points is defined as the difference in their coordinates. For simplicity, let us assume that the two cameras are **rectified**, which means that the image planes of the two cameras are coplanar, and the line joining their centers of projection is parallel to the scanlines (rows) of both images. In other words, the camera positions are related only by a translation parallel to the scanlines, with no rotation between them. When cameras are rectified, as shown in Figure 13.4, the disparity is defined as the x coordinate of a point in the left image minus the x coordinate of its corresponding point in the right image. From Equation (2.1), it is easy to show that the disparity is inversely proportional to depth:

$$d = x_L - x_R = f\frac{x_w + b}{z_w} - f\frac{x_w}{z_w} = \frac{fb}{z_w} \tag{13.1}$$

where b is the distance between the two focal points, called the **baseline**; x_w is the horizontal offset from the right focal point to the world point; and z_w is the depth, or distance, to the point from the focal point along the optical axis. As a result, a disparity of zero indicates a point that is an infinite distance away (e.g., a star in the sky), and increasing disparity means decreasing depth. Note that the disparity is always non-negative: $d = x_L - x_R \geq 0$.

Given a point (x_L, y_L) in the left image, where can its corresponding point (x_R, y_R) in the right image be? At first glance, it might seem that the corresponding point can be anywhere in the image. In fact, however, it is constrained to lie along a line, called the **epipolar line**. This restriction is called the **epipolar constraint**. While the epipolar constraint is

Figure 13.4 Rectified stereo geometry. TOP: A world point is imaged at point (x_L, y) in the left image and (x_R, y) in the right image, with respect to coordinate systems aligned with each image and placed in the top-left corner, as usual. BOTTOM LEFT: The same scene viewed in 2D. (The y axis, going into the page, is not shown.) BOTTOM RIGHT: Overlapping the two imaging rays onto a single (virtual) sensor, the distance $x_L - x_R$ between the two coordinates is the disparity d.

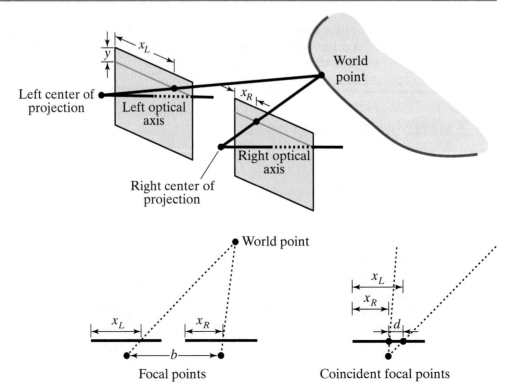

applicable for any configuration of binocular cameras, as we shall explore later in the chapter, for rectified images it leads to the simple constraint that the corresponding point must lie along the same scanline, $y_L = y_R$, because the scanlines are the epipolar lines. Looking back at the images in Figure 13.1 through Figure 13.3 (all of which are rectified), notice that corresponding points always lie on the same row as each other.

To better visualize the correspondence problem, let us define a cuboid with the reflected left image comprising one face and the right image comprising an adjacent face, such that the left borders of both images share an edge in the cuboid, shown in Figure 13.5. The cuboid, called the **matching space**, is usually divided into a discrete grid defined at pixel resolution, with each cell in the grid indicating whether the two pixels indexed by the cell's row and column correspond to each other: a 1 in the cell means that the pixels correspond, while a 0 means that they do not correspond. A slice through the matching space for a pair of scanlines is also shown in the figure, where the origin is placed at the top-left corner, each column indicates the x coordinate of the pixel in the left scanline, and each row indicates the x coordinate of the pixel in the right scanline. From the correspondences between two images, a **disparity map** can be constructed by projecting the matches in the matching space onto the horizontal axis (for the *left* disparity map) or onto the vertical axis (for the *right* disparity map). Occluded pixels are assigned the value of the smaller of the two neighbors.

In a rectified system, an object is said to be **frontoparallel** if it is parallel to both image planes, in which case its depth, and hence disparity, is constant. Now, the disparity for any given cell is determined by its Manhattan distance[†] in the grid to the main diagonal. Therefore, the cells along any given diagonal represent the matches that would occur for a frontoparallel object at the related depth, because $x_L - x_R$ is constant along such a diagonal. Notice that the gray cells in the lower triangle of the grid cannot possibly be matches, since $x_R > x_L$, that is, they yield negative depth. Similarly, for computational efficiency it is customary to put a limit on the maximum allowable disparity $d_{\max}$, which explains the gray cells in the upper triangle of the grid, similar to Panum's fusional area.[‡]

13.2.2 Stereo Constraints

Like most problems in computer vision, establishing stereo correspondence is an underconstrained problem because the nearby pixel values are insufficient to make the decision locally. As a result, a number of constraints have been articulated over the years to make the problem more tractable. While some of these constraints arise from the geometry of the scene and the imaging process, others are more heuristic in nature.

Figure 13.5 Left: The matching space for a pair of rectified stereo images. Right: A horizontal slice through the matching space. (The y axis, going into the page, is not shown.) Each discrete cell in the matching space indicates whether the given pair of pixels (x_L, y) and (x_R, y) correspond to each other. Here the green cell represents a match between the pixel $x_L = 7$ and the pixel $x_R = 4$, so that the disparity is $d = 3$. The maximum disparity, bounded by the shaded region in the upper-right corner, is $d_{\max} = 7$.

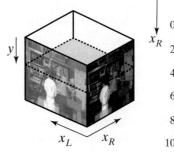
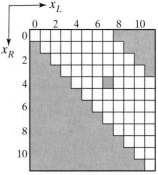

[†] Section 4.3.1 (p. 164).
[‡] Section 2.1.2 (p. 21).

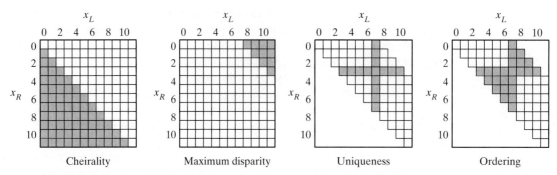

Figure 13.6 Stereo constraints. The gray cells indicate matches that are simply not allowed (left two grids) or that are illegal when the green cell indicates a match (right two grids). Cheirality precludes matches with $x_L < x_R$, which would refer to points behind the camera. Maximum disparity precludes matches whose disparity exceeds a threshold. Uniqueness prevents a pixel in either scanline from matching more than one pixel. Ordering ensures that the pixel coordinates of the matches are monotonically increasing as the pixels along either scanline are traversed. Note that the gray cells in the right grid are the forbidden zone.

Figure 13.6 illustrates some of the more common constraints. The **cheirality constraint**[†] requires $x_L \geq x_R$ for matching pixels, as mentioned above, since only objects in front of the camera can be visible. The **maximum disparity constraint** forbids matches whose disparity exceeds a certain amount, which effectively enforces a minimum distance from the camera to the surface being viewed; this is related to Panum's fusional area, as mentioned above. The **uniqueness constraint** says that if $x_L \leftrightarrow x_R$ is a match, then there is no other match $x_L \leftrightarrow x$ where $x \neq x_R$, and there is no other match $x \leftrightarrow x_R$ where $x \neq x_L$. That is, a pixel in either image can only match at most one other pixel, which is true as long as the surfaces in the world are opaque and the general position assumption with regard to viewpoint is not violated (that is, no imaging ray is aligned with a surface in the world). Furthermore, although not directly illustrated in the figure, the very possibility of drawing the matching process on a 2D grid (as opposed to 3D) owes itself to the epipolar constraint mentioned above, which requires that corresponding pixels share the same y coordinate when the images are rectified.

Another constraint arises from the fact that, as illustrated in Figure 13.7, when a point on a continuous surface is viewed by both cameras, it is not physically possible for another point on the same surface to also be visible in both cameras if it lies within the region defined by two lines passing through the centers of projection and the point. This hourglass-shaped region is called the **forbidden zone**, and denying matches in the forbidden zone is known as the **ordering constraint** or the **monotonicity constraint**. Mathematically, the constraint can be stated thus: If $x_L \leftrightarrow x_R$ is a match, then there is no other match $x \leftrightarrow x'$ such that $x \geq x_L$ and $x' \leq x_R$, and there is no other match $x \leftrightarrow x'$ such that $x \leq x_L$ and $x' \geq x_R$. Equivalently, if $x_L^{(1)} \leftrightarrow x_R^{(1)}$ and $x_L^{(2)} \leftrightarrow x_R^{(2)}$ are both matches, then $x_L^{(1)} < x_L^{(2)}$ implies $x_R^{(1)} < x_R^{(2)}$. Note that the constraint ensures that if the pixels in one scanline are considered in a certain order, then the corresponding pixels in the other scanline are encountered in the same order.

To better understand the ordering constraint, note that it is violated when a thin, opaque object (such as a pole) is close to the camera, and there is a considerable distance between this object and the background behind it. This situation is illustrated in Figure 13.8, where a thin frontoparallel surface labeled $\mathbf{f}$ is in front of a frontoparallel background surface

---

[†] The cheirality constraint derives its name from the Greek word meaning "hand," the idea being that if you look on edge at the imaging plane of a camera pointed to the right, the camera would be able to see all the points on your right hand, and vice versa.

Figure 13.7 The forbidden zone is an hourglass-shaped region defined by the two centers of projection and a particular match (which implies a particular world point). No other matches within the zone are possible, as long as there is a single, opaque surface in the world in the nearby vicinity. The matches in the forbidden zone are exactly those not allowed by the ordering constraint.

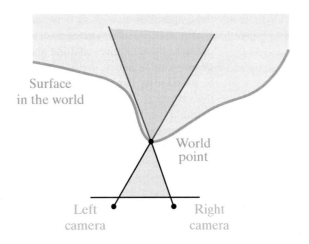

whose individual regions (defined by the visibility rays of the two cameras) are labeled a through e. Notice that if we were to examine the pixels in the left image, from left to right, we would encounter projections of regions a, b, c, f, and e, in that order, since region d is occluded. Similarly, the right image contains, from left to right, projections of regions a, f, c, d, and e, because region b is occluded. The ordering constraint is therefore violated in this case because the order of the regions is different in the two images: whereas c is to the left of f in one image, it is to the right in the other image.

A related concept is the **disparity gradient limit**, which says that surfaces cannot recede too quickly from the viewer. This constraint is inspired by psychophysical experiments showing that the human visual system is incapable of fusing points when the slope of the surface exceeds a maximum value; since disparity is the inverse of depth, this slope is related to the disparity of the gradient, and therefore the human visual system is incapable of fusing points when the disparity gradient exceeds a certain value. To better understand the disparity gradient, imagine holding a flat piece of paper oriented vertically in front of you, so that the disparity of all the points on the paper are the same, and therefore the disparity gradient is zero. However, as you rotate the paper around a vertical axis, the disparity gradient increases until it eventually reaches a point where the paper recedes into the background so much that the visual system is no longer able to fuse the images.

The disparity gradient is defined in terms of **Cyclopean coordinates**. That is, the disparity gradient is given by the derivative $\partial d/\partial x$ of the Cyclopean disparity function with respect to the Cyclopean coordinate, where the Cyclopean disparity function $d(x)$ is

Figure 13.8 A thin, opaque object close to the camera causes the ordering constraint to be violated. LEFT: Two rectified cameras viewing a thin pole in front of a fronto-parallel background. The visibility rays from the cameras to the edges of the pole divide the background into 5 different regions labeled a through e. RIGHT: The matching space with matches shown in green. The letters on the cells indicate the surface to which the match belongs. Regions b and d produce no match due to the occlusion from f.

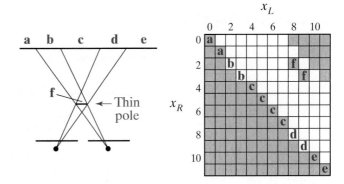

the disparity $d = x_L - x_R$ between two matching pixels, and the Cyclopean coordinate x of a match is the average of the two coordinates: $x \equiv \frac{1}{2}(x_L + x_R)$. The disparity gradient limit requires that the absolute value of the disparity gradient never exceeds a given value—that is, $|\partial d/\partial x| \leq \kappa$—where κ is the limit. Two special numbers should be noted: The human visual system imposes a disparity gradient limit of $\kappa = 1$, while the value $\kappa = 2$ is equivalent to the ordering constraint. One way to visualize this is to define $d^*(x) \equiv \max_a(d(x + a) - \kappa|a|)$, so that graphically d^* is determined by casting shadows from d at an angle of $\arctan(\kappa)$ and taking the maximum of the original value and all the shadows created by neighboring values. Then $d^*(x) = d(x)$ for all x for which d satisfies the disparity gradient limit of κ; and if $\kappa = 2$, then the set of points x such that $d^*(x) > d(x)$ are the occluded points.

As another visualization aid, Figure 13.9 shows the Cyclopean coordinates and absolute disparity gradient for a number of possible matches with respect to the green cell. Whereas the left coordinate axis x_L is horizontal, and the right coordinate axis x_R is vertical, the Cyclopean coordinate axis x is diagonal. Therefore, whereas diagonals parallel to the main diagonal (down and to the right) contain cells with constant disparity, as mentioned earlier, diagonals perpendicular to the main diagonal (that is, up and to the right) contain cells with constant Cyclopean coordinates. As a result, if we were to overlay a black-and-white checkerboard pattern on the matching space, all cells of one color would have integer Cyclopean coordinates, while all the cells of the other color would have fractional (odd multiples of 0.5) Cyclopean coordinates. By comparing the figure with the previous figure, it is clear that a gradient limit of $\kappa \geq 2$ implies the ordering constraint, with $\kappa = 2$ being equivalent to the ordering constraint.

EXAMPLE 13.1	Compute the disparity gradient between the green match and the match just below it in Figure 13.9.
Solution	The match shown is $x_L = 7$ and $x_R = 3$, so the disparity is $7 - 3 = 4$. The Cyclopean coordinate of the match is $x = \frac{1}{2}(x_L + x_R) = 5$. The absolute value of the (discrete approximation to the) disparity gradient between two matches $x_L^{(1)} \leftrightarrow x_R^{(1)}$ and $x_L^{(2)} \leftrightarrow x_R^{(2)}$ is

$$\left|\frac{\partial d}{\partial x}\right| = \frac{d_2 - d_1}{\frac{1}{2}(x_L^{(2)} + x_R^{(2)}) - \frac{1}{2}(x_L^{(1)} + x_R^{(1)})} = \frac{2(x_L^{(2)} - x_R^{(2)} - x_L^{(1)} + x_R^{(1)})}{x_L^{(2)} + x_R^{(2)} - x_L^{(1)} - x_R^{(1)}} \qquad (13.2)$$

where $d_1 = x_L^{(1)} - x_R^{(1)}$ and $d_2 = x_L^{(2)} - x_R^{(2)}$ are the two disparities. Since the match just below the green match is at $x_L = 7$ and $x_R = 4$, we have $x_L^{(1)} = x_L^{(2)} = 7$, $x_R^{(1)} = 3$, and $x_R^{(2)} = 4$. Plugging into Equation (13.2), the absolute disparity gradient is computed as $|\partial d/\partial x| = |2(7 - 4 - 7 + 3)/(7 + 4 - 7 - 3)| = 2$.

13.2.3 Block Matching

The simplest algorithm to compute dense correspondence between a pair of stereo images is **block matching**. Block matching is an *area-based* approach that relies upon a statistical correlation between local intensity regions. For each pixel (x, y) in the left image, the right image is searched for the best match among all possible disparities $0 \leq d \leq d_{max}$:

$$d_L(x, y) = \arg \min_{0 \leq d \leq d_{max}} dissim(I_L(x, y), I_R(x - d, y)) \qquad (13.3)$$

where $dissim(I_L(\mathbf{x}_L), I_R(\mathbf{x}_R))$ is the dissimilarity between the pixel $\mathbf{x}_L = (x_L, y_L)$ in the left image and pixel $\mathbf{x}_R = (x_R, y_R)$ in the right image, d_L is the left disparity map (i.e., the

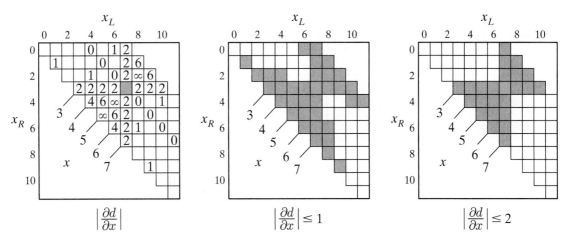

Figure 13.9 Disparity gradient constraint. Left: The number in each cell shows the magnitude of the disparity gradient (only integer values and infinity are shown; empty cells have a non-integral disparity gradient). Middle: The gray cells violate the disparity gradient limit $\kappa = 1$. Right: The gray cells violate the disparity gradient limit $\kappa = 2$. In all plots, the numbers 3, 4, ..., 7 along the main diagonal indicate the Cyclopean coordinates.

disparity map with respect to the left image), and d_{max} is the maximum allowed disparity, which improves computation time significantly. Choices for dissimilarity will be considered in a moment, but usually some sort of absolute difference in intensity (perhaps after applying a preprocessing filter) is used.

The most straightforward way to perform block matching is shown in Algorithm 13.1, BLOCKMATCH1. For each pixel in the left image, a search is conducted for the best disparity, where the best disparity is defined as the one that yields the lowest sum of the dissimilarities over a window around the pixel. The $w \times w$ sized window $\mathcal{W}$ is defined so that $(\tilde{x}, \tilde{y}) \in \mathcal{W}$ means $\tilde{x} = -\tilde{w}, \ldots, \tilde{w}$ and $\tilde{y} = -\tilde{w}, \ldots, \tilde{w}$, where $\tilde{w} \equiv \lfloor \frac{w}{2} \rfloor$ is the half-width. The algorithm uses two scalars: $\hat{d}$ which keeps track of the best disparity seen so far, and $\hat{g}$ which holds the score (i.e., the sum of the dissimilarities) of the best disparity.

ALGORITHM 13.1 Stereo correspondence by block matching

BLOCKMATCH1 $\left(I_L, I_R, d_{min}, d_{max} \right)$

Input: left I_L and right I_R images from a stereo pair, with minimum and maximum disparities
Output: left disparity map

1 **for** $(x, y) \in I_L$ **do**
2      $\hat{g} \leftarrow \infty$
3   **for** $d \leftarrow d_{min}$ **to** d_{max} **do**
4          $g \leftarrow 0$
5   **for** $(\tilde{x}, \tilde{y}) \in \mathcal{W}$ **do**
6              $g \leftarrow g + dissim(I_L(x + \tilde{x}, y + \tilde{y}), I_R(x + \tilde{x} - d, y + \tilde{y})$
7   **if** $g < \hat{g}$ **then**
8              $\hat{g} \leftarrow g$
9              $\hat{d} \leftarrow d$
10     $d_L(x, y) \leftarrow \hat{d}$
11 **return** d_L

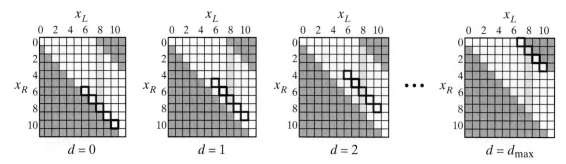

$$d = 0 \qquad\qquad d = 1 \qquad\qquad d = 2 \qquad\qquad d = d_{\max}$$

Figure 13.10 Block matching algorithm. For every pixel in the left image, a search is performed to find the disparity yielding the lowest cost. The red cells indicate the dissimilarities that are aggregated in Lines 5–6 of BLOCKMATCH1, while the yellow cells indicate the matches considered during the search. Shown is the pixel $x_L = 8$ with a window size of $w = 5$.

Not shown in the code is the additional logic to ensure that out-of-bounds errors do not occur in Line 6 near the left border of the right image. The procedure is illustrated in Figure 13.10.

BLOCKMATCH1 is a time-consuming algorithm because it contains five nested **for** loops (note that the loops in Lines 1 and 5 are each double loops). The computing time is therefore $O(n^2 w^2 d_{\max})$, where n is the width or height of the image. This computing time can be reduced substantially using the running-sum techniques that we encountered earlier,[†] which take advantage of redundant computations. This leads to BLOCKMATCH2, which produces exactly the same results as BLOCKMATCH1 but with a running time of only $O((n^2 + w)d_{\max})$. The algorithm uses a 3D array called Δ whose dimensions are the dimensions of the image by $d_{\max} + 1$, and it uses a procedure called COMPUTESUMMEDDISSIMILARITIES to populate

ALGORITHM 13.2 A more efficient version of stereo correspondence by block matching

BLOCKMATCH2$(I_L, I_R, d_{\min}, d_{\max})$

Input: left I_L and right I_R images from a stereo pair, with minimum and maximum disparities
Output: left disparity map

1    $\Delta \leftarrow$ COMPUTESUMMEDDISSIMILARITIES$(I_L, I_R, d_{\min}, d_{\max})$
2   **for** $(x, y) \in I_L$ **do**
3        $d_L(x, y) \leftarrow \arg\min_d \Delta(x, y, d)$
4   **return** d_L

COMPUTESUMMEDDISSIMILARITIES $(I_L, I_R, d_{\min}, d_{\max})$

Input: left I_L and right I_R images from a stereo pair, with minimum and maximum disparities
Output: 3D array Δ containing summed dissimilarities

1   **for** $d \leftarrow d_{\min}$ **to** $d_{\max}$ **do**
2   **for** $(x, y) \in I_L$ **do**
3            $\Delta(x, y, d) \leftarrow dissim(I_L(x, y), I_R(x - d, y))$
4        $\Delta(:, :, d) \leftarrow$ CONVOLVE$(\Delta(:, :, d), \mathbf{1}_{w \times w})$
5   **return** Δ

[†] Section 3.3.2 (p. 87).

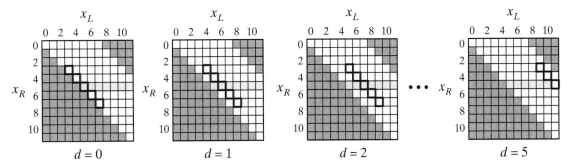

Figure 13.11 Block matching algorithm from the right image. For every pixel in the right image, a search is performed to find the disparity yielding the lowest cost. The red cells indicate the dissimilarities that are aggregated when computing the cost of the disparity. Shown is the pixel $x_R = 5$ with a window size of 5. The maximum disparity is not reached, because the window is near the image border.

the array. Here the dissimilarity between any two given pixels is computed only once, with the aggregation over the window performed afterward by recognizing that summing over a window is the same as convolving with a box kernel of the same size. The notation uses $\mathbf{1}_{w \times w}$ to represent a square window of all ones, but note that in a real implementation the convolution would be split into two 1D convolutions, since the box kernel is separable. Here we use slice notation, so that $\Delta(x, y, :)$ means the $(d_{\max} + 1)$-dimensional vector whose elements are $\Delta(x, y, d)$, $d = 0, \ldots, d_{\max}$, and $\Delta(:, :, d)$ is the dissimilarity slice (same size as the image) at disparity d.

Our choice of the left image as the reference was arbitrary. Alternatively, we could just as easily have performed the search with respect to the right image, as shown in Figure 13.11. In a perfect world, both answers would agree, and it would not matter which image we used as the reference. In practice, however, the pixels will not always agree, due to occlusion, specular reflections, and other phenomena. A convenient trick, called the **left-right disparity check**, is to perform block matching twice, once using the left image as the reference, and once using the right image as the reference. Wherever the pixels agree on their answer, the match is accepted as reliable, but when they disagree, the match is discarded as unreliable, thus yielding a number of pixels for which an answer is simply not determined, as illustrated in Figure 13.12.

Conceptually, the code for performing the left-right consistency check is as follows. First the search is conducted from left to right to compute the disparity map with respect to the left image, and then the search is conducted from right to left to compute the disparity map

Figure 13.12 LEFT TWO COLUMNS: Block matching from the left image finds a disparity (green), while block matching from the right image (using the pixel found by the match) finds a different disparity (green). Since the two matches disagree, they are unreliable. RIGHT TWO COLUMNS: Both left and right images agree on the disparity, leading to a reliable match.

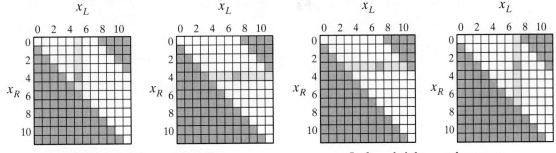

Left and right matches disagree   Left and right matches agree

with respect to the right image. Following these steps, only those answers that are consistent between the two disparity maps are retained. Note in Line 4 that the pixel (x,y) in the left disparity map is not simply compared with pixel (x,y) in the right disparity map. Rather, for a pixel (x,y) in the left disparity map, we must consider its corresponding pixel according to the disparity computed for that pixel: namely, $(x - d, y)$, where d is the disparity computed at pixel (x,y).

The code in BLOCKMATCHWITHLEFTRIGHTCHECK1 is inefficient. To optimize the algorithm, note that both block match searches use the same matching space. As a result, Δ only needs to be computed once, thereby eliminating redundant computations and yielding the more efficient algorithm of BLOCKMATCHWITHLEFTRIGHTCHECK2, whose asymptotic running time is the same as that of BLOCKMATCH2. First Δ is computed. Then, for every pixel (x,y) in the left image, its disparity using the left image as the reference is computed. Then the disparity computed for the corresponding pixel using the right image as the reference is checked to see whether it is the same. If both answers agree, then the solution is retained; otherwise, ignorance is declared. Results of applying block matching, both with and without the left-right disparity check, are shown in Figure 13.13.

13.2.4 Dissimilarity Measures

How should two pixels $\mathbf{x}_L = (x_L, y_L)$ and $\mathbf{x}_R = (x_R, y_R)$ be compared? The most straightforward approach is to compute either the absolute or squared difference in intensity between the two pixels. When the information is aggregated over a window, these two choices are known as the **sum of absolute differences (SAD)** and the **sum of squared differences (SSD)**:

$$dissim(I_L(\mathbf{x}_L), I_R(\mathbf{x}_R)) = |I_L(\mathbf{x}_L) - I_R(\mathbf{x}_R)| \quad (\text{SAD}) \qquad (13.4)$$

$$dissim(I_L(\mathbf{x}_L), I_R(\mathbf{x}_R)) = (I_L(\mathbf{x}_L) - I_R(\mathbf{x}_R))^2 \quad (\text{SSD}) \qquad (13.5)$$

Treating the window of intensities as a vector in a high-dimensional space (e.g., 49 dimensions for a 7×7 window), the SAD and SSD are equivalent to the L^1 norm and the square of the L^2 norm, respectively, between the two vectors.[†] Generally speaking, SAD is the preferred method due to several advantages: It is faster to compute, is less sensitive to outliers, and its output has the same bit depth as the input. On the other hand, the advantage of SSD is that it is differentiable, which makes it more commonly used for computing optical flow, as we shall see later in the chapter.

Figure 13.13 Results of block matching on a stereo image pair. From left to right: the image, the left disparity map, and the disparity map after the left-right consistency check.

[†] Recall from Section 4.3.1 (p. 164) that the L^1 norm is just the Manhattan distance, and the L^2 norm is the Euclidean distance.

ALGORITHM 13.3 Stereo correspondence by block matching with left-right disparity check

BLOCKMATCHWITHLEFTRIGHTCHECK1 (I_L, I_R, d_{max})

Input: left I_L and right I_R images from a stereo pair, with maximum disparity
Output: left disparity map

1 $d_L \leftarrow$ BLOCKMATCH2$(I_L, I_R, 0, d_{max})$
2 $d_R \leftarrow$ BLOCKMATCH2$(I_R, I_L, -d_{max}, 0)$
3 **for** $(x, y) \in I_L$ **do**
4   **if** $d_L(x, y) \neq -d_R(x - d_L(x, y), y)$ **then**
5          $d_L(x, y) \leftarrow$ NOT-MATCHED
6 **return** d_L

ALGORITHM 13.4 More efficient version of stereo correspondence by block matching with left-right disparity check

BLOCKMATCHWITHLEFTRIGHTCHECK2 (I_L, I_R, d_{max})

Input: left I_L and right I_R images from a stereo pair, with maximum disparity
Output: left disparity map

1 $\Delta \leftarrow$ COMPUTESUMMEDDISSIMILARITIES$(I_L, I_R, 0, d_{max})$
2 **for** $(x, y) \in I_L$ **do**
3      $\delta_L \leftarrow$ arg min$\{\Delta(x, y, 0), \Delta(x, y, 1), \ldots, \Delta(x, y, d_{max})\}$
4      $\delta_R \leftarrow$ arg min$\{\Delta(x - \delta_L, y, 0), \Delta(x - \delta_L + 1, y, 1), \ldots, \Delta(x - \delta_L + d_{max}, y, d_{max})\}$
5   **if** $\delta_L = \delta_R$ **then**
6          $d_L(x, y) \leftarrow \delta_L$
7   **else**
8          $d_L(x, y) \leftarrow$ NOT-MATCHED
9 **return** d_L

Another option is to compute the *similarity* between two pixels as the product of their intensities:

$$dissim(I_L(\mathbf{x}_L), I_R(\mathbf{x}_R)) = -I_L(\mathbf{x}_L)I_R(\mathbf{x}_R) \qquad \text{(cross correlation)}$$

where the negative sign is needed to convert to *dissimilarity*. When aggregated over a window, this option is just the negative **cross correlation** between the two signals. There is a fundamental connection between SSD and cross correlation, so that minimizing the SSD is approximately the same as maximizing the cross correlation.[†]

Comparing the raw intensities, as we have just done, is rather simplistic. Not only does this approach assume that the surfaces in the world are Lambertian,[‡] but it also assumes that the cameras have similar photometric characteristics. In practice, however, two cameras rarely generate the same gray level, even when receiving the same irradiance. Perhaps the simplest photometric model to account for these differences is to assume that the two

---

[†] Problem 13.13.
[‡] Section 2.5.4 (p. 61).

cameras have different gains (multiplicative factors) and biases (additive factors), so that the gray levels of corresponding pixels are related by

$$I_L(\mathbf{x}_L) = \alpha I_R(\mathbf{x}_R) + \beta \tag{13.6}$$

where α is the relative gain and β is the relative bias between the two cameras. Ideally, $\alpha = 1$ and $\beta = 0$, but with real cameras the values will often be significantly different from ideal. One approach is to estimate α and β, then adjust gray levels of the pixels in I_R accordingly, then compare the resulting values directly using SAD, SSD, or cross correlation. Removing the effect of the gain and bias is known as photometric (or radiometric) calibration. Taking this concept a step further, we can adjust for gain and bias adaptively across the image, so that different gains and biases are determined for different image locations, leading to **normalized cross correlation**. Alternatively, the Laplacian of Gaussian (LoG)[†] can be used as a prefilter to remove the relative bias between the cameras, although it does not remove the relative gain.

Another source of error is the corruption in the pixel values themselves that occurs near occluding boundaries. Traditional methods such as SAD, SSD, and correlation are based on standard statistical assumptions such as additive Gaussian noise, which model the underlying values as arising from a single statistical population. However, when a window straddles an occluding boundary, some pixels in the window capture intensities from the nearby surface, while other pixels capture intensities from the farther surface. In addition, some of the pixels that are visible in one view are occluded in the other view, which corrupts the pixel values in a way that is difficult to model directly. As a result, the intensities in the window are caused by two distinct subpopulations, a phenomenon known as **factionalism**.

To better tolerate factionalism, we must turn to nonparametric local transforms that look not at the pixel values themselves, but rather at the ordering of the values. One such transform is the **rank transform**, which computes the number of pixels in the neighborhood whose gray level is less than that of the central pixel of the window:

$$\tilde{I}(\mathbf{x}) = |\{\mathbf{x}' \in \mathcal{N}(\mathbf{x}) : I(\mathbf{x}') < I(\mathbf{x})\}| \tag{13.7}$$

where $|\cdot|$ denotes the cardinality of the set, and $\mathcal{N}(\mathbf{x})$ is the neighborhood of $\mathbf{x}$. This is a preprocessing step which, when applied to both images, results in new pixel values that can be used in one of the dissimilarity equations above, such as SAD. Another transform is the **census transform**, which defines a bit string for each pixel, where each bit is 0 or 1 depending upon whether the pixel in the neighborhood has a smaller value than that of the central pixel:

$$\tilde{I}(\mathbf{x}) = [h(I(\mathbf{x}) - I(\mathbf{x}'_1)), \ldots, h(I(\mathbf{x}) - I(\mathbf{x}'_n))]^\mathsf{T} \tag{13.8}$$

where h is the Heaviside operator mentioned earlier,[‡] $\mathbf{x}'_i \in \mathcal{N}(\mathbf{x})$, and $n = |\mathcal{N}(\mathbf{x})|$ is the number of pixels in the neighborhood. Similar to the rank transform, this operator results in a new value for each pixel that can then be used in the SAD computation.

13.2.5 Dynamic Programming

The fundamental limitation of block matching is that the disparity for each window is computed independently of all the other windows. To enforce a global consistency on the solution, the stereo matching problem can be formulated as the minimization of an

---

[†] Section 5.4.1 (p. 242).
[‡] Section 10.2.5 (p. 469).

energy functional.[†] As a result of the epipolar constraint, the matching problem can be viewed as a 1D problem whose goal is to compute the disparity for all pixels along the epipolar line (scanline). Typically, the functional is composed of two terms, one corresponding to the data and the other to the smoothness:

$$E(d_L) = E_D(d_L) + \lambda E_S(d_L) \tag{13.9}$$

where $d_L(\cdot, \cdot)$ is the left disparity map, and λ governs the relative importance of the two terms. This formulation can be tied directly to Bayes' rule, as we have seen before. For the data term, the sum of the dissimilarities of the matches is computed. For the smoothness term, each pair of neighboring pixels is penalized with different disparities according to a potential function V:

$$E_D(d_L) = \sum_{x_L \in I_L} dissim(I_L(x_L, y), I_R(x_R + d_L(x_L, y), y)) \tag{13.10}$$

$$E_S(d_L) = \sum_{x_L \in I_L} V(d_L(x_L, y), d_L(x_L + 1, y)) \tag{13.11}$$

where $V(d, d) = 0$ and $V(d_1, d_2) = V(d_2, d_1)$, so V is at least a semimetric.[‡]

This functional can be minimized using the Viterbi algorithm we saw earlier.[§] To understand the approach, let us consider the problem of computing the **edit distance** between two strings of characters. The edit distance, also known as the **Levenshtein distance**, is defined as the minimum number of operations to change one string into another, where an operation is either an insertion, deletion, or substitution of a single character. The Levenshtein distance is a generalization of the Hamming distance. Suppose we have two strings s_1 and s_2, where each string is a sequence of characters, and let $d(s_1, s_2)$ be the edit distance between them. Dynamic programming relies on *recursive relations*, which for the edit-distance problem are as follows:

$d(\emptyset, \emptyset) = 0$, where $\emptyset$ is the empty string;
$d(s, \emptyset) = d(\emptyset, s) = \kappa_o |s|$, where $|s|$ is the length of the string;
$d(s_1 + c_1, s_2 + c_2) = \min\{d(s_1, s_2) + \gamma(c_1, c_2), d(s_1 + c_1, s_2)$
$\qquad\qquad\qquad + \kappa_o, d(s_1, s_2 + c_2) + \kappa_o\}$

In the second relation, κ_o is the cost of not matching a letter (which, as we shall see, is analogous to a pixel being occluded), while $\gamma(c_1, c_2)$ is the cost of associating letter c_1 with c_2 (which is similar to the dissimilarity between two pixels). In string matching, we normally set $\kappa_o = 1$ and

$$\gamma(c_1, c_2) = \begin{cases} 0 & \text{if } c_1 = c_2 \\ 1 & \text{otherwise} \end{cases} \tag{13.12}$$

A commonly used algorithm for computing the edit distance, shown in Algorithm 13.5, involves the use of an $(|s_1| + 1) \times (|s_2| + 1)$ array, which we shall call φ. The $(i, j)^{\text{th}}$ element $\varphi(i, j)$ is the edit distance between the substrings containing the first i characters of s_1 and the first j characters of s_2. The important property of this algorithm for our purposes is that it not only computes the edit distance, but it also solves the **string matching problem**. That is, once φ has been constructed, it contains information that associates the characters in the two strings. Therefore, the correspondence between the characters can be determined by traversing the array in reverse, beginning with $\varphi(|s_1| - 1, |s_2| - 1)$, keeping track of the previous cell that gave rise to the minimum.

[†] Functionals are defined in Section 10.2.1 (p. 453).
[‡] Section 4.3.1 (p. 164).
[§] Section 10.2.1 (p. 453).

ALGORITHM 13.5 Compute the edit distance between two strings of characters

STRINGMATCHING(s_1, s_2)

Input: strings s_1 and s_2
Output: cost of matching the two strings, and the matching function between them

1   **for** $x_1 \leftarrow 0$ **to** $|s_1| - 1$ **do**
2       $\varphi(x_1, 0) \leftarrow x_1 * \kappa_o$
3   **for** $x_2 \leftarrow 0$ **to** $|s_2| - 1$ **do**
4       $\varphi(0, x_2) \leftarrow x_2 * \kappa_o$
5   **for** $x_1 \leftarrow 1$ **to** $|s_1| - 1$ **do**
6   **for** $x_2 \leftarrow 1$ **to** $|s_2| - 1$ **do**
7           $\varphi(x_1, x_2) \leftarrow \min\{\varphi(x_1 - 1, x_2) + \kappa_o, \varphi(x_1, x_2 - 1) + \kappa_o, \varphi(x_1 - 1, x_2 - 1) + \gamma(s_1(x_1), s_2(x_2))\}$
8   $match \leftarrow$ EXTRACTPATH(φ)
9   **return** $\varphi(|s_1| - 1, |s_2| - 1), match$

EXTRACTPATH (φ)

Input: 2D array of costs computed between two strings
Output: list of matches between the strings

1   $x_1 \leftarrow |s_1|$
2   $x_2 \leftarrow |s_2|$
3   **while** $x_1 \geq 0$ AND $x_2 \geq 0$ **do**
4   **if** $\varphi(x_1, x_2) == \varphi(x_1 - 1, x_2) + \kappa_o$ **then**
5           $x_1 \leftarrow x_1 - 1$
6   **elseif** $\varphi(x_1, x_2) == \varphi(x_1, x_2 - 1) + \kappa_o$ **then**
7           $x_2 \leftarrow_- 1$
8   **else**
9           $x_1 \leftarrow_- 1$
10          $x_2 \leftarrow_- 1$
11          $match$: PUSH(x_1, x_2)
12 **return** $match$

The stereo correspondence problem for two scanlines is strikingly similar to the string matching problem, in which each pixel is a letter, the cost to change a pixel is the dissimilarity between the two pixels, and the cost to insert or delete a pixel is the penalty for occlusion. Pseudocode for the approach is shown in Algorithm 13.6, where EXTRACTPATHSTEREO is similar to EXTRACTPATH. For a real implementation, it is helpful to keep track of which previous cell gave rise to the minimum in Line 3 to make the path extraction easier. Another improvement is to model occlusions explicitly by treating the two scanlines symmetrically, but this involves slightly more bookkeeping.

13.2.6 Energy Minimization in 2D

While dynamic programming is able to find the global minimum of a 1D energy functional along a given scanline, images are inherently 2D. Applying dynamic programming to each scanline independently therefore leads to horizontal streaks in the resulting disparity maps.

ALGORITHM 13.6 Stereo correspondence by dynamic programming

$\textsc{StereoDynamicProgramming}\left(I_L, I_R, d_{\max}\right)$

Input: left I_L and right I_R images from a stereo pair, with maximum disparity
Output: left disparity map

```
1   for y ← 0 to height − 1 do
2       for x_L ← 1 to width − 1 do
3           φ_prev ← min{φ(x_L − 1, 0), . . . , φ(x_L − 1, d_max)}
4           for d ← 0 to d_max do
5               φ(x_L, d) ← min{φ_prev + κ, φ(x_L − 1, d)} + dissim(I_L(x_L, y), I_R(x_L + d, y))
6           d_L(:, y) ← ExtractPathStereo(φ)
7   return d_L
```

To overcome this limitation, one approach, known as **semi-global matching**, repeatedly performs dynamic programming along lines in the image that are not necessarily the scanlines. In contrast, another class of approaches explicitly minimize a 2D energy functional, such as

$$E(d_L) = \sum_{(x,y)} dissim\left(I_L(x, y), I_R(x + d_L(x, y), y)\right) + \lambda \sum_{(x,y)} \sum_{(x', y') \in \mathcal{N}(x, y)} V\left(d(x, y), d(x', y')\right) \quad (13.13)$$

where the **Potts model** is a popular way to enforce piecewise smoothness: $V(d, d') = 0$ if $d = d'$, or 1 otherwise. Most algorithms for minimizing such a 2D energy functional are based on either multiway cuts, which we saw earlier,[†] or **Bayesian belief propagation**. Belief propagation is an inference technique for graphs that involves maintaining probability distributions at the nodes and passing messages between nodes to update these distributions. When the graph contains cycles (as is the case with images), the approach is known as *loopy belief propagation*.

13.2.7 Active Stereo

So far we have discussed algorithms for passive stereo vision, an approach that most closely mimics the way that the human visual system estimates depth. Another passive approach to depth estimation is **photometric stereo**, in which two images are taken by the same camera at the same location but under different lighting conditions—mild assumptions on the surface albedo then allow the recovery of depth. Alternatively, **depth from defocus** uses multiple images taken with different focal lengths to estimate depth from the amount of energy in the derivative of the graylevel signal (since blurring the image reduces the values in the derivative). Such passive systems have struggled for many years to produce dense depth measurements that have sufficient accuracy and resolution for real-world applications.

As a result, most commercially successful depth sensors rely on **active sensing**, in which light is projected onto the scene, and the light is sensed in some way. Active systems overcome the most vexing problem for passive stereo matching—namely, the lack of visual texture. In other words, when the scene contains areas without significant visual texture, then there is not enough local information to reliably estimate the disparity in those regions by simply matching patches of pixel values. This problem is particularly acute in indoor scenes, which often consist of large uniformly colored regions.

[†] Section 10.4.4 (p. 506)

Figure 13.14 An example of an RGB image (left) and a depth image (right) captured by the Kinect active stereo system.

One such active sensor is the **laser range finder**, which measures the depth of a single point. A laser beam is shone on the scene, and the beam's reflection from a surface in the world is sensed by a photodetector (such as a photodiode) placed near the laser. The distance to the surface is then measured as half the travel time multiplied by the speed of light. Combining a laser range finder with one or more rotating mirrors to change the orientation of the emitted ray yields a **scanning lidar** ("light radar"), also known as a **laser scanner**. Laser scanners can include either a single rotating mirror, in which case depth is measured only within a plane in the scene, or a pair of coupled rotating mirrors, in which the depth is measured at all points visible from the sensor.

While laser scanners are widely used in applications that can justify their high cost, the consumer market has only recently begun to be penetrated by depth sensors that are based on more inexpensive technology. One such approach, known as a **time-of-flight (TOF) camera**, involves shining light via either an LED (light-emitting diode) or laser diode and measuring the time required for the light to reflect off surfaces in the scene. As a result, a time-of-flight camera can be thought of as a scannerless laser scanner, since the entire scene is captured simultaneously without any moving parts. An alternate approach, known as **structured light**, uses a video projector that emits specific patterns of noncoherent light. These patterns are then detected by a camera and matched with the known emitted patterns to recover depth via triangulation, in a manner similar to stereo. An example depth image captured by a structured light depth sensor is shown in Figure 13.14.

13.3 Computing Optical Flow

Having considered the problem of estimating the one-dimensional displacement (disparity) between corresponding pixels in a pair of stereo images, in this section we address a closely related problem, namely to estimate the two-dimensional displacement (velocity) between corresponding pixels in a pair of consecutive image frames in a video sequence.

13.3.1 Motion Field

Suppose a video camera is viewing a scene in which either the objects and/or camera are moving. Associated with each point in the scene is a 3D velocity vector that captures the velocity of the point relative to the camera. The **motion field** is defined as the projection onto the 2D image plane of the 3D velocity vectors of all the points in the scene. For example, if you are driving a car in a straight line while looking forward, the 2D vectors in the motion field emanate from a single point in the image, which is known as the **focus of expansion**, as illustrated in Figure 13.15. On the other hand, if you are looking backward

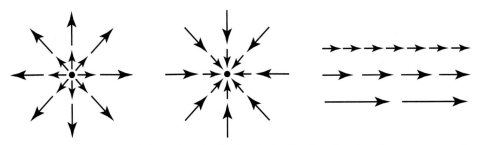

Figure 13.15 Examples of motion field. From left to right: When looking in the direction of motion, all vectors emanate from the focus of expansion; when looking in the direction opposite that of the motion, all vectors terminate on the focus of contraction; when looking parallel to the direction of motion, all vectors are parallel, and their magnitude is inversely proportional to their distance from the camera.

(and hopefully someone else is driving!), then the 2D vectors all aim toward a single point, called the **focus of contraction**. Finally, if you look to the side of the vehicle, then all the vectors are parallel in the image, analogous to a rectified stereo pair of images.

To quantify the motion field, let $\mathbf{p} = [p_x \quad p_y \quad p_z]^T$ be a world point in the camera coordinate frame. Assuming a pinhole camera model with the z axis aligned with the optical axis, the projection of the point onto the image plane, from Equations (2.1)–(2.2), is $(x, y) = \left(\frac{fp_x}{p_z}, \frac{fp_y}{p_z} \right)$. If the relative motion between $\mathbf{p}$ and the camera is given by the translation $\mathbf{t} = [t_x \quad t_y \quad t_z]^T$ and rotation $\omega = [\omega_x \quad \omega_y \quad \omega_z]^T$ velocities, then the velocity of the point in 3D is given by the translation minus the vector cross product of the rotation and the point:

$$\frac{d\mathbf{p}}{dt} = \mathbf{t} - \omega \times \mathbf{p} = \begin{bmatrix} -t_x - \omega_y p_z + \omega_z p_y \\ -t_y - \omega_z p_x + \omega_x p_z \\ -t_z - \omega_x p_y + \omega_y p_x \end{bmatrix} \tag{13.14}$$

Differentiating the projected coordinates (x, y) with respect to time t yields the image velocity $\mathbf{u} \equiv [u \quad v]^T$, where

$$u \equiv \frac{dx}{dt} = \frac{f}{p_z^2} \left(p_z \frac{dp_x}{dt} - p_x \frac{dp_z}{dt} \right) = \frac{xt_z - ft_x}{p_z} - f\omega_y + y\omega_z + \frac{xy\omega_x}{f} - \frac{x^2\omega_y}{f} \tag{13.15}$$

$$v \equiv \frac{dy}{dt} = \frac{f}{p_z^2} \left(p_z \frac{dp_y}{dt} - p_y \frac{dp_z}{dt} \right) = \frac{yt_z - ft_y}{p_z} + f\omega_x - x\omega_z - \frac{xy\omega_y}{f} + \frac{y^2\omega_x}{f} \tag{13.16}$$

Note that in both equations the depth p_z only appears in the first term, which involves translation, whereas the other terms involve rotation. That is, the depth and rotation are decoupled, because rotation of the camera about its focal point yields no **parallax** (the apparent displacement as seen from two different points of view) and hence no depth information.

Let us consider a few special cases. If the motion is pure translation, then $\omega = \mathbf{0}$, leading to

$$u = \frac{xt_z - ft_x}{p_z} = \frac{t_z}{p_z} \left(x - \frac{ft_x}{t_z} \right) \tag{13.17}$$

$$v = \frac{yt_z - ft_y}{p_z} = \frac{t_z}{p_z} \left(y - \frac{ft_y}{t_z} \right) \tag{13.18}$$

This motion field is radial, emanating from the point $\left(\frac{ft_x}{t_z}, \frac{ft_y}{t_z}\right)$, which is the intersection of the image plane with the translation vector. This point is either the focus of expansion (if $t_z < 0$), or the focus of contraction (if $t_z > 0$); it is also known as the **instantaneous epipole**.

If, in addition to $\omega = \mathbf{0}$, there is no motion along the optical axis, $t_z = 0$, then

$$u = -\frac{ft_x}{p_z} \tag{13.19}$$

$$v = -\frac{ft_y}{p_z} \tag{13.20}$$

In this case, all motion vectors are parallel, and the epipoles (which we shall define more precisely later) are at infinity. If, in addition, $t_y = 0$, the equations reduce to those of rectified stereo:

$$u = -\frac{ft_x}{p_z} \tag{13.21}$$

$$v = 0 \tag{13.22}$$

where t_x is the baseline, or translation between the two locations of the focal point, relative to the scene; and the displacement (or disparity, in this case) is inversely proportional to depth, as we saw earlier. Keep in mind that if the scene is static and the camera is moving, then all the equations of this section are applicable to the entire image, or if a single rigid object is moving, then they are applicable only to the pixels on that object.

13.3.2 Optical Flow

Unlike the motion field, which is the *actual* projected motion, **optical flow**[†] refers to the *apparent* motion of the brightness patterns in the image plane. In an ideal world, these two would be equivalent, but there are some obvious pathological cases in which the two are very different. For example, the famous barberpole illusion arises because the diagonal stripes appear to be moving vertically in the image, when in reality they are moving horizontally. Similarly, a rotating ping pong ball with no texture will appear not to be moving even though it is; and a specular reflection on a shiny surface will appear to be moving as the viewpoint (or lighting source) location is changed, even though in reality the surface itself is not changing.

The motion field and optical flow are identical when the projected pixel values of points in the scene remain the same throughout the image sequence. This is known as the **brightness constancy assumption**. Consider a point in the world that projects onto the image plane at the 2D location $(x(t), y(t))$ at time t, so that the trajectory of the projection over time is described by the functions $x(\cdot)$ and $y(\cdot)$. At some small time Δt later, therefore, the point projects to $(x(t + \Delta t), y(t + \Delta t))$. If we assume that the brightness (or, more accurately, gray level) of the projected point remains constant, then we have

$$I(x(t + \Delta t), y(t + \Delta t), t + \Delta t) = I(x(t), y(t), t) \tag{13.23}$$

where we have introduced a third parameter, t, to the function $I(\cdot, \cdot)$ to specify which image in the sequence is meant. This equation says that every pixel in one image (the one captured at time t) has the same gray level as the corresponding pixel in the other image (the one

---

[†] Referred to as *optic flow* in the vision science community.

captured at time $t + \Delta t$), where the correspondence is determined by the projection of the actual motion of the point in space.

You may recall that a continuously differentiable 1D function $f(x)$ can be approximated around a point x_0, using the Taylor series expansion:

$$f(x) \approx f(x_0) + (x - x_0)\frac{df(x_0)}{dx} + \cdots \tag{13.24}$$

where only the linear term is shown. For a function of multiple variables, the differentiation must be performed with respect to each variable. Therefore, the Taylor series expansion of the left-hand side of Equation (13.23) around the point $(x(t), y(t), t)$, ignoring all higher-order terms, is

$$
\begin{aligned}
I(x(t + \Delta t), y(t + \Delta t), t + \Delta t) \approx\ & I(x(t), y(t), t) \\
& + (x(t + \Delta t) - x(t))\frac{\partial I}{\partial x} \\
& + (y(t + \Delta t) - y(t))\frac{\partial I}{\partial y} \\
& + \Delta t\frac{\partial I}{\partial t} \tag{13.25}
\end{aligned}
$$

which, due to the constant-brightness assumption in Equation (13.23) leads to

$$(x(t + \Delta t) - x(t))\frac{\partial I}{\partial x} + (y(t + \Delta t) - y(t))\frac{\partial I}{\partial y} + \Delta t\frac{\partial I}{\partial t} \approx 0 \tag{13.26}$$

Dividing both sides by Δt and taking the limit as $\Delta t \to 0$ yields

$$\lim_{\Delta t \to 0}\frac{(x(t + \Delta t) - x(t))}{\Delta t}\frac{\partial I}{\partial x} + \lim_{\Delta t \to 0}\frac{(y(t + \Delta t) - y(t))}{\Delta t}\frac{\partial I}{\partial y} + \frac{\partial I}{\partial t} \approx 0 \tag{13.27}$$

or, according to the definition of derivatives,[†]

$$\frac{dx}{dt}\frac{\partial I}{\partial x} + \frac{dy}{dt}\frac{\partial I}{\partial y} \approx -\frac{\partial I}{\partial t} \tag{13.28}$$

This is the standard **optical flow constraint equation**, which, as you can see, is somewhat of a misnomer, since it is not an equation at all but rather an approximation. Nevertheless, we shall retain the conventional terminology while leaving the approximation intact to remind us that it is based on the linearized Taylor series. The importance of this reminder will become apparent in a moment. In the meantime, we note that the equation can be written more compactly as the dot product between two vectors:

$$(\nabla I)^\mathsf{T}\mathbf{u} \approx -I_t \tag{13.29}$$

[†] The derivatives of x and y are total derivatives because the functions depend on only one variable, whereas the derivatives of I are partial derivatives because it depends on multiple variables; although the distinction is not important for our purposes. The derivation of this equation is sometimes attributed to the chain rule of differentiation applied to the total derivative of the image: $\frac{dI}{dt} = 0$, but the total derivative of a function of several variables is its gradient, which is not a scalar but a vector.

where $\nabla I \equiv \begin{bmatrix} I_x & I_y \end{bmatrix}^\mathsf{T} \equiv \begin{bmatrix} \partial I/\partial x & \partial I/\partial y \end{bmatrix}^\mathsf{T}$ is the gradient of the image intensity function, $\mathbf{u} \equiv \begin{bmatrix} u & v \end{bmatrix}^\mathsf{T} = \begin{bmatrix} dx/dt & dy/dt \end{bmatrix}^\mathsf{T}$ is the unknown image velocity, and $I_t \equiv \partial I/\partial t$ is the partial derivative of the image with respect to time. In practice, we often use two consecutive images in a video sequence, so that $\Delta t = 1$, and the continuous derivatives are approximated by finite differences:

$$u = \frac{dx}{dt} \approx x(t+1) - x(t) \quad v = \frac{dy}{dt} \approx y(t+1) - y(t) \quad \frac{\partial I(x, y, t)}{\partial t} \approx I(x, y, t+1) - I(x, y, t) \quad (13.30)$$

in which case the unknown velocities u and v are actually displacements along the x and y axes between consecutive images.

The single scalar equation in Equation (13.28) has two unknowns, meaning that the problem of estimating the motion of a single pixel is underconstrained. This is known as the **aperture problem**, which receives its name from the idea that viewing a small part of the image (as if through a small aperture) would not enable the resolution of both components of the velocity vector. Instead, only the component of motion in the direction of the gradient, or, equivalently, normal to the graylevel isocontour, can be computed:

$$u_n = \frac{-I_t}{\|\nabla I\|} = \frac{(\nabla I)^\mathsf{T} \mathbf{u}}{\|\nabla I\|} = \left(\frac{\nabla I}{\|\nabla I\|} \right)^\mathsf{T} \mathbf{u} \qquad (13.31)$$

To better visualize the optical flow constraint equation, it may be helpful to consider an object moving horizontally in the image at approximately constant velocity, as shown in Figure 13.16. In 1D, the augmented displacement vector, given by $\tilde{\mathbf{u}} = (\Delta x, \Delta t)$, is tangent to the graylevel isocontour, which itself is perpendicular to the gradient of $I(x, t)$, so that $\tilde{\mathbf{u}}^\mathsf{T} \nabla I = \begin{bmatrix} \Delta x & \Delta t \end{bmatrix} \begin{bmatrix} I_x & I_t \end{bmatrix}^\mathsf{T} = I_x \Delta x + I_t \Delta t = 0$. Assuming that $\Delta t = 1$ (i.e., consecutive image frames), this yields $I_x \Delta x = -I_t$. Extending to 2D, similar reasoning leads to $I_x \Delta x + I_y \Delta y = -I_t$, which is the optical flow constraint equation above.

13.3.3 Lucas-Kanade Algorithm

One way to overcome the aperture problem is to assume that all pixels in a small region share the same image motion. This assumption leads to the well-known **Lucas-Kanade**[†] method. Recall that for a single pixel at location $\mathbf{x} = \begin{bmatrix} x & y \end{bmatrix}^\mathsf{T}$, the optical flow constraint equation is $I_x(\mathbf{x})u + I_y(\mathbf{x})v \approx -I_t(\mathbf{x})$, where $\mathbf{u} = \begin{bmatrix} u & v \end{bmatrix}^\mathsf{T}$. Let us now reinterpret this approximation as the following equation:

$$I_x(\mathbf{x})u_\Delta + I_y(\mathbf{x})v_\Delta = -I_t(\mathbf{x}) \qquad (13.32)$$

Figure 13.16 Visualization of the optical flow constraint equation (OFCE) using a 1D image, showing that the augmented displacement vector $\tilde{\mathbf{u}}$ of a point is parallel to the graylevel isocontour at that point, and therefore perpendicular to the augmented gradient vector. Shown is an illustration (left) and an actual slice (right) through a spatiotemporal volume of an image sequence, in which a thin object moves to the right faster than the rest of the scene.

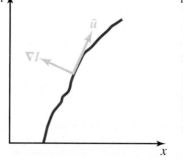

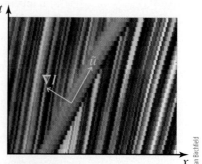

Stan Birchfield

† pronounced kah-NAH-deh.

where $\mathbf{u}_\Delta \equiv \begin{bmatrix} u_\Delta & v_\Delta \end{bmatrix}^T$ is the incremental displacement of the pixel about the current estimate, and the approximation has become an equation because $\mathbf{u}_\Delta$ describes the displacement captured by the linearized equation, rather than the $\mathbf{u}$ that we are ultimately trying to estimate, that is,

$$\mathbf{u} = \mathbf{u}_0 + \mathbf{u}_\Delta^{(1)} + \mathbf{u}_\Delta^{(2)} + \mathbf{u}_\Delta^{(3)} + \cdots \tag{13.33}$$

where $\mathbf{u}_0$ is the initial motion estimate, and $\mathbf{u}_\Delta^{(i)}$ is the incremental displacement computed in the ith iteration.

If we now assume that all the pixels $\mathbf{x}_1, \ldots, \mathbf{x}_n$ in a region $\mathcal{R} = \{\mathbf{x}_i\}_{i=1}^n$ share the same image motion, then we can stack their equations to yield an overdetermined linear system:

$$\underbrace{\begin{bmatrix} I_x(\mathbf{x}_1) & I_y(\mathbf{x}_1) \\ I_x(\mathbf{x}_2) & I_y(\mathbf{x}_2) \\ \vdots & \vdots \\ I_x(\mathbf{x}_n) & I_y(\mathbf{x}_n) \end{bmatrix}}_{\mathbf{A}} \begin{bmatrix} u_\Delta \\ v_\Delta \end{bmatrix} = -\underbrace{\begin{bmatrix} I_t(\mathbf{x}_1) \\ I_t(\mathbf{x}_2) \\ \vdots \\ I_t(\mathbf{x}_n) \end{bmatrix}}_{\mathbf{b}} \tag{13.34}$$

where n is the number of pixels in the region, $\mathbf{A}$ is the $n \times 2$ matrix containing the spatial derivatives of the image at the pixels, and $\mathbf{b}$ is the $n \times 1$ vector containing the temporal derivatives. The temporal derivative is usually approximated by simply subtracting the two images from each other: $I_t(\mathbf{x}) \approx J(\mathbf{x}) - I(\mathbf{x})$, where I is the current image, and J is the next image in the video sequence; such a finite difference is a reasonable approximation to the derivative, and it is easy to compute.

The system of equations can be made more compact by multiplying both sides by $\mathbf{A}^T$:

$$\underbrace{\begin{bmatrix} I_x(\mathbf{x}_1) & \cdots & I_x(\mathbf{x}_n) \\ I_y(\mathbf{x}_1) & \cdots & I_y(\mathbf{x}_n) \end{bmatrix}}_{\mathbf{A}^T} \underbrace{\begin{bmatrix} I_x(\mathbf{x}_1) & I_y(\mathbf{x}_1) \\ \vdots & \vdots \\ I_x(\mathbf{x}_n) & I_y(\mathbf{x}_n) \end{bmatrix}}_{\mathbf{A}} \begin{bmatrix} u_\Delta \\ v_\Delta \end{bmatrix} = -\underbrace{\begin{bmatrix} I_x(\mathbf{x}_1) & \cdots & I_x(\mathbf{x}_n) \\ I_y(\mathbf{x}_1) & \cdots & I_y(\mathbf{x}_n) \end{bmatrix}}_{\mathbf{A}^T} \begin{bmatrix} I_t(\mathbf{x}_1) \\ I_t(\mathbf{x}_2) \\ \vdots \\ I_t(\mathbf{x}_n) \end{bmatrix} \tag{13.35}$$

or

$$\underbrace{\sum_{\mathbf{x} \in \mathcal{R}} \begin{bmatrix} I_x^2(\mathbf{x}) & I_x(\mathbf{x})I_y(\mathbf{x}) \\ I_x(\mathbf{x})I_y(\mathbf{x}) & I_y^2(\mathbf{x}) \end{bmatrix}}_{\mathbf{Z}} \begin{bmatrix} u_\Delta \\ v_\Delta \end{bmatrix} = -\underbrace{\sum_{\mathbf{x} \in \mathcal{R}} \begin{bmatrix} I_x(\mathbf{x})I_t(\mathbf{x}) \\ I_y(\mathbf{x})I_t(\mathbf{x}) \end{bmatrix}}_{\mathbf{e}} \tag{13.36}$$

where $\mathbf{Z} = \mathbf{A}^T\mathbf{A}$ is the 2×2 gradient covariance matrix that we saw in Equation (7.29), and $\mathbf{e} = \mathbf{A}^T\mathbf{b}$ is a 2×1 vector. Therefore, solving the linear system $\mathbf{A}\mathbf{u}_\Delta = \mathbf{b}$ is equivalent to solving $\mathbf{A}^T\mathbf{A}\mathbf{u}_\Delta = \mathbf{A}^T\mathbf{b}$, or $\mathbf{Z}\mathbf{u}_\Delta = \mathbf{e}$. If we let z_x, z_y, and z_{xy} be the elements of $\mathbf{Z}$, as in Equation (7.29), and e_x and e_y be the elements of $\mathbf{e}$, then the equation is easily solved as

$$\mathbf{u}_\Delta = \begin{bmatrix} u_\Delta \\ v_\Delta \end{bmatrix} = \frac{1}{\det(\mathbf{Z})} \begin{bmatrix} z_y e_x - z_{xy} e_y \\ z_x e_y - z_{xy} e_x \end{bmatrix} \tag{13.37}$$

where $\det(\mathbf{Z}) = z_x z_y - z_{xy}^2$ is the determinant of $\mathbf{Z}$.

Now let us consider the difference between $\mathbf{u}$ and $\mathbf{u}_\Delta$. The preceding analysis assumes that the image intensity function is well represented by a linear approximation in the neighborhood of the current estimate. This is known as the **small motion assumption**, and it causes the Taylor series to be truncated to the linear term. To obtain the final result, therefore, we need to iteratively apply this technique, each time linearizing about the new estimate. This approach is known as the **Gauss-Newton method**, and it is can be thought of as a simplification of Newton's method that is applicable in the case of a squared error function, as here. In both techniques the equation is repeatedly linearized about the new estimate to find the root of a function, but Gauss-Newton obviates the need for second derivatives due to the special structure of the problem being solved.

To visualize an iteration of the procedure, Figure 13.17 shows the one-dimensional intensity profile of an image I, along with the profile of the next image J in the video sequence. It is obvious that the brightness constancy assumption holds because the shape of the two profiles is identical, the only difference being the relative shift between them. A right triangle is drawn from the value of the current estimate $I(x_0)$ to the value in the next image $J(x_0)$; from that point, a diagonal line is drawn using $I_x(x_0)$, which is the slope of I computed at x_0. A horizontal line from $I(x_0)$ completes the triangle. It should be clear from the drawing that the length of the horizontal base of the triangle is the unknown u_Δ, whereas the vertical distance is simply the temporal derivative estimate $J(x_0) - I(x_0) \approx I_t$, and the slope is the spatial derivative. Since I_x is the slope ("rise over run"), we have $I_x(x_0) = -I_t(x_0)/u_\Delta$, or

$$u_\Delta = -\frac{I_t(x_0)}{I_x(x_0)} \tag{13.38}$$

which is simply the one-dimensional version of the optical flow constraint equation.

Once u_Δ has been computed, the images are shifted by u_Δ to bring them closer together. Then the process is repeated until convergence. Conceptually, it does not matter whether the shift is applied to the first or second image—that is, whether I is moved (to the right, in our example) or J is moved (to the left), since either way produces essentially the same result. However, for computational efficiency it is better *not* to shift the image for which the spatial derivatives were computed so that those derivatives can be reused in subsequent iterations. As a result, we adopt the convention of shifting J instead of I, a process sometimes known as the **inverse warp**.

We are now ready to provide the pseudocode for Lucas-Kanade in Algorithm 13.7, which shows the steps for computing the displacement of a single set of pixels $\mathcal{R}$, which is usually a square window around the central pixel. First the spatial gradient of the image I is computed, then $\mathbf{u}_\Delta$ is solved for iteratively, effectively shifting J each time. The computation terminates when the minimization converges, as evidenced by the norm of $\mathbf{u}_\Delta$ falling below a threshold, or when a maximum number of iterations has been reached. Typically no more than 10 iterations are needed for good accuracy, and oftentimes as few as three to five are

Figure 13.17 One-dimensional illustration of the small motion assumption made by the optical flow constraint equation.

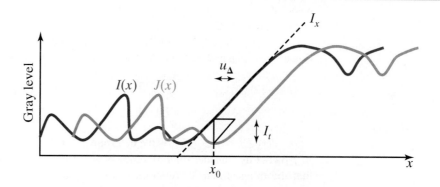

ALGORITHM 13.7 Lucas-Kanade optical flow for a single region

LucasKanade($I, J, \mathcal{R}$)

Input: images I and J, and region $\mathcal{R}$
Output: the 2D displacement $\mathbf{u}$ of the pixel represented by $\mathcal{R}$

1 $\mathbf{u} \leftarrow 0$
2 $iter \leftarrow 0$
3 $G_x, G_y \leftarrow$ Gradient(I)
4 **repeat**
5      $\mathbf{Z} \leftarrow$ Compute2x2GradientMatrix($G_x, G_y, \mathcal{R}$)
6      $\mathbf{e} \leftarrow$ Compute2x1ErrorVector($I, J, G_x, G_y, \mathcal{R}, \mathbf{u}$)
7      $\mathbf{u}_\Delta \leftarrow$ Solve2x2LinearSystem($\mathbf{Z}, \mathbf{e}$)
8      $\mathbf{u} \leftarrow \mathbf{u} + \mathbf{u}_\Delta$
9      $iter \leftarrow_+ 1$
10 **until** $\|\mathbf{u}_\Delta\| <$ *threshold* OR *iter* $\geq$ *max-iter*
11 **return** $\mathbf{u}$

Compute2x2GradientMatrix($G_x, G_y, \mathcal{R}$)

Input: gradient images G_x and G_y, and region $\mathcal{R}$
Output: elements of 2×2 gradient covariance matrix $\mathbf{Z} = \begin{bmatrix} z_x & z_{xy} \\ z_{xy} & z_y \end{bmatrix}$

1 $z_x \leftarrow z_{xy} \leftarrow z_y \leftarrow 0$
2 **for** $(x, y) \in \mathcal{R}$ **do**
3      $z_x \leftarrow_+ G_x(x, y) * G_x(x, y)$
4      $z_{xy} \leftarrow_+ G_x(x, y) * G_y(x, y)$
5      $z_y \leftarrow_+ G_y(x, y) * G_y(x, y)$
6 **return** z_x, z_{xy}, z_y

Compute2x1ErrorVector($I, J, G_x, G_y, \mathcal{R}, \mathbf{u}$)

Input: images I and J, gradient images G_x and G_y, region $\mathcal{R}$, and initial displacement $\mathbf{u} = \begin{bmatrix} u & v \end{bmatrix}^\mathsf{T}$
Output: elements of 2×1 error vector $\mathbf{e} = \begin{bmatrix} e_x & e_y \end{bmatrix}^\mathsf{T}$

1 $e_x \leftarrow e_y \leftarrow 0$
2 **for** $(x, y) \in \mathcal{R}$ **do**
3      $e_x \leftarrow_+ G_x(x, y) * (I(x, y) - J(x + u, y + v))$ ▸ $\partial I/\partial t \approx J - I$, so $-\partial I/\partial t \approx I - J$.
4      $e_y \leftarrow_+ G_y(x, y) * (I(x, y) - J(x + u, y + v))$
5 **return** e_x, e_y

Solve2x2LinearSystem($\mathbf{Z}, \mathbf{e}$)

Input: 2×2 matrix $\mathbf{Z}$ and 2×1 error vector $\mathbf{e} = \begin{bmatrix} e_x & e_y \end{bmatrix}^\mathsf{T}$
Output: displacement $\begin{bmatrix} u_\Delta & v_\Delta \end{bmatrix}^\mathsf{T}$ that solves the equation

1 $det \leftarrow z_x * z_y - z_{xy} * z_{xy}$
2 $u_\Delta \leftarrow (1/det) * (z_y * e_x - z_{xy} * e_y)$
3 $v_\Delta \leftarrow (1/det) * (z_x * e_y - z_{xy} * e_x)$
4 **return** u_Δ, v_Δ

ALGORITHM 13.8 Lucas-Kanade tracking of features throughout a video sequence

$\textsc{LucasKanadeSequence}(I_{0:m}, \textit{window-width})$

Input: sequence of 2D images $I_0, I_1, \ldots, I_m$ and width of square feature window
Output: 2D positions of all features in all frames

1    $features(:, 0) \leftarrow \textsc{SelectGoodFeatures}(I_0)$
2   **for** $i \leftarrow 1$ **to** m **do**
3   **for** $j \leftarrow 1$ $features.\textsc{Size}$ **do**
4            $\mathbf{u} \leftarrow \textsc{LucasKanade}(I_{i-1}, I_i, \textsc{SquareRegion}(features(j, i - 1), \textit{window-width}))$
5            $features(j, i) \leftarrow features(j, i - 1) + \mathbf{u}$
6   **return** $features$

needed. The astute reader will notice that the $\textsc{Compute2x2GradientMatrix}$ call in Line 5 depends only upon parameters that do not change each iteration and can therefore be pulled out of the loop to reduce computation.

Lucas-Kanade is usually used to track a sparse number of feature points throughout a video sequence, as shown in Algorithm 13.8. These features are selected in the first image frame by computing the Harris or Tomasi-Kanade cornerness measure[†] using $\mathbf{Z}$ over all pixels in the image, then performing nonmaximal suppression and selecting the most prominent features. A rectangular window, usually 7×7 or larger, of pixels is associated with each feature, and Lucas-Kanade is used to find the displacement between each consecutive pair of image frames for each feature. One detail to keep in mind is that, after solving the 2×2 system in the first iteration, the value for $\mathbf{u}$ is no longer an integer; as a result, the function $\textsc{Compute2x1ErrorVector}$ needs to access non-integral pixel values, which is usually performed with bilinear interpolation.[‡] Similarly, after the first pair of image frames, the feature coordinates will not be integers either, so that I, J, G_x, and G_y need to access non-integral pixel values in both $\textsc{Compute2x2GradientMatrix}$ and $\textsc{Compute2x1ErrorVector}$. The result of Lucas-Kanade tracking on a pair of frames from a video sequence is shown in Figure 13.18.

Figure 13.18 Two frames of a video sequence, with Lucas-Kanade features overlaid as red dots.

Stan Birchfield

[†] Section 7.4 (p. 342).
[‡] Section 3.8.2 (p. 110).

13.3.4 Generalized Lucas-Kanade

One of the advantages of the Lucas-Kanade approach is that it naturally generalizes to other motion models. To understand how this works, let us derive the equation $\mathbf{Zu}_\Delta = \mathbf{e}$ in an alternate manner using the Taylor series expansion, beginning with the same translation model that we have been considering. Then we will show how this alternate approach generalizes to other models.

Consider a region of pixels $\mathcal{R}$ in an image I, and a translation $\mathbf{u}$ that offsets the region in another image J. The sum-of-squared differences (SSD) between the intensities of the two sets is given by

$$error = \sum_{\mathbf{x} \in \mathcal{R}} (I(\mathbf{x} - \mathbf{u}) - J(\mathbf{x}))^2 \tag{13.39}$$

where $\mathbf{u}$ is the unknown displacement that we are trying to find. Although $J - I$ was considered earlier to be an approximation to I_t, this alternate formulation makes it clear that the two images do not have to be adjacent frames in a video sequence but rather can be any two image patches, such as the current image and a reference template.

Given our previous experience with the Taylor series expansion of a function of multiple variables, such as Equation (7.24), it should not be hard to see that the linearized expansion of I about the point $(\mathbf{x} - \mathbf{u})$ is given by

$$I(\mathbf{x} - \mathbf{u}) \approx I(\mathbf{x}) - u_\Delta \frac{\partial I}{\partial x}(\mathbf{x}) - v_\Delta \frac{\partial I}{\partial y}(\mathbf{x}) \tag{13.40}$$

where the notation is slightly abused by replacing $\mathbf{u}$ with $\mathbf{u}_\Delta \equiv [u_\Delta \quad v_\Delta]^\mathsf{T}$. Substituting back into Equation (13.39) yields

$$error = \sum_{\mathbf{x} \in \mathcal{R}} \left(I(\mathbf{x}) - J(\mathbf{x}) - u_\Delta \frac{\partial I}{\partial x}(\mathbf{x}) - v_\Delta \frac{\partial I}{\partial y}(\mathbf{x}) \right)^2 \tag{13.41}$$

Note that the expression inside the parentheses is none other than the left side of the optical flow constraint equation, if we let $J - I \approx I_t$. In other words, just as the left side of the optical flow constraint equation should be equal to zero under the constant brightness assumption, the linear approximation of the SSD dissimilarity of two intensity patches also seeks to minimize this value.

To find the motion, simply differentiate with respect to the unknowns and set the result to zero:

$$\frac{\partial\, error}{\partial \mathbf{u}_\Delta} = 2 \sum_{\mathbf{x} \in \mathcal{R}} (I(\mathbf{x}) - J(\mathbf{x}) - (\nabla I(\mathbf{x}))^\mathsf{T} \mathbf{u}_\Delta) \nabla I(\mathbf{x}) = 0 \tag{13.42}$$

which, after rearranging terms, yields the same formula that we derived before:

$$\underbrace{\left(\sum_{\mathbf{x} \in \mathcal{R}} \nabla I(\mathbf{x}) (\nabla I(\mathbf{x}))^\mathsf{T} \right)}_{\mathbf{z}} \mathbf{u}_\Delta = \underbrace{\sum_{\mathbf{x} \in \mathcal{R}} \nabla I(\mathbf{x}) (I(\mathbf{x}) - J(\mathbf{x}))}_{\mathbf{e}} \tag{13.43}$$

Although it may not yet look like we have accomplished much, the power of this technique becomes apparent when we want to generalize to other motion models. For example, consider the SSD error using an affine transformation which, as explained earlier,[†] is an approximation to perspective projection:

$$error = \sum_{\mathbf{x} \in \mathcal{R}} (I(\mathbf{Dx} + \mathbf{d}) - J(\mathbf{x}))^2 \tag{13.44}$$

where

$$\mathbf{D} \equiv \begin{bmatrix} 1 + d_1 & d_2 \\ d_3 & 1 + d_4 \end{bmatrix} \tag{13.45}$$

so that $d_1 = d_2 = d_3 = d_4 = 0$ causes $\mathbf{D}$ to reduce to the identity matrix, and $\mathbf{d} \equiv [d_x \quad d_y]^\mathsf{T}$. The linearized Taylor series approximation is similar to before:

$$I(\mathbf{Dx} + \mathbf{d}) \approx I(\mathbf{x}) + (d_1 x + d_2 y + d_x)\frac{\partial I}{\partial x}(\mathbf{x}) + (d_3 x + d_4 y + d_y)\frac{\partial I}{\partial y}(\mathbf{x}) \tag{13.46}$$

If we collect the unknowns into a single vector $\mathbf{p} \equiv [d_1 \quad d_2 \quad d_x \quad d_3 \quad d_4 \quad d_y]^\mathsf{T}$, and if we define $\mathbf{g}(\mathbf{x}) \equiv [x I_x \quad y I_x \quad I_x \quad x I_y \quad y I_y \quad I_y]^\mathsf{T}$, then it is not difficult to show that

$$\frac{\partial\,error}{\partial \mathbf{p}} = 2 \sum_{\mathbf{x} \in \mathcal{R}} (I(\mathbf{Dx} + \mathbf{d}) - J(\mathbf{x})) \mathbf{g}(\mathbf{x}) = 0_{\{6 \times 1\}} \tag{13.47}$$

which leads to the following 6×6 linear set of equations:

$$\left(\sum_{\mathbf{x} \in \mathcal{R}} \mathbf{g}(\mathbf{x}) \mathbf{g}^\mathsf{T}(\mathbf{x}) \right) \mathbf{p}_\Delta = \sum_{\mathbf{x} \in \mathcal{R}} (I(\mathbf{x}) - J(\mathbf{x})) \mathbf{g}(\mathbf{x}) \tag{13.48}$$

The similarity between Equations (13.43) and (13.48) should be obvious, with the 6×1 vector $\mathbf{g}(\mathbf{x})$ playing the role of the image gradient, and the 6×1 vector $\mathbf{p}_\Delta$ substituting for $\mathbf{u}_\Delta$.

The result can be generalized even further. For the case of an arbitrary warp function $\mathbf{x}' = W(\mathbf{x}; \mathbf{p})$ that maps pixels in one image to the corresponding pixels in the other image, where $\mathbf{p}$ are the parameters of the warp function, the sum-of-squared differences (SSD) is given as

$$error = \sum_{\mathbf{x} \in \mathcal{R}} (I(W(\mathbf{x}; \hat{\mathbf{p}} + \mathbf{p}_\Delta)) - J(\mathbf{x}))^2 \tag{13.49}$$

where $\hat{\mathbf{p}}$ is the current estimate of the parameters, and $\mathbf{p}_\Delta$ represents the incremental warp that will be found by the algorithm in the current iteration. Our goal is to find the best warp to align the two images so as to minimize the SSD. Using the first-order Taylor series approximation, we have

$$I(W(\mathbf{x}; \hat{\mathbf{p}} + \mathbf{p}_\Delta)) \approx I(W(\mathbf{x}; \hat{\mathbf{p}})) + (W(\mathbf{x}; \hat{\mathbf{p}} + \mathbf{p}_\Delta) - W(\mathbf{x}; \hat{\mathbf{p}}))^\mathsf{T}\frac{\partial I}{\partial \mathbf{x}} \tag{13.50}$$

$$\approx I(W(\mathbf{x}; \hat{\mathbf{p}})) + \mathbf{p}_\Delta^\mathsf{T} \underbrace{\left(\frac{\partial W}{\partial \mathbf{p}}\right)^\mathsf{T} \frac{\partial I}{\partial \mathbf{x}}}_{\mathbf{g}(\mathbf{x})} \tag{13.51}$$

[†] Section 3.9.4 (p. 124).

where the second equation comes from the Taylor series expansion of the warp:

$$W(\mathbf{x}; \hat{\mathbf{p}} + \mathbf{p}_\Delta) \approx W(\mathbf{x}; \hat{\mathbf{p}}) + \frac{\partial W}{\partial \mathbf{p}} \mathbf{p}_\Delta \tag{13.52}$$

where $I(\cdot)$ is a scalar, $\nabla I = \partial I / \partial \mathbf{x}$ and $W(\cdot)$ are 2×1, $\partial W / \partial \mathbf{p}$ is the $2 \times m$ **Jacobian**, and $\mathbf{p}_\Delta$ and $\mathbf{g}(\mathbf{x})$ are $m \times 1$, where m is the number of parameters in the warp. Differentiating with respect to the unknown parameters and setting the result equal to zero yields

$$\frac{\partial \, error}{\partial \mathbf{p}_\Delta} = \sum_{\mathbf{x} \in \mathcal{R}} 2(I(W(\mathbf{x}; \hat{\mathbf{p}})) - J(\mathbf{x}) + \mathbf{p}_\Delta^\mathsf{T} \mathbf{g}(\mathbf{x})) \mathbf{g}(\mathbf{x}) = \mathbf{0} \tag{13.53}$$

which, after rearranging terms, yields a linear system of equations in the unknown parameters $\mathbf{p}_\Delta$:

$$\underbrace{\left(\sum_{\mathbf{x} \in \mathcal{R}} \mathbf{g}(\mathbf{x}) \mathbf{g}^\mathsf{T}(\mathbf{x}) \right)}_{\mathbf{H}} \mathbf{p}_\Delta = \underbrace{\sum_{\mathbf{x} \in \mathcal{R}} (I(\mathbf{x}) - J(W^{-1}(\mathbf{x}; \hat{\mathbf{p}}))) \, \mathbf{g}(\mathbf{x})}_{\mathbf{e}} \tag{13.54}$$

or $\mathbf{H} \mathbf{p}_\Delta = \mathbf{e}$, where the inverse warp is applied to J rather than the forward warp to I to avoid having to recompute the image gradient at each iteration.

Pseudocode for the generalized version of Lucas-Kanade, provided in Algorithm 13.9, is not substantially different from the specific cases considered above. First, the gradient of the image is precomputed and used to construct m steepest descent images $\mathbf{g}(\mathbf{x})$, along with an outer product matrix $\mathbf{H}$. At each iteration, the image J is warped (using the inverse warp) to construct the vector $\mathbf{e}$. Solving the linear system yields the incremental warp $\mathbf{p}_\Delta$, which is composed with the current warp estimate to yield a new warp estimate. Because the inverse warp is used[†] and the incremental warps are composed rather than added, this pseudocode

ALGORITHM 13.9 Lucas-Kanade alignment of two image patches using an arbitrary warp function

$\textsc{LucasKanadeGeneral}(I, J, \mathcal{R}, \hat{\mathbf{p}})$

Input: images I and J, region $\mathcal{R}$, and current estimate $\hat{\mathbf{p}}$ of the warp parameters
Output: updated estimate $\hat{\mathbf{p}}$ of the warp parameters

1	$\mathbf{p}_\Delta \leftarrow 0$	
2	$iter \leftarrow 0$	
3	$\nabla I \leftarrow \textsc{Gradient}(I)$	Compute gradient of image that is not being warped.
4	compute $\mathbf{g}(\mathbf{x}) \leftarrow \left(\frac{\partial W}{\partial \mathbf{p}} \right)^\mathsf{T} \nabla I$ for all $\mathbf{x} \in \mathcal{R}$	Compute steepest descent images.
5	$\mathbf{H} \leftarrow \sum_{\mathbf{x} \in \mathcal{R}} \mathbf{g}(\mathbf{x}) \mathbf{g}^\mathsf{T}(\mathbf{x})$	Compute outer product matrix.
6	**repeat**	
7	$\quad \mathbf{e} \leftarrow \sum_{\mathbf{x} \in \mathcal{R}} (I(\mathbf{x}) - J(W^{-1}(\mathbf{x}; \hat{\mathbf{p}}))) \mathbf{g}(\mathbf{x})$	Compute error vector.
8	$\quad \mathbf{p}_\Delta \leftarrow \textsc{Solve}(\mathbf{H} \mathbf{p}_\Delta = \mathbf{e})$	Solve Equation (13.54) for $\mathbf{p}_\Delta$
9	$\quad W(\mathbf{x}; \hat{\mathbf{p}}) \leftarrow W(W(\mathbf{x}; \hat{\mathbf{p}}); \mathbf{p}_\Delta)$	Update the warp by composition.
10	**until** $\|\mathbf{p}_\Delta\| < threshold$ **OR** $iter \geq max\text{-}iter$	
11	**return** $\hat{\mathbf{p}}$	

[†] Note that the roles of the images have been reversed, so that the image I for which the gradients are computed is different from the image J to which the inverse warp is applied; this inversion of roles is the key to computational efficiency.

implements what is known as the **inverse compositional method.**[†] Note that in this code a warp estimate $\hat{\mathbf{p}}$ is explicitly passed in, updated, and returned; for the previous version, an initial estimate could have been passed in as well but was implicitly assumed to be zero.

It is easy to show that this general case handles both the translation and affine models already considered. In the case of translation we have $W(\mathbf{x}; \hat{\mathbf{p}} + \mathbf{p}_\Delta) = \mathbf{x} - \mathbf{u}$, so that the parameters are the translation values, $\mathbf{p} = -\mathbf{u}$, where the sign arises from our convention of swapping the roles of I and J. The partial derivative of the warp with respect to the parameters is

$$\frac{\partial W}{\partial \mathbf{p}} = \frac{\partial(\mathbf{x} - \mathbf{u})}{\partial(-\mathbf{u})} = \begin{bmatrix} 1 & 0 \\ 0 & 1 \end{bmatrix} \tag{13.55}$$

so that

$$\left(\frac{\partial W}{\partial \mathbf{p}}\right)^{\mathsf{T}} \frac{\partial I}{\partial \mathbf{x}} = \frac{\partial I}{\partial \mathbf{x}} = \nabla I \tag{13.56}$$

leading to the same result we obtained before:

$$\sum_{\mathbf{x} \in \mathcal{R}} (\nabla I(\mathbf{x}))(\nabla I(\mathbf{x}))^{\mathsf{T}} \mathbf{p}_\Delta = \sum_{\mathbf{x} \in \mathcal{R}} (I(\mathbf{x} - \hat{\mathbf{u}}) - J(\mathbf{x}))) \nabla I(\mathbf{x}) \tag{13.57}$$

where $W(\mathbf{x}; \hat{\mathbf{p}}) = \hat{\mathbf{u}}$ is the current estimate. For the affine transformation, we have

$$W(\mathbf{x}; \hat{\mathbf{p}}) = \begin{bmatrix} 1 + d_1 & d_2 & d_x \\ d_3 & 1 + d_4 & d_y \end{bmatrix} \begin{bmatrix} x \\ y \\ 1 \end{bmatrix} = \begin{bmatrix} (1 + d_1)x + d_2 y + d_x \\ d_3 x + (1 + d_4)y + d_y \end{bmatrix} \tag{13.58}$$

so that

$$\frac{\partial W}{\partial \mathbf{p}} = \begin{bmatrix} x & y & 1 & 0 & 0 & 0 \\ 0 & 0 & 0 & x & y & 1 \end{bmatrix} \tag{13.59}$$

leading to

$$\mathbf{g}(\mathbf{x}) = \left(\frac{\partial W}{\partial \mathbf{p}}\right)^{\mathsf{T}} \frac{\partial I}{\partial \mathbf{x}} = \begin{bmatrix} xg_x & yg_x & g_x & 0 & 0 & 0 \\ 0 & 0 & 0 & xg_y & yg_y & g_y \end{bmatrix}^{\mathsf{T}} \tag{13.60}$$

where $\nabla I = \partial I / \partial \mathbf{x} = [g_x \quad g_y]^{\mathsf{T}}$, and therefore

$$\sum_{\mathbf{x} \in \mathcal{R}} (\mathbf{g}(\mathbf{x}))(\mathbf{g}(\mathbf{x}))^{\mathsf{T}} \mathbf{p}_\Delta = \sum_{\mathbf{x} \in \mathcal{R}} (I(\mathbf{x}) - J(\mathbf{x})) \mathbf{g}(\mathbf{x}) \tag{13.61}$$

13.3.5 Horn-Schunck Algorithm

Returning to the standard Lucas-Kanade algorithm, an alternate approach to computing pixel displacements is the **Horn-Schunck algorithm**, which is based upon the same optical flow constraint equation. Instead of computing the displacement of a sparse number of feature points, where all pixels in a window around the feature point are assumed to have

---

[†] If Line 9 were replaced by $\hat{\mathbf{p}} > \hat{\mathbf{p}} + \mathbf{p}_\Delta$, it would be the *inverse additive method*; in the case of translation only, the two are identical.

the *same* displacement (as in Lucas-Kanade), Horn-Schunck computes a dense optical flow throughout the image by imposing regularization to leverage the assumption that neighboring pixels in the image have *similar* displacements.

Because Horn-Schunck computes a value for every pixel, the displacements u and v are replaced by functions $u(x, y)$ and $v(x, y)$ defined for every pixel. Let us define $E_d(\cdot)$ as the standard left-hand side of the optical flow constraint equation:

$$E_d(x, y, u(x, y), v(x, y)) \equiv I_x(x, y)u(x, y) + I_y(x, y)v(x, y) + I_t(x, y) \qquad (13.62)$$

and let $E_s(\cdot)$ be a smoothness term over the first derivatives of the displacement functions:

$$E_s^2(x, y, u(x, y), v(x, y)) \equiv \left(\frac{\partial u(x, y)}{\partial x}\right)^2 + \left(\frac{\partial u(x, y)}{\partial y}\right)^2 + \left(\frac{\partial v(x, y)}{\partial x}\right)^2 + \left(\frac{\partial v(x, y)}{\partial y}\right)^2 \qquad (13.63)$$

The Horn-Schunck algorithm seeks to solve the following minimization problem:

$$\min_{u, v} \iint \left(E_d^2(x, y, u(x, y), v(x, y)) + \lambda E_s^2(x, y, u(x, y), v(x, y))\right) dx\, dy \qquad (13.64)$$

where λ is a weight governing the relative importance of the two terms.

It is straightforward to solve this minimization problem using the calculus of variations.[†] The two independent variables are x and y, the two dependent variables are $u = \partial x/\partial t$ and $v = \partial y/\partial t$, and the functional to be minimized is $\int E\, dx\, dy$, where $E = E_d^2 + \lambda E_s^2$. The Euler-Lagrange equations to be solved are therefore

$$\frac{\partial E}{\partial u} - \frac{\partial}{\partial x}\left[\frac{\partial E}{\partial(\frac{\partial u}{\partial x})}\right] - \frac{\partial}{\partial y}\left[\frac{\partial E}{\partial(\frac{\partial u}{\partial y})}\right] = 0 \qquad (13.65)$$

$$\frac{\partial E}{\partial v} - \frac{\partial}{\partial x}\left[\frac{\partial E}{\partial(\frac{\partial v}{\partial x})}\right] - \frac{\partial}{\partial y}\left[\frac{\partial E}{\partial(\frac{\partial v}{\partial y})}\right] = 0 \qquad (13.66)$$

Expanding E yields

$$E = E_d^2 + \lambda E_s^2 \qquad (13.67)$$

$$= (I_x u + I_y v + I_t)^2 + \lambda \left[\left(\frac{\partial u}{\partial x}\right)^2 + \left(\frac{\partial u}{\partial y}\right)^2 + \left(\frac{\partial v}{\partial x}\right)^2 + \left(\frac{\partial v}{\partial y}\right)^2\right] \qquad (13.68)$$

$$= I_x^2 u^2 + I_y^2 v^2 + I_t^2 + 2I_x I_y uv + 2I_x I_t u + 2I_y I_t v \qquad (13.69)$$

$$+ \lambda \left[\left(\frac{\partial u}{\partial x}\right)^2 + \left(\frac{\partial u}{\partial y}\right)^2 + \left(\frac{\partial v}{\partial x}\right)^2 + \left(\frac{\partial v}{\partial y}\right)^2\right] \qquad (13.70)$$

leading to the following derivatives:

$$\frac{\partial E}{\partial u} = 2I_x^2 u + 2I_x I_y v + 2I_x I_t \qquad (13.71)$$

$$\frac{\partial}{\partial x}\left[\frac{\partial E}{\partial(\frac{\partial u}{\partial x})}\right] = \frac{\partial}{\partial x}\left[2\lambda\left(\frac{\partial u}{\partial x}\right)\right] = 2\lambda\left(\frac{\partial^2 u}{\partial x^2}\right) \qquad (13.72)$$

[†] Section 10.2.1 (p. 453).

and similarly for $\partial E/\partial v$ and the other derivatives. Plugging these expressions back into Equations (13.65)–(13.66) yields

$$I_x^2 u + I_x I_y v + I_x I_t - \lambda \nabla^2 u = 0 \tag{13.73}$$

$$I_y^2 v + I_x I_y u + I_y I_t - \lambda \nabla^2 v = 0 \tag{13.74}$$

where $\nabla^2 u = \frac{\partial^2 u}{\partial x^2} + \frac{\partial^2 u}{\partial y^2}$ and $\nabla^2 v = \frac{\partial^2 v}{\partial x^2} + \frac{\partial^2 v}{\partial y^2}$ are the Laplacian of u and v, respectively. Rearranging into matrix form yields

$$\begin{bmatrix} I_x^2 & I_x I_y \\ I_x I_y & I_y^2 \end{bmatrix} \begin{bmatrix} u \\ v \end{bmatrix} = \begin{bmatrix} \lambda \nabla^2 u - I_x I_t \\ \lambda \nabla^2 v - I_y I_t \end{bmatrix} \tag{13.75}$$

which, if $\lambda = 0$, is the original underconstrained optical flow constraint equation for a single pixel. Recalling that in the discrete formulation, $\nabla^2 u \approx h(\bar{u} - u)$ for some normalization constant h,[†]

$$\begin{bmatrix} I_x^2 & I_x I_y \\ I_x I_y & I_y^2 \end{bmatrix} \begin{bmatrix} u \\ v \end{bmatrix} = \begin{bmatrix} \lambda h(\bar{u} - u) - I_x I_t \\ \lambda h(\bar{v} - v) - I_y I_t \end{bmatrix} \tag{13.76}$$

which, after rearranging and absorbing the normalization constant into λ, becomes

$$\begin{bmatrix} I_x^2 + \lambda & I_x I_y \\ I_x I_y & I_y^2 + \lambda \end{bmatrix} \begin{bmatrix} u \\ v \end{bmatrix} = \begin{bmatrix} \lambda \bar{u} - I_x I_t \\ \lambda \bar{v} - I_y I_t \end{bmatrix} \tag{13.77}$$

Solving this 2×2 system of equations is straightforward:

$$\begin{bmatrix} u \\ v \end{bmatrix} = \frac{1}{(I_x^2 + \lambda)(I_y^2 + \lambda) - I_x^2 I_y^2} \begin{bmatrix} (I_y^2 + \lambda)(\lambda \bar{u} - I_x I_t) - I_x I_y(\lambda \bar{v} - I_y I_t) \\ (I_x^2 + \lambda)(\lambda \bar{v} - I_y I_t) - I_x I_y(\lambda \bar{u} - I_x I_t) \end{bmatrix} \tag{13.78}$$

$$= \frac{1}{\lambda(\lambda + I_x^2 + I_y^2)} \begin{bmatrix} \lambda I_y^2 \bar{u} + \lambda^2 \bar{u} - I_x I_y^2 I_t - \lambda I_x I_t - \lambda I_x I_y \bar{v} + I_x I_y^2 I_t) \\ \lambda I_x^2 \bar{v} + \lambda^2 \bar{v} - I_x^2 I_y I_t - \lambda I_y I_t - \lambda I_x I_y \bar{u} + I_x^2 I_y I_t \end{bmatrix} \tag{13.79}$$

$$= \frac{1}{\lambda + I_x^2 + I_y^2} \begin{bmatrix} (\lambda + I_y^2)\bar{u} - I_x I_y \bar{v} - I_x I_t \\ (\lambda + I_x^2)\bar{v} - I_x I_y \bar{u} - I_y I_t \end{bmatrix} \tag{13.80}$$

$$= \frac{1}{\lambda + I_x^2 + I_y^2} \begin{bmatrix} (\lambda + I_x^2 + I_y^2)\bar{u} - I_x^2 \bar{u} - I_x I_y \bar{v} - I_x I_t \\ (\lambda + I_x^2 + I_y^2)\bar{v} - I_y^2 \bar{v} - I_x I_y \bar{u} - I_y I_t \end{bmatrix} \tag{13.81}$$

$$= \begin{bmatrix} \bar{u} \\ \bar{v} \end{bmatrix} + \frac{1}{\lambda + I_x^2 + I_y^2} \begin{bmatrix} -I_x^2 \bar{u} - I_x I_y \bar{v} - I_x I_t \\ -I_y^2 \bar{v} - I_x I_y \bar{u} - I_y I_t \end{bmatrix} \tag{13.82}$$

$$= \begin{bmatrix} \bar{u} \\ \bar{v} \end{bmatrix} - \frac{I_x \bar{u} + I_y \bar{v} + I_t}{\lambda + I_x^2 + I_y^2} \begin{bmatrix} I_x \\ I_y \end{bmatrix} \tag{13.83}$$

From this final equation, we see that the iterative Horn-Schunck approach involves repeatedly solving the following equations

$$u^{(k+1)}(\mathbf{x}) = \bar{u}^{(k)}(\mathbf{x}) - \gamma(\mathbf{x})I_x(\mathbf{x}) \tag{13.84}$$

$$v^{(k+1)}(\mathbf{x}) = \bar{v}^{(k)}(\mathbf{x}) - \gamma(\mathbf{x})I_y(\mathbf{x}) \tag{13.85}$$

[†] Section 5.4.1 (p. 242).

for all $\mathbf{x} = (x, y)$, where

$$\gamma(\mathbf{x}) \equiv \frac{I_x \bar{u}^{(k)}(\mathbf{x}) + I_y \bar{v}^{(k)}(\mathbf{x}) + I_t(\mathbf{x})}{\lambda + I_x^2(\mathbf{x}) + I_y^2(\mathbf{x})} \tag{13.86}$$

The pseudocode for Horn-Schunck is provided in Algorithm 13.10. The algorithm essentially solves a sparse $2n \times 2n$ linear system, where n is the number of pixels in the image, using one of several variations. If, at each iteration, the values for the next iteration are carefully stored in a different memory location, that is, if $u^{(k)}$ and $u^{(k+1)}$ are distinct arrays, and similarly $v^{(k)}$ and $v^{(k+1)}$, then the pseudocode implements what is known as the **Jacobi method**. On the other hand, if we are less careful and instead allow the new values to overwrite the old values as soon as they are computed, then the pseudocode implements the **Gauss-Seidel method**. Interestingly, although Gauss-Seidel can be thought of as "sloppy Jacobi," it is actually preferred in practice because it leads to faster convergence. Even faster convergence is obtained by making a copy of the values, running Gauss-Seidel, and combining the values from the previous iteration, $u^{(k)}$ and $v^{(k)}$, with the values resulting from Gauss-Seidel in the current iteration, $u_{GS}^{(k+1)}$ and $v_{GS}^{(k+1)}$, using a linear weighting scheme:

$$u^{(k+1)} = \omega u_{GS}^{(k+1)} + (1 - \omega) u^{(k)} \tag{13.87}$$

$$v^{(k+1)} = \omega v_{GS}^{(k+1)} + (1 - \omega) v^{(k)} \tag{13.88}$$

where $0 < \omega < 2$ to guarantee convergence; this approach is known as **successive over-relaxation (SOR)**, and it is easy to implement. Some results from a modern approach to optical flow that extends the basic Horn-Schunck algorithm to yield increased robustness can be found in Figure 13.19.

ALGORITHM 13.10 Horn-Schunck dense optical flow

HORN-SCHUNCK(I, λ)

Input: grayscale image I, weighting parameter λ
Output: dense optical flow images u and v

1 $I_x, I_y \leftarrow$ GRADIENT(I)
2 **for** $(x, y) \in I$ **do**
3      $u(x, y) \leftarrow 0$
4      $v(x, y) \leftarrow 0$
5 **repeat**
6   **for** $(x, y) \in I$ **do**
7          $\bar{u} \leftarrow \frac{1}{4}(u(x - 1, y) + u(x + 1, y) + u(x, y - 1) + u(x, y + 1))$
8          $\bar{v} \leftarrow \frac{1}{4}(v(x - 1, y) + v(x + 1, y) + v(x, y - 1) + v(x, y + 1))$
9          $\gamma \leftarrow \frac{I_x(x, y)\bar{u} + I_y(x, y)\bar{v} + I_t(x, y)}{\lambda + I_x(x, y)^2 + I_y(x, y)^2}$
10         $u(x, y) \leftarrow \bar{u} - \gamma I_x(x, y)$
11         $v(x, y) \leftarrow \bar{v} - \gamma I_y(x, y)$
12 **until** convergence

T. Brox and J. Malik. Large displacement optical flow: Descriptor matching in variational motion estimation. IEEE Transactions on Pattern Analysis and Machine Intelligence, 33(3):500–513, Mar. 2011

Figure 13.19 Optical flow computed on several image sequences, using a modern version of Horn-Schunck. The top row shows an image from each sequence, while the bottom row shows the optical flow displayed using pseudocolors.

13.4 Projective Geometry

In this section we briefly introduce projective geometry, which is a branch of mathematics that is particularly useful for 3D computer vision. This material provides more complete understanding of the concepts underlying the Euclidean, similarity, affine, and projective transformations introduced earlier;[†] and it provides a foundation for performing camera calibration and 3D reconstruction described later in the chapter.

13.4.1 Homogeneous Coordinates

At the heart of projective geometry is the concept of **homogeneous coordinates**. To understand homogeneous coordinates, two principles must be grasped:

- A point is represented in homogeneous coordinates by appending a 1 to the end.
- Once in homogeneous coordinates, scaling does not matter (as long as the scaling is nonzero).

To see these principles at work, suppose, for example, that (x, y) is a point in 2D. The homogeneous coordinates of this point are obtained by appending a 1 to the end, leading to $(x, y, 1)$. Then, since scaling does not matter, $(2x, 2y, 2)$ represents the exact same point as $(x, y, 1)$. In fact, so do both $(19.6x, 19.6y, 19.6)$ and $(\pi x, \pi y, \pi)$. More generally, $(x, y, 1) = (\lambda x, \lambda y, \lambda)$ for any $\lambda \neq 0$. These relationships can be captured using the proportionality symbol:

$$\begin{bmatrix} x \\ y \\ 1 \end{bmatrix} \propto \begin{bmatrix} 2x \\ 2y \\ 2 \end{bmatrix} \propto \begin{bmatrix} 19.6x \\ 19.6y \\ 19.6 \end{bmatrix} \propto \begin{bmatrix} \pi x \\ \pi y \\ \pi \end{bmatrix} \tag{13.89}$$

[†] Section 3.9 (p. 120).

Once in homogeneous coordinates, a **projective transformation** can be applied, which is simply a matrix multiplied by the point to yield new homogeneous coordinates. If we let (u, v, w), where $u = \lambda x$, $v = \lambda y$, and $w = \lambda$, represent the homogeneous coordinates of the original point (x, y), then the homogeneous coordinates of the transformed point are (u', v', w'), where

$$\begin{bmatrix} u' \\ v' \\ w' \end{bmatrix} = \mathbf{H}_{\{3 \times 3\}} \begin{bmatrix} u \\ v \\ w \end{bmatrix} \tag{13.90}$$

and $\mathbf{H}$ is the 3×3 projective transformation matrix known as a **homography**. Whereas the point (x, y) is in the **Euclidean plane**, both (u, v, w) and (u', v', w') are points in the **projective plane**. And just as the standard coordinates[†] in the Euclidean plane can be obtained from the homogeneous coordinates in the projective plane by dividing by the third coordinate, $(u/w, v/w) = (x, y)$, so the transformed point (x', y') is obtained by dividing by the third coordinate: $(u'/w', v'/w')$. These steps were considered earlier in Example 3.12, which we reproduce here for convenience.

| **EXAMPLE 13.2** | Apply the following projective transformation to the point $(x, y) = (1, 2)$: |

$$\mathbf{H} = \begin{bmatrix} 7 & 3 & 2 \\ 2 & 4 & 8 \\ 1 & 3 & 2 \end{bmatrix} \tag{13.91}$$

Solution

The homogeneous coordinates of the point are obtained by appending a 1 to $(1, 2)$, leading to $\begin{bmatrix} 1 & 2 & 1 \end{bmatrix}^{\mathsf{T}}$. Multiplying the matrix $\mathbf{H}$ by this vector yields:

$$\begin{bmatrix} u' \\ v' \\ w' \end{bmatrix} = \begin{bmatrix} 7 & 3 & 2 \\ 2 & 4 & 8 \\ 1 & 3 & 2 \end{bmatrix} \begin{bmatrix} 1 \\ 2 \\ 1 \end{bmatrix} = \begin{bmatrix} 15 \\ 18 \\ 9 \end{bmatrix} \propto \begin{bmatrix} \frac{15}{9} \\ \frac{18}{9} \\ 1 \end{bmatrix} = \begin{bmatrix} \frac{5}{3} \\ 2 \\ 1 \end{bmatrix} \tag{13.92}$$

so the transformed point is $(x', y') = \left(\frac{15}{9}, \frac{18}{9} \right) = \left(\frac{5}{3}, 2 \right)$. Note the proportionality symbol, which is used since scaling is unimportant.

These three steps are illustrated in Figure 13.20. The point $(x, y) = (1, 2)$ is represented as $(u, v, w) = (1, 2, 1)$, which is then transformed to $(u', v', w') = (15, 18, 9)$, which is then converted back to $(x', y') = (15/9, 2)$. As seen in the figure, the homogeneous coordinates of a 2D point can be visualized as a 3D point lying on the plane $w = 1$, where the coordinate axes are u, v, and w. Equivalently, since scaling does not matter, the coordinates can be normalized so that $u^2 + v^2 + w^2 = 1$, in which case the homogeneous coordinates are visualized as a point lying on the surface of a unit hemisphere.

Why would we ever want to use three coordinates to represent a 2D point? Before delving into the complicated world of projective geometry, let us motivate the approach by first considering three situations in which homogeneous coordinates are useful. First, as we saw earlier,[‡] homogeneous coordinates facilitate more compact mathematical notation for

---

[†] Also known as *inhomogeneous coordinates*.
[‡] Section 3.9.2 (p. 122).

Figure 13.20 Using projective geometry involves 3 steps: 1) augmenting the coordinates of a point by appending a 1 to the end; 2) applying a projective transformation, by multiplying a matrix by the augmented point; and 3) dividing by the final coordinate to yield the transformed point.

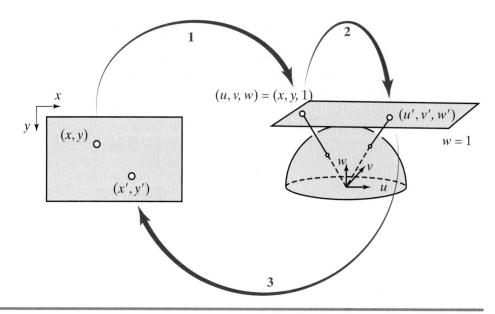

Euclidean transformations, which is especially handy when composing or inverting transformations. For example, if $\mathbf{p} \in \mathbb{R}^2$, $\mathbf{R}_1$ and $\mathbf{R}_2$ are 2×2 rotation matrices, and $\mathbf{t}_1$ and $\mathbf{t}_2$ are 2×1 translation vectors, then instead of writing the complicated expression $\mathbf{p}' = \mathbf{R}_2(\mathbf{R}_1\mathbf{p} + \mathbf{t}_1) + \mathbf{t}_2$, we can simply write $\mathbf{x}' = \mathbf{T}_2\mathbf{T}_1\mathbf{x}$, where $\mathbf{x}$ and $\mathbf{x}'$ are the homogeneous coordinates of $\mathbf{p}$ and $\mathbf{p}'$, respectively, and $\mathbf{T}_1$ and $\mathbf{T}_2$ are 3×3 linear transforms:

$$\mathbf{T}_1 \equiv \begin{bmatrix} \mathbf{R}_1 & \mathbf{t}_1 \\ \mathbf{0}_{\{2\times1\}}^{\mathsf{T}} & 1 \end{bmatrix} \quad \mathbf{T}_2 \equiv \begin{bmatrix} \mathbf{R}_2 & \mathbf{t}_2 \\ \mathbf{0}_{\{2\times1\}}^{\mathsf{T}} & 1 \end{bmatrix} \quad \mathbf{x} \equiv \begin{bmatrix} \mathbf{p} \\ 1 \end{bmatrix} \quad \mathbf{x}' \equiv \begin{bmatrix} \mathbf{p}' \\ 1 \end{bmatrix} \quad (13.93)$$

where $\mathbf{0}_{\{2\times1\}}^{\mathsf{T}}$ is a horizontal vector of two zeros in the bottom row. Here the scaling does not change, because as long as the bottom row consists of 0s followed by a single 1, the final coordinate will always be 1. Such transforms are known as *affine transforms*, of which Euclidean is a special case.

Secondly, homogeneous coordinates simplify common everyday computations. Consider, for example, a line with slope m and y-intercept k, so that any point (x, y) on the line satisfies $y = mx + k$. The problem with this representation is that it is not valid for vertical lines, which is easily solved by representing the point using homogeneous coordinates, $\mathbf{x} = (x, y, 1)$, and parameterizing the line using the triple $\ell = (a, b, c)$, so that any point on the line satisfies $\mathbf{x}^{\mathsf{T}}\ell = \ell^{\mathsf{T}}\mathbf{x} = ax + by + c = 0$. It is easy to see that scaling the point or the line does not affect the result, and that $m = -a/b$ and $k = -c/b$ if the line is not vertical. This representation is sufficient to handle any line, with $a = 0$ indicating a horizontal line, and $b = 0$ indicating one that is vertical.

Similarly, the parameters of a line passing through two points is computed easily as $\mathbf{x}_1 \times \mathbf{x}_2$, where $\times$ is the standard cross-product between two vectors in 3D. (Note that since scaling is unimportant, $\mathbf{x}_2 \times \mathbf{x}_1$ yields the same result, so order does not matter.) Similarly, computing the homogeneous coordinates of a point at the intersection of two lines is $\ell_1 \times \ell_2$ (or $\ell_2 \times \ell_1$). While these expressions may not be obvious at first glance, they are easy to verify. Moreover, the similarity between the expressions here reveals that there exists a **duality**[†] between points and lines in 2D projective geometry (and between points and planes in 3D), so that for any property that holds for a 3×1 vector representing the

---

[†] Recall the concept of duality from Section 4.1.2 (p. 134).

homogeneous coordinates of a point, there is a related property that holds for the same vector interpreted as the parameters of a line.

Finally, homogeneous coordinates simplify the mathematical representation of perspective projection. Recall from Equations (2.1)–(2.2)[†] that in the case of perspective projection, $x = \frac{f x_w}{z_w}$ and $y = \frac{f y_w}{z_w}$, where (x_w, y_w, z_w) are the coordinates of a world point in 3D, and (x, y) are the 2D coordinates of the projection onto the image plane. These are nonlinear equations, making them inconvenient for complicated mathematical analysis. A more compact notation is achieved by representing the points in 3D and 2D projective geometry, respectively, leading to the following linear transformation:

$$\lambda \begin{bmatrix} x \\ y \\ 1 \end{bmatrix} = \begin{bmatrix} f & 0 & 0 & 0 \\ 0 & f & 0 & 0 \\ 0 & 0 & 1 & 0 \end{bmatrix} \begin{bmatrix} x_w \\ y_w \\ z_w \\ 1 \end{bmatrix} \tag{13.94}$$

so that $\lambda x = f x_w$, $\lambda y = f y_w$, and $\lambda = z_w$. The problem of perspective projection will be revisited later in the chapter when we consider the camera parameters.

13.4.2 Points at Infinity (Ideal Points)

Since the projective transformation can be any invertible 3×3 matrix, for some input points the output will have 0 for the 3rd coordinate: that is, $w' = 0$. For example, plugging $(1, -1)$ into Equation (13.92) yields

$$\begin{bmatrix} u' \\ v' \\ w' \end{bmatrix} = \begin{bmatrix} 7 & 3 & 2 \\ 2 & 4 & 8 \\ 1 & 3 & 2 \end{bmatrix} \begin{bmatrix} 1 \\ -1 \\ 1 \end{bmatrix} = \begin{bmatrix} 6 \\ 6 \\ 0 \end{bmatrix} \tag{13.95}$$

When trying to convert such a point back to standard coordinates, a divide-by-zero error occurs: $(u'/0, v'/0)$. What this tells us is that some points that are represented by homogeneous coordinates cannot be represented by standard coordinates; in other words, the projective plane contains some points that are not in the Euclidean plane. Such points are known as **points at infinity** (or *ideal points*). Examining Figure 13.20 again, note that points at infinity satisfy $w = 0$ and therefore lie on the equator of the unit hemisphere, which means that the ray through them and the origin never intersects the $w = 1$ plane.

Since $\mathbf{x}^T \ell$ indicates whether a point lies on a line, it is easy to show that every point at infinity, $(x, y, 0)$, lies on a special line called the **line at infinity**, represented by $\ell_\infty = (0, 0, 1)$. As a result, the projective plane can also be visualized as the standard Euclidean plane augmented with a 1D ring around it representing ℓ_∞. As shown in Figure 13.21, the points at infinity can, in some sense, be considered "beyond infinity" because the plane inside the ring itself is infinite. One advantage of this visualization is that it clearly reveals that points at infinity represent *directions* in the plane; that is, the point $(x, y, 0)$ represents the direction from the origin through the point (x, y), so that the point $(1, m, 0)$ represents slope m. This is easily seen because horizontal lines are given by $(0, b, c)$, so that $\begin{bmatrix} 0 & b & c \end{bmatrix} \begin{bmatrix} 1 & 0 & 0 \end{bmatrix}^T = 0$, and vertical lines are given by $(a, 0, c)$, so $\begin{bmatrix} a & 0 & c \end{bmatrix} \begin{bmatrix} 0 & 1 & 0 \end{bmatrix}^T = 0$. Similarly, the intersection of two parallel lines $\ell_1 = (a, b, c_1)$ and $\ell_2 = (a, b, c_2)$ is given by $(1, -\frac{a}{b}, 0)$, which is the point at infinity associated with their direction.

[†] Section 2.2.3 (p. 35).

Figure 13.21 The projective plane can
be visualized as the Euclidean plane
augmented with ℓ_∞, the line at infinity.
Two parallel lines intersect at the point at
infinity (which lies on the line at infinity)
associated with the direction of the lines.

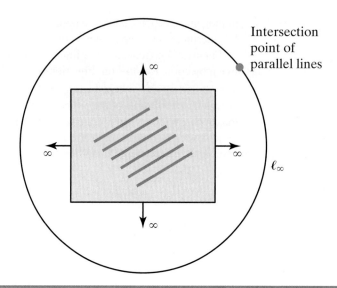

13.4.3 Conics

One of the first principles learned in a drawing class is that a circle in the world is drawn
as an ellipse on the canvas because the perspective projection of a circle is an ellipse. In
fact, it can be shown that all *conic sections* (circles, ellipses, parabolas, and hyperbolas) are
projectively equivalent, meaning that each one can be obtained from the other by a projective transformation. In contrast, the Euclidean transformation of a circle is always a circle,
since Euclidean transformations can only perform rotation and translation.

A conic section is parameterized by a 3×3 symmetric matrix $\mathbf{C}$, as we saw in
Equation (11.55):

$$[x \; y \; 1] \underbrace{\begin{bmatrix} 2a & b & d \\ b & 2c & e \\ d & e & 2f \end{bmatrix}}_{\mathbf{C}} \begin{bmatrix} x \\ y \\ 1 \end{bmatrix} = 0 \qquad (13.96)$$

which represents the conic section $ax^2 + bxy + cy^2 + dx + ey + f = 0$. In other words,
$\mathbf{x}^T \mathbf{C} \mathbf{x} = 0$, where $\mathbf{x} = (x, y, 1)$ are the homogeneous coordinates of the point (x, y). In
projective geometry, a curve represented by Equation (13.96) is simply called a **conic**.

Geometrically, a conic can be considered as a locus of points, as we have just
done, or as the envelope of tangent lines (that is, the set of lines that are tangent to the
conic). Given a point $\mathbf{x}$ on a conic $\mathbf{C}$, the tangent line to $\mathbf{C}$ at $\mathbf{x}$ is given by $\ell = \mathbf{C}\mathbf{x}$.
Plugging $\mathbf{x} = \mathbf{C}^{-1}\ell$ into $\mathbf{x}^T\mathbf{C}\mathbf{x} = 0$ yields the expression for this **dual conic $\mathbf{C}^*$**:
$(\mathbf{C}^{-1}\ell)^T\mathbf{C}(\mathbf{C}^{-1}\ell) = \ell^T\mathbf{C}^{-T}\mathbf{C}\mathbf{C}^{-1}\ell = \ell^T\mathbf{C}^{-1}\ell = 0$, since the matrix is symmetric.
Therefore, the dual conic of a full-rank matrix is given by $\mathbf{C}^* = \mathbf{C}^{-1}$. (On the other hand,
if $\mathbf{C}$ is singular—and therefore the conic is degenerate—then $\mathbf{C}^* = \text{adj}(\mathbf{C})$, the *adjoint*
of the matrix. It is well-known that $\text{adj}(\mathbf{C}) = \det(\mathbf{C})\mathbf{C}^{-T}$ when $\mathbf{C}$ is full rank, which is
equivalent to $\mathbf{C}^{-T}$ for our purposes, since the matrix is symmetric and scaling is unimportant.
Regardless of rank, the adjoint always satisfies $\text{adj}(\mathbf{C})\mathbf{C} = \mathbf{C}\,\text{adj}(\mathbf{C}) = \det(\mathbf{C})\mathbf{I}_{\{3\times3\}}$,
where $\mathbf{I}_{\{3\times3\}}$ is the 3×3 identity matrix.)

13.4.4 Transformations of Lines and Conics

If a matrix $\mathbf{H}$ transforms the point $\mathbf{x}$ into $\mathbf{x}' = \mathbf{H}\mathbf{x}$, then the same matrix also transforms lines and conics in a similar way:

$$\ell' = \mathbf{H}^{-\mathsf{T}}\ell \tag{13.97}$$

$$\mathbf{C}' = \mathbf{H}^{-\mathsf{T}}\mathbf{C}\mathbf{H}^{-1} \tag{13.98}$$

These relationships are easy to derive. If $\mathbf{x}$ lies on a line ℓ, then $\mathbf{x}^{\mathsf{T}}\ell = 0$. Substituting $\mathbf{x} = \mathbf{H}^{-1}\mathbf{x}'$ yields $(\mathbf{H}^{-1}\mathbf{x}')^{\mathsf{T}}\ell = \mathbf{x}'^{\mathsf{T}}(\mathbf{H}^{-\mathsf{T}}\ell) = 0$, which means that $\mathbf{x}'$ lies on the line given by $\mathbf{H}^{-\mathsf{T}}\ell$. Similarly, substituting $\mathbf{x} = \mathbf{H}^{-1}\mathbf{x}'$ into $\mathbf{x}^{\mathsf{T}}\mathbf{C}\mathbf{x} = 0$ yields $(\mathbf{H}^{-1}\mathbf{x}')^{\mathsf{T}}\mathbf{C}(\mathbf{H}^{-1}\mathbf{x}') = \mathbf{x}'^{\mathsf{T}}(\mathbf{H}^{-\mathsf{T}}\mathbf{C}\mathbf{H}^{-1})\mathbf{x}' = 0$.

13.4.5 Hierarchy of Transformations

A hierarchy of transformations exists. As we have seen, a general invertible matrix $\mathbf{H}$ is a **projective transformation**. If the bottom row of $\mathbf{H}$ consists of all zeros followed by a single 1, then it is also an **affine transformation**, which is special case of projective. Furthermore, if the top-left corner consists of a rotation matrix with an overall scaling, then it is known as a **similarity transformation**, which is a special case of affine. Finally, if this scaling is 1, then it is a **Euclidean transformation**, which is a special case of similarity. These transformations, which we saw earlier,[†] are summarized as follows for the case of 2D:

$$\begin{bmatrix} h_{11} & h_{12} & h_{13} \\ h_{21} & h_{22} & h_{23} \\ h_{31} & h_{32} & h_{33} \end{bmatrix} \quad \begin{bmatrix} h_{11} & h_{12} & h_{13} \\ h_{21} & h_{22} & h_{23} \\ 0 & 0 & 1 \end{bmatrix} \quad \begin{bmatrix} kc_\theta & -ks_\theta & t_x \\ ks_\theta & kc_\theta & t_y \\ 0 & 0 & 1 \end{bmatrix} \quad \begin{bmatrix} c_\theta & -s_\theta & t_x \\ s_\theta & c_\theta & t_y \\ 0 & 0 & 1 \end{bmatrix}$$
$$\underbrace{}_{\text{projective}} \qquad \underbrace{}_{\text{affine}} \qquad \underbrace{}_{\text{similarity}} \qquad \underbrace{}_{\text{Euclidean}}$$

where h_{ij} is an arbitrary scalar, $c_\theta = \cos\theta$, $s_\theta = \sin\theta$, t_x and t_y are translation parameters, and k is a nonzero scaling factor.

As the number of allowed transformations increases, the number of invariants decreases, as summarized in Table 13.1. For example, a Euclidean transformation preserves not only the overall shape of an object, but also its size, whereas a similarity transformation preserves

	transformations							invariants					
	rotation	translation	uniform scaling	nonuniform scaling	shear	perspective projection	projective transform	length	angle	length ratio	parallelism	incidence	cross ratio
Euclidean	•	•						•	•	•	•	•	•
similarity	•	•	•						•	•	•	•	•
affine	•	•	•	•	•						•	•	•
projective	•	•	•	•	•	•	•					•	•

TABLE 13.1 Hierarchy of transformations, showing that as the number of transformations increases, the number of invariants decreases. (*Incidence* refers to whether a point lies on a line.)

[†] Section 3.9 (p. 120).

the shape but not necessarily the size, and an affine or projective transformation does not even preserve the shape. More specifically, Euclidean transformations preserve the length of each line segment, similarity transformations preserve the ratio between lengths of any pair of line segments, and affine and projective transformations preserve the **cross ratio**, which is a ratio of ratios of lengths. As a result, the Euclidean transformation of a square is another square of the same size, the similarity transformation is another square not necessarily of the same size, the affine transformation is a parallelogram, and the projective transformation is a quadrilateral.

13.4.6 Absolute Points

An interesting property of conics is that every circle intersects ℓ_∞ at two fixed points, known as the **absolute points**. Recalling the equation for a circle in Equation (11.42), $(x - h)^2 + (y - k)^2 = r^2$, and that a point on ℓ_∞ is parameterized by $(x, y, 0)$, it is easy to see that the absolute points satisfy

$$\begin{bmatrix} x & y & 0 \end{bmatrix} \begin{bmatrix} 1 & 0 & -h \\ 0 & 1 & -k \\ -h & -k & h^2 + k^2 - r^2 \end{bmatrix} \begin{bmatrix} x \\ y \\ 0 \end{bmatrix} = x^2 + y^2 = 0 \qquad (13.99)$$

from which we note that the absolute points are $(1, \pm j, 0),$[†] where $j \equiv \sqrt{-1}$. The absolute points solve the curious phenomenon illustrated in Figure 13.22, namely, that although two overlapping ellipses intersect at 4 points, two overlapping circles intersect at only 2 points, even though a circle is an ellipse. In other words, where did the other two intersection points go? The answer is that every pair of circles also intersect at the absolute points, since the absolute points lie on every circle.

The absolute points can be visualized by plotting them in the complex plane, as in Figure 13.23, from which it is immediately obvious that they are orthogonal to and the same length as each other. As a result, they play the role of orthogonal unit axes (except their overall scaling is not fixed), which explains why they are important in converting from projective to Euclidean (that is, to similarity, since there remains an unknown scaling factor). It is easy to show that the absolute points are invariant to similarity transforms:

$$\begin{bmatrix} kc_\theta & -ks_\theta & t_x \\ ks_\theta & kc_\theta & t_y \\ 0 & 0 & 1 \end{bmatrix} \begin{bmatrix} 1 \\ \pm j \\ 0 \end{bmatrix} = \begin{bmatrix} kc_\theta \mp jks_\theta \\ ks_\theta \pm jkc_\theta \\ 0 \end{bmatrix} \propto \begin{bmatrix} 1 \\ \pm j \\ 0 \end{bmatrix} \qquad (13.100)$$

Figure 13.22 Two ellipses can intersect at four points (at most), but two circles can only intersect at two points (at most). Where are the other two intersection points?

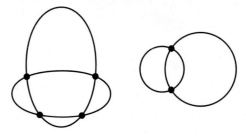

[†] Note that the absolute points could equivalently be represented as $(\mp j, 1, 0)$ by scaling the coordinates.

Figure 13.23 When plotted in the complex plane, the geometric properties of the absolute points are manifest; they are orthogonal to one another and share the same length. As a result, they capture the notions of angle and relative length.

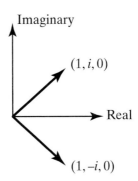

where the proportionality follows from the fact that scaling is not important, so we can divide the second coordinate by the first coordinate:

$$\frac{ks_\theta \pm jkc_\theta}{kc_\theta \mp jks_\theta} = \frac{(s_\theta \pm jc_\theta)(c_\theta \pm js_\theta)}{(c_\theta \mp js_\theta)(c_\theta \pm js_\theta)} = \frac{s_\theta c_\theta \pm jc_\theta^2 \pm js_\theta^2 - s_\theta c_\theta}{c_\theta^2 + s_\theta^2} = \pm j \quad (13.101)$$

13.4.7 Projective Geometry in Other Dimensions

Although we have concentrated on the 2D case (projective plane) for ease of presentation, projective geometry is applicable in any number of dimensions. That is, a point in $\mathbb{R}^d$ is represented in homogeneous coordinates using $d + 1$ elements. Therefore, as we have just seen, a projective transformation in 2D is represented by a 3×3 transformation matrix, and points and lines are duals, whereas a projective transformation in 3D is represented by a 4×4 matrix, and points and planes are duals. Projecting from 3D to 2D, as described in the next section, involves a 3×4 matrix.

13.4.8 Perspective Imaging

Assuming a pinhole camera with no diffraction, light rays travel in straight lines, in which case a point (x_w, y_w, z_w) in the 3D world projects onto a point (x, y) on the image plane by following the ray that connects the point with the center of projection, also known as the focal point. In Equation (13.94) we saw an equation for this projection that is oversimplified because it ignores two facts: (1) the coordinate system for measuring world points is usually not attached to the camera, and (2) the units on the image are different from those in the world.

As shown in Figure 13.24, three coordinate systems are involved in the process of imaging: one attached to the world, one attached to the camera, and one associated with the image. Points in the world are described in the *world coordinate system*, in which lengths are measured in meters. In the *camera coordinate system*, measurements are also in meters (or micrometers), since the camera and world coordinate systems are related by a Euclidean transformation. The camera coordinate system is, by convention, centered at the focal point, with the x and y axes aligned with the rightward horizontal and downward vertical directions, respectively, of the image plane, and the positive z axis pointing along the optical axis toward the world.[†] The *image coordinate system* is centered at the top left corner of the image, with the positive x and y axes pointing along the rows and columns, respectively, of the imaging sensor. In the image coordinate system, measurements are made in pixels.

[†] Recall from Section 2.2.3 (p. 35) that the optical axis is defined as the line passing through the focal point that is perpendicular to the image plane.

Figure 13.24 The projection of a world point (x_w, y_w, z_w) onto an image plane at point (x, y), assuming a pinhole camera model with no diffraction. The three coordinate systems are the world coordinate system (W), the camera coordinate system (C), and the image coordinate system (I).

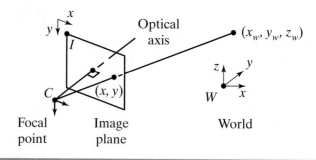

Using homogeneous coordinates, the imaging process is captured mathematically as

$$\begin{bmatrix} x \\ y \\ 1 \end{bmatrix} \propto \underbrace{\begin{bmatrix} p_{11} & p_{12} & p_{13} & p_{14} \\ p_{21} & p_{22} & p_{23} & p_{24} \\ p_{31} & p_{32} & p_{33} & p_{34} \end{bmatrix}}_{\mathbf{P}_{\{3 \times 4\}}} \begin{bmatrix} x_w \\ y_w \\ z_w \\ 1 \end{bmatrix} \qquad (13.102)$$

where $\mathbf{P}$ is a 3×4 projection matrix that itself is composed of two parts:

$$\mathbf{P}_{\{3 \times 4\}} = \underbrace{\begin{bmatrix} \alpha & \gamma & u_0 \\ 0 & \beta & v_0 \\ 0 & 0 & 1 \end{bmatrix}}_{\mathbf{K}_{\{3 \times 3\}}} \begin{bmatrix} \mathbf{R}_{\{3 \times 3\}} & \mathbf{t}_{\{3 \times 1\}} \end{bmatrix} \qquad (13.103)$$

where the rotation matrix $\mathbf{R}$ and translation vector $\mathbf{t}$ relate the world and camera coordinate systems, and the internal calibration matrix $\mathbf{K}$ of the camera not only performs perspective projection but also converts meters to pixels. The coordinates (u_0, v_0) specify the **principal point** (the intersection of the optical axis and the image plane); and α, β, and γ are related to the focal length f_x along the x axis, the focal length f_y along the y axis, and the **skew** θ (the angle between the two axes) as follows:

$$\alpha = f_x$$
$$\beta = f_y / \sin \theta$$
$$\gamma = -f_x / \tan \theta$$

For a real, physical camera, it is safe to assume $\theta = \frac{\pi}{2}$, so that $\beta = f_y$, and $\gamma = 0$, leading to

$$\mathbf{K} = \begin{bmatrix} f_x & 0 & u_0 \\ 0 & f_y & v_0 \\ 0 & 0 & 1 \end{bmatrix} \qquad (13.104)$$

The focal lengths f_x and f_y can alternatively be replaced by a single focal length f and the aspect ratio f_y / f_x. For most cameras, $f_x \approx f_y$ within a tolerance of about 5%. One might be tempted to think that the principal point is in the center of the image, but because of the camera manufacturing process, it can often be as much as 10% away from the center, which can be as much as dozens (or even hundreds) of pixels, depending upon the image dimensions. Nevertheless, assuming that the principal point is in the center does not cause much damage because reprojection equations tend to be insensitive to the location of the principal point.

13.4.9 Lens Distortion

Because real cameras have lenses, light does not travel in straight lines but rather bends due to the curvature in the lens. By far, the dominant distortion of a typical lens is **radial distortion**, so named because it is a function only of the radial distance from the center of the image. Let (x_u, y_u) be the undistorted coordinates of a pixel in the image (that is, the coordinates at which the ray of light would have pierced the image plane had there been no lens distortion) and let (x_d, y_d) be the distorted coordinates (that is, the actual coordinates). Define $r_d \equiv \sqrt{\bar{x}_d^2 + \bar{y}_d^2}$ as the radial distance to the pixel, where $\bar{x}_d \equiv x_d - u_0$ and $\bar{y}_d \equiv y_d - v_0$. Then the transformation from the distorted coordinates to the undistorted coordinates can be modeled by the following approximation:

$$x_u = x_d + \bar{x}_d f(r_d) \tag{13.105}$$

$$y_u = y_d + \bar{y}_d f(r_d) \tag{13.106}$$

where

$$f(r_d) = k_1 r_d^2 + k_2 r_d^4 + k_3 r_d^6 + \cdots \tag{13.107}$$

and only even terms are retained due to symmetry in the distortion. Once the coefficients $k_1, k_2, k_3, \ldots$, have been determined, the image can be unwarped by computing, for each pixel in the undistorted image, the corresponding pixel in the distorted image from which to grab the pixel data.[†]

If a more accurate model is desired, **tangential distortion** (or *decentering distortion*) can be included:

$$x_u = x_d + \bar{x}_d f(r_d) + \left(p_1(r_d^2 + 2\bar{x}_d^2) + 2p_2 \bar{x}_d \bar{y}_d\right)(1 + p_3 r_d^2 + \cdots) \tag{13.108}$$

$$y_u = y_d + \bar{y}_d f(r_d) + \left(2p_1 \bar{x}_d \bar{y}_d + p_2(r_d^2 + 2\bar{y}_d^2)\right)(1 + p_3 r_d^2 + \cdots) \tag{13.109}$$

in which case the parameters include not only k_i but also p_i, $i = 1, 2, 3, \ldots$. However, keep in mind that the larger the number of parameters, the more difficult it is to accurately estimate those parameters, so oftentimes better results will be achieved with a simpler model. In fact, many lenses can be well modeled just using k_1 and k_2.

13.5 Camera Calibration

Calibrating a pinhole camera involves estimating 11 parameters because the matrix $\mathbf{P}$ contains 12 elements, but it is unique only up to an unknown nonzero scaling factor. These numbers can be further broken down as follows: 6 values for the Euclidean rotation and translation (3 values each), and 5 values for the internal calibration matrix $\mathbf{K}$. These are called the **extrinsic** and **intrinsic parameters**, respectively. For a physical camera, $\gamma = 0$ and $f_x \approx f_y$, so that the intrinsic parameters can be reduced to just 3, of which the focal length is by far the most important. Nevertheless, the additional parameters are useful for the sake of completeness, and they are important when studying projections of projections. Whereas the extrinsic parameters depend upon the pose of the camera with respect to the world, the intrinsic parameters do not change as the camera moves, unless the lens has a variable focal length.

[†] This is an inverse mapping, as in Section 3.1.1 (p. 69).

13.5.1 Normalized Direct Linear Transform (DLT) Algorithm

Before we delve into calibration *per se*, let us first address an important and related problem: namely, estimating the homography, given corresponding points on two planes. Usually one of these planes is the image plane and the other is some plane in the world. The most straightforward and widely used method for homography estimation is known as the **direct linear transformation (DLT)**, which solves a least squares solution given the point coordinates. Let $(x_i, y_i) \leftrightarrow (x_i', y_i')$ be the ith corresponding pair of points between the two planes, and let $\mathbf{H}$ be the 3×3 homography matrix. Then $\mathbf{x}_i' \propto \mathbf{H}\mathbf{x}_i$, where $\mathbf{x}_i = \begin{bmatrix} x_i & y_i & 1 \end{bmatrix}^T$ and $\mathbf{x}_i' = \begin{bmatrix} x_i' & y_i' & 1 \end{bmatrix}^T$. Let us define $\mathbf{h}_j^T$ as the jth row of $\mathbf{H}$, $j = 1, 2, 3$:

$$\mathbf{H} = \begin{bmatrix} \mathbf{h}_1^T \\ \mathbf{h}_2^T \\ \mathbf{h}_3^T \end{bmatrix} = \begin{bmatrix} h_{11} & h_{12} & h_{13} \\ h_{21} & h_{22} & h_{23} \\ h_{31} & h_{32} & h_{33} \end{bmatrix} \tag{13.110}$$

With this definition, we can rewrite $\mathbf{x}_i' \propto \mathbf{H}\mathbf{x}_i$ as

$$\mathbf{x}_i' = \begin{bmatrix} x_i' \\ y_i' \\ 1 \end{bmatrix} \propto \begin{bmatrix} \mathbf{h}_1^T \mathbf{x}_i \\ \mathbf{h}_2^T \mathbf{x}_i \\ \mathbf{h}_3^T \mathbf{x}_i \end{bmatrix} \tag{13.111}$$

Of these three equations, only two are independent, because of the scaling ambiguity of homogeneous coordinates. Assuming that $\mathbf{x}_i'$ is a real point, i.e., $\mathbf{h}_3^T \mathbf{x}_i \neq 0$, these two equations are

$$x_i' = \frac{\mathbf{h}_1^T \mathbf{x}_i}{\mathbf{h}_3^T \mathbf{x}_i} = \frac{\mathbf{x}_i^T \mathbf{h}_1}{\mathbf{x}_i^T \mathbf{h}_3} \tag{13.112}$$

$$y_i' = \frac{\mathbf{h}_2^T \mathbf{x}_i}{\mathbf{h}_3^T \mathbf{x}_i} = \frac{\mathbf{x}_i^T \mathbf{h}_2}{\mathbf{x}_i^T \mathbf{h}_3} \tag{13.113}$$

where we have taken advantage of the fact that the transpose of a scalar does not change the scalar. Rearranging terms, these two equations can be rewritten as

$$x_i' \mathbf{x}_i^T \mathbf{h}_3 - \mathbf{x}_i^T \mathbf{h}_1 = 0 \tag{13.114}$$

$$y_i' \mathbf{x}_i^T \mathbf{h}_3 - \mathbf{x}_i^T \mathbf{h}_2 = 0 \tag{13.115}$$

or, in matrix form,

$$\begin{bmatrix} -\mathbf{x}_i^T & \mathbf{0}_{\{3 \times 1\}}^T & x_i' \mathbf{x}_i^T \\ \mathbf{0}_{\{3 \times 1\}}^T & -\mathbf{x}_i^T & y_i' \mathbf{x}_i^T \end{bmatrix} \begin{bmatrix} \mathbf{h}_1 \\ \mathbf{h}_2 \\ \mathbf{h}_3 \end{bmatrix} = \mathbf{0}_{\{9 \times 1\}} \tag{13.116}$$

where $\mathbf{0}_{\{3 \times 1\}}^T$ is the transpose of a vector of 3 zeros, and so forth. There are eight unknowns in these two equations, since $\mathbf{H}$ is only unique up to a scale factor. To solve for $\mathbf{H}$, then, we need at least four corresponding points, with additional points increasing robustness to measurement noise. Stacking these expressions into the rows of a matrix, we have

$$
\underbrace{\begin{bmatrix}
-\mathbf{x}_1^\mathsf{T} & \mathbf{0}_{\{3\times 1\}}^\mathsf{T} & x_1'\mathbf{x}_1^\mathsf{T} \\
\mathbf{0}_{\{3\times 1\}}^\mathsf{T} & -\mathbf{x}_1^\mathsf{T} & y_1'\mathbf{x}_1^\mathsf{T} \\
 & \vdots & \\
-\mathbf{x}_n^\mathsf{T} & \mathbf{0}_{\{3\times 1\}}^\mathsf{T} & x_n'\mathbf{x}_n^\mathsf{T} \\
\mathbf{0}_{\{3\times 1\}}^\mathsf{T} & -\mathbf{x}_n^\mathsf{T} & y_n'\mathbf{x}_n^\mathsf{T}
\end{bmatrix}}_{\mathbf{A}_{\{2n\times 9\}}}
\begin{bmatrix} \mathbf{h}_1 \\ \mathbf{h}_2 \\ \mathbf{h}_3 \end{bmatrix} = \mathbf{0}_{\{2n\times 1\}} \tag{13.117}
$$

where n is the number of correspondences. As we have seen already,[†] this equation can be solved for the elements of $\mathbf{H}$ by selecting the eigenvector of $\mathbf{A}^\mathsf{T}\mathbf{A}$ associated with the smallest eigenvalue, then rearranging into matrix form.

The procedure just described is not adequate for real-world use, because the matrix $\mathbf{A}^\mathsf{T}\mathbf{A}$ is nearly always ill-conditioned. Therefore, just as we saw earlier,[‡] normalization is required to achieve accurate results. Normalization involves applying a similarity transform to both point sets by shifting their respective centroids to the origin and normalizing so that the average distance of a point from the origin is $\sqrt{2}$. The reason for this number is that we want the typical point $\mathbf{x}_i$ or $\mathbf{x}_i'$ to have equal weight in its elements, which is achieved with the homogeneous vector $\begin{bmatrix} 1 & 1 & 1 \end{bmatrix}^\mathsf{T}$ that is $\sqrt{2}$ from the origin at $\begin{bmatrix} 0 & 0 & 1 \end{bmatrix}^\mathsf{T}$.

The similarity transforms are therefore given by

$$
\mathbf{T} = \begin{bmatrix} s & 0 & -sx_0 \\ 0 & s & -sy_0 \\ 0 & 0 & 1 \end{bmatrix} \quad \text{and} \quad \mathbf{T}' = \begin{bmatrix} s' & 0 & -sx_0' \\ 0 & s' & -sy_0' \\ 0 & 0 & 1 \end{bmatrix} \tag{13.118}
$$

where

$$
(x_0, y_0) = \frac{1}{n}\sum_{i=1}^{n}(x_i, y_i) \qquad (x_0', y_0') = \frac{1}{n}\sum_{i=1}^{n}(x_i', y_i') \tag{13.119}
$$

$$
s = \sqrt{2}/d_{avg}, \qquad s' = \sqrt{2}/d_{avg}' \tag{13.120}
$$

$$
d_{avg} = \frac{1}{n}\sum_{i=1}^{n}\sqrt{(x_i - x_0)^2 + (y_i - y_0)^2} \quad d_{avg}' = \frac{1}{n}\sum_{i=1}^{n}\sqrt{(x_i' - x_0')^2 + (y_i' - y_0')^2} \tag{13.121}
$$

Applying these transforms leads to the normalized coordinates $\check{\mathbf{x}} = \mathbf{T}\mathbf{x}$ and $\check{\mathbf{x}}' = \mathbf{T}'\mathbf{x}'$. Substituting into $\mathbf{x}' = \mathbf{H}\mathbf{x}$ reveals

$$
\mathbf{x}' = \mathbf{H}\mathbf{x} \tag{13.122}
$$

$$
\mathbf{T}'^{-1}\check{\mathbf{x}}' = \mathbf{H}\mathbf{T}^{-1}\check{\mathbf{x}} \tag{13.123}
$$

$$
\check{\mathbf{x}}' = \underbrace{\mathbf{T}'\mathbf{H}\mathbf{T}^{-1}}_{\check{\mathbf{H}}}\check{\mathbf{x}} \tag{13.124}
$$

which means that the homography using the normalized coordinates is related to the desired homography by $\check{\mathbf{H}} = \mathbf{T}'\mathbf{H}\mathbf{T}^{-1}$. In other words, after the transforms are applied to the points to yield the normalized points, and after the homography of the normalized points has been

---

[†] Section 11.1.3 (p. 516).

[‡] Section 11.1.2 (p. 515).

ALGORITHM 13.11 Normalized DLT algorithm for estimating a homography

NORMALIZEDDIRECTLINEARTRANSFORM $\left(\{\mathbf{x}_i \leftrightarrow \mathbf{x}_i'\}_{i=1}^n\right)$

Input: n corresponding pairs of points $(x_i, y_i) \leftrightarrow (x_i', y_i')$ between two planes
Output: The 3×3 homography matrix that best satisfies $\mathbf{x}_i' \propto \mathbf{H}\mathbf{x}_i$ for all i

1 Compute $\mathbf{T}$ and $\mathbf{T}'$ using Equation (13.118)
2 Normalize the points: $\check{\mathbf{x}}_i \leftarrow \mathbf{T}\mathbf{x}_i$, $\check{\mathbf{x}}_i' \leftarrow \mathbf{T}'\mathbf{x}_i'$, $i = 1, \ldots, n$
3 Construct the $2n \times 9$ matrix $\check{\mathbf{A}}$ in Equation (13.117) using the normalized points
4 Solve the linear system $\check{\mathbf{A}}\check{\mathbf{h}} = \mathbf{0}$ for $\check{\mathbf{h}}$
5 Rearrange the values in $\check{\mathbf{h}}$ to create $\check{\mathbf{H}}$
6 Unnormalize the result to yield the desired homography: $\mathbf{H} \leftarrow \mathbf{T}'^{-1}\check{\mathbf{H}}\mathbf{T}$.
7 **return H**

found via a least-squares solution, the desired homography is then computed as $\mathbf{H} = \mathbf{T}'^{-1}\mathbf{H}\mathbf{T}$. The pseudocode is provided in Algorithm 13.11. Note that since the least-squares approach minimizes the algebraic error, it is often a good idea to follow this entire procedure with a nonlinear minimization to minimize the geometric error.[†]

13.5.2 Mosaicking

One application of the normalized DLT is to stitch together multiple photographs of a scene into a single larger image. The resulting image is known as a **mosaic**, and the process itself is known as **mosaicking**,[‡] a term borrowed from the traditional art form in which individual pieces of stone, tile, or glass are arranged to create a larger image. For the normalized DLT algorithm to be applicable, the pictures must have been taken from approximately the same vantage point (relative to the distance to the object of interest), so that there is little parallax between the images.

Consider, for example, the pictures in Figure 13.25 taken of a building by a person standing far away from the building. The procedure for creating a mosaic is the following. First, corresponding points are found between pairs of images, by either manually clicking

Figure 13.25 Collection of 12 images of a building from nearly the same viewpoint.

Stan Birchfield

[†] The distinction between algebraic error and geometric error is explained in Section 11.2 (p. 523).
[‡] Or *mosaicing*.

on the images, or using an automated process such as SIFT feature point detection and matching.[†] These correspondences are then fed to the normalized DLT algorithm to determine the homographies between the images, which are then warped according to the homographies to yield images that are all in the same coordinate system. Finally, the images are blended together in the locations where they overlap by weighting the colors according to their distance from the image border, a process known as **feathering**, which reduces the effects of the seams between images. The result of a semi-automated process that is easily implemented using these steps is shown in Figure 13.26, but note that fully automatic mosaicking in a robustness manner requires attention to many details, and homographies alone may not be sufficient when the field of view is large.

13.5.3 Zhang's Calibration Algorithm

Returning to the topic of calibration, the most popular technique, by far, for calibrating the intrinsic camera parameters is **Zhang's algorithm**. Compared with alternate approaches, this technique has the advantage that it only requires a planar calibration target with known coordinates on the plane (e.g., a chessboard, or checkerboard), as shown in Figure 13.27. Multiple images (6 to 10 are usually sufficient) are captured of the target at different positions and orientations. The positions of the points in the images, along with their known locations on the target, are then used to estimate the intrinsic calibration matrix $\mathbf{K}$ using the technique explained in this section. (The extrinsic parameters—namely, rotation and translation of the camera relative to the target for each capture—are also estimated, but these are usually not of interest.)

The world coordinate system is affixed to the planar target so that the plane is at $z_w = 0$, and the image coordinate system is attached to the image plane as usual. As illustrated in Figure 13.28, the projection of a target point $(x_w, y_w, 0)$ onto the image plane is given by (x, y), where

$$\begin{bmatrix} x \\ y \\ 1 \end{bmatrix} \propto \underbrace{\begin{bmatrix} \alpha & \gamma & u_0 \\ 0 & \beta & v_0 \\ 0 & 0 & 1 \end{bmatrix}}_{\mathbf{K}_{\{3 \times 3\}}} \underbrace{\begin{bmatrix} \mathbf{r}_1 & \mathbf{r}_2 & \mathbf{r}_3 & \mathbf{t} \end{bmatrix}}_{\begin{bmatrix} \mathbf{R}_{\{3 \times 3\}} & \mathbf{t}_{\{3 \times 1\}} \end{bmatrix}} \begin{bmatrix} x_w \\ y_w \\ 0 \\ 1 \end{bmatrix} = \underbrace{\mathbf{K} \begin{bmatrix} \mathbf{r}_1 & \mathbf{r}_2 & \mathbf{t} \end{bmatrix}}_{\mathbf{H}_{\{3 \times 3\}}} \begin{bmatrix} x_w \\ y_w \\ 1 \end{bmatrix} \qquad (13.125)$$

Figure 13.26 Mosaic created by warping and stitching the images from the previous figure together.

Stan Birchfield

[†] Section 7.4.6 (p. 347) and Section 7.5.1 (p. 348).

Figure 13.27 A chessboard calibration target captured straight ahead (left) and at an angle (right).

Stan Birchfield

where **K** contains the intrinsic calibration parameters, **R** and **t** contain the extrinsic rotation and translation, respectively, and $\mathbf{r}_i$ is the ith column of **R**. Therefore, the mapping from the target at any particular position and orientation to the image is the homography $\mathbf{H} \propto \mathbf{K}[\mathbf{r}_1 \quad \mathbf{r}_2 \quad \mathbf{t}]$, which is a 3×3 matrix relating the coordinates on one plane to the coordinates on the other plane.

Now suppose that an image of the target at a particular position and orientation has been captured, and the homography has been estimated (using the normalized DLT, for example). A homography has 8 degrees of freedom (because there are 9 elements, but scaling is unimportant), but 6 of these are due to Euclidean geometry (3 rotation, 3 translation). As a result, such a homography provides only 2 constraints on the intrinsic parameters:

$$\mathbf{r}_1^{\mathsf{T}}\mathbf{r}_2 = 0 \quad (\text{orthogonal vectors}) \tag{13.126}$$

$$\mathbf{r}_1^{\mathsf{T}}\mathbf{r}_1 = \mathbf{r}_2^{\mathsf{T}}\mathbf{r}_2 \quad (\text{unit norm vectors}) \tag{13.127}$$

which arise because the columns of **R** must be orthonormal, that is, they must be orthogonal to each other and have unit norm. Note that, in Equation (13.127), the vectors are only ensured to have the *same* norm rather than *unit* norm, due to the overall scaling ambiguity.

If we let $\mathbf{h}_i$ be the ith column of **H**,[†] so that $\mathbf{H} = [\mathbf{h}_1 \quad \mathbf{h}_2 \quad \mathbf{h}_3]$, then $\mathbf{h}_1 \propto \mathbf{K}\mathbf{r}_1$ and $\mathbf{h}_2 \propto \mathbf{K}\mathbf{r}_2$. Plugging into Equations (13.126)–(13.127) yields the 2 constraints in terms of the intrinsic matrix:

$$\mathbf{h}_1^{\mathsf{T}}\mathbf{K}^{-\mathsf{T}}\mathbf{K}^{-1}\mathbf{h}_2 = 0 \tag{13.128}$$

$$\mathbf{h}_1^{\mathsf{T}}\mathbf{K}^{-\mathsf{T}}\mathbf{K}^{-1}\mathbf{h}_1 = \mathbf{h}_2^{\mathsf{T}}\mathbf{K}^{-\mathsf{T}}\mathbf{K}^{-1}\mathbf{h}_2 \tag{13.129}$$

Figure 13.28 The mapping from a point $(x_w, y_w, 0)$ on a world plane to the point (x, y) on the image plane.

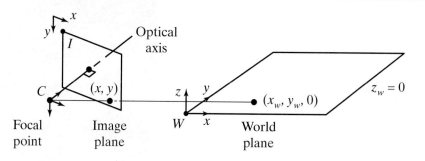

[†] Do not be confused by this notation: $\mathbf{h}_i$ is here used as the ith column, even though it was used as the transpose of the ith row in an earlier subsection.

These equations can be written more compactly as

$$\mathbf{h}_1^\mathsf{T} \boldsymbol{\omega}_\infty \mathbf{h}_2 = 0 \tag{13.130}$$

$$\mathbf{h}_1^\mathsf{T} \boldsymbol{\omega}_\infty \mathbf{h}_1 = \mathbf{h}_2^\mathsf{T} \boldsymbol{\omega}_\infty \mathbf{h}_2 \tag{13.131}$$

by defining the following 3×3 matrix $\boldsymbol{\omega}_\infty$ known as the **image of the absolute conic (IAC)**, which is a fundamental quantity tied to the intrinsic calibration parameters of a pinhole camera:

$$\boldsymbol{\omega}_\infty \equiv (\mathbf{K}\mathbf{K}^\mathsf{T})^{-1} = \mathbf{K}^{-\mathsf{T}}\mathbf{K}^{-1} = \frac{1}{\alpha^2\beta^2} \begin{bmatrix} \beta^2 & -\beta\gamma & \beta\xi \\ -\beta\gamma & \alpha^2 + \gamma^2 & -\alpha^2 v_0 - \gamma\xi \\ \beta\xi & -\alpha^2 v_0 - \gamma\xi & \alpha^2 v_0^2 + \xi^2 + \alpha^2\beta^2 \end{bmatrix} \tag{13.132}$$

where $\xi \equiv \gamma v_0 - \beta u_0$, and

$$\mathbf{K}^{-1} = \frac{1}{\alpha\beta} \begin{bmatrix} \beta & -\gamma & \xi \\ 0 & \alpha & -\alpha v_0 \\ 0 & 0 & \alpha\beta \end{bmatrix} \tag{13.133}$$

Noting by inspection that $\boldsymbol{\omega}_\infty$ is symmetric, let the entries of the matrices be given by

$$\boldsymbol{\omega}_\infty = \begin{bmatrix} w_{11} & w_{12} & w_{13} \\ w_{12} & w_{22} & w_{23} \\ w_{13} & w_{23} & w_{33} \end{bmatrix} \quad \mathbf{H} = \begin{bmatrix} \mathbf{h}_1 & \mathbf{h}_2 & \mathbf{h}_3 \end{bmatrix} = \begin{bmatrix} h_{11} & h_{12} & h_{13} \\ h_{21} & h_{22} & h_{23} \\ h_{31} & h_{32} & h_{33} \end{bmatrix} \tag{13.134}$$

Multiplying the terms, it is easy to show that

$$\mathbf{h}_i^\mathsf{T} \boldsymbol{\omega}_\infty \mathbf{h}_j = \mathbf{a}_{ij}^\mathsf{T} \mathbf{w}, \quad i, j \in \{1, 2, 3\} \tag{13.135}$$

where $\mathbf{a}_{ij}$ is a vector of knowns, and $\mathbf{w}$ is the vector of unknowns:

$$\mathbf{a}_{ij} \equiv \begin{bmatrix} h_{i1}h_{j1} & h_{i1}h_{j2} + h_{i2}h_{j1} & h_{i1}h_{j3} + h_{i3}h_{j1} & h_{i2}h_{j2} & h_{i2}h_{j3} + h_{i3}h_{j2} & h_{i3}h_{j3} \end{bmatrix}^\mathsf{T} \tag{13.136}$$

$$\mathbf{w} \equiv \begin{bmatrix} w_{11} & w_{12} & w_{13} & w_{22} & w_{23} & w_{33} \end{bmatrix}^\mathsf{T} \tag{13.137}$$

With this rearrangement, the two Euclidean constraints in Equations (13.130)–(13.131) become

$$\mathbf{a}_{12}^\mathsf{T} \mathbf{w} = 0 \tag{13.138}$$

$$(\mathbf{a}_{11} - \mathbf{a}_{22})^\mathsf{T} \mathbf{w} = 0 \tag{13.139}$$

which, recall, are the constraints arising from a single image of the target. When multiple images are taken from different viewpoints, there are multiple homographies, which taken together lead to a linear system of equations for determining $\boldsymbol{\omega}_\infty$. The expressions in Equations (13.138)–(13.139) are stacked into the rows of a $2n \times 6$ matrix $\mathbf{A}$, where n is the number of images, yielding a linear system $\mathbf{A}_{\{2n \times 6\}} \mathbf{w}_{\{6 \times 1\}} = \mathbf{0}_{\{2n \times 1\}}$ that is solved for $\mathbf{w}$. (Note that $n > 3$ for the system to be overdetermined.) Then $\boldsymbol{\omega}_\infty$ is obtained by simply rearranging the entries of $\mathbf{w}$ into matrix form.

The only remaining problem is to estimate the five intrinsic parameters (α, β, γ, u_0, and v_0) from $\boldsymbol{\omega}_\infty$. There are two ways to do this. One way is to use the Cholesky decomposition to factorize the symmetric, positive definite matrix $\boldsymbol{\omega}_\infty = \mathbf{L}\mathbf{L}^\mathsf{T}$ into the product of a lower triangular matrix $\mathbf{L} = \mathbf{K}^{-\mathsf{T}}$ and an upper triangular matrix $\mathbf{L}^{\mathsf{T}} = \mathbf{K}^{-1}$. Once $\mathbf{L}$ has been computed, the intrinsic matrix is given by $\mathbf{K} = \mathbf{L}^{-\mathsf{T}}$, and the parameters α, β, γ, u_0, and v_0 can be found by inspection from $\mathbf{K}$.

Alternatively, recall that $\boldsymbol{\omega}_\infty = \lambda \mathbf{K}^{-\mathsf{T}} \mathbf{K}^{-1}$, where $\lambda \neq 0$ is an unknown scale factor. Therefore, from Equation (13.132) the elements of $\boldsymbol{\omega}_\infty$ are related to the parameters as follows:

$$w_{11} = \frac{\lambda}{\alpha^2} \qquad\qquad w_{12} = -\frac{\lambda\gamma}{\alpha^2\beta} \qquad\qquad w_{13} = \frac{\lambda\xi}{\alpha^2\beta} \qquad (13.140)$$

$$w_{22} = \lambda\left(\frac{1}{\beta^2} + \frac{\gamma^2}{\alpha^2\beta^2}\right) \quad w_{23} = \lambda\left(\frac{-v_0}{\beta^2} - \frac{\gamma\xi}{\alpha^2\beta^2}\right) \quad w_{33} = \lambda\left(\frac{v_0^2}{\beta^2} + \frac{\xi^2}{\alpha^2\beta^2} + 1\right) \quad (13.141)$$

These equations are used to compute the scale factor:

$$\begin{aligned}
\lambda &= w_{33} - \lambda\left(\frac{v_0^2}{\beta^2} + \frac{\xi^2}{\alpha^2\beta^2}\right) \\
&= w_{33} - \frac{1}{w_{11}}\left(\frac{\lambda^2 v_0^2}{\alpha^2\beta^2} + \frac{\lambda^2 \xi^2}{\alpha^4\beta^2}\right) \\
&= w_{33} - \frac{1}{w_{11}}\left(\frac{\lambda^2 v_0^2}{\alpha^2\beta^2} + w_{13}^2\right) \\
&= w_{33} - \frac{1}{w_{11}}\left((w_{12}w_{13} - w_{11}w_{23})\, v_0 + w_{13}^2\right) \qquad (13.142)
\end{aligned}$$

where v_0 is determined by a ratio of elements:

$$\frac{w_{12}w_{13} - w_{11}w_{23}}{w_{11}w_{22} - w_{12}^2} = \frac{\lambda^2\left(-\beta^2\gamma\xi + \alpha^2\beta^2 v_0 + \beta^2\gamma\xi\right)}{\lambda^2\left(\alpha^2\beta^2 + \beta^2\gamma^2 - \beta^2\gamma^2\right)} = \frac{\alpha^2\beta^2 v_0}{\alpha^2\beta^2} = v_0 \qquad (13.143)$$

With the scale factor, the remaining quantities are given by solving the equations above:

$$\alpha = \sqrt{\frac{\lambda}{w_{11}}} \qquad (13.144)$$

$$\beta = \sqrt{\frac{\lambda w_{11}}{w_{11}w_{22} - w_{12}^2}} \qquad (13.145)$$

$$\gamma = \frac{-\alpha^2\beta w_{12}}{\lambda} \qquad (13.146)$$

$$u_0 = \frac{\gamma v_0}{\alpha} - \frac{\alpha^2 w_{13}}{\lambda} \qquad (13.147)$$

Either way, once $\mathbf{K}$ has been determined, the extrinsic parameters (rotation and translation) for any given image with homography $\mathbf{H} = \begin{bmatrix} \mathbf{h}_1 & \mathbf{h}_2 & \mathbf{h}_3 \end{bmatrix}$ are easily found as

$$\mathbf{r}_1 = \lambda\mathbf{K}^{-1}\mathbf{h}_1 \qquad (13.148)$$

$$\mathbf{r}_2 = \lambda\mathbf{K}^{-1}\mathbf{h}_2 \qquad (13.149)$$

$$\mathbf{r}_3 = \mathbf{r}_1 \times \mathbf{r}_2 \qquad (13.150)$$

$$\mathbf{t} = \lambda\mathbf{K}^{-1}\mathbf{h}_3 \qquad (13.151)$$

One final step remains to be described before we summarize the approach in Algorithm 13.12. The procedure so far has estimated the intrinsic (and extrinsic, if desired) parameters by minimizing the algebraic error. However, as we saw earlier,[†] although algebraic error is mathematically convenient, it is not closely tied to the actual problem at hand. More accurate values can be estimated by minimizing the **reprojection error**:

$$\sum_{i=1}^{n} \sum_{j=1}^{m} \|\mathbf{x}_{ij} - g(\mathbf{K}, \mathbf{R}_i, \mathbf{t}_i, \mathbf{x}_w^{(j)}, \phi)\|^2 \tag{13.152}$$

where $\mathbf{x}_w^{(j)} = (x_w^{(j)}, y_w^{(j)}, 0)$ is the jth world point, $\mathbf{x}_{ij} = (x_{ij}, x_{ij})$ is the corresponding point in the ith image, and g is the image projection function, taking lens distortion parameters ϕ into account. Using the parameters resulting from the above procedure, Equation (13.152) is minimized in an iterative manner, using a nonlinear minimization routine such as Levenberg-Marquardt, which is a variation of Newton's method. This iterative minimization, which adjusts all the parameters, is known as **bundle adjustment**.

13.5.4 Image of the Absolute Conic (IAC)

Just as the line at infinity ℓ_∞ surrounds the 2D plane, the **plane at infinity** π_∞ surrounds 3D space. This plane is visualized as an invisible sphere surrounding the 3D world in which we live, similar to the way ℓ_∞ is visualized as an invisible ring enveloping the plane. Each point on π_∞ is associated with a direction in 3D space, in the same way that each point on ℓ_∞ is associated with a direction in the plane. Just as there are two special points on ℓ_∞ that remain invariant under similarity transformations (known as the absolute points), there is also a locus of points on π_∞ that remains invariant under similarity transformations. This locus is called the **absolute conic**. Recall that the absolute points are those points $(x, y, 0)$ such that $x^2 + y^2 = 0$. Similarly, the absolute conic is the locus of points $(x, y, z, 0)$ satisfying $x^2 + y^2 + z^2 = 0$.

ALGORITHM 13.12 Zhang's camera calibration routine

CALIBRATECAMERAZHANG

Input: $n \geq 3$ images of a known planar target at different orientations and directions
Output: intrinsic parameters $\mathbf{K}$, lens distortion parameters ϕ, and extrinsic parameters

1   **for** each image **do**
2   Detect image features corresponding to known points on target.
3   Compute the homography using the normalized DLT algorithm.
4   Stack the entries from the homographies into matrix $\mathbf{A}$.
5   Solve $\mathbf{Aw} = \mathbf{0}$ for $\mathbf{w}$, then reshape into ω_∞.
6   Compute the five intrinsic parameters of $\mathbf{K}$ from ω_∞ using either
          Cholesky decomposition, or
          Equation (13.143) for v_0, then Equation (13.142) for λ, then Equations (13.144)–(13.147).
7   Compute extrinsic parameters $\mathbf{R}_i$ and $\mathbf{t}_i$ for each image $i = 1, \ldots, n$ using Equations (13.148)–(13.151).
8   Using these results as a starting point, minimize Equation (13.152) to perform bundle adjustment.

[†] Section 11.2 (p. 523).

From Equation (13.103), a point at infinity $(x, y, z, 0)$ projects onto the image plane at

$$
\begin{bmatrix} x \\ y \\ 1 \end{bmatrix} \propto \mathbf{K} \begin{bmatrix} \mathbf{R} & \mathbf{t} \end{bmatrix} \begin{bmatrix} x_w \\ y_w \\ z_w \\ 0 \end{bmatrix} = \underbrace{\mathbf{KR}}_{\mathbf{H}} \begin{bmatrix} x_w \\ y_w \\ z_w \end{bmatrix} \tag{13.153}
$$

where $\mathbf{H} = \mathbf{KR}$ is the homography that maps π_∞ to the image plane. Any point $(x_w, y_w, z_w, 0)$ on π_∞ that is also on the absolute conic satisfies $x_w^2 + y_w^2 + z_w^2 = 0$, or

$$
\begin{bmatrix} x_w & y_w & z_w \end{bmatrix} \begin{bmatrix} 1 & 0 & 0 \\ 0 & 1 & 0 \\ 0 & 0 & 1 \end{bmatrix} \begin{bmatrix} x_w \\ y_w \\ z_w \end{bmatrix} = 0 \tag{13.154}
$$

In other words, $\mathbf{I}_{\{3 \times 3\}}$ is the mathematical expression of the absolute conic in π_∞. Since from Equation (13.98), conics transform according to $\mathbf{C'} = \mathbf{H}^{-\top} \mathbf{C} \mathbf{H}^{-1}$, the projection of the absolute conic onto the image plane is therefore

$$
\begin{aligned}
\mathbf{H}^{-\top} \mathbf{I}_{\{3 \times 3\}} \mathbf{H}^{-1} &= (\mathbf{KR})^{-\top} \mathbf{I}_{\{3 \times 3\}} (\mathbf{KR})^{-1} & (13.155) \\
&= \mathbf{K}^{-\top} \mathbf{R}^{-\top} \mathbf{R}^{-1} \mathbf{K}^{-1} & (13.156) \\
&= \mathbf{K}^{-\top} \mathbf{K}^{-1} & (13.157) \\
&= \omega_\infty & (13.158)
\end{aligned}
$$

which is the image of the absolute conic (IAC) of Equation (13.132), as we expected.

The absolute points in any given plane are given by $(1, \pm j, 0)$ in the plane's coordinate system. Therefore, when an image is captured of a planar calibration target, these points project onto the image at locations $\mathbf{H}\begin{bmatrix} 1 & \pm j & 0 \end{bmatrix}^\top = \mathbf{h}_1 \pm j\mathbf{h}_2$, where $\mathbf{H} = \begin{bmatrix} \mathbf{h}_1 & \mathbf{h}_2 & \mathbf{h}_3 \end{bmatrix}$ is the homography between the calibration plane and the image plane at that pose. The absolute points, of course, must lie on the absolute conic, and their projections must therefore lie on the image of the absolute conic. As a result, we have

$$
(\mathbf{h}_1 \pm j\mathbf{h}_2)^\top \omega_\infty (\mathbf{h}_1 \pm j\mathbf{h}_2) = 0 \tag{13.159}
$$

Splitting into real and imaginary parts yields two equations:

$$
\begin{aligned}
\mathbf{h}_1^\top \omega_\infty \mathbf{h}_1 &= \mathbf{h}_2^\top \omega_\infty \mathbf{h}_2 & (13.160) \\
\mathbf{h}_1^\top \omega_\infty \mathbf{h}_2 &= 0 & (13.161)
\end{aligned}
$$

which are the same two constraints derived in Equations (13.128) and (13.129).

To summarize, whenever you take a picture with a pinhole camera, you not only capture a visible representation of the scene in terms of colored pixels that you can see, but you also capture an invisible picture of this imaginary locus of points called the absolute conic. Because the absolute conic is unaffected by similarity (and therefore Euclidean) transformations, the image of the absolute conic (IAC) is also unaffected by them, meaning that ω_∞ is the same no matter the pose of the camera. Therefore, since the IAC is only dependent upon the camera's intrinsic (rather than extrinsic) parameters, calibrating the camera is equivalent to discovering the IAC. This is the geometric intuition behind Zhang's algorithm.

13.6 Geometry of Multiple Views

In this section we apply projective geometry to several problems involving multiple images taken of a static scene. In order to make the math easier, we shall assume pinhole cameras throughout, assuming that lens distortion either is insignificant or has been calibrated away using the principles discussed earlier in the chapter.[†] To help motivate this section, Figure 13.29 shows some 3D reconstructions of scenes using a state-of-the-art system. While the details of the system are not discussed here, the basic procedure of matching points from multiple images, estimating camera intrinsic parameters, estimating relative camera pose, and recovering the 3D coordinates of the scene points is based on the fundamental principles covered in this section.

13.6.1 Epipolar Geometry

Suppose two cameras are viewing the same scene, as shown in Figure 13.30, and the two cameras take a picture simultaneously. (The situation is mathematically equivalent if a single camera takes two pictures in sequence, as long as the scene is static.) Let $\mathbf{c} \in \mathbb{R}^3$ be the center of projection (or focal point) of one camera, and $\mathbf{c}' \in \mathbb{R}^3$ be the center of projection of the other camera. The line joining the distance $|\mathbf{c}' - \mathbf{c}|$ between these two centers of projection, $\mathbf{c}' - \mathbf{c}$, is the **baseline**.[‡]

Now suppose the cameras can both see a point $\mathbf{p} \in \mathbb{R}^3$ in the world, which projects onto the first image plane at $\mathbf{q} \in \mathbb{R}^3$ and onto the second image plane at $\mathbf{q}' \in \mathbb{R}^3$. Assuming light travels in straight lines, it is easy to see that $\mathbf{c}$, $\mathbf{c}'$, $\mathbf{p}$, $\mathbf{q}$, and $\mathbf{q}'$ all lie in the same plane, as illustrated in Figure 13.30. This plane is known as the **epipolar plane**, because it is attached to the two poles (centers of projection). The epipolar plane intersects the two image planes along two **epipolar lines**.

Moreover, the baseline intersects the two image planes at the **two epipoles**. In other words, the left epipole is the projection of the right focal point onto the left image plane, and vice versa for the right epipole. These epipoles are at infinity if the image planes are parallel to the baseline. Furthermore, if the image planes are coplanar and rotated so that

Figure 13.29 Results of a 3D reconstruction system that matches points in multiple images, determines the camera positions and orientations, and estimates the 3D coordinates of the scene points.

From Agarwal et al. [2009], "Building Rome in a Day."

[†] Section 13.4.9 (p. 663).

[‡] The term *baseline* can also be used to refer to the distance $\|\mathbf{c}' - \mathbf{c}\|$ between the centers, as we saw in Section 13.2.1 (p. 624).

Figure 13.30 Epipolar geometry of two cameras viewing the same scene. The epipolar lines are shown on each of the image planes, while the epipoles, in this example, are outside the finite extent of the image plane.

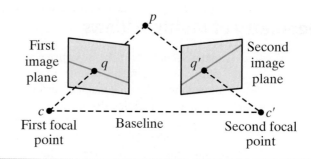

their scanlines are parallel to the baseline, then the epipoles are not only at infinity but they are along the x-axis, so that the epipolar lines are the scanlines; this is the case of rectified images that we considered earlier.

13.6.2 Fundamental Matrix

Suppose the image coordinates (in pixels) of the projections of the world point onto the stereo pair of cameras just described are (x, y) and (x', y'), respectively. Let us represent these points in homogeneous coordinates as $\mathbf{x} = [x \quad y \quad 1]^\mathsf{T}$ and $\mathbf{x}' = [x' \quad y' \quad 1]^\mathsf{T}$. Because these two points are projections of the same world point, we say that they are *corresponding points*, which we denote by $\mathbf{x} \leftrightarrow \mathbf{x}'$. Let ℓ denote the epipolar line in the first image, and ℓ' the epipolar line in the second image, associated with the point. Since $\mathbf{x}$ must lie on the first epipolar line, and $\mathbf{x}'$ must lie on the second epipolar line, we know $\mathbf{x}^\mathsf{T}\ell = \mathbf{x}'^\mathsf{T}\ell' = 0$.

Given the point $\mathbf{x}$, how do we determine its epipolar line ℓ' in the other image? Well, since the imaging process is a linear mapping (when homogeneous coordinates are used), the answer to this question is also a linear function: $\ell' = \mathbf{F}\mathbf{x}$, where $\mathbf{F}$ is the 3×3 **fundamental matrix** that captures the relative geometry between the cameras. Since $\mathbf{x}'^\mathsf{T}\ell' = 0$, this leads to the following equation:

$$\mathbf{x}'^\mathsf{T}\mathbf{F}\mathbf{x} = 0 \qquad (13.162)$$

and since this is a scalar equation, the transpose yields the relationship in the opposite direction:

$$\mathbf{x}^\mathsf{T}\mathbf{F}^\mathsf{T}\mathbf{x}' = 0 \qquad (13.163)$$

which reveals that $\ell = \mathbf{F}^\mathsf{T}\mathbf{x}'$ is the epipolar line in the first image. The result below summarizes what we have learned so far.

Result 1. *Given a fundamental matrix $\boldsymbol{F}$, the epipolar line ℓ' in the second image associated with the point $\boldsymbol{x}$ in the first image is given by $\ell' = \boldsymbol{F}\boldsymbol{x}$. Similarly, the epipolar line ℓ in the first image associated with the point $\boldsymbol{x}'$ in the second image is given by $\ell = \boldsymbol{F}^\mathsf{T}\boldsymbol{x}'$.*

Recall that with rectified cameras, the vertical coordinates of any two corresponding points are the same: that is, $y = y'$. Therefore, when the cameras are rectified, the fundamental matrix takes the special form

$$\tilde{\mathbf{F}} = \begin{bmatrix} 0 & 0 & 0 \\ 0 & 0 & -1 \\ 0 & 1 & 0 \end{bmatrix} \qquad (13.164)$$

which is easily seen by substituting:

$$\mathbf{x}'^{\mathsf{T}}\tilde{\mathbf{F}}\mathbf{x} = \begin{bmatrix} x' & y' & 1 \end{bmatrix} \begin{bmatrix} 0 & 0 & 0 \\ 0 & 0 & -1 \\ 0 & 1 & 0 \end{bmatrix} \begin{bmatrix} x \\ y \\ 1 \end{bmatrix} = \begin{bmatrix} 0 & 1 & -y' \end{bmatrix} \begin{bmatrix} x \\ y \\ 1 \end{bmatrix} = y - y' = 0 \qquad (13.165)$$

or $y = y'$. Note that, since the fundamental matrix, as with all homogeneous quantities, is only defined up to a nonzero scale factor, the choice of location of the negative sign in Equation (13.164) is arbitrary.

EXAMPLE 13.3

Suppose we have two nonrectified cameras, and their fundamental matrix is

$$\mathbf{F} = \begin{bmatrix} 1 & 2 & 1 \\ -2 & -4 & -2 \\ 3 & 3 & 1 \end{bmatrix} \qquad (13.166)$$

Is it possible for $(x, y) = (100, 50)$ and $(x', y') = (78, 40)$ to be corresponding points?

Solution

We answer the question by noting that

$$\ell' = \mathbf{F}\mathbf{x} = \begin{bmatrix} 1 & 2 & 1 \\ -2 & -4 & -2 \\ 3 & 3 & 1 \end{bmatrix} \begin{bmatrix} 100 \\ 50 \\ 1 \end{bmatrix} = \begin{bmatrix} 201 \\ -402 \\ 451 \end{bmatrix} \qquad (13.167)$$

which means that the epipolar line in the second image associated with $\mathbf{x}$ is $201x - 402y + 451 = 0$. Substituting $\mathbf{x}'$ yields $201(78) - 402(40) + 451 = 49 \neq 0$, which tells us that the point $\mathbf{x}'$ does not lie on the epipolar line ℓ associated with $\mathbf{x}$, and therefore these are not corresponding points.

But wait a minute. In the real world there will always be some noise, so we should never expect the result to be exactly zero, and although the residue of 49 may seem large at first glance, overall values are misleading when dealing with homogeneous coordinates because scaling is unimportant. To find the Euclidean distance from the point to the line, we first have to normalize the line parameters of $\ell' = (a', b', c')$ so that its normal vector $\begin{bmatrix} a' & b' \end{bmatrix}^{\mathsf{T}}$ has unit norm:

$$\ell' = \frac{\begin{bmatrix} 201 & -402 & 451 \end{bmatrix}^{\mathsf{T}}}{\sqrt{201^2 + (-402)^2}} = \frac{\begin{bmatrix} 201 & -402 & 451 \end{bmatrix}^{\mathsf{T}}}{449.45} = \begin{bmatrix} 0.447 & -0.894 & 1.003 \end{bmatrix}^{\mathsf{T}} \qquad (13.168)$$

Substituting $\mathbf{x}'$ into this expression yields $0.447(78) - 0.894(40) + 1.003 = 0.109$, so the point is about a tenth of a pixel away from the line, which is well within the range of typical real-world noise. So while the point $(78, 40)$ does not lie exactly on the epipolar line in a mathematical sense, from a practical point of view these points might very well correspond to each other.

Because the fundamental matrix captures the relative geometry between two cameras, it also encodes the epipoles. For any point $\mathbf{x}'$ in the second image, its epipolar line ℓ in the first image passes through the first epipole $\mathbf{e}$, so that $\ell^{\mathsf{T}}\mathbf{e} = 0$. Since $\ell = \mathbf{F}^{\mathsf{T}}\mathbf{x}'$, this means that $\mathbf{x}'^{\mathsf{T}}\mathbf{F}\mathbf{e} = 0$. Since this must be true for any $\mathbf{x}'$, we have $\mathbf{F}\mathbf{e} = \mathbf{0}_{\{3 \times 1\}}$. As a result, the epipole $\mathbf{e}$ in the first image is the right null vector of $\mathbf{F}$. By similar reasoning, $\mathbf{F}^{\mathsf{T}}\mathbf{e}' = \mathbf{0}_{\{3 \times 1\}}$, so the epipole $\mathbf{e}'$ in the second image is the left null vector of $\mathbf{F}$. These are easily computed using the SVD.[†]

Result 2. *Given a fundamental matrix* $\mathbf{F}$, *the epipole* $\mathbf{e}$ *in the first image is the rightmost right singular vector of* $\mathbf{F}$, *and the epipole* $\mathbf{e}'$ *in the second image is the rightmost left singular vector of* $\mathbf{F}$. *That is,* $\mathbf{e} = \mathbf{v}_3$ *and* $\mathbf{e}' = \mathbf{u}_3$, *where* $\mathbf{F} = \mathbf{U}\Sigma\mathbf{V}^\mathsf{T}$, $\mathbf{U} = \begin{bmatrix} \mathbf{u}_1 & \mathbf{u}_2 & \mathbf{u}_3 \end{bmatrix}$, *and* $\mathbf{V} = \begin{bmatrix} \mathbf{v}_1 & \mathbf{v}_2 & \mathbf{v}_3 \end{bmatrix}$.

Since $\mathbf{Fe} = \mathbf{0}$ and $\mathbf{F}^\mathsf{T}\mathbf{e}' = \mathbf{0}$, the fundamental matrix has a null space. As a result, it is a singular matrix, which means that its determinant is zero $(\det \mathbf{F} = 0)$, and therefore it is rank deficient. In fact, it can be shown that no matter the geometry between the cameras, its rank is always 2. This leads to the following important result, allowing us to test whether a given matrix is in fact a fundamental matrix.

Result 3. *A real* 3×3 *matrix* $\mathbf{F}$ *is a fundamental matrix for some camera geometry if and only if* $\mathrm{rank}(\mathbf{F}) = 2$. *That is, its singular values are such that* $\sigma_1 \neq 0$, $\sigma_2 \neq 0$, *and* $\sigma_3 = 0$.

13.6.3 Essential Matrix

Now suppose the cameras have been calibrated, and let $\mathbf{K}$ and $\mathbf{K}'$ be the 3×3 intrinsic calibration matrices of the two cameras. These matrices transform the *metric coordinates* $\overline{\mathbf{x}}$ and $\overline{\mathbf{x}}'$ (which are given in meters, or micrometers, in the camera coordinate system) into *image coordinates* $\mathbf{x}$ and $\mathbf{x}'$ (which are given in pixels in the image coordinate system):[‡]

$$\mathbf{x} = \mathbf{K}\overline{\mathbf{x}} \tag{13.169}$$

$$\mathbf{x}' = \mathbf{K}'\overline{\mathbf{x}}' \tag{13.170}$$

Recall from Equation (13.103) that the 3×1 vectors $\overline{\mathbf{x}}$ and $\overline{\mathbf{x}}'$ are just Euclidean transformations of the world point, so they are actual points in 3D Euclidean space: $\overline{\mathbf{x}} = (\overline{x}, \overline{y}, \overline{z})$ and $\overline{\mathbf{x}}' = (\overline{x}', \overline{y}', \overline{z}')$. On the other hand, the 3×1 vectors $\mathbf{x}$ and $\mathbf{x}'$ are homogeneous coordinates of points on the 2D image plane: $\mathbf{x} = (u, v, w)$ and $\mathbf{x}' = (u', v', w')$.

With calibrated cameras, it is possible to show how the fundamental matrix relates to the geometry of the cameras. The epipolar constraint says that in the Euclidean world, the vector $\overline{\mathbf{x}}$ from the first camera's optical center to the first imaged point, the vector $\overline{\mathbf{x}}'$ from the second optical center to the second imaged point, and the vector $\mathbf{t}$ from one optical center to the other are all coplanar. Recall that if three vectors are coplanar, then the cross product of any two of them is perpendicular to the third, which is represented mathematically as the inner product of the cross product being equal to zero. As a result, if $\mathbf{R}$ and $\mathbf{t}$ are the rotation and translation between the two cameras' coordinate frames (so that $\mathbf{t}$ is along the baseline), then the epipolar constraint can be expressed simply as

$$\overline{\mathbf{x}}'^\mathsf{T}(\mathbf{t} \times \mathbf{R}\overline{\mathbf{x}}) = 0 \tag{13.171}$$

where the multiplication by $\mathbf{R}$ is necessary to transform $\overline{\mathbf{x}}$ into the second camera's coordinate frame. By defining $[\mathbf{t}]_\times$ as the corresponding 3×3 skew-symmetric matrix[§] for $\mathbf{t}$ that captures the cross product:

$$[\mathbf{t}]_\times \equiv \begin{bmatrix} 0 & -t_z & t_y \\ t_z & 0 & -t_x \\ -t_y & t_x & 0 \end{bmatrix}, \quad \text{where} \quad \mathbf{t} = \begin{bmatrix} t_x \\ t_y \\ t_z \end{bmatrix} \tag{13.172}$$

[†] Section 11.1.5 (p. 520).

[‡] Although the term *normalized coordinates* is sometimes used instead of *metric coordinates*, we prefer the latter to avoid confusion with the normalized algorithms presented elsewhere in this and previous chapters.

[§] A skew-symmetric matrix $\mathbf{S}$ satisfies $\mathbf{S} = -\mathbf{S}^\mathsf{T}$.

so that $[\mathbf{t}]_\times \mathbf{y} = \mathbf{t} \times \mathbf{y}$ for any 3×1 vector $\mathbf{y}$, we can rewrite Equation (13.171) as a linear equation:

$$\bar{\mathbf{x}}'^{\mathsf{T}} \mathbf{E} \bar{\mathbf{x}} = 0 \tag{13.173}$$

where $\mathbf{E} \equiv [\mathbf{t}]_\times \mathbf{R}$ is called the **essential matrix** between the two cameras.

The essential and fundamental matrices are cousins. Like the fundamental matrix, the essential matrix captures the geometric relationship (i.e., the epipolar constraint) between the cameras in a compact matrix form. The fundamental matrix relates *uncalibrated* cameras (and therefore transforms points in the image coordinate systems), while the essential matrix relates *calibrated* cameras (and therefore transforms points in the camera coordinate systems). The relationship between the two matrices is easily determined by substituting Equations (13.169)–(13.170) into Equation (13.173):

$$\bar{\mathbf{x}}'^{\mathsf{T}} \mathbf{E} \bar{\mathbf{x}} = 0 \tag{13.174}$$

$$(\mathbf{K}'^{-1}\mathbf{x}')^{\mathsf{T}} \mathbf{E} (\mathbf{K}^{-1}\mathbf{x}) = 0 \tag{13.175}$$

$$\mathbf{x}'^{\mathsf{T}} \underbrace{(\mathbf{K}'^{-\mathsf{T}} \mathbf{E} \mathbf{K}^{-1})}_{\mathbf{F}} \mathbf{x} = 0 \tag{13.176}$$

which reveals $\mathbf{F} = \mathbf{K}'^{-\mathsf{T}} \mathbf{E} \mathbf{K}^{-1}$, or $\mathbf{E} = \mathbf{K}'^{\mathsf{T}} \mathbf{F} \mathbf{K}$.

As defined, the essential matrix appears to have six parameters: 3 for the rotation angles between the two cameras, and 3 for the translation vector. However, the essential matrix is used exclusively in the context of Equation (13.173), which is a homogeneous equation. As a result, multiplying $\mathbf{E}$ by any nonzero scalar results in another 3×3 matrix with the exact same behavior. Another way to look at this is to recognize that only the direction of $\mathbf{t}$ in Equation (13.171) is important, and therefore scaling the translation vector by any nonzero amount does not affect the result. This ambiguity matches experience, for in practice we often do not know the overall magnitude of $\mathbf{t}$ but rather only the direction. Therefore, in practice the essential matrix contains just five parameters: 3 for rotation, and 2 for the direction of translation.

As seen in the analysis above, every essential matrix has a special structure: It is the product of a skew-symmetric matrix and an orthogonal matrix. Since an arbitrary 3×3 matrix has 9 parameters, but an essential matrix has only 5 parameters, it is natural to ask under what conditions a given matrix is also an essential matrix. The answer is that a 3×3 matrix is an essential matrix if and only if two of its singular values are equal, and the third is zero. This is given in the following result.

Result 4. *A real 3×3 matrix $\mathbf{E}$ is an essential matrix if and only if its three singular values are such that $\sigma_1 = \sigma_2 \neq 0$ and $\sigma_3 = 0$. Therefore, $\text{rank}(\mathbf{E}) = 2$.*

If there is no rotation between the cameras, and the only translation is in the x direction, then the essential matrix is

$$\mathbf{E} = [\mathbf{t}]_\times \mathbf{R} = \begin{bmatrix} 0 & 0 & 0 \\ 0 & 0 & -t_x \\ 0 & t_x & 0 \end{bmatrix} \begin{bmatrix} 1 & 0 & 0 \\ 0 & 1 & 0 \\ 0 & 0 & 1 \end{bmatrix} \propto \begin{bmatrix} 0 & 0 & 0 \\ 0 & 0 & -1 \\ 0 & 1 & 0 \end{bmatrix} \tag{13.177}$$

which is the same as the fundamental matrix of rectified cameras if the principal points and focal lengths are identical.

From Equation (13.103), the two epipoles are given by

$$\mathbf{e} = \mathbf{K}\mathbf{R}^{\mathsf{T}}\mathbf{t} \tag{13.178}$$

$$\mathbf{e}' = \mathbf{K}'\mathbf{t} \tag{13.179}$$

which are just the projections of the point at infinity associated with $\mathbf{t}$, that is, $[\mathbf{t}^\mathsf{T} \quad 0]^\mathsf{T}$, onto the two image planes. That is, since $\mathbf{t}$ is already represented in the coordinate system of the second camera, its rotation and translation are zero, so that the projection onto the second image plane is simply $\mathbf{K}'\,[\mathbf{I} \quad \mathbf{0}]\,[\mathbf{t}^\mathsf{T} \quad 0]^\mathsf{T} = \mathbf{K}'\mathbf{t}$. For the first image coordinate system, however, $\mathbf{t}$ must first be rotated by $\mathbf{R}^\mathsf{T}$, leading to $\mathbf{K}\,[\mathbf{R}^\mathsf{T} \quad -\mathbf{t}]\,[\mathbf{t}^\mathsf{T} \quad 0]^\mathsf{T} = \mathbf{K}\mathbf{R}^\mathsf{T}\mathbf{t}$. These epipoles are easily verified using $\mathbf{F}\mathbf{e} = \mathbf{0}$ and $\mathbf{F}^\mathsf{T}\mathbf{e}' = \mathbf{0}$. That is, $\mathbf{F}\mathbf{e} = (\mathbf{K}'^{-\mathsf{T}}\mathbf{E}\mathbf{K}^{-1})\mathbf{e} = \mathbf{K}'^{-\mathsf{T}}[\mathbf{t}]_\times\mathbf{R}\mathbf{K}^{-1}(\mathbf{K}\mathbf{R}^\mathsf{T}\mathbf{t}) = \mathbf{K}'^{-\mathsf{T}}[\mathbf{t}]_\times\mathbf{t} = \mathbf{K}'^{-\mathsf{T}}(\mathbf{t} \times \mathbf{t}) = \mathbf{0}$, where we have used the property that the cross product of a vector with itself is zero; and similarly, $\mathbf{F}^\mathsf{T}\mathbf{e}' = (\mathbf{K}'^{-\mathsf{T}}\mathbf{E}\mathbf{K}^{-1})^\mathsf{T}\mathbf{e}' = \mathbf{K}^{-\mathsf{T}}\mathbf{E}^\mathsf{T}\mathbf{K}'^{-1}(\mathbf{K}'\mathbf{t}) = \mathbf{K}^{-\mathsf{T}}\mathbf{E}^\mathsf{T}\mathbf{t} = \mathbf{K}^{-\mathsf{T}}([\mathbf{t}]_\times\mathbf{R})^\mathsf{T}\mathbf{t}$ $= K^{-\mathsf{T}}\mathbf{R}^\mathsf{T}\,[\mathbf{t}]_\times^\mathsf{T}\mathbf{t} = \mathbf{0}$.

One property of any essential matrix is that $\mathbf{E}^\mathsf{T}\mathbf{t} = \mathbf{0}_{\{3\times 1\}}$. This is easy to see by substitution, similar to above: $\mathbf{E}^\mathsf{T}\mathbf{t} = ([\mathbf{t}]_\times\mathbf{R})^\mathsf{T}\mathbf{t} = \mathbf{R}^\mathsf{T}\,[\mathbf{t}]_\times^\mathsf{T}\mathbf{t} = \mathbf{0}$. In other words, since $\mathbf{t}$ points along the baseline, the point at infinity associated with $\mathbf{t}$ is the epipole in the second image, and therefore $\mathbf{t}$ does not have an associated epipolar line. Another property is that $\mathbf{E}\mathbf{E}^\mathsf{T}$ depends only on the translation: $\mathbf{E}\mathbf{E}^\mathsf{T} = [\mathbf{t}]_\times\mathbf{R}([\mathbf{t}]_\times\mathbf{R})^\mathsf{T} = [\mathbf{t}]_\times\mathbf{R}\mathbf{R}^\mathsf{T}[\mathbf{t}]_\times^\mathsf{T} = [\mathbf{t}]_\times[\mathbf{t}]_\times^\mathsf{T}$. Additional properties can be found in the exercises.[†]

13.6.4 Relationship with Camera Projection Matrices

Now let us consider the relationship between the fundamental matrix and the camera projection matrices. Given a world point at (x_w, y_w, z_w), the homogeneous coordinates of its projection onto the first image are given by $\mathbf{x} = \mathbf{P}\mathbf{w}$, where $\mathbf{P}$ is the 3×4 projection matrix of the first camera, as in Equation (13.103), and $\mathbf{w} \equiv [x_w \quad y_w \quad z_w \quad 1]^\mathsf{T}$ are the homogeneous coordinates of the world point. Without additional information about the depth, we cannot compute the world coordinates of the point from $\mathbf{x}$, but nevertheless we know that $\mathbf{P}^+\mathbf{x}$ is one possible world point, where $\mathbf{P}^+$ is the pseudoinverse of $\mathbf{P}$. That is, $\mathbf{P}^+\mathbf{x}$ lies on the ray passing through the first camera center and the point $\mathbf{x}$. The projection of this point onto the second image is $\mathbf{x}' = \mathbf{P}'\mathbf{P}^+\mathbf{x}$, where $\mathbf{P}'$ is the 3×4 projection matrix of the second camera. Now we know that the epipolar line ℓ' in the second image associated with $\mathbf{x}$ in the first image joins $\mathbf{x}'$ and the second epipole $\mathbf{e}'$. Since, as we have already seen, the line joining two points is the cross product of the two points, this leads to $\ell' = \mathbf{e}' \times \mathbf{P}'\mathbf{P}^+\mathbf{x} = [\mathbf{e}']_\times\mathbf{P}'\mathbf{P}^+\mathbf{x}$. Since $\mathbf{x}'$ lies on ℓ', this yields

$$\mathbf{x}'^\mathsf{T}\ell' = \mathbf{x}'^\mathsf{T}\underbrace{[\mathbf{e}']_\times\mathbf{P}'\mathbf{P}^+}_{\mathbf{F}}\mathbf{x} = 0 \tag{13.180}$$

or $\mathbf{F} = [\mathbf{e}']_\times\mathbf{P}'\mathbf{P}^+$. By similar reasoning, $\mathbf{F}^\mathsf{T} = [\mathbf{e}]_\times\mathbf{P}\mathbf{P}'^+$.

Now suppose that $\mathbf{R}$ and $\mathbf{t}$ are the rotation and translation, respectively, between the two cameras, so that $\mathbf{P} = \mathbf{K}\,[\mathbf{I}_{\{3\times 3\}} \quad \mathbf{0}_{\{3\times 1\}}]$ and $\mathbf{P}' = \mathbf{K}'\,[\mathbf{R} \quad \mathbf{t}]$. Then, straightforward substitution reveals:

$$\mathbf{F} = [\mathbf{e}']_\times\mathbf{P}'\mathbf{P}^+ = [\mathbf{e}']_\times\mathbf{K}'\,[\mathbf{R} \quad \mathbf{t}]\,(\mathbf{K}\,[\mathbf{I} \quad \mathbf{0}])^+ = [\mathbf{e}']_\times\mathbf{K}'\,[\mathbf{R} \quad \mathbf{t}]\begin{bmatrix}\mathbf{I}\\\mathbf{0}^\mathsf{T}\end{bmatrix}\mathbf{K}^{-1} = [\mathbf{e}']_\times\mathbf{K}'\mathbf{R}\mathbf{K}^{-1} \tag{13.181}$$

Although the projection matrices uniquely determine $\mathbf{F}$, for any given $\mathbf{F}$ there is a family of projection matrices compatible with it. If we set $\mathbf{P} = [\mathbf{I}_{\{3\times 3\}} \quad \mathbf{0}_{\{3\times 1\}}]$, then this family is given by $\mathbf{P}' = [[\mathbf{e}']_\times\mathbf{F} + \mathbf{e}'\mathbf{v}^\mathsf{T} \quad \lambda\mathbf{e}']$, where $\mathbf{v}$ is any 3×1 vector, and λ is any nonzero scalar.

[†] Problem 13.39.

For reference, we now summarize the equations for the essential and fundamental matrices. Four of these equations have already been shown, while one is left as an exercise:[†]

$$\mathbf{E} = [\mathbf{t}]_\times \mathbf{R} = \mathbf{R} [\mathbf{R}^\mathsf{T} \mathbf{t}]_\times \tag{13.182}$$

$$\mathbf{F} = \mathbf{K}'^{-\mathsf{T}} \mathbf{E} \mathbf{K}^{-1} = [\mathbf{e}']_\times \mathbf{K}' \mathbf{R} \mathbf{K}^{-1} = [\mathbf{e}']_\times \mathbf{P}' \mathbf{P}^+ \tag{13.183}$$

13.6.5 Estimating the Essential and Fundamental Matrices

Now that we have laid the groundwork for understanding how the fundamental and essential matrices capture the relative geometry between a pair of cameras, we are ready to show how to reconstruct the metric geometry of a scene from point correspondences taken by a pair of calibrated cameras. The general procedure is as follows: (1) estimate the fundamental matrix from the correspondences; (2) construct the essential matrix using the intrinsic camera calibration parameters; (3) decompose the essential matrix into rotation and translation (up to scale); and (4) estimate the 3D coordinates of the world points. These steps are covered in this subsection and the next several subsections.

Since the matrix $\mathbf{E}$ has only five degrees of freedom, in theory only five corresponding points are needed to solve for it. (Note that each corresponding pair yields a single equation, because $\mathbf{x}'^\mathsf{T} \mathbf{E} \mathbf{x} = 0$ is a scalar equation, unlike the homography mapping $\mathbf{x}' \propto \mathbf{H}\mathbf{x}$, which yields two linearly independent equations per corresponding pair.) The so-called **five-point algorithm** takes advantage of this property, but we shall not spend time discussing it because it is quite complicated mathematically, and we usually want to use as many correspondences as possible anyway in order to overcome the effects of noise.

Instead, the most popular approach is the **eight-point algorithm**. Since the essential and fundamental matrices consist of nine elements but are unique only up to an unknown scaling factor, each has eight unique elements. Since each corresponding pair yields one equation, eight pairs of correspondences yield eight equations for the eight unknowns, thereby leading to a linear solution. The beauty of the eight-point algorithm is its simplicity, because it computes a least squares solution if additional correspondences are available, with no modification to the algorithm. The eight-point algorithm is applicable to estimating either $\mathbf{E}$ or $\mathbf{F}$ from corresponding pairs, depending upon whether the measurements are in pixels or meters. We will describe the procedure for computing $\mathbf{F}$, but keep in mind that the same procedure can be used to compute $\mathbf{E}$, with just a very slight modification at the end that we shall mention. Alternatively, once $\mathbf{F}$ has been computed, $\mathbf{E}$ can be found by simply multiplying $\mathbf{F}$ by the intrinsic calibration matrices if they are available, as in Equation (13.176).

If we let $\mathbf{x}$ and $\mathbf{x}'$ be points in the two images, then we can rewrite the fundamental matrix equation by explicitly listing the individual elements of the matrix and vectors as follows:

$$\mathbf{x}'^\mathsf{T} \mathbf{F} \mathbf{x} = \begin{bmatrix} x' & y' & 1 \end{bmatrix} \underbrace{\begin{bmatrix} f_{11} & f_{12} & f_{13} \\ f_{21} & f_{22} & f_{23} \\ f_{31} & f_{32} & f_{33} \end{bmatrix}}_{\mathbf{F}} \begin{bmatrix} x \\ y \\ 1 \end{bmatrix} = 0 \tag{13.184}$$

Multiplying the elements reveals a single scalar equation:

$$f_{11}xx' + f_{12}x'y + f_{13}x' + f_{21}xy' + f_{22}yy' + f_{23}y' + f_{31}x + f_{32}y + f_{33} = 0 \tag{13.185}$$

[†] Problem 13.40.

which, when rearranged, can be written as the dot product of two vectors, one containing known quantities and the other containing unknown quantities:

$$
\begin{bmatrix} xx' & x'y & x' & xy' & yy' & y' & x & y & 1 \end{bmatrix}
\begin{bmatrix} f_{11} \\ f_{12} \\ f_{13} \\ f_{21} \\ \vdots \\ f_{33} \end{bmatrix} = 0
\tag{13.186}
$$

where the vector of unknowns contains the elements of $\mathbf{F}$ is row-major order. This is one equation with eight unknowns, due to the unknown scaling factor. With additional correspondences, the values are stacked to yield additional rows on the left:

$$
\underbrace{\begin{bmatrix}
x_1 x_1' & x_1' y_1 & x_1' & x_1 y_1' & y_1 y_1' & y_1' & x_1 & y_1 & 1 \\
x_2 x_2' & x_2' y_2 & x_2' & x_2 y_2' & y_2 y_2' & y_2' & x_2 & y_2 & 1 \\
x_3 x_3' & x_3' y_3 & x_3' & x_3 y_3' & y_3 y_3' & y_3' & x_3 & y_3 & 1 \\
\vdots & & & & & & & & \\
x_n x_n' & x_n' y_n & x_n' & x_n y_n' & y_n y_n' & y_n' & x_n & y_n & 1
\end{bmatrix}}_{\mathbf{A}_{\{n \times 9\}}}
\underbrace{\begin{bmatrix} f_{11} \\ f_{12} \\ f_{13} \\ f_{21} \\ \vdots \\ f_{33} \end{bmatrix}}_{\mathbf{f}_{\{9 \times 1\}}} = \mathbf{0}_{\{9 \times 1\}}
\tag{13.187}
$$

where $(x_i, y_i) \leftrightarrow (x_i', y_i')$ is the ith corresponding pair, and $n \geq 8$ is the number of correspondences. This equation can be solved by selecting the eigenvector of $\mathbf{A}^\mathsf{T}\mathbf{A}$ associated with the smallest eigenvalue, or equivalently, the right singular vector associated with the smallest singular value of $\mathbf{A}$, as we have already seen. Then, the elements of $\mathbf{f}$ can be rearranged to construct the 3×3 matrix $\mathbf{F}$.

Note, however, that this result will not be a valid fundamental matrix, because image noise will cause it to be full rank. To ensure that the resulting matrix is rank 2, we must compute the SVD, $\mathbf{F} = \mathbf{U\Sigma V}^\mathsf{T}$, then set the smallest singular value to zero, then put the matrix back together: $\mathbf{U} \operatorname{diag}(\sigma_1, \sigma_2, 0)\,\mathbf{V}^\mathsf{T}$, which yields the closest matrix in the Frobenius norm sense that has rank 2 and is therefore a valid fundamental matrix.[†] If, instead, the essential matrix is desired, the same procedure is followed except that in this last step we set $\sigma_1 = \sigma_2 = 1$, leading to $\mathbf{U} \operatorname{diag}(1, 1, 0)\mathbf{V}^\mathsf{T}$.

Just as we did for the DLT, the coordinates must first be normalized to achieve robust results using the matrices $\mathbf{T}$ and $\mathbf{T}'$ which are computed in the same manner as before. If we let $\check{\mathbf{x}} = \mathbf{Tx}$ and $\check{\mathbf{x}}' = \mathbf{T}'\mathbf{x}'$, then substituting into $\mathbf{x}'^\mathsf{T}\mathbf{Fx} = 0$ reveals

$$
\left(\mathbf{T}'^{-1}\check{\mathbf{x}}'\right)^\mathsf{T}\mathbf{F}\left(\mathbf{T}^{-1}\check{\mathbf{x}}\right) = 0
\tag{13.188}
$$

$$
\check{\mathbf{x}}'^\mathsf{T}\underbrace{\mathbf{T}'^{-\mathsf{T}}\mathbf{F}\mathbf{T}^{-1}}_{\check{\mathbf{F}}}\check{\mathbf{x}} = 0
\tag{13.189}
$$

which means $\check{\mathbf{F}} = \mathbf{T}'^{-\mathsf{T}}\mathbf{F}\mathbf{T}^{-1}$, or $\mathbf{F} = \mathbf{T}'^\mathsf{T}\check{\mathbf{F}}\mathbf{T}$. The pseudocode is provided in Algorithm 13.13. As before, the result can be considered a starting point for a nonlinear minimization routine, if more accurate results are desired. An example of the epipolar lines obtained by estimating $\mathbf{F}$ in this manner is shown in Figure 13.31.

[†] Section 11.1.5 (p. 520).

ALGORITHM 13.13 Eight-point algorithm for estimating the fundamental matrix

EightPointFundamental$\left(\{\mathbf{x}_i \leftrightarrow \mathbf{x}_i'\}_{i=1}^n\right)$

Input: n corresponding pairs of points $(x_i, y_i) \leftrightarrow (x_i', y_i')$ between two images
Output: the 3×3 fundamental matrix that best satisfies $(\mathbf{x}_i')^\mathsf{T}\mathbf{F}\mathbf{x}_i = 0$ for all i

1   Compute $\mathbf{T}$ and $\mathbf{T}'$ using Equation (13.118)
2   Normalize the points: $\check{\mathbf{x}}_i \leftarrow \mathbf{T}\mathbf{x}_i$, $\check{\mathbf{x}}_i' \leftarrow \mathbf{T}'\mathbf{x}_i'$, $i = 1, \ldots, n$
3   Construct the $n \times 9$ matrix $\check{\mathbf{A}}$ in Equation (13.187) using the normalized points
4   Solve the linear system $\check{\mathbf{A}}\,\check{\mathbf{f}} = \mathbf{0}$ for $\check{\mathbf{f}}$
5   Rearrange the values in $\check{\mathbf{f}}$ to create $\check{\mathbf{F}}$
6   Compute $\check{\mathbf{U}}, \check{\mathbf{\Sigma}}, \check{\mathbf{V}}^\mathsf{T} \leftarrow \text{Svd}(\check{\mathbf{F}})$
7   Set the smallest singular value to zero: $\check{\mathbf{F}} \leftarrow \check{\mathbf{U}}\, diag(\check{\sigma}_1, \check{\sigma}_2, 0)\check{\mathbf{V}}^\mathsf{T}$
8   Unnormalize the result to yield the desired fundamental matrix: $\mathbf{F} \leftarrow \mathbf{T}'^\mathsf{T}\check{\mathbf{F}}\mathbf{T}$.
9   **return F**

13.6.6 Decomposing the Essential Matrix

Given an essential matrix $\mathbf{E}$, we would like to be able to extract the translation vector $\mathbf{t}$ and rotation matrix $\mathbf{R}$. Since $\mathbf{E} = [\mathbf{t}]_\times \mathbf{R}$, this can be done by decomposing the matrix into the product of a skew-symmetric matrix $[\mathbf{t}]_\times$ and an orthogonal matrix $\mathbf{R}$. To do this, let us define two special matrices:

$$\hat{\mathbf{\Sigma}} \equiv \begin{bmatrix} 1 & 0 & 0 \\ 0 & 1 & 0 \\ 0 & 0 & 0 \end{bmatrix} \quad \mathbf{W} \equiv \begin{bmatrix} 0 & -1 & 0 \\ 1 & 0 & 0 \\ 0 & 0 & 1 \end{bmatrix} \tag{13.190}$$

Since $\mathbf{E}$, by definition, has two nonzero identical singular values and one zero singular value, the former matrix is simply the singular value matrix that results from computing the SVD of $\mathbf{E}$, that is, $\mathbf{E} = \mathbf{U}\hat{\mathbf{\Sigma}}\mathbf{V}^\mathsf{T}$ for some $\mathbf{U}$ and $\mathbf{V}$. The latter matrix is special because it satisfies two important properties, both of which are easy to verify by inspection: (1) $\mathbf{W}\hat{\mathbf{\Sigma}}$ is skew-symmetric, that is, $\mathbf{W}\hat{\mathbf{\Sigma}} = -(\mathbf{W}\hat{\mathbf{\Sigma}})^\mathsf{T}$; and (2) $\mathbf{W}\hat{\mathbf{\Sigma}}\mathbf{W} \propto \hat{\mathbf{\Sigma}}$.

Figure 13.31 Two images of a static scene, with some of the epipolar lines overlaid, whose intersection reveals the epipoles. Note that corresponding points on the two images lie on corresponding epipolar lines.

Stan Birchfield

Decomposing $\mathbf{E}$ is then straightforward by computing its SVD and applying these properties:

$$\mathbf{E} = \mathbf{U}\hat{\boldsymbol{\Sigma}}\mathbf{V}^{\mathsf{T}} \propto \mathbf{U}(\mathbf{W}\hat{\boldsymbol{\Sigma}}\mathbf{W})\mathbf{V}^{\mathsf{T}} = \mathbf{U}\mathbf{W}\hat{\boldsymbol{\Sigma}}(\mathbf{U}^{\mathsf{T}}\mathbf{U})\mathbf{W}\mathbf{V}^{\mathsf{T}} = \underbrace{(\mathbf{U}\mathbf{W}\hat{\boldsymbol{\Sigma}}\mathbf{U}^{\mathsf{T}})}_{\mathbf{S}}\underbrace{(\mathbf{U}\mathbf{W}\mathbf{V}^{\mathsf{T}})}_{\mathbf{R}} \qquad (13.191)$$

where $\mathbf{S} = [\mathbf{t}]_\times$ is a skew-symmetric matrix, and $\mathbf{R}$ is an orthogonal matrix. Therefore, the rotation matrix is computed as $\mathbf{R} = \mathbf{U}\mathbf{W}\mathbf{V}^{\mathsf{T}}$. To obtain the translation vector $\mathbf{t}$, recall that $\mathbf{E}^{\mathsf{T}}\mathbf{t} = \mathbf{V}\hat{\boldsymbol{\Sigma}}\mathbf{U}^{\mathsf{T}}\mathbf{t} = \mathbf{0}_{\{3\times 1\}}$, from which it follows that $\mathbf{t} = \mathbf{U}\begin{bmatrix} 0 & 0 & 1 \end{bmatrix}^{\mathsf{T}} = \mathbf{u}_3$, which is the rightmost column of $\mathbf{U} = \begin{bmatrix} \mathbf{u}_1 & \mathbf{u}_2 & \mathbf{u}_3 \end{bmatrix}$. However, our choice of $\mathbf{W}$ rather than $\mathbf{W}^{\mathsf{T}}$ was arbitrary (a case of the scale ambiguity), and we also know $\mathbf{t}$ only up to an unknown scale factor. So actually there are four possible solutions, namely, $\mathbf{R} = \mathbf{U}\mathbf{W}\mathbf{V}^{\mathsf{T}}$ or $\mathbf{R} = \mathbf{U}\mathbf{W}^{\mathsf{T}}\mathbf{V}^{\mathsf{T}}$, and $\mathbf{t} = \pm\mathbf{u}_3$. Thankfully, exactly one of these four solutions is physically plausible because the other three solutions place world points behind one or both of the cameras, leading to the pseudocode in Algorithm 13.14.

It is now possible to describe a procedure for estimating the projection matrices of two cameras with overlapping fields of view, presented in Algorithm 13.15. Before calling this procedure, the cameras must first be calibrated internally using Algorithm 13.12. Then, a number of corresponding points are determined between the images, either manually or automatically. These points are used to estimate the fundamental matrix, from which the essential matrix is computed. The essential matrix is decomposed into rotation and translation, and the camera projection matrices are constructed. Note that this procedure assumes a particular world coordinate system; therefore, other solutions are possible, as long as the relative rotation and translation between the cameras remains the same.

13.6.7 Computing 3D Point Coordinates

Given the camera projection matrices $\mathbf{P}_{\{3\times 4\}}$ and $\mathbf{P}'_{\{3\times 4\}}$, the 3D coordinates of a world point can be estimated from its projections onto the two images. Let $\mathbf{w}_{\{4\times 1\}}$ be the homogeneous coordinates of the world point, and let $\mathbf{x} = \mathbf{P}\mathbf{w}$ and $\mathbf{x}' = \mathbf{P}'\mathbf{w}$ be its projection onto the two image planes. Note that $\mathbf{x} \leftrightarrow \mathbf{x}'$ are the two corresponding points. Let $\mathbf{p}_i^{\mathsf{T}}$ be the ith row of $\mathbf{P}$, and similarly for $\mathbf{p}_i'^{\mathsf{T}}$:

$$\mathbf{P} = \begin{bmatrix} \mathbf{p}_1^{\mathsf{T}} \\ \mathbf{p}_2^{\mathsf{T}} \\ \mathbf{p}_3^{\mathsf{T}} \end{bmatrix} \quad \text{and} \quad \mathbf{P}' = \begin{bmatrix} \mathbf{p}_1'^{\mathsf{T}} \\ \mathbf{p}_2'^{\mathsf{T}} \\ \mathbf{p}_3'^{\mathsf{T}} \end{bmatrix} \qquad (13.192)$$

ALGORITHM 13.14 Decompose the essential matrix to reveal rotation and translation

DECOMPOSEESSENTIALMATRIX($\mathbf{E}$)

Input: essential matrix $\mathbf{E}$
Output: the 3×3 rotation matrix $\mathbf{R}$ and 3×1 translation vector $\mathbf{t}$ such that $\mathbf{E} = [\mathbf{t}]_\times \mathbf{R}$

1	$\mathbf{U}, \hat{\boldsymbol{\Sigma}}, \mathbf{V}^{\mathsf{T}} \leftarrow \text{SVD}(\mathbf{E})$	➤ Compute the SVD of $\mathbf{E}$, noting that $\sigma_1 = \sigma_2 = 1$ and $\sigma_3 = 0$.
2	$\mathbf{R}_1 \leftarrow \mathbf{U}\mathbf{W}\mathbf{V}^{\mathsf{T}}$	➤ Construct 2 possible rotation matrices,
3	$\mathbf{R}_2 \leftarrow \mathbf{U}\mathbf{W}^{\mathsf{T}}\mathbf{V}^{\mathsf{T}}$	and 2 possible translation vectors, where $\mathbf{U} = \begin{bmatrix} \mathbf{u_1} & \mathbf{u_2} & \mathbf{u_3} \end{bmatrix}$.
4	$\mathbf{t}_1 \leftarrow \mathbf{u}_3$	➤ Then test the four solutions on an arbitrary world point;
5	$\mathbf{t}_2 \leftarrow \mathbf{u}_3$	the solution which places the point
6	**return** either $\mathbf{R}_1, \mathbf{t}_1$ or $\mathbf{R}_1, \mathbf{t}_2$ or $\mathbf{R}_2, \mathbf{t}_1$ or $\mathbf{R}_2, \mathbf{t}_2$	in front of both cameras is the correct one.

ALGORITHM 13.15 Estimate the projection matrices of a pair of internally calibrated cameras

ESTIMATECAMERAPROJECTIONMATRICES$\left(\{\mathbf{x}_i \leftrightarrow \mathbf{x}_i'\}_{i=1}^n, \mathbf{K}, \mathbf{K}'\right)$

Input: n corresponding pairs of points $(x_i, y_i) \leftrightarrow (x_i', y_i')$ between two images intrinsic camera parameters
$\mathbf{K}$ and $\mathbf{K}'$

Output: the 3×4 camera projection matrices $\mathbf{P}$ and $\mathbf{P}'$

1   $\mathbf{F} \leftarrow$ EIGHTPOINTFUNDAMENTAL$\left(\{\mathbf{x}_i \leftrightarrow \mathbf{x}_i'\}_{i=1}^n\right)$
2   $\mathbf{E} \leftarrow \mathbf{K}'^\mathsf{T}\mathbf{F}\mathbf{K}$
3   $\mathbf{R}, \mathbf{t} \leftarrow$ DECOMPOSEESSENTIALMATRIX$(\mathbf{E})$
4   $\mathbf{P} \leftarrow \mathbf{K}\left[\mathbf{I}_{\{3 \times 3\}} \quad \mathbf{0}_{\{3 \times 1\}}\right]$
5   $\mathbf{P}' \leftarrow \mathbf{K}'\left[\mathbf{R} \quad \mathbf{t}\right]$
6   **return** $\mathbf{P}, \mathbf{P}'$

The coordinates of the image point (in pixels) are then given by

$$(x, y) = \left(\frac{\mathbf{p}_1^\mathsf{T}\mathbf{w}}{\mathbf{p}_3^\mathsf{T}\mathbf{w}}, \frac{\mathbf{p}_2^\mathsf{T}\mathbf{w}}{\mathbf{p}_3^\mathsf{T}\mathbf{w}}\right) \quad \text{and} \quad (x', y') = \left(\frac{\mathbf{p}_1'^\mathsf{T}\mathbf{w}}{\mathbf{p}_3'^\mathsf{T}\mathbf{w}}, \frac{\mathbf{p}_2'^\mathsf{T}\mathbf{w}}{\mathbf{p}_3'^\mathsf{T}\mathbf{w}}\right) \tag{13.193}$$

Cross-multiplying yields four equations for the four unknowns:

$$x\mathbf{p}_3^\mathsf{T}\mathbf{w} - \mathbf{p}_1^\mathsf{T}\mathbf{w} = 0 \quad \text{and} \quad y\mathbf{p}_3^\mathsf{T}\mathbf{w} - \mathbf{p}_2^\mathsf{T}\mathbf{w} = 0 \tag{13.194}$$

$$x'\mathbf{p}_3'^\mathsf{T}\mathbf{w} - \mathbf{p}_1'^\mathsf{T}\mathbf{w} = 0 \quad \text{and} \quad y'\mathbf{p}_3'^\mathsf{T}\mathbf{w} - \mathbf{p}_2'^\mathsf{T}\mathbf{w} = 0 \tag{13.195}$$

Since these equations are linear in the unknown $\mathbf{w}$, we can stack them into matrix form:

$$\underbrace{\begin{bmatrix} x\mathbf{p}_3^\mathsf{T} - \mathbf{p}_1^\mathsf{T} \\ y\mathbf{p}_3^\mathsf{T} - \mathbf{p}_2^\mathsf{T} \\ x'\mathbf{p}_3'^\mathsf{T} - \mathbf{p}_1'^\mathsf{T} \\ y'\mathbf{p}_3'^\mathsf{T} - \mathbf{p}_2'^\mathsf{T} \end{bmatrix}}_{\mathbf{A}_{\{4 \times 4\}}} \mathbf{w} = \mathbf{0}_{\{4 \times 1\}} \tag{13.196}$$

and solve for $\mathbf{w}$ by selecting the eigenvector of $\mathbf{A}^\mathsf{T}\mathbf{A}$ associated with the smallest eigenvalue. This **triangulation** method is easy to implement, will usually give acceptable results, and generalizes well to more than two images. Keep in mind, however, that it minimizes the algebraic error, not the geometric error.

13.6.8 Homography Resulting From Two Stereo Images of a Plane

We have covered a lot of ground in this chapter, and the mathematics has been fairly dense. Even if all the details have not been easy to follow, hopefully at least some of the basic procedures and fundamental principles have been grasped. We close this chapter by considering the special case that arises when a stereo pair of cameras views a plane in the world. As we saw earlier, $\boldsymbol{\ell} = \begin{bmatrix} a & b & c \end{bmatrix}^\mathsf{T}$ represents the line $ax + by + c = 0$ in the plane, so that a point $\mathbf{x} = \begin{bmatrix} x & y & 1 \end{bmatrix}^\mathsf{T}$ lies on the line if and only if $\mathbf{x}^\mathsf{T}\boldsymbol{\ell} = 0$. Similarly, $\boldsymbol{\pi} = \begin{bmatrix} a & b & c & d \end{bmatrix}^\mathsf{T}$ represents the plane $ax + by + cz + d = 0$ in 3D space, so that a point $\mathbf{x} = \begin{bmatrix} x & y & z & 1 \end{bmatrix}^\mathsf{T}$ lies on the plane if and only if $\mathbf{x}^\mathsf{T}\boldsymbol{\pi} = 0$. Now, since the projective transform between two planes is a 3×3 homography, if we view a plane $\boldsymbol{\pi}$

in the world with two different cameras, then the transform between this world plane and each image plane is a homography, and therefore the transform between the two image planes is also a homography (that is, for all image points that are projections of points on the world plane).

To relate this homography to the epipolar geometry between two calibrated cameras, let $\mathbf{R}$ and $\mathbf{t}$ be the rotation and translation, respectively, between the two cameras, and let $\mathbf{K}$ and $\mathbf{K}'$ be the intrinsic calibration matrices of the cameras. The homogeneous coordinates of the projection of the world point (x_w, y_w, z_w) onto the first image plane are given by

$$\mathbf{x} \propto \mathbf{P}\mathbf{w} = \mathbf{K}\left[I_{\{3 \times 3\}} \quad \mathbf{0}_{\{3 \times 1\}}\right]\mathbf{w} = \mathbf{K}\begin{bmatrix} x_w \\ y_w \\ z_w \end{bmatrix} \tag{13.197}$$

where $\mathbf{w} \equiv \begin{bmatrix} x_w & y_w & z_w & 1 \end{bmatrix}^\mathsf{T}$. Inverting reveals

$$\begin{bmatrix} x_w \\ y_w \\ z_w \end{bmatrix} \propto \mathbf{K}^{-1}\mathbf{x}, \quad \text{or} \quad \mathbf{w} = \begin{bmatrix} x_w \\ y_w \\ z_w \\ 1 \end{bmatrix} \propto \begin{bmatrix} \mathbf{K}^{-1}\mathbf{x} \\ \lambda \end{bmatrix} \tag{13.198}$$

where $\lambda \neq 0$ is an unknown scale factor.

Let us represent the world plane in the first camera's coordinate system as $\boldsymbol{\pi}_{\{4 \times 1\}} = \begin{bmatrix} \mathbf{n}^\mathsf{T}_{\{3 \times 1\}} & d \end{bmatrix}^\mathsf{T}$, where $\mathbf{n}$ is the normal to the plane, and d is the distance from the plane to the origin (first center of projection). Since the world point lies on the world plane, we have $\mathbf{w}^\mathsf{T}\boldsymbol{\pi} = \boldsymbol{\pi}^\mathsf{T}\mathbf{w} = 0$. Substituting reveals

$$\begin{bmatrix} \mathbf{n}^\mathsf{T} & d \end{bmatrix}\mathbf{w} \propto \begin{bmatrix} \mathbf{n}^\mathsf{T} & d \end{bmatrix}\begin{bmatrix} \mathbf{K}^{-1}\mathbf{x} \\ \lambda \end{bmatrix} = 0 \tag{13.199}$$

which implies $\lambda = -\frac{1}{d}\mathbf{n}^\mathsf{T}\mathbf{K}^{-1}\mathbf{x}$. Applying these results, we see that the projection of the world point onto the other image is given by

$$\mathbf{x}' \propto \mathbf{P}'\mathbf{w} \propto \mathbf{K}'\begin{bmatrix} \mathbf{R} & \mathbf{t} \end{bmatrix}\begin{bmatrix} \mathbf{K}^{-1}\mathbf{x} \\ -\frac{1}{d}\mathbf{n}^\mathsf{T}\mathbf{K}^{-1}\mathbf{x} \end{bmatrix} = \mathbf{K}'\left(\mathbf{R}\mathbf{K}^{-1}\mathbf{x} - \frac{1}{d}\mathbf{t}\mathbf{n}^\mathsf{T}\mathbf{K}^{-1}\mathbf{x}\right) \tag{13.200}$$

or, in other words,

$$\mathbf{x}' \propto \underbrace{\mathbf{K}'\left(\mathbf{R} - \frac{1}{d}\mathbf{t}\mathbf{n}^\mathsf{T}\right)\mathbf{K}^{-1}}_{\mathbf{H}_\pi}\mathbf{x} \tag{13.201}$$

where $\mathbf{H}_\pi = \mathbf{K}'\left(\mathbf{R} - \frac{1}{d}\mathbf{t}\mathbf{n}^\mathsf{T}\right)\mathbf{K}^{-1}$ is the homography between the images induced by the plane $\boldsymbol{\pi}$.

Since the coordinates of the points also satisfy the epipolar constraint, we have $\mathbf{x}'^\mathsf{T}\mathbf{F}\mathbf{x} = 0$. Substituting the homography above yields $(\mathbf{H}_\pi\mathbf{x})^\mathsf{T}\mathbf{F}\mathbf{x} = \mathbf{x}^\mathsf{T}\mathbf{H}_\pi^\mathsf{T}\mathbf{F}\mathbf{x} = 0$. It can be shown that a matrix $\mathbf{A}$ is skew-symmetric, that is, $\mathbf{A} = -\mathbf{A}^\mathsf{T}$, if and only if $\mathbf{x}^\mathsf{T}\mathbf{A}\mathbf{x} = 0$ for all $\mathbf{x}$. Therefore, since this result must hold for all $\mathbf{x}$, it must be the case that $\mathbf{H}_\pi^\mathsf{T}\mathbf{F}$ is skew-symmetric. It follows, then, that a given homography $\mathbf{H}$ and a given fundamental matrix $\mathbf{F}$ are compatible if and only if $\mathbf{H}^\mathsf{T}\mathbf{F} = -\mathbf{F}^\mathsf{T}\mathbf{H}$.

13.7 Further Reading

Stereo perception is covered in the classic book by Julesz [1971], the random dot stereogram is due to Julesz [1960], the autostereogram is due to Tyler and Clarke [1990], and early algorithms for stereo matching include those of Marr and Poggio [1976], Marr et al. [1978], and Marr and Poggio [1979]. The disparity gradient limit was first put into practice by Pollard et al. [1985], citing the earlier discussions of Burt and Julesz [1980], and it was rediscovered independently by Little and Gillett [1990], who also introduced the forbidden zone. The term *matching space* comes from Geiger et al. [1995], and it is similar to the disparity space image of Intille and Bobick [1994]. The concept of casting shadows on the Cyclopean disparity function is due to Belhumeur [1996], which also describes a straightforward dynamic programming algorithm for matching based on the classic dynamic programming approach of Ohta and Kanade [1985]. The more recent dynamic programming approach known as semiglobal matching is due to Hirschmuller [2008].

Regarding block matching, the left-right disparity check is due to Fua [1991], adaptive window sizes are addressed in the classic work of Kanade and Okutomi [1994], the advantage of using more than two cameras is shown by Okutomi and Kanade [1993], and an effective block-matching algorithm with reduced border errors is described by Hirschmuller et al. [2002]. The rank and census transforms for dealing with factionalism are due to Zabih and Woodfill [1994], while a similarity measure that addresses the problem of image sampling can be found in the work of Birchfield and Tomasi [1998]. Another approach to aligning images is to use mutual information, as explained in Viola and Wells [1997]. The well-known study comparing dense two-frame stereo algorithms, along with the accompanying Middlebury stereo website,[†] is that of Scharstein and Szeliski [2002].

Space did not permit us to cover alternative approaches to depth perception in any detail. Nevertheless, those interested in photometric stereo should consult the classic work of Woodham [1980]. An interesting discussion of depth from defocus and how it relates to triangulation-based stereo can be found in Schechner [2000]. A popular survey of structured light techniques is the one by Salvi et al. [2004], and a well-known approach to wide-baseline stereo matching is that of Matas et al. [2004]. We have also not had the space to cover mosaicking, for which a good overview has been provided by Szeliski [2006], or the Perspective-n-Point (PnP) problem, solutions for which can be found in the works of Haralick et al. [1994],

Dementhon and Davis [1995], Lu et al. [2000], and Lepetit et al. [2009]. Another omission due to lack of space is the factorization method of Tomasi and Kanade [1992].

The Lucas-Kanade algorithm was originally presented in the context of stereo by Lucas and Kanade [1981]. It was later extended to feature tracking by Tomasi and Kanade [1991] and augmented with an affine warp for determining lost features by Shi and Tomasi [1994]. It was also modified to track large regions using precomputed templates viewed under variable lighting conditions by Hager and Belhumeur [1998]. A thorough overview of the algorithm, along with variations such as the inverse compositional warp, can be found in Baker and Matthews [2004]. The classic Horn-Schunck algorithm for dense optical flow estimation is described in the classic paper by Horn and Schunck [1981]. Another important problem in optical flow is the estimation of multiple motions as described in Black and Anandan [1996]. More recent approaches to optical flow estimation can be found in the papers by Brox et al. [2004] and Brox and Malik [2011].

Zhang's calibration algorithm, including a description of the image of the absolute conic, can be found in the classic paper by Zhang [2000]; a nearly identical algorithm was discovered simultaneously by Sturm and Maybank [1999]. The lens distortion model presented here is the same as that of Brown [1971]. Several subsequent authors have reversed the roles of the distorted and undistorted coordinates, and some have argued that the same equation works in both directions. In support of this view, de Villiers et al. [2008] show that with enough coefficients it is possible to model the equation in either direction.

The authoritative source on projective geometry for computer vision, and multiple-view geometry in particular, is the book by Hartley and Zisserman [2003], which contains a useful description of the normalized DLT algorithm, along with details about the properties of the fundamental and essential matrices. The DLT algorithm is originally due to Aziz and Karara [1971]. The eight-point algorithm is due to Longuet-Higgins [1981], but for years researchers complained that results were not reliable; these problems largely disappeared once the importance of normalization was realized, as championed by Hartley [1997]. The essential matrix is described thoroughly in the classic text of Faugeras [1993], which was followed by a detailed analysis of the fundamental matrix in Luong and Faugeras [1996]. The relationship between the fundamental matrix and homography of a world plane is due to Luong and Faugeras [1993].

[†] http://vision.middlebury.edu/stereo

PROBLEMS

13-1 Define retinal disparity. Suppose your two eyes are fixated on a small object at some distance away. What is the retinal disparity of the object?

13-2 Suppose two cameras are positioned such that both image planes are parallel, but one plane is slightly in front of the other. Are the cameras rectified? Why or why not?

13-3 Suppose an object 2 m away is viewed by a rectified pair of stereo cameras with a baseline of 50 mm, and the lens of each camera has a focal length of 35 mm. What is the disparity?

13-4 Which of the following pixels in the right image could possibly match the pixel (52,3) in the left image, assuming the images are rectified, and the maximum disparity is 20? (a) (26,3), (b) (48,13), (c) (64,3), (d) (48,3), and (e) (59,6).

13-5 Explain the relationship between the epipolar constraint and rectified cameras.

13-6 Consider the ordering constraint. (a) What other constraint is implied by it? (b) What zone describes the set of matches that it forbids? (c) What is another name for the constraint? (d) Give an example when it is violated.

13-7 Consider a thin, opaque pole just thicker than the interpupillary distance. How close does the pole have to be to the cameras in order to violate the ordering constraint?

13-8 The function $d^*(x) \equiv \max_a(d(x + a) - \kappa|a|)$ is equivalent to grayscale dilation of the function d with what 1D structuring element?

13-9 Given a constant κ, a function f is said to be *Lipschitz continuous* if and only if

$$|f(x + h) - f(x)| \le \kappa|h| \tag{13.202}$$

for all x and h. The smallest such κ is called the Lipschitz constant of the function, and the function is called a Lipschitz function. Lipschitz continuity is a smoothness condition on functions which is stronger than regular continuity. Show that, if the disparity gradient limit is satisfied, then the Cyclopean disparity function is Lipschitz continuous with the same constant κ.

13-10 Suppose the left disparity map of a scanline from a pair of rectified images is given by $0,0,0,2,2,2,0,0,0$ for the pixels $x_L = 0$ through $x_L = 8$.

(a) Compute the right disparity map for $x_R = 0$ through $x_R = 8$, assuming that occluded pixels are part of the background.

(b) Compute the Cyclopean disparity function over the coordinates associated with the discrete matches.

13-11 Compute the disparity gradient (a) between the matches labeled a and b in the figure below, and (b) between the matches labeled c and d.

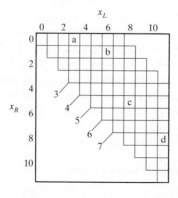

13-12 Implement a basic algorithm for matching two rectified stereo images.

(a)   Implement BLOCKMATCH1 of Algorithm 13.1.

(b)   Implement BLOCKMATCH2 of Algorithm 13.2, and compare running times for different values of w.

(c)   Add the left-right check using Algorithm 13.4, and compare results with those of (a) and (b). What do you notice?

13-13 Show that minimizing the SSD is nearly the same as maximizing the cross-correlation. Under what circumstances are they identical?

13-14 Compute the rank and census transforms of the following window of grayscale pixel values:

$$\begin{bmatrix} 8 & 5 & 3 \\ 7 & 6 & 9 \\ 1 & 1 & 2 \end{bmatrix}$$

13-15 Apply Algorithm 13.5 to compute the edit distance, as well as the matching function, between the strings "cart" and "earth."

13-16 Explain the key idea behind semi-global matching.

13-17 Explain the difference between motion field and optical flow. Given an example where a nonzero motion field results in zero optical flow, and vice versa.

13-18 List some situations that would cause the brightness constancy assumption to fail.

13-19 Explain (a) why the optical flow constraint equation in Equation (13.28) is actually an approximation, not an equation. Also explain (b) why it contains a mixture of partial and total derivatives, and (c) how $\partial I/\partial t$ is usually computed.

13-20 Implement the Lucas-Kanade method.

13-21 What problem is generalized Lucas-Kanade attempting to solve?

13-22 Explain how the Horn-Schunck algorithm relates to the Lucas-Kanade method. What are some similarities and differences?

13-23 Suppose we set $\lambda = 0$ in the Horn-Schunck algorithm. (a) What is the rank of the matrix on the left-hand side of Equation (13.77), and what are the implications for solving for $[u \quad v]^T$? (b) State what direction $[u \quad v]^T$ will be shifted relative to $[\bar{u} \quad \bar{v}]^T$, and relate this answer with the finding of Equation (13.31).

13-24 Given the point (5,2) in the 2D Euclidean plane, (a) write the homogeneous coordinates of the point, (b) write the coordinates of the equivalent point on the $w = 1$ plane, and (c) write the coordinates of the equivalent point on the positive unit hemisphere.

13-25 Convert the homogeneous coordinates (27,18,3) to inhomogeneous coordinates.

13-26 Apply the following homogeneous transformation to the point (3,7), converting the result back to the Euclidean plane:

$$\begin{bmatrix} 8 & 2 & 4 \\ 3 & 9 & 1 \\ 6 & 7 & 2 \end{bmatrix}$$

13-27 Use homogeneous coordinates to simplify the computation of

(a)   The point at the intersection of the lines $7x + 3y - 6 = 0$ and $y = -2x + 16$

(b)   The line joining the points $(6,2)$ and $(1,9)$

13-28 Use homogeneous coordinates to compute the intersection of the lines $8x + 2y - 3 = 0$ and $y = -4x + 12$. What do you notice about the result?

13-29 Given the ellipse $4x^2 + 3y^2 = 100$,

(a)   Draw the ellipse.

(b)   Construct the 3×3 matrix $\mathbf{C}$ representing the conic in homogeneous coordinates.

(c)   Compute the dual conic $\mathbf{C}^*$.

(d)   Use $\mathbf{C}$ to compute three different points on the conic, and add these points to your drawing.

(e)   Use $\mathbf{C}^*$ to compute three different tangent lines to the conic, and add these lines to your drawing.

13-30 Which of the following statements is true?

(a)   A similarity transformation is always an affine transformation.

(b)   A projective transformation is always a similarity transformation.

(c)   A similarity transformation is a Euclidean transformation if the scaling is 1.

13-31 Apply the following similarity transformation to the absolute points:

$$\begin{bmatrix} 5.196 & -3.000 & 6.789 \\ 3.000 & 5.196 & -8.312 \\ 0 & 0 & 1 \end{bmatrix}$$

13-32 How many parameters specify the imaging process of a single pinhole camera? List these parameters, and specify for each whether it is extrinsic or intrinsic.

13-33 List the two most common types of lens distortion.

13-34 Apply the normalized DLT algorithm to the following set of corresponding point pairs:

x	(45,45)	(340,45)	(45,125)	(340,125)	(45,205)	(340,205)
x'	(99,43)	(287,111)	(98,128)	(311,184)	(96,203)	(333,241)

13-35 Suppose you are given the following matrix representing the inverse of the image of the absolute conic (IAC). Compute the camera intrinsic parameters.

$$\omega^{-1} = \mathbf{K}\mathbf{K}^{\mathsf{T}} = \begin{bmatrix} 231664.36 & 76045.34 & 335.00 \\ 76045.34 & 168561.41 & 227.00 \\ 335.00 & 227.00 & 1 \end{bmatrix}$$

13-36 Calibrate an actual camera using Zhang's algorithm. That is, obtain access to a camera, download an implementation of a camera calibration routine (nearly all implementations are some variation of Zhang's algorithm), print out a chessboard pattern and tape it to a hard flat surface, capture several images, obtain coordinates of the corners either manually or automatically, and feed the coordinates into the code to compute the intrinsic parameters. (Ignore the extrinsic parameters.)

13-37 Define the following terms: epipole, epipolar line, epipolar plane, and epipolar constraint.

13-38 Explain the relationship between the fundamental matrix and the essential matrix.

13-39 Prove the following additional properties of essential matrices. For simplicity, just prove the "only if" part—that is, show that these properties hold for any essential matrix:

(a) A real 3×3 matrix $\mathbf{E}$ is an essential matrix if and only if $\det(\mathbf{E}) = 0$ and $\frac{1}{2}tr^2(\mathbf{EE}^\mathsf{T}) = tr((\mathbf{EE}^\mathsf{T})^2)$.

(b) A real 3×3 matrix $\mathbf{E}$ is an essential matrix if and only if $\mathbf{EE}^\mathsf{T}\mathbf{E} = \frac{1}{2}tr(\mathbf{EE}^\mathsf{T})\mathbf{E}$.

13-40 Prove that the essential matrix is given by $\mathbf{E} = \mathbf{R}[\mathbf{R}^\mathsf{T}\mathbf{t}]_\times$, as stated in Equation (13.182).

13-41 (a) Estimate the fundamental matrix from the following set of correspondences, (b) compute the epipolar line for each point, and (c) display the epipolar lines on a pair of plots:

$\mathbf{x}$	(190,155)	(420,114)	(252,29)	(150,111)	(35,228)	(443,230)	(149,240)	(276,324)
$\mathbf{x}'$	(231,138)	(442,98)	(272,9)	(169,91)	(58,209)	(460,217)	(184,225)	(312,310)

13-42 Decompose the following essential matrix into rotation and translation.

$$\begin{bmatrix} -0.8497 & -0.9437 & 9.0579 \\ 0.9437 & -0.8497 & -8.1472 \\ -1.2798 & 12.1155 & 0 \end{bmatrix}$$

13-43 Implement block-based matching of a pair of rectified stereo images, using the SAD dissimilarity measure. For efficiency, your code should precompute the 3D array of dissimilarities, followed by a series of separable convolutions (one pair of convolutions per disparity). Implement the left-to-right consistency check, retaining a value in the left disparity map only if the corresponding point in the right disparity map agrees in its disparity. The resulting disparity map should be valid only at the pixels that pass the consistency check; set other pixels to zero. (*Note:* For simplicity, do not worry about setting the values of pixels along the left border of the left image.)

13-44 Implement the detection and tracking of sparse features points throughout a video sequence. To detect good features in the first frame, use either the Harris corner detector or the Tomasi-Kanade method, as explained in Chapter 7. Then, for each pair of consecutive frames, perform Lucas-Kanade tracking of all the features to update their 2D image positions. Remember to keep the feature coordinates as floating point values throughout the tracking process, only rounding for display purposes; to handle noninteger values, use bilinear interpolation. Do not worry about declaring features lost, but simply allow them to continue tracking throughout the sequence, even if they drift to a neighboring surface in the world due to occlusion. Nevertheless, be sure to perform bounds checking so that features that reach the image border do not cause the program to crash due to out-of-bounds memory access.

BIBLIOGRAPHY

I. E. Abdou and W. Pratt. Quantitative design and evaluation of enhancement/thresholding edge detectors. *Proceedings of the IEEE*, 67(5): 753–763, May 1979.

D. Adalsteinsson and J. A. Sethian. A fast level set method for propagating interfaces. *Journal of Computational Physics*, 118(2): 269–277, May 1995.

A. Adams, J. Baek, and A. Davis. Fast high-dimensional filtering using the permutohedral lattice. In *Eurographics*, 2010.

R. Adams and L. Bischof. Seeded region growing. *IEEE Trans. on Pattern Analysis and Machine Intelligence*, 16(6): 641–647, June 1994.

E. H. Adelson and J. R. Bergen. The plenoptic function and the elements of early vision. In M. S. Landy and A. J. Movshon, editors, *Computational Models of Visual Processing*, pages 3–20. MIT Press, Cambridge, MA, 1991.

S. Agarwal, N. Snavely, I. Simon, S. M. Seitz, and R. Szeliski. Building Rome in a day. In *Proceedings of the International Conference on Computer Vision (ICCV)*, Sept. 2009.

N. Ahmed. How I came up with the discrete cosine transform. *Digital Signal Processing*, 1(4-5): 4–9, Jan. 1991.

N. Ahmed, T. Natarajan, and K. R. Rao. Discrete cosine transform. *IEEE Trans. on Computers*, 23(1): 90–93, Jan. 1974.

H. Akaike. A new look at the statistical model identification. *IEEE Trans. on Automatic Control*, 19(6): 716–723, Dec. 1974.

A. Al-Sharadqah and N. Chernov. Error analysis for circle fitting algorithms. *Electronic Journal of Statistics*, 3: 886–911, 2009.

A. A. Amini, T. E. Weymouth, and R. C. Jain. Using dynamic programming for solving variational problems in vision. *IEEE Tran. on Pattern Analysis and Machine Intelligence*, 12(9): 855–867, Sept. 1990.

P. Arbeláez. Boundary extraction in natural images using ultrametric contour maps. In *Proceedings 5th IEEE Workshop on Perceptual Organization in Computer Vision*, June 2006.

P. Arbeláez, M. Maire, C. Fowlkes, and J. Malik. Contour detection and hierarchical image segmentation. *IEEE Transactions on Pattern Analysis and Machine Intelligence*, 33(5): 898–916, May 2011.

D. Arthur and S. Vassilvitskii. K–means++: The advantages of careful seeding. In *Proceedings of the ACM-SIAM Symposium on Discrete Algorithms (SODA)*, pages 1027–1035, 2007.

K. S. Arun, T. S. Huang, and S. D. Blostein. Least-squares fitting of two 3-D point sets. *IEEE Transactions on Pattern Analysis and Machine Intelligence*, 9(5): 698–700, Sept. 1987.

C. G. Atkeson, A. W. Moore, and S. Schaal. Locally weighted learning. *Artificial Intelligence Review*, 11(1–5): 11–73, Feb. 1997.

F. Attneave. Some informational aspects of visual perception. *Psychological Review*, 61(3): 183–193, 1954.

V. Aurich and J. Weule. Non-linear Gaussian filters performing edge preserving diffusion. In *Proceedings of the DAGM Symposium*, pages 538–545, 1995.

S. Avidan. Ensemble tracking. In *Proceedings of the IEEE Conference on Computer Vision and Pattern Recognition (CVPR)*, 2005.

A. Y. I. Aziz and H. M. Karara. Direct linear transformation into object space coordinates in close-range photogrammetry. In *Proc. of the Symposium on Close-Range Photogrammetry*, pages 1–18, Jan. 1971.

J. Babaud, A. P. Witkin, M. Baudin, and R. O. Duda. Uniqueness of the Gaussian kernel for scale-space filtering. *IEEE Transactions on Pattern Analysis and Machine Intelligence*, 8(1): 26–33, 1986.

S. Baker and I. Matthews. Lucas-Kanade 20 years on: A unifying framework. *International Journal of Computer Vision*, 56(3): 221–255, 2004.

D. Ballard. Generalizing the Hough transform to detect arbitrary shapes. *Pattern Recognition*, 13(2), Apr. 1981.

D. H. Ballard and C. M. Brown. *Computer Vision*. Englewood Cliffs, New Jersey: Prentice-Hall, 1982.

D. Barash. A fundamental relationship between bilateral filtering, adaptive smoothing, and the nonlinear diffusion equation. *IEEE Transactions on Pattern Analysis and Machine Intelligence*, 24(6): 844–847, June 2002.

D. Barash and D. Comaniciu. A common framework for nonlinear diffusion, adaptive smoothing, bilateral filtering and mean shift. *Journal of Image and Vision Computing*, 22(1): 73–81, 2004.

J. Barraquand and J. C. Latombe. Robot motion planning: A distributed representation approach. *International Journal of Robotics Research*, 10(6): 628–649, 1991.

H. G. Barrow and J. M. Tenenbaum. Recovering intrinsic scene characteristics from images. In A. Hanson and E. Riseman, editors, *Computer Vision Systems*, pages 3–26. New York: Academic Press, 1978.

R. H. Bartels, J. C. Beatty, and B. A. Barsky. *An Introduction to Splines for Use in Computer Graphics and Geometric Modeling*. Los Altos, CA: Morgan Kauffman, 1987.

H. Bay, A. Ess, T. Tuytelaars, and L. V. Gool. SURF: Speeded up robust features. *Computer Vision and Image Understanding*, 110(3): 346–359, June 2008.

P. R. Beaudet. Rotationally invariant image operators. In *Proceedings of the International Joint Conference on Pattern Recognition*, pages 579–583, 1978.

M. J. Behe. *Darwin's Black Box: The Biochemical Challenge to Evolution*. Free Press, 1996.

P. N. Belhumeur. A Bayesian approach to binocular steropsis. *International Journal of Computer Vision*, 19(3): 237–260, 1996.

P. N. Belhumeur, J. P. Hespanha, and D. J. Kriegman. Eigenfaces vs. Fisherfaces: Recognition using class specific linear projection. *IEEE Transactions on Pattern Analysis and Machine Intelligence*, 19(7): 711–720, July 1997.

S. Belongie, J. Malik, and J. Puzicha. Shape matching and object recognition using shape contexts. *IEEE Transactions on Pattern Analysis and Machine Intelligence*, 24(4): 509–522, Apr. 2002.

B. Berlin and P. Kay. *Basic Color Terms: Their Universality and Evolution*. Berkeley: University of California Press, 1969.

P. J. Besl and R. C. Jain. Segmentation through variable-order surface fitting. *IEEE Transactions on Pattern Analysis and Machine Intelligence*, 10(2): 167–192, Mar. 1988.

P. J. Besl and N. D. McKay. A method for registration of 3-D shapes. *IEEE Transactions on Pattern Analysis and Machine Intelligence*, 14(2): 239–256, Feb. 1992.

S. Beucher and C. Lantuéjoul. Use of watersheds in contour detection. In *International Workshop on Image Processing: Real-time Edge and Motion Detection/Estimation*, Sept. 1979.

S. Beucher and F. Meyer. The morphological approach to segmentation: The watershed transformation. In E. R. Dougherty, editor, *Mathematical Morphology in Image Processing*, pages 433–481. CRC Press, 1992.

S. Birchfield and C. Tomasi. Depth discontinuities by pixel-to-pixel stereo. In *Proceedings of the Sixth International Conference on Computer Vision (ICCV)*, pages 1073–1080, Jan. 1998a.

S. Birchfield and C. Tomasi. A pixel dissimilarity measure that is insensitive to image sampling. *IEEE Transactions on Pattern Analysis and Machine Intelligence*, 20(4): 401–406, Apr. 1998b.

S. T. Birchfield and S. Rangarajan. Spatiograms versus histograms for region-based tracking. In *Proceedings of the IEEE Conference on Computer Vision and Pattern Recognition (CVPR)*, pages 1158–1163, June 2005.

C. M. Bishop. *Pattern Recognition and Machine Learning*. Springer, 2006.

M. J. Black and P. Anandan. The robust estimation of multiple motions: Parametric and piecewise-smooth flow fields. *Computer Vision and Image Understanding*, 63(1): 75–104, Jan. 1996.

I. Blayvas, A. Bruckstein, and R. Kimmel. Efficient computation of adaptive threshold surfaces for image binarization. *Pattern Recognition*, 39: 89–101, 2006.

H. Blum. A transformation for extracting new descriptors of shape. In W. Wathen-Dunn, editor, *Models for the Perception of Speech and Visual Forms*, pages 362–380. Cambridge, Mass.: MIT Press, 1967.

O. Boiman, E. Shechtman, and M. Irani. In defense of nearest-neighbor based image classification. In *Proceedings of the IEEE Conference on Computer Vision and Pattern Recognition (CVPR)*, June 2008.

M. V. Boland, M. K. Markey, and R. F. Murphy. Automated recognition of patterns characteristic of subcellular structures in fluorescence microscopy images. *Cytometry*, 33: 366–375, 1998.

H. Bond. Making unsharp masks for black and white negatives. *Photo Techniques*, pages 63–66, Jan/Feb 1996.

H. Bond. Unsharp masking update. *Photo Techniques*, pages 36–39, Sep/Oct 1997.

F. L. Bookstein. Fitting conic sections to scattered data. *Computer Graphics and Image Processing*, 9(1): 56–71, Jan. 1979.

G. Borgefors. Distance transformations in digital images. *Computer Vision, Graphics, and Image Processing*, 34(3): 344–371, 1986.

D. Borland and R. M. Taylor, II. Rainbow color map (still) considered harmful. *IEEE Computer Graphics and Applications*, 27(2): 14–17, March-April 2007.

E. Boros and P. L. Hammer. Pseudo-boolean optimization. *Discrete applied mathematics*, 123(1): 155–225, 2002.

K. W. Bowyer and P. J. Phillips. *Empirical Evaluation Techniques in Computer Vision*. Wiley-IEEE Computer Society Press, 1998.

S. Boyd and L. Vandenberghe. *Convex Optimization*. Cambridge University Press, 2004.

Y. Boykov and G. Funka-Lea. Graph cuts and efficient N-D image segmentation. *International Journal of Computer Vision*, 70(2): 109–131, 2006.

Y. Boykov and M.-P. Jolly. Interactive graph cuts for optimal boundary and region segmentation of objects in N-D images. In *Proceedings of the International Conference on Computer Vision*, pages 105–112, July 2001.

Y. Boykov and V. Kolmogorov. An experimental comparison of min-cut/max-flow algorithms for energy minimization in vision. *IEEE Transactions on Pattern Analysis and Machine Intelligence*, 26(9): 1124–1137, 2004.

Y. Boykov, O. Veksler, and R. Zabih. Fast approximate energy minimization via graph cuts. *IEEE Transactions on Pattern Analysis and Machine Intelligence*, 23(11): 1222–1239, 2001.

D. Bradley and G. Roth. Adaptive thresholding using the integral image. *ACM Journal of Graphics Tools*, 12(2): 13–21, 2007.

J. N. Bradley and C. M. Brislawn. The wavelet/scalar quantization compression standard for digital fingerprint images. In *IEEE International Symposium on Circuits and Systems*, pages 205–208, May 1994.

J. N. Bradley, C. M. Brislawn, and T. Hopper. FBI wavelet/scalar quantization standard for gray-scale fingerprint image compression. In *Proceedings of the SPIE 1961, Visual Information Processing II*, Aug. 1993.

D. H. Brainard and A. Stockman. Colorimetry. In M. Bass, editor, *OSA Handbook of Optics*. New York: McGraw-Hill, third edition, 2009.

L. Breiman. Bagging predictors. *Machine Learning*, 24(2): 123–140, Aug. 1996.

L. Breiman. Random forests. *Machine Learning*, 45(1): 5–32, Oct. 2001.

L. Breiman, J. Friedman, C. J. Stone, and R. A. Olshen. *Classification and Regression Trees*. Belmont, California: Wadsworth International Group, 1984.

H. Breu, J. Gil, D. Kirkpatrick, and M. Werman. Linear-time Euclidean distance transform algorithms. *IEEE Transactions on Pattern Analysis and Machine Intelligence*, 17(5): 529–533, May 1995.

R. Brinkmann. *The Art and Science of Digital Compositing*. Morgan Kaufmann, second edition, 2008.

D. C. Brown. Close-range camera calibration. *Photogrammetric Engineering*, 37(8): 855–866, 1971.

T. Brox and J. Malik. Large displacement optical flow: Descriptor matching in variational motion estimation. *IEEE Transactions on Pattern Analysis and Machine Intelligence*, 33(3): 500–513, Mar. 2011.

T. Brox, A. Bruhn, N. Papenberg, and J. Weickert. High accuracy optical flow estimation based on a theory for warping. In *Proceedings of the European Conference on Computer Vision (ECCV)*, volume 4, pages 25–36, May 2004.

R. Brunelli and T. Poggio. Face recognition: Features versus templates. *IEEE Transactions on Pattern Analysis and Machine Intelligence*, 15(10): 1042–1052, Oct. 1993.

A. Buades, B. Coll, and J.-M. Morel. A non-local algorithm for image denoising. In *Proceedings of the IEEE Conference on Computer Vision and Pattern Recognition (CVPR)*, volume 2, pages 60–65, June 2005.

W. Burger and M. J. Burge. *Digital Image Processing: An Algorithmic Introduction Using Java*. Springer, first edition, 2008.

C. J. C. Burges. A tutorial on support vector machines for pattern recognition. *Data Mining and Knowledge Discovery*, 2: 121–167, 1998.

P. Burt and B. Julesz. A disparity gradient limit for binocular fusion. *Science*, 208(4444): 615–617, May 1980.

P. J. Burt. Fast filter transforms for image processing. *Computer Graphics and Image Processing*, 16(1): 20–51, May 1981.

P. J. Burt and E. H. Adelson. The Laplacian pyramid as a compact image code. *IEEE Transactions on Communications*, 31(4): 532–540, 1983.

J. B. Campbell and R. H. Wynne. *Introduction to Remote Sensing*. New York: The Guilford Press, fifth edition, 2011.

J. F. Canny. A computational approach to edge detection. *IEEE Transactions on Pattern Analysis and Machine Intelligence*, 8(6): 679–698, Nov. 1986.

C. Carson, S. Belongie, H. Greenspan, and J. Malik. Blobworld: Image segmentation using expectation-maximization and its application to image querying. *IEEE Transactions on Pattern Analysis and Machine Intelligence*, 24(8): 1026–1038, Aug. 2002.

V. Caselles, F. Catté, T. Coll, and F. Dibos. A geometric model for active contours in image processing. *Numerische Mathematik*, 66(1): 1–31, 1993.

V. Caselles, R. Kimmel, and G. Sapiro. Geodesic active contours. *International Journal of Computer Vision*, 22(1): 61–79, Feb. 1997.

K. R. Castleman. *Digital Image Processing*. Prentice Hall, second edition, 1995.

R. Cattell. The scree test for the number of factors. *Multivariate Behavioral Research*, 1(2): 245–276, Apr. 1966.

T. F. Chan and L. A. Vese. Active contours without edges. *IEEE Transactions on Image Processing*, 10(2): 266–277, Feb. 2001.

F. Chang, C.-J. Chen, and C.-J. Lu. A linear-time component-labeling algorithm using contour tracing technique. *Computer Vision and Image Understanding*, 93(2): 206–220, Feb. 2004.

K. N. Chaudhury. Constant-time filtering using shiftable kernels. *IEEE Signal Processing Letters*, 18(11): 651–654, Nov. 2011.

K. N. Chaudhury, D. Sage, and M. Unser. Fast O(1) bilateral filtering using trigonometric range kernels. *IEEE Transactions on Image Processing*, 20(12): 3376–3382, Dec. 2011.

Chen et al. [2007] {chen2007siggraph}.

Y. Chen and G. Medioni. Object modelling by registration of multiple range images. *Image and Vision Computing*, 10(3): 145–155, Apr. 1992.

D. K. Cheng. *Field and Wave Electromagnetics*. Addison Wesley, second edition, 1989.

Y. Z. Cheng. Mean shift, mode seeking, and clustering. *IEEE Transactions on Pattern Analysis and Machine Intelligence*, 17(8): 790–799, Aug. 1995.

N. Chernov and C. Lesort. Least squares fitting of circles. *Journal of Mathematical Imaging and Vision*, 23(3): 239–252, Nov. 2005.

P. Chockalingam, N. Pradeep, and S. T. Birchfield. Adaptive fragments-based tracking of non-rigid objects using level sets. In *Proceedings of the International Conference on Computer Vision (ICCV)*, Oct. 2009.

C. K. Chow and T. Kaneko. Automatic boundary detection of the left-ventricle from cineangiograms. *Computers and Biomedical Research*, 5(4): 388–410, Aug. 1972.

O. Chum, J. Matas, and J. Kittler. Locally optimized RANSAC. In *DAGM-Symposium*, pages 236–243, 2003.

M. B. Clowes. On seeing things. *Artificial Intelligence*, 2(1): 79–116, 1971.

A. Cohen, I. Daubechies, and J.-C. Feauveau. Biorthogonal bases of compactly supported wavelets. *Communications on Pure and Applied Mathematics*, 45(5): 485–560, June 1992.

D. Comaniciu and P. Meer. Mean shift: A robust approach toward feature space analysis. *IEEE Transactions on Pattern Analysis and Machine Intelligence*, 24(5): 603–619, May 2002.

D. Comaniciu, V. Ramesh, and P. Meer. Kernel-based object tracking. *IEEE Transactions on Pattern Analysis and Machine Intelligence*, 25(5): 564–577, May 2003.

J. W. Cooley and J. W. Tukey. An algorithm for the machine calculation of complex Fourier series. *Mathematics of Computation*, 19(90): 297–301, Apr. 1965.

T. F. Cootes, C. J. Taylor, D. H. Cooper, and J. Graham. Active shape models – their training and application. *Computer Vision and Image Understanding*, 61(1): 38–59, Jan. 1995.

T. F. Cootes, G. J. Edwards, and C. J. Taylor. Active appearance models. *IEEE Transactions on Pattern Analysis and Machine Intelligence*, 23(6): 681–685, June 2001.

T. H. Cormen, C. E. Leiserson, and R. L. Rivest. *Introduction to Algorithms*. Cambridge, Massachusetts: The MIT Press, 1990.

C. Cortes and V. Vapnik. Support-vector networks. *Machine Learning*, 20(3): 273–297, Sept. 1995.

T. M. Cover and J. A. Thomas. *Elements of Information Theory*. Wiley, second edition, 1991.

I. J. Cox, J. Kilian, F. T. Leighton, and T. Shamoon. Secure spread spectrum watermarking for multimedia. *IEEE Transactions on Image Processing*, 6(12): 1673 – 1687, Dec. 1997.

J. F. Crawford. A non-iterative method for fitting circular arcs to measured points. *Nuclear Instruments and Methods In Physics Research*, 211(1): 223–225, June 1983.

F. C. Crow. Summed-area tables for texture mapping. *Computer Graphics (SIGGRAPH)*, 18(3): 207–212, July 1984.

J. L. Crowley. A representation for visual information. Technical Report CMU-RI-TR-82-07, Robotics Institute, Nov. 1981.

J. L. Crowley, O. Riff, and J. H. Piater. Fast computation of characteristic scale using a half-octave pyramid. In *Proceedings of the International Workshop on Cognitive Computing (CogVis)*, Oct. 2002.

N. Dalal and B. Triggs. Histograms of oriented gradients for human detection. In *Proceedings of the IEEE Conference on Computer Vision and Pattern Recognition (CVPR)*, June 2005.

I. Daubechies. Orthonormal bases of compactly supported wavelets. *Communications on Pure and Applied Mathematics*, 41(7): 909–996, Oct. 1988.

D. L. Davies and D. W. Bouldin. A cluster separation measure. *IEEE Transactions on Pattern Analysis and Machine Intelligence*, 1(2): 224–227, Apr. 1979.

E. R. Davies. *Machine Vision: Theory, Algorithms, Practicalities*. Morgan Kaufmann, third edition, 2005.

J. Davis and M. Goadrich. The relationship between precision-recall and ROC curves. In *Proceedings of the 23rd International Conference on Machine Learning (ICML)*, 2006.

J. de Villiers, F. Leuschner, and R. Geldenhuys. Centi-pixel accurate real-time inverse distortion correction. In *Proceedings of the International Symposium on Optomechatronic Technologies (ISOT)*, Nov. 2008.

P. E. Debevec and J. Malik. Recovering high dynamic range radiance maps from photographs. In *Proceedings of SIGGRAPH*, pages 369–378, 1997.

D. Dementhon and L. S. Davis. Model-based object pose in 25 lines of code. *International Journal of Computer Vision*, 15(1–2): 123–141, June 1995.

A. P. Dempster, N. M. Laird, and D. B. Rubin. Maximum likelihood from incomplete data via the EM algorithm. *Journal of the Royal Statistical Society, Series B (Methodological)*, 39(1): 1–22, 1977.

L. Deng and D. Yu. Deep learning: Methods and applications. *Foundations and Trends in Signal Processing*, 7(3–4): 197–387, 2013.

R. Deriche. Separable recursive filtering for efficient multi-scale edge detection. In *Proceedings of the International Conference on Computer Vision (ICCV)*, pages 18–23, Feb. 1987.

R. Deriche. Fast algorithms for low-level vision. *IEEE Transactions on Pattern Analysis and Machine Intelligence*, 12(1): 78–87, Jan. 1990.

R. Deriche. Recursively implementing the Gaussian and its derivatives. Technical Report 1893, INRIA Sophia-Antipolis, Apr. 1993.

I. S. Dhillon, Y. Guan, and B. Kulis. Weighted graph cuts without eigenvectors: A multilevel approach. *IEEE Transactions on Pattern Analysis and Machine Intelligence*, 29(11): 1944–1957, Nov. 2007.

M. B. Dillencourt, H. Samet, and M. Tamminen. A general approach to connected-component labeling for arbitrary image representations. *Journal of the ACM*, 39(2): 253–280, Apr. 1992.

D. Douglas and T. Peucker. Algorithms for the reduction of the number of points required to represent a digitized line or its caricature. *The Canadian Cartographer*, 10(2): 112–122, 1973.

C. E. Duchon. Lanczos filtering in one and two dimensions. *Journal of Applied Meteorology*, 18(8): 1016–1022, Aug. 1979.

R. O. Duda and P. E. Hart. Use of the Hough transformation to detect lines and curves in pictures. *Communications of the ACM*, 15: 11–15, Jan. 1972.

R. O. Duda, P. E. Hart, and D. G. Stork. *Pattern Classification*. Wiley Interscience, second edition, 2001.

F. Durand and J. Dorsey. Fast bilateral filtering for the display of high-dynamic-range images. *ACM Transactions on Graphics (SIGGRAPH)*, 21(3): 257–266, July 2002.

A. A. Efros and T. Leung. Texture synthesis by non-parametric sampling. In *Proceedings of the International Conference on Computer Vision (ICCV)*, pages 1033–1038, Sept. 1999.

D. Eggert, A. Lorusso, and R. B. Fisher. Estimating 3-D rigid body transformations: A comparison of four major algorithms. *Machine Vision and Applications*, 9(5-6): 272–290, Mar. 1997.

M. Enzweiler and D. M. Gavrila. Monocular pedestrian detection: Survey and experiments. *IEEE Transactions on Pattern Analysis and Machine Intelligence*, 31(12): 2179–2195, Dec. 2009.

M. D. Fairchild. *The Color Curiosity Shop*. Honeoye Falls, New York: MDF Publications, 2011.

A. X. Falcão, J. Stolfi, and R. de Alencar Lotufo. The image foresting transform: Theory, algorithms, and applications. *IEEE Transactions on Pattern Analysis and Machine Intelligence*, 26(1): 19–29, Jan. 2004.

O. Faugeras. *Three-Dimensional Computer Vision*. Cambridge, MA: MIT Press, 1993.

P. Felzenszwalb and D. Huttenlocher. Distance transforms of sampled functions. Technical Report TR2004-1963, Cornell Computing and Information Science, Sept. 2004.

P. Felzenszwalb, R. Girshick, D. McAllester, and D. Ramanan. Object detection with discriminatively trained part based models. *IEEE Transactions on Pattern Analysis and Machine Intelligence*, 32(9): 1627–1645, Sept. 2010.

P. F. Felzenszwalb and D. P. Huttenlocher. Pictorial structures for object recognition. *International Journal of Computer Vision*, 61(1): 55–79, Jan. 2005.

M. Fiedler. Algebraic connectivity of graphs. *Czechoslovak Mathematical Journal*, 23(2): 298–305, 1973.

D. J. Field. Relations between the statistics of natural images and the response properties of cortical cells. *Journal of the Optical Society of America A: Optics, Image Science, and Vision*, 4(12): 2379–2394, Dec. 1987.

G. D. Finlayson and S. Süsstrunk. Spectral sharpening and the Bradford Transform. In *Proceedings of Color Imaging Symposium (CIS)*, pages 236–243, 2000.

M. A. Fischler and R. C. Bolles. Random sample consensus: A paradigm for model fitting with applications to image analysis and automated cartography. *Communications of the ACM*, 24(6): 381–395, 1981.

M. A. Fischler and R. A. Elschlager. The representation and matching of pictorial structures. *IEEE Transactions on Computers*, 22(1): 67–92, Jan. 1973.

R. A. Fisher. The use of multiple measurements in taxonomic problems. *Annals of Eugenics*, 7(2): 179–188, Sept. 1936.

A. Fitzgibbon, M. Pilu, and R. B. Fisher. Direct least square fitting of ellipses. *IEEE Transactions on Pattern Analysis and Machine Intelligence*, 21(5): 476–480, May 1999.

E. W. Forgy. Cluster analysis of multivariate data: Efficiency versus interpretability of classifications. *Biometrics*, 21: 768–769, 1965.

D. Forsyth and J. Ponce. *Computer Vision: A Modern Approach*. Prentice-Hall, second edition, 2012.

H. Freeman. Techniques for the digital computer analysis of chain-encoded arbitrary plane curves. In *Proceedings of the National Electronics Conference*, volume 17, pages 421–432, Oct. 1961.

W. Freeman and E. H. Adelson. The design and use of steerable filters. *IEEE Transactions on Pattern Analysis and Machine Intelligence*, 13(9): 891–906, Sept. 1991.

Y. Freund and R. E. Schapire. A short introduction to boosting. *Journal of Japanese Society for Artificial Intelligence*, 14(5): 771–780, Sept. 1999.

P. Fua. Combining stereo and monocular information to compute dense depth maps that preserve depth discontinuities. In *Proceedings of the 12th International Joint Conference on Artificial Intelligence*, pages 1292–1298, Aug. 1991.

K. Fukunaga and L. Hostetler. The estimation of the gradient of a density function, with applications in pattern recognition. *IEEE Transactions on Information Theory*, 21(1): 32–40, Jan. 1975.

P. Gabbur, H. Hua, and K. Barnard. A fast connected components labeling algorithm and its application to real-time pupil detection. *Machine Vision and Applications*, 21(5): 779–787, Aug. 2010.

D. Gabor. Theory of communication. Part 1: The analysis of information. *Journal of the Institution of Electrical Engineers - Part III: Radio and Communication Engineering*, 93(26): 429–441, Nov. 1946.

S. I. Gallant. Perceptron-based learning algorithms. *IEEE Transactions on Neural Networks*, 1(2): 179–191, June 1990.

W. Gander, G. H. Golub, and R. Strebel. Least-squares fitting of circles and ellipses. *BIT Numerical Mathematics*, 34(4): 558–578, 1994.

D. Geiger, B. Ladendorf, and A. Yuille. Occlusions and binocular stereo. *International Journal of Computer Vision*, 14(3): 211–226, Apr. 1995.

S. Geman and D. Geman. Stochastic relaxation, Gibbs distributions, and the Bayesian restoration of images. *IEEE Transactions on Pattern Analysis and Machine Intelligence*, 6(6): 721–741, 1984.

S. Geman and D. McClure. Statistical methods for tomographic image reconstruction. *Bulletin of the International Statistical Institute*, LII(4–5), 1987.

A. S. Georghiades, P. N. Belhumeur, and D. J. Kriegman. From few to many: Illumination cone models for face recognition under variable lighting and pose. *IEEE Transactions on Pattern Analysis and Machine Intelligence*, 23(6): 643–660, June 2001.

A. Gershun. The light field. *Journal of Mathematics and Physics*, 18: 51–151, 1939. (Translated from the 1936 original by P. Moon and G. Timoshenko).

W. E. Gettys, F. J. Keller, and M. J. Skove. *Physics: Classical and Modern*. McGraw-Hill, 1989.

P. J. Giblin and B. B. Kimia. On the local form and transitions of symmetry sets, medial axes, and shocks. *International Journal of Computer Vision*, 54(1-3): 143–157, 2003.

J. J. Gibson. *The senses considered as perceptual systems*. Boston: Houghton-Mifflin Co., 1966.

A. S. Glassner. Frame buffers and color maps. In A. S. Glassner, editor, *Graphics Gems*, pages 215–218. San Diego: Academic Press, 1990.

J. Goldberger, S. Roweis, G. Hinton, and R. Salakhutdinov. Neighbourhood components analysis. In *Advances in Neural Information Processing Systems*, pages 513–520, Dec. 2004.

D. B. Goldman. Vignette and exposure calibration and compensation. *IEEE Transactions on Pattern Analysis and Machine Intelligence*, 32(12): 2276–2288, Dec. 2010.

R. C. Gonzalez and R. E. Woods. *Digital Image Processing*. New Jersey: Prentice Hall, third edition, 2008.

S. J. Gortler, R. Grzeszczuk, R. Szeliski, and M. Cohen. The lumigraph. In *Proceedings of SIGGRAPH*, pages 43–54, 1996.

V. K. Goyal. Theoretical foundations of transform coding. *IEEE Signal Processing Magazine*, 18(5): 9–21, Sept. 2001.

H. Grassman. On the theory of compound colours. *Philosophical Magazine and Journal*, 4(7): 254–264, 1854. (Translated from the original, Zur theorie der farbenmischung, *Poggendorff's Annalen der Physik und Chemie*, 89: 69–84, 1853.).

S. B. Gray. Local properties of binary images in two dimensions. *IEEE Transactions on Computers*, 20(5): 551–561, May 1971.

P. Green. *Understanding Digital Color*. Graphic Arts Technical Foundation, second edition, 1999.

D. M. Greig, B. T. Porteous, and A. H. Seheult. Exact maximum a posteriori estimation for binary images. *Journal of the Royal Statistical Society, Series B (Methodological)*, 51(2): 271–279, 1989.

R. Grosse, M. K. Johnson, E. H. Adelson, and W. T. Freeman. Ground truth dataset and baseline evaluations for intrinsic image algorithms. In *Proceedings of the International Conference on Computer Vision*, pages 2335–2342, Oct. 2009.

J. Guild. The colorimetric properties of the spectrum. *Philosophical Transactions of the Royal Society of London. Series A, Containing Papers of a Mathematical or Physical Character*, 230: 149–187, June 1931.

P. Gurney. Is our 'inverted' retina really 'bad design'? *Journal of Creation*, 13(1): 37–44, Apr. 1999.

A. Haar. Zur theorie der orthogonalen funktionensysteme: Erste mitteilung. *Mathematische Annalen*, 69(3): 331–371, July 1910.

A. Haar. Zur theorie der orthogonalen funktionensysteme: Zweite mitteilung. *Mathematische Annalen*, 71(1): 38–53, July 1911.

Y. HaCohen, R. Fattal, and D. Lischinski. Image upsampling via texture hallucination. In *IEEE International Conference on Computational Photography (ICCP)*, Mar. 2010.

H. Hadwiger. Minkowskische addition und subtraktion beliebiger punktmengen und die theoreme von erhard schmidt. *Mathematische Zeitschrift*, 53(3): 210–218, 1950.

G. D. Hager and P. N. Belhumeur. Efficient region tracking with parametric models of geometry and illumination. *IEEE Transactions on Pattern Analysis and Machine Intelligence*, 20(10): 1025–1039, Oct. 1998.

F. R. Hampel, E. M. Ronchetti, P. J. Rousseeuw, and W. A. Stahel. *Robust Statistics: The Approach Based on Influence Functions*. New York: John Wiley and Sons, 1986.

A. Hanbury. Constructing cylindrical coordinate colour spaces. *Pattern Recognition Letters*, 29(4): 494–500, Mar. 2008.

B. M. Haralick, C.-N. Lee, K. Ottenberg, and M. Nolle. Review and analysis of solutions of the three point perspective pose estimation problem. *International Journal of Computer Vision*, 13(3): 331–356, Dec. 1994.

R. M. Haralick, S. R. Sternberg, and X. Zhuang. Image analysis using mathematical morphology. *IEEE Transactions on Pattern Analysis and Machine Intelligence*, 9(4): 532–550, July 1987.

R. W. Harold. An introduction to appearance analysis. *A reprint from GATFWorld, the magazine of the Graphic Arts Technical Foundation*, SS(84), 2001.

C. G. Harris and M. Stephens. A combined corner and edge detector. In *Proceedings of the 4th Alvey Vision Conference*, pages 147–151, Sept. 1988.

P. E. Hart. How the Hough transform was invented. *IEEE Signal Processing Magazine*, 26(6): 18–22, Nov. 2009.

R. Hartley and A. Zisserman. *Multiple View Geometry in Computer Vision*. Cambridge University Press, second edition, 2003.

R. I. Hartley. In defense of the eight-point algorithm. *IEEE Transactions on Pattern Analysis and Machine Intelligence*, 19(6): 580–593, June 1997.

K. He, J. Sun, and X. Tang. Single image haze removal using dark channel prior. *IEEE Transactions on Pattern Analysis and Machine Intelligence*, 33(12): 2341–2353, Dec. 2011.

X. He, S. Yan, Y. Hu, P. Niyogi, and H.-J. Zhang. Face recognition using Laplacianfaces. *IEEE Transactions on Pattern Analysis and Machine Intelligence*, 27(3): 328–340, Mar. 2005.

P. S. Heckbert. Filtering by repeated integration. *Computer Graphics (SIGGRAPH)*, 20(4): 315–321, Aug. 1986.

M. Heikkilä and M. Pietikäinen. A texture-based method for modeling the background and detecting moving objects. *IEEE Transactions on Pattern Analysis and Machine Intelligence*, 28(4): 657–662, Apr. 2006.

J. Hershberger and J. Snoeyink. Speeding up the Douglas-Peucker line-simplification algorithm. In *Proceedings of the 5th International Symposium on Spatial Data Handling*, pages 134–143, 1992.

H. Hirschmüller. Stereo processing by semi-global matching and mutual information. *IEEE Transactions on Pattern Analysis and Machine Intelligence*, 30(2): 328–341, Feb. 2008.

H. Hirschmüller, P. R. Innocent, and J. M. Garibaldi. Real-time correlation-based stereo vision with reduced border errors. *International Journal of Computer Vision*, 47(1): 229–246, 2002.

H. Hofer, J. Carroll, J. Neitz, M. Neitz, and D. R. Williams. Organization of the human trichromatic cone mosaic. *The Journal of Neuroscience*, 25(42): 9669–9679, Oct. 2005.

B. K. P. Horn. *Robot Vision*. Cambridge, Mass.: MIT Press, 1986.

B. K. P. Horn. Closed-form solution of absolute orientation using unit quaternions. *Journal of the Optical Society of America*, 4: 629–642, Apr. 1987.

B. K. P. Horn and B. G. Schunck. Determining optical flow. *Artificial Intelligence*, 17(185): 185–203, 1981.

S. L. Horowitz and T. Pavlidis. Picture segmentation by a tree traversal algorithm. *Journal of the ACM*, 23(2): 368–388, 1976.

P. V. C. Hough. Machine analysis of bubble chamber pictures. In *International Conference On High-Energy Accelerators (HEACC)*, pages 554–558, Sept. 1959.

M. K. Hu. Visual pattern recognition by moment invariants. *IRE Transactions on Information Theory*, 8(2): 179–187, Feb. 1962.

T. S. Huang, G. J. Yang, and G. Y. Tang. A fast two-dimensional median filtering algorithm. *IEEE Transactions on Acoustics, Speech, and Signal Processing*, 27(1): 13–18, Feb. 1979.

D. H. Hubel. *Eye, Brain, and Vision*. W. H. Freeman and Company, 1988.

D. H. Hubel and T. N. Wiesel. Receptive fields, binocular interaction and functional architecture in the cat's visual cortex. *The Journal of Physiology*, 160(1): 106–154, 1962.

P. J. Huber. Robust estimation of a location parameter. *The Annals of Mathematical Statistics*, 35(1): 73–101, 1964.

P. J. Huber. *Robust Statistics*. New York: John Wiley and Sons, 1981.

D. A. Huffman. A method for the construction of minimum-redundancy codes. *Proceedings of the I. R. E.*, 40(9): 1098–1101, Sept. 1952.

D. A. Huffman. Impossible objects as nonsense sentences. *Machine Intelligence*, 6: 295–323, 1971.

J. Imber, J.-Y. Guillemaut, and A. Hilton. Intrinsic textures for relightable free-viewpoint video. In *Proceedings of the European Conference on Computer Vision (ECCV)*, 2014.

S. S. Intille and A. F. Bobick. Disparity-space images and large occlusion stereo. In *Proceedings of the 3rd European Conference on Computer Vision (ECCV)*, pages 179–186, May 1994.

A. K. Jain. *Fundamentals of Digital Image Processing*. Englewood Cliffs, New Jersey: Prentice-Hall, 1989.

A. K. Jain and F. Farrokhnia. Unsupervised texture segmentation using Gabor filters. *Pattern Recognition*, 24(12): 1167–1186, 1991.

A. K. Jain, R. P. W. Duin, and J. C. Mao. Statistical pattern recognition: A review. *IEEE Transactions on Pattern Analysis and Machine Intelligence*, 22(1): 4–37, Jan. 2000.

R. Jain and H.-H. Nagel. On the analysis of accumulative difference pictures from image sequences of real world scenes. *IEEE Transactions on Pattern Analysis and Machine Intelligence*, 1(2): 206–214, Apr. 1979.

R. Jain, R. Kasturi, and B. G. Schunck. *Machine Vision*. Boston: McGraw-Hill, 1995.

O. Javed, K. Shafique, and M. Shah. A hierarchical approach to robust background subtraction using color and gradient information. In *Proceedings of the Workshop on Motion and Video Computing*, pages 22–27, Dec. 2002.

S. C. Johnson. Hierarchical clustering schemes. *Psychometrika*, 32(3): 241–254, Sept. 1967.

M. J. Jones and J. M. Rehg. Statistical color models with application to skin detection. *International Journal of Computer Vision*, 46(1): 81–96, Jan. 2002.

M. J. Jones and P. Viola. Fast multi-view face detection. Technical Report TR-20003-96, Mitsubishi Electric Research Lab, July 2003.

D. B. Judd. Report of U.S. Secretariat Committee on Colorimetry and Artificial Daylight. In *Proceedings of the Twelfth Session of the CIE*, volume 1, 1951.

B. Julesz. Binocular depth perception of computer-generated patterns. *Bell System Technical Journal*, 39(5): 1125–1162, 1960.

B. Julesz. *Foundations of Cyclopean Perception*. Chicago: The University of Chicago Press, 1971.

B. Julesz. Textons, the elements of texture perception, and their interactions. *Nature*, 290: 91–97, Mar. 1981.

B. Julesz and J. R. Bergen. Textons, the fundamental elements in preattentive vision and perception of textures. *The Bell System Technical Journal*, 62(6): 1619–1645, July-August 1983.

W. Kabsch. A solution for the best rotation to relate two sets of vectors. *Acta Crystallographica Section A*, 32(5): 922–923, Sept. 1976.

H. F. Kaiser. The application of electronic computers to factor analysis. *Educational and Psychological Measurement*, 20: 141–151, 1960.

Y. Kameda and M. Minoh. A human motion estimation method using 3-successive video frames. In *International Conference on Virtual Systems and Multimedia (VSMM)*, pages 135–140, Sept. 1996.

T. Kanade and M. Okutomi. A stereo matching algorithm with an adaptive window: Theory and experiment. *IEEE Transactions on Pattern Analysis and Machine Intelligence*, 16(9): 920–932, Sept. 1994.

T. Kanungo, D. M. Mount, N. S. Netanyahu, C. D. Piatko, R. Silverman, and A. Y. Wu. An efficient k-means clustering algorithm: Analysis and implementation. *IEEE Transactions on Pattern Analysis and Machine Intelligence*, 24(7): 881–892, July 2002.

I. Kåsa. A circle fitting procedure and its error analysis. *IEEE Transactions on Instrumentation and Measurement*, 25(1): 8–14, Mar. 1976.

M. Kass, A. Witkin, and D. Terzopoulos. Snakes: Active contour models. *International Journal of Computer Vision*, 1(4): 321–331, 1988.

R. J. Kauth and G. S. Thomas. The tasseled cap – A graphic description of the spectral-temporal development of agricultural crops as seen by LANDSAT. In *Proceedings of the Symposium on Machine Processing of Remotely Sensed Data*, pages 4B–41–4B–51, June 1976.

C. S. Kenney, M. Zuliani, and B. S. Manjunath. An axiomatic approach to corner detection. In *Proceedings of the IEEE Conference on Computer Vision and Pattern Recognition (CVPR)*, pages 191–197, 2005.

R. G. Keys. Cubic convolution interpolation for digital image processing. *IEEE Transactions on Acoustics, Speech, and Signal Processing*, 29(6): 1153–1160, Dec. 1981.

K. Kimura, S. Kikuchi, and S. Yamasaki. Accurate root length measurement by image analysis. *Plant and Soil*, 216(1): 117–127, 1999.

M. Kirby and L. Sirovich. Application of the Karhunen-Loève procedure for the characterization of human faces. *IEEE Transactions on Pattern Analysis and Machine Intelligence*, 12(1): 103–108, 1990.

L. Kitchen and A. Rosenfeld. Gray-level corner detection. *Pattern Recognition Letters*, 1(2): 95–102, Dec. 1982.

J. Kittler, M. Hatef, R. Duin, and J. Matas. On combining classifiers. *IEEE Transactions on Pattern Analysis and Machine Intelligence*, 20(3): 226–239, 1998.

J. J. Koenderink. The structure of images. *Biological Cybernetics*, 50(5): 363–370, Aug. 1984.

J. J. Koenderink and A. J. van Doorn. Representation of local geometry in the visual system. *Biological Cybernetics*, 55(6): 367–375, Mar. 1987.

R. Kohavi. A study of cross-validation and bootstrap for accuracy estimation and model selection. In *Proceedings of the International Joint Conference on Artificial Intelligence*, pages 1137–1143, Aug. 1995.

P. Kohli, L. Ladický, and P. H. S. Torr. Robust higher order potentials for enforcing label consistency. *International Journal of Computer Vision*, 82(3): 302–324, May 2009.

V. Kolmogorov and R. Zabih. What energy functions can be minimized via graph cuts? *IEEE Transactions on Pattern Analysis and Machine Intelligence*, 26(2): 147–159, 2004.

J. Kopf, M. Uyttendaele, O. Deussen, and M. Cohen. Capturing and viewing gigapixel images. *ACM Transactions on Graphics (SIGGRAPH)*, 26(3), July 2007.

A. Krizhevsky, I. Sutskever, and G. E. Hinton. ImageNet classification with deep convolutional neural networks. In *Advances in Neural Information Processing Systems (NIPS)*, pages 1097–1105, 2012.

N. Krüger, P. Janssen, S. Kalkan, M. Lappe, A. Leonardis, J. Piater, A. Rodriguez-Sanchez, and L. Wiskott. Deep hierarchies in the primate visual cortex: What can we learn for computer vision? *IEEE Transactions on Pattern Analysis and Machine Intelligence*, 35(8): 1847–1871, 2013.

S. Kumar and M. Hebert. Discriminative random fields. *International Journal of Computer Vision*, 68(2): 179–201, June 2006.

J. D. Lafferty, A. McCallum, and F. C. N. Pereira. Conditional random fields: Probabilistic models for segmenting and labeling sequence data. In *Proceedings of the International Conference on Machine Learning (ICML)*, pages 282–289, June 2001.

K. M. Lam. *Metamerism and Colour Constancy*. PhD thesis, University of Bradford, 1985.

Y. Lamdan and H. J. Wolfson. Geometric hashing: A general and efficient model-based recognition scheme. In *Proceedings of the International Conference on Computer Vision (ICCV)*, Dec. 1988.

K. I. Laws. Rapid texture identification. In *Proceedings of the SPIE 0238: Image Processing for Missile Guidance*, pages 376–381, July 1980.

H.-C. Lee. *Introduction to Color Imaging Science*. Cambridge University Press, 2005.

T. S. Lee. Image representation using 2D Gabor wavelets. *IEEE Transactions on Pattern Analysis and Machine Intelligence*, 18(10): 1–13, Oct. 1996.

V. Lepetit, F. Moreno-Noguer, and P. Fua. EPnP: An accurate O(n) solution to the PnP problem. *International Journal of Computer Vision*, 81(2): 155–166, Feb. 2009.

M. Levoy and P. Hanrahan. Light field rendering. In *Proceedings of SIGGRAPH*, pages 31–42, 1996.

C. Li, C. Xu, C. Gui, and M. D. Fox. Level set evolution without re-initialization: A new variational formulation. In *Proceedings of the IEEE Conference on Computer Vision and Pattern Recognition (CVPR)*, pages 430–436, June 2005.

P.-S. Liao, T.-S. Chen, and P.-C. Chung. A fast algorithm for multilevel thresholding. *Journal of Information Science and Engineering*, 17: 713–727, 2001.

A. Likas, N. Vlassis, and J. J. Verbeek. The global k-means clustering algorithm. *Pattern Recognition*, 36(2): 451–461, 2003.

T. Lillesand, R. W. Kiefer, and J. Chipman. *Remote Sensing and Image Interpretation*. Wiley, sixth edition, 2007.

T. Lindeberg. Scale-space for discrete signals. *IEEE Transactions on Pattern Analysis and Machine Intelligence*, 12(3): 234–254, Mar. 1990.

T. Lindeberg. *Scale-Space Theory in Computer Vision*. Springer, 1993.

T. Lindeberg. Scale-space theory: A basic tool for analysing structures at different scales. *Journal of Applied Statistics*, 21(2): 224–270, 1994.

T. Lindeberg. Feature detection with automatic scale selection. *International Journal of Computer Vision*, 30(2): 79–116, Nov. 1998a.

T. Lindeberg. Edge detection and ridge detection with automatic scale selection. *International Journal of Computer Vision*, 30(2): 117–156, Nov. 1998b.

J. J. Little and W. E. Gillett. Direct evidence for occlusion in stereo and motion. *Image and Vision Computing*, 8(4): 328–340, Nov. 1990.

S. P. Lloyd. Least squares quantization in PCM. *IEEE Transactions on Information Theory*, 28(2): 129–137, Mar. 1982.

H. C. Longuet-Higgins. A computer algorithm for reconstructing a scene from two projections. *Nature*, 293(5828): 133–135, Sept. 1981.

D. G. Lowe. Distinctive image features from scale-invariant keypoints. *International Journal of Computer Vision*, 60(2): 91–110, Nov. 2004.

C.-P. Lu, G. D. Hager, and E. Mjolsness. Fast and globally convergent pose estimation from video images. *IEEE Transactions on Pattern Analysis and Machine Intelligence*, 22(6): 610–622, June 2000.

B. D. Lucas and T. Kanade. An iterative image registration technique with an application to stereo vision. In *Proceedings of the 7th International Joint Conference on Artificial Intelligence*, pages 674–679, Aug. 1981.

M. A. Luengo-Oroz, D. Pastor-Escuredo, C. Castro-Gonzalez, E. Faure, T. Savy, B. Lombardot, J. L. Rubio-Guivernau, L. Duloquin, M. J. Ledesma-Carbayo, P. Bourgine, N. Peyrieras, and A. Santos. 3Dt+ morphological processing: Applications to embryogenesis image analysis. *IEEE Transactions on Image Processing*, 21(8): 3518–3530, Aug. 2012.

B. Luo and E. R. Hancock. Structural graph matching using the EM algorithm and singular value decomposition. *IEEE Transactions on Pattern Analysis and Machine Intelligence*, 23(10): 1120–1136, Oct. 2001.

Q.-T. Luong and O. D. Faugeras. Determining the fundamental matrix with planes: Instability and new algorithms. In *Proceedings of the IEEE Conference on Computer Vision and Pattern Recognition (CVPR)*, pages 489–494, June 1993.

Q.-T. Luong and O. D. Faugeras. Fundamental matrix: Theory, algorithms, and stability analysis. *International Journal of Computer Vision*, 17(1): 43–75, Jan. 1996.

J. MacQueen. Some methods for classification and analysis of multivariate observations. In *Proceedings of the 5th Berkeley Symposium on Mathematical Statistics and Probability*, pages 281–297, 1967.

R. Malladi, J. A. Sethian, and B. C. Vemuri. Shape modeling with front propagation: A level set approach. *IEEE Transactions on Pattern Analysis and Machine Intelligence*, 17(2): 158–175, Feb. 1995.

S. Mallat and S. Zhong. Characterization of signals from multiscale edges. *IEEE Transactions on Pattern Analysis and Machine Intelligence*, 14(7): 710–732, July 1992.

S. G. Mallat. A theory for multiresolution signal decomposition: The wavelet representation. *IEEE Transactions on Pattern Analysis and Machine Intelligence*, 11(7): 674–693, July 1989.

D. Marr. *Vision*. San Francisco: W. H. Freeman and Company, 1982.

D. Marr and E. Hildreth. Theory of edge detection. *Proceedings of the Royal Society of London, Series B*, 207: 187–217, 1980.

D. Marr and T. Poggio. Cooperative computation of stereo disparity. *Science*, 194: 283–287, Oct. 1976.

D. Marr and T. Poggio. A computational theory of human stereo vision. *Proceedings of the Royal Society of London, Series B*, 204: 301–328, 1979.

D. Marr, G. Palm, and T. Poggio. Analysis of a cooperative stereo algorithm. *Biological Cybernetics*, 28: 223–239, 1978.

D. Martin, C. Fowlkes, D. Tal, and J. Malik. A database of human segmented natural images and its application to evaluating segmentation algorithms and measuring ecological statistics. In *Proceedings of the International Conference on Computer Vision (ICCV)*, volume 2, pages 416–423, July 2001.

D. R. Martin, C. C. Fowlkes, and J. Malik. Learning to detect natural image boundaries using local brightness, color, and texture cues. *IEEE Transactions on Pattern Analysis and Machine Intelligence*, 26(5): 530–549, May 2004.

J. Matas, O. Chum, M. Urban, and T. Pajdla. Robust wide-baseline stereo from maximally stable extremal regions. *Image and Vision Computing*, 22(10): 761–767, Sept. 2004.

C. R. Maurer, Jr., R. Qi, and V. Raghavan. A linear time algorithm for computing exact Euclidean distance transforms of binary images in arbitrary dimensions. *IEEE Transactions on Pattern Analysis and Machine Intelligence*, 25(2): 265–270, Feb. 2003.

J. C. Maxwell. On the theory of compound colours, and the relations of the colours of the spectrum. *Philosophical Transactions of the Royal Society of London*, 150: 57–84, 1860.

C. S. McCamy, H. Marcus, and J. G. Davidson. A color-rendition chart. *Journal of Applied Photographic Engineering*, 2(3): 95–99, Summer 1976.

F. Meyer. *Cytologie Quantitative et Morphologie Mathématique*. PhD thesis, Paris School of Mines, May 1979.

F. Meyer and S. Beucher. Morphological segmentation. *Journal of Visual Communication and Image Representation*, 1(1): 21–46, Sept. 1990.

K. Mikolajczyk and C. Schmid. A performance evaluation of local descriptors. *IEEE Transactions on Pattern Analysis and Machine Intelligence*, 27(10): 1615–1630, 2005.

H. Minkowski. Über die begriffe länge, oberfläche und volumen. *Jahresbericht der Deutschen Mathematiker Vereiningung*, 9: 115–121, 1901.

M. Minsky and S. Papert. *Perceptrons: An Introduction to Computational Geometry*. The MIT Press, 1969.

D. P. Mitchell and A. N. Netravali. Reconstruction filters in computer graphics. *Computer Graphics (SIGGRAPH)*, 22(4): 221–228, June 1988.

T. Mitchell. *Machine Learning*. McGraw Hill, 1997.

U. Montanari. A method for obtaining skeletons using a quasi-Euclidean distance. *Journal of the ACM*, 15(4): 600–624, Oct. 1968.

G. A. Moore. Automatic scanning and computer processes for the quantitative analysis of micrographs and equivalent subjects. In G. C. Cheng, R. S. Ledley, D. K. Pollock, and A. Rosenfeld, editors, *Pictorial Pattern Recognition*, pages 275–326. Thomson, 1968.

H. P. Moravec. Towards automatic visual obstacle avoidance. In *Proceedings of the International Joint Conference on Artificial Intelligence*, page 584, 1977.

E. N. Mortensen and W. A. Barrett. Intelligent scissors for image composition. *Proceedings of SIGGRAPH*, pages 191–198, 1995.

M. E. Mortenson. *Geometric Modeling*. New York: John Wiley and Sons, second edition, 1997.

D. Mumford and J. Shah. Boundary detection by minimizing functionals. In *Proceedings of the IEEE Conference on Computer Vision and Pattern Recognition (CVPR)*, pages 22–26, 1985.

D. Mumford and J. Shah. Optimal approximation by piecewise smooth functions and associated variational problems. *Communications on Pure and Applied Mathematics*, 42: 577–685, 1989.

A. H. Munsell. *A Color Notation: A Measured Color System, Based on the Three Qualities Hue, Value, and Chroma*. Boston: G. H. Ellis Co., 1905.

A. H. Munsell. *Atlas of the Munsell Color System*. Malden, Mass.: Wadsworth-Howland and Company, 1915.

H. Murase and S. K. Nayar. Visual learning and recognition of 3-D objects from appearance. *International Journal of Computer Vision*, 14(1): 5–24, Jan. 1995.

V. Nair and G. Hinton. Rectified linear units improve restricted Boltzmann machines. In *International Conference on Machine Learning (ICML)*, 2010.

V. S. Nalwa. *A Guided Tour of Computer Vision*. Reading, MA: Addison-Wesley, 1993.

I. Newton. *Opticks, or a Treatise of the Reflexions, Refractions, Inflexions, and Colours of Light*. London, 1704.

A. Y. Ng, M. I. Jordan, and Y. Weiss. On spectral clustering: Analysis and an algorithm. In *Advances in Neural Information Processing Systems (NIPS)*, pages 849–856, 2001.

R. Ng. *Digital Light Field Photography*. PhD thesis, Stanford University, July 2006.

W. Niblack. *An Introduction to Digital Image Processing*. Englewood Cliffs, NJ: Prentice Hall, 1986.

H. K. Nishihara. Practical real-time imaging stereo matcher. *Optical Engineering*, 23(5): 536–545, 1984.

F. O'Gorman and M. B. Clowes. Finding picture edges through collinearity of feature points. *IEEE Transactions on Computers*, 25(4): 449–456, 1976.

Y. Ohta and T. Kanade. Stereo by intra- and inter-scanline search using dynamic programming. *IEEE Transactions on Pattern Analysis and Machine Intelligence*, 7(2): 139–154, Mar. 1985.

M. Okutomi and T. Kanade. A multiple-baseline stereo. *IEEE Transactions on Pattern Analysis and Machine Intelligence*, 15(4): 353–363, Apr. 1993.

B. A. Olshausen and D. J. Field. Emergence of simple-cell receptive field properties by learning a sparse code for natural images. *Nature*, 381(6583): 607–609, June 1996.

A. V. Oppenheim and R. W. Schafer. *Discrete-Time Signal Processing*. New Jersey: Prentice Hall, second edition, 1999.

S. J. Osher and J. A. Sethian. Fronts propagating with curvature dependent speed: Algorithms based on Hamilton-Jacobi formulations. *Journal of Computational Physics*, 79(1): 12–49, Nov. 1988.

N. Otsu. A threshold selection method from gray-level histograms. *IEEE Transactions on Systems, Man, and Cybernetics*, 9(1): 62–66, Jan. 1979.

M. Ozuysal, P. Fua, and V. Lepetit. Fast keypoint recognition in ten lines of code. In *Proceedings of the IEEE Conference on Computer Vision and Pattern Recognition (CVPR)*, June 2007.

A. W. Paeth. Image file compression made easy. In J. Arvo, editor, *Graphics Gems II*, pages 93–100. San Diego: Academic Press, 1991.

N. R. Pal and S. K. Pal. Entropic thresholding. *Signal Processing*, 16(2): 97–108, Feb. 1989.

S. E. Palmer. *Vision Science: Photons to Phenomenology*. Cambridge, Mass.: The MIT Press, 1999.

N. Paragios and R. Deriche. Geodesic active contours and level sets for the detection and tracking of moving objects. *IEEE Transactions on Pattern Analysis and Machine Intelligence*, 22(3): 266–280, Mar. 2000.

S. Paris, P. Kornprobst, J. Tumblin, and F. Durand. Bilateral filtering: Theory and applications. *Foundations and Trends in Computer Graphics and Vision*, 4(1): 1–73, 2009.

E. Parzen. On estimation of a probability density function and mode. *Annals of Mathematical Statistics*, 33(3): 1065–1076, Sept. 1962.

J. B. Pawley. Points, pixels, and gray levels: Digitizing image data. In J. B. Pawley, editor, *Handbook of Biological Confocal Microscopy*. New York: Springer Science, third edition, 2006.

B. Peasley and S. Birchfield. Real-time obstacle detection and avoidance in the presence of specular surfaces using an active 3D sensor. In *IEEE Workshop on Robot Vision (WoRV)*, Jan. 2013.

P. Perona and J. Malik. Scale-space and edge detection using anisotropic diffusion. *IEEE Transactions on Pattern Analysis and Machine Intelligence*, 12(7): 629–639, July 1990.

S. Perreault and P. Hébert. Median filtering in constant time. *IEEE Transactions on Image Processing*, 16(9): 2389–2394, Sept. 2007.

G. Petschnigg, R. Szeliski, M. Agrawala, M. Cohen, H. Hoppe, and K. Toyama. Digital photography with flash and no-flash image pairs. *ACM Transactions on Graphics (SIGGRAPH)*, 23(3): 664–672, Aug. 2004.

P. J. Phillips, H. Moon, S. A. Rizvi, and P. J. Rauss. The FERET evaluation methodology for face-recognition algorithms. *IEEE Transactions on Pattern Analysis and Machine Intelligence*, 22(10): 1090–1104, Oct. 2000.

K. K. Pingle. Visual perception by a computer. In A. Grasselli, editor, *Automatic Interpretation and Classification of Images*, pages 277–284. New York: Academic Press, 1969.

J. C. Platt. Fast training of support vector machines using sequential minimal optimization. In *Advances in Kernel Methods - Support Vector Learning*. MIT Press, Jan. 1998.

B. W. Pogue, M. A. Mycek, and D. Harper. Image analysis for discrimination of cervical neoplasia. *Journal of Biomedical Optics*, 5(1): 72–82, Jan. 2000.

S. B. Pollard, J. E. W. Mayhew, and J. P. Frisby. PMF: A stereo correspondence algorithm using a disparity gradient limit. *Perception*, 14: 449–470, 1985.

F. Porikli. Constant time O(1) bilateral filtering. In *Proceedings of the IEEE Conference on Computer Vision and Pattern Recognition (CVPR)*, June 2008.

T. Porter and T. Duff. Compositing digital images. *Computer Graphics (SIGGRAPH)*, 18(3): 253–259, July 1984.

C. Poynton. The rehabilitation of gamma. In B. E. Rogowitz and T. N. Pappas, editors, *Human Vision and Electronic Imaging III, Proceedings of SPIE/IS&T Conference 3299*, pages 26–30, Jan. 1998.

C. Poynton. *Digital Video and HDTV: Algorithms and Interfaces*. Morgan Kaufmann, 2003.

W. K. Pratt. *Digital Image Processing*. Wiley Interscience, second edition, 1991.

J. M. S. Prewitt. Object enhancement and extraction. In B. S. Lipkin and A. Rosenfeld, editors, *Picture Processing and Psychopictorics*, pages 75–149. New York: Academic Press, 1970.

U. Ramer. An iterative procedure for the polygonal approximation of plane curves. *Computer Graphics and Image Processing*, 1(3): 244–256, Nov. 1972.

S. Rao, A. de Medeiros Martins, and J. C. Príncipe. Mean shift: An information theoretic perspective. *Pattern Recognition Letters*, 30(3): 222–230, Feb. 2009.

R. Rau and J. H. McClellan. Efficient approximation of Gaussian filters. *IEEE Transactions on Signal Processing*, 45(2): 468–471, Feb. 1997.

S. S. Reddi, S. F. Rudin, and H. R. Keshavan. An optimal multiple threshold scheme for image segmentation. *IEEE Transactions on Systems, Man, and Cybernetics*, 14(4): 661–665, Jul-Aug 1984.

X. Ren and J. Malik. Learning a classification model for segmentation. In *Proceedings of the International Conference on Computer Vision (ICCV)*, pages 10–17, Oct. 2003.

T. W. Ridler and S. Calvard. Picture thresholding using an iterative selection method. *IEEE Transactions on Systems, Man, and Cybernetics*, 8(8): 630–632, Aug. 1978.

R. Rithe, P. Raina, N. Ickes, S. V. Tenneti, and A. P. Chandrakasan. Reconfigurable processor for energy-scalable computational photography. In *Proceedings of the IEEE International Solid-State Circuits Conference (ISSCC)*, pages 164–166, Feb. 2013.

J.-F. Rivest, P. Soille, and S. Beucher. Morphological gradients. *Journal of Electronic Imaging*, 2(4): 326–336, Oct. 1993.

L. G. Roberts. *Machine Perception of Three-Dimensional Solids*. PhD thesis, Dept. of Electrical Engineering, M.I.T., June 1963.

F. Rodriguez, E. Maire, P. Courjault-Radé, and J. Darrozes. The black top hat function applied to a DEM: A tool to estimate recent incision in a mountainous watershed. *Geophysical Research Letters*, 29(6): 1–4, Mar. 2002.

F. Rosenblatt. The perceptron: A probabilistic model for information storage and organization in the brain. *Psychological Review*, 65(6): 386–408, Nov. 1958.

A. Rosenfeld and A. C. Kak. *Digital Picture Processing*. San Diego: Academic Press, second edition, 1982.

A. Rosenfeld and J. L. Pfaltz. Sequential operations in digital picture processing. *Journal of the ACM*, 13(4): 471–494, Oct. 1966.

A. Rosenfeld and J. L. Pfaltz. Distance functions on digital pictures. *Pattern Recognition*, 1(1): 33–61, July 1968.

P. Rosin. Thresholding for change detection. In *Proceedings of the International Conference on Computer Vision (ICCV)*, pages 274–279, Jan. 1998.

P. L. Rosin and E. Ioannidis. Evaluation of global image thresholding for change detection. *Pattern Recognition Letters*, 24(14): 2345–2356, 2003.

E. Rosten and T. Drummond. Machine learning for high-speed corner detection. In *Proceedings of the European Conference on Computer Vision (ECCV)*, pages 430–443, May 2006.

C. Rother, V. Kolmogorov, and A. Blake. GrabCut: Interactive foreground extraction using iterated graph cuts. *ACM Transactions on Graphics (SIGGRAPH)*, 23(3): 309–314, Aug. 2004.

C. Rother, V. Kolmogorov, V. Lempitsky, and M. Szummer. Optimizing binary MRFs via extended roof duality. In *Proceedings of the IEEE Conference on Computer Vision and Pattern Recognition (CVPR)*, June 2007.

P. J. Rousseeuw. Least median of squares regression. *Journal of the American Statistical Association*, 79: 871–880, 1984.

P. J. Rousseeuw and A. M. Leroy. *Robust Regression and Outlier Detection*. New York: John Wiley and Sons, 1987.

H. A. Rowley, S. Baluja, and T. Kanade. Neural network-based face detection. *IEEE Transactions on Pattern Analysis and Machine Intelligence*, 20(1): 23–38, Jan. 1998.

S. Roy and I. J. Cox. A maximum-flow formulation of the N-camera stereo correspondence problem. In *Proceedings of the 6th International Conference on Computer Vision (ICCV)*, pages 492–499, Jan. 1998.

Y. Rubner, C. Tomasi, and L. J. Guibas. A metric for distributions with applications to image databases. In *Proceedings of the International Conference on Computer Vision (ICCV)*, Jan. 1998.

D. E. Rumelhart, G. E. Hinton, and R. J. Williams. Learning representations by back-propagating errors. *Nature*, 323: 533–536, Oct. 1986.

S. Rusinkiewicz and M. Levoy. Efficient variants of the ICP algorithm. In *Third International Conference on 3D Digital Imaging and Modeling (3DIM)*, June 2001.

P. K. Sahoo, S. Soltani, A. K. C. Wong, and Y. C. Chen. A survey of thresholding techniques. *Computer Vision, Graphics, and Image Processing*, 41(2): 233–260, Feb. 1988.

P. Saint-Marc, J.-S. Chen, and G. Medioni. Adaptive smoothing: A general tool for early vision. *IEEE Transactions on Pattern Analysis and Machine Intelligence*, 13(6): 514–529, June 1991.

J. Salvi, J. Pagès, and J. Batlle. Pattern codification strategies in structured light systems. *Pattern Recognition*, 37(4): 827–849, Apr. 2004.

J. Sauvola and M. Pietikäinen. Adaptive document image binarization. *Pattern Recognition*, 33(2): 225–236, Feb. 2000.

H. Scharr. *Optimal Operators in Digital Image Processing*. PhD thesis, Rupertus Carola University, May 2000.

D. Scharstein and R. Szeliski. A taxonomy and evaluation of dense two-frame stereo correspondence algorithms. *International Journal of Computer Vision*, 47(1): 7–42, Apr. 2002.

Y. Y. Schechner. Depth from defocus vs. stereo: How different really are they? *International Journal of Computer Vision*, 39(2): 141–162, Sept. 2000.

C. Schmid, R. Mohr, and C. Bauckhage. Evaluation of interest point detectors. *International Journal of Computer Vision*, 37(2): 151–172, June 2000.

H. Schneiderman and T. Kanade. A statistical method for 3D object detection applied to faces and cars. In *Proceedings of the IEEE Conference on Computer Vision and Pattern Recognition (CVPR)*, June 2000.

I. J. Schoenberg. On equidistant cubic spline interpolation. *Bulletin of the American Mathematical Society*, 77(6): 1039–1044, Nov. 1971.

P. H. Schönemann. A generalized solution of the orthogonal Procrustes problem. *Psychometrika*, 31(1): 110, Mar. 1966.

E. F. Schumacher. *Small is Beautiful: A Study of Economics as if People Mattered*. Blond and Briggs, 1973.

G. E. Schwarz. Estimating the dimension of a model. *Annals of Statistics*, 6(2): 461–464, 1978.

S. M. Seitz and S. Baker. Filter flow. In *Proceedings of the International Conference on Computer Vision (ICCV)*, pages 143–150, Oct. 2009.

J. Serra. *Image Analysis and Mathematical Morphology*, volume 1. London: Academic Press, 1982.

J. Sethian. *Level Set Methods and Fast Marching Methods*. Cambridge University Press, 1999.

M. Sezgin and B. Sankur. Survey over image thresholding techniques and quantitative performance evaluation. *Journal of Electronic Imaging*, 13(1): 146–165, Jan. 2004.

F. Shafait, D. Keysers, and T. M. Breuel. Efficient implementation of local adaptive thresholding techniques using integral images. In *IS&T/SPIE Electronic Imaging: Document Recognition and Retrieval XV*, volume 6815, 2008.

Q. Shan, Z. Li, J. Jia, and C.-K. Tang. Fast image/video upsampling. *ACM Transactions on Graphics (SIGGRAPH ASIA)*, 27(5): 153:1–15:7, Dec. 2008.

C. E. Shannon. A mathematical theory of communication. *Bell System Technical Journal*, 27(3): 379–423, July / October 1948.

L. G. Shapiro and G. C. Stockman. *Computer Vision*. New Jersey: Prentice-Hall, 2001.

H. R. Sheikh, A. C. Bovik, and G. de Veciana. An information fidelity criterion for image quality assessment using natural scene statistics. *IEEE Transactions on Image Processing*, 14(12): 2117–2128, Dec. 2005.

J. Shi and J. Malik. Normalized cuts and image segmentation. *IEEE Transactions on Pattern Analysis and Machine Intelligence*, 22(8): 888–905, Aug. 2000.

J. Shi and C. Tomasi. Good features to track. In *Proceedings of the IEEE Conference on Computer Vision and Pattern Recognition (CVPR)*, pages 593–600, June 1994.

P. Simard, L. Bottou, P. Haffner, and Y. LeCun. Boxlets: A fast convolution algorithm for signal processing and neural networks. In *Advances in Neural Information Processing Systems (NIPS)*, pages 571–577, Dec. 1998.

J. Sklansky, R. L. Chazin, and B. J. Hansen. Minimum-perimeter polygons of digitized silhouettes. *IEEE Transactions on Computers*, 21(3): 260–268, Mar. 1972.

A. R. Smith. Color gamut transform pairs. *Computer Graphics*, 12(3): 12–19, Aug. 1978.

S. M. Smith and J. M. Brady. SUSAN - A new approach to low level image processing. *International Journal of Computer Vision*, 23(1): 45–78, May 1997.

L. Snidaro and G. L. Foresti. Real-time thresholding with Euler numbers. *Pattern Recognition Letters*, 24: 1533–1544, 2003.

P. Soille. *Morphological Image Analysis: Principles and Applications*. Berlin: Springer, second edition, 2003.

M. Sonka, V. Hlavac, and R. Boyle. *Image Processing, Analysis, and Machine Vision*. Thomson, third edition, 2008.

N. I. Speranskaya. Determination of spectrum color coordinates for twenty-seven normal observers. *Optics and Spectroscopy*, 7: 424–428, Nov. 1959.

S. N. Srihari. Document image understanding. In *Proceedings of the ACM Fall Joint Computer Conference*, pages 87–96, 1986.

N. Srivastava, G. Hinton, A. Krizhevsky, I. Sutskever, and R. Salakhutdinov. Dropout: A simple way to prevent neural networks from overfitting. *Journal of Machine Learning Research*, 15(1): 1929–1958, Jan. 2014.

P. Stathis, E. Kavallieratou, and N. Papamarkos. An evaluation technique for binarization algorithms. *Journal of Universal Computer Science*, 14(18): 3011–3030, 2008.

C. Stauffer and E. Grimson. Learning patterns of activity using real-time tracking. *IEEE Transactions on Pattern Analysis and Machine Intelligence*, 22(8): 747–757, Aug. 2000.

W. S. Stiles and J. M. Burch. Interim report to the Commission Internationale de l'Éclairage, Zurich, 1955, on the National Physical Laboratory's investigation of colour-matching. *Optica acta: International Journal of Optics*, 2(4): 168–181, 1955.

W. S. Stiles and J. M. Burch. N.P.L. colour-matching investigation: Final report. *Optica acta: International Journal of Optics*, 6(1): 1–26, 1959.

A. Stockman and L. T. Sharpe. Cone spectral sensitivities and color matching. In K. Gegenfurtner and L. T. Sharpe, editors, *Color Vision: From Genes to Perception*, pages 51–85. Cambridge: Cambridge University Press, 1999.

A. Stockman and L. T. Sharpe. Spectral sensitivities of the middle- and long-wavelength sensitive cones derived from measurements in observers of known genotype. *Vision Research*, 40: 1711–1737, 2000.

M. Stokes, M. Anderson, S. Chandrasekar, and R. Motta. A standard default color space for the internet -sRGB. Technical report, Hewlett-Packard and Microsoft, Nov. 1996.

P. F. Sturm and S. J. Maybank. On plane-based camera calibration: A general algorithm, singularities, applications. In *Proceedings of the IEEE Conference on Computer Vision and Pattern Recognition (CVPR)*, pages 432–437, June 1999.

K.-K. Sung and T. Poggio. Example-based learning for view-based human face detection. *IEEE Transactions on Pattern Analysis and Machine Intelligence*, 20(1): 39–51, 1998.

G. Svaetichin. Spectral response curves from single cones. *Acta Physiologica Scandinavica*, 39, Supplement 134, 1956.

M. Swain and D. Ballard. Color indexing. *International Journal of Computer Vision*, 7(1): 11–32, 1991.

R. Szeliski. Image alignment and stitching: A tutorial. *Foundations and Trends in Computer Graphics and Vision*, 2(1): 1–104, 2006.

R. Szeliski. *Computer Vision: Algorithms and Applications*. Springer, 2010.

Y. Taigman, M. Yang, M. Ranzato, and L. Wolf. DeepFace: Closing the gap to human-level performance in face verification. In *Proceedings of the IEEE Conference on Computer Vision and Pattern Recognition (CVPR)*, June 2014.

M. F. Tappen, W. T. Freeman, and E. H. Adelson. Recovering intrinsic images from a single image. *IEEE Transactions on Pattern Analysis and Machine Intelligence*, 27(9): 1459–1472, Sept. 2005.

R. E. Tarjan. Efficiency of a good but not linear set union algorithm. *Journal of the ACM*, 22(2): 215–225, Apr. 1975.

G. Taubin. Estimation of planar curves, surfaces, and nonplanar space curves defined by implicit equations with applications to edge and range image segmentation. *IEEE Transactions on Pattern Analysis and Machine Intelligence*, 13(11): 1115–1138, Nov. 1991.

M. R. Teague. Image analysis via the general theory of moments. *Journal of the Optical Society of America*, 70(8): 920–930, Aug. 1980.

C.-H. Teh and R. T. Chin. On image analysis by the methods of moments. *IEEE Transactions on Pattern Analysis and Machine Intelligence*, 10(4): 496–513, July 1988.

K. Thompson. Alpha blending. In A. S. Glassner, editor, *Graphics Gems*, pages 210–211. San Diego: Academic Press, 1990.

E. Tola, V. Lepetit, and P. Fua. DAISY: An efficient dense descriptor applied to wide baseline stereo. *IEEE Transactions on Pattern Analysis and Machine Intelligence*, 32(5): 815–830, May 2010.

C. Tomasi and T. Kanade. Detection and tracking of point features. Technical Report CMU-CS-91-132, Carnegie Mellon University, Apr. 1991.

C. Tomasi and R. Manduchi. Bilateral filtering for gray and color images. In *Proceedings of the International Conference on Computer Vision (ICCV)*, pages 839–846, Jan. 1998.

C. Tomasi and T. Kanade. Shape and motion from image streams under orthography: A factorization method. *International Journal of Computer Vision*, 9(2):137–154, Nov. 1992.

P. H. S. Torr and A. Zisserman. MLESAC: A new robust estimator with application to estimating image geometry. *Computer Vision and Image Understanding*, 78(1): 138–156, Apr. 2000.

K. Toyama, J. Krumm, B. Brumitt, and B. Meyers. Wallflower: Principles and practice of background maintenance. In *Proceedings of the 7th International Conference on Computer Vision (ICCV)*, pages 255–261, Sept. 1999.

E. Trucco and A. Verri. *Introductory Techniques for 3D Computer Vision*. Upper Saddle River, NJ: Prentice Hall, 1998.

G. V. Trunk. A problem of dimensionality: A simple example. *IEEE Transactions on Pattern Analysis and Machine Intelligence*, 1(3): 306–307, July 1979.

M. Tuceryan and A. K. Jain. Texture analysis. In C. H. Chen, L. F. Pau, and P. S. P. Wang, editors, *Handbook of Pattern Recognition and Computer Vision*, pages 235–276. World Scientific Publishing, New Jersey, 1993.

J. W. Tukey. A survey of sampling from contaminated distributions. In Olkin, editor, *Contributions to Probability and Statistics*. Stanford University Press, 1960.

M. Turk and A. Pentland. Eigenfaces for recognition. *Journal of Cognitive Neuroscience*, 3(1): 71–86, Jan. 1991.

K. Turkowski and S. Gabriel. Filters for common resampling tasks. In A. S. Glassner, editor, *Graphics Gems*, pages 147–165. San Diego: Academic Press, 1990.

C. W. Tyler and M. B. Clarke. The autostereogram. In *Stereoscopic Displays and ApplicationsProceedings of the SPIE*, volume 1256, pages 182–197, Jan. 1990.

S. E. Umbaugh. *Digital Image Processing and Analysis: Human and Computer Vision Applications with CVIPtools*. CRC Press, second edition, 2010.

S. Umeyama. Least-squares estimation of transformation parameters between two point patterns. *IEEE Transactions on Pattern Analysis and Machine Intelligence*, 13(4): 376–380, Apr. 1991.

M. Unser and T. Blu. Mathematical properties of the JPEG2000 wavelet filters. *IEEE Transactions on Image Processing*, 12(9): 1080–1090, Sept. 2003.

B. E. Usevitch. A tutorial on modern lossy wavelet image compression: Foundations of JPEG 2000. *IEEE Signal Processing Magazine*, 18(5): 22–35, Sept. 2001.

V. Vapnik. *The Nature of Statistical Learning Theory*. New York: Springer-Verlag, 1995.

V. Vapnik and A. Chervonenkis. On the uniform convergence of relative frequencies of events to their probabilities. *Theory of Probability and its Applications*, 16(2): 264–280, 1971.

L. Vincent and P. Soille. Watersheds in digital spaces: An efficient algorithm based on immersion simulations. *IEEE Transactions on Pattern Analysis and Machine Intelligence*, 13(6): 583–598, June 1991.

P. Viola and M. J. Jones. Robust real-time face detection. *International Journal of Computer Vision*, 57(2): 137–154, 2004.

P. Viola and W. M. Wells, III. Alignment by maximization of mutual information. *International Journal of Computer Vision*, 24(2): 137–154, Sept. 1997.

P. A. Viola, M. J. Jones, and D. Snow. Detecting pedestrians using patterns of motion and appearance. *International Journal of Computer Vision*, 63(2): 153–161, 2005.

M. Visvalingam and J. D. Whyatt. Line generalisation by repeated elimination of the smallest area. Cartographic Information Systems Research Group (CISRG) Discussion Paper 10, University of Hull, July 1992.

J. J. Vos. Colorimetric and photometric properties of a 2° fundamental observer. *Color Research and Application*, 3(3): 125–128, Autumn 1978.

P. F. Wainwright. Unsharp masks offer benefits beyond sharpness. *Photo Techniques*, 1, Sep/Oct 2004.

D. B. Walther, B. Chai, E. Caddigan, D. M. Beck, and L. Fei-Fei. Simple line drawings suffice for functional MRI decoding of natural scene categories. *Proceedings of the National Academy of Sciences (PNAS)*, 108(23): 9661–9666, 2011.

B. A. Wandell. *Foundations of Vision*. Sunderland, Mass.: Sinauer Associates, Inc., 1995.

Z. Wang, A. C. Bovik, H. R. Sheikh, and E. P. Simoncelli. Image quality assessment: From error visibility to structural similarity. *IEEE Transactions on Image Processing*, 13(4): 600–612, Apr. 2004.

J. Weickert. *Anisotropic Diffusion in Image Processing*. Stuttgart: B. G. Teubner, 1998.

M. J. Weinberger, G. Seroussi, and G. Sapiro. The LOCO-I lossless image compression algorithm: Principles and standardization into JPEG-LS. *IEEE Transactions on Image Processing*, 9(8): 1309–1324, Aug. 2000.

B. Weiss. Fast median and bilateral filtering. *ACM Transactions on Graphics (SIGGRAPH)*, 25(3): 519–526, July 2006.

Y. Weiss. Segmentation using eigenvectors: A unifying view. In *Proceedings of the International Conference on Computer Vision (ICCV)*, Sept. 1999.

Y. Weiss. Deriving intrinsic images from image sequences. In *Proceedings of the International Conference on Computer Vision (ICCV)*, pages 68–75, July 2001.

T. Welch. A technique for high-performance data compression. *Computer*, 17(6): 8–19, June 1984.

W. M. Wells, III. Efficient synthesis of Gaussian filters by cascaded uniform filters. *IEEE Transactions on Pattern Analysis and Machine Intelligence*, 8(2): 234–239, Mar. 1986.

M. Wertheimer. Laws of organization in perceptual forms. In W. D. Ellis, editor, *A Source Book of Gestalt Psychology*, pages 71–88. New York: Harcourt, Brace and Co., 1938.

L. Wiskott, J.-M. Fellous, N. Krüger, and C. von der Malsburg. Face recognition by elastic bunch graph matching. *IEEE Transactions on Pattern Analysis and Machine Intelligence*, 19(7): 775–779, July 1997.

A. Witkin. Scale space filtering. In *Proceedings of the International Joint Conference on Artificial Intelligence*, Aug. 1983.

R. J. Woodham. Photometric method for determining surface orientation from multiple images. *Optical Engineering*, 19(1): 139–144, Jan/Feb 1980.

W. D. Wright. A re-determination of the trichromatic coefficients of the spectral colours. *Transactions of the Optical Society*, 30(4): 141–164, Mar. 1929.

B. Wu and R. Nevatia. Detection and tracking of multiple, partially occluded humans by Bayesian combination of edgelet based part detectors. *International Journal of Computer Vision*, 75(2): 247–266, Nov. 2007.

B. Wu, H. Ai, C. Huang, and S. Lao. Fast rotation invariant multi-view face detection based on real Adaboost. In *IEEE International Conference on Automatic Face and Gesture Recognition (FGR)*, pages 79–84, May 2004.

Z. Wu and R. Leahy. An optimal graph theoretic approach to data clustering: Theory and its application to image segmentation. *IEEE Transactions on Pattern Analysis and Machine Intelligence*, 15(11): 1101–1113, Nov. 1993.

G. Wyszecki and W. S. Stiles. *Color Science: Concepts and Methods, Quantitative Data and Formulae*. New York: Wiley, second edition, 1982.

C. Xu and J. L. Prince. Gradient vector flow: A new external force for snakes. In *Proceedings of the IEEE Conference on Computer Vision and Pattern Recognition (CVPR)*, pages 66–71, June 1997.

C. Xu and J. L. Prince. Snakes, shapes, and gradient vector flow. *IEEE Transactions on Image Processing*, 7(3): 359–369, Mar. 1998.

C. Xu, D. L. Pham, M. E. Rettmann, D. N. Yu, and J. L. Prince. Reconstruction of the human cerebral cortex from magnetic resonance images. *IEEE Transactions on Medical Imaging*, 18(6): 467–480, June 1999.

R. Xu and D. Wunsch, II. Survey of clustering algorithms. *IEEE Transactions on Neural Networks*, 16(3): 645–678, May 2005.

M. H. Yang, D. J. Kriegman, and N. Ahuja. Detecting faces in images: A survey. *IEEE Transactions on Pattern Analysis and Machine Intelligence*, 24(1): 34–58, Jan. 2002.

A. Yilmaz, X. Li, and M. Shah. Contour-based object tracking with occlusion handling in video acquired using mobile cameras. *IEEE Transactions on Pattern Analysis and Machine Intelligence*, 26(11): 1531–1536, Nov. 2004.

I. T. Young and L. J. van Vliet. Recursive implementation of the Gaussian filter. *Signal Processing*, 44(2): 139–151, June 1995.

R. Zabih and J. Woodfill. Non-parametric local transforms for computing visual correspondence. In *Proceedings of the 3rd European Conference on Computer Vision (ECCV)*, pages 151–158, May 1994.

T. Y. Zhang and C. Y. Suen. A fast parallel algorithm for thinning digital patterns. *Communications of the ACM*, 27(3): 236–239, Mar. 1984.

Z. Zhang. Iterative point matching for registration of free-form curves and surfaces. *International Journal of Computer Vision*, 13(2): 119–152, Oct. 1994.

Z. Zhang. A flexible new technique for camera calibration. *IEEE Transactions on Pattern Analysis and Machine Intelligence*, 22(11): 1330–1334, Nov. 2000.

S. C. Zhu and A. L. Yuille. Region competition: Unifying snakes, region growing, and Bayes/MDL for multiband image segmentation. *IEEE Transactions on Pattern Analysis and Machine Intelligence*, 18(9): 884–900, Sept. 1996.

J. Ziv and A. Lempel. A universal algorithm for sequential data compression. *IEEE Transactions on Information Theory*, 23(3): 337–343, May 1977.

J. Ziv and A. Lempel. Compression of individual sequences via variable-rate coding. *IEEE Transactions on Information Theory*, 24(5): 530–536, Sept. 1978.

Z. Zivkovic and F. van der Heijden. Efficient adaptive density estimation per image pixel for the task of background subtraction. *Pattern Recognition Letters*, 27(7): 773–780, May 2006.

INDEX